T0188986

Textbook for Transcultural Health Care: A Population Approach

Larry D. Purnell • Eric A. Fenkl
Editors

Textbook for Transcultural Health Care: A Population Approach

Cultural Competence Concepts in Nursing Care

Fifth Edition

Editors
Larry D. Purnell
College of Health Sciences
University of Delaware/Newark
Sudlersville, MD
USA

Eric A. Fenkl
College of Nursing and Health Sciences
Florida International University
Miami, FL
USA

ISBN 978-3-030-51398-6 ISBN 978-3-030-51399-3 (eBook)
https://doi.org/10.1007/978-3-030-51399-3

© Springer Nature Switzerland AG 2013, 2021
This work is subject to copyright. All rights are reserved by the Publisher, whether the whole or part of the material is concerned, specifically the rights of translation, reprinting, reuse of illustrations, recitation, broadcasting, reproduction on microfilms or in any other physical way, and transmission or information storage and retrieval, electronic adaptation, computer software, or by similar or dissimilar methodology now known or hereafter developed.
The use of general descriptive names, registered names, trademarks, service marks, etc. in this publication does not imply, even in the absence of a specific statement, that such names are exempt from the relevant protective laws and regulations and therefore free for general use.
The publisher, the authors, and the editors are safe to assume that the advice and information in this book are believed to be true and accurate at the date of publication. Neither the publisher nor the authors or the editors give a warranty, expressed or implied, with respect to the material contained herein or for any errors or omissions that may have been made. The publisher remains neutral with regard to jurisdictional claims in published maps and institutional affiliations.

This Springer imprint is published by the registered company Springer Nature Switzerland AG
The registered company address is: Gewerbestrasse 11, 6330 Cham, Switzerland

Preface

With an increase in global migration, the need for culturally competent and congruent health-care providers is a priority. This need is not only driven by the actual numbers of immigrants and migrants, but other dimensions such as the mounting rise of multiculturalism and the waning of the social ideology that includes the idea of the "melting pot."

From within health care, the advocacy for culturally competent approaches is driven in part by recognizing the danger to patient safety, overall inadequacy in the quality of outcomes essential to health care, and costs that are driven up by health care that is not culturally competent and congruent. These concerns include the health disparities experienced by individuals who may or may not be socially and economically disadvantaged. Some of the excess expenditures are associated with poor communication in patient assessment data. As health-care systems move to more consumer and individual responsiveness, the systems are recognizing that a cultural perspective is essential for providing services that earn high levels of consumer satisfaction.

This book, *Textbook for Transcultural Health Care: A Population Approach: Cultural Competence Concepts in Nursing Care*, provides critical lessons to introduce students and practitioners to how different cultures construct their social world and the dramatic impact that culture has on how health-care providers, the community, and their families interact. These insights into the rich variety of human culture are only small steps toward developing real wisdom regarding culture competence and culturally congruent care.

Much of what is done in health care follows "procedures," which implies that there are predetermined steps by which anyone receiving the care or service must fit the intervention rather than vice versa. For an increasingly large part of the population, nothing could be further from the truth.

A culturally competent practitioner must have a sense of comfort with the *experiential* process of engaging others from different cultures. This is perhaps the most difficult of all skills to teach and may only be learned through the practice of engaging others and being able to reflect critically on the experience and its impact on the patient and on the provider. This process is a familiar one as is the core of clinical education. Care providers must learn how to adapt their clinical expertise to different cultures and the individual unique development in a multicultural context.

This textbook is based on the Purnell Model for Cultural Competence and its accompanying organizing framework that have been used in education, clinical practice, administration, and research, giving credence to its usefulness for health-care providers. The model has been translated into Arabic, Danish, Flemish, French, Korean, German, Portuguese, Spanish, and Turkish. Health-care organizations have adapted the organizing framework as a cultural assessment tool, and numerous students have used the Model to guide research for theses and dissertations in the United States and other countries. Thus, the Model's usefulness has been established in the global arena, recognizing and including the clients' cultures in assessment, health-care planning, interventions, and evaluation.

The introductory chapters lay the foundation for individual and organizational cultural competence with five chapters: (a) *Transcultural diversity and health care* that includes essential terminology and concepts related to culture, (b) a comprehensive description of *The Purnell Model for Cultural Competence*, (c) *Individual Competence and Evidence-based Practice* with the inclusion of international standards, (d) *Organizational Cultural Competence*, and (e) *Perspectives on Nursing and Health Care in a Global Context.*

Section two of the textbook includes aggregate data for population-specific groups. We have made a concerted effort to use non-stereotypical language when describing cultural attributes of specific cultures, recognizing that there are exceptions to every description provided and that the differences within a cultural group may be greater than the diversity between and among different cultural groups. We have also tried to include both the sociological and anthropological perspectives of culture. Each chapter in this section has reflexive exercises and case studies.

Specific criteria were used for identifying the groups represented. Groups included in the book were selected based on any of the criteria that follow.

- The group has a large population in North America such as people of Appalachian, Mexican, German, and African American heritage.
- The group is relatively new in its migration status such as people of Asian, Cuban, and Middle Eastern heritage.
- The group is widely dispersed throughout the world such as people of Iranian, Korean, and Filipino heritage.
- The group has little written about it in the health-care literature such as people of Guatemalan, Russian, and Thai heritage.
- The group holds significant disenfranchised status such as people of American Indian and Alaskan native heritage.
- The group is of particular interest to readers such as people of Amish heritage, a group that exists only in the United States.

We have strived to portray each culture positively and without stereotyping with an emphasis on the variant characteristics of culture portrayed on the Purnell Model for Cultural Competence.

Caring for culturally diverse patients and families has become commonplace throughout most of North America and worldwide. Areas that previously were largely homogenous, except for possibly one or two culturally diverse groups, are no longer the norm. The mosaic of diverse ethnicities and cultures being created is adding richness to our communities and societies at large.

Many cultural groups fall into the category of vulnerable populations with increased health and health-care disparities. Even when disparities are not a concern, patient satisfaction may be of concern. If we are to decrease health and health-care related disparities, more attention must be given to cultural values, beliefs, and practices, which includes spirituality and complementary and alternative therapies. If health-care providers and the health-care delivery system do not adapt and include diverse practices, patient and family care are compromised, resulting in not only poorer health outcomes but also an increase in cost.

Recent efforts to promote global recognition of multiculturalism have encouraged health-care providers to become more aware of their own cultures and to cultural differences in order to provide effective, acceptable, and safe health care. Each provider and organization must make a strong commitment to providing culturally competent care, regardless of the setting in which the care is provided and the backgrounds of the participants in care. As a result of this commitment, health-care organizations and educational programs have begun to recognize the need to prepare health-care professionals with the knowledge, skills, and resources essential to the provision of culturally competent care.

Culturally competent providers value diversity and respect individual differences regardless of one's own race, religious beliefs, or ethnocultural background. The major goal of this textbook is to provide a framework that promotes culturally competent care, respecting a person's right to be understood and treated as a unique individual. Foreign works are in italics. A glossary is included. Key words in each chapter are in boldfaced type and are included in the glossary for easy retrieval.

Sudlersville, MD, USA — Larry D. Purnell
Miami, FL, USA — Eric A. Fenkl

Contents

Part I

Foundations for Cultural Competence

1 Transcultural Diversity and Health Care: Individual and Organizational

Larry D. Purnell and Eric A. Fenkl

1.1 Introduction

Cultural competence is one of the most important initiatives in health care in the United States and throughout most of the world. Diversity has increased in many countries due to wars, discrimination, political strife, worldwide socioeconomic conditions, and the creation of the European Union. Due to an increase in the number of migrants across the world over the last two decades (the number reached 258 million in 2017) (U.N. 2017), populations of many countries are becoming more culturally and ethnically diverse. Working with patients of different cultural backgrounds can be challenging due to differences in opinions, beliefs, thoughts, norms, customs, and traditions. If poorly managed, cultural diversity can result in miscommunication, maladaptive behaviors, and interpersonal conflicts. Thus, it has been widely suggested that native and foreign-born health care professionals need to be equipped with cross-cultural competence.

1.2 The Need for Culturally Competent Health Care

Health ideology and health-care providers have learned that it is just as important to understand the patient's culture as it is to understand the physiological responses in illness, disease, and injury. The health-care provider may be very knowledgeable about laboratory values and standard treatments and interventions for diabetes mellitus, heart disease, and asthma, but if the recommendations are not compatible with the patient's own health beliefs, dietary practices, and views toward wellness, the treatment plan is less likely to be followed (Giger et al. 2007). To this end, a number of worldwide initiatives have addressed cultural competence as a means for improving health and health care, decreasing disparities, and increasing patient satisfaction. These initiatives come from the U.S. Office of Minority Health, the Institute of Medicine, Healthy People 2020, the National Quality Forum, the Joint Commission, The American Medical Association, the American Association of Colleges of Nursing, and other professional organizations. Educational institutions—from elementary schools to colleges and universities—

This chapter is an update from a previous edition written by Larry Purnell.

L. D. Purnell
University of Delaware, Newark, DE, USA
e-mail: lpurnell@udel.edu

E. A. Fenkl (✉)
Nicole Wertheim College of Nursing and Health Sciences, Florida International University, Miami, FL, USA
e-mail: efenkl@fiu.edu

© Springer Nature Switzerland AG 2021

L. D. Purnell, E. A. Fenkl (eds.), *Textbook for Transcultural Health Care: A Population Approach*, https://doi.org/10.1007/978-3-030-51399-3_1

are also addressing cultural diversity and cultural competency as they relate to disparities; health promotion and wellness; illness, disease, and injury prevention; and health maintenance and restoration.

Many countries are now recognizing the need for addressing the diversity of their societies. Societies that were largely homogeneous, such as Portugal, Norway, Sweden, Korea, and selected areas in the United States and the United Kingdom, are now facing significant internal and external migration, resulting in ethnic and cultural diversities that did not previously exist, at least not to the degree they do now. Several European countries, such as Denmark, Italy, Poland, the Czech Republic, Latvia, the United Kingdom, Sweden, Norway, Finland, Italy, Spain, Portugal, Hungary, Belgium, Greece, Germany, the Netherlands, and France, either have in place or are developing national programs to address the value of cultural competence in reducing health disparities (Judge et al. 2006).

Whether people are internal migrants, immigrants, or vacationers, they have the right to expect the health-care system to respect their personal beliefs, values, and health-care practices. Culturally competent health care from providers and the system, regardless of the setting in which care is delivered, is becoming a concern and expectation among consumers. Diversity also includes having a diverse workforce that more closely represents the population the organization serves. Health-care personnel provide care to people of diverse cultures in long-term-care facilities, acute-care facilities, clinics, communities, and patients' homes. All health-care providers—physicians, nurses, nutritionists, therapists, technicians, home health aides, and other caregivers—need similar culturally specific information. For example, all health-care providers communicate, both verbally and nonverbally; therefore, all health-care providers and ancillary staff need to have similar information and skill development to communicate effectively with diverse populations. The manner in which the information is used may differ significantly based on the discipline, individual experiences, and specific circumstances of the patient, provider, and organization. If providers and the system are competent, most patients will access the health-care system when problems are first recognized, thereby reducing the length of stay, decreasing complications, and reducing overall costs.

A lack of knowledge of patients' language abilities and cultural beliefs and values can result in serious threats to life and quality of care for all individuals (Joint Commission 2010). Organizations and individuals who understand their patients' cultural values, beliefs, and practices are in a better position to be co-participants with their patients in providing culturally acceptable care. Having ethnocultural-specific knowledge, understanding, and assessment skills to work with culturally diverse patients ensures that the health-care provider can conduct a more targeted assessment. Providers who know culturally specific aggregate data are less likely to demonstrate negative attitudes, behaviors, ethnocentrism, stereotyping, and racism. The onus for cultural competence is health-care providers and the delivery system in which care is provided. To this end, health-care providers need both general and specific cultural knowledge when conducting assessments, planning care, and teaching patients about their treatments and prescriptions.

1.3 World Diversity and Migration

The current world population is 7.8 billion as of February 2020 according to the most recent United Nations estimates with a median age of 29.6 years. The population is expected to approach 9.8 billion by 2050. The estimated population growth rate remains relatively stable at 1.03%, with 18.10 births per 1000 population; 7.7 deaths per 1000 population; and an infant mortality rate of 30.80 per 1000 population, down from 44.13 in 2010. Worldwide life expectancy at birth is currently 70.50 years, up from 66.12 years in 2010 (CIA World Factbook 2020). The ten largest urban populations where significant migration occurs are Tokyo, Japan with 37.3 million; Mumbai, India with 25.97 million; Delhi, India with 25.83 million; Dhaka,

Bangladesh with 22.04 million; Mexico City, Mexico with 21.81 million; São Paulo, Brazil with 21.57 million; Lagos, Nigeria with 21.51 million; Jakarta, Indonesia with 20.77 million; New York, United States with 20.4 million; and Karachi, Pakistan with 18.94 million (CIA World Factbook 2020).

As a first language, Mandarin Chinese is the most popular, spoken by 1.1 billion people; followed by English spoken by 983 million people; Hindustani (544 million speakers); Spanish (527 million speakers); Arabic (422 million speakers); Malay (281 million speakers); Russian (267 million speakers); and Bengali (261 million speakers). On a global scale, 86.3% of the world population is literate. When technology is examined, more people have a cell phone than a landline—with a ratio of 3:1. The internet has been one of the most transformative and fast-growing technologies. Globally the number of internet users increased from only 413 million in 2000 to over 3.4 billion in 2016 (CIA World Factbook 2020). Language literacy has serious implications for immigration. Over two-thirds of the world's 785 million illiterate adults are found in only eight countries: Bangladesh, China, Egypt, Ethiopia, India, Indonesia, Nigeria, and Pakistan. Of all the illiterate adults in the world, two-thirds are women; extremely low literacy rates are concentrated in three regions: the Arab states, South and West Asia, and Sub-Saharan Africa, where around one-third of the men and half of all women are illiterate (CIA World Factbook 2020).

The number of refugee status individuals has increased dramatically in the last several years. As of 2019, an unprecedented 70.8 million people have been forcibly displaced worldwide, and 37,000 people are forced to flee their homes every day due to conflict or persecution. The global number of migrants worldwide reached an estimated 272 million in 2019, an increase of 51 million since 2010. Currently, international migrants comprise 3.5% of the global population, compared to 2.8% in the year 2000 (U.N. Refugee Agency 2009).

In 1997, the International Organization for Migration studied the costs and benefits of international migration. A comprehensive update has not been undertaken since that time. According to the report, ample evidence exists that migration brings both costs and benefits for sending and receiving countries, although these are not shared equally. Trends suggest a greater movement toward circular migration with substantial benefits to both home and host countries. Migration boosts the working-age population. Migrants arrive with skills and contribute to human capital development of receiving countries. Migrants also contribute to technological progress. Understanding these impacts is important if our societies are to usefully debate the role of migration.

The perception that migrants are more of a burden on than a benefit to the host country is not substantiated by research. For example, in the Home Office Study (2002) in the United Kingdom, migrants contributed US$4 billion more in taxes than they received in benefits. In the United States, the National Research Council (1998) estimated that national income had expanded by US$8 billion because of immigration. Thus, because migrants pay taxes, they are not likely to put a greater burden on health and welfare services than the host population. However, undocumented migrants run the highest health risks because they are less likely to seek health care. This not only poses risks for migrants but also fuels sentiments of xenophobia and discrimination against all migrants.

- *What evidence do you see in your community that migrants have added to the economic base of the community? Who would be doing their work if they were not available? If migrants (legal or undocumented) were not picking vegetables (just one example), how much more do you think you would pay for the vegetables?*

1.4 U.S. Population and Census Data

As of 2020, the U.S. population is estimated at 331 million people according to UN data. The United States population is equivalent to

4.25% of the total world population. This is an increase of 15 million since the 2010 census. The 2010 census data included changes designed to more clearly distinguish Hispanic ethnicity as not being a race. In addition, the Hispanic terms have been modified to include *Hispanic* (used more heavily on the East Coast), *Latino* (used more heavily in California and the West Coast), and *Spanish*. The most recent census data estimate that the White, non-Hispanic or Latino population make up 61.3% of the nation's total, 17.8% are Hispanic/Latino, 13.0% are Black, 5.0% are Asian, 2.0% are American Indian or Alaskan Native, and 0.2% are Native Hawaiian or other Pacific Islander. These groupings will be more specifically reported as the census data are analyzed. The categories to be used in the 2020 U.S. Census will remain the same as those used in the 2010 census and are as follows:

1. *White* refers to people having origins in any of the original peoples of Europe and includes Middle Easterners, Irish, German, Italian, Lebanese, Turkish, Arab, and Polish.
2. *Black,* or *African American,* refers to people having origins in any of the black racial groups of Africa and includes Nigerians and Haitians or any person who self-designates this category regardless of origin.
3. *American Indian* and *Alaskan Native* refer to people having origins in any of the original peoples of North, South, or Central America and who maintain tribal affiliation or community attachment.
4. *Asian* refers to people having origins in any of the original peoples of the Far East, Southeast Asia, or the Indian subcontinent. This category includes the terms *Asian Indian*, *Chinese*, *Filipino*, *Korean*, *Japanese*, *Vietnamese*, *Burmese*, *Hmong*, *Pakistani*, and *Thai*.
5. *Native Hawaiian* and *other Pacific Islander* refer to people having origins in any of the original peoples of Hawaii, Guam, Samoa, Tahiti, the Mariana Islands, and Chuuk.
6. *Some other race* was included for people who are unable to identify with the other categories.
7. In addition, the respondent could identify, as a write-in, with two races (U.S. Census Bureau 2020).

The Hispanic/Latino and Asian populations continue to rise in numbers and in percentage of the overall population; however, although the Black/African American, Native Hawaiian and Pacific Islanders, and American Indian and Alaskan Natives groups continue to increase in overall numbers, their percentage of the population has decreased. Of the Hispanic/Latino population, most are Mexicans, followed by Puerto Ricans, Cubans, Central Americans, South Americans, and Dominicans. Salvadorans are the largest group from Central America. Three-quarters of Hispanics live in the West or South, with 50% of the Hispanics living in just two states: California and Texas. The median age for the entire U.S. population is 38.0 years, and the median age for Hispanics is 30.2 years (U.S. Census Bureau 2020). The young age of Hispanics in the United States makes them ideal candidates for recruitment into the health professions, an area with crisis-level shortages of personnel, especially of minority representation.

Before 1940, most immigrants to the United States came from Europe, especially Germany, the United Kingdom, Ireland, the former Union of Soviet Socialist Republics, Latvia, Austria, and Hungary. Since 1940, immigration patterns to the United States have changed: Most are from Mexico, the Philippines, China, India, Brazil, Russia, Pakistan, Japan, Turkey, Egypt, Thailand, and countries of the Middle East. People from each of these countries bring their own culture with them and increase the cultural mosaic of the United States. Many of these groups have strong ethnic identities and maintain their values, beliefs, practices, and languages long after their arrival. Individuals who speak only their indigenous language are more likely to adhere to traditional practices and live in ethnic enclaves and are less likely to assimilate into their new society. The inability of immigrants to speak the language of their new country creates additional challenges for health-care providers working with these populations. Other countries in the world face similar immigration challenges and opportunities for diversity enrichment. However, space does not permit a comprehensive analysis of migration patterns.

What changes in ethnic and cultural diversity have you seen in your community over the last 5 years? Over the last 10 years? Have you had the opportunity to interact with these newer groups?

1.5 Racial and Ethnic Disparities in Health Care

A number of organizations have developed documents addressing the need for cultural competence as one strategy for eliminating racial and ethnic disparities. The Agency for Healthcare Research and Quality (AHRQ) released in 2018, the National Healthcare Quality and Disparities Report" (AHRQ 2018), which provided a comprehensive overview of health disparities in ethnic, racial, and socioeconomic groups in the United States. This report was a combined document with the National Healthcare Quality Report released initially (2006), which was an overview of quality health care in the United States. Healthy People 2010's (www.healthypeople.gov) goals were to increase the quality and the length of a healthy life and to eliminate health disparities. Healthy People provided science-based, 10-year national objectives for improving the health of all Americans. For three decades, Healthy People has established benchmarks and monitored progress over time in order to (a) encourage collaborations across communities and sectors, (b) empower individuals toward making informed health decisions, and (c) measure the impact of prevention activities (http://www.healthypeople.gov/2020/about/default.aspx).

The Healthy People 2020 (www.healthypeople2020.gov) report had a renewed focus on identifying, measuring, tracking, and reducing health disparities through determinants of health such as the social and economic environment, the physical environment, and the person's individual characteristics and behaviors.

Although the term *disparities* is often interpreted to mean racial or ethnic disparities, many dimensions of disparity exist in the United States, particularly in health. If a health outcome is seen in a greater or lesser extent among different populations, a disparity exists. Race or ethnicity, sex, sexual identity, age, disability, socioeconomic status, and geographic location all contribute to an individual's ability to achieve good health. During the past two decades, one of *Healthy People's* overarching goals focused on disparities. Indeed, in *Healthy People 2000,* the goal was to reduce health disparities among Americans; in *Healthy People 2010,* it was to completely eliminate, not just reduce, health disparities; and in Healthy People 2020, the goal was expanded to achieve health equity, eliminate disparities, and improve the health of all groups.

Healthy People 2020 defines a *health disparity* as "a particular type of health difference that is closely linked with social, economic, and/or environmental disadvantage." Health disparities adversely affect groups of people who have systematically experienced greater obstacles to health based on their racial or ethnic group; religion; socioeconomic status; gender; age; mental health; cognitive, sensory, or physical disability; sexual orientation or gender identity; geographic location; or other characteristics historically linked to discrimination or exclusion. In addition, powerful, complex relationships exist among health and biology, genetics, and individual behavior, and among health and health services, socioeconomic status, the physical environment, discrimination, racism, literacy levels, and legislative policies. These factors, which influence an individual's or population's health, are known as determinants of health (Healthy People 2020).

What health disparities have you observed in your community? To what do you attribute these disparities? What can you do as a professional to help decrease these disparities?

More specific data on ethnic and cultural groups are included in individual chapters. As can be seen by the overwhelming data, much more work needs to be done to improve the health of the nation.

1.6 Culture and Essential Terminology

1.6.1 Culture Defined

Anthropologists and sociologists have proposed many definitions of *culture*. For the purposes of this book, which is primarily focused on individual cultural competence instead of the culturally competent organization, **culture** is defined as the totality of socially transmitted behavioral patterns, arts, beliefs, values, customs, lifeways, and all other products of human work and thought characteristics of a population of people that guide their worldview and decision making. Health and health-care beliefs and values are assumed in this definition. These patterns may be explicit or implicit, are primarily learned and transmitted within the family, are shared by most (but not all) members of the culture, and are emergent phenomena that change in response to global phenomena. Culture refers to a large and diverse set of mostly intangible aspects of social life consisting of the values, beliefs, systems of language, communication, and practices that people share in common and that can be used to define them as a collective (Cole 2020).

Culture, a combined anthropological and social construct, can be seen as having three levels: (a) a tertiary level that is visible to outsiders, such as things that can be seen, worn, or otherwise observed; (b) a secondary level, in which only members know the rules of behavior and can articulate them; and (c) a primary level that represents the deepest level in which rules are known by all, observed by all, implicit, and taken for granted (Koffman 2006). Culture is largely unconscious and has powerful influences on health and illness.

An important concept to understand is that cultural beliefs, values, and practices are learned from birth: first at home, then in other places where people congregate such as educational settings and places of worship. Therefore, a 3-month-old female child from Russian of Ashkenazi Jewish heritage who is adopted by a European American family and reared in a dominant European American environment will have a European American worldview. However, if that child's heritage has a tendency toward genetic/hereditary conditions, they would come from her Russian Jewish ancestry, not from European American genetics.

Who in your family had the most influence in teaching you cultural values and practices? Outside the family, where else did you learn about your cultural values and beliefs? What cultural practices did you learn in your family that you no longer practice?

When individuals of dissimilar cultural orientations meet in a work or a therapeutic environment, the likelihood for developing a mutually satisfying relationship is improved if both parties attempt to learn about one another's culture. Moreover, race and culture are not synonymous and should not be confused. For example, most people who self-identify as African American have varying degrees of dark skin, but some may have white skin. However, as a cultural term, *African American* means that the person takes pride in having ancestry from both Africa and the United States; thus, a person with whiter skin tones could self-identify as African American.

1.6.2 Important Terms Related to Culture

Attitude is a state of mind or feeling about some aspect of a culture. Attitudes are learned; for example, some people think that one culture is better than another. No one culture is "better" than another; they are just different, and many different cultures share the same customs. A **belief** is something that is accepted as true, especially as a tenet or a body of tenets accepted by people in an ethnocultural group. A belief among some cultures is that if you go outside in the cold weather with wet hair, you will catch a cold. Attitudes and beliefs do not have to be proven; they are unconsciously accepted as truths. **Ideology** consists of the thoughts and beliefs that reflect the social needs and aspirations of an individual or an ethnocultural group. For example, some people believe that health care is the right of all people, whereas others see health care as a privilege.

The literature reports many definitions of the terms *cultural knowledge*, *cultural awareness, cultural sensitivity,* and *cultural competence*. Sometimes, these definitions are used interchangeably, but each has a distinct meaning. **Cultural knowledge** is all we know that characterize a particular **culture**. It can include descriptions such as those known as **cultural** dimensions and can also include other information that may explain why people conduct themselves in a particular way. **Cultural awareness** has to do with an appreciation of the external signs of diversity, such as the arts, music, dress, foods, and physical characteristics. **Cultural sensitivity** has to do with personal attitudes and not saying things that might be offensive to someone from a cultural or ethnic background different from that of the health-care provider's cultural or ethnic background. **Cultural competence** in health care is having the knowledge, abilities, and skills to deliver care more congruent with the patient's cultural beliefs and practices. Increasing one's consciousness of cultural diversity improves the possibilities for health-care practitioners to provide culturally competent care. Cultural competence should be a part of health-care provider basic training and based on cultural knowledge and experiential learning methods as well as having the opportunity to be exposed to different cultures (Khatib and Hadid 2019).

What activities have you done to increase your cultural awareness and competence? How do you demonstrate that you are culturally sensitive?

One progresses from unconscious incompetence (not being aware that one is lacking knowledge about another culture), to conscious incompetence (being aware that one is lacking knowledge about another culture), to conscious competence (learning about the patient's culture, verifying generalizations about the patient's culture, and providing cultural-specific interventions), and, finally, to unconscious competence (automatically providing culturally congruent care to patients of diverse cultures). Unconscious competence is difficult to achieve and failure to do so is potentially dangerous as individual differences exist within cultural groups. To be even minimally effective, culturally competent care must have the assurance of continuation after the original impetus is withdrawn; it must be integrated into, and valued by, the culture that is to benefit from the interventions.

Developing mutually satisfying relationships with diverse cultural groups involves good interpersonal skills and the application of knowledge and techniques learned from the physical, biological, and social sciences as well as the humanities. An understanding of one's own culture and personal values and the ability to detach oneself from presumptions and biases associated with personal worldviews are essential for cultural competence. Even then, traces of ethnocentrism may unconsciously pervade one's attitudes and behavior. **Ethnocentrism**—the universal tendency of human beings to think that their ways of thinking, acting, and believing are the only right, proper, and natural ways (which most people practice to some degree)—can be a major barrier to providing culturally competent care. Ethnocentrism perpetuates an attitude in which beliefs that differ greatly from one's own are strange, bizarre, or unenlightened and, therefore, wrong. Values are principles and standards that are important and have meaning and worth to an individual, family, group, or community. For example, the dominant U.S. culture places high value on youth, technology, and money. The extent to which one's cultural values are internalized influences the tendency toward ethnocentrism. The more one's values are internalized, the more difficult it is to avoid the tendency toward ethnocentrism.

Given that everyone is ethnocentric to some degree, what do you do to become less ethnocentric? With which groups are you more ethnocentric? If you were to rate yourself on a scale of 1 to 10, with 1 being only a little ethnocentric and 10 being very ethnocentric, what score would you give yourself? What score would your friends give you? What score would you give your closest friends?

The Human Genome Project (2003) determined that 99.9% of all humans share the same genes. One-tenth percent of genetic variations account for the differences among humans, although these differences may be significant when conducting health assessments and prescribing medications and treatments. Ignoring this small difference, however, is ignoring the beliefs, practices, and values of a small ethnic or cultural population to whom one provides care. The controversial term *race* must still be addressed when learning about culture. A race is a grouping of humans based on shared physical or social qualities into categories generally viewed as distinct by society. The word "**race**" itself, intended to define an identifiable group of people who share a common descent, was introduced into English in about 1580, from the Old French "rasse" (1512) and from Italian "razza." The term was first used to refer to speakers of a common language and then to denote national affiliations. By the seventeenth century the term began to refer to physical (phenotypical) traits. Race is genetic in origin and includes physical characteristics that are similar among members of the group, such as skin color, blood type, and hair and eye color (Giger et al. 2007). People from a given racial group may, but do not necessarily, share a common culture. Race as a social concept is sometimes more important than race as a biological concept. Race has social meaning, assigns status, limits or increases opportunities, and influences interactions between patients and clinicians. Some believe that race terminology was invented to assign low status to some and privilege, power, and wealth to others (American Anthropological Association 1998). Thus, perhaps the most significant aspect of race is social in origin. Moreover, one must remember that even though one might have a racist attitude, it is not always recognized because it is ingrained during socialization and leads to ethnocentrism.

How do you define race? What other terms do you use besides race to describe people? In what category did you classify yourself on the last census? What categories would you add to the current census classifications?

Worldview is the way individuals or groups of people look at the universe to form basic assumptions and values about their lives and the world around them. Worldview includes cosmology, relationships with nature, moral and ethical reasoning, social relationships, magico-religious beliefs, and aesthetics.

Any **generalization**—reducing numerous characteristics of an individual or group of people to a general form that renders them indistinguishable—made about the behaviors of any individual or large group of people is almost certain to be an oversimplification. When a generalization relates less to the actual observed behavior than to the motives thought to underlie the behavior (i.e., the *why* of the behavior), it is likely to be oversimplified. However, generalizations can lead to **stereotyping**, an oversimplified conception, opinion, or belief about some aspect of an individual or group. Although generalization and stereotyping are similar, functionally, they are very different. Generalization is a starting point, whereas stereotyping is an endpoint. The health-care provider must specifically ask questions to determine these values and avoid stereotypical views of patients. See the section on Variant Characteristics of Culture in this chapter.

Everyone engages in stereotypical behavior to some degree. We could not function otherwise. If someone asks you to think of a nurse, what image do you have? Is the nurse male or female? How old is the nurse? How is the nurse dressed? Is the nurse wearing a hat? How do you distinguish a stereotype from a generalization?

Within all cultures are subcultures and ethnic groups whose values/experiences differ from those of the dominant culture with which they identify. Indeed, subcultures share beliefs according to the *variant characteristics* of culture, as described later in this chapter. In sociology, anthropology, and cultural studies, a **subculture** is defined as a group of people with a culture that differentiates them from the larger culture of which they are a part. Subcultures may be distinct or hidden (e.g., gay, lesbian, bisexual, and transgendered populations). If the subculture is characterized by a systematic opposition to the dominant culture, then it may be described as a counterculture. Examples of subcultures are Goths, punks, and stoners, although popular lay literature might call these groups cultures instead of subcultures. **Countercultures**, on the other hand, are cultures with values and mores that run counter to those of established society and whose norms and values may be incompatible with prevailing cultural norms. A counterculture might include religious cults or other groups whose behaviors contradict societal norms and expectations (Merriam Webster Online Dictionary 2020).

The terms *transcultural* versus *cross-cultural* have been hotly debated among experts in several countries but especially in the United States. Specific definitions of these terms vary. Some attest that they are the same, whereas others say they are different. Historically, nursing seems to favor the word *transcultural*. Indeed, the term has been credited to a nurse anthropologist, Madeleine Leininger, in the 1950s (Leininger and McFarland 2006), and it continues to be popular in the United States, the United Kingdom, and many European countries. The term *cross-cultural* can be traced to anthropologist George Murdock in the 1930s and is still a popular term used in the social sciences, although the health sciences have used it as well. The term implies comparative interactivity among cultures.

Cultural humility, another term found in cultural literature, focuses on the process of intercultural exchange, paying explicit attention to clarifying the professional's values and beliefs through self-reflection and incorporating the cultural characteristics of the health care professional and the patient into a mutually beneficial and balanced relationship (Trevalon and Murray-Garcia 1998). This term appears to be most popular with physicians and some professionals from the social sciences.

Cultural safety is a popular term in Australia, New Zealand, and Canada, although it is used elsewhere. Cultural safety expresses the diversity that exists within cultural groups and includes the social determinants of health, religion, and gender, in addition to ethnicity (*Guidelines for Cultural Safety* 2005). **Cultural leverage** is a process whereby the principles of cultural competence are deliberately invoked to develop interventions. It is a focused strategy for improving the health of racial and ethnic communities by using their cultural practices, products, philosophies, or environments to facilitate behavioral changes of the patient and professional (Fisher et al. 2007).

Acculturation occurs when a person gives up the traits of his or her culture of origin as a result of contact with another culture. Acculturation is not an absolute, and it has varying degrees. Traditional people hold onto the majority of cultural traits from their culture of origin, which is frequently seen when people live in ethnic enclaves and can get most of their needs met without mixing with the outside world. Bicultural acculturation occurs when an individual is able to function equally in the dominant culture and in one's own culture. People who are comfortable working in the dominant culture and return to their ethnic enclave without taking on most of the dominant culture's traits are usually bicultural. Marginalized individuals are not comfortable in their new culture or their culture of origin. **Assimilation** is the gradual adoption and incorporation of characteristics of the prevailing culture (Portes 2007).

Enculturation is a natural conscious and unconscious conditioning process of learning accepted cultural norms, values, and roles in society and achieving competence in one's culture through socialization. Enculturation is facilitated by growing up in a particular culture, and it can be through formal education, apprenticeships, mentorships, and role modeling (Clarke and Hofsess 1998).

1.6.3 Individualism, Collectivism, and Individuality

All cultures worldwide vary along an individualism and collectivism scale and are subsets of broad worldviews. A continuum of values for individualistic and collectivistic cultures includes orientation to self or group, decision making, knowledge transmission, individual choice and personal responsibility, the concept of progress, competitiveness, shame and guilt, help-seeking, expression of identity, and interaction/communication style (Hofstede 1991; Hofstede and Hofstede 2005).

Elements and the degree of individualism and collectivism exist in every culture. People from an individualist culture will more strongly identify with the values at the individualistic end of the scale. Moreover, individualism and collectivism fall along a continuum, and some people from an individualistic culture will, to some degree, align themselves toward the collectivistic end of the scale. Some people from a collectivist culture will, to some degree, hold values along the individualistic end of the scale. Acculturation is a key component of adopting individualistic and collectivistic values. Those who live in ethnic enclaves usually, but not always, adhere more strongly to their dominant cultural values, sometimes to such a degree that they are more traditional than people in their home country. Acculturation and the variant characteristics of culture determine the degree of adherence to traditional individualistic and collectivist cultural values, beliefs, and practices (Hofstede 1991; Hofstede and Hofstede 2005).

Communicating, assessing, counseling, and educating a person from an individualistic culture, where the most important person in society is the individual, may require different techniques than for a person in a collectivist culture where the group is seen as more important than the individual (Hofstede and Hofstede 2005). The professional must not confuse individualism with individuality—the degree that varies by culture and is usually more prevalent in individualistic countries. **Individuality** is the sense that each person has a separate and equal place in the community and where individuals who are considered "eccentrics or local characters" are tolerated (Purnell 2010).

Some highly individualistic cultures include traditional European American (in the United States), British, Canadian, German, Norwegian, and Swedish, to name a few. Some examples of collectivist cultures include traditional Arabic, Amish, Chinese, Filipino, Korean, Japanese, Latin American, Mexican, American Indians (and most other indigenous Indian groups), Taiwanese, Thai, Turkish, and Vietnamese. Far more world cultures are collectivistic than are individualistic. It may be difficult for a nurse who is from a highly collectivist culture to communicate with patients and staff in highly individualistic cultures, such as the United States and Germany (Hofstede and Hofstede 2005).

Cultures differ in the extent to which health and information are explicit or implicit. In low-context cultures, great emphasis is placed on the verbal mode, and many words are used to express a thought. Low-context cultures are individualistic. In high-context cultures, much of the information is implicit where fewer words are used to express a thought, resulting in more of the message being in the nonverbal mode. Great emphasis is placed on personal relationships. High-context cultures are collectivistic (Hofstede 1991, 2001).

Consistent with individualism, individualistic cultures encourage self-expression. Adherents to individualism freely express personal opinions, share many personal issues, and ask personal questions of others to a degree that may be seen as offensive to those who come from a collectivistic culture. Direct, straightforward questioning is usually appreciated with individualism. However, the professional should take cues from the patient before this intrusive approach is initiated. Small talk before getting down to business is not always appreciated. Individualistic cultures usually tend to be more informal and frequently use first names. Ask the patient by what name he or she prefers to be called. Questions that require a "yes" or "no" answer are usually answered truthfully from the patient's perspective. In individualistic cultures that value autonomy and productivity, one is expected to be a productive member of society. Among collectivistic cultures, people with a mental or physical disability are *more likely* to be hidden from society to "save face," and the cultural norms and values of the family unit mean that the family provides care in the home (Purnell 2001).

Indeed, it is absolutely imperative to include the family, and sometimes the community, in health care for effective counseling; otherwise, the treatment plan may falter. However, among many Middle Eastern and other collectivistic cultures, family members with mental or physical disabilities are hidden from the community for fear that children in the family might not be able to obtain a spouse if the condition is known. For other impairments, such as HIV, the condition may be kept from public view, not because of confidentiality rights but for fear that news of the condition will spread to other family members and the community.

The greater the perceived cultural stigma, the more likely the delay in seeking counseling, resulting in the condition being more severe at the time of treatment. Individualistic cultures socialize their members to view themselves as *independent*, separate, distinct individuals, where the most important person in society is self. A person feels free to change alliances and not feel bound by any particular group (shared identity). Although they are part of a group, they are still free to act independently within the group and less likely to engage in "groupthink." In individualism, competition, whether individual or group, permeates every aspect of life. Separateness, independence, and the capacity to express one's own views and opinions are both explicitly valued and implicitly assumed.

In individualistic cultures, a person's identity is based mainly on one's personal accomplishments, career, and challenges. A high standard of living supports self-efficiency, self-direction, self-advocacy, and independent living. Decisions made by elders and people in hierarchal positions may be questioned or not followed because the ideal is that all people expect to, and are expected to, make their own decisions about their lives. Moreover, people are personally responsible and held accountable for their decisions. Improving self, doing "better" than others (frequently focused on material gains), and making progress on a community or national level are expected. If one fails, the blame and shame are on the individual alone.

In collectivistic cultures, people are socialized to view themselves as members of a larger group, family, school, church, educational setting, workplace, and so on. They are bound through the expectations of loyalty and personal and familial lifetime protective ties. Children are socialized where priority is given to connections and interrelationship with others as the basis of psychological well-being. Older people and those in hierarchical positions are respected, and people are less likely to openly disagree with them. Parents and elders may have the final say in their children's careers and life partners. The focus is not on the individual but on the group.

Collectivism is characterized by not drawing attention to oneself, and people are not encouraged to ask controversial questions about themselves or others. When one fails, shame may be extended to the family, and external explanations, spiritual, superiors, or fate may be given. To avoid offending someone, people are expected to practice smooth interpersonal communication by not openly disagreeing with anyone and being evasive about negative issues. Among most collectivist cultures, disagreeing with or saying "no" to a health-care professional is considered rude. In fact, in some languages, there is no word for "no." If you ask a collectivist patient if she knows what you are asking, if she understands you, and if she knows how to do something, she will always answer "yes." But "yes" could mean (a) I hear you, but I do not understand you; (b) I understand you, but I do not agree with what you are saying; and (c) I know how to do that, but I might not do it. Repeating what has been prescribed does not ensure understanding; instead, ask for a demonstration or some other response that is more likely to determine understanding.

Reflective Exercise 1.1

Does your cultural heritage primarily have a collectivistic or individualistic cultural worldview? Rate your culture on a scale of 1 to 10 with 1 = collectivistic and 10 = individualistic. Is your culture tolerant of individuality? Are you consistent with your cultural heritage? Provide some specific behaviors to support your answer.

Collectivism					Individualism				
1	2	3	4	5	6	7	8	9	10

1.6.4 Variant Characteristics of Culture

Great diversity exists within a cultural group. Major influences that shape people's worldviews and the degree to which they identify with their cultural group of origin are called the "variant characteristics of culture." Some variant characteristics cannot be changed, while others can. They include but are not limited to the following:

- *Nationality:* One cannot change his or her nationality, but over time many people have changed their names to better fit into society or to decrease discrimination. For example, many Jews changed the spelling of their last names during and after World War II to avoid discrimination.
- *Race:* Race cannot be changed, but people can and do make changes in their appearance, such as with of cosmetic surgery.
- *Color:* Skin color cannot usually be changed on a permanent basis.
- *Age:* Age cannot be changed, but many people go to extensive lengths to make themselves look younger. One's worldview changes with age. In some cultures, older people are looked upon with reverence and increased respect. Age difference with the accompanying worldview is frequently called the *generation gap*.
- *Religious affiliation:* People can and do change their religious affiliations or self-identify as atheists. However, if someone changes his or her religious affiliation—for example, from Judaism to Pentecostal or Baptist to Islam—a significant stigma may occur within their family or community.
- *Educational status:* As education increases, people's worldview changes and increases their knowledge base for decision making.
- *Socioeconomic status:* Socioeconomic status can change either up or down and can be a major determinant for access to and use of health care.
- *Occupation:* One's occupation can change. Of course, an occupation can be a health risk if employment is in a coal mine, on a farm, or in a high-stress position. In addition, someone who is educated in the health professions would not have as much difficulty with health literacy.
- *Military experience:* People who have military experience may be more accustomed to hierarchical decision making and rules of authority.
- *Political beliefs:* Political affiliation can change according to one's ideology. One of the major reasons for migration is ideological and political beliefs.
- *Urban versus rural residence:* People can change their residence with concomitant changes in ideology with different health risks and access to health care.
- *Enclave identity:* For people who primarily live and work in an ethnic enclave where they can get their needs met without mixing with the world outside, they may be more traditional than people in their home country.
- *Marital status:* Married people and people with partners frequently have a different worldview than those without partners.
- *Parental status:* Often, when people become parents—having children, adopting, or taking responsibility for raising a child—their worldview changes, and they usually become more futuristic.
- *Sexual orientation:* Sexual orientation is usually stable over time, but some people are bisexual. In addition, people who are incarcerated may engage in same-sex activity but return to a heterosexual lifestyle when released from prison. Gender reassignment is now a possibility for some, although a significant stigma may occur.
- *Sex and Gender*: Sex is not necessarily binary and can be fluid. Sex identity may include males, females, transgender man or transgender woman and/or other gender non-conforming individuals. Not all individuals identify with the gender assigned to them at birth. Additionally, men and women may have different concerns in regards to type of work and work hours, pay scales, and health inequalities.
- *Physical characteristics:* One's physical characteristics may have an effect on how people see themselves and how others see

them and can include such characteristics as height, weight, hair color and style, and skin color.

- *Immigration status (sojourner, immigrant, or undocumented status):* Immigration status and length of time away from the country of origin also affect one's worldview. People who voluntarily immigrate generally acculturate and assimilate more easily. Sojourners who immigrate with the intention of remaining in their new homeland for only a short time on work assignments or refugees who think they may return to their home country may not have the need or desire to acculturate or assimilate. Additionally, undocumented individuals (illegal immigrants) may have a different worldview from those who have arrived legally. Many in this group remained hidden in society so they will not be discovered and returned to their home country.
- *Length of time away from the country of origin:* Usually, the longer people are away from their culture of origin, the less traditional they become as they acculturate and assimilate into their new culture.

Reflective Exercise 1.2
What are your variant characteristics of culture? How has each one influenced you and your worldview? How has your worldview changed as your variant characteristics have changed? How is each of these a culture or a subculture?

Some examples of how variant cultural characteristics change one's worldview follow.

Consider two people with the following variant characteristics. One is a 75-year-old devout Islamic female from Saudi Arabia, and the other is a 19-year-old African American fundamentalist Baptist male from Louisiana. Obviously, the two do not look alike, and they probably have very different worldviews and beliefs, many of which come from their religious tenets and country of origin.

The variant cultural characteristics of being a single transsexual woman urban business executive will most likely have a different worldview from that of a married heterosexual rural woman secretary who has two teenagers. In another case, a migrant farm worker from the highlands of Guatemala with an undocumented status has a different perspective than an immigrant from Mexico who has lived in New York City for 10 years.

1.7 Ethics across Cultures

As globalization grows and population diversity with nations increases, health-care providers are increasingly confronted with ethical issues related to cultural diversity. At the extremes stand those who favor multiculturalism and postmodernism versus those who favor humanism. Internationally, multiculturalism asserts that no common moral principles are shared by all cultures. Postmodernism asserts a similar claim against all universal standards, both moral and immoral. The concern is that universal standards provide a disguise, whereas dominant cultures destroy or eradicate traditional cultures.

Humanism asserts that all human beings are equal in worth, that they have common resources and problems, and that they are alike in fundamental ways (Macklin, 1999). Humanism does not put aside the many circumstances that make individuals' lives different around the world. Many similarities exist as to what people need to live well. Humanism says that certain human rights should not be violated. Macklin (1998) asserts that universal applicability of moral principles is required, not universal acceptability. Beaucamp (1998) concurs that fundamental principles of morality and human rights allow for cross-cultural judgments of immoral conduct. Of course, there is a middle ground.

Throughout the world, practices are claimed to be cultural, traditional, and beneficial, even when they are exploitative and harmful. For example, female circumcision, a traditional cultural practice, is seen by some as exploiting women. In many cases, the practice is harmful and can even lead to death. Although empirical, anthropological research has shown that different cultures and historical eras contain different moral beliefs and practices, it is far from certain that what is right or wrong can be determined only by the beliefs and practices within a particular culture or subculture. Slavery and apartheid are examples of civil rights violations.

Accordingly, codes of ethics are open to interpretation and are not value-free. Furthermore, ethics belong to the society, not to professional groups. Ethics and ethical decision making are culturally bound. The Western ethical principles of patient autonomy, self-determination, justice, do no harm, truth telling, and promise-keeping are highly valued, but not all cultures—non-Western societies—place such high regard on these values. For example, in Russia, the truth is optional, people are expected to break their promises, and most students cheat on examinations. Cheating on a business deal is not necessarily considered dishonorable (Birch 2006).

In health organizations in the United States, advance directives give patients the opportunity to decide about their care, and staff members are required to ask patients about this upon admission to a health-care facility. Western ethics, with its stress on individualism, asks this question directly of the patient. However, in collectivist societies, such as among some ethnic Chinese and Japanese, the preferred person to ask may be a family member. In addition, translating health forms into other languages can be troublesome because a direct translation can be confusing. For example, "informed consent" may be translated to mean that the person relinquishes his or her right to decision making.

Some cultural situations occur that raise legal issues. For instance, in Western societies, a competent person (or an alternative such as the spouse, if the person is married) is supposed to sign her or his own consent for medical procedures. However, in some cultures, the eldest son is expected to sign consent forms, not the spouse. In this case, both the organization and the family can be satisfied if both the spouse and the son sign the informed consent.

Instead of Western ethics prevailing, some authorities advocate for universal ethics. Each culture has its own definition of what is right or wrong and what is good or bad. Accordingly, some health-care providers encourage international codes of ethics, such as those developed by the International Council of Nurses (2010). These codes are intended to reflect the patient's culture and whether the value is placed on individualism or collectivism. Most Western codes of ethics have interpretative statements based on the Western value of individualism. International codes of ethics do not contain interpretative statements but, rather, let each society interpret them according to its culture. As our multicultural society increases its diversity, health-care providers need to rely upon ethics committees that include members from the cultures they serve.

As the globalization of health-care services increases, providers must also address very crucial issues, such as cultural imperialism, cultural relativism, and cultural imposition. **Cultural imperialism** is the practice of extending the policies and practices of one group (usually the dominant one) to disenfranchised and minority groups. An example is the U.S. government's forced migration of Native American tribes to reservations with individual allotments of lands (instead of group ownership), as well as forced attendance of their children at boarding schools attended by white people. Proponents of cultural imperialism appeal to universal human rights values and standards (Purnell 2001).

Cultural relativism is the belief that the behaviors and practices of people should be judged only from the context of their cultural system. Proponents of cultural relativism argue that issues such as abortion, euthanasia, female circumcision, and physical punishment in child rearing should be accepted as cultural values without judgment from the outside world. Opponents argue that cultural relativism may undermine condemnation of human rights violations, and family violence cannot be justified or excused on a cultural basis (Purnell 2001).

Cultural imposition is the intrusive application of the majority group's cultural view upon individuals and families (Universal Declaration of Human Rights 2001). Prescription of special diets without regard to patients' cultures and limiting visitors to immediate family, a practice of many acute-care facilities, border on cultural imposition (Purnell 2001).

What practices have you seen that might be considered a cultural imposition?

What practices have you seen that might be considered cultural imperialism?

What practices have you seen that might be considered cultural relativism?

What have you done to address them when you have seen them occurring?

Health-care providers must be cautious about forcefully imposing their values regarding genetic testing and counseling. No group is spared from genetic disease. Advances in technology and genetics have found that many diseases, such as Huntington's chorea, have a genetic basis. Some forms of breast and colon cancers, adult-onset diabetes, Alzheimer's disease, and hypertension are some of the newest additions. Currently, only the well-to-do can afford broad testing. Advances in technology will provide the means for access to screening that will challenge genetic testing and counseling. The relationship of genetics to disability, individuals with a disability, and those with a potential disability will create moral dilemmas of new complexity and magnitude.

Many questions surround genetic testing. Should health-care providers encourage genetic testing? What is, or should be, done with the results? How do we approach testing for genes that lead to disease or disability? How do we maximize health and well-being without creating a eugenic devaluation of those who have a disability? Should employers and third-party payers be allowed to discriminate based on genetic potential for illness? What is the purpose of prenatal screening and genetic testing? What are the assumptions for state-mandated testing programs? Should parents and individuals be allowed to "opt out" of testing? What if the individual does not want to know the results? What if the results could have a deleterious outcome to the infant or the mother? What if the results got into the hands of insurance companies that then denied payment or refused to provide coverage? Should public policy support genetic testing, which may improve health and health care for the masses of society? Should multiple births from fertility drugs be restricted because of the burden of cost, education, and health of the family? Should public policy encourage limiting family size in the contexts of the mother's health, religious and personal preferences, and the availability of sufficient natural resources (such as water and food) for future survival? What effect do these issues have on a nation with an aging population, a decrease in family size, and decreases in the numbers and percentages of younger people? What effect will these issues have on the ability of countries to provide health care for their citizens? Health-care providers must understand these three concepts and the ethical issues involved because they will increasingly encounter situations in which they must balance the patient's cultural practices and behaviors with health promotion and wellness, as well as illness, disease, and injury prevention activities for the good of the patient, the family, and society. Other international issues that may be less controversial include sustainable environments, pacification, and poverty (Purnell 2001; Centers for Disease Control 1985; National Standards of Cultural and Linguistic Services 2007).

References

Agency for Healthcare Research and Quality (AHRQ) (2018) AHRQ national healthcare quality and disparities report. Content last reviewed October 2019. Agency for Healthcare Research and Quality, Rockville. https://www.ahrq.gov/research/findings/nhqrdr/nhqdr18/index.html

American Anthropological Association (1998) Statement on race: position paper. http://www.aaanet.org/stmts/racepp.htm

Beaucamp T (1998) The mettle of moral fundamentalism: a reply to Robert baker. Kennedy Inst Ethics J 8(4):389–401

Birch D (2006, October 27) In Russia, the truth is optional. Baltimore Sun, pp 2F, 6F

Centers for Disease Control (1985) Perspectives on disease prevention and health promotion. www.cdc.gov/mmwr/preview/mmwrthml/00000688.htm

CIA World Factbook (2020) World. https://www.cia.gov/library/publications/the-world-factbook/geos/xx.html

Clarke L, Hofsess L (1998). Acculturation) In: Loue S (ed) Handbook of immigrant health. Plenum Press, New York, pp 37–59

Cole NL (2020) So what is culture, exactly? https://www.thoughtco.com/culture-definition-4135409

Council of New Zealand (2005) Guidelines for cultural safety: the treaty of Waitangi and Maori health. http://www.nursingcouncil.org.nz
Fisher TL, Burnet DL, Huang ES, Chin TL, Cagney KA (2007) Cultural leverage: interventions using culture to narrow racial disparities in health care. Med Care Res Rev (5 Suppl):64, 243S–282S
Giger J, Davidhizar R, Purnell L, Taylor Harden J, Phillips J, Strickland O (2007) American Academy of Nursing expert panel Report: developing cultural competence to eliminate health disparities in ethnic minorities and other vulnerable populations. J Transcult Nurs 18(2):95–102
Healthy People (2010) 2001 Companion document for lesbian, gay, bisexual, and transgender health. http://www.lgbthealth.net
Healthy People (2020). www.healthypeople2020.gov
Hofstede G (1991) Cultures and organizations: software of the mind. McGraw-Hill, New York
Hofstede G (2001) Culture's consequences: comparing values, behaviors, institutions, and organizations across nations, 2nd edn. Sage, Thousand Oaks
Hofstede G, Hofstede J (2005) Cultures and organizations: software of the mind, 2nd edn. McGraw-Hill, New York
Home Office Study (2002). http://news.google.com/news?q=Home+Office+Study++United+Kingdom&hl=en&lr=&sa=Xoi=news&ct=title
Human Genome Project (2003). http://www.ornl.gov/sci/techresources/Human_Genome/home.shtml
International Council of Nurses (2010) Ethics. http://www.icn.ch/publications/ethics/
Joint Commission (2010) Effective communication, cultural competence and family centered care: a roadmap for hospitals
Judge K, Platt S, Costongs C, Jurczak K (2006) Health inequalities: a challenge for Europe. http://eurohealthnet.eu/sites/eurohealthnet.eu/files/publications/pu_2.pdf
Khatib M, Hadid S (2019) Developing cultural competence as part of nursing studies: language, customs and health issues. Int J Stud Nurs 4(1):63–71
Koffman J (2006) Transcultural and ethical issues at the end of life. In: Cooper J (ed) Stepping into palliative care. Radcliffe Publishing Ltd, Abington, pp 171–186
Leininger M, McFarland M (2006) Culture care diversity and universality: a worldwide theory. Jones and Bartlett, Burlington
Macklin R (1998) A defense of fundamental principles and human rights: a reply to Robert baker. Kennedy Inst Ethics J 8(4):389–401
Macklin R (1999) Against relativism: cultural diversity and the search for ethical universals in medicine. Oxford University Press, New York
Merriam Webster Online Dictionary (2020) Counterculture. http://www.merriam-webster.com/dictionary/counterculture
National Research Council (1998). www.nationalacademies.org/news.nsf/isbn/0309063566?OpenDocument
National Standards of Cultural and Linguistic Services (2007). http://minorityhealth.hhs.gov/templates/browse.aspx?lvl=2&lvlID=15
Portes A (2007) Migration, development and segmented assimilation: a conceptual review of evidence. Ann Am Acad Polit Soc Sci 610:73–97
Purnell L (2001) Cultural competence in a changing health-care environment. In: Chaska NL (ed) The nursing profession: tomorrow and beyond. Sage Publications, Thousand Oaks, pp 451–461
Purnell L (2010) Cultural rituals in health and nursing care. In: Esterhuizen P, Kuckert A (eds) Diversiteit in de verpleeg-kunde [Diversity in nursing]. Bohn Stafleu van Loghum, Amsterdam, pp 130–196
Trevalon M, Murray-Garcia J (1998) Cultural humility versus cultural competence. J Health Care Poor Underserved 9(2):117–125
U.N. Migration Report (2017). https://www.un.org/en/development/desa/population/migration/publications/migrationreport/docs/MigrationReport2017_Highlights.pdf
U.N. Refugee Agency (2009). http://www.unhcr.org
U.S. Census Bureau (2020) Demographic profiles. 2020.census.gov/news/press-kits/demographic-profiles.html
Universal Declaration of Human Rights (2001). www.amnesty.org

2 The Purnell Model and Theory for Cultural Competence

Larry D. Purnell

2.1 Introduction

Purnell model for cultural competence has been classified as holographic complexity grand theory because it is applicable to all health professionals, has a graphic display of the model, and an extensive assessment framework. The model has been translated into Arabic, Czechoslovakian, Danish, French, German, Korean, Portuguese, Spanish, and Turkish. The assumptions upon which the model is based. The model provides a comprehensive and systematic framework for learning and understanding culture and for use in the clinical practice setting. The empirical framework of the model provides a basis for health-care providers, educators, researchers, managers, and administrators in all health disciplines to provide holistic, culturally competent, therapeutic interventions; health promotion and wellness; illness, disease, and injury prevention; health maintenance and restoration; and health teaching across educational and practice settings. Reflective exercises for the health-care provider are included in each section of the organizing framework.

L. D. Purnell (✉)
University of Delaware, Newark, DE, USA
e-mail: lpurnell@udel.edu

2.1.1 The Purposes of This Model Are the Following

- Provide a framework for all health-care providers to learn concepts and characteristics of culture.
- Define circumstances that affect a person's cultural worldview in the context of historical perspectives.
- Provide a model that links the most central relationships of culture.
- Interrelate characteristics of culture to promote congruence and to facilitate the delivery of consciously sensitive and competent health care.
- Provide a framework that reflects human characteristics such as motivation, intentionality, and meaning.
- Provide a structure for assessing and analyzing cultural data.
- View the individual, family, or group within their unique ethnocultural environment.

2.1.2 The Major Explicit Assumptions upon Which the Model Is Based Are as Follows

- All health-care professions need similar information about cultural diversity and culturally congruent care.

© Springer Nature Switzerland AG 2021
L. D. Purnell, E. A. Fenkl (eds.), *Textbook for Transcultural Health Care: A Population Approach*,
https://doi.org/10.1007/978-3-030-51399-3_2

- All health-care professions share the metaparadigm concepts of global society, family, person, and health.
- One culture is not better than another culture; they are just different.
- Core similarities are shared by all cultures.
- Differences exist within, between, and among cultures.
- Cultures change slowly over time.
- The variant cultural characteristics (see Fig. 2.1) determine the degree to which one varies from the dominant culture and should be considered to prevent stereotyping.
- If patients are coparticipants in their care and have a choice in health-related goals, plans, and interventions, their adherence and health outcomes will be improved.
- Culture has a powerful influence on one's interpretation of and responses to health care.
- Individuals and families may belong to several subcultures.
- Each individual has the right to be respected for his or her culture heritage and individuality.
- Caregivers need both culture-general and population specific information in order to provide culturally competent and congruent care.
- Caregivers who can assess, plan, intervene, and evaluate in a culturally congruent manner will improve the care of patients for whom they care.
- Learning culture is an ongoing process that increases by working with diverse encounters.
- Prejudices, biases, and stereotyping can be minimized with cultural understanding.
- To be effective, health care must reflect the unique understanding of the values, beliefs, attitudes, lifeways, and worldviews of diverse populations and individual acculturation patterns.
- Differences in race, ethnicity, and culture often require adaptations to standard interventions.
- Cultural awareness improves the caregiver's self-awareness.
- Professions, organizations, and associations have their own culture, which can be analyzed using a grand theory of culture.
- Every patient contact is a cultural encounter.

2.2 Overview of the Theory, the Model, and the Organizing Framework

The model is a circle: the outer rim represents global society, the second rim represents community, the third rim represents family, and the inner rim represents the person (Fig. 2.1). The interior of the circle is divided into 12 pie-shaped wedges depicting cultural domains and their concepts. The dark center of the circle represents unknown phenomena. Along the bottom of the model, a jagged line represents the nonlinear concept of cultural consciousness. The 12 cultural domains (constructs) provide the organizing framework of the model. Following the discussion of each domain, a table provides statements that can be adapted as a guide for assessing patients in various settings. Accordingly, health-care providers can use these same questions to better understand their own cultural beliefs, attitudes, values, practices, and behaviors.

2.2.1 Macroaspects of the Model

The macro aspects of this interactional model include the metaparadigm concepts of a global society, community, family, person, and conscious competence. The theory and model are conceptualized from biology, anthropology, sociology, economics, geography, history, ecology, physiology, psychology, political science, pharmacology, and nutrition, as well as theories from communication, family development, and social support. The model can be used in clinical practice, education, research, and the administration and management of health-care services or to analyze organizational culture.

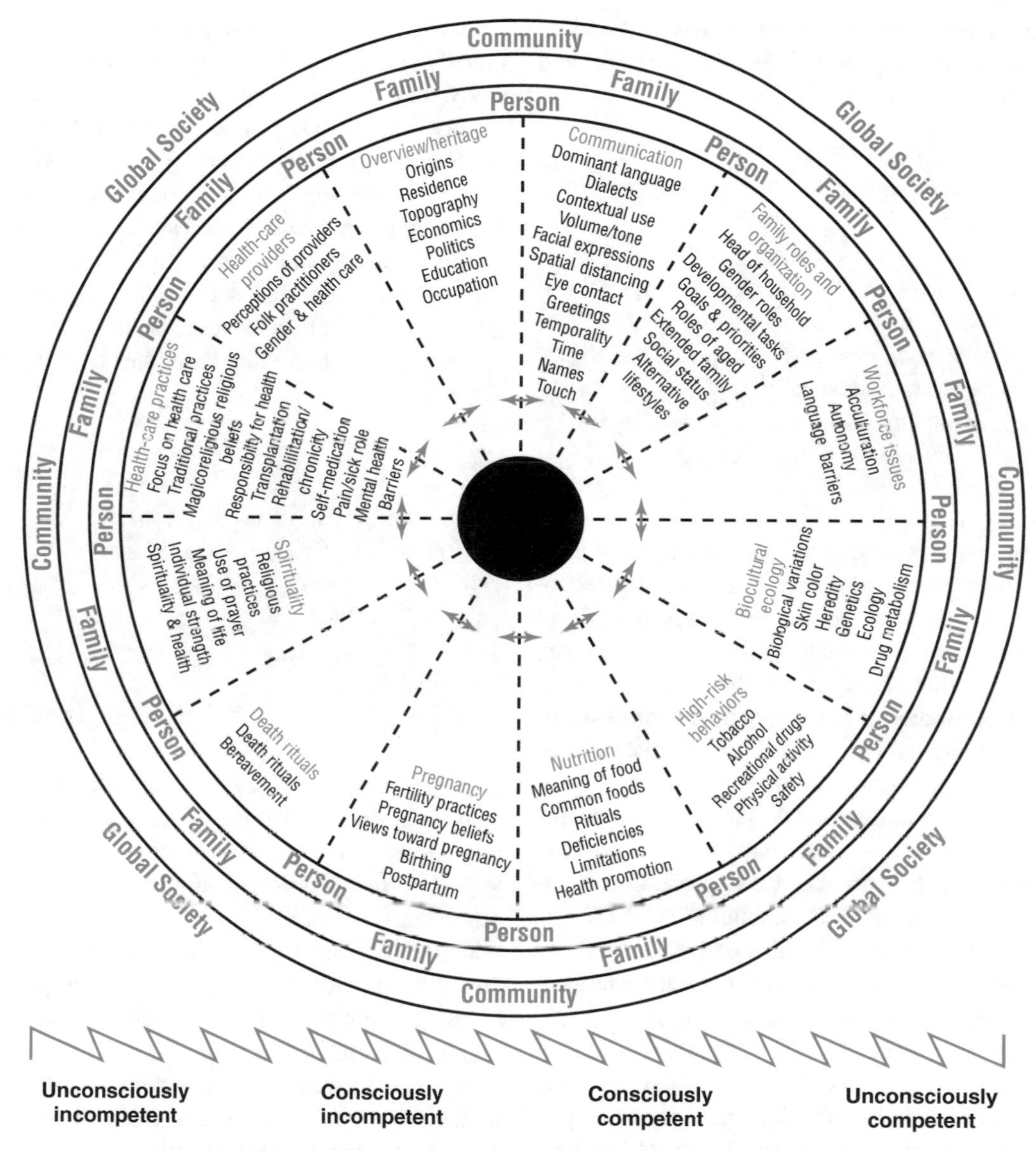

Fig. 2.1 The Purnell model for cultural competence. Variant cultural characteristics: age, generation, nationality, race, color, gender, religion, educational status, socioeconomic status, occupation, military status, political beliefs, urban versus rural residence, enclave identity, marital status, parental status, physical characteristics, sexual orientation, gender issues, and reason for migration (sojourner, immigrant, undocumented status). Unconsciously incompetent: not being aware that one is lacking knowledge about another culture. Consciously incompetent: being aware that one is lacking knowledge about another culture. Consciously competent: learning about the client's culture, verifying generalizations about the client's culture, and providing culturally specific interventions. Unconsciously competent: automatically providing culturally congruent care to clients of diverse cultures

Phenomena related to a global society include world communication and politics; conflicts and warfare; natural disasters and famines; international exchanges in education, business, commerce, and information technology; advances in health science; space exploration; and the expanded opportunities for people to travel around the world and interact with diverse societies. Global events that are widely disseminated by television, radio, satellite transmission, newsprint, and information technology affect all societies, either directly or indirectly. Such events may create chaos while consciously and unconsciously forcing people to alter their lifeways and worldviews.

Reflective Exercise
Think of a recent event that has affected global society, such as a conflict or war, health advances in technology, possible environmental exposure causing health problems, or volcanic eruptions. How did you become aware of this event? How has this event altered your views and other people's views of worldwide cultures?

In the broadest definition, community is a group of people who have a common interest or identity that goes beyond the physical environment. Community includes the physical, social, and symbolic characteristics that cause people to connect. Bodies of water, mountains, rural versus urban living, and even railroad tracks help people define their physical concept of community. Today, however, technology and the Internet allow people to expand their community beyond physical boundaries through social and professional networking sites. Economics, religion, politics, age, generation, and marital status delineate the social concepts of community. Symbolic characteristics of a community include sharing a specific language or dialect, lifestyle, history, dress, art, or musical interest. People actively and passively interact with the community, necessitating adaptation and assimilation for equilibrium and homeostasis in their worldview. Individuals may willingly change their physical, social, and symbolic community when it no longer meets their needs.

Reflective Exercise
1. How do you define your community in terms of objective and subjective cultural characteristics?
2. How has your community changed over the last 5–10 years? The last 15 years? The last 20 years?
3. If you have changed communities, think of the community in which you were raised.

A family is two or more people who are emotionally connected. They may, but do not necessarily, live in close proximity to one another. Family may include physically and emotionally close and distant consanguineous relatives, as well as physically and emotionally connected and distant non–blood-related significant others. Family structure and roles change according to age, generation, marital status, relocation, immigration status, and socioeconomic status, requiring each person to rethink individual beliefs and lifeways.

Reflective Exercise
1. Whom do you consider family?
2. Are they all blood related?
3. How have they influenced your culture and worldview?
4. Who else has helped instill your cultural values?

A person is a biopsychosociocultural being who is constantly adapting to her or his community and environment. Human beings adapt biologically and physiologically with the aging process; psychologically in the context of social relationships, stress, and relaxation; socially as they interact with the changing community; and culturally within the broader global society. In highly individualistic cultures (see Chap. 1), a person is a separate physical and unique psychological being and a singular member of society. The self is separate from others.

However, in highly collectivistic cultures, the individual is defined in relation to the family or other group rather than a basic unit of nature.

Reflective Exercise
In what ways have you adapted (a) biologically and physiologically to the aging process, (b) psychologically in the context of social relationships, (c) socially in your community, and (d) culturally within the broader society and culture?

Health, as used in this book, is a state of wellness as defined by the individual within his or her cultural group. Health generally includes physical, mental, and spiritual states because group members interact with the family, community, and global society. The concept of health, which permeates all metaparadigm concepts of culture, is defined globally, nationally, regionally, locally, and individually. Thus, people can speak about their personal health status or the health status of the community, state, or nation. Health can also be subjective or objective in nature.

Reflective Exercise

1. How do you define health? Is health the absence of illness, disease, injury, and/or disability?
2. How does your profession define health? How does your community or nation define health?
3. How do these definitions compare with your original cultural heritage?

On a micro level, the model's organizing framework consists of 12 domains and their concepts, which are common to all cultures. These 12 domains are interconnected and have implications for health and well-being. The utility of this organizing framework can be used in any setting and applied to a broad range of empirical experiences and can foster inductive and deductive reasoning in the assessment of cultural domains. Once cultural data are analyzed, the practitioner can fully adopt, modify, or reject health-care interventions and treatment regimens in a manner that respects patients' cultural differences. Such adaptations improve the quality of the patients' health-care experiences and personal existence.

2.2.2 The Twelve Domains of Culture

These are the 12 domains that are essential for assessing the ethnocultural attributes of an individual, family, or group:

1. Overview, inhabited localities, and topography
2. Communication
3. Family roles and organization
4. Workforce issues
5. Biocultural ecology
6. High-risk behaviors
7. Nutrition
8. Pregnancy and childbearing practices
9. Death rituals
10. Spirituality
11. Health-care practices
12. Health-care providers

2.3 Overview, Inhabited Localities and Typography

This domain includes concepts related to the country of origin, the current residence, and the effects of the topography of both the country of origin and the current residence on health, economics, politics, reasons for migration, educational status, and occupations. Learning about a culture includes becoming familiar with the heritage of its people and understanding how discrimination, prejudice, and oppression influence value systems and beliefs used in everyday life.

2.3.1 Heritage and Residence

Heritage and residence include not only the country of origin and ancestry, but other places where the person has lived. For example, one's ancestry might be German as the country of origin but born in the United States and lived or worked in Asia or Central American, where that person might have been exposed to tropical diseases unknown in the United States or Germany. Likewise, the topography and physical environment of one's residence may increase one's chances of being inflicted with and leave with an infection. There is a difference between a disease and an illness and we have both here. Illnesses such as malaria from swampy areas, asthma from polluted inner-city environments, or cancer if exposed to radioactive fallout. Regardless of one's environment and lifestyle, one's heritage may be an increased risk for genetic and hereditary diseases that are common among the Amish and Ashkenazi Jews (see specific chapters on these cultural groups).

One's occupation can also have deleterious effects on health if exposed to asbestos, working in farming with pesticides, or in textile factories with increased risk for respiratory, eye, and ear infections. A complete health history may be required because people might have worked in several occupations over a lifetime.

2.3.2 Reasons for Migration and Associated Economic Factors

The social, economic, religious, and political forces of the country of origin play an important role in the development of the ideologies and the worldview of individuals, families, and groups and are often a major motivating force for emigration. People emigrate for better economic opportunities; because of religious or political oppression, and ethnic cleansing; as a result of environmental disasters, such as earthquakes and hurricanes in their home countries; and by forced relocation, such as with slaves, indentured servants, and human trafficking. Others have emigrated for educational opportunities and personal ideologies or a combination of factors. Most people emigrate in the hope of a better life, but the individual or group personally defines this ideology. A common practice for many immigrants is to relocate to an area that has an established population with similar ideologies that can provide initial support, serve as cultural brokers, and orient them to their new culture and health-care system. When immigrants settle and work exclusively in predominantly ethnic communities, primary social support is enhanced. However, acculturation and assimilation into the wider society may be hindered. Although ethnic enclaves to an extent assist with acculturation, they may need extra help in adjusting to their new homeland's language as well as securing access to health-care services, living accommodations, and employment opportunities. Further, people who move voluntarily are likely to experience less difficulty with acculturation than people who are forced to migrate. Some individuals migrate with the intention of remaining in the selected country only a short time, making money, continuing their education, and returning home. Others migrate with the intention of relocating permanently.

Reflective Exercise

1. What is your cultural heritage?
2. How might you find out more about it?
3. How does your cultural heritage influence your current beliefs and values about health and wellness?
4. What brought you/your ancestors to your current country of residence?
5. Why did you/your ancestors emigrate?

2.3.3 Educational Status and Occupations

The value placed on formal education differs among cultural and ethnic groups and is often related to their socioeconomic status in their homeland and their abilities and reasons for emigrating. Some people place a high value on formal education; however, some do not stress formal education because it is not needed for employment in their homeland. Consequently, they may become engulfed in poverty, isolation, and enclave identity, which may further limit their potential for formal educational opportunities and planning for the future. In regard to learning styles, the Western system places a high value on the ability to categorize information using linear, sequential thought processes (West et al. 2007).

Reflective Exercise

1. How strongly do you believe in the value of education?
2. Who in your life has been responsible for instilling this value?
3. Do you consider yourself to be a more linear/sequential learner or a random-patterned learner?

However, not everyone adheres to this pattern of thinking. Others have spiral and circular thought patterns that move from concept to concept without being linear or sequential; therefore, they may have difficulty placing information in a stepwise methodology, which is common in individualistic cultures. When someone is unaware of the value given to such behaviors, the person may seem disorganized, scattered, and faulty in their cognitive patterns, resulting in increased difficulty with written and verbal communications.

Some educational systems stress application of content over theory. Most European educational programs emphasize theory over practical application, and Arab education emphasizes theory with little attention given to practical application. As a result, Arab students are more proficient at tests requiring rote learning than at those requiring conceptualization and analysis. Being familiar with the individual's personal educational values and learning modes allows health-care providers, educators, and employers to adjust teaching strategies for patients, students, and employees. Educational materials and explanations must be presented at a level consistent with the patient's educational capabilities and within their cultural framework and beliefs.

Immigrants bring job skills from their homelands and traditionally seek employment in the same or similar trades. Sometimes, these job skills are inadequate for the available jobs in the new society; thus, immigrants are forced to take low-paying jobs and join the ranks of the working poor and economically disadvantaged. Immigrants may be employed in a broad variety of occupations and professions; however, limited experiential, educational, and language abilities of more recent immigrants often restrict employment possibilities. More importantly, experiential backgrounds sometimes encourage employment choices that are identified as high risk for chronic diseases, such as exposure to pesticides and chemicals. Others may work in factories that manufacture hepatotoxic chemicals, in industries with pollutants that increase the risk for pulmonary diseases, and in crowded conditions with poor ventilation that increase the risk for tuberculosis or other respiratory diseases.

Understanding patients' current and previous work background is essential for health screening. For example, newer immigrants who worked in malaria-infested areas in their native country, such as Egypt, Italy, Turkey, and Vietnam may need health screening for insect borne diseases. Those who worked in mining, such as in Ireland, Poland, and parts of the United States may need screening for respiratory diseases. Those who lived in overcrowded and unsanitary conditions, such as with refugees and migrant workers, may need to be screened for such infectious diseases as tuberculosis, parasitosis, and respiratory diseases (see Table 2.1 Section on Overview/Heritage).

2.4 Communication

Perhaps no other domain has the complexities as that of communication. Communication is interrelated with all other domains and depends on verbal language skills that include the dominant language, dialects, and contextual use of the language, as well as paralanguage variations such as voice volume, tone, intonations, reflections, and willingness to share thoughts and feelings. Other important communication characteristics include nonverbal communications, such as eye contact, facial expressions, use of touch, body language, spatial distancing practices, and acceptable greetings; temporality in terms of past, present, or future orientation of worldview; clock versus social time; and the name format and the degree of formality in the use of names. Communication styles may vary among insiders (family and close friends) and outsiders (strangers and unknown health-care providers). Hierarchical relationships, gender, and some religious beliefs affect communication.

2.4.1 Dominant Language and Dialects

The health-care provider must be aware of the dominant language and the difficulties that dialects may cause when communicating in a

Table 2.1 Organizing framework and cultural assessment guide

Cultural assessment tool	Sample rationale/example
Overview, inhabited localities, and topography	
Where do you currently live?	Someone living in a wooded area with deer and other wildlife are at an increased risk for Lyme disease
What is your ancestry?	Ashkenazi Jewish population has a high incidence of genetic and hereditary health conditions
Where were you born?	Immigrants from Eastern Europe near Chernobyl have an increased risk for genetic mutations and hereditary defects related to radioactive contamination
How many years have you lived in the United States (or other country, as appropriate)?	Length of time away from the home country may determine the degree of assimilation and acculturation
Were your parents born in the United States (or other country, as appropriate)?	Generation may determine the degree of assimilation and acculturation
What brought you (your parents/ancestors) to the United States (or other country, as appropriate)?	The reason for migration may determine the degree of assimilation and acculturation
Describe the land or countryside where you live. Is it mountainous, swampy, etc.?	The physical environment has a host of infectious diseases
What was the land or countryside like when you lived there?	People immigrating from or who have recently visited parts of central American may be at increased for and need to be assessed for arthropod-borne diseases
What is your income level?	Income level has implications for affording medications, dressings, and prescriptive devices
Does your income allow you to afford the essentials of life?	Determines the ability to afford healthy diets
Do you have health insurance?	Refer to social services or other source for financial support if no insurance
Are you able to afford health insurance on your salary?	The working poor cannot afford health insurance
What is your educational level (formal/informal/self-taught)?	Educational level may determine ability to understand health prescriptions
What is your current occupation? If retired, ask about previous occupations.	People working in home remodeling may be at risk for asbestosis
Have you worked in other occupations? What were they?	A person may currently be retired or may now work as a salesperson but previously worked as a coal miner, increasing the risk of black lung disease
Are there (were there) any particular health hazards *associated* with your job(s)	Previous employment may have health concerns long-term
Have you been in the military? If so, in what foreign countries were you stationed?	People who served in the military may suffer from post-traumatic stress syndrome or diseases contracted in their military experiences
Communications	
What is your full name?	Part of a standard assessment
What is your legal name?	Part of a standard assessment and medical record keeping
By what name do you wish to be called?	Most people respond better if called by a preferred name
What is your primary language?	Determining the preferred language for consent forms and discharge instructions
Do you speak a specific dialect?	A dialect-specific interpreter is preferred
What other languages do you speak?	Sometimes a second or third language may be helpful for interpretation if a preferred language interpreter is not available
Do you find it difficult to share your thoughts, feelings, and ideas with family? Friends? Health-care providers?	Additional time may be needed to establish trust and get full disclosure, especially with sensitive topics
Do you mind being touched by friends? Strangers? Health-care workers?	Asking permission and explaining the rationale before touching reinforces the trust relationship
How do wish to be greeted? Handshake? Nod of the head, etc.?	Demonstrates respect and helps establish trust

Table 2.1 (continued)

Cultural assessment tool	Sample rationale/example
Are you usually on time for appointments?	Explain rationale for the expectation of timeliness: Might not be seen and have to reschedule and may still be charged for the appointment
Are you usually on time for social engagements?	Ask only if question is pertinent
Observe the patient's speech pattern. Is the speech pattern high- or low-context? Remember, patients from highly contexted cultures place greater value on silence	Patients from high-context cultures place greater value on silence and implicit communication and may take more time to give a response
Observe the patient when physical contact is made. Does he/she withdraw from the touch or become tense?	Being aware of the patient's level of comfort helps establish trust. Reinforce the necessity and ask permission before touching
How close does the patient stand when talking with family members? With health-care providers?	Spatial distancing is culture bound. Do not take offense if a patient stands closer or farther away than what you are accustomed to
Does the patient maintain eye contact when talking with the nurse/physician/etc.?	Some avoid eye contact with people in hierarchal positions as a sign of respect. The health-care provider is in a hierarchal position
Family roles and organization	
What is your *marital*/partner status?	Part of a standard assessment
How many children do you have?	Part of a standard assessment
Who makes most of the decisions in your family?	If the decision maker is not accessed, no decision will be made and time will be wasted
What types of decisions do(es) the female(s) in your family make?	In many traditional families, the female usually makes decision about the household and child care
What types of decisions do(es) the male(s) in your family make?	In many traditional families, the male usually makes decisions about affairs outside the household, but not always
What are the duties of the women in the family?	Determining the division of labor can become important when illness occurs
What are the duties of the men in the family?	Determining the division of labor can become important when illness occurs
What should children do to make a good impression for themselves and for the family?	A child's behavior in the Appalachian and Greek cultures can bring shame upon the family
What should children not do to make a good impression for themselves and for the family?	Among traditional Koreans, children are expected to do well in school, or shame may come to the family
What are children forbidden to do?	Among traditional Germans, taboo behaviors include talking back to elders and touching another person's possessions
What should adolescents do to make a good impression for themselves and for the family?	Among Somalis, young adults are expected to marry and assist older family members
What should adolescents not do to make a good impression for themselves and for the family?	Among traditional Mexican families, young adults should not dress in a provocative manner; otherwise, shame can come to them or their family
What are adolescents forbidden to do?	A taboo behavior for young female adults in Haiti is engaging in sexual activity before marriage
What are the priorities for your family?	For lower socioeconomic families, the priority may be having adequate food and shelter, with stress on the present
What are the roles of the older people in your family? Are they sought for their advice?	Among traditional Turks, no decision is made until after seeking the advice of older adults
Are there extended family members in your household? Who else lives in your household?	Most traditional Asian cultures live in extended family arrangements in their home country
What are the roles of extended family members in this household? What gives you and your family status?	Extended family members provide significant financial and social support and are important sources for child care

(continued)

Table 2.1 (continued)

Cultural assessment tool	Sample rationale/example
Is it acceptable to you for people to live together and not be married?	Among traditional Arab families, shame may occur if a pregnancy occurs outside of marriage
Is it acceptable to you for people to admit being gay or lesbian?	Among many Asian cultures, a man and woman living together without being married may cause them to be rejected by their family
What is your sexual preference/orientation? (if appropriate, and then later in the assessment after a modicum of trust has been established)	Not all cultures or individuals are accepting of gay, lesbian, or gender reassignment individuals
Workforce issues	
Do you usually report to work on time?	Not all cultures espouse timeliness in reporting to work. If timeliness is important, this must be explicitly explained along with consequences
Do you usually report to meetings on time?	If timeliness is important for attendance at meetings, this must be explicitly explained
What concerns do you have about working with someone of the opposite gender?	Strict orthodox separation of the sexes may cause disharmony if men and women are expected to work in close proximity
Do you consider yourself a "loyal" employee? How long do you expect to remain in your position?	Among the Japanese, an employer may expect absolute loyalty, and employees remain with the same company their entire lives
What do you do when you do not know how to do something related to your job?	Among many traditional Koreans and Filipinos, when an employee does not know how to do something, rather than admitting it, they may go to a co-worker of the same nationality (if available)
Do you consider yourself to be assertive in your job?	Traditional Asians are sometimes not seen as assertive as some American employers would like. Most professionals are assertive, but in a different way from assertiveness in individualistic cultures
What difficulty does English (or another language) give you in the workforce?	Low verbal and written literacy may have implications for accuracy in fulfilling job requirements
What difficulties do you have working with people older (younger) than you?	Some professionals and some non-professionals are hesitant to follow directions from someone younger than they are
What difficulty do you have in taking directions from someone younger/older than you?	Some professionals and some non-professionals are hesitant to follow directions from someone younger than they are
What difficulty do you have working with people whose religions are different from yours?	Some professionals and some non-professionals prefer to only work with people from their religion
What difficulty do you have working with people whose sexual orientation is different from yours?	Some professionals and some non-professionals do not like to work with people whose sexual orientation is different from theirs
What difficulty do you have working with someone whose race or ethnicity is different from yours?	Some professionals and some non-professionals do not like to work with people whose race or ethnicity is different from theirs
Do you consider yourself to be an independent decision maker?	Independent think is required in many positions
Biocultural ecology	
Are you allergic to any medications?	Part of a standard assessment
What problems did you have when you took over-the-counter medications?	Part of a standard assessment
What problems did you have when you took prescription medications?	Part of a standard assessment
What are the major illnesses and diseases in your family?	Part of a standard assessment

Table 2.1 (continued)

Cultural assessment tool	Sample rationale/example
Are you aware of any genetic diseases in your family?	Part of a standard assessment
What are the major health problems in the country from which you come (if appropriate)?	Migrants may bring new infections from their home country. Should be part of a standard assessment
With what race do you identify?	Part of a standard assessment
Observe skin coloration and physical characteristics	To assess for rashes on people with dark skin, the health-care provider may need to palpate rather than rely on visual cues
Observe for and document physical handicaps and disabilities	Many people do not disclose handicaps or disabilities upon initial encounter unless specifically asked, especially learning disabilities
High-risk health behaviors	
How many cigarettes a day do you smoke?	Part of a standard assessment
Do you smoke a pipe (or cigars)?	Part of a standard assessment
Do you chew tobacco?	Part of a standard assessment
For how many years have you smoked/chewed tobacco?	Part of a standard assessment
How much do you drink each day? Ask about wine, beer, spirits?	Part of a standard assessment
How many energy drinks do you consume each day?	Part of a standard assessment
What recreational drugs do you use?	Recreational drug use is part of a standard assessment. In order for the patient to disclose this sensitive information, ask in a non-judgmental manner
How often do you use recreational drugs?	Part of a standard assessment
What type of exercise do you do each day?	Part of a standard assessment. Physical activity or lack thereof is part of a standard assessment for health promotion and wellness
Do you use seat belts?	Part of a standard assessment
What precautions do you take to prevent getting a sexually transmitted infections or HIV/AIDS?	Part of a standard assessment
Nutrition	
Are you on a special diet?	Part of a standard assessment
Are you satisfied with your weight?	Not all cultures adhere to or believe in the U.S. weight recommendations
Which foods do you eat to maintain your health?	Food choices are seen as a means for promoting health
Do you avoid certain foods to maintain your health?	Each culture has certain foods people avoid for maintaining their health. E.g.
Why do you avoid these foods?	Kosher Jews do not eat pork or pork products
Which foods do you eat when you are ill?	Common foods eaten when ill among many cultures include toast and tea or ginger ale when ill
Which foods do you avoid when you are ill?	If the health-care provider recommends a food that the person culturally or personally avoids, it may not be followed
Why do you avoid these foods (if appropriate)?	This may be a culturally learned practice
For what illnesses do you eat certain foods?	A common practice is to drink a "hot toddy" for a cold or minor illness. The ingredients vary, but generally include tea, lemon or lime, sugar or honey, and some type of alcohol, such as whiskey or rum
Which foods do you eat to balance your diet?	Many cultures adhere to specific foods for balancing a diet; frequently related to opposite qualities of food such as the hot-and-cold theories
Which foods do you eat every day?	Incorporating these foods into dietary prescriptions will increase compliance with dietary instructions

(continued)

Table 2.1 (continued)

Cultural assessment tool	Sample rationale/example
Which foods do you eat every week?	Incorporating these foods into dietary prescriptions will increase compliance with dietary instructions
Which foods do you eat that are part of your cultural heritage?	Including culturally preferred foods into nutritional recommendations increases adherence
Which foods are high-status foods in your family/culture?	High-status foods vary according to cost and availability
Which foods are eaten only by men? Women? Children? Teenagers? Older people?	Among some Guatemalan highland indigenous populations, men primarily eat eggs for the added protein value. The belief is that because men do heavy labor, they need more protein. However, they are supposed to share the protein foods on their plates with children
How many meals do you eat each day?	Many Turks eat 4–6 times a day, but in smaller amounts than most European Americans do
What time do you eat each meal?	May have implications for medication administration
Do you snack between meals?	Snacks can be a significant source of added calories
What foods do you eat when you snack?	Many snacks are not considered healthy food choices
What holidays do you celebrate?	Holidays are a time for special meals and a time when many people overconsume calories
Who usually buys the food in your household?	Many times, it is just as important to talk with the person who purchases the food as it is with the person who prepares the meals. In migrant worker camps, the person who prepares the meals is not the person who purchases food for the group. If one member of the group needs a special diet, such as with a diabetic, the purchaser of the food needs to be included in nutritional education
Who does the cooking in your household?	The person who does the cooking should be included in dietary counseling and education for special diets
Do you have a refrigerator?	For the homeless and those in severe poverty, proper food storage must be taken into consideration
How do you cook your food?	Preparation practices with butter, lard, etc., can add significant calories to meals
How do you prepare meat?	Preparation practices can add significant calories
How do you prepare vegetables?	Preparation practices add significant calories to meals, such as adding butter or bacon fat to vegetables
What do you drink with your meals?	Beverages can add significant calories to meals. Be sure to ask if sugar is added to beverages, including natural juices.
Do you drink special teas?	Teas are used by many people for health promotion and wellness and in times of illness
Do you have any food allergies or intolerances?	Many American Indians and Asians have lactose intolerance
Are there certain foods that cause you problems when you eat them?	Looking for allergies or side effects of specific foods to avoid in dietary counseling
How does your diet change with each season?	For those who live in colder climates, fresh fruits and vegetables may be too expensive in the colder months
Are your food habits different on days you work versus when you are not working?	This is a common practice with working adults
Pregnancy and childbearing practices	
How many children do you have?	Part of a standard OB/GYN assessment
What do you use for birth control?	Each cultural and religious group has acceptable and unacceptable methods of birth control
What special foods do you eat when you are pregnant?	Although there are no specifically prescribed foods for a pregnant Polish woman, she is expected to eat for two
What foods do you avoid when you are pregnant?	Chinese women are reluctant to take iron because they believe it will make delivery more difficult

Table 2.1 (continued)

Cultural assessment tool	Sample rationale/example
What activities do you avoid when you are pregnant?	A belief among many traditional Panamanians is that a pregnant woman should not walk in the moonlight for fear the baby will be born with a cleft lip or palate
Do you do anything special when you are pregnant?	Korean women are expected to work hard during pregnancy to help ensure having a smaller baby
Do you eat non-food substances when you are pregnant?	Eating non-food substances, pica, is common among many cultural groups. One example is clay, which can interfere with iron absorption
Who do you want with you when you deliver your baby?	Some women prefer their mothers or another female family member rather than their husbands
In what position do you want to be when you deliver your baby?	Traditional Indian women in Guatemala prefer to deliver in a squatting position rather than in the supine position. Negotiating for the position during delivery may be necessary in some organizations
What special foods do you eat after delivery?	Hindu women may be restricted to liquids, rice, gruel, and bread
What foods do you avoid after delivery?	Guatemalan women avoid eating spicy foods because the milk will cause irritability in the baby
What activities do you avoid after you deliver?	Russian women should do no strenuous activity after delivery to prevent any complications
Do you do anything special after delivery?	Traditional Japanese women should not wash their hair for several days postpartum
Who will help you with the baby after delivery?	Looking for home support for the mother
What bathing restrictions do you have after you deliver?	Many Egyptian women may be reluctant to bathe postpartum because air may get into the mother and cause illness. However, a sponge bath is acceptable
Do you want to keep the placenta?	Some American Indians bury the placenta outside their home to keep away evil spirits
What do you do to care for the baby's umbilical cord?	A common practice among Mayans is to place a coin or metal object, held on with an abdominal binder, to prevent the umbilicus from protruding when the baby cries
Death rituals	
What special activities need to be performed to prepare for death?	When death is impending, Muslims want the bed to face toward Mecca
Would you want to know about your impending death?	A belief among traditional Somalis is that a person might give up hope if impending death is made known
What is your preferred burial practice? Interment, cremation?	Patient's wishes should be granted
How soon after death does burial occur?	For traditional Jews, burial is before sundown the next day
How do men grieve?	In some cultures, men are expected to be stoical and maintain control of their emotions. Expressions of grief have a wide variation
How do women grieve?	In some cultures, women are expected to be histrionic with their grief to demonstrate their care for the deceased loved one. Expressions of grief have a wide variation
What does death mean to you?	Among Hindus, death means rebirth
Do you believe in an afterlife?	Many Christians believe that there is a better life after death
Are children included in death rituals?	The Amish include children in all aspects of dying and burial
Spirituality	
What is your religion?	Part of a standard assessment

(continued)

Table 2.1 (continued)

Cultural assessment tool	Sample rationale/example
Do you consider yourself deeply religious?	Religion may have more influence than the culture
How many times a day do you pray?	Islam requires prayer five times a day
What do you need in order to say your prayers?	If possible, Muslims need a prayer rug
Do you meditate?	Meditation can be used for relaxation and for pain control
What gives strength and meaning to your life?	For some, the most important thing in their life is family
In what spiritual practices do you engage for your physical and emotional health?	Prayer, meditation, yoga, and quiet time are some examples
Health-care practices	
In what prevention activities do you engage to maintain your health?	A strong value in the dominant European American culture is to have regularly scheduled health check-ups, including self breast examinations, mammograms, and colonoscopies
Who in your family takes responsibility for your health?	Among Arabs, family, not the individual, has the primary responsibility for a person's health-seeking care
What over-the-counter medicines do you use?	All cultural groups and individuals use over-the-counter medication; some use them to the exclusion of prescription medicines
What herbal teas and folk medicines do you use?	Many Hispanic/Latino and other populations, use a wide variety of herbal teas for many health conditions
For what conditions do you use herbal medicines?	Iranians use a variety of berries, leaves, seeds, and dried flowers steeped in hot or cold water and drunk for digestive problems
What do you usually do when you are in pain?	Some African Americans may see pain and suffering as inevitable and something that is to be endured
How do you express your pain?	Among Mexicans, being able to endure pain is seen as a sign of strength
How are people in your culture viewed or treated when they have a mental illness?	Having a mental illness in many Arab cultures is seen as a stigma; therefore, the person with a mental illness may be well cared for but kept hidden from society
How are people with physical disabilities treated in your culture?	The Amish approach disability as a community responsibility, and those with a disability are incorporated into all family and social activities
What do you do when you are sick? Stay in bed, continue your normal activities, etc.?	For many from the European American culture, a belief is "if you are not dead," take something for relief and continue with your daily routines
What are your beliefs about rehabilitation?	Studies demonstrate that for Germans, if rehabilitation is needed to function at maximum capacity, then all rehab exercises are done
How are people with chronic illnesses viewed or treated in your culture?	For most Arabs, if a chronic illness is debilitating, family members readily assume that person's responsibilities
Are you averse to blood transfusions?	Besides a religious prohibition for a Jehovah's witness to receive blood, many people do not want a blood transfusion for fear of contracting HIV/AIDS
Is organ donation acceptable to you?	Jewish law views organ transplants for four perspectives: The recipient, the living donor, the cadaver donor, and the dying donor. Because life is sacred, if the recipient's life can be prolonged without considerable risk, then the transplant is favorably viewed
Are you an organ donor?	Part of a standard assessment
Would you consider having an organ transplant if needed?	Organ donation and transplantation among Muslims are individual decisions
Are health-care services readily available to you?	The health-care provider needs to be aware of access problems for health care and make attempts to improve access
Do you have transportation problems accessing needed health-care services?	Many organizations have vouchers for public transportation

Table 2.1 (continued)

Cultural assessment tool	Sample rationale/example
What traditional health-care practices do you use? Acupuncture, acupressure, *cai gao*, *moxibustion*, aromatherapy, coining, etc.?	If the health-care provider is familiar with traditional practices within the culture, more specific information can be obtained
Health-care practitioners	
What health-care providers do you see when you are ill? Physicians, nurses?	Not all patients see Western allopathic practitioners for illnesses, at least not as first access. Some use Western providers and traditional providers simultaneously
Do you prefer a same-sex health-care provider for routine health problems? For intimate care?	Among orthodox Jewish and Islamic patients, a same-sex provider should be assigned unless it is an emergency
What healers do you use beside physicians and nurses?	If the health-care provider is familiar with the specific culture, better/more pointed questions can be asked. Among many Hispanics/Latinos, folk practitioners are consulted for the evil eye and other conditions
For what conditions do you use healers?	Many American Indians use a variety of traditional healers. Being able to integrate traditional healers with allopathic professionals will increase compliance with recommendations

An extensive cultural assessment is rarely completed in the clinical setting because of time and other circumstances. A seasoned clinical practitioner will know when further assessment is required. Thus, this tool should be used as a guide. Some i*tems are part of any standard assessment. Other items may also be part of a standard assessment, depending on the organization, setting, and clinical area*

patient's native language. For example, English is a monochromic, low-contextual language in which most of the message is in the verbal mode. Verbal communication is frequently seen as being more important than nonverbal communication. Thus, people for whom English is the dominant language are more likely to miss the more subtle nuances of communication. Accordingly, if a misunderstanding occurs, both the sender and the receiver of the message take responsibility for the miscommunication.

Reflective Exercise

1. What is your dominant language?
2. Do you have difficulty understanding other dialects of your dominant language?
3. Have you traveled abroad where you had difficulty understanding the dialect or accent?
4. What other languages beside your dominant language do you speak?

English differs somewhat in its pronunciation, spelling, and choice of words from English spoken in Great Britain, Australia, and other English-speaking countries. Within each country, several dialects can exist, but generally the differences do not cause a major concern with communications. However, accents and dialects within a country, region, or local area can cause misunderstanding; for example, the "Elizabethan English" that is spoken in parts of the United States and the English spoken in Glasgow, Scotland, are both different from the English spoken in Central London. The Spanish spoken in Spain differs from the versions spoken in Puerto Rico, Argentina, Panama, or Mexico, which has as many as 50 different dialects. In such cases, dialects that vary widely may pose substantial problems for health-care providers and interpreters in obtaining accurate health data and increasing the difficulty of making an accurate diagnosis.

When speaking in a non-native language, health-care providers must select words that have relatively pure meanings, be certain of the voice intonation, and avoid the use of regional slang and jargon to avoid being misunderstood. Minor variations in pronunciation may change the entire meaning of a word or a phrase and result in inappropriate interventions.

Given the difficulty of obtaining the precise meaning of words in a language, it is best for health-care providers to obtain someone who can interpret the meaning and message, not just translate the individual words. Remember, translation refers to the written word. Interpretation refers to the spoken word. Children should never be used as interpreters for sensitive topics for their family members. Not only can it have a negative bearing on family dynamics, but sensitive information may not be transmitted. However, when discussing dietary concerns, in family centered care family members can be used (see Chap. 3 for guidelines for using interpreters).

Those with limited language ability may have inadequate vocabulary skills to communicate in situations in which strong or abstract levels of verbal skills are required, such as in the psychiatric setting (Purnell 2011). Helpful communication techniques with diverse patients include displaying tact, consideration, and respect; gaining trust by listening attentively; addressing the patient by preferred name; and showing genuine warmth and openness to facilitate full information sharing. When giving directions, be explicit. Give directions in sequential procedural steps (e.g., first, second, third). Do not use complex sentences with conjunctions or contractions because many languages do not use them.

Reflective Exercise

1. Give some examples of problems communicating with patients who did not speak or understand English or the dominant language.
2. What did you do to promote effective communication?

Before trying to engage in more sensitive areas of the health interview, the health-care provider may need to start with social exchanges to establish trust if time permits, use an open-ended format rather than yes or no closed-response questions, elicit opinions and beliefs about health and symptom management, and focus on facts rather than feelings. An awareness of nonverbal behaviors is essential to establishing a mutually satisfying relationship.

The context within which a language is spoken is an important aspect of communication. Some languages are low in context, and most of the message is explicit, requiring many words to express a thought (English and Spanish, and other romance languages). Other languages are highly contextual (Chinese and Korean), with most of the information either in the physical context or internalized, resulting in the use of fewer words with more emphasis on unspoken understandings.

Reflective Exercise

1. On a scale of 1 to 10, with 1 being low and 10 being high, where do you place yourself in the scale of high-contextual versus low-contextual communication?
2. Do you tend to use a lot of words to express a thought?
3. Do you know family members/friends/acquaintances who are your opposite in terms of low-contextual versus high-contextual communication?
4. Does this sometimes cause concerns in communication?
5. Do you think biomedical language is high or low context?

Voice volume and tone are important paralanguage aspects of communication. A loud voice volume may be interpreted as reflecting anger, when in fact a loud voice is merely being used to express their thoughts in a dynamic manner. Thus, health-care providers must be cautious about voice volume and tones when interacting with diverse cultural groups so that their intentions are not misunderstood.

2.4.2 Cultural Communication Patterns

Communication includes the willingness of individuals to share their thoughts and feelings. Some cultures encourage people to disclose very personal information about themselves, such as information about sex, drugs, and family problems that is more common in individualistic cultures. In some cultures, having well-developed verbal skills is seen as important, whereas in other cultures, the person who has very highly developed verbal skills is seen as having suspicious intentions. Some cultures willingly share their thoughts and feelings among family members and close friends, but they may not easily share thoughts, feelings, and health information with "outsiders" (i.e., health-care providers) until they get to know and trust them that is more common with collectivistic cultures. If the situation permits, engaging in small talk and inquiring about family members before addressing the patient's health concerns, health-care providers can help establish trust and, in turn, encourage more open communication and sharing of important health information.

Touch, a method of nonverbal communication, has substantial variations in meaning among cultures. For the most part, individualistic cultures are low-touch cultures, which are reinforced by sexual harassment guidelines and policies.

Reflective Exercise

1. How willing are you to share personal information about yourself?
2. How does it differ with family, friends, or strangers?
3. Do you tend to speak faster, slower, or about the same rate as the people around you?
4. What happens when you meet someone who speaks much more rapidly or much more slowly than you do?
5. Do you normally speak in a loud or low voice volume?
6. How do you respond when someone speaks louder or softer than you do?

For many, even casual touching may be seen as a sexual overture and should be avoided whenever possible. In individualistic cultures, people of the same sex (especially men) or opposite sex do not generally touch each other unless they are family members of close friends. It is recognized that the low-touch individualistic culture has variations according to age and location. However, among most collectivist cultures, two people of the same gender can touch each other without it having a sexual connotation, although modesty remains important. Always explain the necessity and ask permission before touching a patient for a health examination. Being aware of individual practices regarding touch is essential for effective health assessments.

Reflective Exercise

1. How comfortable are you with being touched on the arm or shoulder by friends? By people who know you well?
2. Do you consider yourself to be a "person who touches frequently" or do you rarely touch friends?
3. Can you think of groups in the clinical setting for whom therapeutic touch is not appropriate?

Personal space needs to be respected when working with multicultural patients and staff. Among more individualistic cultures, conversants tend to place at least 18 inches of space between themselves and the person with whom they are talking. Most collectivist cultures require less personal space when talking with each other (Anasthasia 2015). They are quite comfortable standing closer to each other than are people from individualistic cultures; in fact, they interpret physical proximity as a valued sign of emotional closeness (Anasthasia 2015). Patients who stand very close and stare during a conversation may offend some health-care practitioners. These patients may interpret health-care providers as being cold because they stand so far away, perhaps appearing as being "standoffish". Thus, an understanding of personal space and distancing characteristics can enhance the quality of communication among individuals.

Regardless of the class or social standing of the conversants, people from individualistic cultures are expected to maintain direct eye contact without staring. A person who does not maintain eye contact may be perceived as not listening, not being trustworthy, not caring, or being less than truthful. Among some traditional collectivist cultures, sustained eye contact can be seen as offensive; further, a person of lower social class or status is expected to avoid eye contact with superiors or those with a higher educational status. Thus, eye contact must be interpreted within its cultural context to optimize relationships and health assessments.

Reflective Exercise

1. What are your spatial distancing practices?
2. How close do you stand to family? Friends? Strangers?
3. Does this distancing remain the same with the opposite gender?
4. Do you maintain eye contact when speaking with people?
5. Is it intense?
6. Does it vary with the age or gender of the person with whom you are conversing?
7. What does it mean when someone does not maintain eye contact with you?
8. How do you feel in this situation?

The use of gestures and facial expressions varies among cultures. Most European Americans gesture moderately when conversing and smile easily as a sign of pleasantness or happiness, although one can smile as a sign of sarcasm. A lack of gesturing can mean that the person is too stiff, too formal, or too polite.

Preferred greetings and acceptable body language also vary among cultural groups. An expected practice for many cultures in business is to extend the right hand when greeting someone for the first time. More elaborate greeting rituals occur in Asian, Arab, and Latin American countries and are covered in individual chapters.

Although many people consider it impolite or offensive to point with one's finger, many do so and do not see it as impolite. In other cultures, beckoning is done by waving the fingers with the palm down, whereas extending the thumb, like thumbs-up, is considered a vulgar sign. Among some cultures, signaling for someone to come by using an upturned finger is a provocation, usually done to a dog. Among the Navajo, it is considered rude to point; rather, the Navajo shift their lips toward the desired direction.

Reflective Exercise

1. Do you tend to use your hands a lot when speaking?
2. Can people tell your emotional state by your facial expressions?

2.4.3 Temporal Relationships

Temporal relationships—people's worldview in terms of past, present, and future orientation—vary among individuals and among cultural groups. Some cultures, usually highly individualistic ones, are future-oriented, and people are encouraged to sacrifice for today and work to save and invest in the future. The future is important in that people can influence it. Fatalism, the belief that powers greater than humans are in control, may be seen as negative; however, to many others, it is seen as a fact of life not to be judged. Other cultures are regarded as a past-oriented society, in which laying a proper foundation by providing historical background information can enhance communication. However, for people in many societies, temporality is balanced among past, present, and future in the sense of respecting the past, valuing and enjoying the present, and saving for the future.

Differences in temporal orientation can cause concern or misunderstanding among health-care providers. For example, in a future-oriented culture, a person is expected to delay purchase of nonessential items to afford prescription medications. However, in less future-oriented cultures, the person may purchase the nonessential item because it is readily available and defers purchasing the prescription medication. The attitude is, why not purchase it now; the prescription medication can be purchased later.

Most people from individualistic cultures see time as a highly valued resource and do not like to be delayed because it "wastes time." When visiting friends or meeting for strictly social engagements, punctuality is less important, but one is still expected to appear within a "reasonable" time frame. In the health-care setting, if an appointment is made for 9 a.m., the person is expected to be there at 8:45 a.m. so she or he is ready for the appointment and does not delay the health-care provider, although the health-care provider may be late. For immigrants from rural settings, time may be even less important. Although it is rare, these individuals may not even own a timepiece or be able to tell time. Expectations for punctuality can cause conflicts between health-care providers and patients, even if one is cognizant of these differences. These details must be carefully explained to individuals when such situations occur. Being late for appointments should not be misconstrued as a sign of irresponsibility or not valuing one's health.

Reflective Exercise

1. How timely are you with professional appointments?
2. With social engagements?
3. What does it mean to you when people are chronically late?
4. Can you give examples indicating that you are past oriented? Present oriented? Future oriented?
5. Do you consider yourself more one than the other?

2.4.4 Format for Names

Names are important to people, and name formats differ among cultures. The most common Western system is to have a first or given name, a middle name, and then the family surname. The person would usually write the name in that order. In formal situations, the person would be addressed with a title of Mr., Mrs., Ms., Miss, or Doctor and the last name. Friends and acquaintances would call the person by the first name or perhaps a nickname. Married women may take their husband's last name, keep their maiden name, or use both their maiden and married names. However, in some cultures, the family or surname name comes first, followed by the given name and then the middle name. The person would usually write and introduce himself or herself in that order. Married women usually keep their maiden name. Other name formats are even more complex and may include a given name, a middle name, the father's family name, and the mother's maiden name. When a woman marries, she may keep all these names plus add the surname of her husband. She may choose any name she wants for legal purposes. When in doubt, the health-care provider needs to ask which name is used for legal purposes. Such extensive naming formats can create a challenge for health-care workers keeping a medical record when they are unaware of differences in ethnic recording of names. See individual chapters for name formats (see Table 2.1: Section on Communication).

Reflective Exercise

1. How do you prefer to be addressed or greeted?
2. Does this change with the situation?
3. How do you normally address and greet people?
4. Do your responses change with the situation?

2.5 Family Roles and Organization

The cultural domain of family roles and organization affects all other domains and defines relationships among insiders and outsiders. This domain includes concepts related to the head of the household, gender roles, family goals and priorities, developmental tasks of children and adolescents, roles of the aged and extended family members, individual and family social status in the community, and acceptance of alternative lifestyles such as single parenting, non-traditional sexual orientations, childless marriages, and divorce. Family structure in the context of the larger society determines acceptable roles, priorities, and the behavioral norms for its members.

2.5.1 Head of Household and Gender Roles

Reflective Exercise

1. How would you classify the decision-making process in your family—patriarchal, matriarchal, or egalitarian?
2. Does it vary by what decision has to be made?
3. Are gender roles prescribed in your family?
4. Who makes the decisions about health and health care?
5. Who would you want to make health-care decisions if you are unable to do so?

An awareness of family decision-making patterns (i.e., patriarchal, matriarchal, or egalitarian) is important for determining with whom to speak when health-care decisions have to be made. Among many cultures, it is acceptable for women to have a career and for men to assist with child care, household domestic chores, and cooking responsibilities. Both parents work in many families, necessitating placing children in child-care facilities. In some families, fathers are responsible for deciding when to seek health care for family members, but mothers may have significant influence on final decisions.

Among many, the decisions may be egalitarian, but the male's role in the family is to be the spokesperson for the family. The health-care provider, when speaking with parents, should maintain eye contact and direct questions about a child's illness to both parents.

2.5.2 Prescriptive, Restrictive, and Taboo Behaviors for Children and Adolescents

Every society has prescriptive, restrictive, and taboo practices for children and adolescents. Prescriptive beliefs are things that children or teenagers *should do* to have harmony with the family and a good outcome in society. Restrictive practices are things that children and teenagers *should not do* to have a positive outcome. Taboo practices are those things that, if done, are likely to cause significant concern or negative outcomes for the child, teenager, family, or community at large.

For some Western cultures, a child's individual achievement is valued over the family's financial status. This is different from some non-Western cultures in which attachment to family may be *more important* than the need for children to excel individually. At younger ages, rather than having group toys, each child has his or her own toys and is taught to share them with others. Individualistic cultures encourage autonomy in children, and after completing homework assignments (with which parents are expected to help), children are expected to contribute to the family by doing chores, such as taking out the garbage, washing dishes, cleaning their own room, feeding and caring for pets, and helping with cooking. They are not expected to help with heavy labor at home, except in rural farm communities.

In Western cultures, children are allowed and encouraged to make their own choices, including managing their own allowance money and deciding who their friends might be—although parents may gently suggest one friend as a better choice than another. Children and teenagers are permitted and encouraged to have friends of both the same and opposite genders. They are expected to be well behaved, especially in public. They are taught to stand in line—first come, first served—and to wait their turn. As they reach the teenage years, they are expected to refrain from premarital sex, smoking, using recreational drugs, and drinking alcohol until they leave the home. However, this does not always occur, and teenage pregnancy and the use of recreational alcohol and drugs remain high. When children become teenagers, most are expected to get a job, such as babysitting, delivering newspapers, or doing yard work to make their own spending money, which they manage as a way of learning independence. The teenage years are also seen as a time of natural rebellion.

In Western cultures, when young adults become 18 or complete their education, they usually move out of their parents' home (unless they are in college) and live independently or share living arrangements with nonfamily members. If the young adult chooses to remain in the parents' home, then she or he might be expected to pay rent. However, young adults are generally allowed to return home, as needed, for financial or other purposes. Individuals over the age of 18 are expected to be self-reliant and independent, which are virtues in the Western cultures. This differs from most collectivist cultures in which children are expected to live at home with their parents until they marry because dependence, not independence, is the virtue.

Adolescents have their own subculture with its own values, beliefs, and practices that may not be in harmony with those of their dominant culture. Being in harmony with peers and conforming to the prevalent choice of music, clothing, hairstyles, and adornment may be especially important to adolescents. Thus, role conflicts can become considerable sources of family strain in many more traditional families who may not agree with the values of individuality, independence, self-assertion, and egalitarian relationships. Many teens may experience a cultural dilemma with exposure outside the home and family.

Reflective Exercise

1. Were you taught to be independent and autonomous or dependent in your family?
2. Was there more emphasis on the individual or on the group?

2.6 Family Goals and Priorities

In most cultures, family goals and priorities are centered on raising and educating children. During this stage, young adults make a personal commitment to a spouse or significant other and seek satisfaction through productivity in career, family, and civic interests. In most societies, young adulthood is the time when individuals work on Erikson's developmental tasks of *intimacy versus isolation* and *generativity versus stagnation.*

Western cultures place a high value on children, and many laws have been enacted to protect children who are seen as the "future of the society." In most collectivist cultures, children are desirable and highly valued as a source of family strength; family members are expected to care for one another more so than in Western cultures.

Collectivistic cultures have great reverence for the wisdom of older people; families eagerly make space for them to live with extended families. Children are expected to care for elders when they are unable to care for themselves. A great embarrassment may occur to family members when they cannot take care of their older family members.

The concept of extended family membership varies among societies. The extended family is extremely important, especially in collectivist cultures. Health-care decisions are often postponed until the entire family is consulted. The extended family may include biological relatives and nonbiological members who are considered brothers, sisters, aunts, or uncles. In some cultures, the influence of grandparents in decision making is considered more important than that of the parents.

Individualistic cultures also place a high value on egalitarianism, non-hierarchical relationships, and equal treatment regardless of their race, color, religion, ethnicity, educational or economic status, sexual orientation, or country of origin. However, these beliefs are theoretical and not always seen in practice. For example, throughout the world, women usually have a lower status than men, especially when it comes to prestigious positions and salaries, although progress is being made. Most top-level politicians and corporate executive officers are White men, although some progress is being made. Subtle classism does exist, as evidenced by comments referring to "working-class men and women." Many Western cultures are known for their informality and for treating everyone the same. They call people by their first names very soon after meeting them, whether in the workplace, in social situations, in classrooms, in restaurants, or in places of business. Some readily talk with waitstaff and store clerks and call them by their first names, considering this respectful behavior.

Formality can be communicated by using the person's last (family) name and title such as Mr., Mrs., Miss, Ms., or Dr. To this end, achieved status is more important than ascribed status. What one has accumulated in material possessions, where one went to school, and one's job position and title are more important than one's family background and lineage. However, in some families ascribed status has equal importance to achieved status. Without a caste or class system, theoretically one can move readily from one socioeconomic position to another. To some, if formality is maintained, it may be seen as pompous or arrogant, and some even deride the person who is very formal. However, formality is a sign of respect in many other cultures, especially collectivistic cultures.

Reflective Exercise

1. Do you consider your family nuclear or extended?
2. How close are you to your extended family?
3. How is status measured in your family?
4. By money or by some other attribute?
5. What are your personal views of two people of the same gender living together in a physical relationship?
6. What about heterosexual couples?
7. Does divorce cause a stigma in your culture? In your family?

Reflective Exercise

1. What were prescriptive behaviors for you as a child? As a teenager? As a young adult?
2. What were restrictive behaviors for you as a child? As a teenager? As a young adult?
3. What were taboo behaviors for you as a child? As a teenager? As a young adult?
4. How are elders regarded in your culture? In your family?

2.6.1 Alternative Lifestyles

The traditional family is nuclear, with a married man and woman living together with one or more unmarried children. This concept of family is becoming a more varied community, including unmarried people, both women and men, living alone; single people of the same or different genders living together with or without children; single parents with children; and blended families consisting of two parents who have remarried, with children from their previous marriages and additional children from their current marriage. However, in some cultures, the traditional family is extended, with parents, unmarried children, married children with their children, and grandparents all sharing the same living space or at least living in very close proximity.

Social attitudes toward homosexual activity vary widely, and homosexual behavior occurs in societies that deny its presence. Homosexual behavior carries a severe stigma in some societies. Discovering that one's son or daughter is gay is akin to a catastrophic event for some, whether it is a collectivistic or individualistic culture. In the last 10 years, sex marriage has become more accepted in both individualistic and collectivistic cultures and they are also able to adopt children.

When the health-care provider needs to provide assistance and make a referral for a person who is gay, lesbian, bisexual, or transgender, a number of options are available. Some referral agencies are local, whereas others are national, with local or regional chapters. Many are ethnically or religiously specific identifies guidelines for assessing the cultural domain (see Table 2.1: Section on Family Roles and Organization.

2.7 Workforce Issues

2.7.1 Culture in the Workplace

A fourth domain of culture is workforce issues. Differences and conflicts that exist in a homogeneous culture may be intensified in a multicultural workforce. Factors that affect these issues include language barriers, degree of assimilation and acculturation, and issues related to autonomy. Moreover, such concepts as gender roles, cultural communication styles, health-care practices of the country of origin, and selected concepts from all other domains affect workforce issues in a multicultural work environment.

Timeliness and punctuality are two culturally based attitudes that can create serious problems in the multicultural workforce. In most Western cultures, people are expected to be punctual on their job, with formal meetings, and with appointments. With social engagements, punctuality is not as important. However, in many cultures, punctuality is not stressed unless one is meeting with officials or it is required for transportation schedules, such as for trains or air travel. Timeliness for social engagements may not be taken seriously and may simply begin when most of the people arrive. The lack of adherence to meeting time demands in other countries is often in direct opposition to the Western concept and the ethic for punctuality.

Reflective Exercise

1. How timely are you in reporting to work?
2. Do you see people in the workforce who do not report to work on time?
3. What problems does it cause if they are not on time?
4. What would you do as a supervisor to encourage people to report to work on time?

Clinical professionals trained in their home countries now occupy a significant share of technical and laboratory positions in health-care facilities in many counties throughout the world. Service employees, such as food preparation workers, nursing assistants, orderlies, housekeepers, and janitors represent the most culturally diverse component of hospitals workforce. These unskilled and semiskilled positions are among the most attainable for new immigrants.

Reflective Exercise

1. How important are technical skills and verbal skills in your work environment?
2. Does your organization encourage more formal or more informal communication? Why?
3. Do you believe that everything needs to be proven scientifically?
4. Do you value a more direct or indirect style of communication?
5. Does your workforce (class) reflect the ethnic and racial diversity of the community? Why? Why not? What might you do to increase this diversity?

2.7.2 Issues Related to Autonomy

Cultural differences related to assertiveness influence how health-care providers view one another. In most Western individualistic cultures, professionals are expected to be assertive with other professionals for the benefit of the patient. However, in some collectivist and patriarchal societies, women, for example, may be unprepared for the level of sophistication and autonomy expected in individualistic cultures. Educational training for health-care providers varies significantly throughout the world.

Language ability in a new country may not meet the standards expected in the workforce, especially in the health-care environment and in positions where highly developed verbal skills are required. Thus, the newer immigrant—for whom the language of the host country is new—may need extra time in translating messages and formulating replies.

Reflective Exercise

1. How many generations are in your work group (class)?
2. Are their beliefs and practices similar to or different from what is reported in the literature?
3. Do the generational differences cause conflict?
4. Which generation takes the lead in resolving conflicts when they arise?

Reflective Exercise

1. Does your profession encourage autonomy in the workforce?
2. Does your current work (class) encourage autonomy and independence?
3. Do you see any cultural or gender differences in autonomy?
4. Do people speak different languages at work?
5. What difficulty does this cause?

When individuals speak in their native language at work, it may become a source of contention for both patients and health-care providers. Some organizations prohibit this, even in social situations at work. Most employees do not want to exclude or offend others, but it is easier to speak in their native language to articulate ideas, feelings, and humor among themselves. Negative interpretations of behaviors can be detrimental to working relationships in the health-care environment. Some foreign graduates, with limited aural language abilities, may need to have care instructions written or procedures demonstrated.

2.7.3 Generational Differences in the Workforce

Not only is the workforce becoming more multicultural in most countries, but over the last decade, increased interest has been found in the professional literature regarding generational differences in the workforce. Most of the literature on generational differences describes the dominant culture of the United States, with little mention as to how these differences might coincide with the multi-ethnic workforce. However, these descriptions do not always "fit" the generalizations as well as they do for the dominant, nonethnic, non-immigrant populations (see Table 2.1: Section Workforce Issues).

2.8 Biocultural Ecology

The domain *biocultural ecology* identifies specific physical, biological, and physiological variations in ethnic and racial origins. These variations include skin color and physical differences in body habitus; genetic, hereditary, endemic, and topographic diseases; psychological makeup of individuals; and differences in the way drugs are metabolized by the body. No attempt is made here to explain or justify any of the numerous, conflicting, and highly controversial views and research about racial variations in drug metabolism and genetics. More research needs to be completed.

2.8.1 Skin Color and Other Biological Variations

Reflective Exercise

1. Do you have difficulty assessing rashes, bruises, and sunburn in people with a skin color different from yours?
2. Do you have difficulty assessing jaundice and oxygenation in people with a skin color different from yours?
3. How does your assessment of skin differ between patients with light versus dark skin?
4. Do you take precautions and protect yourself against the sun? Why? Why not?

Skin coloration is an important consideration for health-care providers because anemia, jaundice, and rashes require different assessment skills in dark-skinned people than in light-skinned people. To assess for oxygenation and cyanosis in dark-skinned patients, the practitioner must examine the sclera, buccal mucosa, tongue, lips, nail beds, palms of the hands, and soles of the feet rather than relying on skin tone alone. Jaundice is more easily determined in Asians by assessing the sclera rather than relying on the overall change in skin color. Health-care providers may need to establish a baseline skin color (by asking a family member or someone known to the individual), use direct sunlight (if possible), observe areas with the least amount of pigmentation, palpate for rashes, and compare skin in corresponding areas. With people who are generally fair-skinned, prolonged exposure to the sun places them at an increased risk for skin cancer.

Reflective Exercise

1. What are the most common illnesses and diseases in your family? In your community?
2. What might you do to decrease the incidence of illness and diseases in your family? In your community?
3. Are you aware of any outbreaks of new illnesses or diseases in your community? In other parts of the world?
4. How might these outbreaks have been prevented?
5. What are the most common illnesses and diseases in your family? In your community?
6. What might you do to decrease the incidence of illness and diseases in your family? In your community?
7. Are you aware of any outbreaks of new illnesses or diseases in your community? In other parts of the world?
8. How might these outbreaks have been prevented?

Variations in body habits occur among ethnic and racially diverse individuals in terms of bone density, length of long bones, and shoulder and hip width, but do not usually cause a concern for health-care providers. However, bone density is greater in Whites than in Asian and Pacific Islanders; osteoporosis is lowest in Black males and highest in White females (Peacock et al. (2009). Given diverse gene pools, this type of information is often difficult to obtain, and much of the research is inconclusive.

2.8.2 Diseases and Health Conditions

Some diseases are more prevalent and some are even endemic in certain racial or ethnic groups, especially with migration. Specific health problems are covered in individual chapters. In general, many adverse health conditions are a result of genetics, lifestyle, and the environment. Genetic conditions occur among families in all races, but some conditions, such as Tay-Sachs disease, hemophilia, and cystic fibrosis are more common among particular ethnic and racial groups. Lifestyle causes include cultural practices and behaviors that can generally be controlled—for example, smoking, diet, and stress. Environmental causes refer to factors (e.g., air and water pollution) and situations over which the individual has little or no control (e.g., presence of malarial or dengue mosquitoes, exposure to chemicals and pesticides, access to care, and associated diseases).

Reflective Exercise

1. Why is it important for health-care providers to be aware of variations in drug metabolism in the body?
2. What conditions besides genetics have an influence on drug metabolism?

Information regarding drug metabolism among racial and ethnic groups has important implications for health-care providers when prescribing medications. Besides the effects of (a) smoking, which accelerates drug metabolism; (b) malnutrition, which affects drug response; (c) a high-fat diet, which increases absorption of antifungal medication, whereas a low-fat diet renders the drug less effective; (d) cultural attitudes and beliefs about taking medication; and (e) stress, which affects catecholamine and cortisol levels on drug metabolism, studies have identified some specific alterations in drug metabolism among diverse racial and ethnic groups (Burroughs Valentine et al. 2002). Information for specific groups is included in each chapter. Health-care providers need to investigate the literature for ethnic-specific studies regarding variations in drug metabolism, communicate these findings to other colleagues, and educate their patients regarding these side effects.

Medication administration is one area in which health-care providers see the importance of culture, ethnicity, and race (see Table 2.1, section Biocultural Ecology).

2.9 High-Risk Behaviors

High-risk behaviors include use of tobacco, alcohol, or recreational drugs; lack of physical activity; increased calorie consumption; unsafe driving practices (speeding, driving and texting); failure to use seat belts and helmets; failure to take precautions against human immunodeficiency virus (HIV) and sexually transmitted infections (STIs); and high-risk recreational activities. High-risk behaviors occur in all ethnocultural groups, with the degree and types of high-risk behaviors varying.

Alcohol consumption crosses all cultural and socioeconomic groups. Enormous differences exist among ethnic and cultural groups around the use of and response to alcohol. Even in cultures in which alcohol consumption is taboo, it is not ignored. However, alcohol problems are not simply a result of how much people drink. When drinking is culturally approved, it is typically done more by men than women and is more often a social, rather than a solitary, act. The group in which drinking is most frequently practiced is usually composed of same-age social peers (Caetano et al. 1998). Studies on increasing controls on the availability of alcohol to decrease alcohol consumption, with the premise that alcohol-related problems occur in proportion to per capita consumption, have not been supported. Furthermore, countries with temperance movements have greater alcohol-related behavior problems than do countries without temperance movements (Purnell and Foster 2003a, b).

Countries in which drinking alcoholic beverages is integrated into rites and social customs, and in which one is expected to have self-control and sociability, have lower rates of alcohol-related problems than those of countries and cultures in which ambivalent attitudes toward drinking prevail (Purnell and Foster 2003a, b). Hilton's (1987) study demonstrated a clear and distinct difference in the alcohol abuse rate by socioeconomic status. The conclusion of many studies suggests that alcohol-related violence is a learned behavior, not an inevitable result of alcohol consumption (Purnell and Foster 2003a, b).

2.9.1 Health-Care Practices

Obesity and being overweight are a result of an imbalance between food consumed and physical activity. National data have shown an increase in the calorie consumption of adults and no change in physical activity patterns. However, obesity is a complex issue related to lifestyle, environment, and genes. Many underlying factors have been linked to the increase in obesity, such as increased portion sizes; eating out more often; increased consumption of sugar-sweetened drinks; increased television viewing, computer, electronic gaming time; and fear of crime, which prevents outdoor exercise. In some cultures, what is seen as overweight or obese according to actuarial tables, is seen as positive and means that one could afford to lose weight if one is ill, the individual has good socioeconomic status, the person is more desirable to the other gender.

The practice of self-care by using folk and magico-religious practices before seeking professional care may also have a negative impact on the health status for some individuals. Overreliance on these practices may mean that the health problem is in a more advanced stage when a consultation is sought. Such delays make treatment more difficult and prolonged. Selected complementary and alternative health-care practices are addressed in this chapter under the domain *health-care practices* and in each population specific chapter.

Reflective Exercise

1. In which high-risk health behaviors do you engage?
2. What do you do to control or reduce your risk?
3. Which high-risk health behaviors do you see most frequently in your family? In your community?
4. What might you do to help decrease these high-risk behaviors?

The cultural domain of *high-risk behaviors* is one area in which health-care providers can make a significant impact on patients' health status. High-risk health behaviors can be controlled through ethnic-specific interventions aimed at health promotion and health-risk prevention. This can be accomplished through educational programs in schools, business organizations, churches, and recreational and community centers, as well as through one-on-one and family counseling techniques. Taking advantage of public communication technology can enhance participation in these programs if they are geared to the unique needs of the individual, family, or community (see Table 2.1, Section High-risk Behaviors).

2.10 Nutrition

The cultural domain of *nutrition* includes much more than merely having adequate food for satisfying hunger. This domain also comprises the meaning of food to the culture; common foods and rituals; nutritional deficiencies and food limitations; and the use of food for health promotion and wellness, illness and disease prevention, and health maintenance and restoration. Understanding a patient's food choices and preparation practices is essential for providing culturally competent dietary counseling. Health-care providers may be considered professionally negligent when prescribing, for example, an American diet to a Hispanic or an Asian patient whose food choices and mealtimes may be different from American food patterns.

2.10.1 Meaning of Food

Food and the absence of food—hunger—have diverse meanings among cultures and individuals. Cultural beliefs, values, and the types of foods available influence what people eat, avoid, or alter to make food congruent with cultural lifeways. Food offers cultural security and acceptance. Food plays a significant role in socialization and can denote caring or lack of caring or closeness.

Reflective Exercise

1. What are your personal beliefs about weight and health?
2. Do you agree with the dominant American belief that thinness correlates with desirability and beauty?
3. What does food mean in your culture besides satisfying hunger?

2.10.2 Common Foods and Food Rituals

Traditional food habits are basic to satisfactory nutrition to most people. Perhaps a traditional diet does not really exist for some people; rather, they have favorite foods and preparation practices that health-care providers need to assess for effective dietary recommendations for illness and disease prevention and health promotion and wellness. Most immigrants bring their favorite foods with them when they relocate, including preferred mealtimes. Food choices may vary according to the region of the country, urban versus rural residence, and weekdays versus weekends. In addition, food choices vary by marital status, economic status, climate changes, religion, ancestry, availability, and personal preferences.

Many older people and people living alone, regardless of cultural background, frequently do not eat balanced meals. They state that they do not take the time to prepare a meal, even though most homes have labor-saving devices such as stoves, microwave ovens, refrigerators, and dishwashers. For those who are unable to prepare their own meals because of disability or illness, most communities have a Meals on Wheels program through which community and church organizations deliver, usually once a day, a hot meal along with a cold meal for later and food for the following morning's breakfast. Socioeconomic status may dictate food selections—for example, hamburger instead of steak, canned or frozen vegetables and fruit rather than fresh produce, and fish instead of shrimp or lobster. Special occasions and holidays are frequently associated with ethnic-specific foods. Many religious groups are required to fast during specific holiday seasons. However, health-care providers may need to remind patients that fasting is not required during times of illness or pregnancy.

Reflective Exercise

1. What do you eat to maintain your health?
2. What does a healthy diet mean to you?
3. Do you agree with the U.S. Department of Agriculture Food Pyramid? Why? Why not?
4. What do you eat when you are ill?

Given the intraethnic variations of diet, it is important for health professionals to inquire about the specific diets of their patients. Expecting the patient to eat according to a set mealtime schedule and to select foods from an exchange list may be unrealistic for patients of different cultural backgrounds. Counseling about food-group requirements, intake restrictions, and exercise must respect cultural behaviors and individual lifeways. Culturally congruent dietary counseling, such as changing amounts and preparation practices while including preferred ethnic food choices, can reduce the risk for obesity, cardiovascular disease, and cancer. Whenever possible, determining a patient's dietary practices should be started during the intake interview.

Reflective Exercise

1. In what food rituals does your family engage?
2. Do you have specific foods and rituals for holidays?
3. What would happen if you changed these rituals?
4. Do food patterns change for you by the season?
5. During the week versus the weekend?

2.10.3 Dietary Practices for Health Promotion

The nutritional balance of a diet is recognized by most cultures throughout the world. Most cultures have their own distinct theories of nutritional practices for health promotion and wellness, illness and disease prevention, and health maintenance and restoration. Common folk practices and selected diets are recommended during periods of illness and for prevention of illness or disease. For example, cultures subscribe to the hot-and-cold (opposites) theory of food selection to prevent illness and maintain health. Although each of these cultural groups has its own specific name for the hot-and-cold theory of foods, the overall belief is that the body needs a balance of opposing foods. These practices are covered in culture-specific chapters.

2.10.4 Nutritional Deficiencies and Food Limitations

Because of limited socioeconomic resources or limited availability of their native foods, immigrants may eat foods that were not available in their home country. These dietary changes may result in health problems when they arrive in a new environment. This is more likely to occur when individuals immigrate to a country where they do not have native foods readily available and do not know which new foods contain the necessary and comparable nutritional ingredients. Consequently, they do not know which foods to select for balancing their diet.

Reflective Exercise

1. In what food rituals does your family engage?
2. Do you have specific foods and rituals for holidays?
3. What would happen if you changed these rituals?
4. Do food patterns change for you by the season?
5. During the week versus the weekend?

Enzyme deficiencies exist among some ethnic and racial groups. For example, many people are lactose-intolerant and are unable to drink milk or eat dairy products (unless cooked) to maintain their calcium needs. Thus, the health-care provider may need to assist patients and their families in identifying foods high in calcium when they are unable to purchase their native foods. In general, the wide availability of foods reduces the risks of these disorders as long as people have the means to obtain culturally nutritious foods. Recent emphasis on cultural foods has resulted in larger grocery stores having sections designated for ethnic goods and in small businesses selling ethnic foods and spices to the general public. The health-care provider's task is to determine how to assist the patient and identify alternative foods to supplement the diet when these stores are not financially or geographically accessible (see Table 2.1, Section on Nutrition).

Reflective Exercise

1. What enzyme deficiencies run in your family?
2. Do you have any difficulty getting your preferred foods?
3. What other food limitations do you have?

2.11 Pregnancy and Childbearing Practices

The cultural domain *pregnancy and childbearing practices* includes culturally sanctioned and unsanctioned fertility practices; views toward pregnancy; and prescriptive, restrictive, and taboo practices related to pregnancy, birthing, and the postpartum period.

Many traditional, folk, and magico-religious beliefs surround fertility control, pregnancy, childbearing, and postpartum practices. The reason may be the mystique that surrounds the processes of conception, pregnancy, and birthing. Ideas about conception, pregnancy, and childbearing practices are handed down from generation to generation and are accepted without validation or being completely understood. For some, the success of modern technology in inducing pregnancy in postmenopausal women and others who desire children through in vitro fertilization and the ability to select a child's gender raises serious ethical questions.

2.11.1 Fertility Practices and Views toward Pregnancy

Commonly used methods of fertility control include natural ovulation methods, birth control pills, foams, Norplant, the morning-after pill, intrauterine devices, tubal ligation or sterilization, vasectomy, prophylactics, and abortion.

Although not all of these methods are acceptable to all people, many women use a combination of fertility control methods. The most extreme examples of fertility control are sterilization and abortion. Sterilization in the United States and most other countries is now strictly voluntary. Abortion remains a controversial issue in many countries and religions. For example, in some countries, women are encouraged to have as many children as possible, and abortion is illegal. However, in other countries, abortion is commonly used as a means of limiting family size for a variety of reasons. The "morning-after pill" also continues to be controversial to some.

Reflective Exercise

1. Does pregnancy have a special meaning in your culture?
2. Is fertility control acceptable in your culture?
3. Do most people adhere to fertility control practices in your culture?
4. What types of fertility control are acceptable? Unacceptable?

Fertility practices and sexual activity, sensitive topics for many, is one area in which "outside" health-care providers may be more effective than health-care providers known to the patient because of the concern about providing intimate information to someone they know. Some of the ways health-care providers can promote a better understanding of practices related to family planning include using videos in the native language and videos and pictures of native ethnic people, using material written at the individual's level of education, and providing written instructions in both English and the native language. Health-care providers should avoid family planning discussions on the first encounter; such information may be better received on subsequent visits when some trust has developed. Approaching the subject of family planning obliquely may make it possible to discuss these topics more successfully.

2.11.2 Prescriptive, Restrictive, and Taboo Practices in the Childbearing Family

Most societies have prescriptive, restrictive, and taboo beliefs for maternal behaviors and the delivery of a healthy baby. Such beliefs affect sexual and lifestyle behaviors during pregnancy, birthing, and the immediate postpartum period. Prescriptive practices are things that the mother should do to have a good outcome (healthy baby and pregnancy). Restrictive practices are those things that the mother should not do to have a positive outcome (healthy baby and delivery). Taboo practices are those things that, if done, are likely to harm the baby or mother.

One prescriptive belief is that women are expected to seek preventive care, eat a well-balanced diet, and get adequate rest to have a healthy pregnancy and baby. A restrictive belief is that pregnant women should refrain from being around loud noises for prolonged periods of time. Taboo behaviors during pregnancy include smoking, drinking alcohol, drinking large amounts of caffeine, and taking recreational drugs—practices that are sure to cause harm to the mother or baby.

A taboo belief common among many cultures is that a pregnant woman should not reach over her head because the baby may be born with the umbilical cord around its neck. A restrictive belief among others is that permitting the father to be present in the delivery room and seeing the mother or baby before they have been cleaned can cause harm to the baby or mother. If the father is absent from the delivery room or does not want to see the mother or baby during birthing or immediately after birth, it does not mean that he does not care about them. However, in many cultures, the father is often encouraged to take prenatal classes with the expectant mother and provide a supportive role in the delivery process; fathers with opposing beliefs may feel guilty if they do not comply. The woman's female relatives provide assistance to the new mother until she is able to care for herself and the baby. Additional cultural beliefs carried over from cultural migration and diversity include the following:

- If you wear an opal ring during pregnancy, it will harm the baby.
- Birthmarks are caused by eating strawberries or seeing a snake and being frightened.
- Congenital anomalies can occur if the mother sees or experiences a tragedy during her pregnancy.
- Nursing mothers should eat a bland diet to avoid upsetting the baby.
- The infant should wear a band around the abdomen to prevent the umbilicus from protruding and becoming herniated.
- A coin, key, or other metal object should be put on the umbilicus to flatten it.
- Cutting a baby's hair before baptism can cause blindness.
- Raising your hands over your head while pregnant may cause the cord to wrap around the baby's neck.
- Moving heavy items can cause your "insides" to fall out.
- If the baby is physically or mentally abnormal, God is punishing the parents.

Reflective Exercise

1. What are some prescriptive practices for pregnant women in your culture?
2. What are some restrictive practices for pregnant women in your culture?
3. What are some taboo practices for pregnant women in your culture?
4. What special foods should a woman eat to have a healthy baby in your culture?
5. What foods should be avoided? What foods should a nursing mother eat postpartum?
6. What foods should she avoid?

In some other cultures, the postpartum woman is prescribed a prolonged period of recuperation in the hospital or at home, something that may not be feasible income countries because of the shortened length of confinement in the hospital after delivery. he health-care provider must respect cultural beliefs associated with pregnancy and the birthing process when making decisions related to the health care of pregnant women, especially those practices that do not cause harm to the mother or baby. Most cultural practices can be integrated into preventive teaching in a manner that promotes compliance (see Table 2.1: Section Pregnancy and Childbearing Families).

2.12 Death Rituals

The cultural domain *death rituals* includes how the individual and the society view death, euthanasia, rituals to prepare for death, burial practices, and bereavement practices. Death rituals of ethnic and cultural groups are the least likely to change over time and may cause concerns among health-care personnel. Some staff may not understand the value of customs with which they are not familiar, such as the ritual washing of the body. Death practices, beliefs, and rituals vary significantly among cultural and religious groups. To avoid cultural taboos, health-care providers must become knowledgeable about unique practices related to death, dying, and bereavement.

2.12.1 Death Rituals and Expectations

For many health-care providers educated in a culture of mastery over the environment, death is seen as one more disease to conquer, and when this does not happen, death becomes a personal failure. Thus, for many, death does not take a natural course because it is "managed" or "prolonged," making it difficult for some to die with dignity. Moreover, death and responses to death are not easy topics for many to verbalize. Instead, many euphemisms are used rather than verbalizing that the person died—for example, "passed away," "no longer with us," and "was visited by the Grim Reaper." The individualistic cultural belief in self-determination and autonomy extends to people making their own decisions about end-of-life care. Mentally competent adults have the right to refuse or decide what medical treatment and interventions they wish to extend life, such as artificial life support, artificial feeding, and hydration.

Reflective Exercise

1. What terms do you use when referring to death?
2. Why do you use these terms?
3. What specific burial practices do you have in your family/culture?

Among most Westerners, the belief is that a dying person should not be left alone, and accommodations are usually made for a family member to be with the dying person at all times. Health-care personnel are expected to care for the family as much as for the patient during this time. Most people are buried or cremated within 3 days of the death, but extenuating circumstances may lengthen this period to accommodate family and friends who must travel a long distance to attend a funeral or memorial service or where frozen land prevents the burial. The family can decide whether to have an open casket—so family and friends can view the deceased—or a closed casket. Cremation is common among many groups. Significant variations in burial practices occur with other ethnocultural groups throughout the world.

2.12.2 Responses to Death and Grief

Reflective Exercise

1. How do men grieve in your culture?
2. How do women grieve in your culture?
3. Do you have a living will or advance directive? Why? Why not?
4. Are you an organ donor? Why? Why not?
5. Is there a specific time frame for bereavement?

Numerous countries have been launching major initiatives to help patients die as comfortably as possible without pain. As a result, more people are choosing to remain at home or to choose palliative care or hospice for end-of-life care where their comfort needs are better met. When death does occur, some people conservatively control their grief, although women are usually more expressive than men. For many, especially men, they are expected to be stoic in their reactions to death, at least in public. Generally, tears are shed, but loud wailing and uncontrollable sobbing rarely occur. The belief is that the person has moved on to a better existence and does not have to undergo the pressures of life on earth. Regardless of the gender or culture, bereavement is a very private issue, and there are no norms; people grieve in their own way.

Variations in the grieving process may cause confusion for health-care providers, who may perceive some patients as overreacting and others as not caring. The behaviors associated with the grieving process must be placed in the context of the specific cultural belief system in order to provide culturally competent care. Caregivers should accept and encourage ethnically specific bereavement practices when providing support to family and friends. Bereavement support strategies include being physically present, encouraging a reality orientation, openly acknowledging the family's right to grieve, accepting varied behavioral responses to grief, acknowledging the patient's pain, assisting them to express their feelings, encouraging interpersonal relationships, promoting interest in a new life, and making referrals to other resources, such as a priest, minister, rabbi, or pastoral care (Table 2.1: Section Death and Dying).

2.13 Spirituality

The domain *spirituality* involves more than formal religious beliefs related to faith and affiliation and the use of prayer. For some people, religion has a strong influence over and shapes nutrition practices, health-care practices, and other cultural domains. Spirituality includes all behaviors that give meaning to life and provide strength to the individual. Furthermore, it is difficult to distinguish religious beliefs from cultural beliefs because for some, especially the very devout, religion guides the dominant beliefs, values, and practices even more than their culture does.

Spirituality, a component of health related to the essence of life, is a vital human experience that is shared by all humans. Spirituality helps provide balance among the mind, body, and spirit. Trained, lay, and traditional religious leaders may provide comfort to both the patient and the family. Spirituality does not have to be scientifically proven and is patterned unconsciously from a person's worldview. Accordingly, an individual may deviate somewhat from the majority view or position of the formally recognized religion.

2.13.1 Dominant Religion and Use of Prayer

Of the major religions in the world, 31.4% of people are Christians, 23.2% are Muslim; 15% are Hindu; 7.1% are Buddhists; 5% practice folk religions, 0.2% Jewish, 0.8% other, and unaffiliated, 16.4% (CIA World Factbook 2019).

Many people have migrated for religious acceptance or freedom. Furthermore, specific religious groups are concentrated regionally within a country. Unlike in some countries that support a specific church or religion and in which people discuss their religion frequently and openly, religion is not an everyday topic of conversation for many. The health-care provider who is aware of the patient's religious practices and spiritual needs is in a better position to promote culturally congruent health care. The health-care provider must demonstrate an appreciation of and respect for the dignity and spiritual beliefs of patients by avoiding negative comments about religious beliefs and practices. Patients may find considerable comfort in speaking with religious leaders in times of crisis and serious illness.

Reflective Exercise

1. Do identify with a certain religion?
2. Do you consider yourself devout?
3. Do you need anything special to pray?
4. When do you pray?
5. Do you pray for good health?
6. Do religiosity and spirituality differ for you?
7. What gives meaning to your life?
8. How are spirituality, religiosity, and health connected for you?

Prayer takes different forms and different meanings. Some people pray daily and may have altars in their homes. Others may consider themselves devoutly religious and say prayers only on special occasions or in times of crisis or illness. Health-care providers may need to make special arrangements for individuals to say prayers in accordance with their belief systems. A religious leader may be helpful to staff.

2.13.2 Meaning of Life and Individual Sources of Strength

What gives meaning to life varies among and within cultural groups and among individuals. To some people, their formal religion may be the most important facet of fulfilling their spirituality needs, whereas for others, religion may be replaced as a driving force by other life forces and worldviews. For others, family is the most important social entity and is extremely important in helping meet their spiritual needs. For many, what gives meaning to life is good health and well-being. For a few, spirituality may include work or money.

A person's inner strength comes from different sources. For some, inner self is dependent on being in harmony with one's surroundings, whereas for others, a belief in a supreme being may give personal strength. For most people, spirituality includes a combination of these factors. Knowing these beliefs allows health-care providers to assist individuals and families in their quest for strength and self-fulfilment.

2.13.3 Spiritual Beliefs and Health-Care Practices

Spiritual wellness brings fulfillment from a lifestyle of purposeful and pleasurable living that embraces free choices, meaning in life, satisfaction in life, and self-esteem. For some, ritual dancing and herbal treatments (combined with prayers and songs) are performed for total body healing and the return of spirits to the body. Practices that interfere with a person's spiritual life can hinder physical recovery and promote physical illness.

Health-care providers should inquire whether the person wants to see a member of the clergy even if she or he has not been active in church. Religious emblems should not be removed because they provide solace to the person, and removing them may increase or cause anxiety. A thorough assessment of spiritual life is essential for the identification of solutions and resources that can support other treatments (see Table 2.1: Section Spirituality).

2.14 Health-Care Practices

Another domain of culture is *health-care practices*. The focus of health care includes traditional, magico-religious, and biomedical beliefs; individual responsibility for health; self-medicating practices; and views toward mental illness, chronicity, rehabilitation, and organ donation and transplantation. In addition, responses to pain and the sick role are shaped by specific ethnocultural beliefs. Significant barriers to health care may be shared among cultural and ethnic groups.

2.14.1 Health-Seeking Beliefs and Behaviors

For centuries, people's health has been maintained by a wide variety of healing and medical practices. Currently, most of the world is undergoing a paradigm shift from one that places high value on curative and restorative medical practices with sophisticated technological care to one of health promotion and wellness; illness, disease, and injury prevention; health maintenance and restoration; and increased personal responsibility. Most believe that the individual, the family, and the community have the ability to influence their health. However, among other populations, good health may be seen as a divine gift from a superior being, with individuals having little control over health and illness.

The primacy of patient autonomy is generally accepted as an enlightened perspective in individualistic cultures. To this end, advance directives, such as "durable power of attorney" or a "living will" are an important part of medical care and is common in hospitals and long-term care in the United States. Accordingly, patients can specify their wishes concerning life and death decisions before or upon entering an inpatient facility. The durable power of attorney for health care allows the patient to name a family member or significant other to speak for the patient and make decisions when or if the patient is unable to do so. The patient can also have a living will that outlines the person's wishes in terms of life-sustaining procedures in the event of a terminal illness. Most inpatient facilities have forms that patients may sign, or they can elect to bring their own forms, many of which are available on the Internet. Most countries and cultural groups engage in preventive vaccines for children. Guidelines for vaccines were developed largely as a result of the influence of the World Health Organization. Specific vaccine schedules and the ages at which they are prescribed vary widely among countries and can be obtained from the WHO website (World Health Organization: Immunizations 2019). However, some religious groups, such as Christian Scientists and ultra-conservative Jews, do not believe in vaccinations. Beliefs like this, which restrict optimal child health, have resulted in court battles with various outcomes.

2.14.2 Responsibility for Health Care

The world is moving to a paradigm in which people take increased responsibility for their health. In a society in which individualism is valued, people are expected to be self-reliant. In fact, people are expected to exercise some control over disease, including controlling the amount of stress in their lives. If someone does not maintain a healthy lifestyle and then gets sick, some believe it is the person's own fault. Unless someone is very ill, she or he should not neglect social and work obligations.

Reflective Exercise

1. What do you do to take responsibility for your health?
2. Do you take vaccines to prevent the flu or other illnesses?
3. Do you have adequate health insurance?
4. Do you have regular checkups with your health-care provider?

The health-care delivery system of the country of origin and degree of individualism and collectivism may shape patients' beliefs regarding personal responsibility for health care. Most countries in the world have some kind of basic universal coverage for their citizens, although access and quality may vary significantly from rural and urban settings and for vulnerable populations.

A potential high-risk behavior in the self-care context includes self-medicating practices. Self-medicating behavior in itself may not be harmful, but when combined with or used to the exclusion of prescription medications, it may be detrimental to the person's health. A common practice with prescription medications is for people to take medicine until the symptoms disappear and then discontinue the medicine prematurely. This practice commonly occurs with antihypertensive medications and antibiotics. No culture is immune to self-medicating practices; almost everyone engages in it to some extent.

Each country has some type of control over the purchase and use of medications. The United States is more restrictive than many countries and provides warning labels and directions for the use of over-the-counter medications. In many countries, pharmacists may be consulted before physicians for fever-reducing and pain-reducing medicines. In parts of Central America, a person can purchase antibiotics, intravenous fluids, and a variety of medications over the counter; most stores sell medications and vendors sell drugs in street-corner shops and on public transportation systems. People who are accustomed to purchasing medications over the counter in their native country frequently see no problem in sharing their medications with family and friends. To help prevent contradictory or exacerbated effects of prescription medications and treatment regimens, health-care providers should ask about patients' self-medicating practices. One cannot ignore the ample supply of over-the-counter medications in pharmacies worldwide, the numerous television advertisements for self-medication, and media campaigns for new medications, encouraging viewers to ask their doctor or health-care provider about a particular medication.

Reflective Exercise

1. In what self-medicating practices do you engage?
2. What makes you decide when to see your health-care provider when you have an illness?

2.14.3 Folk and Traditional Practices

Some cultures and individuals favor traditional, folk, or magico-religious health-care practices over biomedical practices and use some or all of them simultaneously. For many, what are considered alternative or complementary health-care practices in one country may be mainstream medicine in another society or culture. In the United States, interest has increased in alternative and complementary health practices (National Center for Complementary and Integrative Health 2019) to bridge the gap between traditional and non-traditional therapies.

Reflective Exercise

1. In complementary and alternative practices have you practiced?
2. For what conditions have you used them?
3. Were they helpful?
4. How willingly do you accept other people's traditional practices?

As an adjunct to biomedical treatments, many people use acupuncture, acupressure, acumassage, herbal therapies, and other traditional treatments. Some cultural groups and individuals commonly visit traditional healers because modern medicine is viewed as inadequate. Examples of folk medicines include covering a boil with axle grease, wearing copper bracelets for arthritis pain, taking wild turnip root and honey for a sore throat, and drinking herbal teas. The Chinese subscribe to the yin-and-yang theory of treating illnesses, and Hispanic groups believe in the hot-and-cold theory of foods for treating illnesses and disease (see specific population-based chapters. Traditional schools of pharmacy in many countries sell folk remedies. Most people practice folk medicine in some form; they may use family remedies passed down from previous generations.

An awareness of combined practices when treating or providing health education to individuals and families helps ensure that therapies do not contradict one another, intensify the treatment regimen, or cause an overdose. At other times, they may be harmful, conflict with, or potentiate the effects of prescription medications. Many times, these traditional, folk, and magico-religious practices are and should be incorporated into the plans of care for patients resulting in integrative medicine. Inquiring about the full range of therapies being used, such as food items, teas, herbal remedies, non-food substances, over-the-counter medications, and medications prescribed or loaned by others, is essential so that conflicting treatment modalities are not used. If patients perceive that the health-care provider does not accept their beliefs, they may be less compliant with prescriptive treatment and less likely to reveal their use of these practices.

2.14.4 Barriers to Health Care

For people to receive adequate health care, a number of considerations must be addressed. For example, a lack of fluency in language, verbal or written, can be a barrier to receiving adequate health care. Other barriers include the following:

- *Availability:* Is the service available and at a time when needed? For example, no services exist after 6 p.m. for someone who needs suturing of a minor laceration. Clinic hours coincide with patients' work hours, making it difficult to schedule appointments for fear of work reprisals.
- *Accessibility:* Transportation services may not be available or rivers and mountains may make it difficult for people to obtain needed health-care services when no health-care provider is available in their immediate region. For example, it can be difficult for a single parent with four children to make three bus transfers to get one child immunized.
- *Affordability:* The service is available but the patient does not have financial resources.
- *Appropriateness:* Maternal and child services are available but what might be needed are geriatric and psychiatric services.
- *Accountability:* Are health-care providers accountable for their own education and do they learn about the cultures of the people they serve? Are they culturally aware, sensitive, and competent?
- *Adaptability:* A mother brings her child to the clinic for a vaccine. Can she get a mammogram at the same time or must she make another appointment?
- *Acceptability:* Are services and patient education offered in a language preferred by the patient?

- *Awareness:* Is the patient aware that needed services exist in the community? The service may be available but if patients are not aware of it the service will not be used.
- *Attitudes:* Adverse subjective beliefs and attitudes from caregivers mean that the patient will not return for needed services until the condition is more compromised. Do health-care providers have negative attitudes about patients' home-based traditional practices?
- *Approachability:* Do patients feel welcomed? Do health-care providers and receptionists greet patients in the manner in which they prefer? This includes greeting patients with their preferred names.
- *Alternative practices and practitioners:* Do biomedical providers incorporate patients' alternative or complementary practices into treatment plans?
- *Additional services:* Are child- and adult-care services available if a parent must bring children or an aging parent to the appointment with them?
- *Literacy:* Language has been identified as the biggest barrier to health care and not just for those for whom English is a second language.

Reflective Exercise

1. Looking at the list of barriers to health care, which apply to you?
2. How can you decrease these barriers?
3. What are the barriers to health care in your community?
4. Looking at the list of barriers to health care, which apply to you?
5. How can you decrease these barriers?
6. What are the barriers to health care in your community?

Health-care providers can help reduce some of these barriers by calling an area ethnic agency or church for assistance, establishing an advocacy role, involving professionals and laypeople from the same ethnic group as the patient, using cultural brokers, and organizationally providing culturally congruent and linguistically appropriate services. If all of these elements are in place and used appropriately, they have the potential of generating culturally congruent and responsive care.

2.14.5 Cultural Responses to Health and Illness

Significant research has been conducted on patients' responses to pain, which has been called the "fifth vital sign." Most health-care professionals believe that patients should be made comfortable and not have to tolerate high levels of pain. Accrediting bodies survey organizations to ensure that patients' pain levels are assessed and that appropriate interventions are instituted.

A number of studies related to pain and the ethnicity/culture of the patient have been completed. Most of the studies have come from end-of-life care.

- Communication between patient and health-care provider influences pain diagnosis and treatment.
- The brain's pain-processing and pain-killing systems vary by race and ethnicity.
- Few patients are told in advance about possible side effects of pain medicine and how to manage them.
- African American, Hispanic, and other groups with severe pain are less likely than White patients to be able to obtain needed pain medicine because they live in communities that are crime ridden and the pharmacies do not carry the medicines.
- African Americans are less likely to have their pain recorded (see population specific chapters).
- Inadequate education of pain and analgesia expectations may contribute to poor pain relief in the Asian populations.
- Disparities in pain management and quality care at end of life exist among African American women in general and, specifically, those with breast cancer.

- Hispanic patients are more likely to describe pain as "suffering," the emotional component. African Americans are more likely to describe pain as "hurts," the sensory component.
- Socioeconomic factors negatively influence prescribing pain medicine.
- Pain does not have the same debilitating effect for patients from Eastern cultures as it does for patients from Western cultures.
- Stoicism, fatalism, family, and spirituality have a positive impact on Hispanics and pain control.
- Most Chinese, Korean, and Vietnamese patients do not favor taking pain medicine over a long period of time.
- Vietnamese Canadians prefer herbal therapies over prescription pain medicine (Voyer et al. 2005).
- Many Haitians, Haitian Americans, and Haitian Canadians combine herbal therapies with prescription medicine without telling the health-care provider (Voyer et al. 2005).
- Black, Hispanic, and Asian women receive less epidural analgesia than do White women.
- Cultural background, worldview, and variant characteristics of culture influence the pain experience.
- The greater the language differences, the poorer the pain control.
- For Asians, tolerating pain may be a way of atoning for past sins.

Pain scales are in different languages and with faces appropriate to the language and ethnicity of the patient. Additional resources for pain are the American Pain Foundation, The American Pain Society, the Boston Cancer Pain Education Center (in 11 languages), and the OUCHER Pain scale for children (OUCHER!, n.d.), all of which are available on the Internet. Health-care practitioners must investigate the meaning of pain for each person within a cultural explanatory framework to interpret diverse behavioral responses and provide culturally congruent care. The health-care provider may need to offer and encourage pain medication and explain that it can help the healing progress.

Reflective Exercise

1. What is your first line of intervention when you are having pain?
2. When do you decide to see a health-care practitioner when you are in pain?
3. What differences do you see between yourself and others when they are in pain?
4. Where did you learn your response to pain?
5. Do you see any difference in the clinical setting in response to pain among ethnic and cultural groups?
6. Between men and women?

The manner in which mental illness is perceived and expressed by a cultural group has a direct effect on how individuals present themselves and, consequently, on how health-care providers interact with them. In some societies, mental illness may be seen by many as not being as important as physical illness. Mental illness is culture-bound; what may be perceived as a mental illness in one society may not be considered a mental illness in another. For some, mental illness and severe physical handicaps are considered a disgrace and taboo. As a result, the family is likely to keep the mentally ill or handicapped person at home as long as they can. This practice may be reinforced by the belief that all individuals are expected to contribute to the household for the common good of the family, and when a person is unable to contribute, further disgrace occurs. In some cultures, children with a mental disability are stigmatized. The lack of supportive services may cause families to abandon their loved ones because of the cost of long-term care and the family's desire and desperate need for support. Such children may be kept from the public eye in hope of saving the family from stigmatization.

Reflective Exercise

1. What are your perceptions about mental illness?

2. Does mental illness have the same value as physical illness and disease?
3. When you are having emotional difficulties, what is your first line of defense?
4. Have you observed different attitudes/responses from providers regarding physical and mental illnesses?

Rehabilitation and occupational health services focus on returning individuals with handicaps to productive lifestyles in society as soon as possible. The goal of the health-care system is to rehabilitate everyone: convicted individuals, people with alcohol and drug problems, as well as those with physical conditions. To establish rapport, health-care practitioners working with patients suffering from chronic disease must avoid assumptions regarding health beliefs and provide rehabilitative health interventions within the scope of cultural customs and beliefs. Failure to respect and accept patients' values and beliefs can lead to misdiagnosis, lack of cooperation, and alienation of patients from the health-care system.

Reflective Exercise

1. Do you see physically challenged individuals as important as non-physically challenged individuals in terms of their worth to society?
2. What are your beliefs about rehabilitation?
3. Should everyone have the opportunity for rehabilitation?

Sick role behaviors are culturally prescribed and vary among ethnic societies. Traditional individualistic cultural practice calls for fully disclosing the health condition to the patient. However, traditional collectivistic families may prefer to be informed of the bad news first, and then slowly break the news to the sick family member. Given the ethnocultural acceptance of the sick role, health-care providers must assess each patient and family individually and incorporate culturally congruent therapeutic interventions to return the patient to an optimal level of functioning.

Reflective Exercise

1. What do you normally do when you have a minor illness?
2. Do you go to work (class) anyway?
3. What would make you decide not to go to work or class?
4. Does the sick role have a specific meaning in your culture?

2.14.6 Blood Transfusions and Organ Donation

Most religions favor organ donation and transplantation and transfusion of blood or blood products. Jehovah's Witnesses do not believe in blood transfusions. Some individuals and cultures choose not to participate in organ donation or autopsy because of their belief that they will suffer in the afterlife or that the body will not be whole on resurrection. Health-care providers may need to assist patients in obtaining a religious leader to support them in making decisions regarding organ donation or transplantation.

Reflective Exercise

1. Are you averse to receiving blood or blood products? Why? Why not?
2. Are you an organ donor? Why? Why not?

Some people do not sign donor cards because the concept of organ donation and transplantation is not customary in their homelands. Health-care providers should supply information regarding organ donation on an individual basis, be sensitive to individual and family concerns, explain procedures involved with organ donation and procurement, answer questions factually, and explain involved risks. A key to successful marketing approaches for organ donation is cultural awareness (see Table 2.1: Section Health-care Practices).

2.15 Health-Care Providers

The domain *health-care providers* includes the status, use, and perceptions of traditional, magico-religious, and biomedical health-care providers. This domain is interconnected with communications, family roles and organization, and spirituality. In addition, the gender of the health-care provider may be significant for some people.

2.15.1 Traditional Versus Biomedical Providers

Most people combine the use of biomedical health-care providers with traditional practices, folk healers, and/or magico-religious healers. The health-care system abounds with individual and family folk practices for curing or treating specific illnesses. A significant percentage of all care is delivered outside the perimeter of the formal health-care arena. Many times, herbalist-prescribed therapies are handed down from family members and may have their roots in religious and cultural beliefs. Traditional and folk practices often contain elements of historically rooted beliefs.

Reflective Exercise

1. What alternative health-care providers do you see for your health-care needs besides traditional allopathic-care providers?
2. For what conditions do you use nonallopathic providers?
3. Do you think traditional health-care providers are as valuable as allopathic health-care providers?

The traditional practice in the United States and other countries is to assign staff to patients regardless of gender differences, although often an attempt is made to provide a same-gender health-care provider when intimate care is involved, especially when the patient and caregiver are of the same age. However, health-care providers should recognize and respect differences in gender relationships when providing culturally competent care because not all ethnocultural groups accept care from someone of the opposite gender. Health-care providers need to respect patients' modesty by providing adequate privacy and assigning a same-gender caregiver whenever possible.

Reflective Exercise

1. Do you prefer a same-gender health-care provider for your general health care?
2. Do you mind having an opposite-gender provider for intimate care? Why? Why not?
3. Do you prefer Western-trained health-care providers or does it not make any difference?

2.15.2 Status of Health-Care Providers

Health-care providers are perceived differently among ethnic and cultural groups. Individual perceptions of selected health-care providers may be closely associated with previous contact and experiences with health-care providers. In many Western societies, health-care providers, especially physicians, are viewed with great respect,

Reflective Exercise

1. Does one type of health-care provider have increased status over another type?

2. Should all health-care providers receive equal respect, regardless of educational requirements?
3. Does the ethnicity or race of a provider make any difference to you? Why? Why not?

although recent studies show that this is declining among some groups.

Although many nurses in the United States do not believe they are respected, public opinion polls usually place patients' respect of nurses higher than that of physicians. The advanced practice role of registered nurses is gaining respect as more of them have successful careers and the public sees them as equal or preferable to physicians and physician assistants in many cases.

Depending on the country of origin and experience of working with professional nurses, some physicians may misunderstand the assertive behavior of Western-educated nurses because in their home country, nurses were not expected to be assertive. Some patients perceive older male physicians as being of higher rank and more trustworthy than younger health professionals, especially for patients who come from a collectivist culture where they are taught from a very early age to respect elders and to show deference to nurses and physicians, regardless of gender or age.

Evidence suggests that respect for professionals is correlated with their educational level. In some cultures, the nurse is expected to defer to physicians. In many countries, the nurse is viewed more as a domestic than as a professional person, and only the physician commands respect. Health beliefs are not border bound. People bring their beliefs with them upon migration.

In some cultures, folk and magico-religious health-care providers may be deemed superior to biomedically educated physicians and nurses. It may be that folk, traditional, and magico-religious health-care providers are well known to the family and provide more individualized care. In such cultures, health-care providers take time to get to know patients as individuals and engage in small talk totally unrelated to the health-care problem to accomplish their objectives. Establishing satisfactory interpersonal relationships is essential for improving health care and education in these ethnic groups (see Table 2.1: Section Health-care Practitioners).

References

Anasthasia (2015) Understanding culture and people with Hofstede dimensions. https://www.cleverism.com/understanding-cultures-people-hofstede-dimensions

Burroughs Valentine J, Randall W, Levy RA (2002) Racial and ethnic differences in response to medicines: towards individualized pharmaceutical treatment. J Natl Med Assoc 2002(10):1–26

Caetano R, Clark CL, Tam T (1998) Alcohol consumption among racial/ethnic minorities. Alcohol Health Res World 22(4):233–242

CIA World Factbook (2019) The World. https://www.cia.gov/library/publications/the-world-factbook/geos/xx.html

Hilton M (1987) Demographic characteristics and the frequency of heavy drinking as predictors of drinking problems. Br J Addict 82:913–925

National Center for Complementary and Integrative Health (2019). https://www.nih.gov/about-nih/what-we-do/nih-almanac/national-center-complementary-integrative-health-nccih

Peacock M, Buckwalter KA, Persohn S, Hangatner TN, Econs MJ, Hui S (2009) Race and sex differences in bone mineral density and geometry at the femur. Bone 45(2):218–225

Purnell L (2011) Application of transcultural theory to mental health-substance use in an international context. In: Cooper DB (ed) Interventions in mental health-substance use. Radcliffe Publishing, London

Purnell L, Foster J (2003a) Cultural aspects of alcohol use: part I. Drug Alcohol Professional 3(3):17–23

Purnell L, Foster J (2003b) Cultural aspects of alcohol use: part II. Drug Alcohol Professional 2(3):3–8

Voyer P, Rail G, Laberge S, Purnell L (2005) Cultural minority older women's attitudes toward medication and implications for adherence to a drug regimen. J Divers Health Soc Care 2(1):47–61

West CR, Kahn JH, Nauta MM (2007) Learning styles as predictors of self-efficacy and interest in research: implications for graduate research training. Train Educ Prof Psychol 1(3):174–183

World Health Organization: Immunizations (2019). https://www.who.int/topics/immunization/en/

3 Individual Competence and Evidence-Based Practice (with Inclusion of the International Standards)

Susan W. Salmond

3.1 Introduction

In general, cultural competence is a journey involving the willingness and ability of an individual to deliver culturally congruent and acceptable health and nursing care to the patients to whom one provides care. To this author, individual cultural competence can be arbitrarily divided among cultural general approaches, the clinical encounter, and language.

3.2 Individual Cultural Competence

3.2.1 Self-Awareness and Health Professions

Culture has a powerful unconscious impact on patients and health professionals. Culture in health-care settings is extremely complex for the following reasons:

- Each patient has a culture.
- There is diversity within cultural groups stemming from dimensions such as race, ethnicity, gender, class background, education, and immigration status.
- Each health-care provider has a culture that may be different from that of the patient.
- Each profession, nursing, medicine, physical therapy, occupational therapy, and social work, to name a few, has a culture.
- Each specialty such as medicine, surgery, gerontology, psychiatry, hospice/palliative care, pediatrics, rehabilitation, to name a few, has a subculture of the dominant professional culture.
- Each organization has a culture with subcultures within each organization.

When all these competing cultures and subcultures are combined, a significant mismatch may occur, increasing the complexity of providing culturally competent care. The way health-care providers perceive themselves as competent providers is often reflected in the way they communicate with patients. Thus, it is essential for health-care providers to think about their cultures, their behaviors, and their communication styles in relation to their perceptions of cultural differences. They should also examine the impact their beliefs have on others, including patients and co-workers, who are culturally diverse. Before addressing the multicultural backgrounds and unique individual perspectives of each patient, health-care providers must first address their own personal and professional knowledge, values, beliefs, ethics, biases, and life experiences in a manner that optimizes interactions and assessment of culturally diverse individuals.

Self-knowledge and understanding promote strong professional perceptions that increase prejudice awareness and allow health care providers

S. W. Salmond (✉)
Rutgers School of Nursing, Newark, NJ, USA
e-mail: susan.salmond@rutgers.edu

© Springer Nature Switzerland AG 2021

L. D. Purnell, E. A. Fenkl (eds.), *Textbook for Transcultural Health Care: A Population Approach*,
https://doi.org/10.1007/978-3-030-51399-3_3

to interact with others in a manner that preserves personal integrity and respects uniqueness and differences among individual patients. The process of professional development and diversity competence begins with *self-exploration* or critical reflection. Although the literature provides numerous definitions of *self-awareness*, discussion of research integrating the concept of self-awareness with multicultural competence is minimal. Many theorists and diversity trainers imply that self-examination or awareness of personal prejudices and biases is an important step in the cognitive process of developing cultural competence (Boyle and Andrews 2011; Calvillo et al. 2009; Giger et al. 2007). Complicating this self-examination is the fact that some bias or stereotypes may not be consciously endorsed but may negatively affect communication, compassion, and equitable care (Chapman et al. 2013; Gatewood et al. 2019; Narayan 2019). These unintended biases, or implicit bias, can negatively impact culturally competent care and patient outcomes. Discussions of emotional feelings elicited by this cognitive awareness are somewhat limited, given the potential impact of emotions and conscious feelings on behavioral outcomes.

- *In your opinion, why is there conflict about working with culturally diverse patients? What attitudes are necessary to deliver quality care to patients whose culture is different from yours?*
- *Consider a time when you made an assumption about a patient that affected your ability to care for this patient. How did this make you feel?*

Self-awareness in cultural competence is a deliberate and conscious cognitive and emotional process of getting to know yourself: your personality, your values, your beliefs, your biases, your professional knowledge standards, your ethics, and the impact of these factors on the various roles you play when interacting with individuals different from yourself. Critically analyzing our own values and beliefs in terms of how we see differences enables us to be less fearful of others whose values and beliefs are different from our own (Calvillo et al. 2009). The ability to understand oneself sets the stage for integrating new knowledge related to cultural differences into the health-care provider's knowledge base, perceptions of health, interventions, and the impact these factors have on the various roles of professionals when interacting with multicultural patients.

- *What have you done in the last 5–10 years to increase your self-awareness? Has increasing your self-awareness resulted in an increased appreciation for cultural diversity? How might you increase your knowledge about the diversity in your community? In your school?*

3.2.2 Measuring Individual Cultural Competence

Loftin and colleagues (2013) integrative review of measures of cultural competence provides an overview of 11 instruments measuring cultural competence capturing select domains of awareness, knowledge, sensitivity, attitudes, desire, and skills. They identity 4 cultural general tools that have undergone thorough psychometric testing including the Transcultural Self-Efficacy Scale (Jeffreys 2000), Campinha-Bacote's Inventory for Assessing the Process of Cultural Competence among Healthcare Professionals Revised (Campinha-Bacote 2002), the Cultural Competence Assessment (Schim et al. 2003) and the Nurse Cultural Competence Scale (Perng and Watson 2012). A limitation of the tools is that they are self-report and these responses may reflect socially acceptable responses rather than the most accurate responses. Using an internet search engine such as scholar.google.com and entering "cultural competence measurement" or "cultural competence assessment tools" in the search field will provide an array of tools that one can select from.

In addition to assessment tools, standards for cultural competence provide a framework from which to judge cultural competence. The Office of Minority Health's Standards for Culturally and Linguistically Appropriate Services in Health and Health Care (CLAS Standards) establish a framework to advance health equity, improve quality, and help to eliminate health care disparities by providing a path for health facilitates to achieve culturally and linguistically appropriate services (thinkculturalhealth.hhs.gov/clas). Cultural Competence Standards (www.omhrc.gov).

The American Academy of Nursing with representatives from the Transcultural Nursing Society has developed Standards of Practice for Culturally Competent Nursing Care based on social justice (Douglas et al. 2011). Box 3.1 lists the 12 standards. These standards were incorporated into guidelines for implementing culturally competent nursing from both the caregivers and health care organizations Leaders/Managers perspectives (Douglas et al. 2014).

Box 3.1 Standards of Practice for Culturally Competent Nursing Care

- **Standard 1: Social Justice**

Professional nurses shall promote social justice for all. The applied principles of social justice guide decisions of nurses related to the patient, family, community, and other health-care professionals. Nurses will develop leadership skills to advocate for socially just policies.

- **Standard 2: Critical Reflection**

Nurses shall engage in critical reflection of their own values, beliefs, and cultural heritage in order to have an awareness of how these qualities and issues can impact culturally congruent nursing care.

- **Standard 3: Knowledge of Cultures**

Nurses shall gain an understanding of the perspectives, traditions, values, practices, and family systems of culturally diverse individuals, families, communities, and populations for whom they care, as well as knowledge of the complex variables that affect the achievement of health and well-being.

- **Standard 4: Culturally Competent Practice**

Nurses shall use cross-cultural knowledge and culturally sensitive skills in implementing culturally congruent nursing care.

- **Standard 5: Cultural Competence in Healthcare Systems and Organizations**

Healthcare organizations should provide the structure and resources necessary to evaluate and meet the cultural and language needs of their diverse patients.

- **Standard 6: Patient Advocacy and Empowerment**

Nurses shall recognize the effect of healthcare policies, delivery systems, and resources on their patient populations, and shall empower and advocate for their patients as indicated. Nurses shall advocate for the inclusion of their patients' cultural beliefs and practices in all dimensions of their health care when possible.

- **Standard 7: Multicultural Workforce**

Nurses shall actively engage in the effort to ensure a multicultural workforce in health-care settings. One measure to achieve a multicultural workforce is through strengthening of recruitment and retention effort in the hospital and academic setting.

- **Standard 8: Education and Training in Culturally Competent Care**

Nurses shall be educationally prepared to promote and provide culturally congruent health care. Knowledge and skills necessary for ensuring that nursing care is culturally congruent shall be included in global health-care agendas that mandate formal education and clinical training, as well as required ongoing, continuing education for all practicing nurses.

- **Standard 9: Cross-Cultural Communication**

Nurses shall use culturally competent verbal and nonverbal communication skills to identify patient's values, beliefs, practices, perceptions, and unique health-care needs.

- **Standard 10: Cross-Cultural Leadership**

Nurses shall have the ability to influence individuals, groups, and systems to achieve positive outcomes of culturally competent care for diverse populations.

- **Standard 11: Policy Development**

Nurses shall have the knowledge and

skills to work with public and private organizations, professional associations, and communities to establish policies and standards for comprehensive implementation and evaluation of culturally competent care.

• **Standard 12: Evidence-Based Practice and Research**

Nurses shall base their practice on interventions that have been systematically tested and shown to be the most effective for the culturally diverse populations that they serve. In areas where there is a lack of evidence of efficacy, nurse researchers shall investigate and test interventions that may be the most effective in reducing the disparities in health outcomes.

3.2.3 Cultural General Approaches

A number of general approaches exist to help health-care providers achieve cultural competence, including the following:

1. Developing an awareness of oneself as a racial/cultural being and of the biases, stereotypes, and assumptions that influence worldviews.
2. Developing an awareness of the worldviews of culturally diverse patients and staff.
3. Continuing to learn cultures of patients to whom one provides care.
4. Demonstrating knowledge and understanding of the patient's culture, health-related needs, and meanings of health and illness.
5. Accepting and respecting cultural differences in a manner that facilitates the patient's and the family's ability to make decisions to meet their needs and beliefs.
6. Recognizing that the health-care provider's beliefs and values may not be the same as the patient's.
7. Resisting judgmental attitudes such as "different is not as good."
8. Being open to new cultural encounters.
9. Recognizing that variant cultural characteristics determine the degree to which patients adhere to the beliefs, values, and practices of their dominant culture.
10. Having contact and experience with the communities from which patients come.
11. Being willing to work with patients of diverse cultures and subcultures.
12. Accepting responsibility for one's own education in cultural competence by attending conferences, reading literature, and observing cultural practices.
13. Promoting respect for individuals by discouraging racial and ethnic microaggressions (brief and commonplace verbal, behavioral or environmental indignities that may be intentional or unintentional that communicate hostile or negative racial slights and insults toward people of color (Sue et al. 2009).
14. Intervening with staff behavior that is insensitive, lacks cultural understanding, or reflects prejudice.
15. Having a cultural general framework for assessment as well as having cultural-specific knowledge about the patients to whom care is provided.

3.2.4 The Clinical Encounter

The clinical encounter is a rich area for learning about and becoming more culturally competent. Cultural competence requires that we recognize heterogeneity within cultural groups surrounding communication, socio-cultural issues, behavioral patterns and perceptions of health and illness and their explanatory model and preferences for treatment. Clinical assessment and rounds should not just focus on pathophysiology and management, but the patients cultural background and its impact on disease and heath behavior. Some specific approaches that are helpful in becoming more culturally competent are:

- Adapting care to be congruent with the patient's culture and worldviews.

- Responding respectively to all patients and their families (includes addressing patients and family members as they prefer, formally or informally).
- Collecting cultural data on assessments to capture the individual view versus the stereotypical views of cultural groups.
- Forming generalizations as a method for formulating questions rather than stereotyping.
- Recognizing culturally based health-care beliefs and practices.
- Knowing the most common diseases and illnesses affecting the unique population to whom care is provided.
- Individualizing care plans to be consistent with the patient's cultural beliefs.
- Having knowledge of the communication styles of patients to whom you provide care.
- Accepting varied gender roles and childrearing practices from patients to whom you provide care.
- Having a working knowledge of the religious and spirituality practices of patients to whom you provide care.
- Having an understanding of the family dynamics of patients to whom you provide care.
- Using faces and language pain scales in the ethnicity and preferred languages of the patients.
- Recognizing and accepting traditional, complementary, and alternative practices of patients to whom you provide care.
- Incorporating patient's cultural food choices and dietary practices into care plans and
- Incorporating patient's health literacy into care plans and health education initiatives.

3.3 Language Interpretation, Health Literacy, and Translation

3.3.1 Language

Language ability, as mentioned previously, is the biggest barrier to effective health-care access, diagnosis, assessment, and comprehension of medication and health prescription instructions. Strategies for improving language ability for effective communication with patients and family follow:

- Explaining things in a way that patients and families can understand–avoid medical jargon, speak slowly and use teach back when indicated.
- Developing skills and using interpreters (includes sign language) trained in cultural competence and health literacy with patients and families who have limited English proficiency.
- Ensuring that patient education materials are formatted simply and clearly, written in simple language, and use images and infographics to reinforce clinical concepts.
- Providing patients with educational documents that are culturally congruent, inclusive, and translated into their preferred language.
- Providing discharge instructions at a level the patient and the family understand and in the language the patient and the family prefer.
- Providing medication and treatment instructions in the language the patient prefers.
- Using pain scales in the preferred language of the patient.

Look at the list of activities that promote individual cultural competence.

- *Which of these activities have you used to increase your cultural competence?*
- *Which ones can you easily add to increase your cultural competence?*
- *Which ones are the most difficult for you to incorporate?*

Provider cross-cultural skills and interpreter services, and written patient materials in different languages and at a low level of health literacy, ensure that patients understand their options, choices, costs, and benefits. The health-care provider and the organization should become familiar with the Code of Ethics for Medical Interpreters (1987) and make the National Standards on Interpreting in Health Care (2007) best practices in your organization.

Some recommendations for interpretation are shown in Box 3.2.

Box 3.2 Recommendations for Working with an Interpreter

Use interpreters who can decode the words and provide the meaning behind the message.
Use dialect-specific interpreters whenever possible.
Use interpreters trained in the health-care field.
Give the interpreter time alone with the patient.
Provide time for translation and interpretation.
Use same-gender interpreters whenever possible.
Maintain eye contact with both the patient and the interpreter to elicit feedback: Read nonverbal cues.
Speak slowly without exaggerated mouthing, allow time for translation, use the active rather than the passive tense, wait for feedback, and restate the message. Do not rush; do not speak loudly.
Use as many words as possible in the patient's language and nonverbal communication when unable to understand the language.
Use phrase charts and picture cards if available.
During the assessment, direct your questions to the patient, not to the interpreter.
Ask one question at a time, and allow interpretation and a response before asking another question.
Be aware that interpreters may affect the reporting of symptoms, insert their own ideas, or omit information.
Remember that patients can usually understand more than they can express; thus, they need time to think in their own language. They are alert to the health-care provider's body language, and they may forget some or all of their English in times of stress.
Avoid using children as interpreters, especially with sensitive topics.
Avoid the use of relatives, who may distort information or not be objective.
Avoid idiomatic expressions and medical jargon.
If a certified interpreter is unavailable, the use of an uncertified interpreter may be acceptable; however, the difficulty might be omission of parts of the message or distortion of the message, including transmission of information not given by the speaker and messages not being fully understood.
If available, use an interpreter who is older than the patient.
Review responses with the patient and interpreter at the end of a session.
Be aware that social class differences between the interpreter and the patient may result in the interpreter's not reporting information that he or she perceives as superstitious or unimportant.

3.3.2 Health Literacy

Health literacy is the ability to find, understand, and use information about health and services needed to make appropriate health decisions (Mabachi et al. 2016). Over 40% of adults have significant literary challenges, and 88% of adults have less than "proficient" health literacy skills. Health literacy deficits can be a significant contributor to poor health outcomes and disparities. Low health literacy and cultural barriers may be a reason for delay in seeking care and engaging in a care regime. Nurses need to participate in conducting health literacy and cultural and linguistic competence audits at a unit and organizational level, examine written and spoken communication and examine difficulties in navigating facilities and complex systems in order to contribute to changes needed to become a health literate organization.

The health care provider and the facility should adopt a 'universal precautions' approach and assume that all patients have gaps in health literacy. The Agency for Healthcare Research and Quality has a Health Literacy Universal Precautions Toolkit available at https://www.ahrq.gov. This guide provides 21 tools addressing spoken communication, written communication, self-management and empowerment, and supportive systems. The Centers for Medicare & Medicaid Services also has a comprehensive toolkit for making written material clear and effective that is available on-line at cms.gov. In interacting with patients be alert to "red flag" statements that ms suggest literacy problems such as "I will read it later as I forgot my glasses" or "I'll let my daughter fill that out for me when I get home." Health care providers should ask the patient if he or she needs assistance in completing forms and offer to assist the patient.

3.3.3 Translation

Whereas interpretation is verbal, translation is written. Sometimes a patient may have adequate verbal skills for understanding health-related information but not have adequate reading skills. To help ensure health literacy of written materials, present materials at a fifth-grade or lower reading level, use bulleted points for crucial information, and translate materials into the patient's preferred language. Health education materials should be reviewed for literacy level, quality of translation, and cultural appropriateness. The National Network of Libraries of Medicine (nnlm.gov) has a range of resources on health information, health literacy, and translation.

3.4 Evidence-Based Practice and Culturally Congruent Best Practices

3.4.1 Health Professions Education

The 2003 IOM report, *Health Professions Education: A Bridge to Quality* called for an overhaul of health professions education surrounding 5 core areas: patient-centered care, interdisciplinary teamwork, evidence-based practice, quality improvement, and informatics. In response to this call, the Robert Wood Johnson Foundation supported a 10-year Quality and Safety in Nursing Education (QSEN) project to define nursing competencies in the five core areas. A sixth competency, safety, was added by the QSEN group. After the knowledge, skills, and attitudes were identified for each competency, the American Association of Colleges of Nursing became a partner to develop teaching strategies that address identified competencies at both the baccalaureate and graduate level (www.aacn.nche.edu). As described in Box 3.3, these core areas are important to evidence-based practice and culturally congruent care.

Box 3.3 Quality and Safety Education for Nurses

Patient-Centered Care

Patient-centered care accounts for and recognizes that the patient and family are co-participants to ensure culturally competent care and that their cultural preferences, values, and needs are addressed (see cultural general approaches and the clinical encounter in this chapter). In order for preferences to be adequately addressed, culturally acceptable communication is a core requirement for developing trust with full disclosure (Institute for Patient and Family Centered Care: www.ipfcc.org). Shared decision-making bridges patient-centered care and evidence based practice. Accomplishing this requires that health-care providers respect patients' expertise and recognize that beliefs and values of the health-care providers may not be the same as those of the patients.

Teamwork and Cultural Care

Well functioning multidisciplinary teams improve health outcomes and limit adverse event (Epstein 2014). Cultural competence has been linked to better teamwork (Kumra et al. 2020). Given the complexity of the culture of the patient, the cultures of individual team members, the cultures of the professions and specialties, and the culture of the organization (see Chap. 4), to function effectively, mutual respect and collaboration are essential to coordinated, reliable, quality care. Awareness of individual and other team member roles and interacting without professional power dominance increases team satisfaction and team success. The Interprofessional Education Collaborative Expert Panel published core competencies for interprofessional collaborative practice (2011).

Evidence-Based Practice

Evidence-based practice (EBP) is a key to promoting best practices and avoiding the underuse, misuse, and overuse of care. EBP requires not only an ability to access and appraise best evidence but to use this evidence as only one factor in clinical decision making along with clinical expertise, clinical context, and patient values and wishes. Primary research evidence that is culturally congruent with the patient being served is often lacking. This requires adaptation of best practices or clinical guidelines to accommodate cultural preferences and values.

Quality Improvement

Quality improvement in cultural care is a combination of individual and organizational cultural competence (see Chap. 4). Quality cultural care cannot occur without the support of the organization in which care is delivered and should include expertise from the community it serves.

Safety

Safety and minimizing risk are concerns on multiple levels, but specific to cultural aspects health literacy and interpretation and translation services (see language interpretation, translation, and health literacy in this chapter) take the forefront. Health-care providers in all professions must discuss safety issues with their patients in both home and environmental contexts.

Informatics

Informatics and information technology are key to providing solutions to improve the efficacy of health systems. Application of informatics to administrative and financial management, an explosion of health-related information available via the Internet and health related apps designed to promote wellness and assist individuals to manage chronic illness, and gains in making evidence syntheses, practice guidelines and health services research more accessible are key ways informatics is changing care.

3.4.2 Evidence-Based Practice

The mandate for evidence-based practice (EBP) to reduce the "know-do" gap (Antes et al. 2006; Knighton et al. 2019) between known science and implementation in practice has been driven by the demand for improved safety and quality outcomes for clients. This has necessitated a culture shift from an opinion-based culture grounded in intuition, clinical experience/expertise, and pathophysiological rationale (Swanson et al. 2010) to a culture of EBP in which there is conscientious, explicit, and judicious use of current best evidence in making decisions about the care of individuals or groups of patients. The aim of EBP is to deliver appropriate care in an efficient manner to patients and to reduce illogical variation in care that may lead to quality concerns and unpredictable health outcomes (Stevens 2013).

Appropriate care stems from best evidence; however, EBP is more than evidence alone. The four components of EBP, all of which contribute to the best patient outcomes, are best research evidence, clinical expertise, patient preferences and values, and clinical context. Evidence-based practice is operationalized when evidence, clinical expertise, clinical context, and patient preferences and values inform one another in a positive way (Jylhä et al. 2017). Although a greater emphasis is frequently given to the evidence component of EBP, no one component is the most important; rather, the weight given to each component varies according to the clinical situation (Melnyk et al. 2009). Table 3.1 summarizes the components of EBP and the actions and resources needed to facilitate its implementation.

Table 3.1 portrays the dynamic nature of the process as it combines the four core components contributing to clinical decision making. It summarizes the components of EBP and the actions and resources needed to facilitate its implementation.

3.4.3 Best Research Evidence

Best evidence refers to approaches to treatment or prevention that have shown to be effective or that provides information about care based on qualitative evidence of the patient experience across multiple research studies.

Table 3.1 The evidence-based practice process

	Components	Resources/change needed
Identify best evidence	*Clinical inquiry:* What knowledge is needed? *Informed skepticism:* Why are we doing it this way? Is there a better way to do it? What is the evidence for what we do? Would doing this be as effective as doing that? (Salmond 2007)	Shift from "know how" and doing to "know why" Reflect on what information is needed to provide "best" care Generate questions about practice and care Role model clinical inquiry at report, rounds, conferences Use interdisciplinary case reviews to evaluate actual care Include clinical librarians as members of teams participating in clinical rounds and conferences
	Convert information needs from practice into focused, searchable questions (patient intervention-comparison-outcome [PICO] framework)	Consider recurring clinical issues, need for information, negative incidents/events or QI concerns as sources for questions or information needs Identify clinical issues sensitive to nursing interventions Narrow broad clinical issues/questions into searchable, focused questions Use the mnemonic PICO to frame questions Specify population of interest by using the specific cultural group identifier or broader terms such as multiethnic, multicultural
	Search databases for highest level of evidence in a timely manner	Use evidence-searching skills to target relevant focused evidence Begin search with filtered resources- systematic reviews, clinical guidelines, and appraised single studies Understand the match between question and design Search strategies: Key terms, multiple databases, point-of-care data Use assistance of clinical librarian
	EB practitioner leaders appraise unfiltered literature to determine strength and validity of evidence and relevance to one's practice, summarize findings, and assist clinicians with understanding research design	Demonstrate knowledge of research design Demonstrate knowledge of statistics Use critical appraisal tools to guide process of research critique Utilize journal clubs Summarize findings from evaluation, resolving conflicting evidence
Clinical experience and expertise	Use clinical expertise to determine how to use evidence in care of patient and how to manage patient in absence of evidence or presence of conflicting evidence	Consider evidence in relation to own patient population Consider cost-benefit ratio Consider multidimensionality of patient and clinical situation in relation to evidence that is often reductionistic Ensure holistic assessment and planning inclusive of the social and cultural context
Patient values and preferences	Practice from a patient-centered paradigm Demonstrate ability to perform a culture assessment and identify patient preferences and values that inform the clinical decision	Understand culture-general and culture-specific knowledge to guide interactions with patient Use interview skills to avoid culture imposition and seek client's true preferences Communicate evidence and treatment options considering patient values and preferences using decision aids when available Involve patient and family in both information giving and decision making
Translation of evidence from total process into clinical decisions and strategies for best patient outcomes	Use all four components in clinical decision-making process and implementation of clinical decision	Provide plan of care based on evidence, clinical judgment, patient preferences, and organizational context Use implementation frameworks for translating evidence into practice throughout an organization or care site including the community Evaluate EBP guidelines and evidence that meets the clinical and cultural needs of the community Make cultural modifications in guidelines or best practice based on input of providers and community members
Monitor patient outcomes	Use outcome tools to track patient outcomes	Develop audit systems to track patient outcomes Make clinical outcomes accessible electronically for analysis Analyze outcomes and effectiveness of "evidence-based" clinical intervention

3.4.3.1 Moving to "As Needed" Evidence

When one considers the volume of evidence, more than 1800 new articles and 75 new clinical trials per day, the impossibility of staying current in all of the conditions and situations that patients present with becomes apparent. Practicing from an evidence-based (EB) perspective requires the clinician to recognize these knowledge limitations and to reflect on their ongoing practice to determine what evidence they are relying on and when they need evidence. Ask "why" things are being done as they are, "whether" there is evidence supporting the approach, or "what" the evidence suggests may be best in this clinical situation and "whether" there are likely to be contextual or cultural considerations that necessitate examining evidence specific to your cultural group (Salmond 2007). With this clinical inquiry approach, the clinician can pose questions in search of best evidence.

3.4.3.2 Asking Clinical Questions to Gather Evidence

Practicing from an EB perspective requires asking clinical questions and searching for the evidence to guide practice. There is a technique to asking clinical questions so that the evidence can be retrieved quickly and efficiently. The key is to ask focused questions. Who is the population of interest? What is the intervention or phenomenon of interest? What is the outcome of interest? If the broad interest is duration of breastfeeding, it should be narrowed further. Determine if evidence for a particular cultural or social group is needed, and specify this population by using key words as outlined in Table 3.2. For effectiveness questions, narrow down what the intervention of interest is—for example, barriers to breastfeeding, interventions to support breastfeeding, or educational and support programs to encourage breastfeeding. Then select an outcome that defines how success of the program will be measured—for example, initiation of breastfeeding, exclusive breastfeeding, or duration of breastfeeding. Putting it all together, the question might be "Does a structured doula program (support program) affect the initiation of breastfeeding among urban minority women?" If the interest is to gain greater awareness of the patient experience, a qualitative question may be asked such as "What is the experience of breastfeeding for teenage mothers?" Table 3.2 provides a quick reference to asking clinical questions.

3.4.3.3 Locating Best Evidence from the Literature

The question has been asked, and the "best" evidence needs to be located. For the busy clinician, the key is to be able to find this evidence quickly for use in clinical decision-making. This means, that primary research studies, that need to be appraised and summarized with other similar studies, are not the go-to source. Rather, the starting place should be sources where evidence has been pre-processed or filtered, meaning it has already been critically appraised and determined to be of sufficient rigor to be considered for application into practice (Lehane et al. 2019). These sources include systematic reviews, critically appraised topics (guidelines), and critically appraised individual articles.

3.4.3.4 Systematic Reviews

As one study is not enough to change practice, the best evidence often comes from systematic reviews that have pooled the primary research data for assessment and summarization. By using a systematic review, the clinician can generally rely on the fact that a comprehensive search for all available information was done and the information was systematically reviewed for rigor and relevance to the clinical question. The high-quality studies actually included in the final summarization or synthesis are pooled together, providing more precise, powerful, and convincing conclusions. The end product is synthesized, trustworthy evidence that the clinician can then use as a component in the EB decision making process (Jordan 2019). Sources for systematic reviews (see Table 3.2) include the AHRQ, Cochrane Collaboration, Campbell Collaboration, and Joanna Briggs libraries. Bibliographic databases should also be searched. The online journal, *JBI Evidence Synthesis*, is a publication dedicated to systematic reviews—full reports and protocols.

Table 3.2 Locating evidence sources

Asking Clinical Questions and Search Terms to Locate Practices with Cultural Variation	1. Include your specific phenomena of interest or intervention of interest: (breastfeeding, literacy, patient education, type of drug, exercise regimen) 2. Include your population and/or group interest: Consider who you are interested in, age, gender, diagnosis (middle-aged obese women, adolescents, elders residing in the community), cultural group that could include: Culture (or specific culture group, i.e., Hispanic, Muslim), cross-cultural, transcultural Ethnocultural, multicultural, multiethnic groups Minority, ethnic minority groups Immigrants, newcomers Country of interest (i.e., Canada, Vietnam, Tanzania) 3. Include outcome of interest (breastfeeding duration, breastfeeding satisfaction, weight loss, quality of life, HbA1c levels, adherence to low-sodium diet) 4. Include type of information desired 5. Systematic review 6. Clinical practice guidelines 7. Research Method (i.e., randomized controlled trial, qualitative)
Systematic Reviews	1. Cochrane Collaboration Library: cochranclibrary.com 2. Campbell Collaboration Library: campbellcollaboration.org 3. Joanna Briggs Institute Library: journals.lww.com/jbisrir 4. Agency for Healthcare Research & Quality: ahrq.gov/clinic/epcindex.htm
Evidence-Based Guidelines	1. AHRQ https://www.ahrq.gov/ 2. Guidelines International Network (G-I-N): http://www.openclinical.org/prj_gin.html 3. CPG Infobase: Clinical Practice Guidelines https://joulecma.ca/cpg/homepage 4. Professional Organization Websites
Bibliographic Databases	1. Academic Search Premiere PsychARTICLES 2. CINAHL PsychINFO 3. Embase PubMed 4. Medline REHABDATA 5. Proquest Social Science Journals
Grey Literature	1. New York Academy of Medicine Grey Literature: http://www.nyam.org/library/online-resources/grey-literature-report/current-grey-literature.html 2. WorldWideScience.org: http://worldwidescience.org/ 3. World Health Organization: http://www.who.int/topics/ 4. Pan American Health Organization: http://new.paho.org/ 5. United Nations Educational, Scientific and Cultural Organization: http://portal.unesco.org/culture 6. Kaiser Permanente Institute for Health Research: http://www.kpco-ihr.org/ 7. Kaiser Family Foundation: http://www.kff.org/ 8. Culture Link Network: http://www.culturelink.org/
Journals Specific to Evidence-Based Care	Clinical Evidence Evidence-Based Nursing, Evidence-Based Mental Health, Evidence-Based Healthcare Electronic Journals on Evidence-Based Practice: http://www.wcpt.org/node/29660
Patient Decision Aids	1. Ottawa Hospital Research Institute: http://decisionaid.ohri.ca/index.html 2. Dartmouth Hitchcock Center for Shared Decision Making:https://med.dartmouth-hitchcock.org/csdm_toolkits.html 3. Foundation for Informed Medical Decision Making: https://www.healthwise.org/solutions/care-transformation.aspx 4. Mayo Clinic Shared Decision-Making National Resource Center: http://shareddecisions.mayoclinic.org/

3.4.3.5 Practice Guidelines

Practice guidelines (preferably based on systematic reviews) translate research findings into systematically developed statements to assist health-care providers and patients in making decisions about appropriate health care for specific clinical circumstances. Guidelines are just what the name implies: a guide to inform health-care providers in applying best practice, not a cookbook where a recipe must be followed. A limitation to guidelines is that they often target one diagnosis and do not take into consideration cultural and contextual realities (Greenhalgh et al. 2014). As many of our patients have more than one diagnosis, multiple comorbidities and varying cultural and contextual experiences, guidelines can't be used as a recipe to be followed by must be adapted and contextualized to the setting and individual. Sources for guidelines include AHRQ, the International Guideline Library maintained by the Guidelines International Network (GIN) and the CPG Infobase, a Canadian guideline repository (see Table 3.2). Guidelines are also developed by numerous professional organizations crossing specialties and disciplines.

3.4.3.6 Filtered and Non-filtered Single Studies

In the absence of systematic reviews or practice guidelines, search for single studies using bibliographic databases (see Table 3.2). Many EB journals, such as *Evidence-Based Nursing*, *Evidence-Based Medicine*, and *Evidence-Based Mental Health*, provide filtered literature synopses of primary research providing the reader with a summary, an appraisal, and recommendations for translation into practice. To review filtered literature, a single-paper search should be done in these journals by adding the journal in the key word search.

If the question is still unanswered after searching the filtered literature, search for primary studies (nonfiltered) by searching the bibliographic databases. If there are journals that commonly carry articles related to the topic, include the journal title in the electronic search or hand-search the journal. The *Journal of Transcultural Nursing* may be a helpful source for information on culture.

Articles retrieved from nonfiltered sources need to be critically appraised to determine scientific rigor prior to use in practice. This requires determining whether the best or strongest design was used for the particular questions and whether the study design was rigorous. For questions of intervention (e.g., "In African Americans with newly diagnosed hypertension, what is the best diuretic treatment in lowering blood pressure?"), randomized controlled trials are the best type of design, followed by cohort studies, case-controlled studies, case series, and descriptive studies. For questions about meaning or understanding an experience (e.g., "What is the experience of marginalization in new African immigrants?"), qualitative studies are the best design. Rigor is assessed by reviewing the article for its adherence to design principles. There are many tools to assist in this process. One good source is the CASP International (www.caspinternational.org), which has a tool for the different study design types.

3.4.3.7 Grey Literature

Another source of evidence that can provide valuable information, especially in the area of culture, is ***grey literature***, which consists of material that is not formally published by commercial publishers or peer-reviewed journals. It includes technical reports, fact sheets, state-of-the-art reports, conference proceedings, dissertations, and other documents from institutions, organizations, government agencies, Internet-based materials and sites, and other forms of media (newspapers, films, published photographs) and the like. The "grey" non-peer-reviewed literature is an important source of information on culture because there are few peer-reviewed publications on specific diseases and cultural implications or diseases and management among the culturally and linguistically diverse. It is important that grey literature be authenticated as reliable and accurate as far as this can be assessed.

The New York Academy of Medicine has catalogued grey literature from 1999 to 2016 highlighting health services research, health aging, prevention, disparities, and urban health topics.

Although the database has not being updated since January 2017, the resources are all accessible. Other key grey literature sites are listed in Table 3.2. Examples of valuable grey literature reports relevant to understanding the impact of culture on health decision making or reporting on cultural health issues include Cross-Cultural Considerations in Promoting Advance Care Planning in Canada (Con 2008); Culture and Mental Health in Haiti: A Literature Review (WHO/PAHO 2010); 2018 National Healthcare Quality and Disparities Report (Agency for Healthcare Research and Quality 2018).

3.5 Best Clinical Expertise

Although locating and drawing from best evidence is important, by itself it cannot direct practice. One reason for this is the lack of quality evidence on topics of interest. This is especially true for evidence on social and cultural influences on health and health outcomes. Very few studies of effectiveness have been devoted to ethnic minority groups. Although legislation was passed in the early 1990s requiring NIH-funded researchers to include ethnic minorities as subjects to allow for valid subgroup analyses of differences in effect by ethnic group, still only about 10% of patients enrolled in clinical trials are minorities (Levine and Greenberg (2016) with reports as low as 4% in cancer clinical trials (Nazha et al. 2019). Similar to the problem with clinical guidelines, another reason one cannot simply rely on evidence is that there is poor fit between our patients in actual practice and those studied in research studies. Most studies control for one or two variables, whereas in practice our patients' problems can be very complex and their value systems very different. Mosley (2009) articulates that "we are living in a spectrum from good evidence at one end to no evidence at the other. We spend most of our lives in the grey area in the middle with somewhat adequate evidence, and we are often not really sure what is good evidence and what is not for the findings to be put into practice." Navigating this grey area requires clinicians to use their tacit, professional-craft knowledge, or practical know-how (Rycroft-Malone et al. 2004; Greenhalgh et al. 2014).

Evidence informed clinicians consider evidence along with their clinical judgment and expertise to thoroughly assess the patient and differentiate nuances that influence treatment perspectives. Too often in health care, significant emphasis is placed on assessment of the biophysical domain, with much less attention paid to psychological, cultural, and social factors that clearly affect health behaviors and outcomes. Clinical expertise must be holistic and recognize the psychological, social and cultural determinants of health (McMurray 2004) and use this in an inclusive skill set for cultural assessment and culturally competent interaction. The clinician must be able to evaluate and adapt research evidence and clinical guidelines in light of not only the clinical presentation but in response to social and cultural values of the client. They must use their clinical acumen to question why the client is or is not responding to treatment and use a holistic framework to determine whether there are intervening biophysical, psychological, social, or cultural considerations that have not been accounted for that could be influencing outcomes and then make necessary adjustments (Shah and Chung 2009).

3.5.1 Patient Values and Preferences

It is insufficient to simply blend expertise and evidence because at the heart of the issue is the patient—their clinical state, clinical setting, circumstances, values and preferences. Although critics of EBP have indicated that there is too great an emphasis on empirical evidence and clinical expertise, the reality is that EBP that integrates all four components is both contextually based and patient-centered (Donnelly 2014). Practicing from an EB perspective requires the clinician to recognize the uniqueness of the patient and family and to value the patient as a co–decision maker in selection of interventions or approaches toward his or her improved health.

The individual's or group's beliefs about

health and illness must be understood if one is to design interventions that are likely to have an impact on health behavior. Their definitions of health and their perceptions of the importance of health states such as mobility, freedom from pain, prolonged life expectancy, and preservation of faculties are important to define because they are valued differently, and these values influence both clinician recommendations and patient decisions. Failure to consider these patient preferences and practicing from a medical model value system leads to unintentional bias toward a professional's view of the world. If EBP is to be value-added, it is critical to ensure that the users of the knowledge—the health-care providers—become active shapers of knowledge and action (Clough 2005). Recognizing that there may be a monocultural bias in evidence, health-care providers must be prepared to make "real-time" adjustments to their approach to care based on patient feedback.

Culture embodies a way of living, a worldview targeting our beliefs about human nature, interpersonal relationships, relationships of people to nature, time or the temporal focus of life, and ways of living one's life (activity). Health-care providers armed with this culture-general knowledge are more open to multiple ways of being, and it serves as a framework for building culture-specific knowledge. It is important to understand the factors that an individual from one cultural group believes cause different types of illnesses and the culture-specific remedies to treat those illnesses. Although not easy to assess, Hulme (2010) provides questions to determine an individual's explanatory model for health conditions. Questions address the patients' perception of what they think caused the problem, why they think it started, and when it started; what they think their sickness does to them; how they perceive the severity and duration of the illness; what they expect from the treatment; the main problems the illness has caused them; and their fears about the illness.

When designing counseling and prevention programs for communities and populations, it is important to be guided by the notion that best practices in counseling and prevention programs do not automatically translate intact across cultural lines (Gone 2015). Adaptations of best practices to integrate culture-specific approaches can become an active ingredient in enhancing outcomes. A systematic review examining interventions to improve cultural competency in healthcare similarly found that these interventions resulted in better disease control, patient satisfaction and trust (Truong et al. 2014).

Cultural translation is the process of adapting EB guidelines or best practice to be congruent with select populations of interest and should be undertaken when there is "variability across groups, when cultural or contextual processes influence risk or protection from target problems, or when the external validity of evidence-based interventions is jeopardized by differences in engagement (e.g., participation rates, attrition, and compliance)" (DePue et al. 2010). A valuable resource in modifying practices to be more culturally congruent is the *Toolkit for Modifying Evidence-Based Practices to Increase Cultural Competence* (Samuels et al. 2009).

Understanding and integrating patient values require attention to salient ethnocultural factors, such as beliefs, language, and traditions. Developing a relationship with the patient; listening to the patient's expectations, concerns, and beliefs; and two-way information sharing between the patient and clinician about the evidence is the beginning to making the patient central to the decision-making process. Shared decision-making begins with finding out what matters to the patient. With this base it is possible for a professional's perspective as health-care provider and the patient's preferences and characteristics to be weighed equally in the decision-making process, a process known as shared decision making. In shared decision making there is a bi-directional respectful, collaborative relationship where the clinician contributes technical expertise (evidence and clinical expertise), while the patient is the expert on his or her own needs, situations, and preferences (Truglio-Londigran and Slyer 2018). Bringing the two together advances the goal of the decision-making process to match care with patient preferences and to shift the locus of decision making from solely the clinician to be inclusive of the patient (Johnson et al. 2010).

Shared decision making is called for when there is no clearly indicated "best" therapeutic option or in preference-sensitive situations or situations where the best choice depends on the patients' values or their preferences for the benefits, harms, and scientific uncertainties of each option (Godolphin 2009). It is the process of interacting with patients who wish to be involved in arriving at an informed, values-based choice among two or more medically reasonable alternatives. Examples of preference-sensitive situations include the following:

1. Should I have knee replacement surgery for my arthritis?
2. Should I take warfarin (or other anticoagulant) to prevent a stroke?
3. Should I take allergy shots?
4. Should I have an MRI for low back pain?
5. Should I stop taking my antidepressants and take herbal treatments?

Facilitating shared decision making involves communicating individualized information on treatment options, treatment outcomes, and probabilities of the benefits and risks and having patients reflect and discuss their personal values or the importance they place on benefits versus harms so that a decision on the best strategy can be reached. Hulme (2010) emphasizes that not involving patients in shared decision making because of perceptions of inability to pay or different cultural models is paternalistic at best and an example of institutional racism at worst. Clinicians need to partake in shared decision making with all patients and be willing to discuss options congruent with culture.

To assist clinicians with shared decision-making and patient-centered care, EB patient decision aids (PtDAs) have been developed. PtDAs are tools that help people participate in decision making by providing information about the options and outcomes and by clarifying personal values (Ankolekar et al. 2018). Decision aids are different from traditional patient education material because they present balanced, personalized information about options in enough detail for patients to make informed judgments about the personal value of the options (O'Connor et al. 2007). They are designed to complement, rather than replace, counseling from a health-care provider and assist people in making specific and deliberative choices among options.

In a systematic review on decision aids for people facing health treatment or screening decisions, Stacey et al. (2017) found that people exposed to decision aids feel more knowledgeable, better informed, and clearer about their values, and they probably have a more active role in decision making and more accurate risk perceptions.

The National Quality Forum (2016) provides multi-stakeholder guidance on national standards and a sustainable process for the certification of patient decision aids. They identified seven criteria that were necessary for an aid to be eligible for certification. These criteria include:

1. Describes the health condition or problem for which the index decision is required.
2. Identifies the target user of the patient decision aid.
3. Explicitly states the index decision under consideration.
4. Describes the options available for the decision, including nontreatment.
5. Describes the positive features of each option.
6. Describes the negative features of each option.
7. Clarifies patient values for outcomes of options by: a. asking patients to consider or rate which positive and negative features matter most to them; and/or b. describing the features of options to help patients imagine the physical and/or social and/or psychological effects.

In accomplishing these criteria, decision aids provide tailored information, exercises that support values clarification, and guidance in how to arrive at decisions. For a patient, a decision-making aid can help to clarify what he or she wants in a treatment, to weigh the pros and cons of different options, and to understand how the options would affect her or him personally (Edwards and

Elwyn 2009). With the use of a decision-making aid, patients can feel confident that they have the information necessary to make a decision. For clinicians, decision aids can promote more effective counseling by providing the clinician with more accurate, structured, and complete information; reducing the need to memorize information; and helping ensure compliance with standards (Elwyn et al. 2010). Further guidance in judging the quality of patient decision aids is available through the International Patient Decision Aids Standards (IPDAS) Collaboration and includes Patient Decision Aid Checklists for Users (Volk et al. 2013).

The Ottawa Hospital Research Institute has as one of its primary missions' practice-changing research with an emphasis on knowledge translation, clinical decision rules, and patient decision aids. It is a leader in shared decision-making research and has a Personal Decision Guide that can be customized for any health or social decision, as well as over 300 decision aids on specific treatment topics. Other developers of PtDAs include the Foundation for Informed Medical Decision Making and its commercial partner Health Dialog, Healthwise, the Mayo Clinic, and the Dartmouth Hitchcock Center for Shared Decision Making. The National Cancer Institute and the Centers for Disease Control and Prevention are compiling and managing clearinghouses of decision aids. Searches for decision aids can be done by keying in the terms *decision aid, patient decision aid, decision guide,* or *patient decision guide* and the condition of interest.

3.5.2 Clinical Context

The clinical context encompasses the setting in which health care is provided or the environment in which the proposed change is to be implemented (McCormack et al. 2002). Drennan (1992) argues that culture, or "the way things are done around here," at the individual, team, and organizational levels creates the context for practice and change. Organizational culture is a paradigm—a way of thinking about the organization, comprising a linkage of basic assumptions, values, and artifacts (Schein 1992) and having its own belief system, paradigms, customs, and language. In addition to organizational culture, the medical culture also has a powerful influence on treatment approaches and modalities. The medical culture values objectivity, cause and effect, biophysical care, and, in many cases, the power of their own expertise and status. These cultures may be resistant to new paradigms calling for EBP or to new evidence-based processes being proposed to replace old ways. These organizational values may be in opposition to patient values, creating clashes between providers and patients.

A largely unexplored disparity exists between the beliefs and expectations of health-care providers and patients, particularly when there is also a disparity between the cultures/ethnicity of the two (Asthma and Allergy Foundation of America 2005; Enarson and Ait-Khaled 1999; Walker et al. 2005). Viewing biomedicine as a culture in itself, such that interactions between patients and health-care providers become a communication between cultures or transactions between worldviews, appears to be a necessary process in establishing a trusting and effective partnership and thus improving the health outcomes of patients. Moreover, clinicians must have an understanding of the environment in which health behaviors are enacted. This goes beyond the traditional health-care organizational setting to include the home, residential care, the neighborhood, and the broader community.

The importance of context is becoming more apparent, with a greater emphasis on knowledge translation and implementation science or the study of methods to promote the transfer of research findings into routine health-care policy and practice (Squires et al. 2019). Evidence must be adapted to the clinical context and the clinical context must be understood as it may either enable or block goals for improved clinical practice. Strifler et al.' 2018 scoping review identified 159 knowledge translation theories, models and frameworks used in evidence implementation studies. These guides typically call for an assessment of barriers and supports specific to the evidence itself, the knowledge users, and the practice environment. With an understanding of the enablers and barriers, an implementation plan tailored to the context to enhance the translation of knowledge to action.

The abundance of new evidence that has *not* been successfully translated into practice is a critical reminder of the importance of context and the strength of the existing culture. Difficult questions remain to be addressed. What should be done with health-care providers who cannot or will not adapt to EBP? How will lack of interdisciplinary collaboration be approached? How will it be handled if long-standing treatment approaches show no evidence of fostering improvement? What is the individual's responsibility compared with the organization's and community's responsibilities in ensuring readiness for EBP? What are the best approaches for facilitating knowledge translation in different contexts? Knowing the answer to some of these questions will influence outcomes of getting knowledge into practice.

Evidence-based practice is not research-directing practice but evidence-informing practice. The clinician evaluates best evidence in light of his or her clinical expertise, patient preferences, and clinical context to make decisions about patient or program management. Health-care providers practicing from an EB perspective need the skills to acquire and evaluate evidence, make decisions about adaptation of evidence, or plan to be congruent with the patient values and clinical context.

References

Agency for Healthcare Research and Quality (2018) 2018 National healthcare disparities report. U.S. Department of Health and Human Services, Rockville. https://www.ahrq.gov/research/findings/nhqrdr/nhqdr18/index.html

Ankolekar A, Dekker A, Fijten R, Berlanga A (2018) The benefits and challenges of using patient decision aids to support shared decision making in health care. JCO Clin Cancer Inform 2:1–10

Antes G, Sauerland S, Seiler CM (2006) Evidence-based medicine—from best research evidence to a better surgical practice and health care. Arch Surg 391:61–67

Asthma and Allergy Foundation and the National Pharmaceutical Council (2005) Asthma and allergy Foundation of America. Ethnic disparities in the burden and treatment of asthma. National Pharmaceutical Council

Boyle J, Andrews M (2011) Transcultural concepts in nursing, 5th edn. Lippincott, Philadelphia

Calvillo E, Clark L, Ballantyne J, Pacquiao D, Purnell L, Villarruel A (2009) Cultural competency in baccalaureate education. J Transcult Nurs 20(2):137–145

Campinha-Bacote J (2002) The process of cultural competence in the delivery of healthcare services: a model of care. J Transcult Nurs 13(3):181–184

Chapman EN, Kaatz A, Carnes M (2013) Physicians and implicit bias: how doctors may unwittingly perpetuate health care disparities. J Gen Intern Med 28(11):1504–1510

Clough E (2005) Foreword. In: Burr J, Nicholson P (eds) Researching health care consumers, critical approaches. Palgrave (MacMillan), Basingstoke, pp ix–xi

Con A (2008) Cross-cultural considerations in promoting advance care planning in Canada. http://www.bccancer.bc.ca/NR/rdonlyres/E17D408A-C0DB-40FA-9682-9DD914BB771F/28582/COLOUR030408_Con.pdf

DePue JD, Rosen RK, Batts-Turner M, Bereolos N, House M, Held RF, Nu'usolia O, Tuitele J, Goldstein MG, McGarvey ST (2010) Cultural translation of interventions: diabetes care in American Samoa. Am J Public Health 100(11):2085–2093

Donnelly G (2014) Evidence-based practice and contextual care. Holist Nurs Pract 28(3):159–159. https://doi.org/10.1097/HNP.0000000000000028

Douglas M, Pierce J, Rosenkoetter M, Clark Callister L, Hattar-Pollara M, Lauderdale J, Milstead J, Nardi D, Pacquiao D, Purnell L (2011) Standards of practice for culturally competent nursing care. J Transcult Nurs 22(4):317–334

Douglas M, Rosenkoetter M, Pacquiao DF, Callister LC, Hattar-Pollara M, Lauderdale J, Milstead J, Nardi D, Purnell L (2014) Guidelines for implementing culturally competent nursing care. J Transcult Nurs 25(2):109–121

Drennan D (1992) Transforming company culture. McGraw-Hill, London

Edwards A, Elwyn G (eds) (2009) Shared decision-making. Achieving evidence-based patient choice. Oxford University Press, Oxford

Elwyn G, Frosch D, Volandes AE, Edwards A, Montori VM (2010) Investing in deliberation: a definition and classification of decision support interventions for people facing difficult health decisions. Med Decis Mak 30:701–711

Enarson DA, Ait-Khaled N (1999) Cultural barriers to asthma management [commentary]. Pediatr Pulmonol 28:297–300

Epstein NE (2014) Multidisciplinary in-hospital teams improve patient outcomes: a review. Surg Neuro Intern 5(Suppl 7):S295

Gatewood E, Broholm C, Herman J, Yingling C (2019) Making the invisible visible: implementing an

implicit bias activity in nursing education. J Prof Nurs 35(6):447–451
Giger J, Davidhizar R, Purnell L, Taylor Harden J, Phillips J, Strickland O (2007) American Academy of Nursing expert panel report: developing cultural competence to eliminate health disparities in ethnic minorities and other vulnerable populations. J Transcult Nurs 18(2):95–102
Godolphin W (2009) Shared decision making. Healthc Q 12:186–190
Gone JP (2015) Reconciling evidence-based practice and cultural competence in mental health services. Transcult Psychiatry 52(2):139–149
Greenhalgh T, Howick J, Maskrey N (2014) Evidence based medicine: a movement in crisis? BMJ 348:g3725
Hulme PA (2010) Cultural considerations in evidence-based practice. J Transcult Nurs 21(3):271–280
International Medical Interpreters Association (1987) Code of ethics for medical interpreters. http://www.imiaweb.org/uploads/pages/376.pdf
Jeffreys MR (2000) Development and psychometric evaluation of the transcultural self-efficacy tool: a synthesis of findings. J Transcul Nurs 11(2):127–136
Johnson SL, Kim YW, Church K (2010) Towards client-centered counseling: development and testing of the WHO decision-making tool. Patient Educ Couns 81:355–361
Jordan Z (2019) Focus is the new intelligence quotient for evidence-based practice. Evid-Based Healthc 17(4):189–190
Jylhä V, Oikarainen A, Perälä ML, Holopainen A (2017) Facilitating evidence-based practice in nursing and midwifery in the WHO European region. WHO, Geneva. http://www.euro.who.int/__data/assets/pdf_file/0017/348020/WH06_EBP_report_complete.pdf
Knighton AJ, McLaughlin M, Blackburn R, Wolfe D, Andrews S, Hellewell JL et al (2019) Increasing adherence to evidence-based clinical practice. Qual Manage Healthc 28(1):65–67
Kumra T, Hsu YJ, Cheng TL, Marsteller JA, McGuire M, Cooper LA et al (2020) The association between cultural competence and teamwork climate in a network of primary care practices. Health Care Manage review
Lehane E, Leahy-Warren P, O'Riordan C, Savage E, Drennan J, O'Tuathaigh C et al (2019) Evidence-based practice education for healthcare professions: an expert view. BMJ Evid-Based Med 24(3):103–108
Levine D, Greenberg R (2016) More minorities needed in clinical trials to make research relevant. https://www.aamc.org/news-insights/more-minorities-needed-clinical-trials-make-research-relevant-all
Loftin C, Hartin V, Branson M, Reyes H (2013) Measures of cultural competence in nurses: an integrative review. Sci World J
Mabachi NM, Cifuentes M, Barnard J, Brega AG, Albright K, Weiss BD et al (2016) Demonstration of the health literacy universal precautions toolkit: Lessons for quality improvement. J Ambul Care Manage 39(3):199
McCormack B, Kitson A, Harvey G, Rycroft-Malone J, Titchen A, Seers K (2002) Getting evidence into practice: the meaning of "context". J Adv Nurs 38(1):94–104
McMurray A (2004) Culturally sensitive evidence-based practice. Collegian 11:14–18
Melnyk BM, Fineout-Overholt E, Stillwell SB, Williamson KM (2009) Igniting a spirit of inquiry: an essential foundation for evidence-based practice. Am J Nurs 109(11):49–52
Mosley C (2009) Evidence-based medicine: the dark side. J Pediatr Orthop 29(8):839–843
Narayan MC (2019) CE: addressing implicit Bias in nursing: a review. Am J Nurs 119(7):36–43
National Council on Interpreting in Health Care (2007). www.ncihc.org/
National Quality Forum (2016, December) National standards for the certification of patient decision aids. National Quality Forum, Washington, DC
Nazha B, Mishra M, Pentz R, Owonikoko TK (2019) Enrollment of racial minorities in clinical trials: old problem assumes new urgency in the age of immunotherapy. Am Soc Clin Oncol Educ Book 39:3–10
O'Connor AM, Wennberg JE, Legare F, Llewellyn-Thomas HA, Mouton BW, Sepucha KR, Sodano AG, King JS (2007) Toward the "tipping point": decision aids and informed patient choice. Health Aff 26(3):716–725
Perng SJ and Watson R (2012) Construct validation of the nurse cultural competence scale: a hierarchy of abilities. J Clinic Nurs 21(11–12):1678–1684
Rycroft-Malone J, Seers K, Titchen A, Harvey G, Kitson A, McCormack B (2004) What counts as evidence in evidence-based practice? J Adv Nurs 47(1):81–90
Salmond S (2007) Advancing evidence-based practice: a primer. Orthop Nurs 26(2):114–123
Samuels J, Schudrich W, Altschul D (2009) Toolkit for modifying evidence-based practice to increase cultural competence. Research Foundation for Mental Health, Orangeburg
Schein EH (1992) Organizational culture and leadership, 2nd edn. Jossey-Bass, San Francisco
Schim SM, Doorenbos AZ, Miller J, Benkert R (2003) Development of a cultural competence assessment instrument. J Nurs Measure 11(1):29–40
Shah HM, Chung KC (2009) Archie Cochrane and his vision for evidence-based medicine. Plast Reconstr Surg 124(3):982–988
Squires JE, Aloisio LD, Grimshaw JM, Bashir K, Dorrance K, Coughlin M et al (2019) Attributes of context relevant to healthcare professionals' use of research evidence in clinical practice: a multi-study analysis. Implement Sci 14(1):52
Stacey D, Légaré F, Lewis K, Barry MJ, Bennett CL, Eden KB et al (2017) Decision aids for people facing health treatment or screening decisions. Cochrane Database Syst Rev 4:CD001431
Stevens K (2013) The impact of evidence-based practice in nursing and the next big ideas. Online J Issues Nurs 18(2):4
Strifler L, Cardoso R, McGowan J, Cogo E, Nincic V, Khan PA et al (2018) Scoping review identifies sig-

nificant number of knowledge translation theories, models, and frameworks with limited use. J Clin Epidemiol 100:92–102

Sue DW, Lin AI, Torino GC, Capodilupo CM, Rivera DP (2009) Racial microaggressions and difficult dialogues on race in the classroom. Cultu Div Ethnic Minor Psychol 15(2):183

Swanson JA, Schmitz D, Chung KC (2010) How to practice evidence-based medicine. Plast Reconstr Surg 126(1):286–294

Truglio-Londrigan M, Slyer JT (2018) Shared decision-making for nursing practice: an integrative review. Open Nurs J 12:1–14

Truong M, Paradies Y, Priest N (2014) Interventions to improve cultural competency in healthcare: a systematic review of reviews. BMC Health Scrv Rcs 14(1):99

Volk RJ, Llewellyn-Thomas H, Stacey D, Elwyn G (2013) Ten years of the international patient decision aid standards collaboration: evolution of the core dimensions for assessing the quality of patient decision aids. BMC Med Inform Decision 13(Suppl 2):S1

Walker C, Weeks A, McAvoy B, Demetriou E (2005) Exploring the role of self-management programs in caring for people from culturally and linguistically diverse backgrounds in Melbourne, Australia. Health Expect 8:315–323

World Health Organization/Pan American Health Organization (HO/PAHO) (2010) Culture and mental health in Haiti: a literature review. WHO, Geneva. http://www.who.int/mental_health/emergencies/culture_mental_health_haiti_eng.pdf

4 Organizational Cultural Competence

Larry D. Purnell

4.1 Introduction

Worldwide, an emerging consensus is that culturally competent and congruent care are essential components of social and ethical responsive quality health care. Key components include not only direct health-care providers, but also leadership support from the organization. Moreover, providing culturally competent care to some extent has been hindered by a dearth of systematic approaches and organizational support.

Cultural competence does not relate solely to the care of patients, families, and the community; it is also applicable to educational, health-care, and professional organizations. As described in the Purnell Model (see Chap. 2), the workforce issues domain can be used to assess organizational culture and cultural issues among staff at all levels in the organization.

The purpose of this chapter, therefore, is to provide an overview of the requisite organizational infrastructure designed to create and sustain cultural competency. Major resources include (a) The National CLAS Standards from the U.S. Department of Health and Human Services Office of Minority Health (2019); (b) The National Academies of Sciences, Engineering, and Medicine (2019), (c) The National Library of Medicine (2003), (d) Intercultural Competence for All: Preparation for Living in a Heterogeneous World (2012), (e) the Cultural Competence Assessment Profile (Health Resources Service Administration 2002), and (f) the Purnell Model for Cultural Competence. Together, resources provide the organizing framework for this chapter.

This chapter is a revision made of the original chapter written by Stephen Marrone in the previous edition of the book.

L. D. Purnell (✉)
University of Delaware, Newark, DE, USA
e-mail: lpurnell@udel.edu

4.2 Health Disparities

A growing realization among health-care researchers, clinicians, and advocates is that a focus on health-care disparities is an important aspect of improving health-care outcomes. Activities toward improvement bring together many elements of a health-care delivery system. Health disparities are difference in which disadvantaged social groups such as the poor, racial/ethnic minorities, women, and other groups who have persistently experienced social disadvantage or discrimination systematically experiencing worse health or greater health risks than more advantaged social groups. The term describes an increased presence and severity of certain diseases, poorer health outcomes, and greater difficulty in obtaining health-care services for these races and ethnicities. When sys-

© Springer Nature Switzerland AG 2021

L. D. Purnell, E. A. Fenkl (eds.), *Textbook for Transcultural Health Care: A Population Approach*,
https://doi.org/10.1007/978-3-030-51399-3_4

temic barriers to good health are avoidable, yet still remain, they are often referred to as health inequities (Virginia Department of Health 2019). Moreover, race, ethnicity, geography, education, and income have demonstrated valuable insights for health policy experts and advocates. Learning about disparities can help lessen these inequalities. Racial and ethnic disparities are not only morally wrong and fiscally unwise, but stress the health infrastructure, including reimbursement for health-care costs.

Racial and ethnic diversity among health-care providers has, in some instances, correlated with the delivery of quality care to diverse patient populations (Betancourt et al. 2003). Evidence demonstrates that racial concordance between minority patients and minority physicians is associated with greater patient satisfaction and higher self-rated quality of care (Saha et al. 1999). Evidence also supports that minority patients prefer minority physicians regardless of practice location or other geographic issues (Saha et al. 2000). Spanish-speaking patients in the US, for example, report greater satisfaction when care is provided by Spanish-speaking providers than by non–Spanish-speaking providers (Morales et al. 1999). Likewise, African-American patients report more satisfaction with care when their physician uses an inclusive, participatory decision-making approach to health care (Cooper-Patrick et al. 1999). However, other surveys report that not all patients prefer a health-care provider of the same background (Robert Wood Johnson Foundation 2011).

A plethora of data from around the world suggest that minority groups experience a disproportionately higher rate of illness, more severe complications, and increased mortality and morbidity related to cardiovascular disease, diabetes, asthma, and cancer. Multiple factors external to the health-care system influence health disparities—namely, lower socioeconomic status of minorities, hazardous jobs with increased incidence of injury, lower educational and literacy levels, lack of or inadequate health insurance, fear of the health-care system, overuse of over-the-counter medications and home remedies, and the use of the emergency department for care. A systemic review of the literature entitled *Unequal Treatment: Confronting Racial/Ethnic Disparities in Health Care*, conducted by the Institute of Medicine, reported the findings of more than 175 studies that illustrated racial and ethnic disparities in the diagnosis and treatment of multiple medical conditions, even when analyses were controlled for socioeconomic status, insurance status, site of care, stage of disease, comorbidities, and age (Institute of Medicine 2002).

The root causes of health disparities relate to a disconnect between patients' health beliefs, values, preferences, and behaviors and those of the dominant health-care system. This lack of fit includes variations in patient recognition of symptoms, thresholds for seeking care the ability to communicate symptoms to a provider who understands their meaning; the ability to understand the prescribed treatment plan, including use of medications; expectations of care access to and utilization of diagnostic and therapeutic procedures and adherence to preventive measures (Einbinder and Schulman 2000). These core factors are some of the primary influencers for decision making among patients and health-care providers, physicians in particular, and the degree to which patients access and interact with the health-care delivery system (Pacquiao 2018a). Emphasis on cultural competency in health care and culturally competent health-care organizations has emerged as a result of these findings.

4.3 Culturally Competent Health-Care Organizations

Cultural competence in health care has been defined as "a set of congruent behaviors, attitudes, and policies that come together in a system, agency, or among professionals and enable that system, agency, or those professionals to work effectively in cross-cultural situations" (Health Resource Services Administration 2002, p. 3). The tenets of cultural competency are not specific to one health-care discipline and must be inclusive of all professional disciplines, as well as clerical, technical, and unlicensed assistive personnel. Hence, the provision of culturally safe care relies on all members of the health-care team receiving consistent and comparable information about the needs of the diverse patients, families, and communities they serve. It is important to understand that cultural competency is a process and not a result (Purnell et al. 2011). To be effective, health care must reflect the unique understanding of the values, beliefs, attitudes, lifeways, and worldviews of diverse populations and individual acculturation patterns (Purnell and Fenkl 2019).

A culturally competent health-care organization incorporates culture at all levels to meet culturally unique needs (Purnell et al. 2011). Culturally competent health-care organizations out perform their competitors by achieving and sustaining greater performance and outcomes measures and increased market share as evidenced by improved consumer access to care. Enhanced quality of care reduces health disparities and improves health outcomes for vulnerable and underserved populations. Thus, greater patient and staff satisfaction leads to an increased consumer market share to and secures financial sustainability of the organization (Marrone, 2010).

Reflection: *Examining where you work/go to school, what evidence can you find that the mission and philosophy of the organization include statements on diversity and inclusion? Is the board of trustees reflective of the diversity of the community and patient population?*

Culturally competent health-care delivery organizations provide consumers with effective, understandable, and respectful care provided in ways that fit with their cultural values and beliefs and in the consumer's preferred language. To achieve this goal, organizations develop, implement, and promote a written strategic plan that outlines clear goals, policies, operational plans, and management accountability/oversight mechanisms to provide culturally and linguistically appropriate services. Consequently, to ensure the design of an evidence-based strategic plan, the organization conducts initial and ongoing organizational self-assessments of diversity-related activities. It integrates cultural and linguistic competence-related measures into internal audits, performance improvement programs, patient satisfaction assessments, and outcomes-based evaluations. Finally, it collects and updates information related to consumers' race, ethnicity, and spoken and written language(s) and integrates this information into the organization's data management system. These data help maintain a current demographic, cultural, and epidemiological profile of the community to plan for and implement services that respond to its cultural and linguistic characteristics (U.S. Department of Health and Human Services Office of Minority Health (2019).

Reflection: *Does the organization where you work/go to school have a strategic plan that reflects the needs of the community? What would you recommend to improve organizational competence?*

To determine the need for diversity-related services, it is essential that organizations create a community demographic needs assessment tool to assess the cultural beliefs and language needs of the people who live there. To provide effective patient education, it is necessary to review the language literacy level and the use of culturally respectful images in written and visual (i.e., television or video) patient education materials. Organizations should develop systems that indicate whether language assistance is needed prior to or at the point of entry into the organization (Purnell et al. 2011).

A key ingredient to creating and sustaining organizational cultural competency is to designate *diversity champions* who have acquired the requisite knowledge and skills to provide culturally congruent care at all levels within the organization. Champions can mentor other health-care providers within their discipline and/or department to expand their influence on consumer care and services. Furthermore, culturally competent human resources departments promote patient-centered care by including patient satisfaction measures in employee performance appraisals (Purnell et al. 2011) by establishing diversity-related sentinel events and by completing a root cause analysis. Culturally competent organizations create their own or revise standardized consumer satisfaction tools to include items related to the provision of culturally and linguistically appropriate care and monitor data at all levels of committee and council meetings throughout the organization. Developing a data bank with best practices and lessons learned need not be complex. The following items have been used successfully:

1. "Did you receive care that was respectful to your cultural and religious beliefs?"
2. "Did you receive care in your preferred language?"
3. "Did you receive care in a language that helped you to make informed decisions?"

To be effective, written satisfaction surveys should be translated into the languages that represent the catchment area for the organization (Purnell et al. 2011).

Andrews (1998) provided the following six-step framework for ensuring organizational cultural competency:

1. Collect demographic and descriptive data of the prevalent cultural, ethnic, linguistic, and spiritual groups represented among patients, families, visitors, the community, and the staff in the service area.
2. Describe the effectiveness of current systems and processes in meeting diverse needs.
3. Assess the organization's strengths and limitations by examining the institution's ethos toward cultural diversity and the presence or absence of a corporate culture that promotes accord among its constituents.
4. Determine organizational need and readiness for change through dialogue with key stakeholders aimed at discovering foci of anticipated support and recognizing areas of potential resistance.
5. Implement strategic plans, policies, and procedures that include measurable benchmarks of success and an ongoing process to ensure that change is maintained.
6. Evaluate actual outcomes against established benchmarks utilizing performance improvement, quality, and customer satisfaction data.

Culturally competent health-care organizations implement strategies to recruit, retain, and promote at all levels of the organization a diverse staff and leadership team that are representative of the demographics of the service area. The goal of recruiting and retaining a diverse workforce that matches the demographics of the service area is to reduce health disparities among vulnerable and underserved populations that often result from discordant consumer–provider relationships. Culture and language discordance can lead to decreased access to care, decreased quality of care, increased cost of care, decreased patient satisfaction, recidivism, discrimination, and poor health outcomes (American Association of Critical Care Nurses 2008; Europa 2010).

Reflection: *In the organization where you work/go to school, are pictures, posters, and calendars representing the diversity of the patient and staff posted throughout the organization? What additional pictures or posters would you include?*

To reduce discordance between consumers and providers, an organization needs to integrate diversity into the organization's mission statement, strategic plans, and goals (Purnell et al. 2011). A diverse workforce program should include mentoring programs, community-based internships, and collaborations with academic partners such as universities, local schools, training programs, and faith-based organizations. To expand the recruitment base, organizations should recruit at minority health and recruitment fairs, advertise in multiple languages, and list job opportunities in minority publications such as local newspapers, community newsletters, and local businesses (Purnell et al. 2011).

4.4 CLAS Standards

The National Standards on Culturally and Linguistically Appropriate Services in Health Care (CLAS Standards) were developed via national consensus by the U.S. Department of Health and Human Services—Office of Minority

Health (OMH 2019). The CLAS Standards were intended to guide health-care organizations in the provision of safe care and services that were culturally, ethnically, linguistically, and spiritually appropriate and effective. The guiding principles and associated actions and interventions of culturally and linguistically appropriate health-care services are intended to be integrated throughout the organization and designed, implemented, and evaluated in partnership with the communities being served (Office of Minority Health 2019).

4.4.1 The 15 CLAS Standards Are Organized According to the Following Themes

4.4.1.1 Principal Standard

1. Provide effective, equitable, understandable, and respectful quality care and services that are responsive to diverse cultural health beliefs and practices, preferred languages, health literacy and other communication needs.

4.4.1.2 Governance, Leadership and Workforce

2. Advance and sustain organizational governance and leadership that promotes CLAS and health equity through policy, practices, and allocated resources.
3. Recruit, promote, and support a culturally and linguistically diverse governance, leadership, and workforce that are responsive to the population in the service area.
4. Educate and train governance, leadership, and workforce in culturally and linguistically appropriate policies and practices on an ongoing basis.

4.4.1.3 Communication and Language Assistance

5. Offer language assistance to individuals who have limited language proficiency and/or other communication needs at no cost to facilitate timely access to all health care and services.
6. Inform all individuals of the availability of language assistance services clearly and in their preferred language verbally and in writing.
7. Ensure the competence of individuals providing language assistance, recognizing that the use of untrained individuals and/or minors as interpreters should be avoided.
8. Provide easy-to-understand print and multimedia materials and signage in the languages commonly used by the populations in the service area.

4.4.1.4 Engagement, Continuous Improvement, and Accountability

9. Establish culturally and linguistically appropriate goals, policies, and management accountability and infuse them throughout the organizations' planning and operations.
10. Conduct ongoing assessments of the organization's CLAS-related activities and integrate CLAS-related measures into assessment measurement and continuous quality improvement activities.
11. Collect and maintain accurate and reliable demographic data to monitor and evaluate the impact of CLAS on health equity and outcomes and to inform service delivery.
12. Conduct regular assessments of community health assets and needs and use the results to plan and implement services that respond to the cultural and linguistic diversity of populations in the service area.
13. Partner with the community to design, implement, and evaluate policies, practices, and services to ensure cultural and linguistic appropriateness.
14. Create conflict- and grievance-resolution processes that are culturally and linguistically appropriate to identify, prevent, and resolve conflicts or complaints.
15. Communicate the organization's progress in implementing and sustaining CLAS to all stakeholders, constituents, and the general public.

4.5 Cultural Competence Assessment Profile

The Cultural Competence Assessment Profile, funded by the U.S. Department of Health and Human Services—Health Resources and Services Administration (HRSA), was developed to answer the question "How do we know cultural competence when we see it?" (Health Resources Service Administration 2002). The Profile is based on evidence from foundational work in organizational cultural competency such as the CLAS Standards and provides the infrastructure to conceptualize how to assess cultural competence at the organizational level. Essentially, the profile is intended to gather information based on the specific performance and outcomes characteristics that should be evident across the health-care continuum in a culturally competent organization. The Profile can assist organizations by providing a framework to organize activities related to the cultural competence and quality monitoring for compliance with cultural competence standards (Health Resources Service Administration 2002).

The Assessment Profile was built on the following assumptions:

- Organizational cultural competence is an integral component of patient-centered care and can contribute to improving access to care, quality of care, and health outcomes. The authors of this chapter would include family-centered care because most people live within a family structure.
- Health-care organizations drive the development and maintenance of individual provider cultural competence and the environment of care.
- Cultural competence is a business imperative that supports organizational branding and increases the organization's market share among diverse cultural groups, thereby leading to continuous service and process improvements (Health Resources Services Administration 2002).

The performance areas of the Cultural Competence Assessment Profile include organizational values, governance, planning and monitoring/evaluation, communication, staff development, organizational infrastructure, and services/interventions.

1. Organizational values refer to the organization's viewpoint regarding cultural competence and its commitment to provide culturally congruent care.
2. Governance relates to goal-setting, policy-making, and oversight methods used to help ensure the delivery of culturally congruent care.
3. Planning, monitoring, and evaluating include the use of internal and external stakeholders in short- and long-term planning for the delivery of culturally congruent health-care services.
4. Communication focuses on the schema through which information is exchanged, vertically and horizontally, with internal and external consumers, executives, and members of the health-care team in order to promote cultural competence.
5. Staff development underscores the need for organizations to ensure that staff at all levels of the organization acquire the attitudes, knowledge, and skills for delivering culturally congruent care.
6. Organizational infrastructure refers to organizational resources required to hardwire the delivery of culturally congruent care and services throughout the continuum of care. Services and interventions relate to an organization's delivery of clinical and community health services that reflect the needs of the diverse consumer groups within the organization's service area (Health Resource Services Administration 2002).

4.6 The Purnell Model

Theories and conceptual models are essential in scientific disciplines; they enable health-care providers to describe, explain, and predict concepts and phenomena. Nursing and health-care theories guide practice, influence decisions and interventions, and provide a framework for evaluating outcomes. Theory serves as the foundation for the provision of culturally congruent care by culturally competent health-care providers.

The Purnell Model is based on theories and evidence derived from organizational, administrative, communication, and family development theories, in addition to anthropology, sociology, psychology, anatomy and physiology, biology, ecology, nutrition, pharmacology, religion, history, economics, political science, and linguistics. Hence, the major assumptions of the Purnell Model that affect organizational cultural competency were developed from a broad perspective, allowing their use across practice disciplines and organizational/environmental contexts. See Chap. 2 for a complete description of the Purnell Model.

The Purnell Model is germane to all health-care disciplines in a variety of environmental contexts. The 12 domains of the Purnell Model provide an organizing framework for organizational cultural competency as it highlights the importance of interprofessional collaboration with emphasis on patient- and family-centered care, managed care, and case management across the health-care continuum. The model can guide the development of assessment instruments, planning strategies, and individualized patient, family, and community interventions.

As outlined in Table 4.1, triangulation among the Purnell Model, National CLAS Standards (2018), and the Cultural Assessment Profile (HRSA 2002) can be used to frame the design of an organizational cultural competency program. The Purnell domain, overview and heritage, supports that culturally competent organizations articulate mission and vision statements, values, a strategic plan, and standard operating procedures (policies and procedures) that reflect the value of diversity. The mission, vision, and values drive the shared governance model that provides for a dedicated chief diversity officer to oversee the activities of a transdisciplinary diversity council. A cultural health assessment should be included in patient history and assessment data at all points of entry into the system. Traditional or folk healing practices that are integrated into the plan of care should be identified.

The Purnell domain, communication, supports organizational efforts to ensure that signage is placed in all areas in multiple languages related to directions within the organization. It also requires that the availability of language assistance services include sign language, the translation of critical documents such as consents and patient education materials, pain scales, patient evaluation surveys, and communication boards/aids for patients who are not able to speak or understand the dominant language.

4.7 Knowledge and Skill Acquisition

Health-care organizations should ensure that staff at all levels and across all disciplines receive ongoing education related to culturally and linguistically appropriate service delivery. To ensure the successful acquisition and maintenance of culturally and linguistically appropriate knowledge and skills, organizations must allocate fiscal resources to educate staff at all levels in order to develop the requisite role-specific competencies for the provision of culturally congruent care. In addition, if bilingual staff express an interest in, and are able to provide the service, internal and/or external funding sources should be made available to support the training of staff as medical interpreters and translators.

Educational programming and learning outcomes related to cultural competence need to include, principally, the cognitive and affective domains of learning, with, to a lesser degree, the psychomotor domain. The curriculum should follow the educational design principle of *simple-to-complex* and *general-to-specific*. The learning objectives and educational content should be evidence-based and address definitions of cultural competence; discrimination, prejudice, and stereotyping; role-specific performance criteria for the provision of culturally congruent care;

Table 4.1 Crosswalk: Purnell Model, CLAS standards, and cultural assessment

Purnell Model domains	CLAS standards	Cultural assessment profile	Indicators
Overview/Heritage Advocates for the formation of a transdisciplinary diversity council, ethics committee, and patient education committee that include community members and give attention to the health literacy of the service area. A strong community partnership with key individuals and agencies reflective of the demographics in the service area, and the health-care team should include community members of governing board.	**Standard 1** Effective, respectful care in the preferred language **Standard 2**: Cultural, demographic, and epidemiological community profile	**Domains** Organizational values Governance Staff development Organizational infrastructure	• Mission, vision, values, strategic plan, and standard operating procedures (policies and procedures) that reflect value of diversity • Shared governance model • Dedicated chief diversity officer • Diversity competence program for all levels of staff • History and assessment databases that include traditional and folk healing practices • Cultural health assessment performed by clinical staff • Code of professionalism that outlines diverse behavioral expectations • Breaches trigger a "culture code" for immediate intervention • Transdisciplinary diversity council, ethics committee, and patient education committee that include community members • Community partnerships with key individuals and agencies reflective of the demographics in the service area and the health-care team • Community members of the governing board and decision-making groups • Dynamic demographic data collection and management systems
Communication Communication supports organizational efforts to ensure that signage is placed in all areas in multiple languages related to directions within the organization. Requires that the availability of language assistance services and includes sign language, the translation of critical documents such as consents and patient education materials, pain scales, and communication boards/aids for patients who are not able to speak or understand the dominant language.	**Language assistance services** Standard 5: Notice of language assistance services Standard 6: Competence of language assistance services provided by interpreters Standard 7: Translated signage and patient materials	**Domain** Governance	• Signage in multiple languages • Translation of critical documents such as consent forms, patient education materials, pain scales, communication boards • Patient education programs that address diversity of the service area, such as meal planning for diabetes management of Caribbean Americans • Customer service initiatives that address diversity • Maintain an organizational language

Family roles and organization Can guide the development of organizational visitation policies, including open visiting hours that address cultural norms and visitor role responsibilities; policies related to diverse decision-making practices, such as informed consent; and the purchase and utilization of racially ethnically and age-appropriate toys for infants and children and high-fidelity manikins used in simulation education.	**Standard 1:** Effective, respectful care in preferred language	**Domains** Organizational values Planning Monitoring, and evaluation Staff development Organizational infrastructure Services and interventions	• Visitation policies, including open visiting hours that address cultural norms and visitor role responsibilities • Protocols for diverse decision-making practices, such as informed consent • Racially ethnically, and age-appropriate toys and manikins used in simulation education
Workforce issues Supports the development of diversity-related education content and competencies integrated into orientation, in-service education, staff development, and continuing education programs for all levels of staff and the integration and utilization of a standardized communication method such as SBAR (Situation, Background, Assessment, and Recommendation) among the interprofessional healthcare team	**Standard 2** Recruit, retain, and promote diversity Standard 3: Ongoing staff education Standard 8: Written strategic plan Standard 9: Conduct organizational self-assessment standard 10: Collect demographics and integrate into data management system Standard 11: Implement services based on demographic profile Standard 12: Collaborative community partnerships Standard 13: Culturally relevant conflict management processes Standard 14: Inform public of diversity- related initiatives	**Domains** Organizational values Governance Communication Staff development Organizational infrastructure	• Diversity content and competencies integrated into orientation, in-service education, staff development, and continuing education programs for all levels of staff • Standardized communication methods among caregivers, such as using SBAR or other shared mental and communication models

(continued)

Table 4.1 (continued)

Purnell Model domains	CLAS standards	Cultural assessment profile	Indicators
Biocultural ecology Gives direction to the development of educational programs and clinical practices that recognize physical and genetic variations that have an impact on assessment and treatment plans including preventive services and early screening, and to the design of patient education, counseling, and screening related to known genetic predispositions to diseases and biological variations regarding the pharamcokinetics of medications and other substances.	**Standard 1** Effective, respectful care in preferred language	***Domains*** Organizational values Planning Monitoring and evaluation Staff development Organizational infrastructure Services and interventions	• Educational programs and clinical practices that recognize physical and genetic variations that affect assessment and treatment plans including preventive services and early screening. • Patient education, counseling, and screening related to known genetic predispositions to diseases and biological variations regarding drug metabolism.
High-risk behavior identifies the need for assessing known high-risk behaviors in the service area and including these behaviors on the risk assessment portions of patient admission histories and health screenings for inclusion in patients' plans of care.	**Standard 1** Effective, respectful care in preferred language	**Domains** Organizational values Planning Monitoring and evaluation Staff development Organizational infrastructure Services and interventions	• Assessment for known high risk behaviors in the service area included on admission histories and health screenings to be included in the patients plan of care. • Patient and community education programs that address high-risk behaviors in the catchment area, i.e., smoking cessation, alcohol and drug abuse, teen pregnancy. • Community outreach programs and community partnerships that address high risk behaviors related to morbid and mortality in the service area.
Nutrition Lends importance to creating menu plans that reflect the demographics of the services area, such as offering kosher or halal meals or other ethnic meal choices; formulating policies and procedures that address bringing ethnic foods from home; using hot versus cold foods and drinks during illness and recovery; and establishing flexible meal times to accommodate culturally driven meal time preferences, for example, offering dinner at sundown for fasting during Ramadan for Muslim patients.	**Standard 1** Effective, respectful care in preferred language.	**Domains** Organizational values Planning Monitoring and evaluation Staff development Organizational infrastructure Services and interventions	• Menu plans that reflect the demographics of the services area, i.e., Kosher; Halal, or other ethnic meal choices. • Policies and procedures that address bring food from home. • Use of hot versus cold foods during illness and recovery. • Flexible meal times to accommodate culturally driven meal time preferences i.e., day time fasting for Muslim patients during Ramadan.

Pregnancy and childbearing practices Supports the development of perinatal practices, policies, and procedures that address diverse birthing practices, including gender roles and responsibilities before, during, and after birth; hot versus cold; and views of birth as a sickness/ illness or natural experience.	**Standard 1** Effective, respectful care in preferred language.	**Domains** Organizational values Planning Monitoring and evaluation Staff development Organizational infrastructure Services and interventions	Labor and delivery and postpartum policies and procedures that address diverse birthing practices, including gender roles and responsibilities, hot versus cold, and views of birth as a sickness/ illness or natural experience
Death rituals Requires that organizations develop plans for the provision of culturally relevant palliative care services, policies, procedures, and staff competencies that address diverse dying and bereavement practices across the lifespan, and culturally relevant counseling services related to death and dying, organ donation, and care of the body at the time of death.	**Standard 1** Effective, respectful care in preferred language.	**Domains** Organizational values Planning Monitoring and evaluation Staff development Organizational infrastructure Services and interventions	• Culturally relevant palliative care services. • Policies, procedures, and competencies that address diverse dying and bereavement practices across the lifespan. • Culturally relevant counseling services related to death and dying, organ donation, and care of the body.
Spirituality Requires organizations to make certain that chaplain services are available for each religion that is represented within the service area. Partnerships with key community religious leaders, including traditional and folk healers should be established in response to the cultural and spiritual needs of consumers and staff.	**Standard 1** Effective, respectful care in preferred language.	**Domains** Organizational values Planning Monitoring and evaluation Staff development Organizational infrastructure Services and interventions	• Chaplain services for all dominant religions. • Partnerships with community leaders. • Policies and procedures that address work schedule, meal and medication times, and hours of operation (i.e., clinic visits) in response to cultural and spiritual needs of consumers and staff.
Health-care practices Underscores the need for culturally competent education for all levels of clinical and administrative staff that addresses cultural perspectives of the sick role; medical management, including folk and traditional healing practices. Cultural emphasis on preventive versus acute care can guide the provision of community health and wellness programs to increase access to care, enhance quality of care, improve health outcomes at the individual and community levels, and reduce health disparities.	**Standard 1** Effective, respectful care in preferred language	**Domains** Organizational values Staff development	Cultural competency education that addresses cultural perspectives of the sick role, medical management including folk and traditional healing practices, genetic implications for care across the continuum, and cultural emphasis on preventive versus acute.

(continued)

Table 4.1 (continued)

Purnell Model domains	CLAS standards	Cultural assessment profile	Indicators
Health-care providers Highlights the need for a dedicated chief diversity officer to oversee diversity- related initiatives such as the diversity council, affirmative action program, discrimination events, culture codes, and culturally related conflict management and grievance procedures. Additionally, this domain supports the creation of job descriptions that outline role-related cultural competencies, job requirements that specify role specific language proficiency and cultural competencies, and initial and ongoing performance appraisals that include cultural competency requirements.	**Standard 1** Effective, respectful care in preferred language	**Domains** Organizational values, Communication, Organizational infrastructure	• Chief Diversity Officer position to oversee diversity-related initiatives. affirmative action, discrimination, culture codes, conflict management, and grievance procedures. • Job descriptions that outline cultural competency. • Job requirements that specify language proficiency and cultural competency. • Initial and ongoing performance appraisals that include cultural competency requirements.

and the completion of a cultural health assessment, in general, and specific culture care needs of the most commonly encountered demographics of the service area in particular. Moreover, the education should also include self-reflection, critical thinking, and cross-cultural communication, including the appropriate use of medical interpreters and translators. Generational diversity and the diversity that exists among the health-care team should also be addressed.

Diversity-related education must start in orientation and continue through unit/department-based, population-specific orientation programs. Additionally, diversity education should be woven into annual educational initiatives and performance appraisals using evidence-based assessment instruments to ensure the initial and ongoing maintenance of competency, including the proficiency of trained medical interpreters and translators. To accommodate the variety of learning styles that exist within the health-care team, a variety of educational venues, such as face-to-face classroom interaction, online, Web-based programs, and online and/or hard-copy resources at the point of care should be available. Informal venues such as *lunch-and-learn*, and including diversity-related topics on staff meeting agenda has been helpful to keep cultural competency visible in daily operations.

Transdisciplinary, interprofessional team learning approaches have demonstrated improved communication within the health-care team. Onsite consultation and conferences and workshops conducted by experts in the fields of transcultural nursing, cultural competency, and organizational culture have been reported to help sustain diversity initiatives in fast-paced health-care delivery systems (IOM 2002; Marrone 2008). Other successful strategic initiatives within culturally competent organizations that support staff knowledge and skill acquisition include providing staff with incentives such as reward and recognition ceremonies, pins, acknowledgment in organizational newsletters or Web sites, preference to attend external conferences and workshops for staff who have completed initial and ongoing cultural competency education and competency requirements, and incentives for staff to volunteer in the community to learn about community members and the cultures represented within the service area, including health fairs where specific populations of the catchment area are present.

4.8 Language Assistance Services

Culturally competent health-care organizations must provide language assistance services, including interpreter and translation services, at no cost to the customer with limited or dominant language proficiency at all points of contact and in a timely manner during all hours of operation. In addition, organizations must provide consumers in their preferred language both verbal and written notices informing them of their rights to receive language assistance services. Likewise, organizations must ensure the competence of language assistance to those with limited proficiency of dominant language. Patient-related materials and post signage in the languages of the commonly encountered groups and/or groups represented in the service area are additional requirements.

According to multiple sources, language discordance can lead to decreased access to care, decreased quality of care, increased cost of care, decreased patient satisfaction, recidivism, discrimination, and poor health outcomes (American Association of Critical Care Nurses 2008; Europa 2010). Patients who speak little or none of the dominant language are at greater risk of medical errors or misdiagnosis. If they are not provided with an interpreter, they are less likely to use preventive care services and are more likely to use emergency rooms than the dominant language (Cornelio 2004).

Culturally competent health-care organizations must provide language assistance at all points of entry and care and during all hours of operation. Incorrect interpretation and/or translation can result in patient confusion, threaten patient safety, and cause emotional distress, resulting in increased costs to the organization. Resources for interpretation and translation services include interpretation agencies, community language banks, telephonic services at the point of care, and interactive video- or computer-based services for the deaf or hearing impaired (Tang 2010). Frequently used techniques to identify the languages spoken within the community and service area—as well as the need for sign language—include information from the local community; community organizations; directly from consumers; national, regional, and community census data; and community needs assessment (Tang 2010). Critical documents such as informed consent forms, patient education materials, and pain assessment tools, as well as signage and directions within the organization, should be translated into the major languages spoken by the consumers in the service area (Purnell et al. 2011).

An emerging body of knowledge related to health literacy and its relationship to patient health, compliance with medical treatment plans and access to care exists. Health literacy is the degree to which an individual is able to read, understand, and use information to make health-care decisions. Low health literacy can have a negative impact on health outcomes and potentially increase the risk of medical errors. Negative sequelae can be minimized by the use of correctly interpreted and simple printed information with pictures and diagrams. Written policies and procedures for the development and purchase of written and/or video/audio patient education materials in the dominant and nondominant languages should be evaluated by a transdisciplinary patient education team, the members of which should include community partners and organizational constituents who are knowledgeable in health literacy and cultural competency (Marrone 2010; Tang 2010).

4.9 Community Resources and Partnerships

Socially responsible educational, health-care, and professional organizations, must be forged that use a variety of formal and informal mechanisms to facilitate community and consumer involvement in designing, implementing, and evaluating diversity-related initiatives. Active, bidirectional, and mutually beneficial partnerships with formal and informal community leaders and key informant interviews and focus group meetings with cultural and spiritual leaders, political and regulatory leaders and accrediting agencies, natural and lay healers, and community elders are all mechanisms to ensure that organizations provide culturally relevant services. Health-care organizations, in particular, need to negotiate with managed care organizations for culturally relevant health services across the continuum of care and advocate on behalf of the vulnerable and underserved populations within the organization's service sector.

4.10 Advocacy

Advocacy on behalf of consumer and health-care team diversity is a critical element within the schema of organizational cultural competency. Culturally competent health-care organizations ensure that conflict and grievance resolution processes are culturally and linguistically sensitive and capable of identifying, preventing, and resolving cross-cultural conflicts or complaints by patients/consumers (Pacquiao 2018b). Furthermore, culturally competent health-care organizations establish an environment committed to diversity non-discrimination through clearly articulated behavioral and performance expectations that are communicated to all levels of staff both verbally and in writing. In essence, culturally competent organizations create and sustain a culture of *zero tolerance* for discrimination in all sectors of the work environment.

Reflection: *Does the organization where you work/go to school have cultural brokering/mentoring programs for new employees? What strategies would you use to initiate such a program?*

Best practice supports the development of the chief diversity officer role to lead all diversity-related organizational initiatives, such as a diversity council. The chief diversity office and diversity council would have the following roles:

- Advise the chief executive team in the development of strategies that support diversifying the organization's workforce.

- Review organizational policies, recruitment practices, patient education materials, and care practices that may have an adverse impact on one or more consumer groups within the service area (Thornicroft et al. 2008).
- Establish written, evidence-based criteria for hiring external and promoting internal candidates and apply policies consistently to all candidates.
- Revise job standards for job performance to reasonably accommodate individuals with disabilities (Tartaglia et al. 2007).
- Develop policies that address discrimination, conflict management, and grievance resolution processes and incorporate them into the patient bill of rights.
- Educate staff as mediators in cross-cultural conflicts.
- Train or hire patient advocates.
- Post signage that notifies patients and families that a grievance process exists.

Reflection: *Does the organization where you work/go to school have culturally appropriate toys available on pediatric units, in the Emergency Department, and in reception areas where children are likely to be? Where might you go to obtain such toys?*

4.11 Transparency

Health-care organizations are encouraged to regularly make available to the public information about their progress and successful innovations in implementing diversity initiatives and to provide public notice in their communities about the availability of this information (OMH 2019). Strategies that have proven beneficial include the following (Marrone 2010; Purnell et al. 2011):

- Create and distribute brochures to patients/families and include in admissions packets that highlight the attention to and respect for diversity within the organization.
- Include diversity services on patient/family education television channels and in the organization's Web site and brochures.
- Publish articles in professional health-care journals and local/neighborhood periodicals to market diversity services and to share success stories.
- Inform community agencies and local advocacy groups regarding the diversity services that are offered by the organization and the benefits thereof.
- Reach out to professional associations to present and publish diversity-related initiatives and outcomes in the association's publications, on their Web sites, and at local, regional, national, and international conferences.
- Partner with case managers and discharge planners regarding patients with needs related to culture, health literacy, and the ability of ambulatory/community services to provide culturally relevant care following hospitalization.
- Use legislative representative as a vehicle for constituents who need health-care providers and who are sensitive to cultural issues.

4.12 Outcomes Metrics to Assess Cultural Competence

Several methods have been developed to assess organizational cultural competence. However, to date, consensus has not been achieved regarding which data elements to be measured. Few reliable data collection instruments are available to assess organizational cultural competence. In spite of this limitation, a review of the literature reveals that indicators of cultural competence in health-care delivery organizations typically include organizational values and governance structures, quality monitoring and evaluation, communication, education services, community involvement, access, health outcomes, financial stability/viability, and data management/data-driven decisions (Joint Commission 2011; Joint Commission International 2017).

Outcomes metrics for organizational cultural competency include the following:

1. Organizational Values and Governance Structures

(a) Philosophy, mission, vision, values, and strategic plan that reflect responsiveness to internal and external diversity.
(b) Annual Report Cards that demonstrate accomplishments related to diversity-related initiatives for patients, families, and the community.

2. Quality Monitoring and Evaluation
(a) Increased consumer satisfaction with care and services particularly related to
Providers rated as being actively engaged with the consumer/practitioner partnership, and,
Increased time with providers during primary care visits.
(b) Improvements in health and wellness status in the service area, particularly related to underserved populations.
(c) Improved public safety in the service area.
(d) Increased compliance with treatment plans.
(e) Decreased medical errors related to informed consent and wrong patient, wrong site/side surgeries or procedures.

3. Communication
(a) Documented use of trained medical interpreters and/or use of language telephone interpretation services for patients who do not understand the primary language of care used within the organization.
(b) Documented use of patient education materials that are culturally sensitive and reflect the health literacy level and language proficiency of the patient/family.
(c) Documented use of critical documents such as consent forms and patient education materials that have been translated into the most frequent languages represented by the demographic data of the service area.
(d) Documentation systems that include cultural health assessment data and the integration of assessment findings into a transdisciplinary plan of care.

4. Education
(a) Cultural competency performance criteria that are integrated into performance appraisals for all levels of job descriptions.
(b) Learning outcomes that reflect culturally, ethnically, linguistically, and spiritually relevant health-care interventions, plans of care, and clinical evaluation strategies.

5. Services
(a) Increased access to services by diverse, underserved, and vulnerable populations.
(b) Increase in cultural- and language-related services and programs provided in response to community needs assessment date.

Reflection: *In the organization where you work/go to school, are food pyramids and food selections available and reflective of the patients' and staffs' languages and culture?*

6. Community Involvement
Community leaders represented among key organizational decision making and stakeholder groups, such as Board of Trustees, Governing Board, Diversity Council, Patient Education Committee, and Ethics Committee.

7. Access
Increased use of preventive services with resultant decreased use in Emergency Department visits for nonemergent health needs.

8. Health Outcomes
Decrease in racial and ethnic health disparities in the service area.

9. Financial Stability/Viability
(a) Decreased health-care costs chiefly related to decreased length of stay (LOS), decreased complications, issues of "noncompliance," decreased Emergency Department visits for nonemergent health issues, and decreased readmissions and recidivism.
(b) Increased revenue principally related to increased use of services by underserved populations and increased throughout related to decreased LOS.
(c) Decreased litigation and malpractice.

10. Data Management
    (a) Decreased disparities related to the access and use of health-care services or in received and/or recommended treatment.
    (b) Review and analysis of consumer satisfaction with care and services data and evidence that information has been used to influence the design/redesign of strategic initiatives and services.

4.13 Resources to Support Culturally Competent Health-Care Organizations

Many governmental, regulatory, and professional agencies provide information that can assist with the design of an organizational infrastructure that supports cultural competency. The World Health Organization provides technical support to assist countries and regions in addressing priority health issues and engages in partnerships to establish health and care norms and standards; policy, program, and human resources development; and the prevention and control of major communicable diseases (WHO 2010).

The U.S. Office of Minority Health (U.S. Department of Health and Human Services Office of Minority Health (2019) develops health policies and programs that are aimed at protecting the health of minority populations through the elimination of health disparities among vulnerable populations. The OMH developed the CLAS Standards in Health Care to guide health-care delivery systems toward meeting the culture care needs of consumers within their respective service areas. The Institute of Medicine (IOM) serves as an independent advisor for health and science policy development. The IOM established Core Competencies for Health Care Professionals that encourage patient-centered care, collaborative interprofessional teams, evidence-based practice, quality improvement, and informatics (IOM 2002).

The Joint Commission is a U.S. agency with an international affiliate that is aimed at continuously improving the safety and quality of care provided to the public through the accreditation of health-care facilities. The Joint Commission developed evaluation strategies in support of culturally competent organizations. Organizational cultural competency evaluation strategies include the following:

1. Allocating resources to initial and ongoing team-building, cultural training, and educational programs such as the development of interdisciplinary cultural diversity committees and outreach programs to minority nursing and medical organizations.
2. Integrating cultural diversity initiatives into and cited cultural competency standards within all levels of the organization such as vision and mission statements, strategic objectives, learning outcomes, clinical performance criteria, policies and procedures, documentation systems, and research.
3. Assessing the cultural composition of the staff as compared with the demographics of the community.
4. Developing hiring practices, promotion strategies, and outreach programs that underscore diversity as a priority.
5. Assessing and integrating patient satisfaction, staff satisfaction, quality improvement, and health outcomes data related to cultural, spiritual, and linguistic diversity into all levels of the organizational strategic planning.
6. Using only trained medical interpreters and translators.
7. Integrating patients' health and illness values and traditions into written plans of care and progress notes.
8. Assessing patient/family understanding of teaching and discharge instructions (Joint Commission 2011).

The Transcultural Nursing Society (2012) is a professional nursing organization that is committed to enhancing the quality of culturally congruent care provided by nurses and other health-care providers prepared in transcultural nursing and health care that supports improved health and wellness for people worldwide. The Society developed a certification process for nurses and other health-care professionals that ensures competency in providing culturally competent care. The International Council of Nurses (ICN) (2010) promotes healthy lifestyles, healthy workplaces, and healthy communities by working closely with the national nursing associations representing 130 countries. The Council supports programs that mitigate poverty, pollution, and other underlying causes of illness, includes care strategies that address meeting spiritual and emotional needs, and advocates that prevention, care, and cure are the rights of every human being.

The American Organization of Nurse Executives (AONE) (2011) developed the AONE Guiding Principles for Diversity in Health Care Organizations (2007) and the AONE Diversity for Health Care Organizations Toolkit to assist health-care organizations to establish a healthy practice and work environment that reflects the diversity through a commitment to inclusivity, tolerance, and governance structures.

4.14 The Future of Culturally Competent Health-Care Organizations

Global geographic migrations resulting in wide-reaching demographic variations are anticipated to grow over the next several decades. These demographic changes amplify the importance of addressing cultural, ethnic, racial, linguistic, and spiritual health disparities. Minority populations who are currently experiencing poorer health status are expected to grow more rapidly, particularly in industrialized countries. Governments and health-care organizations are focusing more on reducing health disparities by ensuring the cultural competency of the health-care system and its providers.

Successful culturally competent health-care organizations use evidence-based organizing frameworks that articulate well with one another to ensure the integrity and reliability of diversity-related initiatives. The triangulated framework used in this chapter to illustrate the critical elements of organizational cultural competency included the CLAS Standards, the Purnell Model, and the Cultural Assessment Profile. Collectively, the components provide the road map and guideposts for the delivery of culturally and linguistically appropriate health-care services to diverse consumers.

In summary, hallmark characteristics of culturally competent health-care organizations include the following (Marrone 2010):

- An organizational infrastructure that respects and celebrates diversity as reflected in the vision, mission, values, strategic plan, standard operating procedures (policies and procedures), quality initiatives, education plan, employee competencies, and clinical and operational performance outcomes
- The design, implementation, and evaluation of a well-structured strategic plan that ensures the provision of culturally, spiritually, and linguistically appropriate services consistent with the demographics of the service area
- Retention, recruitment, and promotion strategies that attract and retain health-care professionals prepared in transcultural concepts at all levels of the organization that are representative of the consumer demographics in the service area
- Availability of diversity-related resources at the point of care
- Education programs that ensure that all levels of staff receive initial and ongoing learning and skill acquisition related to the culture care needs of the consumer demographics in the service area
- Evidence-based assessment strategies that reliably measure access to care, quality of care, health outcomes, and patient and staff satisfaction and stratify data related to the demographics of the service area
- A network of active partnerships with community leaders and consumer groups that reflect the demographics of the service area who assist the organization in the assessment, planning, implementation, and evaluation of diversity-related initiatives

Finally, essential to the provision of safe, quality language services, culturally competent health-care organizations must develop systems and processes that ensure ongoing self-assessments to determine if consumers are receiving care in their preferred language, maintain a current organizational language data bank that reflects the dynamic changes in languages spoken within the service area, employ trained and validated medical interpreters and translators to provide language assistance services at all points of care and across the continuum of care at no cost to the consumers, and have in place retention and recruitment strategies for diverse staff at all levels of the organization represented in the service area.

Culturally competent organizations must provide patients and visitors with written notices of their right to receive language assistance services; maintain signage in the major languages spoken by consumers, including sign language for the deaf/hearing impaired and Braille for the blind/visually impaired; establish and maintain partnerships with key formal and informal community leaders; safeguard that interpretation and translation services are efficient and accessible at the point of care throughout the continuum of care; and use subject matter experts and specialists in transcultural care and cultural competency to guide the language services product development.

References

American Association of Critical Care Nurses (2008) AACN standards for establishing and sustaining healthy work environments. http://www.aacn.org/wd/hwe/content/resources.pcms?menu=practice

American Organization of Nurse Executives (2011) Guidelines and toolkit for diversity in health care organizations. http://www.nursezone.com/nursing-news-events/more-news/AONE-Releases-Online-Tool-Kit-To-Support-Diversity_32546.aspx

Andrews MM (1998) A model for cultural change: nurse leaders must realize the importance of transculturally based administrative practices. Nurs Manag 29:62–66

AONE Guiding principles for diversity in health care organizations (2007) AONE Guiding principles for diversity in health care organizations. Nurse Lead 5(5):17–24. https://doi.org/10.1016/j.mnl.2007.08.001

Betancourt JR, Green AR, Carrillo JE, Ila OA (2003) Defining cultural competence: a practical framework for addressing racial/ethnic disparities in health and health care. Public Health Rep 118:293–302

Cooper-Patrick L, Gallo JJ, Powe NR, Steinwachs DM, Eaton WW, Ford DE (1999) Mental health service utilization by African Americans and whites: the Baltimore epidemiological catchment area follow-up. Med Care 37(10):1034–1045

Cornelio M (2004) Quality translation in health care: Kaiser Permanente—meeting the challenge. Apuntes 12(4):14–15

Einbinder LC, Schulman KA (2000) The effect of race on the referral process for invasive cardiac procedures. Med Care Res Rev 1:162–177

Europa (2010) Commission takes steps to promote patient safety in Europe [Press release]. http://europa.eu/rapid/press ReleasesAction.do?reference=IP/08/1973&format=HTML&aged=0&language=EN&guiLanguage=en

Health Resources and Services Administration (2002) Indicators of cultural competence in health care delivery organizations: an organizational cultural competence assessment profile. U.S. Department of Health and Human Resources. http://www.hrsa.gov/CulturalCompetence/healthdlvr.pdf

Institute of Medicine (2002) Unequal treatment: confronting racial and ethnic disparities in health care. National Academies Press, Washington

Intercultural competence for all: preparation for living in a heterogeneous world (2012)

International Council of Nurses (2010) Vision statement. http://www.icn.ch/visionstatement.htm

Joint Commission (2011) Patient safety: About hospitals, language, and culture. www.jointcommission.org/assets/1/6/hlc_paper.pdf

Joint Commission International Accreditation Standards for Hospitals (2017). https://www.jointcommissioninternational.org/-/media/jci/jci-documents/accreditation/hospital-and-amc/jci-standards-only_6th-ed-hospital.pdf

Marrone SR (2008) Factors that influence critical care nurses' intentions to provide culturally congruent care to Arab Muslims. J Transcult Nurs 19:8–15

Marrone SR (2010) Organizational cultural competency. In: Douglas M, Pacquiao D (eds) Core curriculum in transcultural nursing and health care. Sage, Thousand Oaks

Morales LS, Cunningham WE, Brown JA, Liu H, Hays RD (1999) Are Latinos less satisfied with communication by healthcare providers? J Gen Intern Med 14:409–417

National CLAS Standards (2018). https://minorityhealth.hhs.gov/omh/browse.aspx?lvl=2&lvlid=53

Office of Minority Health (2019) Eliminating racial and ethnic disparities in health: overview. http://www.raceandhealth.hhs.gov/sidebars/sbinitover.htm

Pacquiao D (2018a) Conceptual framework for culturally competent health care. In: Global applications of applications of culturally competent health care: guidelines for practice. Springer, Cham, pp 1–27

Pacquiao D (2018b) Advocacy and empowerment of individuals, families, and communities. Conceptual framework for culturally competent health care. In: Global applications of applications of culturally competent health care: guidelines for practice. Springer, Cham, pp 239–253

Purnell L, Fenkl EA (2019) The need for culturally competent health care. In: Purnell L, Fenkl E (eds). Springer, ChamHandbook for culturally competent care

Purnell L, Davidhizar R, Giger G, Fishman D, Strickland O, Allison D (2011) A guide to developing a culturally competent organization. J Transcult Nurs 1(22):5–14

Robert Wood Johnson Foundation (2011) Getting minority patients' points of view about cultural barriers to health care. http://www.rwjf.org/reports/grr/055258.htm

Saha S, Komaromy M, Koepsell TD, Bindman AB (1999) Patient-physician racial concordance and the perceived quality and use of health care. Arch Internal Med 159:997–1004

Saha S, Taggart SH, Komaromy M, Bindman AB (2000) Do patients choose physicians of their own race? Health Aff 19:76–83

Tang G (2010) Organizational cultural competency: interpretation services. In: Douglas M, Pacquiao D (eds) Core curriculum in transcultural nursing and health care. Sage, Thousand Oaks

Tartaglia A, McMahon BT, West SL, Belongia L, Shier Beach L (2007) Workplace discrimination and healthcare: the national EEOC ADA research project. J Vocat Rehabil 27:163–169

The National Academies of Sciences, Engineering, and Medicine (2019). https://www.nationalacademies.org/

The National Library of Medicine (2003) Health professions education: a bridge to quality. The Core competencies needed for health care professionals. https://www.ncbi.nlm.nih.gov/books/NBK221519/

Thornicroft G, Brohan E, Kassam A, Lewis-Holmes E (2008) Reducing stigma and discrimination: candidate interventions. Int J Ment Heal Syst 2:1–7

Transcultural Nursing Society (2012) Mission, vision, philosophy, and values. http://www.tcns.org

U.S. Department of Health and Human Services Office of Minority Health (2019). https://minorityhealth.hhs.gov/omh/browse.aspx?lvl=2&lvlid=53

Virginia Department of Health (2019). http://www.vdh.virginia.gov/health-equity/

World Health Organization (2010) Role of WHO in public health. http://www.who.int/en/

Part II

Aggregate Data for Cultural-Specific Groups

5 People of African American Heritage

Josepha Campinha-Bacote and Rebecca C. Lee

5.1 Introduction

African Americans are the second largest minority population in the United States. They are at greater risk for many diseases, especially those associated with low-income, stressful life conditions, lack of access to primary health care, and negating health behaviors such as violence, poor dietary habits, lack of exercise, and lack of importance placed on seeking primary health care early. In addition, African Americans carry the burden of several diseases, which is reflected in higher rates of diabetes, obesity, hypertension, stroke, hepatitis, cancer, asthma, and human immunodeficiency virus infection and acquired immune deficiency syndrome (HIV/AIDS). This chapter provides insight and knowledge regarding the health status, traditions, values, beliefs, and practices of African Americans and implications for culturally specific patient care.

The current chapter is a revision written by Josepha Campinha-Bacote of the original chapter in the previous edition of the book.

J. Campinha-Bacote (✉)
Transcultural C.A.R.E. Associates,
Blue Ash, OH, USA

R. C. Lee
University of Cincinnati, Cincinnati, OH, USA
e-mail: Rebecca.lee@uc.edu

5.2 Overview, Inhabited Localities, and Topography

5.2.1 Overview

African Americans are the second largest minority population in the United States, with the Hispanic/Latino population being the largest. In July 2017, the population of African Americans, including those of more than one race, was estimated at 41.4 million, comprising 12.7% of the total population (Office of Minority Health 2019f). This number is projected to rise to 65.7 million (15%) of the total population by the year 2050.

African Americans are mainly of African ancestry, but many have non-African ancestors due to the fact that the slave trade resulted in a diaspora from West and Central Africa to many parts of the world, including the West Indies, South America, Central America, and the United States. Over the centuries, in all parts of the world, the African has mixed with other local ethnic groups. In America this intermixing has largely been with American Indians and European Americans. Although African Americans and African immigrants in the United States share some similarities in respect to phenotype features and experience with racism, discrimination, and prejudices, there is considerable heterogeneity among people of African descent. Even among Africans, there is not just one African culture, for

© Springer Nature Switzerland AG 2021
L. D. Purnell, E. A. Fenkl (eds.), *Textbook for Transcultural Health Care: A Population Approach*,
https://doi.org/10.1007/978-3-030-51399-3_5

Africa has 54 countries with varying colonial histories and with people from a variety of tribes who speak different languages and have different customs (Abdullah and Brown 2011). African Americans are also a diverse group. There are marked historical differences, as well as differences in degree of identification with their African heritage, socio-economic status, and level of education. Because of the significant diversity that exists among African Americans, health-care providers must be aware of the intracultural variations that exist within this ethnic group.

African Americans have been identified as "Negro," "colored," "black," "black American," "Afro-American," and "people of color." Depending on their cohort group, some African Americans may prefer to identify themselves differently. For example, younger African Americans may prefer the term African American, whereas elderly African Americans may use the terms Negro and Colored. In contrast, middle-aged African Americans refer to themselves as black or black American. Although the term Negro is not a commonly used today, and many African Americans are offended by its use, the U.S. Census Bureau still included the term Negro on the 2010 U.S. Census, along with the terms black and African American. These different descriptors can cause confusion for those who are attempting to use the politically correct term for this ethnic group. In addition, organizational titles, such as the National Black Nurses Association, National Center for the Advancement of Blacks in the Health Professions, the National Association for the Advancement of Colored People, and the United College Negro Fund still exist, which clearly depict the differences in how African Americans prefer to be identified. Therefore, it is culturally responsive to ask African Americans what they prefer to be called.

5.2.2 Heritage and Residence

African Americans are largely the descendants of Africans who were brought forcibly to this country as slaves between 1619 and 1860. The literature contains many conflicting reports of the exact number of slaves who arrived in this country. Varying estimates reveal that from 3.5 to 24 million slaves landed in the Americas during the slave trade era. Many slaves who were brought to the American colonies and early United States came from the west coast of Africa, from the Kwa- and Bantu-speaking people. The legacy of African American heritage and history of slavery is often passed on from generation to generation through African American folktales and lived experiences (Mitchem 2007).

African American slaves were settled mostly in southern states. In 2017, 58% of African Americans still lived in the South. The ten states with the largest African American population were New York, Florida, Texas, Georgia, California, North Carolina, Illinois, New Jersey, Virginia, and Louisiana (Office of Minority Health 2019f).

5.2.3 Reasons for Migration and Associated Economic Factors

The Civil War in the US ended slavery in 1865, and particularly in the state of South Carolina, the Reconstruction Act allowed African Americans the right to vote and participate in state government. However, most African Americans in the South were denied their civil rights and were segregated. They were unable to get out of debt and support their families despite being successful farmers. Thus, African Americans lived in poverty and encountered many hardships. After the Civil War, more African Americans migrated from southern rural areas to northern urban areas. African Americans migrated because of a lack of security for life and property. World War II was also a major catalyst in fostering migration to urban and northern areas, which provided greater economic opportunities and brought African Americans and European Americans into close contact for the first time. Jaynes and Williams (1989) reported that during the 1940s, a net outmigration from the South totalled approximately 1.5 million

African Americans (15% of the South's African American population). Although the migration was viewed as a positive move, many African Americans encountered all the problems of fragmented urban life, racism, poverty, and covert segregation.

5.2.4 Educational Status and Occupations

Before 1954, educational opportunities for African Americans were compromised. School systems were segregated, and African Americans were victims of inferior facilities. In fact, in 1910, almost one-third of all African Americans were illiterate (Washington 2002). Conant (1961) described the plight of African Americans in segregated schools and, to some extent, predicted the long-term social consequences of such a system. His predictions have been borne out as inadequate job opportunities and poor wages, resulting in poverty. Despite the 1954 Supreme Court decision in *Brown v. Board of Education of Topeka* that ruled against the segregation of African Americans and European Americans in the public-school systems, poverty has continued to have a ripple effect on African American communities, often leading to poor health outcomes and increased morbidity and mortality (Noonan et al. 2016).

While Braithwaite et al. (2000) have reported past dropout rates among African Americans as high as 61%, data from the Current Population Survey (CPS) show that from 2000 to 2016, the African American dropout rate decreased from 13.1 to 6.2% (National Center for Education Statistics 2019). This is in comparison to the 2016 European American dropout rates of 5.2% and the Hispanic dropout rate of 8.6%. The decrease in dropout rate among African Americans is partially related to the high value that most African American families place on education. In 2017, 86% of African Americans 25 years old and over had earned at least a high school diploma (as compared to 92.9% European Americans), 21.4% had a bachelor's degree or higher (as compared to 35.8% European Americans), and 8.1% had a graduate or advanced professional degree (as compared to 13.8% of the European American population) (Office of Minority Health 2019f).

The African American family views education as the process most likely to ensure work security and social mobility. Families often make great sacrifices so at least one child can go to college. In African American families, it is not uncommon to see cooperative efforts among siblings to assist one another financially to obtain a college education. For example, as the older child graduates and becomes employed, that child then assists the next sibling, who, in turn, assists the next one. This continues until all of the children who attend college have graduated. Before the civil rights movement, a major emphasis for African Americans in higher education was vocational. The thinking was that if African Americans could learn a trade or vocation, they could become self-sufficient and improve their economic well-being. Preparation for vocational careers is evidenced in the name, mission, and goals of two of the renowned, historically Black institutions, Hampton University and Tuskegee University, formerly known as Hampton Institute and Tuskegee Normal and Industrial Institute.

African Americans make up 11.6% of the U.S. workforce; however, representation of many African Americans and other ethnic groups in the health professions is far below their representation in the general population. African Americans are underrepresented in the Life, Physical, and Social Sciences and Health Diagnosing and Treating Practitioners occupations (National Center for Health Workforce Analysis 2017). Specifically, African Americans are underrepresented in all occupations, except among Dieticians and Nutritionists (15.0%), and Respiratory Therapists (12.8%). African Americans are also underrepresented in many of the occupations in the Health Technologists and Technicians category (Dental Hygienists, 3.1%; Dispensing Opticians, 5.5%; EMT paramedics, 6.3%; and Diagnostic Related Technologists and Technicians, 7.8%) (National Center for Health Workforce Analysis 2017). In contrast to other occupational groups, African Americans are well represented in occupations belonging to the Healthcare Support and Personal Care and Services occupational categories. African Americans have their highest representation in Nursing, Psychiatric and Home Health Aides (32.0%) and have nearly twice their representation in the occupations of Counselor (18.8%) and Social Workers (21.5%) (National Center for Health Workforce Analysis 2017). Increasing racial and ethnic diversity among health professionals is critical because evidence indicates that diversity is associated with improved access to care for racial and ethnic minority patients, greater patient choice and satisfaction, and better educational experiences for all students (Institute of Medicine 2004).

African Americans continue to be approximately twice as likely as European Americans and Asians to be among the working poor (Bureau of Labor Statistics 2018). One reason for this disproportionate representation in professional and managerial positions is believed to be discrimination in employment and job advancement. In 1961, President John F. Kennedy established the Committee on Equal Employment Opportunity to protect minorities from discrimination in employment. However, most African Americans still believe that job discrimination is a major variable contributing to problems they encounter in obtaining better jobs or successful career mobility. With the dismantling of affirmative-action programs, based on misinterpretation of their purpose, this view will, perhaps, continue to gain support.

Most working-class African Americans do not typically advance to the higher socioeconomic levels. The occupational categories in which the greatest proportion of African American men are employed include machine operators, fabricators, and laborers. Because African Americans are overrepresented in the working class, they are more likely to be employed in hazardous occupations, resulting in an increased risk for occupation-related diseases and illnesses (Seabury et al. 2017). For example, African Americans are at a higher risk to occupational exposures to lung cancer causing agents in the workplace. These cancer-causing agents include diesel exhaust, asbestos, nickel, chromium, arsenic, second-hand smoke, dust particulate matter, and other particulate matter pollutions in the workplace setting (Alberg and Nonemaker 2008; Cooley and Jennings-Dozier 1998; Hicks 2010). Implications are that health-care providers must not only assess African American patients for occupation and stress-related diseases such as hypertension, but must also be familiar with the government's *Healthy People 2020* goals for the health and safety of individuals in the work environment (Office of Disease Prevention and Health Promotion 2019).

5.3 Communication

5.3.1 Dominant Language and Dialects

The dominant language spoken among African Americans is English. Some African Americans use a language that sociolinguists refer to as *African American Vernacular English (AAVE)* (Zienkiewicz 2008). Over the years, a number of names have been used to describe the different varieties or dialects of AAVE. Some of the more common terms are *Black Dialect, Black Folk Speech, Black English, African American English (AAE), Ebonics, and Black Vernacular English (BVE)*.

African American Vernacular English is rooted in history (Rickford et al. 2015). The two main hypotheses about the origin of AAVE are the dialect hypothesis and the Creole hypothesis. The dialect hypothesis supports the position that African slaves, upon arriving in the United States, picked up English very slowly and learned it incorrectly. In turn, these inaccuracies have been passed down through generations. The Creole hypothesis maintains that AAVE is the result of a Creole derived from English and various West African Languages.

The major problem that AAVE speakers face is prejudice, as most people believe that AAVE is inferior to Standard American English (SAE). It has often been considered a language deficit, based on a lack of understanding about the AAVE variety (Pearson and Jackson 2013). At times, African Americans who use AAVE are misinterpreted as being uneducated. However, it is common for educated African Americans who are extremely articulate in SAE to use AAVE when conversing with one another. Thompson et al. (2004) referred to this ability as *dialect shifting* (also known as *code switching*). The literature suggests that AAVE provides African Americans with a framework for communicating unique cultural ideas and also serves as a way to symbolize racial pride and identity (Allender and Spradley 2001; Murray and Zentner 2001; Rickford et al. 2015).

5.3.2 Cultural Communication Patterns

African American communication has been described as high-context (Cokley et al. 2005). Nonverbal cues, such as facial expressions, eye movement, and the tone of voice are very important in understanding a high-context communicator. The volume of African Americans' voices is often louder than those in some other cultures; therefore, health-care providers may misunderstand this attribute and automatically assume this increase in tone is reflecting anger and aggression.

While some ethnic groups, such as Native American and Asian American cultures, may value silence, African Americans place a high value on verbal skills and their members tend to speak more. African American speech is generally more dynamic, animated, expressive, and emotionally textured (Campinha-Bacote 2017) and they communicate more interactively than European Americans. Wood (2009) states that this may explain why some African Americans shout out responses such as "Tell it," "All right," and "Keep talking" during speeches or church sermons. While many European Americans may consider these comments as an interruption, some African Americans regard it as complimentary participation in communication.

Face-to-face communication is very important in the African American culture. They also tend to rely on the use of non-verbal messages. Their non-verbal communication is usually accompanied by physical expression (e.g., body movements such as gesturing of arms and hands, head nodding, and demonstrative facial expressions), varying vocal qualities (typically louder), and frankness of expression (Monroe 2006). African Americans are reported to be comfortable with a closer personal space than other ethnic groups. Touch is another form of nonverbal communication seen when African Americans are interacting with relatives and extended family members. When interacting with African Americans, the power of touching should not be underestimated for its healing powers (Cokley et al. 2005).

In communicating among themselves, African Americans place a strong value on oral tradition. Oral tradition is the face-to-face transmittal of elements of the African American culture from one generation to another by the spoken word. One example of this form of communication is storytelling. Storytelling has been a method of intergenerational communication and connection for a number of years in the African American community (Fabius 2016). These stories convey important values and morals on how to live life. Storytelling has emerged as a culturally specific intervention for health promotion in African American patients. Houston et al. (2011) conducted a study investigating the use of storytelling as an effective way to teach African Americans about hypertension and improve blood pressure control. The storytelling intervention produced significant improvements in blood pressure for patients with baseline uncontrolled hypertension.

Another oral tradition is the use of humor within the African American community. Humor is a universal phenomenon but is also culturally tinted (Jiang et al. 2019). From a historical context, humor was necessary for survival, for it provided a space where African Americans could express their unjust treatment as slaves. Today, humor continues to serve as a tool to release angry feelings and to reduce stress and ease racial tension. The *dozens*, a social game in which African Americans use humor, is a joking relationship between two African Americans in which each in turn is, by custom, permitted to tease or make fun of the other. Frequently, humor is used among the African American population as a preventive mechanism to ward off an anticipated attack. Often, the joking is loud and can be mistaken for aggressive communication if not understood within the context of the African American culture. Being aware of and understanding the function that humor serves in the African American culture can assist health-care providers to formulate culturally responsive health-care interventions. For example, Campinha-Bacote (1993, 1997) documented the effective use of culturally specific humor groups with African American patients with psychiatric disorders.

Many African Americans mistrust health-care providers and express their feelings only to trusted friends or family. What transpires within the family is viewed as private and not appropriate for discussion with strangers. A common phrase that reflects this perspective is "Don't air your dirty laundry in public." Health-care providers must be sensitive to this form of communication in that older and more traditional African Americans may not embrace "talk therapy."

5.3.3 Temporal Relationships

In general, African Americans tend to be more present-oriented than past- or future-oriented. African Americans with a present-time orientation may not see the need to take preventive medication or to complete a course of antibiotics when symptoms disappear. Therefore, it is critical for health-care providers to provide detailed explanations of the need to complete medication regimes. Additionally, some African Americans may delay seeing a health-care provider until symptoms are severe or begin interfering with their work or life. However, the mental health literature suggests a positive aspect of a present-time orientation. Williams et al. (2012) contend that an African American worldview focusing on a present-time orientation may in fact contribute to reducing anxiety. Anxiety is characterized as an attentive bias toward future possible threat over present moment experiences. Interventions for anxiety disorders help relieve anxiety by teaching patients to shift their focus to the present. Therefore, health-care providers can utilize an African American's focus on present-time orientation as an asset, not always a challenge, when developing culturally specific interventions for specific mental health conditions.

Due to the intercultural variation that exits among African Americans, the past or future may also be valued in specific subgroups of African Americans. For example, the elderly, tend to place greater emphasis on the past than on the present. In contrast, younger and middle-aged African Americans are more present oriented, with evidence of becoming more future oriented, as indicated by the value placed on education.

Linear time is not adopted to the same extent it is in the dominant society. African Americans tend to be more relationship-oriented and therefore more relaxed about time. Within this context, they may not be prompt for their appointments. It is more important for them to show up for an appointment than to be on time. What they see as significant is the fact that they are there, even though they may arrive 1 or 2 h late. Therefore, flexibility in timing appointments may be necessary for African Americans, who have a circular sense of time rather than the dominant culture's categorically imperative linear sense of time (Murray and Zentner 2001).

5.3.4 Format for Names

Most African Americans prefer to be greeted formally as Dr., Reverend, Pastor, Mr., Mrs., Ms., or Miss. They prefer their surname because the family name is highly respected and connotes pride in their family heritage. However, African Americans do not use such formal names when they interact among themselves. An African American youth commonly addresses an unrelated African American who lives in the community as Uncle, Aunt, or Cousin. Adult African Americans may also be called names different from their legal names. In addition, it is not uncommon for African Americans to describe members of their church community using kinship terms in which fellow congregation members are called "Brother" or "Sister" (Coe et al. 2015). Since most African Americans prefer to be greeted formally as Dr., Reverend, Pastor, Mr., Mrs., Ms., or Miss., until invited to do otherwise, greet African American patients by using their last name and appropriate title.

Reflective Exercise

Mr. Williams, a 45-year African American male, presents to an outpatient mental clinic for a psychiatric evaluation. He is 45 min late for his appointment and doesn't offer any explanation for his tardiness. The nurse conducts a mental status examine and reports that he appears to be a very guarded individual who is extremely suspicious and mistrustful of all health-care providers. She adds that Mr. Williams began the interview by strongly stating, "I do not trust you and will not take your medication." The nurse states that he appears angry, which she maintains is clearly reflected in his loud voice, continuous raising of his hand and arms, facial grimaces, and situating himself extremely close to her. The nurse also notes that at one brief point during the examination his speech became extremely incoherent and he appeared to be experiencing audio and visual hallucinations with delusions of grandeur. For example, when asked to whom he was talking, Mr. Williams responded that he was talking and praying to the "One most powerful who gives him power." The nurse reports that his insight and judgment are poor, as evidenced by his frank denial that he has any mental health problems.

1. How might Mr. William's tardiness be explained from a temporal relationship perspective.
2. What could explain Mr. Williams' mistrust of health-care providers?
3. From a cultural perspective, how can the documentation of audio and visual hallucinations and delusions of grandeur be explained?
4. The nurse interpreted Mr. Williams' loud voice, continuous raising of his hand and arms, and his facial grimaces as being agitated. What could be another interpretation of these behaviors?

5.4 Family Roles and Organization

5.4.1 Head of Household and Gender Roles

The history of African American families in the United States is complex and the experiences of these families are incredibly different than those of other cultural groups. While in current society it is common to find a patriarchal system present in African American families, nevertheless, a high percentage of families still have a matriarchal system in which the head of the household can be a single mother, grandmother, or aunt. Therefore, a single head of household is accepted without associated stigma in African American families. When single mothers are unable to provide emotional and physical support for their children, grandmothers, aunts, the church, and extended or augmented families readily provide assistance or take responsibility for the children. Nationally, 4.9 million children under the age of 18 live in grandparent-headed households (U.S. Census Bureau 2014). Reasons for grandchildren residing with grandparents include maternal substance abuse (18.9%), paternal substance abuse (14.7%), paternal incarceration (14.4%), and maternal abuse (12.3%) (Radel et al. 2016). Since 2000, there has been a steady decline in the number of African American grandparents living in the same home as their children, decreasing from 8.2% in 2000 to 5.6% in 2014 (U.S. Census Bureau 2014).

Structural barriers are one contributor to the current trends in the African American family structure. Ladner and Gourdine (1992) state, "Single parenting and poverty are viewed as the causal factors in destabilizing the African American family" (p. 208). According to recent U.S. Census data reports (U.S. Census Bureau 2018), 21% of African American women lived in poverty, and nearly 1 in 3 (30.8%) of African American children lived in poverty. More than half of all children living in poverty lived in families headed by women. Nearly two in five (39%) of single-mother African American families with children are living in poverty.

An additional structural barrier that has historically undermined the African American family is the absence of African American males due to high unemployment rates, low life expectancy, and high incarceration rates. Gender roles and child-rearing practices in the African American family vary widely depending on ethnicity, socioeconomic class, rural versus urban location, and educational achievement. The diverse family structure extends the care of family members beyond the nuclear family to include relatives and nonrelatives (Wilson 1995). Similar to the pattern in the general society, dual employment of many middle-class African American families requires cooperative teamwork. Many family tasks such as cooking, cleaning, childcare, and shopping are shared, requiring flexibility and adaptability of roles.

A recent review of research regarding African American gender roles (Jones et al. 2018), revealed that gendered phenomena have a significant impact on indices of mental wellness for African American men and women. Overall, less traditional gender role attitudes and/or androgyny were associated with mental wellness, enhanced relationship quality, and positive vocational and educational outcomes (Jones et al. 2018). Despite this fact, the authors of the study did not locate any literature detailing how gender roles affect clinical practice, nor how to support individuals in the adoption of adaptive gender role attitudes. This supports the need for future research to examine the effectiveness of interventions such as, parent workshops, couples counseling, and/or gender-specific therapy groups, in shifting maladaptive gender role beliefs.

Because many African American families, especially those with a single head of household, are matrifocal in nature, the health-care provider must recognize women's importance in decision making and disseminating health information. Also, the health-care provider must focus on, and work with, the strengths of African American families, especially single-parent families. In addition, it is important that health-care providers not neglect the important role that African American fathers can and do play in the nurturance of their families, and identify opportunities for strengthening these bonds. In the past, research has focused on the detrimental impact of non-residential African American fathers on the lives of their children. However, more recently, research has focused instead on the concept of father involvement—how much time African American fathers spend engaged with or accessible to their children (Brown et al. 2018). While residential fathers are typically more involved with their children, positive involvement with either residential or non-residential fathers has a positive influence on the children's development, including heightened social and emotional well-being and greater academic achievement (Adamsons and Johnson 2013; Thomas et al. 2007).

5.4.2 Prescriptive, Restrictive, and Taboo Roles for Children and Adolescents

Given African Americans' strong work and achievement orientation, they value self-reliance and education for their children. A dichotomy might exist here because many parents do not expect to get full benefit from their efforts because of discrimination. Thus, families tend to be more protective of their children and act as a buffer between their children and the outside world. Research finds that especially in low-income African American female-headed households, parental involvement can promote positive educational outcomes and that parental involvement is influenced by parents' own histories of involvement (Jarrett and Coba-Rodriguez 2015). It is important for health-care providers to note that previous research with African American families in the United States has taken a deficit perspective focused on single parents living in poverty (Cross-Barnet and McDonald 2015). Nevertheless, there is considerable diversity among African American families as a whole, with variation found within and among social and demographic groups (Cross-Barnet and McDonald 2015). Research indicates that no one social identity determines attitudes or values regarding childrearing or parenting; rather, race, class, and gender interact. Health-care providers must consider these intersections rather than evaluate each category alone (Cross-Barnet and McDonald 2015).

Respectfulness, obedience, conformity to parent-defined rules, and good behavior are stressed for African American children. The belief is that a firm parenting style, structure, and discipline are necessary to protect the child from danger outside of the home. In violence-ridden communities, mothers try to keep young children off the streets and encourage them to engage in productive activities. Adolescents are assigned household chores as part of their family responsibility or seek employment for pay when they are old enough, thus learning "survival skills."

The importance of role-modeling strength and resilience in the face of adversity was found to be an important cultural value for African American mothers of Appalachian ethnicity in one ethnonursing study of family homelessness (Lee 2012). All participants, both African American Appalachian and European American Appalachian, shared stories of their experience of becoming and being homeless while maintaining their mothering role, a role that provided an important source of identity despite living in the shelter. Participant observation and interviews with mothers revealed that some shelter policies were regarded by African American mothers as interfering with this important role. One particularly negative experience was associated with required attendance at weekly parenting classes, which were based on general assumptions of deficits in parenting as a result of their homeless status. For all mothers in the study, this was seen as a non-caring policy that failed to acknowledge the mothering strengths they possessed. As a result of this important finding, shelter administrators asked the nurse researcher to work with the families to develop a replacement educational offering. In partnership, the mothers and researcher co-created a new class entitled Family Development Class. This class was built around a strength-based approach which, according to mothers, recognized that all families, even strong ones, can develop additional skills (Lee 2012).

The United States has made remarkable progress in reducing both teen pregnancy and racial and ethnic differences (Centers for Disease Control and Prevention 2016). Nationally, the teen birth rate (number of births per 1000 15 to 19-year-old females) declined over 40% from 2006 to 2014, with rates for African American teens declining 44% (Hamilton 2015). While much progress has been made in reducing teen pregnancy rates, the reality is, it continues to be a problem in the African American community because of poor pregnancy outcomes such as premature and low-birth-weight infants and obstetric complications. Furthermore, the teenage mother is expected to assume primary responsibility for her child, whereas the extended family becomes a strong support system. While teen pregnancy is not encouraged in African American families, it is generally accepted after the fact, with support provided by extended family networks. In some instances, the infant may be informally adopted, and someone other than the mother may become the primary caregiver.

By better understanding the many factors that contribute to teen pregnancy health-care providers can better design, implement, evaluate, and improve prevention interventions and further reduce disparities. Certain social determinants, such as high unemployment, low education, and low income, have been associated with higher teen birth rates. Interventions that address socioeconomic conditions like these can play a critical role in addressing disparities observed in U.S. teen births rates (Centers for Disease Control and Prevention 2016). The U.S. Department of Health and Human Services Office of Adolescent Health (OAH) has identified 44 evidence-based teen pregnancy prevention interventions (Office of Adolescent Health 2017). One drawback of many of these interventions is that they are time and staff intensive. Therefore, adapting existing interventions to specifically target at-risk African American adolescents is one promising approach. Plant et al. (2019) used entertainment education (EE) to develop an educational video to reach this group. EE is a process of creating and implementing an entertainment program to increase knowledge and change attitudes and behaviors regarding a social or health issue (Plant et al. 2019). One important strategy used by this research team was the co-creation of the video with input from diverse stakeholders, including African American and Latina adolescent teenaged girls. Brief, low-resource interventions such as this are promising as they can be broadly disseminated and easily implemented to have widespread impact.

5.4.3 Family Roles and Priorities

African American families share a wide range of characteristics, family values, goals, and priorities. One example of a strong family value is the level of respect bestowed upon the elders in the African American community. African American grandparents tend to play vital roles in their grandchildren's lives. Often grandparents reside in the same household as their grandchildren, with multigenerational households being common. Partly due to earlier mortality among African American men, grandchildren are more likely to have substantive relationships with their grandmothers. Because of this, the role of the grandmother is one of the most central roles in the African American family. These grandmothers are respected for their insight and wisdom. In addition, grandmothers are frequently the economic support of African American families, while also often playing a critical role in childcare.

An understanding of the role and importance of the extended family in the lives of African Americans is essential for health care providers. Extended family and church-based social networks are important resources for African Americans (Krause and Bastida 2011; Taylor et al. 2004) because they provide social support to their members in the form of instrumental, emotional, social, and psychological assistance and resources. Several African American extended-family models exist. Billingsley (1968) divided them into four major types: subfamilies, families with secondary members, augmented families, and nonblood relatives. Subfamily members include nieces, nephews, cousins, aunts, and uncles. Secondary members consist of peers of the primary parents, older relatives of the primary parents, and parents of the primary parents. In an augmented family, the head of the household raises children who are not his or her own relatives. Nonblood relatives are individuals who are unrelated by blood ties but who are closely involved with the family functioning. Nonblood relatives are also referred to as "fictive kin." As a result of long-standing relationships with the family, fictive kin may be serving as the primary caregivers or even as the substitute decision makers and sometimes may be more involved than the related family members (Hargrave 2010).

Studies have found that African American families exhibit about 70 diverse structural formations versus about 40 among European American families (Barbarin 1983). Barbarin adds that this comparison points to the variability of the African American family structure and to the flexibility of family roles. Nguyen et al. (2016) examined social network typologies among African American adults as well as their sociodemographic correlates. Network types were derived from indicators of the family and church networks. Results indicated four distinct network types: ambivalent, optimal, family centered, and strained. These four types were distinguished by (a) degree of social integration, (b) network composition, and (c) level of negative interactions.

Social status is important within the African American community. Certain occupations receive higher esteem than others. For example, African American physicians and dentists tend to have privileged positions. Ministers and clergy also receive respect in the African American community. Individuals in these professions have historically held a high status in African American communities and are critical "First Responders" (Cokley et al. 2005). African Americans who move up the socioeconomic ladder often find themselves caught between two worlds. They have their roots in the African American community, but at times they find themselves interacting more within the European American community. Other African Americans refer to these individuals as "oreos"—a derogatory term that means "black on the outside, but white on the inside." In Frazier's (1957) seminal and controversial publication *Black Bourgeoisie,* he highly criticized middle-class African Americans. He argued that African American families who achieve upper-middle-class and middle-class status—the so-called black bourgeoisie—perpetuate a myth of "Negro society." According to Frazier, this term describes behavior, attitudes, and values of a make-believe world created by middle- and upper-class African Americans in order to escape feelings of inferiority in American society.

5.4.4 Alternative Lifestyles

African Americans are, and have always been, represented among the lesbian, gay, bisexual, transgender, queer, intersex, asexual, and others community (LGBTQIA+) (Gold, 2018). The term is an inclusive way to describe a diverse group of people who identify as having different characteristics, including biological sex, gender identity, and sexual attraction. For many years the term LGBT was widely used to describe people who do not identify as exclusively heterosexual. However, terms LGBTQIA and LGBTQIA+ have become more widespread, in recognition of the fact that there are a wide range of differences between how people identify. Unfortunately, reporting of health statistics has not kept pace with this expanding term, with much of the literature citing information on LGBT African American populations. According to the Williams Institute (2013), there are more than 1 million LGBT African Americans currently living in the United States, with approximately 3.7% of all African American people identifying as LGBT (Williams Institute 2013). LGBT African Americans are disproportionately young and disproportionately female, and nearly one-third of all African American same-sex couples are raising children (Williams Institute). Lesbian and gay relationships undoubtedly occur as frequently among African Americans as in other ethnic groups.

A review of the literature reveals that African Americans are less supportive of homosexuality than other racial and ethnic groups, but the reasons have more to do with religion than race. African Americans are markedly more religious on a variety of measures than the U.S. population as a whole, including level of affiliation with a religion, attendance at religious services, frequency of prayer, and religion's importance in life (Pew Research Center 2014). Negy and Eisenman (2005) reported that while initial results of their study suggested that African Americans had modestly higher homophobia and homonegativity scores than European Americans, these differences did not hold after controlling for frequency of church attendance, religious commitment, and socioeconomic status. For both ethnic groups, religiosity significantly predicted homophobia and homonegativity.

Some of the important issues facing LGBTQ African Americans that have the potential to impact health include economic insecurity, risk for violence and harassment, human immunodeficiency virus infection (HIV) and health inequity, religious intolerance, and criminal injustice. For example, LGBTQ African Americans continue to be economically disadvantaged because of persistent discrimination, housing insecurity, a lack of quality, affordable healthcare and fewer educational opportunities.

5.5 Workforce Issues

5.5.1 Culture in the Workplace

Research reveals that African Americans have a long history of workforce disadvantage. Among race and ethnicity, African Americans have the lowest labor force participation (Bureau of Labor Statistics 2017). Despite their low level of participation, they are disproportionately exposed to increased work stressors including shift work, low control and high demands, poor-to-none health insurance, longer hours, and job insecurity (Archibald 2019). Their survival is often met with ethnic or racial tension. *Ethnic* or *racial tension* can be defined as a negative workplace atmosphere motivated by prejudicial attitudes about cultural background and/or skin color. Race-based discrimination at work is associated with poor job quality, reduced organizational productivity, commitment, trust, satisfaction and morale as well as increased cynicism, absenteeism and staff turnover (Trenerry and Paradies 2012). Racism is also considered a fundamental cause of adverse health outcomes for racial/ethnic minorities and racial/ethnic inequities in health (Williams et al. 2019, p. 105). While some work organizations are making strides in their efforts to fight against discrimination, African Americans continue to encounter challenges imposed on them through the multifaceted interactions of racially motivated negative attitudes and actions of individual and organizational policies and practices that are not encountered by European Americans.

Discrimination in the workplace can covertly take the form of microaggressions. Racial microaggressions are brief and commonplace daily verbal, behavioral, and environmental indignities, whether intentional or unintentional, that communicate hostile, derogatory, or negative racial slights and insults to the target person or group (Sue et al. 2008). They are not limited to human encounters alone but may also be environmental in nature. Racial microaggressions and discrimination in the workplace have been documented to cause considerable psychological and physical distress among African Americans (DeCuir-Gunby and Gunby 2016; Goosby et al. 2015; Hollingsworth et al. 2017; Pittman 2012; Sue et al. 2008). It is necessary for health-care providers to have an enhanced understanding of how discrimination combines with other stressors to shape health and racial/ethnic inequities in health (Williams et al. 2019).

African Americans are underrepresented in highly skilled and managerial positions and overrepresented in low-status positions. Middle-class African Americans who hold higher-paying jobs often experience the "glass ceiling" effect, in which access to higher positions is blocked. Airen (2017) asserts that African Americans are still fighting for equity and equality in the workforce and thus experience a different form of "glass ceiling." Airen proposes the term, "Color Ceiling," to reflect an origin that is based on research focused directly on the workforce barriers uniquely faced by African Americans. The Color Ceiling refers to the "invisible barriers that impede financial equity, employment equity, and promotional advancement for African Americans in the workforce" (Airen 2017, p. 3). Health-care providers must increase their sensitivity and awareness of cultural nuances and issues that create ethnic or racial tension in the workplace environment, for these factors can have an impact on such stress-related conditions as mental health disorders and hypertension.

5.5.2 Issues Related to Autonomy

African Americans typically work in jobs characterized by more routinization and less autonomy when compared to their European American counterparts (Sloan et al. 2013). In addition, they have been routinely exposed to employment discrimination that includes the lack of autonomy (Archibald 2019). Research demonstrates that compared to European Americans, African Americans are less likely than their European American counterparts to perceive workplace autonomy (Campos-Castillo and Ewoodzi 2014; Petrie and Roman 2004).

Lowenstein and Glanville (1995) found that along with historical circumstances, culture and politics affect the employment of African Americans in the health-care industry, often relegating African Americans to nonskilled roles. Many African Americans continue to be frustrated at their lower-level positions and the absence of African American leadership and autonomy in many workplaces. Having opportunities for advancement, recognition for one's work, decision freedom, and autonomy, strongly predict positive perceptions of one's work environment (McCluney et al. 2018).

African American men may experience a difficult time in taking direction from European American supervisors or bosses. This difficulty is thought to have stemmed from the era of slavery when African Americans were considered the property of their masters. African Americans desire for equality and autonomy formed their definition of freedom. Thus, if the professional health-care provider who directs and supervises nonprofessional workers lacks cultural sensitivity toward other ethnic groups, the stage is set for cultural conflict.

Because the dominant language of African Americans is English, they usually have no difficulty communicating verbally with others in the workforce. However, some people may inaccurately view African Americans who exclusively speak AAVE as poorly educated or unintelligent. This misinterpretation may affect employment and job promotion where verbal skills are more valued. In addition, the nonverbal communication style (e.g., strong intonation and animated body movements) of some African Americans is often misunderstood and labelled as more aggressive than assertive in comparison with that of other cultural groups.

5.6 Biocultural Ecology

5.6.1 Skin Color and Other Biological Variations

African Americans encompass a gene pool of over 100 racial strains. Therefore, skin color among African Americans can vary from light to very dark. As health-care providers, we are trained in the art of using alterations in skin color and deviations from an individual's normal skin tone to aid in our diagnoses. For example, jaundice is a sign of a liver disorder; pink and blue skin changes are associated with pulmonary disease; ashen or gray color signals possible cardiac disease; copper skin tone indicates Addison's disease; and a nonblanchable erythema response signifies the presence of a stage I pressure ulcer (Salcido 2002). We commonly use these alterations in skin color as potential signals of pathology because we can visualize changes such as the increased blood flow (erythema) that signals such problems as inflammation. However, these acquired assessment skills are based on a Eurocentric rather than a melanocentric approach to skin assessment (Campinha-Bacote 2007). Sommers (2011) urges health-care providers to cultivate color awareness in regard to assessing the skin of African Americans. Color awareness recognizes that skin color is relevant to health and should not be ignored. Furthermore, by applying color awareness to health assessment, health-care providers can more appropriately manage skin conditions among patients of all skin colors and help reduce disparities in health-care delivery.

Assessing the skin of most African American patients requires clinical skills different from those for assessing people with white skin. For example, pallor in dark-skinned African Americans can be observed by the absence of the underlying red tones that give the brown and black skin its "glow" or "living color." Lighter-skinned African Americans appear more yellowish-brown, whereas darker-skinned African Americans appear ashen. Assessing such conditions as inflammation, cyanosis, jaundice, and petechiae in African Americans may require natural light and the use of different assessment skills. African Americans exhibiting inflammation or petechiae must be assessed by palpation of the skin for warmth, edema, tightness, or induration. If feasible, do not to wear gloves to perform the skin assessment, because they have a tendency to diminish sensitivity to skin temperature changes. To assess for cyanosis in dark-skinned African Americans, the health-care provider needs to observe the oral mucosa or conjunctiva. Jaundice is assessed more accurately in dark-skinned persons by observing the sclera of the eyes, the palms of the hands, and the soles of the feet, which may have a yellow discoloration. In performing a skin assessment, it may also be helpful to ask the patient, family, significant other, or caregiver to point out an area of normal skin color, temperature, and texture to serve as a baseline (Sommers 2011).

Researchers studying forensic sexual assault examinations found data suggesting African American women had a lower incidence of genital injury after rape when compared to European American women (Sommers et al. 2008). However, they maintain that the difference in reported injury prevalence was not related to race or ethnicity but rather due to reduced visibility of injury in dark-skinned women. Their research demonstrated that skin color explained more of the differences in the numbers of genital injuries than race or ethnicity, concluding that the prevalence of genital injuries in dark-skinned women has likely been underreported because of difficulty seeing the injuries. Sommers (2011) argues that these findings are important given the role of forensic evidence in the criminal justice system; women whose injuries are documented during the forensic examination have better judicial outcomes than women without documented injuries.

The literature also confirms that health-care providers are not doing an adequate job of detecting and reducing pressure ulcer risk in African Americans. According to recent studies, African Americans are at higher risk for developing more severe pressure ulcers and associated mortality and morbidity (Bauer et al. 2016; Saladin and Krause 2009). Salcido (2002) asserted that it may be due to our lack of ability to make an early diagnosis of skin in jeopardy of breaking down. Currently, researchers are testing a variety of devices that could be used to detect and diagnose alterations in blood flow, regardless of the color of the patient's skin. These devices include visible and near-infrared spectroscopy, pulse oximetry, laser Doppler, and ultrasound (Matas et al. 2001; Salcido 2002; Sowa et al. 2002).

Several skin disorders are found among the African American population. The major skin disorder is postinflammatory hyperpigmentation, which is the darkening of the skin after resolution of skin trauma, lesions of a dermatosis, or as a result of treatments administered for skin disorders. Hypopigmentary changes have also been noted in these instances. African Americans also have a tendency toward the overgrowth of connective tissue associated with the protection against infection and repair after injury. Keloid formation is one example of this tendency. Diseases such as lymphoma and systemic lupus erythematosus occur in African Americans secondary to this overgrowth of connective tissue. Dermatosis papulosa nigra is a common benign skin condition that occurs predominately among dark-skinned individuals, with the highest prevalence in African Americans (Kundu and Patterson 2013). It is usually characterized by multiple, small, hyperpigmented, asymptomatic papules that are typically located on the face, neck, upper back and chest. The lesions resemble tiny pigmented seborrheic keratoses and occur usually as multiple black papules with a diameter of 1–5 mm.

Certain skin conditions are gender-specific among some African Americans. Pseudofolliculitis barbae ("razor bumps") is more common among African American males. This skin condition results from curved hairs growing back into the skin, causing itchy and painful bumps. It is recognized by follicularly-based erythematous and hyperpigmented papules and pustules. African American males should be counselled regarding the best shaving method to keep this disorder to a minimum, for although the pustules are frequently sterile, secondary infections can occur. Suggestions include the use of electric clippers, shave in the direction of hair growth, use clippers, a single-blade razor, depilatories, or laser therapy. Dissecting cellulitis of the scalp (also known as *perifolliculitis capitis abscedens et suffodiens*) is an uncommon inflammatory scalp condition that mostly affects African American men 20–40 years of age. Symptoms include the development of pus-filled spots and lumps, often leading to scarring and patchy areas of permanent hair loss. Acne keloidalis nuchae, a skin condition found almost exclusively among African American men, is a progressive chronic folliculitis resulting in keloid-like papules and plaques on the occipital scalp.

Certain hair care practices and hairstyles are unique among women of African descent, and may contribute to specific types of hair loss seen in this population (Lawson et al. 2015; Ogunleye et al. 2014). These conditions include traction alopecia and central centrifugal cicatricial alopecia. Traction alopecia (hair loss from excessive pulling) is usually observed in African American girls and women who wear their hair in styles that put continuous traction on hair follicles (such as tight braids). This tension results in a gradual, symmetrical thinning and loss of hair in the frontal and temporal areas (McMichael 2003). Traction alopecia can lead to scarring and permanent hair loss. Central centrifugal cicatricial alopecia includes the scarring alopecia formerly referred to by the terms "follicular degeneration syndrome," "hot comb alopecia," and "pseudopelade" in African Americans (Olsen et al. 2003). This condition begins as an asymptomatic, skin-colored, noninflammatory scarring alopecia of the central scalp that gradually enlarges centrifugally. Treatment suggestions include avoiding frequent use of heat and blow dryers, using chemical relaxer touch-ups no more frequently than every 8–10 weeks; using gentle combing and grooming methods; and the use conditioners after shampooing. It is key for health-care providers to have a better understanding of the common skin and hair disorders afflicting African American women for more satisfactory outcomes in this patient population.

African Americans, in general, also experience a disproportionate amount of pigment discoloration, with vertiligo (white patches) being the most common. This autoimmune disease manifests as European American patches on the skin and causes skin discoloration and is also associated with diabetes and thyroid disorders. Birthmarks are also more prevalent in African Americans, occurring in 20% of the African American population compared with 1 to 3% in other ethnic groups. One example is mongolian spots, which are found more often in African American newborns but disappear over time. Another pigment discoloration common among African Americans is melasma. Familial predisposition is the most important risk factor for its development. It is also more common among darker-skinned African American females during pregnancy, and thus often referred to as "the mask of pregnancy." This condition is characterized by brown spots or patches on the face. Melasma has a significant impact on appearance, causing psychosocial and emotional distress, and reducing the quality of life for some patients (Handel et al. 2014). Health-care providers must be attuned to the potential need for psychological support for certain skin conditions that occur among the African American population.

African Americans must also be screened for skin cancer. Although skin cancer occurs with increased frequency among European Americans, African Americans suffer from a higher morbidity and mortality when diagnosed with skin cancer (Higgins et al. 2018). Skin cancer comprises 1–2% of all cancers in African Americans (Gloster and Neal 2006). Whereas squamous cell carcinoma is the second most common type of skin cancer in European American patients, it is the most common type in patients of African and Asian Indian descent. The peak occurrence in African Americans is from ages 40 to 49. Basal cell cancer is the second most common skin cancer of African Americans and is associated with chronic sun exposure. This type of skin cancer is more aggressive in African Americans than in European Americans. While melanoma is uncommon in African Americans, it is often terminal. The overall melanoma survival rate for African Americans is only 77%, as compared with 91% for European Americans (Ries et al. 2008). Many African Americans believe that they are not at risk for skin cancer because of their higher concentration of melanin; however, health-care providers must help to dispel this myth and educate African Americans regarding skin cancer protection.

5.6.2 Diseases and Health Conditions

Health disparities are preventable differences in the burden of disease, injury, violence, or opportunities to achieve optimal health that are experienced by certain racial and ethnic groups and socially disadvantaged populations (CDC 2008;

Kaiser Family Foundation 2018). Reducing health disparities is an important initiative to improve health outcomes for patients from diverse backgrounds. In 2002, a landmark study conducted by the Institute of Medicine (IOM) provided healthcare providers with overwhelming evidence documenting the severity of health disparities among African Americans (Smedley et al. 2002). Whereas previous research attributed the problem of health disparities among African Americans and other minority groups to access-related factors, income, age, comorbid conditions, insurance coverage, socioeconomic status, and expressions of symptoms, the IOM's report also cited racial prejudice and differences in the quality of health care as possible reasons for increased disparities (Burroughs et al. 2002).

According to Census Bureau projections, the 2015 life expectancy for African American men is 72.9 years as compared with 77.5 years for European American men (Office of Minority Health 2019f). African American women's life expectancy is 78.9 years compared with 82.0 years for European American women. Death rates for all major causes of death are higher for African Americans than for European Americans, contributing in part to a lower life expectancy for both African American men and African American women.

Health disparities also reach to the youngest members of African American culture. African American infants have 2.2 times the infant mortality rate as European American infants, are 3.2 times as likely to die from complications related to low birthweight, and had over twice the sudden infant death syndrome mortality rate as European American infants in 2014 (Office of Minority Health 2019f). While African American mothers were 2.2 times more likely than European American mothers to receive late or no prenatal care, research suggests this may not be the only factor contributing to disparities seen in infant mortality. It has been suggested that institutionalized and interpersonal racism, including poverty, unemployment, and residential segregation, may make African American women more vulnerable to disparate sexual and reproductive health outcomes, such as infant mortality (Prather et al. 2018).

While suicide rates for adolescents and young adults are higher in European Americans than in African Americans, these rates are not consistent with elementary-aged children. An analysis of suicide trends in United States children (ages 5–11 years between 1993–1997 and 2008–2012) revealed a significant increase in the suicide rate among African American children and a significant decrease for European American children (Bridge et al. 2015). Moreover, Sheftall's et al. (2016) study confirmed that children who died by suicide were more commonly male, African American, died by hanging/strangulation/suffocation, and died at home.

In examining the relationship of social characteristics such as education, income, and occupation to health indicators, African Americans have worse indicators when compared with those of European Americans. African Americans are at greater risk for many diseases, especially those associated with low-income, stressful life conditions, lack of access to primary health care, and negating health behaviors such as violence, poor dietary habits, lack of exercise, and lack of importance placed on seeking primary health care early. They carry the burden of several diseases, which is reflected in higher rates of diabetes, obesity, hypertension, stroke, hepatitis, cancer, asthma, and human immunodeficiency virus infection and acquired immune deficiency syndrome (HIV/AIDS). Moreover, lower vaccine uptake rates among African Americans, coupled with a disproportionate burden of chronic diseases, place many African Americans at high risk for complications, hospitalizations and premature mortality (Quinn 2018).

African Americans are almost twice as likely to be diagnosed with diabetes as European Americans and are twice as likely as European Americans to die from diabetes (Office of Minority Health 2019b). Reasons for this disparity are that they are less likely to seek preventive care or screening tests and are more likely to rely on the emergency department for routine health care. In addition, African Americans are more likely to suffer complications from diabetes, such as end-stage renal disease and lower extremity amputations, eye disease, and higher rates of hospitalization for diabetes when compared with European Americans.

African Americans also have a higher rate of obesity and tend to carry upper-body obesity, which puts them at additional risk for diabetes. African American women have the highest rates of being overweight or obese when compared to other groups in the U.S. About four out of five African American women are overweight or obese. In 2015, African Americans were 1.4 times as likely to be obese as European Americans (Office of Minority Health 2019e). Data from 2011 to 2014, reports that African American girls were 50% more likely to be overweight than their European American counterparts (Office of Minority Health 2019e).

The racial disparity in hypertension and hypertension-related outcomes has been recognized for decades with African Americans having greater risks than European Americans. Hypertension prevalence among African Americans is the highest in the world at 44% (Mozaffarian et al. 2016). African American adults are less as likely than their European American counterparts to have their blood pressure under control (Office of Minority Health 2019c). The higher blood pressure levels for African Americans are associated with higher rates of stroke, end-stage renal disease and congestive heart failure. Specifically, end-stage renal disease is five times more common for African American men and women and stroke risk among African Americans is 2–3 times higher than that of European Americans (Mozaffarian et al. 2016). Moreover, a Center for Disease Control (CDC) health interview survey reported that African American stroke survivors were more likely to become disabled and have difficulty with activities of daily living than their European American counterparts (Blackwell et al. 2014). In addition, perceived discrimination among African Americans is generally linked to increased levels of inflammation and systolic and diastolic blood pressure (Goosby et al. 2015).

The literature suggests that the pathophysiology of hypertension in African Americans is related to volume expansion, decreased renin, and increased intracellular concentration of sodium and calcium. Genetic cardiovascular researchers have hypothesized that there might be a "hypertensive-heart failure genotype" (Moore 2005). However, it is more likely that the etiology of hypertension among African Americans is multifaceted, including genetics, diet, lifestyle, stress, environment, medication nonadherence, and socioeconomic status (Moore 2005; Ritchey et al. 2016). Due to theses aforementioned etiological factors, self-management is complex. Disparities in hypertension self-management disproportionately affect African Americans. Wright, Still, Jones, and Moss (2018) recommend that health-care providers partner with African Americans to cocreate culturally specific interventions to address perceived stress in the self-management of their hypertension.

For hepatitis B and C infections, African Americans disproportionately carry a large burden of disease (Forde et al. 2014). African Americans have higher rates of infection and hepatitis C-related death compared with the overall population. In 2016, African Americans were almost twice as likely to die from hepatitis C as compared to the European American population and 2.5 times more likely to die from hepatitis B than European Americans (Office of Minority Health 2019c). Furthermore, it has been well documented that African ancestry individuals respond more poorly to hepatitis C virus drug treatment than European American and Asian individuals (Lu et al. 2014).

African Americans have the highest mortality rate and shortest survival rate of any racial and ethnic group for most cancers (American Cancer Society 2019). Moreover, African Americans experience higher incidence and mortality rates from many cancers that are amenable to early diagnosis and treatment. Reasons for late diagnosis include such factors as lack of knowledge or awareness of cancer screenings, lack of access to general preventive health care services, institutional or system barriers, socioeconomic status, language barriers, immigrant status, and cultural beliefs. Health-care providers must recognize the critical need for a focus on prevention and early diagnosis of cancer among this population.

Prostate cancer is most commonly diagnosed in African American men, and breast cancer the most common among African American women. The four most common cause of cancers among African Americans are lung, breast, prostate and colorectal. According to the American Cancer Society, lung cancer accounts for the largest number of cancer deaths among African Americans. In a study conducted by Lin et al. (2014), it was reported that negative surgical beliefs, fatalism, and mistrust explained almost one-third of the observed disparities in lung cancer treatment among African American patients. These authors called for interventions targeting cultural factors to help reduce undertreatment of minorities.

The burden of asthma also falls on the African American population. The National Health Interview Survey Data of 2015, found that almost 2.6 million African Americans reported that they currently have asthma (Office of Minority Health 2019a). African Americans are almost three times more likely to die from asthma related causes than the European American population. Asthma is a leading health problem for African American children. In 2015, African American children were 4 times more likely to be admitted to the hospital for asthma, as compared to European American children and had a death rate ten times that of European American children (Office of Minority Health 2019a). Causes of asthma in the African American population are exposure to secondhand tobacco smoke, poverty, lack of education, not being able to get to a doctor, and residing in urban industrial or substandard housing. It is noted that African Americans are more likely to reside near sources of air pollution and a greater distance from air quality monitoring sites (Noonan et al. 2016).

Acquired immune deficiency syndrome (AIDS) contributes to lower life expectancy of African Americans when compared with European Americans. HIV/AIDS continues to be a devastating epidemic with African American communities carrying the brunt of the impact (Williams et al. 2010). Although African Americans represent 13% of the U.S. population, they account for 44% of HIV infection cases in 2016. African American males have 8.6 times the AIDS rate as European American males, while African American females have 18.6 times the AIDS rate as compared to European American females (Office of Minority Health 2019d). Mortality rates for African Americans living with HIV/AIDS are equally disproportionate. African American men are almost 6 times as likely to die from HIV/AIDS as European American men and African American women are almost 18 times as likely to die from HIV/AIDS as European American women. In 2016, African Americans were 8.4 times more likely to be diagnosed with HIV infection, as compared to the European American population (Office of Minority Health 2019d). In addition, African Americans also continue to experience higher rates of sexually transmitted infections (STIs) than any other race/ethnicity in the United States.

African Americans are at higher risk for certain health-threatening environmental conditions. Lead exposure is an environmental threat for poorer African American communities, with African American children having the highest prevalence of elevated blood lead levels in the United States. Despite this disparity, little is known when the course of this disproportionate burden of lead exposure first emerges (Cassidy-Bushrow et al. 2017). Cassidy-Bushrow et al. (2017) conducted a study to determine if there were racial disparities in lead levels during fetal development and early childhood. Findings revealed the disproportionate burden of lead exposure is vertically transmitted (i.e., mother-to-child) to African American children before they are born and persists into early childhood. These findings suggest the need of testing African American women for lead during pregnancy to identify the risk to their future offspring.

In addition to the exposure to harmful environmental conditions, African Americans suffer from certain genetic conditions. Sickle cell disease is the most common genetic disorder among the African American population, affecting 1 in every 500 African Americans, and represents several hemoglobinopathies including sickle cell anemia, sickle cell hemoglobin C disease, and sickle cell thalassemia. Sickle cell disease is also found among people from geographic areas in which malaria is endemic, such as the Caribbean, the Middle East, the Mediterranean region, and Asia. In addition to sickle cell disease, glucose-6-phosphate dehydrogenase deficiency, which interferes with glucose metabolism, is another genetic disease found among African Americans.

Health is a state of complete physical, mental and social well-being and not merely the absence of disease or infirmity (World Health Organization 1946). Therefore, a health-care providers' knowledge of the aforementioned diseases and health conditions prevalent among African Americans is only one facet in promoting more equitable health outcomes for this population. Healthy People 2020 highlights the importance of addressing the social determinants of health as a means of creating social and physical environments that promote good health for all (Office of Disease Prevention and Health Promotion 2019). Social determinants of health, such as socioeconomic status, poverty, location, environment, unemployment, violence, and criminal justice, can negatively impact on health disparities found among African Americans.

African Americans experience the highest poverty rates. In 2017, the poverty rate for African Americans was 21.2% representing 9.0 million people in poverty (Fontenot et al. 2018). According to the Census Bureau in 2017, the average African American median household income was $40,165 in comparison to $65,845 for European American households (Office of Minority Health 2019f). Higher education often has been touted as the "great equalizer," as a mechanism to reduce the wealth gap between European Americans and African Americans (Darity et al. 2018, p. 5). However, at every level of educational attainment, the median wealth among African American families is significantly lower than European American families. Darity et al. (2018) reported that on average, an African American household with a college-educated head has less wealth than a European American family whose head did not obtain a high school diploma. Additional studies show that it requires a postgraduate education for an African American to have comparable levels of wealth to a European American household with some college education or an associate degree (Meschede et al. 2017).

Neighborhood poverty can result in environmental inequalities such as the lack of green space (an area of grass, trees, or other vegetation set apart for recreational or aesthetic purposes in an urban environment). Wen et al. (2013) examined the relationship between neighborhood social disadvantages and access to parks and green spaces and found that areas with higher levels of poverty and greater concentrations of African Americans had less green space. The lack of green space has been noted to adversely affect health status, while positive well-being has been documented in individuals living in areas with more green space. Specifically, individuals have less mental distress, less anxiety and depression, greater wellbeing and healthier cortisol profiles when living in areas with more green space compared with less green space (Barton and Rogerson 2017).

African Americans tend to live in the poorest neighborhoods with the highest rates of violence and homicide. Violence is a major cause of injury, disability and premature death among African Americans, with homicide being the leading cause of death among young African American males between the ages of 15 and 34. In 2016, African American men were nearly 10.4 times more likely than European American men to die by homicide in the United States (Centers for Disease Control and Prevention 2017). Moreover, African American females were four times more likely to be murdered by a boyfriend or girlfriend than their European American counterparts (Frazer et al. 2018). A study conducted by Sheats et al. (2018) found that a disproportionate exposure to violence for African Americans may contribute to disparities in physical injury and long-term mental and physical health. Sheats et al. assert that a health-care provider's understanding of the violence experiences of African Americans and the social context surrounding these experiences can improve the health status for this population.

African Americans experience the highest unemployment rates. In 2017, the unemployment rate for African Americans was twice that of European Americans (9.5% and 4.2%, respectively) (Office of Minority Health 2019f). Unemployment has been linked to poor physical and mental health. Individuals who report being unemployed experience significantly worse perceived mental health, are less likely to have access to health care coverage, and are more likely to delay medical treatment due to cost when compared to employed people (Pharr et al. 2012). In addition, African Americans face a higher unemployment rate than European Americans at every level of education (Jones and Schmitt 2014).

African Americans are over-represented in the criminal justice systems. While imprisonment rate of sentenced African American adults declined by 4% from 2016 to 2017 and by 31% from 2007 to 2017; at year-end 2017, the imprisonment rate for sentenced African American males was still almost six times that of sentenced European American males (Bronson and Carson 2019). It is well documented that the prison population suffers from infectious and chronic disease at rates that are four to ten times higher than for the total population and often receive inadequate healthcare (Noonan et al. 2016).

In further understanding the social determinants of health among African Americans, a vexing concern is the health status of the middle-class African American. While middle-class Americans are healthier than those living in poverty, this financial advantage does not hold true for African Americans. Studies report that middle-class status provides restricted health returns to upward mobility for African-Americans, with gains in family income resulting in significantly smaller improvements in health status (Colen et al. 2018a, b; Woolf et al. 2015). Competing explanations for this disparity include early life exposures to socioeconomic disadvantage that shape subsequent health trajectories, and racial barriers that continue to operate at structural (institutional), interpersonal, and intrapersonal levels. Therefore, health-care providers must be aware that middle-class African Americans not living in poverty are also subjected to poorer health status as compared to their European American counterparts.

The awareness of underlying causes of poor health has the potential to reduce health disparities that continue to exist within the African American population. Educating health-care providers to identify and address social determinants of health that negatively impact on the health status of African Americans is considered one of the key principles in promoting more equitable health outcomes for this population (Andermann 2016). In addition, community-based participatory research (CBPR) is a culturally conscious and responsive approach that health-care providers can utilize when conducting research about environmental health and social justice issues within the African American community (Fitzgerald and Campinha-Bacote 2019). The principles of CBPR include promotion of 'equitable engagement' with community partners throughout the research process (Cacari-Stone et al. 2014, p. 1615).

5.6.3 Variations in Drug Metabolism

African Americans have been treated as a representative population for African ancestry for many purposes, including pharmacogenomic studies. However, it must be noted that genetic diversity is greater in Africa than in other continental populations (Rajman et al. 2017). Rajman et al. (2017) conducted a review of the published literature on *CYP* polymorphisms which demonstrated and confirmed that genetic variation is greater in African populations than in Asian and European populations. This finding is important for cytochrome P450 (CYP) enzymes play a major role in drug metabolism. For example, ethnic differences in the dosing of the drug Warfarin are well documented in the literature, with African Americans requiring a higher initial dosing than Asians and Europeans (Johnson 2008). However, this difference is often ignored when an African American patient begins anticoagulant therapy on Warfarin. Other examples of drugs that African Americans respond to or metabolize differently are psychotropic drugs, immunosuppressants, antihypertensives, cardiovascular drugs, and antiretroviral medications.

From a prescribing viewpoint, studies have clearly documented that African American patients with psychotic disorders are more likely than their European American counterparts to receive excessive doses of typical antipsychotics; more likely than European American patients to be treated with older, high-potency first-generation antipsychotics (FGA); and less likely to receive second-generation antipsychotics (SGA) (Chaudhry et al. 2008; Cook et al. 2015; Herbeck et al. 2004). These racial disparities continue to exist in the prescribing of FGA for African Americans despite the introduction of SGA over 30 years ago. Research confirms that African American patients continue to disproportionately receive FGA regardless of psychiatric diagnosis, comorbidities, SGA with equal efficacy and lower extrapyramidal side effects, metabolic factors, treatment setting, and payer source (Cook et al. 2015). These findings negatively impact the incidence and severity of side effects from the over use of high-potency medications and underuse of new generation antipsychotics among African Americans.

Studies report that African Americans are twice as likely to develop tardive dyskinesia than their European American counterparts when placed on specific neuroleptics (Alblowi and Alosaimi 2015; Cook et al. 2015; Woerner et al. 2011). African American psychiatric patients experience a higher incidence of extrapyramidal effects with haloperidol decanoate than that found in European Americans. African Americans are also more susceptible to tricyclic antidepressant (TCA) delirium than are European Americans. Strickland et al. (1995) reported that for a given dose of a TCA, African Americans show higher blood levels and a faster therapeutic response. As a result, African Americans experience more toxic side effects from a TCA than do European Americans. In addition, African Americans have a higher risk of lithium toxicity and side effects related to less efficient cell membrane lithium-sodium transport and increased lithium red blood cell to plasma ratio (Henderson et al. 2010). Some African Americans have a lower baseline leukocyte count (benign leukopenia), which puts them at risk for side effects of specific antipsychotic drugs, such as clozapine, which can cause agranulocytosis. Health-care providers must make extended efforts to observe African American patients for side effects related to TCAs and other psychotropic medications.

Dirks et al. (2004) reported ethnic differences in the pharmacokinetics of immunosuppressants among African Americans and European Americans. They found that the oral bioavailability of these drugs in African Americans was 20 and 50% lower than in non-African Americans. This finding suggests that there is a need for higher dose requirements in African Americans to maintain average concentrations of specific immunosuppressants. Dirk et al. (2004) maintained that recognition of these findings has the potential to improve post-transplant immunosuppressant therapy among African Americans.

African Americans may differ in their response to beta-blockers, angiotensin-converting enzyme (ACE) inhibitors, angiotensin receptor blocking agents, and diuretics used either alone or in combination for the treatment of hypertension (Burroughs et al. 2002). In 0.1–0.5% of patients, ACE inhibitors induce a rapid swelling in the nose, throat, larynx, mouth, glottis, lips, and/or tongue (angioedema), but African Americans have a 4.5 times greater risk of ACE inhibitor-induced angioedema (Brunton et al. 2008). Studies report that African Americans do not respond as readily to the beta-blocker propanolol as European Americans do. However, their response to the diuretic hydrochlorothiazide is greater when taken alone or with a calcium channel blocker. Diuretics, alone or in combination with another antihypertensive agent, are reported to counteract increases in salt retention noted among African Americans. Although there has been much discussion about the best type of antihypertensive drug to administer in African Americans, health-care providers must remember "There is no specific class of antihypertensive drugs that categorically should not be used based on race" (Burroughs et al. 2002, p. 18).

In 2005, the Food and Drug Administration (FDA) approved the drug BiDil (Nitro Med) as adjunct standard therapy in self-identified African American patients for heart failure. This drug is based on the chemical nitric oxide, found naturally in the body, which dilates the blood vessels, allowing the blood to flow more easily and thus easing the burden on the heart. Although this drug was initially considered a drug failure in 2003, when the results were re-examined by race, it was found that a significantpercentage of the 400 African American patients in the trial seemed to respond. It was postulated that heart failure in African Americans is somehow associated with how they produce and metabolize nitric oxide (Maglo et al. 2014). Specifically, African Americans may produce less nitric oxide and destroy it too quickly. BiDil became the first drug approved by the FDA and marketed for a single racial-ethnic group, African Americans, in the treatment of congestive heart failure.

The approval of BiDil for "blacks only" is a highly controversial subject (Brody and Hunt 2006; Ellison et al. 2008; Maglo et al. 2014; Minority Nurse 2013; Mitchell et al. 2011). There are ongoing debates over whether and how to use race and ethnicity as categories in biomedical research. This position maintains that racial categories are more a societal construct than a scientific one and therefore, BiDil biologizes racial groups. An obvious question is, in a world of mixed heritages, how do health-care providers determine a person's race? Others call for the need for further testing of BiDil across a broader range of ethnicities. In addition, the cost of this drug is expensive. The controversial nature of BiDil and its prohibited cost for the average African American requiring this drug, has resulted in its underutilization.

Health-care providers must be cautious in promoting drugs for specific ethnic groups, since it could easily lead to stereotyping and discrimination. Whereas race and ethnicity are important for public health issues, they are not true biological or genetic categories. One solution is the designing of drugs that target specific genes, eliminating the need to rely on race. The field of personalized medicine holds promise of incorporating an individual's genomic profile, family history, and social and other health details into clinical decision-making (Ortega and Meyer 2014).

Research has identified the possibility that a genetic mutation may make antiretroviral treatment less effective in Africans and African Americans (Schaeffeler et al. 2001). The P-glycoprotein (PGP) membrane protein appears to transport antiretroviral drugs out of cells, thus making the drugs less effective. A double mutation of the gene that encodes this protein (C/C genotype) leads to an increased amount of the PGP protein. Schaeffeler et al. (2001) examined the frequency of the C/C genotype in 537 Whites, 142 Ghanaians (from West Africa), 50 Japanese, and 41 African Americans. The C/C genotype was found in 83% of the Ghanaians and 61% of the African Americans, and only 34% of the Japanese and 26% of the Whites. It was hypothesized that certain antiretroviral drugs may not be as effective in people with the C/C genotype. Considering that African Americans account for half of the diagnosed HIV/AIDS cases, this finding has serious implications in efforts to treat the AIDS epidemic among the African American population.

Clinician bias is one of several contributors to racial inequalities in health care and outcomes. Research studies cite the persistence of these patterns and expose that higher implicit bias scores among physicians are associated with biased treatment recommendations in the care of African American patients (van Ryn et al. 2011). Studies have found that African Americans are more likely to be over diagnosed with having a psychotic disorder and more liable to be treated with antipsychotic drugs, regardless of diagnosis. DelBello et al. (2000) found that in a study with adolescents, although there were no differences in psychotic symptoms (14% of the African Americans and 18% of the European Americans were diagnosed as having psychotic symptoms), those who were African American, despite not being more psychotic, received more antipsychotic medications. There are several possible explanations; however, DelBello and colleagues contend that one plausible explanation is that clinicians perceived African Americans to be more aggressive and, thus, more psychotic, and prescribed the antipsychotics.

Implicit bias among health-care providers is not only noted in the treatment of African Americans, but also in the diagnosing of African Americans. Decades of research confirms that African Americans are at a higher risk of misdiagnosis for psychiatric disorders (specifically, schizophrenia) and, therefore, inappropriately treated with antipsychotics (Lawson 1999; Strakowski et al. 1996; Strickland et al. 1995). Current studies continue to document this incidence (Gara et al. 2019). Moreover, the findings of Gara et al. (2019) support that racial differences in the diagnosis of schizophrenia among African Americans result in part from clinicians underemphasizing the relevance of mood symptoms among African Americans compared with other racial-ethnic groups. Chapman et al. (2011) maintain that the contribution of implicit bias to health care disparities could decrease if health-care providers acknowledge their susceptibility to it, and intentionally practice perspective-taking and individuation when providing patient care.

There have been racial/ethnic disparities in the assessment and treatment of pain (Hoffman et al. 2016; Meghani and Keane 2017). African Americans are less likely to receive guideline-recommended analgesia for pain when compared to non-minority patients. In addition, access to pain medication is an issue for African Americans and other minority groups. African Americans with severe pain are less likely than European Americans to be able to obtain commonly prescribed pain medication because pharmacies in predominantly non- European American communities do not sufficiently stock opiates (Burroughs et al. 2002). Morrison, Wallenstein, Natale, Senzel, and Huang (2000) examined the percentage of pharmacies in New York City stocked with adequate opioid medications and found that pharmacies in predominantly minority neighborhoods were much less likely to stock opioid medications. Only 25% of the pharmacies in minority neighborhoods had an ample supply of opioid medications to treat severe pain, compared with 72% of pharmacies in predominantly European American neighborhoods.

Malnutrition and eye color can also influence drug response. Protein, vitamin, and mineral deficiencies can hinder the function of metabolic enzymes and alter the body's ability to absorb or eliminate a psychotherapeutic drug. This may pose a problem for newly arriving refugees from Ethiopia/Eritrea and other East African countries where malnutrition is considered a major medical problem. In addition, psychotherapeutic medications, such as antidepressants, that require fat in order to be absorbed are not as effective in patients with exceptionally low body fat or differing fat metabolism (Wandler 2003). This is a factor to consider when caring for Ghanaians, who may differ in fat metabolism as compared to European Americans (Banini et al. 2003). When there are unexplained variations in a patient's response to a medication, it is imperative for the health-care provider to assess the patient's dietary habits (Campinha-Bacote 2007). Eye color is a genetic variation and is related to difference in response to specific drugs. For example, light eyes dilate wider in response to mydriatic drugs than do dark eyes. This difference in response to a mydriatic drug must be taken into consideration when treating African Americans who have a significant higher incidence of dark eyes.

Reflective Exercise

Mr. Townsen, a 57-year old African American male with a history of diabetes mellitus and hypertension, is admitted to the hospital in a hypertensive crisis. During the nursing assessment, Mr. Townsen shares that he has been experiencing a lot of stress and racial tension in his workplace. The nurse ascertains that he could benefit from a mental health consultation and shares her assessment with the medical team. A psychiatric consultation is ordered and to the nurse's surprise, the psychiatric consultation reveals a diagnosis of Schizophrenia. A high-potency first-generation antipsychotic is prescribed.

1. What concerns should you have regarding the prescribing of a high-potency

first-generation antipsychotic to an African American patient?
2. What issues should be raised regarding a diagnosis of Schizophrenia in an African American patient?
3. What are some culturally sensitive discharge planning strategies that should be discussed regarding Mr. Townsen's diagnose of hypertension and diabetes?

5.7 High-Risk Behaviors

Tobacco use is the leading cause of preventable disease, disability, and death in the United States. As of 2017, about 34 million US adults smoke cigarettes (Office on Smoking and Health 2019). Current smoking prevalence among African Americans declined from 21.5% in 2005 to 14.9% in 2017, with 22.4% of African American men and 15% of African American women 18 years and over reporting smoking (Wang et al. 2018). Although African Americans usually smoke fewer cigarettes and start smoking cigarettes at an older age, they are more likely to die from smoking-related diseases than European Americans (CDC 2018). Compared with European Americans, African Americans are at increased risk for lung cancer even though they smoke approximately the same amount. In addition, African American children and adults are more likely to be exposed to secondhand smoke than any other racial or ethnic group (Tsai et al. 2018).

Cultural tailoring of health promotion messages is a strategy being implemented in African American communities to reduce health-risk behaviors such as smoking (Webb Hooper et al. 2018). Culturally tailored or culturally specific interventions are framed within a cultural context and are based on theory, empirical evidence, and characteristics of the target population (Resnicow et al. 2000). Webb Hooper et al. are currently implementing a tobacco cessation intervention project for African Americans to test the effects of a culturally specific video intervention combined with standard quitline counseling. This study represents a core population-based intervention with the potential to reduce barriers to help-seeking as well as reducing tobacco cessation disparities.

Other high-risk health behaviors among African Americans can be inferred from the high incidences of HIV/AIDS and other STIs, teenage pregnancy, violence, unintentional injuries, smoking, alcoholism, drug abuse, sedentary lifestyle, and delayed seeking of health care. Community health-care providers can have a significant impact on these detrimental practices by providing health education at community events located in African American communities. The goals of health education are to change high-risk health behaviors and improve decision making (World Health Organization 2016). Health-care providers must understand influential factors affecting decision making regarding health behaviors. These factors include values, attitudes, beliefs, religion, previous experiences with the health-care system, and life goals. Yarnell et al. (2018) examined multiple risk behaviors (violence, delinquency, and substance use) among 240 African American and 262 Hispanic preadolescent boys from urban schools in the Midwest United States. Personal involvement in school and community action among parents were highlighted as protective factors. Suggestions for prevention programming based on results included early intervention, addressing criminal justice involvement, providing academic enrichment programs, and promoting community action among parents.

5.7.1 Health-Care Practices

Because a significant proportion of African Americans are poor and live in inner cities, they tend to concentrate on day-to-day survival. Health care often takes second place to the basic needs of the family, such as food and shelter. In addition, the role of the family has an impact on the health-seeking behaviors of African Americans. African Americans have strong family ties; when an individual becomes ill, that individual is frequently taught to seek health care from the family rather than from health-care professionals. This cultural practice may contribute to the failure of African Americans to seek treatment at an early stage, especially for older adults who may seek treatment from home remedies, prayer, spiritual healers, and advice from family and friends. Therefore, screening programs located in the community and church activities in which the entire family is present hold increased potential for effectiveness.

African American self-care practices during illness continue to be influenced by the overriding historical struggle for survival in the face of racism and oppression. Thus, those self-care practices that do exist emerge from strategies for survival and long-term efforts to overcome adversity (Becker et al. 2004). Self-care practices are thus regulatory, preventive, reactive, and restorative. Important self-care practices emerging from research with African Americans include spirituality, social support and advice, and non-biomedical healing traditions (Becker et al. 2004). These cultural factors were present regardless of socioeconomic status and encompassed a diverse range of activities.

Mistrust of the health care system by African Americans continues to be a major influence on health-care practices. This major problem must be addressed in order to improve health outcomes for African Americans, reduce, and eliminate health disparities for this population. Mistrust of the health care system is especially prevalent among older African Americans who grew up experiencing segregation and discrimination in health care and social service systems (Hargrave 2010). Additionally, in the past, health-care research was conducted in the community without the community being an active partner in all phases of the research process (Fitzgerald and Campinha-Bacote 2019). Perhaps the most infamous research study conducted behind this veil of secrecy was the United States Public Health Service Tuskegee Study (Tuskegee University 2019).

The Tuskegee Study, which ran from 1932 to 1972 was undertaken in Macon County, Alabama from 1932 to 1972 for the purpose of recording the natural history of syphilis in African American men. Numerous atrocities and violations of human rights took place during the course of this research involving 600 participants. African American men were recruited to the study without being told the nature of the investigation of the disease referred to simply as "bad blood." In exchange for enrollment, the men were provided medical care and survivors insurance far beyond what was the norm for African Americans at that time (Tuskegee 2019). At the initiation of the research, there were no proven treatments for syphilis; however, penicillin became the standard treatment for the disease in 1947. Nevertheless, the antibiotic was withheld from all study participants, including both experimental and control groups (Tuskegee 2019). Only after a news reporter for the Associated Press broke the story in 1972 was action taken to close the Tuskegee Study. The legacy of Tuskegee was part of the evidence leading to landmark legislation regarding the conduct of human subjects' research and the creation of Institutional Review Boards to uphold the protection of human rights. Nevertheless, this legacy also lives on in the memories of African Americans, leading to mistrust of health-care providers, health-care researchers and the lower rates of participation in research studies.

5.8 Nutrition

5.8.1 Meaning of Food

Historically, food was used to exert power over the African slaves that were brought to this country. Rations were often used as a powerful form of control on many plantations. By supervising food, slave-owners could regularly establish their authority over enslaved people. Thus, eating foods previously identified with slavery has provided many African Americans with a sense of their identity and tradition, and allowed them to reclaim power through food traditions.

Special meaning is attached to the soul food diet, a southern tradition handed down from generation to generation. The term *soul food* comes from the need for African Americans to express the group feeling of soul, and as a result, soul foods are seen to nourish not only the body but also the spirit. This style of cooking originated during American slavery, when African slaves were given only the "leftover" and "undesirable" cuts of meat from their masters. Although African Americans have incorporated soul foods into their diets, these foods are more commonly consumed for occasions such as special events, holidays, and birthdays. Therefore, the everyday diet of African Americans may more closely resemble the "American" diet, based on convenience and cost.

5.8.2 Common Foods and Food Rituals

Southern food, often perceived as the quintessential American cuisine, is actually derived from a complex blend of European, Native American, and African origins that found realization in the hands of enslaved people. While Southern food has evolved from sources and cultures of diverse regions, classes, races, and ethnicities, African and African American slaves have one of the strongest yet least recognized roles in its history. Chitterlings (pig intestines often either fried or boiled with hot peppers, onions, and spices), okra, ham hocks, corn, cornbread, barbeque, fried chicken pork fat, and sweet potato pie are foods uniquely identified as Southern African American foods. Common ways for African Americans to prepare food include frying, barbecuing, and using gravy and sauces. African American women in particular are encouraged in the preparation of customary dishes and ways of eating learned from their mothers and grandmothers, even though cooking methods such as frying of foods aren't considered healthy today (Potter et al. 2016).

African American diets are typically high in fat, cholesterol, and sodium. African Americans eat more animal fat, less fiber, and fewer fruits and vegetables than the rest of American society. Traditional breads of Southern African Americans are cornbread and biscuits, and the most popular vegetables are greens such as mustard, collard, or kale. Vegetables are preferred cooked rather than raw, with some type of fat, such as salt pork, fatback, and bacon or fat meat. Salt pork is a key ingredient in the diet of many African Americans. Salt pork is inexpensive and, therefore, more frequently purchased.

Infant feeding methods may vary among African American parents. While breast milk is considered the gold standard of infant feeding, a majority of African American mothers are not exclusively breastfeeding their newborn infants (Asiodu et al. 2017). Rather than providing support for breastfeeding, new mothers may be encouraged by their elders to begin feeding solid foods, such as cereal, at an early age (usually before 2 months). The cereal is mixed with the formula and given to the infant in a bottle. African Americans believe that giving only formula is starving the baby and that the infant needs "real food" to sleep through the night. Cultural-specific interventions are needed to educate African American parents regarding the potential harmful effects of giving infants solid foods at an early age.

Black et al. (2001) conducted a study with first-time African American adolescent mothers living in multigenerational households. The intervention focused on reducing the cultural barriers to the acceptance of the recommendations of the American Academy of Pediatrics and World Health Organization on complementary feeding. Culturally specific interventions included non-food strategies for managing infant behavior and mother–grandmother negotiation strategies. Based on a more recent study of first-time African American mothers, researchers suggest that health- care providers develop health promotion interventions geared towards this population that include social media interventions, culturally inclusive messaging around breastfeeding and lactation, and increased education for identified social support persons (Asiodu et al. 2017).

5.8.3 Dietary Practices for Health Promotion

Obesity continues to be a public health threat and an ongoing concern within our society. African American women have the highest rates of being overweight or obese compared to all other groups in the U.S. (Office of Minority Health 2019e). Some African Americans believe that a healthy person is one who has a good appetite. Foods such as milk, vegetables, and meat are referred to as *strength foods*. In the African American community, individuals who are at an ideal body weight are commonly viewed as "not having enough meat on their bones" and, therefore, unhealthy. African Americans believe that it is important to carry additional weight in order to be able to afford to lose weight during times of sickness. Therefore, being slightly overweight is seen as a sign of good health. Researchers examining lay knowledge about eating practices in two African American communities found that familial members reminded participants about their shared cultural repertoire of food, rather than exhorting them to engage in new or different eating practices, including those that were healthier (Potter et al. 2016).

One common belief among Southern African Americans is the concept of "high blood" and "low blood." The healthy state is when the blood is in balance—neither too high nor too low. High blood is viewed as more serious than low blood. High blood is often interchangeable with high blood pressure. High blood is believed to be a condition in which the blood expands in volume or moves higher in the body, usually to the head. Some African Americans believe that rich foods or foods red in color, especially red meat, are considered the primary cause of high blood. Another common African American folk belief is that excess blood will travel to the head when one eats large amounts of pork, thereby causing hypertension. Some African Americans believe that the treatment of high blood is to drink vinegar or eat pickles to "thin" the blood. Garlic is also seen as a health food. Garlic water is consumed to treat hypertension as well as hyperlipidemia in the African American population. In contrast, low blood is believed to be caused by eating too many acidic foods. Low blood is believed to be the cause of anemia. Treatment is aimed at trying to thicken the blood by eating rich foods and red meats. Another treatment for anemia, as well as for malnutrition, is to drink "pot liquor," the liquid that remains after a pot of greens has been cooked.

5.8.4 Nutritional Deficiencies and Food Limitations

While dietary guidelines have been in place for years, adherence to many food groups has been low and racial/ethnic differences have been observed. The National Health and Nutrition Examination Surveys (NHANES II) reported that intake levels of vitamins A, C, B6, folate, thiamin, riboflavin, iron, zinc, calcium, magnesium, and potassium were lowest in African Americans than all other groups (Malek et al. 2019). One factor that may explain the low calcium intake among African Americans is the lack of awareness of the health risks associated with this deficiency. Another factor is the high level of lactose intolerance in this population. Lactose intolerance occurs in 75% of the African American population (Malek et al. 2019). Low levels of thiamine, riboflavin, vitamins A and C, and iron are also noted among African Americans and are mostly associated with a poor diet secondary to a low socioeconomic status.

Many African Americans are Protestant and have no specific food restrictions. However, a significant number of African Americans are members of religious groups who have dietary restrictions. These may include Seventh-Day Adventists, Muslims, and Jehovah's Witnesses. For example, a Muslim *halal* diet forbids pork or pork products. Muslims also refuse pork-based insulin. They consider these products to be filthy. In addition, some African Americans, especially those from Jamaica and other parts of the Caribbean, may be Rastafarians. Their religious beliefs mandate that they follow a clear dietary restriction, which includes eating fresh foods of vegetable origin and avoiding meat, salt, and alcohol. The health-care provider must always ask about any religious or cultural prohibitions on types of food consumed.

Health promoting nutritional choices for African Americans have been encouraged in recent years through educational programming in various settings (schools, churches, etc.) and through food supplement programs such as Women, Infant, & Children (WIC) and Supplemental Nutrition Assistance Program (SNAP). In one of these community-based participatory programs, the GoodNEWS (Genes, Nutrition, Exercise, Wellness, & Spiritual Growth) program, researchers partnered with eighteen African American churches to deliver an educational intervention focusing on nutrition and exercise to reduce cardiovascular disease risk factors (DeHaven et al. 2011). This program built upon recognition that the church plays a leadership role on issues in the African American community ranging from education, health, and social justice. In another study, researchers delivered eight 2-h sessions focused on nutrition and physical activity in the form of soul line dancing as part of the Nice to Your Heart community-based wellness venture (Carter et al. 2016).

A wide range of media sources including television and the Internet have broadly disseminated messages regarding what is considered healthy nutrition. As a result, society has created an image of what "healthy" eating looks like. Those who are "privileged socially and economically are able to position health as a priority in their lives and have the economic and educational resources to do so" (Lupton 2013, p. 297). However, individuals who are economically disadvantaged may have no choice but to eat what is available at the lowest cost. Access to adequate nutritional resources can be limited by many of the social determinants of health, such as lack of transportation to travel to grocery stores, lack of economic resources to purchase food, inadequate housing for safe storage and preparation of nutritious foods, and finally, lack of available environmental resources, such as areas designated as food deserts (U.S. Department of Agriculture 2019). Food deserts are defined as parts of the country vapid of fresh fruit, vegetables, and other healthful whole foods, usually found in impoverished areas (United States Department of Agriculture, 2019). This is largely due to a lack of grocery stores, farmers' markets, and healthy food providers.

5.9 Pregnancy and Childbearing Practices

5.9.1 Fertility Practices and Views Toward Pregnancy

Historically, African American families have been large, especially in rural areas. A large family was viewed as an economic necessity, and African American parents depended on their children to support them when they could no longer work. However, as families moved to cities, they soon found that large families could become an economic burden. A review of demographic data from 1988 to 2014 shows that the number of African American families with four or more children has decreased from 30% in 1988 to 18% in 2014, while increases have been seen in families with 1–3 children have been noted (Livingston et al. 2015; Pew Research Center 2014).

Sustained efforts have not reduced racial and ethnic disparities in unintended pregnancy and effective contraceptive use in the United States. Despite similar patterns of sexual activity, socioeconomically disadvantaged African American adolescents and young adults (AYA) have disproportionately higher rates of unintended pregnancies than other racial/ethnic groups (Finer and Zolna 2014). Among youth who reported having sex, Martinez et al. (2011) found that a higher percentage of African American and Latina youth, as compared with European American youth, reported not using contraception at the first sex and the last time they had sex.

Methods of birth control among African American women ages 15–49 include female sterilization (22.9%), long-acting reversible contraceptives (9.6%), oral contraceptives (8.3%), and condoms (6.8%) (Daniels and Abma 2018). Religious beliefs may also play a role in selection of a contraceptive. For example, African American Catholics may choose the rhythm method over other forms of birth control. African American communities also hold many views on the issue of pregnancy versus abortion. Many African Americans who oppose abortion do so because of religious or moral beliefs. Others oppose abortion because of moral, cultural, or

Afrocentric beliefs. Such beliefs may cause a delay in making a decision so that having an abortion is no longer safe.

5.9.2 Prescriptive, Restrictive, and Taboo Practices in the Childbearing Family

African American women usually respond to pregnancy in the same manner as women in other ethnic groups, based on their satisfaction with self, economic status, and career goals. The elders in the family provide advice and counseling about what should and should not be done during pregnancy. The African American family network guides many of the practices and beliefs of the pregnant woman, including *pica*. Pica is a general term referring to the craving and purposive consumption of nonfood items, such as earth (geophagy), raw starch (amylophagy), large quantities of ice (pagophagy), charcoal, ash, and chalk (Young 2010). It has been suggested that pica is an adaptive behavior, with potential benefits including provision of iron and detoxification of harmful dietary components (Young et al. 2011). Women have reported that these items reduce nausea and cause an easy birth. However, geophagia can also lead to a potassium deficiency, constipation, and anemia. Although it is a common practice among many African Americans, independent of socioeconomic or educational level, some are unaware that the practice exists. Knowing that some view this behaviour negatively can lead mothers to hide this behaviour from family members and not reveal it to health-care providers.

Other practices are believed to be taboo during pregnancy. For example, some African Americans believe that pregnant women should not take pictures because it may cause a stillbirth, nor should they have their picture taken because it captures their soul. Some also believe that it is not wise to reach over their heads if they are pregnant because the umbilical cord will wrap around the baby's neck. Another taboo concerns the purchase of clothing for the infant prior to birth.

The impact of obesity and excess weight gain on pregnancy outcomes has become more salient in recent years, as the prevalence of overweight and obesity in our society has grown. 74% of African American women of childbearing age are classified as overweight or obese (Allison et al. 2012). Many African American women expect to experience cravings during pregnancy. Several beliefs related to the failure to satisfy this food craving exist. Some African Americans claim that if the mother does not consume the specific food craving, the child can be birthmarked, or, more seriously, it can result in a stillbirth. Caribbean food beliefs during pregnancy focus on pregnancy outcomes and eating specific food groups. For example, consuming milk, eggs, tomatoes, and green vegetables is believed to result in a large baby, whereas drinking too many liquids will drown the baby.

Snow (1993) reported several home practices related to initiating labor in pregnant African American women. Taking a ride over a bumpy road, ingesting castor oil, eating a heavy meal, or sniffing pepper are all thought to induce labor. If a baby is born with the amniotic sac (referred to as a "veil") over its head or face, the neonate is thought to have special powers. In addition, certain children are thought to have received special powers from God: those born after a set of twins, those born with a physical problem or disability, or a child who is the seventh son in a family.

Some negative pregnancy-related experiences for African American women during pregnancy are related to stressful environmental conditions such as living in chaotic and high crime neighborhoods (Giurgescu et al. 2015). Health-care providers must recognize that neighborhood conditions may increase women's stress during pregnancy, and be prepared to recommend relaxation techniques and refer those women to social service agency for additional support. In an analysis of delivery-related experiences across racial and ethnic groups, African American women were most likely to report preferring and experiencing low-intervention births, and desiring the support of a doula during labor. They also reported a lack of support during the labor from a spouse or partner (Declercq et al. 2014).

The postpartum period for the African American woman can be greatly extended. African American postpartum care is a rite of passage that restores strength to the mother and protects the newborn against illness, by caring for the mother, loving the newborn, and working with the extended family. This model of care is built on public health, traditions and rituals, and the respect of elder wisdom (Wenzel 2016). Some believe that during the postpartum period, the mother is at greater risk than the baby. She is cautioned to avoid cold air and is encouraged to get adequate rest to restore the body to normal. Postpartum practices for childcare can involve the use of a bellyband or a coin. When placed on top of the infant's umbilical area, these objects are believed to prevent the umbilical area from protruding outward.

5.9.3 Death Rituals and Expectations

Death rituals for African Americans may vary owing to the diversity in their religious affiliations, geographic location, educational level, and socioeconomic background. African Americans are very family oriented, and it is important that family members and extended family stay at the bedside of the dying patient in the hospital. They desire to hold on to their loved ones for as long as possible, and as a result may avoid signing Do Not Resuscitate (DNR) orders or making preparations for death (Lobar et al. 2006, p. 47).

Studies report that African Americans prefer more aggressive therapies at the end of life (LoPresti et al. 2016; Welch et al. 2005). Johnson et al. (2005, p. 711) maintain that spirituality is an important part of African American culture and is often the rationale for more aggressive treatment preferences at the end of life. African Americans believe that death is God's will, and may tend to believe that life support should be continued as long as necessary.

African Americans do not believe in rushing to bury the deceased. Therefore, it is common to see the burial service held 5–7 days after death. Allowing time for relatives who live far away to attend the funeral services is important. Visual display of the body is also important. Southern and rural African Americans observe the custom of having the deceased's body remain in the house the evening before the funeral (Lobar et al. 2006). This practice allows the extended family time to "pay respect" to their deceased loved one.

African Americans believe that the body must be kept intact after death and it is common to hear an African American say, "I came into this world with all my body parts, and I'll leave this world with all my body parts!" Based on this belief, African Americans are less likely to donate organs or consent to an autopsy. Health-care providers must be aware that talking about organ donation may be considered an insult to the family.

5.9.4 Responses to Death and Grief

The grieving and death rituals of African Americans are strongly influenced by religion. One of the ways that African Americans respond to death involves the experience of an ongoing spiritual connection with the deceased (Laurie and Neimeyer 2008, p. 176). One African American religious belief regarding death is the belief that death is not a final ending, but rather part of the continuum of life (Barrett 1998). For most African Americans, death does not end the connection among people, especially family. They believe the deceased is in God's hands and that they will be reunited in heaven after death. Relatives communicating with the deceased's spirit is one example of this endless connection.

A response to hearing about a death of a family member or close member in the African American culture is "falling out," which is manifested by sudden collapse, paralysis, and the inability to see or speak. However, the individual's hearing and understanding remain intact. Health-care providers must understand the African American culture to recognize this condition as a cultural response to the death of a family member and not as a medical condition requiring emergency intervention. Some African Americans are less likely to express grief openly and publicly. Nevertheless, they do express their feelings openly during the funeral. There is often strong, loud and unrelieved weeping and wailing during the funeral ceremony (Brooten et al. 2017).

Many African Americans and other underrepresented groups tend to utilize hospice less often than their European American counterparts. Current data from the National Hospice and Palliative Care Organization (NHPCO) documents persistent racial and ethnically based disparities in hospice use between African Americans and European Americans. NHPCO reports that African American Medicare beneficiaries who received hospice care in 2017 was 8.2%, as comparted to 82.5% European Americans (National Hospice and Palliative Care Organization [NHPCO] 2019). Washington et al. (2008) conducted an in-depth review of the literature regarding the underuse of hospice services by African Americans and found the following six factors:

1. Lack of awareness of hospice services
2. Mistrust of the health-care system
3. Anticipated lack of ethnic minority employees in hospice agencies
4. Personal or cultural values in conflict with hospice philosophy
5. Concerns about burdening the family
6. Economic factors

Research suggests that African Americans are also less likely to complete advance directives (Carr 2011). Spiritual beliefs that may influence the treatment decisions of African Americans at the end of life include the belief that only God has power to decide life and death; religious restrictions against physician-assisted death or advance directives limiting life-sustaining treatments; and divine intervention and miracles can occur (Johnson et al. 2005). Many African Americans believe that God is in ultimate control of the timing of death. Implications of these findings suggest a need for culturally relevant discussion and education in the African American community regarding advance directives and end-of-life services.

Reflective Exercise

Ms. Gordon, an African American woman of 81 years of age, is admitted to the hospital with the diagnosis of end-stage renal disease. She is not a candidate for dialysis or a kidney transplant. Ms. Gordon's condition rapidly deteriorates over the course of three days and death is imminent. The family does not want Ms. Gordon to be placed on a Do Not Resuscitate (DNR) order and are in constant prayer that "a miracle" will happen.

1. What are some of the cultural reasons for an African Americans' preference for more aggressive therapies at the end of life?
2. What should the nurse's approach be regarding the discussion of organ donation?
3. How would you approach the topic of Hospice care with Ms. Gordon and her family?
4. What can the nurse expect regarding the grieving of an African American family member?

5.10 Spirituality

5.10.1 Dominant Religion and Use of Prayer

Faith, church and strong religious affiliations are often central to family and community life in the African American culture. Many African Americans attend weekly religious services as well as pray daily. This community also tends to believe in God with absolute certainty and believe in miracles (Johnson et al. 2005). For many African Americans, religious organizations are important vehicles for social support and for receiving information on healthy lifestyles and preventive care (Campbell et al. 2007; Ellison et al. 2010). Moreover, African American churches have played a major role in the development and survival of African Americans.

African Americans expect to take an active part in religious activities. In reviewing the literature, Johnson et al. (2005, p. 712) found that African Americans "participate more often in organizational (attendance at religious services) and nonorganizational (prayer or religious study) religious activities and endure higher levels of intrinsic religiosity (personal religious commitment) than do Caucasians." In addition, research has noted that church attendance was an important correlate of positive health-care practices (Aaron et al. 2003). For most African Americans, there is a relationship between religious involvement and health. This relationship is important, for African Americans are disproportionately impacted by health conditions. Holt et al. (2014) conducted a study of 2370 African Americans to test a model of the religion-health connection to determine whether religious coping plays a mediating role in health behaviors in a national sample of African Americans. Findings revealed that religious coping was in fact a mediator of that relationship.

The vast majority of African Americans (79%) self-identify as Christian, according to Pew Research Center's 2014 Religious Landscape Study (Pew Center 2015). Most African American Christians and are associated with historically black Protestant churches (Baptist, Methodist, Pentecostal), according to the study. Smaller percentages of African Americans identify with evangelical Protestantism (14%), Catholicism (5%), mainline Protestantism (4%) and Islam (2%). Many other denominations and distinct religious groups are also represented in African American communities in the United States. These include African Methodist, Episcopalian, Jehovah's Witnesses, Church of God in Christ, Seventh-Day Adventists, Pentecostal, Apostolic, Presbyterian, Lutheran, Roman Catholic, Nation of Islam, and Islamic sects, as well as nondenominational and evangelical churches.

African Americans strongly believe in the use of prayer for all situations they may encounter. They also pray for the sake of others who are experiencing problems. Due to the high levels of spirituality among African Americans, when one has a reason to seek mental health services many African Americans prefer to "take it to the Lord in prayer." Faith in God is preferable to faith in mental health providers, who are often not trusted by African Americans (Gary 2005). Many who are religiously or spiritually focused indicate that fasting and praying will handle all difficulties.

Spirituality is an important multidimensional cultural resource and coping strategy used by many African Americans for managing many chronic health conditions (Clark et al. 2018; Harvey and Cook 2010; Spruil et al. 2015). It is believed that the high degree of spirituality among African Americans is grounded in a long history of oppression and mistreatment (Parks 2003). Maliski et al. (2010) studied the role of the use of faith by low-income, uninsured African American men in coping with prostate cancer and its treatment and adverse effects. They found that faith took on a much broader context than religion alone. Findings revealed that faith was a major resource for African American men in assisting them to move from a perception of cancer as a death sentence to the integration of their cancer into their lives. Based on these findings, it is important for health-care providers to recognize that faith is multifaceted and not necessarily limited to religious practice.

African Americans believe in the laying on of hands while praying. The belief is that certain individuals have the power to heal the sick by placing hands on them. African Americans may pray in a language that is not understood by anyone but the person reciting the prayer. This expression of prayer is referred to as *speaking in tongues* or *the gift of tongues*. Pentecostals introduced speaking in tongues as an indication that a person had truly received the Spirit of God (Appiah 2019). If health-care providers are not aware of this common spiritual practice among African Americans, an assessment of their speech could be considered incoherent and potentially lead to the misdiagnosis of a mental disorder (Campinha-Bacote 2017).

5.10.2 Meaning of Life and Individual Sources of Strength

Most African Americans' inner strength comes from trusting in God and maintaining a biblical worldview of health and illness. Some African Americans believe that whatever happens is "God's will." Because of this belief, African Americans may be perceived to have a fatalistic view of life. Mishra et al. (2009) reported that African Americans perceive illnesses as reflecting someone who is spiritually flawed or has a lack of faith. Therefore, having faith in God is a major source of inner strength for many African Americans. Frameworks such as Campinha-Bacote's (2005) Biblically Based Model of Cultural Competence in the Delivery of Healthcare Services can provide health-care providers with strategies for implementing culturally specific interventions for African Americans who share a biblical worldview of health and illness.

5.10.3 Spiritual Beliefs and Health

Spiritual beliefs strongly direct many African Americans as they cope with illness and the end of life. In a review of the literature on spiritual beliefs and practices of African Americans, Johnson et al. (2005) noted the following recurrent themes: spiritual beliefs and practices are a source of comfort, coping, and support and are the most effective way to influence healing; God is responsible for physical and spiritual health; and the doctor is God's instrument. African Americans consider themselves spiritual beings, and God is thought to be the supreme healer. Health-care practices center on religious and spiritual activities such as going to church, praying daily, laying on of hands, and speaking in tongues.

Drayton-Brooks and White (2004) conducted a qualitative study to explore health-promoting behaviors among African American women with faith-based support. They concluded that "health beliefs, attitudes, and behaviors are not developed outside of social systems; therefore, the facilitation of healthy lifestyle behaviors may be best addressed and influenced within a context of reciprocal social interaction such as a church." As health-care providers develop culturally specific interventions for African Americans, it is important to understand that the church community can serve as a viable support system in developing health-promoting behaviors. Underwood and Powell (2006) further added that considerable improvements can occur in the health status of African Americans if health education and outreach efforts are presented and promoted through religious, spiritual and faith-based efforts. However, Musgrave et al. (2002) cautioned public health not to "use" faith communities or the spirituality of individuals to its own end. Instead, there must be a partnership between public health and faith communities in which the central undertaking of faith is respected.

5.11 Health-Care Practices

5.11.1 Health-Seeking Beliefs and Behaviors

Spirituality, communalism, oral tradition, internal strength, resolve, and respect for elders are central values that guide the health-seeking beliefs and behaviors among the traditional African American culture (Carroll 2014; Coffman et al., 2010). Spirituality depicts an inner strength that comes from trusting in God for good health. Communalism reflects African cosmology that values the interconnected essence of all living entities. This belief system has forged a strong history of collective group orientation that includes personal relationships, social support systems, and shared resources over individualism in maintaining health. Oral tradition is an important tool for African Americans in sharing knowledge about health behaviors and practices. Internal strength and resolve originate from survival skills learned through challenging conditions and slavery. Respect for elders refers to African American elders who are revered for their experience and wisdom in areas concerning health and well-being.

According to Snow (1974), many African Americans are pessimistic about human relationships and believe that it is more natural to do evil than to do good. Snow concluded that some African Americans' belief systems emphasize three major themes:

1. The world is a very hostile and dangerous place to live.
2. The individual is open to attack from external forces.
3. The individual is considered to be a helpless person who has no internal resources to combat such an attack and, therefore, needs outside assistance.

Some African Americans, particularly those of Haitian background, may believe in sympathetic magic. *Sympathetic magic* assumes everything is interconnected and includes the practice of imitative and contagious magic. *Contagious magic* is the belief that once an entity is physically connected to another, it can never be separated; what one does to a specific part, they also do to the whole. This type of belief is seen in the practice of voodoo. An individual will take a piece of the victim's hair or fingernail and place a hex, which they believe will cause the person to become ill (voodoo illness). *Imitative magic* is the belief that "like follows like" (Campinha-Bacote 1993). For example, a pregnant woman may sleep with a knife under her pillow to "cut" the pains of labor. Another example is the use of a doll or a picture of an individual to inflict harm on that person. Whatever harm is done to the picture is also simultaneously done to the person.

5.11.2 Responsibility for Health Care

The African American population believes in natural and unnatural illnesses. *Natural illness* occurs in response to normal forces from which individuals have not protected themselves. *Unnatural illness* is the belief that harm or sickness can come to you via a person or spirit. In treating an unnatural illness, African Americans seek clergy or a folk healer or pray directly to God. In general, health is viewed as harmony with nature, whereas illness is seen as a disruption in this harmonic state owing to demons, "bad spirits," or both.

African Americans may use home remedies to maintain their health and treat specific health conditions as well as seek health care from Western health-care providers. When taking prescribed medications, African Americans commonly take the medications differently from the way prescribed. For example, in treating hypertension, African Americans may take their antihypertensive medication on an "as-needed" basis (Sowers et al. 2002). To provide services that are effective and culturally acceptable to African Americans, health-care providers must conduct thorough culturally specific medication assessments. Gaw's (2001) "Clinician's Inquiry into the Meaning of Taking Medications" assessment guide offers health-care providers a culturally sensitive guide to explore the patient's worldview on taking medication. Without assessing an African American's worldview on taking medication, prescribed treatment can result in nonadherence to the drug treatment, misinterpretation of dosage and/or timing of taking the drug and ultimately, poor treatment outcomes. Finally, health-care providers must become partners with the African American community and be open to listening to their voices regarding how the care they wish to receive. Strategies such as focus groups and community forums can provide health-care providers with insight into health-care practices acceptable to African Americans.

5.11.3 Folk and Traditional Practices

African Americans, like most ethnic groups, engage in folk medicine. The history of African American folk medicine has its origin in slavery. Slaves had a limited range of choices in obtaining health care. Although they were expected to inform their masters immediately when they were ill, slaves were reluctant to submit themselves to the harsh prescriptions and treatments of eighteenth- and nineteenth-century European American physicians (Savitt 1978). They preferred self-treatment or treatment by friends, older relatives, or "folk doctors." This led to a dual system: "white medicine" and "black medicine" (Savitt 1978).

Snow (1993) studied hundreds of folk practices used by African Americans. One example is the belief that drinking a mixture of an alcoholic beverage and fish blood can cure alcoholism. This is believed to give an undesirable taste and cause nausea and vomiting when subsequent alcoholic drinks are taken. A secondary analysis conducted of a nationally representative cross-sectional sample of 2107 adult African Americans living in the United States in 1979 and 1980 found that 69.6% reported that their families used home remedies and 35.4% reported that they used home remedies themselves (Boyd et al. 2000).

African American traditional medicine can be traced back beyond enslavement in the United States to their native cultures in Africa (Fett 2002; Savitt 1978). One class used for the treatment of illness and promotion of health by African Americans are botanicals. A botanical is a plant or plant part valued for its medicinal or therapeutic properties, flavor, and/or scent. Some research suggests that the use of botanicals among African Americans has decreased (Gunn and Davis 2011; Kelly et al. 2006). In a more recent study of African American and European American urban adults as part of the Healthy Aging in Neighborhoods of Diversity, African American participants showed lower use of botanicals, but greater use of dietary supplements such as vitamins and minerals, than their European American counterparts (Duffy et al. 2017). African Americans also used magical and herbal cures from their homelands, but over time they borrowed additional herbal lore and curative practices from Native Americans and adopted colonial European approaches. These approaches include purgatives, bleedings, and preventive measures based on classical humoral pathology, leading to an amalgamated ethnomedical system with many regional variants (Puckrein 1981). This system reflects spiritual power in action and is part of a sacred worldview (Smith 1994).

5.11.4 Barriers to Health Care

Healthy People 2020 defines health literacy as "the degree to which individuals have the capacity to obtain, process, and understand basic health information and services needed to make appropriate health decisions" (U.S. Department of Health and Human Services [USDHHS] 2010). Research shows that health literacy is the single best predictor of health status. Low health literacy affects older people, immigrants, the impoverished, and minorities. Low health literacy affects 40% of African Americans and is considered a barrier to receiving optimal health care. Low health literacy is also driven by poor patient-provider communication (USDHHS 2010). Health-care providers can reduce low health literacy by limiting the amount of information provided at each visit, avoiding medical jargon, using pictures or models to explain important health concepts, ensuring understanding with the "show-me" technique, and encouraging patients to ask questions.

Negative attitudes from health-care professionals can greatly affect African Americans' decision to seek medical attention (McNeil et al. 2002). McNeil and colleagues maintained that the attitude of the health-care provider is one of the most significant barriers to the care of African Americans (p. 132). One study reported that 12% of African Americans, compared with 1% of European Americans, felt that health-care providers treated them unfairly or disrespectfully because of their race (Kaiser Family Foundation 2001). Kennedy et al. (2007) reported that African Americans feel that just receiving health care is very often a demeaning and humiliating experience.

Numerous recent studies have focused on the harmful effects of the unconscious, negative attitudes of health-care providers toward certain populations. Implicit bias refers to any unconsciously held set of assumption about a social group. These assumptions can result in stereotyping. Evidence suggests that implicit bias held by health-care providers is one possible contributor to disparities in health outcomes for a number of stigmatized groups, including African Americans

(Zestcott et al. 2016). Despite their explicit commitment to providing equal care, some studies suggest that implicit prejudice and stereotyping can impact the judgment and behavior of health-care providers when they interact with stigmatized patients (Chapman et al. 2011; Green et al. 2007). Consistent with other populations, health-care providers demonstrate implicit biases indicative of more negative attitudes toward African Americans than European Americans (Blair et al. 2013; Cooper et al. 2012; Haider et al. 2015a; Haider et al. 2015b; Hausmann et al. 2015; Oliver et al. 2014; Schaa et al. 2015). This research also suggests that contemporary approaches to teaching cultural competence and minority health are generally insufficient to reduce implicit bias among health-care providers. Some potential strategies suggested for reducing implicit bias among health-care providers include training in perspective-taking, bias awareness, and promoting bias reduction at an institutional level (Zestcott et al. 2016).

Some research suggests that the emphasis on Western biomedical beliefs and practices in medical institutions, coupled with little recognition of alternative beliefs and healing practices, may foster sentiments of cultural mistrust among African Americans (Blank et al. 2002; Chandler 2010). Fear of racial discrimination in medical facilities has also been found to deter African Americans from utilizing available services (Lee et al. 2009; Shavers et al. 2012). Further, when compared to European Americans, African Americans report lower levels of trust in both their physician and in the health care system, and express greater concerns regarding the quality of care (Pullen et al. 2014; Wyn et al. 2004). In all, racial attitudes and experiences may have a greater and more measurable impact on the use of some preventative care services, like an annual physical, which are perceived as less urgent or necessary than other types of health care services.

African Americans may also experience economic and geographic barriers to health-care services. Needed health-care services may not be accessible or affordable for African Americans in lower socioeconomic groups. Although some services may be available, accessible, and affordable for other African Americans, they may not be culturally relevant. For example, a health-care provider may prescribe a strict American Diabetic Association diet to a newly diagnosed diabetic African American patient without taking into consideration this person's dietary habits. Therefore, therapeutic interventions developed by health-care providers may be underused or ignored.

Underrepresentation of ethnic minority health-care providers is an additional barrier to health care for many minorities and an issue addressed in numerous studies on the topic of race concordance. Race concordance occurs when the race of a patient matches the race of his/her physician and discordance occurs when races do not match. In the absence of adequate representation, minority populations are less likely to access and use health-care services. A systematic review of literature exploring doctor–patient race concordance and its impact on predicting greater health-care utilization and satisfaction among minorities between 1995 and 2016 revealed that African American patients consistently experienced poorer communication quality, information-giving, patient participation, and participatory decision-making than European American patients. Across these studies, racial concordance was more clearly associated with better communication across all domains, except quality (Shen et al. 2018). These results continue to support the need to increase the number of minority physicians, as well as improve the ability of physicians to interact with patients who are not of their own race. These findings are relevant for all health-care providers.

Another barrier that many African Americans face in obtaining health care is inadequate health insurance coverage. In 2009, 19% of African Americans did not have health insurance. Having access to health insurance is a critical factor in reducing the current health disparities that exit among African Americans. In an attempt to reduce these disparities, the Affordable Care Act (ACA) was passed in 2010. This health insurance reform legislation included a series of measures to guarantee that insurance companies would no longer be able to deny coverage to anyone with pre-existing conditions, a significant benefit for the many African Americans who are plagued with higher rates of chronic diseases, illnesses, and comorbidity (Jefferson 2010). The Affordable Care Act also expanded access to preventive care, a needed service to reduce health disparities for millions of African Americans by helping to prevent many diseases that have a disproportionate impact on this group.

Since the passage of the ACA, there have been arguments that the insurance coverage afforded under the ACA provides insufficient protection against high costs or offers limited networks that prevent accessing intended care (Glied et al. 2017). Despite the efforts of the ACA, current data still reflects a disparity in health insurance coverage for African Americans. In 2017, 55.5% of African Americans in comparison to 75.4% of European Americans used private health insurance. Also, in 2017, 43.9% of African Americans in comparison to 33.7% of European Americans relied on Medicaid or public health insurance. Finally, 9.9% African Americans in comparison to 5.9% of European Americans were uninsured (Office of Minority Health 2019f).

5.11.5 Cultural Responses to Health and Illness

To understand the African American responses to health and illness, it is important to first understand their worldview. The literature discusses an Afrocentric, or African-centered, worldview held by some African Americans (Carroll 2010, 2014; Dixon 1971). Within an Afrocentric worldview, African Americans place the highest on interpersonal relationships among people and the collective group. Therefore, it is key to establish rapport early on in the patient–health care provider interaction. Time invested early on in creating this relationship is crucial in order to establish trust and open communication between patient and health care provider.

African American culture argues that knowledge is acquired through the five senses and beyond, while truth is interconnected, interrelated, and interdependent on the universe (Carroll 2014). Therefore, this worldview maintains that one can discover knowledge and truth through feelings or emotions. It is not uncommon for an African American patient to say, "It doesn't *feel* right" when asked questions regarding compliance issues. Afrocentric logic highlights seeing the union of opposites (diunital logic). For example, an African American patient may be both optimistic and pessimistic about the future at the same time and see no conflict in this view. An Afrocentric worldview asserts that one should live in harmony with nature, and spirituality must hold the most significant place in life. Cooperation, collective responsibility, and interdependence are the central values to which all should aspire. The Afrocentric worldview is a circular one, in which all events are tied together with one another. Therefore, it may be challenging for health-care providers to isolate specific health problems when taking a patient history. Crucial skills needed for enhancing the holistic assessment of African American patients include active listening, adaptive questioning, empathy, validation, and reassurance.

African Americans often perceive pain as a sign of illness or disease. Therefore, it is possible that if they are not experiencing severe and/or immediate pain, a regimen of regularly prescribed medicine may not be followed. For example, African Americans may take their antihypertensive drugs or diuretics only when they experience head or neck pain. This cultural practice interferes with successful and effective treatment of hypertension. In other cases, some African Americans believe, as part of their spiritual and religious foundation, that suffering and pain are inevitable and must be endured, thus contributing to their high tolerance levels for pain. Prayers and the laying on of hands are thought to free the person from all suffering and pain, and people who still experience pain are considered to have little faith.

Given that stigma is a social construction, it is influenced by socially important categories such as culture and ethnicity (Corrigan et al. 2014). Therefore, it is not surprising that some African Americans hold a stigma against mental illness. In addition to religious beliefs, low educational levels among African Americans from lower socioeconomic levels may also limit their access to information about the etiology and treatment of mental illness. Cultural values of communalism, kinship, and group identity may lead African Americans to distance themselves from the person with mental illness in order to protect the integrity of their kin (Abdullah and Brown 2011).

Close family and spiritual ties within the African American family allow one to enter the sick role with ease. Extended and nuclear family members willingly care for sick individuals and assume their role responsibilities without hesitation. Sickness and tragedy bring African American families together, even in the presence of family conflict. Health-care providers must take into consideration the strength and closeness of African American families when planning care. Ensuring that members of the family are included, when desired by the patient, in decision-making is vital. However, it is important to also recognize the need for referrals of patients and their families to additional community support agencies in order to avoid overtaxing personal and fiscal resources of the family.

5.11.6 Blood Transfusions and Organ Donation

Blood transfusions are generally accepted in the African American patient. However, some religious groups, such as Jehovah's Witnesses, do not permit this practice. In addition, Jehovah's Witnesses believe that any blood that leaves the body must be destroyed, so they do not approve of an individual storing her or his own blood for a later autologous transfusion.

Organ and tissue transplantation have contributed to marked improvements in the ability to extend and save lives; however, a low level of organ donation among African Americans has been cited (Robinson et al. 2014). This is particularly concerning since African Americans make up the largest group of minorities in need of an organ transplant (Office of Minority Health 2016). This disparity in need for organ transplants among African Americans is related to the fact that they are disproportionately impacted by chronic conditions such as diabetes, heart disease, and hypertension, which often creates the need for life-saving organ transplant (Office of Minority Health 2016). This reluctance is associated with a lack of information about organ donation, religious fears and beliefs, distrust of health-care providers, fear that organs will be taken before the patient is dead, and concern that proper medical attention will not be given to patients if they are organ donors.

In regard to kidney donations, it should be noted that African Americans donate in proportion to their share of the population. African Americans, for example, represent about 13% of the population and account for 12% of kidney donors (Office of Minority Health 2016). It may appear that there is a low level of kidney donation among the African American community because they are disproportionately represented (35%) on the kidney waiting list (Office of Minority Health 2016). Their rate of organ donation does not keep pace with the number of those needing transplants. This increased need for organ donors led the Congress of National Black Churches to make organ and tissue donation a top-priority health issue. As a result of efforts such as this as well as recent improvements in national kidney transplant policy, the rates at which African American transplant candidates receive kidneys from deceased donors has now equalled the rates of transplant for Hispanics and Caucasians (United Network for Organ Sharing 2017).

5.12 Health-Care Providers

5.12.1 Traditional Versus Biomedical Providers

Physicians are recognized as heads of the health-care team, with nurses having lesser importance. However, as nurses are becoming more educated and operating in advanced practice roles, both African American males and female are holding them in higher regard. For example, Wehbe-Alamah et al. (2011) found that African American men receiving primary care in a nurse-managed clinic reported that that they felt nurse practitioners (NPs) spent more time with patients and demonstrated more caring behaviors. Similar studies revealed that African Americans reported that NPs provided non-judgmental care, showed more care than physicians, spent more time with them than physicians to explain things, were trusted more, and rendered a holistic approach to find the best treatment (Benkert and Tate 2008; Gunn and Davis 2011; Wehbe-Alamah et al. 2011). These findings suggest that some African Americans are able to develop a strong and trusting relationship with health-care providers despite their long history of distrust.

Whereas some African Americans may prefer a health-care provider of the same gender for urological and gynecological conditions, generally gender is not a major concern in the selection of health-care provider. Men and women can provide personal care to the opposite sex. On occasion, young men may prefer that another man or an older woman give personal care. With the current emphasis on women's health and the responses of women to illness and treatment regimens, some African American women prefer female primary-care providers. Health-care providers should respect these wishes when possible.

African American folk healing sees sickness as arising from situations that break "relational connections" of the unborn, born, and the dead, which are all intertwined. All healing links in with and emphasizes African American identity and culture (Mitchem 2007). Among the African American community, traditional/folk practitioners can be spiritual leaders, grandparents, elders of the community, voodoo doctors, or priests. For example, the pastor in the African American church is noted to be "a healer of the sick" (Drayton-Brooks and White 2004, p. 86).

5.12.2 Status of Health-Care Providers

Western health-care providers do not generally regard folk practitioners with high esteem. However, as homeopathic and alternative medicine increases in importance in preventive health, these practitioners are gaining more recognition, respect, and utilization. Folk practitioners are respected and valued in the African American community and frequently used by African Americans of all socioeconomic levels. In a recent study of older African Americans health care encounters, participants reported greater comfort with the use of alternative health-care providers, particularly Chinese medicine providers and herbalists when compared to traditional approaches to health care (Hansen et al. 2016).

Many African Americans perceive biomedical health-care providers as outsiders, and they resent them for telling them what their problems are or telling them how to solve them (Hansen et al. 2016). Generally, most African Americans are suspicious and cautious of health-care providers they have not heard of or do not know. In a recent study, older African Americans also reported experiencing the added insult of ageism from health-care providers, displayed through impatience with their slow movement, lack of money, and decreased hearing. Participants went further, stating that this disrespectful behaviour toward older adults on the part of health-care providers was not dependent on race. Thus, they perceived ageism as cutting across other sociocultural and racial divisions, and the preeminent defining barrier to health care communication (Hansen et al. 2016).

Because interpersonal relationships are highly valued in this group, it is important to initially focus on developing a sound, trusting relationship. Providers must be especially vigilant and mindful of any potential implicit bias when caring for older African American patients. When patients actively engage in their own care, this advantage reduces anxiety, increases knowledge, and motivates them to ask specific questions. Awareness, knowledge, communication, and training may reduce negative behavior or bias of health-care providers and encourage patient participation. Therefore, patient-provider communication is essential to the development of a trusting relationship for best possible outcomes.

References

Aaron K, Levine D, Burstin H (2003) African American church participation and health care practices. J Gen Intern Med 18(11):908–913

Abdullah T, Brown T (2011) Mental illness stigma and ethnocultural beliefs, values, and norms: an integrative review. Clin Psychol Rev 31:934–948

Adamsons K, Johnson SK (2013) An updated and expanded meta-analysis of nonresident fathering and child well-being. J Fam Psychol 27(4):589–599

Airen O (2017) The color ceiling: African Americans still fighting for equity and equality. J Hum Serv Train Res Prac 2(1). http://scholarworks.sfasu.edu/jhstrp/vol2/iss1/1

Alberg A, Nonemaker J (2008) Who is at high risk for lung cancer? Population-level and individual-level perspectives. Semin Respir Crit Care Med 29(3):223–232

Alblowi M, Alosaimi D (2015) Tardive dyskinesia occurring in a young woman after withdrawal of an atypical antipsychotic drug. Neuroscience (Riyadh) 20(4):376–379

Allender J, Spradley B (2001) Community health nursing: concepts and practice. Lippincott, Philadelphia

Allison KC, Wrotniak BH, Paré E, Sarwer DB (2012) Psychosocial characteristics and gestational weight change among overweight, African American pregnant women. Obstet Gynecol Int 2012:878607

American Cancer Society (2019) Cancer facts and figures for African Americans 2019–2021. Author, Atlanta. https://www.cancer.org/content/dam/cancer-org/research/cancer-facts-and-statistics/cancer-facts-and-figures-for-african-americans/cancer-facts-and-figures-for-african-americans-2019-2021.pdf

Andermann A (2016) Taking action on the social determinants of health in clinical practice: a framework for health professionals. Can Am J (CMAJ) 188(17–18):E474–E483

Appiah C (2019) Capitalizing on my African American Christian heritage in the cultivation of spiritual formation and contemplative spiritual disciplines. George Fox University, Portland. Doctor of Ministry, p 288. https://digitalcommons.georgefox.edu/cgi/viewcontent.cgi?article=1287&context=dmin

Archibald A (2019) Perspective: work-related stress and mortality among black men. Ethn Dis 29(1):21–22

Asiodu IV, Waters CM, Dailey DE, Lyndon A (2017) Infant feeding decision-making and the influences of social support persons among first-time African American mothers. Matern Child Health J *21*(4):863–872

Banini A, Allen J, Boyd L, Lartey A (2003) Fatty acids, diet, and body indices of type II diabetic American Whites and Blacks and Ghanaians. Nutrition 19(9):722–726

Barbarin OA (1983) Coping with ecological transitions by black families: a psychosocial model. J Community Psychol 11(4):308–322

Barrett R (1998) Sociocultural considerations for working with Blacks experiencing loss and grief. In: Doka KJ, Davidson JD (eds) Living with grief: who we are, how we grieve, pp 83–96

Barton J, Rogerson M (2017) The importance of greenspace for mental health. BJPsych Int *14*(4):79–81

Bauer K, Rock K, Nazzal M, Jones O, Qu W (2016) Pressure ulcers in the United States' inpatient population from 2008 to 2012: results of a retrospective nationwide study. Ostomy Wound Manage 62(11):30–38

Becker G, Gates RJ, Newsom E (2004) Self-care among chronically ill African Americans: culture, health disparities, and health insurance status. Am J Public Health *94*(12):2066–2073

Benkert R, Tate N (2008) Trust of nurse provider and physicians among African Americans with hypertension. J Am Acad Nurse Provider *20*(5):273–280

Billingsley A (1968) Black families in White America. Prentice Hall, Upper Saddle River

Black M, Siegel E, Abel Y, Bentley M (2001) Home and videotape intervention delays early complementary feeding among adolescent mothers. Pediatrics 107(5):E67

Blackwell D, Lucas J, Clarke T (2014) Summary health statistics for U.S. adults: National Health Interview Survey, 2012. Vital Health Stat 10(260):i-161. https://www.cdc.gov/nchs/data/series/sr_10/sr10_260.pdf

Blair IV, Havaranek EP, Price DW, Hanratty R, Fairclough DL, Farley T et al (2013) Assessment of biases against Latinos and African Americans among primary care providers and community members. Am J Public Health 103:92–98

Blank M, Mahmood M, Fox J, Guterbock T (2002) Alternative mental health services: the role of the black church in the south. Am J Public Health 92:1668–1672

Boyd E, Taylor S, Shimp L, Semler C (2000) An assessment of home remedy use by African Americans. J Natl Med Assoc *92*(7):341–353

Braithwaite R, Taylor S, Austin J (2000) Building health coalitions in the black community. Sage Publications, Thousand Oaks

Bridge J, Asti L, Horowitz L, Greenhouse J, Fontanella C, Sheftall A et al (2015) Suicide trends among elementary school-aged children in the United States from 1993 to 2012. J Am Med Assoc Pediatr 169(7):673–677

Brody H, Hunt L (2006) BiDil: assessing a race-based pharmaceutical. Ann Fam Med 4(6):556–560

Bronson J, Carson A (2019) Prisoners in 2017. U.S. Department of Justice. https://www.bjs.gov/index.cfm?ty=pbdetail&iid=6546

Brooten D, Youngblut J, Charles D, Roche R, Hidalgo M, Malkawi M (2017) Death rituals reported by White, Black, and Hispanic parents following the ICU death of an infant or child. J Pediatr Nurs *31*(12):132–140

Brown GL, Kogan SM, Kim J (2018) From fathers to sons: the intergenerational transmission of parenting behavior among African American young men. Fam Process *57*(1):165–180

Brunton L, Goodman L, Blumenthal D, Buxman L (2008) Goodman & Gillman's manual of pharmacology and therapeutics. McGraw-Hill Companies, New York

Bureau of Labor Statistics (2017) *Labor force characteristics by race and ethnicity, 2016*. Department of Labor, Washington, D.C. https://www.bls.gov/opub/reports/race-and-ethnicity/2016/home.htm

Bureau of Labor Statistics (2018) A profile of the working poor, 2016. BLS reports. Report 1074. U.S. Department of Labor, Washington, D.C. https://www.bls.gov/opub/reports/working-poor/2016/pdf/home.pdf

Burroughs V, Mackey M, Levy R (2002) Racial and ethnic differences in response to medicines: towards individualized pharmaceutical treatment. J Natl Med Assoc 94(10):1–20

Cacari-Stone L, Wallerstein N, Garcia A, Minkler M (2014) The promise of community-based participatory research for health equity: a conceptual model for bridging evidence with policy. Am J Public Health 104(9):116–123

Campbell M, Hudson M, Resnicow K, Blakeney N, Paxton A, Baskin M (2007) Church-based health promotion interventions: evidence and lessons learned. Annu Rev Public Health 28(33):213–234

Campinha-Bacote J (1993) Soul therapy: humor and music with African American patients. J Christ Nurs 10(2):23–26

Campinha-Bacote J (1997) Humor therapy for culturally diverse psychiatric patients. J Nurs Jocularity 7(1):38–40

Campinha-Bacote J (2005) A biblically based model of cultural competence in the delivery of healthcare services. Transcultural C.A.R.E. Associates, Cincinnati

Campinha-Bacote J (2007) The process of cultural competence in the delivery of healthcare services: the journey continues. Transcultural C.A.R.E. Associates, Cincinnati

Campinha-Bacote J (2017) Cultural considerations in forensic psychiatry: the issue of forced medication. Int J Forensic Psychiatry 50:1–8

Campos-Castillo C, Ewoodzi K (2014) Relational trustworthiness: how status affects intra-organizational inequality in job autonomy. Soc Sci Res 44:60–74

Carr D (2011) Racial differences in end-of-life planning: why don't Blacks and Latinos prepare for the inevitable? Omega (Westport) 63(1):1–20

Carroll K (2010) A genealogical analysis of the worldview framework in African-centered psychology. J Pan African Stud *3*(8):109–143

Carroll KK (2014) An introduction to African-centered sociology: worldview, epistemology, and social theory. Crit Sociol 40(2):257–270

Carter SR, Walker A, Abdul-Latif S, Maurer L, Masunungure D, Tedaldi E, Patterson F (2016) Nice to your heart: a pilot community-based intervention to improve heart health behaviors in urban residents. Health Educ J 75(3):306–317

Cassidy-Bushrow A, Sitarik A, Havstad S, Park S, Bielak L, Austin C et al (2017) Burden of higher lead exposure in African-Americans starts in utero and persists into childhood. Environ Int 108:221–227

Centers for Disease Control (2016) Reduced disparities in birth rates among teens aged 15–19 Years—United States, 2006–2007 and 2013–2014. MMWR. https://www.cdc.gov/mmwr/volumes/65/wr/mm6516a1.htm

Centers for Disease Control and Prevention (2008) Community health and program services (CHAPS): health disparities among racial/ethnic populations. U.S. Department of Health and Human Services, Atlanta

Centers for Disease Control and Prevention (2017) About underlying cause of death, 1999–2016. http://wonder.cdc.gov/ucd-icd10.html

Centers for Disease Control and Prevention (2018) *African Americans and tobacco use*. Office on Smoking and Health. https://www.cdc.gov/tobacco/disparities/african-americans/index.htm#prevalence

Chandler D (2010) The underutilization of health services in the black community: an examination of causes and effects. J Black Stud 40(5):915–931

Chapman E, Kaatz A, Carnes M (2011) Physicians and implicit bias: how doctors may unwittingly perpetuate health care disparities. J Gen Intern Med 28(11):1504–1510

Chaudhry I, Neelam K, Duddu V, Husain N (2008) Ethnicity and psychopharmacology. J Psychopharmacol 22(6):673–680

Clark E, Williams B, Huang J, Roth D, Holt C (2018) A longitudinal study of religiosity, spiritual health locus of control, and health behaviors in a national sample of African Americans. J Relig Health 57(6):2258–2278

Coe K, Keller C, Walker J (2015) Religion, kinship and health behaviors of African American Women. J Relig Health 54:46–60

Coffman K, Aten JD, Denney RM, Futch T (2010) African-American spirituality. In: Leeming DA, Madden K, Marlan S (eds) Encyclopedia of psychology and religion. Springer, Boston

Cokley K, Cooke B, Nobles W (2005) Guidelines for providing culturally appropriate services for people of African ancestry exposed to the trauma of Hurricane Katrina. The Association of Black Psychologists, Washington. http://cretscmhd.psych.ucla.edu/nola/Video/PTSD/materials/abpsi_article1.pdf

Colen C, Krueger P, Boettner B (2018a) Do rising tides lift all boats? Racial disparities in health across the lifecourse among middle-class African-Americans and Whites. SSM Popul Health 6:125–135

Colen C, Ramey D, Cooksey E, Williams D (2018b) Racial disparities in health among nonpoor African Americans and Hispanics: the role of acute and chronic discrimination. Soc Sci Med 199:167–180

Conant J (1961) Slums and suburbs. New American Library Publishers, New York

Cook T, Reeves G, Teufel J, Postolache T (2015) Persistence of racial disparities in prescription of first-generation antipsychotics in the USA. Pharmacoepidemiol Drug Saf. https://www.

ctsi.umn.edu/sites/ctsi.umn.edu/files/journal_club-urp-article1.pdf
Cooley M, Jennings-Dozier K (1998) Lung cancer in African Americans: a call for action. Cancer Pract 6(2):99–106
Cooper LA, Roter DL, Carson KA, Beach MC, Sabin JA, Greenwald AG, Inui TS (2012) The associations of clinicians' implicit attitudes about race with medical visit communication and patient ratings of interpersonal care. Am J Public Health 102:979–987
Corrigan PW, Druss BG, Perlick DA (2014) The impact of mental illness stigma on seeking and participating in mental health care. Psychol Sci Public Interest 15(2):37–70
Cross-Barnet C, McDonald K (2015) It's all about the children: an intersectional perspective on parenting values among black married couples in the United States. Societies 5(4):855–871
Daniels K, Abma JC, National Center for Health Statistics (U.S.) (2018) Current contraceptive status among women aged 15–49: United states, 2015–2017. U.S. Department of Health and Human Services, Centers for Disease Control and Prevention, National Center for Health Statistics, Hyattsville
Darity W, Hamilton D, Paul M, Aja A, Price A, Moore A, Chiopris C (2018) What we get wrong about closing the racial gap. Samuel DuBois Cook Center on Social Equity, Insight Center for Community Economic Development. https://insightcced.org/wp-content/uploads/2018/07/Where-We-Went-Wrong-COMPLETE-REPORT-July-2018.pdf
Declercq ER, Sakala C, Corry MP, Applebaum S, Herrlich A (2014) Major survey findings of listening to MothersSM III: new mothers speak out: report of national surveys of Women's childbearing experiences. J Perinat Educ 23(1):17–24
DeCuir-Gunby J, Gunby N (2016) Racial microaggressions in the workplace: a critical race analysis of the experiences of African American educators. Urban Educ 51(4):390–414
DeHaven MJ, Ramos-Roman MA, Gimpel N, Carson J, DeLemos J, Pickens S et al (2011) The GoodNEWS (genes, nutrition, exercise, wellness, and spiritual growth) trial: a community-based participatory research (CBPR) trial with African-American church congregations for reducing cardiovascular disease risk factors—recruitment, measurement, and randomization. Contemp Clin Trials 32(5):630–640
Delbello MP, Soutullo CA, Strakowski SM (2000) Racial differences in treatment of adolescents with bipolar disorder. Am J Psychiatry 157(5):837–838
Dirks N, Huth B, Yates C, Melbohm B (2004) Pharmacokinetics of immunosuppressants: a perspective on ethnic difference. Int J Clin Pharmacol Ther 42(12):719–723
Dixon V (1971) African-oriented and Euro-American-oriented worldviews: research methodologies and economics. Rev Black Polit Econ 7(2):119–156
Drayton-Brooks S, White N (2004) Health promoting behaviors among African American women with faith-based support. ABNF J 15(5):84–90
Duffy GF, Shupe ES, Kuczmarski MF, Zonderman AB, Evans MK (2017) Motivations for botanical use by socioeconomically diverse, urban adults: does evidence support motivation? J Altern Complement Med 23(10):812–818
Ellison G, Kaufman J, Head R, Martin P, Kahn J (2008) Flaws in the U.S. Food and Drug Administration's rationale for supporting the development and approval of BiDil as a treatment for heart failure only in black patients. J Law Med Ethics 36(3):449–457
Ellison C, Hummer R, Burdette A, Benjamins M (2010) Race, religious involvement, and health: the case of African Americans. In: Ellison C, Hummer R (eds) Religion, families and health: population-based research in the United States. Rutgers University Press, New Brunswick, pp 321–348
Fabius C (2016) Toward an integration of narrative identity, generationitivity, and storytelling in African American Elders. J Black Stud 47:423–434
Fett SM (2002) Working cures: healing, health, and power on Southern slave plantations. University of North Carolina Press, Chapel Hill
Finer LB, Zolna MR (2014) Shifts in intended and unintended pregnancies in the United States, 2001–2008. Am J Public Health 104(S1):S43–S48
Fitzgerald E, Campinha-Bacote J (2019) Cultural competemility Part II: an intersectionality approach to the process of competemility. Online J Issues Nurs 24(2). https://ojin.nursingworld.org/MainMenuCategories/ANAMarketplace/ANAPeriodicals/OJIN/TableofContents/Vol-24-2019/No2-May-2019/Articles-Previous-Topics/Intersectionality-Approach-to-Cultural-Competemility.html
Fontenot K, Semega J, Kollar M (2018) Income and poverty in the United States: 2017 current population reports. U.S. Department of Commence Economics and Statistics Administration, U.S. Census. https://www.census.gov/content/dam/Census/library/publications/2018/demo/p60-263.pdf
Forde K, Tanapanpanit O, Reddy K (2014) Hepatitis B and C in African Americans: current status and continued challenges. Clin Gastroenterol Hepatol 2(5):738–748
Frazer E, Mitchell R, Nesbitt L, Williams M, Mitchel E, Williams R, Browne D (2018) The violence epidemic in the African American community: a call by the National Medical Association for comprehensive reform. J Natl Med Assoc 110(1):4–15
Frazier E (1957) Black bourgeoisie. Collier Books, New York
Gara M, Minsky S, Silverstein S, Miskimen T, Strakowski S (2019) A naturalistic study of racial disparities in diagnoses at an outpatient behavioral clinic. Psychiatr Serv 70(2):130–134
Gary F (2005) Stigma: barrier to mental health care among ethnic minorities. Issues Ment Health Nurs 26(10):979–999

Gaw A (2001) Concise guide to cross-cultural psychiatry. American Psychiatric Publishing Inc., Washington, DC, p 158

Giurgescu C, Misra DP, Sealy-Jefferson S, Caldwell CH, Templin TN, Slaughter-Acey JC, Osypuk TL (2015) The impact of neighborhood quality, perceived stress, and social support on depressive symptoms during pregnancy in African American women. Soc Sci Med 130:172–180

Glied S, Ma S, Borja A (2017) Effect of the Affordable Care Act on health care access. The Commonwealth Fund. https://www.commonwealthfund.org/publications/issue-briefs/2017/may/effect-affordable-care-act-health-care-access

Gloster H, Neal K (2006) Skin cancer in skin of color. J Am Acad Dermatol 55(5):741–760

Goosby J, Malone S, Richardson E, Cheadle J, Williams D (2015) Perceived discrimination and markers of cardiovascular risk among low-income African American youth. Am J Hum Biol 27(4):546–552

Green AR, Carney DR, Pallin DJ, Ngo LH, Raymond KL, Iezzoni LI, Banaji MR (2007) Implicit bias among physicians and its prediction of thrombolysis decisions for Black and White patients. J Gen Intern Med 22:1231–1238

Gunn J, Davis S (2011) Beliefs, meanings, and practices of healing with botanicals re-called by elder African American women in the Mississippi Delta. Online J Cult Care Nurs Healthc 1(1):37–49

Haider AH, Schneider EB, Sriram N, Dossick DS, Scott VK, Swoboda SM et al (2015a) Unconscious race and social class bias among acute care surgical clinicians and clinical treatment decision. J Am Med Assoc Surg 150:457–464

Haider AH, Schneider EB, Sriram N, Scott VK, Swoboda SM, Zogg CK et al (2015b) Unconscious race and social class bias among registered nurses: vignette-based study using implicit association testing. J Am Coll Surg 220:1077–1086

Hamilton B (2015) Births: final data for 2014. Nat Vital Stat Rep 65(12):1–64

Handel A, Miot L, Miot H (2014) Melasma: a clinical and epidemiological review. An Bras Dermatol 89(5):771–782

Hansen BR, Hodgson NA, Gitlin LN (2016) It's a matter of trust: older African Americans speak about their health care encounters. J Appl Gerontol 35(10):1058–1076

Hargrave R (2010) In: Periyakoil VS (ed) Health and health care of African American older adults. eCampus-Geriatrics, Stanford. https://geriatrics.stanford.edu/ethnomed/african_american.html

Harvey I, Cook L (2010) Exploring the role of spirituality in self-management practices among older African-American and non-Hispanic White women with chronic conditions. Chronic Illn 6:111–124

Hausmann LRM, Myaskovsky L, Niyonkuru C, Oyster ML, Switzer GE, Burkitt KH et al (2015) Examining implicit bias of physicians who care for individuals with spinal cord injury: a pilot study and future directions. J Spinal Cord Med 38:102–110

Henderson D, Wills M, Fricchione G (2010) Culture and psychiatry. In: Stern T, Fricchione G (eds) Handbook of general Hospital. Saunders, Philadelphia, pp 629–638

Herbeck D, West J, Ruditis I, Duffy F, Fitek D, Bell C, Snowden L (2004) Variations in use of second-generation antipsychotic medication by race among adult psychiatric patients. Psychiatr Serv 55(6):677–684

Hicks W (2010) Too many Cases, too many deaths: lung cancer in African Americans. American Lung Association, Washington, DC

Higgins S, Azadeh N, Chow M, Wysong A (2018) Review of nonmelanoma skin cancer in African Americans, Hispanics, and Asians. Dermatol Surg 44(7):903–910

Hoffman K, Trawalter S, Axt J, Oliver M (2016) Racial bias in pain assessment and treatment recommendations, and false beliefs about biological differences between Blacks and Whites. Proc Natl Acad Sci U S A 13(16):4296–4301

Hollingsworth D, Cole A, O'Keefe V, Tucker R, Story C, Wingate L (2017) Experiencing racial microaggressions influences suicide ideation through perceived burdensomeness in African Americans. J Couns Psychol 64(1):104–111

Holt C, Clark E, Debnam K, Roth D (2014) Religion and health in African Americans: the role of religious coping. Am J Health Behav 38(2):190–199

Houston T, Allison J, Sussman M, Horn W, Holt C, Trobaugh J et al (2011) Culturally appropriate storytelling to improve blood pressure: a randomized trial. Ann Intern Med 154:77–84

Institute of Medicine (2004) In the nation's compelling interest: ensuring diversity in the health care workforce. Author, Washington, DC

Jarrett RL, Coba-Rodriguez S (2015) "My mother didn't play about education": low-income, African American mothers' early school experiences and their impact on school involvement for preschoolers transitioning to kindergarten. J Negro Educ 84(3):457–472

Jaynes D, Williams R (1989) A common destiny: Blacks and American society. National Academy Press, Washington, DC

Jefferson C (2010) What health-care reform means for African Americans. The Root. https://www.theroot.com/what-health-care-reform-means-for-african-americans-1790881318

Jiang T, Li H, Hou Y (2019) Cultural differences in humor perception, usage, and implications. Front Psychol 10:123

Johnson J (2008) Ethnic differences in cardiovascular drug response: potential contribution of pharmacogenetics. Circulation 118(3):1383–1393

Johnson K, Elbert-Avila K, Tulsky J (2005) The influence of spiritual beliefs and practices on the treatment preferences of African-Americans. J Am Geriatr Soc 53(4):711–719

Jones J, Schmitt A (2014) A college degree is no guarantee. Center for Economic and Policy Research,

Washington, DC. http://cepr.net/documents/black-coll-grads-2014-05.pdf
Jones MK, Buque M, Miville ML (2018) African American gender roles: a content analysis of empirical research from 1981 to 2017. J Black Psychol 44(5):450–486
Kaiser Family Foundation (2001) *African Americans view of the HIV/AIDS epidemic at 20 years*. Author, Menlo Park
Kaiser Family Foundation (2018) Disparities in health and health care: five key questions and answers. https://www.kff.org/disparities-policy/issue-brief/disparities-in-health-and-health-care-five-key-questions-and-answers/
Kelly J, Kaufman D, Kelley K, Rosenberg L, Mitchell A (2006) Use of herbal/natural supplements according to racial/ethnic group. J Altern Complement Med 12(6):555–561
Kennedy B, Mathis C, Woods A (2007) African Americans and their distrust of the health care system: healthcare for diverse populations. J Cult Divers 14(2):56–60
Krause N, Bastida E (2011) Social relationships in the church during late life: assessing differences between African Americans, whites, and Mexican Americans. Rev Relig Res 53(1):41–63
Kundu R, Patterson S (2013) Dermatologic conditions in skin of color: Part II. Disorders occurring predominantly in skin of color. Am Fam Physician *87*(12):859–865
Ladner J, Gourdine R (1992) Adolescent pregnancy in the African-American community. In: Braithwaite R, Taylor S (eds) Health issues in the black community. Josey-Bass, San Francisco, pp 206–221
Laurie A, Neimeyer R (2008) African Americans in bereavement: grief as a function of ethnicity. Omega 57(2):173–193
Lawson W (1999, May 19) Ethnicity and treatment of bipolar disorder. Presented at the 152nd annual meeting of the American Psychiatric Association. Washington, DC
Lawson C, Hollinger J, Sethi S, Rodney I, Sarkar R, Dlova N, Callender V (2015) Updates in the understanding and treatments of skin & hair disorders in women of color. Int J Women's Dermatol *1*:59–75
Lee RC (2012) Family homelessness viewed through the lens of health and human rights. Adv Nurs Sci 35(2):E47–E59
Lee C, Ayers S, Kronenfeld J (2009) The association between perceived provider discrimination, healthcare utilization and health status in racial and ethnic minorities. Ethn Dis 19(3):330–337
Lin J, Mhango G, Wall M, Lurslurchachai L, Bond K, Nelson J et al (2014) Cultural factors associated with racial disparities in lung cancer care. Ann Am Thorac Soc 11(4):489–495
Livingston G, Parker K, Rohal M (2015) Childlessness falls, family size grows among highly educated women. Pew Research Center. https://www.pewsocialtrends.org/2015/05/07/childlessness-falls-family-size-grows-among-highly-educated-women/
Lobar S, Youngblut J, Brooten D (2006) Cross-cultural beliefs, ceremonies, and rituals surrounding death of a loved one. Pediatr Nurs 32(1):44–55
LoPresti M, Dement F, Gold H (2016) End-of-life care for people with cancer from ethnic minority groups: a systematic review. Am J Hosp Palliat Care 33(3):291–305
Lowenstein A, Glanville CL (1995) Cultural diversity and conflict in the health care workplace. Nurs Econ *13*(4):203–209
Lu Y, Goldstein D, Angrist M, Cavalleri G (2014) Personalized medicine and human genetic diversity. Cold Spring Harb Perspect Med 4(9):a008581
Lupton D (2013) Fat. Routledge, New York
Maglo K, Rubinstein J, Huang B, Ittenbach R (2014) BiDil in the clinic: an interdisciplinary investigation of physicians' prescription patterns of a race-based therapy. AJOB Empirical Bioethics 5(4):37–52
Malek A, Newman J, Hunt K, Marriott B (2019) Race/ethnicity, enrichment/fortification, and dietary supplementation in the US population, NHANES 2009–2012. Nutrients 11(5):1005
Maliski S, Connor S, Williams L, Litwin M (2010) Faith among low-income, African American/Black men treated for prostate cancer. Cancer Nurs 33(6):470–478
Martinez G, Copen C, Abma J (2011) Teenagers in the United States: sexual activity, contraception use, and childbearing, 2006–2010 National Survey of Family growth. National Center Health Stat Vital Health Stat 23(31):1–5
Matas A, Sowa M, Taylor V, Taylor G, Schattka B, Mantsch H (2001) Eliminating the issue of skin color in assessment of the blanch response. Adv Skin Wound Care 14:180–188
McCluney C, Schmitz L, Hicken M, Sonnega A (2018) Structural racism in the workplace: does perception matter for health inequalities? Soc Sci Med 199:106–114
McMichael A (2003) Hair and scalp disorders in ethnic populations. Dermatol Clin 21(4):629–644
McNeil J, Campinha-Bacote J, Tapscott E, Vample G (2002) BESAFE: National minority AIDS education and training center cultural competency model. Howard University Medical School, Washington, DC
Meghani S, Keane A (2017) Preference for analgesic treatment for cancer pain Among African Americans. J Pain Symptom Manage 34(2):136–147
Meschede T, Taylor J, Mann A, Shapiro T (2017) "Family achievements?": how a college degree accumulates wealth for Whites and not for Blacks. Federal Reserve Bank St Louis Rev 99(1):121–137
Minority Nurse (2013, March 30) BiDil controversy continues as FDA approves first "race-specific" drug. https://minoritynurse.com/bidil-controversy-continues-as-fda-approves-first-race-specific-drug/
Mishra S, Lucksted A, Gioia D, Barnet B, Baquet C (2009) Needs and preferences for receiving mental health information in an African American focus groups. Community Mental Health 45(2):117–126

Mitchell J, Ferdinand K, Watson C, Wenger N, Watkins L, Flack J et al (2011) Treatment of heart failure in African Americans—a call to action. J Natl Med Assoc 103(2):86–98

Mitchem S (2007) African American folk healing. NYU Press, New York

Monroe C (2006) Misbehavior or misinterpretation? Closing the discipline gap through cultural synchronization. Kappa Delta Pi Record 42(4):161–165. http://files.eric.ed.gov/fulltext/EJ738081.pdf

Moore J (2005) Hypertension: catching the silent killer Nurse Provider 30(10):16–18, 23–24, 26–27

Morrison R, Wallenstein S, Natale D, Senzel R, Huang L (2000) "We don't carry that"—failure of pharmacies in predominantly nonwhite neighborhoods to stock opioid analgesics. N Engl J Med 342:1023–1026

Mozaffarian D, Benjamin E, Go A, Arnett D, Blaha M, Cushman M et al (2016) Heart disease and stroke statistics—2016 update. Circulation 133:38–60

Murray R, Zentner J (2001) Health promotion strategies through the lifespan. Prentice Hall, Upper Saddle River

Musgrave C, Allen C, Allen G (2002) Rural health and women of color: spirituality and health for women of color. Am J Public Health 92(4):557–560

National Center for Education Statistics (2019) Status and trends in the education of racial and ethnic groups. Indicator 17: High school status dropout rates. https://nces.ed.gov/programs/raceindicators/indicator_RDC.asp#info

National Center for Health Workforce Analysis (2017) Sex, race, and ethnic diversity of U.S. health occupations (2011–2015). Department of Health and Human Services, Health Resources and Services Administration, Rockville. https://bhw.hrsa.gov/sites/default/files/bhw/nchwa/diversityushealthoccupations.pdf

National Health Interview Survey (NHIS) Data (2015) Current asthma prevalence percents by age. National Health Interview Survey. http://www.cdc.gov/asthma/nhis/2015/table4-1.htm

National Hospice and Palliative Care Organization (2019). NHPCO facts and figures 2018 edition (revision 7-2-2019). Am J Hosp Palliat Care 35(3):431–439. Retrieved from https://39k5cm1a9u1968hg74aj3x51-wpengine.netdna-ssl.com/wp-content/uploads/2019/07/2018_NHPCO_Facts_Figures.pdf

Negy C, Eisenman R (2005) A comparison of African American and white college students' affective and attitudinal reactions to lesbian, gay, and bisexual individuals: an exploratory study. J Sex Res 42(4):291–298

Nguyen AW, Chatters LM, Taylor RJ, Mouzon DM (2016) Social support from family and friends and subjective well-being of older African Americans. J Happiness Stud 17(3):959–979

Noonan A, Velasco-Mondragon H, Wagner F (2016) Improving the health of African Americans in the USA: an overdue opportunity for social justice. Public Health Rev 37:12

Office of Adolescent Health (2017) About the teen pregnancy prevention (TPP) program. https://www.hhs.gov/ash/oah/grant-programs/teen-pregnancy-prevention-program-tpp/about/index.html

Office of Disease Prevention and Health Promotion (2019) History & development of healthy people. https://www.healthypeople.gov/2020/About-Healthy-People/History-Development-Healthy-People-2020

Office of Minority Health (2016) Organ donation and African Americans. https://minorityhealth.hhs.gov/omh/browse.aspx?lvl=4&lvlid=27.

Office of Minority Health (2019a) Asthma and African Americans. U.S. Department of Mental Health and Human Resources, Washington, DC. https://www.minorityhealth.hhs.gov/omh/browse.aspx?lvl=4&lvlid=15

Office of Minority Health (2019b) Diabetes and African Americans. U.S. Department of Mental Health and Human Resources, Washington, DC. https://www.minorityhealth.hhs.gov/omh/browse.aspx?lvl=4&lvlid=18

Office of Minority Health (2019c) Heart disease in African Americans. U.S. Department of Mental Health and Human Resources, Washington, DC. https://www.minorityhealth.hhs.gov/omh/browse.aspx?lvl=4&lvlid=19

Office of Minority Health (2019d) HIV/AIDS and African Americans. U.S. Department of Mental Health and Human Resources, Washington, DC. https://www.minorityhealth.hhs.gov/omh/browse.aspx?lvl=4&lvlid=21

Office of Minority Health (2019e) Obesity and African Americans. U.S. Department of Mental Health and Human Resources, Washington, DC. https://www.minorityhealth.hhs.gov/omh/browse.aspx?lvl=4&lvlid=25

Office of Minority Health (2019f) Profile: Black/African Americans. U.S. Department of Mental Health and Human Resources, Washington, DC. https://www.minorityhealth.hhs.gov/omh/browse.aspx?lvl=3&lvlid=61

Office on Smoking and Health (2019) Smoking & tobacco use. National Center for Chronic Disease Prevention and Health Promotion

Ogunleye A, McMichael A, Olsen E (2014) Central centrifugal cicatricial alopecia: what has been achieved, current clues for future research. Dermatol Clin 32(2):173–181

Oliver MN, Wells KM, Joy-Gaba JA, Hawk-ins CB, Nosek BA (2014) Do physicians' implicit views of African Americans affect clinical decision making? J Am Board Fam Med 27:177–188

Olsen E, Bergfeld W, Cotsarelis G, Price V, Shapiro J, Sinclair R et al (2003) Summary of North American Hair Research Society (NAHRS)-sponsored workshop in cicatricial alopecia. J Am Acad Dermatol 48(10):103–110

Ortega V, Meyer D (2014) Pharmacogenetics: implications of race and ethnicity on defining genetic profiles for personalized medicine. J Allergy Clin Immunol 133(1):16–26

Parks F (2003) The role of African American folk beliefs in the modern therapeutic process. Clin Psychol Sci Pract 10:456–467

Pearson B, Jackson C (2013) Removing obstacles for African American English-speaking children through greater understanding of language difference. Dev Psychol 49(1):31–44

Petrie M, Roman P (2004) Race and gender differences workplace autonomy: a research note. Sociol Inquiry 74(4):590–603

Pew Research Center (2014) Childlessness falls, family size grows among highly educated women. Paw Research Center, Washington, DC

Pew Research Center (2015) America's changing religious landscape. Pew Center, Washington, DC. https://www.pewforum.org/2015/05/12/americas-changing-religious-landscape/

Pharr J, Moonie S, Bungum T (2012) The impact of unemployment on mental and physical health, access to health care and health risk behaviors. International Scholarly Research Notices (ISRN) Public Health

Pittman C (2012) Racial Microaggressions: the narratives of African American faculty at a predominantly White university. J Negro Educ 81(1):82–92

Plant A, Montoya JA, Snow EG, Coyle K, Rietmeijer C (2019) Developing a video intervention to prevent unplanned pregnancies and sexually transmitted infections among older adolescents. Health Promot Pract 20(4):593–599

Potter DA, Markowitz LB, Smith SE, Rajack-Talley TA, D'Silva MU, Della LJ, Best LE, Carthan Q (2016) Healthicization and lay knowledge about eating practices in two African American communities. Qual Health Res 26(14):1961–1974

Prather C, Fuller TR, Jeffries WL 4th, Marshall KJ, Howell AV, Belyue-Umole A, King W (2018) Racism, African American women, and their sexual and reproductive health: a review of historical and contemporary evidence and implications for health equity. Health Equity 2(1):249–259

Puckrein GA (1981) Humoralism and social development in colonial America. JAMA 245:1755–1757

Pullen E, Perry B, Oser C (2014) African American women's preventative care usage: the role of social support and racial experiences and attitudes. Sociol Health Illn 36:1037–1053

Quinn S (2018) African American adults and seasonal influenza vaccination: changing our approach can move the needle. Hum Vaccin Immunother 14(3):719–723

Radel L, Bramlett M, Chow K, Waters A (2016) Children living apart from their parents: highlights from the National Survey of Children in Nonparental Care. National Survey of Children's Health, 2016. Retrieved from https://aspe.hhs.gov/system/files/pdf/203352/NSCNC.pdf

Rajman I, Knapp L, Morgan T, Masimirembwa C (2017) African genetic diversity: implications for cytochrome P450-mediated drug metabolism and drug development. EBioMedicine 17:67–74

Resnicow K, Soler R, Braithwaite RL, Ahluwalia JS, Butler J (2000) Cultural sensitivity in substance use prevention. J Community Psychol 28(3):271–290

Rickford J, Duncan G, Gennetian L, Gou R, Greene R, Katz L et al (2015) Neighborhood effects on use of African-American Vernacular English. Proc Natl Acad Sci U S A 112(38):1817–1822

Ries L, Melbert D, Krapcho M, Mariotto A, Miller B, Feuer E (eds) (2008) SEER cancer statistics review, 1975–2004. National Cancer Institute. Bethesda, MD. https://seer.cancer.gov/csr/1975_2005/, based on November 2007 SEER data submission, posted to the SEER web site

Ritchey M, Chang A, Powers A, Loustalot F, Schieb L, Ketcham M et al (2016) Vital signs: disparities n antihypertensive medication nonadherence among medicare part d beneficiaries—United States, 2014. Morbidity and Mortality Weekly Report (MMWR) 65:967–976

Robinson DHZ, Gerbensky Klammer SM, Perryman JP, Thompson NJ, Arriola KRJ (2014) Understanding African American's religious beliefs and organ donation intentions. J Relig Health 53(6):1857–1872

van Ryn M, Burgess D, Dovidio J, Phelan S, Saha S, Malat J et al (2011) The impact of racism on clinician cognition, behavior, and clinical decision making. Du Bois Rev 8(1):99–218

Saladin L, Krause J (2009) Pressure ulcer prevalence and barriers to treatment after spinal cord injury: comparisons of four groups based on race-ethnicity. NeuroRehabilitation 24(1):57–66

Salcido S (2002) Finding a window into the skin. Adv Skin Wound Care 15(3):100

Savitt T (1978) Medicine and slavery. University of Illinois Press, Chicago

Schaa KL, Roter DL, Biesecker BB, Cooper LA, Erby LH (2015) Genetic counselors' implicit racial attitudes and their relationship to communication. Health Psychol 34:111–119

Schaeffeler E, Eichelbaum M, Brinkmann U, Penger A, Asante-Poku S, Zanger U, Schwab M (2001) Frequency of C3435T polymorphism of gene in African people. The Lancet 358(9279):383–384

Seabury S, Terp S, Boden L (2017) Racial and ethnic differences in the frequency of workplace injuries and prevalence of work-related disability. Health Aff 36(2):266–273

Shavers V, Fagan P, Jones D, Klein W et al (2012) The state of research on racial/ethnic discrimination in the receipt of health care. Am J Public Health 102(5):953–966

Sheats K, Irving S, Mercy J, Simon T, Crosby A, Ford D et al (2018) Violence-related disparities experienced by Black youth and young adults: opportunities for prevention. Am J Prev Med 55(4):462–469

Sheftall A, Asti L, Horowitz L, Felts A, Fontanella C, Campo J, Bridge J (2016) Suicide in elementary school-aged children and early adolescents. Pediatrics 138(4):e20160436. https://doi.org/10.1542/peds.2016-0436

Shen MJ, Peterson EB, Costas-Muñiz R, Hernandez MH, Jewell ST, Matsoukas K, Bylund CL (2018) The effects of race and racial concordance on patient-physician communication: a systematic review of the literature. J Racial Ethn Health Disparities 5(1):117–140

Sloan M, Newhouse R, Thompson A (2013) Counting on coworkers: race, social support, and emotional experiences on the job. Soc Psychol Q 76(4):343–372

Smedley B, Stith A, Nelson A (2002) Unequal treatment: confronting racial and ethnic disparities in healthcare. National Academy Press, Washington, DC

Smith TH (1994) Conjuring culture: biblical formations of Black America. Oxford University Press, New York

Snow L (1974) Folk medical beliefs and their implications for care of patients. Ann Intern Med 81:82–96

Snow L (1993) Walkin' over medicine. Westview Press, Boulder

Sommers M (2011) Color awareness: a must for patient assessment. Am Nurse Today 6(1). https://www.americannursetoday.com/color-awareness-a-must-for-patient-assessment/

Sommers M, Zink T, Fargo J, Baker RB, Buschur C, Shambley-Ebron DZ, Fisher BS (2008) Forensic sexual assault examination and genital injury: is skin color a source of health disparity? Am J Emerg Med 26:857–866

Sowa M, Matas A, Schattka B, Mantsch H (2002) Spectroscopic assessment of cutaneous hemodynamics in the presence of high epidermal melanin concentration. Clin Chim Acta 317:203–212

Sowers J, Ferdinand K, Bakris G, Douglas J (2002) Hypertension-related disease in African Americans. Postgrad Med 112(4):24–48

Spruil I, Magwood G, Nemeth L, Williams T (2015) African Americans' culturally specific approaches to the management of diabetes. Global Qualitative Nursing Research 2

Strakowski S, McElroy S, Keck P, West S (1996) Racial influence on diagnosis in psychotic mania. J Affect Disord 39(2):157–162

Strickland T, Lin K, Fu P, Anderson D, Zheng Y (1995) Comparison of lithium ratio between African-American and Caucasian bipolar patients. Biol Psychiatry 37(5):325–330

Sue DW, Capodilupo CM, Holder AMB (2008) Racial microaggressions in the life experience of Black Americans. Prof Psychol Res Pract 39(3):329–336

Taylor RJ, Chatters LM, Levin J (2004) Religion in the lives of African Americans: social, psychological, and health perspectives. Sage, Thousand Oaks

Thomas PA, Krampe EM, Newton RR (2007) Father's presence, family structure, and feelings of closeness to the father among adult African American children. J Black Stud 38(4):541

Thompson C, Craig H, Washington J (2004) Variable production of African American English across oracy and literacy contexts. Lang Speech Hear Serv Sch 35(3):269–282

Trenerry B, Paradies Y (2012) Organizational assessment: an overlooked approach to managing diversity and addressing racism in the workplace. J Divers Manage 7(1):11–26

Tsai J, Homa DM, Gentzke AS, Mahoney M, Sharapova SR, Sosnoff CS et al (2018) Exposure to secondhand smoke among nonsmokers—United States, 1988–2014. MMWR Morb Mortal Wkly Rep 67(48):1342–1346

Tuskegee University (2019) National Center for Bioethics in Research and Health Care. https://www.tuskegee.edu/about-us/centers-of-excellence/bioethics-center

U.S. Census Bureau (2014) Coresident grandparents and their grandchildren: 2012. https://www.census.gov/content/dam/Census/library/publications/2014/demo/p20-576.pdf

U.S. Census Bureau (2018, September 13) Poverty rate in the United States in 2017, by age and gender [Graph]. In Statista. https://www.statista.com/statistics/233154/us-poverty-rate-by-gender/

U.S. Department of Agriculture (2019) Food access research atlas. Economic Research Service. https://www.ers.usda.gov/data-products/food-access-research-atlas/documentation

U.S. Department of Health and Human Services (2010) Office of Disease Prevention and Health Promotion. National action plan to improve health literacy. Author, Washington, DC

Underwood S, Powell R (2006) Religion and spirituality: Influence on health/risk behavior and cancer screening behavior of African Americans. Assoc Black Nursing Faculty (ABNF) 17(10):20–31

United Network for Organ Sharing (2017) Odds for receiving a kidney transplant now equal for black, white, and Hispanic candidates. https://unos.org/news/odds-equal-of-kidney-transplant-for-minorities/

Wandler K (2003) Psychopharmacology of patients with eating disorders. Remuda Rev 2(1):1–7

Wang TW, Asman K, Gentzke AS, Cullen KA, Holder-Hayes E, Reyes-Guzman C et al (2018) Tobacco product use among adults—United States, 2017. Morb Mortal Wkly Rep 67(44):1225–1232

Washington L (2002) Commentary on the history of the African immigrants and their American descendants. iUniverse Publishers, Lincoln

Washington K, Bickel-Swenson D, Stephens N (2008) Barriers to hospice use among African Americans: a systematic review. Health Soc Work 33(4):267–274

Webb Hooper M, Carpenter K, Payne M, Resnicow K (2018) Effects of a culturally specific tobacco cessation intervention among African American Quitline enrollees: a randomized controlled trial. BMC Public Health 18(1):123

Wehbe-Alamah H, McFarland M, Macklin J, Riggs N (2011) The lived experiences of African American women receiving care from nurse provider leave as practitioner in an urban nurse-managed clinic. Online J Cult Competence Nurs Healthcare 1(1):15–26

Welch L, Teno J, Mor V (2005) End-of-life care in Black and White: race matters for medical care of dying patients and their families. J Am Geriatr Soc 53(7):1145–1153

Wen M, Zhang X, Harris C, Holt J, Croft J (2013) Spatial disparities in the distribution of parks and green spaces in the USA. Ann Behav Med 45(Suppl 1):18–27

Wenzel A (ed) (2016) The Oxford handbook of perinatal psychology. Oxford University Press, New York

Williams Institute (2013) LGBT African-Americans and African-American same—sex couples. http://williamsinstitute.law.ucla.edu/wp-content/uploads/Census-AFAMER-Oct-2013.pdf

Williams J, Wyatt G, Wingood G (2010) The four Cs of HIV prevention with African Americans: crisis, condoms, culture, and community. Curr HIV/AIDS Rep 7(4):185–193

Williams M, Chapman L, Wong J, Turkheimer R (2012) The role of ethnic identity in symptoms of anxiety and depression in African Americans. Psychiatry Res 199(1):31–36

Williams D, Lawrence J, Davis B (2019) Racism and health: evidence and needed research. (2019). Annu Rev Public Health 40:105–125

Wilson M (1995) African American family life: its structural and ecological aspects. Jossey-Bass, San Francisco

Woerner M, Correll C, Alvir J, Greenwald B, Delman H, Kane J (2011) Incidence of tardive dyskinesia with risperidone or olanzapine in the elderly: results from a 2-year, prospective study in antipsychotic-naïve patients. Neuropsychopharmacology 36(8):1738–1746

Wood J (2009) Communication in our lives. Wadsworth Cengage Learning, Boston

Woolf S, Aron L, Dubay L, Simon S, Zimmerman E, Luk K (2015) How are income and wealth linked to health and longevity? Urban Institute, Washington, DC. https://www.urban.org/sites/default/files/publication/49116/2000178-How-are-Income-and-Wealth-Linked-to-Health-and-Longevity.pdf

World Health Organization (1946) Preamble to the Constitution of the World Health Organization as adopted by the International Health Conference. World Health Organization, New York. http://www.who.int/suggestions/faq/en/

World Health Organization (2016) What is health promotion? https://www.who.int/features/qa/health-promotion/en/

Wright K, Still C, Jones L, Moss K (2018) Designing a cocreated intervention with African American older adults for hypertension self-management. Int J Hypertens 2018:7591289

Wyn, R., Ojeda, V., Ranji, U., & Salganicoff, A. (2004). Racial and ethnic disparities in women's health coverage and access to care: findings from the 2001 Kaiser women's health survey. Report. Menlo Park: Henry J. Kaiser Family Foundation.

Yarnell LM, Pasch KE, Perry CL, Komro KA (2018) Multiple risk behaviors among African American and Hispanic boys. J Early Adolesc 38(5):681–713

Young SL (2010) Pica in pregnancy: new ideas about an old condition. Annu Rev Nutr 30(40):3–22

Young SL, Sherman PW, Lucks JB, Peltro GH (2011) Why on earth?: Evaluation hypotheses about the physiological functions of human geophagy. Q Rev Biol 86:97–120

Zestcott CA, Blair IV, Stone J (2016) Examining the presence, consequences, and reduction of implicit bias in health care: a narrative review. Group Process Intergroup Relat 19(4):528–542

Zienkiewicz S (2008) African American Vernacular English (AAVE). Multicultural Topics in Communications and Disorders. Portland State University, Speech and Hearing Sciences. https://www.pdx.edu/multicultural-topics-communication-sciences-disorders/african-american-vernacular-english-aave

6 Indigenous American Indians and Alaska Natives

Kathy Prue-Owens

6.1 Introduction

American Indians and Alaska Natives are the original inhabitants of North America. Although these groups are referred to as *Native Americans* and *Alaskan Natives* (AI/AN), many prefer to be called *American Indians* or *Alaska Natives* or names more specific to their cultural heritage, self-identity, distinct language, distinct social, economic and political systems. A more recent name used by AI/AN is the term "indigenous" people which is estimated to be more than 370 million spread across 70 countries worldwide.

6.2 Overview, Inhabited Localities, and Topography

6.2.1 Overview

Among the indigenous people are those who inhabited the country or a geographical region at the time when others from different cultures or ethnic origins arrived and dominated native nations (Anderson 2016; United Nations Office of the High Commissioner for Human Rights 2013). The amount of Indian blood necessary to be considered a tribal member varies with each tribe (Table 6.1). Navajos and Cherokees claim the distinction of the largest tribes. Navajos require a blood quantum of 25%, whereas the Cherokees have no minimum eligibility requirements (Navajo Tribal Policy n.d.; Department of Interior BIA Federal Register 2019). Cherokee Tribal Citizenship depends on the provision of documents that connect one to an enrolled lineal ancestor who is listed on the Dawes Roll, final rolls of Citizens and Freedmen of the Five Civilized Tribes (Mihesuah 2015; National Archives and Records Administration 2019). This roll was taken between 1899 and 1906 of Citizens and Freedmen residing in Indian Territory [Northeast Oklahoma (Cherokee Tribal Policy n.d.)]. Identity as an AI/AN tribal member has been a continuous debate because unlike other ethnic groups, particularly, American Indians must constantly prove their identity. This forces tribal groups to question their cultural heritage and blood quantum membership requirements (Schmidt 2011; Blood Quantum American Indian Enterprise and Business Council 2019; Indian Ancestry 2004). Health care professionals can help limit barriers to ancestry linkage issues (Blanchard et al. 2017). Efforts to guard religious freedom have fallen short to protect sacred places and a way of life for AI/AN (Navajo Nation Human Rights Commission 2014; Shadow

This chapter is a revision made of the original chapter written by Olivia Hodgins and David Hodgins in the previous edition of the book.

K. Prue-Owens (✉)
University Colorado Colorado Springs, Colorado Springs, CO, USA
e-mail: kprueowe@uccs.edu

© Springer Nature Switzerland AG 2021
L. D. Purnell, E. A. Fenkl (eds.), *Textbook for Transcultural Health Care: A Population Approach*, https://doi.org/10.1007/978-3-030-51399-3_6

Table 6.1 Tribal blood quantum

Tribes requiring 1/2-degree blood quantum for citizenship (equivalent to one parent)
Kialegee Tribal Town
Miccosukee Tribe of Indians, Florida
Mississippi Band of Choctaw Indians, Mississippi
St. Croix Chippewa Indians, Wisconsin
White Mountain Apache Tribe, Arizona
Yomba Shoshone Tribe, Nevada
Tribes requiring 1/4-degree blood quantum for citizenship (equivalent to one grandparent)
Absentee-Shawnee Tribe of Indians, Oklahoma
Cheyenne and Arapaho Tribes, Oklahoma
Confederated Tribes and Bands of the Yakama Nation, Washington
Ho-Chunk Nation, Wisconsin
Kickapoo Tribe, Oklahoma
Kiowa Tribe, Oklahoma
Fort McDowell Yavapai Nation, Arizona
Fort Peck Assiniboine and Sioux Tribes, Montana
Navajo Nation, Arizona, Utah and New Mexico
Oneida Tribe of Indians, Wisconsin
Pascua Yaqui Tribe, Arizona
Prairie Band Potawatomi Nation, Kansas
Shoshone Tribe of the Wind River Reservation, Wyoming
Standing Rock Sioux Tribe, North and South Dakota
United Keetoowah Band of Cherokee Indians, Oklahoma
Utu Utu Gwaitu Paiute Tribe, California
Yavapai-Prescott Tribe, Arizona
Tribes requiring 1/8-degree blood quantum for citizenship (equivalent to one great-grandparent)
Apache Tribe of Oklahoma
Comanche Nation, Oklahoma
Delaware Nation, Oklahoma
Confederated Tribes of the Siletz Reservation, Oregon
Fort Sill Apache Tribe of Oklahoma
Karuk Tribe of California
Muckleshoot Indian Tribe of the Muckleshoot Reservation, Washington
Northwestern Band of Shoshoni Nation of Utah (Washakie)
Otoe-Missouria Tribe of Indians, Oklahoma
Pawnee Nation, Oklahoma
Ponca Nation, Oklahoma
Sac and Fox Nation, Oklahoma
Sac & Fox Nation of Missouri in Kansas and Nebraska
Squaxin Island Tribe of the Squaxin Island Reservation, Washington
Suquamish Indian Tribe of the Port Madison Reservation, Washington
Three Affiliated Tribes of the Fort Berthold Reservation
Upper Skagit Indian Tribe of Washington
Wichita and Affiliated Tribes (Wichita, Keechi, Waco and Tawakonie)
Tribes requiring 1/16-degree blood quantum for citizenship (equivalent to one great-great-grandparent)
Caddo Nation
Confederated Tribes of the Grand Ronde Community of Oregon
Fort Sill Apache Tribe
Iowa Tribe, Oklahoma
Sac and Fox Nation, Oklahoma
Eastern Band of Cherokee Indians, North Carolina
Confederated Tribes of Siletz Indians

Table 6.1 (continued)

Tribes determining membership by lineal descent
These tribes do not have a minimum blood quantum requirement; however, this does not mean anyone with any amount of Indian blood can enroll. Members must be direct descendants of original enrollees.
Alabama-Quassarte Tribal Town
Cherokee Nation
Chickasaw Nation
Choctaw Nation
Citizen Potawatomi Nation
Delaware Tribe of Indians
Eastern Shawnee Tribe
Kaw Nation
Mashantucket Pequot Tribe, Connecticut
Miami Tribe, Oklahoma
Modoc Tribe
Muscogee Creek Nation
Osage Nation
Ottawa Tribe, Oklahoma
Peoria Tribe of Indians
Quapaw Tribe, Oklahoma
Sault Ste. Marie Tribe of Chippewa Indians, Michigan
Seminole Nation
Seneca-Cayuga Tribe, Oklahoma
Shawnee Tribe
Thlopthlocco Tribal Town
Tonkawa Tribe
Wyandotte Nation

Report; American Indian Religious Freedom Act 1978). This chapter describes AI/AN's tribal customs and traditions to inform and promote culturally congruent care by health-care providers.

6.2.2 Inhabited Localities and Topography

The Bureau of Indian Affairs (BIA) (BIA 2019) became a resource to allow more rapid westward expansion of European Americans (Reyhner n.d.). President Andrew Jackson did not continue the peaceful policy of the previous administration and the government established the Indian Territory in present-day Oklahoma (Sturgis 2007). Through what is now termed "ethnic cleansing," the federal government forced the five civilized, plains, and northern tribes to leave their ancestral homes and walk to a new homeland where they were promised they could have forever, Oklahoma (Yellow-Horse-Brave Heart and DeBruyn 2004). However, the westward expansion included Oklahoma that was opened in 1889 to pioneers despite the federal government's promises. The federal government policy at that time was if Indians or Natives could not be eradicated or isolated in an Indian Territory, they would have to be civilized (Yellow-Horse-Brave Heart and DeBruyn 2004). Of some 400 treaties negotiated between tribes and the government before such treaty-making ended in 1871, 120 contained educational provisions to move Indians toward "civilization" through education (Sturgis 2007).

For over 200 years, the BIA has been the treaty negotiator between the US and tribes. The agency has been involved in the implementation of federal laws that have significantly impacted the ways of AI/AN. The Indian Citizenship Act of 1887 granted AI/AN citizenship and the right to vote. This led to the Indian Reorganization Act of 1934 where tribal governments were established (Fig. 6.1). Relocation after World War II and into the 1960s and 1970s witnessed the takeover of the BIA headquarters which resulted in the development of the Indian Self-Determination and Education Assistance Act of 1975 (Public

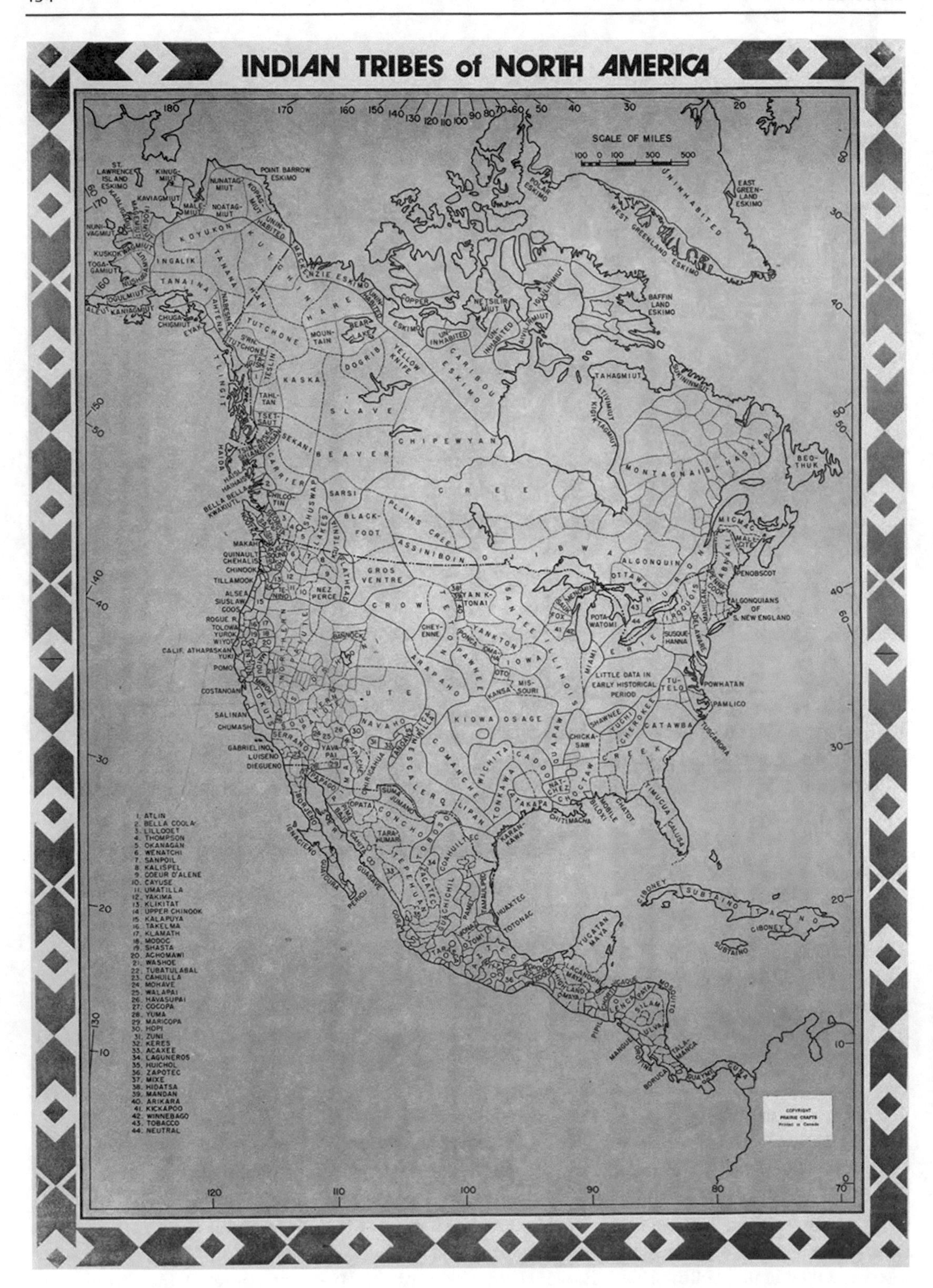

Fig. 6.1 Map of Indian Tribes of North America

Law 93-638 1975). Along came the Tribal Self-Governance Act of 1994 which fundamentally changed how tribes would operate with each other and non-Indians off the reservations or tribal offices (BIA 2019). During the time when the federal government promised health-care services and education to AI/AN tribes in exchange for land for which they could not pay full market value. The Snyder Act of 1921 (Public Law 67-85 n.d.), also known Indian Citizenship Act (1924) codified federal engagement in land management and oversight of health care. The Indian Health Care Improvement Act of 1976 was a further delineation of the government's responsibility in providing health care. Variations of this act have been approved yearly since this time continuing the federal government's responsibility to American Indian and Alaska Natives (Halabi 2019).

6.2.3 Heritage and Residence

The Bureau of Indian Affairs (BIA 2019) recognizes 573 different American Indian tribes and Alaska Natives that extend throughout Alaska and Canada, from Maine to Florida, and from the east coast to the west coast. Subdivisions of American Indians include the Algonquian, Siouan, Dhegiha, Chiwere, Winnebago, Mandan, Hidatus, Caddoan, Pueblos, Navajo, Apache, and the Five Civilized Tribes: Creek, Choctaw, Chickasaw, Cherokee, and Seminole.

Alaskans are divided into several main groups: the Southeastern Coastal Indians (the Tlingit, Haida, and Tsimshian), the Athabascans, the Aleut, and the two groups of Eskimos, the Inupiat and the Yup'ik. Alaskan natives comprise 14.8% up from the 2018 census of 7% (U.S. Bureau of Census 2018). Of the 573 tribes, there are over 230 federally recognized Alaskan villages; Alaska Natives is the preferred name or Inuits. Native Alaskan tribes are diverse with 20 different languages in five geographic areas, and they are organized under 13 Alaska Native Regional Corporations (GAO-13-121 Regional Alaska Native Corporations 2012).

Most AI/AN tribes attribute the present-day world to a Creator, and their stories were told by traditional priests or shamans who had the privilege to recite and explain it. These creation stories deal with the origins and wanderings of the people from the beginning of the world to the final settlement of the tribe in its home (Johnson 2004). According to tribal cosmology, human beings came into being by a creator and after the creation of the animals. The creation story of the Salinan Indians of southern California states that "When the world was finished, the Bald Eagle, chief of the animals, saw that the world was incomplete and decided to make some human beings" (Johnson 2004, para 4). First, he made man and decided he should not be alone so he created woman from one of his feathers. The Pima Indians of southern Arizona believe that Earth Maker took some clay and mixed it with his own sweat, forming it into two figures, a man and a woman. He breathed life into them; they began to walk around; they lived and had children and occupied the land and built villages. Understanding creation within the framework of delivering care allows for the acceptance of family emergence when a family member is ill.

6.2.4 Reasons for Migration and Associated Economic Factors

Forced migration of AI/ANs was driven by colonization of the US by Europeans from England in the east, French and Spanish in the southeast, Russians in Alaska, and Spanish in the southwest. President Andrew Jackson and Congress forced natives from all over the US to travel to the Indian Territory, present-day Oklahoma (Indian Removal Act of 1830 n.d.). The U.S. Army waged a brutal war in the western part of the United States marching through Indian territory to destroy their culture including languages and traditions, and starved them in an attempt at ethnic annihilation (Yellow-Horse-Brave Heart and DeBruyn 2004). The Indian Removal Act of 1830 grant Jackson the authority to move AI west of the Mississippi in exchange for land. Evidence of these attempts are documented in tribal history (Carlson and Roberts 2006; Resch 2005). Troops, led by Christopher (Kit) Carson, captured and killed Navajos throughout northern New Mexico and Arizona. In 1864, most Navajos surrendered to Carson; he forced nearly 8000 Navajo men, women, and children to take what later came to be called the *Long Walk*, 300 miles from Arizona to Fort Sumner, New Mexico (Remini 1998). Cherokees were forced out of their native lands and documented in the Trail of Tears drama. Thousands of tribal members died from starvation, cold, diarrhea, measles, and cholera (Sturgis 2007).

Reservations were created to contain the AI/ANs on undesirable lands that European American people did not view as valuable. Because of severe economic conditions and high unemployment rates, significant migration occurs into and out of the reservations. Commercial activities on reservations are limited to businesses owned by AI/ANs. On some reservations, the business owner must be from the same tribe, while on other reservations any AI/AN tribe can own a business. Such restrictions severely limit growth and employment on the reservations. Occasionally, big businesses contract with a reservation for mining, timbering, and electrical power services (Indian Affairs Manual 2019).

In Alaska, discovery of oil and gold brought people from the lower 48 states, Europe, and Asia. The strikingly different culture and economy of *bush* Alaska creates further complications. Isolated communities have survived in a harsh wilderness for centuries by heavy reliance on the animals, fish, and birds they hunted or vegetable resources they gathered in subsistence activities. In the past, a cash economy was unnecessary as it was primarily survival oriented. Alaska is a land of extremely cold temperatures, small concentrations of population, rugged terrain, and distance. In the 1970s, AN groups started to form tribal villages and corporations. Communication, transportation, and housing difficulties make education costlier and harder to administer. Many parts of northern Alaska are locked in by ice; supplies must arrive by a bush plane or other means or rural mainland reservations can be isolated from resources (Pacheco et al. 2015).

In the lower 48 states, south-western and northern tribes tend to remain on designated reservations. Many who leave the reservation experience culture shock and some return as the result of a lack of social support systems and loss of identity and self-esteem. Those who work in cities often return to their reservations on a regular basis to refresh and renew themselves with ceremonies and cultural activities such as traditional pow wow (Fig. 6.2) or sun dances; where teaching of heritage to the youth is important (Figs. 6.3 and 6.4).

AI/AN tribes have fought to maintain tribal sovereignty and are a cornerstone of jurisprudence within tribal law (Garrow 2015). Self-determination is a huge component of past treaties and tribes have prevailed in legal battles to ensure its continuation. In contrast to the poor economic environment of some tribes, the Osage tribe in Oklahoma has oil leases. Each tribal member has head rights and receives an income from these leases. Gambling has become a leading revenue producer for many tribes. Depending on the tribe, oil production and lease agreements provides a slight economic advantage. Land can have various uses among the tribes, such as grazing for cattle and water use for fishing or crop growing (Office of Program Planning and Development Navajo Area Indian Health Service 1988).

6.2.5 Educational Status and Occupations

In *Light of the Feather: Pathways Through Contemporary Indian America,* author Mick Fedullo (1992) writes about his experiences as an educational consultant in AI/AN schools. Fedullo (1992) quotes an older Apache who says that the students' parents had been to school in their day and that usually meant a bad BIA boarding school. Older Apaches remember that school was taught by these Anglos [White people] trying to destroy their way of life. Teachers were trying to make them turn against their parents, telling them

Fig. 6.2 A little Native American girl, dressed in full pow wow apparel, holds the hand of one of the lady elders, as many elders stand, dressed in full pow wow regalia such as brightly colored dresses with intricate beading. They are caring beautiful shawls and fans made of feathers. Each is adorned in hand-made Native American jewelry (Intertribal Pow Wow)

that Indian ways were evil. As a result, many students came to believe that their teachers were the evil ones, and so anything that had to do with "education" was also evil, including books. This conflict resulted in rejecting the school's language and culture and created an environment that did not value education. If students rejected their language and culture, the result was an inability to communicate with their parents and other extended family members. In Alaska Native communities, schools were operated by the federal government and a variety of church mission schools. Alaska Natives were allowed to attend territorial school only if they were at least *mixed breed* and lived a civilized life (Getches 1977). Cultural response teaching as pedagogy has the potential to no longer make students feel isolated or misunderstood and promote a positive learning environment. This response can help students to express cultural beliefs/values improving academic achievements (Morgan 2010; Stowe 2017). The use of social cultural competence can illicit the help from computer technology/telehealth (Fraser et al. 2017; Carroll et al. 2011) where comprehensive integration and implementation of pedagogy conditions can serve to promote competence in both students, providers, system (Cunningham et al. 2014); even among care to older adults (Rakhimova et al. 2017; Cuellar 2015).

A larger percentage of AI/AN's eighth-grade students report absences from school more than any other racial group (DeVoe et al. 2008). In 2006, while 87.7% Caucasians graduated from high school, only 74.7% of AI/AN students graduated, lower than any other minority group. Today, only 14% of American Indians have a college degree—less than half the national average (American Indian College Fund 2019). Consequentially, educational levels for AI/ANs

Fig. 6.3 A youth Lakota dancer at the grand entry of the Rosebud Pow Wow 2019 (photo courtesy of Kernit Grimshaw)

are lower than those of similar populations, creating another barrier to employment. AI/ANs have been identified as the most underrepresented of all minority groups in colleges and universities, especially in nursing (Simi and Matusitz 2016). Before the 1970s, an increase occurred in educational achievement among American Indians; however, since then, there has been a downward curve (Office of Indian Education 2006).

In contrast to other cultures, competitiveness is generally discouraged among AI/AN populations. Group achievements are more important than individual achievement. Occupations selected by AI/ANs include social work, teaching, and artistry. Math, science, and health-care fields have the lowest number of AI/ANs (DeVoe et al. 2008). Recruitment efforts concentrate on these fields, but recruitment does not mean they graduate. In an attempt to improve the success rate for nursing schools, resilience is being studied within the context of AI/AN culture. Additionally, adding programs to support AI/AN nursing students to understand cultural norms has the potential to keep students retained (Katigbak et al. 2015). The Indian Health Service (IHS) as a federal health program for American Indians and Alaska Natives; under the Title V of the Indian Healthcare Improvement Act partnered with the Arizona State University nursing program to provide programs to help AI/AN nursing students successfully complete their nursing degrees, increase the number of American Indian nurses, and increase the number of nurses providing care to American Indians (Indian Health Service 2019).

Historically, art as a decorative concept did not exist among indigenous people (Hendrix 1999). Objects were utilitarian, although decorated in ways that conveyed images of spiritual or physical activity. After Europeans first made contact with the native people in the seventeenth century, nonutilitarian art objects began to be traded in exchange for metal implements, cloth, and foodstuffs such as tea, flour, or sugar. Northwest Tribes, including Alaska Natives, are known for their ceremonial masks, totem poles, baskets, beadwork, ivory, and wood carving. Art takes such forms as rug weaving, basket weaving, pot-

Fig. 6.4 A young Lakota girl dancing during a competition at the annual Rosebud Pow Wow 2019 (photo courtesy of Kernit Grimshaw)

tery making, and beadwork. Rug weaving and silversmithing are the most common forms of art among the Navajo. Sand paintings were traditionally used in healing ceremonies by medicine people and were not originally intended for sale. The Pueblos are known for their pottery, basketry, kachinas, and fetishes.

6.3 Communication

6.3.1 Dominant Languages and Dialects

Tribal groups in Alaska have four distinct Yuipik dialects and Athabascan, which is the basis of the Navajo and Apache languages. The Navajo language was not reduced to writing until the 1970s; consequently, many older Navajo speak only their native language and few are literate in the English language. Navajos who traded with the Spanish have since died so Spanish is no longer being spoken by the elders. The younger populations are becoming more acculturated; thus, many understand their language but are not able to speak it fluently. Many tribes had assistance from Europeans to develop a written dialect. Cherokees were most notably the first to develop their written language.

Health-care providers must be extremely careful when attempting to speak an AI/AN language because minor variations in pronunciation may change the entire meaning of a word or phrase. Today, trained interpreters in health care communicate with the older patient populations. Health-care workers must remember not to use family members or other health-care workers unless these have been trained as interpreters. On one

Fig. 6.5 Navajo Code Talkers, Saipan, June 1944

such occasion, a patient had cancer and the interpreter was supposed to be interpreting the physician's discussion. A family member who was present had some medical background, and the interpreter was not correctly interpreting the diagnosis and discussion. A valuable lesson was learned, and incidents such as this led to the establishment of interpreter-training programs on reservations.

AI/ANs have served with distinction in each World War including the Spanish American War. During World War II, Navajos were used to develop codes to prevent enemies from determining the locations of ships and tactics. The Japanese could not break these codes and eventually were defeated. Choctaws serving in the U.S. Army during World War I pioneered "Code Talking." Other American Indian code talkers were deployed by the U.S. Army during World War II, including Cherokee, Choctaw, Lakota· Meskwaki, and Comanche soldiers (Stanley 1997) (Fig. 6.5). Code Talkers were top secret until the 1980s. The Code Talkers received numerous awards for their participation in World War II. Marines were assigned to guard the talkers. It has been said that these Marines were under orders to kill the talkers to protect the code from falling into the hands of the Japanese, but the federal government has denied that allegation. Native Americans has the highest number of individuals who have served our nation (Figs. 6.6 and 6.7).

6.3.2 Cultural Communication Patterns

Nonverbal communication styles have different connotations within each tribe. For example, the willingness of AI/ANs to share their thoughts and feelings varies from group to group and from individual to individual. In addition, no set pattern exists regarding their willingness to share tribal ceremonies. However, suspicion always exists because earlier government and church groups banned tribal ceremonies and events. The Pueblo groups, especially the Hopi, exclude outsiders from viewing most of their ceremonies including the famous Snake Dance.

Fig. 6.6 Lakota Sioux Color Guard, Rosebud, South Dakota (photo courtesy of Kernit Grimshaw)

Fig. 6.7 Native American Veterans, Francis White Hat, US Army Medic, Vietnam (photo courtesy of Kernit Grimshaw)

AI/ANs are comfortable with periods of silence. Interest in what an individual says is shown through attentive listening skills. For example, when a group of nursing students from Iowa visited an Indian reservation, a native elderly woman was asked what is the most important thing a nurse can do for their patient. The elderly women responded "just listen, just listen." Chisholm (1983) reported that to establish a positive social relationship, the rule of silence is considered a serious matter that calls for caution, careful judgment, and plenty of time to reflect. Failing to allow adequate time for information processing may result in an inaccurate response or no response in older AI/ANs. One may be considered immature if answers are given too quickly or if one interrupts another who is forming a response.

Tribal members tend to be somber and less likely to laugh aloud except in family settings. For example, a nurse complained, "I don't know why the family comes in here, for they only sit there. They don't even say a word to the person, and then they get up and walk out." What this nurse failed to realize was that the visitors were supporting the individual, not through talking but by visiting and quite listening. The nurse's view reflects his or her cultural bias that saying something is essential to demonstrate support; in reality, for AI/ANs silence is being supportive. An awareness of this nonverbal communication support is extremely helpful for health-care providers who wish to establish mutually satisfying relationships.

In some tribes, touch is very important because many forms of traditional medicine involve massaging and rubbing by the traditional healer, a family member, or both. However, if a Western health-care provider were to do this, even within the context of treatment, it would not be permissible. If contact is made, it is in the form of a handshake. Close observation of body language is very important for determining cues related to the permissibility of touch.

Questions should be modified to age and acculturation level; health-care providers must be careful not to make assumptions based on appearance or age. Health-care providers must slow down when communicating with an AI/AN elder, especially during initial encounters and when explanations of treatments/medications/health-care decisions are being given. Questions should be carefully structured to convey caring and not indicate idle curiosity about the culture or cultural practices. AI/ANs are part of a shared society that promotes reliance upon a close bond with family members, community, and tribe. The individual is not the focus; the group is the focus. Elders transmit the ancestral knowledge to the youth of their tribe, the community at large, and specifically their family. Understanding the cultural values of AI/AN will help health care professional focus on what is important to this group of people (Table 6.2) (Sando 1976).

Table 6.2 Comparison of cultural value systems

Tribal traditional cultural values	European american values
Group, clan, or tribal emphasis	Individual emphasis
Present oriented	Future oriented
Time, always with us	Time, use every minute
Age	Youth
Cooperation	Competition
Harmony with nature	Conquest of nature
Giving, sharing	Saving
Pragmatic	Theoretical
Mythological	Scientific
Patience	Impatience
Mystical	Skeptical
Shame	Guilt
Permissiveness	Social coercion
Extended family and clan	Immediate family
Nonaggressive	Aggressive
Modest	Overconfident
Silence	Noise
Respect others' religion	Convert others to religion
Religion, way of life	Religion, a segment of life
Land, water, forest	Land, etc.
Belong to all	Private domains
Beneficial, reasonable	Avarice, greedy
Use of resource	Use of resource

Source: Sando, J. S. (1976). Pueblo Indians. San Francisco, CA: Indian Historian Press

6.3.3 Temporal Relationships

The time sequences of importance for AI/ANs are present, past, and future, in comparison with the time sequences important to most European Americans, which are present, future, and past (Burke et al. 1989). Most AI/AN tribes are not future oriented. Very little planning is done for the future because their view is that many things are outside of one's control and may affect or change the future. In fact, when the Navajo language was written it did not include a future-tense verb. Time is not viewed as a constant or something that one can control, but rather as something that is always with the individual. Planning for the future is viewed as foolish. Past events are an important part of the AI/AN heritage as evidenced in verbal histories passed down from generation to generation; the present is addressed as a here-and-now issue.

The term *Indian time* has little meaning in a European American worldview, but it assumes particular importance for those who supervise an AI/AN employee in which time has no meaning or importance and the concept of controlling time is foreign. Events do not always start on time; rather, events start when the group gathers. To help prevent frustration in scheduling events, time factors need to be taken into consideration and the speaker made aware of these unique time perceptions. Appointments may not always be kept, especially if someone else in the community needs help. Many do not have transportation and must wait for a ride from family or community member such as the community resource.

6.3.4 Addressing Family Members

Older people are addressed as grandmother, grandfather, mother, or father, or by members in his or her tribe/band/nation/pueblo/village/community. Otherwise, they are called by a nickname. A health-care provider can call an older Navajo client grandmother or grandfather as a sign of respect. In the Apache Tribe, children often breastfeed with other mothers, consequently children address most women as mother.

6.4 Family Roles and Organization

6.4.1 Head of Household and Gender Roles

The roles vary greatly from region to region and from tribe to tribe; in some cases, even from band to band within a tribe or people. Over the years the roles have changed but largely remain intact. The Algonquians and Cherokee have an egalitarian society and women serve on tribal counsels. The Dene tribe (Athapaskans) of Athabascan-speaking people of Alaska, Pacific Northwest, and Siouan tribes are patriarchal and patrilineal. Gender roles vary widely among the Inuit. Early Inuit had few preconceptions concerning "appropriate" male and female roles so gender roles varied widely. The Iroquois, Navajo, Pueblos, and Haida are matriarchal societies.

6.4.2 Prescriptive, Restrictive, and Taboo Behavior for Children and Adolescents

AI/AN naming traditions vary greatly from tribe to tribe and are frequently determined by nature, animals, or character. The Southwest Hopis have a tradition of placing an ear of corn, representing Mother Earth, close to a newborn baby. Twenty days after the birth of a baby, corn is rubbed over its body, the baby is held facing the rising sun; the baby is named when the first ray of sun hits its forehead. Anthropologists believe that the tradition of not naming the infant until this time came about because of high infant mortality in the past.

In the past, AI/ANs women practiced breastfeeding exclusively. Since the early to mid-1980s, the use of formula has become popular. As a result, an increased incidence of bottle caries has been observed among AI/AN children, who as babies often went to bed with a bottle of juice or soda pop (IHS Fact Sheet n.d.). This practice causes children to lose their teeth by the age of 4 years. One health-promotion priority is to educate parents about "bottle mouth" to prevent dental caries and to encourage a return to breastfeeding.

A primary social premise is that no person has the right to speak for another. Parents tend to be more silent, noninterfering, and permissive in their childrearing practices. Children are allowed to make decisions that other cultures may consider irresponsible. For example, children may be allowed to decide whether they want to take their medicine and allow them to live with extended family members. This practice may be perceived as neglect by some health-care providers. Children who do not listen to their parents or older people accept the consequences regardless of their age. If children do not want to attend school, they stay at home.

Ceremony plays a vital, essential role in AI/ANs everyday life. Indians were eager to embrace ceremonies or portions of ceremonies that provided power to conquer the difficulties of life. One life cycle ceremonial ritual in AI/AN societies includes the onset of menarche, which is celebrated with special foods that symbolize passage into adulthood. Other ceremonies are conducted to heal those with pain (Gone et al. 2017; Greensky et al. 2014).

> The most important thing to remember about ceremony is that it is a way for humans to give back to the Creation some of the energy that they are always receiving. The Earth Mother constantly gives us two-leggeds a surface on which to place our two feet; Father Sun warms us, and Grandmother moon brings dreams. The element of Earth gives us a place to grow food and the ability to make homes and tools. The water keeps us alive. The fire warms our homes and cooks our food. The air gives us the sacred breath of life. Through ceremony, we learn how to give back.—Sun Bear, Anishinaabe Nation, Cited in McFadden 1994, p. 30

6.4.3 Family Goals and Priorities

Family goals are a priority in the American Indian culture; family remains strong, even if marriage joins the couple with another family. When a couple marries in the Pueblo tribes, the man goes to live in the woman's house. In Navajo tradition, families have separate dwellings but are grouped by familial relationships. The Navajo family unit consists of the nuclear family and relatives such as sisters, aunts, and their female descendants. For American Indian and Alaska Natives, extended family members play an important role in the infants' life.

Older adults are looked on with clear deference (Hendrix 1999). Elders play an important role in keeping rituals and in instructing children and grandchildren. It is not uncommon for grandparents to be raising grandchildren from two generations (Dennis and Brewer 2017; Lewis et al. 2018). Even though older people are respected, elder neglect is on the rise, possibly as a result of the loss of middle-aged individuals due to chronic disease (Graves et al. 2004). Younger adults are faced with the responsibility of caring for elders over a longer period. In addition, there are few nursing homes, and hospitals are forced to keep patients until nursing-home placement is found. When nursing-home placement is found, it may be a great distance from the family, making family visits difficult. Hence, older people are often taken from the nursing home, even though the family is not in a position to care for their needs. Patients are often readmitted to the hospital; a revolving-door syndrome develops.

Social status is determined by age and life experiences. However, among the Pueblos, governors are chosen from a particular clan; unless one is born in the clan that person cannot run for tribal governor. Generally, individuals are discouraged from having more possessions than their peers; those who display more material wealth are ignored. Standing out is not encouraged among the different tribal groups.

6.4.4 Alternative Lifestyles

Two-spirit is a contemporary term, adopted in 1990 from the Northern Algonquin word *niizh manitoag*, meaning "two-spirits," signifying the personification of both feminine and masculine spirits within one person (Anguksuar 1997). This term is used contemporarily to connote diverse gender and sexual identities among the AI/AN population (Balsam et al. 2004). Early in AI/AN history, many two-spirit tribal members had sacred or ceremonial roles in their communities. Colonization and compulsory Christianity led to the suppression of two-spirit roles in many AI/AN communities. Today most two-spirits face homophobic oppression from both mainstream U.S. society and their own tribes and communities. This may be particularly true for AI/ANs who live off their reservation and in urban areas (Walters 1997).

6.5 Workforce Issues

6.5.1 Culture in the Workplace

Many AI/ANs remain traditional in their practice of religious activities. Family matters are more important than work, resulting in high rates of absenteeism. Additionally, tribal ceremonies are seen as necessary and they often must take time from work or school. The needs of the individual must be weighed against organizational requirements in the development of a reasonable solution. Tribal members in the community function informally as cultural brokers and assist by helping non–American Indian staff to understand important cultural issues. Cultural sensitivity by the employer is of utmost importance. Local tribal policies or even federal IHS or BIA policies are in place to address these needs, employees use annual leave, or they may earn religious compensatory time for use at a later date. This cultural necessity is conveyed to non-Indian staff when they are in orientation so that they are sensitive to the local population (The Navajo Nation Personnel Policies Manual 2017; The U.S. Equal Employment Opportunity Commission, Policy Statement on Indian Preference 1988; Indian Health Manual 2019).

European Americans who are upset or in conflict often want to talk through the issue that caused the conflict. AI/AN wisdom teaches that one must be mindful of a person's emotional damage. Supervisors should emphasize a need for conflict resolution and strive to heal broken relationships and rebuild personal self-esteem and confidence. Addressing these dimensions directly provides an opportunity for both resolution and closure. Sometimes AI/AN's methods of resolving anger and conflict among themselves is much different. Conflict is addressed indirectly through third parties in some tribes such as the Pueblos and in some conflict is handled with violence without the involvement of police.

6.5.2 Issues Related to Autonomy

Group activities are an important norm in AI/AN cultures. One individual should not be singled out to answer a question because the student's mistakes are generally not forgotten by the group. For example, if an individual is quick to answer and is wrong, the entire group laughs about the mistake. Conversely, when remarks are made concerning an individual without group participation, revenge may be sought in a passive-aggressive manner. In AI/AN societies transgressions are usually not forgotten from tribe to tribe. The Pueblos remember the Navajos raided their feasts and stole food. Today conflict still occurs as some may not want other tribes to receive care at their facility.

This concept is also true in the classroom. The instructor who allows adequate time for observation has a greater chance of success and gains respect with AI/AN students. Improved success is achieved if the AI/ANs are allowed to observe the task several times before being asked to demonstrate it. Their first effort at completing the task in front of a group should occur without error. This is especially true when delivering care to their people. Mistakes that have been made are discussed in the community and mistrust can develop. Alpers (2018) conducted a qualitative study on distrust and patients in the intercultural healthcare and found that what is important is that in order to create trust, health care providers must require cultural competency.

Issues of superior-subordinate roles exist and are related to age. Younger supervisors may not be respected because they are perceived as not possessing the life experiences necessary to lead. In like situations, major decisions are made by the group with the assistance of the group leader, who is generally the senior woman. Thus, a young male manager on the reservation may face resistance when attempting to direct a work group composed of older women.

Aside from Veterans Administration, the IHS is an organization allowed to initiate preference hiring practices; it is required to hire an AI/AN when possible. This law is referred to as the *Indian Preference Law*. Leadership problems have been identified on recruitment and retention surveys of staff members leaving the IHS. This creates a problem with retaining staff, and there are not enough AI/ANs to care for their population. Examination of Emotional intelligence (EI) in the European American population has been identified as a positive trait for a leader. Traits associated with EI in supervisors demonstrate low EI scores. There was a direct correlation on the Emotional Quotient (EQ) Map with staff's perception of the individual leadership abilities (Hodgins 2004). The higher the score, the more the staff viewed the supervisor as having positive leadership traits on a Health Resources Management Index (HRMI). The research opened many more questions than it answered. Some of the reasons expressed to the author were that boarding schools negatively impacted their lives. More research is needed to determine the causal relationships between the scores on the EQ Map and whether the relationship is statistically significant. One must allow extra time for a verbal response from an AI/AN because English may not be their primary language. When translating from the native Indian language into English, adjectives and adverbs sometimes follow the noun or verb, making it appear that the person is speaking backward.

Many AI/AN students have difficulty with examinations, even if they are fluent in their language. Although the individual may have the knowledge necessary to complete an examination, the translation of test questions to select an answer is very difficult. When tested verbally on the same material, students pass the examination (Hodgins 2004). To the extent possible, examinations that consist of return demonstration and that do not use abstract terminology are preferred. Instructors, who demonstrate caring, respect, and holistic attitudes, will be very effective in assisting their AI/AN students to be successful.

6.6 Biocultural Ecology

6.6.1 Skin Color and Other Biological Variations

Skin color among AI/ANs varies from light to very dark brown, depending on the tribe. To assess for oxygenation in darker-skinned people, health-care professionals must examine the client's mucous membranes and nailbeds for capillary refill. Anemia is detected by examining the mucous membranes for pallor and the skin for a grayish hue. To assess for jaundice, it is necessary to examine the sclera rather than to rely on skin hue. Newborns and infants commonly have Mongolian spots on the sacral area. Health-care professionals unfamiliar with this trait may mistake these spots for bruises and suspect child abuse.

Each of the American Indian tribes has varying degrees of Asian traits. In facial appearance, the Athabascan tribes such as the Navajo appear Asian, with epithelial folds over the eyes. In the southwest Navajo are generally taller and thinner than other AI/ANs. In the northwest, the Seneca are tall, generally males are over 6 ft. Pueblo tribes are generally much shorter with heights of 5 ft to 5 ft 5 in. tall. Health-care professionals must be careful not to generalize these traits to draw conclusions about an AI/AN patient. Never assume that an AI/AN patient is from a particular tribe; if wrong, he or she will be offended.

6.6.2 Diseases and Health Conditions

Historically, most diseases affecting AI/ANs were infectious, such as tuberculosis, smallpox, and influenza. In the past, contact with settlers who had communicable diseases eliminated entire tribes because they had no acquired immunity. Common diseases related to living in close contact with others include upper respiratory illnesses and pneumonia (Indian Health Service 2002). Diseases of the heart, malignant neoplasm, unintentional injuries, diabetes mellitus, and cerebrovascular disease are the top five leading causes of AI/AN deaths (IHS "Fact Sheet" 2004–2006). American Indians/Alaska Natives have greater risk factor prevalence and mortality rates for cardiovascular disease and that these disparities are poorly addressed in research (Struthers et al. 2006; Brooks et al. 2019; Indian Health Service 2019; Poudel et al. 2018; Hutchinson and Shin 2014; Phillip et al. 2017; Adakai et al. 2018; Lewis and Myhra 2018).

Hantavirus-associated disease was first reported in the United States in 1993 during an outbreak in the Four Corners region (southwestern corner of Colorado, south-eastern corner of Utah, north-eastern corner of Arizona, and northwestern corner of New Mexico). Centers for Disease Control and Indian Health Service physicians determined the vector was rodents (Chapman and Khabbaz 1994). Many of these illnesses are due to the area's rodent population, consisting of prairie dogs and deer mice. Older tribal members believed that the hantavirus was caused by a plentiful crop of pinion nuts and spirits would return for the young people. This disease has actually been recorded for centuries within the Navajo and Zuni tribes but had not been documented in modern medicine.

Type 1 diabetes mellitus is almost non-existent in AI/ANs, but type 2 diabetes mellitus is one of the top three mortality disparities for AI/AN (Indian Health Service 2019). The incidence of diabetes varies among tribes but has steadily increased with poor control and dietary compliance associated with major long-term complications such as retinopathy, cardiovascular disease, and kidney failure.

Congress has delegated funds exclusively for diabetes research and education in this population because of the huge number of complications (HRSA n.d.). The National Institutes for Health and the diabetic teams on several reservations are engaged in research with high school students to determine the presence of insulin resistance. The IHS director has changed the initiatives of the IHS to case management of the chronic disease population, health promotion and disease prevention, and behavioral health.

Other conditions include a high incidence of severe combined immunodeficiency syndrome (SCIDS), an immunodeficiency syndrome, unrelated to AIDS, which results in a failure of the antibody response and cell-mediated immunity (Marcaigh et al. 2006) among the Athbascans. Factors being examined include space, time, pedigree, and immunologic status. Affected infants who survive initially are sent to tertiary-care facilities. Survivors must receive gamma globulin on a regular basis until a bone marrow transplant can be performed. Thus far, studies indicate that SCIDS is unique to this Navajo population. Lanying et al. (1998) researched SCIDS among Athabascan-speaking AI/ANs and found linkage analysis of 300 microsatellite markers in 14 affected families. It is an autosomal recessive trait and creates a deficiency in T cell and B cell immunity.

Navajo neurohepatopathy (NNH) has been researched since 1974, and in 2006 it was discovered to be the result of a mutation of the MPV17 gene (Charalampos et al. 2006). Through homozygosity mapping, the genetic defect causing NNH was determined, showing that a single mutation in exon 2 of *MVP17* accounts for the different forms of the disease. The identification of *MPV17* as the disease-causing gene will provide a definitive pre- and postnatal diagnosis of NNH and will shed light on the pathogenic mechanism of the disorder. Characteristics of this disease include poor weight gain, short stature, sexual infantilism, serious systemic infections, and liver derangement. Manifestations include weakness, hypotonia, areflexia, loss of sensation in the extremities, corneal ulcerations, acral mutilation, and painless fractures. Serial nerve biopsies show a nearly complete absence of myelinated fibers, which is different from other neuropathies that present a gradual demyelination process (Johnsen et al. 1993). Individuals, who survive have many complications, are ventilator dependent and none have survived past 24 years old.

Albinism occurs in the Navajo and Pueblo tribes. Navajos who lived in Rainbow Grand Canyon are genetically prone to blindness that develops in individuals during their late teens and early 20s. Mulligan et al. (2004) attribute a low level of genetic diversity in AI/AN groups as a possible reason for albinism.

The Zunis have an incidence of cystic fibrosis of approximately one in 333 or seven and one-half times that found for Caucasians. Cystic fibrosis has not been documented in any other tribal group. Kessler et al. (2008) found the carrier frequency in the general population to be 6.7%, a significant number when compared with the carrier frequency for all cystic fibrosis mutations in the Caucasian population of approximately 4%.

Families with satisfactory environmental conditions in their homes, which include safe water and sewerage systems, require appreciably fewer medical services and place fewer demands on the Indian Health Service (IHS) and tribal primary health-care delivery system. The Indian Sanitation Facilities Act, Pub Law 86-121 (1959), authorizes the IHS to provide essential sanitation facilities, such as safe drinking water and adequate sewerage systems to AI/AN homes and communities. The Sanitation Facilities Construction Program (SFC) is a preventative health program that yields positive benefits. A recent cost-benefit analysis indicated that for every dollar IHS spends on sanitation facilities to serve eligible existing homes; at least a 20-fold return in health benefits is achieved. The IHS SFC Program has been the primary provider of these services since 1960. Safe and adequate water supply and waste disposal facilities are lacking in approximately 9% of American Indian and Alaska Native homes, compared to 1% of homes for the U.S. general population (Public Law 86-121 Annual Report The Sanitation Facilities Construction Program of the Indian Health Service 2014; Indian Health Services "Fact Sheets" n.d.; Mitchell 2019).

Researchers and many AI/ANs believe that their traumatic history and current underfunding has made them susceptible to post-traumatic stress syndrome disorder and many other mental illnesses (Yellow Horse-Brave Heart & Debruyn 2004; Bassett et al. 2014). Methamphetamine (meth) abuse and suicide are two top concerns in Indian country. Suicide rates among American Indians and Alaska Natives (AI/ANs) are 1.6 times higher than the national average. Suicide is the second leading cause of death behind unintentional injuries for Indian youth ages 15–24 residing in IHS service areas and is 3.5 times higher than the national average (IHS "Fact Sheet" Behavioral Health, n.d.; Centers for Disease Control and Prevention Morbidity and Mortality Weekly Report 2018; O'Keefe and Reger 2017; Pollock et al. 2018; Leavitt et al. 2018). Methamphetamines (meth) use is increasing 3.1% among youth. People who are addicted to meth are a new challenge for health-care professionals and for community recovery groups. Because of the need for mental-health professionals, the IHS has instituted telepsychiatry (telehealth), which evaluation of the effectiveness is unknown and/or limited. Leading health care systems within the IHS explore the idea of research in this area (Savin et al. 2006).

6.6.3 Variations in Drug Metabolism

Research has documented adverse reactions to medication in Navajo populations (Hodgins and Still 1989). Lidocaine reactions occur in 29% of the Navajo population, compared with 11–15% of European Americans. Little research has been completed that distinguishes absorption differences of specific medications in AI/AN populations.

6.7 High-Risk Behaviors

Most AI/AN tribes exhibit high-risk behaviors related to alcohol misuse with its subsequent morbidity and mortality. Alcohol use is more prevalent than any other form of chemical misuse. Health problems related to alcoholism include motor vehicle accidents, homicides, suicides, and cirrhosis (IHS 2019; Substance Abuse and Mental Health Services Administration [SAMHSA], Park-Lee et al. 2018). Many accidents are attributed to driving while under the influence of alcohol. Although alcohol is illegal on most reservations, many purchase it off the reservation, and bootleggers make money selling it on reservations at grossly inflated prices. The northern tribes living on the Rosebud Sioux reservation have a higher alcoholism rate than that of most other AI/AN tribes. Often, this rate is attributed to an unemployment rate of 50% among these tribes. Another disparity among AI/AN that link to high risk behaviors is the issue of mental health (Payne et al. 2018). The HIS and Substance Abuse and Mental Health Services Administration (SAMHSA) under the guidance of the Department of Health and Human Services (HHS) are responsible for substance abuse prevention and services to eligible applicants (SAMHSA 2019). Payne and colleague clearly identified gaps in reducing mental health disparities, first is funding, second is practice based on principles of self-determination and finally, support diverse interventions (Payne et al. 2018). This translates to the efforts of infusion of cultural competency among health care professionals into the healing and wellness of AI/AN (Goodkind et al. 2015).

Spousal abuse is common and frequently related to alcohol use. The wife is the usual recipient of the abuse, but occasionally, the husband is abused. Emergency rooms have documented cases of husbands being beaten with baseball bats by their wives in response to their drinking. The effects of alcohol abuse are also evidenced in newborns as fetal alcohol syndrome, in teenagers as pregnancies and sexually transmitted infections, and in adults as liver failure. One research study on school achievement among AI/ANs linked weekly and daily alcohol abuse with low high school graduating rates of only 33–40%. Incarcerated Native American youths in major urban centers began drinking at an earlier age than other youths, had more binge-drinking episodes, and used more illegal drugs (Social Issues Research Center 2002). Although smoking is not as prevalent as in some other cultures, the use of smokeless tobacco has steadily increased among teenagers and those in their early 20s (Boudreau et al. 2016). Cocaine is rarely used by AI/ANs because of its high cost and limited accessibility.

6.7.1 Healthcare Practices

A number of programs among AI/ANs promote public awareness and encouragement for positive health-seeking behaviors. These programs encourage seatbelt and helmet use for those who ride bicycles or motorcycles. Unfortunately, the success of these programs has been limited. Seatbelts are required by state law as well as by tribal law-enforcement agencies. Although many adults comply with these laws, noncompliance is high among younger Indians. Therefore, it is not uncommon to see children without seatbelts on. Unfortunately, children are often permitted to ride in the back of open trucks, resulting in serious and sometimes fatal injuries when they fall off or are thrown from vehicles.

Health promotion programs have been initiated in an attempt to re-establish healthy behaviors. These programs also promote runs, relay races, and acrobics classes. However, before programs can be effective, the need for cultural competency in health care must be at the forefront (Abrishami 2018).

6.8 Nutrition

6.8.1 Meaning of Food

Food has major significance beyond nourishment in AI/AN populations. Food is offered to family and friends or may be burned to feed higher powers and those who have died. Life events, dances, healing, and religious ceremonies evolve around food. The importance of food is evident when a family sponsors a dance. Sponsors of this event are expected to feed the participants and their entire families. Food preparation takes several days. Women cook large amounts of food, which may include green chili stew, mutton, and fry bread cooked over an outdoor fireplace.

6.8.2 Common Foods and Food Rituals

Foods like cornbread, turkey, cranberry, blueberry, squash, hominy, and mush are a few of the foods European American adapted from AI/ANs. Corn is an important staple in the diet of American Indians. Rituals such as the green corn dance of the Cherokees and harvest-time rituals for the Zuni surround the use of corn. Corn pollen is used in the Blessingway ceremony and many other ceremonies by the Navajo (Nava et al. 2015). In the Midwest, tribes gathered wild rice from local areas.

In the Pacific Northwest and for Alaska Natives, a few of the dishes include dried meats such as smoked salmon strips, salted salmon, an Inuit dish of brined salmon in a heavy concentration of salt water, left for months to soak up salts. Stink fish is an Inuit dish of dried fish kept underground until ripe for later consumption. Akutaq, also called "Eskimo ice cream," is made from caribou or moose tallow and meat, berries, seal oil, and sometimes fish that whipped together with snow or water. Walrus flipper soup, an Inuit dish, is made from walrus flippers.

Feasts are common with the Pueblo and are held throughout the summertime, and one of the favorite dishes is green chili stew. Southwestern U.S. tribal groups make use of wild ginger, miners' lettuce, and juniper in various dishes. Sheep are a major source of meat for these groups and Navajo consider sheep brains a delicacy. Other delicacies include prairie dog, rabbits, and deer. Each tribe has its own version of fry bread. Access to fresh fruits and vegetables is minimal during wintertime. Among some of the tribes such as the Navajo, foods once taboo are eaten today. Chicken was once taboo; now it is an integral part of their diet. Southeastern American Indian meats came from the hunting of native game such as venison, rabbits, squirrels, opossums, and raccoons. Additionally, they ate organ meats such as liver, brains, and intestines. This tradition remains today in Northeast Oklahoma among the Cherokee because each spring they serve brains and eggs with wild onions as a treat. Additionally, hog fries are common at church gatherings or ceremonies among the Cherokee.

6.8.3 Dietary Practices for Health Promotion

The establishment of diabetic projects within the IHS has prompted teaching that integrates the optimal selection, preparation, and quantities of native foods to encourage good health habits. This is especially important for older tribal members who are less likely to change their diets but may be willing to change methods of preparation or amounts eaten. This can be important for students who attend college and need strategies for healthy choices (Keith et al. 2018; Walters et al. 2018). Herbs are used in the treatment of many illnesses to cleanse the body of ill spirits or poisons. Corn is used in many rituals for health and wellness.

6.8.4 Nutritional Deficiencies and Food Limitations

AI/AN diets may be deficient in vitamin D because many members suffer from lactose intolerance or do not drink milk. Many individuals in the Navajo tribe and other isolated tribes lack electricity for refrigeration. Therefore, they have difficulty storing fresh vegetables or milk. After some ceremonies, individuals may not eat salt or particular foods. For example, during initiation into the religion, young Hopi boys have a restricted diet. Cultural assessment is important to determine if ceremonies have been done recently and ask whether there are specific food restrictions. After certain ceremonies there may be ashes smeared on the body so one should ask if it is okay to wash the individual. Cultural ceremonies offer thanks in order to restore healing between the mind, body and spirit (Fig. 6.8) (Garrett et al. 2011; Nadeau et al. 2012; Novins et al. 2012); honouring ceremonies are special events to honour and pay tribute to AI/AN veterans who have served or continue to serve our country (Goss et al. 2019; Greenleaf et al. 2019).

Dietary changes have had a major impact on the health of many tribes. The mid-1800s saw many southwestern tribes forced onto military reservation where the diet changed to military

Fig. 6.8 Inupiat drummers at Eskimo dance

rations consisting of salt pork, bacon, beans, white flour, potatoes, coffee, sugar, tea, and lard (Teufel 1996; Americans at War 2005). A report from the early part of the twentieth century concluded that the most important item affecting the health of AI/ANs was the food supply. AI/ANs health care was responsibility of the Bureau of Indian Affairs (BIA) and under the control of the Department of the Interior. Ninety-five areas of jurisdiction had been established and 55 of these were surveyed and found the dietary staples were meat or fish, bread, beans, sugar, and coffee or tea. Milk and eggs were almost totally lacking. Distances from settlements or villages to larger off-reservation towns limited the availability of fresh food items because of shipping and storage issues. These issues remained with some tribes into the mid-twentieth century leading to the introduction of bills in congress to combat hunger among AI/ANs (Targeted News Service 2018; Chino et al. 2009).

Among the Navajo, nutritional surveys from the 1950s reported fewer than 10% of Navajo families reporting intake of fruits or green vegetables (Jackson 1986). Data from the 1991–1992 Navajo Health and Nutritional Survey (Mendlien et al. 1997) reported less than daily consumption of fruits and vegetables. The study also reported 41% of the energy and 15–46% of the macronutrients were obtained from tortillas, fry bread, home fried potatoes, coffee, tea, soft drinks, mutton, and mixed food items containing processed meats. Subsequent studies using data from the Navajo Health and Nutritional Survey, reported increasing numbers of patients with hypertension (Ballew et al. 1997; Percy et al. 1997) and higher prevalence of risk factors for coronary artery disease including obesity, hypertension, and type II diabetes (Hutchinson and Shin 2014; Poudel et al. 2018; Brooks et al. 2019; Phillip et al. 2017).

From this review of Navajo dietary history, the impact of change and the adoption of a Westernized diet resulted in an increase in heart disease and diabetes, conclusions supported by Byers and Hubbard (1997), Trude et al. (2015), and Nava et al. (2015). Even with attempts to introduce the long-ago diet (Ornelas et al. 2018), the health implications of obesity and the dietary changes that occurred over two generation continue to be serious health problems. Food insecurities and dietary patterns across both rural and urban American Indians and Navajos with children is a challenge for those in a lower educational level, single household and those who use the Women, Infant and Children (WIC) programs (Ornelas et al. 2018; Pardilla et al. 2014; Tomayko et al. 2017).

Reflective Exercise #1

A 65-year-old American Indian man is brought to the emergency room by his granddaughter after she found him lying on the floor of his house. The granddaughter tells you that he has lost 15 pounds and has not been eating good the last month. His diet consists mainly of mountain dew, chips and soup with fry bread when relatives bring him food. The granddaughter does not know the medications and medical history of her grandfather.

(a) In planning the care for this patient, what would you tell the granddaughter is the priority for the patient?
(b) Describe any cultural values that might be a barrier to delivery of care for this patient.
(c) What community resources could be involved in the care of this patient?

6.9 Pregnancy and Childbearing Practices

6.9.1 Fertility Practices and Views toward Pregnancy

The definitive goal for a woman in AI/AN societies is being a mother and rearing a healthy family. Infant mortality was very high among early AI/ANs women and many died in childbirth. Early Plains Indian women relied on herbal medicines, myths, and taboos to guide them during their pregnancies. Older tribal women instructed first-time mothers to avoid certain foods and observe these taboos because both were believed to be the cause of a difficult delivery or defect in the unborn baby. Traditional AI/ANs do not practice birth control and often do not limit family size. Historical birth rate among AI/ANs is 96% higher than the birth rate in the U.S. population. Survival rates of AI/AN children have greatly improved since the 1970s; they are at 8.9 compared with 7.2 in the general population (IHS 2002). However, trends in disparity in health status, health care, income, and incarceration continued to be well documented (Chang et al. 2015; Dickman et al. 2017; Bailey et al. 2017; Bor et al. 2017; Gaffney and McCormick 2017; Wildeman and Wang 2017).

Historical perspective informs the reader that in the past, many traditional women did not seek prenatal care because pregnancy was not considered a physical state requiring health care. During labor, women usually gathered together for a birth and the prospective mother was assisted by female relatives or other midwives with knowledge of birth customs. In the Northern tribes, it was the custom for the expectant mother to leave camp and deliver her baby alone. In general, the AI/AN cultures expect both mother and baby remain alone for a period. Using a cultural intervention tool (Box 6.1) can assist with the cultural expectations of an expecting AI/AN women.

Box 6.1 Nursing Care and Beliefs of Expectant AI/AN Women: Cultural Intervention Tool

Question	Yes or No	Comments
1. Do you want blood, urine, or other specimens returned?		
2. Do you wish to use herbs during the labor and delivery?		
3. What foods are you allowed to eat or must you avoid?		
4. During labor would you like a medicine woman or man present to perform a ceremony if necessary?		
5. Are you a member of the Native American Church?		
6. Is there anyone that you want excluded from the labor room?		
7. What position do you prefer to deliver in?		

Question	Yes or No	Comments
8. Do you prefer to deliver without family members present?		
9. After delivery, do you want a medicine man or special person to complete a ceremony or ritual?		
10. Do you want the afterbirth, meconium, or umbilical cord?		
11. Do you want to breastfeed immediately or do you want to wait until a specified time?		
12. What things would you like the nurses to do or not to do?		

Adapted from Ursula Wilson, July 1987, IHS In-service in Tuba City, Arizona

In Apache and Navajo tribes, twins are not looked on favorably and are frequently believed to be the work of a witch, in which case one of the babies must die. Recent observations reveal that this no longer happens, but sometimes the mother may have difficulty caring for two infants. For example, twins may be readmitted to the hospital for neglect and failure to thrive. In such instances, culturally sensitive counseling assists adoption. Despite a social worker's assistance, some mothers may not be able to cope with twins, and eventually, other members adopt the children.

Plains Indian women relied on natural materials for newborns; one item was dried buffalo manure. The women pounded the manure into a fine powder and used it as an absorbent in which to swaddle the baby; when the child was soiled the swaddling was replaced with a fresh supply of powder. Plains Indians parents' rituals were performed to welcome the newborn member into their community after a specified period.

6.9.2 Prescriptive, Restrictive, and Taboo Practices in the Childbearing Family

In the AI/AN society the bonds between mothers and daughters are very special but much of training of young girls fell to the grandmothers. Grandmothers taught them to sew and cook, to tan hides to make their clothing, to fashion and decorate items, instilling the tribe's moral values and traditions in their granddaughters. Instructions on proper conduct intensified as a girl approached puberty. The mother and grandmother would increase the number of tasks assigned to her to prepare for later life. Girls and boys had certain rites of passage into adulthood. Ceremonial rites of initiating adolescent girls into womanhood were usually performed by their grandmothers, with the assistance of their mothers. To provide culturally competent care, many facilities are reinstituting traditional AI/AN methods of birthing into the hospital setting. Blending Western traditional methods with AI/AN practices has greatly improved the health care of AI/AN women.

Reflective Exercise #2

A 24-year-old American Indian woman comes to the family practice clinic and is not sure if she is 2 or 3 months pregnant. This is her first perinatal appointment; she expresses concerns about not feeling movement of the baby. She states that she has a history of alcohol intake of two drinks per day and her friend told her that drinking can harm the baby. Her final concern is trusting the provider.

(a) Describe nursing strategies that can be implemented to reassure the competency of the provider.
(b) How do you explain to the woman that alcohol consumption can harm the baby?
(c) Discuss the prenatal resources available to this patient.

6.10 Death Rituals

6.10.1 Death Rituals and Expectations

Death rituals vary among tribes, maintaining and adapting them to their regional environments. Most AI/AN tribes believe that the souls of the dead pass into a spirit world and became part of the spiritual forces that influence every aspect of their lives. Early in history in Alaska, natives left their dead in isolated areas for animals. Southeastern tribes dug up their corpses, cleansed the bones, and reburied them. The Iroquois, in the seventeenth century, saved skeletons in a mass burial that included furs and ornaments for the dead spirits' use. Northwest coastal tribes put their dead in canoes fastened to poles. In western mountain areas, tribes left their dead in caves or fissures in the rocks. Plains tribes either buried their dead or left them on tree platforms or on scaffolds in special areas; no one was allowed in these areas except for certain tribal members.

Today some tribes maintain their traditional practices but use a mortuary or use the IHS morgue to prepare their dead. The Pueblo tribes prepare their own dead and only certain family members are allowed to prepare the body. Hopis bury their dead before the next setting of the sun and bury them in upright sitting positions with food and goods in the grave with the person. After the Zuni burial ceremony, the members must take off 3 days from work for a cleansing ceremony.

Rites among AI/AN tribes tended to focus on aiding the deceased in their afterlife. Some tribes left food and possessions of the dead person in or near the gravesite such as the Plains tribes. Among many tribes, mourners, especially widows, cut their hair or injured themselves in some manner to demonstrate their grief such as the Lakota Sioux did. The body must go into the afterlife as whole as possible. In some tribes, amputated limbs are given to the family for a separate burial and later the limb is buried with the body.

In some tribes, family members are reluctant to deal with the body because those who work with the dead must have a ceremony to protect them from the deceased's spirit. In the Navajo, if the person dies at home, the body must be taken out of the north side of the hogan and a ceremony conducted to cleanse the hogan or it must be abandoned. Tribal members who choose mortuary care as a profession are rare and people tend to avoid the area where the dead are prepared. Before a tribal member is buried, a ring is placed on the index finger of each hand and their shoes are placed on the wrong feet. This allows the living to recognize the individual if he or she comes back and attends ceremonial dances. Individuals who return are called "Skin Walkers." Health-care providers should not wear rings on their index fingers, as older people will not want to be around them. Individuals who speak the name of a dead person may be cursed, and bad things will happen to them as a consequence.

Advance directives, required by law, should be handled with sensitivity. Older adults are reluctant to discuss advance directives once they discover what it means. Health-care providers must be careful when discussing end-of-life issues with tribal members. Effective discussions require that the issue be discussed in the third person, as if the illness or disorder is happening to someone else. The health-care provider must never suggest that the client is dying. If a provider makes a statement such as "if you don't get medical care, you will die," this implies that the provider wishes the client dead. If the patient does die or is extremely ill, the provider might be considered a *witch*.

6.10.2 Responses to Death and Grief

The fear of the power of the dead is very real; excessive displays of emotion are not looked on favorably among some tribes. The Navajo are not generally open in their expression of grief; they often will not touch or pick up the body or prepare the body for burial. Grief among the Pueblo and Plains Tribes is expressed openly and involves much crying among extended family members. Even if the deceased is a distant relative and has not been seen in years, much grief is expressed. An all-night wake is held and the body is prepared at the morgue and buried by the family members. Health-care providers must support survivors and permit family bereavement and grieving in a culturally congruent and sensitive manner that respects the beliefs of each tribe.

6.11 Spirituality

6.11.1 Dominant Religion and Use of Prayer

AI/AN religion predominates in many tribes. Missionaries continue their efforts to convert AI/ANs to Christian religions such as the Roman Catholics, Church of Jesus Christ of Latter-Day Saints, Jehovah's Witnesses, and to a lesser extent, Evangelical groups. In the past many tribes integrated their religion with Christianity. When illnesses are severe, consultations with appropriate religious organizations are sought. Sometimes, hospital admissions are accompanied by traditional ceremonies and consultation with a pastor. Even if people are strong in their adopted beliefs, they honor their parents and families by having a traditional healing ceremony. Each director of the IHS has issued memorandums reaffirming the rights of AI/ANs to conduct ceremonies in health-care facilities. This memorandum also directs IHS health-care providers to be attuned to the total needs of patients in order to provide culturally competent and congruent care. AI/AN tribal traditional members start the day with prayer, meditation, and corn pollen. Prayers ask for harmony with nature and for health and invite blessings to help the person exist in harmony with the earth and sky. Along with certain ceremonies, prayer using smudge sticks (Fig. 6.9) helps the AI/ANs to attain fulfil-

Fig. 6.9 Smudge sticks

ment and inner peace with themselves and their environment.

6.11.2 Meaning of Life and Individual Sources of Strength

Spirituality for most AI/ANs is based on harmony with nature. The meaning of life for AI/ANs is derived from being in harmony with nature. The individual's source of strength comes from the inner self and depends on being in harmony with one's surroundings. Some tribal shamans or medicine men believe that the current proliferation of *New Age* imitations of traditional AI/AN spiritual practices are genocidal. Insensitive and opportunistic non-Indian "healers" corrupt and attempt to profit from stereotypic distortions of traditional ceremonies such as the sweat lodge. For example, such attitudes toward the sacred pipes and ceremonies are reminiscent of the entitlement and subsequent aggressive actions inherent in European American cultures.

6.11.3 Spiritual Beliefs and Healthcare Practices

Spirituality cannot be separated from the healing process in ceremonies. Illnesses, especially mental illnesses, result from not being in harmony with nature, from the spirits of evil persons such as a witch, or through violation of taboos. Healing ceremonies restore an individual's balance mentally, physically, and spiritually. The following are core concepts to traditional Indian medicine:

- AI/ANs believe in a Supreme Creator.
- Each person is a threefold being composed of mind, body, and spirit.
- All physical things, living and nonliving, are a part of the spiritual world.
- The spirit existed before it came into the body and it will exist after it leaves the body.
- Illness affects the mind and the spirit as well as the body.
- Wellness is harmony with nature and spirits.
- Natural unwellness is caused by violation of a taboo.
- Unnatural wellness is caused by witchcraft.
- Each of us is responsible for our own health.

6.12 Health Care Practices

6.12.1 Health-Seeking Beliefs and Behaviors

Traditional AI/AN beliefs influence health care decisions. For example, for many older people the germ theory is nearly impossible to comprehend. In addition, asking patients questions to make a diagnosis fosters mistrust. This approach is in conflict with the practice of traditional medicine men who tell people their problem without their having to say anything. AI/ANs must feel *bad* to seek treatment and with high blood pressure the patient does not feel bad. The patient is reluctant to take high blood pressure, diabetic, and renal failure medications. What healthcare providers assume is that the patient must feel bad when in stage V kidney failure and often they do not and delay dialysis until it is an emergency. Efforts to evaluate training among providers about pain management and opioid substance use indicate knowledge, self-efficacy and attitude are key elements to caring for this population (Katzman et al. 2016; Haozous et al. 2016).

6.12.2 Responsibility for HealthCare

Through existing treaties, the federal government assumed responsibility for the healthcare needs of AI/ANs. However, with the Indian Self-Determination and Education Assistance Act of 1975 (Public Law 93-638 1975) many tribes have contracted for this money to operate their own health-care systems. Few tribal members have traditional health insurance; recent efforts at healthcare reform have caused many tribes to fear that the government will not continue to honor its obligations under existing treaties. Inequalities within the native health-care systems have created wider gaps to access to care, economic, social, behavioural and health care outcomes (Dickman et al. 2017).

Indian Health Services has attempted to shift its focus from acute care to health promotion, disease prevention, and chronic health conditions. In 2020, DHHS revealed the nation's new 10-year goals and objectives for health promotion and disease prevention (Healthy People 2020). The Healthy People initiative is grounded in the principle that setting national objectives and monitoring progress can motivate action for improved health; in just the last decade, preliminary analyses indicate that the country has either progressed toward or met 71% of its Healthy People targets for diabetes among AI/ANs.

Since the early 1980s, an increase has occurred in wellness-promotion activities and a return to past traditions such as running for health, avoiding alcohol, and using purification ceremonies. Mental-health programs are not well funded and understaffed in IHS (Ross et al. 2015). Craig (2006) stated that culturally competent healthcare can be fostered through creating partnerships that connect diverse cultures with technology/telehealth (Fraser et al. 2017; Rakhimova et al. 2017). Technology can influence these partnerships but one must understand the health disparities (Chang et al. 2015) in the AI/AN populations where heart disease, unintentional injuries and diabetes are leading causes of death where the people of federally recognized tribes die at a substantially higher rates than any other ethnic and racial groups in the United States (Brooks et al. 2019; Indian Health Service 2019). This results in high percentage rates for chronic liver disease and cirrhosis (368% higher), diabetes (177% higher), unintentional injuries (138%), assault/homicide (82% higher), and intentional self-harm-suicide (65% higher) (Ross et al. 2015).

AI/ANs healthcare focus is curative and based on promoting harmony with Mother Earth. Physicians are oriented to traditional healing practices but if patients perceive reluctance to accept these practices, they do not reveal their use. This is especially true among older people who seek hospital or clinic treatments only when their conditions become life-threatening. Younger generations seek treatment sooner and use the healthcare system more readily than do older people. However, if their parents are traditional, they may combine native traditional medicine with Western medicine.

6.12.3 Folk and Traditional Practices

Tribal traditional practices involve wellness and a state of harmony. Illnesses are believed to be a state of disharmony. Medicine men, diagnosticians, crystal gazers, and shamans tell them how to restore harmony. The medicine man is expected to diagnose the problem and prescribe necessary treatments for regaining health. In Western health care, the practitioner asks the client what he or she thinks is wrong and prescribes a treatment. This practice is sometimes interpreted by AI/ANs as a lack of knowledge by the European American practitioner. The Joint Commission (2019) (https://www.jointcommission.org/) requires that each time the nurse identify the patient by asking his or her name and date of birth. In clinical nursing practice where a student often asks the patient his/her name and date of birth can create suspicious especially in a task such as medication administration by the student nurse. In a clinical practice scenario where the student left the room and the instructor asked the patient if she had any concerns. The patient voiced that the student was not competent because he had to ask her name all the time and was asking her if she knew her medications. The instructor explained the process for identification and patient family teaching on medications. This intervention relieved the patient and she was much more receptive to the student. Health care professionals need to communicate clearly the meaning of tools such as the pain scale to patients within the Native American culture; this scale can be difficult to interpret in elderly patients.

6.12.4 Folk and Traditional Practices Barriers to Health Care

The U.S. Commission on Civil Rights (2004) examined the efforts of the Department of Health and Human Services and concluded after 35 years of equal access to health care was not accorded the same federal protections as equal opportunity in housing, education, and employment. The Commission reported that discrimination against minority populations manifests itself in a variety of ways, including

- Differential treatment of health services
- Inability to access health services because of lack of financial resources
- Culturally incompetent providers, language barriers, and the unavailability of services
- Exclusion from health-related research

On some reservations, hospitals are some distance away and tribal members must find rides to receive care. Many families do not have adequate transportation and must wait for others to transport them to their appointments. Dependency on community health workers for home visits and/or this transportation is unreliable (Gampa et al. 2017). The challenge can also be in which Urban facilities provide care to the AI/AN members but funding and care is limited to Urban Indians.

Immunizations may be missed because parents do not have transportation. Close attention should be paid to the immunization status of patients on arrival to the emergency department or clinic. If the client is not current with immunizations, scheduling an appointment may be a waste of time because they may not be able to return until a ride is found. Healthcare providers might have better success by taking the time to administer the immunization on the spot or by making a referral to the public health nursing office.

Reflective Exercise #3

An elderly 90-year-old American Indian woman who lives alone is brought to the community clinic for a follow up appointment related to her hypertension and diabetes. The patient request to have an AI/AN provider take care of her. At the time, there is not a Native American provider available; the community health nursing department is asked to visit the patient along with the provider. The elderly woman converses in her native language and wishes to have her family in the room.

(a) How would you explain to the elderly woman about how the care can be provided to her?
(b) After seeking assistance from the community health nursing department, what would be the priority for this elderly patient?
(c) How would you accommodate the language conversation and the patient's wishes to have her family in the room?

6.12.5 Cultural Responses to Health and Illness

AI/ANs receiving care within the context of Western medicine are concerned about obtaining adequate pain control. Frequently, pain control is ineffective because the intensity of their pain is not obvious to the healthcare provider because patients do not request pain medication. Older adult AI/AN patients view pain as something that is to be endured and may not ask for analgesics or may not understand that pain medication is available. At other times, herbal medicines are preferred and used without the knowledge of the healthcare provider. Not sharing the use of herbal medicine is a carryover from times when individuals were not allowed to practice their native medicine.

Mental illness is perceived as resulting from witches or *witching* (placing a curse) on a person. In these instances, a healer who deals with dreams or a crystal gazer is consulted. Individuals may wear turquoise or other items such as a medicine bag to ward off evil. However, a person who wears too much turquoise is sometimes believed to be an evil person and someone to avoid. In some tribes, mental illness may mean that the affected person has special powers.

The concept of rehabilitation is relatively new to AI/ANs because in years past, they did not survive to old age to which chronic diseases became an issue. Because life expectancy is increasing, additional stress is placed on families as they are expected to care for older relatives. Many families do not have the resources to assume this responsibility. Home health care is occasionally available, but this is a recent development that tribes are just beginning to accept. Federal public health nursing is also available to assist with some home care. Physical or mental handicaps are not seen as different; rather, the limitation is accepted and a role is found for them within the society.

Cultural perceptions of the sick role for AI/ANs are based on the ideal of maintaining harmony with nature and with others. Ill people have violated a taboo and created disharmony or have had a curse placed on them. In either case, support of the sick role is not generally accepted, but rather, support is directed at assisting the person with regaining harmony. Older people frequently work even when seriously ill and often must be encouraged to rest.

6.12.6 Blood Transfusions and Organ Donation

Autopsy and organ donation are becoming a little more accepted among traditional AI/ANs. The concepts of organ transplant and organ donation may result in a major cultural dilemma. For example, in one case, a woman needing a kidney transplant consulted a medicine man, who advised against having the transplant performed. She elected to have the transplant done against the medicine man's advice, which created a cultural dilemma for her family. Increasing use of Western medical practices on reservations must be accompanied by attempting to incorporate culturally congruent traditional care into Western practices to improve patient care outcomes for AI/AN people.

6.13 Health-Care Providers Practitioners

6.13.1 Traditional Versus Biomedical Providers Practitioners

AI/AN healers are divided primarily into three categories: those working with the power of good, the power of evil, or both. Generally, medicine people are from specific clans promote activities that encourage self-discipline and self-control and involve acute body awareness. Within these three categories are several types of practitioners. Some are endowed with supernatural powers, whereas others have knowledge of herbs and specific manipulations to "suck" out the evil spirits. The first group is people who can use their power only for good, transform themselves into other forms of life, *shapeshift,* and maintain cultural integration in times of stress. The second group can use their powers for both evil and good and are expected to do evil against someone's enemies. People in this group know witchcraft, poisons, and ceremonies designed to afflict the enemy. The third group is the diviner diagnostician, such as a *crystal gazer* or a *hand trembler* who can see what caused the problem but not implement a treatment. Hand tremblers practice trembling over the sick person to determine the cause of an illness. A fourth group are the specialist medicine people and they treat the disease after it has been diagnosed and may specialize in the use of herbs, massage, or midwifery. A fifth group cares for the soul and sends guardian spirits to restore a lost soul. A sixth group, considered to be the most special, are singers, who cure through the power of their song. These healers are involved in the laying on of hands and usually remove objects or draw disease-causing objects from the body while singing.

Healthcare providers must be careful not to open medicine bags or remove them from the patient. These objects contribute to patients' mental well-being, and their removal creates undue stress. Treatment regimens prescribed by a medicine man not only cure the body but also restore the mind. An example of a health-maintenance practice among the Navajo is the Blessingway ceremony, in which prayers and songs are offered. Individuals living off reservations frequently return to participate in this ceremony, which returns them to harmony and restores a sense of well-being.

Acceptance of Western medicine is variable with a blending of traditional healthcare beliefs. Experienced IHS providers understand the concepts of holistic health for AI/ANs, and behavioral health specialists are beginning to make referrals to the medicine man. Few physicians possess this level of cultural experience and there are very few AI/AN physicians. This is also true of the nursing professionals. Most registered nurses in the IHS are not AI/ANs, and careers in the healthcare field often go against traditional beliefs. Male health-care providers are generally limited in the care provided to women, especially during their menses or delivery unit. Older women are generally modest and wear several layers of slips but not necessarily underpants.

6.13.2 Status of Health-Care Providers

If an AI/AN becomes a physician the conclusion is drawn that the physician must not be traditional. The factors that influence acceptance of AI/AN health-care providers have not been adequately researched. Lack of respect for AI/AN's beliefs by some Western health-care providers has contributed to mistrust. Many health concerns of AI/ANs can be treated by both traditional and Western healers in a culturally competent manner, when these practitioners are willing to work together and respect each other's differences. Each AI/AN culture is unique; but many share similar views regarding cosmology, medicine, and family organization. Health-care providers can gain respect through sincerity and a desire to learn about the cultures and traditions of the people they serve, rather than have preconceived ideas about cultural aspects (Cunningham et al. 2014; Katigbak et al. 2015; Lewis and Prunuske 2017). Trust, engagement, acknowledgment, awareness, and a genuine interest of cultural competency is paramount for the care of AI/AN people who value a nation of individuals, community, and systems. Health care professionals who listen attentively can learn to sew like Eskimo children did many years ago (Fig. 6.10).

Fig. 6.10 Eskimo children learning to sew

References

Abrishami D (2018) The need for cultural competency in health care. Radiol Technol 89(5):441–448

Adakai M, Sandoval-Rosario M, Xu F, Aseret-Manygoats T, Allison M, Greenlund KJ, Barbour KE (2018) Health disparities among American Indians/Alaska natives—Arizona, 2017. Morb Mortal Wkly Rep 67(47):1314–1318

Alpers L-M (2018) Distrust and patients in intercultural healthcare: a qualitative interview study. Nurs Ethics 25(3):313–323. https://doi.org/10.1177/0969733016652449

American Indian College Fund (2019). https://standwithnativestudents.org/

American Indian Religious Freedom Act of 1978 (AIRFA) (1978) 42 U.S.C. § 1996

Anderson RW (2016) Native American reservation constitutions. Const Polit Econ 27:377–398. https://doi.org/10.1007/s10602-016-9225-7

Anguksuar L (1997) A post-colonial perspective on Western misconceptions of the cosmos and the restoration of indigenous taxonomies. In: Jacobs SE, Thomas W, Lang S (eds) Two spirit people: native American gender identity, sexuality, spirituality. University of Illinois Press, Chicago

Bailey ZD, Krieger N, Agenor M, Graves J, Linos N, Bassett MT (2017) America: equity and equality in health 3 structural racism and health inequities in the USA: evidence and interventions. Lancet 389:1453–1463

Ballew C, White LL, Strauss KF, Benson LJ, Mendlein JM, Mokdad AH (1997) Intake of nutrients and food sources of nutrients among the Navajo: finding from the Navajo Health and Nutritional Survey. J Nutr 127:2085S–2093S

Balsam K, Huang B, Fieland K, Simoni J, Walters K (2004) Culture, trauma and wellness: a comparison of heterosexual and lesbian, gay, bisexual, and two-spirit native Americans. Cult Divers Ethn Minor Psychol 10(3):287–301

Bassett D, Buchwald D, Manson S (2014) Postraumatic stress disorder and symptoms among American Indians and Alaska Natives: a review of the literature. Sociol Psychiatr Psychiatr Epidemiol 49(3):417–433. https://doi.org/10.1007/s00127-013-0759-y

Blanchard JW, Tallbull G, Wolpert C, Powell J, Foster MW, Royal C (2017) Barriers and strategies related to qualitative research on genetic ancestry testing in indigenous communities. J Empir Res Hum Res Ethics 12(3):169–179. https://doi.org/10.1177/1556264617704542

Blood Quantum (2019) American Indian Enterprise & Business Council-CDC, NGO, 501C3. http://www.aiebc.org

Bor J, Cohen GH, Galea S (2017) America: equity and equality in health 5 population health in an era of rising income inequality: USA, 1980–2015. Lancet 389:1475–1490

Boudreau G, Hernandez C, Hoffer D, Starlight Preuss K, Tibbetts-Barto L, Toves Villaluz N, Scott S (2016) Why the world will never be tobacco-free: reframing "tobacco control" into a traditional tobacco movement. Am J Public Health 106(7):1188–1195. https://doi.org/10.2105/AJPH.2016.303125

Brooks JL, Berry DC, Currin EG, Ledford A, Knafl GJ, Fredrickson BL, Beeber LS, HAPPI Community Partnership Committee, Peden DB, Corbie-Smith GM (2019) A community-engaged approach to investigate cardiovascular-associated inflammation among American Indian women: a research protocol. Res Nurs Health, 42(3), 165–175. https://doi.org/10.1002/nur.21944

Bureau of Indian Affairs (BIA) (2019). https://www.bia.gov/bia

Burke S, Kisilevsky B, Maloney R (1989) Time orientations of Indian mothers and white nurses. Can J Nurs Res 21(4):14–20

Byers T, Hubbard J (1997) The Navajo health and nutrition survey: research that can make a difference. J Nutr 127:2075S–2077S

Carlson LA, Roberts MA (2006) Indian lands, "Squatterism," and slavery: economic interests and the passage of the Indian removal act of 1830. Explor Econ Hist 43:486–504. https://doi.org/10.1016/j.eeh.2005.06.003

Carroll M, Cullen T, Ferguson S, Hogge N, Horton M, Kokesh J (2011) Innovation in Indian healthcare: using health information technology to achieve health equity for American Indians and Alaska native populations. Perspect Health Inf Manage 8:1d

Centers for Disease Control and Prevention Morbidity and Mortality Weekly Report (2018, March 2) Suicides among American Indian/Alaska natives National Violet Death Reporting System, 18 States, 2003–2014

Chang M-H, Moonesinghe R, Athar HM, Truman BI (2015) Trends in disparity by sex and race/ethnicity for the leading causes of death in the United States—1999-2010. J Publ Health Manage Pract 22(1 Suppl):S13–S24. https://doi.org/10.1097/PHH.0000000000000267

Chapman L, Khabbaz R (1994) Etiology and epidemiology of the four corners hantavirus outbreak. Infect Agents Dis 3(5):233–244

Charalampos L, Karadimas C, Vu T, Holve S, Chronopoulou P, Quinzii C, Johnsen S, Kurth J, Eggers E, Palenzuela L, Tanji K, Bonilla E, DeVivo D, DiMauro S, Hirano M (2006) Navajo neurohepatopathy is caused by a mutation in the MPV17 gene. Am J Hum Genet 79(3):544–548

Cherokee Tribal Policy (n.d.). http://www.cherokee.org/Services/146/Page/Default.aspx

Chino M, Haff DR, Dodge Francis C (2009) Patterns of commodity food use among American Indians. Pimatisiwin J Aborig Ind Commun Health 7(2):279–289. https//digitalscholarship.unlv.edu/env_occ_health_fac_articles/51

Chisholm JS (1983) Navajo infancy: an ethnological study of child development. Aldine, New York

Craig G (2006) Native American elders health care series. Comput Inf Nurs 24(3):133–135

Cuellar NG (2015) Providing culturally congruent health care to older adults. J Transcult Nurs 26(2):109. https://doi.org/10.1177/1043659615569540

Cunningham BA, Marsteller JA, Romano MJ, Carson KA, Noronha GJ, McGuire MJ, Yu A, Cooper LA (2014) Perceptions of health system orientation: quality, patient centeredness, and cultural competency. Med Care Res Rev 71(6):559–579. https://doi.org/10.1177/1077558714557891

Dennis MK, Brewer JP (2017) Rearing generations: Lakota Grandparents commitment to family and community. J Across Cult Gerontol 32:95–113. https://doi.org/10.1007/s10823-016-9299-8

Department of Interior Bureau of Indian Affairs Federal Register (2019) Indian entitles recognized by and eligible to receive services from the United States Bureau of Indian Affairs. 84(22). February 1, 2019/Notices

DeVoe J, Darling-Churchhill K, Snyder T (2008) Status and trends in the education of American Indians and Alaska Natives, 2008. U.S. Department of Education

Dickman SL, Himmelstein DU, Woolhandler S (2017) America: equity and equality in health 1 inequality and the health-care system in the USA. Lancet 389:1431–1441

Fedullo M (1992) Light of the feather: pathways through contemporary Indian America. William Morrow, New York

Fraser S, Mackean T, Grant J, Hunter K, Towers K, Ivers R (2017) Use of telehealth for health care of Indigenous peoples with chronic conditions: a systematic review. Rural Remote Health 17:4205. http://www.rrh.org.au

Gaffney A, McCormick D (2017) America: equity and equality in health 2 the affordable care act: implications for health-care equity. Lancet 389:1442–1452

Gampa V, Smith C, Muskett O, King C, Sehn H, Malone J, Curley C, Brown C, Begay M-G, Shin S, Nelson A (2017) Cultural elements underlying the community health representative-client relationship on Navajo Nation. BMC Health Serv Res 17:19. https://doi.org/10.1186/s12913-016-1956-7

GAO-13-121 Regional Alaska Native Corporations (2012) Status 40 years after establishment, and future considerations. United States Government Accountability Office

Garrett MT, Torres-Rivera E, Brubaker M, Portman T, Brotherton D, West-Olatunji C, Conwill W, Grayshield L (2011) Crying for a vision: the native American sweat lodge ceremony as therapeutic intervention. J Counsel Dev 89:318–325

Garrow CE (2015) Tribal criminal law and procedure, 2nd edn. Rowman & Littlefield, London

Getches D (1977) The influence of federal and state legislation upon education of rural Alaska Natives. University of Alaska-Fairbanks, Center for Northern Educational Research. http://www.alaskool.org/

Gone JP, Blumstein KP, Dominic D, Fox N, Jacobs J, Lynn RS, Martinez M, Tuomi A (2017) Teaching tradition: diverse perspectives on the pilot urban Amercian Indian traditional spirituality program. Am J Community Psychol 59(3–4):382–389. https://doi.org/10.1002/ajcp.12144

Goodkind JR, Gorman B, Hess JM, Parker DP, Hough RL (2015) Reconsidering culturally competent approaches to American Indian healing and well-being. Qual Health Res 25(4):486–499. https://doi.org/10.1177/1049732314551056

Goss CW, Richardson WJ, Shore JH (2019) Outcomes and lessons learned from the tribal veterans representative program: a model for system engagement. J Community Health 44:1076–1085. https://doi.org/10.1007/s10900-019-00683-0

Graves K, Smith S, Easley C, Kanaqlak G (2004) Conferences of Alaska native elders: our view of dignified aging. Department of Health and Human Services, Washington, DC

Greenleaf AT, Roessger KM, Williams JM, Motsenbocker J (2019) Effects of a rite of passage ceremony on veterans' well-being. J Couns Dev 97:171–181. https://doi.org/10.1002/jcad.12248

Greensky C, Stapleton MA, Walsh K, Gibbs L, Abrahason J, Finnie DM, Hathaway J, Vickers-Douglas KS, Cronin JB, Townsend CO, Hooten WM (2014) A qualitative study of traditional healing practices among American Indians with chronic pain. Pain Med 15:1795–1802

Halabi S (2019) The role of provinces, states, and territories in shaping federal policy for indigenous peoples health. Am Rev Can Stud 49(2):231–246. https://doi.org/10.1080/02722011.2019.1613797

Haozous EA, Doorenbos AZ, Stoner S (2016) Pain management experiences and the acceptability of cognitive behavioral strategies among American Indians and Alaska natives. J Transcult Nurs 27(3):233–240. https://doi.org/10.1177/1043659614558454

Healthy People (2020). https://www.healthypeople.gov/2020/

Hendrix L (1999) Health and Health Care of American Indian and Alaska Natives Elders. Ethnogeriatric Curriclum Module. http://www.stanford.edu/group/ethnoger/americanindian.html

Hodgins O (2004) Emotional intelligence: implications in native Americans. Unpublished doctoral dissertation, Kennedy Western University, Thousand Oaks, CA

Hodgins D, Still O (1989) Lidocaine reactions in the American Indian population: quality assurance study. Unpublished research, Tuba City Indian Health Center, Tuba City, AZ

HRSA Office of Minority Health and Bureau of Primary Health Care (n.d.) American Indians and Alaska Natives and diabetes. The Providers Guide to Quality and Culture. http://erc.msh.org/quality&culture

Hutchinson R, Shin S (2014) Systematic review of health disparities for cardiovasuclar diseases and associated factors among American Indian and Alaska native populations. PLoS One 9(1):e90973–e80973. https://doi.org/10.1371/journal.pone.0080973

Indian Affairs Manual (2019). https://www.bia.gov/policy-forms/manual

Indian Ancestry and How to enroll or register in a Federally Recognized Tribe (2004). http://www.native-american-online.org/tribal-enrollment.html
Indian Health Service (2002) Mortality disparities rates. http://info.ihs.gov/Files/DisparitiesFacts-Jan2006.pdf
Indian Health Service (2019). https://www.ihs.gov
Indian Health Service "Fact Sheets" (n.d.). https://www.ihs.gov/newsroom/factsheets/
Indian Removal Act: 1830 (n.d.) Library of Congress. http://www.loc.gov/rr/program/bib/ourdocs/Indian.html
Jackson M (1986) Nutrition in American Indian health: past, present and future. J Am Diet Assoc 86:1561–1565
Johnsen SD, Johnson P, Stein SR (1993) Familial sensory autonomic neuropathy with arthropathy in Navajo children. Neurology 43:1120–1125
Johnson B (2004) American genesis: the cosmological beliefs of the Indians. Institute for Creation Research. http://www.icr.org. Accessed 22 Mar 2011
Katigbak C, Van Devanter N, Islam N, Trinh-Shevrin C (2015) Partners in Health: a conceptual framework for the role of community health workers in facilitating patients' adoption of healthy behaviors. Am J Public Health 105(5):872–880. https://doi.org/10.2105/AJPH.2014.302411
Katzman JG, Fore C, Bhatt S, Greenberg N, Salvador J, Comerci GC, Camarata C, Marr L, Monette R, Arora S, Bradford A, Taylor D, Dillow J, Karol S (2016) Evaluation of American Indian Health Service training in pain management and opioid substance use disorder. Publ Health Pract 106(7):1427–1429. https://doi.org/10.2105/AJPH.2016.303193
Keith JF, Stastny S, Brunt A, Agnew W (2018) Barriers and strategies for healthy food choices among American Indian tribal college students: a qualitative analysis. J Acad Nutr Diet 118(6):1017–1026. https://doi.org/10.1016/j.jand.2017.08.003
Kessler D, Moehlenkamp C, Kaplan G (2008) Determination of cystic fibrosis carrier frequency for Zuni Native Americans. Clin Genet 49(2):95–97
Lanying L, Dennis D, Hu D, Hayward A, Gahagan S, Pabst H, Cowan M (1998) The gene for severe combined immunodeficiency disease in Athabascan-speaking Native Americans is located on chromosome 10p. Am J Hum Genet 62:136–144
Leavitt RA, Ertl A, Sheats K, Petrosky E, Ivey-Stephenson A, Fowler KA (2018) Suicides among American Indian/Alaska natives—National Violent Death Reporting System, 18 states, 2003–2014. Centers Dis Control Prev Morb Mortal Wkly Rep 67(8):237–242
Lewis ME, Myhra LL (2018) Integrated care with indigenous populations: considering the role of health care systems in health disparities. J Health Care Poor Underserved 29:1083–1107. https://doi.org/10.1007/s10823-018-9350-z
Lewis M, Prunuske A (2017) The development of an indigenous health curriculum for medical students. Acad Med 92(5):641–648. https://doi.org/10.1097/ACM.0000000000001482
Lewis JP, Boyd K, Allen J (2018) "We raise our grandchildren as our own:" Alaska native grandparents raising grandchildren in Southwest Alaska. J Cross Cult Gerontol 33:265–286. https://doi.org/10.1007/s10823-018-9350-z
Marcaigh ASO, Baker M, Savik K (2006) Severe combined immunodeficiency. J Obstet Gynecol Neonatal Nurs 27(7):703–709
McFadden S (ed) (1994) The little book of native American wisdom. Element, Rockport
Mendlien J, Freedman D, Peter D, Allen B, Percy C, Ballew C (1997) Risk factors for coronary heart disease among the Navajo Indians: findings from the Navajo Health and Nutritional Survey. J Nutr 127:2009–2105
Mihesuah DA (2015) Sustenance and health among the five tribes in Indian territory, postremoval to statehood. Ethnohistory 62:263–284. https://doi.org/10.1215/00141801-2854317
Mitchell FM (2019) Water (in)security and American Indian Health: social and environmental justice implications for policy, practice, and research. The Royal Society for Public Health 176:98–105. https://doi.org/10.1016/j.puhe.2018.10.010
Morgan H (2010) Teaching native American students: what every teacher should know. Educ Dig Essent Read Condens Quick Rev 75(6):44–47
Mulligan C, Hunley K, Cole S, Long J (2004) Populations and genetics, history, and health patterns in Native Americans. Annu Rev Genomics Hum Genet 5:295–315
Nadeau M, Blake N, Poupart J, Rhodes K, Forster JL (2012) Circles of tobacco wisdom learning about traditional and commercial tobacco with native elders. Am J Prev Med 43(5S3):S222–S228. https://doi.org/10.1016/j.amepre.2012.08.003
National Archives and Records Administration (2019). U.S. Department of the Interior. A guide to tracing American Indian & Alaska Native Ancestry. http://www.archives.gov/research/native-americans/dawes/intro.html
Nava LT, Zambrano JM, Arviso KP, Brochetti D, Becker KL (2015) Nutrition-based interventions to address metabolic syndrome in the Navajo: a systematic review. J Clin Nurs 24:3024–3045
Navajo Nation Human Rights Commission 2014 Shadow report, regarding the United States of America report submitted by states parties under article 9 of the international convention on the elimination of all forms of racial discrimination. Navajo Nation, AZ
Navajo Tribal Policy (n.d.). http://www.cherokee.org/Services/146/Page/Default.aspx
Novins DK, Boyd ML, Brotherton DT, Fickenscher A, Moore L, Spicer P (2012) Walking on: celebrating the journeys of native American adolescents with substance use problems on the winding road to healing. J Psychoactive Drugs 44(2):153–159. https://doi.org/10.1080/02791072.2012.684628
O'Keefe VM, Reger GM (2017) Suicide among American Indian/Alaska native military Servie members and

veterans. Psychol Serv 14(3):289–294. https://doi.org/10.1037/ser0000117
Office of Indian Education (2006) Academic Achievement: Indian Education. http://www.ade.state.az.us/asd/indianed/
Office of Program Planning and Development Navajo Area Indian Health Service: Area Profile (1988) Navajo Area Indian Health Service
Ornelas IJ, Osterbauer K, Woo L, Bishop SK, Deschenie D, Beresford SAA, Lombard K (2018) Gardening for health: patterns of gardening and fruit and vegetable consumption among the Navajo. J Community Health 43:1053–1060. https://doi.org/10.1007/s10900-018-0521-1
Pacheco JA, Pacheco CM, Lewis C, Williams C, Barnes C, Rosenwasser L, Choi WS, Daley C (2015) Ensuring healthy American Indian generations for tomorrow through safe and healthy indoor environments. Int J Environ Res Public Health 12:2810–2822. https://doi.org/10.3390/ijerph120302810
Pardilla X, Prasad X, Suratkar X, Gittelsohn X (2014) High levels of household food insecurity on the Navajo Nation. Public Health Nutr 17(1):58–65. https://doi.org/10.1017/S1368980012005630
Park-Lee E, Lipari RN, Bose J, Hughes A, Greenway K, Glasheen C, Herman-Stahl M, Penne M, Pemberton M, Cajka J (2018). Substance use and mental health issues among U.S.-Born American Indians or Alaska natives residing on and off tribal lands. Substance Abuse and Mental Health Services Administration (SAMHSA) CBHSQ Data Review, pp 1–24
Payne HE, Steele M, Bingham JL, Sloan CD (2018) Identifying and reducing disparities in mental health outcomes among American Indians and Alaskan natives using public health, mental healthcare and legal perspectives. Adm Policy Ment Health Ment Health Serv Res 45:5–14. https://doi.org/10.1007/s10488-016-0777-7
Percy C, Freedman D, Gilbert T, White L, Ballew C, Mokdad A (1997) Prevalence of hypertension among Navajo Indians: findings from the Navajo Health and Nutritional Survey. J Nutr 127:2111–2119
Phillip J, Ryman TK, Hopkins SE, O'Brien DM, Bersamin A, Pomeroy J, Thummel KE, Austin MA, Boyer BB, Dombrowski K (2017) Bi-cultural dynamics for risk and protective factors for cardiometabolic health in an Alaska Native (Yup'ik) population. PLoS One 12(11):1–24. https://doi.org/10.1371/journal.pone.0183451
Pollock NJ, Naicker K, Loro A, Mulay S, Colman I (2018) Global incidence of suicide among indigenous peoples: a systematic review. BMC Med 16(1):145–152. https://doi.org/10.1186/s12916-018-1115-6
Poudel A, Yi Zhou J, Story D, Li L (2018) Diabetes and associated cardiovascular complications in American Indians/Alaskan Natives: a review of risks and prevention strategies. J Diab Res 2018:1–8. https://doi.org/10.1155/2018/2742565
Public Law 67-85 The Snyder Act of 1921 (n.d.). https://www.bia.gov/
Public Law 86-121 Annual Report (2014) The Sanitation Facilities Construction Program of the Indian Health Service. http://www.dsfc.ihs.gov
Public Law 93-638 (1975) Indian Self-Determination and Education Assistance Act of 1975. https://www.bia.gov/
Rakhimova AE, Yashina ME, Mukhamadiarova AF, Sharipova AV (2017) The development of sociocultural competence with the help of computer technology. Interchange 48:55–70. https://doi.org/10.1007/s10780-016-9279-5
Remini R (1998) The reform begins. Andrew Jackson: History Book Club. HarperCollins, New York
Resch JP (2005) Americans at war. Gale Publisher, Detroit
Reyhner J (n.d.) American Indian/Alaska Native education: an overview. http://jan.ucc.nau.edu/jar/AIE/Ind_Ed.html
Ross RE, Garfield LD, Brown DS, Raghavan R (2015) The affordable care act and implications for health care services for Ameican Indian and Alaska Native Individuals. J Health Care Poor Underserved 26(4):1081–1088. https://doi.org/10.1353/hpu.2015.0129
Sando JS (1976) Pueblo Indians. Indian Historian Press, San Francisco
Savin D, Garry M, Zuccaro P, Novins D (2006) Telepsychiatry for treating rural American Indian youth. J Am Acad Child Adolesc Psychiatry 45(4):484–488
Schmidt RW (2011) American Indian identity and blood quantum in the 21st century: a critical review. J Anthropol 2011:1–9. https://doi.org/10.1155/2011/549521
Simi D, Matusitz J (2016) Native American students in U.S. higher education: a look from attachment theory. Interchange 47:91–108. https://doi.org/10.1007/s10780-015-9256-4
Social Issues Research Center (2002) Social and cultural aspects of drinking. http://www.sirc.org/news/sirc_in_the_news_2002.html. Accessed 22 Mar 2011
Stanley J (1997) Personal experience of a Battalion Commander and Brigade Signal Officer. In: 105th Field Signal Battalion in the Somme Offensive, September 29–October, U.S. Army, 1932
Stowe R (2017) Culturally responsive teaching in an Oglala Lakota classroom. Soc Stud 108(6):242–248. https://doi.org/10.1080/00377996.2017.1360241
Struthers M, Baker M, Savik K (2006) Cardiovascular risk factors among Native American Women Inter-Tribal Heart Project participants. J Obstet Gynecol Neonatal Nurs 35(4):482–490
Sturgis A (2007) The trail of tears and Indian removal. Greenwood Press, Westport
Substance Abuse and Mental Health Services Administration (2019) About us. http://www.samhsa.gov/about-us
Targeted News Service (2018) Sen. Tina Smith introduces bill to allow tribal governments to determine how programs to combat hunger best serve Indian country. Targeted News Service. https://

advance-lexis-com.libproxy.uccs.edu/api/document?collection=news&id=urn:contentItem:5SJC-DFR1-JC11-1264-00000-00&context=1516831
Teufel N (1996) Nutrient-health associations in the historic and contemporary diets of southwest native Americans. J Nutr Environ Med 6(2):179–189
The Joint Commission (2019). https://www.jointcommission.org/
The Navajo Nation Personnel Policies Manual CAP-30-17 (2017, April 20) Resolution of the Navajo Nation Council, 23rd Navajo Nation Council—Third Year, 2017
The U.S. Equal Employment Opportunity Commission. Policy Statement on Indian Preference (1988) Number 915.027
Tomayko EJ, Mosso KL, Cronin KA, Carmichael L, Kim K, Parker T, Yaroch AL, Adams AK (2017) Household food insecurity and dietary patterns in rural and urban American Indian families with young children. BMC Public Health 17:611. https://doi.org/10.1186/s12889-017-4498-y
Trude ACB, Kharmats A, Jock B, Liu D, Lee K, Martins PA, Pardilla M, Swartz J, Gittelsohn J (2015) Patterns of food consumption are associated with obesity, self-reported diabetes and cardiovascular disease in five American Indian Communities. Ecol Food Nutr 54(5):437–454. https://doi.org/10.1080/03670244.2014.922070
U.S. Bureau of Census (2018) Census Report for 2018. http://www.census.gov
U.S. Commission on Civil Rights (2004) Native American health care disparities briefing, executive summary. Office of General Counsel, Washington, DC
United Nations Office of the High Commissioner for Human Rights (OHCHR) (2013) Fact Sheet No. 9, Rev. 2, Indigenous Peoples and the United States Human Rights System. https://www.refworld.org/docid/5289d7ac4.html
Walters K (1997) Urban lesbian and gay American Indian identity: implications for mental health service delivery. J Gay Lesbian Soc Serv 6(2):43–65
Walters KL, Johnson-Jennings M, Stroud S, Rasmus S, Charles B, John S et al (2018) Growing from our roots: strategies for developing culturally grounded health promotion interventions in American Indians, Alaska Native, and Native Hawaiian Communities. Prev Sci:1–11. https://doi.org/10.1007/s11121-018-0952-z
Wildeman C, Wang EA (2017) America: equity and equality in health 4 mass incarceration, public health, and widening inequality in the USA. Lancet 389:1464–1474
Yellow-Horse-Brave Heart M, DeBruyn L (2004) The American Indian holocaust: healing historical unresolved grief. J Natl Center Am Indian Alaska Native Prog 8(2):60–82

7 People of Amish Heritage

Christine Nelson-Tuttle

7.1 Introduction

Today's Amish live in rural areas in a band of over 30 states stretching westward from Vermont and Maine, through the most populous states of Pennsylvania, Ohio, and Indiana, and reaching as far as Montana. There are additional settlements as far south as Florida and Texas. While the majority of Canadian settlements are found in Ontario, there are Amish settlements in the Canadian provinces of Manitoba, New Brunswick, and Prince Edward Island. A few settlements can be found in Honduras, Belize, and Mexico. The **Old Order Amish**, so-called for their strict observance of traditional ways that distinguishes them from other, more progressive "plain folk," are the largest and most notable group among the Amish.

This chapter is a revision made of the original chapter written by Anna Frances Wenger † in the previous edition of the book.

C. Nelson-Tuttle (✉)
School of Nursing, D'youville College,
Buffalo, NY, USA
e-mail: tuttlec@dyc.edu

7.2 Overview, Inhabited Localities, and Topography

7.2.1 Overview

As dusk gathers on the hospital parking lot, a man first ties his horse to the hitching rack and then helps down from the carriage a matronly figure who is wrapped in a shawl as dark as his own greatcoat. On their mother's heels, a flurry of children dressed like undersized replicas of their parents turn their wide eyes toward the fluorescent-lit glass façade of the reception area, a glimmering beacon from the world of high-technology health care. Their excitement is muted by their father's soft-spoken rebuke in a language more akin to German than English, and in a hush, the Amish family crosses a cultural threshold—into the world of health-care providers.

This Amish family appears to come from another time and place. Those familiar with the health-care needs of the Amish know the profound cultural distance they have bridged in seeking professional help. Others, only marginally acquainted with Amish ways, may ask why this group dresses, acts, and talks like visitors to the North American cultural landscape of the twenty-first century. Amish are "different" by intention and by conviction. For most of the ways in which they depart from the norm for contemporary American culture, they cite a reason

© Springer Nature Switzerland AG 2021

L. D. Purnell, E. A. Fenkl (eds.), *Textbook for Transcultural Health Care: A Population Approach*,
https://doi.org/10.1007/978-3-030-51399-3_7

related to their understanding of the biblical mandate to live a life separated from a world they see as unregenerate or sinful.

As noted in the variant cultural characteristics in the introduction to cultural diversity in Chap. 2, dissimilar appearance, behavior, or both may signal deeper underlying differences in the Amish culture. Noting these differences does not, of necessity, lead to better acceptance or deeper understanding of attitudes and behaviors. Appearances can be misleading. For example, the Amish family's arrival at the hospital by horse and carriage might suggest a general taboo against modern technological conveniences. In fact, while most Amish homes are not furnished with electric and electronic laborsaving devices and appliances, many have propane generated equipment that adapts to similar needs answered by electricity. Lack of electricity and internet connections in their homes does not preclude the Amish's openness to using state-of-the-art medical technology if it is perceived as necessary to promoting their health (Granville and Gilbertson 2017).

This minority group's unique features of dress and language may disguise true motivations regarding health-seeking behaviors, which they share with the larger, or majority, culture. To enable such patients to attain their own standard of health and well-being, health-care providers need to look beyond the superficial appearance and to listen more carefully to the cues they provide.

7.2.2 Heritage and Residence

It is as important to locate the Amish topographically according to cultural and religious coordinates as well as to the geographical areas they inhabit. The hospital visit scene just portrayed could have taken place in any one of a number of towns spanning the many states in which Amish currently reside. The basic circumstances surrounding the interaction with professional caregivers and the cultural assumptions underlying it are basically similar. For the Amish, seeking help from health-care providers requires them to go outside their own people, and in so doing, cross over a significant "permeable boundary" that delimits their community in cultural-geographic terms.

Today's Amish live in rural areas in a band of over 30 states stretching westward from Vermont and Maine, through the most populous states of Pennsylvania, Ohio, and Indiana, and reaching as far as Montana. There are additional settlements as far south as Florida and Texas. While the majority of Canadian settlements are found in Ontario, there are Amish settlements in the Canadian provinces of Manitoba, New Brunswick, and Prince Edward Island (Young Center 2019). A few settlements can be found in Honduras, Belize, and Mexico. About 63% of their estimated US total population of over 335,000 is concentrated in Pennsylvania, Ohio, and Indiana (Young Center 2019). The largest increase in concentration of Amish population is in New York State. The Old Order Amish, so-called for their strict observance of traditional ways that distinguishes them from other, more progressive "plain folk," are the largest and most notable group among the Amish. As such, they constitute an ethnoreligious cultural group in modern America with roots in Reformation-era Europe.

7.2.3 Reasons for Migration and Associated Economic Factors

The Amish emerged after 1693 as a variant of one stream of the **Anabaptist** movement that originated in Switzerland in 1525 and spread to neighboring lands to the east, north, and northwest, especially along the Rhine River, to the Netherlands. The Amish embraced, among other essential Anabaptist tenets of faith, the baptism of adult believers as an outward sign of membership in a voluntary community with an inner commitment to live peaceably with all. The Amish parted ways with the larger Anabaptist group, now known as *Mennonites*, over the Amish propensity to strictly avoid community members whom they excluded from fellowship in their church (Hostetler 1993). The Amish name is derived from the surname of Jacob Ammann, a seventeenth-century Anabaptist who led the Amish division from the Anabaptists in 1693 (Hüppi 2000). Similarly, the name Mennonite is derived from the given name of Menno Simons, a former Catholic priest, who was a key leader of the Anabaptist movement in Europe.

Anabaptists were disenfranchised and deported, and their goods expropriated for their refusal to bear arms as a civic service and to accept the authority of the state church in matters of faith and practice. Their attempts at radical discipleship in a "free church," following the guidelines of the early church as set forth in the New Testament, resulted in conflict with Catholic and Protestant leaders. After experiencing severe persecution and martyrdom in Europe, the Amish and related groups emigrated to America in the 17th and 18th centuries. No Old Order Amish live in Europe today, the last survivors having been assimilated into other religious groups (Hostetler 1993). There are settlements of Beachy Amish found in Ukraine, Ireland and Belgium (Nolt 2015). As a result, the Amish, unlike many other ethnic groups in the United States, have no larger reference group in their former homeland to which their customs, language, and lifeways can be compared.

Denied the right to hold property in their homelands, the Amish sought not only religious freedom but also the opportunity to buy farmland where they could live out their beliefs in peace. In their communities, the Amish have transplanted and preserved a way of life that bears the outward dress of preindustrial European peasantry. In modern industrial America, they have persisted in general social isolation based on religious principles, a paradoxically separated life of Christian altruism. Living for others entails a caring concern for members of their in-group, a community of mutuality, but it also calls them to reach out to others in need outside their immediate Amish household of faith (Hostetler 1993).

Although the Amish value inner harmony, mutual caring, and a peaceable life in the country, it would be a mistake to see Amish society as an idyllic, pastoral folk culture, frozen in time and serenely detached from the dynamic developments all around them. Since the mid-nineteenth century, Amish communities have experienced inner conflicts and dissension as well as outside pressures to conform and modernize. Over time, the Amish have continued to adapt and change, but at their own pace, accepting innovations selectively (Johnson-Weiner 2014).

One cost of controlled, deliberate change has been the loss of some members through factional divisions over "progressive" motivations, both religious and material. The influence of revivalism led to religious reform variants, which introduced Sunday schools, missions, and worship in meetinghouses instead of homes. Others who were impatient to use modern technology, such as gasoline-powered farm machinery, telephones, electricity, electronic devices, and automobiles, also split off from the main body of the most conservative traditionalists, now called the Old Order Amish. Some variant groups were named after their factional leaders (e.g., Egli and Beachy Amish); some were called Conservative Amish Mennonites, and others The New Order Amish. Today, these progressives stand somewhere between the parent body, the Mennonites, and the Old Order Amish (technically Old Order Amish Mennonites), hereafter simply referred to as the Amish (Hostetler 1993). This latter group, the (Old Order) Amish, which has been widely researched and reported on, provides the observational basis for this present culture study.

7.2.4 Educational Status and Occupations

The controversy over schooling of Amish children is a good example of a policy issue that attracts public attention. Amish parents assume primary responsibility for child rearing, with the constant support of the extended family and the church community to reinforce their teachings. On the family farm, parents and older siblings model work roles for younger siblings. Corporate worship and community religious practices nurture and shape their faith. Learning how to live and to prepare for death is more important in the Amish tradition than acquiring special skills or knowledge through formal education or training (Hostetler 1993).

The mixed-grade, one-room schoolhouses, typical of rural America before 1945, were acceptable to the Amish because the schools were more amenable to local control. With the introduction of consolidated high schools, however, the Amish resisted secondary education, particularly compulsory schooling mandated by state and federal agencies, and raised objections both on principle and on scale. To illustrate the latter, the amount of time required by secondary education and the distances required to bus students out of their home communities were cited as problems. But probably more crucial, was the understanding that the high school promised to socialize and instruct the young in a value system that was not consistent with the Amish way of life. For example, in the high school, individual achievement and competition were promoted, rather than mutuality and caring for others in a communal spirit. On pragmatic grounds, Amish parents objected to "unnecessary" courses in science, advanced math, and computer technology, which seemed to have no place and little relevance in their tradition (Johnson-Weiner 2015).

The Amish response to this perceived threat to their culture was to build and operate their own private elementary schools. Their right to do so was litigated but finally upheld in the U.S. Supreme Court in the 1972 *Wisconsin v. Yoder* ruling (U.S. Supreme Court Base Syllabus, 406 US 205, Johnson-Weiner 2015). Today, school-age children are encouraged to attend only eight grades, but Amish parents actively support local private and public elementary schools.

The Amish rejection of higher learning for their children means that only the rare individual may pursue professional training and still remain Amish. Health-care providers, by definition, are seen as outsiders who mediate information on health promotion, make diagnoses, and propose therapies across cultural boundaries. To the extent that they do so with sensitivity and respect for Amish cultural ways, they are respected, in turn, and valued as an important resource by the Amish.

As the twentieth century drew to a close, important changes were underway among the Amish in North America, whose principal and preferred occupations have long been agricultural work and farm-related enterprises. They had typically settled on productive farmland from their earliest immigration some 250 years ago. As cultivatable land at an affordable price became an increasingly scarce commodity near centers of Amish settlements, the trend toward other work away from home led to a reshaping of the Amish family (Lutz 2017). Income from goods and services once delivered for internal domestic consumption came increasingly from cottage industry production for the retail market and wage-earning with nearby employers. The alternative, seeking new farmland at a distance, has led to community resettlements far away from the concentration of Amish in Pennsylvania, Ohio and Indiana (Moledina et al. 2014).

Young women who have learned quantity cookery at the many church and family get-togethers may find jobs in restaurants and catering, or skills learned in household chores may be exchanged for wages in child care or housecleaning. Young men who bring skills from the farm may practice carpentry or cabinetmaking in the trades and construction industry. This, in turn, brings a change in family patterns, since "lunch-box daddies" are absent during daylight workday hours and the burden for parenting is borne more by stay-at-home mothers. The bonds of family and church have proved resilient but are clearly experiencing more tension in the current generation.

In summary, jobs away from home, an established majority culture pattern, and increased contacts with non-Amish people test the strength of sociocultural bonds that tie young people to the Amish culture. Given the enticements of the majority culture to change and to acculturate, it is noteworthy that most young Amish seek full membership (baptism) in the ethnoreligious culture that nurtured them.

7.3 Communication

7.3.1 Dominant Languages and Dialects

Like most people, the Amish vary their language usage depending on the situation and the individuals being addressed. American English is only one of three language varieties in their repertoire. For the Amish, English is the language of school, of written and printed communications, and, above all, the language used in contacts with most non-Amish outsiders, especially business contacts. Because English serves a useful function as the contact language with the outside world, Amish schools all use English as the language of instruction, with the strong support of parents, because elementary schooling offers the best opportunity for Amish children to master the language. But within Amish homes and communities, use of English is discouraged in favor of the vernacular ***Deitsch***, or Pennsylvania German. Because all Amish except preschool children are literate in their second language, American English language usage helps to define their cultural space (Hostetler 1993).

The first language of most Amish is *Deitsch*, an amalgamation of several upland German dialects that emerged from the interaction of immigrants from the Palatinate and Upper Rhine areas of modern France, Germany, and Switzerland. Their regional linguistic differences were resolved in an immigrant language better known in English as "Pennsylvania German" (also known as "Pennsylvania Dutch"). Amish immigrants who later moved more directly from the Swiss Jura and environs to midwestern states (with minimal mixing in transit with *Deitsch*-speakers) call their home language *Düütsch*, a related variety with marked Upper Alemannic features. Today, *Deitsch* and *Düütsch* both show a strong admixture of vocabulary borrowed from English, whereas the basic structure remains clearly nonstandard German. Both dialects have practically the same functional distribution (Meyers and Nolt 2005; Wenger 1970).

Deitsch is spoken in the home and in conversation with fellow Amish and relatives, especially during visiting, a popular social activity by which news is disseminated orally. It is important to note that *Deitsch* is primarily a spoken language. Some written material has been printed in Pennsylvania German, but Amish seldom encounter it in this form. Even Amish publications urging the use of *Deitsch* in the family circle are printed in English, by default the print replacement for the vernacular, the spoken language (What is in a language? 1986).

Health-care providers can expect all their Amish patients of school age and older to be fluently bilingual. They can readily understand spoken and written directions and answer questions presented in English, although their own terms for some symptoms and illnesses may not have exact equivalents in *Deitsch* and English. Amish patients may be more comfortable consulting among themselves in *Deitsch*, but generally they intend no disrespect for those who do not understand their mother tongue.

Although of limited immediate relevance for health-care considerations, the third language used by the Amish deserves mention in this cultural profile to complete the scope of their linguistic repertoire. Amish proficiency in English varies according to the type and frequency of contact with non-Amish, but it is increasing. The use of Pennsylvania German is in decline outside the Old Order Amish community. Its retention by Amish, despite the inroads of English, has been related to their religious communities' persistent recourse to *Hochdeitsch*, or Amish High German, their so-called third language, as a sacred language (Huffines 1994).

Amish do not use Standard Modern High German, but an approximation, which gives access to texts printed in an archaic German with some regional variations. Rote memorization and recitation for certain ceremonial and devotional functions, and for selected printed texts from the Bible, from the venerable "Ausbund" hymnbook, and from devotional literature are a part of public and private prayer and worship among the Amish. Such restricted and non-productive use of a third language hardly justifies the term "trilingual" because it does not encompass a fully developed range of discourse. However, Amish High

German does provide a situational-functional complement to their other two languages (Enninger and Wandt 1982). Its retention is one more symbol of a consciously separated way of life that reaches back to its European heritage.

Within a highly contextual subculture like the Amish, the base of shared information and experience is proportionately larger. As a result, less overt verbal communication is required than in the relatively low-contextual American culture, and more reliance is placed on implicit, often unspoken understandings. Amish children and youth may learn adult roles in their society more through modeling, for example, than through explicit teaching. The many and diverse kinds of multigenerational social activities on the family farm provide the optimal framework for this kind of enculturation. Although this may facilitate the transmission of traditional, or accepted, knowledge and values within a high-context culture, this same information network may also impede new information imparted from the outside, which entails some behavior changes. Wenger (1988, 1991c) suggested that nurses and other health-care providers should consider role modeling as a teaching strategy when working with Amish patients. Later, a brief example of the promotion of inoculation is presented to illustrate how public-health workers can use culture-appropriate information systems to achieve fuller cooperation among the Amish.

In a final note on language and the flow of verbal information, health-care providers should be aware that much of what passes for "general knowledge" in our information-rich popular culture is screened, or filtered, out of Amish awareness. The Amish have severely restricted their own access to print media, permitting only a few newspapers and periodicals. Most have also rejected the electronic media, beginning with radio and television, but also including entertainment and information applications of film and computers. Conversely, the Amish are openly curious about the world beyond their own cultural horizons, particularly regarding a variety of literature that deals with health and quality-of-life issues. They especially value the oral and written personal testimonial as a mark of the efficacy of a particular treatment or health-enhancing product or process. Wenger (1988, 1994) identified testimonials from Amish friends and relatives as a key source of information in making choices about health-care providers and products.

7.3.2 Cultural Communication Patterns

Fondness and love for family members are held deeply but privately. Some nurses have observed the cool, almost aloof behavior of Amish husbands who accompany their wives to maternity centers, but it would be presumptuous to think that it reflects a lack of concern. The expression of joy and suffering is not entirely subdued by dour or stoic silence, but Amish are clearly not outwardly demonstrative or exuberant. Amish children, who can be as delightfully animated as any other children at play, are taught to remain quiet throughout a worship service lasting more than 2 h. They grow up in an atmosphere of restraint and respect for adults and elders. But privately, Amish are not so sober as to lack a sense of humor and appreciation of wit.

Beyond language, much of the nonverbal behavior of Amish is also symbolic. Many of the details of Amish garb and customs were once general characteristics without any particular religious significance in Europe, but in the American setting, they are closely regulated and serve to distinguish the Amish from the dominant culture as a self-consciously separate ethnoreligious group (Kraybill 2001).

It is precisely in the domain of ideas held to be normative for the religious aspects of Amish life that they find their English vocabulary lacking. The key source texts in *Hochdeitsch* and the oral interpretation of them in *Deitsch* are crucial to an understanding of two German values, which have an important impact on Amish nonverbal behavior. *Demut*, German for "humility," is a priority value, the effects of which may be seen in details such as the height of the crown of an Amish man's hat, as well as in very general features such as the modest and unassuming bearing and demeanor usually shown by Amish people in public. This behavior is reinforced by frequent verbal warnings against its opposite, *hochmut*, which means "pride" or "arrogance," and should be avoided (Hostetler 1993).

The second term, *gelassenheit*, is embodied in behavior more than it is verbalized. *Gelassenheit* is treasured not so much for its contemporary German connotations of passiveness, even of resignation, as it is for its earlier religious meanings, denoting quiet acceptance and reassurance, encapsulated in the biblical formula "godliness with contentment" (1 Tim. 6:5). The following Amish paradigm for the good life flows from the calm assurance found through inner yielding and forgoing one's ego for the good of others:

1. One's life rests secure in the hands of a higher power.
2. A life so divinely ordained is therefore a good gift.
3. A godly life of obedience and submission will be rewarded in the life hereafter (Kraybill 2001).

A combination of these inner qualities; an unpretentious, quiet manner; and modest outward dress in plain colors lacking any ornament, jewelry, or cosmetics presents a striking contrast to contemporary fashions, both in clothing styles and in personal self-actualization. Amish public behavior is consequently seen as deliberate rather than rash, deferring to others instead of being assertive or aggressive, avoiding confrontational speech styles and public displays of emotion in general.

Health-care providers should greet Amish patients with a handshake and a smile. Amish use the same greeting both among themselves and with outsiders, but little touching follows the handshake. Younger children are touched and held with affection, but older adults seldom touch socially in public. Therapeutic touch, conversely, appeals to many Amish and is practiced informally by some individuals who find communal affirmation for their gift of warm hands. This concept is discussed further in the section on health-care practices.

In public, the avoidance of eye contact with non-Amish may be seen as an extension, on a smaller scale, of the general reserve and measured larger body movements related to a modest and humble being. But in one-on-one clinical contacts, Amish patients can be expected to express openness and candor with unhesitating eye contact.

Among their own, Amish personal space may be collapsed on occasions of crowding together for group meetings or travel. In fact, Amish are seldom found alone, and a solitary Amish person or family is the exception rather than the rule. But Amish are also pragmatic, and in larger families, physical intimacy cannot be avoided in the home, where childbearing and care of the ill and dying are accepted as normal parts of life. Once health-care providers recognize that Amish prefer to have such caregiving within the home and family circle, providers will want to protect modest Amish patients who feel exposed in the clinical setting.

7.3.3 Temporal Relationships

So much of current Amish life and practice has a traditional dimension reminiscent of a rural American past that it is tempting to view the Amish culture as "backward-looking." In actuality, Amish self-perception is very much grounded in the present, and historical antecedents or reasons for current consensus have often been lost to common memory. Conversely, the Amish existential expression of Christianity focused on today is clearly seen as a preparation for the afterlife. One may say that Amish are also future-oriented, at least in a metaphysical sense, although not as it relates to modern, progressive, or futuristic thought.

After generations of rural life guided by the natural rhythms of daylight and seasons, the Amish manage the demands of clock time in the dominant culture. They are generally punctual and conscientious about keeping appointments, although they may seem somewhat inconvenienced by not owning a telephone or car. These communication conveniences, deemed essential by the dominant American culture, are viewed by the most conservative Amish as technological advances that could erode the deeply held value of community, in which face-to-face contacts are easily made. Therefore, telephones and automobiles are generally owned by nearby non-Amish neighbors and used by Amish only when it is deemed essential, such as for reaching healthcare facilities (Nolt 2016).

Because the predominant mode of transportation for the Amish is horse and carriage, travel to a doctor's office, a clinic, or a hospital requires the same adjustment as any other travel outside their rural community to shop, trade, or attend a wedding or funeral. The latter three reasons for travel are important means of reinforcing relationship ties, and on these occasions, the Amish may use hired or public transportation, excluding flying. Taking time out of normal routines for extended trips related to medical treatments is not uncommon, such as a visit to radioactive mines in the Rocky Mountains or to a laetrile clinic in Mexico to cope with cancer (Wenger 1988). Some Amish participate in types of "medical tourism" trips to locations for care, renting a bus with other Amish, for transportation (King 2017).

7.3.4 Format for Names

Using first names with Amish people is appropriate, particularly because generations of intermarriage have resulted in a large number of Amish who share only a limited number of surnames. It is preferable to use first names during personal contacts rather than titles such as Mr. or Mrs. Miller. In fact, within Amish communities with so many Millers, Lapps, Yoders, and Zooks, given names like Mary and John are overused to the extent that individuals have to be identified further by nicknames, residence, a spouse's given name, or a patronymic, which may reflect three or more generations of patrilineal descent. For example, a particular John Miller may be known as "Red John," or "Gap John," or "Annie's John," or "Sam's Eli's Roman's John" (Hostetler 1993).

During an interview with an Amish mother and her 5-year-old son, Wenger (1988) asked the child where he was going that day. The boy replied that he was going to play with Joe Elam John Dave Paul, identifying his age-mate Paul with four preceding generations. This little boy was giving useful everyday information, while at the same time, unknown to him, keeping oral history alive. The patronymics also illustrate the cultural value placed on intergenerational relationships and help to create a sense of belonging that embraces several generations and a broad consanguinity. Thus, one can see that medical record keeping can be a challenge when serving an extensive number of Amish patients.

7.4 Family Roles and Organization

7.4.1 Head of Household and Gender Roles

From the time of marriage, the young Amish man's role as husband is defined by the religious community to which he belongs. Titular patriarchy is derived from the Bible: Man is the head of the woman as Christ is the head of the church (I Cor. 3). This patriarchal role in Amish society is balanced or tempered by realities within the family in which the wife is accorded high status and respect for her vital contributions to the success of the family. Practically speaking, husband and wife may share equally in decisions regarding the family farming business. In public, the wife may assume a retiring role, deferring to her husband, but in private, they are typically partners. However, it is best to listen to the voices of Amish women themselves as they reflect on their values and roles within Amish family and their shared ethnoreligious cultural community.

Traditionally, the highest priority for the parents is child rearing, an ethnoreligious expectation in the Amish culture. With a completed family averaging at least 5 children, the Amish mother contributes physically and emotionally to the burgeoning growth in the Amish population (Kraybill et al. 2013). She also has an important role in providing family food and clothing needs, as well as a major share in child nurturing. Amish society expects the husband and father to contribute guidance, serve as a role model, and discipline the children. This shared task of parenting takes precedence over other needs, including economic or financial success in the family business. On the family farm, all must help as needed, but in general, field and barn work and animal husbandry are primarily the work of men and boys, whereas food production and preservation, clothing production and care, and management of the household are mainly the province of women.

7.4.2 Prescriptive, Restrictive, and Taboo Behaviors for Children and Adolescents

Children and youth represent a key to the vitality of the Amish culture. Babies are welcomed as a gift from God, and the high birth rate is one factor in their population growth. Another is the surprisingly high retention of youth, an estimated 75% or more, who choose as adults to remain in the Amish way (Kraybill et al. 2013). Before and during elementary school years, parents are more directive as they guide and train their children to assume responsible, productive roles in Amish society.

Young people over 16 years of age may be encouraged to work away from home to gain experience or because of insufficient work at home or on the family farm, but their wages are still usually sent home to the parental household because of the cultural value that the whole family contributes to the welfare of the family. Some experimentation with non-Amish dress and behavior among Amish teenagers is tolerated during Rumspringa (Stevick 2014). This is a period of relative leniency, but the expectation is that an adult decision to be baptized before marriage will call young people back to the discipline of the church, as they assume adult roles (Mendez Ruiz 2017).

In recent years, the media have been fascinated with this period of Amish teenage life as Americans in general have learned more about the Amish as a distinctive culture. Meyers and Nolt (2005) contended that although some Amish teenagers do experiment with behaviors that are incongruent with Amish beliefs and values, they do so in a distinctive Amish way. Amish teenagers are aware of the dominant American culture, and when they choose to participate in behaviors, some of which may involve the legal system, they do so in ways that are not common to American teenagers in general. For example, Amish youth will usually experiment with other Amish youth, rather than with non-Amish teenagers.

7.4.3 Family Goals and Priorities

The Amish family pattern is referred to as the ***freindschaft***, the dialectical term used for the three-generational family structure. This kinship network includes consanguine relatives consisting of the parental unit and the households of married children and their offspring. All members of the family personally know their grandparents, aunts, uncles, and cousins, with many Amish knowing their second and third cousins as well.

Individuals are identified by their family affiliation. Children and young adults may introduce themselves by giving their father's first name or both parents' names so they can be placed geographically and genealogically. Families are the units that make up church districts; the size of a church district is measured by the number of families rather than by the number of church members. This extended family pattern has many functions. Families visit together frequently, thus learning to anticipate caring needs and preferences. Health-care information often circulates through the family network, even though families may be geographically dispersed. Wenger (1988) found that informants referred to *freindschaft* when discussing the factors influencing the selection of health-care options. "The functions of family care include maintaining *freindschaft* ties, bonding family members together intergenerationally, and living according to God's will by fulfilling the parental mandate to prepare the family for eternal life" (Wenger 1988, p. 134).

As grandparents turn over the primary responsibility for the family farm to their children, they continue to enjoy respected status as elders, providing valuable advice and sometimes material support and services to the younger generation. Many nuclear families live on a farm with an adjacent grandparent's cottage or a smaller apartment attached to the main farmhouse, This promotes frequent interactions across generations. Grandparents provide child care and help in rearing grandchildren and, in return, enjoy the respect generally paid by the next generations. This emotional and physical proximity to older adults also facilitates elder care within the family setting. In an ethnonursing study on care in an Amish community, Wenger (1988) reported that an informant discussed the reciprocal benefits of having her grandparents living in the attached *daadihaus* and her own parents living in a house across the road. Her 3-year-old daughter could go across the hall to spend time with her great-grandfather, which the mother reported, was good for him in that he was needed, whereas the small child benefited from learning to know her great-grandfather, and the young mother gained some time to do chores. There is no set retirement age among the Amish, and grandmothers also continue in active roles as advisers and assistants to younger mothers.

Assuming full adult membership and responsibility means the willingness to put group harmony ahead of personal desire. In financial terms, it also means an obligation to help others in the brotherhood who are in need. This mutual aid commitment also provides a safety net, which allows Amish to rely on others for help in emergencies. Consequently, the Amish do not seek federal pension or retirement support; they have their own informal "social security" plan. Amish of varying degrees of affluence enjoy approximately the same social status, and extremes of poverty and wealth are uncommon. Property damage or loss and unusual health-care expenses are also covered to a large extent by an informal brotherhood alternative to commercial insurance coverage (Kraybill 2019). The costs of high-technology medical care present a new and severe test of the principle of mutual aid or "helping out," which is almost synonymous with the Amish way of life.

7.4.4 Alternative Lifestyles

There is little variation from the culturally sanctioned expectations for parents and their unmarried children to live together in the same household while maintaining frequent contact with the extended family. Unmarried children live in the parents' home until marriage, which usually takes place between the ages of 20 and 30 years of age. Some young adults may move to a different community to work and live as a boarder with another Amish family. Being single is not stigmatized, although almost all Amish do marry. Single adults are included in the social fabric of the community with the expectation that they will want to be involved in family-oriented social events.

Individuals of the same gender do not live together except in situations in which their work may make it more convenient. For example, two female schoolteachers may live together in an apartment or home close to the Amish school where they teach. There are no available statistics on the incidence of homosexuality in Amish culture. Isolated incidents of homosexual practice may come to the attention of health providers, but homosexual lifestyles do not fit with the deeply held values of Amish family life and procreation.

Pregnancy before marriage does not usually occur, and it is viewed as a situation to be avoided. When it does occur, in most Amish families, the couple would be encouraged to consider marriage. If they are not yet members of the church, they need to be baptized and to join the church before being married. Although not condoning pregnancy before marriage, the families and the Amish community support the young couple about to have a child. If the couple chooses not to marry, the young girl is encouraged to keep the baby and her family helps raise the child. Abortion is an unacceptable option. Adoption by an Amish family is an acceptable alternative.

7.5 Workforce Issues

7.5.1 Culture in the Workplace

In every generation except the present one, the Amish have worked almost exclusively in agriculture and farm-related tasks. Their large families were ideally suited to labor-intensive work on the family farm. As the number of family farms has been drastically reduced because of competition from agribusinesses that use mechanized and electronically controlled production methods, few options are available for Amish youth.

Traditionally, the Amish have placed a high value on hard work, with little time off for leisure or recreation. Productive employment for all is the ideal, and the intergenerational family provides work roles appropriate to the age and abilities of each person. But prospects began to narrow with the increased concentration of family farms in densely settled Amish communities as their population increased.

In addition, several cultural factors combine to limit the opportunities for young Amish to adapt to new work patterns. Amish children, who are encouraged to attend school through only eight grades, have a limited basis for vocational training in many work areas other than agriculture. Amish avoidance of compromising associations with "worldly" organizations, such as labor unions, restricts them to non-union work, which often pays lower hourly rates. Work off the family farm, at one time a good option for unmarried youth, has become an economic necessity for some parents, although it is considered less acceptable for social reasons. Fathers who "work away," sometimes called "lunch pail daddies," have less contact with children during the workday, which in turn has an impact on the traditional father's modeling role and places more of the responsibility for child rearing on stay-at-home mothers. This shift in traditional parental roles is a source of some concern, although the effects are not yet clear.

Another concern for the Amish culture in relation to the workplace is the use of technologies that may be of concern for them. Hurst and McConnell (2010) describe survey results of Amish in Holmes County, Ohio, where 9 out of 10 persons "believe there are some technologies that are harmful to the stability and integrity of Amish culture, regardless of how they are used" (p. 210). Computers, Internet, and TV were mentioned the most. These technologies seem to be so pervasive in non-Amish lifestyles and workplaces. For the Amish, their concern is the difficulty in using these technologies in healthy ways that uphold Amish beliefs and values.

7.5.2 Issues Related to Autonomy

As described previously, external and internal factors have converged in the early twenty-first century to cause doubt about the continued viability of compact Amish farming communities. Exorbitant land prices triggered group movement and resettlement in states to the west and south. The declining availability of affordable prime farm land in and around the centers of highest Amish population density is due in part to their non-Amish neighbors' land-use practices, especially in areas of suburban sprawl. A powerful internal force is at work as well in the population growth rate among the Amish, now well above the national average. So, contrary to popular notions that such a "backward" subculture is bound to die out, the Amish today are thriving.

Population growth continues even without a steady influx of new immigrants from the European homeland or significant numbers of new converts to their religion or way of life (Kraybill 2001). The Young Center for Anabaptist and Pietist Studies at Elizabethtown College (2019) reports that the estimated population of the Amish of North America (adults and children) as of June 2018 is 330,270. This is an increase of approximately 11,880 since 2017, a growth rate of 3.73%. In the 20-year period from 1991 to 2018, the Amish in North America (adults and children) increased from 123,500 in 1991 to 330,270 in 2017, an overall growth of 260%. This population growth has been attributed largely to the size of families and the retention rate of young adults.

The resulting pressures to control the changes in their way of life while maintaining its religious basis, particularly the high value placed on in-group harmony, have challenged the Amish to develop adaptive strategies. One outcome is an increasingly diversified employment base, with a trend toward cottage industries and related retail sales, as well as toward wage labor to generate cash needed for higher taxes and increasing medical costs. Another recent development includes a shift from traditional multigenerational farmsteads, as some retirees and crafts workers employed off the farm have begun to relocate to the edges of country towns. In summary, pressures to secure a livelihood within the Amish tradition have heightened awareness of the tension field within which the Amish coexist with the surrounding majority American culture.

Because English is the language of instruction in schools and is used with business contacts in the outside world, there is generally no language barrier for the Amish in the workplace. English vocabulary that is lacking in their normative ideas for religious aspects of Amish life is rarely a concern in the workplace.

7.6 Biocultural Ecology

7.6.1 Skin Color and Other Biological Variations

Most Amish are descendants of eighteenth-century Southern German and Swiss immigrants; therefore, their physical characteristics vary, as do those of most Europeans, with skin variations ranging from light to olive tones. Hair and eye colors vary accordingly. No specific health-care precautions are relevant to this group.

7.6.2 Diseases and Health Conditions

Since 1962, several hereditary diseases have been identified among the Amish. The earliest findings of the genetic studies have been published by Dr. Victor McKusick (1978) of the Johns Hopkins University. The Clinic for Special Children in Strasburg, PA was started in 1989 by Dr. Holmes Morton and continues to provide medical care for Amish and Mennonite children with genetic disorders (Strauss 2015). Because Amish tend to live in settlements with relatively little domiciliary mobility, and because they keep extensive genealogical and family records, genetic studies are more easily done than with more mobile cultural groups (Bass and Waggoner 2018). Many years of collaboration between the Amish and geneticists from the Johns Hopkins Hospital and the Clinic for Special Children have resulted in mutually beneficial projects. The Amish received printed community directories, and geneticists compiled computerized genealogies for the study of genetic diseases that continue to benefit society in general (Strauss et al. 2012).

The Amish are essentially a closed population with exogamy occurring very rarely. However, they are not a singular genetically closed population. The larger and older communities are consanguineous, meaning that within the community the people are related through bloodlines by common ancestors. Several consanguine groups have been identified in which relatively little intermarriage occurs between the groups. "The separateness of these groups is supported by the history of the immigration into each area, by the uniqueness of the family names in each community, by the distribution of blood groups, and by the different hereditary diseases that occur in each of these groups" (Hostetler 1993, p. 328). These diseases are one of the indicators of distinctiveness among the groups.

Hostetler (1993) cautioned that although inbreeding is more prevalent in Amish communities than in the general population, it does not inevitably result in hereditary defects. Through the centuries in some societies, marriages between first and second cousins were relatively common without major adverse effects. However, in the Amish gene pool are several recessive tendencies that in some cases are limited to specific Amish communities in which the consanguinity coefficient (degree of relatedness) is high for the specific genes (Hou et al. 2017). Of at least 12 recessive diseases, 4 should be noted here (Hostetler 1993; McKusick 1978; Troyer 1994).

Dwarfism has long been recognized as obvious in several Amish communities. Ellis–van Creveld syndrome, known in Europe and named for Scottish and Dutch physicians, is especially prevalent among the Lancaster County, Pennsylvania, Amish (McKusick et al. 1964). This syndrome is characterized by short stature and an extra digit on each hand, with some individuals having a congenital heart defect and nervous system involvement, resulting in a degree of mental deficiency. The Lancaster County Amish community, the second largest Amish settlement in the United States, is the only one in which Ellis–van Creveld syndrome is found (Baby et al. 2016). The lineage of all affected people has been traced to a single ancestor, Samuel King, who immigrated in 1744 (Troyer 1994).

Cartilage hair hypoplasia, also a dwarfism syndrome, has been found in nearly all Amish communities in the United States and Canada and is not unique to the Amish (Riley et al. 2015). This syndrome is characterized by short stature and fine, silky hair. There is no central nervous system involvement and, therefore, no mental deficiency. However, most affected individuals have deficient cell-mediated immunity, thus increasing their susceptibility to viral infections (Troyer 1994).

Pyruvate kinase anemia, a rare blood cell disease, was described by Bowman and Procopio in 1963. The lineage of all affected individuals can be traced to Jacob Yoder (known as "Strong Jacob"), who immigrated to Mifflin County, Pennsylvania, in 1792 (Hostetler 1993; Troyer 1994). This same genetic disorder was found later in the Geauga County, Ohio, Amish community. Notably, the families of all those who were affected had migrated from Mifflin County, Pennsylvania, and were from the "Strong Jacob" lineage. Symptoms usually appear soon after birth, with the presence of jaundice and anemia. Transfusions during the first few years of life and eventual removal of the spleen can be considered cures.

Hemophilia B, another blood disorder, is disproportionately high among the Amish, especially in Ohio. Ratnoff (1958) reported on an Amish man who was treated for a ruptured spleen. It was discovered that he had grandparents and 10 cousins who were hemophiliacs; 5 of the cousins had died from hemophilia. Research studies on causative mutations indicated a strong probability that a specific mutation may account for much of the mild hemophilia B in the Amish population (Ketterling et al. 1991).

Through the vigilant and astute observations of some public-health nurses, a major health-care problem was noted in a northern Indiana Amish community. A high prevalence of phenylketonuria (PKU) was found in the Elkhart-Lagrange Amish settlement (Martin et al. 1965; Bass and Waggoner 2018). Those affected are unable to metabolize the amino acid phenylalanine, resulting in high blood levels of the substance and, eventually, severe brain damage if the disorder is untreated. Through epidemiological studies, the health department found that 1 in 62 Amish were affected, whereas the ratio in the general population was 1 in 25,000 at that time. Through the leadership of these nurses, the county and the state improved case funding for PKU and health-care services for affected families throughout Indiana, which was followed by improved health services in Amish communities in other states as well.

In recent years, a biochemical disorder called glutaric aciduria has been studied by Dr. Holmes Morton, a Harvard-educated physician who has chosen to live and work among the Amish in Lancaster County, Pennsylvania. Morton made house calls, conducted research at his own expense because funding was not forthcoming, and established a clinic in the Amish community to screen, diagnose, and educate people to care for individuals afflicted with the disease (Allen 1989). By observing the natural history of glutaric aciduria type I, the researchers postulated that the onset or progression of neurological disease in Amish patients can be prevented by screening individuals at risk; restricting dietary protein; and thus, limiting protein catabolism, dehydration, and acidosis during illness episodes.

Dr. Morton was well received in the Amish community, with many people referring friends and relatives to him. When he noted the rapid onset of the symptoms and the high incidence among the Amish, he did not wait for them to come to his office. He went to their homes and spent evenings and weekends driving from farm to farm, talking with families, running tests, and compiling genealogical information (Wolkomir and Wolkomir 1991). In 1991, he built a clinic with the help of donations, in part the result of an article in the *Wall Street Journal* about the need for this non-profit clinic. Hewlett-Packard donated the needed spectrometer that cost $80,000; local companies provided building materials, and an Amish couple donated the building site. Although volunteers helped to build the clinic, a local hospital provided temporary clinic space lease-free because the community recognized the very important contribution Morton was making, not only to the Amish and the advancement of medical science but also to the public health of the community.

A countywide screening program is now in place. Health-care providers are able to recognize the onset of symptoms. Research continues on this metabolic disorder, its relationship to cerebral palsy in the Amish population, and the biochemical causes and methods of preventing spastic paralysis in the general population. However, education remains a highly significant feature of any community health program. Nurses and physicians need to plan for family and community education about genetic counseling, screening of newborns, recognition of symptoms during aciduric crises in affected children, and treatment protocols. In The New York Times Magazine (Belkin 2005), Dr. Morton was called "a doctor for the future"' because he practices what is now referred to as genetic medicine, which recognizes genetics as part of all medicine. But to the Amish, he is their friend who cares about their children, knows their families by name, and comes to their homes to see how they are able to cope with the manifestations of these genetically informed diseases. Although he has retired from full time work in the clinic he established, he personally trained the clinicians and researchers that carry on his work.

Similar to the views of non-Amish parents, there is great variability in parental willingness to obtain genetic counselling (Teapole 2019). As in most health care decisions, the intent to obtain genetic counselling remains with the individual. Family members, friends, and religious leaders may give advice. Genetic counselling may or may not be done before pregnancy and may be requested prior to delivery. This information may be helpful in childbirth decisions made in collaboration with health care providers.

Extensive studies of manic-depressive illnesses have been conducted in the Amish population. Comparative studies have been done on both Amish (Gill et al. 2016) and non-English families (Foroud et al. 2000) to determine the genetic basis of mental disorders. While new information on the genome has been obtained, the specific locus for bipolar disorders had not yet been found. Attempts have been made to gain knowledge about the affective response the Amish have to their ethnoreligious cultural identity and experience. Lantz (2019) studied the protective factors of the Amish lifestyle in mental health disorders.

The incidence of alcohol and drug abuse, which can complicate psychiatric diagnoses, is much lower among the Amish than in the general North American population, thus contributing to the importance of the Amish sample. Although the incidence of bipolar affective disorder is not found to be higher in the Amish, some large families with several affected members continue to contribute to medical science by being subjects in the genetic studies. Because the Old Order Amish descend from 30 pioneer couples whose descendants have remained genetically isolated in North America, have relatively large kindred groups with multiple living generations, and generally live in close geographic proximity, they are an ideal population for genetic studies.

7.6.3 Variations in Drug Metabolism

No drug studies specifically related to the Amish were found in the literature. Given the genetic disorders common among selected populations of Amish, astute clinicians are tailoring interventions to provide the safest medical care of Amish patients (Strauss 2015; Weller 2017). The impact of genetic variance in pharmacogenetics, specifically for the Amish, is one area in which more research needs to be conducted.

Reflective Exercise 1

You are working with a young married Amish couple who are pregnant with their first child. They both have a significant family history of genetic disorders. They have presented to your care as they would like to have genetic testing. They want "to be prepared".

- How would you proceed?
- Are there any other professionals you would include in preparing the best care?
- Are there any other friends, family or others significant to this young couple you anticipate would be participating? How would you prepare for this?

Reflective Exercise 2

You are a nurse practitioner in a small rural clinic that is one of the few clinics that will provide "cash" services for the Amish patients in the area. You are seeing a set of twins for an Amish family who has a history of significant genetic diseases that already resulted in deaths of two of their children. This condition results in neurological decline but outcomes can be improved with the addition of a specially made formula. This formula is expensive.

- How would you approach this appointment with the parents?
- What are the care priorities?
- What resources would you research to provide the best care?

Reflective Exercise 3

Mary and Elmer's fifth child, Melvin, was born 6 weeks prematurely and is now 1 month old. The consulting physician has recommended that Mary have no more children because the difficulties she had with this pregnancy and delivery. Their other children Sarah, age 13, Martin, age 12, and Wayne, age 8, attend the Amish elementary school located 1 mile from their home. Lucille, age 4, is staying with Mary's sister and her family for a week because baby Melvin has been having respiratory problems, and their physician told the family he will need to be hospitalized if he does not get better within 2 days. They are seeking further advice because they believe the amoxicillin for the baby is not helping. They do not want the baby to be hospitalized.

- How would you proceed? What questions would you be asking to gather more assessment information to proceed the best plan of care?
- What concerns do you have for each family member?
- Address the priorities for care for the immediate concerns for baby Melvin.
- What are the priorities addressing further reproductive care for Mary? How would you proceed?

Reflective Exercise 4
Aaron and Annie Schlabach, aged 70 and 68 years, live in the attached grandparents' cottage of their youngest daughter, Mary. Most of Aaron's and Annie's children live in the area or in another Amish community about 20 miles away. Aaron continues to help with the farm work, despite increasing pain in his hip, which the doctor advises should be replaced. He has tried numerous options to try to manage the pain but comes today for assistance in planning for a hip replacement.

- What additional history would you obtain to provide the best care?
- What are the care priorities for Aaron currently and post operatively (if he decides to have the surgery)?
- What are the care priorities if he decides to not have the surgery?
- What stressors would you anticipate for the best outcome for Aaron and Annie?

Reflective Exercise 5
You are the primary community health nurse for a small Amish community. A few of the elders have asked you about the medications they have been prescribed by a local nurse practitioner that you feel is a competent provider. They would like your opinion on using their own remedies with these prescribed medications. They ask if you will work with the person from whom they obtain their herbal and alternative remedies from.

- How do you assess this?
- Would you communicate any of this information with the prescriber? If so, how would you do that?
- Would you work with their preferred herbalist?
- What additional information would you desire before answering this request

7.7 High-Risk Behaviors

Amish are traditionally agrarian and prefer a lifestyle that provides intergenerational and community support systems to promote health and mitigate against the prevalence of high-risk behaviors. Genetic studies using Amish populations are seldom confounded by the use of alcohol and other substances. However, health providers should be alert to potential alcohol and recreational drug use in some Amish communities, especially among young, unmarried men. When young adult men exhibit such behavior as straying from the Amish way of life and "sowing their wild oats" before becoming baptized church members and before marriage, it is tolerated. Although this may be considered a high-risk behavior, it is not prevalent in all communities, nor is it promoted in any. Parents confide in each other and sometimes in trusted outsiders that this errant behavior causes many heartaches, although at the same time, they try to be patient and keep contact with the youth so the latter may choose to espouse the Amish lifeways.

Another lifestyle pattern that poses potential health risks is nutrition. Amish tend to eat high-carbohydrate and high-fat foods with a relatively high intake of refined sugar. Wenger (1994) reported that in an ethnonursing study on health and health-care perceptions, informants talked about their diet being too high in "sweets and starches" and knowing they should eat more vegetables. The prevalence of obesity was found to be greater among Amish women than for women in general in the state of Ohio (Fuchs et al. 1990). In this major health-risk survey of 400 Amish adults and 773 non-Amish adults in Ohio, the authors found that the pattern of obesity in Amish women begins in the 25-year-old and older cohort, with the concentration occurring between the ages of 45 and 64. An explanation for the propensity for weight gain among the Amish may be related to the central place assigned to the consumption of food in their culture and the higher rates of pregnancy throughout their childbearing years (Wenger 1994). However, in recent studies related to eating behaviors, obesity, and diabetes, the Old Order Amish cohorts showed some significant differences from other Whites in the majority culture. Hsueh and colleagues (2002) reported in the Third National Health and Nutrition Examination Survey that the Old Order Amish sample evidenced diabetes approximately half as frequently as did other whites in the survey. There has been additional research done on linkage of familial hypercholesterolemia and diabetes among related Amish families (Xu et al. 2017). Another important difference was the level of daily physical activity, which was reported to be higher among both Amish men and women than among other white cohorts (Katz et al. 2012).

7.7.1 Health-Care Practices

Most Amish are physically active, largely owing to their chosen agrarian lifestyle and farming as a preferred occupation. Physical labor is valued, and men as well as women and children help with farm work. Household chores and gardening, generally considered to be women's work, require physical exertion, particularly because the Amish do not choose to use electrically operated appliances in the home or machinery that conserve human energy. The impact of increased physical activity is found in comparison studies of Amish and non-Amish children (Hairston et al. 2013). Nevertheless, many women do contend with a tendency to be overweight. In recent years, it is not uncommon to find Amish women seeking help for weight control.

Farm and traffic accidents are an increasing health concern in communities with a dense Amish population. In states such as Indiana and other states as well with relatively high concentrations of Amish who drive horse-drawn vehicles, blinking red lights and large red triangles are required by law to be attached to their vehicles. Research done on both adult and pediatric injuries show trends in trauma with accidents involving farming equipment, animal injuries and buggy accidents (Strotmeyer et al. 2019). Transportation-related injuries (buggy vs. car) continue to be the biggest threat for all ages. Amish families need to be encouraged to monitor their children who operate farm equipment and transportation vehicles and to teach them about safety factors. Concern about accidents is evident in Amish newsletters, many of which have a regular column reporting accidents and asking for prayers or expressing gratitude that the injuries were not more severe, that God had spared the person, or that the community had responded in caring ways (Wenger 1988).

7.8 Nutrition

7.8.1 Meaning of Food

Among the Amish, food is recognized for its nutritional value. Most Amish prefer to grow their own produce for economic reasons and because for generations they have been aware of their connections with the earth. They believe that God expects people to be the caretakers of the earth and to make it flourish.

The Amish serve food in most social situations because food also has a significant social meaning. Because visiting has a highly valued cultural function, occasions occur during most weeks for Amish to visit family, neighbors, and friends, especially those within their church district. Some of these visits are planned when snacks or meals are shared, sometimes with the guests helping to provide the food. Even if guests come unexpectedly, it is customary in most Amish communities for snacks and drinks to be offered.

7.8.2 Common Foods and Food Rituals

Typical Amish meals include meat, potatoes, noodles, or both; a cooked vegetable; bread; something pickled (e.g., pickles, red beets); cake or pudding; and coffee. Beef is usually butchered by the family and then kept in the local commercially owned freezer for which they pay a rental storage fee. However, some more progressive Amish have freezer capabilities that are gasoline powered. Some families also preserve beef by canning, and most families have chickens and other fowl, such as ducks or geese, which they raise for eggs and for meat. Amish families still value growing their own foods and usually have large gardens. A generation ago, this was an unquestioned way of life, but an increasing number of families living in small towns and working in factories and construction own insufficient land to plant enough food for the family's consumption.

Snacks and meals in general tend to be high in fat and carbohydrates. A common snack is large, home-baked cookies about 3 in. in diameter. Commercial non-Amish companies have recognized large soft cookies as a marketable commodity and have advertised their commercially made products as "Amish" cookies, even though no Amish are involved in the production. Other common snacks are ice cream (purchased or homemade), pretzels, and popcorn.

When Amish gather for celebrations such as weddings, birthdays, work bees, or quiltings, the tables are usually laden with a large variety of foods. The selection, usually provided by many people, includes several casseroles, noodle dishes, white and sweet potatoes, some cooked vegetables, few salads, pickled dishes, pies, cakes, puddings, and cookies. Hostetler (1993) provided a detailed ethnographic description of the meaning and practices surrounding an Amish wedding, including the food preparation, the wedding dinner and supper, and the roles and functions of various key individuals in this most important rite of passage that includes serving food.

In communities in which tourists flock to learn about the Amish, many entrepreneurs have used the Amish love of wholesome, simple foods to market their version of Amish cookbooks, food products, and restaurants that more aptly reflect the Pennsylvania German, commonly referred to as *Pennsylvania Dutch*, influence of communities such as Lancaster County, Pennsylvania. Many of these bear little resemblance to authentic Amish foods, and some even venture to sell "Amish highballs" or "Amish sodas" (Hostetler 1993). Some Amish families help to satisfy the public interest in their way of life by serving meals in their homes for tourists and local non-Amish. But most Amish view their foods and food preparation as commonplace and functional, not something to be displayed in magazines and newspapers. Because many Amish are wary of outsiders' undue interest, health-care providers need to discuss nutrition and food as a part of their lifeways to promote healthy nutritional lifestyles.

In Amish homes, a "place at the table" is symbolic of belonging (Hostetler 1993). Seating is traditionally arranged with the father at the head and boys seated youngest to oldest to his right. The mother sits to her husband's left, with the girls also seated youngest to oldest or placed so that an older child can help a younger one. The table is the place where work, behavior, school, and other family concerns are discussed. During the busy harvesting season, preference is given to the men and boys who eat and return to the fields or barn. At mealtimes, all members of the household are expected to be present unless they are working away from home or visiting at a distance, making it difficult to return home.

Sunday church services, which for the Old Order Amish are held in their homes or barns, are followed by a simple meal for all who attended church. The church benches, which are transported from home to home wherever the church service is to be held, are set up with long tables for serving the food. In many communities, some of the benches are built so they can quickly be converted into tables. Meals become ritualized so the focus is not on what is being served but rather on the opportunity to visit together over a simple meal. In one community, an Amish informant who had not attended services because of a complicated pregnancy told the researcher that she missed the meal, which in that community consisted of bread, butter, peanut butter mixed with marshmallow creme and honey, apple butter, pickles, pickled red beets, soft sugar cookies, and coffee (Wenger 1988).

7.9 Pregnancy and Childbearing Practices

7.9.1 Fertility Practices and Views toward Pregnancy

Children are viewed as a gift from God and are welcomed into Amish families. Estimates place the average number of live births per family at seven (Elajami et al. 2016). The Amish fertility pattern has remained constant during the past 100 years, while many others have declined. Household size varies from families with no children to couples with 15 or more children* (Meyers and Nolt 2005). Even in large families, the birth of another child brings joy because of the core belief that children are "a heritage from the Lord," and another member of the family and community means another person to help with the chores (Hostetler 1993).

Having children has a different meaning in Old Order Amish culture than in the dominant European American culture. In a study on women's roles and family production, the authors suggested that women in Amish culture enjoy high status despite the apparent patriarchal ideology because of their childbearing role and their role as producers of food (Lipon 1985). A large number of children benefit small labor-intensive farms, and with large families comes an apparent need for large quantities of food. Interpretation of this pragmatic view of fertility should always be moderated with recognition of the moral and ethical core cultural belief that children are a gift from God, given to a family and community to nurture in preparation for eternal life.

Scholars and researchers of long-term acquaintance with Old Order Amish agree that the pervasive Amish perception of birth control is that it interferes with God's will and thus should be avoided (Kraybill 2001). Nevertheless, fertility control does exist, although the patterns are not well known and very few studies have been reported. Wenger (1980) discussed childbearing with two Amish couples in a group interview and they conceded that some couples do use the rhythm method. In referring to birth control, one Amish father stated, "It is not discussed here, really. I think Amish just know they shouldn't use the pill" (Wenger 1980, p. 5). Three physicians and three nurses were interviewed; they reported that some Amish do ask about birth control methods, especially those with a history of difficult perinatal histories and those with large families. Some Amish women do use intrauterine devices, but this practice is uncommon. Most Amish women are reluctant to ask physicians and nurses and, therefore, should be counseled with utmost care and respect because this is a topic that generally is not discussed, even among themselves. Approaching the subject obliquely may make it possible for the Amish woman or man to sense the health provider's respect for Amish values and thus encourage discussion. "When you want to learn more about birth control, I would be glad to talk to you" is a suggested approach.

7.9.2 Prescriptive, Restrictive, and Taboo Practices in the Childbearing Family

Amish tend to have their first child later than do non-Amish. A retrospective chart review examining pregnancy outcomes of 39 Amish and 145 non-Amish women at a rural hospital in southern New York found that Amish had their first child an average of 1 year later than non-Amish couples (Lucas et al. 1991). The Amish had a narrower range of maternal ages and had proportionately fewer teenage pregnancies. All subjects received prenatal care, with the Amish receiving prenatal care from Amish lay midwives during the first trimester.

In some communities, Amish have been reputed to be reluctant to seek prenatal health care. Providers who gain the trust of the Amish learn that they want the best perinatal care, which fits with their view of children being a blessing (Miller 1997). However, they may choose to use Amish and non-Amish lay midwives who promote childbearing as a natural part of the life cycle. In a study of childbearing practices as described by Amish women in Michigan, Miller (1997) learned that they prefer home births, they had "limited formal knowledge of the childbirth process" (p. 65), and health-care providers were usually consulted only when there were perceived complications. Although many may express privately their preference for perinatal care that promotes the use of nurse-midwifery and lay midwifery services, home deliveries, and limited use of high technology, they tend to use the perinatal services available in their community. In ethnographic interviews with informants, Wenger (1988) found that grandmothers and older women reported greater preference for hospital deliveries than did younger women. The younger women tend to have been influenced by the increasing general interest in childbirth as a natural part of the life cycle and the deemphasis on the medicalization of childbirth. Some Amish communities, especially those in Ohio and Pennsylvania, have a long-standing tradition of using both lay midwifery and professional obstetric services, often simultaneously.

In Ohio, the Mt. Eaton Care Center developed as a community effort in response to retirement of an Amish lay midwife known as Bill Barb (identified by her spouse's name, as discussed in the section on communication). She provided perinatal services, including labor and birth, with the collaborative services of a local Mennonite physician who believed in providing culturally congruent and safe health-care services for this Amish population. At one point in Bill Barb Hochstetler's 30-year practice, the physician moved a trailer with a telephone onto Hochstetler's farm so that he could be called in case of an emergency (Huntington 1993). Other sympathetic physicians also delivered babies at Bill Barb's home. After state investigation, which coincided with her intended retirement, Hochstetler's practice was recognized to be in a legal grey area. The Mt. Eaton Care Center became a reality in 1985 after careful negotiation with the Amish community, Wayne County Board of Health, Ohio Department of Health, and local physicians and nurses. Physicians and professional nurses and nurse-midwives, who are interested in Amish cultural values and health-care preferences, provide low-cost, safe, low-technology perinatal care in a homelike atmosphere. In 1997, the New Eden Care Center, modeled after the Mt. Eaton Care Center, was built in LaGrange County in northern Indiana and, in recent years, has had more than 400 births per year (Meyers and Nolt 2005).

Because the Amish want family involvement in perinatal care, outsiders may infer that they are open in their discussion of pregnancy and childbirth. In actuality, most Amish women do not discuss their pregnancies openly and make an effort to keep others from knowing about them until physical changes are obvious. Mothers do not inform their other children of the impending birth of a sibling, preferring for the children to learn of it as "the time comes naturally" (Wenger 1988). This fits with the Amish cultural pattern of learning through observation that assumes intergenerational involvement in life's major events. Anecdotal accounts exist of children being in the house, though not physically present, during birth. Fathers are expected to be present and involved, although some may opt to do farm chores that cannot be delayed, such as milking cows.

Amish women do participate in prenatal classes, often with their husbands. The women are interested in learning about all aspects of perinatal care but may choose not to participate in sessions when videos are used. Prenatal class instructors should inform them ahead of time when videos or films will be used so they can decide whether to attend. For some Amish in which the *Ordnung* (the set of unwritten rules prescribed for the church district) is more prescriptive and stricter, the individuals may be concerned about being disobedient to the will of the community. Even though the information on the videos may be acceptable, the type of media is considered unacceptable.

Amish have no major taboos or requirements for birthing. Men may be present; most husbands choose to be involved. However, they are likely not to be demonstrative in showing affection verbally nor physically. This does not mean they do not care; it is culturally inappropriate to show affection openly in public. The laboring woman cooperates quietly, seldom audibly expressing discomfort. Because many women tend to be stoical with pain, the health-care provider needs to assess vital signs that may indicate the need for pain medicine.

Given the Amish acceptance of a wide spectrum of health-care modalities, the nurse or physician should be aware that the woman in labor might be using herbal remedies to promote labor. Knowledge about and a respect for Amish health-care practices alert the physician or nurse to a discussion about simultaneous treatments that may be harmful or helpful. It is always better if these discussions can take place in a low-stress setting before labor and birth.

As in other hospitalizations, the family may want to spend the least allowable time in the hospital. This is generally related to the belief that birth is not a medical condition and because most Amish do not carry health insurance. In their three-generational family, and as a result of their cultural expectations for caring to take place in the community, many people are willing and able to assist the new mother during the postpartum period. Visiting families with new babies are expected and generally welcomed. Older siblings are expected to help care for the younger children and to learn how to care for the newborn. The postpartum mother resumes her family role managing, if not doing, all the housework, cooking, and child care within a few days after childbirth. For a primiparous mother, her mother often comes to stay with the new family for several days to help with care of the infant and give support to the new mother. It also is not uncommon for a teenage female relative (who has completed schooling) to come stay with the family to provide assistance in household chores and care of the children. She may stay a few weeks to a few months.

The day the new baby is first taken to church services is considered special. People who had not visited the baby in the family's home want to see the new member of the community. The baby is often passed among the women to hold as they become acquainted and admire the newcomer.

7.10 Death Rituals

7.10.1 Death Rituals and Expectations

Amish customs related to death and dying have dual dimensions. On the one hand, they may be seen as holdovers from an earlier time when, for most Americans, major life events such as birth and death occurred in the home. On the other hand, Amish retention of such largely outdated patterns is due to distinctively Amish understandings of the individual within and as an integral part of the family and community. Today, when 70% of elderly Americans die in hospitals and nursing homes, some still reflect nostalgically on death as it should be and as, in fact, it used to be, in the circle of family and friends, a farewell with familiarity and dignity. In Amish society today, in most cases, this is still a reality. As physical strength declines, the expectation is that the family will care for the aging and the ill in the home (Farrar et al. 2018). Hostetler's (1993) brief observation that Amish prefer to die at home is borne out by research findings.

Clearly, these preferences are motivated by more than a wish to dwell in the past or an unwillingness to change with the times. The obligation to help others, in illness as in health, provides the social network that supports Amish practices in the passage from life to death. In effect, it is a natural extension of caregiving embraced as a social duty with religious motivation. The Amish accept literally the biblical admonition to "bear one another's burdens," and this finds expression in communal support for the individual, whether suffering, dying, or bereaved. Life's most intensely personal and private act becomes transformed into a community event.

Visiting in others' homes is, for the Amish, a normal and frequent reinforcement of the bonds that tie individuals to extended family and community. As a natural extension of this social interaction, visiting the ill takes on an added poignancy, especially during an illness believed to be terminal. Members of the immediate family are offered not only verbal condolences but many supportive acts of kindness as well. Others close to them prepare their food and take over other routine household chores to allow them to focus their attention and energy on the comfort of the ailing family member.

7.10.2 Responses to Death and Grief

Ties across generations, as well as across kinship and geographic lines, are reinforced around death as children witness the passing of a loved one in the intimacy of the home. Death brings many more visitors into the home of the bereaved, and the church community takes care of accommodations for visitors from a distance as well as funeral arrangements. The immediate family is thus relieved of responsibility for decision making, which otherwise may add distraction to grief. In some Amish settlements, a wake-like "sitting up" through the night provides an exception to normal visiting patterns. The verbal communication with the bereaved may be sparse, but the constant presence of supportive others is tangible proof of the Amish commitment to community. The return to normal life is eased through these visits by the resumption of conversations.

Both Amish and non-Amish friends and relatives come to pay their respects to both the deceased and the family. The funeral ceremony is as simple and unadorned as the rest of Amish life. A local Amish cabinetmaker frequently builds a plain wooden coffin. In the past, interment was in private plots on Amish farms, contrasting with the general pattern of burial in a cemetery in the churchyard of a rural church. Because Amish worship in their homes and have no church buildings, they also have no adjoining cemeteries. An emerging pattern is burial in a community cemetery, sometimes together with other Mennonites. Burial restrictions may also be the result of town and county ordinances restricting burials to traditional cemeteries.

Grief and loss are keenly felt, although verbal expression may seem muted, as if to indicate stoic acceptance of suffering. In fact, the meaning of death as a normal transition is embedded in the meaning of life from the Amish perspective. Parents are exhorted to nurture their children's faith because life in this world is seen as a preparation for eternal life.

7.11 Spirituality

7.11.1 Dominant Religion and Use of Prayer

Amish religious and cultural values include honesty; order; personal responsibility; community welfare; obedience to parents, church, and God; nonresistance or nonviolence; humility; and the perception of the human body as a temple of God. Amish settlements are subdivided into church districts similar to rural parishes with 30–50 families in each district. Local leaders are chosen from their own religious community and are generally untrained and unpaid. Authority patterns are congregationalist, with local consensus directed by local leadership, designated as bishops, preachers, and deacons, all of whom are male. No regional or national church hierarchy exists to govern internal church affairs, although a national committee may be convened to address external institutions of government regarding issues affecting the broader Amish population.

In addition to prayer in church services, silent prayer is always observed at the beginning of a meal, and in many families, a prayer also ends the meal. Children are taught to memorize prayers from a German prayer book for beginning and ending meals and for silent prayer. The father may say an audible "amen" or merely lift his bowed head to signal the time to begin eating.

7.11.2 Meaning of Life and Individual Sources of Strength

Outsiders, who are aware of the Amish detachment from the trappings of our modern materialistic culture, may be disappointed to discover in their "otherworldliness" something less than a lofty spirituality. Amish share the earthy vitality of many rural peasant cultures and a pragmatism born of immediate life experiences, not distilled from intellectual pursuits such as philosophy or theology. Amish simplicity is intentional, but even in austerity, there is a relish of life's simpler joys rather than a grim asceticism.

If death is a part of life and a portal to a better life, then individuals are well advised to consider how their lives prepare them for life after death. Amish share the general Christian view that salvation is ultimately individual, preconditioned on one's confession of faith, repentance, and baptism. These public acts are undertaken in the Amish context as part of preparing to fully assume one's adult role in a community of faith. In contrast with the ideals of American individualism, however, the Amish surrender much of their individuality as the price of full acceptance as members of a community. In practical, everyday terms, the religiously defined community is inextricably intertwined with a social reality, which gives it its distinctive shape.

For the Amish, the importance of conformity to the will of the group can hardly be exaggerated. To maintain harmony within the group, individuals often forgo their own wishes. In terms of faith-related behavior, outsiders sometimes criticize this "going along with" the local congregational group as an expression of religiosity, rather than spirituality. The frequent practice of corporate worship, including prayer and singing, helps to build this conformity. It is regularly tested in "counsel" sessions in the congregational assembly in which each individual's commitment to the corporate religious contract is reviewed before taking communion (Kraybill 2001).

Non-Amish occasionally are baffled at reports of the Amish response to grave injury or even loss of life at the hands of others. Owing to deeply held community values, and especially constrained by love for others, Amish often eschew retaliatory or vengeful attitudes and actions when the majority culture might justify such means. Amish are socialized to sustain such injuries, grieve, and move on without fixing blame or seeking redress or punishment for the perpetrator. The felt need to forgive is for the Amish as strong as others perceive a need to bring wrongdoers to justice. The need to forgive is considered to be "second nature" in the Amish community. It does not indicate moral superiority or a heroic strength of forbearance in the face of adversity, but flows consistently from a biblical mandate to express love, even for an apparent adversary, as a practical application of the "The Golden Rule" (Matt. 7:12). A current example, claiming both national and international attention, was the Amish response of forgiveness in the face of the Nickel Mines, Pennsylvania, tragedy when 10 Amish schoolgirls were held hostage and 5 of the girls were shot to death on October 2, 2006 (Complete Coverage of Nickel Mines Tragedy Web site at http://local.lancasteronline.com/1/91). Forgiveness in such situations may not come easily for many persons. Krabill et al. (2007) contend that for the Amish, forgiveness is part of the Anabaptist "habits" begun in the sixteenth century that continue to undergird Old Order Amish culture even today. Amish "values incorporate a willingness to place tragedy in God's hands without demanding divine explanation for injustice" (p. 71).

7.11.3 Spiritual Beliefs and Health-Care Practices

As seen in earlier sections on communication among Amish and their socioreligious provenance, many symbols of Amish faith point to the separated life, which they live in accordance with God's will. Over time, they have chosen to embody their faith rather than verbalize it. As a result, they seldom proselytize among non-Amish and nurture among themselves a non-creedal, often primitive form of Christianity that emphasizes "right living." Their untrained religious leaders offer unsophisticated views of what that entails based on their interpretation of the Bible. Most members are content to submit to the congregational consensus on what right living means, with the assumption that it is based on submission to the will of a loving, benevolent God, an aspect of their spirituality that is seldom articulated (Kraybill 2001).

Although the directives of religious leaders are normative for many types of decisions, this appears not to be the case for health-care choices (Wenger 1991a). When choosing among health-care options, families usually seek counsel from religious leaders, friends, and extended family, but the final decision resides with the immediate family. Unfortunately, the sources of healthcare information sought by many Amish, may not provide evidence-based guidance. Health-care providers need to be aware of the Amish cultural context and may need to adjust the normal routines of diagnosis and therapy to fit Amish patients' socioreligious context.

7.12 Health-Care Practices

7.12.1 Health-Seeking Beliefs and Behaviors

The Amish believe that the body is the temple of God and that human beings are the stewards of their bodies. This fundamental belief is based on the Genesis account of creation. Medicine and health care should always be used with the understanding that it is God who heals. Nothing in the Amish understanding of the Bible forbids them from using preventive or curative medical services. A prevalent myth among health-care providers in Amish communities is that Amish are not interested in preventive services. Although it is true that many times the Amish do not use mainstream health services at the onset of recognized symptoms, they are highly involved in the practices of health promotion and illness prevention.

Although the Amish, as a people, have a reputation for honesty and forthrightness, they may withhold important medical information from medical providers by neglecting to mention folk and alternative care being pursued at the same time. When questioned, some Amish admit to being less than candid about using multiple therapies, including herbal and chiropractic remedies, because they believe that "the doctor wouldn't be interested in them." Making choices among folk, complementary, and professional health-care options does not necessarily indicate a lack of confidence or respect for the latter, but rather reflects the belief that one must be actively involved in seeking the best health care available (Wenger 1994).

7.12.2 Responsibility for Health Care

The Amish believe that it is their responsibility to be personally involved in promoting health. As in most cultures, health-care knowledge is passed from one generation to the next through women. In the Amish culture, men are involved in major health-care decisions and often accompany the family to the chiropractor, physician, or hospital. Grandparents are frequently consulted about treatment options. In one situation, a scheduled consultation for a 4-year-old was postponed until the maternal grandmother was well enough after a cholecystectomy to make the three-hour automobile trip to the medical center.

A usual concern regarding responsibility for health care is payment for services. Many Amish do not carry any insurance, including health insurance. However, in most communities, there is some form of agreement for sharing losses caused by natural disasters as well as catastrophic illnesses. Some have formalized mutual aid, such as the Amish Aid Society. Wenger (1988) found that her informants were opposed to such formalized agreements and wanted to do all they could to live healthy and safe lives, which they believed would benefit their community in keeping with their Christian calling. Many hospitals have been astounded by the Amish practice of paying their bills despite financial hardship. Because of this generally positive community reputation, hospitals have been willing to set up payment plans for the larger bills.

Active participation was found to be a major theme in Wenger's (1991a, 1994, 1995) studies on cultural context, health, and care. The Amish want to be actively involved in health-care decision making, which is a part of daily living. "To do all one can to help oneself" involves seeking advice from family and friends, using herbs and other home remedies, and then choosing from a broad array of folk, alternative, and professional health-care services. One informant, who visited an Amish healer while considering her physician's recommendation that she have a computerized axial tomography (CAT) scan to provide more data on her continuing vertigo, told the researcher, "I will probably have the CAT scan, but I am not done helping myself, and this [meaning the healer's treatment] may help and it won't hurt." In this study, health-care decision making was found to be influenced by three factors: type of health problem, accessibility of health-care services, and perceived cost of the service. When the Amish use professional health-care services, they want to be partners in their health care and want to retain their right to choose from all culturally sanctioned health-care options.

Caring within the Amish culture is synonymous with being Amish. "It's the Amish way" translates into the expectation that members of the culture be aware of the needs of others and thus fulfill the biblical injunction to bear one another's burdens. Caring is a core value related to health and well-being. Care is expressed in culturally encoded expectations that they can best describe in their dialect as *abwaarde*, meaning "to minister to someone by being present and serving when someone is sick in bed." A more frequently used term for helping is *achtgewwe*, which means "to serve by becoming aware of someone's needs and then to act by doing things to help." Helping others is expressed in gender-related and age-related roles, *freindschaft* (the three-generational family), church district, community (including non-Amish), Amish settlements, and worldwide. No outsiders or health-care providers can be expected to fully understand this complex, caring network, but health-care providers can learn about it in the local setting by establishing trust in relationships with their Amish patients.

When catastrophic illness occurs, the Amish community responds by being present, helping with chores, and relieving family members so that they can be with the afflicted person in the acute care hospital. Some do opt to accept medical advice regarding the need for high-technology treatment, such as transplants or other high-cost interventions. The patient's family seeks prayers and advice from the bishop and deacons of their church and their family and friends, but the decision is generally a personal or family one.

Amish engage in self-medication. Although most Amish regularly visit physicians and use prescription drugs, as indicated previously, they also use herbs and other non-prescription remedies, often simultaneously. When discussing the meaning of health and illness, Wenger (1988, 1994) found that her Amish informants considered it their responsibility to investigate their treatment options and to stay personally involved in the treatment process rather than to relegate their care to the judgment of the professional physician or nurse. Consequently, they seek testimonials from other family members and friends about what treatments work best. They may also seek care from Amish healers and other alternative-care practitioners, who may suggest nutritional supplements. One informant told how she would take "blue cohosh" pills with her to the hospital when she was in labor because she believed they would speed up the labor.

Because of the Amish practice of self-medication, it is essential that health-care providers inquire about the full range of remedies being used. For the Amish patient to be candid, the provider must develop a context of mutual trust and respect. Within this context, the Amish patient can feel assured that the provider wants to consider and negotiate the most advantageous yet culturally congruent care.

7.12.3 Folk and Traditional Practices

The Amish, like many other cultures, have an elaborate health-care belief system that includes traditional remedies passed from one generation to the next (King 2017). They also use alternative health care that is shared by other Americans, though often not sanctioned by medical and other health-care providers. Although the prevalence of specific health-care beliefs and practices, such as use of chiropractic, Western medical and health-care science, reflexology, iridology, osteopathy, homeopathy, and folklore, is influenced mainly by *freindschaft* (Wenger 1991b), variations depend on geographic region and the conservatism of the Amish community.

Herbal remedies include those handed down by successive generations of mothers and daughters. One elderly grandmother showed the researcher the cupboard where she kept some cloths soaked in a herbal remedy and shared the recipe for it. She stated that the cupboard was where she remembered her grandmother keeping those same remedies when her grandmother lived in the *daadihaus,* the grandparents' cottage attached to the family farmhouse where her daughter and son-in-law lived. She also confided that, although she prepared the herb-soaked cloths for her daughters when they married, she thought they opted for more modern treatments, such as herb pills and prescription drugs. This is a poignant example of the effect of modern health care on a highly contextual culture.

"Of all Amish folk health care, *brauche* has claimed the most interest of outsiders, who are often puzzled by its historical origins and contemporary application" (Wenger 1991b, p. 87). *Brauche* is a folk-healing art that was practiced in Europe around the time of the Amish immigration to North America and is not unique to the Amish, but is a common healing art used among Pennsylvania Germans. As with some other European practices, the Amish have retained *brauche* in some communities. In other communities, the practice is considered suspect, and it has been the focus of some church divisions.

Brauche is sometimes referred to as sympathy curing or pow-wowing. It is unrelated to American Indian pow-wowing, and the use of this English term to refer to the German term *brauche* is unclear. In most literary descriptions of sympathy curing, it refers to the use of words, charms, and physical manipulations for treating some human and animal maladies. In some communities, the Amish refer to *brauche* as "warm hands," the ability to feel when a person has a headache or a baby has colic. Informants describe situations in which some individuals can "take" the stomach ache from the baby into their own bodies in what is described by researchers as *transference*. Wenger (1991a, 1994) stated that all informant families volunteered information about *brauche,* using that term or "warm hands" to describe folk healing. One informant asked the author if she could "feel" it, too.

A few folk illnesses have no Western scientific equivalents. The first is *abnemme*, which refers to a condition in which the child fails to thrive and appears undernourished. Specific treatments given to the child may include incantations. Some of the older people remember these treatments, and some informants remember having been taken to a healer for the ailment. The second is *aagwachse*, or *livergrown,* meaning "hidebound" or "grown together," once a common ailment among Pennsylvania Germans (Hostetler 1993). Symptoms include crying and abdominal discomfort that is believed to be caused by jostling in rough buggy rides. Wenger (1988) reported accompanying an informant with her newborn baby to an Amish healer, and the woman carried the baby on a pillow because she believed the baby to be suffering from *aagwachse.* As stated previously, Amish patients are more likely to discuss folk beliefs and practices with providers if the nurse or physician gives cues that it is acceptable to do so.

7.12.4 Barriers to Health Care

Barriers to health care include delay in seeking professional health care at the onset of symptoms, occasional overuse of home remedies, and a prevailing perception that health-care providers are not interested in, or may disapprove of, the use of home remedies and other alternative treatment modalities (Rohr et al. 2019). In addition, some families may live far from professional health-care services, making travel by horse and buggy difficult or inadvisable. Because in some Amish communities, such as the Old Order Amish, telephones are not permitted in the home, there may be delays in communication with Amish patients (Elms 2014, 2019). Finally, the cost of health care without health insurance can deter early access to professional care, which could result in more complex treatment regimens.

7.12.5 Cultural Responses to Health and Illness

The Amish are unlikely to display pain and physical discomfort. The health-care provider may need to check changes in vital signs for pain and remind the Amish patient that medication is available for pain relief if they choose to accept it.

Community for the Amish means inclusion of people who are chronically ill or "physically or mentally different." Amish culture approaches these differences as a community responsibility. Children with mental or physical differences are sometimes referred to as "hard learners," who are expected to go to school and be incorporated into the classes with assistance from other student "scholars" and parents. A culturally congruent approach is for the family and others to help engage those with differences in work activities, rather than to leave them sitting around and getting more anxious or depressed.

Hostetler (1993) stated that "Amish themselves have developed little explicit therapeutic knowledge to deal with cases of extreme anxiety" (p. 332). They do seek help from trusted physicians, and some are admitted to mental health centers or clinics (Cates 2014). However, the mentally ill are generally cared for at home whenever possible. Studies of clinical depression and manic-depressive illness were discussed in the section on biocultural ecology.

As previously mentioned, when individuals are sick, other family members take on additional responsibilities. Little ceremony is associated with being sick, and members know that to be healthy means to assume one's role within the family and community. Caring for the sick is highly valued, but at the same time, receiving help is accompanied by feelings of humility. Amish newsletters abound with notices of thanks from individuals who were ill. A common expression is "I am not worthy of it all." A care set identified in one research study is that "giving care involves privilege and obligation, and receiving care involves expectation and humility" (Wenger 1991a). The sick role is mediated by very strong values related to giving and receiving care.

The Amish culture also sanctions time out for illness when the sick are relieved of their responsibilities by others who minister to their needs. A good analogy to the communal care of the ill is found in the support offered by family and church members at the time of bereavement, as noted in the section on dying. The informal social support network is an important factor in the individual's sense of well-being. An underlying expectation, however, is that healthy individuals will want to resume active work and social roles as soon as their recovery permits. With reasonable adjustments for age and physical ability, it is understood that a healthy person is actively engaged in work, worship, and social life of the family and community (Wenger 1994). Work and rest are kept in balance, but for the Amish, the accumulation of days or weeks of free time or time off for vacation outside the framework of normal routines and social interactions is a foreign idea.

In a study of Amish women's construction of health narratives, Nelson (1999) found that the "collective descriptions [of] health included a sense of feeling well and the physical ability to complete one's daily work responsibilities" (p. vi). Women's health traditions included the use of herbal and other home remedies and consulting lay practitioners. In general, health values and beliefs are influenced by cultural group membership and personal developmental history.

7.12.6 Blood Transfusions and Organ Donation

No cultural or religious rules or taboos prohibit Amish from accepting blood transfusions or organ transplantation and donation. In fact, with the genetic presence of hemophilia, blood transfusion has been a necessity for some families. Anecdotal evidence is available regarding individuals who have received heart and kidney transplants, although no research reports or other written accounts were found. Thus, some Amish may opt for organ transplantation after the family seeks advice from church officials, extended family, and friends, but the patient or immediate family generally makes the final decision.

7.13 Health-Care Providers

7.13.1 Traditional Versus Biomedical Providers

Amish usually refer to their own healers by name rather than by title, although some say *brauch-doktor* or ***braucher***. In some communities, both men and women provide these services. They may even specialize, with some being especially good with bed-wetting, nervousness, women's problems, or livergrown. Some set up treatment rooms, and people come early in the morning and wait long hours to be seen. They do not charge fees but do accept donations. A few also treat non-Amish patients. In some communities, Amish folk healers use a combination of treatment modalities, including physical manipulation, massage, *brauche,* herbs and teas, and reflexology. A few have taken short courses in reflexology, iridology, and various types of therapeutic massage. In a few cases, their practice has been reported to the legal authorities by individuals in the medical profession or others who were concerned about the potential for illegal practice of medicine. Huntington (1993) chronicled several cases, including those of Solomon Wickey and Joseph Helmuth, both in Indiana. Both men continue to practice with some carefully designed restrictions.

7.13.2 Status of Health-Care Providers

For the Old Order Amish, health-care providers are always outsiders because, thus far, this sect has been unwilling to allow their members to attend medical, nursing, or other health-related professional schools or to seek higher education in general. Therefore, the Old Order Amish must learn to trust individuals outside their culture for health care and medically related scientific knowledge. Hostetler (1993) contended that the Amish live in a state of flux when securing health-care services. They rely on their own tradition to diagnose and sometimes treat illnesses, while simultaneously seeking technical and scientific services from health-care providers.

Most Amish consult within their community to learn about physicians, dentists, and nurses with whom they can develop trusting relationships. For more information on this practice, see the Amish informants' perceptions of caring physicians and nurses in Wenger's (1994, 1995) chapter and article on health and health-care decision making. Amish prefer health-care providers who discuss their health-care options, giving consideration to cost, need for transportation, family influences, and scientific information. They also like to discuss the efficacy of alternative methods of treatment, including folk care. When asked, many Amish, like others from diverse cultures, claim that health-care providers do not want to hear about non-traditional health-care modalities that do not reflect dominant American health-care values.

Amish hold all health-care providers in high regard. Health is integral to their religious beliefs, and care is central to their worldview. They tend to place trust in people of authority when they fit their values and beliefs. Because Amish are not sophisticated in their knowledge of physiology and scientific health care, the health-care provider who gains their trust should consider that because the Amish respect authority, they may unquestioningly follow orders. Therefore, health-care providers should make sure that their patients understand instructions. Role modeling and other concrete teaching strategies are recommended to enhance understanding.

Amish obtaining gasoline for home use

References

Allen F (1989, September 20) Country doctor: how a physician solved the riddle of rare disease in children of Amish. Wall Street J 1:A16

Baby T, Pillai R, Bindhu PR, Thomas P (2016) Ellis-van Creveld syndrome: a case report of two brothers. Oral Maxillofac Pathol J 7(1):698–701

Bass P, Waggoner D (2018) Red flags for genetic disorders. Contemp Pediatr 35(10):22–25

Belkin L (2005, November 6) A doctor for the future. New York Times WSJ Magazine, pp 68–115

Bowman HS, Procopio J (1963) Hereditary non-sperocytic hemolytic anemia of the pyruvate kinase deficient type. Ann Intern Med. 58:561–591

Cates J (2014) Serving the Amish. Johns Hopkins University Press, Baltimore

Elajami T, Giuseffi J, Avila M, Hovnanians N, Mukamal K et al (2016) Parity, coronary heart disease and mortality in the old order Amish. Atherosclerosis 254:14–19

Elizabeth Town College (2019) The Young Center for Anabaptist and Pietist Studies. http://www.etown.edu/centers/young-center/

Ems L (2014) Amish workarounds: towards a dynamic, contextualized view of technology use. J Amish Plain Anabaptist Stud 2(1):42–58

Ems L (2019) Amish philosophies on information communication technology design and use. In: CHI'19 extended abstracts: standing on the shoulders of giants: exploring the intersection of philosophy and HCI, May 04–09, 2019, Glasgow, Scotland. ACM, New York, 6 p

Enninger W, Wandt K-H (1982) Pennsylvania German in the context of an old order Amish settlement. Yearbook German-American Stud 17:123–143

Farrar H, Kulig J, Sullivan-Wilson J (2018) Older adult caregiving in the Amish: an integrative review. J Cult Divers 25(2):54–65

Foroud T, Casteluccio P, Kollar D, Edenberg H, Miller M, Boman L (2000) Suggestive evidence of a locus on chromosome 10p using the NIMH genetics initiative bipolar affective disorder pedigrees. Am J Med Genet 96(1):18–23

Fuchs JA, Levinson R, Stoddard R, Mullet M, Jones D (1990) Health risk factors among Amish: results of a survey. Health Educ Q 17(2):197–211

Gill K, Cardenas S, Kassem L, Schulze T, McMahon F (2016) Symptom profiles and illness course among Anabaptist and non-Anabaptist adults with major mood disorders. Int J Bipolar Disord 4(1):1–7

Granville K, Gilbertson A (2017) In Amish Country, the future is calling. New York Times, (166) 57723, 9/17/2017. Sunday Business. pp 6–7

Hairston K, Ducharme J, Treuth M, Hsueh W, Jastreboff A et al (2013) Comparison of BMI and physical activity between old order Amish children and non-Amish children. Diabetes 36(4):873–878

Hostetler JA (1993) Amish society, 4th edn. Johns Hopkins University Press, Baltimore

Hou L, Kember RL, Roach JC et al (2017) A population-specific reference panel empowers genetic studies of Anabaptist populations. Sci Rep 7:6079. https://doi.org/10.1038/s41598-017-05445-3

Hsueh WC, Mitchell BD, Aburomia R, Pollin T, Sakul H, Gelder EM, Michelsen BK, Wagner MJ, St. Jean PL, Knowler WC, Burns DK, Bell CJ, Shuldine AR (2002) Diabetes in the old order Amish: characterization and heritability analysis of the Amish family diabetes study. Am J Clin Nutr 75(6):1098–1106

Huffines ML (1994) Amish languages. In: Dow JR, Enninger W, Raith J (eds) Internal and external perspectives on Amish and Mennonite life. 4: Old and new world Anabaptist studies on the language, culture, society and health of Amish and Mennonites. University of Essen, Essen, pp 21–32

Huntington GE (1993) Health care. In: Kraybill DB (ed) The Amish and the state. Johns Hopkins University Press, Baltimore

Hüppi J (2000) Research note: identifying Jacob Ammann. Mennonite Quart Rev 74(10):329–339

Hurst CE, McConnell DL (2010) An Amish paradox: diversity and change in the world's largest Amish community. The Johns Hopkins University Press, Baltimore

Johnson-Weiner K (2014) Technological diversity and cultural change among contemporary Amish groups. Mennonite Quart Rev 88(1):5

Johnson-Weiner K (2015) Old order Amish education: the Yoder decision in the 21st century. J Amish Plain Anabaptist Stud 3(1):25–44

Katz ML, Ferketich AK, Broder-Oldach B et al (2012) Physical activity among Amish and non-Amish adults living in Ohio Appalachia. J Community Health 37:434–440. https://doi.org/10.1007/s10900-011-9460-9

Ketterling RP, Bottema CD, Koberl DD, Setsuko I, Sommer SS (1991) $T^{296}M$, a common mutation causing mild hemophilia B in the Amish and others: Founder effect, variability in factor IX activity assays, and rapid carrier detection. Hum Genet 87:333–337

King M (2017) Crafting an Amish biomedical landscape. Med Anthropol Theory 4(1):105–122

Kraybill D (2019) The Amish of Lancaster County, 2nd edn. Stackpole Books, Guilford

Kraybill DB, Nolt SM, Weaver-Zercher DL (2007) Amish grace: how forgiveness transcended tragedy. Wiley, San Francisco

Kraybill DB, Johnson-Weiner K, Nolt S (2013) The Amish. John Hopkins University Press, Baltimore

Lantz G (2019) Perceptions of lifestyle as mental health protective factors among Midwestern Amish, Unpublished doctoral dissertation. Walden University, Minneapolis

Lipon T (1985) Husband and wife work roles and the organization and operation of family farms. J Marriage Fam 47(3):759–764

Lucas CA, O'Shea RM, Zielezny MA, Freudenheim JL, Wold JF (1991) Rural medicine and the closed society. N Y State J Med 91(2):49–52

Lutz M (2017) The Amish in the market: competing against the odds? Am Stud J 63:1–7

Martin PH, Davis L, Askew D (1965) High incidence of phenylketonuria in an isolated Indiana community. J Indiana State Med Assoc 56:997–999

McKusick VA (1978) Medical genetics studies of the Amish: selected papers assembled with commentary. Johns Hopkins University Press, Baltimore

McKusick VA, Egeland JA, Eldridge D, Krusen EE (1964) Dwarfism in the Amish I. The Ellis-van Creveld syndrome. Bull Johns Hopkins Hosp 115:306–330

Mendez Ruiz A (2017) The Amish rule of order: conformity and deviance among Amish youth. Arcadia University ScholarWorks, Arcadia University, Glenside

Meyers TJ, Nolt SM (2005) An Amish patchwork: Indiana's old orders in a modern world. Quarry Books, Indiana University Press, Bloomington
Miller NL (1997) Childbearing practices as described by old order Amish women. Doctoral dissertation, Michigan State University. Dissertation Abstracts International, UMI 1388555
Moledina A, McConnell D, Sugars S, Connor B (2014) Amish economic transformations: new forms of income and wealth distribution in a traditionally 'flat' community. J Amish Plan Anabaptist Stud 2(1):1–22
Nelson WA (1999) A study of Amish women's construction of health narratives. Unpublished doctoral dissertation, Kent State University, Kent, OH
Nolt S (2015) A history of the Amish. Goodbooks Publishing, New York
Nolt S (2016) The Amish: a concise introduction. Johns Hopkins University Press, Baltimore
Ratnoff OD (1958) Hereditary defects in clotting mechanisms. Adv Intern Med 9:107–179
Riley P, Weiner D, Leighley B, Jonah D, Morton DH et al (2015) Cartilage hair hypoplasia: characteristics and orthopaedic manifestations. J Child Orthop 9(2):145–152
Rohr JM, Spears KL, Geske J et al (2019) Utilization of Health Care Resources by the Amish of a Rural County in Nebraska. J Community Health 44:1090–1097. https://doi.org/10.1007/s10900-019-00696-9
Stevick R (2014) Growing up Amish: The Rumspringa years (Young Center Books in Anabaptist and Pietist Studies), 2nd edn. John Hopkins University Press, Baltimore
Strauss K (2015) Genomics for the people. Sci Am 313(6):66–73
Strauss K, Puffenberger E, Morton D (2012) One community's effort to control genetic disease. Am J Public Health 102(7):1300–1306
Strotmeyer S, Koff A, Honeyman JN et al (2019) Injuries among Amish children: opportunities for prevention. Inj Epidemiol 6:49. https://doi.org/10.1186/s40621-019-0223-x
Teapole B(2019) Amish perspectives of the genetic counseling process. Unpublished master's thesis. University of South Carolina, Columbia. https://scholarcommons.sc.edu/etd/5208
Troyer H (1994) Medical considerations of the Amish. In: Dow JR, Enninger W, Raith J (eds) Internal and external perspectives on Amish and Mennonite life 4: old and new world Anabaptist studies on the language, culture, society and health of the Amish and Mennonites. University of Essen, Essen, pp 68–87
Weller G (2017) Caring for the Amish: what every anesthesiologist should know. Anesth Analg 124(5):1520–1528
Wenger MR (1970) A Swiss German dialect study: three linguistic islands in Midwestern USA. University Microfilms, Ann Arbor
Wenger AF (1980, October) Acceptability of perinatal services among the Amish. Paper presented at a March of Dimes symposium, future directions in perinatal care, Baltimore, MD
Wenger AFZ (1988) The phenomenon of care in a high-context culture: the old order Amish. Doctoral dissertation, Wayne State University. Dissertation Abstracts International, 50/02B
Wenger AFZ (1991a) The culture care theory and the old order Amish. In: Leininger MM (ed) Cultural care diversity and universality: a theory of nursing. National League for Nursing, New York, pp 147–178
Wenger AFZ (1991b) Culture-specific care and the old order Amish. Imprint 38(2):81–82, 84, 87, 93
Wenger AFZ (1991c) The role of context in culture-specific care. In: Chinn PL (ed) Anthology of caring. National League for Nursing, New York, pp 95–110
Wenger AFZ (1994) Health and health-care decision-making: the old order Amish. In: Dow JR, Enninger W, Raith J (eds) Internal and external perspectives on Amish and Mennonite life 4: old and New World Anabaptists studies on the language, culture, society and health of the Amish and Mennonites. University of Essen, Essen, pp 88–110
Wenger AFZ (1995) Cultural context, health and health-care decision making. J Transcult Nurs 7(1):3–14
What is in a language? (1986) Family Life, p 12
Wolkomir R, Wolkomir J (1991, July) The doctor who conquered a killer. Reader's Digest 139:161–166
Xu H, Ryan K, Jaworek T, Southam L, Reid J et al (2017) Familial hypercholesterolemia and type 2 diabetes in the old order Amish. Diabetes 66(7):2054–2058

8 People of Appalachian Heritage

Sandra J. Mixer and Mary Lou Clark Fornehed

8.1 Introduction

The Appalachian region includes 13 states and 420 counties that span 205,000 square miles and is called home by approximately 25 million people, 8% of the U.S. population (Appalachian Regional Commission [ARC] 2017a). This vast region is one of contrasts, starting at the southernmost mountains of New York and continuing south to northern Mississippi encompassing all of West Virginia and parts of Alabama, Georgia, Kentucky, Maryland, Mississippi, New York, North Carolina, Ohio, Pennsylvania, South Carolina, Tennessee, and Virginia.

This chapter is a revision made of the original chapter written by Kathleen Huntlinger in the previous edition of the book.

S. J. Mixer (✉)
University of Tennessee-Knoxville, Knoxville, TN, USA
e-mail: smixer@utk.edu

M. L. C. Fornehed
Tennessee Technological University, Cookeville, TN, USA

8.2 Overview, Inhabited Localities, and Topography

8.2.1 Overview

Appalachia is characterized by a rolling topography with rugged ridges and hilltops, some extending 4000–6000 ft high with remote valleys between them. The surrounding valleys give one a sense of isolation, peacefulness, and separateness from the more heavily travelled urban areas. Even though the Appalachian region includes several large cities, many people live in small settlements and in limited-accessible hollows or "hollers". This isolation and rough topography have contributed to the development of secluded communities where people, over time, have developed a strong sense of independence and family and community cohesiveness. Persons who do not have access to private transportation experience accessibility issues because public transportation is not widely available, and even then only in the larger, more urbanized areas.

8.2.2 Heritage and Residence

The Appalachian culture is dynamic, complex, diverse, and rich with distinctive music, art, and literature. American Indian tribes lived in the region for centuries before European settlers arrived. Germans, Scots-Irish, Welsh, French, and

© Springer Nature Switzerland AG 2021

L. D. Purnell, E. A. Fenkl (eds.), *Textbook for Transcultural Health Care: A Population Approach*,
https://doi.org/10.1007/978-3-030-51399-3_8

British constitute the primary groups who settled the region between the seventeenth and nineteenth centuries. It was thought that people wanting to escape South Carolina's mosquitos and malaria were drawn to the easternmost part of Tennessee. There was a sense that the area's isolation and ruggedness gave them a feeling of being free from the persecution they experienced in their previous homes over religious, social, and economic differences (Gobble 2009; Stephenson 1984).

Although predominantly White, many settlers maintained a strong family identity with Native American Indians. For example, the Cherokee Indians once called vast portions of Kentucky and Tennessee and northern portions of Alabama, Georgia, and South Carolina home. As White settlers demanded more land, the Indian removal act of 1830 forced over 16,000 Cherokee Indians to walk to Indian Territory, now Oklahoma. Approximately 4000 died enroute. Some Cherokee owned land and were allowed to stay, some hid in the mountains, and some later walked back. In the 1850s the Eastern Band Cherokee Indians (EBCI) was formed as a sovereign nation. They purchased 57,000 acres of their land in Western North Carolina from the federal government and today number over 14,000 members (EBCI 2020). Many Appalachians in the region trace family heritage to Cherokee. Cherokee and regional Appalachian culture is intertwined, e.g. folk medicine practices.

African Americans have been in Appalachia since the 1500s. Early Spanish and French explorers brought with them African slaves; free persons of color were among the earliest settlers. While a plantation economy never developed in the region, many wealthy mountain people owned slaves who worked in stores and inns, logging, and mining, as well as on farms. After the Civil War, many freed African Americans bought land to farm and lived rural lifestyles similar to those of their White neighbors. By 1860, African Americans comprised 13% of the Appalachian population (Burns et al. 2006; Webb 2013).

Melungeons are a group of mixed-race persons living in the mountains of rural Appalachia. For many years, there was mystery surrounding their origins. However, a 2012 DNA study found Melungeons are the offspring of sub-Saharan African men and White women of northern or central European origin. In many regions, people endured prejudice and segregation: therefore, Melungeon heritage was often denied. However, there is a resurgence of identification of Melungeon ancestry and many persons gather yearly at the Melungeon Heritage Association annual conference (Neal 2015; Webb 2013).

Although predominately White, Appalachia has always had a racially and ethnically diverse population. With recent immigration in the United States, the Appalachian region has become more ethnically diverse including people of Asian, African, eastern European, and Hispanic heritage (Campbell 2006).

Many of the people who live in Appalachia can trace their family roots back 200 years or more and it is common to find whole communities comprising extended, related families (ARC 2019b). It is important to remember that simply taking up residence in Appalachia does not make one an "Appalachian," since a significant value is held for those whose roots are well identified within the region. A medical student doing his rotation in rural Appalachia made a poignant observation: "My Yankee accent and my formidable shirt and tie made it clear I was an outsider. It took more than one coal miner covered in soot in the exam room, more than one duct-taped puncture wound before I realized that such sights were commonplace… One day… I offered up to a patient that I had married a girl from their county and from that time on, it all changed. I was welcomed into the fold as family… To be an insider in Appalachia is to be tied to the people. For them, community is everything" (Carter 2009, p. 470).

Like many marginalised groups, the people of Appalachia have been described in stereotypically negative terms (e.g., "poor White trash") that in no way represent the people or the culture as a whole. Rural Appalachians generally are distrustful of outsiders and often are misunderstood. They have been stereotyped as being 'stupid,' 'hillbillies,' and 'rednecks' (Coyne et al. 2006; Mixer et al. 2014). During the past 60 years, the media have perpetuated the stereotypes with cartoon strips such as "Li'l Abner" and television programs such as the *Dukes of Hazzard*, and stories of the feuding Hatfields and McCoys. However, Appalachian persons cherish the solitude of their environment, stay close to family and community and maintain a faith in God. They describe cultural values of a strong work ethic, loyalty, caring, hardiness, independence, humility, honesty, patriotism, and resourcefulness. (Coyne et al. 2006; Mixer et al. 2014; Salyers and Ritchie 2006).

8.2.3 Reasons for Migration and Associated Economic Factors

Approximately 300 years ago, people came to Appalachia to seek religious freedom, land for themselves, and control over social interactions with the outside world. Over the years, mining and timber resources have become depleted, farmland has eroded, and jobs have become scarce, which has resulted in an out-migration of working age people to larger urban areas. For example, North to cities such as Detroit, Cleveland, and Cincinnati for industrial jobs; to the South in cities like Charlotte and Atlanta; and even to states as far away as Nevada and California (Pollard and Jacobsen 2011). This migration began after World War II and has remained constant ever since (Obermiller and Brown 2002). Those who moved to urban areas often feel alone and sometimes become depressed as they are separated from family and friends. Many of those who have remained in urban settings have become bicultural, adapting to the culture of urban life while retaining, as much as possible, their traditional Appalachian culture. In recent times, Appalachia has seen migration both into and out of the mountains and small communities (DeYoung et al. 2006; Phipps 2006). Out-migration continues for those seeking a better life through more education and better employment. The increase of returning rural Appalachians may be due to persons choosing to return home to live, work, and serve while others retire and live close to what they consider their home.

Initially dependent on mining, forestry, and agriculture, Appalachia has become more diverse over the past 50 years to include manufacturing and service industries. However, because of its vastness, the economic status of the Appalachian region varies from state to state. The eastern portion of rural Appalachia is remote and continues to have widespread poverty and limited access to health care, which are in stark contrast to other regions of these states.

Between 1969 and 2017, the population in the Appalachian region and employment both declined 2% compared to the rest of the nation, evidence that fewer people live and work in Appalachia today (ARC 2019b). The per capita income in the Appalachian region is approximately $41,000/year versus $51,500 in the U.S. The three-year average unemployment rate 2015–2017 was 5.4% compared to 4.8% in the U.S. The poverty rate from 2013 to 2017 was 16.3% compared to 14.6% in the nation. Kentucky has the worst poverty rate in the region with a 25.4% rate in the Appalachian portion versus 18.9% rate for the rest of the state (ARC 2017b). Using U.S. census bureau data, annually ARC compares each Appalachian county unemployment rate, per capita market income, and poverty rates to the national average. Counties are ranked as distressed if median family income is less than 67% of U.S. average and poverty rate is 150% of U.S. average. Eighty-one Appalachian counties were considered economically distressed in 2019. Kentucky far exceeds other Appalachian states with 38 counties designated as distressed. West Virginia is second with 16 distressed counties and Tennessee follows with nine. However, every Appalachian state has numerous counties designated as economically at risk and transitional (ARC 2020b, d).

Employment growth was slower in all industries across Appalachia from 2002 to 2017 and fewer jobs were available per person in Appalachia than the rest of the country. In fact, the economic growth rate was one half the rate of the U.S. However, the economic decline over the past 15 years has been much slower, yet still in a downward trend. While there was a decline in manufacturing jobs, there has been growth in service economy jobs. Interestingly, coal, gas, and mining ranks 14th out of 15 job industries, contributing only 1.1% of the regions earnings. While farming ranks 12th contributing only 0.7% of earnings. In the South-Central region, Alabama, Georgia, and South Carolina experienced the greatest economic growth while Kentucky, New York, and Virginia Appalachian regions experienced declines, despite growth in each state's non-Appalachian regions. As one might expect, Appalachian metropolitan counties outperformed rural counties (ARC 2019b). The Appalachian Regional Commission (ARC) was created in 1965 as a federal, state, and local partnership to promote economic growth. Today, goals include business and workforce development, critical infrastructure, natural and cultural assets, and leadership and community capacity (ARC 2019c). The ARC Appalachian states local leaders and private investors have partnered to create or retain more than 17,300 jobs and train and educate 51,000 students and workers (ARC 2019c).

8.2.4 Educational Status and Occupations

Although many of the original immigrants to this area were highly educated when they arrived, limited access to more formal education resulted in the isolation of later generations with fewer educational opportunities. The education received in Appalachia in the twenty-first century is guided by policies and procedures established at the state and federal level (DeYoung et al. 2006). These policies have presented problems, as few are culturally sensitive or address the actual needs of some of the rural Appalachian counties. Rural Appalachia is in crucial need of policy reform to address the educational needs of its students (The Rural School Community and Trust 2014). Without policy reform, rural Appalachian instructional salaries and expenditures per student will continue to be low. There also has been much debate about consolidating some schools and using bus transportation, which could mean very long travel times for children.

Rural Appalachian cultural marginalization has persisted because of old stereotypes of being a "hillbilly" and "stupid," due to the region's economic distress and poverty. The cultural marginalization of this area has had a significant impact on the type of careers rural Appalachian students aspire to and choose (Ali and McWhirter 2006; Ali and Saunders 2009). Rural Appalachian adolescents receive mixed cultural messages from their peers regarding their educational advancement. Additionally, the historical presence of mining and manufacturing employment in these small communities has often led to the belief that education is not necessary.

With the decrease in manufacturing and mining jobs, some adolescents lack confidence for a better life in rural Appalachia. For some families, education beyond high school is viewed as not as important as earning a living to help support the family. Some Appalachian parents, and several who belong to more conservative and secular religious sects, do not want their children influenced by mainstream middle-class American behaviors and actions. These influences can leave adolescents poorly prepared to join an evolving workforce that demands education beyond a high school diploma (Ali and Saunders 2009; Hendrickson 2012). Young people who are unprepared and unambitious contribute to the cycle of poverty in this area.

Many programs exist to combat educational challenges and help parents, students, and teachers in the region. For example, beginning in the summer of 1990, ARC and the Oak Ridge National Laboratory partnered to offer an all-expenses paid yearly residential summer math and science technology institute and science academy targeting high school (2 weeks) and middle school (1 week) students and teachers. Students participate in hands on math, science, and computer science research projects while teachers work with science practitioners to develop science, technology, engineering, and math (STEM) curriculum. Additional interactive, physically active, team building and fun activities such as high ropes, zipline, swimming, and a trip to Dollywood are included. Program goals are to inspire students to (a) continue education beyond high school (b) consider a career in STEM; learning about STEM jobs available in Appalachia and

(c) improve STEM instruction in Appalachian schools (ARC 2020a).

Varying data provide evidence that Appalachian educational outcomes trail non-Appalachians. In 2018, the states of Tennessee, Kentucky, Alabama, and Mississippi ranked within the 10 least educated states. In a different ranking, only West Virginia and Alabama ranked so low (CATO Institute 2018). For the period 2013–2017, bachelor degree completion rates for Appalachia were 23.7% compared to 30.9% for the U.S. and high school graduation rates were 76.8% (Appalachians) compared to 80.4% U.S. (ARC 2020e).

A variety of issues contribute to educational challenges in the region. Access to colleges and universities, community colleges, and technical training has improved; however, there remains limitations in the number and placement of educational facilities accessible throughout the region. Despite U.S. representatives work to ensure rural Appalachians have access to the internet, grave disparity exists today. Without affordable, reliable, high-speed internet access, rural Appalachians are not able to access online education, educational and informational resources, nor numerous economic opportunities. The federal government program, Connect America, only requires companies to deliver 10 mb download, while the FCC considers broadband to be at least 25 mb and many cities are getting 1000 mb. To address such lack of service, many Appalachians are working to solve the problem locally using creative solutions such as cooperatives (Institute for Local Self-Reliance 2020). Additionally, the mountainous terrain often limits those services that require "line of sight" such as cell phone, further complicating Appalachians' internet access.

8.3 Communication

8.3.1 Dominant Language and Dialects

The dominant language of the Appalachian region is English. The seemingly archaic nature of the language and phrases of rural Appalachian people have long been ridiculed by those living outside the community (Montgomery 2006). While the unique language patterns are considered regional, there is no single dialect, which can make communication with outsiders even more difficult. There has been much debate regarding the linguistic ancestry of the language, but with the influx of European immigrants and the American Indians, the speech is felt to be of mixed origin.

Some of the more isolated groups in Appalachia speak Elizabethan English, which has its own distinct vocabulary and syntax and can cause communication difficulties for those who are not familiar with it. Some examples of variations in pronunciation for words are *allus* for "always" and *fit* for "fight." Word meanings may be different for example, *sass* for "vegetables." The Appalachian region is also noted for its use of strong preterits such as *drug* for "dragged" and *swelled* for "swollen." Plural forms of monosyllabic words are formed like Chaucerian English, which adds *es* to the word—for example, "post" becomes *postes*, "beast" becomes *beastes*. Many people, especially in the non-academic environment, drop the *g* on words ending in *ing*. For example, "writing" becomes *writin'* and "reading" becomes *readin'*. In addition, vowels may be pronounced with a diphthong that can cause difficulty to one unfamiliar with this dialect—hence, *poosh* for "push," *warsh* for "wash," *deef* for "deaf," *welks* for "welts," *whar* for "where and *your'n* for "your." However, when a word is written, the meaning is apparent (Wilson 1989).

Language clashes between rural Appalachians and outsiders have led to frustration when communicating needs. It can take a while for persons to get to the crux of what they are trying to communicate. To effectively communicate, health care providers need to allow adequate time to clarify words and meanings. Clinicians may ask the person to write the words (if the person has writing skills) to help prevent miscommunication, incorrect diagnoses and to improve outcomes and following directions with health prescriptions and treatments. (Carter 2009;

Diddle and Denham 2010; Lozier and Althouse 1975; Rowles 1991; Salyers and Ritchie 2006). The introduction of foreign healthcare providers in the region to offset the health labor shortage (ARC 2020c) further complicates communication and meeting Appalachians' healthcare needs in a culturally satisfactory way.

8.3.2 Cultural Communication Patterns

Appalachians practice the ethic of neutrality, which helps shape communication styles, their worldview, and other aspects of the Appalachian culture. Four dominant themes affect communication patterns in the Appalachian culture: avoiding aggression and assertiveness, not interfering with others' lives unless asked to do so, avoiding dominance over others, and avoiding arguments and seeking agreement (Smith and Tessaro 2005).

In general, Appalachians are a very private people who do not want to offend others, nor do they easily trust or share their thoughts and feelings with *outsiders*. They are more likely to say what they think the listener wants to hear rather than what the listener needs to hear. Many Appalachians dislike authority figures and institutions that attempt to control their behavior. Individualism and self-reliant behavior are idealized.

Appalachians may be sensitive to direct questions about personal issues. Sensitive topics are best approached with indirect questions and suggestions and without critical innuendo. Appalachians are taught to deny anger and not complain. Information should be gathered in the context of broader relationships with respect for the ethic of equality, which implies more horizontal than hierarchical relationships, allowing cordiality to precede information sharing. Starting with sensitive issues may invite ineffectiveness; thus, the health-care provider may need to "sit a spell" and "chat" before getting down to the business of collecting health information. To establish trust, the health-care provider must show interest in the community, the patient's family, and other personal matters; drop hints instead of give orders; and solicit the patient's opinions and advice. These actions increase the patient's self-worth and self-esteem and help to establish the trust needed for an effective working relationship. A health-care provider may need to use more open-ended questions when obtaining health information and eliciting opinions and beliefs about health-care practices, such as "What do you believe might be causing your illness?" (Huttlinger 2013).

To communicate effectively with Appalachian patients, nonverbal behaviors must be assessed within the contextual framework of the culture. Traditional Appalachians value personal physical space so they are more likely to stand at a distance when talking with people in both social and health-care situations (Coyne et al. 2006). Therefore, many people may perceive direct eye contact, especially from strangers, as an aggressive or hostile act. Staring is considered bad manners.

Many Appalachians are comfortable with silence and when talking with health-care providers who are outsiders, they are likely to speak without emotion, facial expression, or gestures and avoid telling unpleasant news to avoid hurting someone's feelings. Health-care providers who are unfamiliar with the culture may interpret these nonverbal communication patterns as not caring. Within this context, the health-care provider needs to allow sufficient time to develop rapport by dropping hints instead of giving orders (Coyne et al. 2006). In addition, to communicate effectively with traditional Appalachians, health-care providers must not ignore speech patterns; they must clarify any differences in word meanings, translate medical terminology into everyday language using concrete terms, explain not only what is to be done but also why, and ask patients to repeat or demonstrate instructions to ensure understanding. Adopting an attitude of respect and flexibility demonstrates interest and helps bridge barriers imposed by health-care providers' personal ideologies and cultural values. Throughout history, Appalachians have enjoyed storytelling, a practice that still continues; accordingly, some individuals may respond better to verbal instructions and education, with reinforcement from videos rather than printed communications.

An example of a typical interaction in an Appalachian community may be illustrated by this statement from a key informant in the Counts and Boyle (1987) Genesis Project that took place from 1985 to 1994. Miss Ruth, a 94-year-old native Appalachian, was interviewed in the house in which she was born. In fact, she had her appendix removed in the living room of this same house by a traveling nurse. After returning from a trip to Africa (she had a doctorate and liked to travel, but always returned home), Miss Ruth described the concept of "neighboring" as a double-edged sword. The positive side is that when you are sick, everyone comes around to take care of you; however, on the negative side, when you try to do something quietly, everyone knows about it. This example continues to reflect the experiences of many in Appalachia today.

The following research study highlights the critical nature of culturally appropriate health provider communication. The incidence of cervical cancer rates in central Appalachia is 35% above the national average. In Eastern Kentucky, the estimated rate of invasive cervical cancer exceeds 67% the national average. Researchers undertook evaluating the effectiveness of health-care provider communication to help rural Appalachian women understand changes in recommended cervical cancer screening guidelines. Forty-one Appalachian women ages 24–65 participated. Health-care provider communication both aggravated and alleviated women's uncertainty about updated cervical cancer screening guidelines (Cohen et al. 2016).

Generally, women understood they were to be screened yearly throughout their lives via a PAP smear and expressed many concerns about the new guidelines. For example, they wondered if women after age 65 years weren't screened, perhaps they didn't matter anymore? Were the new guidelines in the best interest of women; were they credible? Participants spoke about trusting screening recommendations from their provider whom they believed looked out for their needs. Appalachian women have verbalized that the new guidelines for cervical cancer screening have not been adequately communicated to them, thus leaving them uncomfortable with the new guidelines and with many questions regarding the schedule for screening. The study demonstrated that effective communication among rural Appalachian women includes building trust, deep listening, thorough, culturally sensitive assessment, and managing uncertainty. These women were then armed with the knowledge needed to improve cervical cancer screening behavior (Cohen et al. 2016).

8.3.3 Temporal Relationships

The traditional Appalachian culture is present oriented, living for today, as compared with future-oriented, planning for the future. Living for today may mean ignoring expert advice and "accepting one's lot in life." Other problems may be more pressing, and "just getting by" may be the most important activity. Health-care providers must realize that the emphasis on illness prevention in our current society may not be a focus for some Appalachians. For those living in poverty and isolation, the trend is to "live for today" and to rely on more traditional approaches for those things that cannot be controlled. This worldview is common with present-oriented societies in which a higher power is in charge of life and its outcomes resulting in a deterrent to obtaining preventive health services. With a fatalistic view, individuals have little or no control over nature, and the time of death is "predetermined by God." Such beliefs have a foundation in biblical teachings; Jesus taught followers to trust him, not worry about tomorrow, and live for today. One frequently hears expressions such as "I'll be there, God willing and if the crick [creek] don't rise." As communication systems such as televisions, satellite dishes, and the Internet become more commonplace, temporal relationships are becoming more future-oriented.

For the traditional Appalachian, life is unhurried and body rhythms, not the clock, control activities. One may come early or late for an appointment and still expect to be seen. If individuals are not seen because they are late for an appointment and are asked to reschedule, they are likely to not return because they may feel rejected. Many Appalachians are hesitant to make appointments because "somethin' better might come up" or they may not be sure of transportation until the last minute.

8.3.4 Format for Names

Although the format for names in Appalachia follows the U.S. standard given name plus family name, individuals address nonfamily members by their last name. A common practice that denotes neighborliness with respect is to call a person by his or her first name with the title Mr. or Miss (pronounced "miz" similar to "Ms.," when referring to women, whether single or married)—for example, Miss Lillian or Mr. Bill. Miss Lillian may or may not be married. There is also a need to provide a link with both families of origin. Many times Appalachians refer to a married woman as "she was [born] a…," thus linking the families and enhancing the feeling of continuity.

8.4 Family Roles and Organization

8.4.1 Appalachian Family Culture, Head of Household and Gender Roles

The Appalachian worldview has historically included limited contact with the world beyond the mountain hollow and distrust of outsiders (Gobble 2009). This combination of beliefs created strong and complex kinship/community networks that included a heightened sense of caring for and being loyal to family. This perspective shapes how Appalachians view, understand, and express their physical, mental, and spiritual needs in this world.

Rural Appalachian persons often refer to their family members and neighbors as kin. An anthropologist examining semantics and kinship in rural Appalachian communities found that kinship was often referred to as a "set" (Batteau 1982, p. 18). The kin set is used to describe the family units as they are geographically connected within communities. Kin sets include persons who are descendants of an original settler and still live in close proximity to one another. These communities are very often within the valleys or "hollers" of the mountains that often follow a stream or river. These kin sets offer support to one an another in good and bad times, often deeding land to children and grandchildren so additional households can begin, thereby extending the kin set (Burns et al. 2006). Families are more than genetic relationships and are described as including brothers, sisters, aunts, uncles, parents, grandparents, cousins, in-laws, and out-laws (those related by marriage).

Historically, gender roles for Appalachian men and women were more clearly defined and followed traditional family roles. Men engaged in physical work to support the family financially and to provide transportation. Women took care of the house and assumed responsibility for child rearing and care for elders, ill or dying family members (Gobble 2009). Self-made individuals and families, or those who carried out their own subsistence and depended little on outsiders, were idealized. In modern families, the mother may now work outside the home while retaining the position of caregiver and household manager. Many families have had to become flexible within their traditional roles in order to survive. People have out-migrated beyond rural Appalachian borders to survive financially. In such cases, extended family or kin sets often step in to provide the support necessary during temporary job loss or illness (Burns et al. 2006; Lozier and Althouse 1975; Salyers and Ritchie 2006).

The traditional Appalachian household is viewed by outsiders as patriarchal; however, insiders freely share that "mama is the boss". Some families are becoming more egalitarian in their beliefs and practices. Women are generally the providers of emotional strength, with older women having a lot of clout in health-care matters (Mixer et al. 2015). Older women are usually responsible for preparing herbal remedies and folk medicines and are sought out by family members and neighbors for these preparations (Fornehed 2017; Mixer et al. 2014).

Grandparent headed households (GHH) have proven to be one of the fastest growing family units in Appalachia (Hatcher et al. 2018) and have long thought to be a vulnerable population with higher risks of health problems. These units are sometimes referred to as "skipped generation families" (p. 42), conceived out of necessity so kinship ties could remain intact rather than grandchildren being placed into state care.

Multifactorial reasons for a GHH to exist include but not limited to parental death and incarceration resulting from the opioid epidemic. Hatcher et al. (2018) does report positive experiences for the GHH. These included improved health and strong social support systems. There were also barriers to healthy living within this family unit with the most notable being too young to take advantage of older adult services such as retirement benefits that would be helpful to the family unit. Being a GHH could mean the caregiver must be home to care for small children and without a retirement or Social Security benefit, this led to other barriers for healthy living, such as having food insecurities. These GHH also reported a lack of support with the stress of dealing with a drug addicted adult child who was often hostile and interfered with the raising of the grandchild. These barriers place the grandparents at risk for increased health issues. Authors called for interventions that will decrease GHH vulnerability to these health threats (Hatcher et al. 2018).

8.4.2 Prescriptive, Restrictive, and Taboo Behaviors for Children and Adolescents

Children are important to the Appalachian culture (Coyne et al. 2006). Large families are common and children are usually accepted regardless of whether the parents are married. Parents may impose strict social conformity for family members in fear of community censure and their own parental feelings of inferiority. Permissive behavior at home is unacceptable, and hands-on physical punishment to a degree that some perceive as abuse is common. For Appalachian children who have problems with school performance, the most effective approach to increase performance is to provide individualized attention rather than group support or attention, an approach that is congruent with the ethic of neutrality. To be effective in changing negative behavior, it is necessary to emphasize positive points.

As children progress into their teens, mischievous behavior is accepted but not condoned. Continuing formal education may not be stressed because many teens are expected to get a job to help support the family. Children are seen as being important, and to many, having a child, even at an early age (less than 18 years), means fulfilment. Motherhood increases the woman's status in the church and the community. In previous generations, it was not uncommon for teenagers to marry by the age of 15, and some as early as 13. Children, single or married, may return to their parents' home, where they are readily accepted, whenever the need arises.

Teens in Appalachia enter into a cultural dilemma when exposed to other lifestyles outside the home and family. Health-care providers can assist adolescents and their family members in working through these cultural differences by helping them resolve personal conflicts resulting from being exposed to different cultures and lifestyles. Some ways to promote a positive self-awareness that conveys a respect for their culture are discussing personal parenting practices and providing information about health promotion and wellness; disease, illness, and injury prevention; and health restoration and maintenance in a culturally congruent way.

Young children and adolescents in Appalachia and across the U.S. do experience adverse childhood experiences (ACEs) (Centers for Disease Control and Prevention [CDC] 2019a). ACEs are childhood (age 0–17 years) experiences where children have either witnessed or been directly involved in violent acts or abuse. ACEs are reported by at least 60% of adults across 25 states. Children who reside in vulnerable populations who are under-resourced, segregated, and with food insecurities are at the highest risk for ACEs. Children and adolescents in Appalachia who have grown up in a culture of substance misuse such as watching parents use drugs, being offered their drugs, seen parents arrested, or incarcerated are at great risk. Why does this matter? ACEs are linked to chronic health problems like heart disease, diabetes, mental health illness, and suicide. These are costly to the community and have lasting negative effects long into these children's adult life (CDC 2019a).

8.4.3 Family Goals and Priorities

Appalachian families take great pride in being independent and doing things for themselves. Even when economics may permit paying others to do some tasks, great pride is taken in being able to do for oneself. This is an area in which the editor (L. Purnell) can still strongly relate to Appalachian roots. Even though reaching a financial position at which he can pay someone to do chores on home and farm, he continues to take pride in doing them for himself. Traditionally, nuclear and extended families are important in the Appalachian culture, so family members frequently live in close proximity. Relatives help one another with major chores or building projects and are sought for advice on child rearing and most other aspects of daily living.

Elders are respected and honored in the Appalachian family. They are cared for by other family members or kin sets, and usually are treated with great respect (Burns et al. 2006; Lewis et al. 1985). Appalachian adults 65 years and older constitute approximately 18% of the population, compared to 15% in the remainder of the US (Pollard and Jacobsen 2019). Grandparents frequently care for grandchildren, especially if both parents work. This form of child care is readily accepted and is an expectation in large extended families. Elders usually live close to or with their children when they are no longer able to care for themselves. The physical structure of the home is designed to assist aging parents. Many adult children do not consider nursing home placement because it is "the equivalent of a death sentence". Migration of children out of the home area may force many older people to relocate outside their home area to be with their children. A dilemma occurs because they have an equally strong Appalachian value of attachment to place and family (Salyers and Ritchie 2006). As a compromise, some practice "snow birding"—leaving their home in the winter and moving in with their children, then returning to their home in the summer. It is not unusual for adult children to drive 3–5 h on days off work to spend time with and help maintain their aging parents at their homes in Appalachia.

One's obligation to extended family outweighs the obligations to school or work. The family feels a personal responsibility for nieces and nephews and readily takes in relatives when the need arises. This extended family is important regardless of the socioeconomic level. Upon migration to urban areas, the nuclear family becomes dominant because the extended family is usually left behind in Appalachia. This strong sense of family, in which the family distrusts outsiders and values privacy, can be a deterrent to getting involved in community activities or joining self-help group activities.

Social status is gained from having the respect of family and friends. Formal education and position do not gain one respect. Respect has to be earned by proving that one is a good person and "living right." Living right is based on the ethic of neutrality and on being a good "Christian person." Having a job, regardless of what the work might be, is as important as having a prestigious position. Families are very proud of their family members and let the entire community know about their accomplishments. In some instances, migration to the city may result in mixed views toward one's status. Monetary gain does not necessarily improve one's status in the family and community. Rather, skills and character traits that allow one to achieve financial comfort are given high status.

The Appalachian family network can be a rich resource for the health-care provider when health teaching and assistance with personalized care are needed. For programs with Appalachians to be effective, support must begin with the family, specifically the grandmothers, and immediate neighborhood activities. The health-care provider must respect each person as an individual and be nonbureaucratic in nature. The family, rather than the individual, must be considered as the basic unit of care.

8.4.4 Family Decision Making at End of Life

Treatment of life-limiting illness has changed dramatically in the last 20 years (Fornehed et al. 2020). What was once deemed a death sentence is now treated aggressively, leading to very complex decisions for health care to be made by family. Decision making by persons with life-limiting illness can be a complex process that may include decisions to continue aggressive care with or without palliative care, decisions to abandon aggressive care with or without palliative hospice care, and decisions regarding with whom to communicate and disclose their illness and prognosis (Campbell et al. 2010; Wallace 2015). Throughout this process, family plays an integral role in making decisions about caregiving at end of life. Palliative service providers, including hospice care providers, report that the care they provide is not just for the person with life-limiting illness, but also for the family in that person's life (NHPCO 2018).

In rural Appalachia, end of life care decisions are made with family, accepting death as a part of life (Fornehed et al. 2020). Even though Appalachians are realist, they still hold out "hope" that the illness is not severe, and their family member will recover. Communication is essential to rural Appalachians when making decisions at end of life. Loyal family decision makers also voiced they wished to remain true to the severely ill person's wishes at end of life. Knowledge and economics play a large role in the decisions rural Appalachians make when deciding end-of-life care. Limited knowledge about end-of-life care options by the interdisciplinary team and financial hardships sometimes suffered by family were all factors in the decisions families made when loved ones were dying. Finally, rural Appalachians value combining folk care with professional care when they are dying. Persons with life limiting illness wished to be comfortable in their everyday living, even though they knew their time for living was limited. Herbal remedies, being with family, gardening, farming, and watching their favorite television show were all supportive comfort measures families expressed were important to their loved ones at end of life (Fornehed et al. 2020). Supporting and validating difficult and emotional decisions made by families at end of life when a loved one is dying, was found to be essential for building trust between the family and the healthcare team (Cronin et al. 2015; Fornehed et al. 2020).

Reflective Exercise 8.1

Mr. Smith was living alone with a chronic disease. His condition declined and he was worked-up for aggressive treatment at a research-intensive medical center. Mr. Smith declined any further curative treatment and returned home. As his condition worsened, he was hospitalized at the local hospital. Mr. Smith did not have an advance directive naming a surrogate. As the health care team communicated with Mr. Smith and several of his adult children, the decision was made for hospice care so he could die "at home in peace". Several family members, including those who hadn't seen him in many years wanted "everything done." Family shared later they felt that communication should have begun years ago when the life-limiting diagnosis was made and in many healthcare encounters since. Fornehed et al. (2020) identified the need for health care providers to communicate with persons and family early and often using easily understood language, illustrations, and sharing palliative/end-of-life care options well in advance of needing such services.

1. How would you have dealt with the Smith family?
2. What approach do you and/or your interdisciplinary team take when working with persons (of any age) with chronic disease and their families?

3. Health promotion: Do you encourage your patients over the age of 18 to complete an advance directive (a) designating a surrogate decision maker; a person they would trust to honor their wishes, if ever needed and (b) describing how they want to live life, e.g. being able to complete activities of daily living and living independently?
4. As a member of your family and your community, have you engaged in proactive conversations with significant others about what you or your family or your friends/neighbors/faith-community members would want or not want?
5. Examine these links: https://fivewishes.org/https://advancingexpertcare.org/acp and Share widely!

8.4.5 Community as Family

Community has always been an essential part of rural Appalachian culture. Family and kinship groups form communities that support each other economically, socially, emotionally, educationally, and spiritually. Community members gather together at social events to play, worship, visit, make music, pass down cultural values, and share personal and family stories; many that extend back hundreds of years. Traditional gathering places still exist today such as country stores, the family's front porch, barber shops, community centers, and churches. Some of these gatherings, such as quilt making groups are symbolic of this region. These functions preserve Appalachian cultural traditions, identity, and local history that might be otherwise lost, foster relationships among the community members, and provide a mechanism for welcoming and orienting new community members (Burns et al. 2006; DeYoung et al. 2006).

8.4.6 Alternative Lifestyles

Alternative lifestyles such as single and divorced parents are usually readily accepted in the Appalachian culture. Same-sex couples and families living together are accepted and rarely discussed. Such acceptance is congruent with the ethic of neutrality, the Appalachian need for privacy, not interfering with others' lives unless asked to do so, avoiding arguments, and seeking agreement, even though agreement may be implied rather than spoken. Hall et al. (2018) found a stark contrast when exploring the end of life narratives of persons with HIV/AIDs living in rural Appalachia. Narratives revealed spiritual condemnation and dehumanization of the lesbian, gay, bisexual, transgender, queer (LGBTQ) living with HIV/AIDS by clergy, thus shunning of an already marginalized community.

8.5 Workforce Issues

8.5.1 Culture in the Workplace

Because many Appalachians value family above all else, reporting to work may become less of a priority when a family member is ill or other family obligations are pressing. When family illnesses occur, many Appalachian individuals willingly quit their jobs to care for family members. For some, the preferred work pattern is to work for an extended period of time, take some time off, and then return to work. Although work patterns may change, a deep-seated work ethic exists. Liberal leave policies for funerals and family emergencies are seen as a necessary part of the work environment.

Personal space is important; consequently, many Appalachians use a greater distance when communicating in the workplace. Close face-to-face encounters, hugs, and the like are rarely seen. A harmonious environment that fosters cooperation and agreement in decision making is valued and desired. Health-care providers who come from outside the area may have some difficulty establishing rapport in the workplace if they lack an understanding and appreciation for Appalachian workplace etiquette.

8.5.2 Issues Related to Autonomy

In general, a lack of leadership is not uncommon because ascribed status is more important than achieved status and because there is an attempt to keep hierarchal relationships to a minimum (Coyne et al. 2006). The Appalachian ethic of neutrality and the values of individualism and nonassertiveness with a strong people orientation, may pose a dichotomous perception at work for outsiders who may not be familiar with the Appalachian way of life. However, when conflicts occur, mutual collaboration for seeking agreement is consistent with the ethic of neutrality. Because many Appalachians align themselves more closely with horizontal rather than hierarchical relationships, they may be reluctant to take on management roles. When they do accept management roles, they take great pride in their work and in the organization as a whole.

Most middle-class Americans gain self-actualization through work and personal involvement with doing. Appalachians seek fulfilment through kinship and neighborhood activities of being. To foster positive and mutually satisfying working relationships, organizations should capitalize on individual strengths such as independence, sensitivity, and loyalty that are recognized values in the Appalachian culture. Many Appalachians prefer to work at their own pace, devising their own work rules and methods for getting the job done. Some local factories, mines, lumber mills, and health-care facilities that hire managers and administrators from outside the region often provide educational seminars about the Appalachians' worldview, work culture, and way of life in order to foster cultural sensitivity and a general understanding of the people with whom they work.

8.6 Biocultural Ecology

8.6.1 Skin Color and Other Biological Variations

Since its first settlement, the Appalachian region has had a predominantly White European-ancestry population. In particular, those with Scot-Irish heritage have light White pigment, raising concerns with sun exposure. As described earlier, some individuals can trace their heritage to a mixture of White ancestry along with Cherokee, Apalachee, Choctaw, and other indigenous tribes of the region. Intermarriage with American Indians and White settlers was not uncommon (Campbell 2006). A few Blacks, a distinct minority of 9% live in Appalachia (Pollard and Jacobsen 2019). The influence of American Indians and Blacks can be seen in skin color along with pronounced epicanthic eye folds, high cheekbones characteristic for American Indian ancestry, and darker skin tones and black, curly hair that are characteristic in many Blacks.

8.6.2 Diseases and Health Conditions

The Appalachian Regional Commission in partnership with the Robert Wood Johnson Foundation sponsor an innovative multi-part health research initiative that will, over time, document health disparities and health outcomes in rural Appalachia. The first of these reports discussed the most prominent health disparities for the region. Domains studied included mortality, morbidity, behavioral health, child health, community characteristics, lifestyle, health care systems, quality of care, and social determinants (ARC 2017a).

Some key indicators demonstrated the Appalachian region performed much better overall when compared to the United States in the following domains: HIV prevalence, travel time to work, excessive drinking, student–teacher ratio, chlamydia prevalence, uninsured adults less than 65 years of age, and diabetes monitoring for Medicare patients. Remaining indicators studied revealed that the rural Appalachian region performed much worse than the United States overall, including higher mortality rates in 7 of the 10 leading causes of death and drug overdoses or poisoning mortality. Premature mortality was seen in rural Appalachians' with heart disease, cancer, chronic obstructive pulmonary disease (COPD), injury, stroke, and diabetes. In turn, these mortality rates translate to a 25% higher years-of-potential-life-lost for rural Appalachians when compared to the nation as a whole. Rural areas of Appalachia saw a significant increase in mortality for these indicators when compared to the urban counterparts in Appalachia (ARC 2017a).

To impact Appalachian health, nurses and other health-care providers must create and carry forth research designed to honor Appalachian culture, individuals, and families. For example, in West Virginia, 14% of the population over age 65 years suffer from heart failure (HF) and the state reports the highest rate of heart failure (HF) deaths in the nation. A culturally appropriate study is planned to address this rural Appalachian health disparity. Using FamPALcare materials, nurses will coach patients and family caregivers in managing advanced heart failure at home including choosing among conservative end-of-life palliative care options and determining choices for advance directives. Key to conducting the study is using Appalachian community and faith-based nurses to recruit and engage participants in coaching sessions. These nurses will honor the Appalachian cultural values of developing trust through relationship-building, scheduling flexibility, and engaging multiple family caregivers. In addition, rural community leaders and interprofessional health-care providers will participate in focus groups, thus providing additional essential data (Piamjariyakul et al. 2019). Findings will impact individual and family health and quality of life while honoring Appalachian cultural values.

Access to clean water is a primary concern for many residents of rural, isolated Appalachian communities where critical infrastructure for water utilities is minimal to non-existent. Piping municipal water to neighborhoods in remote mountain hollows is expensive due to steep terrain or geographic distance between hollow communities. Those who have no access to tap water in the home use local wells and springs, which are often contaminated by inadequate human waste management or failing septic systems. Water-borne illness is an unfortunate consequence due to human exposure of affected streams during recreational contact like swimming/fishing, and especially when using as a potable water source. A community-academic partnership between local Appalachian community members and a university-based inter-professional team of nursing, architecture, civil and environmental engineering, and law enforcement faculty and students used shared expertise to assess, plan, and implement local solutions. Strong multi-sectoral partnerships with community stakeholders are critical for promoting culturally-driven sustainable solutions to make positive improvements in health and societal outcomes in the Appalachian region. Strengthening the capacity of a community using partners' shared values of health, trust, and innovative vision is the key (Arcipowski et al. 2017).

Cancer, suicide, and accident rates in some parts of Appalachia are significantly greater than the national average. Rural Appalachians experience cancer disparities when compared to the U.S. population, including rural non-Appalachians. Despite a drop in cancer mortality rates across the country, rural Appalachians continue to have the highest incidence of cancer, are diagnosed in later stages of cancer, and experience poorer survivorship. Researchers suggest a holistic approach to address these challenges, improving access to and utilization of quality care and healthy lifestyle interventions (Yao et al. 2017).

Dental care continues to remain a challenge throughout Appalachia. Many adults and children receive little or no dental care and in portions of Appalachia, people experience the worst oral health in the U.S. Poor oral health and dental caries in Appalachia is believed to be linked to genetics, household socioeconomic status, family behavior and stress, and access to dental care (Krause et al. 2016; Center for Oral Health Research in Appalachia 2020).

Huttlinger and colleagues (2004) found when prioritizing health and other needs of the family, dental health often falls last, and most people will simply "do without" or rely upon home remedies and/or home extraction. In a study examining oral health indicators across the United States and Appalachia, researchers found significant oral health disparity for Appalachians (Krause et al. 2016). Only 70% of participants had seen a dentist in the last year; nearly half of adults age 35–44 had lost some teeth; and 25% of adults over age 65 had no remaining teeth. Disparity varied across the region. Oral health was poorer in rural versus urban Appalachia and in central and southern regions. However, oral health outcomes in rural Appalachia was actually better when compared to poor rural areas in other U.S. regions (Krause et al. 2016).

8.6.3 Variations in Drug Metabolism

Current medical and research literature reports no studies specific to the pharmacodynamics of drug interactions among Appalachians. Given the diverse gene pool of many residents, the healthcare provider needs to observe each individual for adverse drug interactions.

8.7 High-Risk Behaviors

Compared with non-Appalachians, Appalachians seem to be less concerned about their overall health and risks associated with smoking (Huttlinger 2013). Their use of tobacco and smokeless tobacco is the highest in the country, and deaths from tobacco-related uses such as cancer, are the highest in the nation (CDC 2004). Underage use of tobacco and alcohol is widespread among teens.

8.7.1 Substance Misuse and Mental Health Disorders

Nearly 20 million Americans (aged 12 and older) battled a substance misuse disorder in 2017. Of those, 38% struggled with illicit drug use, 74% with alcohol, and 8.5 million suffered from a mental health disorder (Substance Abuse and Mental Health Services Administration 2018a, b). Genetics, environmental factors—such as chaotic home, parent's drug use, community influences, and poor academic achievement—and mental health challenges contribute to risk for drug use (National Institute on Drug Abuse [NIDA] 2018a, b). Rural Appalachians have a disproportionately higher rate of substance misuse disorder, specifically, opioid use disorder including fentanyl, heroin, and prescription opioid use. The rate of drug overdose deaths involving opioids in the U.S. was among the highest in Appalachians states (NIDA 2019c). West Virginia had the highest rate: 49.6 deaths per 100,000 persons; 833 total deaths. Ohio had the second highest rate of 39.2 deaths per 100,000 persons and the highest number of total deaths at 4293. Kentucky ranked sixth overall with 27.9 deaths per 100,000 persons; 1160 total deaths. These figures compare to the U.S. rate of 21.7 deaths per 100,000 persons (NIDA 2019c).

Seventy-eight million people need mental health care in the United States and less than half of that number receive care (National Alliance on Mental Illness 2018). Rural Appalachians experience significantly reduced access to mental health care. When compared to other regions in the U.S., the ratio of mental health providers for the population is approximately 50% fewer in rural Appalachia. The Substance Abuse and Mental Health Services Administration reports that "People with mental and substance abuse disorders may die decades earlier than the average person—mostly from untreated and preventable chronic illnesses like hypertension, diabetes, obesity, and cardiovascular disease that are aggravated by poor health habits such as inadequate physical activity, poor nutrition, smoking, and substance abuse (Substance Abuse and Mental Health Services Administration [SAMHSA] 2018a, b)."

Experts believe that substance use disorder (SUD) in rural Appalachia snowballs due to lack of resources in smaller isolated communities. "Everyone knows everyone" and therefore stigma, guilt, and shame contribute to persons hiding their challenges. There are limited services offered in rural communities and if offered, persons are reluctant to access treatment as then others in the community will "know." Regardless, transportation challenges exist for persons to get to services whether local or further away. In rural Appalachia and nationally, addressing stigma was identified as the most important step to prevent and treat SUD; followed by access to care with telehealth viewed as key to reaching persons in rural counties (Davis, S. [personal communication, January 22, 2020]).

Building on resilient rural Appalachian cultural values of strong communities, families, and faith, much hope exists to improve SUD in rural Appalachia. In addition to the example described in the Reflective Exercise 8.3, prevention, treatment, and recovery measures include resiliency training for parents, youth programs in schools, community centers, and churches that address preventing adverse childhood experiences (ACES), take-back drug days, county coalitions, rural opioid specialists teach communities about and distribute narcan, peer-to-peer support, treatment in jails with warm hand-offs to community mentors, housing, and job placement post recovery (ARC 2019a; CDC 2019b; Davis, S. [personal communication, January 22, 2020]).

Reflective Exercise 8.2

Between 1999 and 2016, deaths in Tennessee related to prescription medication overdose more than quadrupled. Efforts to reduce the number of opioids prescribed was effective; however, deaths related to drug overdose and opioids continued to rise in 2017–2018 (Tennessee Department of Health, n.d.) at a rate of 19.3 deaths per 100,000 compared to the national rate of 14.6 deaths per 100,000 persons (NIDA 2019c). The East Tennessee Appalachian region is noted for its vulnerability to opioid use disorder (OUD). Rural communities are particularly vulnerable, as emphasized by extreme rates of neonatal abstinence syndrome and opioid overdose and lack of treatment and recovery services and health-care providers. The Rural Communities Opioid Response Program-East Tennessee Planning Consortium (RCORP-ETC) was established through collaborative efforts between the University of Tennessee College of Nursing and East Tennessee community partners in ten rural Appalachian counties highly affected by OUD. HRSA grant funds were used to develop a comprehensive strategic plan and workforce plan to address OUD prevention, treatment, and recovery services and access to care (Davis 2018). Building upon the RCORP-ETC consortium work, Project HOPE (**H**ealing **O**UD through **P**revention and **E**xpertise), is using HRSA funding and a community-engaged approach with health care providers, youth, and the greater community (e.g., faith communities), to develop and implement research- and theory-based services that enhance OUD prevention, treatment, and recovery. For example, team members are reducing OUD stigma using digital storytelling and youth advocacy. They emphasize the personal impact of OUD across individuals, families, and communities (Davis 2019).

1. Consider your community (use the broad definition of community from Chap. 2); what partnerships exist to enhance the prevention, treatment, and recovery of substance use disorder in a culturally congruent manner?
2. What steps can you take as a health care provider to reduce stigma among colleagues? Within the healthcare system? Your workplace? Your community, family, and friends?
3. Substance use disorder affects everyone. Use this link https://www.drugabuse.gov/related-topics/trends-statistics to examine the latest research, trends, statistics and incidence of opioid-involved death rates in your state. Search current resources to address SUD and determine their cultural acceptability.

Similar to many other regions of the U.S., methamphetamine and marijuana are additional products used in Appalachia. The seclusion of mountain hollows and the number of remote and available barns and sheds have contributed to methamphetamine production. This highly addictive drug is made using common ingredients such as over-the-counter (OTC) cold medications, acetone, and rock salt. Setting up a laboratory does not require a lot of room. Unfortunately, ingredients and recipes are not hard to find. It is cooked up in homemade laboratories using items such as paint thinner, camping fuel, starter fluid, gasoline additives, mason jars, and coffee filters (U.S. Department of Justice Drug Enforcement Administration 2017).

8.8 Nutrition

8.8.1 Meaning of Food

As with most ethnic and cultural groups, food has meaning beyond providing nutritional sustenance. To many Appalachians, wealth means having plenty of food to share with family, friends, and at social gatherings. It is generally believed that one should drink plenty of fluids and eat plenty of good food to have a strong body. A strong body is a healthy body. Food and the sharing of food have broad social implications. Appalachians love to get together with family members, friends, and neighbors for meals. Weekend meals at a family member's home are common and serve as a mechanism to share information, community events and happenings, and gossip. Church suppers are also commonplace, with members contributing favorite dishes.

8.8.2 Common Foods and Food Rituals

Many Appalachians, and especially those living in the more remote areas, include wild game in their diet. Muskrat, groundhog, rabbit, squirrel, duck, turkey, and venison commonly supplement "store-bought" meats. Wild game traditionally has a lower fat content than meat raised for commercial purposes. However, consistent with traditional practices from previous decades, most parts of both wild and domesticated animals are eaten including high-cholesterol organ meats. Additionally, low-fat game meat is usually breaded and fried. Most diets include sweet pre-packaged drinks, Kool-Aid with added sugar, very sweet iced tea, and soda. In fact, "sweet tea" is a year-round favorite, and most people keep a jar of it in the refrigerator or on the porch in cooler weather.

Food preparation practices may increase dietary risk factors for cardiac disease because many recipes contain lard and meats that are preserved with salt. Other common foods in particular regions of Appalachia that may be unfamiliar to non-native Appalachians are sweet potato pie; molasses candy; apple beer; gooseberry pie; pumpkin cake; and pickled beans, fruit, corn, beets, and cabbage, all of which are high in sodium. Fried green tomatoes, biscuits, and thick gravies are ongoing favorites (Huttlinger 2013).

Appalachians celebrate Thanksgiving, Christmas, other national and religious holidays, and many other occasions with food. The value of self-reliance is enhanced during the "cannin" season when foodstuffs are preserved. Canning becomes a social or family occasion and is an excellent avenue for health teaching if the health-care provider is willing to participate and learn. Additional celebrations with food occur during times of death and grieving, when friends and participants bring dishes specifically prepared for the occasion.

8.8.3 Dietary Practices for Health Promotion

Many Appalachians believe that good nutrition has an effect on one's health. In one study with rural Appalachians, young mothers were asked what it meant to eat well for good health. They referred to "taking fluids" and "eating right," but they were unable to describe healthy eating patterns any further (Gainor et al. 2005). Because of health intervention programs, publicity though television, magazines, newspapers, and the Internet, residents of most Appalachian communities are aware of "good foods" and "bad foods" in terms of general health. However, a lack of money, having a meager budget that requires the use of food stamps, and distance to supermarkets may limit choices.

Grow Appalachia is a community food security organization begun in 2009, headquartered in

Berea, Kentucky. Their goal is "to teach families in central Appalachia how to grow food for themselves and their families, cook the produce in heart-healthy ways, and preserve excess for the winter and sell at the farmers market" (Grow Appalachia 2020).

Working with over 60 non-profit partnerships across six states in central Appalachia (Ohio, Kentucky, Tennessee Virginia, West Virginia and North Carolina), the program provides financial, technical, and educational assistance tailored to meet specific family needs. Grow Appalachia has helped over 6000 families grow four million pounds of organic food (Grow Appalachia 2020).

Health-care providers in clinics and school settings have an excellent opportunity to have a positive impact on the nutritional status of individuals and families. School breakfast and lunch programs, Meals on Wheels, and church-sponsored meal plans are some of the ways in which health-care providers can encourage and support families to attain better nutrition practices.

8.8.4 Nutritional Deficiencies and Food Limitations

A common practice for rural and urban Appalachian children is to replace meals with snacks. The most common snacks are candy, salty foods, desserts, and carbonated beverages, which can result in nutritional deficiencies. There are no specific food limitations or enzyme deficiencies associated with the people of Appalachia.

8.9 Pregnancy and Childbearing Practices

8.9.1 Fertility Practices and Views toward Pregnancy

Birth outcomes in the more rural areas of Appalachia are poorer than among middle-class White groups in rural, suburban, and urban populations. In one study that compared birth outcomes among rural, rural-adjacent, and urban women, rural women had the worst birth outcomes overall; rural-adjacent women had the best birth outcomes of the three groups, yet they were the youngest, least educated, least likely to be married and least likely to be privately insured (Gainor et al. 2005). Contraceptive practices of Appalachians follow the general pattern of the U.S. population. As a group, a disproportionate number of teenage pregnancies occur at a younger age compared with non-Appalachians.

Fertility practices and sexual activity, both sensitive topics for many adolescents, are topics in which outsiders unknown to the family may be more effective than health-care practitioners who are known to the family. To be effective, counseling by the health-care provider must be accomplished within the cultural belief patterns of this group and must be approached in a non-hierarchical manner, preferably with a health-care provider of the same gender.

8.9.2 Prescriptive, Restrictive, and Taboo Practices in the Childbearing Family

The literature reports no specific research or studies related to prescriptive, restrictive, or taboo practices during pregnancy or postpartum.

8.9.3 Neonatal Abstinence Syndrome

When pregnant women use opioids, babies are born suffering from neonatal abstinence syndrome (NAS) or neonatal opioid withdrawal syndrome (NAS/NOWS). According to the National Institute on Drug Abuse (2019a, b, c), approximately every 15 minutes a baby is born with NAS and suffers from opioid withdrawal. Between 2004 and 2014, there was a dramatic five-fold increase in NAS. Babies with NAS are more likely to have low birthweight and respiratory complications and an estimated 80% of moms come from lower-income communities with higher rates of unemployment and counties designated as mental health care provider shortage areas (NIDA 2019c). Using Tennessee as an example, in 2016 three times more addicted babies were born in Tennessee compared to other states; the East Tennessee Appalachian region had the highest incidence of NAS within the state (Tennessee Department of Health 2016) demonstrating a six-fold increase in NAS cases 2006–2016. Numerous national, state, and local programs, and research were implemented to address this overwhelming public health concern. Fortunately, after peaking in 2017 at just over 1000, NAS cases are slowly declining and were 769 at the end of 2019 (Tennessee Department of Health 2019).

Unfortunately, babies experiencing NAS drug withdrawal may suffer from high-pitched crying, seizures, temperature instability, tremors, sleep problems, sneezing, and feeding difficulties. Babies frequently are fussy and hard to soothe. Skin-to-skin contact, swaddling, being gentle, and breastfeeding can calm babies. The health-care community is learning about the long-term effects on health and development. Minimally, challenges include developmental delays, behavior and learning challenges, and problems with vision, hearing, speech and language, ear infections, and sleeping (March of Dimes 2020).

A study funded by NIDA (2018a, b) found barriers for pregnant women in Appalachia with opioid use disorder included limited access to medication assisted treatment (buprenorphine) and a limited number (less than half) of health care providers accepted Medicaid or private insurance (versus cash). As communities, families, health-care providers, and law enforcement come together to address the overall opioid crisis (as described above), there is hope and improved outcomes for these mothers and babies.

Decisions to take prescribed medication during pregnancy may present challenges for Appalachian mothers as well as those throughout the U.S. As discussed previously, mental health issues including depression is common in the region. A review of research on fetal exposure to selective serotonin reuptake inhibitors-SSRIs (antidepressant medication) taken by mothers found that children may exhibit later developmental challenges with cognition or behavior. Findings are not conclusive. However, since SSRIs cross the placental barriers, mothers may use caution or avoid SSRIs during pregnancy (Hermansn and Melinder 2015). However, women suffering from depression tend to smoke and use alcohol and drugs more. Trusted health care providers can provide support and help guide Appalachian mothers through making decisions best for their individual circumstances.

8.10 Death Rituals

8.10.1 Death Rituals and Expectations

Death is considered a part of living and, because of their faith in God; Appalachians believe they will spend eternity with their deceased loved ones. One rural Appalachian described her beliefs in these terms, "We're Baptists. We believe in Jesus and that He died on the cross for our sins and when we die, we're going to heaven" and will see our family again (unpublished research findings from study Mixer et al. 2014).

As Appalachians near the end of life, there is diversity in how people and their families approach death when family members have life-limiting illness. Family decision-making at end of life was discussed previously. Additional person, family, and healthcare provider data from several studies the authors have conducted revealed that some families readily accept death while others continue praying for a miracle (Fornehed et al. 2020; Mixer et al. 2014). One nurse described how families struggle, "the good Lord will take me and it will be ok,′ and I think on that same dime, some of 'em think 'No, I need everything done, so I can stay alive…" Another nurse shared, strong faith seemed to help family members be more "settled about the next step in life" and accepting that making end-of-life care decisions was necessary (unpublished research findings from study Fornehed et al. 2020). Campbell and Ash (2007) described Appalachian faith is strong even when facing life-limiting illness. Their connection with God through prayer was a mechanism for receiving symptom relief, feeling hope, staying connected with family and their community, and gaining a source of strength while going through the dying process.

The values of family cohesiveness and being neighborly helped shape Appalachians' view of death. Historically, death was common for Appalachians, with diseases and accidents occurring frequently in rural areas (Hufford 2006). A gathering at the bedside then began and continued until the person either returned to health or died. The women in the family would prepare the body for burial and the men of the community would build the coffins and dig the graves (Lozier and Althouse 1975). These death rituals have changed little over the years in rural Appalachia. The gathering at the bedside may now be at a hospital or nursing home or individual's home. After death, the body now goes to a funeral home to be prepared for burial. Visitation by family and friends still occurs, but at the funeral home. The community stills cares for the family of the deceased in a neighborly way.

There are variations in death rituals across Appalachia. Some individuals design their own funeral services long before their death. A traditional Appalachian death would be handled as follows: The body is shown for 3 days. The deceased is usually buried in her or his best clothes. A common practice is to bury the deceased with personal possessions. At the funeral home, a picture of the deceased, or other personal items may be displayed. Extended family and community members visit to "pay their respect." On the final day, the funeral is held either at the church or funeral home. The preacher conducts a "church service" including hymns and prayers. All process to the cemetery with cars in a line with magnetic funeral flags, lights on, and police escort. Custom dictates traffic pulls over and allows the funeral procession to pass without breaking the line. A graveside mini-service with prayer is held. The body/coffin is lowered into the ground; covered later by the cemetery personnel. Family and friends return to the church or family home for a meal. Food demonstrates caring throughout this entire process and is provided by extended family, friends, and community "by the droves" to the grieving family; many times, for weeks to come (Fornehed, personal communication, January 27, 2020).

Historically and today, there are many family and community burial sites scattered throughout rural Appalachia. For example, the co-author's "daddy" was buried just down the road, where the family and community have been buried for over 200 years. As others move into the Appalachian region, cremation is becoming more popular; however, the practice is rarer among those raised here (Fornehed, personal communication, January, 2020). Cemeteries throughout Appalachia show frequent visitations and give a sense of place and relationship to the land. Plots are carefully tended with displays of flowers, wreaths, and flags. Other beliefs regarding burial practices include placing graveyards on hillsides for fear that graves may be flooded out in low-lying areas.

8.10.2 Responses to Death and Grief

Clergy help families through the grieving process by providing counseling and support to family members. Family members, fellow church members, friends, and community leaders often assist the bereaved. Typically, family members get together and reminisce about their deceased loved one. Friends and neighbors bring food for about a week or more and share memories of the "one who passed on."

A rural Appalachian minister and study participant discussed the need to address grief in the context of meaning for the person, congregation, and community. He had 25 years of experience ministering throughout the central Appalachian region in both White and Black Methodist churches. A synopsis of his perspective is paraphrased: As many experience, a person participates in the life of a congregation and community where he/she shares significant life celebrations and shoulders inevitable suffering. When one of these persons faces death, the whole congregation and community experience the imminent loss of that person and community wide grief. It is critical for both the individual and the community to address death in a mutually supportive manner. End of life is a significant theological rite of passage. It is a sacred event sustaining the dignity of the community and empowering the liturgical witness with hope (unpublished research findings from study Mixer et al. 2014). Thus, there is a critical need to collaborate with people, their families, faith communities, and community members in the region to develop culturally and theologically acceptable methods of educating about end of life and end of life care.

Health-care providers are encouraged to create these partnerships to develop culturally acceptable approaches for coping with death and grief that are tailored to meet specific needs in the Appalachian communities they serve.

8.11 Spirituality

8.11.1 Dominant Religion and Use of Prayer

The dominant religion in Appalachia is Christianity based on Protestant doctrine. Baptist, Methodist, Presbyterian, Holiness, Pentecostal, and Episcopalian are the most common denominations. Many churches are associated with national church organizations; however, because central organization of churches was difficult to retain in rural mountain areas, some people individualized their chosen church structure; there are still many small unaffiliated congregations. There is great diversity in religious practices throughout the region. Roman Catholic, Eastern Orthodox, Lutheran, Jehovah's Witness, Jewish, Muslim, Mormon, Episcopalian, Mennonite, Quaker, and Unitarian as well as people of other faiths live and worship in Appalachia (Gowen 2003; Huttlinger 2013; Mixer et al. 2014).

For most Appalachians, the church is the center for social and community activities. One study participant shared: if you were "raised in Appalachia… you talk about the three priorities: God, family, land. It's always in that priority" (Mixer et al. 2015, p. 901). Prayer for many Appalachians is a primary source of strength and engaged in daily whether or not they formally attend church. People and families described that prayer was a significant part of their lives and also used to ask for healing (Fornehed et al. 2020; Mixer et al. 2014, 2015). One Hispanic mother shared, "[we] pray to God and ask for healing," while a Caucasian mother said, "I don't know where I would be without my faith right now…" (Mixer et al. 2015, p. 901). Religious beliefs may also be of a spiritual nature, not tied to the tenets of any singular faith, and reflect the harmony of the mountains and being at one with life.

8.11.2 Rural Appalachian Faith

In rural Appalachia, there are several characteristics common to most religious practices. The most basic characteristic is the belief that "God is the controlling force in life" (Dorgan 2006, p. 1283). Other characteristics that have remained mostly intact in the area over the years include the belief that clergy are called and trained by God; religion and faith are part of life whether or not a church is attended, every event is in God's hands, people have little control over outcomes (**fatalism**), faith in God goes beyond religion, and faith is the basis for most decisions and actions in life (Campbell and Ash 2007; Diddle and Denham 2010; Dorgan 2006; Fornehed et al. 2020; Rowles 1991).

Many small churches have lay preachers instead of trained ministers. Many of the Baptist faiths believe that baptism must be done in a river, pond, or lake so that the body can be submerged. Another practice, feet washing, is believed to emulate Jesus' model of serving, demonstrating humility. Fundamentalist churches may segregate women and children from men in the seating arrangement within the church; men and older boys sit on one side, and women and children sit on the opposite side. In some churches, men sit on the right side of the church to represent the "right hand of God," while women sit on the left.

Rural Appalachian spirituality is usually based on Pentecostal doctrine. These religious beliefs lead rural Appalachians to take the Bible literally, thereby believing there are both God and the devil; good and evil are not abstract concepts but tangible entities. Rural Appalachians may aspire to live pure lives, believing that the devil causes sickness in impure people (Gobble 2009). Rural Appalachians profess belief in an active Holy Spirit that can be accessed through prayer to heal some illnesses.

In several qualitative studies, rural Appalachians described having deep faith in "God", "Jesus", and "Lord." People described that faith in God gives them comfort and hope and they trust his timing for illness, healing, and death. Faith practices such as bible reading and prayer were cornerstones of their daily lives. Several salient quotes illustrate these values and beliefs: "I've worn that bible out;" [after praying] "God healed me;" "Now, you know, we can do this with God's help" (Fornehed et al. 2020; Mixer et al. 2014).

Fatalism among rural Appalachians has often been identified in the literature as a barrier to both seeking and receiving health care. For example, Behringer and Krishnan (2011) and Shell and Tudiver (2004) discussed the barriers of fatalism to cancer screening and caring for cancer patients. They described that when rural Appalachians profess having a personal relationship with God and the belief that their health is under God's control, they may believe that having cancer likely is a death sentence, regardless of whether they were screened or treated for the disease.

However, fatalism was described differently by a nurse, born and raised in rural Appalachia whose family lived there for generations. She stated, "… when you're given a life-altering disease process… I think the rest of the world outside of Appalachia thinks… you're supposed to crumble… that doesn't happen in Appalachia because… if your world's falling apart, you drop to your knees, you pray to God. You go to your church family and your friends. You have prayer circles and that kind of stuff… and that's why [Appalachians are] viewed as fatalistic… knowing that God is still in control and you willingly and freely submit to His will, whether it's health or disease…" (Mixer et al. 2015, p. 902).

8.11.3 Less Common Spiritual Practices

Some freewill churches—for example, the Holiness Church—preach against attending movies, ball games, and social functions where dancing occurs. Other sects believe in handling poisonous snakes. Although the practice is rare, it is believed that the snake will not bite those who have faith based on Bible verses from Mark 16:17–18. If a person is bitten, a usual course of action is to heal themselves rather than go to a hospital, despite historical data that deaths have occurred following snake-handling rituals.

Another practice, the ingestion of strychnine in small doses during religious services, is believed to increase sensory stimuli. Needless to say, this practice can precipitate convulsions if ingested in large enough amounts. Fire-handling is practiced by some groups, again with the belief that the hot coals will not burn those who have faith.

8.11.4 Meaning of Life and Individual Sources of Strength

Meaning in life comes from the family and "living right," which is defined by each person and usually means living right with God and in the beliefs of a chosen church. Religion tends to be less focused on institutional rituals and ceremonies and consists more of personalized beliefs in God, Christ, and church. Because life in the mountainous regions can often be harsh, religious beliefs and faith make life worth living in a grim situation. The church provides a way of coping with the hurts, pains, and disappointments of a sometimes hostile environment and becomes a source for celebration and a social outlet.

Common themes that give Appalachians strength are family, traditionalism, personalism, self-reliance, religiosity, and a worldview of being and not having undue concern about things that one cannot control, such as nature and the future. Appalachians believe that rewards come in another life, in which God repays one for kind deeds done on earth (Huttlinger 2013). Appalachians' faith is central to everyday life, healthcare decision-making, coping, and making meaning of life (Fornehed et al. 2020; Mixer et al. 2014, 2015).

8.11.5 Spiritual Beliefs and Health-Care Practices

Spiritual care provided by family, neighbors, church family, and health-care providers was highly valued by rural Appalachian study participants. Participants in these studies self-reported ascribing to Christian faith. Nevertheless, not all rural Appalachians are Christians and Christians have varying values and beliefs, especially as Appalachians migrate in and out of the region. Thus, health-care providers are encouraged to assess each Appalachian's spiritual care needs; honor Appalachian similar and diverse faith practices; and help them tap into personal, family, and community spiritual support systems. As one provider study participant simply stated, "you know we do everything we can to honor and respect their beliefs" (Fornehed et al. 2020; Mixer et al. 2014, 2015).

Research demonstrated that forming partnerships between health-care providers and faith-related organizations for health promotion and illness and disease prevention is effective to improve the health status of Appalachians (Simpson and King 1999; Studts et al. 2012). Health-care providers who are aware of Appalachians' religious practices and spirituality needs are in a better position to promote culturally competent health care and to incorporate nonharmful practices into individual care plans. Health-care providers must indicate an appreciation and respect for the dignity and spiritual beliefs of Appalachians without expressing negative comments about differing religious beliefs and practices.

8.12 Health-Care Practices

8.12.1 Health-Seeking Beliefs and Behaviors

The Appalachian definition of health encompasses three levels: body, mind, and spirit. Good health is feeling well and being able to meet one's obligations. This definition precludes viewing disease as a problem unless it interferes with one's functioning. Consequently, many conditions are denied or ignored until they progress to the point of decreasing function. Health-related decisions usually are made at the last minute and for acute problems only. Appalachians may seek medical care from a healthcare professional only as a last resort (Coyne et al. 2006; Shell and Tudiver 2004). Literature points to several cultural reasons for delaying or managing care in rural Appalachia: initial reliance on emic or folk care to treat illness, a fatalistic view of disease, a day-to-day view of living, distrust of healthcare professionals, language barriers, lack of medical knowledge, and the concern that family problems will become community knowledge (Coyne et al. 2006; Lewis et al. 1985; Shell and Tudiver 2004).

Huttlinger and colleagues (2013) surveyed a large sample of Appalachians from southwestern Virginia and northeastern Tennessee to determine access to health care. They also addressed factors related to "good health." Over 75% stated that their health was "God's will," and over half stated that their families, church, and community played a vital role in their overall health and well-being. A 10-step pattern of health-seeking behaviors was identified: (a) self-care practices, (b) advice from female family members, (c) over the counter medications, (d) folk care, (e) talk to a pharmacist (f) see a provider (g) if no resolution of the issue, refer to specialist. (h) specialist treats the condition (i) no resolution, referred to closet tertiary medical center.

These 10 steps may not always follow the sequence presented and some steps may be skipped. Moreover, the time frame may be over the course of several years. It is not unusual that by the time a person is referred for definitive treatment, compensatory reserves have been depleted and one dies at a large medical center. The story is then passed on in the "holler": "So-and-so went to [Hospital X] and died." This pattern leads to a significant mistrust of large medical centers and continued reluctance to use these facilities effectively (Huttlinger 2013). Health-care providers can have a significant impact on improving a patient's health-seeking behaviors by providing information early on. Nurses especially can help to reverse this pattern because they are viewed, by the people they serve in Appalachia, as knowledgeable, non-judgmental, and respectful of Appalachian lifestyles.

8.12.2 Responsibility for Health Care

Self-care is a primary focus of health and perceived as an individual responsibility. Because many Appalachians value the ability to respond to and cope with events of daily life, home remedies, treatments, and active consultation with family members are sought before seeking outside help (Huttlinger 2013). Care within the medical system is used when the condition is perceived as serious, does not respond to self-care, or has a high potential for death. Furthermore, because self-reliance activities and nature predominate over people, many believe that it is best to let nature heal. Health-care providers need to keep this in mind when giving explanations and instructions to make them more acceptable to patients and their families.

When entering the biomedical health-care arena, Appalachians might feel powerless to control their own health. They often abdicate responsibility for their own care and expect that the health-care provider will completely take over their care. Many have high expectations for their health-care provider, with an unrealistic dependence on the system and an abandonment of more self-reliance activities (Coyne et al. 2006).

One major health concern for many Appalachians is the state of the blood, which is described as being thick or thin, good or bad, and high or low; these conditions can be regulated through diet (Huttlinger 2013; Obermiller and Brown 2002). Venereal disease and Rh-negative blood fall into the category of bad blood. Sour foods can also cause bad blood. Appalachian men, in general, report a greater number of backaches, with women reporting a greater number of headaches, than the rest of society (Coyne et al. 2006).

When establishing rapport, a health-care provider can go a long way in achieving trust by using churches, grange halls, and other community places (e.g., libraries, schools) as meeting places to work with entire Appalachian families at the community level.

8.12.3 Folk and Traditional Practices

A strong belief in folk medicine is a traditional part of the Appalachian culture. Folk and traditional practices were learned from the Cherokee and Apalachee Indians living in the region and have been passed down from generation to generation over centuries (Hufford 2006). Folk healers were sought out because they were trusted individuals, known to all within the community, and able to speak the language of rural Appalachians (Lewis et al. 1985; Rowles 1991). Using herbal medicines, poultices, and teas is common practice among individuals of all socioeconomic levels. However, the following factors have led to growing acceptance of a slowly changing health culture that combines professional and folk care: progress in the integration of healthcare professionals into the community, improvement in culturally appropriate communication with care recipients, and increased understanding of the culture by healthcare professionals (Ahijevych et al. 2003; Barish and Snyder 2008; Presley 2013; Rowles 1991; Schoenberg et al. 2015).

There is wide variation in use of folk remedies and some younger Appalachians will use folk remedies to honor their elders. Table 8.1 presents a reference guide complied by Dr. Larry Purnell for the health-care provider with the major ingredients and conditions for which the folk treatments are used. Although many of these home remedies are not harmful, some may have a deleterious effect when used to the exclusion of, or in combination with, prescription medications. Health-care providers need to assess for use of folk remedies simultaneously with prescription medications and treatment regimens. Those who partner with their Appalachian patients and integrate their folk medicine practices into allopathic prescriptions have a greater chance of the plan of care will be used Health-care providers are reminded that today's scientific medicine may be traditional or folk medicine to the next generation.

Reflective Exercise 8.3

The Johnson family presents at your rural clinic with their 18 month daughter Rebekah. They describe Rebekah has been "cutting teeth" and describe symptoms of colic.

1. Describe how you would conduct a culturally sensitive assessment of medications and folk care practices so that the Appalachian family is comfortable sharing honestly and openly.
2. Once you have built trust, you discover the Johnsons have been using whiskey on Rebekah's gums to ease the pain of teething and giving her Calamus root tea in her bottle for colic.
3. Which interdisciplinary team member(s) would be most helpful to consult?
4. How would you approach the family to negotiate alternatives for care and combine professional and folk care?

Table 8.1 Health conditions and Appalachian folk medicine practices

Health condition	Folk medicine practices
Arthritis	Make tea from boiling the roots of ginseng. Drink the tea or rub it on the arthritic joint Mix roots of ginseng and goldenseal in liquor and drink it. Ginseng is used heavily by many Koreans and was exported to Korea in the eighteenth and nineteenth centuries Eat large amounts of raw fruits and vegetables Carry a buckeye around in a pocket Drink tea from the stems of the barbell plant Drink a mixture of honey, vinegar, and moonshine (or other liquor) Drink tea made from alfalfa seeds or leaves Drink tea made from rhubarb and whiskey Place a magnet over the joint to draw the arthritis out of the joint
Asthma	Drink tea from the bark of wild yellow plum trees, mullein leaves, and alum. Take every 12 h Combine gin and heartwood of a pine tree. Take twice a day Suck salty water up the nose Smoke or sniff rabbit tobacco Swallow a handful of spiderwebs Smoke strong tobacco until choking occurs Drink a mixture of honey, lemon juice, and whiskey Inhale smoke from ginseng leaves
Bedbugs/ chiggers	Apply kerosene liberally to all parts of the body. *Caution:* Kerosene can cause significant irritation to sensitive skin, especially when exposed to sunlight
Bleeding	Place a spiderweb across the wound. This is also used in rural Scotland Put kerosene oil on the cut Place soot from the fireplace into a cut. Be sure to wash out the soot after bleeding is stopped, or the area will scar Apply a mixture of honey and turpentine on the bleeding wound Apply a mixture of soot and lard on the wound Place a cigarette paper over the wound Put pine resin over the cut Place kerosene oil on the wound. *Caution:* If used in large doses, kerosene will burn the skin
Blood builders	Drink tea from the bark of a wild cherry tree Combine cherry bark, yellowroot, and whiskey. Take twice each day Eat fried pokeweed leaves
Blood purifiers	Drink tea from burdock root Drink tea from spice wood
Blood tonic	Take a teaspoon of honey and a tiny amount of sulfur Take a teaspoon of molasses and a tiny amount of sulfur Drink tea made from bloodroot Soak nails in a can of water until they become rusty. Drink the rusty water
Boils or sores	Apply a poultice of walnut leaves or the green hulls with salt Apply a poultice of the houseleek plant Apply a poultice of rotten apples Apply a poultice of beeswax, mutton tallow, sweet oil, oil of amber, oil of spike, and resin Apply a poultice of kerosene, turpentine, petroleum jelly, and lye soap Apply a poultice of heart leaves, lard, and turpentine Apply a poultice of bread and milk Apply a poultice of slippery elm and pork fat Apply a poultice of flaxseed meal Apply a poultice of beef tallow, brown sugar, salt, and turpentine
Burns	Apply a poultice of baking soda and water Place castor oil on the burn Apply a poultice of egg white and castor oil Place a potato on the burn Wrap the burn in gauze and keep moist with salt water Place linseed oil on the burn Apply a poultice of lard and flour Put axle grease on the burn. This is also a practice with some Germans in Minnesota

(continued)

Table 8.1 (continued)

Health condition	Folk medicine practices
Chapped hands and lips	Apply lard, grease, or tallow from pork or mutton
Chest congestion	Apply a poultice of kerosene, turpentine, and lard to the chest. Make sure the poultice is not applied directly to the chest but rather on top of a cloth Apply mutton tallow directly to the chest Apply a warm poultice of onions and grease Rub pine tar on the chest Chew leaves and stems of peppermint Drink a combination of ginger and sugar in hot water Make a mixture of rock candy and whiskey. Take several teaspoons several times each day Drink tea made from ginger, honey, and whiskey Drink tea made from pine needles Put goose grease on the chest Drink red pepper tea Eat roasted onions Drink brine from pickles or kraut Make tea from boneset, rosemary, and goldenrod Make tea from butterfly weed
Colic	Make tea from Calamus root and catnip. (Calamus is a suspected carcinogen) Tie an asafetida bag around the neck Drink baking soda and water Chew and swallow the juice of camel root Massage stomach with warm castor oil Drink ginseng tea
Constipation	Take two tablespoons of turpentine Combine castor oil and mayapple roots Take castor oil or Epsom salts
Croup	Have child wear a bib containing pine pitch and tallow Apply cloth to the chest saturated with groundhog fat, turpentine, and lamp oil Drink juice from a roasted onion Apply a poultice of mutton tallow and beeswax to the back Eat a spoonful of sugar with a drop of turpentine Eat honey with lemon or vinegar Eat onion juice and honey
Diarrhea	Drink tea from the ladyslipper plant Place soot in a glass of water, let the soot settle to the bottom of the glass, and drink the water Drink tea made from blackberry roots Drink tea from red oak bark Drink blackberry or strawberry juice Drink tea made from strawberry or blackberry leaves Drink tea made out of willow leaves Drink the juice from the bark of a white oak or a persimmon tree
Earache	Place lukewarm salt water in the ear Put castor oil or sweet oil in the ear Put sewing machine oil in the ear Place a few drops of urine in the ear Place cabbage juice in the ear Blow smoke from tobacco in the ear Place a Vicks VapoRub–soaked cotton ball in the ear

Table 8.1 (continued)

Health condition	Folk medicine practices
Eye ailments	Place a few drops of castor oil in the eye Drop warm, salty water in the eye Drink tea made from rabbit tobacco or snakeroot
Fever	Drink tea made from butterfly weed, wild horsemint, or feverweed Mash garlic bulbs and place in a bag tied around the pulse points Drink water from wild ginger
Headache	Drink tea made of ladyslipper plants Tie warm fried potatoes around the head Take Epsom salts Tie ginseng roots around the head Place crushed onions on the head Rub camphor and whiskey on the head
Heart trouble	Drink tea made from heartleaf leaves or bleeding heart Eat garlic
High blood pressure (not to be mistaken for high blood)	Drink sarsaparilla tea Drink a half cup of vinegar
Kidney trouble	Drink tea made from peach leaves or mullein roots Drink tea made from corn silk or arbutus leaves
Liver trouble	Drink tea made from lion's tongue leaves Drink tea made from the roots of the spinet plant
Poison ivy	Urinate on the affected area Take a bath in salt water and then apply petroleum jelly Wash the area with bleach Wash the area with the juice of the milkweed plant Apply a poultice of gunpowder and buttermilk Apply baking soda to wet skin
Sore throat	Gargle with sap from a red oak tree Eat honey and molasses Eat honey and onions Drink honey and whiskey Tie a poultice of lard of cream with turpentine and Vicks VapoRub to the neck Apply a poultice of cottonseed to the throat Swab the throat with turpentine

8.12.4 Barriers to Health Care

As discussed previously, there are significant health disparities, poor health status indicators, and challenging social determinants of health in Appalachia. Barriers to health care for Appalachians are numerous and center on accessibility, affordability, adaptability, acceptability, appropriateness, and awareness. Many counties in Appalachia are designated medically underserved and/or health provider shortage areas and/or geographic high needs. Provider shortages are in primary care, dental health, and mental health. Health-care facilities (USDHHS 2020a) are closing in some areas of Appalachia. Often, the closings are related to reduced reimbursement and decreasing availability of health-care providers and the ability to pay competitive salaries, especially for registered nurses (Huttlinger 2013; Mixer, personal communication, 2020).

There is an acute shortage of specialty providers and especially those for respiratory and pulmonary diseases, oncology, dental services, and ophthalmology. Those physicians who settle in the region quickly learn that flexible fee schedules, patience, and hands-on treatment approaches work best. Referral to specialty care in the larger urban centers must be made with consideration of travel and other expenses. For example, a referral to a pulmonologist in Charlottesville, Virginia for a person who lives in Wise might require a 3-day trip because 2 days are needed just to travel each way. Add to this the expense of gasoline, a relative taking off work to drive the person, and two nights in a motel, and it becomes something many Appalachians cannot afford. Innovations in telehealth often provide needed services closer to home (University of Virginia 2009; Mixer, personal communication, 2020).

As noted earlier in the chapter, the rugged terrain and distance is a deterrent to accessing health-care services. Even though ARC has sponsored road-building campaigns in the mountainous regions of Appalachia since 1965, transportation problems continue to exist in parts of the region. The author worked with Appalachians in rural Eastern Kentucky where people had to drive 45 minutes one way to fill a prescription. This created numerous challenges: Was a reliable vehicle available for transportation? Could one afford the gas? Could one afford the prescription? Could one get time off from work or family responsibilities (child or elder care) to make the round trip?

The high rate of unemployment in Appalachia means that many people cannot afford basic health care. A disproportionate number of Appalachians, especially those who are self-employed, unemployed, or underemployed are uninsured or underinsured and have limited resources to meet basic family needs. For some who do not believe in owing money, seeing a health-care provider may be postponed until the condition is severe or until they have the money. If services can be offered on a sliding scale, more people may be willing to access them.

Even when services are available, people may feel they are not delivered in an appropriate manner. Outsider health-care providers may be seen as disrespectful of Appalachian ways and self-care practices; patients may see the health-care providers' advice as criticism. If the health-care provider uses language that the patient does not understand, the health-care provider may be perceived as "stuck up." Many Appalachians do not like the impersonal care delivered in large clinics and, therefore, shop around and ask friends and family for suggestions for a private health-care provider. "Sittin' for a spell and engagin' in small talk" with the patient before an examination or treatment will help ensure return visits for follow-up care.

When health-care facilities have limited hours or are not adaptable, patients may not return for scheduled appointments. For example, a mother may bring her child in for an immunization. If the mother has a health problem and perhaps needs a Pap smear, she may be willing to have the test performed while having the child examined. However, if she is given an appointment to return at a later date, she may not keep the appointment because it is too far to travel for a problem she sees as nonurgent. If services are not available during evening hours, people may be afraid of taking time off during regular work hours for fear of losing their job.

Low literacy is widespread in Appalachia. Bureaucratic, written forms foster fear and suspicion of health-care providers, which can lead to confusion, distrust, and negative stereotyping by both parties. Some individuals fear "being cut on" or "going under the knife" and feel that a hospital is a place where you go only to give birth or die.

8.12.5 Cultural Responses to Health and Illnesses

Appalachians take care of their own and accept a person as a "whole individual." Thus, those with mental impairments or physical handicaps are generally accepted into their communities and not turned away. People with intellectual disabilities may be seen as having "bad nerves," "quite turned," or "odd turned." Appalachians may label certain behaviors as "lazy," "mean," "immoral," "criminal," or "psychic" and may recommend punishment by either the social group or the legal system or tolerate these behaviors (Obermiller and Brown 2002).

Traditional Appalachians believe that disability is a natural and inevitable part of the aging process. To establish trust and rapport when working with Appalachian patients with chronic diseases, health-care providers must avoid assumptions regarding health beliefs and provide health maintenance interventions within the scope of cultural customs and beliefs.

Individual responses to pain cannot be classified among Appalachians. The Appalachian background is too varied. A belief among some elders is that if one places a knife or axe under the bed or mattress of a person in pain, the knife will help cut the pain. This practice occurs with childbearing and other conditions that cause pain. The editor (L. Purnell) is aware of an Appalachian woman who requested to have a knife or axe placed under the bed or mattress postoperatively to help cut (or decrease) the pain associated with surgery. He offered a small pocketknife or butter knife to place under the bed. Both were unacceptable as the pocketknife was too small and the butter knife was too dull to be of use. A sharp meat-cutting knife from the kitchen was deemed appropriate because it was both large enough and sharp enough to help cut the pain.

8.12.6 Blood Transfusions and Organ Donation

Appalachians generally do not have any specific rules or taboos about receiving blood, donating organs, or undergoing organ transplantation. These decisions are largely one's own, but advice is usually sought from family and friends.

8.13 Health-Care Providers

8.13.1 Traditional Versus Biomedical Providers

For decades, both lay and trained nurses have provided significant health-care services, including obstetrics. Granny midwives and more formally trained midwives have provided obstetric services throughout the history of Appalachia. Although many practitioners and herbalists are older women, men may also become healers. Grannies and herb doctors are trusted and known to the individual and the community for giving more personalized care.

The entire Appalachian area has a shortage of health personnel and facilities as described above. In recent years, government incentives for graduate health-care provider education has improved the number of primary-care providers, including advanced nurse practitioners, available in the region. Health Resources Service Administration provides scholarships, loans, and loan repayment for primary care, dental, and mental health providers willing to work in health professional shortage areas (USDHHS 2020b).

The Frontier Nursing Service, started by Mary Breckenridge, is the oldest, and one of the most well-known and highly-awarded, nurse-run clinics and educational institutions in the United States and is a notable example of nurses, midwives, and nurse practitioners providing health care in Appalachia in a culturally competent manner. Frontier Nursing Service was started in one of the most rural areas of Appalachia in response to a lack of physicians and the high birth and child mortality rates in the area (Dawley 2003; Frontier Nursing 2020; Jesse and Blue 2004). Many Appalachians prefer to go to *insider* health-care professionals, especially in the more rural areas, because the system of payment for services is accepted on a sliding scale, and in some communities, even an exchange of goods for health services exists. One nurse practitioner in private practice states that the only time she locks her car is when the zucchini are "in." If she does not, when she gets in her car after a clinic session, she has no room to drive because of all the "presents" of the large vegetables (Huttlinger 2013).

Locally respected Appalachians are engaged to facilitate acceptance of outside programs and of the staff who participate at the grassroots level in planning and initiating programs. For Appalachian patients to become more accepting of biomedical care, it is important for health-care providers to approach individuals in an unhurried manner consistent with their relaxed lifestyle, to engage patients in decision making and care planning, and to use locally trained support staff whenever possible.

Health information on the Appalachian patient should be gathered in the context of broader family relationships and cordiality that precedes information sharing, as the family rather than the individual is the basic unit for treatment. Because direct approaches are frowned upon, health-care providers need to learn to approach sensitive topics, such as contraception and alcohol and drug use, indirectly. Many Appalachians expect the health-care provider to establish an advocacy role and to understand and accept their cultural differences.

Educational information presented in a nonjudgmental manner can have a significant impact on the health of Appalachian patients. Patients generally prefer verbal rather than printed material to obtain health-related information. In fact, an effective success strategy used by ALIC is storytelling, a strong tradition in Appalachia and one in which people can relate. Thus, the presentation of health and educational material needs to include the entire family and be linked with improvement in function in order to be taken seriously (Huttlinger 2013; Mixer et al. 2014).

8.13.2 Status of Health-Care Providers

Most herbal and folk practitioners are highly respected for their treatments, mostly because they are well known to their people and trusted by those who need health care. Physicians, nurses, and other health-care providers are frequently seen as outsiders to the Appalachian population and are, therefore, mistrusted. This initial mistrust is rooted in outsider behaviors that exploited the Appalachian people and took their land for timbering and coal mining in earlier generations.

Trust for an outsider is gained slowly. Health-care providers need to approach interactions using a patient- and family-centered care approach; communicate understanding that the person and family are in control of their health-care plan, and a desire to be in partnership as the plan of care is developed. Health-care providers need to demonstrate verbal and non-verbal acceptance of the person/family specific Appalachian cultural values while co-developing the plan of care.

Once the person gets to know and trust the health-care provider, the provider is given much respect. Studies demonstrate that Appalachians prefer to have biomedical health-care provider care combined with traditional folk care (Fornehed et al. 2020; Mixer et al. 2014, 2015).

To obtain full person and family engagement, the health-care provider needs to ask what they consider to be the problem before devising a plan of care. If the provider begins with an immediate diagnosis without considering the person's explanation, there is a good chance that the provider's treatment or recommendation will be ignored. It is important to decrease language barriers by decoding the jargon of the health-care environment. Because educational levels of individuals within the Appalachian regions vary, it is essential for health-care providers to assess the health literacy and basic understanding of health and disease of individuals when providing any kind of intervention. Educational materials and explanations must be presented at literacy levels that are consistent with patients' understanding. If materials are presented at a level that is not understandable to patients, providers may be seen as being "stuck-up," "putting on airs," or "not understanding them and their ways". Further strategies for providing culturally competent care have been discussed throughout the chapter.

Appalachian people have a proud heritage and live in a beautiful land. Their strong family, community, and spiritual cohesiveness and extraordinary resilience has seen them through hundreds of years of change and challenges; this legacy will carry them forward.

References

Ahijevych K, Kuun P, Christman S, Wood T, Browning K, Wewers ME (2003) Beliefs about tobacco among Appalachian current and former users. Appl Nurs Res 16(2):93–102. https://doi.org/10.1016/S0897-1897(03)00009-0

Ali SR, McWhirter EH (2006) Rural Appalachian youth's vocational/educational postsecondary aspirations applying social cognitive career theory. J Career Dev 33(2):87–111. https://doi.org/10.1177/0894845306293347

Ali SR, Saunders JL (2009) The career aspirations of srural Appalachian high school students. J Career Assess 17:172–188. https://doi.org/10.1177/1069072708328897
Appalachian Regional Commission [ARC] (2017a) Census population change, 2000–2010. https://www.arc.gov/reports/custom_report.asp?REPORT_ID=41
Appalachian Regional Commission [ARC] (2017b) Poverty rates, 2013–2017: Appalachian Tennessee. https://www.arc.gov/reports/custom_report.asp?REPORT_ID=77
Appalachian Regional Commission [ARC] (2019a) Appalachian regional commission substance abuse listening sessions: a catalogue of ideas. https://www.arc.gov/images/research/ARC-SubstanceAbuseSessions-Catalogue-of-Ideas.pdf
Appalachian Regional Commission [ARC] (2019b) Industrial make-up of the Appalachian region. https://www.arc.gov/assets/research_reports/IndustrialMakeUpoftheAppalachianRegion2002-2017.pdf
Appalachian Regional Commission [ARC] (2019c) The Appalachian region. http://www.arc.gov/appalachian_region/TheAppalachianRegion.asp
Appalachian Regional Commission [ARC] (2020a) ARC-Oak Ridge 2018 summer STEM programs. https://www.arc.gov/program_areas/ARCOakRidge2018SummerPrograms.asp. Accessed 24 Jan 2020
Appalachian Regional Commission [ARC] (2020b) Fiscal year 2019: Appalachian region fact sheet. https://www.arc.gov/about/index.asp
Appalachian Regional Commission [ARC] (2020c) J-1 Visa Waivers. https://www.arc.gov/program_areas/J1VisaWaivers.asp. Accessed 26 Jan 2020
Appalachian Regional Commission [ARC]. (2020d). Number of population of counties per ecnomic level and state in Appalachia. https://www.arc.gov/appalachian_region/CountyEconomicStatusandDistressedAreasinAppalachia.asp. Accessed 24 Jan 2020
Appalachian Regional Commission [ARC]. (2020e). Socioeconomic data. https://www.arc.gov/reports/socio_report.asp. Accessed 21 Jan 2020
Arcipowski E, Schwartz J, Davenport L, Hayes M, Nolan T (2017) Clean water, clean life: promoting healthier, accessible water in rural Appalachia. J Contemp Water Res Educ 161:1–18
Barish R, Snyder AE (2008) Use of complementary and alternative healthcare practices among persons served by a remote area medical clinic. Fam Community Health 31(3):221–227. https://doi.org/10.1097/01.FCH.0000324479.32836.6b
Batteau A (1982) Mosbys and broomsedge: the semantics of class in an Appalachian kinship system. Am Ethnol 9(3):445–466
Behringer B, Krishnan K (2011) Understanding the role of religion in cancer care in Appalachia. Southern Med J 104(4):295–296. https://doi.org/10.1097/SMJ.0b013e3182084108
Burns SL, Scott SL, Thompson DJ (2006) Family and community. In: Abramson R, Haskell J (eds) Encyclopeida of Appalachia. The University of TN Press, Knoxville, pp 149–197
Campbell RM (2006) Race, ethnicity, and identity: recent immigrants. In: Abramson R, Haskell J (eds) Encyclopedia of Appalachia. The University of Tennessee Press, Knoxville, pp 239–283
Campbell CL, Ash CR (2007) Keeping faith. J Hosp Palliat Nurs 9(1):31–41. https://doi.org/10.1097/00129191-200701000-00008
Campbell CL, Williams IC, Orr T (2010) Factors that impact end-of-life decision making in African Americans with advanced cancer. J Hosp Palliat Nurs 12(4):214–224
Carter J (2009) Reflections on a rotation in Appalachia. Fam Med 41(7):470–471
CATO Institute (2018) Fixing the bias in current state K-12 education rankings. https://www.cato.org/publications/policy-analysis/fixing-bias-current-state-k-12-education-rankings?gclid=EAIaIQobChMI_KDn3J2c5wIVCZezCh1AOQugEAAYAiAAEgI1UvD_BwE. Accessed 24 Jan 2020
Center for Oral Health Research in Appalachia (2020). https://www.dental.pitt.edu/research/center-oral-health-research-appalachia. Accessed 27 Jan 2020
Centers for Disease Control and Prevention [CDC] (2004) CDC report shows cancer death rates in Appalachia higher than national. https://www.scienceblog.com/community
Centers for Disease Control and Prevention [CDC] (2019a) Adverse childhood experiences (ACEs). https://www.cdc.gov/violenceprevention/childabuseandneglect/acestudy/index.html. Accessed 26 Jan 2020
Centers for Disease Control and Prevention [CDC] (2019b) Preventing adverse childhood experiences. https://www.cdc.gov/violenceprevention/childabuse-andneglect/aces/fastfact.html
Cohen EL, Scott AM, Shaunfield S, Jones MG, Collins T (2016) Using communication to manage uncertainty about cervical cancer screening guideline adherence among Appalachian women. J Appl Commun Res 44(1):22–39. https://doi.org/10.1080/00909882.2015.1116703
Counts MM, Boyle JS (1987) Nursing, health, & policy within a community context. Adv Nurs Sci 9(3):12–23. https://doi.org/10.1097/00012272-198704000-00009
Coyne CA, Demian-Popescu C, Friend D (2006) Social and cultural factors influencing health in southern West Virginia: a qualitative study. Prev Chronic Dis 3(4):A124
Cronin J, Arnstein P, Flanagan J (2015) Family members' perceptions of most helpful interventions during end-of-life care. J Hosp Palliat Nurs 17(3):223–228. https://doi.org/10.1097/NJH.0000000000000151
Davis S (Prinicipal Investigator) (2018) Rural Communities Opioid Response Program for East Tennessee (RCORP-ET). (Project No. 1G25RH32484) (HRSA Grant)
Davis S (Prinicipal Investigator) (2019) Project HOPE: Healing OUD through Prevention and Expertise. (Project No. 1GA1RH33552) (HRSA Grant)
Davis S (2020) (personal communication January 22, 2020)

Dawley K (2003) Origins of nurse-midwifery in the United States and its expansion in the 1940s. J Midwifery Womens Health 48(2):86–95

DeYoung AJ, Glover M, Herzog MJR (2006) Education. In: Abramson R, Haskell J (eds) Encyclopedia of Appalachia. The University of Tennessee Press, Knoxville, pp 1517–1561

Diddle G, Denham SA (2010) Spirituality and its relationships with the health and illness of Appalachian people. J Transcult Nurs 21(2):175–182. https://doi.org/10.1177/1043659609357640

Dorgan H (2006) Religion. In: Abramson R, Haskell J (eds) Encyclopedia of Appalachia. The University of Tennessee Press, Knoxville, pp 1281–1359

Eastern Band of Cherokee Indians [EBCI] (2020) Take a journey to the home of the Eastern Band of Cherokee Indians. https://visitcherokeenc.com/eastern-band-of-the-cherokee/. Accessed 21 Jan 2020

Fornehed ML (2017) Rural Appalachian person and family decision making at end of life [unpublished doctoral dissertation]. University of Tennessee, Knoxville

Fornehed ML (2020) (personal communication)

Fornehed ML, Mixer SJ, Lindley LC (2020) Families' decision making at end of life in rural Appalachia. J Hosp Palliat Nurs 22(3):188–195

Frontier Nursing University (2020) History of frontier nursing. https://frontier.edu/about-frontier/. Accessed 26 Jan 2020

Gainor RE, Fitch C, Pollard C (2005) Maternal diabetes and perinatal outcomes in West Virginia Medicaid enrollees. W V Med J 102(1):314–316

Gobble CD (2009) The value of story theory in providing culturally sensitive advanced practice nursing in rural Appalachia. Online J Rural Nurs Health Care 9(1):94–105

Gowen T (2003) Religion in Appalachia. Smithsonian Folklife festival newsletter. https://festival.si.edu/articles/2003/religion-in-appalachia. Accessed 26 Jan 2020

Grow Appalachia (2020) Planting the seeds for a sustainable future. https://growappalachia.berea.edu/

Hall J, Hutson SP, West F (2018) Anticipating needs at end of life in narratives related by people living with HIV/AIDS in Appalachia. Am J Hosp Palliat Med 35(7):985–992

Hatcher J, Voigts K, Culp-Roche A, Adegboyega A, Scott T (2018) Rural grandparent headed households: a qualitative description. Online J Rural Nurs Health Care 18(1):40–62

Hendrickson KA (2012) Student resistance to schooling: disconnections with education in rural Appalachia. High School J 95(4):37–49

Hermansen TK, Melinder A (2015) Prenatal SSRI exposure: effects on later child development. Child Nueropsychol 21(5):543–569. https://doi.org/10.1080/09297049.2014.942727

Hufford M (2006) Folklore and folklife. In: Abramson R, Haskell J (eds) Encyclopedia of Appapachia. The University of Tennessee Press, Knoxville, pp 844–909

Huttlinger K, Schaller-Ayers J, Lawson T (2004) Health care in Appalachia: a population-based approach. Public Health Nurs 21(2):103–110

Huttlinger K (2013) People of Appalachian heritage. In: Purnell LD (ed) Transcultural health care: a culturally competent approach, 4th edn. F.A. Davis, Philadelphia, pp 137–158

Institute for Local Self-Reliance (2020) Community networks. https://muninetworks.org/tags-165. Accessed 24 Jan 2020

Jesse D, Blue C (2004) Mary Breckinridge meets *Healthy People 2010*: a teaching strategy for visioning and building healthy communities. J Midwifery Womens Health 49(2):126–131

Krause D, Cossman J, May WL (2016) Oral health in Appalachia: regional, state, sub state, and national comparisons. J Appalachian Stud 22(1):80–102

Lewis S, Messner R, McDowell WA (1985) An unchanging culture. J Gerontol Nurs 11(8):20–26. https://doi.org/10.3928/0098-9134-19850801-07

Lozier J, Althouse R (1975) Retirement to the porch in rural Appalachia. Int J Aging Hum Dev 6(1):7–15

March of Dimes (2020) Neonatal absinence syndrome (NAS). https://www.marchofdimes.org/complications/neonatal-abstinence-syndrome-(nas).aspx. Accessed 26 Jan 2020

Mixer SJ (2020) (personal communication)

Mixer SJ, Fornehed ML, Varney J, Lindley LC (2014) Culturally congruent end-of-life care for rural Appalachian people and their families. J Hosp Palliat Nurs 16(8):526–535. https://doi.org/10.1097/NJH.0000000000000114

Mixer SJ, Carson E, McArthur PM, Abraham C, Silva K, Davidson R, Sharp D, Chadwick J (2015) Nurses in action: a response to cultural care challenges in a pediatric acute care setting. J Pediatr Nurs 30:896–907. https://doi.org/10.1177/1049909112444592

Montgomery M (2006) Language. In: Abramson R, Haskell J (eds) Encyclopedia of Appalachia. The University of Tennessee Press, Knoxville, pp 999–1033

National Alliance on Mental Illness [NAMI] (2018) Mental health by the numbers. https://www.nami.org/Learn-More/Mental-Health-By-the-Numbers. Accessed 26 Jan 2020.

National Hospice and Palliative Care Organization (NHPCO) (2018) NHPCO's Facts and figures: hospice care in America, 2018 edition. https://www.nhpco.org/research/

National Institute on Drug Abuse [NIDA] (2018a) Pregnant women in Appalachia face barriers to opiod treatment. https://www.drugabuse.gov/news-events/latest-science/pregnant-women-in-appalachia-face-barriers-to-opioid-treatment

National Institute on Drug Abuse [NIDA] (2018b) Drugs, brains, and behavior: the science of addiction. https://www.drugabuse.gov/publications/drugs-brains-behavior-science-addiction/drug-misuse-addiction

National Institute on Drug Abuse [NIDA] (2019a) Dramatic increases in maternal opiod use and neonateal abstinence syndrome. https://www.drugabuse.gov/related-topics/trends-statistics/infographics/dramatic-increases-in-maternal-opioid-use-neonatal-abstinence-syndrome

National Institute on Drug Abuse [NIDA] (2019b) Higher rates of NAS linked with economic conditions. https://www.drugabuse.gov/news-events/news-releases/2019/01/higher-rates-nas-linked-economic-conditions

National Institute on Drug Abuse [NIDA] (2019c) Opiod summaries by state. https://www.drugabuse.gov/drugs-abuse/opioids/opioid-summaries-by-state

Neal D (2015) Melungeons explore mysterious mixed-race origins. Newspaper article Ashville (N.C.). Citizen-Times June 25, 2015. https://www.usatoday.com/story/news/nation/2015/06/24/melungeon-mountaineers-mixed-race/29252839/

Obermiller PJ, Brown MK (2002) Appalachain health status in greater Cincinnati: a research overview. Cincinnati: Urban Appalachian Council Working Paper No. 18, February, 2002

Phipps SR (2006) Settlement and migration. In: Abraham J, Haskell J (eds) Encyclopedia of Appalachia. The University of Tennessee Press, Knoxville, pp 285–345

Piamjariyakul U, Petitte T, Smothers A, Wen S, Morissey E, Young S, Sokos G, Moss AH, Smith CE (2019) Study protocol of coaching end-of-life palliative care for advanced heart failure patients and their family caregivers in rural Appalachia: a randomized controlled trial. BMC Palliat Care 18:119

Pollard K, Jacobsen LA (2011) The Appalachian region in 2010: a census data overview. http://www.arc.gov/assets/research_reports/AppalachianRegion2010CensusReport1.pdf

Pollard K, Jacobsen LA (2019) The Appalachian region: a data overview from the 2013–2017 American community survey chartbook. https://www.arc.gov/appalachian_region/TheAppalachianRegion.asp. Accessed 26 Jan 2020.

Presley C (2013) Cultural awareness: enhancing clinical experiences in rural Appalachia. Nurse Educ 38(5):223–226. https://doi.org/10.1097/NNE.0b013e3182a0e556

Rowles GD (1991) Changing health culture in rural appalachia: implications for serving the elderly. J Aging Stud 5(4):375–389. https://doi.org/10.1016/0890-4065(91)90017-m

Salyers KM, Ritchie MH (2006) Multicultural counseling: an Appalachian perspective. J Multicult Couns Dev 34(3):130–142. https://doi.org/10.1002/j.2161-1912.2006.tb00033.x

Schoenberg NE, Bundy HE, Baeker Bispo JA, Studts CR, Shelton BJ, Fields N (2015) A rural Appalachian faith-placed smoking cessation intervention. J Relig Health 54(2):598–611. https://doi.org/10.1007/s10943-014-9858-7

Shell R, Tudiver F (2004) Barriers to cancer screening by rural Appalachian primary care providers. J Rural Health 20(4):368–373. https://doi.org/10.1111/j.1748-0361.2004.tb00051.x

Simpson MR, King MG (1999) "God brought all these churches together": issues in developing religion-health partnerships in an Appalachian community. Public Health Nurs 16(1):41–49. https://doi.org/10.1046/j.1525-1446.1999.00041.x

Smith SL, Tessaro IA (2005) Cultural perspectives on diabetes in an Appalachian population. Am J Health Behav 29(4):291–301

Stephenson JB (1984) Escape to the periphery—commodifying place in rural Appalachia. Appalachian J 11(3):187–200

Studts CR, Tarasenko YN, Schoenberg NE, Shelton BJ, Hatcher-Keller J, Dignan MB (2012) A community-based randomized trial of a faith-placed intervention to reduce cervical cancer burden in Appalachia. Prev Med 54(6):408–414

Substance Abuse and Mental Health Services (2018a) Can we live longer? https://www.thenationalcouncil.org/integrated-health-coe/about-us/

Substance Abuse and Mental Health Services (2018b) Key substance use and mental health indicators in the United States: results from the 2017 National Survey on Drug Use and Health. https://www.thenationalcouncil.org/integrated-health-coe/about-us/

Tennessee Department of Health (2016) Neonatal abstinence syndrome surveillance annual report 2016. https://www.tn.gov/content/dam/tn/health/documents/nas/NAS_Annual_report_2016_FINAL.pdf

Tennessee Department of Health (2019) Neonatal abstinence syndrome surveillance summary week 52: December 22–December 28, 2019. https://www.tn.gov/content/dam/tn/health/documents/nas/NASsummary_Week_5219.pdf. Accessed 28 Jan 2020

Tennessee Department of Health (n.d.) Welcome to the Tennessee Drug Overdose Dashboard. https://www.tn.gov/health/health-program-areas/pdo/pdo/data-dashboard.html. Accessed 25 Jan 2020

The Rural School Community and Trust (2014) Why rural matters 2013–2014: the condition of rural education in 50 states. http://www.ruraledu.org/user_uploads/file/2013-14-Why-Rural-Matters.pdf

University of Virginia (2009) Health appalachia: improving cancer control through telehealth in Southwestern Virginia. http://www.virginiacpac.org/PDFs/HealthyAppalachia_0909.pdf

U.S. Department of Health and Human Services (USDHHS) (2020a) Administration shortage designation advisor. https://datawarehouse.hrsa.gov/tools/analyzers/hpsafind.aspx. Accessed 20 Jan 2020

U.S. Department of Health and Human Services (USDHHS) (2020b). Loan repayment. https://www.hrsa.gov/loan-scholarships/repayment/index.html. Accessed 27 Jan 2020.

U.S. Department of Justice Drug Enforcement Administration (2017) Drugs of abuse. https://www.dea.gov/factsheets/methamphetamine. Accessed 27 Jan 2020,

Wallace CL (2015) Family communication and decision making at the end of life: a literature review. Palliat Support Care 13(3):815–825. https://doi.org/10.1017/S1478951514000388

Webb A (2013) African Americans in Appalachia. Accessed 26 Jan 2020. http://www.oxfordaasc.com/public/features/archive/0213/essay.jsp

Wilson CM (1989) Elizabethan America. In: Neil WK (ed) Appalachian images in folk and popular cultures. UMI Press, London, pp 205–216

Yao N, Hector EA, Anderson R, Balkrishnan R (2017) Cancer dispartities in rural Appalachia: incidence, early detection and survivorship. J Rural Health 33:375–381

9 People of Arab Heritage

Anahid Dervartanian Kulwicki

9.1 Introduction

Arabs trace their ancestry and traditions to the nomadic desert tribes of the Arabian Peninsula. They share a common language, **Arabic**, and most are united by **Islam**, a major world religion that originated in seventh-century Arabia. Despite these common bonds, Arab residents of a single Arab country are often characterized by diversity in thoughts, attitudes, and behaviours. Indeed, cultural variations may be significant within and across countries and regions. For example, a tradition-bound farmer from rural Yemen may appear to have little in common with a small business owner from Beirut. It is important to note that there are individual differences among Arabs who transcends cultural, religious, and social norms. In general, immigrant Arab populations may exhibit great cultural differences based on such additional factors as religion, country of origin, refugee status, time since arrival, ethnic identity, education, economic status, employment status, social support, and English language skills.

This chapter is an update previously written by Anahid Dervartanian Kulwicki and Suha Ballout.

A. D. Kulwicki (✉)
Lebanese American University, Beirut, Lebanon
e-mail: anahid.kulwicki@lau.edu.lb

9.2 Overview, Inhabited Localities, and Topography

In recent years, Western sentiments due to armed conflict and political turmoil in addition to the incident of September 11, 2001 terrorist attack on the United States (US), there has been an increase in hostility toward Arabs and Arab Americans in US. Health-care providers need to be cognizant that many Arab Americans do not share the ideological anti-Western movement of the terrorist attacks and that individuals must not be stereotyped by their cultural background. A study conducted by Kulwicki and colleagues (Kulwicki and Khalifa 2008) on the effects of 9/11 on Arab American nurses in Detroit revealed that Muslims and Arabs were discriminated against by being called names, intimidation, and verbal attacks about their appearance and religion. They were also subject to suspicious questioning; the media supported this. Martin (2015) reports similar perceptions of Arab Americans related to anti-Muslim sentiments by health-care providers.

The diversity among Arabs makes presenting a representative account of Arab Americans a formidable task because of the variant cultural characteristics (see Chap. 2) and the limited research literature on Arabs in the Americas. The earliest Arab immigrants arrived as part of the great wave of immigrants at the end of the nineteenth century and the beginning of the twentieth century. They were predominantly

© Springer Nature Switzerland AG 2021
L. D. Purnell, E. A. Fenkl (eds.), *Textbook for Transcultural Health Care: A Population Approach*,
https://doi.org/10.1007/978-3-030-51399-3_9

Christians from the region that is present-day Lebanon and Syria and, like most newcomers of the period, they valued assimilation and were rather easily absorbed into mainstream U.S. society. Arab Americans tend to disappear in national studies because they are counted as White in census data rather than as a separate ethnic group. Therefore, to portray Arab Americans as fully as possible, including the large numbers of new arrivals since 1965, literature that describes Arabs in their home countries is used to supplement research completed by groups studying Arab Americans residing in Michigan, Illinois, New York, Ohio, and the San Francisco Bay area of California. An underlying assumption is that the attitudes and behaviours of first-generation immigrants may be similar in some aspects to those of their counterparts in the Arab world. However, readers are cautioned not to make assumptions about Arabs or Arab American clients without thoroughly assessing individual behaviours that may or may not be similar to their counterparts within or outside United States.

Islamic doctrines and practices are included because most post-1965 Arab American immigrants are **Muslims**. Religion, whether Islam, Christianity, or minority faiths, may be an integral part of everyday Arab life. In addition, Islam is the official religion in most Arab countries, Lebanon being a notable exception, and Islamic law is identified as the source of national laws and regulations. Consequently, knowledge of religion is critical to understanding the Arab patient's cultural frame of reference and for providing care that considers specific religious beliefs and practices.

9.2.1 Heritage and Residence

Arab Americans are defined as immigrants from the 22 Arab countries of North Africa and Southwest Asia: Algeria, Bahrain, Comoros, Djibouti, Egypt, Iraq, Jordan, Kuwait, Lebanon, Libya, Mauritania, Morocco, Oman, Palestine, Qatar, Saudi Arabia, Somalia, Sudan, Syria, Tunisia, United Arab Emirates, and Yemen. Some Arabs may originate from neighboring states such as Chad and Iran. The Arab American Institute Foundation (AAIF), 2018 Census Bureau estimates that at least 1.9 million Americans are of Arab descent. The Arab American Institute Foundation, (2018) estimates that the number is closer to 3.7 million Arab Americans live in the United States, with approximately 94 living in metropolitan areas. New York, Detroit, Los Angeles, Chicago, and Washington D.C. are the top five metropolitan areas with Arab American populations, though there are also large numbers of Arab Americans in Florida, Texas, New Jersey, Ohio, Massachusetts, Pennsylvania, and Virginia (AAIF 2018).

9.2.2 Reasons for Migration and Associated Economic Factors

First-wave immigrants came to the US between 1887 and 1913 seeking economic opportunity and perhaps the financial means to return home and buy land or set up a shop in their ancestral villages. Most first-wave Arab Americans worked in unskilled jobs, were male and illiterate (44%), and were from mountain or rural areas (Naff 1980). Today, 32% of Arab Americans are from Lebanon, and 11% come from Egypt (Arab American Institute 2010).

Second-wave immigrants entered the US after World War II; the numbers increased dramatically after the Palestinian-Israeli conflict erupted and the passage of the Immigration Act of 1965 (Naff 1980). Unlike the more economically motivated Lebanese-Syrian Christians, most second-wave immigrants are refugees from nations beset by war and political instability—chiefly, occupied Palestine, Jordan, Iraq, Yemen, Lebanon, and Syria. The latest immigrants from the Arab world (2008–2018) originate from Iraq, Egypt, Somalia, Morocco, and Lebanon (Arab American Institute Foundation 2018). Included in this group are a large number of professionals and individuals seeking degrees who have subsequently remained in the US.

9.2.3 Educational Status and Occupations

Because Arabs favor professional occupations, education, as a prerequisite to white-collar work, is valued. Not surprisingly, both U.S. and foreign-born Arab Americans are more educated than the average American. Over 90% of Arab Americans have a high school education, compared with 32% of the total US population. Approximately 20% have post-graduate degrees (AAIF 2018).

In comparison with European Americans, Arab Americans are more likely to be self-employed and much more likely to be in managerial and professional specialty occupations (AAIF 2018). Nearly 45% are employed in managerial and professional positions, 29% in sales, and 16% in service jobs. Arab American households in the US have a mean annual income of $60,398, compared with $52,029 for all households (AAIF 2018).

9.3 Communication

9.3.1 Dominant Language and Dialects

Arabic is the official language of the Arab world. Modern or classical Arabic is a universal form of Arabic used for all writing and formal situations ranging from radio newscasts to lectures. Dialectal or colloquial Arabic, of which each community has a variety, is used for everyday spoken communication. Arabs often mix Modern Standard Arabic and colloquial Arabic according to the complexity of the subject and the formality of the occasion. The presence of numerous dialects with differences in accent, inflection, and vocabulary may create difficulties in communication between Arab immigrants from Syria and Lebanon and, for example, Arab immigrants from Iraq and Yemen.

An Arab person's speech is likely to be characterized by repetition and gesturing, particularly when involved in serious discussions. Arabs may be loud and expressive when involved in serious discussions to stress their commitment and their sincerity in the subject matter. Observers witnessing such impassioned communication may assume that Arabs are argumentative, confrontational, or aggressive.

English is a common second language in Egypt, Jordan, Lebanon, Yemen, Iraq, and Kuwait; French is a common second language in Lebanon, Algeria and Morocco. In contrast, literacy rates among adults in the Arab world vary from 93% for men and 30% for women (CIA World Factbook 2011a, b, c) in Yemen; 93% for men and 82% for women in Lebanon; and 95% for men and 85% for women in Jordan. More than half speak a language other than English at home, although many have a good command of the English language (Arab American Institute 2010). Despite this, ample evidence indicates that language and communication pose formidable problems in American health-care settings. For example, Kulwicki and Miller (1999) reported that 66% of respondents using a community-based health clinic spoke Arabic at home, and only 30.2% spoke both English and Arabic. Even English-speaking Arab Americans report difficulty in expressing their needs and understanding health-care providers.

Health-care providers have cited numerous interpersonal and communication challenges including erroneous assessments of patient complaints, delayed or failed appointments, reluctance to disclose personal and family health information, and in some cases adherence to medical treatments (Kulwicki 1996; Kulwicki et al. 2000), as well as a tendency to exaggerate when describing complaints (Sullivan 1993). This has been shown to create a barrier to access (Kulwicki et al. 2010).

9.3.2 Cultural Communication Patterns

Arab communication has been described as highly nuanced, with more communication contained in the context of the situation than in the actual words spoken. Arabs value privacy and resist disclosure of personal information to strangers, especially when it relates to familial disease conditions. Conversely, among friends and relatives, Arabs express feelings freely. These patterns of communication become more comprehensible when interpreted within the Arab cultural frame of reference. Many personal needs may be anticipated without the individual having to verbalize them because of close family relationships. The family may rely more on unspoken expectations and nonverbal cues than overt verbal exchange.

Arabs need to develop personal relationships with health-care providers before sharing personal information. Because meaning may be attached to both compliments and indifference, manner and tone are as important as what is said. Arabs are sensitive to the courtesy and respect they are accorded, and good manners are important in evaluating a person's character. Therefore, greetings, inquiries about well-being, pleasantries, and a cup of tea or coffee precede business. Conversants stand close to one another, maintain steady eye contact, and touch (only between members of the same sex) the other's hand or shoulder. Sitting and standing properly is critical, because doing otherwise is taken as a lack of respect. Within the context of personal relationships, verbal agreements are considered more important than written contracts. Keeping promises is considered a matter of honor. However, increasingly, written contracts are considered essential in business relations.

Substantial efforts are directed at maintaining pleasant relationships and preserving dignity and honor. Hostility in response to perceived wrongdoing is warded off by an attitude of ***maalesh***: "Never mind; it doesn't matter." Individuals are protected from bad news for as long as possible and are then informed as gently as possible. For example, they may be protected from being informed about a cancer diagnosis. When disputes arise, Arabs hint at their disagreement or simply fail to follow through. Alternatively, an intermediary, someone with influence, may be used to intervene in disputes or present requests to the person in charge. Mediation saves face if a conflict is not settled in one's favour and reassures the petitioner that maximum influence has been employed (Nydell 1987).

9.3.3 Guidelines for Communicating with Arab Americans Include the Following:

1. Health-care providers should employ an approach that combines expertise with warmth. They should minimize status differences, because Arab Americans report feeling uncomfortable and self-conscious in the presence of authority figures. Also, health-care providers should pay special attention to the person's feelings. Arab Americans perceive themselves as sensitive, with the potential for being easily hurt, belittled, discriminated against, and slighted (Reizian and Meleis 1987, Martin, 2015).
2. Nurses and other health-care providers should take time to get acquainted before delving into business. If sincere interest in the person's home country and adjustment to American life is expressed, he or she is likely to enjoy relating such information, much of which is essential to assessing risk for traumatic immigration experience and understanding the person's cultural frame of reference. Sharing a cup of tea does much to give an initial visit a positive beginning (Kulwicki 1996).
3. Nurses may need to clarify role responsibilities regarding history taking, performing physical examinations, and providing health information for newer immigrants. Although some recent Arab American immigrants may now recognize the higher status of nurses in the US, they are still accustomed to nurses functioning as medical assistants and housekeepers.
4. Nurses will need to perform a comprehensive assessment and explain the relationship of the information needed for physical complaints.
5. Health-care providers should interpret family members' communication patterns within a cultural context. Care providers should recognize that a spokesperson may answer questions directed to the patient and that the family members may edit some information that they feel is inappropriate (Kulwicki 1996). Family members can also be expected to act as the

patient's advocates; they may attempt to resolve problems by taking appeals "to the top" or by seeking the help of an influential intermediary.

6. Health-care providers need to convey hope and optimism. The concept of "false hope" is not meaningful to Arabs because they regard God's power to cure as infinite. The amount and type of information given should be carefully considered.
7. It is important to be mindful of the patient's modesty and dignity. Islamic teachings forbid unnecessary touch (including shaking hands) between unrelated adults of opposite sexes (Al-Shahri 2002). Observation of this teaching is expressed most commonly by female patients with male health-care providers and may cause the patient to be shy or hesitant in allowing the health-care provider to do physical assessments. Health-care providers must make concerted efforts to understand the patient's feelings and to take them into consideration.

9.3.4 Temporal Relationships

First-generation Arab immigrants may believe in predestination—that is, God has predetermined the events of one's life. Accordingly, individuals are expected to make the best of life while acknowledging that God has ultimate control over all that happens. Consequently, plans and intentions are qualified with the phrase "***inshallah***", "if God wills" and blessings and misfortunes are attributed to God rather than to the actions of individuals.

Throughout the Arab world, there is nonchalance about punctuality except in cases of business or professional meetings; otherwise, the pace of life is more leisurely than in the West. Social events and appointments tend not to have a fixed beginning or end. Although certain individuals may arrive on time for appointments, the tendency is to be somewhat late. However, for most Arab Americans who belong to professional occupations or who are in the business field, punctuality and respecting deadlines and appointments are considered important (Kulwicki 2001).

9.3.5 Format for Names

Etiquette requires shaking hands on arrival and departure. However, when an Arab man is introduced to an Arab woman, the man waits for the woman to extend her hand. Traditional Muslims may not shake hands with the opposite sex. Women and men put their hand on their chest as a gesture to replace shaking hands.

Titles are important and are used in combination with the person's first name (e.g., Mr. Khalil or Dr. Ali). Some may prefer to be addressed as mother (*Um*) or father (*Abu*) of the eldest son (e.g., Abu Khalil, "father of Khalil"). Married women usually retain their maiden names.

9.4 Family Roles and Organization

9.4.1 Head of Household and Gender Roles

Arab Muslim families are often characterized by a strong patrilineal tradition (Aswad 1999). Although traditional gender roles still exist in some parts of Arab countries, these roles are being transformed towards more liberal approaches to gender differences. In traditional families, women are seen as subordinate to men, and young people are subordinate to older people. Consequently, within his immediate family, the man is the head of the family and his influence is overt. In public, a wife's interactions with her husband are formal and respectful. However, behind the scenes, she typically wields tremendous influence, particularly in matters pertaining to the home and children. A wife may sometimes be required to hide her power from her husband and children to preserve the husband's view of himself as head of the family.

Within the larger extended family, the older male figure assumes the role of decision maker. Women attain power and status in advancing years, particularly when they have adult children. The bond between mothers and sons is typically strong, and most men make every effort to obey their mother's wishes, and even her whims (Nydell 1987). These traditional norms are changing as nuclear families are replacing extended families.

Gender roles are clearly defined and regarded as a complementary division of labor. Men are breadwinners, protectors, and decision makers, whereas women are responsible for the care and education of children and for maintenance of a successful marriage by tending to their husbands' needs. Although women in more urbanized Arab countries such as Lebanon, Syria, Jordan, and Egypt often have professional careers with some women advocating for women's liberation, the family and marriage remain primary commitments for the majority. Most educated women still consider caring for their children as their primary role after marriage. The authority structure and division of labor within Arab families are often interpreted in the West as creating a subservient role for women, fueling common stereotypes of the overly dominant Arab male and the passive and oppressed Arab female. Thus, by extension, conservative Arab Americans perceive the stereotypical understanding of the subordinate role of women as a criticism of Arab culture and family values (Kulwicki 2000).

Arabs value modesty among both men and women and typically will cover the extremities and avoid revealing garb. Many practicing Muslim women view the ***hijab***—"covering the body except for one's face and hands"—as offering them protection in situations in which the sexes mix because it is a recognized symbol of Muslim identity and good moral character. Ironically, many Americans associate the *hijab* with oppression rather than protection. The *hijab* is not universal, and one may find women within the same family choosing to wear or not wear it as a matter of personal choice. Even if Muslim women don't wear the *hijab* during the day, they wear it during prayer and while reading the Quran.

9.4.2 Prescriptive, Restrictive, and Taboo Behaviors for Children and Adolescents

In the traditional Arab family, the roles of the father and mother as they relate to the children are quite distinct. Typically, the father is the disciplinarian, whereas the mother is an ally and mediator, an unfailing source of love and kindness. Although some fathers feel that it is advantageous to maintain a degree of fear, family relationships are usually characterized by affection and sentimentality. Children are dearly loved, indulged, and included in all family activities.

Among Arabs, raising children so they reflect well on the family is an extremely important responsibility. A child's character and successes (or failures) in life are attributed to upbringing and parental influence. Because of the emphasis on collective familism rather than individualism within the Arab culture, conformity to adult rules is favored. Correspondingly, child-rearing methods are oriented toward accommodation and cooperation. Family reputation is important; children are expected to behave in an honorable manner and not bring shame to the family. Child-rearing patterns also include great respect toward parents and elders. In traditional families, children are raised to not question elders and to be obedient to older brothers and sisters (Kulwicki 1996), although these norms are evolving providing children with greater sphere of self control. Methods of discipline may include physical punishment and shaming. Children are made to feel ashamed because others have seen them misbehave, rather than to experience guilt arising from self-criticism and inward regret.

Whereas adolescence in the West is centered on acquiring a personal identity and completing the separation process from family, Arab adolescents are expected to remain enmeshed in the family system. Family interests and opinions often influence career and marriage decisions. Arab adolescents are pressed to succeed academically, in part because of the connections between professional careers and social status. Conversely, behaviours that would bring family dishonor, such as academic failure, sexual activity, illicit drug use, and juvenile delinquency, are avoided. For girls in particular, chastity and decency are required. Adolescence in North America may provide more opportunities for academic success and more freedom in making career choices than can be accessed by their counterparts in the Arab countries. Cultural conflicts between American values and

Arab values often cause significant conflicts for Arab American families. Arab American parents cite a variety of concerns related to conflicting values regarding dating, after-school activities, drinking, and drug use (Zogby 2002).

Reflective Exercise 9.1

Mr. and Mrs. AbulMuna presented to the clinic with Samah, their 17-year-old daughter, who is not married. Samah was complaining of nausea and a metallic aftertaste. Mrs. AbulMuna also told the nurse that she noticed Samah was pale and seemed weak during that period. Samah also told the nurse that she was having breast tenderness for the past 2 weeks but did not get her menstrual period yet. Mr. and Mrs. AbulMuna were worried because they did not want Samah to be sick for her final exams. During the assessment, the nurse asked Samah if she was sexually active and if she could be pregnant. Mr. and Mrs. AbulMuna were angry and thought the nurse's questions were inappropriate. They argued that because Samah was unmarried, it was inappropriate for the nurse to ask her about sexual activity and pregnancy. After he left the room, the nurse could hear Mr. and Mrs. AbulMuna furiously asking Samah why the nurse would ask her such a question. Samah insisted that she believes her symptoms were related to something she ate and that she cannot possibly be pregnant. Mr. AbulMuna told the nurse they were going to take their daughter to another facility.

1. How should the nurse deal with this situation?
2. Identify culturally appropriate strategies that may be effective in addressing the needs of the AbulMuna family.
3. How might the nurse ensure that the best care is provided to Samah?
4. Should the nurse ask Samah if she might be pregnant while her parents are out of the room?

9.4.3 Family Goals and Priorities

The family is the central socioeconomic unit in Arab society. Family members cooperate to secure livelihood, rear children, and maintain standing and influence within the community. Family members live nearby, sometimes intermarry (first cousins, although this varies in different countries), and expect a great deal from one another regardless of practicality or ability to help. Loyalty to one's family takes precedence over personal needs. Maintenance of family honor is paramount.

Within the hierarchical family structure, older family members are accorded great respect. Children, sons in particular, are held responsible for supporting elderly parents. Therefore, regardless of the sacrifices involved, the elderly parents are almost always cared for within the home, typically until death.

In traditional Arab families, responsibility for family members rests with the older men of the family. In the absence of the father, brothers are responsible for unmarried sisters. In the event of a husband's death, his family provides for his widow and children. In general, family leaders are expected to use influence and render special services and favors to kinsmen. These norms are evolving as nuclear families in some Arab countries are replacing extended family structure and care of the elderly is becoming a burden to married couples to care for their elderly or family members (Al-Olama and Tarazi 2017).

Although educational accomplishments (doctoral degrees), certain occupations (medicine, engineering, law), and acquired wealth contribute to social status, family origin continues to be the primary determinant in traditional Arab cultures. Certain character traits such as piety, generosity, hospitality, and good manners may also enhance social standing.

9.4.4 Alternative Lifestyles

Most adults marry. Although the Islamic right to marry up to four wives is sometimes exercised, particularly if the first wife is chronically ill or infertile; most marriages are monogamous and for life. Recent studies have reported very small percentage of Arab Muslim marriages are polygamous (Kulwicki 2000). Whereas homosexuality occurs in all cultures, it is still stigmatized among Arab cultures. In Michigan, 46% of the Arab HIV/AIDS cases were men having sex with men (Michigan Department of Community Health 2010a, b). In some Arab countries, it is considered a crime. Fearing family disgrace and ostracism, gays and lesbians remain closeted (*Global Gayz* 2006). However, in recent years, Arab American gays and lesbians have been active in gay and lesbian organizations, and some have been outspoken and publicly active in raising community awareness about gay and lesbian rights in Arab American communities. Recently, some Arab countries like Lebanon have Lesbian, Gay, Bisexual, transgender and Questioning/Queer (LGBTQ) organizations that support the rights of this population and are working toward decreasing or removing the legal and cultural barriers to alternative lifestyles. A survey of health care providers by Naal et al. (2019) in Lebanon indicated positive attitudes and behaviors toward LGBT patients by mental health providers when compared by non-mental health providers and when compared to previous research findings. In the US, several LGBTQ Arab organizations are actively involved in educating the Arab communities on the rights of the populations, aiming at removing cultural stereotypes and stigma associated with being a member of LGBTQ population.

9.5 Workforce Issues

9.5.1 Culture in the Workplace

Cultural differences that may have an impact on work life include beliefs and practices regarding family, gender roles, one's ability to control life events, maintaining pleasant personal relationships, guarding dignity and honor, and the importance placed on maintaining one's reputation. Arabs and Americans may also differ in attitudes toward time, instructional methods, patterns of thinking, and the amount of emphasis placed on objectivity. However, because many second-wave professionals were educated in the US and thereby socialized to some extent, differences are probably more characteristic of less-educated, first-generation Arab Americans.

In recent years and more so with the continued political unrest in Arab countries, rising unemployment and poverty in the home countries, stress, depression and mental health conditions has become more prevalent among first-generation immigrants and refugees and for people in the war-torn Arab counties. Sources of stress include separation from family members, difficulty adjusting to American life, marital tension, and intergenerational conflict, specifically coping with adolescents socialized in American values through school activities (Seikaly 1999). In addition, worrying about family members' safety, economic well being of relatives in the home country has increased the Arab immigrants susceptibility to mental health conditions. Issues related to discrimination has been reported as a major source of stress among Arab Americans in their work environment. In a recent study exploring the perceptions and experiences of Arab American nurses in the aftermath of 9/11, the majority of nurses did not experience major episodes of discrimination at work such as termination and physical assaults. However, some did experience other types of discrimination such as intimidation, being treated suspiciously, negative comments about their religious practices, and refusal by some patients to be treated by them (Kulwicki and Khalifa 2008). Arab Americans are keenly aware of the misperceptions Americans hold about Arabs (Martin 2015), such as notions that Arabs are inferior, backward, sinister, and violent. In addition, the American public's ignorance of mainstream Islam and the stereotyping of Muslims as fanatics, extremist, and confrontational burden Muslim Arab Americans. Muslim Arab Americans face a variety of challenges as they practice their faith in a secular American society. For example, Islamic and American civil law differ on matters such as marriage, divorce, banking, and inheritance. Individuals who wish to attend Friday prayer services and observe religious holidays frequently encounter job-related conflicts. Children are often torn between fulfilling Islamic obligations regarding prayer, dietary restrictions, and dress and hiding their religious identity in order to fit into the American public school culture.

9.5.2 Issues Related to Autonomy

Whereas American workplaces tend to be dominated by deadlines, profit margins, and maintaining one's competitive edge, a more relaxed, cordial, and relationship-oriented atmosphere prevails in the Arab world. Friendship and business are mixed over cups of sweet tea to the extent that it is unclear where socializing ends and work begins. Managers promote optimal performance by using personal influence and persuasion, and performance evaluations are based on personality and social behaviour as well as job skills.

Significant differences also exist in workplace norms. In the US and some other countries, position is usually earned, laws are applied equally, work takes precedence over family, honesty is an absolute value, facts and logic prevail, and direct and critical appraisal is regarded as valuable feedback. In the Arab world, position is often attained through one's family and (these practices are changing), rules are bent, family obligations take precedence over the demands of the job, subjective perceptions often dictate actions; criticism is often taken personally as an affront to dignity and family honor (Nydell 1987). In Arab offices, supervisors and managers are expected to praise their employees to assure them that their work is noticed and appreciated. Whereas such direct praise may be somewhat embarrassing for Americans, Arabs expect and want praise when they feel they have earned it (Nydell 1987).

9.6 Biocultural Ecology

9.6.1 Skin Color and Other Biological Variations

Although Arabs are uniformly perceived as swarthy, and whereas many do, in fact have dark or olive skin, they may also have blonde or auburn hair, blue eyes, and fair complexions. Arabs from North Africa, such as Egypt, Morocco, and Tunis, may be black and have African features. Because color changes are more difficult to assess in dark-skinned people, pallor and cyanosis are best detected by examination of the oral mucosa and conjunctiva.

9.6.2 Diseases and Health Conditions

The major public health concerns in the Arab world include trauma related to motor vehicle accidents, cancer, cardiovascular diseases, maternal-child health, suicide, mental health conditions, and control of communicable diseases. The incidence of infectious diseases such as tuberculosis, malaria, trachoma, typhus, hepatitis, typhoid fever, dysentery, and parasitic infestations varies between urban and rural areas and from country to country. For example, disease risks are relatively low in modern urban centers of the Arab world, but are quite high in the countryside where animals such as goats and sheep virtually share living quarters, open toilets are commonplace, and running water is not available. Schistosomiasis (also called bilharzia), with which about one-fifth of Egyptians are infected, has been called Egypt's number-one health problem. Its prevalence is related to an entrenched social habit of using the Nile River for washing, drinking, and urinating. Similarly, outbreaks of cholera and meningitis are continuous concerns in Saudi Arabia during the Muslim pilgrimage season. In Jordan, where contagious diseases have declined sharply, emphasis has shifted to preventing accidental death and controlling non-communicable diseases such as cancer and heart disease. Correspondingly, seat belt use, smoking habits, and pesticide residues in locally grown produce are major issues. Campaigns directed at improving children's and young adults' health include smoking prevention, hepatitis B vaccinations, and dental health programs.

Glucose-6-phosphate dehydrogenase (G-6-PD) deficiency, sickle cell anemia, and the thalassemias are extremely common in the eastern Mediterranean region, probably because carriers enjoy an increased resistance to malaria (Hamamy and Alwan 1994). High consanguinity rates—roughly 30% of marriages in Iraq, Jordan, Kuwait, and Saudi Arabia are between first cousins—and the trend of bearing children up to menopause also contribute to the prevalence of genetically determined disorders in Arab countries (Hamamy and Alwan 1994).

With modernization, acculturation and increased life expectancy, multifactorial disorders—hypertension, diabetes, and coronary heart disease—have also emerged as major problems in Eastern Mediterranean countries and among Arab Americans in US (Al-Dahir et al. 2013; Kulwicki 2001). In many countries, cardiovascular disease is a major cause of death. In Lebanon, the increased frequency of familial hypercholesterolemia is a contributing factor. Individuals of Arabic ancestry are also more likely to inherit familial Mediterranean fever, a disorder characterized by recurrent episodes of fever, peritonitis, or pleurisies, either alone or in some combination.

The extent to which these conditions affect the health of Arab Americans is little understood, most notably because epidemiological studies have primarily originated from southeast Michigan, home of the highest concentration of Arab Americans in the US. A Wayne County Health Department (1994) project, which conducted telephone surveys with Arabs residing in the Detroit, Michigan, area, identified cardiovascular disease as one of two specific risks based on the high prevalence of cigarette smoking, high-cholesterol diets, obesity, and sedentary lifestyles. Although the prevalence of hypertension was lower in the Arab community than in the rest of Wayne County, Arab respondents were less likely to report having their blood pressure checked. In fact, lower rates for appropriate testing and screening, such as cholesterol testing, colorectal cancer screening, and uterine cancer screening, were considered a major risk for this group of Arab Americans. In recent years, the rate of mammography has increased dramatically. The Institute of Medicine's report *Unequal Treatment: Confronting Racial and Ethnic Disparities in Health Care* (2002) indicated that death rates for Arab females, compared with those of other white groups, was higher from heart disease and cancer but lower from strokes. However, the death rate for Arab males from coronary heart disease is higher when compared with that of White males. Metabolic syndrome is also highly prevalent and increased with age in both male and female Arab Americans (Jaber et al. 2004). Lung and colorectal cancer are the two leading causes of death among Arab Americans. For Arab American men, lung cancer is the leading cause of death; breast cancer is the leading cause of death in Arab American women (Schwartz et al. 2005).

The Michigan Department of Community Health (2018a, b), Behavioral Risk Factors Survey (BRFS) of Arab Americans reported higher usage of hookah among Arab Americans when compared to Michigan population (13.5%), lower percentage of cervical (64.2%) and prostate cancer screenings than all adults in Michigan. and only 1 in 12 Arab adults met physical activity guidelines.

The rate of infant mortality in the Arab world is high (26.4), ranging from 45 per 1000 births in Yemen to 6 per 1000 births in United Arab Emirates and Qatar. In Bahrain, the infant mortality is low: 8.5 per 1000 births (World Bank 2019). Although overall infant mortality rates for Arab Americans are slightly lower than the Michigan population (6.6), figures for Michigan show a higher infant mortality rate for Arab Americans (6 per 1000 births) than for White infants (4.5 per 1000 births) (Michigan Department of Community Health 2018a, b).

9.6.3 Variations in Drug Metabolism

Information describing drug disposition and sensitivity in Arabs is limited. Between 1 and 1.4% of Arabs are known to have difficulty metabolizing debrisoquine and substances that are metabolized similarly, such as antiarrhythmics, antidepressants, beta-blockers, neuroleptics, and opioid agents. Consequently, a small number of Arab

Americans may experience elevated blood levels and adverse effects when customary dosages of antidepressants are prescribed. Conversely, typical codeine dosages may prove inadequate because some individuals cannot metabolize codeine to morphine to promote an optimal analgesic effect (Levy 1993).

9.7 High-Risk Health Behaviors

Despite Islamic beliefs discouraging tobacco use, smoking remains deeply ingrained in Arab culture. For many Arabs, offering cigarettes is a sign of hospitality. Consistent with their cultural heritage, Arab Americans are characterized by higher smoking rates and lower quitting rates than European Americans (Darwish-Yassine and Wang 2005; Rice and Kulwicki 1992; WHO 2006). Despite country efforts in reducing tobacco use in many Arab countries, tobacco use continues to be on the rise in some countries. Besides smoking cigarettes, smoking tobacco through a water pipe, commonly known as a hookah, shisha, or narghile, is a common practice among both adults and youths.

According to the *2018 Arab and Chaldean Behavioral Risk Factor Survey*, Arab American reported that they have ever smoked or used hookah had decreased from 38.1% in 2013 to 30.8% in 2018. The majority of smokers believe that hookah use is a healthier choice than tobacco use. The rate of cigarette smoking among Arab adults was reported to be 17.2% compared to all Michigan adults (20.45). Prevalence of smoking is significantly higher among male Arabs than females (Michigan Department of Community Health 2018a, b). The level of acculturation was found to influence nicotine dependence, with less-assimilated Arab Americans smoking more than other Arab Americans who primarily socialized with Americans or behaved like them (Al-Omari and Scheibmeir 2010).

Limited information is available on alcohol use among Arab Americans. However, Islamic prohibitions do appear to influence patterns of alcohol consumption and attitudes toward drug use. In a study of publicly funded treatment centers in Michigan, Arfken et al. (2007) indicated that the number of Arab Americans admitted for substance abuse treatment centers was lower for Arab Americans than others and that most abusers were concentrated in the metropolitan Detroit area. Most common drugs used were alcohol (34.8%), marijuana (17.9%), heroin (17.4%), and crack cocaine (15.6%). The majority of patients admitted to treatment centers were male (76.3%), mostly unemployed (62.1%), and more than half were involved in the criminal justice system (58%). Ninety percent of the Arab respondents in the survey reported that they abstain from drinking alcohol. None reported heavy drinking; a limited number reporting binge drinking (2.2%) and driving under the influence of alcohol (1.4%). All respondents believed that occasional use of cocaine entails "great" risk, with most saying the same about occasional use of marijuana. In 2018, an estimated 29% of Arab adults reported using alcohol consumption compared to 57.2% of the Michigan population. Alcohol us among females were significantly lower than the Arab males (MDCH 2018a, b).

The actual risk for, and incidence of, HIV infection and AIDS in Arab countries and among Arab Americans is low. However, an increase in the rate of infection has been noticed among many Arab countries and among Arab Americans (Centers for Disease Control and Prevention (CDC)). The reported number of individuals having AIDS in the Arab countries varies and may not be an accurate reflection of the real incidence owing to restrictions placed on HIV/AIDS research by some Arab countries. The largest number of AIDS cases is seen in Djibouti (214); the lowest numbers of individuals reported as having AIDS are found in Palestine (1), Kuwait (11), Syria (18), Lebanon (24), Yemen (45), and Egypt (63) (WHO 2004). According to latest reports, the Arab region still has one of the lowest HIV cases in the world (less than 0.1%).

Despite the reported low rate of HIV/AIDS among Arab Americans, 4% of the Arab American respondents surveyed by Kulwicki and Cass in 1994 reported that they were at high risk for AIDS. In addition, the sample demonstrated less knowledge of primary routes of transmission and more misconceptions regarding unlikely modes of transmission than other populations surveyed. In 2010, the Michigan Department of Community Health [MDCH] stated that only 110 Arab Americans have ever been diagnosed with HIV and reported in Michigan. Of these, 83 are living, and 54% have progressed to AIDS (MDCH 2010a, b). Misconceptions of HIV/AIDS by Arab Americans has considerably improved since the first HIV/AIDS discovery. In 2016, an estimated 23.1% of Arab adults reported being tested for HIV which is significantly lower Michigan adults (41.4%) of Michigan adults of same ages.

Cultural norms of modesty for Arab women are also a significant risk related to reproductive health among Arab Americans. For example, the rate of breast cancer screening among Arab women in 2016 was 75.8%, similar to 74.0% of all women aged 40 years or older in Michigan. The recent results indicate a significant improvement in breast screening from previous reports of 50.8%, compared with 71.2% for other women in Michigan. In 2018, the rate of cervical Pap smears was 54.9% lower than the reported rate of 59.9% in previous BRFS report (Kulwicki 2000; MDCH, Michigan, 2018a, b). Arab American women, especially new immigrants, may be at a higher risk for domestic violence because of the higher rates of stress, poverty, poor spiritual and social support, and isolation from family members owing to immigration (Kulwicki et al. 2010).

A systematic review of the health status of Arabs living in the US reports little consensus on the rates of cardiovascular disease among Arab Americans. A cross-sectional study among Arab Americans in Michigan found an overall prevalence of self-reported heart disease of 7.1%; Arab Americans were four times more likely to have heart disease than Black Americans in this sample. The prevalence of Hypercholesterolemia ranged from 24.6 to 44.8% among Arab Americans in various national and community-level non-random samples (Al-Daher et al. 2013).

Many Arab Americans are refugees fleeing war and political and religious conflicts, placing them at greater risk for psychological distress, depression, and other psychiatric illnesses (Hikmet et al. 2002; Kinzie et al. 2002; Kira et al. 2010; Jaber et al. 2015). Psychological distress was also documented among immigrants who themselves were not victims of war and conflict but who worried over family members who were in areas of conflict. Studies conducted with Iraqi refugees and victims of torture in the US identified higher prevalence of post-traumatic stress disorder and depression (Kinzie et al. 2002; Kulwicki et al. 2010). However, Kira et al. (2006) found that although tortured Arab immigrants have multilateral trauma experiences and, thus, a significantly higher trauma dose, they have more post-traumatic growth, they are more resilient, and they practice their religion more. In 2016, an estimated 15.6% of Arab adults reported having been told by a health-care provider that they had a depressive disorder; an estimated 11.4% of Arab adults met the criteria for major depression (MDCH 2018a, b).

According to the Wayne County Health Department (1994), Arab Americans' risk in terms of safety is mixed. Factors enhancing safety include low rates of gun ownership and high recognition of the risks associated with having guns in the house. Conversely, lower rates of fire escape planning and seat belt usage for adults and older children (car seats are generally used for younger children); higher rates of physical assaults threaten their safety.

In most health areas surveyed in Michigan, education and income were important determinants of risk for people of Arab descent. Socioeconomic status was also a strong indicator in accessing health-care services. In 2016, the prevalence of no health-care coverage among Arab adults aged 18–64 years (11.0%) was higher than among White, nonHispanic adults (8.2%) in Michigan. Use of health-care services for prenatal care was, however, higher among Arab American females than other ethnic groups in Michigan (Michigan Department of Community Health 2009). Physical or mental disability among Arab Americans in Michigan was almost equal to that of white Americans.

9.8 Nutrition

9.8.1 Meaning of Food

Sharing meals with family and friends is a favorite pastime. Offering food is also a way of expressing love and friendship, hospitality, and generosity. For traditional Arab woman whose primary role is caring for her husband and children, the preparation and presentation of an elaborate midday meal (replaced by dinner) is taken as an indication of her love and caring. Similarly, in entertaining friends, the types and quantity of food served, often several entrees, are indicators of the level of hospitality and esteem for one's guests. Honor and reputation are based on the manner in which guests are received. In return, family members and guests express appreciation by eating heartily.

9.8.2 Common Foods and Food Rituals

Although cooking and national dishes vary from country to country and seasoning from family to family, Arabic cooking shares many general characteristics. Familiar spices and herbs such as cinnamon, allspice, cloves, ginger, cumin, mint, parsley, bay leaves, garlic, and onions are used frequently along with nutmeg, cardamom, marjoram, thyme, and rosemary. Skewer cooking and slow simmering are typical modes of preparation. Yogurt is used in cooking or served plain. All countries have rice and wheat dishes, stuffed vegetables, nut-filled pastries, and fritters soaked in syrup. Dishes are garnished with raisins, pine nuts, pistachios, and almonds. It is also popular to prepare hot drinks from several herbs such as chamomile.

Favorite fruits and vegetables include dates, figs, apricots, guavas, mangos, melons, apples, papayas, bananas, citrus fruits, carrots, tomatoes, cucumbers, parsley, mint, spinach, and grape leaves. Grains are also an important part of the diet such as fava beans, chickpeas, peas, corn, lentils, kidney beans, and white beans. Lamb and chicken are the most popular meats. Muslims are prohibited from eating pork and pork products (e.g., lard). Arab Christians may eat pork, but few of them do. Similarly, because the consumption of blood is forbidden, Muslims are required to cook meats and poultry until well done. Bread accompanies every meal and is viewed as a gift from God. In many respects, the traditional Arab diet is representative of the U.S. Department of Agriculture's food pyramid. Bread is a mainstay, grains and legumes are often substituted for meats, fresh fruit and juices are especially popular, and olive oil is widely used. In addition, because foods are prepared "from scratch," consumption of preservatives and additives is limited.

Lunch is the main meal in Arab households. However, this practice is changing in the US where dinner is becoming main meal. Encouraging guests to eat is the host's duty. Guests often begin with a ritual refusal and then succumb to the host's insistence. Food is eaten with the right hand because it is regarded as clean. Beverages may not be served until after the meal because some Arabs consider it unhealthy to eat and drink at the same time. Similar concerns may exist regarding mixing hot and cold foods.

Health-care providers should also understand **Ramadan**, the Muslim month of fasting. The fast, which is meant to remind Muslims of their dependence on God and the poor who experience involuntary fasting, involves abstinence from eating, drinking (including water), smoking, and marital intercourse during daylight hours. Although the sick are not required to fast, many pious Muslims insist on fasting while hospitalized, necessitating adjustments in meal times and medications, including medications given by nonoral routes. In outpatient settings, health-care providers need to be alert to potential nonadherence to treatment. Patients may omit or adjust the timing of medications. Of particular concern are medications requiring constant blood levels, adequate hydration, or both (e.g., antibiotics that may crystallize in the kidneys). Health-care providers may need to provide appointment times after sunset during Ramadan for individuals requiring injections (e.g., allergy shots).

9.8.3 Dietary Practices for Health Promotion

Arabs associate good health with eating properly, consuming nutritious foods, and fasting to cure disease. For some, concerns about amounts and balance among food types (hot, cold, dry, moist) may be traced to the prophet Mohammed, who taught that "the stomach is the house of every disease, and abstinence is the head of every remedy" (Al-Akili 1993, p. 7). Within this framework, illness is related to excessive eating, eating before a previously eaten meal is digested, eating nutritionally deficient food, mixing opposing types of foods, and consuming elaborately prepared foods. Conversely, abstinence allows the body to expel disease.

The condition of the alimentary tract has priority over all other body systems in the Arab perception of health (Meleis 2005). Gastrointestinal complaints are often the reason Arab Americans seek care (Meleis 2005). Obesity is a problem for second-generation Arab American women and children, most of whom reported eating American snacks that are high in fat and calories. Most women try to lose weight by reducing caloric intake (Wayne County Health Department 1994).

9.8.4 Nutritional Deficiencies and Food Limitations

In Arab countries, diet is influenced by income, government subsidies for certain foods (e.g., bread, sugar, oil), and seasonal availability. Arab Americans most at risk for nutritional deficiencies include newly arrived immigrants from Yemen and Iraq (Ahmad 2004) and Arab American households below the poverty level. Lactose intolerance sometimes occurs in this population. However, the practice of eating yogurt and cheese, rather than drinking milk, probably limits symptoms in sensitive people.

Many of the most common foods are available in American markets. Some Muslims may refuse to eat meat that is not **halal**—"slaughtered in an Islamic manner." *Halal* meat can be obtained in Arabic grocery stores and through Islamic centers or **mosques**.

Islamic prohibitions against the consumption of alcohol and pork have implications for American health-care providers. Conscientious Muslims are often wary of eating outside the home and may ask many questions about ingredients used in meal preparation: Are the beans vegetarian? Was wine used in the meat sauce or lard in the pastry crust? Muslims are equally concerned about the ingredients and origins of mouthwashes, toothpastes, and medicines (e.g., alcohol-based syrups and elixirs), as well as insulin and capsules (gelatin coating) derived from pigs. However, if no substitutes are available, Muslims are permitted to use these preparations.

Reflective Exercise 9.2

Rida, a 42-year-old man diagnosed with pancreatic cancer, is receiving chemotherapy. As a side effect of his cytotoxic medications, Rida is constantly nauseated with decreased appetite. Some days, Rida would not touch any of the food on his tray. The nurse realizes that Rida is not eating his food and requests that the dietician visit him and follow up with his dietary preferences. Rida tells the dietician that he is not eating his soup and most of the food because it had meat in it, and he was not sure if the meat was *halal*. The nurse tries to explain that the food was brought based on his agreement with the dietician. Rida is uncomfortable that the nurse cannot reassure him that the meat is *halal*. Rida also does not believe that the food is nutritious because it does not have vegetables and salads. The nurse tries to explain that having raw vegetables is not appropriate for his neutropenia. Rida continues to express his discomfort with his dietary management.

1. Based on your readings about the Arab culture, what measures should the nurse have taken?
2. How can the nurse and the dietician enhance Rida's food intake?
3. How can the nurse prevent similar instances from taking place?

9.9 Pregnancy and Childbearing Practices

9.9.1 Fertility Practices and Views toward Pregnancy

Fertility rates in the countries from which most Arab Americans emigrate range from 1.8 in Tunisia and Lebanon, to 2.4 in Morocco, to 5.2 in Yemen (UNICEF 2008). Fertility practices of Arabs are influenced by traditional Bedouin values supporting tribal dominance, popular beliefs that "God decides family size," and "God provides," and Islamic rulings regarding birth control, treatment of infertility, and abortion.

High fertility rates are favored. Procreation is regarded as the purpose of marriage and the means of enhancing family strength. Accordingly, Islamic jurists have ruled that the use of "reversible" forms of birth control is "undesirable but not forbidden." These should be employed only in certain situations, listed in decreasing order of legitimacy: threat to the mother's life, too frequent childbearing, risk of transmitting genetic disease, and financial hardship. Moreover, irreversible forms of birth control such as vasectomy and tubal ligation are ***haram***—"absolutely unlawful." Muslims regard abortion as *haram* except when the mother's health is compromised by pregnancy-induced disease or her life is threatened (Ebrahim 1989). Therefore, unwanted pregnancies are dealt with by hoping one miscarries "by an act of God" or by covertly arranging for an abortion. Recently, great decline in fertility rates has occurred in Arab countries and among Arab Americans. According to Michigan's birth registration data, fertility rates among Arab Americans are highest when compared with those of the total population (Office of Minority Health 2001).

Among Jordanian husbands, religion and the fatalistic belief that "God decides family size" were most often given as reasons why contraceptives were not used. Contraceptives were used by 27% of the husbands, typically urbanites of high socioeconomic status. Although the intrauterine device (IUD) and the pill were most widely favored, 4.9% of females used sterilization despite religious prohibitions (Hashemite Kingdom of Jordan 1985). A survey of a random sample of 295 Arab American women in Michigan indicated that 29.1% of the surveyed women did not use any birth control methods because of their desire to have children, 4.3% did not use any form of contraceptives because of their husband's disapproval, and 6% did not use contraceptive methods because of religious reasons. The use of birth control pills was the highest (33.2%) among the users of contraceptive methods, followed by tubal ligation (12.9%) and IUD (10.7%) (Kulwicki 2000).

Indeed, among Arab women in particular, fertility may be more of a concern than contraception because sterility in a woman could lead to rejection and divorce. Islam condones treatment for infertility, as Allah provides progeny as well as a cure for every disease. However, approved methods for treating infertility are mostly limited to artificial insemination using the husband's sperm and in vitro fertilization involving the fertilization of the wife's ovum by the husband's sperm.

9.9.2 Prescriptive, Restrictive, and Taboo Practices in the Childbearing Family

Because of the emphasis on fertility and the bearing of sons, pregnancy traditionally occurs at a younger age; the fertility rate among women in the Arab world was higher than among Arab American women. However, as educational and economic conditions for Arab women have improved both in the Arab world and in the US, fertility rates have fallen.

The pregnant woman is indulged and her cravings satisfied, lest she develop a birthmark in the shape of the particular food she craves. Because of the preference for male offspring, the sex of the child can be a stressor for mothers without sons. Friends and family often note how the mother is "carrying" the baby as an indicator of the baby's sex (i.e., high for a girl and low for a boy). Although pregnant women are excused from fasting during Ramadan, some Muslim women may be determined to fast and thus suffer potential consequences for glucose metabolism and hydration.

Labor and delivery are women's affairs. In Arab countries, home delivery, with the assistance of ***dayahs*** ("midwives") or neighbors was common because of limited access to hospitals, "shyness," and financial constraints. However, recently, the practice of home delivery has decreased dramatically in Arab countries, and hospital deliveries have become common. During labor, some women openly express pain through facial expressions, verbalizations, and body movements. Nurses and medical staff may mistakenly diagnose Arab women as needing medical intervention and administer pain medications more liberally to alleviate the pain.

The call to prayer is recited in the Muslim newborn's ear. Male circumcision is almost a universal practice; for Muslims, it is a religious requirement. Female circumcision is practiced in some Arab countries like Egypt.

Breastfeeding is often delayed until the second or third day after birth because of beliefs that the mother requires rest, that nursing at birth causes "colic" pain for the mother, and that "colostrum makes the baby dumb" (Cline et al. 1986). Postpartum care also includes special foods such as lentil soup to increase milk production and tea to flush and cleanse the body. The 40 days after delivery are valued for women to rest. Mothers, in-laws, and other female members of the extended family may step in to help. The newly delivered woman expects to receive guests to congratulate her for the birth of the child from all family and friends.

According to the WHO infant feeding indicators, breastfeeding practices were suboptimal in several aspects with a low proportion of children being exclusively breastfed, short breastfeeding duration and early introduction of complementary feeding, despite high socioeconomic status. These findings suggest that there is a need to understand potential barriers towards breastfeeding in order to develop appropriate strategies to promote and support breastfeeding in Abu Dhabi (Taha 2018).

Reflective Exercise 9.3

Mrs. Khairallah is a 32-year-old pregnant woman who arrives at the delivery suite with contractions. While the nurses are getting Mrs. Khairallah to her bed, her husband calls the nurse and tells her that he would prefer that no man be allowed to enter his wife's room. He insists that all nurses, doctors, and other staff be female. He also does not want any male person to enter the room during the night because his wife will have her veil removed while she is sleeping. The nurse is conflicted because the doctor in the delivery suite is a male, and Mrs. Khairallah is fully dilated and ready to deliver at any time. The nurse puts a note next to the Khairallah family's room saying that males are not allowed in the room based on patient preference.

1. Based on your nursing training, evaluate the response of this nurse.
2. Explain the cultural connotations of Mr. Khairallah's behavior.
3. What can the nurse do in this situation?

A Michigan study with 2755 Arab Americans reported the experiences of Arab American mothers and infants as fairly comparable with their White counterparts with regard to adequacy of prenatal care, maternal complications, infant mortality, and birth complications. In addition, fewer Arab American mothers smoke, drink alcohol, or gain too little weight (Kulwicki et al. 2007).

Although these state-wide statistics are quite favorable, it is important to mention that earlier studies revealed an alarming rate of infant mortality among Arab American mothers in Dearborn, Michigan, a particularly disadvantaged community of new immigrants with high rates of unemployment. Factors contributing to poor pregnancy outcomes include poverty; lower levels of educational attainment; inability to communicate in English; personal, family, and cultural stressors; cigarette smoking; and early or closely spaced pregnancies. Fear of being ridiculed by American health-care providers and a limited number of bilingual providers limit access to health-care information.

9.10 Death Rituals

9.10.1 Death Rituals and Expectations

Although Arabs insist on maintaining hope regardless of prognosis, death is accepted as God's will. According to Muslim beliefs, death is foreordained and worldly life is but a preparation for eternal life. Hence, from the Qur'an, Surrah III, v. 185:

> Every soul will taste of death. And ye will be paid on the Day of Resurrection only that which ye have fairly earned. Whoso is removed from the Fire and is made to enter Paradise, he indeed is triumphant. The life of this world is but comfort of illusion. (Pickthall 1977, p. 70)

Muslim death rituals include turning the patient's bed to face the holy city of Mecca and reading from the Qur'an, particularly verses stressing hope and acceptance. After death, the deceased is washed three times by a Muslim of the same sex. The body is then wrapped, preferably in white material, and buried underground as soon as possible, usually the day of or the day after death, in a brick- or cement-lined grave facing Mecca. Prayers for the deceased are recited at home, at the mosque, or at the cemetery. Women dress in black but do not ordinarily attend the burial unless the deceased is a close relative or husband. Instead, they gather at the deceased's home and read the Qur'an. Similar memorials are planned 1 week and 40 days after the death. Cremation is not practiced.

Family members do not generally approve of autopsy because of respect for the dead and feelings that the body should not be mutilated. Islam allows forensic autopsy and autopsy for the sake of medical research and instruction.

Death rituals for Arab Christians are similar to Christian practices in the rest of the world. Arab American Christians may have a Bible next to the patient, expect a visit from the priest, and expect medical means to prolong life if possible. Organ donations and autopsies are acceptable. Wearing black during the mourning period is also common. Widows may wear black for the remainder of their lives. For both Christians and Muslims, patients, especially children, are not told about terminal illness. The family spokesperson is usually the person who should be informed about the impending death. The spokesperson will then communicate news to family members.

9.10.2 Responses to Death and Grief

Mourning periods and practices may vary among Muslims and Christians emigrating from different Arab countries. Extended mourning periods may be practiced if the deceased is a young man, a woman, or a child. However, in some cases, Muslims may perceive extended periods of mourning as defiance of the will of God. Family members are asked to endure with patience and good faith in Allah what befalls them, including death. Whereas friends and relatives are to restrict mourning to 3 days, a wife may mourn for 4 months, and in some special cases, mourning can extend to 1 year. Although weeping is allowed, beating the cheeks or tearing garments is prohibited. For women, wearing black is considered appropriate for the entire period of mourning.

9.11 Spirituality

9.11.1 Religious Practices and Use of Prayer

Not all Arab Americans are Muslims. Prominent Christian groups include the Copts from Egypt, the Chaldeans from Iraq, and the Maronites from

Lebanon. Despite their distinctive practices and liturgies, Christians and Muslims share certain beliefs because of Islam's origin in Judaism and Christianity. Muslims and Christians believe in the same God and many of the same prophets, the Day of Judgment, Satan, heaven, hell, and an afterlife. One major difference from Catholicism and Christian Orthodoxy is that Islam has no priesthood. Islamic scholars or religious **sheikhs**, the most learned individuals in an Islamic community, assume the role of ***imam***, or "leader of the prayer." The imam also performs marriage ceremonies and funeral prayers and acts as a spiritual counselor or reference on Islamic teachings. Obtaining the opinion of the local imam may be a helpful intervention for Arab American Muslims struggling with health-care decisions.

As with any religion, observance of religious practices varies. Some nominally practice their religion, whereas others are devout. However, because Islam is the state religion of most Arab countries and, in Islam, there is no separation of church and state, a certain degree of religious participation is obligatory.

To illustrate, consider a few examples of Islam's impact on Jordanian life. Because of Islamic law, abortion is investigated as a crime, and foster parenting is encouraged, whereas adoption is forbidden. The infertility treatments available are those approved by Islamic jurists. Islamic law courts rule on family matters such as marriage, divorce, guardianship, and inheritance employing *shariah*, or Islamic law. Public schools have classes on Islam and prayer rooms. School and work schedules revolve around Islamic holidays and the weekly prayer. During Ramadan, restaurants remain closed during daylight hours and workdays are shortened to facilitate fasting. Because Muslims gather for communal prayer on Friday afternoons, the workweek runs from Saturday through Thursday. Finally, because of Islamic tradition that adherents of other monotheistic religions be accorded tolerance and protection, Jordan's Christians have separate religious courts and schools, and non-Muslims attending public schools are not required to participate in religious activities. Similar arrangements exist in other Arab countries. In Saudi Arabia, the practice of other religions is officially banned.

For Arab American Christians, church is an important part of everyday life. Most celebrate Catholic and Orthodox Christian holidays with fasting and ceremonial church services. They may display or wear Christian symbols such as a cross or a picture of the Virgin Mary. There are also schools that offer classes on Christianity.

9.11.2 Meaning of Life and Individual Sources of Strength

For Muslims, adherents of the world's second largest religion, Islam means "submission to Allah." Life centers on worshipping Allah and preparing for one's afterlife by fulfilling religious duties as described in the Qur'an and the ***hadith***, the putative sayings of the Prophet Muhammad. The five major pillars, or duties, of Islam are declaration of faith, prayer five times daily, almsgiving, fasting during Ramadan, and completion of a pilgrimage to Mecca.

Despite the dominance of familism in Arab life, religious faith is often regarded as more important. Whether Muslim or Christian, Arabs identify strongly with their respective religious groups, and religious affiliation is as much a part of their identity as family name. God and his power are acknowledged in everyday life.

9.11.3 Spiritual Beliefs and Health-Care Practices

Many Muslims believe in combining spiritual medicine, performance of daily prayers, and reading or listening to the Qur'an with conventional medical treatment. The devout patient may request that her or his chair or bed be turned to face Mecca and that a basin of water be provided for ritual washing or ablution before praying. Providing for cleanliness is particularly important because the Muslim's prayer is not acceptable unless the body, clothing, and place of prayer are clean.

Islamic teachings urge Muslims to eat wholesome food; abstain from pork, alcohol, and illicit drugs; practice moderation in all activities; be conscious of hygiene; and face adversity with faith in Allah's mercy and compassion, hope, and acceptance. Muslims are also advised to care for the needs of the community by visiting and assisting the sick and providing for needy Muslims.

Sometimes, illness is considered punishment for one's sins. Correspondingly, by providing cures, Allah manifests mercy and compassion and supplies a vehicle for repentance and gratitude (Al-Akili 1993). Some emphasize that sickness should not be viewed as punishment, but as a trial or ordeal that brings about expiation of sins and that may strengthen character (Ebrahim 1989). Common responses to illness include patience and endurance of suffering because it has a purpose known only to Allah, unfailing hope that even "irreversible" conditions might be cured "if it be Allah's will," and acceptance of one's fate. Suffering by some devout Muslims may be viewed as a means for greater reward in the afterlife (Lovering 2006). Because of the belief in the sanctity of life, euthanasia and assisted suicide are forbidden (Lawrence and Rozmus 2001).

Arab American Christians have spiritual beliefs related to health care that are similar or the same as Orthodox or Catholic Christians. Caring for the body and burial practices are similar. A priest is always expected to visit the patient; if the patient is Catholic, a priest administers the sacrament of the sick.

9.12 Health-Care Practices

9.12.1 Health-Seeking Beliefs and Behaviors

Good health is seen as the ability to fulfill one's roles. Diseases are attributed to a variety of factors such as inadequate diet, hot and cold shifts, exposure of one's stomach during sleep, emotional or spiritual distress, envy, or the evil eye. Arabs are expected to express and acknowledge their ailments when ill. Muslims often mention that the Prophet urged physicians to perform research and the ill to seek treatment because "Allah has not created a disease without providing a cure for it" (Ebrahim 1989, p. 5), except for the problem of old age.

Despite beliefs that one should care for health and seek treatment when ill, some Arab women are often reluctant to seek care. Because of the cultural emphasis placed on modesty, some women express shyness about disrobing for examination. Similarly, some families object to female family members being examined by male physicians. Because of the fear that a diagnosed illness such as cancer or psychiatric illness may bring shame and influence the marriage ability of the woman and her female relatives, delays in seeking medical care may be common.

Whereas Arab Americans readily seek care for actual symptoms, preventive care is not generally sought among the low income (Kulwicki 1996; Kulwicki et al. 2000). Similarly, pediatric clinics are used primarily for illness and injury rather than for well-child visits (Lipson et al. 1987). Laffrey et al. (1989) and are attributed to Arabs' present orientation and reluctance to plan and to the meaning that Arab Americans attach to preventive care. Whereas American health-care providers focus on screening and managing risks and complications, Arab Americans value information that aids in coping with stress, illness, or treatment protocols. Arab Americans' failure to use preventive care services may be related to other factors such as insurance coverage, the availability of female physicians who accept Medicaid patients, and the novelty of the concept of preventive care for immigrants from developing countries.

9.12.2 Responsibility for Health Care

Dichotomous views regarding individual responsibility and one's control over life's events often cause misunderstanding between Arab Americans and health-care providers (Abu Gharbieh 1993). For example, individualism and an activist approach to life are the underpinnings of the

American health-care system. Accordingly, practices such as informed consent, self-care, advance directives, risk management, and preventive care are valued. Patients are expected to use information seeking and problem solving in preference to faith in God, patience, and acceptance of one's fate as primary coping mechanisms. Similarly, American health-care providers expect that the patient's hope be "realistic" in accordance with medical science.

However, in the Arab culture, quite different values, familism, and reliance on God's will influence health care and responses to illness. Rather than engage in self-care and decision making, patients often allow family members to oversee care. Family members indulge the individual and assume the ill person's responsibilities. Although the patient may seem overly dependent and the family overly protective by American standards, family members' vigilance and "demanding behavior" should be interpreted as a measure of concern. For Muslims, care is a religious obligation associated with individual and collective meanings of honor (Luna 1994). Individuals are seen as expressing care through the performance of gender-specific role responsibilities as delineated in the Qur'an.

Although most American health-care providers consider full disclosure an ethical obligation, most Arab physicians do not believe that it is necessary for a patient to know a serious diagnosis or full details of a surgical procedure. In fact, communicating a grave diagnosis is often viewed as cruel and tactless because it deprives patients of hope. Similarly, preoperative instructions are believed to cause needless anxiety, hypochondriasis, and complications. In Lebanon, a qualitative study revealed that communication with the physician was a means of relieving stress among cancer patients (Doumit and Abu-Saad 2008). However, some patients still prefer the traditional nondisclosure approach, and thus it is best to ask patients what they want to know about their illness. Most Arabs expect physicians because of their expertise to select treatments. The patient's role is to cooperate. The authority of physicians is seldom challenged or questioned. When treatment is successful, the physician's skill is recognized; adverse outcomes are attributed to God's will unless there is evidence of blatant malpractice (Sullivan 1993).

Not all Arabs may be familiar with the American concept of health insurance. Traditionally, the family unit through its communal resources provides insurance. Certain Arab countries, such as Saudi Arabia, Syria, and Kuwait, provide free medical care at the point of entry; in other countries many citizens are government employees and are entitled to low-cost care in government-sector facilities. Private physicians and hospitals are preferred because of the belief that the private sector offers the best care.

Because some medications requiring a prescription in the US are available over the counter in Arab countries, Arabs are accustomed to seeking medical advice from pharmacists.

9.12.3 Folk Practices

Although Islam disapproves of superstition, witchcraft, and magic, concerns about the powers of jealous people, the evil eye, and certain supernatural agents such as the devil and **jinn** are part of the folk beliefs. Those who envy the wealth, success, or beauty of others are believed to cause adversity by a gaze, which brings misfortune to the victim. Beautiful women, healthy-looking babies, and the rich are believed to be particularly susceptible to the evil eye; expressions of congratulations may be interpreted as envy. Protection from the evil eye is afforded by wearing amulets such as blue beads or figures involving the number 5, reciting the Qur'an, or invoking the name of Allah (Kulwicki 1996). Barren women, the poor, and the unfortunate are usually suspects for casting the evil eye. Mental or emotional illnesses may be attributed to possession by evil *jinn*, or demons. Some believe that insanity or **jinaan** ("possession by the jinn") may also be caused by the evil wishes of jealous individuals.

Traditional Islamic medicine is based on the theory of four humors and the spiritual and physical remedies prescribed by the Prophet. Because illness is viewed as an imbalance between the humors—black bile, blood, phlegm, and yellow bile—and the primary attributes of dryness, heat, cold, and moisture, therapy involves treating with the disease's opposite: hot disease, cold remedy. Although methods such as cupping, cautery, and phlebotomy (bloodletting) may be employed, treating with special prayers or simple foods such as dates, honey, salt, and olive oil is preferred (Al-Akili 1993). Yemeni or Saudi Arabian patients may apply heat (cupping, moxibustion) or use cautery in combination with modern medical technology.

9.12.4 Barriers to Health Care

Newly arrived and unskilled refugees from poorer parts of the Arab world are at particular risk for both increased exposure to ill health and inadequate access to health care. Factors such as refugee status, recency of arrival, differences in cultural values and norms, inability to pay for health-care services, and inability to speak English add to the stresses of immigration (Kulwicki 2000; Kulwicki et al. 2010, 2016, b) and affect both health status and responses to health problems. Moreover, these immigrants are less likely to receive adequate health care because of cultural and language barriers, lack of transportation, limited health insurance, poverty, a lack of awareness of existing services, and poor coordination of services (Kulwicki 1996, 2000, 2010).

Although a lack of insurance coverage is a factor for a significant number of Wayne County Health Department respondents, other studies suggest that Arab Americans regard other barriers and services as more significant. For instance, language and communication remain serious barriers for recent Arab American immigrants (Kulwicki 2000, 2010). Transportation to health-care facilities and culturally competent service providers also adds to the problems of accessing health-care services.

9.12.5 Cultural Responses to Health and Illness

Arabs regard pain as unpleasant and something to be controlled (Reizian and Meleis 1986). Because of their confidence in medical science, Arabs may anticipate immediate postoperative relief from their symptoms. This expectation, in combination with a belief in conserving energy for recovery, often contribute to a reluctance to comply with typical postoperative routines such as frequent ambulation. Although expressive, emotional, and vocal responses to pain are usually reserved for the immediate family, under certain circumstances such as childbirth and illnesses accompanied by spasms, Arabs express pain more freely (Reizian and Meleis 1986). The tendency of Arabs to be more expressive with their family and more restrained in the presence of health-care providers may lead to conflicting perceptions regarding the adequacy of pain relief. Whereas the patient and nurse may assess pain relief as adequate, family members may demand that their relative receive additional analgesia.

The attitude that mental illness is a major social stigma is particularly pervasive (Jaber et al. 2013; Dardas and Simmons 2015; Kulwicki 2016, b). Psychiatric symptoms may be denied, attributed to "bad nerves" (Hattar-Pollara et al. 2001), or blamed on evil spirits (Kulwicki 1996, 2016, b). Lebanese families are still denying the presence of mental illness and many individuals choose not to seek professional help out of fear of their communities' reactions. There are still misconceptions and stigma associated with mental illness in the Lebanese population, similar to other Arab countries. Al-Krenawi et al. found that Arab patients with mental illness avoid the negative reactions of the public towards their illness by not disclosing their psychiatric symptoms to others (Kulwicki 2016, b). Under recognition of signs and symptoms may occur because of the somatic orientation of Arab patients and physicians, patients' tolerance of emotional suffering, and relatives' tolerance of behavioral disturbances (El-Islam 1994). Indeed, home management with standard but crucial adjustments within the family may abort or control symptoms until remission occurs. For example, female family members manage the mother's postpartum depression by assuming care of the newborn and/or by telling the mother she needs more help or more rest. Islamic legal prohibitions further confound attempts to estimate the incidence of problems such as alcoholism and suicide, resulting in underreporting of these conditions because of potential for social stigma.

When individuals suffering from mental distress seek medical care, they are likely to present with a variety of vague complaints, such as abdominal pain, lassitude, anorexia, and shortness of breath. Patients often expect and may insist on somatic treatment, at least "vitamins and tonics" (El-Islam 1994). When mental illness is accepted as a diagnosis, treatment by medications rather than by counseling is preferred. In a sample of United Arab Emirates subjects, the main treatment adopted for psychiatric illness was prayer, herbal ingredients, or both; counseling by a psychiatrist was the least preferred due to stigma (Salem et al. 2009). Hospitalization is resisted because such placement is viewed as abandonment (Budman et al. 1992). Although Arab Americans report family and marital stress as well as various mental health symptoms, they often seek family counseling or social services rather than a psychiatrist (Aswad and Gray 1996). A study in 2007 suggested that immigrant Muslim women in the US are at an increased risk for experiencing anxiety and depressive symptoms, as well as stressors such as acculturative stress, discrimination, and trauma (Hassouneh and Kulwicki 2007).

Yousef (1993) described the Arab public's attitude toward the disabled as generally negative, with low expectations for education and rehabilitation. Yousef also related misconceptions about mental retardation to the dearth of Arab literature about disability and the public's lack of experience with the disabled. Because of social stigma, the disabled are often kept from public view. Similarly, although there is a trend toward educating some children with mild mental retardation in regular schools, special education programs are generally institutionally based.

Reiter et al. (1986) found that parents who were most intimately involved with the developmentally disabled held rather positive attitudes. More tolerant views were expressed among Arab-Israeli parents, Muslims, the less educated, and residents of smaller villages than among Christians, the educated, and residents of larger villages with mixed populations. Reiter et al. (1986) linked the less positive attitudes of the latter groups to the process of modernization, which affects a drive toward status and a weakening of family structures and traditions. Traditions include regarding the handicapped as coming from God, accepting the disabled person's dependency, and providing care within the home.

Dependency is accepted. Family members assume the ill person's responsibilities. The ill person is cared for and indulged. From an American frame of reference, the patient may seem overly dependent and the family overly protective.

9.12.6 Blood Transfusions and Organ Donation

Although blood transfusions and organ transplants are widely accepted, organ donation is a controversial issue among Arabs and Arab Americans. Practices of organ donation may vary among Arab Muslims and non-Muslims based on their religious beliefs about death and dying, reincarnation, or their personal feelings about helping others by donating their organs to others or for scientific purposes Organ donations are generally low among Muslim countries and also low among Muslims living in other countries as minorities. Most Muslim scholars have agreed that organ donation is permitted on the conditions that it will help the recipient with certainty, it does not cause harm to the donor, and the donor donates the organ or tissue voluntarily and without financial compensation. The organ transplant rates, especially for deceased-donor transplant, in most Islamic countries were less than expected. Some of the causes of low transplant activity included lack of public education and awareness, lack of approval and support by Islamic scholars, and lack of government infrastructure, and financial resource (Kulwicki 2001; Ghods 2015). Health-care providers should be sensitive to personal, family, and religious practices toward organ donation among Arab Americans and should not make any assumptions about organ donation unless family members are asked.

9.13 Health-Care Providers

9.13.1 Traditional Versus Biomedical Providers

Although Arab Americans combine traditional and biomedical care practices, they are very cognizant of the effective medical treatments in the West and often seek treatment in US, especially for complex health conditions. Because of their profound respect for medicine, Arab Americans seek treatments for physical disorders or ailments. Medical treatments that require surgery, removal of causative agents, or eradicating by intravenous treatments are valued more than therapies aimed at health promotion or disease prevention. Although most Arab Americans have high regard for medicine related to physical disorders, many do not have the same respect or trust for mental or psychological/psychiatric treatment. A pervasive feeling among many Arab Americans and Arabs in general is that psychiatric services or therapies related to mental disorders are not effective and are required only for individuals who have severe mental disorders. Psychiatric services are, therefore, underutilized among Arab Americans despite greater need for such services among distressed immigrant populations.

Gender and, to a lesser extent, age are considerations in matching Arab patients and health-care providers. In Arab societies, unrelated males and females are not accustomed to interacting. Shyness in women is appreciated, and in traditional Arab societies Muslim men may ignore women out of politeness. Health-care settings, patient units, and sometimes waiting rooms are segregated by sex. In some countries, male nurses do not care for female patients.

Given this background, some Arab Americans may find interacting with a health-care provider of the opposite sex quite embarrassing and stressful. Discomfort may be expressed by refusal to discuss personal information and a reluctance to disrobe for physical assessments and hygiene. Arab American women may prefer to be seen by male American health-care providers, excluding or denying men the opportunity to interact or appropriately diagnose health conditions for high-risk Arab American females.

9.13.2 Status of Health-Care Providers

Arab Americans have great respect for science and medicine. Most Arab Americans are aware of the historical contributions of Arabs in the field of medicine and are proud of their accomplishments. Knowledge held by a doctor is believed to convey authority and power. When ill, most Arab American patients who lack English communication skills prefer to see Arabic-speaking doctors because of their feelings of cultural and linguistic affinity toward Arab American doctors. Many Arabic-speaking patients also feel that Arab American doctors understand them better; they feel more at ease speaking with someone from their own culture. However, patients who are able to communicate in English do not usually show preferences for seeing Arab doctors over American doctors. In some cases, these patients prefer to be seen by American doctors because they view American doctors as more professional and more respectful to patients than their Arab American counterparts.

Although medicine is perhaps the most respected prestigious profession in Arab society, nursing is viewed as a less desirable profession that conflicts with societal norms proscribing certain female behaviour. However, many Arab countries are making considerable effort in improving the image of nurses and nursing profession through country efforts and improving nursing work environment for nurses. American nurses are regarded more favourably because of their education, expertise, and performance of roles ascribed solely to Arab physicians (e.g., performing physical examinations). However, younger immigrants, and especially immigrants who come from Lebanon, Iraq, and Jordan have more favorable perceptions about nursing as a profession than the older generation of Arab American immigrants (Kulwicki and Kridli 2001).

Perhaps because Arab physicians tend to be older males and Arab nurses are typically young females, the status and roles of physicians and nurses mirror the hierarchical family structure of Arab society. Physicians require that nurses "know their place" and leave the interpretation of data, decision making, and disclosure of information to them. Nurses conform to the role expectations of physicians and the public, and they function as medical assistants and housekeepers rather than as critical thinkers and health educators. Recently, nursing has established professional organizations in many Arab countries that resemble the American Nurses Association.

References

Abu Gharbieh P (1993) Culture shock. Cultural norms influencing nursing in Jordan. Nurs Health Care 14(10):534–540

Ahmad NM (2004) Arab-American culture and health care. http://www.case.edu/med/epidbio/mphp439/Arab-Americans.html

Al-Akili M (1993) Natural healing with the medicine of the prophet. Pearl Publishing House, Philadelphia

Al-Dahir S, Brakta F, Khalil A, Benrahla M (2013) The impact of acculturation on diabetes risk among Arab Americans in Southeastern Louisiana. J Health Care Poor Underserved 24:47

Al-Olama M, Tarazi F (2017) Middle Eastern cultures treasure the elderly, making Alzheimer's a complex scourge. https://www.weforum.org/agenda/2017/10/alzheimers-MENA/

Al-Omari H, Scheibmeir M (2010) Arab Americans' acculturation and tobacco smoking. J Transcult Nurs 20(2):227–233

Al-Shahri MZ (2002) Culturally sensitive caring for Saudi patients. J Transcult Nurs 13(2):133–138

Arab American Institute (2018) Demographics. http://www.aaiusa.org/arab-americans/22/demographics

Arfken CL, Kubiak SP, Koch AL (2007) Health issues in the Arab American community. Arab Americans in publicly financed substance abuse treatment. Ethnicity Dis 17(2 Suppl 3):S3-72–S3-76

Aswad B (1999) Arabs in America: building a new future. In: Suleiman MW (ed) Attitudes of Arab immigrants toward welfare. Temple University Press, Philadelphia, pp 177–191

Aswad BC, Gray N (1996) Challenges to the Arab-American family and ACCESS (Arab Community Center for Economic and Social Services). In: Aswad BC, Bilgé B (eds) Family and gender among American Muslims: Issues facing Middle Eastern immigrants and their descendants. Temple University Press, Philadelphia, pp 223–240

Budman C, Lipson J, Meleis A (1992) The cultural consultant in mental health care: the case of an Arab adolescent. Am J Orthopsychiatry 62(3):359–370

CIA World Factbook (2011a) Jordan. https://www.cia.gov/library/publications/the-world-factbook/geos/jo.html

CIA World Factbook (2011b) Lebanon. https://www.cia.gov/library/publications/the-world-factbook/geos/le.html

CIA World Factbook (2011c). Yemen. https://www.cia.gov/library/publications/the-world-factbook/geos/ym.html

Cline S, Abuirmeileh N, Roberts A (1986) Woman's life cycle. Fundamentals of health education. Yarmouk University, Yarmouk, pp 48–77

Dardasm LA, Simmons LA (2015) The stigma of mental illness in Arab families: a concept analysis. J Psychiatr Ment Health Nurs 22(9):668–679

Darwish-Yassine M, Wang D (2005) Cancer epidemiology in Arab Americans and Arabs outside the Middle East. Ethn Dis 15:S1–S8

Doumit MA, Abu-Saad HH (2008) Lebanese cancer patients: communication and truth-telling preferences. Contemp Nurse 28:74–82

Ebrahim A (1989) Abortion, birth control and surrogate parenting. An Islamic perspective. American Trust Publications, Indianapolis

El-Islam M (1994) Cultural aspects of morbid fears in Qatari women. Soc Psychiatr Psychiatr Epidemiol 29:137–140

Ghods AJ (2015) Current status of organ transplant in Islamic countries. Exp Clin Transplant 1:13–17

Global Gayz: Muslims (2006). http://www.globalgayz.com/art-index.html#middleeast

Hamamy H, Alwan A (1994) Hereditary disorders in the Eastern Mediterranean region. Bull World Health Organ 72(1):145–154

Hashemite Kingdom of Jordan, Department of Statistics (1985) Jordan's husbands' fertility survey. Author, in collaboration with Division of Reproductive Health, Centers for Disease Control and Prevention, Atlanta, GA, Amman

Hassouneh DM, Kulwicki A (2007) Mental health, discrimination, and trauma in Arab Muslim women living in the U.S.: a pilot study. Ment Health Relig Cult 10(3):257–262

Hattar-Pollara M, Meleis AI, Nagib H (2001) A study of spousal role of Egyptian women in clerical jobs. Health Care Women Int 21(4):305–517

Hikmet J, Hakim-Larson J, Farrag M, Jamil L (2002) A retrospective study of Arab American mental health clients: trauma and the Iraqi refugees. Am J Orthopsychiatry 72:355–361

Institute of Medicine (2002) Unequal treatment: Confronting racial and ethnic disparities in health care. http://www.iom.edu/?id=4475

Jaber LA, Brown MB, Hammad A, Zhu Q, Herman WH (2004) The prevalence of the metabolic syndrome among Arab Americans. Diabetes Care 27:234–238

Jaber RM, Farroukh M, Ismail M, Najda J, Sobh H, Hammad A (2015) Measuring depression and stigma towards depression and mental health treatment among adolescents in an Arab-American community. Int J Cult Ment Health 8(3):247–254

Kinzie J, Boehnlein JK, Riley C, Sparr L (2002) The effects of September 11 on traumatized refugees: reactivation of posttraumatic stress disorder. J Nerv Ment Dis 190:437–441

Kira I, Templin T, Lewandoski L, Clifford D, Wieneck P, Hammad A et al (2006) The effects of torture: two community studies. Peace Conflict J Peace Psychol 12(3):205–228

Kira I, Smith I, Lewandoski L, Templin T (2010) The effects of gender discrimination on refugee torture survivors: a cross-cultural traumatology perspective. J Am Psychiatr Nurs Assoc 16:299–306

Kulwicki A (1996) Health issues among Arab Muslim families. In: Aswad BC, Bilgé B (eds) Family and gender among American Muslims: issues facing Middle Eastern immigrants and their descendants. Temple University Press, Philadelphia, pp 187–207

Kulwicki A (2000) Arab women. In: Julia M (ed) Constructing gender: multicultural perspectives in working with women. Brooks/Cole, Nelson, BC, pp 89–98

Kulwicki A (ed) (2001) Ethnic resource guide. Henry Ford Hospital, Dearborn

Kulwicki A (2016) Domestic violence: cultural determinants. Reducing risks, a enhancing resilience. In: Amer M, Awad G (eds) Handbook of Arab American psychology. Tyler and Francis, pp 206–218

Kulwicki A, Cass P (1994) An assessment of Arab-American knowledge, attitudes, and beliefs about AIDS. Image J Nurs Scholarsh 26(1):13–17

Kulwicki A, Khalifa R (2008) The impact of September 11, 2001, on Arab American nurses in Michigan. J Transcult Nurs 19(2):134–139

Kulwicki A, Kridli S (2001) Health-care perceptions and experiences of Chaldean, Arab Muslim, Arab Christian, and Armenian women in the metropolitan area of Detroit. Unpublished manuscript

Kulwicki A, Miller J (1999) Domestic violence in the Arab American population: transforming environmental conditions through community education. Issues Ment Health Nurs 20(3):199–215

Kulwicki A, Miller J, Schim S (2000) Collaborative partnership for culture care: enhancing health services for the Arab community. J Transcult Nurs 11(1):31–39

Kulwicki A, Smiley K, Devine S (2007) Smoking behavior in pregnant Arab Americans. Am J Matern Child Nurs 32(6):363–367

Kulwicki A, Aswad B, Carmona T, Ballout S (2010) Barriers in the utilization of domestic violence services among Arab immigrant women: perceptions of professionals, service providers & community leaders. J Fam Violence 25(8):727–735

Laffrey S, Meleis A, Lipson J, Solomon M, Omidian P (1989) Assessing Arab-American health care needs. Soc Sci Med 29(7):877–883

Lawrence P, Rozmus C (2001) Culturally sensitive care of the Muslim patient. J Transcult Nurs 12:228–233

Levy R (1993) Ethnic and racial differences in response to medicines: preserving individualized therapy in managed pharmaceutical programmes. Pharm Med 7:139–165

Lipson J, Reizian A, Meleis A (1987) Arab-American patients: a medical record review. Soc Sci Med 24(2):101–107

Lovering S (2006) Cultural attitudes and beliefs about pain. J Transcult Nurs 17(4):389–395

Luna L (1994) Care and cultural context of Lebanese Muslim immigrants: using Leininger's theory. J Transcult Nurs 5(2):12–20

Martin B (2015) Perceived discrimination of Muslims in health care. 10.3998/jmmh.10381607.0009.203

Meleis A (2005) Arabs. In: Lipson J, Dibble S (eds) Culture and clinical care. The Regents, University of California, San Francisco, pp 42–57

Michigan Department of Community Health (2009) National trends and deterrent strategies for prescription and OTC drug abuse. www.deadiversion.usdoj.gov/pubs/presentations/stermidcpgov09.pdf

Michigan Department of Community Health (2010a) 2010 Profile of HIV/AIDS in Michigan—Special populations: Arab-Americans. http://www.michigan.gov/documents/mdch/21arab_335588_7.pdf

Michigan Department of Community Health (2010b) Michigan Department of Community Health 2009 health disparities report. Author, Lansing

Michigan Department of Community Health (2018a) Arab-Chaldean behavioral risk factor survey health risk behaviors among Arab adults within the State of Michigan. www.michigan.gov/brfs www.michigan.gov/minorityhealth.

Michigan Department of Community Health (2018b) Number of Hispanic and Middle Easterner infant deaths by selected county and Michigan residents, 2008–2018. https://www.mdch.state.mi.us/osr/InDxMain/HispInfantDeaths.asp

Naal H, Abboud S, Harfoush O, Mahmoud H (2019) Examining the attitudes and behaviors of healthcare providers toward LGBT patients in Lebanon. J Homosex. https://doi.org/10.1080/00918369.2019.1616431

Naff A (1980) Arabs in America: a historical overview. Harvard Encyclopedia of American Ethnic Groups, Boston, pp 128–136

Nydell M (1987) Understanding Arabs. A guide for Westerners. Intercultural Press, Yarmouth

Office of Minority Health (OMH) (2001) Health facts. http://www.mdch.state.mi.us/pha/omh/aram-ch.htm

Pickthall M (1977) The meaning of the glorious Qur'an. Muslim World League, Mecca

Reiter S, Mar'i S, Rosenberg Y (1986) Parental attitudes toward the developmentally disabled among Arab communities in Israel: a cross-cultural study. Int J Rehabil Res 9(4):335–362

Reizian A, Meleis A (1986) Arab-Americans' perceptions of and responses to pain. Crit Care Nurse 6(6):30–37
Reizian A, Meleis A (1987) Symptoms reported by Arab-American patients on the Cornell Medical Index (CMI). Western J Nurs Res 9(3):368–384
Rice V, Kulwicki A (1992) Cigarette use among Arab Americans in the Detroit metropolitan area. Public Health Rep 107(5):589–594
Salem MO, Saleh B, Yousef S, Sabri S (2009) Help-seeking behavior of patients attending the psychiatric service in a sample of United Arab Emirates population. Int J Soc Psychiatry 55(2):141–148
Schwartz S, Darwish-Yassine M, Wing D (2005) Cancer diagnosis in Arab Americans and Arabs outside the United States. Ethnicity Dis 15:1–8
Seikaly M (1999) Arabs in America: building a new future. In: Suleiman MW (ed) Attachment and identity: The Palestinian community of Detroit. Temple University Press, Philadelphia, pp 25–38
Sullivan S (1993) The patient behind the veil: Medical culture shock in Saudi Arabia. Can Med Assoc J 148(3):444–446
Taha Z (2018) Patterns of breastfeeding practices among infants and young. internationalbreastfeedingjournal.biomedcentral.com
UNICEF (2008) United Nations Population Division. http://www.unicef.org/infobycountry/index.html
Wayne County Health Department (1994) Arab community in Wayne County, Michigan: behavior risk factor survey (BRFS). Michigan State University, Institute for Public Policy and Social Research, East Lansing
World Bank (2019) Infant mortality rate for the Arab world. https://www.worldbank.org/
World Health Organization (WHO) (2004) Regional database on HIV/AIDS. WHO Regional Office for the Eastern Mediterranean. http://www.emro.who.int/asd/
World Health Organization (WHO) (2006) Country and selection for infant deaths. www.who.int/whosis/database/mort/table2.cfm
Yousef J (1993) Education of children with mental retardation in Arab countries. Ment Retard 31(2):117–121
Zogby J (2002) What Arabs think: values, beliefs and concerns. http://www.zogby.com/News/ReadNews.dbm?ID=629

People of Brazilian Heritage

10

De Anne K. Hilfinger Messias

10.1 Introduction

As a result of centuries of racial and ethnic mixing, the native Brazilian population is quite diverse. When Portuguese explorers arrived in the 1500s, they encountered numerous indigenous tribes spread across the vast territory. Subsequent arrivals included French and Dutch settlers as well as African slaves (UNESCO 1999). In the nineteenth and twentieth centuries, the influx of Italian, German, Arab, Chinese, and Japanese immigrants contributed to increasing population diversity in the southern regions of the country (Levy 1974). Widespread intermarriage across these various ethnic groups in the early twentieth century contributed to Brazil's rich mixture of native indigenous peoples, and Portuguese, Africans, French, Dutch, Germans, Italians, Japanese, Chinese, and Arabs (Levy 1974). Thus, there is tremendous variations in health beliefs and practices among Brazilians (Canelas and Gisselquist 2018).

This chapter is a revision made of the original chapter written by Marga Simon Coler † in the previous edition of the book.

D. A. K. Hilfinger Messias (✉)
University of South Carolina, Columbia, SC, USA
e-mail: DKMESSIA@mailbox.sc.edu

10.2 Overview, Inhabited Localities, and Topography

10.2.1 Overview

Brazil is the largest country in South America in both landmass and population. The country extends 2695 miles from north to south and 2691 miles wide east to west, with a landmass of 3,286,487 square miles, which is approximately 400,000 square miles smaller than the continental United States and 600,000 square miles smaller than Canada (CIA World Factbook 2019). Bordered on the east by the Atlantic Ocean, Brazil shares borders with every other South American country except Chile and Ecuador. According to the *Instituto Brasileiro de Geografia e Estatística* (IBGE 2019), the official population estimate in July 2018 was 208,846,892; the vast majority (85%) of the population consists of urban dwellers (The World Bank 2017). There are 15 metropolitan areas with over a million residents, the largest of which are São Paulo, with a population over ten million and Rio de Janeiro, with over six million inhabitants (Population Review 2019).

Brazil is quite diverse in both topography and climate. The sparsely populated equatorial Amazon region valley has little variation in temperature throughout the year in contrast to distinct summer and winter seasons in the southern regions. Higher temperatures and humidity

© Springer Nature Switzerland AG 2021

L. D. Purnell, E. A. Fenkl (eds.), *Textbook for Transcultural Health Care: A Population Approach*,
https://doi.org/10.1007/978-3-030-51399-3_10

characterize the climate along the coastal regions. Many locations also have a distinct rainy season, accompanied by somewhat lower temperatures. However, the vast interior of the country consists of high plateaus traversed with low mountain ranges where the climate varies, with little or no rain much of the year (Central Intelligence Agency 2019; IBGE 2019).

10.2.2 Heritage and Residence

In more recent years, there have been significant increases in domestic mobility, secondary to access to both employment and education, with many Brazilians migrating from the more rural areas in the northeast to southern urban centers. Despite the diverse cultural heritage of the Brazilian population, language is a unifying factor, with Portuguese being the primary language of the country. The diversity of the Brazilian population is reflected in the racial variations among Brazilian living in the United States. It is important to note that Brazilians *do not* consider themselves Hispanics, despite similarities in their Iberian ancestry. Portuguese is the native language spoken throughout the country and most Brazilians do not study or speak Spanish as a foreign language.

In the professional healthcare literature in the United States, information about Brazilian subcultures is often unidentifiable due to the tendency to incorporate Brazilians into aggregate data on Hispanics. Therefore, the exact number of Brazilians residing in the United States is not known (Zong and Batalova 2016). In recent years, factors including economic recession in the United States, stepped-up U.S. immigration enforcement, and improved economic prospects in Brazil have led a significant number of immigrants to return to Brazil, especially those who are unauthorized and/or employed in low-skilled jobs. Meanwhile, the number of Brazilian international students in the United States has more than tripled, from 7000 in 2005 to 24,000 in 2015—in part thanks to increased investment by the Brazilian government in international scholarships. Factors associated with both lower Brazilian immigration and higher return migration of Brazilians include improved economic prospects in Brazil. However, between 2005 and 2015, the number of Brazilians enrolled as international students in American colleges and universities more than tripled, reaching 24,000 in 2015. In part, this increase reflected increased Brazilian investments in international scholarships (Zong and Batalova, July 13, 2016).

The 10 most destinations for people leaving Brazil to settle in another country are the United States with the majority (1,066,500), Japan, Paraguay, Portugal, Spain, United Kingdom, Germany, Italy, France, and Switzerland (Emigration from Brazil 2019). The majority of Brazilians in the US reside in Florida, Massachusetts, New Jersey, California, and Connecticut (Center for Latin American, Caribbean, and Latino Studies 2010). Many of the more established Brazilian immigrant communities have their own churches, spiritualists, beauty shops, travel services, and support services. The number of Brazilians seeking U.S. citizenship in 2010 was 8800, a 125% increase since 2001. In 5 years, the non-immigrant visa issuances have nearly tripled to more than half a million annually (Ministério dos Relações Exteriores 2011).

10.2.3 Reasons for Migration and Associated Economic Factors

Similar to many immigrants, Brazilians emigrated in search of opportunities for improving their economic situations while planning to return to their homeland after having acquired sufficient personal wealth to live comfortably (Margolis 2004, 2010). Many send money home to Brazil to help their families or build their "nest eggs." Toward this end, many subsist in urban slums without privacy and think only of earning money. Others flee family problems, come for educational opportunities, and leave their homeland searching for a more humane life with greater dignity.

Like other immigrants, many Brazilians are underemployed after emigrating, often giving up their professions to earn money as illegal domestic workers, waiters, cab drivers, and in other low-paying positions. Even these low-paying jobs pay more than many professional workers can earn in Brazil, which has a per capita yearly income of US $10,800 (CIA World Factbook 2019). Brazilian immigrants often move to large cities where networks help find "under-the-table" wages. Overall, these individuals represent a wide range of professions—from law, medicine, and academics to the arts—as well as young men and women who have enough money for plane fare and a tourist visa and have the courage to disappear into the fairly accessible underground networks if necessary. Most Brazilians emigrating are between the ages of 20 and 39 years of age. More men come than women, and most are representative of the middle and lower-middle socioeconomic groups (Center for Latin American, Caribbean, and Latino Studies 2010). Children, wives, and family are frequently left behind to become slaves of work in any type of situation. Those who are in the United States legally include those who have married and raised families and those who have been sent to the United States as Brazilian government employees.

There are students, former students, and those who get lost in the "zone" between legal entry as tourists and illegal residence. Others emigrate because they find it difficult to market their skills in their home country, creating a "brain drain" in Brazil. University-educated Brazilians are commonly employed in manual work upon emigration.

Since the visit by Secretary of State Hillary Clinton in March 2010 shortly after the inauguration of Brazil's first female president, Dilma Rousseff, there have been increasing ties between Brazilian and American academic researchers in both the private and governmental sectors. Medical, agricultural, and technological and professional collaboration has escalated, especially in relation to environmentally friendly research. There is an increasing awareness against "**biopiracy**," which involves the unauthorized taking of genetic resources or traditional knowledge of indigenous communities in Brazil by foreign researchers. This visit launched several agreements, including the Defense Cooperation Agreement, the Bi-National Energy Working Group Joint Action Plan, the Tropical Forests Conservation Act, and the General Security of Information Agreement. Other topics discussed were trade and finance, biofuels, non-proliferation and arms control, human rights and trafficking, international crime, and environmental and climate change issues (U.S. Department of State, Bureau of Western Hemisphere Affairs, Diplomacy in Action 2019).

10.2.4 Educational Status and Occupations

According to IBGE (2019), there has been improvement in the educational status of the population in Brazil. They report a decrease of illiteracy levels and an increase in the frequency of schooling. The adult literacy rate in Brazil is estimated to be over 90%, and elementary education for the underprivileged has risen to a fifth-grade level. Still, economic reasons as well as the lack of transportation, accessibility, and time create insurmountable barriers for the poor. The federal government has been trying to upgrade public education to the extent that school buses have become visible, but the great majority of services remains managed by private owners or offered with extra charges.

In spite of the increasing governmental and constitutional intervention, middle socioeconomic families often "stretch" their finances to register their children in private schools, hoping for a better education. In many areas, public schools lack necessary supplies and other resources. Disciplining students and enforcing punctuality are not part of their strengths. Children and adolescents of the upper-lower socioeconomic citizens are often able to attend a parochial or an inexpensive private school. Middle- and upper- socioeconomic students generally do not attend public schools.

Lack of competency in English makes it difficult for professionals to pass required professional examinations in the United States. Children and professionals in Brazil are frequently taught by noncertified individuals who had been abroad and who essentially learned English outside of a formal classroom.

10.3 Communication

10.3.1 Dominant Languages and Dialects

Portuguese is the one official language of Brazil and is the primary language among Brazilian immigrants in the US and other parts of the world. Brazilian Portuguese differs from the mother language in the meanings of certain words, accents, and dialects. As in many countries, dialects vary. One who is well versed in the language can frequently ascertain a compatriot's origin. Language from the interior regions of Brazil is a mixture of aboriginal Indian languages and Portuguese. Brazilians from interior towns; the *sertão*, or the dry regions; and the *matta*, or jungle; tend to abbreviate words and frequently run them together. These groups, however, are rare among immigrants because they are the *pobres*, the poor, and cannot afford to emigrate. This dominant class of the country often leads a hand-to-mouth existence. Their speech appears rapid, is full of *giria*, slang, and is difficult for outsiders to understand. The language from the interior is filled with formal second-person expressions such as the English old fashioned "thou." The Portuguese taught to foreigners no longer emphasizes pronouns and verb endings.

10.3.2 Cultural Communication Patterns

General greetings are different from those of other cultures. Americans use the general greeting "how are you?" without an expectation of obtaining a true response, whereas Brazilians seem to hold a strong desire to truly know the answer. Many Brazilians continue to be of "proper" Old World orientation in which true feelings are not divulged for fear of hurting the feelings of the receiver of the communication. Everything is said to be ***tudo bom***, great, almost in a stoic sense. However, in the intimate circle of family, relatives, and friends, sharing thoughts and feelings is common. Young adult and adolescent Brazilians in the United States and other countries are generally more acculturated because of their desire and need to assimilate into the new culture. Among these groups, intragenerational communication is probably more common than intergenerational and transcultural communication when it comes to sharing thoughts and feelings.

Similar to other Latin Americans, Brazilians frequently use touch and usually maintain eye contact when directly interacting with others. It is common for women to kiss one another on both cheeks when they meet and when they say goodbye. Men shake each other's hands and slap each other on the back with the other hand. This gesture frequently ends in an embrace. Children are kissed and often touched or embraced. Spatial distancing is close. Facial expressions and symbolic gestures are commonplace. People from the northeast tend to be more expressive than their more-Westernized compatriots of Rio de Janeiro, São Paulo, the south, and southeast.

10.3.3 Temporal Relationships

Although most Brazilians in America are future oriented, temporality in Brazil is focused on the present because of an unpredictable future. Therefore, for emotional survival, the time factor must necessarily be oriented toward the present. This is changing as Brazil is obtaining world leadership status. During the decades of inflation greater than 100%, Brazilians learned to spend their money immediately to avoid devaluation of the currency of the moment. Presently, lavish credit card spending is the mode of shopping.

Brazilians, in general, are not concerned with punctuality. They tend to arrive late—from a few minutes to hours—especially for social occasions. Given the expectation of general tardiness, schedules are fairly flexible. Even public events may be delayed as the audience waits for the arrival of a distinguished individual to give the beginning oration. Workers are often away for lunch for longer than the usual 2 h and may use this time for errands. However, punctuality is expected in business and professional environments.

10.3.4 Format for Names

Traditionally, Brazilian names are lengthy, consisting of proper name(s) and both parents' surnames. Traditionally, names appear as the first name, the mother's family name, and the father's family name. When a woman marries, she may opt to drop her maiden name or her father's name, keep her father's name, or she may keep them both. At times, *de*, *da/do*, or *das/dos* is added to a name to denote "of"; this seems to be done out of tradition. Junior is added to a name if the son has been named after the father and *neto* if the son has been named after the grandfather. No rigid protocol is apparent. Children who have no father by marriage of the mother are often given their mother's maiden name or the name *da Silva* may be added, denoting that the line of paternity is unclear. This depends on the subculture.

However, in social situations individuals tend to use only the first and last names and in day-to-day encounters adults are called by their first name, often with the title *Seu* (*Senhor*) preceding the first name of a man or *Dona* preceding the first name of a woman. Physicians, lawyers, and other professionals are addressed as *Doutor* (male) or *Doutora* (female), followed by the first name. Similarly, teachers are *Professor* (male) or *Professora* (female), followed by the first name.

Rather than using the personal pronoun "you," grandmothers and other elder women are addressed as *a Senhora*; fathers, grandfathers, and respected men are called *o Senhor.* In the same vein, God is referred to as *O Senhor*.

10.4 Family Roles and Organization

10.4.1 Head of Household and Gender Roles

Gender roles vary for Brazilians according to socioeconomics and education. Brazilian society had been one of ***machismo***, with the middle and upper classes being patriarchal in structure. Generally, women enjoy equality as is evident by a female president elected in 2010. Lower socioeconomic households tend to be more matriarchal in nature.

Reflective Exercise 10.1

Yara Lima, age 65 years from Brazil, is visiting her sister and brother-in-law, in the United States. She discloses to her sister, a nurse, that she has been experiencing chills, fever, and fatigue for two weeks. Her sister suspects malaria and takes her to the neighborhood clinic for an evaluation by a friend who is a physician.

When the physician greets Mrs. Lima by her first name, she gasps and says nothing. When asked how she is doing she answers *tudo bom.* She finally admits to experiencing chills, fever, and fatigue but says that she is not worried and prefers to wait until she returns to Brazil next month to be seen by a ***curandeiro*** because she does not have the money for a physician.

1. How should the physician have greeted this patient upon meeting her the first time?
2. *Tudo bom* is a common Brazilian phrase. What does it mean?
3. If Mrs. Lima does have malaria, what advice would you give her for further prevention?
4. What is a *curandeiro*?
5. Besides money, what other reasons might Mrs. Lima want to see a *curandeiro*?

Social status is very important in the Brazilian society. This is well demonstrated in the titles that people use with one another and the practice of listing both parents' surnames. Class separation is discretely maintained by literacy status.

Children are important in Brazilian families. A wealthier family may raise the child of a poorer relative. These children often enter the family in a second-class capacity, are sent to public or less-expensive private schools, and are taught to help around the house during their free time. Although no documentation substantiates the state of immigrant Brazilian adolescents, they seem to be vulnerable in their attempt to be accepted.

10.4.2 Family Goals and Priorities

The goals of the family are unity and success. Among middle and upper socioeconomic citizens, a good education for children is sought; whereas among lower socioeconomic citizens, the goal is survival. This is increasingly changing as public education becomes stronger. Night school is a very important asset for individuals who, in the past, had little future. A good example is the household *empregada,* live-in housekeeper, who worked for a family for little money. Increasingly these housekeepers are hard to find and almost nonexistent in Southern Brazil, where similar per diem workers are paid a good hourly wage. For upper and middle (and increasingly lower) socioeconomic Brazilians, the outcome of education is to enter the workforce as a university graduate.

Family members living in the same household pool their money so that priority needs can be met. Priorities may include a new washing machine, a 15th-birthday celebration, or the electric bill. The Brazilian father frequently sets his son up in business. For example, a physician father might buy a farm and set his son up in aquaculture, while holding on to the financial reins until the son becomes self-sufficient. Parents with a business of their own, such as a beauty parlor or bar, frequently train their daughters or sons to take over. Whereas a sense of responsibility and loyalty to family and country is strong, a sense of responsibility to political causes may be weak. In the latter scenario, loyalty can easily be bought.

Older people live with one of their children when self-care becomes a concern; nursing home placement is uncommon. Older adults are respected and are often seen as family counsellors. They are included in family activities such as child care and frequently accompany their children's families on vacation. Older people receive benefits such as free public bus fares and special lines in banks and supermarkets. The waiting lines often have benches. Frequently, designated parking spaces are denoted for older people in shopping centers and other public areas. This respect for older people is displayed by the younger generations who help them secure priority places wherever they are. Younger generations commonly give up their seats to older people on public transportation.

Brazilians are loyal to their extended families and help relatives. The extended family is very important: a ***jeitinho***, knack, is frequently procured for employment or in housing relatives in any situation, which can vary from the government or a bank to helping a relative get into a special university or school. Family businesses are common, even among lower and middle socioeconomic citizens, in which everyone pools their money to live comfortably.

Godparents (***madrinha/padrinho***) are a very important family extension. Poor families frequently ask their ***patron*** and ***patrona***, employer and his wife, to be godparents to their child. Godparent's responsibilities include helping to provide clothing and schooling and caring for the child in case of the parent's death, or in times of need. The godmother is called ***comadre***; ***compadre*** refers to the godfather.

10.4.3 Alternative Lifestyles

Although historically common in the lower socioeconomic classes, middle socioeconomic households with a single female parent are becoming increasingly common among Brazilians. The society has also become more accepting of gay and lesbian relationships. Gay and lesbian newsletters and journals exist. National and regional conferences, videos, and other informational materials publicize their movements. AIDS and safe sex are frequent topics of their seminars and political movements.

10.5 Workforce Issues

10.5.1 Culture in the Workplace

Brazilians value diplomacy over honesty, even when they promise to attend to something the next day, knowing that it will be impossible. This is due, in part, to their fatalistic beliefs and, in part, to "save face." Most Brazilians report on time to work. However, in northern and northeastern Brazil where life often represents a struggle and telephone lines frequently break and collapse, people are more flexible regarding time commitments and accepting of a person who may not appear for work or who leaves work early during lunch or at the end of a day. This flexibility in time fosters early closings of businesses or offices, with employees going home before the day's work is completed. When questioned about when a key person will return, a favorite answer is ***d'aqui a pouco***, a little while from now. This may mean 5 min to a half hour, to the next workday. Thus, immigrant Brazilians may find it difficult to adhere to the rigid time schedules in the workplace with individualistic cultures that expect timeliness. Necessities of immediate and extended family members frequently take priority over work, as exemplified by a son or daughter having to take his or her mother to the physician during working hours.

10.5.2 Issues Related to Autonomy

Some Brazilians may have a difficult time adapting to English if they have not had good instruction before entering the country. English intonation and the pronunciation of certain words are particularly difficult. Many undocumented Brazilians find employment within the Brazilian community where they may never have to learn the new language. Finding regular employment is difficult when one is unsure of the language or aware of one's accent. In addition, categorizing Brazilians under the general category of Hispanics adds to their discomfort.

Brazilians generally respect authority and are frequently more comfortable in employment situations in which rules and job specifications are well defined. Brazilians tend to have a lesser sense of responsibility than that seen in the dominant individualistic American culture. For example, when educated people believe that they can do something more efficiently, they are apt not to ask permission from their supervisor to do what they believe is required to complete the job. Brazilian work culture is not as "rigid" as that of the United States.

Reflective Exercise 10.2

Graduate students went to Northeastern Brazil for a clinical experience in international nursing. Rich, who spoke the language, decided to learn about Brazilian health services by helping a man whose acquaintance he made at a cookout given by the American *dono* (owner) of the *granja* (small farm). Severino de Silva, the sole employee of the farm, filled the position of a caretaker and cared for the animals, and performed farming duties, and repairs. As part of his contract, Severino had his house rent-free with utilities paid for by the *patron* (employer). He lived in the house with his wife and two children, ages 15 and 9 years. Severino worked hard and was honest. Although he loved taking care of the fruit trees and vegetable garden, he did not enjoy taking care of the livestock. He tended the chickens, provided the *dono* and his family with eggs, and sold the rest. On weekends Severino would "go out" and sometimes did not return home. Yet, he reported for work each morning.

Sadly, when the *dono* and his family went away for an extended period, he received reports from his neighbors and friends that the farm was "falling apart"; the dog became emaciated as did the rest of the animals. Although Severino was permitted to sell the harvest and keep the income, it seemed to the neighbors that the reason the animals were becoming so emaciated was because not only did Severino use the money from the harvest, but he also spent some of the animal food allowance for weekend drinking bouts.

Severino's behavior deviated during the patron's absences, in spite of the fact that he was appropriate when the owner returned. Severino's wife and children would not discuss their concerns, although she would burst into tears when asked about her husband's behavior. Severino's decompensation became increasingly visible over the years. Once his wife came crying to the patron, stating that Severino claimed the furniture was being moved by ghosts.

1. How should Rich initially greet Severino?
2. How does the Brazilian culture address high-risk behaviors?
3. Is sharing mental health issues, thoughts, and feelings acceptable among Brazilian families? Among outsiders?
4. Rich is an outsider to this family and the Brazilian culture. How might Rich approach Severino to seek help for his alcohol misuse?
5. What value does the Brazilian culture place on family?
6. Should Rich elicit help from Severino's wife to address his alcohol intake?

10.6 Biocultural Ecology

10.6.1 Skin Color and Other Biological Variations

The "typical" Brazilian is a ***moreno***, characterized by brown skin and eyes and black or brown hair (Telles 2004). However, individuals from the southern states of Brazil may have blond hair and blue eyes due to a strong European heritage. Asian Brazilians, most of whom emigrated from Japan, now total more than 9% of the population and most live in the state of São Paulo (CIA World Factbook 2019). It is not unusual to see a Japanese first name with a Portuguese last name or vice versa. A diverse gene pool of native Indians and a multitude of other nationalities make it impossible to actually describe a typical Brazilian.

10.6.2 Diseases and Health Conditions

The overall infant mortality rate in Brazil is 16.9 per 1000 live births with male infant mortality being 19.9 and female infant mortality significantly lower at 13.8 (CIA World Factbook 2019). Causes of death among children under age 5 years, in descending order, are diarrheal disease, measles, malaria, pneumonia, and injuries. The overall causes of death among adults, in descending order, are ischemic heart disease, cerebrovascular disease, violence, diabetes mellitus, lower respiratory infections, chronic obstructive lung disease, hypertensive heart disease, road traffic accidents, and inflammatory heart disease (Pan American Health Organization 2011). In addition, a number of infectious and parasitic diseases continue to plague Brazil and include tuberculosis, malaria, Chagas disease, leishmaniasis, dengue fever, schistosomiasis, typhoid fever, hepatitis, and cholera (Centers for Disease Control and Prevention 2010). Because intestinal worms are common in Brazilian immigrants, parasitic diseases should be considered when health assessments are taken. No data were found addressing the overall health conditions for Brazilians residing in the United States or other countries.

Interviews with Brazilian Americans have substantiated that the incidence of gastrointestinal illnesses increases when Brazilians first move to the United States. Changes in eating habits from the long and ample midday dinner to fast foods have left Brazilians in America with numerous gastric complaints. Different methods of milk pasteurization, along with a genetic tendency toward lactose intolerance, can contribute to some of these gastric problems. Many Brazilians' stomachs do not tolerate foods served in salad bars. Personal interviews report an increased incidence of allergies, especially in children of Brazilian immigrants.

10.6.3 Variations in Drug Metabolism

Although recent studies and citations note drug-response variations for some ethnic groups from environmental, cultural, psychosocial, and genetic factors, specific studies on the Brazilian population are not available. However, Levy (1993), in his review of ethnic and racial differences, identified poor and rapid drug metabolizers by race and ethnicity. In this process, he identified various classes of medicines and linked the rate of metabolic activity to race and ethnicity. Unfortunately, the typical Brazilian cannot be classified as Black, Hispanic, Chinese, or White because of the racial mix. A study of Brazilians in this respect is indicated.

10.7 High-Risk Behaviors

Because Brazilian immigrants frequently settle in Brazilian enclaves in large cities when they emigrate, they are subject to the same risk factors as any socially vulnerable urban subpopulation. The greatest risks are violence, drugs, and crime. Adolescents run the risk of resolving their adolescent identity crises by either banding together or joining other gangs.

Because cigarette smoking had been a part of the Brazilian culture, smoking is a high-risk behavior among Brazilians when they emigrate. Among men, drinking hard liquor is also prevalent. Accessibility and use of street drugs and an individual's desperate search for quick money are other identifiable high-risk behaviors and often include living in crowded ghetto conditions where rent is less expensive. The undocumented status of Brazilian immigrants places them at high risk for non-assimilation into the culture of the community in which they live.

Another risk factor, especially for adolescents, is that of contracting HIV or other sexually transmitted infections. The only endemic disease following Brazilians to the United States, and for which documentation is found, is HIV. The HIV rate is 0.6% of the (CIA World Factbook 2019).

10.8 Nutrition

10.8.1 Meaning of Food

Food is important in the celebration of all rites among Brazilians. Food and its counterpart, hunger, are often viewed as symbols that determine social relations. Food has symbolic content, is used as a reward or punishment, and establishes and maintains social relations.

10.8.2 Common Foods and Food Rituals

The mainstay of the Brazilian diet continues to be rice, beans, farina, and *cuscus,* a dry, cornmeal mush. Beef, chicken, and seafood are sought when they are not too expensive. *Cafe de manha,* breakfast, typically consists of bread with *cafe com leite,* half coffee and half hot milk. Sometimes, *cuscus* is served with milk. Fruit, fruit juices, and scrambled eggs are common breakfast fares among middle socioeconomic families. Sometimes, sweet potatoes, yams, and *macaxeira,* cassava, grace the table. Cold cereals have become a favorite breakfast in many middle socioeconomic homes.

Reflective Exercise 10.3

Ana, age 27 years from Brazil, has a bachelor degree in social work. With a 3-month tourist visa, her plans were to visit family members, including her *madrinha* and *padrinho*, and to travel throughout the northeastern part of the United States.

Between exciting visits to landmarks and visiting famous universities she knew from textbooks and authors she had read, Ana started thinking about ways to better her own career prospects. She talked with her family to make a more thorough plan. She could stay where she was or she could get a new start, which evinced feelings of leaving friends in Brazil and facing unfamiliar situations. She planned to enroll in a continuing education program during the summer as a way of testing her abilities with the host language. However, she needed to support herself somehow.

Using her 3-month tourist visa, she could enroll at the university, but she did not have a Social Security number or authorization to work. Therefore, there were not many options in terms of jobs. Her family helped her find a job in a demanding, fast-paced restaurant requiring 8–10 h of work each day. After work, she had a 1-hour walk home, often in snow. For the first time since she arrived, she thought about her country of origin—Brazil, with its tropical weather with two seasons instead of four. She missed harbor walks in her hometown but recognized the beauty of snow, which reminded her about Christmas movies she used to watch when she was younger. "I'm here now," she realized, "it's real."

With demanding long workdays and not getting home until midnight, she realized that school would not be a priority. Besides, just a smile and memorized greetings to communicate with customers was not a good assessment tool to measure her proficiency in the host language. Customers strained to understand her as evidenced by wrinkled foreheads and other facial expressions. She realized she had to improve her language skills if she were to remain on her goal to continue her education.

1. What are *madrinha* and *padrinho?*
2. What are the immigration issues facing Ana with a 3-month visa?
3. What might happen if it is discovered that Ana is working and collecting payment "under-the-table"?
4. What are some positive aspects if Ana decides to immigrate?
5. What are some negative aspects if Ana decides to immigrate?
6. What evidence do you see of *familism* in this reflective exercise?
7. Identify some community resources that could facilitate Ana's acculturation.

O almoco, lunch, is eaten around noontime. This heavy meal, consisting of beans, rice, and farina, often includes *puree,* mashed potatoes, and *macarão,* pasta. Desserts such as *pudim de leite,* custard, various cornmeal pastries, fruit, and *doce,* a sweet paste made by boiling sugar and fruit or fruit pulp, are common, especially during late June when the holidays of St. Anthony, St. John, and St. Peter are celebrated. A typical vegetable salad traditionally consists of finely cubed carrots, potatoes, and *shushu,* a summer squash-like plant. A fruit salad with finely cubed fruits is also common. This picture is rapidly changing as various salads, fruit salads, or sliced fruits without sugar appear on the table.

Almoco in a middle socioeconomic home has at least one course of meat, chicken, or fish. Beef is preferred very well done. Here, too, the trend is changing as the Brazilian populace becomes more nutrition conscious; less red meat and green salads and vegetables are more common. Brazilian "self-service" restaurants frequented by many of the working class have tempting salad bars. There is a clear tendency for all meals to become more Westernized with an awareness of good nutrition.

After a heavy, tiring midday meal, a noontime nap is often welcome. The noon lunch break for some workers is from 2 to 2.5 h long. The workday, however, begins early and often lasts until 5:30 or 6:00 p.m. *Jantar,* supper, is light and generally eaten late in the evening.

In Brazil, *goma.* a manioc starch, fills the stomach. In fact, the manioc root may be viewed as the symbolic plant, which, when made into gruel, fills babies' stomachs for mothers who can no longer provide breast milk because of chronic malnutrition. This nutritionally unfortified gruel is used by all socioeconomic groups as a traditional satisfier for hungry babies. Brazilians in the United States and other countries have increased their use of pizza and fast-food places such as McDonalds and Dunkin' Donuts. The food is fast, liked by all in the family, and easy to put on the table by working dads and moms. For the many single Brazilians, it surpasses going home, cooking, and the like, although traditional *cuscus*, which is easily prepared and is culturally satisfying, still graces the Brazilian table at home.

10.8.3 Dietary Practices for Health Promotion

Brazilians have become vitamin and health food conscious. However, this luxury is often not available to those who have emigrated to other countries for fast money. Legal residents generally become health food consumers. The preference, especially among young Brazilian women, is to rely on vitamins instead of a heavy diet to help them remain thin.

10.8.4 Nutritional Deficiencies and Food Limitations

Individuals in lower socioeconomic groups frequently experience nutritional deficiencies. Undocumented Brazilians who immigrate to earn fast money may experience malnutrition. Many native fruits are expensive, as are other special foods that are common to the Brazilian diet. Food limitations are imposed by expense and lack of availability of Brazilian mainstay foods. However, many Brazilian communities in the United States and other countries have ethnic markets and restaurants. Large chain supermarkets often carry a section of ethnic foods, some of which are reasonably priced.

10.9 Pregnancy and Childbearing Practices

10.9.1 Fertility Practices and Views toward Pregnancy

Although Brazil is predominantly a Catholic country, birth control is taught and used. Women are encouraged by their physicians or clinic personnel to have tubal ligations to prevent unwanted pregnancies. Herbal teas are used for bringing on late menstrual periods and for stimulating natural abortions. Brazil is a fatalistic country, so unwanted pregnancies and abortions are, in the end, left in God's hands. Fatalism, however, is mixed with a strong sense of realism. Therefore, immigrants to other countries generally practice birth control so pregnancy will not interfere with their reason for emigrating.

At times, single women try to become pregnant to facilitate their chance of remaining permanently in the United States because the baby is a U.S. citizen by having been born in that country. This opportunity may be somewhat enhanced if the child is born here and has been able to attend school. Thus, fertility practices among immigrant Brazilians are a matter of convenience with a traditional fatalistic overtone.

Brazilians are aware of the overpopulation problem, and modern middle socioeconomic Brazilians like to have a *casal*, a family of one boy and one girl. Pregnancies are generally accepted fatalistically (God's will). Frequent topics of conversation among northeastern Brazilian women in the lower socioeconomic groups are pregnancy, abortion, stillbirths, and child mortality. Pregnancies among immigrants are treated according to the mother's beliefs. Stories tell of pregnant women returning home to their families to receive care and to have their babies in Brazil and of mothers who have expectations that their North American–born children will have dual citizenship.

10.9.2 Prescriptive, Restrictive, and Taboo Practices in the Childbearing Family

Many restrictions are related to pregnancy. Women are encouraged not to do heavy work and not to swim. Taboos also warn against having sexual relations during pregnancy. Some foods are to be avoided, and specific foods are recommended during pregnancy. Taboos generally vary according to geographic region, socioeconomic class, and ethnic background. Thus, a list of taboo foods cannot be listed and the health-care provider needs to specifically ask about taboo foods.

Whenever possible, women in Brazil go to a *maternidade,* a hospital specializing in obstetrics, for prenatal care and to have their children. These maternity hospitals vary in quality and quantity of services. Generally, hospitals for the lower socioeconomic citizens and university hospitals provide excellent, modern services, such as regular prenatal visits with physicians. There are, however, private hospitals with unsanitary conditions and flies due to lack of screening. C-sections are very common. All licensed nurses practice midwifery. Use of sedation is common and preferred by many over natural childbirth. Lay midwives, known as ***parteiras,*** deliver babies at home and often are the provider of choice among women without access to transportation, health insurance, and financial resources.

Since the 1950s, scientific evidence has demonstrated that artificially fed infants have much higher rates of morbidity and mortality than those who are breastfed. Breast milk contains immunoglobulins, phagocytes, T-lymphocytes, enzymes such as lysozymes, and many other factors, including cells, antibodies, hormones, and other important constituents not present in infant formula, that help protect the infant against infection. Yet, many Brazilian mothers prefer to give their babies powdered formula instead of breast milk. Middle socioeconomic women wish to regain their figures as soon as possible. Lower socioeconomic women often feel that their milk is *fraca,* or weak. Even though powdered milk formula exposes babies to risks including contaminated water, over dilution, and contaminated utensils, many working mothers prefer its convenience. New customs continue to evolve as bottle-feeding replaces breastfeeding. Breastfeeding is still linked to a social stigma; a mother who breastfeeds may often be thought of as abandoned or sexually unattractive.

A postpartum woman eats chicken soup to help her body return to normal. She is also advised not to eat spicy foods or *rapadura,* a molasses candy, and not to drink *garapa,* sugar water, or *caldo de cana,* sugarcane juice if she breastfeeds her infant.

10.10 Death Rituals

10.10.1 Death Rituals and Expectations

Death rituals in Brazil frequently follow religious prescriptions. However, in the more interior of the country, it is rare to see a hearse or a funeral home. In those areas, the deceased, especially in the lower socioeconomic groups, are kept at home until the body is buried. A photographer may be called to take a picture of the body in the coffin, which, after some touching up to make it look natural, may be used as a photo to adorn the living room wall, along with photos of other deceased relatives. The deceased frequently appear in visions and dreams to inform intimate survivors of their needs.

Bones of a loved one are sometimes buried in the same plot as other family members to keep the family together. A great fear is to have a body destroyed or mutilated so that all the parts are not together. Those in power, such as police or other oppressors, sometimes take advantage of this belief to subdue believers who are generally in the lower socioeconomic groups.

If possible, the family carries the coffin to the cemetery, which is usually on the outskirts of a community and separated by a solid cement wall. Cemeteries consist of specially purchased family lots containing vaults in which the dead are placed. In addition, unmarked 2-foot graves are provided for the unclaimed and poor. Many Brazilians prefer to be placed in a coffin rather than risk being buried alive in a vault. Everyone's desire is to be buried in his or her own coffin, regardless of whether it is lined with silk or cardboard. Coffins are frequently not nailed shut to facilitate escape.

Coffins may be pink, blue, or white, with specially designed coffins for babies and children. Babies and children are buried with their eyes open so that they may see God and His angels. Frequently, children are buried holding candles to light their way. The death of a baby or an infant historically has been, and continues to be, treated joyfully and without much sadness, for the child died pure and is regarded as an angel. Children are dressed in white with their hair curled, and ribbons or garlands are interwoven. The mouth is fixed into a smile, and the hands are folded. Flowers fill the coffin, and notes to the Virgin Mary or a saint may be tucked into the hands. A festive celebration, the Wake of the Angels, is a mixture of joy and sadness. One may still see children in their best clothes carrying the coffin to the cemetery, representing a procession of the angels. The custom of wrapping the dead body in its personal hammock for burial is still practiced among lower socioeconomic citizens in the interior of Brazil.

If financially possible, families of Brazilians who die in another country personally accompany the body to Brazil for burial in the family vault. If family members cannot come, relatives meet at the airport upon the body's arrival in Brazil.

10.10.2 Responses to Death and Grief

Responses to death and grief depend on the family. To a poor family, a continuously suffering person is rescued. The fatalistic expression, "It was God's will," helps grieving among the rich and the poor. Older people wear black for various lengths of time depending on the relationship of the family member. Frequently, the final portrait is hung in the family chapel or near the family altar, and prayers are recited. An eternal light burns. Relatives are honored on the anniversaries of their death, both at home and at masses. Often, the family places an obituary of remembrance with or without a picture of the deceased in the local newspaper on the anniversary of the death for several years. ***Anojamento*** is the Brazilian term for deep mourning or grief.

10.11 Spirituality

10.11.1 Dominant Religion and Use of Prayer

Seventy-four percent of Brazilians are nominally Roman Catholic, 15% Protestant, 1.3% Spiritualist, 0.3% Bantu/voodoo, 1.8% other, and 7.4% unspecified (CIA World Factbook 2019). Jewish temples and synagogues and structures of various Eastern religions are also present in Brazil. Spiritualism often occurs in the form of Afro-Brazilian sects and the Universal Church of the Reign of God. Spirits and souls are called to intervene for various problems of health, life, and death. Although most traditional religions are represented in Brazil, prayer is an individual matter. The family altar is a common site of prayer. Frequently, saints and "Our Lady" are asked for help.

10.11.2 Meaning of Life and Individual Sources of Strength

The meaning of life for most Brazilians is found in religion, economy, fatalism, and reality. For some, life is *uma luta,* literally, a battle. For others, life is an almost-hedonistic attitude. Women and children, and often men, dance the native dances the minute familiar music is played, often by an impromptu band of three or four playing Brazilian instruments.

Brazilians in general are hard workers during the week while waiting for weekend activities. Social gatherings are the most common way to socialize, meeting in public places such as beaches, shopping malls, parks, and bars/restaurants. This can be seen in other countries where many Brazilians have immigrated. The greatest source of strength for Brazilians is their immediate and extended families. Tradition and folk religion are other sources of strength.

10.11.3 Spiritual Beliefs and Health-Care Practices

Curandeiros (folk healers), or similar special healers, exorcise and pray for the wellness of their patients. Saints are asked for help, and some people wear medals or little pouches of special powders around their necks to ward off bad spirits. ***Rezadeiras***, or spiritual leaders, also have a strong influence on health practices among populations of small towns, especially in the north eastern region of Brazil.

10.12 Health-Care Practices

10.12.1 Health-Seeking Beliefs and Behaviors

Health care in Brazil is provided by both private and government institutions. The Ministry for Health and Ageing administers national health policy. Free health care at the point of entry into the system is provided by the public health system known as *Sistema Unificado da Saúde*. Health-seeking behaviors among Brazilians living in the United States, and most likely other countries as well, are increasing. Information about safe sex is frequently sought to prevent sexually transmitted infections. A paradigm shift from acute care to preventive care is evident among Brazilians.

10.12.2 Responsibility for Health Care

The family is the nucleus of responsibility for health care. Brazilians in Brazil and in the United States are joining the Western approach for taking responsibility for their own health promotion and wellness. Lower socioeconomic citizens seem to value prevention but frequently lack the resources for accessing these services.

Brazilians are familiar with private and public insurance options. In Brazil, national health insurance is mandatory for each salaried person and her or his family. Middle and upper socioeconomic Brazilians frequently select private plans. Society still borders on feudalism in the north and northeast where the *patrão,* employer, assumes responsibility for meeting the person's medical needs. This responsibility frequently extends to the employee's family, wherever they reside.

10.12.3 Blood Transfusions and Organ Donation

Similar to the United States and other parts of the world, acceptance of blood transfusions, organ donation, and organ transplantation depends on religious credence and individual preference. The same is true for blood transfusions.

10.12.4 Self-medication Practices

Because Brazilians tend to self-medicate, the procurement of health care is often avoided or delayed. Consulting with someone who has the same condition or with friends who know someone who has a similar condition may be the first step. A trip to the local pharmacist may be the second. Middle and upper socioeconomic Brazilians frequently select private plans. Society still borders on feudalism in the north and northeast, where the *patrão* (employer) assumes responsibility for meeting the person's medical needs. This responsibility frequently extends to the employee's family, wherever they reside.

Antibiotic, neuroleptic, antiemetic, and most other prescription drugs are easily obtained over the counter in Brazilian pharmacies. Many prefer and use homeopathic medicines and herbs. Once in the United States, it becomes difficult to obtain the many drugs readily available in Brazil so they may family members send them drugs from home. Consequently, incoming Brazilians often bring medicines requested by their friends and, thus, maintain the circulation of medications not available to them living in other countries.

10.12.5 Pain/Sick Role

Brazilians generally do not like to talk about pain. However, once the emotional barrier is removed, they feel relieved to be able to discuss their discomfort. Many pain-relieving medicines are available in Brazil without a prescription. Frequently, a person living in the United States requiring these on a regular basis can request that friends or friends of friends bring a supply from Brazil.

Most Brazilians do not work if they are seriously ill. Sickness is a neutral role and is considered socially exempt—free of guilt, blame, and responsibility. Illness is looked at from a fatalistic point of view. ***Nervos***, an ever-present folk diagnosis, identifies weakness, craziness, and anger associated principally with hunger. Among lower socioeconomic citizens, the term ***doença dos nervos*** refers to an all-encompassing illness. This diagnosis reveals, and simultaneously conceals, the truth of the existence of a still-struggling people.

10.12.6 Mental Health and Disabilities

In Brazil, people with physical and mental handicaps are usually cared for and kept at home. However, people with physical handicaps can be seen begging on street corners. Both physical and emotional rehabilitation facilities are available, but access is difficult. Although the literature contains little data regarding how Brazilians view mental illness in general, mental health care and services are available in the private and public sectors in Brazil. Thus, one might expect at least a minimal acceptance of mental illness among Brazilian immigrants residing elsewhere.

Following the trend of many European and North American health systems, substandard public mental hospitals have been closed or modernized, and the responsibility for treating mental illness has fallen into the realm of the community health system. Drug treatment centers exist for those who are habituated. Some university and private inpatient or day-treatment facilities offer modern psychiatric treatment. A slow trend toward family-based psychiatric services is apparent.

10.12.7 Barriers to Health Care

At times, support services for legal and undocumented Brazilians hard to find for those who do not have language skills or the self-esteem to become assimilated into a new culture of their newly found environments. In fact, language is one of the major problems for these immigrants. They neglect to learn the language of the host country and get by in their enclave community, which may be detrimental to accessing health-care facilities. Those with a good command of the host language can more readily incorporate new technical terms into their vocabulary. Much health-care information is translated into Portuguese, although much more material exists for Hispanics.

Another barrier to health care for immigrant Brazilians is cost. This, combined with lack of knowledge about the health-care system and facilities impedes both legal and undocumented residents. Most Brazilians do not talk about their illnesses unless these are very serious. Generally, illness is discussed only within the family. Many Brazilians feel that talking about an illness, such as cancer, negatively influences their condition. This may be denial or an actual culturally self-imposed restriction, perhaps linked with some fear of prejudice. For example, patients with gastric cancer may insist they are in good health.

Because many Brazilians tend to shun hospitals, the family remains with the patient when hospitalization is a necessity. The patient is often brought food from home. Brazilian families are eager to participate in patient care and, thus, can be taught various procedures and care activities.

10.13 Health-Care Providers

10.13.1 Folk and Traditional Practices

The Brazilian culture is rich in folk practices that vary with geographic region, ethnic background, socioeconomic factors, and generation. Traditional and homeopathic pharmacies are supplemented by ***remedios populares*** (folk medicines), and ***remedios caseiros*** (home remedies). In Brazil, open-air markets have stands that specialize in herbs and home medicines. Traditional schools of pharmacy grow, sell, and teach courses on folk remedies. Home remedies such as herbal teas, mixtures, and syrup with lemon and honey are used frequently to decrease illness symptoms. Prolonged symptoms or more-serious indications of disease are common precedents in the search for medical attention. Folk remedies and traditional health-care practices become intermeshed, such as when a serious illness may be treated best by traditional caretakers. Some patients are prescribed homeopathic *bolinhas* (literally, little balls), tailored for specific ailments. **Curandeiros** (folk healers) often are sought after by the poor, who may have less access to—and less patience with—free public health services, which often require long waiting times. Homemade remedies (e.g., herbs, roots, leaf teas, and ointments) are widely employed for a variety of ailments.

10.13.2 Traditional Healers

There are a wide variety of Brazilian folk healers who exorcise physical, emotional, or spiritual illness. These include spiritually gifted healers known as curandeiros and rezadeiras (literally, praying women) elicited to exorcise illness. Some Brazilians also seek the services of other traditional healers, which include card readers who predict fortunes, **espiritualistas** who summon the assistance of departed souls and spirits, **conselheiros** who provide spiritual counsel and advice, and **catimbozeiros** who practice sorcery. Within the Afro-Brazilian religions of Umbanda and Xango, the head priestess *(mãe de santo)* and priest *(pai de santo)* have spiritual healing powers.

10.13.3 Professional Health-Care Providers

In Brazil, physicians are generally viewed as the primary source of medical care, although the roles of nurses, social workers, physiotherapists, and nutritionists are evolving to include independent professional practice. Education for the health care professions—including medicine and nursing—is primarily provided at the undergraduate level; however, over the last decade, a number of PhD and DNP programs have been initiated. Admission is based on performance on competitive entrance examinations. Given the higher social prestige of the medical profession, demand for medical education is higher than nursing and the medical workforce is much larger than the professional nursing workforce. Despite the elevated cost, the majority of medical and nursing students are enrolled in private institutions of higher education.

In many healthcare institutions and settings, minimally trained nursing assistants *(atendentes de enfermagem)* and practical nurses *(auxiliares de enfermagem)* practice as independent providers without the supervision of a licensed nurse. Nursing personnel employed in private practice or in private physician-owned clinics often are not well-regarded by either employers or patients. The systematic implementation of nursing diagnoses (Oliveira Salgado and Machado Chianca, 2011) has contributed to strengthening professional nursing practice in Brazil.

References

Brasil Ministério da Saúde. [Brazil Ministry of Health] (2011). http://portal.saude.gov.br/portal/saude/profissional/area.cfm?id_area=1483

Canelas C, Gisselquist RM (2018) Horizontal inequality as an outcome. Oxf Dev Stud 46(3):305–324

Centers for Disease Control and Prevention (2010) Emerging infectious diseases. http://www.cdc.gov/ncidod/eid/vol4no1/momen.htm
CIA World Factbook: Brazil (2019). https://www.cia.gov/library/publications/the-world-factbook/geos/br.html
Emigration from Brazil (2019). https://thebrazilbusiness.com/article/emigration-from-brazil
Instituto Brasileiro de Geografia e Estatística [Brazilian Institute of Geography and Statistics [IBGE] (2019) Atlas do censo demográfi co 2019/IBGE. Rio de Janeiro: IBGE. https://www.cia.gov/library/publications/the-world-factbook/geos/br.html,
Levy MSF (1974) O papel da migração internacional na evolução da população Brasileira 1872 a 1972 [the role of international migration in the evolution of the Brazilian population from 1872 to 1972]. Rev Saude Publica 8:49–90
Levy R (1993) Ethnic and racial differences in response to medicines: preserving individualized therapy in managed pharmaceutical programmes. Pharm Med 7:139–165
Margolis M (2004) Brazilians in the United States, Canada, Europe, Japan and Paraguay. In: Ember M, Ember CR, Skoggard I (eds) Encyclopedia of diasporas. Kluwer Academic/Plenum, New York
Margolis M (2010) Brazilians. Encyclopedia of immigration, vol 3. Elliot Barkan (Santa Barbara): ABC-CLIO
Oliveira Salgado P, Couto Machado Chianca T (2011) Identification and mapping of the nursing diagnoses and actions in an intensive care unit. Rev Lat Am Engermagem 19(4):928–935. https://doi.org/10.1590/S0104-11692011000400011
Population of Cities in Brazil 2019 World population review. http://worldpopulationreview.com/countries/brazil-population/cities/
Telles EE (2004) Race in another America: the significance of skin color in Brazil. Princeton University Press, Princeton
UNESCO (1999) The African slave trade from 15th to the 19th centuries. UNESCO reports and papers. https://unesdoc.unesco.org/ark:/48223/pf0000038840
World Bank. Brazil (2017 Urban population. https://data.worldbank.org/indicator/sp.urb.totl.in.zs
World Health Organization (2011) Pan American Health Organization. http://new.paho.org/hq/index.php?option=com_content&task=view&
Zong J, Batalova J (2016) Brazilian immigrants in the United States. Migration policy institute. https://www.migrationpolicy.org/article/brazilian-immigrants-united-states

11 People of Chinese Heritage

Hsiu-Min Tsai, Tsung-Lane Chu, and Wen-Pin Yu

11.1 Introduction

Although some Western health-care providers, including some researchers, categorize all Asians into aggregate data as if they were one group, each nationality is very different. Cultural values differ even among Chinese people according to their geographic location within Taiwan, Hong Kong, China:—north, south, east, west; rural versus urban; interior versus port city—as well as other variant cultural characteristics (see Chap. 2). Chinese immigrants to Western countries are even more diverse, with a mixture of traditional and Western values and beliefs. These differences must be acknowledged and appreciated.

This chapter is an update previously written by Hsiu-Min Tsai.

H.-M. Tsai (✉)
Chang Gung University of Science and Technology, Tao-Yuan Taiwan Chang Gung University, Tao-Yuan, Taiwan

Linkou Chang Gung Memorial Hospital, Tao-Yuan, Taiwan
e-mail: hmtsai@gw.cgust.edu.tw

T.-L. Chu
Linkou Chang Gung Memorial Hospital, Tao-Yuan, Taiwan

W.-P. Yu
Keelung Chang Gung Memorial Hospital, Kee-lung, Taiwan

11.2 Overview, Inhabited Localities, and Topography

11.2.1 Overview

Han Chinese are the principal ethnic group of China, constituting about 91.6 percent of the population of mainland China, especially as distinguished from Manchus, Mongols, Huis, and other minority nationalities. The remaining 8.4 percent are other (includes Zhuang, Hui, Manchu, Uighur, Miao, Yi, Tujia, Tibetan, Mongol, Dong, Buyei, Yao, Bai, Korean, Hani, Li, Kazakh, Dai, and other nationalities) 7.1% (CIA World Factbook 2019a, b). Substantial genetic, linguistic, cultural, and social differences exist among these subgroups. Because of the complexity of their values, it is critical to consider social and cultural contexts to develop appropriate interventions and provide culturally competent care for multi-ethnic Chinese patients. The information included in this chapter is only a beginning point for understanding the Chinese people; it is not meant to be a definitive profile.

Children born to Chinese parents in Western countries tend to adopt Western culture easily, whereas their parents and grandparents tend to maintain their traditional Chinese culture in varying degrees. Chinese who live in the "Chinatowns" of North America and other places outside of China maintain many of their cultural and social beliefs and values and insist that health-care providers respect these values and beliefs with their prescribed interventions.

© Springer Nature Switzerland AG 2021
L. D. Purnell, E. A. Fenkl (eds.), *Textbook for Transcultural Health Care: A Population Approach*, https://doi.org/10.1007/978-3-030-51399-3_11

11.2.2 Heritage and Residence

The Chinese culture is one of the oldest in recorded human history, beginning with the Xia dynasty, dating from 2200 B.C. to the present-day People's Republic of China (PRC). The Chinese name for their country is ***Zhong guo***, which means "middle kingdom" or "center of the earth." Many of the current values and beliefs of the Chinese remain grounded in their history; many believe that the Chinese culture is superior to other Asian cultures. Ideals based on the teachings of Confucius (551–479 B.C.) continue to play an important part in the values and beliefs of the Chinese. These ideals emphasize the importance of accountability to family and neighbors and reinforce the idea that all relationships embody power and rule. Although industrialization, urbanization, and interaction with Western society have affected some Chinese, the ideas and behavioral patterns related to Confucianism are still deep-seated.

During early Communist rule, an attempt was made to break down the values grounded in Confucianism and substitute values consistent with equal social responsibility. This was initially achieved and rank in society was no longer seen as important. During the People's Revolution, feudal rank frequently meant loss of social importance, physical punishment, imprisonment, and even death. Later, during the Cultural Revolution, the young were held responsible for the deaths of many previously esteemed older adults and educated Chinese. Today, many of the Confucian values have reasserted themselves. Families, older adults, and highly educated individuals are again considered important. Research completed by the Chinese Culture Connection, a group of Chinese sociologists, lists 40 important values in modern China, including filial piety, industry, patriotism, paying deference to those in hierarchical status positions, tolerance of others, loyalty to superiors, respect for rites and social rituals, knowledge, benevolent authority, thrift, patience, courtesy, and respect for tradition (Hu and Grove 1991). Since China's economic reforms social development in the late 1990s, the Chinese society has been on the stage of full transformation. According to Chinese sociologists, the main social values consist of richness, democracy, harmony, and innovation (Chen 2009).

The population of China is over 1.38 billion people, with 7.2 million in Hong Kong, over 606,000 in Macau, and over 23.5 million in Taiwan (CIA World Factbook 2019a, b). According to 2010 China National statistics, 50.32 percent live in rural communities, a decrease of approximate 133.2 million persons since the 2000 census. In other words, urban residents increased by 13.46 percentage points compared with the 2000 population census (Wang 2011). The higher level of urbanization is a result of economic and social development.

China is over 9.6 million square kilometers (3.7 million square miles), with 23 provinces; 5 regions, including Tibet, Hong Kong, and Taiwan; and 4 municipalities. Each province, region, and municipality functions independently and in many different ways. The Chinese consider each region as part of greater China and predict that the day will come when all of China is reunited. Tibet has already been re-assimilated, Hong Kong returned to Chinese control in 1997, and Macau in 1999 (CIA World Factbook 2019a, b).

The largest communities of Chinese Americans are in California, New York, Florida, and Texas. Chinese Americans compose the largest subgroup among Asians/Pacific Islanders (APIs), exceeding 4.1 million people (U.S. Census Bureau 2017).

11.2.3 Reasons for Migration and Associated Economic Factors

The Chinese have immigrated to the United States (US) since the 1800s. Chinese immigration was initially fueled by economic needs. Over 100,000 male peasants from Guangdong and Fujian provinces came to the US without their families in the early 1830s to make their fortune on the transcontinental railroad. This immigration continued through the Gold Rush of 1849. Many believed that they could make money in the US to help their families and later return to China. Unfortunately, most found that opportunities were limited to hard labor and other vocations not desired by European Americans. Their culture and physical features made them readily identifiable in the predominantly White American society. They could not simply change their names and blend in with other, primarily European immigrant populations. The Chinese had few rights and were barred from becoming U.S. citizens. Racial violence and prejudice against them were common, and the courts did not punish the violators. Compared with other ethnic groups, their immigration numbers were small until 1952, when the McCarran-Walters Bill relaxed immigration laws and permitted more Chinese to enter the country (U.S. Citizenship and Immigration Services 2011).

The most recent immigrants from Taiwan, Hong Kong, and mainland China are strikingly different from earlier Chinese immigrants in that they are more diverse. In addition, whereas many emigrated to reunite with their families, students, scholars, and professionals flocked to the US to pursue higher education or research. For their safety and the maintenance of their cultural values, most Chinese and Taiwanese settled in closed communities.

Other countries that have significant Chinese immigration are Great Britain, Spain, Bangladesh, Italy, Singapore, Australia, Canada, Japan, and Korea (China Daily: Top 10 immigration destinations 2019). Regardless where Chinese migrate, they bring their culture with them.

11.2.4 Educational Status and Occupations

Influenced by the Confucian principles, the Chinese believe that "to be a scholar is to be at the top of society" (Ho and Lee 2010). In Chinese society, academic achievement is highly valued for increasing one's own career benefits and enhancing a family's reputation and position. Chinese and Taiwanese parents are much more willing to provide their children with the best possible education and invest huge sums of money in supplementary education.

Education is compulsory in China, Hong Kong, and Taiwan. Most children receive the equivalent of a twelfth-grade education, middle school students still have to complete a state examination to determine their eligibility to enter a general high school, vocational high school or college, or to begin their lives as employees. Those who complete either the general or the preparatory high school experience compete academically to continue their education at college and university levels. The Chinese educational system is complex and is not presented here in its entirety; further study is encouraged.

A university education is highly valued; however, few have the opportunity to achieve this life goal because enrollment in better educational institutions are limited. Because competition for top universities is keen, many families select less valued universities to ensure that their child is accepted into a university rather than slated for a technical school education. After their undergraduate or graduate programs, many young adults come to Western countries to attend universities to seek more advanced education or research. A foreign education is considered prestigious in Chinese and Taiwanese society.

Initially, in the West, the Chinese tended to be either highly or poorly educated. This dichotomy may have resulted in health-care providers categorizing patients in a similar manner. Many people believed that Chinese occupations were limited to restaurant work, service employment, and the garment industry. However, this phenomenon has changed since the 1980s. A significant number of Chinese students and scholars from the PRC and Taiwan come to the US to study every year. Because of the competitive educational system in mainland China and Taiwan where only the brightest students go to a university, Chinese immigrants with a college education are often very well educated. Student immigrants are expected to return to China or Taiwan when their education and research are completed. However, many do not return but elect to remain in Western countries, having obtained graduate degrees in the US where many find employment in high-technology companies or educational and research institutes.

Another group of Chinese immigrants are professionals from Hong Kong who moved to North America and other Western countries to avoid repatriation in 1997. These immigrants usually have family connections or close friends who are highly educated and skilled in Western countries. A third group of immigrants consists of uneducated individuals with diverse manual labor skills. Finding employment opportunities for this group may be more difficult. They often settle with family members who are not skilled or highly educated. This arrangement drains family resources for many years until they obtain financial security, learn the language, and become acculturated in other ways.

11.3 Communication

The Chinese speak a variety of different languages and dialects. The official language of China is *Mandarin* (**pu tong hua**), which means "common speech" (Cheung et al. 2005) and is spoken by approximately 70 percent of the population, primarily in northern China, but there are 10 major, distinct dialects, including Cantonese, Fujianese, Shanghainese, Toishanese, and Hunanese. For example, *pu tong hua* is spoken in Beijing, the capital of China in the north, and Shanghainese is spoken in Shanghai. The two cities are only 1462 km (about 665 miles) apart, but because the dialects are so different, the two groups cannot understand one another verbally. Even though people from one part of China cannot understand those from other regions, the written language is the same throughout the country and consists of over 50,000 characters (about 5000 common ones); thus, most children are at least 10–12 years old before they can read the newspaper.

Most Chinese people tend to be more passive and less sharing when explaining or discussing something, whereas European Americans appreciate more direct and clear explanations. Many European Americans might not understand these communication practices and become upset because they consider Chinese communication to be indirect and offensive. In addition, the Chinese are less likely to tell the other party things that may upset him or her (O'Keefe and O'Keefe 1997). Because of this communication style, the Chinese might avoid sharing a health concern such as a mental illness, a chronic disease, or cancer when consulting with unfamiliar health-care providers about a specific health issue. To prevent misunderstanding, American health-care providers have to be aware of these differences.

Although many times the Chinese sound loud when talking with other Chinese, they generally speak in a moderate to low voice volume. Americans are considered loud to most Chinese, and health-care providers must be cautious about their voice volume when interacting with Chinese patients in English so intentions are not misinterpreted.

When possible, health-care providers should use the Chinese language to communicate. Table 11.1 lists some common phrases; be care-

Table 11.1 Frequently used words and phrases

English word or phrase	Chinese pinyin	Phonetic pronunciation
Hello	*Ňi hăo*	Nee how (note tones to be used[a])
Good-bye	*Zài jiàn*	Dzai jee en
How are you?	*Ňi hăo mā*	Nee how mah
Please	*Qing*	Ching
Thank you	*Xīe xie*	Shee eh shee eh
I don't understand	*Ŵo bù dŏng*	Wah boo doong
Yes	*Sh`i `de* or *`dui*	Shur da or doee (no real yes or no comparable saying—this means I agree or okay)
No	*Bú s h`i `de* or *b`u hăo*	Boo shur or boo how
My name is	*Ŵo jiào*	Wah djeeow
Very good	*Đen hăo*	Hun hao
Hurt	*Téng*	Tung
I, you, he/she/it	*Wo, ňi, tā*	Wah, nee, tah
Hot	*Rè*	Ruh
Cold	*Leňg*	Lung
Happy	*Gāo xi Đgu*	Gow shing
Where	*Ňa li*	Na lee
Not have	*ĐeiŶou*	May yo
Doctor	*Đi shēng*	Yee shung
Nurse	*Hù ś hi*	Who shur

[a]Each *pu tong hua* Chinese word is pronounced with five different tones: First tone is high and even across the word (–); Second tone starts low and goes high (–`); Third tone starts neutral, goes low, and then goes high (ˇ); Fourth tone is curt and goes low (.`); Fifth tone is neutral and pronounced very slowly

ful to avoid jargon and to use the simplest terms. Many times, verbs can be omitted because the Chinese language has only a limited number of verbs. The Chinese appreciate any attempt to use their language. They do not mind mistakes and will correct speakers when they believe it will not cause embarrassment. When asked whether they understand what was just said, the Chinese invariably answer yes, even when they do not understand because admitting it is embarrassing to them. Thus, it is better to have Chinese patients or family members repeat the instructions they have been given or ask them to give a demonstration.

Negative queries are difficult for the Chinese to understand. For example, do not say, "You know how to do that, don't you?" Instead say, "Do you know how to do that?" Also, it is easier for them to understand instructions placed in a specific order, such as the following:

1. At 9 o'clock every morning, get the medicine bottle.
2. Take two tablets out of the bottle.
3. Get your water.
4. Swallow the pills with the water.

Do not use complex sentences with *ands* and *buts*. The Chinese have difficulty deciding what to respond to first when the speaker uses compound or complex sentences.

11.3.1 Cultural Communication Patterns

There is a very obvious difference between traditional Chinese communication patterns and American communication patterns. The Chinese have a reputation for not openly displaying emotion. They tend not to discuss their concerns with health-care providers, while most Americans are willing to openly give and accept comments with others. The Chinese often consider their concerns to be personal and to be shared only with the family, friends, and relatives. They might share information freely with health-care providers once a trusting relationship has developed. This is not always easy, because Western health-care providers may not have the patience or time to develop such relationships. In situations in which Chinese people perceive that health-care providers or other people of authority may lose face or be embarrassed, they may choose not to be totally truthful. As a result, they may always give a "no" response when unfamiliar health-care providers ask them questions.

Touching between health-care providers and Chinese patients should be kept to a minimum. Most Chinese maintain a formal distance with one another, which is a form of respect. Some are uncomfortable with face-to-face communications, especially when there is direct eye contact. Because they prefer to sit next to others, the health-care providers may need to rearrange seating to promote positive communication. When touching is necessary, the health-care provider should provide explanations to Chinese patients.

Facial expressions are used extensively among family and friends. The Chinese love to joke and laugh. They use and appreciate smiles when talking with others. However, if the situation is formal, smiles may be limited. In most greeting and communication situations, shaking hands is common; hugs are limited. The health-care provider should watch for cues from their Chinese patients.

11.3.2 Format for Names

Among the Chinese, introductions, either by name card or verbally, are different from those in Western countries. For example, the family name is stated first and then the given name. Calling individuals by any name except their family name is impolite unless they are close friends or relatives. If a person's family name is Li and the given name is Ruiming, then the proper form of address is Li Ruiming. Men are addressed by their family name, such as *Ma*, and a title such as *Ma xian sheng* ("Mister Ma"), *lao Ma* ("respected older Ma"), or *xiao Ma* ("young Ma"). Titles are important and meaningful to the Chinese people, so, when possible, identify the person's title and use it.

Women in China and Taiwan rarely use their husband's last name after they get married and retain their own family last name. Therefore, unless the woman is from Hong Kong, or has lived in a Western country for a long time, do not assume that her last name is the same as her husband's. Her family name comes first, followed by her given names, and finally by her title. Many Chinese living in Western countries take an English name as an additional given name because their name is difficult for Westerners to pronounce. Their English name can be used in many settings. Addressing them as "Miss Millie" or "Mr. Jonathan" rather than simply by their English name is better. Even though they have adopted an English name, some Chinese may give permission to use only the English name. In addition, some Chinese switch the order of their names to be the same as Westerners, with their family name last. This practice can be confusing; therefore, health-care providers should address Chinese patients by their whole name or by their family name and title, and then ask them how they wish to be addressed. For medical record keeping, what is the legal name as well as what the person wishes to be called.

Reflective Exercise 11.1

Mr. Wang is a 75-year-old first-generation Chinese American and lives with his 70-year-old Chinese wife. Mr. Wang speaks very limited English, while his wife speaks only Chinese. Mr. Wang visited a clinic where he was told by his physician that he had lung cancer. The physician wanted Mr. Wang to be hospitalized for further treatment and chemotherapy. Mr. Wang gave no response.

1. What barriers might exist for Mr. Wang in deciding to accept hospitalization and seek treatment?
2. What concerns might the nurse have regarding a support system for Mr. Wang if he decides to have his lung cancer treated?
3. With the limited English-language ability of Mr. Wang and his wife, how can a health-care professional ensure effective communication?
4. If Mr. Wang prefers traditional Chinese medical treatment, how might the nurse respond?

11.4 Family Roles and Organization

11.4.1 Head of Household and Gender Roles

Kinship traditionally has been organized around the male lineage. Fathers, sons, and uncles are the important, recognized relationships between and among families in politics and in business. Each family maintains a recognized head who has great authority and assumes all major responsibilities for the family. A common and desirable domestic traditional structure is to have four generations under one roof. However, with the improvement in the standards of living and other changes, families have gradually gotten smaller. The first generation of one-child families appeared in the 1970s when China introduced a policy of family planning. This phenomenon led to the nuclear family of three (parents and one child) gradually becoming the mainstream family structure in cities. In 2010, of 401.52 million family households in China, the average number of people in each household was 3.10, or 0.34 persons fewer as compared with the 3.44 persons in the 2000 population census (Wang 2011). The shrinking household size might be caused by the decline of fertility, the increase of migration, and the independent living arrangement of young couples after marriage (Wang 2011).

Family life takes on various faces and follows new trends. Because young people face greater and greater pressure from work and want a higher standard of living and spiritual life, the traditional concept of raising children has faded, and more couples are choosing the "double income, no kids" (DINK) way of life.

Because increasing numbers of young people leave home to work in other parts of the country or to study abroad, the number of households consisting of older couples is also rising. Improvements in housing conditions make it possible for the younger generations to move out of the house and live apart from senior members of the family. Longer life spans have resulted in more seniors living alone and those who have lost their spouse living by themselves in one-person households. "Empty nests" will become the norm for seniors as parents of the first generation of single-child families get older. That, in turn, means a switch from an old system in which children looked after their parents to one in which seniors are cared for by society in general through benefits.

Another traditional practice in many rural Chinese families is the submissive role of the daughter-in-law to the mother-in-law. Often the mother-in-law is demanding and hostile to the daughter-in-law and may treat her worse than the servants. This relationship has changed significantly since modern culture was introduced to Chinese society. However, such relationships may continue to influence some Chinese families today to some extent, or mothers-in-law and daughters-in-law may simply not get along with each other. Overseas, Chinese are quite different. The involvement of parents, especially the husband's parents, in the new family's life may have a great impact on families.

A Confucian cosmology of gender roles permeates Chinese society. A woman is characterized as a ***yin*** union, while a man is a ***yang*** union (Shim 2001). As *yang*, man is superior, and as *yin*, woman is inferior (Li 2000). Thus, a woman is expected to be obedient and dependent on a man (Shrestha and Weber 1994). Because of stereotypical roles of men and women, men are largely in control of the country. However, since the founding of the PRC, this has been changing somewhat. In 1949, the Communist Party stated that "women hold up half the sky" and are legally equal to men.

Almost half of the workforce in China are women. Favored professions are education, culture and arts, broadcasting, television and film, finance and insurance, public health, welfare, sports, and social services. In trades requiring higher technical skills and knowledge such as computer science, telecommunications, environmental protection, aviation, engineering design, real estate development, finance and insurance, and the law are preferred. To promote Chinese women's employment, the Chinese government declared "The Program for the Development of Chinese Women" in 2001 (Huangjuan 2009). Since then, the number of women employed increased from 291 million in 1990 to 337.7 million in 2017. In 2017, there were 65.45 million employed women in China urban areas, accounting for 37.1 percent of the total employees in urban sites (China National Bureau of Statistics 2018).

In patriarchal Confucianism, the roles of Chinese women are referred to as *nei* (the internal), while the roles of Chinese men are referred to as *wai* (the external) (Li 2000). In other words, women are socialized to assume domestic roles and are expected to be responsible for household tasks (Chang 2004; Chen 2009). The traditional gender roles of women are changing, but a sense remains that a woman's responsibility is to maintain a happy and efficient home life, especially in rural China. Recently, some Chinese men have begun to include housework, cooking, and cleaning as their responsibilities when their spouses work. Most Chinese believe that the family is most important, and thus each family member assumes changes in roles to achieve this harmony.

11.4.2 Prescriptive, Restrictive, and Taboo Behaviors for Children and Adolescents

The one-child policy was introduced in 1979 by the Chinese Government (Zeng and Hesketh 2018). Because of overpopulation in China, the government had mandated that each married couple may have only one child; however, in some specific situations, family plans for a second child were allowed. For instance, in rural areas, if the firstborn child was female or if the only child of a couple was disabled or killed in an accident, the couple might be permitted to have a second child (Zhuhong 2006).

In 2015, China announced that the iconic one-child policy had finally been replaced by a universal two-child policy. This change is highly significant because anyone in China is restricted to have just one child for the last 36 years (Zeng and Hesketh 2018). Families often wait many years, until they are financially secure, to have a child. After the child is born, many family resources are lavished on the child. Families may be able to afford only to live with relatives in a two-room apartment, but if the family believes that the child will benefit by having a piano, then the resources will be found to provide a piano. Children are well dressed and kept clean and well fed.

In Chinese society, the child is highly protected from birth; independence is not fostered. The entire family makes decisions for the child, even into young adulthood. Children usually depend on the family for everything. Few teens earn money because they are expected to study hard and to help the family with daily chores rather than to seek employment. Children are pressured to succeed and improve the future of the family and the country. Their common goal is to score well on the national examinations when they reach age 18. Most Chinese children and adolescents value studying over playing and peer relationships. They recognize that they are constantly evaluated on having healthy bodies and minds and achieving excellent marks in school.

With the traditional structure of a patriarchal family, girls are valued less than boys in Chinese society. In rural communities, male children are more valued than female children because they continue the family lineage and provide labor. In urban areas, however, female children are valued as highly as male children. Children in China are taught to curb their expression of feelings because individuals who do not stand out are successful. However, this is changing. The young in China today frequently think that their parents are too cautious. The children are becoming even more outspoken as they read more and watch more television, movies, and internet.

From elementary school to university, students take courses in Marxist politics and learn not to question the doctrine of the country. If they do, they may be interrogated and ridiculed for their radical thoughts. Nationalism is important to Chinese children, and they want to help their country continue to be the center of the world. Children are also expected to help their parents in the home. Many times in the cities when children get home from school before their parents, they are expected to do their homework immediately and then do their household chores. They exhibit their independence not so much by expressing their individual views but by performing chores on their own. The children are expected to earn good grades, and household chores are not encouraged; this is exhibited in overseas Chinese families as well. Lin and Fu (1990) studied 138 children—44 Chinese, 46 Chinese Americans, and 48 White Americans—in kindergarten through second grade and found that both Chinese and Chinese American parents expected increased achievement and parental control over their children. One surprising finding was the high expectation for independence in Chinese and Chinese American children.

Boys and girls play together when they are young, but as they get older, they do not because their roles and the corresponding expectations are predetermined by Chinese society. Girls and boys both study hard. Boys are more active and take pride in physical fitness. Girls are not nearly as interested in fitness as boys, preferring reading, art, and music.

Adolescents are expected to determine who they are and what they want to do with their lives. Adolescents maintain their respect for older people even when they disagree with them. Although they may argue with their parents and teachers, they have learned that it seldom does any good. Teens value a strong and happy family life and seldom do things that jeopardize that unanimity. Adolescents question affairs of life and make great efforts to see at least two sides of every issue. They enjoy exploring different views with their peers, and they try to explore them with their parents as well.

Teenage pregnancy is becoming a common issue among the Chinese. According to a government survey, more than 70 percent of 5000 students from 10 universities in Beijing have participated in a one-night stand (Huangjuan 2009). More female teens are willing to have sex to show their affection for their boyfriends, and most of them do not use any contraception, which leads to a high rate of pregnancy among teenage females (Zhuhong 2009a). Young men and women enter the workforce immediately after high school if they are unable to continue their education. Many continue to live with their parents and contribute to the family, even after marriage, into their 20s and, if they have a child, into their 30s.

11.4.3 Family Goals and Priorities

The Chinese perception of family is through the concept of relationships. In Confucian principles, hierarchical relationships exist between father and son, ruler and ruled, husband and wife, elder brother and younger brother, and friend and friend (Cheung et al. 2005). Each person identifies himself or herself in relation to others in the family. The individual is not lost, just defined differently from individuals in Western cultures. Personal independence in Chinese family is not a great favor; rather, Confucian teachings state that true value is in the relationships a person has with others, especially the family.

Older children who experienced the Cultural Revolution may feel some discomfort with their traditional parents. During the Cultural Revolution, the young were encouraged to inform on older people and peers who did not espouse the doctrine of the time. Most of those who were reported were sent to "re-education camps," where they did hard labor and were "taught the correct way to think." As a result, many families have been permanently separated.

Extended families are important to the Chinese and function by providing ways to get ahead. Often, children live with their grandparents or aunts and uncles so individual family members can obtain a better education or reduce financial burdens. Relatives are expected to help one another through connections, called ***guanxi***, which the Chinese society uses in a manner similar to the way other cultures use money. Such connections are perceived as obligations and are placed in a mental bank with deposits and withdrawals. These commitments may remain in the "bank" for years or generations until they are used to get jobs, housing, business contacts, gifts, medical care, or anything that demands a payback.

Filial loyalty to the family is extended to other Chinese. When Chinese immigrants need additional assistance, health-care providers may be able to call on local Chinese organizations to obtain help for patients.

Older people in China are venerated just as they were in earlier years. Chinese government leaders are often older and remain in power until they are in their 70s, 80s, and older. Traditional Chinese people view older people as very wise, a view that communism still required. Chinese children are expected to care for their parents, and in China, this is mandated by law.

Younger Chinese who adopt Western ideas and values may find that the expectations of older people are too demanding. Even though younger Chinese Americans do not live with their older relatives, they maintain respect and visit them frequently. Older Chinese mothers are viewed as central to family feelings, and older fathers retain their roles as leaders. As generations live in areas removed from China and families become more Westernized, family relationships need to be assessed on an individual basis. An extended-family pattern is common and has existed for over 2000 years. The traditional marriage still remains nuclear. Historically in China, marriage was used to strengthen positions of families in society.

Kinship relationships are based on the concept of loyalty and the young experience pressure to improve the family's standing. Many parents give up items of daily living to provide more for their children, thereby increasing opportunities for them to get ahead.

Maintaining reputation is very important to the Chinese and is accomplished by adhering to the rules of society. Because power and control are important to Chinese society, rank is very important. True equality does not exist in the Chinese mind; their history has demonstrated that equality cannot exist. If more than one person is in power, then consensus is important. If the person in power is not present at decision-making meetings, barriers are raised, and any decisions made are negated unless the person in power agrees. Even after negotiations have been concluded and contracts signed, the Chinese continue to negotiate.

The Chinese concept of privacy is even more important than recognized social status, corresponding values, and beliefs. The Chinese word for "privacy" has a negative connotation and means something underhanded, secret, and furtive. People grow up in crowded conditions, they live and work in small areas; their value of group support does not place a high value on privacy. The Chinese may ask many personal questions about salary, life at home, age, and children. Refusal to answer personal questions is accepted as long as it is done with care and feeling. The one subject that is taboo is sex and anything related to sex. This may create a barrier for a Western health-care provider who is trying to assess a Chinese patient with sexual concerns. The patient may feel uncomfortable discussing or answering questions about sex with honesty. Privacy is also limited by territorial boundaries. Some Chinese may enter rooms without knocking or invade privacy by not allowing a person to be alone. The need to be alone is viewed as "not good" to some Chinese; they may not understand when a Westerner wants to be alone. A mutual understanding of these beliefs is necessary for harmonious working relationships.

11.4.4 Alternative Lifestyles

In Chinese society, people have very little acceptance of gays and lesbians. In many provinces, homosexuality is illegal. Divorce is legal but is not encouraged; Although it is evident that divorce is a growing trend in China approaching 30 percent in 2006 (Alexander and Marg 2006). The reasons of divorce are multifaceted. First, society is going through a transitional period, which is greatly affecting the stability of marriages. Second, as living standards improve, people have higher expectations toward marriage and love. Third, the simplification of marriage and divorce procedures has made getting a divorce much easier (Women of China 2006a).

Tradition, consideration for children's feelings, and difficulty in remarrying are some of the reasons many Chinese families would rather stay in an unhealthy marriage than divorce. Remarriage is encouraged, but some difficult relationships may occur in the blended family, especially remarriage with children from previous marriages.

11.5 Workforce Issues

11.5.1 Culture in the Workplace

China is becoming more Westernized with high technology and increased knowledge. The Communist Party is responsible for establishing the ***dan wei***—local Chinese work units—that are responsible for jobs, homes, health, enforcement of governmental regulations, and problem solving for families. Although recent immigrants know that the culture in the workplace is different in the US and other countries, they adapt to it quickly. The Chinese acculturate by learning as much as possible about their new culture in the workplace. They observe people from the culture and listen closely for nuances in language and interpersonal connections. They frequently call on other Chinese people to teach them and to discuss how to fit into the new culture more quickly. Chinese support one another in new settings and help one another find resources and learn to live effectively and efficiently in the new culture. They also watch television, listen to music, and go to movies to learn about Western ways of life. They read about the new culture in magazines, books, newspapers, and on the Internet. They love to travel, and when an opportunity arises to see different aspects of the new culture, they do not hesitate to do so.

The Chinese are accustomed to giving co-workers small gifts of appreciation for helping them acculturate and adapt to the workforce. Often, Americans seek opportunities to reciprocate with a gift, such as at a birthday party, farewell party, or other occasion. Whereas a wide variety of gifts is appropriate, some gifts are not. For example, giving an umbrella means that one wishes to have the recipient's family dispersed; giving a gift that is white in color or wrapped in white could be interpreted as meaning the giver wants the recipient to die; and giving a clock could be interpreted as never wanting to see the person again or wishing the person's life to end (Smith 2002).

On the surface, Chinese immigrants form classic external networks, including groupings by family surname, locality of origin in China, dialect or subdialect spoken, craft practiced, and trust from prior experience or recommendation. Therefore, Chinese immigrants approximate external networks with some characteristic of internal networks (Haley et al. 1998).

Guanxi is a Mandarin term with no exact English translation. This term includes the concept of trust and presenting uprightness to build close relationships and connections. It definitely helps to build networks. This *Guanxi* network can be used in the work-related, decision-making process and is also used with family, friends, and community-related issues in Chinese immigrant communities.

11.5.2 Issues Related to Autonomy

Historically, the Chinese have been autonomous. They had to exhibit this characteristic to survive through difficult times. However, their autonomy is limited and is based on functioning for the good of the group. When a new situation arises that requires independent decision making, many times the Chinese know what should be done but do not take action until the leader or superior gives permission. Because of deferring to authority, Chinese people tend to avoid conflict and will not challenge anyone whom they regard as a leader or expert. For example, if a doctor prescribes an incorrect medication, Chinese people might accept the doctor's prescription without saying anything. However, the Western workforce expects independence, and some Chinese may need to be taught that true autonomy is necessary to advance. Health-care providers should be aware, however, that the training might not be successful because it is foreign to Chinese cultural values. A demonstration is the best alternative, leaving it up to the individuals to determine whether assertiveness can be a part of their lives. After acculturation takes place, Chinese do not differ significantly in assertiveness.

Language may be a barrier for Chinese immigrants seeking assimilation into the Western workforce. Western languages and Chinese have many differences, among them sentence structure and the use of intonation. The Chinese language does not have verbs that denote tense, as in Western languages. Whereas the ordering of the words in a sentence is basically the same, with the subject first and then the verb, the Chinese language places descriptive adjectives in different orders. Intonation in Chinese is in the words themselves rather than in the sentence. Chinese people who have taken English lessons can usually read and write English competently, but they may have a hard time understanding and speaking it. Research has reported that the ability to speak English is a significant factor among Chinese people in accessing health care (Cheung et al. 2005). Bilingual interpreters might be helpful to improve communication gap between Western healthcare providers and Chinese people.

11.6 Biocultural Ecology

11.6.1 Skin Color and Other Biologicsal Variations

The skin color of Chinese is varied. Many have skin color similar to that of Westerners, with pink undertones. Some have a yellow tone, whereas others are very dark. Mongolian spots, dark bluish spots over the lower back and buttocks are present in about 80 percent of infants. Bilirubin levels are usually higher in Chinese newborns with the highest levels occurring on the fifth or sixth day after birth.

Although Chinese are distinctly Mongolian, their Asian characteristics have many variations. China is very large and includes people from many different backgrounds, including Mongols and Tibetans. Generally, men and women are shorter than Westerners, but some Chinese are over 6 feet tall. Differences in bone structure are evidenced in the ulna, which is longer than the radius. Hip measurements are significantly smaller: females are 4.14 cm shorter, and males 7.6 cm shorter than Westerners (Seidel et al. 1994). Not only is overall bone length shorter, but bone density is also less. Chinese have a high hard palate, which may cause them problems with Western dentures. Their hair is generally black and straight, but some have naturally curly hair. Most Chinese men do not have much facial or chest hair. The Rh-negative blood group is rare, and twins are not common in Chinese families, but they are greatly valued, especially since the emergence of China's child laws.

11.6.2 Diseases and Health Conditions

Many Chinese who emigrate settle in large cities so they are at risk for the same problems and diseases experienced by other inner-city populations. For example, crowding in large cities often results in poor sanitation and increases the incidence of infectious diseases, air pollution, and violence.

Non-infectious chronic diseases have become the major threat to Chinese people, claiming 85 percent of deaths in China (Zhang 2011). The Chinese Ministry of Health (2009) reports that cerebrovascular disease, cancer, respiratory disease, and heart disease are the four principal causes of death in China.

In 1949, the average life expectancy was only 35 years (People's Daily 2002). According to the CIA World Factbook (2019a, b), the overall average life expectancy in China was 75.8 years—73.7 years for men and 78.1 years for women. The life expectancy in China has dramatically increased due to the improvement of living conditions and medical facilities, as well as a nationwide fitness campaign. Disease incidence has decreased as well. Tobacco use is a major problem and results in an increased incidence of lung disease. Health-care providers must screen newer immigrants from China for these health-related conditions and provide interventions in a culturally congruent manner.

Many Chinese immigrants have an increased incidence of hepatitis B and tuberculosis. Poor living conditions and overcrowding in some areas of China enhance the development of these diseases, which persist after immigrants settle in other countries.

According to the Chinese American women have higher rate of pancreatic cancer than the non-Hispanic, Whites (Thompson 2016) and all Chinese have higher death rates owing to diabetes (International Diabetes Federation 2019). The incidence of different types of cancer, including cervical, liver, lung, stomach, multiple myeloma, esophageal, pancreatic, and nasopharyngeal cancers is higher among Chinese Americans (Thompson 2016). Overall, the incidence of disease in this population has not been studied sufficiently, and continuing research is desperately needed.

11.6.3 Variations in Drug Metabolism

Studies outlining problems with drug metabolism and sensitivity have been conducted among the Chinese. Results suggest a poor metabolism of mephenytoin (e.g., diazepam) in 15–20 percent of Chinese; sensitivity to beta-blockers, such as propranolol, as evidenced by a decrease in the overall blood levels accompanied by a seemingly more profound response; atropine sensitivity, as evidenced by an increased heart rate; and increased responses to antidepressants and neuroleptics given at lower doses. Analgesics have been found to cause increased gastrointestinal side effects, despite a decreased sensitivity to them. In addition, the Chinese have an increased sensitivity to the effects of alcohol (Levy 1993).

Delineating specific variations in drug metabolism among the Chinese is difficult because various studies tend to group them in aggregate as Asians. Much more research needs to be completed to determine variations between Westerners and Asians, as well as among different groups of Asians.

11.7 High-Risk Behaviors

High-risk behaviors are difficult to determine with accuracy among the Chinese in the United States and elsewhere because most of the data on the Chinese are included in the aggregate called *Asian Americans*. Smoking is a high-risk behavior for many Chinese men and teenagers. Smoking-related diseases kill roughly 1.2 million Chinese people every year, and the death rate is expected to keep climbing in the coming decades (Yang 2011).

Smoking will cause about 20% of all adult male deaths in China during the 2010s. The tobacco-attributed proportion is increasing in men, but low, and decreasing, in women. Although overall adult mortality rates are falling. As the adult population in China grows and the proportion of male deaths due to smoking increases, the number of deaths in China that are caused by tobacco will rise from about 1 million in 2010 to 2 million in 2030 and 3 million in 2050, unless there is widespread cessation (Chen 2015).

Yu et al. (2002) reported that the male prevalence of smoking in Chicago is higher than that reported in California and exceeds the rate for African Americans aged 18 years and older. Most Chinese women do not smoke, but recently, the numbers for women are increasing, especially after immigration. Travelers in China see more cigarette street vendors than any other type. The decrease in smoking in the United States resulted in cigarette manufacturers targeting China as a good market in which to sell their product.

Alcohol consumption has increased significantly in China. In two random samples of 2327 and 2613 people, it was found that 90 percent of men and 55 percent of women drank alcohol in 2005 (Zhang et al. 2008). The greatest increase in alcohol consumption occurred in 18- and 19-year-olds and among older women (Zhang et al. 2008).

11.8 Nutrition

11.8.1 Meaning of Food

Food habits are important to the Chinese who offer food to their guests at any time of the day or night. Most celebrations with family and business events focus on food. Foods served at Chinese meals have a specific order, with the focus on a balance for a healthy body. The importance of food is demonstrated daily in its use to promote good health and to combat disease and injury. Traditional Chinese medicine frequently uses food and food derivatives to prevent and cure diseases and illnesses and to increase the strength of weak and older people.

11.8.2 Common Foods and Food Rituals

The typical Chinese diet is difficult to describe because each region in China has its own traditional foods. Peanuts and soybeans are popular. Common grains include wheat, sorghum, and maize. Rice is usually steamed but can be fried with eggs, vegetables, and meats. Many Chinese eat beans or noodles instead of rice. The Chinese eat steamed and fried rice noodles, which are usually prepared with a broth base and include vegetables and meats. Meat choices include pork (the most common), chicken, beef, duck, shrimp, fish, scallops, and mussels. Tofu, an excellent source of protein, is a staple of the Chinese diet and can be fried or boiled or eaten cold like ice cream. Bean products are another source of protein; many of the desserts or sweets in Chinese diets are prepared with red beans.

At celebrations, before-dinner toasts are usually made to family and business colleagues. The toasts may be interspersed with speeches, or the speeches may be incorporated into the toasts. Cold appetizers often include peanuts and seasonal fruits. Chopsticks, a chopstick holder, a small plate, and a glass are part of the table setting. If the foods are messy, like Beijing duck, then a finger towel may be available. The Chinese use ceramic or porcelain spoons for soup. Knives are unnecessary because the food is usually served in bite-sized pieces. Eating with chopsticks may be difficult for some at first, but the Chinese are good-natured and are pleased by any attempt to use them. Chopsticks should never be stuck in the food upright because that is considered bad luck (Smith 2002). Westerners soon learn that slurping, burping, and other noises are not considered offensive, but are appreciated. The Chinese are very relaxed at meals and commonly rest their elbows on the table.

Fruits and vegetables may be peeled or eaten raw. Some vegetables commonly eaten raw by Westerners are usually cooked by the Chinese. Unpeeled raw fruits and vegetables are sources of contamination owing to unsanitary conditions in China. The Chinese enjoy their vegetables lightly stir-fried in oil with salt and spices. Salt, oil, and oil products are important parts of the Chinese diet.

Drinks with dinner include tea, soft drinks, juice, and beer. Foreign-born Chinese and older Chinese may not like ice in their drinks. They may just not like anything cold while eating or may believe it is damaging to their body and shocks the body systems out of balance. Conversely, hot drinks are enjoyed and believed to be safe for the body. This "goodness" of hot drinks may stem from tradition in which the only safe drinks were made from boiled water. All food is put in the center of the table, arriving all at one time, but usually multiple courses are served. The host either serves the most important guests first or signals everyone to start.

11.8.3 Dietary Practices for Health Promotion

For the Chinese, food is important in maintaining their health. Foods that are considered *yin* and *yang* prevent sudden imbalances and indigestion. A balanced diet is considered essential for physical and emotional harmony. Health-care providers need to provide special instructions regarding risk factors associated with diets that are high in fats and salt. For example, the Chinese may need education regarding the use of salty fish and condiments, which increase the risk for nasopharyngeal, esophageal, and stomach cancers.

11.8.4 Nutritional Deficiencies and Food Limitations

Little information is available about dietary deficiencies in the Chinese diet. The life span of the Chinese is long enough to suggest that severe dietary deficiencies are not common as long as food is available. Periodically, some deficiencies, such as rickets and goiters have occurred. The Chinese government added iodine to water supplies and fish, which is rich in iron to enhance the diets of people with goiters. Native Chinese generally do not drink milk or eat milk products because of a genetic tendency for lactose intolerance. Their healthy selection of green vegetables limits the incidence of calcium deficiencies. Health-care providers may need to screen newer Chinese immigrants for these deficiencies and assist them in planning an adequate diet.

Most Chinese do not eat desserts with a high sugar content. Their desserts are usually peeled or sliced fruits or desserts made of bean and bean curd. The higher death rate from diabetes in Western countries mentioned earlier in this chapter may be due to a change from the typical Chinese diet with few sweets to a Western diet with many sweets.

11.9 Pregnancy and Childbearing Practices

11.9.1 Fertility Practices and Views Toward Pregnancy

From 1979 to 2014, China made efforts to slow the rate of population growth by enforcing a one-child law. The most popular form of birth control is the intrauterine device. Sterilization is common even though oral contraception is available.

Contraception is free in China. Abortion is fairly common, with 13 million abortions performed in China every year (Zhuhong 2009b). Health-care providers working in women's health need to be aware of the abortion issues among newly arrived Chinese immigrants, as well as among Chinese immigrants who may still adhere to premigration practices. It is critical for women's health to reduce the abortion issue among Chinese women.

Most Chinese families see pregnancy as positive and important in the immediate and extended family. Many couples wait a long time to have their first and only child. If a woman does become pregnant before the couple is ready to start a family, she may have an abortion. When the pregnancy is desired, the nuclear and extended families rejoice in the new family member. Overall, pregnancy is a crucial role among Chinese women, although Chinese men are beginning to demonstrate an active interest in pregnancy and the welfare of the mother and baby.

Under one-child policy, the gender imbalance had become a serious issue because many families, especially those in rural areas, prefer boys to girls. China has 119 boys born for every 100 girls, whereas the global ratio is 103–107 boys for every 100 girls (The World Bank 2012). In China, the "Care for Girls" program was initiated in 2003 to promote the social status of women; attempts are being made to decrease gender identification abortions without a medical purpose and the abandonment of newborn girls. Political action is through professional organizations but it is mostly *sub rosa* (XinHua News Agency 2006). As a result, the sex ratio declined from 106.74 in 2000 to 104.81 in 2017. Of the population enumerated in the 2010 census, males accounted for 51.13 percent, while females accounted for 48.87 percent (National Bureau of Statistics of China 2018).

11.9.2 Prescriptive, Restrictive, and Taboo Practices in the Childbearing Family

Because Chinese women are very modest, many insist on a female midwife or obstetrician. Some agree to use a male physician only when an emergency arises. Pregnant women usually add more meat to their diets because their blood needs to be stronger for the fetus. Many women increase the amount of organ meat in their diet, and even during times of severe food shortages, the Chinese government tried to ensure that pregnant women receive adequate nutrition. These traditions are also reflected in Chinese families outside China.

Other dietary restrictions and prescriptions may be practiced by pregnant women, such as avoiding shellfish during the first trimester because it causes allergies. Some mothers may be unwilling to take iron because they believe that it makes the delivery more difficult.

The Chinese government is proud of the fact that since the People's Revolution in 1949, infant mortality has been significantly reduced. In 2017, the mortality rates for infants and children under age 5 were reduced to 6.8 per 1000 from 35.9 per 1000 (UNICEF 2017a). This has been accomplished by providing a three-level system of care for pregnant women in rural and urban populations. Over 99 percent of childbirths take place under sterile conditions by qualified personnel. The infant mortality rate dropped from 20.3 per thousand in 2005 to 6.8 per thousand in 2017 (UNICEF 2017b). Therefore, most Chinese who have immigrated to Western countries are familiar with modern sterile deliveries.

In China, a woman stays in the hospital for a few days after delivery to recover her strength and body balance. Traditional postpartum care includes 1 month of recovery, with the mother eating cooked and warm foods that decrease the *yin* (cold) energy. The Chinese government supports recuperation period through special provisions on labor protection. Working women get full payment for maternity leave for 98 days (China State Council 2012). Women who return to work are allowed time off for breastfeeding, and in many cases, factories provide a special lounge for women to breastfeed. Families who come to Western societies expect the same importance to be placed on motherhood and may be surprised to find that many Western countries do not provide similar benefits.

Traditional prescriptive and restrictive practices continue among many Chinese women during the postpartum period. Drinking and touching cold water are taboo for women in the postpartum period. Raw fruits and vegetables are avoided because they are considered "cold" foods. They must be cooked and be warm. Mothers eat five to six meals a day with high nutritional ingredients, including rice, soups, and seven to eight eggs. Brown sugar is commonly used because it helps rebuild blood loss. Drinking rice wine is encouraged to increase the mother's breast milk production, but mothers need to be cautioned that it may also prolong the bleeding time. Many mothers do not expose themselves to the cold air and do not go outside or bathe for the first month postpartum because the cold air can enter the body and cause health problems, especially for older women. Some women wear many layers of clothes and are covered from head to toe, even in the summer, to keep the air away from their bodies. However, this practice has changed among some young women who live in Western cultures for a long period of time and when there are no older Chinese parents around during the postpartum period.

Adopted Chinese children display a similar pattern of growth and developmental delays and medical problems as seen in other groups of internationally adopted children. An exception is the increased incidence of elevated lead levels (overall 14 percent). Although serious medical and developmental issues were found among Chinese children, overall their health was better than expected based on recent publicity about conditions in the Chinese orphanages. The long-term outcome of these children remains unknown (Miller 2000). Many children adopted from China have antibody titers that do not correlate with those expected from their medical records. These children, unlike children adopted from other countries, have documented evidence of adequate vaccinations. However, they should be tested for antibody concentrations and reimmunized as necessary (Schulpen 2001).

Reflective Exercise 11.2

Mrs. Huang, a 32-year-old woman from China, lives with her husband and his parents. She delivered a healthy baby girl yesterday. This morning, Mrs. Huang says she is constipated. The physician told Mrs. Huang that constipation was a common problem after delivery. The nurse suggested that she eat more fresh fruits and vegetables to facilitate a bowel movement. However, Mrs. Huang rejected fruits and vegetables and stated that a woman was characterized as a *yin* union and eating too much cold food was not good for a postpartum woman. Later in the day, Mrs. Huang told the nurse that she did not think she wanted to experience childbirth again. When the nurse approached Mrs. Huang about possible contraception, she said that her parents-in-law wanted her be become pregnant very soon in the hopes of having a male child.

1. According to Chinese perspectives, foods are divided into yin and yang categories. What information does the nurse need before discussing Mrs. Huang's concerns about constipation?
2. What other culturally congruent dietary instructions might the nurse recommend to resolve Mrs. Huang's concerns with constipation?
3. As a daughter-in-law in a traditional Chinese family, Mrs. Huang's autonomy to make decisions seems to be limited. What concerns might the nurse have when discussing contraception?
4. Describe some traditional postpartum Chinese practices.

11.10 Death Rituals

11.10.1 Death Rituals and Expectations

Chinese death and bereavement traditions are centered on ancestor worship. Ancestor worship is frequently misunderstood; it is not a religion, but rather a form of paying respect. Many Chinese believe that their spirits can never rest unless living descendants provide care for the grave and worship the memory of the deceased. These practices were so important to early Chinese that Chinese pioneers to the West had statements written into their work contract that their ashes or bones be returned to China (Halporn 1992).

The belief that the Chinese greet death with stoicism and fatalism is a myth. In fact, most Chinese fear death, avoid references to it, and teach their children this avoidance. The number 4 is considered unlucky by many Chinese because it is pronounced like the Chinese word for death; this is similar to the bad luck associated with the number 13 in many Western societies. Huang (1992) wrote:

> At a very young age, a child is taught to be very careful with words that are remotely associated with the "misfortune" of death. The word "death" and its synonyms are strictly forbidden on happy occasions, especially during holidays. People's uneasiness about death often is reflected in their emphasis on longevity and everlasting life .... In daily life, the character "Long Life" appears on almost everything: jewelry, clothing, furniture, and so forth. It would be a terrible mistake to give a clock as a gift, simply because the pronunciation of the word "clock" is the same as that of the word "ending." Recently, many people in Taiwan decided to avoid using the number "four" because the number has a similar pronunciation to the word "death." (p. 1)

Many Chinese are hesitant to purchase life insurance because of their fear that it is inviting death. The color white is associated with death and is considered bad luck. Black is also a bad luck color.

Many Chinese believe in ghosts and the fear of death is extended to the fear of ghosts. Some ghosts are good and some are bad, but all have great power. Communism discourages this thinking and sees it as a hindrance to future growth and development of the society, but the ever-pragmatic Chinese believe it is better not to invite trouble with ghosts just in case they might exist.

The dead may be viewed in the hospital or in the family home. Extended family members and friends come together to mourn. The dead are honored by placing objects around the coffin that signify the life of the dead: food, money designated for the dead person's spirit, and other articles made of paper. In China, cremation is preferred by the state because of a lack of wood for coffins and a limited space for burial. The ashes are placed in an urn and then in a vault. As cities grow, even the space for vaults is limited. In rural areas, many families prefer traditional burial and have family burial plots. It is preferable to burying an intact body in a coffin.

11.10.2 Responses to Death and Grief

The Chinese react to death in various ways. Death is viewed as a part of the natural cycle of life, and some believe that something good happens to them after they die. These beliefs foster the impression that Chinese are stoic. In fact, they feel similar emotions to Westerners but do not overtly express those emotions to strangers. During bereavement, a person does not have to go to work, but instead can use this mourning time for remembering the dead and planning for the future. Bereavement time in the larger cities is 1 day to 1 week, depending on the policy of the government agency and the relationship of family members to the deceased. Mourners are recognized by black armbands on their left arm and white strips of cloth tied around their heads.

11.11 Spirituality

11.11.1 Dominant Religion and Use of Prayer

In mainland China, the practice of formal religious services is minimal. The ideals and values of the different religions are practiced alone rather than with people coming together to participate in a formal religious service. In recent years in some parts of China, religion is becoming more popular. The main formal religions in China are Taoism, Buddhism, Christianity (3–4 percent), and Islam (1–2 percent) (CIA World Factbook 2019a, b).

As immigration from China increases, Chinese who practice Christian religions have become more visible on the American landscape. Chinese immigrants from the People's Republic of China may express perspectives on religious beliefs different from those of the Chinese from other countries, or from Hong Kong and Taiwan, where they have been permitted to practice Christianity. At first, they may go to a church attended by other Chinese people; eventually some are baptized, and others continue to attend Bible studies. In cities in the United States, churches are playing a very important role in the local Chinese community in terms of providing support and services to Chinese immigrants, students, scholars, and their families. An understanding of this concept is essential when the health-care provider attempts to obtain religious counseling services for Chinese patients.

Prayer is generally a source of comfort. Some Chinese do not acknowledge a religion such as Buddhism, but if they go to a shrine, they burn incense and offer prayers.

11.11.2 Meaning of Life and Individual Sources of Strength

The Chinese view life in terms of cycles and interrelationships, believing that life gets meaning from the context in which it is lived. Life cannot be broken into simple parts and examined because the parts are interrelated. When the Chinese attempt to explain life and what it means, they speak about what happened to them, what happened to others, and the importance and interrelatedness of those events. They speak not only of the importance of the current phenomena but also about the importance of what occurred many years, maybe even centuries, before their lives. They live and believe in a true systems framework.

"Life forces" are sources of strength to the Chinese. These forces come from within the individual, the environment, the past and future of the individual, and society. Chinese use these forces when they need strength. If one usual source of strength is unsuccessful, they try another. The individual may use many different techniques such as meditation, exercise, massage, and prayer. Drugs, herbs, food, good air, and artistic expression may also be used. Good luck charms are cherished, and traditional and nontraditional medicines are used.

The family is usually one source of strength. Individuals draw on family resources and are expected to return resources to strengthen the family. Resources may be financial, emotional, physical, mental, or spiritual. Calling on ancestors to provide strength as a resource requires giving back to the ancestors when necessary. The interconnectedness of life provides a source of strength for individuals from before birth to death and beyond.

Health-care providers need to understand this multidimensional manner of thinking and believing. Assessments, goal setting, interventions, and evaluations may be different for Chinese patients than for American patients. The context of client problems is the emphasis, and the physical, mental, and spiritual aspects of the person's life are the focal points.

11.12 Health-Care Practices

11.12.1 Health-Seeking Beliefs and Behaviors

Health care in China and Taiwan is provided for most citizens. Every work unit and neighborhood has its own clinic and hospital. Traditional Chinese medicine shops abound. Even department stores and supermarkets have Western medicines and traditional Chinese medicines and herbs.

The focus of health has not changed over the centuries, and it includes having a healthy body, a healthy mind, and a healthy spirit. Preventive health-care practices are a major focus in China and Taiwan today. An additional focus is placed on infectious diseases such as schistosomiasis, tuberculosis, HIV. Chinese Ministry of Health statistics indicate that China had 501,000 HIV-infected people in 2014 (Avert 2019). China accounts for 3% of new HIV infections globally each year. In 2018, a 14% of new infections was reported, with 40,000 in the second quarter alone (Avert 2019). Treatment, care and support challenges prevail in China. In 2014 alone, 21,000 people died from AIDS-related causes (UNAIDS 2013). The number of people living with HIV on treatment has steadily increased, however, progress in reducing mother-to-child transmission rates is still regarded as slow. Progress has also been slow in addressing the high levels of stigma and discrimination people living with HIV experience across the country (China Ministry of Health 2014).

A regulation on AIDS prevention and control (effective January 1, 2007) spells out the plan to administer the free test in areas of the province where the AIDS situation is "grave." HIV carriers and AIDS patients will be asked to inform their spouses or sex partners of the results, or the local disease prevention authorities will do so (Women of China 2006b).

Whereas many Chinese have made the transition to Western medicine, others maintain their roots in traditional Chinese medicine, and still others practice both types. The Chinese are similar to other nationalities in seeking the most effective cure available. Younger Chinese people usually do not hesitate to seek health-care providers when necessary. They generally practice Western medicine unless they feel that it does not work for them; then they use traditional Chinese medicine. Conversely, older people may try traditional Chinese medicine first and only seek Western medicine when traditional medicine does not seem to work.

Among Chinese immigrants, these health-seeking beliefs, practices, and patterns remain the same as the ones in China. This results in sicker older people seeking care from Western health-care providers. Even after seeking Western medical care, many older Chinese continue to practice traditional Chinese medicine in some form. However, some Chinese patients may not tell health-care providers about other forms of treatment they have been using because they are conscious of saving face. Health-care providers need to understand this practice and include it in their care. Members of the health-care team need to develop a trusting relationship with Chinese patients so all information can be disclosed. Health-care providers must impress upon patients the importance of disclosing all treatments because some may have antagonistic effects.

11.12.2 Responsibility for Health Care

Chinese people often self-medicate when they think they know what is wrong or if they have been successfully treated by their traditional medicine or herbs in the past. They share their knowledge about treatments and their medicines with friends and family members. This often happens among Chinese immigrants as well because of the belief that occasional illness can be ameliorated through the use of nonprescription drugs. Many consider seeing Western health-care providers as a waste of time and money. Health-care providers need to recognize that self-medication and sharing medications are accepted practices among the Chinese. Thus, health-care providers should inquire about this practice when making assessments, setting goals, and evaluating the results of treatments. A trusting relationship between members of the health-care team and the patient and family is necessary to enhance the disclosure of all treatments.

11.12.3 Traditional Chinese Medicine Practices

Traditional Chinese medicine is practiced widely, with concrete reasons for the preparation of medications, taking medicine, and the expected outcomes. Western medicine needs to be explained to Chinese patients in equally concrete terms.

Traditional Chinese medicine has many facets, including the five basic substances (*qi*, energy; *xue*, blood; *jing*, essence; *shen*, spirit; and *jing ye*, body fluids). The pulses and vessels for the flow of energetic forces are (*mai*), the energy pathways (*jing*), the channels and collaterals, the 14 meridians for acupuncture, moxibustion, and massage (*jing luo*); the organ systems (*zang fu*); and the tissues of the bones, tendons, flesh, blood vessels, and skin. The scope of traditional Chinese medicine is vast and should be studied carefully by professionals who provide health care to Chinese patients.

Acupuncture and moxibustion are used in many of the treatments. Acupuncture is the insertion of needles into precise points along the channel system of flow of the *qi* called the 14 meridians. The system has over 400 points. Many of the same points can be used in applying pressure and massage to achieve relief from imbalances in the system. The same systems approach is used to produce localized anesthesia.

Moxibustion is the application of heat from different sources to various points. For example, one source, such as garlic, is placed on the distal end of the needle after it is inserted through the skin, and the garlic is set on fire. Sometimes, the substance is burned directly over the point without a needle insertion. Localized erythema occurs with the heat from the burning substance, and the medicine is absorbed through the skin. Cupping is another common practice. A heated cup or glass jar is put on the skin, creating a vacuum, which causes the skin to be drawn into the cup. The heat generated is used to treat joint pain.

The Chinese believe that health and a happy life can be maintained if the two forces, the ***yang*** and the ***yin***, are balanced. This balance is called the ***dao***. Heaven is *yang*, and Earth is *yin*; man is *yang*, and woman is *yin*; the sun is *yang*, and the moon is *yin*; the hollow organs (bladder, intestines, stomach, gallbladder), head, face, back, and lateral parts of the body are *yang*, and the solid viscera (heart, lung, liver, spleen, kidney, and pericardium), abdomen, chest, and the inner parts of the body are *yin*. The *yang* is hot, and the *yin* is cold. Health-care providers need to be aware that the functions of life and the interplay of these functions, rather than the structures, are important to the Chinese people.

Central to traditional medicine is the concept of the ***qi***; it is considered the vital force of life including air, breath, or wind; it is present in all living organisms. Some of the *qi* is inherited, and other parts come from the environment, such as in food. The *qi* circulates through the 14 meridians and organs of the body to give the body nourishment. The channels of flow are also responsible for eliminating the bad *qi*. All channels, the meridians and organs, are interconnected. The results resemble a system in which a change in one part of the system results in a change in other parts, and one part of the system can assist other parts in their total functioning.

Diagnosis is made through close inspection of the outward appearance of the body, the vitality of the person, the color of the person, the appearance of the tongue, and the person's senses. The health-care provider uses listening, smelling, and questioning techniques in the assessment. Palpation is used by feeling the 12 pulses and different parts of the body. Treatments are based on the imbalances that occur. Many are directly related to the obvious problem, but many more are related through the interconnectedness of the body systems. Many of the treatments not only "cure" the problem but are also used to "strengthen" the entire human being. Traditional Chinese medicine cannot be learned quickly because of the interplay of symptoms and diagnoses. Health-care providers take many years to become adept in all phases of diagnosis and treatment.

T'ai chi, practiced by many Chinese, has its roots in the twelfth century. This type of exercise is suitable for all age groups, even the very old. *T'ai chi* involves different forms of exercise, some of which can be used for self-defense. The major focus of the movements is mind and body control. The concepts of *yin* and *yang* are included in the movements, with a *yin* movement following a *yang* movement. Total concentration and controlled breathing are necessary to enable the smoothness and rhythmic quality of movement. The movements resemble a slow-motion battle, with the participant both attacking and retreating. Movements are practiced at least twice a day to bring the internal body, the external body, and the environment into balance (Mayo Clinic 2010). Yoga is also a fashionable exercise among women. Yoga incorporates meditation, relaxation, imagery, controlled breathing, stretching, and other physical movements. Yoga has become increasingly popular in Western cultures as a means of exercise and fitness training. Yoga needs to be better recognized by the health-care community as a complement to conventional medical care.

Herbal therapy is integral to traditional Chinese medicine and is even more difficult to learn than acupuncture and moxibustion. Herbs fall into four categories of energy (cold, hot, warm, and cool), five categories of taste (sour, bitter, sweet, pungent, and salty), and a neutral category. Different methods are used to administer the herbs, including drinking and eating, applying topically, and wearing on the body. Each treatment is specific to the underlying problem or a desire to increase strength and resistance.

11.12.4 Barriers to Health Care

In China, the government is primarily responsible for providing basic health care within a multilevel system. Native Chinese are accustomed to the neighborhood work units called *dan wei*, where they get answers to their questions and health-care services are provided. After transition to the US, Chinese patients face many of the same barriers to health care faced by Westerners, yet they have other special concerns and difficulties that prevent them from accessing health-care services. Ma (2000) summarized these barriers as the following:

- *Language barriers:* This is one of the major reasons that Chinese Americans do not want to see Western health-care providers. They feel uncomfortable and frustrated with not being able to communicate with them freely and not being able to adequately express their pains, concerns, or health problems. Even highly educated Chinese Americans, who have limited knowledge in the medical field and are unfamiliar with medical terminology, have difficulty complying with recommended procedures and health prescriptions.
- *Cultural barriers:* Lack of culturally appropriate and competent health-care services is another key obstacle to health-care service utilization. Many Chinese Americans have different cultural responses to health and illness. Although they respect and accept the Western health-care provider's prescription drugs, they tend to alternate between Western and traditional Chinese physicians.
- *Socioeconomic barriers:* Being unable to afford medical expenses is another barrier to accessing health-care services for some Chinese Americans. However, having health insurance does not always ensure the utilization of the health-care system or the benefits of health insurance. There may be a sense of distrust between patients and health-care providers or between patients and insurance companies. In addition, many do not know the cost of the service when they enter a clinic or hospital. They are frustrated with being caught in the battle between health insurance companies and the clinic or hospital.
- *Systemic barriers:* Not understanding the Western health-care system and feeling inconvenienced by managed-care regulations deter many from seeking Western health-care providers unless they are seriously ill. The complexity of the rules and regulations of public agencies and medical assistance programs such as Medicaid and Medicare block their effective use.

Tan (1992), in a different perspective, summarized barriers for Chinese immigrants seeking health care:

- Many Chinese Americans have great difficulty facing a diagnosis of cancer because families are the main source of support for patients, and many family members are still in China.
- Because many Chinese Americans do not have medical insurance, any serious illness will lead to heavy financial burdens on the family.
- Once the patient responds to initial treatment, the family tends to stop treatment and the patient does not receive follow-up care or becomes nonadherent. Chinese American families may be reluctant to allow autopsies because of their fear of being "cut up."

- The most difficult barrier is frequently the reluctance to disclose the diagnosis to the patient or the family.

In recent years, clinics of Chinese medicine and health-care providers who are originally from China have been significantly visible in the US, especially in the larger cities. These provide opportunities or options for those Chinese Americans who prefer to seek traditional Chinese treatment for certain illness.

11.12.5 Cultural Responses to Health and Illness

Chinese express their pain in ways similar to those of Americans, but their description of pain differs. A study by Moore (1990) included not only the expression of pain but also common treatments used by the Chinese. The Chinese tend to describe their pain in terms of more diverse body symptoms, whereas Westerners tend to describe pain locally. The Western description includes words like "stabbing" and "localized," whereas the Chinese describe pain as "dull" and more "diffuse." They tend to use explanations of pain from the traditional Chinese influence of imbalances in the *yang* and *yin* combined with location and cause. The study determined that the Chinese cope with pain by using externally applied methods, such as oils and massage. They also use warmth, sleeping on the area of pain, relaxation, and aspirin.

The balance between *yin* and *yang* is used to explain mental as well as physical health. This belief, coupled with the influence of Russian theorists such as Pavlov, influence the Chinese view of mental illness. Mental illness results more from metabolic imbalances and organic problems. The effect of social situations, such as stress and crises, on a person's mental well-being is considered inconsequential, but physical imbalances from genetics are the important factors. Because a stigma is associated with having a family member who is mentally ill, many families initially seek the help of a folk healer. Many use a combination of traditional and Western medicine. Many mentally ill clients are treated as outpatients and remain in the home.

Although the Chinese do not readily seek assistance for emotional and nervous disorders, a study of 143 Chinese Americans found that younger, lower socioeconomic, and married Chinese with better language ability seek help more frequently (Ying and Miller 1992). The researchers recommended that new immigrants be taught that help is available when needed for mental disorders within the mental health-care system.

Chinese in larger cities are becoming more supportive of people with disabilities, but for the most part, support services are popular. Because the focus has been on improving the overall economic growth of the country, the needs of the disabled have not had priority. The son of Deng Xiaoping was crippled in the Cultural Revolution and has been active in making the country more aware of the needs of the disabled. The Beijing Paralympic Games were held on September 6–17, 2008, and opened 12 days after the 29th Olympic Games. A successful Paralympics in Beijing promoted the cause of disabled persons in Beijing as well as throughout China. The games urged the whole of society to pay more attention to this special segment of our population and reinforced the importance of building accessible facilities for the disabled and thus enhance efforts to construct a harmonious society in China (Nan 2006). Overall, the Chinese still view mental and physical disabilities as a part of life that should be hidden.

The expression of the sick role depends on the level of education of the patient. Educated Chinese people who have been exposed to Western ideas and culture are more likely to assume a sick role similar to that of Westerners. However, the highly educated and acculturated may exhibit some of the traditional roles associated with illness. Each patient needs to be assessed individually for responses to illness and for expectations of care. Traditionally, the Chinese ill person is viewed to be passive and accepting of illness. To the Chinese, illness is expected as a part of the life cycle. However, they do try to avoid danger and to live as healthy a life as possible. To the Chinese, all of life is interconnected; therefore, they seek explanations and connections for illness and injury in all aspects of life. Their explanations to health-care providers may not make sense, but the health-care provider should try to determine those connections so they can be incorporated into treatment regimens. The Chinese believe that because the illness or injury is caused from an imbalance, there should be a medicine or treatment that can restore the balance. If the medicine or treatment does not seem to do this, they may refuse to use it.

Native Chinese and Chinese immigrants like treatments that are comfortable and do not hurt. Treatments that hurt are physically stressful and drain their energy. Health-care providers who have been ill themselves can appreciate this way of thinking, because sometimes the cure seems worse than the illness. Treatments will be more successful if they are explained in ways that are consistent with the Chinese way of thinking. The Chinese depend on their families and sometimes on their friends to help them while they are sick and provide much of the direct care; health-care providers are expected to manage the care. The family may seem to take over the life of the sick person; the sick person is very passive in allowing them the control. One or two primary people assume this responsibility, usually a spouse. Health-care providers need to include the family members in the plan of care and, in many instances, in the actual delivery of care.

11.12.6 Blood Transfusions and Organ Donation

Modern-day Chinese accept blood transfusions, organ donations, and organ transplants when absolutely essential, as long as they are safe and effective. Chinese Americans have the same concerns as Americans about blood transfusion because of the perceived high incidence of HIV and hepatitis B. No overall ethnic or religious practices prohibit the use of blood transfusions, organ donations, or organ transplants. Of course, some individuals may have religious or personal reasons for denying their use.

11.13 Health Care Providers

11.13.1 Traditional Versus Biomedical Providers

China uses two health-care systems. One is grounded in Western medical care and the other is anchored in traditional Chinese medicine. The educational preparation of physicians, nurses, and pharmacists is similar to Western health-care education. Ancillary workers have responsibility in the health-care system, and the practice of midwifery is widely accepted by the Chinese. Physicians in Chinese medicine are trained in universities, and traditional Chinese pharmacies remain an integral part of health care.

11.13.2 Status of Health-Care Providers

Traditional Chinese medicine providers are shown great respect. In many instances, they are shown equal, if not more respect than Western health-care providers. The Chinese may distrust Western health-care providers because of the pain and invasiveness of their treatments. The hierarchy among Chinese health-care providers is similar to that of Chinese society. Older health-care providers receive respect from the younger providers. Men usually receive more respect than women, but that is beginning to change. Physicians receive the highest respect, followed closely by nurses with a university education. Other nurses with limited education are next in the hierarchy, followed by ancillary personnel.

Health-care providers are usually given the same respect as older people in the family. Chinese children recognize them as authority figures. Physicians and nurses are viewed as individuals who can be trusted with the health of a family member. Nurses are generally perceived as caring individuals who perform treatments and procedures as ordered by the physician. Nursing assistants provide basic care to patients. Adult Chinese respond to health-care providers with respect, but if they disagree with the health-care provider, they may not follow instructions. They may not verbally confront the health-care provider because they fear that either they or the provider will suffer a loss of face.

Reflective Exercise 11.3

Mrs. Cheng brought her 4-year-old son, Justin, to the emergency department early one morning. She stated that her son has had a high fever for 4 days. Her mother-in-law used traditional herbal medicine, but it was ineffective. The advanced practice nurse diagnosed a pulmonary infection and prescribed liquid antibiotics.

1. From a traditional Chinese medical perspective, how might the nurse incorporate Western medical prescriptions while respecting Mrs. Cheng's family, who wishes to continue Chinese herbal treatments?
2. What additional cultural and socioeconomic barriers should the nurse assess to provide culturally competent health-care and nursing services to Justin?
3. Identify and describe traditional, non-herbal Chinese medical practices that are used to treat pulmonary disorders?
4. Describe from the traditional Chinese individual the ideal health-care provider.

The Chinese respect their bodies and are very modest when it comes to touch. Most Chinese women feel uncomfortable being touched by male health-care providers, and most seek female health-care providers.

References

Alexander RM, Marg AM (2006) Chinese women local leaders exchange realities with WEI. Women Environ 74:17–19

Avert (2019) HIV and AIDS in China. https://www.avert.org/professionals/hiv-around-world/asia-pacific/china

Chang JS (2004) Refashioning womanhood in 1990s Taiwan: an analysis of Taiwanese edition of cosmopolitan magazine. Modern China 30:361–397

Chen SE (2009) Mother of the culture: founding a Taiwanese feminist theology. Fem Theol 27:81–88

Chen Z (2015) Contrasting male and female trends in tobacco-attributed mortality in China: evidence from successive nationwide prospective cohort studies. Lancet 358:1447–1456

Cheung R, Nelson W, Advincula L, Young CV, Canham DL (2005) Understanding the culture of Chinese children and families. J Sch Nurs 21(1):3–9

China Daily (2019). https://www.chinadaily.com.cn/business/2014-12/19/content_19121890_10.htm

China Ministry of Health (2014) China country progress report 2014. http://www.unaids.org/sites/default/files/documents/CHN_narrative_report_2014.pdf

China National Bureau of Statistics (2018) 2017 "China Women's Development Outline (2011–2020)" statistical monitoring report. http://www.stats.gov.cn/tjsj/zxfb/201811/t20181109_1632537.html

China State Council (2012) Order No. 619 of the State Council of the People's Republic of China. Female workers have special regulations for labor protection. http://www.gov.cn/zwgk/2012-05/07/content_2131567.htm

Chinese Ministry of Health (2009) Chinese factfile. http://english.gov.cn/about.htm

CIA World Factbook (2019a). https://www.cia.gov/library/publications/the-world-factbook/geos/ch.html

CIA World Factbook (2019b) China. https://www.cia.gov/library/publications/the-world-factbook/geos/ch.html

Haley G, Tan C, Haley U (1998) New Asian emperors: the overseas Chinese, their strategies and competitive advantages. Butterworth-Heinemann, Woburn, MA

Halporn R (1992) Introduction. In: Chen CL, Lowe WC, Ryan D, Kutscher AH, Halporn R, Wang H (eds) Chinese Americans in loss and separation. Foundation of Thanatology, New York, pp v–xii

Ho HF, Lee HL (2010) Great expectations: family educational expenditure in Taiwan vs China. Eur J Soc Sci 17(4):628–637

Hu W, Grove CL (1991) Encountering the Chinese. Intercultural Press, Yarmouth, MA

Huang W (1992) Attitudes toward death: Chinese perspectives from the past. In: Chen CL, Lowe WC, Ryan D, Kutscher AH, Halporn R, Wang H (eds) Chinese Americans in loss and separation. Foundation of Thanatology, New York, pp 1–5

Huangjuan (2009) One-night stands accepted, few practice. http://www.womenofchina.cn/html/node/104673-1.htm

Levy RA (1993) Ethnic and racial differences in response to medicines: preserving individualized therapy in managed pharmaceutical programmes. Pharm Med 7:139–165

Li C (2000) Confucianism and feminist concerns: overcoming the confucian "Gender complex.". J Chin Philos 27:187–199

Lin CC, Fu VR (1990) A comparison of child-rearing practices among Chinese, immigrant Chinese, and Caucasian-American parents. Child Dev 61:429–433
Ma GX (2000) Barriers to the use of health services by Chinese Americans. J Allied Health 29(2):64–70
Mayo Clinic (2010) Tai chi: discover the many possible health benefits. http://www.mayoclinic.com/health/tai-chi/SA00087
Miller LC (2000) Health of children adopted from China. Pediatrics 105(6):76
Moore R (1990) Ethnographic assessment of pain coping perceptions. Psychosom Med 52:171–181
Nan C (2006) Beijing gears up for 2008 paralympic games. http://www.btmbeijing.com
National Bureau of Statistics of China (2018) China's population sex ratio. http://www.data.stats.gov.cn/easyquery.htm?cn=C01
O'Keefe H, O'Keefe W (1997) Chinese and Western behavioral differences: understanding the gaps. Int J Soc Econ 24:190–197
People's Daily (2002) China's life expectancy averaged 71.8 year. http://english.peopledaily.com.cn
Schulpen T (2001) Immunization status of children adopted from China. Lancet 358:2131–2132
Seidel H, Ball J, Dains J, Benedict W (1994) Quick reference to cultural assessment. St. Mosby, Louis, MO
Shim YH (2001) Feminism and the discourse of sexuality in Korea: continuities and changes. Hum Stud 24:133–148
Shrestha M, Weber KE (1994) Reflection of Confucianism, Hinduism, and Buddhism on gender relations and gender specific occupation in Thai society. J Popul Soc Stud 5(1–2):31–54
Smith CS (2002) Beware of cross-cultural faux pas in China. New York Times, New York
Tan CM (1992) Treating life-threatening illness in children. In: Chen CL, Lowe WC, Ryan D, Kutscher AH, Halporn R, Wang H (eds) Chinese Americans in loss and separation. Foundation of Thanatology, New York, pp 26–33
Thompson CA, Gomez SL, Hastings KG, Kapphahn K, Yu P, Shariff-Marco S, Bhatt AS, Wakelee HA, Patel MI, Cullen MR, Palaniappan LP (2016). The burden of cancer in Asian Americans: a report of national mortality trends by Asian ethnicity. Cancer Epidem Biomar. 25(10):1371–1382. https://doi.org/10.1158/1055-9965
International Diabetes Federation (2019) Demographic and geographic outline. https://diabetesatlas.org/en/sections/demographic-and-geographic-outline.html
The World Bank (2012) Gender highlights: 2012 world development report. https://datacatalog.worldbank.org/dataset/gender-highlights-2012-world-development-report
U.S. Census Bureau (2017) Selected population profile in the United States. http://factfinder.census.gov/
U.S. Citizenship and Immigration Services (2011) Immigration and nationality act. http://www.uscis.gov/
UNAIDS (2013) HIV in Asia and the Pacific. http://www.unaids.org/en/resources/documents/2013/20131119_HIV-Asia-Pacific
UNICEF (2017a) Figure 3.7 infant mortality rate, 1991–2017. https://www.unicef.cn/en/figure-37-infant-mortality-rate-19912017
UNICEF (2017b) Figure 3.15 hospital delivery rate and maternal mortality ratio, 1990–2017. https://www.unicef.cn/en/figure-315-hospital-delivery-rate-and-maternal-mortality-ratio-19902017
Wang Y (2011) Press release on major figures of the 2010 National Population Census. http://www.womenofchina.cn
Women of China (2006a) Survey of divorce rates in different countries
Women of China (2006b) HIV test mandatory
care services. http://www.womenofchina.cn
XinHua News Agency (2006) Official calls for more efforts to curb gender imbalance. http://english.peopledaily.com.cn
Yang Z (2011) Chinese smokers not prepared to face hasty smoking ban. http://www.womenofchina.cn/html/node/129421-1.htm
Ying Y, Miller LS (1992) Help-seeking behavior and attitude of Chinese Americans regarding psychological problems. Am J Community Psychol 20(4):549–556
Yu ES, Edwin H, Chen K, Kim K, Sawsan A (2002) Smoking among Chinese Americans: behavior, knowledge, and beliefs. Am J Public Health 92(6):1007–1013
Zeng Y, Hesketh T (2018) The effects of China's universal two-child policy. https://www.ncbi.nlm.nih.gov/pmc/articles/PMC5944611/
Zhang S (2011) Non-infectious chronic diseases become major health threat, claiming 85% of deaths in China. http://www.womenofchina.cn/html/report/130433-1.htm
Zhang J, Casswell S, Cai H (2008) Increased drinking in a metropolitan city in China: a study of alcohol consumption patterns and changes. Addiction 103(3):416–423
Zhuhong A (2006) Population and family planning law of the People's Republic of China. http://www.womenofchina.cn
Zhuhong A (2009a) Survey shows more teens have sexual experience. http://www.womenofchina.cn/html/node/102669-1.htm
Zhuhong A (2009b) Abortion issue causes for concern. http://www.womenofchina.cn/html/node/102775-1.htm

People of Cuban Heritage

12

Jorge A. Valdes and Victor Delgado

12.1 Introduction

Over 1.6 million Cuban live in the United States (U.S. Census Bureau 2010). Cubans in Miami-Dade County, Florida—the dominant center of Cuban settlement—are credited with the area's socioeconomic transformation (Boswell 2002). In this ethnic enclave, Cubans have created businesses and rejuvenated the economy, leading some to speak of the "great Cuban miracle." Over 300,000 Cubans have relocated to Spain, Puerto Rico, Venezuela, Mexico, and other Latin American and Caribbean countries, as well as Canada and European nations such as Germany, Italy, and France (Migration Policy Institute 2017). Little is written about Cubans who have migrated to these countries.

12.2 Overview, Inhabited Localities, and Topography

12.2.1 Overview

Cuba, an island nation originally named "*Juana*" after Queen Isabella's son by Christopher Columbus, is the largest island in the Caribbean (Wilkinson 2019). The native Indians had named the island, Cubanacan which was slowly shorten to Cuba by the Spanish (Wilkinson 2019). The Republic of Cuba, with a population of over 11.1 million people, is located 90 miles south of Key West, Florida (CIA World Factbook 2019). The Capital, Havana, is the largest city. Fidel Castro was president of this communist country from 1959 until 2008, at which time he resigned due to health problems, handing control to his younger brother. Cuba is a multiracial society, with a population of primarily Spanish and African origins; other significant ethnic groups include Chinese, Haitians, and Eastern Europeans (CIA World Factbook 2019).

The distinctive Cuban culture is evidenced by their music, dance, and art. Cubans have made a number of popular dances including the rumba, the cha-cha, the guaracha, the bolero, and the conga. The classical ballerina, Alicia Alonso, was a Cuban dancer famous for, among other things, her portrayal in the ballet *Carmen*. The film *Fresa y Chocolate* (*Strawberries and Chocolate*) won the Silver Bear Award at the Berlin Film Festival in February 1995 (Cultural Orientation Resource Center 2002).

The experience of Cubans in their homeland and in the US and elsewhere is distinct from other Hispanic groups. The history and culture of Cuba and the Cuban people have been heavily influenced by Spain, the US, the Soviet Union, and

This chapter is an update of a chapter in a previous edition written by Larry Purnell and Jorge Gil.

J. A. Valdes (✉) · V. Delgado
Florida International University, Miami, FL, USA
e-mail: jvalde@fiu.edu

© Springer Nature Switzerland AG 2021

L. D. Purnell, E. A. Fenkl (eds.), *Textbook for Transcultural Health Care: A Population Approach*,
https://doi.org/10.1007/978-3-030-51399-3_12

through the slave trade in Cuba's sugar industry by West African groups such as the Yoruba.

Cuba was under Spanish control from 1511 until 1898, making it one of Spain's last colonies in the New World. Control of the sugar industry by Spanish *peninsulares* (individuals born in Spain) was challenged by the growing class of *criollo* landowners (individuals of Spanish ancestry born in Cuba) and the *independentista* movement. This absentee ownership created political turmoil and social imbalances that gave rise to the Cuban national character. The mistrust of government reinforced a strong personalistic tradition resulting in a sense of national identity evolving from family and interpersonal relationships (Szapocznik and Hernandez 1988).

Under the Cuban Adjustment Act of 1966, Cubans were welcomed by the U.S. government and were provided with support from the Cuban Refugee Program begun by the Kennedy administration. There is a common feeling of thankfulness and appreciation among newer Cuban generations in the US and also from the first cohort of Cuban immigrants who arrived in the early 1960s. Cubans in the US are a strong presence, not only economically but also politically. Overwhelmingly, Cuban Americans tend to be conservative, Republican, and anti-Communist, although that trend appears to be reversing among the younger American-born of Cuban descent population. They have demonstrated high voter turnout and tend to vote in blocs during local and national elections (National Council of la Raza 2011).

Cubans have managed to adjust to the mainstream culture while remaining close to their Cuban roots. However, young adults and adolescents who were educated in Cuba with strict Communist ideation and who emigrated with their parents may find the clash in values between Cuba and their new country confusing and negative. The bicultural Cuban American population can help in their adjustment. Many Cubans outside Cuba possess a strong ethnic identity, speak Spanish, and adhere to traditional Cuban values and practices at home while working in the dominant culture of their new homeland.

In 2008, Cuba began to accepted new foreign investments and begun to allow citizens to own property. In 2014, President Barrack Obama, with the help of the Vatican, began a process of communication which culminated with a US—Cuban thaw in 2016. The US re-opened an embassy on the island nation in 2015; the embassy had been closed since 1961 (Alvarez 2017).

12.2.2 Cuban Economy

From the late 1800s to mid-1900s, Cuba was considered one of the most prosperous countries in Latin America. The economy was based on treaties mainly with US, France, and Spain. Taxes were collected from people at high rates compared with those of countries from the "first world." Since the 1959 Communist Revolution the island has based its economy on subsidies from communist countries such as the former Soviet Union and China. Cuban economy today is primarily based on tourism and gastronomy. Travel to Cuba was simplified in 2016 when commercial flights resumed from the US to Cuba. Flights to Cuba and cruise lines to Cuba flourished from 2016 to 2019. However, the US again imposed travel restrictions in June of 2019, making it difficult to travel to the island (U.S. Embassy 2019).

12.2.3 Heritage and Residence

Ethnically, Cubans are 61.1 percent White, 24.8 percent mulatto or ***mestizo***, and 10.1 percent Black (CIA World Factbook 2018). The native Arawak Indian population inhabited the island when Columbus landed in 1492; most died from diseases brought by Spanish settlers in the nineteenth century. The Monroe Doctrine led to a special relationship between Cuba and the US. The US military controlled the island from 1898 to 1902. In 1902, Cuba was a politically independent capitalist state, although the US reserved the right to intervene in the preservation of Cuban independence; this was known as the Platt amendment (Alvarez 2017). During the U.S. prohibition; Cuba's economy flourished as casinos and many mobsters moved their operations to

Havana. By the 1950s many industries were owned by U.S. companies. Fulgencio Batista became president in 1940 with the promise to undo corruption and stand up to the perceived American influence. In 1959, Fidel Castro led a revolution and established a totalitarian Communist government, which still controls the country through the sole party, the Cuban Communist Party (PCC).

In the US, most Cuban Americans reside in four states: Florida, New Jersey, California, and New York. The largest proportion lives in Florida, especially in Miami-Dade County. The Cuban American population is aging, with a median age of 43.6 years and more than 20 percent over 65 years old. By comparison, Mexicans, Puerto Ricans, and Central and South Americans living in the US have median ages between 11 and 16 years younger than the average for Cuban Americans. The higher median age is explained by lower fertility rates of Cuban American women and the older age of those who immigrate from Cuba (Boswell 2002; Martinez 2002). About two-thirds of Cuban Americans residing in the US were born in Cuba, making this group a largely immigrant population.

Based on the 2010 U.S. Census, the estimated Cuban population in the US is almost 1.6 million, which is the third largest Hispanic population in this country (U.S. Census Bureau 2010). The major concentration of Cubans is located in Miami-Dade County, Florida, with 778,389 (U.S. Census Bureau 2010)—almost half of the total estimated number of Cuban Americans. The second concentration of Cubans is located in New York City, followed by Texas (U.S. Census Bureau 2010). Other states like South Dakota and Alaska have smaller Cuban populations (U.S. Census Bureau 2010).

12.2.4 Reasons for Migration and Associated Economic Factors

Most Cubans arrived on the U.S. mainland after the 1959 revolution where Fidel Castro came to power and changed the social, economic, and political landscape of Cuba. The desire for personal freedom, the hope of refuge and political exile, and the promise of economic opportunities have been the main reasons for Cuban immigration. During the early 1960s, some 14,000 Cuban children and teens were flown to the US without their parents through *Operation Pedro Pan*. Triggered by fears their children would be made wards of the state and forced to participate in counterrevolutionary activities, Cuban parents sent their children to the US. The *Pedro Pan* children were placed with foster families and relocated to different parts of the country; some never saw their parents again (Conde 1999). Although a number of children were eventually reunited with their parents, many suffered years of isolation and estrangement from their families (Portes and Bach (1985).

From 1990 to 2000, 191,506 Cuban immigrants entered the US (U.S. Immigration and Naturalization Service 2000). The term *balseros* was derived from *balsa* (raft), denoting the arrival of Cubans in the 1990s using homemade rafts. This wave of migration was preceded by deteriorating living conditions in Cuba with long electric power outages and chronic shortages of food and basic necessities. Of 35,000 *balseros* who were allowed by the Castro government to leave in 1994, only 30,000 were estimated to have arrived in the US. Many did not survive the crossing because of dehydration or boats that capsized. The most-celebrated case was of 5-year-old Elian Gonzalez, who was rescued floating on an inner tube after his mother and others perished when their boat capsized (Skaine 2004). From 2001 until the present, increasing numbers of Cubans have been immigrating by land through Mexico, Canada, Spain, and other countries (Skaine 2004). Two immigration accords signed by the US and Cuba set a limit of 20,000 visas annually for Cuban immigrants and stipulated that any illegal immigrants would be repatriated. In 1995 the US revised the 1966 Cuban Adjustment Act with the "wet foot, dry foot" policy. U.S. law enforced the wet-foot/dry-foot policy from 1995 to 2017. Under this policy if a Cuban refugee reached dry land in the United States, that individual would be awarded legal immigrant status. The policy ended as relations with Cuba changed dramatically in 2017 (Alvarez 2017).

In the four decades of Cuban immigration, significant change has been observed in the waves of immigrants, from the elite classes of the first stage, called the *golden exiles*, to the ***Marielitos*** of the sixth stage and the *balseros* of the 1990s. Each wave is distinct: the earliest waves of immigrants represented higher educational and economic status in Cuba than subsequent waves; the later groups were more representative of the Cuban population. The motivation for immigration also changed from the desire to escape political and religious persecution in the earlier waves to the hope for economic improvement in the later waves (Skaine 2004).

12.2.5 Educational Status and Occupations

The level of educational attainment of Cuban Americans is higher than some other Hispanic groups. About 22 percent of Cuban Americans are college graduates. The educational preparation of Cuban Americans is reflected in their median income, which is also higher than that of other Hispanic groups.

12.3 Communication

12.3.1 Dominant Language and Dialects

Language is often used as an index of assimilation of an immigrant group into the dominant culture. Virtually all first-generation Cubans in the US speak Spanish as their first language, although Cuban Spanish varies somewhat in choice of words and pronunciation from the Spanish spoken in Spain and Central and South America.

Some Cuban Americans consider English to be their dominant language, others consider Spanish to be their dominant language; yet others are completely bilingual. Because many Cubans live and transact business in Spanish-speaking ethnic enclaves, they have little need or motivation to learn English and are less likely to acculturate. Many, like one of the authors, speak ***Spanglish***, a mixture of Spanish and English with phrases such as *Have a buen dia; Hola, donde va today?* (Have a good day; Hello, where are you going?). The large number and variety of Spanish-language media, including newspapers, magazines, and radio programs, also reflect some Cuban immigrants' preference for Spanish over English. A stroll through Little Havana in Miami or Little Havana North along New Jersey's Union City–west New York corridor, as well as other places in the US, reveals that Spanish is reflected in billboard and poster advertisements.

12.3.2 Cultural Communication Patterns

Like other Hispanic groups, Cubans value ***simpatia*** and ***personalismo*** in their interactions with others. *Simpatia* refers to the need for smooth interpersonal relationships and is characterized by courtesy, respect, and the absence of harsh criticism or confrontation. *Personalismo* emphasizes intimate interpersonal relationships over impersonal bureaucratic relationships. ***Choteo***, a light hearted attitude with teasing, bantering, and exaggerating may often be observed in the way Cubans communicate with one another (Bernal 1994).

Conversations among Cubans are characterized by animated facial expressions, direct eye contact, hand gestures, and gesticulations. Voices tend to be loud and the rate of speech faster than may be observed with non-Cuban groups. Linguistically, the use of the second-person form *usted* to address older people and authority figures has fallen into disuse, replaced by the familiar form *tu*; although some older people prefer the formal use of language, especially in hierarchal realtionships such as with health professionals. The use of *tu* in interpersonal situations serves to reduce distance and promotes *personalismo*. Touching, in the form of handshakes or hugs, is acceptable among family, friends, and acquaintances. In the health-care setting, patients and family members may hug or kiss the health-care provider to express gratitude and appreciation.

Cubans feel a sense of "specialness" about themselves and their culture that may be conveyed in communication with others. This sense of specialness arises from pride in their unique culture, a fusion of European and African, the geopolitical importance of Cuba in relation to powerful countries in history, and the exceptional success they have achieved in adapting to their new environment. This sense of specialness, combined with the fast rate and loud volume of speech, may sometimes be interpreted as arrogance or grandiosity in a non-Cuban cultural context (Bernal 1994).

12.3.3 Temporal Relationships

Cubans tend to be present oriented compared with future-oriented European Americans. A greater emphasis is paid to current issues and problems than on projections into the future. In the clinical setting, health-care providers must realize Cuban patients tend to be motivated to seek help in response to crisis situations. Hence, visits to health-care providers for resolution of a crisis must be used as opportunities for teaching and promotion of personal growth.

Reflective Exercise

Pedro is a 12-year-old child who arrived from Cuba 6 months ago. He flew from the island with his parents and a 4-year-old sister. He now lives with his paternal grandparents in the Coral Gables area of Miami. He started school 1 week ago in a local public school after he came from Cuba. Pedrito, as he is called by his family and close friends, used to do very well in school. He had received several accommodations from his teachers, always had good grades, was very friendly and active socially, was always a team player, and had several awards in different sports.

For the past 4 or 5 months, he appears to be lonely and quiet. His mother approached him once and asked him what changed his attitude. Pedrito told her that he does not understand English, the teachers talk too fast, and when he asks questions, several children in the classroom laugh at him. He told his mother that he wants to go back to Cuba, hates his new country, and will no longer go to school.

1. Do you think Pedrito's behavior is a common pattern in all immigrant children during the process of acculturation? Explain your answer.
2. What are some strategies for the parents to take?
3. Describe three consequences of Pedrito's behavior for his future professional and personal development if his parents do not take early measures to help him.

Hora cubana (Cuban time) refers to a flexible time period that stretches from 1 to 2 h beyond the designated clock time. A Cuban understands that when a party starts at 8 p.m., the socially acceptable time to arrive is between 9 and 10 p.m. However, families who have acculturated to European American individualistic values may adhere to a more rigid clock time. When setting up appointments for clinic visits, the health-care provider must determine the patient's level of acculturation with respect to time and make arrangements for flexible scheduling, if necessary.

12.3.4 Format for Names

Modelled after Spanish and other Latin American societies, Cubans use two surnames representing the mother's and the father's sides of the family. For example, a woman may use the name Regina Morales Colon, indicating that her patrilineal surname is Morales and her matrilineal surname is Colon. When a Cuban woman marries, she adds *de* and her husband's name after her father's surname and drops her mother's surname. In the previous example, if Regina marries Mr. Ordonez, her name will be Regina Morales de Ordonez

(Skaine 2004). When addressing Cuban patients, especially the elderly, the health-care provider should use the formal rather than the familiar form, unless told otherwise. In the previous example, the appropriate appellation would be Señora Morales, or Mrs. Morales, instead of Regina.

Cubans translate English, Russian, or any other language in their own "Cuban way," and this is true for names as well. Some examples of common terms that have been adapted from other languages are *Naivy* (like U.S. Navy), *Yusimi* (You see me), and *Yeneisy* (Yeah, nein—German for "no" and "see").

12.4 Family Roles and Organization

12.4.1 Head of Household and Gender Roles

As among most Hispanic/Latino populations, family is the most important social unit among Cubans and Cuban Americans. The traditional Cuban family structure is patriarchal, characterized by a dominant and aggressive male and a passive, dependent female, although the more acculturated families have become more egalitarian. *La casa*, the house, is considered the province of the woman, and *la calle*, the street, the domain of the man. *La calle* includes everything outside the home, which is considered a proper testing ground for masculinity but dangerous and inappropriate for women. Traditionally, Cuban wives are expected to stay at home, manage the household, and care for the children. Husbands are expected to work, provide, and make major decisions for the family. However, with acculturation and more women working outside the home, egalitarian decision making prevails.

Cultural values acquired through four centuries of Spanish domination influence the behavior of Cuban men and women toward one another. The concept of ***honor*** is described as personal goodness or virtue, which can be lost or diminished by an immoral or unworthy act. Honor is maintained mainly by fulfilling family obligations and by treating others with ***respeto*** (respect). ***Verguenza***, a consciousness of public opinion and the judgment of the entire community is considered more important for women than men. ***Machismo*** dictates that men display physical strength, bravery, and virility.

Women's right to vote was granted in 1933 (History of Cuba 2019) In Cuba, the transition from an agricultural to an industrial economy, the rising educational attainment of women, the increased participation of women in the workforce, and the passage of the Family Code of 1975 resulted in more gender equality and parity between men and women with respect to marriage, divorce, property relations, and sharing household responsibilities (Skaine 2004). Women participate in government at higher rates than in the US' in 2013, 48.9% of parliamentary seats in the Cuban National Assembly where held by a Women (AAUW 2013).

Since the massive migration from Cuba in 1959, the traditional Cuban family has undergone a transition to a less male-dominated, less segregated, and more egalitarian structure. Cuban women who arrived in the United States were frequently the first in the family to find jobs and contribute to the survival of the family. According to Gallagher (1980), Cuban immigrant women were more receptive to life in the US, more flexible, and more readily hired for jobs than were men. Eventually, as their contributions to the family's economic well-being increased, women's' power to make decisions was enhanced. Cuban American women have the highest rate of labor participation when compared with all other groups of women in the US (Suarez 1993). Thus, contemporary Cuban families from the 1980s to the present may demonstrate greater gender equality in decision making for the family.

12.4.2 Prescriptive, Restrictive, and Taboo Practices for Children and Adolescents

Cuban parents tend to pamper and overprotect their children, showering them with love and attention. Among Cubans, the expectation is that children study and respect their parents and older people. Children are encouraged to acquire knowledge and learning *porque eso no te lo puede quitar nadie* (because no one can take that away from you) (Bernal 1994).

When a Cuban daughter reaches the age of 15 years, a ***quince***, a birthday party is typically held to celebrate this rite of passage. Socially, the *quince* is indicative of the young woman's readiness for courting by a *novio* (boyfriend). In Cuba, as among many families in the US, the *quince* is celebrated with food, music, and dancing among family and friends. Parents may save up for years to prepare for a daughter's *quince*, which has today evolved into a large, extravagant party.

Many Cuban adolescents may undergo an identity crisis, not knowing whether they are fully Cuban or American. During this time, they may reject traditional cultural values; parents may feel threatened when their authority is being challenged. The opposing values and demands of their Cuban heritage and new society create a potential for tension and conflict between Cuban adolescents and their parents. Some examples are the Cuban practice of chaperoning unmarried couples when they date. Unmarried daughters are expected to live at home with the family until they marry.

12.4.3 Family Goals and Priorities

Cubans have tightly knit nuclear families that allow for inclusion of relatives and ***padrinos*** (godparents). *La familia* (the family) is the most important source of emotional and physical support for its members. Extended, multigenerational households are common, with grandparents often being part of the nuclear family. Compared with other Hispanic ethnic or cultural groups, Cubans have the lowest proportion of families with children. Cubans also have the highest proportion of people aged 65 and older who live with their relatives. The high proportion of older people living with family members has led to the typical three-generation Cuban family (Perez 2002).

A system of personal relationships known as ***compadrazgo*** is also typical. A set of godparents, or ***compadres***, is selected for each child who is baptized and confirmed. *Compadres* tend to be close friends or relatives of the child's natural parents and may be counted on for moral or financial assistance. *Compadres* are usually considered part of the Cuban family, whether or not a true blood relationship exists.

In recent years, as Cubans have become more acculturated and as the children of Cuban immigrants have become more Americanized and more economically successful than their parents, family dynamics, expectations, and behaviors are changing. Multigenerational living arrangements are markedly declining, with increased numbers of older adults becoming more independent and living alone. Despite these trends, the need and desire for frequent family contact through daily telephone calls and frequent visits are still predominant. Although more likely now to be living alone, older Cuban adults have close interactions not only with their children but also with grandchildren, siblings, cousins, and other relatives (Martinez 2002).

12.4.4 Alternative Lifestyles

There is a high proportion of divorced women among Cuban Americans compared with other Hispanic and non-Hispanic groups. In spite of this, Cubans have the highest percentage of children under 18 years living with both parents, a low percentage of families headed by women with no husbands present, and the lowest rate of mothers and children living within a larger family unit. One explanation for these patterns may be that divorced Cuban women return to their parents' home, but because they typically have fewer children, they do not tend to be accompanied by children (Perez 2002).

In dealing with some Cuban, health-care providers may hear the term *Marielito* used in a derogatory manner to refer to the estimated 4 percent of the 125,000 Cubans who arrived during the Mariel boatlift. Little or no data are available on the occurrence of homosexuality among

Cuban Americans, although the gay lifestyle would be contradictory to the prevailing *machismo* orientation of Cuban culture. Same-sex couples living together may be alienated from their families, especially among first-generation Cubans who adhere closely to traditional gender roles and family values. Undoubtedly, gay and lesbian films such as *Gay Cuba*, *Strawberry and Chocolate*, and *La Carne de Rey* and the Miami Gay and Lesbian Film Festival are attempts at making alternative lifestyles more acceptable. Given the stigma associated with homosexuality, a matter-of-fact, non-judgmental approach must be used by health-care providers when questioning Cuban patients regarding sexual orientation or sexual practices. LGBT rights have improved in the last 50 years, prior to 1979 being gay was a crime in Cuba. Today the government has a state run National Center for Sex Education (CENESEX) as a voice advocating for LGBT community (Rohrlich 2014) Progress of LGBT rights is slow due to the *machismo* dogma and year of abuse by the Castro regime whom at one point sent homosexuals to labor camps.

Reflective Exercise

Alberto Gonzaga is a 43-year-old Cuban American male. He migrated 18 years ago with his wife and 3-year-old son, Alberto. Before he emigrated to the US, Mr. Gonzaga was imprisoned in Cuba for political reasons. He did not complete the 20-year sentence imposed by the Castro regime and was released from jail after 12 years for good behavior. He was immediately granted a U.S. visa for himself and his family. Since his arrival, he has been an active member of the Republican Party and has participated in the local Miami area in the anticommunist movement. His son is now 21, and he still lives with his parents, helping them financially. He works full-time and pursues a law degree in a local college. There have been several confrontations between Mr. Gonzaga and his son. The new, more liberal "open" era has brought Cuban musicians from the island to perform in public concerts in Miami. Mr. Gonzaga is totally opposed to this. He claims that "these Communist musicians will take our money and our taxes and give it to Castro" and prohibits his son from going to the concerts. Alberto states that music has nothing to do with politics and that many of these musicians are opposed to the regime, but this is a way to travel outside Cuba and earn some money. He confronts his father, telling him that he is an adult and he will go to the concerts, despite his father's opposition.

1. How different is the Cuban population that migrated 20 years ago compared with those who arrived 5 or 6 years ago?
2. Are the children of the Cubans who arrived in the 1980s maintaining their traditions? Is their way of thinking the same as their parents?
3. What could be the consequences of the confrontations between Mr. Gonzaga and his son?
4. Is the Castro government still separating the family even outside Cuba, or is this situation just a matter of character?

12.5 Workforce Issues

12.5.1 Culture in the Workplace

Cubans have enjoyed enormous economic success in the US. Twenty percent of first-generation Cuban Americans and 43 percent of second-generation Cuban Americans are college graduates. The high educational achievement is reflected in the large proportions—53 percent for first-generation Cuban Americans and 75 percent for second-generation Cuban Americans—who are employed in managerial and technical jobs, the two highest-paying occupational categories (Boswell 2002). Their strong entrepreneurial abilities tend to be concentrated in construction, transportation, textiles, wholesale, and retail trades. The existence of several Cuban ethnic enclaves with a familiar language and culture has created numerous employment opportunities for recent Cuban immigrants.

A frequent source of tension in the workplace is the tendency of Cubans to speak Spanish with other Cuban or Hispanic co-workers. Speaking the same language allows them to form a common bond, relieve anxieties at work, and feel comfortable with one another. In Blank and Slipp (1994) study, one Cuban supervisor asserted, "Others should know that we tend to go back and forth in language—Spanish when we're talking personally and English when it's professional."

12.5.2 Issues Related to Autonomy

Traditional Cubans tend to be hierarchical in their relationships, recognizing supervisors or superiors as authority figures and treating them with respect and deference. In mainstream American culture, collegial relationships, in which workers can exercise initiative, question the supervisor, and participate in decision making, may make Cubans uncomfortable. Cubans value a structure characterized by *personalismo*—that is, one that is oriented around people rather than around concepts or ideas. For Cubans, personal relationships at work are considered an extension of family relationships. Cuban workers may function best in a working environment that is warm and friendly and fosters *personalismo*. Because of the emphasis on the job or task in the individualistic workplace, many Cubans view this workplace as being too individualistic, business-like, and detached. In the past, the language barrier may have insulated Cuban Americans from the dominant culture, retarded acculturation, and fostered some interethnic tensions. As the ability to speak English and acculturation increases, Cuban Americans have fewer interethnic tensions.

12.6 Biocultural Ecology

12.6.1 Skin Color and Other Biological Variations

Most Cuban Americans are White. Because of their predominantly European ancestry, Cuban Americans have skin, hair, and eye colors that vary from light to dark. A minority, who are of African Cuban extraction, are dark-skinned and may have physical features similar to those of African Americans.

Reflective Exercise

Pablo Perez is a 42-year-old Cuban immigrant. He arrived in the United States 2 years ago with his wife and a 10-year-old daughter. He was a successful physician in Cuba. In 1991, he was selected to travel outside the island to Spain to assist in an international congress. He was chosen to provide medical services in Senegal in 2002. Both times, he returned to Cuba—the first time because his mother was in bad health (she died a short time later), and the second time because he had a wife and daughter waiting for him. In 2006, he and his wife were selected to go to Venezuela. After 2 years there where he demonstrated an excellent professional and communist attitude, the consulate allowed their daughter to join them.

This was the chance Dr. Perez was waiting for. He and his family immediately headed north, crossed the Mexican border into the US, and settled in Las Vegas. Nevada. Both Dr. and Mrs. Perez each have two full-time jobs. They have had problems learning the language. Dr. Perez has taken the medical boards twice, but he failed both times. He has told his wife that if they moved to Miami, it will be better for them because there are more Hispanic people there, the weather is better, and the support system is greater.

1. Do you think that Dr. and Mrs. Perez's behavior in Venezuela is a common pattern among Cuban professionals who are trying to get into the US?
2. Compare and contrast the Cuban health system with that of the US?
3. What options do Dr. and Mrs. Perez have for joining the health-care field again in the US?
4. Do you think that moving to Miami will solve Dr. and Mrs. Perez's problems?

12.6.2 Diseases and Health Conditions

Nath's (2005) analysis of data from the Hispanic Health and Nutrition Examination Survey (HHANES) reported that among Cuban Americans, major health conditions include a high prevalence of coronary heart disease, hypertension, overweight or obesity, type 2 diabetes mellitus, and depression. Twenty-nine percent of Cuban American men and 34 percent of Cuban American women are overweight according to actuarial tables.

In a comparison of hypertension-related mortality among Hispanic groups, Cuban Americans were found to have the lowest death rate and Puerto Ricans had the highest death rate. In addition, age-standardized hypertension-related mortality rates in Cuban Americans were 39 percent lower than those for non-Hispanic whites (Centers for Disease Control and Prevention 2006).

12.6.3 Variations in Drug Metabolism

Although some studies have reported differences in drug metabolism among Hispanics, little or no data specific to Cuban Americans are available.

12.7 High-Risk Behaviors

Devieux et al. (2005) at Florida International University conducted an assessment of HIV risk behaviors of adolescents participating in an HIV risk-reduction intervention. Of the 137 participants in the interview, 81 were African American teens and 57 were Cuban American teens. Cuban American teens reported more unprotected sex acts and more anal sex acts in the 6 months prior to the interview than did African American teens. The groups were similar on the total number of sexual partners and sex acts reported. Regarding drug use, a greater proportion of Cuban American teens reported using drugs in the prior 6 months than did African American teens, and more Cuban American teens reported engaging in unprotected sex while using drugs than did African American teens. The authors speculate that higher acculturation of Cuban American teens and accompanying family conflict may account for the relatively riskier behaviors among Cuban American teens. More research is needed to clarify the processes leading teens of different backgrounds to initiate and maintain risky behaviors and to identify the most effective ways to intervene to reduce risk (Devieux et al. 2005).

The HHANES findings also revealed that drinking alcohol was significantly more common among Cuban males than females and among younger versus older Cuban groups, a pattern that was similar to that in Mexicans and Puerto Ricans. Among middle-aged and older Cuban males who tend to be relatively well educated and have higher incomes compared with the younger, more recent Cuban immigrants, control of intoxication is important. Among Cuban women, the proportion of lifelong abstainers increased significantly from the younger to the older groups (Black and Markides 1994).

Smoking is responsible for 87 percent of the lung cancer deaths in the US. Overall, lung cancer is the leading cause of cancer deaths among Hispanics. Lung cancer deaths are about three times higher for Hispanic men (23.1 per 100,000) than for Hispanic women (7.7 per 100,000).

12.7.1 Health-Care Practices

An obstacle to good nutritional practices is the Cuban cultural perspective of the "healthy body." A healthy and beautiful Cuban infant is fat. Even among adults, a little heaviness is considered attractive. *Que gordo estas*! (How fat you are!) is considered a compliment. The traditional Cuban diet—high in calories, starches, and saturated fats—predisposes individuals to the development of obesity. In Cuba, health care is viewed as a basic human right and occupies a prominent place in the Cuban government's domestic and foreign policies. Polyclinics in communities are the basic unit of health care. Physician–nurse teams attend patients in these polyclinics, as well as in the home, school, day-care center, and the workplace.

In the United States, Cubans exhibit high levels of preventive health behaviors, as evidenced by routine physical examinations within the last 2 years. The utilization of preventive services was usually associated with accessibility, which, in turn, was significantly influenced by education, annual income, and age (Solis et al. 1990).

Lopez and Masse (1993) found that unmarried Cuban American women who had little recreational activity tended to have a higher mean weight. Body mass in Cuban American women was not significantly associated with income (Lopez and Masse 1993).

12.8 Nutrition

12.8.1 Meaning of Food

Besides satisfying hunger, food has a powerful social meaning among Cubans, allowing families to reaffirm kinship ties, promote a sense of community, and perpetuate their customs and heritage. To grasp this fully, one needs only to observe multigenerational families assembled for dinner on a Saturday or Sunday evening in a Cuban restaurant in Miami's Little Havana or Cuban friends sharing a cup of *cafe cubano* and *pastelitos* at a stand-up sidewalk counter. In Miami alone, the demand for Cuban food and food products has resulted in the establishment of about 400 Latin restaurants, mostly Cuban, and some 700 *bodegas*, or grocery stores. Other Cuban enclaves paint a similar picture.

12.8.2 Common Foods and Food Rituals

Cuban foods reflect the environmental influences of Cuba's tropical climate and agriculture, the historical influences of Spanish colonial rule, the African slave trade, and the Arawak Indians' cultivation methods. Typical staple foods are root crops like yams, yucca, malanga, and boniato; plantains; and grains. Traditional Spanish dishes like *arroz con pollo* and *paella* are frequently served. Many dishes are prepared with olive oil, garlic, tomato sauce, vinegar, wine, lime juice (called *sofrito*), and spices. Meat is usually marinated in lemon, lime, sour orange, or grapefruit juice before cooking (Kittler & Sucher, 2008).

The main course in Cuban meals is meat, usually pork or chicken. Some popular entrees are roast pork (*lechon*), fried pork chunks (*masas de puerco*), sirloin steak (*palomilla*), shredded beef (*ropa vieja*), pot roast (*boliche*), and roasted chicken (*pollo asado*). A roasted suckling pig is traditionally served on Christmas Eve, New Year's Day, and other festive celebrations. Black beans are prepared with a sauce containing fat, pork, and spices. Ripe plantains (*platanos maduros*) or green plantains (*platanos verdes*) are served fried. Fried green plantains (*tostones* or *mariquita*) may also be smashed between a brown paper bag and the fist (*un cartucho y el puno*), giving them the familiar name *platanos a punetazo*. Desserts are rich and very sweet, such as custard (*flan*), egg pudding (*natilla*), rice pudding (*arroz con leche*), coconut pudding (*pudin de coco*), or bread pudding (*pudin de pan*) (Kittler & Sucher, 2008).

Beverages may include sugar cane juice (*guarapo*), iced coconut milk (*coco frio*), milkshakes (*batidos*), Cuban soft drinks such as Iron Beer or Materva, sangria, or beer. The strong and bittersweet coffee called *cafe cubano* is a standard drink after meals and throughout the day, whether at home, in restaurants, or in other social situations. In the US, Cubans may drink the *cafe cubano* as *cortadito* or with a dash of milk to cut the strength and bittersweet taste. A traditional Cuban meal includes a generous helping of white rice with black beans or black bean soup, fried plantains, roasted pork or fried chicken, a tuber such as malanga or yucca, followed by dessert and espresso. Thus, the typical diet is high in calories, starches, and saturated fats. A leisurely noon meal (*almuerzo*) and a late evening dinner (*comida*), sometimes as late as 10 or 11 p.m., are customary.

12.8.3 Nutritional Deficiencies and Food Limitations

The major food groups are well represented in the Cuban diet; however, leafy green vegetables may be lacking in the average Cuban meal. Therefore, when assessing the nutritional adequacy of a Cuban patient's diet, the health-care provider must ensure sufficient fiber content.

12.9 Pregnancy and Childbearing Practices

12.9.1 Fertility Practices and Views Toward Pregnancy

The low fertility rate of Cuban women, which is consistent in every maternal age group, has been attributed to three factors (Perez 2002):

1. Cuban American women have a high rate of labor force participation.
2. Before the revolution, Cuba had the lowest birth rate in Latin America.
3. Cuba's current reproductive rate is among the lowest in the developing world.

In an analysis of HHANES data, Stroup-Benham and Trevino (1991) found that only 9 percent of Cuban American women took oral contraceptives. In the same study, hysterectomies, oophorectomies, and tubal ligations were not common among Cuban American women. Based on these data, Cuban American women appear to be at greatest risk for unintended pregnancies. Many Cuban folk beliefs and practices surround pregnancy. For example, some Cuban women believe that they have to eat for two during the pregnancy and end up gaining excessive weight. Some believe that morning sickness is cured by eating coffee grounds, that eating a lot of fruit ensures that the baby will be born with a smooth complexion, and that wearing necklaces during pregnancy causes the umbilical cord to be wrapped around the baby's neck.

Among Cubans, childbirth is a time for celebration. Family members and friends congregate in the hospital, awaiting the delivery of the baby. Although traditionally it was not acceptable for men to attend the birth of their children, the younger and more acculturated fathers tend to be present to support their wives during labor and delivery. In the postpartum period, it is believed that ambulation, exposure to cold, and going barefoot place the mother at risk for infection. Because of this, family and relatives often care for the mother and baby for about 4 weeks postpartum.

12.9.2 Prescriptive, Restrictive, and Taboo Practices in the Childbearing Family

Cuban Americans participate in prenatal care if it is affordable. Rest is encouraged. Abstaining from strenuous activities and loud noises is recommended. Fresh fruits are encouraged for the health of the mother and the fetus. More acculturated fathers participate in prenatal classes and support the mother in the delivery room. Breastfeeding among Cuban women is becoming more popular than in the past. Most do a combination of breast- and bottle-feeding (Varela 2005).

Thomas and DeSantis (1995) related the early introduction of solid foods and prolonged bottle-feeding of Cuban children to the traditional Cuban beliefs that "a fat child is a healthy child" and that breastfeeding may contribute to a deformity or asymmetry of the breasts. In the same study, 97 percent of Cuban mothers indicated that they administer vitamin preparations to promote the healthy development of their children. Cuban mothers also used advice about child health given by their spouses, mothers, mothers-in-law, and clerks and pharmacists who sold them over-the-counter drugs (Thomas and DeSantis 1995).

Traditionally, postpartum mothers and their infants are not supposed to leave the house for 41 days. This initial postpartum period is a time for mothers to rest and devote their energies to caring for the baby. The new mother's immediate family—mother and sisters—help care for the new mother and baby. The mother is sheltered from bad news and any stress that could harm her or her baby. She is also encouraged to eat more to foster milk production (Varela 2005).

12.10 Death Rituals

12.10.1 Death Rituals and Expectations

In death, as in life, the support of the extended family network is important. Whether in the hospital or at home, the dying person is typically surrounded by a large gathering of relatives and friends. In Catholic families, individual and group prayers are offered for the dying to provide a peaceful passage to the hereafter. Religious artifacts such as rosary beads, crucifixes, and ***estampitas*** (little statues of saints) are placed in the dying person's room.

Depending on the dying person's religious beliefs, a Catholic priest, a Protestant minister, a rabbi, or a **santero** may be summoned to the deathbed to perform appropriate death rites. For adherents of **Santería**, death rites may include animal sacrifice, chants, and ceremonial gestures. Health-care providers need to be open-minded and responsive to both the physical and the psychosocial needs of the dying and the bereaved and, regardless of religious beliefs, accord them the utmost respect and privacy.

After a person's death, candles are lighted to illuminate the path of the spirit to the afterlife. A wake, or ***velorio***, is usually held at a funeral parlor, where friends and relatives gather to support the bereaved family. The wake lasts for 2–3 days until the funeral. Burial in a cemetery is the common practice for Cuban Catholics, although some may choose cremation.

12.10.2 Responses to Death and Grief

Bereavement is expressed openly among Cubans, with loud crying and other physical manifestations of grief considered socially acceptable. Death is an occasion for relatives living far away to visit and commiserate with the bereaved family. Women from the immediate family usually dress in black during the period of mourning. Visitors make offerings of candles and floral wreaths (*coronas*), provide assistance with household chores, and attend to visitors or funeral arrangements. Cubans customarily remember and honor the deceased on their birthdays or death anniversaries by lighting candles, offering prayers or masses, bringing flowers to the grave, or gathering with family members at the grave site.

12.11 Spirituality

12.11.1 Dominant Religion and Use of Prayer

Most Cubans are Roman Catholics followed by Protestants, Jews, and believers in the African Cuban practice of *Santería*. The original habitants in Cuba were the Guanatayabes Indians, located mainly in the center-west area of the island, and the Taino Indians, mainly in the east side. When Spaniards arrived, they not only abused and killed the Indians, but they also imposed their Catholic religion. This religion continued in Cuba for many years and was combined with some Christian practices during the colonization period.

The Roman Catholic Church has been an important source of support, especially for first-generation Cuban immigrants. A number of predominantly Cuban parishes with Cuban clergy are located in cities where large Cuban populations reside. The Roman Catholic Church has exerted an important influence on Cuban families by providing educational opportunities in Catholic schools. Many Cuban parents, especially the upper middle class, prefer to have their children educated in private Catholic schools.

Roman Catholicism as practiced by Cubans is personal rather than institutional in nature. The religious practice of Cuban Catholics is characterized by devotion and intimate, confiding relationships with the Virgin Mary, Jesus, and the saints.

Some families may have shrines dedicated to *La Caridad del Cobre* (the patron saint of Cuba) or other saints at the entrance to their homes, in their yards, or in commercial establishments. The three favorite saints are Santa Barbara, San Lazaro, and *La Caridad del Cobre*. Inside the home, crucifixes and pictures or statues depicting images of saints may be found. When someone is ill, small pictures of saints, called ***estampitas***, may be placed under the pillow or at the sick person's bedside.

Significant religious holidays for Cubans include Christmas, *Los Tres Reyes Magos* (Three Kings' Day), and the festivals of the *La Caridad del Cobre* and Santa Barbara. The Cuban community usually celebrates the feast of *La Caridad del Cobre* (September 8) by transporting the statue of the patron saint on a boat to a specific location, where a mass is held in her honor. Cuban families also celebrate Christmas Eve (*Noche Buena*) with a traditional Cuban meal. Typically, a pig is cooked all day in a wooden box lined in metal (*una caja china*) and set in the backyard. The pig is placed at the bottom of the box and is covered with charcoal. The meat is served with black beans and rice, yucca, and *turones* (Spanish dessert). The evening concludes with the family attending Midnight Mass (*Misa de Gallo*).

With the arrival of slaves from Africa in the late 1700s and early 1800s, a new type of religious practice emerged in Cuba. One group from the Bantu tribe in the Congo were called *palos* (sticks) by the Spaniards because they used sticks for their religious practices. Today, the *paleros* (plural for people who practice the religion with sticks and who perform black magic) are viewed in a negative way, representing a "bad" type of African Cuban religion. Another group of slaves came from the Carabali tribe from South Nigeria (also known as the Abakua tribe). They still exist today, but they do not have a negative reputation like the *paleros*. In addition, other groups of slaves mixed with the *criollos* (native Cubans born from Spaniards and Indians), Spaniards, French, and other ethnicities who were already residing in Cuba in the mid-1800s. Because of this extensive mixture of races and ethnicities, Cubans say that *en Cuba el que no tiene de Congo tiene de Carabali* (in Cuba if you don't have it from Congo, you will have it from Carabali), denoting that wonderful combination of cultures.

Another African Cuban religion is the *Yoruba/Lucumi*. Yoruba is an African dialect also known as Lucumi. The minister is known as "Olorisha" or owner of Orisa or Orisha, the saints. The different types of Orishas are Eleggua, Ogun, Oshun, Babalu-Aye, Chango, Oya, Obatala, Yemaya, and Orula. When this priest initiates other priests, they are known as *babalorishas* (the father of Orishas) or *Iyalorishas* (the mother of Orishas). The Supreme Priest is known as *Ifa*, the Father who knows the secret. *Ifas* are commonly known among the general population as *Babalaos* or, correctly said, *Babalawo*. They are the most widely seen *santeros* in Cuba and the ones from whom people seek help with their health or a better economy.

Santeria, or *Regla de Ocha*, is a 300-year-old African Cuban religious system that combines elements of Roman Catholicism with ancient Yoruba tribal beliefs and practices. *Santeria* originated among the Yoruba people of Nigeria, who brought their beliefs with them when they arrived in the New World as slaves. As a condition of their entry into the West Indies, slaves were required to be baptized as Roman Catholics (Perez y Pena 1998). In the process of adapting to their new non-African environment, the slaves altered their beliefs to incorporate those of their predominantly Catholic masters. *Santeria* evolved from two main cultural antecedents: the worship of the *orishas* among the Yoruba tribe of Nigeria and the cult of saints from the Roman Catholicism of Spain. Through their exposure to the Catholic religion, the slaves came to associate their African gods, called *orishas*, with the Roman Catholic saints, or *santos*. The worship of the *orishas* and the associated beliefs, rituals, incantations, magic, and spirit possession are central to *Santeria*.

Table 12.1 displays the seven African powers, or main *orishas* (Martinez and Wetli 1982). The Yoruba deity of fire and thunder, called *Chango*, became identified with Santa Barbara, the patron saint of the Spanish artillery, who appeared in Catholic lithographs in red, the color of the *orisha* (Sandoval 1979). *Chango*, the most popular god in *Santeria*, controls thunder, violent storms, lightning, and fire. The six other *orishas*, the Catholic saints with whom they are identified, and their corresponding functions and powers are also shown in Table 12.1.

When people decide to practice *Santeria*, their *orishas* become known to them and must be worshipped throughout their lives. Followers of *Santeria* believe in the magical and medicinal properties of flowers, herbs, weeds, twigs, and leaves. Sweet herbs such as *manzanilla*, *verbena*, and *mejorana* are used for attracting good luck, love, money, and prosperity. Bitter herbs such as *apasote*, *zarzaparilla*, and *yerba bruja* are used to banish evil and negative energies. Adherents of *Santeria* also believe in the power of consecrated objects such as stones (*otanes*) in which the *orishas* reside. Necklaces, bracelets, and charms may be given by *santeros* to their patients to protect them from evil and strengthen their well-being.

Sacrifice, or *ebo* (pronounced "egbo" or "igbo"), is a central ritual in *Santeria*. The main purpose of *ebo* is to establish communication between the spirits and human beings. The initiation of a *santero* involves the sacrifice of a four-legged animal and a series of rites lasting 7 days. Transition through major life events such as birth, death, and marriage require ritual sacrifices to appease the gods and solicit their support.

Sacrificial objects in *Santeria* include plants, foods, and animals. Plants and foods include

Table 12.1 Seven African powers or main *Orishas*

Orisha	Christian saint	Function/power	Punishment	Propitiation
Eleggua	Holy Child of Atocha	Guardian of entrances, roads, and paths; Trickster	Blindness, paralysis, and birth deformities	Blood of goats; black rooster; smoked fish; smoked junia; yams; sugar cane
Obatala	Our Lady of Mercy	Father of all human beings; gives advice; is source of energy, wisdom, purity, and peace	Death, suicide by fire	White pigeons; white canaries; female goat; plums; yam puree
Chango	Saint Barbara	Warrior deity; controls thunder and violent storms, lightning, and fire	Abdominal distress, social and domestic strife	Roosters; goats; lambs; apples; bananas
Oshun	Our Lady of Charity	Deity that controls money and love, makes marriages, protects genitals	Respiratory distress	Female goat; white chickens; sheep; honey
Yemaya	Our Lady of Regla	Primary mother of the Santos, protects womanhood, owns seas	Leprosy, gangrene, skin diseases	Ducks; lambs; female goats; watermelons; black-eyed peas
Babaluaye	Saint Lazarus	Patron of the sick, especially diseases of the skin	Violent death (such as an automobile accident)	Spotted rooster; snakes; cigars; pennies; glasses of water
Ogun	Saint Peter	Warrior deity, owns all metals and weapons		Blood and feathers; young bulls; roosters; steel knife; railroad tracks

Source: Adapted from Martinez, R., & Wetli, C. (1982). Santeria: A magicoreligious system of Afro-Cuban origin. *American Journal of Social Psychiatry*, 2(3), 34

plantains, malanga, yam, okra, flour, gourds, and ground black-eyed peas wrapped in plantain leaves. Animals used for sacrifice, such as hens, birds, lambs, or goats, are killed by wringing the head or severing the carotid arteries with a knife. The animal's blood is offered as a type of communion with the deities. In 1993, the Supreme Court struck down anti–animal sacrifice laws in Hialeah, Florida and recognized the right of a *Santeria* sanctuary, the Church of Chango Eyife, to offer an animal sacrifice as a religious sacrament (Gonzalez 1995).

Santeria, viewed as a link to the past, is used among Cubans to cope with physical and emotional problems. When someone is sick, that person's physical complaints may be diagnosed and treated by a physician, but the *santero* may be summoned to assist in balancing and neutralizing the various aspects of the illness. *Santeria* is actively practiced in Miami, New York, New Jersey, and California where Cubans reside. Sanitaria is also practiced in other parts of the world where Cubans migrate.

In eliciting a complete history from patients, health-care providers must include information regarding the type of religion being practiced, if any. Patients' religious beliefs and practices must be viewed in an open, sincere, and nonjudgmental manner. In the hospital setting, maintaining privacy is important if patients and families need to perform certain rituals or prayers. A visit from a priest, rabbi, or *santero* may provide a sense of psychological support and spiritual well-being. At times, *santeros* have been known to make sacrificial offerings at the patient's hospital bedside. As long as standards of safety and sanitation are maintained, families must be allowed space and privacy to be able to engage in specific religious ceremonies.

In Cuba today, four decades of Fidel Castro's revolution have significantly affected religious beliefs and practices. Only about 30–40 percent of Cubans are Catholic, whereas *Santeria* has about 55–60 percent adherents. The multiple groups that follow syncretic *Santeria* practices include *Abakua*, *Yoruba*, *Regla Conga*, *Regla Ocha*, *Regla Arara*, *Regla Arada*, and *Yebbe*. Thus, compared with their peers from previous migration waves, recent Cuban immigrants may be less likely to be Catholic. Further, large numbers of Cubans consider themselves adherents of Catholicism and *Santeria* simultaneously (Ramos 2002).

12.11.2 Meaning of Life and Individual Sources of Strength

As in other Latin American communities, the family is the most important source of strength, identity, and emotional security. Cubans usually rely on a network of family members and relatives for assistance in times of need. The sense of specialness Cubans feel, stemming from pride in their culture and their remarkable success in adapting to their new country, is likewise a source of self-esteem and self-identity. For many Cubans, deeply held religious beliefs have provided guidance and strength during the long and difficult process of migration and adaptation and continue to play an important role in their day-to-day lives.

12.11.3 Spiritual Beliefs and Health-Care Practices

Many Cubans tend to be fatalistic, feeling that they lack control over circumstances influencing their lives. The belief in a higher power is evident in a variety of practices—such as using magical herbs, special prayers or chants, ritual cleansing, and sacrificial offerings—that Cubans may engage in for the purpose of maintaining health and well-being or curing illness.

When Cuban patients consult health-care providers, in all likelihood they have already tried some folk remedies primarily advised by older women in their family or obtained from a *botanica*. Most folk remedies are harmless and do not interfere with biomedical treatment. In most cases, patients may be encouraged to continue using these remedies, such as herbal teas.

Encourage patients to report the use of specific teas and herbs. For example, chamomile tea may increase bleeding time, while jaborandi may decrease bleeding time. Other teas and herbs may increase or decrease glucose metabolism. Health-care providers should be alert to the frequent practice of sharing prescription medications in families and among relatives. A family member who found an antibiotic effective in curing an ailment may share the medication with another relative suffering from the same symptoms. The health history must always include assessment of past or present medication use, whether traditional, over-the-counter, or prescription. Appropriate explanations must be given regarding the actions and adverse effects of drugs and the reasons why they cannot be shared with other family members.

12.12 Health-Care Practices

12.12.1 Health-Seeking Beliefs and Behaviors

Cubans rely on the family as the primary source of health advice. Typically, older women in the family are sought for information such as traditional home remedies for common ailments. Herbal teas or mixtures may be prepared to relieve mild or moderate symptoms. Concurrently or alternatively, a *santero* may be consulted, or a trip to the botanica may be warranted to obtain treatment.

Socialized into a strong health ideology and successful primary-care system in Cuba, Cubans are able to use biomedical services as primary or secondary sources of care. Cuba has a regionalized, hierarchically organized, national health system that provides universal coverage and standardization of services. An innovative family practice program assigns physicians and nurses to city blocks and remote communities to promote physical fitness, detect risk factors for disease, and cure disease. In the US, many Cuban clinics have evolved into health maintenance organizations (HMOs).

12.12.2 Responsibility for Health Care

Most Cuban access the health-care system for preventive care and health screenings. Cubans with a more recent history of immigration to are accustomed to preventive health activities as part of the Cuban governmental services under Castro. Practices for healthy living, including avoiding stress and bad news and avoiding extremes of hot and cold are important for health maintenance. Most take full advantage of vaccinations.

12.12.3 Folk and Traditional Practices

Cubans may use traditional medicinal plants in the form of teas, potions, salves, or poultices. As noted above, in Cuban communities like Little Havana in Miami, stores called **botanicas** sell a variety of herbs, ointments, oils, powders, incenses, and religious figurines to relieve maladies, bring luck, drive away evil spirits, or break curses. In addition, *Santeria* necklaces and animals used for ritual sacrifice are available at botanicas.

Herbal teas that may be used to treat common ailments include the following:

Cosimiento de anis (anise): to relieve stomachaches, flatulence, and baby colic; and to calm nerves.

Cosimiento de limon con miel de abeja (lemon and honey): to relieve cough and respiratory congestion.

Cosimiento de apasote (pumpkin seed): to treat gastrointestinal worms.

Cosimiento de canela (cinnamon): to relieve cough, respiratory congestion, and menstrual cramps.

Cosimiento de manzanilla (chamomile): to relieve stomachaches.

Cosimiento de naranja agria (sour orange): to relieve cough and respiratory congestion.

Cosimiento de savila (aloe vera): to relieve stomachaches.

Cosimiento de tilo (linden leaves): to calm nerves.

Cosimiento de yerba buena (spearmint leaves): to relieve stomachaches and calm nerves.

Chamomile tea: to calm nerves and calm babies with colic.

Fruits and vegetables, abundant in the natural tropical environment of Cuba, may include the following:

Chayote (vegetable): to calm nerves.

Zanaoria (carrots): to help problems with vision.

Toronja y ajo (grapefruit and garlic): to lower blood pressure.

Papaya y toronja y pina (papaya, grapefruit, and pineapple): to eliminate gastrointestinal parasites.

Remolacha (beets): to treat influenza and anemia.

Cascara de mandarina (fruit): to relieve cough.

Other home remedies may include the following:

Agua con sal (salt water): to relieve sore throat.

Agua de coco (coconut water): to relieve kidney problems and infections.

Agua raja (turpentine): to relieve pain in sore muscles and joints.

Bicarbonato, limon, y agua (baking soda, lemon, and water): to relieve stomach upset or heartburn.

Cebo de carnero (fat of lamb): to treat contusions and swelling; applied directly on the skin.

Mantequilla (butter): to soothe pain; applied directly on burns.

Clara de huevos (egg white): to promote hair growth; applied directly over scalp.

Cuban families may use an ***azabache, la manito de coral***, or *ojitos de Santa Lucia* for various protective purposes. The *azabache* is a black stone placed on infants and children as a bracelet or pin to protect them from the evil eye. *La manito de coral*, symbolic of the hand of God protecting a person, may also be worn as a necklace or bracelet. *Los ojitos de Santa Lucia*, or the eyes of Saint Lucy, may be hung on a bracelet or necklace for prevention of blindness and protection from the evil eye.

12.12.4 Barriers to Health Care

Poverty and lack of financial resources may be barriers to health care for Cuban families. Other barriers include language, time lag, and transportation, especially if they do not live in an urban environment. Others indicate that the red tape and paperwork required by health-care facilities are deterrents to accessing care, especially preventive care and health wellness checkups. For some, overdependence on family and folk practices may also be a barrier to accessing care.

12.12.5 Cultural Responses to Health and Illness

Because of the many losses they experienced in leaving their homeland and the difficulties associated with adaptation to a new culture and environment, Cuban immigrants may suffer from loneliness, depression, anger, anxiety, insecurity, and health problems. In evaluating Cuban families, Bernal (1994) suggested that health-care providers assess the following:

1. *Migration phase associated with the family.* It is important to know how long the family has lived in their new country and the reasons for migration. Information about political and social pressures that prompted the move should be elicited. Because family members acculturate at different rates, the level of acculturation should also be determined.
2. *Degree of connectedness to the culture of origin.* Conflicts in value orientations must be identified when assessing Cuban families. For example, the varying expectations between mainstream and Cuban cultures with respect to dependence and independence may give rise to tension and conflict.
3. *Differentiation between stresses of migration, differences in cultural values, and family developmental conflicts.* In a clinical situation, health-care providers must be able to recognize whether the patients' responses are due to

migration-related problems, value orientation conflicts, or dysfunctional family development.

Among Cuban Americans, dependency is a culturally acceptable sick role. Sick family members are showered with attention and support. Frequently, a hospitalized Cuban patient will have a room full of flower arrangements and visitors. Favorite dishes may be brought to the hospital from home. The extended family network is relied on to temporarily assume the household chores and other tasks usually performed by the sick person. Family members are consulted and typically participate in decision making relative to the patient's treatment.

Cuban tend to seek help in response to crisis situations. The experience of pain constitutes a signal of a physical disturbance that warrants consultation with a traditional or a biomedical healer. Cuban tend to express their pain and discomfort. Verbal complaints, moaning, crying, and groaning are culturally appropriate ways of dealing with pain. The expression of pain itself may serve a pain-relieving function and may not necessarily signify a need for administration of pain medication.

African Cubans may seek biomedical care for organic diseases but consult a *santero* for spiritual or emotional crises. Conditions such as ***decensos*** (fainting spells) or ***barrenillos*** (obsessions) may be treated solely by a *santero* or simultaneously with a physician. The trance state achieved through *Santeria* enables the patient to act out emotional problems in a manner that is nonthreatening to the person's self-esteem.

Reflective Exercise

Consuegro Luna is a retired 72-year-old Cuban American. She arrived in the US in the early 1960s after Castro's Cuban Revolution. Her entire family lives in Miami; she has no family connections in Cuba. However, she has deep emotional roots to her beloved island. Lately, she has become depressed because she is afraid that she will die without ever seeing Cuba again. She sees her primary physician, but apparently is not adherent to the therapy.

1. What would be her primary physician's best approach for Mrs. Luna?
2. Based on her Cuban cultural background, how would you involve Mrs. Luna's family in her treatment?
3. Besides her nutrition and prescribed/over-the-counter medications, what other information could be relevant for the treatment of Mrs. Luna?

12.12.6 Blood Transfusions and Organ Donation

Receiving blood transfusions and organ donations is usually acceptable for Cubans. This is probably due to their experience with the sophisticated, high-technology medical-care system in Cuba.

12.13 Health-Care Providers

12.13.1 Traditional Versus Biomedical Providers

Cubans use both traditional and biomedical care. Initially, folk remedies may be used at home to treat an ailment or illness. If the condition persists, folk practitioners such as *santeros* and biomedical practitioners may be used either simultaneously or successively. When seeing Cuban patients, health-care providers must always ask about the use of folk remedies and consultations with folk practitioners to prevent conflicting therapeutic regimens.

Although *Santeria* was once associated with the lower, uneducated classes in Cuba, it has emerged as a viable and dynamic religious and health system among middle-class Cubans. The *santero* may prescribe treatment or perform the appropriate rituals or ceremonies to enable ill people to recover. The *santero* may invoke various types of supernatural deities to intervene in their lives and make them well. Often, the *santero* is seen simultaneously with allopathic practitioners, sometimes without the knowledge of the other one.

Many Cubans consult a family physician for primary care. Before the revolution, Cuba had an organized, government-supported health program that provided medical care to most citizens. Since the 1959 revolution, the Cuban government has articulated a fundamental principle that health care is a right of all and a responsibility of the state. Thus, a national health-care system provides universal coverage, equitable geographic distribution of health-care facilities, and standardization of health services.

Cuban families in Miami gained access to primary health-care services predominantly through private health practitioners and private clinics, whereas in Union City, the main sources of health care were private health practitioners. An extensive network of privately owned and operated health clinics exists in Dade County, mainly located in Miami's Cuban ethnic enclaves: Little Havana and Hialeah. The private health clinics are believed to be popular among the Cubans because they provide services that are culturally sensitive to Cuban needs, such as emphasis on the family, use of the Spanish language, focus on preventive health-care behaviors, and low cost.

12.13.2 Status of Health-Care Providers

Although Hispanics, including Cubans, represent 13 percent of the U.S. population, they are seriously underrepresented in the health occupations. In the National Sample Survey of Registered Nurses (RNs) (U.S. Department of Health and Human Services 2008), of over 3 million registered nurses, only 3.6 percent are Hispanic. Cubans generally have respect for all health professionals, including nurses. Respect and trust are increased if the nurse know some Spanish.

References

AAUW (2013) Gender equality and the role of women in Cuban society. https://www.aauw.org/files/2013/01/Cuba_whitepaper.pdf

Alvarez CF (2017) History of Cuba Cuba libre! Cuban history from Christopher Columbus to Fidel Castro. CreateSpace Independent Publishing Platform, Scotts Valley, CA

Bernal G (1994) Cuban families. In: Uriarte-Gaston M, Canas-Martinez J (eds) Cubans in the United States. Center for the Study of the Cuban Community, Boston, pp 135–156

Black SA, Markides KS (1994) Aging and generational patterns of alcohol consumption among Mexican Americans, Cuban Americans and mainland Puerto Ricans. Int J Aging Hum Dev 39(2): 97–103

Blank R, Slipp S (1994) Voices of diversity. American Management Association, New York, pp 63–64

Boswell TD (2002) A demographic profile of Cuban Americans. Cuban American National Council, Miami, FL

Centers for Disease Control and Prevention (2006) Hypertension-related mortality among Hispanic subpopulations—United States, 1995–2002. MMWR Morb Mortal Wkly Rep 55(7):177–180

CIA World FactBook (2018) Retrieved from https://www.cia.gov/library/publications/the-world-factbook/geos/cu.html

CIA World Factbook (2019) Cuba. https://www.cia.gov/library/publications/the-world-factbook/geos/cu.html

Conde YM (1999) Operation Pedro Pan: the untold exodus of 14,048 Cuban children. Routledge, New York

Cultural Orientation Resource Center (2002) The Cubans: their history and culture. http://www.culturalorientation.net.cubans/index.htm

Devieux JG, Malow RM, Ergon-Perez E, Samuels D, Rojas P, Kushal SR, Jean-Gilles M (2005) Research findings—research on behavioral and combined treatments for drug abuse. J Soc Work Pract Addict 2(1):69–83

Gallagher PL (1980) The cuban exile: A socio-political analysis. New York: Arno Press

Gonzalez AM (1995) Santeria still shrouded in secrecy. The Miami Herald, p 1B–5B

History of Cuba (2019). www.historyofcuba.com

Kittler P, Sucher K (2008) Food and culture in America (3rd ed.). Belmont, CA: Wadsworth Publishing Co.

Lopez LM, Masse BR (1993) Income, body fatness, and fat patterns in Hispanic women from the Hispanic Health and Nutrition Examination Survey. Health Care Women Int 14:117–128

Martinez IL (2002) The elder in the Cuban family. Making sense of the real and ideal. J Comp Fam Stud 33(3):359–375

Martinez R, Wetli C (1982) Santeria: a magico-religious system of Afro-Cuban origin. Am J Soc Psychiatry 2(3):496–503

Migration Policy Institute (2017). https://www.migrationpolicy.org/article/cuban-migration-postrevolution-exodus-ebbs-and-flows

Nath SD (2005) Coronary heart disease risk factors among Cuban Americans. Ethn Dis 15:607–614

National Council of la Raza (2011). http://www.nclr.org/

Perez L (2002) Cuban families in the United States. In: Taylor RL (ed) Minority families in the United States:

a multicultural perspective. Prentice Hall, Englewood Cliffs, NJ, pp 95–112
Perez y Pena A (1998) Cuban Santeria, Haitian Vodun, Puerto Rican spiritualism: a multiculturalist inquiry into syncretism. J Sci Study Relig 37(1):15–27
Portes A, Bach RL (1985) Latin journey: Cuban and Mexican immigrants in the United States. University of California Press, Berkeley
Ramos MA (2002) Religion and religiosity in Cuba: past, present, and future. Cuba occasional paper series. Trinity College, Washington, DC
Rohrlich J (2014) Cuba wants you to think it's a gay paradise. It's not. FP. https://foreignpolicy.com/2014/07/03/cuba-wants-you-to-think-its-a-gay-paradise-its-not/
Sandoval M (1979) Santeria as a mental health care system: an historical overview. Soc Sci Med 13:137–151
Skaine R (2004) The Cuban family. Custom and change in an era of hardship. McFarland & Co., Jefferson, NC
Solis JM, Marks G, Garcia M, Shelton D (1990) Acculturation, access to care, and use of preventive services by Hispanics: findings from HHANES 1982–84. Am J Public Health 80(Suppl):11–19
Stroup-Benham CA, Trevino FM (1991) Reproductive characteristics of Mexican-American, mainland Puerto Rican, and Cuban-American women. J Am Med Assoc 265(2):222–226
Suarez ZE (1993) Cuban Americans. From golden exiles to social undesirables. In: McAdoo HP (ed) Family ethnicity: strength in diversity. Sage, Newbury Park, CA, pp 164–176
Szapocznik J, Hernandez R (1988) The Cuban American family. In: Mindel CH, Habenstein RW, Wright R (eds) Ethnic families in America, 3rd edn. Elsevier, New York, pp 160–172
Thomas JT, DeSantis L (1995) Feeding and weaning practices of Cuban and Haitian immigrant mothers. J Transcult Nurs 6(2):34–42
U.S. Census Bureau (2010) Hispanic or Latino origin by specific origin. Retrieved from https://data.census.gov/cedsci/table?q=Hispanic%20or%20Latino%20origin%20by%20specific%20origin%20&hidePreview=false&tid=ACSDT1Y2018.B03001&t=Hispanic%20or%20Latino&vintage=2018
U.S. Department of Health and Human Services (2008) The registered nurse population: national sample survey of registered nurses. U.S. Government Printing Office, Washington, DC
U.S. Embassy in Cuba (2019) Brief Diplomatic in Cuba. https://cu.usembassy.gov/our-relationship/policy-history/
U.S. Immigration and Naturalization Service (2000) U.S. department of justice immigration and naturalization service. Statistical Yearbook of Immigration and Naturalization Service, p 1–67
Varela L (2005) Cubans. In: Lipson J, Dibble SL (eds) Culture and clinical care, 2nd edn. UCSF Nursing Press, San Francisco, CA, pp 121–131
Wilkinson J (2019) History of Cuba. http://www.keyshistory.org/cuba.html

13 People of European American Heritage

Larry D. Purnell

13.1 Introduction

The United States of America, noted for its individualism, is the third largest country in the world. Thus, the cultural values, beliefs, and practices vary tremendously according to the variant cultural characteristics as noted in Chap. 2. The chapter follows the 12 domains of the Purnell Model of Cultural Competence: overview/heritage, communication, family roles and organization, workforce issues, biocultural ecology, high-risk behaviors, nutrition, pregnancy, death rituals, spirituality, health-care practices, and health-care practitioners.

13.2 Overview, Inhabited Localities, and Topography

13.2.1 Overview

This chapter presents dominant European American cultural values, practices, and beliefs, but may not be consistent with variant cultural characteristics as depicted on the Purnell Model (see Chap. 2). Early immigrants to the United States, primarily Caucasians from Europe, adapted and adopted each other's forming a distinct, new culture. Many other groups have assimilated and now self-identify with the European American culture as well. *European American* in this chapter refers to the dominant middle-class values of citizens of mainland United States, and the term *European American* is shortened to American. This chapter does not address the objective culture—arts, literature, humanities, and so on—but rather with the subjective culture.

This chapter emphasises political and culture values starting with the constitution. Given the size, population density, and diversity of the United States makes every generalization in this chapter is subject to exceptions. Moreover, the descriptions about the dominant American culture are aggregate data for White middle-class Americans who still hold the majority of prestigious positions in the United States. The degree to which people conform to or agree with the American culture depends on their variant cultural characteristics as discussed in Chap. 2, as well as individual personality differences. While foreigners believe that all Americans are rich, live in fancy apartments or houses with little or no poverty, crime is rampant, drive expensive gasoline-inefficient cars, these are misconceptions that come from the media and wealthier Americans.

Overall, the dominant culture values and beliefs include

L. D. Purnell (✉)
College of Health Sciences, University of Delaware/Newark, Sudlersville, MD, USA
e-mail: lpurnell@udel.edu

© Springer Nature Switzerland AG 2021
L. D. Purnell, E. A. Fenkl (eds.), *Textbook for Transcultural Health Care: A Population Approach*, https://doi.org/10.1007/978-3-030-51399-3_13

1. **Achievement and Success:** American is a competitive society and stresses personal achievement. Achievement and success are most often measured by how much money one makes, how famous you are, or by how powerful you have become. Failure to achieve is often seen as a defect in your character. Success is often equated with "bigness" and "newness".
2. **Activity and Work**: Americans also value busyness, speed, bustle, and action. The frontier idea of work for survival still exists as is the Puritan ethic of work before play. A person's worth is often measured by his or her work performance and success.
3. **Morality**: Americans think in terms of good and bad, right and wrong. Early Puritan ideas of working hard, leading an orderly life, having a reputation for integrity and fair dealing, avoiding a reckless lifestyle, and carrying out one's purpose in life are still ideals that hold weight today.
4. **Humanitarianism**: In America, emphasis is placed on concern, helpfulness, personal kindness, aid and comfort, spontaneous aid in mass disasters, as well as on charity. This emphasis is related to equality in a republic, but often clashes with the values of rugged individualism.
5. **Efficiency and Practicality**: Most Americans like modern innovation, expediency, mass production, industry, and getting things done! The culture values science, research, and new inventions that make lives easier.
6. **Progress**: Most Americans look forward more to the future than back to the past. Many reject old-fashioned, outmoded ideals. Progress is often identified with Darwin's idea of survival of the fittest and with the free enterprise system.
7. **Material Comfort**: Americans enjoy being content, gratified, and comfortable. Many long for material possessions that provide comfort as well as entertainment. Activities, work, and daily lives are centered around the beliefs in the comforts of life.
8. **Equality**: History has stressed the importance of opportunity, especially economic opportunity. Most Americans fundamentally believe that everyone has the same and equal opportunity to achieve. While discrimination exists much attention is paid to formal civil and legal rights of citizens.
9. **Freedom**: Americans also seek freedom and have confidence in individuals making own choices, which is consistent with individualism. Freedom enters into free enterprise, progress, individual choice, and equality. It does not mean the absence of social or governmental control.
10. **External Conformity**: Americans also believe in adherence to a group, especially for success and are often economically, politically, and socially dependent upon each other. Interdependence results in some conformity.
11. **Science**: Most Americans have faith in science and its tools because its rational, functional, and active and adds to material comfort and progress.
12. **Nationalism and Patriotism**: Most Americans feel some sense of loyalty to their country, national symbols and its history, and believe that America is the greatest country in the world.
13. **Democracy**: Americans have grown to accept majority rule, representative government, and the rights of the individual and reject monarchies and dictatorships.
14. **Individuality**: America protects individualism by laws and by the belief in one's own worth. Americans pride themselves on being self-sufficient, strong, and rugged individualists.
15. **Racism and Group Superiority**: Although a more negative value, many Americans believe that they (or their individual group) are the best. Unfortunately, this belief can take its form in racial, religious, political, ethnic, and sexual discrimination and crosses all boundaries of American belief systems. Individualistic cultures are not as accepting of difference as are collectivistic cultures.

16. **Order and Security**: Americans believe in a structured religious, political, and social system that keeps them from physical harm and chaos (Dominant American Values 2019; Kohl 2019).
17. **Individualism:** For most Americans, dominant cultural values and beliefs include individualism, free speech, rights of choice, independence and self-reliance, confidence, "doing" rather than "being," egalitarian relationships, non-hierarchical status of individuals, achievement status over ascribed status, "volunteerism," friendliness, openness, futuristic temporality, ability to control the environment, and an emphasis on material things and physical comfort.
18. **Violence**: The US is often seen as a violent country with a high suicide rate. When compared to 22 other high-income nations, the United States' gun-related murder rate is 25 times higher. The United States' suicide rate is similar to other countries; however, the nation's gun-related suicide rate is eight times higher than other high-income countries (How U.S. Gun Deaths Compare with other countries (2019).

13.2.2 Heritage and Residence

The United States comprises 3.5 million square miles and a population of 328,199,644, making it the world's third most populous country (U.S. Census Bureau 2019). The United States is mostly temperate but tropical in Hawaii and Florida, arctic in Alaska, semiarid in the Great Plains west of the Mississippi River, and arid in the Great Basin of the southwest. Low winter temperatures in the northwest are ameliorated in January and February by warm Chinook winds from the eastern slopes of the Rocky Mountains. There is a vast central plain; mountains in the west; hills and low mountains in the east; rugged mountains and broad river valleys in Alaska; and rugged, volcanic topography in Hawaii (CIA World Factbook Online 2019).

When Europeans began settling the United States in the sixteenth century American Indians, who mostly lived in geographically isolated tribes, populated the land. The first permanent European settlement in the United States was St. Augustine, Florida, which was settled by the Spanish in 1565. The first English settlement was Jamestown, Virginia, in 1607 (American Chronology 2019). By 1610, the non-native population in the United States amounted to 350 people. By 1700, the population increased to 250,900; by 1800, to 5.3 million; and by 1900, to 75.9 million (American Chronology 2019). From 1607 until 1890, most immigrants to the United States came from Europe and essentially shared a common European culture. Britain's American colonies broke with the mother country in 1776 and were recognized as the new nation of the United States of America following the Treaty of Paris in 1783.

During the nineteenth and twentieth centuries, 37 new states were added to the original 13 as the nation expanded across the North American continent and acquired a number of overseas possessions. The two most traumatic experiences in the nation's history were the Civil War (1861–1865), in which a northern Union of states defeated a secessionist Confederacy of 11 southern slave states, and the Great Depression of the 1930s, an economic downturn during which about a quarter of the labor force lost its jobs. Buoyed by victories in World Wars I and II and the end of the Cold War in 1991, the United States remains the world's most powerful nation state. Over a span of more than five decades, the economy has achieved steady growth, low unemployment and inflation, and rapid advances in technology (CIA World Factbook Online 2019).

The Constitution of the United States was ratified in 1789 and included seven articles, which laid the foundation for an independent nation. The Bill of Rights, the first 10 amendments to the Constitution, guarantees freedom of religion, speech, and the press; the right to petition; the right to bear arms; and the right to a speedy trial. Only 17 additional amendments have been made to the Constitution. The 13th Amendment in 1865 prohibited slavery; the 14th Amendment in 1868 defined citizenship and privileges of citizens; the 15th Amendment in 1870 gave suffrage rights regardless of race or color; and the 19th Amendment in 1920 gave women the right to vote. The United States is the world's oldest Federal Republic with strong democratic traditions and has three branches of government: the executive branch, which includes the Office of the President and the administrative departments; the legislative branch, Congress, which includes both the Senate and the House of Representatives; and the judicial branch, which includes the Supreme Court and the lesser federal courts. The Supreme Court has nine members appointed by the president and approved by Congress. The justices serve a life term if they so choose. The president serves a 4-year term and can be re-elected only one time. The president is the commander in chief of the armed forces and oversees the executive departments (Lewis et al. 2019). The members of the House of Representatives are divided among the states based on the population of each state. Members of the House of Representatives serve 2-year terms. Each state has two senators, regardless of the population of the state. Senators serve 6-year terms. Each of the 50 states has its own constitution establishing, for the most part, a parallel structure to the federal government, with the executive branch headed by a governor, a state congress with representatives and senators, and a state court system (What is the Length of Term for the Senate and House of Representatives? 2019).

No limitations were placed on immigrants from Europe until the late 1800s, with the majority coming from Germany, Ireland, and England. From 1892 to 1952, most European immigrants to America came through Ellis Island, New York, where they had to prove to officials that they were financially independent. More severe restrictions were placed on other immigrant groups, particularly those from Asia. In the 1960s, immigration policy changed to allow immigrants from all parts of the world without favoritism to or restrictions on ethnicity. Today, the United States includes immigrants or descendants from immigrants from almost every nation and culture of the world and is the world's premier international nation. The United States admitted 72,646 refugees in fiscal year 2016 but only 18,936 in fiscal year 2018. The most recent data on asylees, 26,568 were admitted (Migration Policy Institute 2019).

The United States has the largest and most technologically powerful economy in the world, with a per capita gross domestic product (GDP) of $59,500 (CIA World Factbook Online 2019). In this market-oriented economy, private individuals and business firms make most of the decisions, and the federal and state governments buy needed goods and services predominantly in the private marketplace. U.S. firms are at or near the forefront in technological advances, especially in computer technology and in medical advancements, aerospace technology, and military equipment. Their advantage has narrowed since the end of World War II. The on-rush of technology largely explains the gradual development of a "two-tier labor market," in which those at the bottom lack the education and the professional/technical skills of those at the top and, more and more, fail to get comparable pay raises, health insurance coverage, and other benefits. People have been attracted to the United States because of its vast resources and economic and personal freedoms, particularly the dogma that "all men are created equal." Immigrants and their descendants achieved enormous material success, which further encouraged immigration.

13.2.3 Reasons for Migration and Associated Economic Factors

The ethnic and racial distribution in the United States is 62% White, 16.9% Hispanic, 12.6% Black, 5.2% Asian, 2.3% mixed, and 1% other (Race and Ethnicity in the United States 2019). The United States has a very large middle-class population and a small, but growing, wealthy population. Approximately 13.4 percent of the population lives in poverty, with higher rates among children, older persons, Blacks, and non-White Hispanics (Race and Ethnicity in the United States 2019).

The earlier settlers in the United States came for better economic opportunities because of religious and political oppression; environmental disasters, such as earthquakes and hurricanes in their home countries; and by forced relocation, such as with slaves and indentured servants. Others have immigrated for educational opportunities and personal ideologies or a combination of factors. Most people immigrate in the hope of a better life; however, the individual or group personally defines this ideology.

13.2.4 Educational Status and Occupations

In the United States, preparation in elementary and secondary education varies widely. There is no national curriculum that each school is expected to follow, although there is standardized testing at a national level, which is used in the selection process for admission to institutions of higher education. Most states require children to attend school until the age of 16, although the child can drop out of school at a younger age with parents' signed permission. Overall, the United States has the goal of producing a well-rounded individual with a variety of courses and 100 percent literacy. Theoretically, people have the freedom to choose a profession, regardless of gender and background. The American educational system stresses application of content over theory. The United States' dominant system places a high value on the student's ability to categorize information using linear, sequential thought processes, which is common in individualistic cultures.

13.3 Communication

13.3.1 Dominant Language and Dialects

Over 82 percent of the U.S. population speaks English (CIA World Factbook 2019), mostly *American English,* which differs somewhat in its pronunciation, spelling, and choice of words from English spoken in Great Britain, Australia, and other English-speaking countries. Within the United States, several dialects exist, but generally the differences do not cause a major concern with communications. Aside from people with foreign accents, in certain areas of the United States people speak with a dialect; these include the South and Northeast, in addition to local dialects such as "Elizabethan English" in some parts of Appalachia, and "western drawl." In such cases, dialects that vary widely may pose substantial problems for health-care providers and interpreters in performing health assessments and in obtaining accurate health data, in turn increasing the difficulty of making accurate diagnoses.

American English is a monochromic, low-contextual language in which most of the message is in the verbal mode, and verbal communication is frequently seen as being more important than nonverbal communication. Other common languages spoken in the United States include Spanish (10 percent), other Indo-European languages (3.8 percent), Asian and Pacific Islander languages (2.7 percent), and other (0.7 percent) (CIA World Factbook 2019). In addition, the speed at which people speak varies by region; for example, in parts of Appalachia and the South, people speak more slowly than do people in the northeastern part of the United States. Americans may be perceived as being loud and boisterous because their volume carries to those nearby.

13.3.2 Cultural Communication Patterns

Many Americans are willing to disclose very personal information about themselves, including information about sex, drugs, and family problems. In fact, personal sharing is encouraged in a wide variety of topics, but not religion. In the United States, having well-developed verbal skills is seen as important.

For the most part, America is a low-touch society, which has recently been reinforced by sexual harassment guidelines and policies. For many, even casual touching may be seen as a sexual overture and should be avoided whenever possible until people get to know each other. People of the same sex (especially men) or opposite sex do not generally touch each other unless they are close friends.

American conversants tend to place at least 18 inches of space between themselves and the person with whom they are talking. Clients may interpret American health-care providers as being cold because they stand so far away. An understanding of personal space and distancing characteristics can enhance the quality of communication among individuals.

Regardless of class or social standing of the conversants, Americans are expected to maintain direct eye contact without staring. A person who does not maintain eye contact may be perceived as not listening, not being trustworthy, not caring, or being less than truthful.

Most Americans gesture moderately when conversing and smile easily as a sign of pleasantness or happiness, although one can smile as a sign of sarcasm. A lack of gesturing can mean that the person is too stiff, too formal, or too polite. However, when gesturing to make, emphasize, or clarify a point, one should not raise one's elbows above the head unless saying hello or good-bye.

For American men and women in business, the practice is to extend the right hand with a firm handshake when greeting someone for the first time. Confidence and competence are associated with a relaxed posture. Although many people consider it impolite or offensive to point with one's finger, many Americans do so, and do not see it as impolite.

13.3.3 Temporal Relationships

The American culture is future oriented and people are encouraged to sacrifice for today and work to save and invest in the future. The future is important in that people can influence it. Americans generally see fatalism, the belief that powers greater than humans are in control, as negative; to many others, however, it is seen as a fact of life not to be judged. For many, temporality is balanced among past, present, and future in the sense of respecting the past, valuing and enjoying the present, and saving for the future.

Americans see time as a highly valued resource and do not like to be delayed because it "wastes time." When visiting friends or meeting for strictly social engagements, punctuality is less important, but one is still expected to appear within a "reasonable" time frame. In the health-care setting, if an appointment is made for 8 a.m., the person is expected to be there at 7:45 a.m. so she or he is ready for the appointment and does not delay the healthcare provider. Some organizations refuse to see the patient if he or she is more than 15–30 min late for an appointment; a few charge a fee, even though the patient was not seen, giving the impression that money is more important than the person.

13.3.4 Format for Names

The American name David Thomas Jones denotes a man whose first name is David, middle name is Thomas, and family surname is Jones. Friends would call him by his first name, David. In the formal setting, he would be called Mr. Jones. In addition, he could also have a nickname that would be used by family and close friends—for example, Davy from his first name or Tom or Tommy from his middle name. When women marry, they may drop their maiden name and adopt their husband's last name, or they may keep both their maiden and husband's names. In this case, they may or may not hyphenate it, as in Elizabeth Parker-Jones or Elizabeth Parker Jones. Their children usually, but not always, take the husband's last name.

13.4 Family Roles and Organization

13.4.1 Head of Household and Gender Roles

Among Americans, it is acceptable for women to have a career and for men to assist with child care, household domestic chores, and cooking responsibilities. Both parents work in many families, necessitating placing children in child-care facilities. In some families, fathers are responsible for deciding when to seek health care for family members, but mothers may have significant influence on final decisions.

13.4.2 Prescriptive, Restrictive, and Taboo Behaviors for Children and Adolescents

For most Americans, a child's individual achievement is valued over the family's financial status. In many middle- and upper-class American families, children have their own room, television, and telephone, and even their own computer and mobile phones. At younger ages, rather than having group toys, each child has his or her own toys and is taught to share them with others. Americans encourage autonomy in children, and after completing homework assignments (with which parents are expected to help), children are expected to contribute to the family by doing chores, such as taking out the garbage, washing dishes, cleaning their own rooms, feeding and caring for pets, and helping with cooking. They are not expected to help with heavy labor except in rural farm communities.

Children are allowed and encouraged to make their own choices, including managing their own money allowance and deciding who their friends might be, although parents may gently suggest one friend as a better choice than another. American children and teenagers are permitted and encouraged to have friends of both the same and opposite genders. They are expected to be well behaved, especially in public. They are taught to stand in line—first come, first served—and to wait their turn. As they reach the teenage years, they are expected to refrain from premarital sex, smoking, using recreational drugs, and drinking alcohol until they leave home. However, this does not always occur, and teenage pregnancy and use of recreational alcohol and drugs remain high, although some statistics report that teen pregnancy is on the decline. When children become teenagers, most are expected to get a job, such as babysitting, delivering newspapers, or doing yard work to make their own spending money, which they manage as a way of learning independence. The teenage years are also seen as a time of natural rebellion.

In American society, when young adults become 18 or complete their education, they usually move out of their parents' home (unless they are in college) and live independently or share living arrangements with nonfamily members. If the young adult chooses to remain in the parents' home, then she or he might be expected to pay room and board. However, young adults are generally allowed to return home when they are needed or for financial or other purposes. Individuals over the age of 18 are expected to be self-reliant and independent, which are virtues in the American culture.

Adolescents have their own subculture, with its own values, beliefs, and practices that may not be in harmony with those of their parents' wishes. Being in harmony with peers and conforming to the prevalent choice of music, clothing, hairstyles, and adornments may be especially important to adolescents. Thus, role conflicts can become considerable sources of family strain in many more traditional families who may not agree with the American values of individuality, independence, self-assertion, and egalitarian relationships. Many teens may experience a cultural dilemma with exposure outside the home and family in a multi-cultural society.

13.4.3 Family Goals and Priorities

American family goals and priorities are centered on raising and educating children. During this stage in the American culture, young adults make a personal commitment to a spouse or significant other and seek satisfaction through productivity in career, family, and civic interests.

The median age at first marriage in the United States has gradually increased over the last 10 years from 26.8 to 27.7 years for men; for women, the median age of first marriage has gradually increased from 25.1 to 26 years (In 1900, the divorce rate was 0.7 per year per 1000 marriages; this increased to 5.3 per year by 1981 and has gradually declined since then to 3.6 per year per 1000 (American Chronology 2019).

The United States has seen an explosion in its older population. Those over the age of 65 comprise 16% of the population (CIA World Factbook Online 2019). The American culture, which emphasizes youth, beauty, thinness (although 36.2% of the population is obese), independence, and productivity, contributes to some societal views of the aged as less important members and tends to minimize the problems of older people. A contrasting view among some emphasizes the importance of older people in society because of their wisdom.

Americans also place a high value on egalitarianism, non-hierarchical relationships, and equal treatment regardless of their race, color, religion, ethnicity, educational or economic status, sexual orientation, or country of origin. However, these beliefs are theoretical and not always seen in practice. For example, women still have lower status than men, especially when it comes to prestigious positions and salaries. Most top-level politicians and corporate executive officers are White men. Subtle classism does exist, as evidenced by comments referring to "working-class men and women." Despite the current inequities, Most Americans value equal opportunities for all, and most would agree that significant progress is being made.

Americans are known worldwide for their informality and for treating everyone the same. They call people by their first names very soon after meeting them, whether in the workplace, in social situations, in classrooms, in restaurants, or in places of business. Americans readily talk with waitstaff and store clerks and call them by their first names. In fact, most of the waitstaff, store clerks, and nurses introduce themselves by their first names. Most Americans consider this respectful behavior. Formality can be communicated by using the person's last (family) name and title such as Mr., Mrs., Miss, Ms., or Dr. To this end, achieved status is more important than ascribed status. What one has accumulated in material possessions, where one went to school, and one's job position and title are more important than one's family background and lineage. However, in some families especially in the South and the Northeast, one's ascribed status has equal importance to achieved status. Theoretically, the United States does not have a caste or class system and one can move readily from one socioeconomic position to another. To many Americans, if formality is maintained, it may be seen as pompous or arrogant, and some even deride the person who is very formal. However, formality is a sign of respect and is valued by most older Americans.

13.4.4 Alternative Lifestyles

The American family is becoming a more varied community, including unmarried people, both women and men, living alone; single people of the same or different genders living together with or without children; single parents with children; and blended families consisting of two parents who have remarried and have children from their previous marriages and additional children from their current marriage. Over the last decade, the lesbian, gay, bisexual, and transgender populations have become more readily accepted my many, especially the younger generations. Many laws exist discrimination of any kind.

The newest category of family, domestic partnerships, is sanctioned by many cities or counties and grants some of the rights of traditional married couples to unmarried heterosexual, homosexual, older people, and disabled couples who share the traditional bond of the family. Some states allow gay and lesbian couples to marry and to adopt children. The last 10 years have seen many hotly debated issues regarding same-sex marriages and civil unions. Among more rural subcultures, same-sex couples living together may not be as accepted or recognized in the community as they are in larger cities. As gay parents have become more visible, lesbian and gay parenting groups have started in many cities across the United States to offer information, support, and guidance, resulting in more lesbians and gay men considering parenthood through adoption and artificial insemination.

13.5 Workforce Issues

13.5.1 Culture in the Workplace

Americans are expected to be punctual on their jobs, with formal meetings, and with appointments. If one is more than a minute or two late, an apology is expected, and if one is late by more than 5 or 10 min, a more elaborate apology is expected. When people know they are going to be late for a meeting, the expectation is that they call or send a message indicating that they will be late. The convener of the meeting or teacher in a classroom is expected to start and stop on time out of respect for the other people in attendance. However, in social situations, a person can be 15 or more minutes late, depending on the importance of the gathering. In this instance, an apology is not really necessary or expected; however, most Americans will politely provide a reason for the tardiness.

The American workforce stresses efficiency (time is money), operational procedures on how to get things done, task accomplishment, and proactive problem solving. Intuitive abilities are not usually valued as much as technical abilities. The scientific method is valued and everything has to be proven. Americans want to know *why*, not *what*, and will search for a single factor that is the cause of the problem and the reason why something is to be done in a specific way. Many are obsessed with collecting facts and figures before they make decisions. Pragmatism is valued. In the United States, everyone is expected to have a job description, meetings are to have a predetermined agenda, although items can be added at the beginning of the meeting. The agenda is followed throughout the meeting. Americans prefer to vote on almost every item on an agenda, including approving the agenda itself. Everything is given a time frame and deadlines are expected to be respected. In these situations, American values expect that the needs of individuals are subservient to the needs of the organization. However, with the postmodernist movement, where there are no absolute truths and most aspects of a person's worldview are based on perceptions and social contexts, greater credibility and recognition have been given to approaches other than the scientific method.

13.5.2 Issues Related to Autonomy

Most Americans place a high value on "fairness" and rely heavily on procedures and policies in the decision-making process. However, Americans' value for individualism, in which the individual is seen as the most important element in society, favors a person's decision to further her or his own career over the needs or wants of the employer. Therefore, individuals frequently demonstrate little loyalty to the organization and leave one position to take a position with another company for a better opportunity or higher salary. In organizations in which people generally conform because of the fear of failure, there is a hierarchical order for decision making. The person who succeeds is the one with strong verbal skills and conforms to the hierarchy's expectations. This person is well liked and does not stand out too much from the crowd. Frequently, others view as a threat the person with a high level of competence and who stands out. Thus, to be successful in the highly technical American workforce, the individual must get the facts, control feelings, have precise and technical communication skills, be informal and direct, and clearly and explicitly state his or her conclusions.

During workforce shortages, American healthcare facilities rely on emigrating nurses and physicians from the Philippines, Canada, England, Ireland, India, and other countries to supplement their numbers. Some foreign nurses, such as British and Australian, culturally assimilate into the workforce more easily than others but still have difficulty with *defensive* charting as is required in the United States. Some may have difficulty with the assertiveness expected from American nurses.

Reflective Exercise 13.1

Melanie Hilliard, a European American Roman Catholic age 42 years, was diagnosed with breast cancer 8 months ago and had bilateral mastectomies. She underwent chemotherapy and radiation therapy, but the cancer has metastasized to her lungs, ribs, and brain. She has decided to forgo additional allopathic medical treatments, but is open to herbal therapies. Because of intense pain, she has been admitted to a home hospice service. Bonnie, her daughter, has never been married and currently lives with her life partner of 22 years, Florence Walbert, who has durable power of attorney for health-care decision making.

On the intake home hospice assessment, the nurse asked who was the next of kin and was told that it was Florence. The nurses then asked what blood relative would make decisions about health care if she was unable to make decisions. Bonnie again gave Florence's name and explained their relationship. The nurse stated she was unsure if this was acceptable and would ask her supervisor. Melanie explained that both her parents were deceased and that she was estranged from her two older brothers and younger sister, at which point the nurse responded, "I can understand why!"

1. What is the incidence of breast cancer among White women in the United States?
2. Is there a difference in breast cancer among different racial and ethnic groups?
3. Is there a difference between breast cancer rates in lesbians and heterosexual women?
4. What do you know about advance directives and durable power of attorney in healthcare decision making? Can Florence legally be the decision maker for Bonnie?
5. What family support systems does Melanie have?
6. Would you ask Melanie if her family is supportive of their relationship?
7. What resources besides family might be available for Melanie and Florence?
8. What is your response to Melanie's deciding to forgo allopathic treatments but be willing to entertain herbal therapies?
9. How do you feel about same-sex intimate partnerships? What differences do you see between providing care to patients with same-sex partnerships and to patients from heterosexual partnerships?
10. If Melanie and Florence were to seek support from the Catholic Church, how do you think church members would respond? What is the Roman Catholic Church's view on same-sex intimate partner relationships?

13.6 Biocultural Ecology

13.6.1 Skin Color and Other Biological Variations

The majority of European Americans are generally fair-skinned, and thus prolonged exposure to the sun places them at an increased risk for skin cancer. Oxygenation determination can easily be determined from skin color as well as the nail beds.

13.6.2 Diseases and Health Conditions

The top death-causing chronic diseases in the US includes heart disease, cancer, respiratory disease, Alzheimer's, and diabetes (Centers of Disease Control and Prevention 2019). Additional conditions are kidney disease, mental illness, liver disease, hypertension, and Parkinson's. These chronic diseases account for 10 of the 15 leading causes of death in the

US. Furthermore, arthritis and diabetes are two leading causes of disability (Centers of Disease Control and Prevention 2019). Lack of physical activity, poor diet, tobacco use and alcohol misuse are the four factors that the CDC cites as major contributors to the prevalence of chronic diseases. While no habits can negate the risk of chronic diseases, preventative measures include avoiding tobacco, exercising for at least 30 min multiple times a week, avoiding overindulging in saturated fats, and getting plenty of fruits and vegetables, quality sleep, and regular health checkups.

In 2017, 38,739 people received an HIV diagnosis in the US. The annual number of new HIV diagnoses has remained stable (Centers for Disease Control and Prevention 2017). More than 1 million people are living with HIV in the US. One in five (21 percent) of those people living with HIV is unaware of the infection. The most affected populations high to low are Black male-to-male, Hispanic/Latino male-to-male, Black female heterosexual contact, Black men heterosexual contact, Hispanic/Latina female heterosexual contact, and White female heterosexual contact (Centers for Disease Control and Prevention 2017).

Sexually transmitted infections (STIs) remain a major public health challenge in the United States. CDC's surveillance report includes data on the three STIs that physicians are required to report to the agency—chlamydia, gonorrhea, and syphilis—which represent only a fraction of the true burden of STIs. Some common STIs, such as human papillomavirus (HPV) and genital herpes, are not reported to CDC. In total, CDC estimates that there are approximately 19 million new STIs each year (Centers of Disease Control and Prevention 2019).

Despite the continued high burden of STIs, the latest CDC data show some signs of progress:

1. The national gonorrhea rate is at the lowest level ever recorded.
2. Continuing increases in chlamydia diagnoses likely reflect expanded screening efforts and not necessarily a true increase in disease burden.
3. For the first time in 10 years, reported syphilis cases did not increase among women overall.
4. Likewise, cases of congenital syphilis (transmitted from mother to infant) did not increase for (Centers for Disease Control and Prevention 2017).

13.6.3 Variations in Drug Metabolism

Information regarding drug metabolism among racial and ethnic groups has important implications for healthcare practitioners when prescribing medications. Besides the effects of smoking, which accelerates drug metabolism; malnutrition, which affects drug response; a high-fat diet, which increases absorption of antifungal medication, whereas a low-fat diet renders the drug less effective; cultural attitudes and beliefs about taking medication; and stress, which affects catecholamine and cortisol levels on drug metabolism, studies have identified some specific alterations in drug metabolism among diverse racial and ethnic groups (Prows and Prows 2004). The studies on drug metabolism across ethnicity and race have used White ethnic groups/race as the control. Thus, differences in drug metabolism variations among Whites are not reported. Health-care providers need to investigate the literature for ethnic-specific studies regarding variations in drug metabolism, communicate these findings to other colleagues, and educate their clients regarding these side effects.

13.7 High-Risk Behaviors

The steady decline in smoking prevalence has been observed nationally; however, in certain segments of the population, the incidence remains high, thus highlighting the need for expanded interventions that can better reach persons of low socioeconomic status and populations living in poverty. In the US, tobacco use is responsible for about one in five deaths annually and includes tobacco-related deaths as a result of second hand smoke exposure. On average, smokers die 13–14 years earlier than non-smokers; the percentage varies by race and ethnicity, as evidenced by the following statistics:

- 23.2 percent of American Indian/Alaska Native adults
- 22.1 percent of White adults
- 21.3 percent of African American adults
- 14.5 percent of Hispanic adults
- 12 percent of Asian American adults (excluding Native Hawaiians and other Pacific Islanders) (Centers for Disease Control and Prevention 2016).

Alcohol use is very common in the United States and has immediate effects that can increase the risk of many harmful health conditions. Excessive alcohol use, either in the form of *heavy drinking* (drinking more than two drinks per day on average for men or more than one drink per day on average for women) or *binge drinking* (drinking five or more drinks during a single occasion for men or four or more drinks during a single occasion for women), can lead to increased risk of health problems, such as liver disease or unintentional injuries.

According to the National Institute on Alcohol Facts and Statistics (2018), 56% drank alcohol in the last 6 months, approximately 7 percent of the total population drank heavily and 26 percent of the population drank in the last year. The conclusion of many studies suggests that alcohol-related violence is a learned behavior, not an inevitable result of alcohol consumption (Purnell and Foster 2003a, b).

13.7.1 Health-Care Practices

Much of American society has become "obesogenic," characterized by environments that promote increased food intake, nonhealthful foods, and physical inactivity; however, some progress has been made but food desserts are still much more common in inner city poor neighborhoods. Policy and environmental change initiatives that make healthy choices in nutrition and physical activity available, affordable, and easy will likely prove most effective in combating obesity. The Division of Nutrition, Physical Activity, and Obesity (DNPAO) is working to reduce obesity and obesity-related conditions through state programs, technical assistance and training, leadership, surveillance and research, intervention development and evaluation, translation of practice-based evidence and research findings, and partnership development. Although the obesity rate appears to be declining or (at least) levelling, obesity remains high among adults and children (Centers for Disease Control and Prevention 2019). Obesity is a complex issue related to lifestyle, environment, and genes. Many underlying factors have been linked to the increase in obesity, such as increased portion sizes; eating out more often; increased consumption of sugar-sweetened drinks; increased television, computer, electronic gaming time; changing labor markets; and fear of crime, which prevents outdoor exercise. Almost all neighborhoods have a plethora of fast food restaurants, especially in low income neighborhoods. Healthcare providers can assist overweight clients in reducing calorie consumption by identifying healthy choices among culturally preferred foods, altering preparation practices, and reducing portion size.

The practice of self-care using folk and magico-religious practices before seeking professional care may also have a negative impact on the health status of some individuals. Overreliance on these practices may mean that the health problem is in a more advanced stage when a consultation is sought. Such delays make treatment more difficult and prolonged.

Reflective Exercise 13.2

Sharon Shorts, a 54-year-old patient care assistant, has worked on the same acute-care teenage pediatric unit for 12 years. She tells everyone that she loves her job and the teenagers. The unit got a new male nurse manager, Mr. Wilbuer, 2 years ago. During that time, Mrs. Shorts has had six counseling sessions for coming to work more than 10 min late and leaving early.

She insists that she still gets her work completed, despite coming in late and leaving early. Besides, what is the big deal about 10 or 15 min? The nurse manager has informed her that if she is late or leaves early one more time, she will be suspended without pay. Mrs. Shorts believes she is being treated unfairly and, unknown to Mr. Wilbuer, has gone to his supervisor and complained that he has touched her on the shoulder unnecessarily and that he is singling her out because she is female and he is male. In addition, he calls her "Sharon" instead of "Mrs. Shorts."

1. Is the expectation of punctuality on reporting on time consistent with the European American culture?
2. In the American workforce, when employees have a complaint or concern about their supervisor, who is the first person they should speak to about it?
3. What should Mrs. Shorts first response be to being touched unnecessarily?
4. What should Mrs. Shorts first response be if she is offended by Mr. Wilbuer calling her by her first name? Is this common with the European American culture?
5. What is the first thing you would do to resolve this situation?
6. If your first action does not resolve the situation, what is your second action?

13.8 Nutrition

13.8.1 Meaning of Food

When Americans invite a guest to dinner for the first time, the guest frequently brings a gift, although this is not required Some of the choices are often food, beer, and wine. There are no specific rules as to what type of food to bring, but wine, cheese baskets, and candy are usually appropriate. Bread (unless it is a very special bread) is not usually appropriate unless specifically requested.

13.8.2 Common Foods and Food Rituals

American food and preparation practices reflect traditional food habits of early settlers who brought their unique cuisines with them. Accordingly, the "typical American diet" has been brought from elsewhere. Americans vary their mealtimes and food choices according to the region of the country, urban versus rural residence, and weekdays versus weekends. In addition, food choices vary by marital status, economic status, climate changes, religion, ancestry, availability, and personal preferences.

Many older people and people living alone do not eat balanced meals, stating they do not take the time to prepare a meal, even though most American homes have labor saving devices such as stoves, microwave ovens, refrigerators, and dishwashers. For those who are unable to prepare their own meals because of disability or illness, most communities have a Meals on Wheels program through which community and church organizations deliver, usually once a day, a hot meal along with a cold meal for later and food for the following morning's breakfast. Other community and church agencies prepare meals for the homeless or collect food, which is delivered to those who have none. When people are ill, they generally prefer toast, tea, juice, and other easily digested foods. Apple sauce and jello are also common.

Given the size of the United States and its varied terrain, food choices differ by region: beef in the Midwest, fish and seafood in coastal areas, and poultry in the South and along the Eastern Seaboard. Vegetables vary by season, climate, and altitude, although larger grocery stores have a wide variety of all types of American and international meats, fruits, and vegetables. Many television stations and major newspapers have large sections devoted to foods and preparation practices, a testament to the value that Americans place on food and diversity in food preparation.

Special occasions and holidays are frequently associated with specific foods and may vary according to the ethnicity of the family. For example, hot dogs are consumed at sports events, and turkey is served at Thanksgiving. Goose, if available, is common on New Year's day.

13.8.3 Dietary Practices for Health Promotion

Overall, the typical American diet is high in fats and cholesterol and low in fiber (Food-Based Dietary Guidelines—United States of America 2019). Daily recommendations include 6–11 servings of bread, cereal, rice, or pasta; 3–5 servings of vegetables; 2–4 servings of fruit; 2–3 servings of milk, yogurt, or cheese; 2–3 servings of meat, poultry, fish, dry beans, eggs, and nuts; and limited use of fats, oils, and sweets.

13.8.4 Nutritional Deficiencies and Food Limitations

Socioeconomic status may dictate food selections—for example, hamburger instead of steak; canned or frozen vegetables and fruit rather than fresh produce; and fish instead of shrimp or lobster. Most grocery stores have an adequate supply of frozen and canned fruits and vegetables, although they may be high in sodium and sugars. Given the size of the United States and proximity of farms to urban areas, there are few overall limitations for food choices for most Americans. Food desserts for health choices continue to be problematic in many lower socioeconomic and culturally diverse neighborhoods.

13.9 Pregnancy and Childbearing Practices

13.9.1 Fertility Practices and Views Toward Pregnancy

Commonly used methods of birth control among Americans include natural ovulation methods, birth control pills and patches, foams, Norplant, the morning-after pill, intrauterine devices, sterilization, vasectomy, prophylactics, vaginal rings, and abortion (Guttmacher Institute: Contraceptive Use in the United States 2019). These data are not broken down by race or ethnicity. Although not all of these methods are acceptable to all people, many women use a combination of fertility control methods. The most extreme examples of fertility control are sterilization and abortion. Sterilization in the United States is strictly voluntary. Abortion remains a controversial issue in the United States, as it is in other countries. The "morning-after pill", regardless of gender, continues to be controversial to some. The pill no longer requires a prescription there is no age limitation. An estimated 527,476 vasectomies were performed in the United States in 2015. From 2007 to 2015 there was a decrease in the proportion of vasectomies performed in all age groups and in all locations of the country. The end of the year and the month of March are when the most vasectomies are performed (Ostrowski et al. 2018).

Reflective Exercise 13.3

Lawrence Hilliard, a White 64-year-old manager in a large shoe store, presented at a neighborhood clinic with a chief complaint of difficulty urinating, especially in the morning. He has been divorced for over 20 years, lives alone, and has no children. He admits to weekend-only binge drinking when he is out with friends. During the week, he works 10-h days and does not have time to go out with friends in the evening. His diet consists of frozen TV dinners, prepared fast foods, sandwiches made from deli meats and cheese, and snacks that include potato chips, pretzels, cheese twists, soft drinks, and energy drinks. Mr. Hilliard's past medical history includes skin cancer, hemorrhoids, varicose vein surgery, and hypertension and hypercholesterolemia, for which he is on medication.

1. Name one occupation-related condition for which Mr. Hilliard is at risk.

2. What lifestyle changes would you recommend to Mr. Northrop?
3. Do you think giving him the Food-based dietary guidelines—United States of America will help him change his dietary habits?
4. Can you suggest an alternative for Mr. Hilliard for his weekend binge drinking?
5. Would you use low- or high-contexted communication with Mr. Hilliard?

Fertility practices and sexual activity, sensitive topics for many, are two areas where "outside" healthcare providers may be more effective than healthcare providers known to the client because of the concern about providing intimate information to someone they know.

13.9.2 Prescriptive, Restrictive, and Taboo Practices in the Childbearing Family

A prescriptive belief among Americans is that women are expected to seek preventive care, eat a well-balanced diet, and get adequate rest to have a healthy pregnancy and baby. The American healthcare system encourages women to breastfeed, and many places of employment have made arrangements so women can breastfeed while at work.

A restrictive belief among Americans is that pregnant women should refrain from being around loud noises for prolonged periods of time, for example a motorcycle or loud construction sites. Taboo behaviors during pregnancy among Americans include smoking, drinking alcohol, drinking large amounts of caffeine, and taking recreational drugs—practices that are sure to cause harm to the mother and/or baby.

In the American culture, in which the father is often encouraged to take prenatal classes with the expectant mother and provide a supportive role in the delivery process, fathers with opposing beliefs may feel guilty if they do not comply. The woman's female relatives may provide assistance to the new mother until she is able to care for herself and the baby. The current trend is for men to attend prenatal classes and be in the delivery suite.

Additional cultural beliefs carried over from cultural migration and shared among other cultures as well as the European American culture include the following:

- A pregnant woman should not reach over her head because the baby may be born with the umbilical cord around its neck.
- If you wear an opal ring during pregnancy, it will harm the baby.
- Birthmarks are caused by eating strawberries or seeing a snake and being frightened.
- Congenital anomalies can occur if the mother sees or experiences a tragedy during her pregnancy.
- Nursing mothers should eat a bland diet to avoid upsetting the baby.
- The infant should wear a band around the abdomen to prevent the umbilicus from protruding and becoming herniated.
- A coin, key, or other metal object should be put on the umbilicus to flatten it.
- Cutting a baby's hair before baptism can cause blindness.
- Moving heavy items can cause your "insides" to fall out.
- If the baby is physically or mentally abnormal, God is punishing the parents.

In the past, postpartum women were prescribed a prolonged period of recuperation in the hospital or at home, something that is no longer as feasible the shortened length of confinement in the hospital after delivery.

The healthcare provider must respect nonharmful cultural beliefs associated with pregnancy and the birthing process when making decisions related to the health care of pregnant women, especially those practices that do not cause harm to the mother or the baby. Most cultural practices can be integrated into preventive teaching in a manner that promotes compliance.

Reflective Exercise 13.4

Cindy Roberts, age 15 years, is 3 months pregnant and making her first visit to the maternal child nurse practitioner. She is accompanied by her 17-year-old boyfriend, Tommy, who is the father. Tommy declines to give his last name but says he will take care of Cindy and the baby. Cindy and Tommy currently live with Cindy's mother and three younger siblings. Cindy says she has no health problems and wants to keep the baby. She has made this appointment because she just wants to have a healthy baby. She is particularly concerned because her classmates tell her that she cannot lift heavy packages or carry her book bag because she might have a miscarriage. Her mother told her she could eat anything she wants, but Cindy is afraid that if she does, she will gain lots of weight like her mother, her boyfriend will leave her, and the baby might be born with a birthmark.

1. What is your overall first impression of this case?
2. Why might Tommy not be willing to provide his last name?
3. What kind of support does Cindy have? Do you think this support will continue after the baby is born?
4. What is your response to Cindy's concern about lifting heavy objects?
5. What is your response to Cindy's concern about eating whatever she wants?
6. What is your response to Cindy's concern that the baby might be born with a birthmark?
7. What overall advice do you have for Cindy's at this time?
8. Do you think you should call Cindy's mother and discuss the pregnancy?

13.10 Death Rituals

13.10.1 Death Rituals and Expectations

For many American healthcare providers educated in a culture of mastery over the environment, death is seen as one more disease to conquer, and when this does not happen, death becomes a personal failure. Thus, for many, death does not take a natural course because it is "managed" or "prolonged," making it difficult for some to die with dignity. Moreover, death and responses to death are not easy topics for many Americans to verbalize. Instead, many euphemisms are used rather than verbalizing that the person died: "passed away," "no longer with us," and "went to heaven."

The American cultural belief in self-determination and autonomy extends to people making their own decisions about end-of-life care. Mentally competent adults have the right to refuse or decide what medical treatment and interventions they wish to extend life, such as artificial life support and artificial feeding and hydration. Euthanasia is illegal in the US but assisted suicide/assisted death is legal in Washington, D.C., California, Colorado, Oregon, Vermont, Maine, New Jersey, Hawaii, and Washington (States with Legal Physician-Assisted Suicide 2019).

Most Americans believe that a dying person should not be left alone; accommodations are usually made for a family member to be with the dying person at all times. Healthcare providers are expected to care for the family as much as for the patient during this time.

Most people are buried or cremated within 3 days of the death, but extenuating circumstances may lengthen this period to accommodate family and friends who must travel a long distance to attend a funeral or memorial service. In some extremely cold regions, burial may be prolonged due to the frozen terrain. The family can decide whether to have an open or closed casket at the viewing (or wake). Cremation appears to be gaining in popularity and is usually decided by the person while they are still mentally competent; however, the family can also make that decision on cremation after death. A few elect to have burial at sea.

13.10.2 Responses to Death and Grief

American society has been launching major initiatives to help patients die with dignity and as comfortably as possible without pain. Once such initiative is the National Hospice and Palliative Organization (2013). As a result, more people are choosing to remain at home or to enter a hospice for end-of-life care where their comfort needs are better met. Some hospitals and long-term care facilities have started specialized units in recent years.

One of the requirements for entering a hospice in the United States is that the patient must sign documents indicating that he or she does not want extensive life-saving measures performed. Bereavement support strategies include being physically present, encouraging a reality orientation, openly acknowledging the family's right to grieve, accepting varied behavioral responses to grief, acknowledging the patient's pain, assisting them to express their feelings, encouraging interpersonal relationships, promoting interest in a new life, and making referrals to other resources such as a priest, minister, rabbi, or pastoral care.

13.11 Spirituality

13.11.1 Dominant Religion and Use of Prayer

In the United States, major religious groups include Protestant (46.5 percent), Roman Catholic (20.8 percent), Mormon (1.6 percent), other Christian (0.9 percent), Jewish (1.9 percent), Muslim (0.9 percent), Jehovah Witness (0.8 percent), Hindu (0.7 percent), Buddhist (0.7 percent), other or unspecified (22.8 percent), and none (4 percent) (CIA World Factbook 2019). These numbers are not broken down by race or ethnicity.

Many immigrants settled in America for religious freedom. Furthermore, specific religious groups are concentrated regionally in the United States, with Baptists in the South, Lutherans in the North and Midwest, and Catholics in the Northeast, East, and Southwest. Within this context, there is a separation of church and state, and the U.S. government cannot support a particular religion or prevent people from practicing their chosen religion. However, this does not include cults or extremist groups. Examples of cults and extremist groups examples Heaven's Gate, Aum Shunrikyo, and others. Cults have become associated with mind control, strict discipline, and extreme group mentality. And usually devote themselves to esoteric ideals and fads. They are usually small and tightly-knit, whose beliefs and religious practices operate outside the mainstream and are usually led by highly charismatic, controlling men, and subscribe to radical variants of Christianity.

Even though there is a separation of church and state in the United States, many public events and ceremonies open with a prayer, and phrases such as "one nation under God" are often included, something that is becoming more contentious by some. American money still has the phrase "in God we trust" printed on it. Most people see these religious symbols as harmless rituals. Instead of speaking to "religious values," politicians speak to "family values" as a way of getting around religious principles. However, these issues are subject to debate from time to time. Unlike many countries that support a specific church or religion and in which people discuss their religion frequently and openly, religion is not an everyday topic of conversation for most Americans.

The healthcare provider who is aware of the client's religious practices and spiritual needs is in a better position to promote culturally congruent health care. The practitioner must demonstrate an appreciation of and respect for the dignity and spiritual beliefs of clients by avoiding negative comments about religious beliefs and practices. Clients may find considerable comfort in speaking with religious leaders in times of crisis and serious illness.

13.11.2 Meaning of Life and Individual Sources of Strength

When Americans are asked what gives their lives meaning and where they find strength, a variety of answers are offered. Formal religion is one, but other common responses include family, work, self-improvement, friends, music, dance, hobbies, sports events, and meditation. For some, sources and meaning of life come from having a pet: dogs and cats are the most common.

13.11.3 Spiritual Beliefs and Healthcare Practices

Spiritual wellness brings fulfillment from a lifestyle of purposeful and pleasurable living that embraces free choices, meaning in life, satisfaction in life, and self-esteem. Practices that interfere with a person's spiritual life can hinder physical recovery and promote physical illness.

Healthcare providers should inquire whether the person wants to see a member of the clergy even if she or he has not been active in church. Religious emblems should not be removed without the patient's permission because they provide solace to the person. Removing them may increase or cause anxiety. A thorough assessment of spiritual life is essential for the identification of solutions and resources that can support health treatments.

13.12 HealthCare Practices

13.12.1 Health-Seeking Beliefs and Behaviors

The United States has undergone a paradigm shift from one that places high value on curative and restorative healthcare practices with sophisticated technological care to one of health promotion and wellness; illness, disease, and injury prevention; health maintenance and restoration; and increased personal responsibility. Most believe that the individual, the family, and the community have the ability to influence their health. For a few, good health may be seen as a divine gift from God, with individuals having little control over health and illness.

The primacy of patient autonomy is generally accepted as an enlightened perspective in American society. To this end, advance directives such as "durable power of attorney" or a "living will" are an important part of medical care. Accordingly, patients can specify their wishes concerning life and death decisions before entering an inpatient facility. The durable power of attorney for health care allows the patient to name a family member or significant other to speak for the patient and make decisions when or if the patient is unable to do so. The patient can also have a living will that outlines the person's wishes in terms of life-sustaining procedures in the event of a terminal illness. Each inpatient facility has these forms available and will ask the patient what his or her wishes are. Patients may sign these forms at the hospital or elect to bring their own forms, many of which are on the Internet.

Guidelines for immunizations were developed largely as a result of the influence of the World Health Organization's Immunization Standards (2019). Specific immunization schedules and the ages at which they are prescribed vary widely among countries and can be obtained from the WHO Web site (World Health Organization: Immunizations Standards 2019). Recently, controversy has arisen when some facilities have made the requirement that all employees must be immunized against the flu. Most employees comply, but a few see it as an infringement of their individual rights. If they do not comply, they may be required to wear a mask when on duty. In addition, some religious groups, such as Christian Scientists, and Ultra-traditional Jews do not believe in vaccinations. Beliefs like this, which restrict optimal child health, have resulted in court battles with various outcomes.

13.12.2 Responsibility for Health Care

The United States is moving to a paradigm in which people take increased responsibility for their own health. In a society in which individualism is valued, people are expected to be self-reliant. In fact, people are expected to exercise some control over disease, including controlling the amount of stress in their lives. If someone does not maintain a healthy lifestyle and then gets sick, some believe it is the person's own fault. Unless someone is very ill, she or he should not neglect social and work obligations.

In the United States, everyone, regardless of socioeconomic or immigration status, can receive acute-care services. However, they will be charged a fee for the service, and they may not be able to get nonacute follow-up care unless they can prove they are able to pay for the service, unless they go to one of the "free" outpatient services which are rare. Even if they are covered by health insurance, an insurance company representative may need to approve the visit and then have a list of procedures, medicines, and treatments for which it will pay.

Healthcare providers should not assume that clients who do not have health insurance or practice health prevention do not care about their health. Many are included in the working poor where they make minimum wage, and they cannot afford it.

Self-medicating behavior in itself may not be harmful, but when combined with or used to the exclusion of prescription medications, it may be detrimental to the person's health. A common practice with prescription medications is for people to take medicine until the symptoms disappear and then discontinue the medicine prematurely. This practice commonly occurs with antihypertensive medications and antibiotics. No culture is immune to self-medicating practices; almost everyone engages in it to some extent.

One cannot ignore the ample supply of over-the-counter medications in American pharmacies and grocery stores, the numerous television advertisements for self-medication, and media campaigns for new medications, encouraging viewers to ask their doctor or healthcare provider about a particular medication for diabetes and other conditions, erectile dysfunction, and a host of other conditions.

13.12.3 Folk and Traditional Practices

Some Americans favor traditional, folk, or magico-religious healthcare practices over biomedical practices and use some or all of them simultaneously. For many, what are considered alternative or complementary healthcare practices may be mainstream medicine for another person. In the United States, interest has increased in alternative and complementary health practices. The National Institutes of Health: National Center for Complementary and Integrative Medicine (2019).

As an adjunct to biomedical treatments, many people use acupuncture, acupressure, acumassage, herbal therapies, and other traditional treatments. Examples of folk medicines include covering a boil with axle grease, wearing copper bracelets for arthritis pain, mixing wild turnip root and honey for a sore throat, and drinking herbal teas. Most Americans practice folk medicine in some form; they may use family remedies passed down from previous generations.

An awareness of combined practices when treating or providing health education helps ensure that therapies do not contradict each other, intensify the treatment regimen, or cause an overdose. At other times, they may be harmful, conflict with, or potentiate the effects of prescription medications. Many times, these traditional, folk, and magico-religious practices are and should be incorporated into the plans of care for clients. Inquiring about the full range of therapies being used, such as food items, teas, herbal remedies, non-food substances, over-the-counter medications, and medications prescribed or loaned by others, is essential so that conflicting treatment modalities are not used. If clients perceive that the healthcare provider does not accept their beliefs, they may be less compliant with prescriptive treatment and less likely to reveal their use of these practices.

13.12.4 Barriers to Health Care

Barriers to health care for European Americans are essentially the same for any culture and include availability, accessibility, affordability, appropriateness, accountability, adaptability, acceptability, awareness, attitudes, approachability, alternative and complementary practices and providers, and health literacy (see Chap. 2). In order for people to receive adequate health care, a number of considerations need to be addressed. Several studies have identified that a lack of fluency in language is the primary barrier to receiving adequate health care in the United States (see Chap. 1).

Healthcare providers can help reduce some of these barriers by calling an area ethnic agency or church for assistance, establishing an advocacy role, involving professionals and laypeople from the same ethnic group as the client, using cultural brokers, and organizationally providing culturally congruent and linguistically appropriate services. If all of these elements are in place and used appropriately, they have the potential of generating culturally responsive and congruent care.

13.12.5 Cultural Responses to Health and Illness

Significant research has been conducted on patients' responses to pain, which has been called the "fifth vital sign", although this phrase is not as popular as previously because of the ongoing opioid epidemic. Most Americans believe that patients should be made comfortable and not have to tolerate high levels of pain. Cultural backgrounds, worldviews, and the variant cultural characteristics profoundly influence the pain experience.

Additional resources for pain are the U.S. Pain Foundation (2019), the American Pain Society that addresses the opioid epidemic (2019), and the OUCHER Pain scale (OUCHER! n.d.) for children. These pain scales go across racial and ethnicity populations. The healthcare provider may need to offer and encourage pain medication and explain that it will help the healing to progress. Research continues to be conducted in the areas of ethnic pain experiences and management of pain.

For some Americans, mental illness may be seen as being as important as physical illness. For some, mental illness and severe physical handicaps are considered a disgrace or cause a stigma. As a result, some families might keep the mentally ill or handicapped person at home as long as they can. This practice may be reinforced by the belief that all individuals are expected to contribute to the household for the common good of the family, and when a person is unable to contribute, further disgrace occurs.

In previous decades, physically handicapped individuals in the United States were seen as less desirable than those who did not have a handicap. If the handicap was severe, the person was sometimes hidden from the public's view. In 1992, the Americans with Disabilities Act went into effect, protecting handicapped individuals from discrimination (ADAgov n.d.).

Rehabilitation and occupational health services focus on returning individuals with handicaps to productive lifestyles in society as soon as possible. The goal of the American healthcare system is to rehabilitate everyone: people with criminal conviction, people with alcohol and drug problems, as well as those with physical conditions. Rehabilitation seems now seems be well established in the United States.

13.12.6 Blood Transfusions and Organ Donation

Most Americans and most, but not all, religions favor organ donation and transplantation and transfusion of blood or blood products. Jehovah's Witnesses do not believe in blood transfusions. This is a religious issue rather than a medical one. Both the Old and New Testaments clearly command to abstain from blood (Jehovah's Witnesses 2019). Recent developments have seen an increase in bloodless medicine and surgery, an alternative to blood transfusion that among other benefits, has been shown to reduce infections and help patients recover faster. Healthcare providers may need to assist clients in obtaining a religious leader to support them in making decisions regarding organ donation or transplantation.

13.13 HealthCare Providers

13.13.1 Traditional Versus Biomedical Providers

Most Americans combine the use of biomedical healthcare practitioners with traditional practices, folk healers, and magico-religious healers. The healthcare system abounds with individual and family folk practices for curing or treating specific illnesses. A significant percentage of all care is delivered outside the perimeter of the formal healthcare arena. Many times, folk and traditional therapies are handed down from family members and may have their roots in religious beliefs. Traditional and folk practices often contain elements of historically rooted beliefs.

The American practice is to assign staff to patients regardless of gender differences, although often an attempt is made to provide a same-gender healthcare provider when intimate care is involved, especially when the patient and caregiver are of the same age. However, healthcare providers should recognize and respect differences in gender relationships when providing culturally congruent care because not all people accept care from someone of the opposite gender.

13.13.2 Status of Healthcare Providers

Individual perceptions concerning competence and acceptability of providers may be closely associated with previous contact and experiences with healthcare providers. In general, healthcare providers, especially physicians, are viewed with great respect, although recent studies show that this is declining among some groups. Although many nurses in the United States do not believe they have respect, public opinion polls usually place patients' respect of nurses higher than that of physicians. For many years, surveys report that nursing is the most trusted profession. The advanced practice role of registered nurses (nurse practitioners) is gaining popularity and respect and the public sees them as equal or preferable to physicians in many cases. Evidence suggests that respect for professionals is correlated with their educational level, including baccalaureate, masters, and doctoral-level programs of study, and the impact of nursing interventions on healthcare outcomes. In the United States, approximately 10 percent of nurses are men, with very active campaigns in some areas of the country to recruit men and other underrepresented groups into nursing. The data do not have a breakdown for race and ethnicity (Data Brief Update: Current Trends of Men in Nursing 2017; Munoz and Hilgenberg 2005; Office of Disease Prevention and Promotion 2019).

References

American Chronology (2019). http://www.usahistory.info/timeline/

American Pain Society (2019). http://americanpainsociety.org/

Americans with Disabilities Act (n.d.). https://www.ada.gov/

Centers for Disease Control and Prevention (2016). https://www.cdc.gov/media/releases/2016/p0804-ethnic-groups-smoking.html

Centers for Disease Control and Prevention (2017). https://www.cdc.gov/hiv/statistics/overview/ataglance.html

Centers of Disease Control and Prevention (2019). https://www.cdc.gov/

CIA World Factbook (2019) United States. https://www.cia.gov/library/publications/the-world-factbook/geos/us.html

Data Brief Update: Current Trends of Men in Nursing (2017). http://healthworkforcestudies.com/publications-data/data_brief_update_current_trends_of_men_in_nursing.html

Dominant American Values (2019). http://whs.wsd.wednet.edu/Faculty/Cloke/documents/DominantAmericanValues.pdf

Food-Based Dietary Guidelines—United States of America (2019) Food and Agricultural Organization

of the United Nations. http://www.fao.org/nutrition/education/food-dietary-guidelines/regions/countries/united-states-of-america/en/
Guttmacher Institute: Contraceptive Use in the United States (2019). https://www.guttmacher.org/fact-sheet/contraceptive-use-united-states
How U.S. Gun Deaths Compare with Other Countries (2019). https://www.cbsnews.com/news/how-u-s-gun-deaths-compare-to-other-countries/
Jehovah's Witnesses (2019). https://www.jw.org/en/jehovahs-witnesses/faq/jehovahs-witnesses-why-no-blood-transfusions/
Kohl LR (2019) Values in American culture. http://www.bu.edu/isso/files/pdf/AmericanValues.pdf
Lewis PF, Zelinsky W, Beeman RR, Naisbitt J, Donald D, Pessen E (2019) History of the United States. https://www.britannica.com/place/United-States
Migration Policy Institute (2019). https://www.migrationpolicy.org/article/refugees-and-asylees-united-states
Munoz C, Hilgenberg C (2005) Ethnopharmacology. Am J Nurs 105(8):40–49
National Hospice and Palliative Organization (2013). https://www.nhpco.org/
National Institute on Alcohol Facts and Statistics (2018). https://www.niaaa.nih.gov/alcohol-health/overview-alcohol-consumption/alcohol-facts-and-statistics
Office of Disease Prevention and Promotion (2019). https://health.gov/DietaryGuidelines/
Ostrowski KA, Holt SK, Haynes B, Davies BJ, Walsh TJ (2018) Evaluation of vasectomy trends in the United States. Urology 118(10):76–79. https://www.ncbi.nlm.nih.gov/pubmed/29578040
OUCHER! (n.d.). http://www.oucher.org/the_scales.html
Prows CA, Prows DR (2004) Tailoring drug therapy with pharmacogenetics. Am J Nurs 104(5):60–71
Purnell L, Foster J (2003a) Cultural aspects of alcohol use: part I. Drug Alcohol Professional 3(3):17–23
Purnell L, Foster J (2003b) Cultural aspects of alcohol use: part II. Drug Alcohol Professional 2(3):3–8
Race and Ethnicity in the United States (2019). https://statisticalatlas.com/United-States/Race-and-Ethnicity
States with Legal Physician-Assisted Suicide (2019). https://euthanasia.procon.org/view.resource.php?resourceID=000132
The National Institutes of Health: National Center for Complementary and Integrative Medicine (2019). https://nccih.nih.gov/
U.S. Census Bureau (2019). http://factfinder2.census.gov/faces/tableservices/jsf/pages/productview.xhtml?pid=ACS_10_5YR_S1701&prodType=table
U.S. Pain Foundation (2019). https://uspainfoundation.org/
What is the Length of Term for the Senate and House of Representatives? (2019). https://www.reference.com/government-politics/length-term-senate-house-representatives-8c9b5deb7e2eefb2
World Health Organization: Immunization Standards (2019). https://www.who.int/immunization_standards/en/

People of Filipino Heritage

14

Nelson Tuazon

14.1 Introduction

Generational differences within families are associated with age and time of migration from the Philippines. Other factors influencing diversity include pre- and postmigration level of education, occupation, and intermarriage, as well as other variant cultural characteristics (see Chap. 2). This chapter discusses the major characteristics of mainstream Filipino culture, offering some insights into some differences among groups. The reader should avoid using this information as a universal template for every Filipino.

14.2 Overview, Inhabited Localities, and Topography

14.2.1 Overview

The Philippines is located in Southeast Asia in the Western Pacific Ocean. To its north lies Taiwan, and to its west is Vietnam. The tropical climate makes the Philippines prone to earthquakes and typhoons but has also endowed the country with natural resources and made it one of the richest areas of biodiversity in the world. The Philippines is an archipelago consisting of 7107 islands, with three main geographical divisions: Luzon, Visayas, and Mindanao (CIA World Factbook 2012). With a landmass of 300,000 square kilometers (115,830 square miles), it is slightly larger than the state of Arizona. The terrain is mostly mountainous with narrow-to-extensive coastal lowlands. The tropical climate consists of dry and rainy seasons suitable for year-round agriculture and fishing, but it is affected by the seasonal northeastern and southwestern monsoons. With an estimated population of about 104 million, the population has decreased since 2005 (CIA World Factbook 2012). Although the country is rich in natural resources and has a mixed economy of agriculture, light industry, and support services, almost 33 percent of the population lives below the poverty level (CIA World Factbook 2012). According to the Philippine Statistics Authority (2019), the proportion of poor Filipinos in 2018 was at 16.6%, down from 23.3% in 2015.

The Spaniards colonized the country for over three centuries from 1565 to 1898. Following the Spanish-American War, the islands were ceded to the United States and given the anglicized name "the Philippines." Filipinas (Pilipinas) and Philippines are used interchangeably today. Native speakers refer to the country as Filipinas or Pilipinas and use Philippines when speaking to outsiders or writing in English. In 1946, when the Philippines gained its independence from the

This chapter is an update from a previous edition written by Cora Munoz.

N. Tuazon (✉)
University Health System, San Antonio, TX, USA
e-mail: Nelson.tuazon@uhs-sa.com

© Springer Nature Switzerland AG 2021

L. D. Purnell, E. A. Fenkl (eds.), *Textbook for Transcultural Health Care: A Population Approach*,
https://doi.org/10.1007/978-3-030-51399-3_14

United States, it adopted the Tagalog-based *Pilipino* as its national language. In 1959, Pilipino was officially declared the national language. In 1986, however, the national assembly declared the national language as *Filipino*, based on existing Philippine and other languages. Generally, *Filipino* is used interchangeably with *Filipino American*. The term *Pilipino* is generally used to distinguish indigenous identity and nationalistic empowerment. Filipino Americans are a diverse group because of regional variations in the Philippines, which influence the dialect spoken, food preferences, religion, and traditions.

14.2.2 Heritage and Residence

The Filipino way of life is a tapestry of multicultural influences superimposed on indigenous tribal origins. The people are predominantly of Malayan ancestry with overlays of Chinese, Japanese, East Indian, Indonesian, Malaysian, and Islamic cultures (CIA World Factbook 2012). The Philippine culture is distinct from its Asian neighbors largely because of the major influences from Spanish and American colonization.

The Filipino sense of morality and justice evolved from tribal times. Close-knit, kin-based groups known as *barangays* emerged to protect communities from outside atrocities. Communal values of collective welfare and solidarity fostered security of its members in an unstable environment. Outsiders to the culture recognize these values in the Filipino traits of collective loyalty, generosity, hospitality, and humility. These basic values are strong components of childhood socialization in the family. Filipinos inculcate a strong sense of family loyalty beyond the nuclear family. Family obligations extend to cousins, in-laws, and others who are intimately linked with the family by ceremonies such as serving as sponsors of marriage or baptisms (Bautista 2002).

Most Filipinos in North America and elsewhere were born in the Philippines. The majority of Filipino Americans reside in the states of California, Hawaii, Illinois, New Jersey, New York, Washington, and Texas. Filipinos make up the second largest foreign-born population after Mexicans in the United States (Reeves and Bennett 2004). In 2016, more than 1.9 million Filipinos lived in the US, representing about 4% of the 44 million immigrants in the US. The Philippines has ranked the fourth largest origin country for immigrants since 2010, down from being the second-largest in 1990 (Zong and Batalova 2018).

14.2.3 Reasons for Migration and Associated Economic Factors

Many Filipinos migrate to America and elsewhere primarily for economic prosperity (Asis 2006). The economic challenges in the Philippines are exaggerated by its unemployment rate, foreign debt, population growth, and social inequities. These factors contribute to the migration of Filipinos, including physicians, nurses, engineers, information technology experts, and other laborers. American corporations, trade companies, and financial institutions partnering with Philippine companies have invested in promoting economic development and exchanges with the country. These larger political and economic influences, along with social and cultural exchanges, have helped create transnational community networks and established multiple migration streams over time (Espiritu 2003).

These early migrants were ineligible for citizenship and were denied privileges such as employment requiring citizenship, union membership, the right to own land, and the right to marry in states with antimiscegenation laws (laws prohibiting cohabitation, sexual relations, or marriage between people of different races). The Great Depression in the United States heightened racial animosity toward Filipino workers, and passage of the Tydings-McDuffie Act (Philippines Independence Act) in 1934 virtually ended immigration (Ceniza-Choy 2003).

In 1946, immigration restrictions for Filipinos were eased and they were granted naturalization rights. Between 1946 and 1965, 33,000 immigrants entered the United States and contributed to a 44 percent increase in the Filipino population. The Immigration Act of 1965 initiated a period of renewed mass immigration by promoting family reunification and recruitment of occupational immigrants. Since the passage of the 1965 Act, the Philippines has become the largest source of immigrants from Asia. A search for better economic and educational opportunities and reunification with family members in the United States continue to be the primary motivating factors for emigration. Working adult children sponsor their older relatives to come to the United States to care for their young children. In turn, older people facilitate the subsequent immigration of other children.

Because the Philippine economy has been unable to provide jobs for college graduates, an estimated 6 million Filipino professionals work overseas. Export of professional and skilled labor is one of the biggest industries in the Philippines. Remittances sent home by Filipinos overseas contribute as much as 10 percent to the country's gross domestic product, estimated at between 11 and 13 billion pesos (approximately US$278 million) in 2006 (IBON Foundation 2007).

14.2.4 Educational Status and Occupations

Around 1900, Americans introduced public education in the Philippines. Early training of schoolteachers was provided by the Thomasites, forerunners of the U.S. Peace Corps. The development of educational programs in the Philippines was highly influenced and patterned after those in the United States, as in the case of nursing and medicine. Early missionaries and philanthropic organizations such as the Daughters of the American Revolution, the Catholic Scholarship Fund, and the Rockefeller Foundation were instrumental in the Westernization of health-care education and practice in the Philippines. American nursing educators went to the Philippines, and Filipino nurses were sent to the United States for training. They subsequently returned to the Philippines and assumed leadership positions in nursing schools and hospitals. Since 1970, all nursing curricula have converted to a 4-year degree program leading toward a BSN (Pacquiao 2004). The BSN curriculum in the Philippines complies with international standards used for eligibility requirements for nursing practice (Reblando 2018).

The Philippines has one of the highest literacy rates in Asia at 96 percent. Schools are either publicly or privately funded. Formal education starts at the age of 7 years, with 6 years of primary education. Nursery school and kindergarten are offered in most private schools. Students get 4 years of secondary education in either a vocational-technical or an academic school. A high school graduate is 2 years younger than those graduating from U.S. high schools because of the omission of middle school years.

Filipinos view educational achievement as a pathway to economic success, status, and prestige for both the individual and the family. A person's profession is always identified when introducing, addressing, or writing about the person (e.g., Doctor, *Magpantay*, or Engineer, *Paredes*). A family's status in the community is enhanced by the educational achievement of the child, and a child's education is considered an investment for the whole family. Both male and female children are expected to do well in school, and parents do their best to provide for their children's full-time education. Adolescents who closely identify with their families are found to be concerned with the potential effect of their scholastic achievement on their families' reputation (Salazar et al. 2000). Family members and other relatives commonly contribute toward the education of their kin. Among lower- and middle-class families, siblings take turns going to college in order to maximize resources for one member to finish school, who can then contribute to the education of her or his siblings. One's choice of profession is generally a family decision and is based on potential economic return to the group. Hence, increased demand for nurses abroad attracts higher enrolment in nursing, as families view this occupation as a pathway to economic improvement.

Filipinos appear to be assimilated and successful and tend to blend into American society, which gives them a reputation as a "model minority." In reality, high educational attainment of American-born and immigrant Filipinos does not guarantee their entry into well-paying or high-status jobs. Significant discrimination confronts native-born and immigrant Filipinos in the American labor market linked with factors such as ethnicity, gender, region of residence, and level of education (Yamane 2002). As is the experience of many foreign graduates, Filipinos' education and experience are rarely matched with a suitable job because of the restricted labor market, resulting in many individuals competing for low-level jobs for which many are overqualified. Only those who are educated in health-care fields tend to find jobs consistent with their education.

Whereas American nursing education stresses critical thinking, in the Philippines, mastery of facts and rote learning are emphasized. A defined hierarchy exists in schools, with the teacher as the expert authority. This hierarchy is congruent with the social organization in the broader society, in which age and position are markers of status and power. The younger generations are rewarded for accepting the ideas and counsel of older people and teachers. Challenging authority and asserting one's creative ideas are unnatural predispositions, especially for the young. Nursing faculty have identified the tendency of Filipino students to take things at face value, avoid conflict, communicate nonassertively, and learn by rote memorization. Students' traditional values at home were in conflict with values in school and teacher expectations (Pacquiao 1996). Facilitating understanding of the dominant cultural values and norms in school, in addition to teaching the subject matter, is essential to facilitate these students' academic success.

14.3 Communication

14.3.1 Dominant Language and Dialects

Filipinos were influenced by American language and culture beyond the period when the United States recognized its independence in 1946. *Tagalog* is the primary language spoken in the Philippines. The two other official languages are English and Spanish. Within the Philippines there are approximately 75 ethnolinguistic groups who speak more than 100 dialects. For the sake of simplicity, the Philippine government has given all of these languages the collective name of Pilipino (Munoz and Luckmann 2005).

English is the official language used for business and legal transactions, as well as taught in secondary schools and universities. This has an impact on how Filipinos adjust in the United States and other English speaking countries, making it easier and faster to navigate the systems as compared with other Asian immigrant groups. Business and social interactions commonly use a hybrid of both Tagalog and English (*Tag-Lish*) in the same sentence. Tag-Lish is often used in health education. Most Filipinos speak the national language, Filipino, which is based on Tagalog (Tatak Pilipino 2007). This has created some regional tensions because the other dialects are not well represented in Tagalog as the national language, making it difficult for those in the Central and Southern regions of the country to learn and speak Tagalog.

Many Spanish words are found in the Filipino language such as *sopa* (soup), *calle* (street), *hija/hijo* (daughter/son), and *respeto* (respect). The influence of indigenous Filipino and Spanish languages produces distinct characteristics when Filipinos speak English. There is absence of certain sounds in the Filipino language such as short *i*, long *a*, and long *o*. Hence, *liver* may be enunciated as *lever*, *make* as *mik*, and *flow* as *flaw*. Many Filipinos are unable to differentiate *s* from *sh* (*physiology* as *fishiology*), *u* from short *o* or short *a* sounds (*cut* as *cot or cat*; *church* as *charts*). They have a tendency to place emphasis on the second syllable of a multisyllabic word (in ter'fe rence, pen ni'cill in, Ro bi'tus sin(TM)).

Filipino social hierarchy is evident in the language. Specific nouns rather than pronouns are used to denote a person's age, gender, and position in the social hierarchy. For instance, *Manang* and *Manong* are used to refer to or address an older woman and man, respectively. These nouns are used to address the person or when speaking about her or him. There is absence of the "she/

he" in the Filipino language. Rather, generic and gender-neutral pronouns *siya* (singular "she/he") and *sila* (plural "they/them") are used. Hence, many Filipinos may unconsciously use "she" and "he" interchangeably in reference to the same individual.

Although many Filipinos speak English, their ethnic language or dialect, knowledge and use of the English language, and age of migration often influence enunciation, pronunciation, and accentuation. Older Filipinos who originated from *non–Tagalog*-speaking regions may understand and speak better English than other Filipinos. In multigenerational Filipino American families, different languages may be used to communicate with family members and friends. Although many Filipinos speak and write fluently in English, they may have difficulty understanding American idiomatic expressions. For example, to a new immigrant, "How are you?" may be interpreted as a question about the person's well-being, requiring an elaboration of one's situation, rather than a mere greeting. Filipinos may have difficulty communicating their lack of understanding to others and may use ritualistic language and euphemistic behavior that appear to be the opposite of how they actually perceive the situation. Saving face, or concealment (Pasco et al. 2004), is a characteristic pattern of behavior employed to protect the integrity of both parties, which is a consequence of the cultural value on maintaining smooth interpersonal relations. Desirous of group approval, the individual becomes sensitive to the feelings of others and, in turn, develops a high sense of sensitivity to personal insults.

Traditional Filipino communication is highly contextual. It is basically formal, addressing individuals by their academic titles such as Dr. or Mr. and Mrs. The communication pattern is also rooted in the past and is hierarchical, which is consistent with respect for older adults or those with known accomplishment and achievement (Munoz and Luckmann 2005).

The individual is enculturated to attend to the context of the interaction and to adopt appropriate behaviors. Many Filipinos are keenly observant, displaying an intuitive feeling about the other person and the contextual environment during interactions. Contextual variables include the presence of *ibang tao* (outsiders) versus *hindi ibang tao* (insiders) and the age, social position, and gender of the other individual. In the company of insiders, such as one's family, each member develops an intuitive knowledge of the other so that words are unnecessary to convey a message and meanings are embedded in nonverbal communication.

In the presence of outsiders, a child's emotional outburst may be met with adults' stern silence, indifference, or euphemistic grins. These behaviors imply to insiders that emotional outbursts are inappropriate in front of outsiders. One may not disagree, talk loudly, or look directly at a person who is older and who occupies a higher position in the social hierarchy. Honorific terms of address denoting an individual's status within the hierarchy exist in all dialects. In Tagalog, when communicating to an older person or a person of status, he or she is addressed using gender and age-specific honorific nouns such as *Lolo/Lola* (Grandpa/Grandma), and *ate/kuya* (older sister/older brother).

Filipino interpersonal and social life operates to maintain smooth interpersonal relationships; communication tends to be indirect and ambiguous to prevent the risk of offending others. Filipinos may sacrifice clear communication to avoid stressful interpersonal conflicts and confrontations. As saying no to a superior is considered disrespectful, it predisposes an individual to make an ambiguous positive response. Filipinos are often puzzled, and sometimes offended, by the precision and exactness of American communication. Newly recruited Filipino nurses are stunned by their American coworkers' abrasiveness and open expressions of anger toward one another and their subsequent behavior of sitting down at coffee "as if nothing happened."

To many traditional Filipinos, actions speak louder than words. They value respect and might find questions like "Do you understand?" or "Do you follow?" disrespectful. It is preferable for the speaker to say, "Please let me know if I understood you correctly." When speakers occupy different positions in the social hierarchy, an informal and familiar manner of speaking by the subordinate may be perceived as impolite and disrespectful. Allowing time for a Filipino to respond not only communicates respect but also gives time for translating the dialect into English. Speaking clearly and slowly facilitates appreciation of varying pronunciation and accentuation of the English language across cultures.

14.3.2 Cultural Communication Patterns

Relational orientation has been suggested as the essence of Asian social psychology. Enriquez (1994) posited that the Filipino core values of shame (*hiya*), yielding to the leader or majority (*pakikisama*), gratitude (*utang na loob*), and sensitivity to personal affront (*amor propio*) emphasize a strong sense of human relatedness. These values originate from the central concept of ***kapwa***, which arises from the awareness of shared identity with others. *Kapwa* embraces the insider-outsider categories of human relations and prescribes different levels of interrelatedness or involvement with others. *Pakakikipagkapwa,* being one with others, implies accepting and dealing with the other individual as a fellow human being. *Kapwa* is grounded in the fundamental value of shared inner perception or feeling for another, from which all other attributes for human relations are made possible.

Eight levels of social interactions were identified by Enriquez within the core concept of *kapwa*. These levels demonstrate a hierarchy of human relatedness within the Filipino language and context of meanings. The contextual axis of interactions is conceptualized within a continuum of how the "other" is categorized—whether as an insider or outsider. The degree of sharing and involvement with outsiders may progress from levels 1 to 5, whereas interactions at levels 6 to 8 are observed with insiders. The eight levels are *pakikitungo* (civility, level 1), *pakikisalimuha* (interacting, level 2), *pakikihalok* (participating, level 3), *pakikibagay* (conforming, level 4), *pakikisama* (adjusting, level 5), *pakikipagpalagayang loob* (understanding and accepting, level 6), *pakikisangkot* (getting involved, level 7), and *pakikiisa* (being one with, level 8).

Developing working relationships with Filipinos requires an understanding of where one is situated within the insider-outsider continuum. Outsiders can move toward higher levels of interactions by observing cultural norms of communication, using trusted gatekeepers to mediate conflicts, seeking validation of perceptions of behaviors from more acculturated members of the group, and allowing face-saving opportunities to prevent embarrassment and personal denigration. When confronting a Filipino co-worker, provide privacy, and point out positive attributes as well as the problem. Observing nonverbal behaviors and interpreting them within the Filipino cultural context help promote culturally congruent interactions. Accommodating differential sharing and involvement between insiders and with outsiders shows cultural understanding that enhances development of intercultural relationships. For example, a Filipino speaking Tagalog with another reinforces the value of being one with others. Learning and using some Filipino greetings and honorific terms of address facilitate movement of the relationship to higher levels of involvement. Defining work situations in which Filipino dialects may be spoken demonstrates cultural sensitivity and accommodation. The insider and outsider delineations may be less important to some Filipinos who are highly educated and take pride in their global outlook. Unlike other immigrants who settle in ethnic enclaves, more recent Filipino immigrants acculturate and relate well with people from various cultures.

Smiling and giggling are often observed, especially among young Filipino women. The meanings of these spontaneous and highly unconscious behaviors are embedded in the context of the situation and may range from glee, genuine interest, and agreement, to discomfort, politeness, or indifference. It is helpful to point out how the behavior can be misinterpreted by patients and others, if inappropriate to the situation. Behavior change can be expected if correction is done in a timely, respectful, and sincere manner.

Having a heightened sensitivity to personal insults, Filipinos have a remarkable ability to maintain a proper front to protect their self-esteem when threatened. Conflict-avoidance behaviors to conceal discomfort or distress are evident in euphemistic denial of anger, minimization of pain, and silence. However, pent-up emotions and accumulated resentment may result in explosive anger, depression, and somatization. Health-care providers should be sensitive to these behaviors and explore the underlying causes by establishing trust and maintaining respectful relationships. Offering pain medications and attending to nonverbal behaviors, rather than waiting for the patient to verbalize his or her needs, are culturally congruent approaches.

First-generation Filipinos in North America have high regard for health-care providers (Abe-Kim et al. 2004) and present themselves in therapy sessions as polite, cooperative, verbal, and engaging. However, agreement with health-care providers does not ensure that patients will follow through with the recommendations. Health-care providers should be comfortable with patients' deferential attitudes without resorting to authoritarian approaches, which may be perceived as oppressive and may encourage euphemistic complaint behaviors. Once trust is developed, expression of authentic feelings is possible. Filipinos who are accustomed to indirect communication may perceive focusing on action-oriented strategies and outcomes as intrusive and coercive.

Direct eye contact varies among Filipinos depending on the degree of acculturation, length of time in emigration, age, and education. Some individuals may avoid prolonged eye contact with authority figures and older people as a form of respect. Older men may refrain from maintaining eye contact with young women because it may be interpreted as flirtation or a sexual advance. Filipinos are comfortable with silence and may allow the other person to initiate verbal interaction as a sign of respect. During a teaching session, a Filipino patient's nod may have several meanings that can range from "Yes, I hear you," "Yes, we are interacting," "Yes, I can see the instructions," or some other message that may be difficult for the patient to disclose. Validating a patient's response in a sensitive and respectful manner, as well as observing her or his behaviors, can prevent miscommunication.

Touch is used freely, especially with insiders. Greater distance is observed when interacting with outsiders and people in positions of authority. Same-gender closeness and touching, which may be perceived as homosexual adult behavior to some, are considered normal. Young adults of the same gender may hold hands, put one arm over another's shoulder, or walk arm-in-arm. As they become more acculturated, many Filipinos become aware of the differences and adapt to the new culture.

The implicit rules of the social hierarchy are observed when conflicts arise. A subordinate does not confront his or her superiors. Rather, a mediator who is likely to be a trusted individual at the same level of hierarchy as the superior may be employed to mediate and approach the superior on behalf of the subordinate. This behavior may be interpreted as dishonest by those who value direct and assertive communication.

14.3.3 Temporal Relationships

Filipinos have a relaxed temporal outlook. They have a healthy respect for the past, an ability to enjoy the present, and hope for the future. Past orientation is evident in their respect for older people and dead ancestors (*galang*), and a sense of gratitude and obligation to kin (*utang na loob*). Future orientation is manifested in the family's commitment to provide for the education of the young, parental participation in the care of their children and grandchildren, and a strong work ethic. A strong present orientation is associated with the cultural emphasis on maintaining positive relationships with others. Permanent social bonds with kin and significant others outside of kin are nurtured. Filipinos enjoy their families, fiestas, and life. They spend generously to make family events memorable and enjoyable. Although most Filipinos have adapted to American punctuality in the business sphere, promptness for social events is situationally determined. "Filipino time" means arriving much later than the scheduled appointment, which can be from 1 to several hours. The focus is on the gathering rather than on the schedule. A Filipino host may invite American guests at least 1 h later than the Filipino guests in the hope that both will arrive at the same time.

Reflective Exercise 14.1

Marianita de la Fuente, a 95-year-old female, was admitted to a general medical unit for chest pain, generalized weakness, and dizziness. She has diabetes mellitus and hypertension for which she is on medication. With 8 children and 23 grandchildren, she proudly tells all the health-care providers that all her children finished college and her grandchildren are doing very well in school. The patient had several diagnostic tests in the first 2 days of her hospitalization.

On the third day of her hospitalization, her condition worsened. Because she is a devout Catholic, the family requested a Catholic priest to provide the sacrament for the sick. With a large and extended family, Mrs. de la Fuente's room is always crowded with visitors. The staff are complaining about the number of visitors and how there is not adequate room when providing care for the patient.

1. What cultural value regarding family does the staff need to consider in this situation?
2. What are some culturally sensitive strategies that nurses can do to provide care to this patient and address the number of visitors in the patient's room?
3. Describe the Filipino family kinship system and the roles of older parents and adult children.
4. How does Mrs. de la Fuente express her value of education?

14.3.4 Format for Names

The Filipino family is bilineally extended to several generations. Kinship and family affinity can be legally and spiritually claimed equally from both sets of families, giving the child the identity of the extended family. This bilineal kinship is reflected in their names. Children carry the surnames of both parents. For example, Jose Romagos Lopez and Leticia Romagos Lopez are the children of Maria Romagos and Eduardo Lopez. The middle name or initial (R) is the mother's maiden name, Romagos. After marriage, Jose keeps the same name, whereas his sister's name becomes Leticia L. Lukban (her husband being Ernesto Lukban). Leticia's maiden name, Lopez, is abbreviated as her middle initial.

Many Filipino names are of Spanish origin. Symbolic of Filipinos' Catholic faith, saint names are often used with first names. Filipino females may have a Ma. (for Maria) before their given names—for example, Ma. Luisa stands for Maria Luisa. Although the name Maria is often given to girls, some males may use Maria as a first or second name—hence, Ma. Jose Romagos Lopez and Jose Ma. Paredes Castro. The saint name is an integral part of the first name, so an individual will use both first names: Maria Luisa or Jose Maria. Few Filipino American women keep their own surname after marriage, although this may increase among second- and third-generation Filipinos.

Adults use first names to address young children. Nicknames symbolizing affectionate regard for the person (*Nini*, Baby, *Bongbong*) are commonly used instead of the first name. These nicknames may indicate special meanings, positions, and/or outstanding characteristics of the child. First names are avoided when addressing older adults and those occupying higher positions in the hierarchy. In formal business transactions, prefixes such as Mr., Mrs., Miss, or Ms. or the person's professional degree are used before the person's last names (Dr. Abaya or Attorney Abaya).

14.4 Family Roles and Organization

14.4.1 Head of Household and Gender Roles

Since the pre-Spanish era, Filipino women have been held in high regard, having rights equal to those of men (Agoncillo and Guerrero 1987). In contemporary Filipino families, although the father is the acknowledged head of the household, authority in the family is considered egalitarian. The mother plays an equal, and often major, role in decisions regarding health, children, and finances.

Traditional female roles include caring for the sick and children, maintaining kinship ties, and managing the home. Parents and older siblings are involved in the care and discipline of younger children. In extended family households, older relatives and grandparents share much authority and responsibility for the care and discipline of younger members. Traditional Filipino families may not expect female children to engage in activities that are considered appropriate for men, such as driving, bicycling, and other functions requiring mechanical or technical skills. Blurring of roles between men and women occurs with increased education, urbanization, and emigration to a new culture.

Filipino families predominantly consist of married couples with both spouses working. Filipino womanhood has evolved from the Spanish construct of modesty, demureness, and femininity to a contemporary image of a woman who is educated, working, and adept at balancing traditional roles and career demands. Traditional Filipino parents expect their male and female children to pursue college education and economically productive careers and also to have a family. Family members and Filipino friends or acquaintances are the preferred caregivers of young children when parents are working. Older adults, especially grandmothers, emigrate in time for the birth of their grandchildren and are expected to take care of them on behalf of their working adult children (Pacquiao 1993).

14.4.2 Prescriptive, Restrictive, and Taboo Behaviors for Children and Adolescents

Like other cultural groups, Filipinos highly value their children. They are seen as gifts from God and therefore are considered special blessings to the family. The strong in-group consciousness of Filipinos is rooted in the centrality of family and kin, to the exclusion of others, in the socialization of individuals. As the strongest unit of society, the family demands the deepest loyalties and significantly influences an individual's social interactions. Ascriptive and particularistic personal ties with kin are significant in the allocation of rank, authority, and power to individuals. Generational position conditions the status as well as the role performance of individuals. The family and one's familial role define and order authority, rights, obligations, and modes of interaction. Younger generations are taught to be respectful and heed the authority of older siblings and relatives, parents, and grandparents. Respect is manifested in both speech and actions by using honorific terms of address, avoiding confrontation and offensive language, keeping a low tone of voice, greeting older people by kissing their forehead or back of their hand, avoiding direct eye contact when being admonished, offering food, touching, and so forth. Husbands and wives address each other using the honorific terms that they wish to model for their children. In front of the children, a husband will address his wife as *Inay* (Mother) and the wife correspondingly refers to her husband as *Itay* (Father). Under no circumstance are children permitted to call their parents by their first names. Friends of Filipino children are expected to show respect to adult members of the family when they visit.

Reciprocal obligations among kin are embodied in the value of *utang na loob*, a personal sense of indebtedness and loyalty to kin, which carries an obligation to repay or perform services for one another. Filial respect and obligation for caring for one's parents is the ultimate confluence of generational respect and reciprocal obligation. Childhood socialization to the mechanism of shame (*hiya*) reinforces the value of *utang na loob* and generational respect. Failure to perform or recognize reciprocal obligations, as well as disrespect of older people or people of authority, results in the loss of one's self-esteem and status, as well as incurs shame to one's family.

Conditions such as mental illness, divorce, terminal illness, criminal offenses, unwanted pregnancy, homosexuality, and HIV/AIDS are not readily shared with outsiders until trust is established. The extent to which a Filipino patient may disclose personal information is contextualized. Family presence may act as a barrier to full disclosure of conditions that may be perceived as putting the family at risk for shame.

Dating at an early age is discouraged for young daughters who are advised that a short courtship period may suggest that they are "easy to get." Young men with sincere intent must strive to get on the good side of the family and have patience with a long courtship. Open demonstrations of affection with sexual undertones are to be avoided by the young couple. Ideally, the groom's parents formally ask for the bride's parents' consent for the marriage of their children. Traditional families desire that their daughters remain chaste before marriage. Pregnancy out of wedlock brings shame to the whole family. Modernization and urbanization have changed the social mores in the Philippines; yet, many Filipino American families are still perceived by younger family members as having an overly protective attitude toward children in matters of "hanging out" with friends, dating, and courtship. Girls are subjected to greater limitations than boys, which contributes to higher reports of contemplating suicide by Filipino girls. Studies of second-generation Filipino students in high schools revealed greater parental control over daughters, with more latitude allowed for sons. For many Filipinas, high school achievement was met by parental control over their choice of colleges and pressure to remain close to home and family supervision (Wolf 1997).

14.4.3 Family Goals and Priorities

In addition to blood relatives, fictive kinship is established through the *compadrazgo* system in which friends and associates are invited to become godparents or surrogate parents in religious ceremonies, such as baptism and marriage. Fictive kinship is a significant support system for Filipino Americans who left families or relatives in the home country. In times of illness, the extended family provides support and assistance. Sometimes, a family visit to the hospital takes on the semblance of a family reunion.

The family is the basic social and economic unit of Filipino kinship. Family relations strongly influence individual decisions and actions. Relatives and family constitute the reference group for individuals, determining their behavior as well as that of their relatives in any social exchange. Family loyalties and obligations supersede individual interests and residential migration. This is evident in migration patterns of adult children and aged parents, which are planned to maximize the economic welfare and support for group members.

Family emphasis on communal values and generational respect is highly institutionalized. Community activities generally center on the family. Fiestas, weddings, baptisms, illnesses, and funerals are occasions for reinvigorating relations with kin and rekindling local connections, in which the presence and, more importantly, the absence of relatives are viewed as highly significant. Early child-rearing practices are permissive, with emphasis on providing an emotionally secure environment for the child. Priority is placed on promoting the child's well-being and social acceptance. The child is introduced early into various mechanisms designed to impose compliance with family values. A family's prestige is measured by the upbringing of their children, judged by their adherence to traditional cultural values.

The family emphasis on faithfulness to religious obligations is tied with the cultural values of generational respect and reciprocal obligation. Child-rearing practices stress entire family participation in the religious education and adherence to rituals by young members. Older generations share the responsibility for reinforcing these values. Religious sacraments, such as marriage, are embedded in the age-grading activities of the extended family.

As the basic economic unit of society, the family defines the economic obligations of kin to one another. Children are looked upon as economic assets and as sources of support for parents in old age. Thus, educating young members becomes a family priority. The socioeconomic status of the aged is closely linked with the family's wealth; if resources are limited, older people rely on children and relatives. Older parents and grandparents are integrated within the family, thus lessening the impact of advancing age. Traditional Filipinos consider institutionalization of aged parents tantamount to abandonment of filial obligation and respect for older people. Many older people aspire to return to the Philippines to spend their remaining years with loving kin.

The development of *pakiramdam* and *kapwa* is the defining goal of the family. Group cohesiveness, loyalty, and faithfulness to shared obligation are expectations that transcend distant migration, marriage, and adulthood. Students who feel obligated to maintain their family's reputations believe that effort and interest, rather than ability, can result in school success (Salazar et al. 2000).

Filipino American older people have reported experiencing conflict between the maintenance of family obligations, such as babysitting for their grandchildren and their desire to be more independent from their adult children. Family obligations may result in their inability to meet medical appointments, obtain needed medications, and make meaningful social connections because of lack of independent transportation. Depression has been associated with loneliness, feelings of isolation, and financial difficulty (McBride and Parreno 1996). Older Filipino Americans identified integration in the family of their adult children, participation in community activities with family and close friends, and maintaining religious functions as highly important (Pacquiao 1993).

Diversity exists in the degree to which Filipino Americans adhere to the traditional cultural values. Some middle-aged immigrant Filipino parents do not expect to live with their children in old age. Diversity in family member roles and priorities exists as a result of the financial resources of the family. Reciprocal obligations with kin are expressed differentially based on the capacity of older people and adult children to meet them and include economic, physical, emotional, and social support dimensions.

14.4.4 Alternative Lifestyles

Traditional Filipino parents seldom provide sex education; sex is not discussed openly at home. Homosexuality may be recognized and considered an aberrant behavior, but it is not openly practiced in order to save face and prevent shame for the family. However, in recent years, younger gay, lesbian, bisexual, and transgender Filipinos in the Philippines and in North America are taking a more active role in being recognized and expressing their rights.

Although the tenets of the Catholic Church have a direct bearing on sexual mores for older generations of Filipinos, they have less influence on younger generations, as is seen in the high incidence of HIV/AIDS among Filipinos compared with that of other Asian/Pacific Islanders (APIs). The family may not be the primary source of support for individuals, who may be isolated to prevent stigma to the family. The nuclear family may protect the affected member from outsiders and intentionally remove them from a network of friends and extended family. Providing an atmosphere that fosters the much-needed sense of belonging should be the goal of culturally congruent services.

Divorce can carry a stigma for older and more traditional Filipinos, especially those who are devout Catholics. The stigma may be worse for Filipinos in the Philippines for whom divorces are not allowed and are considered a religious taboo. Divorces among Filipino Americans and in other countries generally result from failed marital duties, lack of mutual support between partners, and marital infidelity.

14.5 Workforce Issues

14.5.1 Culture in the Workforce

Experience with racism is a continuing theme voiced by Filipino nurses working with White American nurses (Spangler 1992). Among Filipino American female nurses and nurse's aides, longer residence in the United States is associated with increased stress, evidenced by higher levels of serum norepinephrine, and higher diastolic pressure and lower dips in blood pressure readings during sleep (Brown and James 2000). The requirement by the American Nurses Association for equal pay for the same job transformed foreign nurse recruitment into a competitive enterprise, in which employers and existing staff expected recruited nurses to be functionally competent on the job as soon as they received their American RN license because they will receive pay comparable with that of other RNs. In reality, providing transitional support for foreign nurses requires a significant commitment of time and financial investment and a prolonged acculturation process (Pacquiao 2004).

Filipino nurses have been recruited in large numbers to staff mostly evening and night shifts in which acute shortages of American trained nurses exist. This has reinforced the cultural tendency toward collective solidarity by defining the context of interactions within the insider-outsider continuum. American nurses and administrators of health organizations with large contingents of Filipino nurses are becoming aware of the need for special knowledge and skills in understanding and managing a diverse workforce and in developing culturally specific staff development programs.

Cultural conflicts in the workplace stem from different communication patterns: the dominant norm of assertiveness versus the highly contextual Filipino communication. The cultural concept of shared identity with other Filipinos creates a propensity among Filipino nurses to speak in their own dialect with one another to the exclusion of non-Filipino co-workers and patients. A lack of fluency in speaking and in enunciating English words results in anxiety when interacting with outsiders. Assertive communication is difficult for Filipinos, who have been enculturated to avoid conflict. Filipino nurses may consider it impolite and disrespectful to confront or challenge the authority of a superior. When a problem with a manager occurs, a Filipino nurse may communicate through a mediator, usually another Filipino nurse, who is in the same level within the hierarchy as the manager. Communicating disagreement with a physician is difficult for many Filipino nurses. Conversely, Filipino registered nurses expect their subordinates to be deferential toward them.

Conflict can result from different cultural values about caring. Coming from a highly collective orientation, Filipinos define caring in terms of active caring for others. This perspective differs from the American value of self-care. Filipino nurses feel comfortable performing what they perceive as caring tasks for patients that American nurses expect patients to do for themselves. Initially, they may not be inclined to teach and demonstrate procedures to patients because of their traditional belief in doing the caring tasks for patients. Outsiders may misconstrue Filipino nurses' preoccupation with caring tasks as disorganization or lack of assertiveness.

Different views about a valued co-worker may be another source of conflict. The Filipino values of shared perception and being one with others create a cooperative, rather than a competitive, outlook. A valued individual produces for the group and puts the group above her or his own personal gain. Humility, hard work, loyalty, and generosity are admired. The business like and competitive perspectives of Americans, in which behavior is internally motivated by individual gain, may be interpreted as selfish and uncaring. Self-proclamations of accomplishments are viewed as cocky and offensive. Instead, it is up to the group to recognize a member's achievement, which is assessed in terms of how the action benefited the group.

Health-care organizations are cultural entities defined by norms that reflect the dominant values of the host society. Professional schools mirror these dominant societal norms, which are congruent with those of health-care organizations. Among outsiders to the dominant culture, the experience in nursing schools and health-care organizations is dissonant with previous life experiences, which require an understanding of both cultural and occupational role differences. Bicultural development of Filipino and non-Filipino staff should be the goal of occupational orientation and training. Biculturalism requires awareness of self and others and the ability to adapt behaviors that build positive relationships with others who may be different from oneself (Pacquiao 2003). Understanding cultural differences and similarities allows for the development of intercultural understanding and skills that promote teamwork. Bicultural mentors who can teach cultural norms of the organization and work with diverse patients and staff will foster the individual's ability to adapt behaviors. Staff development requires training in frame switching—using different frameworks to understand behaviors of others and commitment to the belief that other perspectives are equally sound in explaining our experiences. Impression management is a bicultural skill that is grounded in the ability to interpret behaviors of others within their own cultural context and manifest behaviors that promote relationship and intercultural understanding (Pacquiao 2001).

14.5.2 Issues Related to Autonomy

A core Filipino cultural concept is *bahala na*, which consists of the belief and predisposition to trust the divine providence and social hierarchy to resolve problems. Filipinos may avoid taking an active role in managing problems because of their fatalistic belief that a "greater power" will prevail. Outsiders may interpret this behavior as a lack of initiative or responsibility. Many Filipino nurses are hesitant to assume leadership roles and assert their points of view, especially with outsiders. After an initial effort, further attempts to resolve the problem are generally left to the leader or hierarchy. Providing support and acting as a role model help nurses assert themselves and feel confident in problem solving and conflict resolution. Filipinos are proud people who place importance on maintaining self-esteem and dignity by saving face and avoiding shame. Their sensitivity and attention to other people's feelings are often exhibited as indecisiveness, which many interpret as lack of assertiveness.

The Filipino hierarchy and emphasis on collectivism bring a consequent group-oriented sense of responsibility and accountability. The leader is respected, followed, and expected to make decisions on behalf of members. The leader is trusted to act in the best interests of the group. The concept of individual accountability and responsibility in a highly litigious society, such as the United States, may initially be difficult for Filipino nurses to understand. Supportive role modeling in assuming individual accountability is important for Filipino-educated nurses.

14.6 Biocultural Ecology

14.6.1 Skin Color and Other Biological Variations

Variations in anthropomorphic physical and biophysiological characteristics of Filipinos exist as a result of ethnic and racial intermingling. The people of one of the Filipino aboriginal tribes, the Aeta, are negroid and petite in stature. They are believed to have migrated from Africa through land bridges during the Ice Age. However, like other tribal groups in the Philippines, they are now a minority. The typical native-born or immigrant Filipino may be of Malay stock (brown complexion) with a multiracial genetic background.

The youthful features of Filipinos make it difficult for some to assess their age. Common Filipino physical features may include jet black to brunette or light brown hair, dark to light brown pupils with eyes set in almond-shaped eyelids, deep brown to very light tan skin tones, and mildly flared nostrils and slightly low to flat nose bridges. The eye structure may challenge health-care providers in assessments such as observing pupillary reactions for increased intracranial pressure, measuring ocular tension, and evaluating peripheral vision. The flat nose bridge may be overlooked by opticians when fitting and dispensing eyeglasses.

The high-melanin content of the skin and mucosa may pose problems when assessing signs of jaundice, cyanosis, and pallor. This feature also poses difficulty in diagnosing retinal, gum-related, and oral tissue abnormalities. When performing skin assessments, practitioners should consider the complexion and skin tone of the Filipino patient. The usual manifestations of anemia (pallor and jaundice) should be assessed in the conjunctiva. Newborns may have Mongolian spots—bluish-green discolorations on the buttocks—that are physiological and eventually disappear.

Filipinos range in height from under 5 feet to the height of average Americans. Body weight varies according to nativity and other factors such as nutrition, physical activity, and heredity. Filipinos commonly gain weight when they migrate. There are no definitive studies relating nutrition with standard height and weight measures for this population; therefore, it is essential to assess for weight changes on an individual basis.

Filipinos have a small thoracic capacity. Approximately 40 percent have blood type B and a low incidence of the Rh-negative factor (Anderson 1983). As more interracial families emerge in Filipino communities, changes in their serologic profile will likely occur.

14.6.2 Diseases and Health Conditions

Filipino men and women have the highest prevalence of hypertension compared with Whites (Ma et al. 2017; Ryan et al. 2000), compared to other racial and ethnic minority populations (Ursua et al. 2013), and compared with other Asians characterized by sodium sensitivity (Garde et al. 1994). Despite the high prevalence of hypertension among Filipino Americans, the rate of controlled hypertension is lower than the rates for Whites and other Asians. Filipino Americans are less aware of their hypertension compared to other minority populations (Ursua et al. 2014). Ursua et al. (2018) have proposed that a community-based program using community health workers can improve blood pressure control among Filipino Americans.

Filipinos also had the highest rate and risk of type 2 diabetes at 32.1 percent, compared with 5.8 percent in Whites and 12.1 percent in African American women. Filipino women were found to have higher visceral adipose tissues compared with non-Hispanic Whites and African American women. Filipino Americans have a disproportionately high prevalence of diabetes in a non-obese population (Bateman et al. 2009). Family history of diabetes and higher body mass index (BMI) are related to an increased perceived risk of diabetes (Fukuoka et al. 2015).

High incidence of hyperuricemia is attributed to a shift from a Filipino to an American diet (McBride et al. 1995). Liver cancer tends to be diagnosed in the late stages of the disease and appears to be associated with the presence of the hepatitis B virus. Silent carriers of the virus are common among Asians; its presence is detected only when other problems are being evaluated. Health-care providers should routinely screen for hepatitis B virus, especially among recent immigrants. A high incidence of glucose-6-phosphate dehydrogenase (G-6-PD), thalassemias, lactose intolerance, and malabsorption exist among the Filipino population (Anderson 1983).

Compared with other APIs and White males, Filipinos are more likely to be diagnosed with advanced-stage colorectal and prostatic cancers. They have the worst survival rates from these cancers (Lim et al. 2002). Like other APIs, Filipinos underuse cancer screening tests (Kagawa-Singer and Pourat 2000). Filipino Americans are at increased risk for type 2 diabetes and have higher visceral adipose tissue (VAT) than Whites and African Americans (Araneta and Barrett-Connor 2005; Bender et al. 2018;

Maglalang et al. 2017). The three leading causes of mortality among Filipino Americans are cardiovascular disorders (Bayog and Waters 2017; Domingo et al. 2018) followed by stroke (Hastings et al. 2015) and cancer (Chawla et al. 2015; Cuaresma et al. 2018; Tran et al. 2018).

Lack of insurance, low income, and limited access to care were found to have a significant impact on APIs' use of health services (Coughlan and Uhler 2000; Yu et al. 2004). A Canadian study using the 2001 Community Health Survey revealed that minorities, including Filipinos, were less likely to be admitted in the hospital, tested for prostate-specific antigen (PSA), or given a mammogram or Pap test, despite the fact that they had more contact with a general practitioner than White Canadians (Quan et al. 2006). Among older Filipinas, length of residence in the United States and having had a checkup when no symptoms were present were associated with adherence to cancer screening (Maxwell et al. 2000).

Compared with White European Americans, Filipinos have higher levels of depression. In contrast, strong ethnic identity characterized by sense of ethnic pride, involvement in ethnic practices, and cultural commitment to one's racial and ethnic identity were significant factors in mitigating depressive symptoms among Filipino Americans (3Mossakowski 2003). Strong bonds with members of the community and access to culturally congruent health services promoted commitment of older Filipinas to planned physical activity (Maxwell et al. 2002).

14.6.3 Variations in Drug Metabolism

Studying how ethnicity affects drug response is challenging, in part because of the tremendous variations that exists within each ethnic group. Many studies have used broad categories when classifying participants without differentiating among subgroups—for example, using the term *Asians* to refer to Filipinos, Korean, and Chinese, among others (Munoz and Hilgenberg 2006). Ethnographic research has uncovered significant differences in how people of color metabolize drugs differently. There are variations in both pharmacodynamics and pharmacokinetics mechanisms of action. Also, certain ethnic groups have more of these variations than others (Lin and Smith 2000).

Compared with White Americans, Asians require lower doses of central nervous system depressants such as haloperidol, have a lower tolerance for alcohol, and are more sensitive to the adverse effects of alcohol (Levy 1993). Owing to the sodium-sensitive nature of hypertension affecting Filipinos and the high-sodium content of their diet, use of diuretics should be considered. Culturally congruent stress management in addition to dietary modifications and physical activity should be included in the treatment plan to control high blood pressure.

Because of availability of over-the-counter antibiotics and lack of adequate medical monitoring of these drugs in the Philippines, Filipino immigrants may be insensitive to the effects of some anti-infectives. A positive reaction to tuberculin or the Mantoux test is observed because of the practice of giving bacille Calmette-Guérin (BCG) vaccinations in childhood. Chest X-rays and sputum cultures are recommended for screening and diagnosis of tuberculosis. More research is needed to determine pharmacodynamics among Filipinos, including gender differences. Health-care providers, as with all patients, need to assess Filipino patients individually when administering and monitoring medication effects.

14.7 High-Risk Behaviors

Gender differences are evident in the Filipino tolerance and acceptance of high-risk health behaviors related to alcohol, drugs, cigarettes, and safe sex. More Filipino men than women are heavy drinkers. Most Filipino Americans report drinking socially, with a small number reporting having three or more drinks per day (Garde et al. 1994). Because denial is closely associated with alcoholism, the frequency and amount of alcohol taken are generally underreported.

Cigarette smoking is more prevalent among Filipino men than women. Smoking rates have been positively correlated with lower educational levels and income and a tendency to think or speak in a Filipino language, and, for women, being born in the United States. Most Filipino youths reported living with an adult who smoked, and their first substance of choice was cigarettes, followed by alcohol and inhalants (Maxwell et al. 2007). When compared with immigrant Filipinos, second-generation Filipinos were more likely to smoke cigarettes (Bayog and Waters 2018).

Filipinos constitute the largest number of reported HIV/AIDS cases among APIs in the United States (Reeves and Bennett 2004). Low knowledge scores on information about HIV transmission and unprotected sex with multiple partners underscore the urgency of HIV and AIDS education and prevention.

14.7.1 Health-Care Practices

Cultural, social, and economic factors are implicated as reasons for Filipino Americans' underutilization of health services. Typical of the ethnically underserved, older people in the United States may be unaware of available services and are reluctant to access social and health services, particularly when culturally sensitive and bilingual providers are unavailable. Lack of transportation, fear of going to the area where services are located, and inappropriate program design are some of the other reasons for low utilization of services by this group. More recent Filipino immigrants differ significantly from their earlier counterparts in their access and utilization of health services. This group is highly educated and accesses many of the health-care services in the United States.

A study of the experiences of Filipino women with breast cancer screening services identified a pattern of avoidance. Factors contributing to this behavior included cultural beliefs, lack of health insurance, and lack of a familiar source of care (Wu and Bancroft 2006). Some believe that undergoing the test and attempting to know one's condition could tempt faith, which can bring bad luck. Avoidance of an unpleasant diagnosis and concealment of serious illnesses are consequent behaviors of this belief. Many Filipinos seek a familiar and consistent health practitioner who has established a relationship with them. Gender-congruent health-care providers are preferred for conditions specific to women's or men's health. Preference for culturally congruent services and practitioners and the presence of supportive social connections increased participation and commitment among older Filipinas for health promotion (Maxwell et al. 2002).

14.8 Nutrition

14.8.1 Meaning of Food

Food to any group is symbolic and is associated with the affective state of the individual. It is a source of nourishment to the body, as well as a source of pleasure and satisfaction, depending on one's emotional and psychological state. To the Filipino, food is a fundamental form of socialization. Food and meal patterns are integral to the cultural emphasis on generosity, hospitality, and thoughtfulness that support group cohesiveness. No social gathering of Filipinos occurs without food. Food is offered as a token of gratitude and caring, to welcome others, to celebrate accomplishments and important events, to offer support in times of sickness or crisis, and to reinforce social bonds in everyday interactions. Sharing food with others, or at the very least inviting others to share one's food, is expected of Filipinos and considered a sign of good upbringing. The insider versus outsider context influences the choice of food offered (Enriquez 1994).

In the Philippines, traditional Filipino meals are labor intensive, requiring the participation of several family members. Meats are costly, so small amounts are cut in pieces and expanded using vegetables and starches to feed an entire family. All family members, regardless of age, attend social gatherings at which a variety of dishes are prepared to accommodate individual choices. The hosting family serves large amounts of food to accommodate invited guests and those who happen to be around. Guests customarily linger for several meals because the focus is on the gathering. Latecomers are welcomed and expected to fully participate in the entire meal and the company of other guests. Dishes are served all at once from appetizers to desserts so guests are free to eat their courses without waiting for everyone to arrive. Guests are encouraged to return to the table to join arriving guests. More food means more portions for each one and vice versa.

14.8.2 Common Foods and Food Rituals

Indigenous Filipino cooking is characterized by boiling, steaming, roasting, broiling, marinating, or sour-stewing to preserve the fresh and natural taste of food. Spanish, Chinese, and American influences are integrated into contemporary Filipino cuisine. Foods may be sautéed, fried, or served with a sauce. Because of the tropical climate of the Philippines, many types of plants and animals flourish. Seafood (fish and shellfish) forms the bulk of the Filipino diet. Fresh, dried, and marinated fish are abundant in the diet.

In the Philippines, animal sources of protein are chicken and pork because cows and water buffalo are primarily used for farming. Because protein-rich foods are costly, meals generally consist of larger portions of carbohydrates, primarily rice. Plants are the second most important food source and include a variety of seaweeds, edible roots, delicate leaves, tendrils, tropical fruits, seeds, and some flowers. Fruits and vegetables are consumed in large quantities in a variety of ways. Rice is eaten at every meal—steamed, fried, or as a dessert. Less acculturated Filipinos tend to prepare and serve more traditional Filipino foods at home (De la Cruz et al. 2000). Filipino and Asian food stores are abundant in regions where many APIs reside.

Except for babies and young children, milk is almost absent in the Filipino diet, partly due to lactose intolerance. However, milk in desserts such as egg custard (flan) and ice cream seems to be tolerated. In the Filipino food pyramid, milk and dairy products are incorporated in the major protein groups rather than as a separate category. Dietary calcium is derived from green leafy vegetables and seafood. Coconut milk is a common cooking additive among the Bicolanos of southern Luzon. Salty (soy sauce, fish sauce/*patis*, salted shrimp fry, or fermented fish/*bagoong*) and spicy sauces known as *sawsawan* complement meals. These sauces are distinct from the salt added during cooking.

In the Philippines, breakfast consists of rice, meat or fish, and vegetable dishes or dinner leftovers. The breakfast beverage may be coffee, chocolate, or juice. In urban areas, Western-style meals are more common. For many Filipinos, breakfast, lunch, and dinner are not complete without steamed or fried rice served with fish, meat (especially pork), and vegetables. Snacks of bananas, yams, rice cakes, and rice-flour cakes are served as midday snacks, between meals, and before bedtime. The midday meal is the heaviest meal of the day, although this pattern is becoming more difficult among urban dwellers who cannot go home during lunchtime.

14.8.3 Dietary Practices for Health Promotion

Filipinos believe health is maintained by moderation. Although Filipinos enjoy food and love to eat, they adhere to the wisdom that too much of a good thing can be harmful. In some parts of the Philippines, it is considered polite to leave food on one's plate. For many Filipino Americans, moderation in food intake is a special challenge because of the abundance and great variety of quality products at reasonable costs. Significant increases in weight patterns among new immigrants are associated with changes in dietary habits.

The principle of hot and cold is observed by many traditional Filipinos to promote health. A warm beverage is served first at breakfast after a long evening fast, and hot soups are served as the first course to enhance digestion. Cold drinks may be avoided when one has a cold or fever to restore balance and promote harmony between the body and its environment. Eating rice is considered to be essential to a healthy life. *Arroz caldo,* chicken and rice soup, is generally offered to promote recovery after an illness. Chicken soup with *malunggay* leaves is believed to cleanse the blood.

Garlic and onions are believed to thin the blood and combat hypertension. Ginger root is boiled and served as a beverage to relieve sore throats and promote digestion. Guava shoots are eaten to treat diarrhea. Drinking coconut juice and water from boiled fresh corn silk promotes diuresis. Bitter melon is eaten as a vegetable to prevent diabetes. Greens such as *malunggay* and *ampalaya* leaves are used in stews to regain stamina for someone believed to be anemic or run-down.

14.8.4 Nutritional Deficiencies and Food Limitations

In the Philippines, nutrition is greatly affected by socioeconomic factors. Malnutrition persists in the country, especially among the poor and less educated, and is one of the leading causes of infant mortality. In the United States and elsewhere, Filipino immigrants may be at risk for nutritional deficiencies during their adjustment period, especially when they come with limited resources and without a support network of family and friends. Postmenopausal and pregnant women may be vulnerable to calcium deficiency owing to lactose intolerance and decreased intake of seafood and green leafy vegetables that were plentiful in the Philippines but limited in availability and variety in food stores in other countries. Changing food patterns and lifestyle is associated with migration and acculturation. Afable et al. (2016) have reported a significant and adverse association between the risk of being overweight the length of time living in the United States. In the study of Battie et al. (2016), a high percentage of Filipino women who were born in the Philippines but have resided in the US for a period of time, were overweight. Filipino Americans experience similar problems such as obesity, hyperlipidemia, and diabetes seen in the general population. Changes in the dietary intake of Filipino immigrants have been shown in the literature (Bhimla et al. 2017; Vargas 2018; Vargas and Jurado 2015). Knowledge of indigenous food sources and meal patterns, nutritional content of foods, changes in nutritional patterns, and accessibility of traditional ingredients is important for nutritional assessment and counselling.

14.9 Pregnancy and Childbearing Practices

14.9.1 Fertility Practices and Views Toward Pregnancy

The Roman Catholic Church and Filipino family values significantly influence childbearing and fertility practices. In marriage, the only acceptable method of contraception is the rhythm method. Abortion is considered a sin and is generally not acceptable. Whereas these beliefs remain strong among many Filipinos, education, global communication, and modernization are causing changes, particularly in metropolitan cities such as Manila. Recent Filipino immigrants who come from large urban areas are more educated and less committed to the Church's position on birth control and premarital sex.

Filipino culture is child-centered, and abortion evokes strong reactions, even among liberal Filipinos. Though some may support the right to abortion, they may have difficulty having one themselves and feel guilty for considering this option. Pregnancy is considered normal and is a time when a woman can demand attention and pampering from her husband and family members. Health-care providers who do not understand this special period for the pregnant Filipino woman may feel that the patient is "lazy and spoiled." Pregnancy and childbirth are times for the family to draw closer together. Everyone assists in anticipation of the new baby, especially the pregnant woman's mother, who has a strong influence during this period. For mother and daughter, this is a special event in which the bond between them becomes stronger.

In the Filipino American community, women openly give advice to pregnant women, share their own birthing experiences, and ask personal questions that may be considered rather intrusive by outsiders. Elaborate baby showers are hosted by family members and friends, and it is customary to invite male spouses, relatives, and friends as well as children. Male guests do not join in the activities and congregate separately from the women.

14.9.2 Prescriptive, Restrictive, and Taboo Practices in the Childbearing Family

Childbearing is widely celebrated by Filipino families. Children are perceived to be God's blessings and therefore to be accepted and be grateful. Filipino practices surrounding pregnancy are influenced by indigenous beliefs, Western practices, and socioeconomic factors. In the Philippines, most mothers receive prenatal care from a doctor, nurse, or midwife, although two-thirds of births are delivered at home. Traditional birth attendants (***hilots***) use massage and are consulted for physical, spiritual, and psychological advice and guidance (National Statistics Office, Philippines 2010).

After childbirth, the new mother continues to be pampered. Relatives help with the new baby and in running the household. Eighty-eight percent of Filipino babies are breastfed for some time, with a median duration of 13 months. However, supplementation of breastfeeding with other liquids and foods occurs too early, with 19 percent of newborns less than 2 months of age receiving supplemental foods or liquids other than water (National Statistics Office, Philippines 2010). Lactating mothers are encouraged to take plenty of hot soups (chicken with papaya) to promote milk production (Hawaii Community College 2005).

Some Filipino American women refuse to take vitamins during pregnancy for fear that these could deform the fetus. Some believe that when pregnant women crave certain foods, especially during the first trimester, the craving should be satisfied to avoid harm to the baby. Some women believe that the baby takes on the appearance of the craved food. Thus, if the mother craves dark-skinned fruit or dark-colored food, the infant's skin will be dark. Pregnant women are protected from sudden fright or stress because of the belief that this may harm the developing fetus. Table 14.1 provides a summary of traditional

Table 14.1 Traditional Filipino beliefs and practices surrounding pregnancy and childbirth

Prenatal	Postpartum
Eating blackberries will make the baby have black spots.	Use warm water to drink and bathe for a month
Eating black plums will give the baby dark skin.	Don't name the baby before it is born.
Eating twin bananas will result in twin births.	Don't name the baby after a dead person.
Eating apples will give the baby red lips.	Give money to charity or the needy when a baby comes to your house the first time.
When a woman's stomach is not round, the baby will be a boy.	Eating sour or ice-cold foods may cause abdominal cramps.
If a woman's face is blemished, the baby will be a boy.	Wrap the baby's abdomen with a cloth until the umbilical cord falls off.
Going outside during a lunar eclipse is harmful to the baby.	The mother and baby should not go out for a month except to visit a doctor.
Going out in the morning dew is bad for the baby because evil spirits are present.	Putting garlic, salt, or a rosary near the baby's crib will keep evil spirits away.
Funerals are avoided because the spirit of the dead person may affect the baby.	
Wearing necklaces may cause the umbilical cord to wrap around the baby's neck.	
Sitting by a doorway will make the delivery difficult.	
Sitting by a window when it is dark may let evil spirits come to the pregnant woman.	
Sweeping at night may sweep away the good spirits.	
Knitting might tangle the baby's intestines at birth.	

Source: Adapted from Hawaii Community College (2005)

beliefs and practices observed among some Filipinos in Hawaii. Becoming aware of the pregnant Filipino woman's network of family and community health advisers, whose opinions she respects, is important for building trust and rapport in the patient–health provider relationship.

Some women prefer to have their mothers rather than their husbands in the delivery room. Mothers of pregnant women serve as coaches and teachers and are often respected over health-care providers for their experience and knowledge. This may be puzzling to health-care providers who view pregnancy as an emancipating event. Conflicts are likely to occur if the coach and teacher believe in practices that are contrary to Western childbearing practices.

During postpartum, exposure to cold is avoided. Showers are prohibited because these may cause an imbalance and predispose illness. However, the mother is given a sponge bath with aromatic oils and herbs, or a *hilot* gives an aromatic herbal steam bath followed by full body massage, including the abdominal muscles, stimulating a physiological reaction that has both physical and psychological benefits.

Childbirth experiences of Filipino women immigrants in a hospital in Australia revealed language and communication problems as barriers to seeking antenatal care, perceived discrimination by the hospital staff, and conflicting expectations of delivery practices between the mothers and the health-care providers. The women preferred to be examined by female health-care providers and assume a squatting position for birthing. Contrary to their birthing practices, health-care providers expected the husbands to be with them during delivery. The women felt that they were not consulted about their care and preferred to deliver at home (Hoang 2008).

14.10 Death Rituals

14.10.1 Death Rituals and Expectations

Death for Filipinos as a spiritual event is based on the Roman Catholic belief system and doctrine. Illness and death may be attributed to supernatural and magico-religious causes such as punishment from God, angry spirits, or sorcery. Religiosity and fatalism contribute to stoicism in the face of pain or distress as a way of accepting one's fate (Lipson and Dibble 2005). Planning for one's death is taboo and may be considered tempting fate. Hence, many traditional Filipinos are averse to discussing advance directives or living wills (Pacquiao 2001). When death is imminent, contacting a priest is important if the family is Catholic. Religious medallions, rosary beads, scapulars, and religious figures may be found on the patient or at the bedside. Family members generally wish to provide the most intimate care to the patient.

After death, a wake is planned. In the Philippines, the wake may last 3 days or longer to allow time for relatives to arrive from distant places. In the United States, the wake is much shorter because it is costly. Although a wake is generally held in the home in the rural regions, funeral parlors are used in urban areas and in the United States. Families and friends gather to give support and recall the special traits of the deceased. Food is provided to all guests throughout the wake and after the burial.

The burial rites are consistent with the religious traditions of the family, which may be Judeo-Christian, Muslim, Buddhist, or other religions. Among Catholics, 9 days of novenas are held in the home or in the church. These special prayers ask God's blessing for the deceased. Depending on the economic resources of the family, food and refreshments are served after each prayer day. Sometimes, the last day of the novena takes on the atmosphere of a *fiesta* or a celebration. Filipino families in the United States and elsewhere follow variations of this ritual according to their social and economic circumstances. Funerals in the Philippines can be simple or elaborate, with a band accompaniment, several priests officiating, and a large throng of mourners. Reciprocal obligation continues in death through the performance of rituals such as the wake, novenas, and establishing a burial site acceptable for the entire family.

On the 1-year anniversary of the death, family and friends are reunited in prayer to celebrate this memorable event. Most Filipino women wear black clothing for months or up to a year after the death of a spouse or close family member. The 1-year anniversary ends the ritual mourning. Before this period, family members postpone weddings and other celebrations in deference to the memory of the deceased. Memories and love for the deceased are shown on All Soul's Day, a Catholic feast day celebrated in November, when families visit and decorate the graves of their loved ones. Filipino American families may continue these traditions, particularly when strong kinship is present and the clan lives in close proximity. Many who die in the United States and elsewhere are buried in the Philippines; the family in that country continues the tradition.

Beliefs related to cremation vary according to individual preference. Ordinarily, bodies are buried, but cremation is acceptable to avoid the spread of disease and limit the high costs of burial plots. Since the process of cremation has been accepted by the Roman Catholic Church in the Philippines, there has been an increase in this option. In America and elsewhere, some Filipinos who wish to return their deceased family members to the Philippines may choose cremation for practical and economic reasons.

14.10.2 Responses to Death and Grief

Most Filipinos believe in life after death. Caring for the spiritual needs of the dying is one way of ensuring peaceful rest of the soul or one's spirit. Family presence around the dying and immediate period after death to pray for the soul of the departed is considered a priority. If the patient is Catholic, the priest anoints the patient and gives Holy Communion if the patient is able to participate. Caring is shown by providing a peaceful environment, speaking in low tones, and praying with the ill person.

After death, grief reaction varies. Women generally show emotions openly by crying, fainting, or wailing. Men are expected to be more stoic and grieve silently. Young children are admonished for behaving inappropriately because this is considered disrespectful to the deceased. Family members gather together and provide physical and emotional support for each other. Praying for the deceased and following the implicit guidelines of behavior during mourning are ways of demonstrating grief appropriately. Wearing black or subdued colors (gray, white, navy blue, brown); avoiding parties and playing loud, distracting music; postponing weddings; or devoting time to one's studies to honor the dead are some of the acceptable ways of expressing grief. Honoring the memory of the deceased is a continuing obligation among close kin.

14.11 Spirituality

14.11.1 Dominant Religion and the Use of Prayer

The Philippines is the only predominantly Christian country in the Far East. In 2000, Roman Catholics accounted for 80.9 percent of the total population. Other religious groups include Muslims, other Christians, Evangelicals, Iglesia ni Kristo, Aglipay, and others (CIA World Factbook 2012). The spread of the fundamentalist movement within Roman Catholicism is becoming more evident. Christianity in the Philippines is a blend of Spanish Catholicism, American Christianity, and surviving indigenous animistic traditions.

Although Filipinos seek medical care, they believe that part of the efficacy of a cure is in God's hands or by some mystical power. Novenas and prayers are often said on behalf of the sick person. Families may bring religious items such as rosaries, medals, scapulars, and talismans for the sick person to wear. Talismans and amulets are believed to protect one from the forces of darkness, one's enemies, and sickness. Blessed holy water or oil is used to rub on the area in the body believed to be the source of distress. Performance of religious obligations and sacraments and daily prayers are some of the ways many Filipinos believe health and peaceful death are achieved. Providing for spiritual needs of Filipino patients requires accommodation to their various ways of practicing beliefs.

14.11.2 Meaning of Life and Individual Sources of Strength

Filipinos consider a meaningful existence to be a healthy and appropriate relationship with nature, God, and kin. Indigenous Filipino beliefs are embedded in the relationship between humans within the cosmology of the universe. This concept is demonstrated by the integration of supernatural, magico-religious, and natural phenomena in the belief system and practices toward health and illness. Filipinos do not see themselves as victims but rather as part of the larger cosmos, subject to both the controllable and the uncontrollable forces of nature. To the traditional Filipino, strength comes from an intimate relationship with God, family, friends, neighbors, and nature. The concept of self is formed from the relationship with a divine being and the social collective.

Many Filipinos find religion to be a source of strength in their daily lives. Some Filipinos are considered fatalistic in that they tend to accept fate easily, especially when they feel they cannot change a situation. Moreover, the acceptance of fate or destiny comes from their close relationship and healthy respect for nature. The acceptance of events they cannot change is tied to their religious faith. A common expression uttered by Filipinos is *bahala na*, originating from *bathala na* (it is up to God). *Bahala na* is often used when the person has used all resources to deal with a problem, and it is up to a higher power to take care of the rest (Enriquez 1994). Nevertheless, an element of self-reliance exists among Filipinos, manifested by their confidence that the situation is within their sphere of influence through education and hard work.

14.11.3 Spiritual Beliefs and Health-Care Practices

Holism and integration characterize Filipino health-care beliefs and practices. Religious and spiritual dimensions are important components in health promotion. The importance of harmony between humans and nature and the role of natural and supernatural forces in health and illness are included in their beliefs about causes of illness and healing modalities. Prayers, religious offerings, appeasing natural spirits, and witchcraft may be practiced simultaneously along with biomedical interventions. Despite increasing notoriety and scandal associated with Filipino faith healers, this healing modality is widely sought in the Philippines. Many Filipinos seek biomedical and integrative ways of healing and do not subscribe to the competitive reductionism of the West. They believe in the synergistic relationship of differing modalities and have no problem subscribing to both ways of healing. Many Filipino American health-care providers participate in religious pilgrimages to Lourdes, France, and the shrine of Fatima in Portugal to pray for good health and healing.

14.12 Health-Care Practices

14.12.1 Health-Seeking Beliefs and Behaviors

Filipinos seek out family and close kin first for help when they are ill. When illness is more defined, mobilization of support occurs within the family. Decisions about when, where, and from whom to seek help are largely influenced by the intimate circle of family. Among Filipino older people, the choice of health-care providers is based on accessibility and availability to their working adult children (Pacquiao 1993). Linguistically and ethnically congruent health-care providers are preferred. A dual system of personal health care exists for many Filipinos, including those who are established in American communities. Filipinos may accept and adhere to medical recommendations and may use alternative sources of care suggested by trusted friends and family members. Often, they adhere to Western and indigenous medicine simultaneously, creating more choices to deal with their own or their family's health issues.

Reflective Exercise 14.2

Araceli Montemayor is a 41-year-old single mother with a 2-year-old son. She was recently diagnosed with breast cancer for which she has initiated chemotherapy. She continues to work but has periods of fatigue and other side effects of the treatment, making it very difficult for her to take care of her energetic 2-year-old. She told the nurse that she believes that her breast cancer resulted from her son accidentally kicking her in the breast, which was when she first felt the lump in her breast. She had not had mammograms previously because of lack of insurance and reluctance to take time off from work. She believes that she will be cured through miraculous healing by praying to Mary, the mother of Jesus. She is contemplating whether she should continue with chemotherapy and experience the side effects or ask her community to pray and intercede for her healing.

1. How would you describe Mrs. Montemayor's belief system regarding the cause of illness?
2. How should the nurse respond to the patient's decision not to continue treatment for her breast cancer?
3. Who should be included in helping Mrs. Montemayor's decision-making process?
4. Do you think factual information about chemotherapy would be beneficial?
5. If available, would you consider contacting a *hilot*?

Many Filipinos consult an informal network of friends and family members, including physicians, nurses, pharmacists, or neighbors, who have had similar symptoms. Once the person finds the brand name of the "effective" medicine, the person can easily purchase the drug by asking family or friends to purchase medication in the Philippines. Hoarding prescription drugs and sharing medicine may be practiced by Filipinos in the United States. Those who do not believe in wastefulness or who believe that office visits are expensive may practice these behaviors.

When educating Filipino patients about medication, health-care providers should stress that medications need to be taken as prescribed; medications are ordered specifically for each ailment; unused drugs should be discarded; and the use of medications by individuals other than the intended patient may have serious consequences. Assessing these behaviors and delivering the message in a respectful, courteous, and unhurried manner may enhance the patient–health-care provider relationship, especially for traditional Filipino patients.

Health-care practices stress balance and moderation for the Filipino. Health is the result of balance, and illness is the consequence of imbalance. Imbalances that threaten health are brought about by personal irresponsibility or immorality. Care of the body through adequate sleep, rest, nutrition, and exercise is essential for health. A high value is also placed on personal cleanliness. Keeping oneself clean and free of unpleasant body odors is viewed as essential to health and social acceptance. To be slovenly and disorderly is to be shamelessly irresponsible. Aromatic baths are taken both for pleasure and to restore balance.

14.12.2 Responsibility for Health Care

Parents may seek all possible assistance that they can personally generate from family, friends, the church, the community, and the formal health-care system (often in that order) for a child with a serious illness such as cancer, eventually accepting the inevitability of death. From a Western perspective, the outcome may be slightly different than if formal services were accessed as early as possible. Adult children, especially those working in the United States, are responsible for the health care of their aged parents and extended kin. Responsibility may be in different forms, such as decision making, accepting financial responsibility, providing supportive presence, performing caretaking tasks, or negotiating with the health-care provider and the system.

In general, older adult women provide direct care for younger members. Older men participate in caring tasks such as driving the patient to the clinic. Decisions and financial support are relegated to family members who are deemed qualified and able. The family acts as a unit, and the individualistic paradigm commonly used by American caregivers is replaced by a social ethic of care. Before the decision is made to inform the patient about his or her terminal condition, a discussion among family members occurs, and they may request that the physician not divulge the truth to protect the patient. The ethical principles of beneficence and nonmaleficence take precedence over patient autonomy (Pacquiao 2003).

Filipino family hierarchy may require consulting with family members before decisions are made. This may pose a problem to Western practitioners who believe in the adult patients' autonomy to make decisions about their own lives. The same perspective of Filipinos may result in their inability to question and assert ideas with physicians, who are regarded to be in a higher position of authority. Major decisions may be delegated to the physician rather than the patient or family taking an active collaborative role in decision making. Failure to develop a trusting relationship with the health-care provider can lead to noncompliance with prescribed regimens because of lack of participation in the decision-making process.

14.12.3 Folk and Traditional Practices

Supernatural and magico-religious beliefs about health and illness are integrated with scientific medicine. Mental illness may be attributed to an external cause such as witchcraft, soul loss, or spirit intrusion. Illness in infancy and childhood may be attributed to the evil eye. This belief system is consistent with the variety of Filipino folk healers. Healing rituals may involve religious rites (prayers and exorcism), sacrifices to appease the spirits, use of herbs, and massage.

Balance and moderation are embedded in the hot-and-cold theory of healing. The ideal environment is warm, moderate, and balanced. The underlying principle is that change should be introduced gradually. Sudden changes from hot to cold, from activity to inactivity, from fasting to overeating, and so forth introduce undue bodily stresses, which can cause illness. After strenuous physical activity, a rest should precede a shower; otherwise, the person could develop arthritis. Cold drinks or foods such as orange juice or fresh tomatoes are not served for breakfast to prevent stomach upset. Exposure to sudden cold drafts may induce colds, fever, rheumatism, pneumonia, or other respiratory ailments. Some Filipinos avoid hand washing with cold water after ironing or heavy labor. Exposure to cold such as showers is avoided during menstruation and the postpartum period.

The Department of Health in the Philippines (2005), through its Traditional Health Program, has endorsed 10 herbs that have been thoroughly tested and clinically proven to have medicinal value in the relief and treatment of various ailments (Table 14.2). The Philippine government has encouraged production of these herbal medicines to provide affordable medicines for the populations who have limited or no access to Western health care. Widespread acceptance of these herbal medicines is evident among educated and higher-income groups.

14.12.4 Barriers to Health Care

Studies of Filipinos in the United States show report, for many reasons, Filipinos generally do not seek care for illness until it is quite advanced. Some take minor ailments stoically and consider them natural imbalances that will run their normal course and disappear. Others claim to watch the progress of their illness so that the appropriate health-care provider can be consulted. Still others may not seek help because of economic reasons, lack of insurance, distrust of the health-care system, religious reasons, lack of knowledge, or an inability to articulate their needs (McBride et al. 1995).

Table 14.2 Herbal medicines approved by the Department of Health in the Philippines

Filipino name/ generic name	English name	Uses
Akapulko (*Cassia alata*) "bayas-bayasan"	Ringworm bush	Ringworms and skin fungal infections
Ampalaya (*Momordica charantia*)	Bitter gourd or bitter melon	Non-insulin-dependent diabetes
Bawang (*Allium sativum*)	Garlic	Cholesterol reduction Blood pressure control
Bayabas (*Psidium guajava*)	Guava	Antiseptic to disinfect wounds mouthwash to treat tooth decay and gum infection
Lagundi (*Vitex negundo*)	Five-leaf chaste tree	Relief of coughs and asthma
Niyog-niyogan (*Quisqualis indica*)	Chinese honeysuckle	Dried matured seeds to eliminate intestinal worms, particularly *Ascaris* and *Trichina*
Sambong (*Blumea balsamifera*)	Blumea camphora	Diuretic, helps in the excretion of urinary stones and treatment of edema
Tsaang gubat (*Ehretia microphylla lam*)		Taken as tea; used in treating intestinal motility and as a mouthwash because leaves have a high fluoride content
Ulasimang bato (*Pepperomia pellucida*) "pansit-pansitan"		Arthritis and gout; may be prepared as tea or eaten as a salad
Yerba Buena (*Clinopodium douglasii*)	Peppermint	Analgesic to relieve body aches and pain; may be taken internally or applied locally

Source: Adapted from Department of Health (2005)

Many Filipinos are reluctant to participate in health-promotion programs such as cancer screening (Maxwell et al. 2014; Tsoh et al. 2018) and health education (Ghimire et al. 2018). Cuaresma et al. (2018) have reported that colorectal screening rates among Filipinos are below the Healthy People 2020 goals. Aging Filipino veterans may be denied health services because of lack of insurance and consequently referred to various non-profit community clinics. Older Filipino émigrés did not have adequate health benefits through their place of employment. Thus, they may have been used to postponing seeking care until the illness was quite advanced. Health-care providers should expect wide variations in health behaviors among Filipino American patients. A non-judgmental history taking should be well documented (Francisco et al. 2014). Turning on the "multicultural ear" and listening with care to the context of these actions can provide insight for a health-care provider, particularly when the health-care provider is under time pressure.

14.12.5 Cultural Responses to Health and Illness

Filipinos view pain as part of living an honorable life, as well as part of the process of spiritual purification while still on earth. Some view this as an opportunity to reach a fuller spiritual life or to atone for past transgressions. Thus, they may appear stoic and tolerate a high degree of pain. Health-care providers may need to offer and even encourage pain relief interventions for patients who do not complain of pain despite physiological indicators. Others may have a strong sensitivity to the "busyness" of health-care providers, quietly diminishing their own need for attention so that others can receive care, or they may simply have little knowledge of how pain management can be maximized.

Minimal expression of psychological and emotional discomfort may be observed. The discomfort in discussing negative emotions with outsiders may be manifested by somatic complaints or ritualistic behaviors, such as praying. Exploring the underlying meaning of somatization (loss of appetite, inability to sleep) and observing the patient's interactions with others can provide valuable information. Filipino patients may display visible evidence of their religion, such as religious medals, prayer cards, and rosary beads, to manage anxiety and pain. These artifacts should be incorporated into their treatment regimen. Using cultural mediators or brokers to probe innermost feelings of patients may be helpful if used appropriately. Pain assessment can include the role of prayer by the patient and members of the support network. Questions such as "Do you have someone praying for you?" or "Is there a special prayer to help you deal with pain?" may provide vital information for individualizing care.

Most Filipinos believe that mental illness carries a certain amount of stigma. The first choice is caring by family members, friends, and relatives rather than seeking health professionals (Gong et al. 2003) to minimize exposing the problem to outsiders.

Among rural residents and less-educated Filipinos in the Philippines, mental illness is generally attributed to external causes such as sorcery, soul loss, or spirit intrusion. Witch doctors, fortune-tellers, and faith healers are often sought. Filipinos in the United States seek professional interventions when symptoms are advanced. Psychiatric symptoms are precipitated by a loss in self-esteem, loss of status, and shame related to the stresses of immigration. Separation from family, inability to find suitable employment, uncertainty, lack of money, and other relocation stressors create serious psychological reactions among Filipinos. Among Filipino Americans, religiosity was correlated with seeking help from the religious clergy, whereas spirituality was associated with less help-seeking from professional mental health practitioners (Abe-Kim et al. 2004). Using sociocultural behaviors learned early in life, Filipinos have a remarkable ability to maintain a proper front to protect their self-esteem and self-image. Mental health-care providers should recognize that despite the possibility of a Filipino patient's refusing professional mental health services, involving a trusted family member or friends; initiating contact with a Filipino mental health worker, especially a Filipino physician; or using both practices may increase the odds of getting the person into a culturally compatible treatment program. Deference to authority may successfully bring the Filipino patient into treatment, with the patient's expectation that the authority figure will fix the problem.

14.12.6 Blood Transfusions and Organ Donation

The value of blood transfusion is recognized and accepted by Filipinos. However, organ donation may be less acceptable, except perhaps in cases in which a close family member is involved. Many Filipinos who follow Catholic traditions believe that keeping the body intact as much as possible until death is a reasonable preparation for the afterlife. Asian Americans, including Filipinos, hold more negative attitudes toward organ donation than European Americans. They are less likely to participate in large, urban organ donor program (Alden and Cheung 2000).

14.13 Health-Care Providers

14.13.1 Traditional Versus Biomedical Providers

Western medicine is familiar and acceptable to most Filipinos. Many recent Filipino immigrants are educated in the health-care fields. Some Filipinos accept the efficacy of folk medicine and may consult both Western-trained and indigenous healers. Traditional healers are sought more in the rural areas of the Philippines. Folk healers are less common in the United States, with the exceptions of the West Coast and Hawaii. When available, they contribute by facilitating cultural rapport between health-care providers and the patient and by increasing utilization of needed health-care services. For example, the ***hilot*** is often willing to be included in the counselling session and provide support for the patient's adherence to the medical treatment. The *hilot* may provide a special prayer to be incorporated into the medically prescribed treatment plan to increase the patient's sense that all available resources are being used. In some areas on the West Coast, the *hilot* has a distinct role and function in the Filipino community. A few Filipino health professionals have learned the *hilot*'s art, skills, and spiritual approach, which they blend into their professional practice.

A health-care provider of the same gender and the same culture may encourage more Filipinos to take advantage of disease prevention services. The availability of Filipino primary-care providers and, whenever possible, a bilingual person are critical to improving health care for older Filipinos.

14.13.2 Status of Health-Care Providers

Filipinos generally consider the physician as the primary leader of the health-care team, and other providers are expected to defer to the physician. As Filipino families become more acculturated and aware of how health-care services are accessed in the United States, changes in attitude and behavior may be expected.

When ill, Filipinos may first consult a family member or a friend who is a physician or other professional before arranging a medical appointment. Some prefer physicians from their own region, when possible, whereas others indicate preference for physicians who are knowledgeable and competent and have good bedside manners regardless of culture or ethnic background. Factors considered in choosing health-care providers by middle-aged immigrant Filipino women were concern for privacy, feelings of modesty, approval from family members (especially the spouse), and, most important, the overall caring environment in the system.

Reflective Exercise 14.3

Concepcion Miraflor had major depression for many years. Becoming progressively depressed for the past 6 months, she has expressed to her daughter that everyone would be better off if she were dead. Her daughter brought her to the mental health clinic for evaluation, where she was assessed to be a threat to herself and was admitted to the psychiatric unit and started on an antidepressant. She is 51 years old, has completed a sixth-grade education, is unemployed, and has no health insurance. Although she can speak limited English, she is unable to respond in English since she became depressed. She lives with her youngest daughter and her family and takes care of her two young grandchildren at home. Her son-in-law, who is from Pakistan, is a practicing Muslim. As a devout Catholic, Mrs. Miraflor was not happy with this inter-religious marriage. She is very concerned because her two grandchildren were not baptized in the Catholic Church.

1. How is the patient's level of spirituality influencing her relationship with her family and the possible area of conflict?
2. What issues can the nurse assist the patient in addressing while she is in the hospital?
3. Would you suggest a visit from a spiritual counselor?
4. Would you suggest involvement with a traditional healer?
5. What are some nursing implications related to ethnopharmacology that the nurse needs to consider while the patient is on antidepressant therapy?

Interactions of Filipinos with Canadian nurses in the hospital reflected their *kapwa*-oriented worldview, which categorized nursing approaches and interactions within the insider-outsider continuum. Patients based their preferences for which nurses to perform their personal and private tasks or receive information on the nurses' ability to provide spontaneous and unsolicited care and monitoring of their condition. Organizational policies and protocols, in addition to short hospital stays, were identified as barriers toward moving the patient–nurse relationship toward higher intimacy and trust (Pasco et al. 2004).

References

Abe-Kim J, Gong F, Takeuchi D (2004) Religiosity, spirituality, and help-seeking among Filipino Americans: religious clergy or mental health professionals? J Community Psychol 32:675–689

Afable A, Ursua R, Wyatt LC, Aguilar D, Kwon SC, Islam NS, Trinh-Shevrin C (2016) Duration of US residence is associated with overweight risk in Filipino immigrants living in NY metro area. Fam Community Health 39(1):13–23. https://doi.org/10.1097/FCH.0000000000000086

Agoncillo T, Guerrero M (1987) History of the Filipino people, 7th edn. Garcia Publishing, Quezon City

Alden DL, Cheung AHS (2000) Organ donation and culture: a comparison of Asian American and European American beliefs, attitudes, and behaviors. J Appl Soc Psychol 30(2):293–314

Anderson JN (1983) Health and illness in Pilipino immigrants. West J Med 139(6):811–819

Araneta MRG, Barrett-Connor E (2005) Ethnic differences in visceral adipose tissue and type 2 diabetes: Filipino, African-American, and White women. Obes Res 13:1458–1465

Asis M (2006) The Philippines' culture of migration. The Online Journal of the Migration Policy Institute. https://www.migrationpolicy.org/article/philippines-culture-migration

Bateman W, Abesamis-Mendoza N, Ho-Asjoe H (2009) Praeger handbook of Asian American health: taking notice and taking action. ABC-CLIO, Simi Valley, CA

Battie CA, Borja-Hart N, Ancheta IB, Flores R, Rao G, Palaniappan L (2016) Comparison of body mass index, waist circumference, and waist to height ratio in the prediction of hypertension and diabetes mellitus: Filipino-American women cardiovascular study. Prev Med Rep 4:608–613. https://doi.org/10.1016/j.pmedr.2016.10.003

Bautista V (2002) The Filipino Americans (1763–present): their history, culture, and tradition, 2nd edn. Bookhaus, Naperville, IL

Bayog ML, Waters CM (2017) Cardiometabolic risks, lifestyle health behaviors, and heart disease in Filipino Americans. Eur J Cardiovasc Nurs 16(6):522–529. https://doi.org/10.1177/1474515117697886

Bayog ML, Waters CM (2018) Nativity, chronic health conditions, and health behaviors in Filipino Americans. J Transcult Nurs 29(3):249–257. https://doi.org/10.1177/1043659617703164

Bender MS, Cooper BA, Park LG, Padash S, Arai S (2018) Correction: a feasible and efficacious mobile-phone-based lifestyle intervention for Filipino Americans with type 2 diabetes: randomized controlled trial. JMIR Diabetes 3(4):e12784. https://doi.org/10.2196/12784

Bhimla A, Yap L, Lee M, Seals B, Aczon H, Ma G (2017) Addressing the health needs of high-risk Filipino Americans in the greater Philadelphia region. J Community Health 42(2):269–277. https://doi.org/10.1007/s10900-016-0252-0

Brown DE, James GD (2000) Physiological stress responses in Filipino-American immigrant nurses: the effects of residence time, life-style, and job strain. Psychosom Med 62:394–400

Ceniza-Choy C (2003) Empire of care: nursing migration in Filipino American history. Duke University Press, Durham and London

Chawla N, Breen N, Liu B, Lee R, Kagawa-Singer M (2015) Asian American women in California: a pooled analysis of predictors for breast and cervical cancer screening. Am J Public Health 105(2):e98–e109. https://doi.org/10.2105/AJPH.2014.302250

CIA World Factbook (2012) Philippines. https://www.cia.gov/library/publications/the-world-factbook/geos/rp.html

Coughlan SS, Uhler RJ (2000) Breast and cervical cancer screening practices among Asian and Pacific islander women in the US, 1994–1997. Cancer Epidemiol Biomarkers Prev 9:597–603

Cuaresma CF, Sy AU, Nguyen TT, Ho RCS, Gildengorin GL, Tsoh JY, Jo AM, Tong EK, Kagawa-Singer M, Stewart SL (2018) Results of a lay health education intervention to increase colorectal cancer screening among Filipino Americans: a cluster randomized controlled trial. Cancer 124(7):1535–1542. https://doi.org/10.1002/cncr.31116

De la Cruz FA, Padilla GV, Agustin EO (2000) Adapting a measure of acculturation for cross-cultural research. J Transcult Nurs 11(3):191–198

Department of Health (2005) Ten herbal medicines approved by DOH. www.philippineherbalmedicine.org/doh_herbs.htm

Domingo JB, Gavero G, Braun KL (2018) Strategies to increase Filipino American participation in cardiovascular health promotion: a systematic review. Prev Chronic Dis 15:170,294. https://doi.org/10.5888/pcd15.170294

Enriquez VG (1994) From colonial to liberation psychology: the Philippine experience. De La Salle University Press, Manila

Espiritu YL (2003) Homebound: Filipino American lives across cultures, communities and countries. University of California Press, Berkeley, Los Angeles

Francisco D, Rankin L, Kim S (2014) Adherence to colorectal cancer and polyps screening recommendations among Filipino-Americans. Gastroenterol Nurs 37(6):384–390. https://doi.org/10.1097/SGA.0000000000000071

Fukuoka Y, Choi J, Bender MS, Gonzalez P, Arai S (2015) Family history and body mass index predict perceived risks of diabetes and heart attack among community-dwelling Caucasian, Filipino, Korean, and Latino Americans—DiLH survey. Diabetes Res Clin Pract 109(1):157–163. https://doi.org/10.1016/j.diabres.2015.04.015

Garde P, Spangler Z, Miranda B (1994) Filipino-Americans in New Jersey: a health study. Final report of the Philippine Nurses' Association of America to the State of New Jersey Department of Health, Office of Minority Health

Ghimire S, Cheong P, Sagadraca L, Chien L, Sy FS (2018) A health needs assessment of the Filipino American community in the greater Las Vegas area. Health Equity 2(1):34–312. https://doi.org/10.1089/heq.2018.0042

Gong F, Gage S-JL, Tacata LA Jr (2003) Help seeking behavior among Filipino Americans: a cultural analysis of face and language. J Community Psychol 31(5):469–488

Hastings K, Jose P, Kapphahn K, Frank A, Goldstein B, Thompson C, Palaniappan L (2015) Leading causes of death among Asian American subgroups (2003-2011). PLoS One 10(4):e0124341

Hawaii Community College (2005) Traditional Filipino health beliefs. http://www.hawcc.hawaii.edu/nursing/tradfil2.htm

Hoang H (2008) Language and cultural barriers of Asian migrants in accessing maternal care in Australia. Int J Lang Soci Cult 26(6):55–61. http://www.educ.utas.edu.au/users/tle/JOURNAL/issues/2008/26-6.pdf

IBON Foundation (2007). http://www.ibon.org

Kagawa-Singer M, Pourat N (2000) Asian American and Pacific islander breast and cervical carcinoma screening rates and Healthy People 2000 objectives. Cancer 89(3):696–705

Levy R (1993) Ethnic and racial differences in response to medicines: preserving individualized therapy in managed pharmaceutical programmes. Pharm Med 7:139–165

Lim S, Clarke CA, Prehn AW, Glaser SL, West DW, O'Malley CD (2002) Survival differences among Asian subpopulations in the United States after prostate, colorectal, breast, and cervical carcinomas. Cancer 94(4):1175–1182

Lin KM, Smith MW (2000) Psychopharmacology in the context of culture and ethnicity. In: Ruiz P (ed) Ethnicity & psychopharmacology. American Psychiatric Press, Washington

Lipson JG, Dibble SL (2005) Culture and clinical care. University of California San Francisco Nursing Press, San Francisco, CA

Ma GX, Lee M, Bhimla A, Tan Y, Gadegbeku CA, Yeh MC, Aczon H (2017) Risk assessment and prevention of hypertension in Filipino Americans. J Community Health 42(4):797–805. https://doi.org/10.1007/s10900-017-0320-0

Maglalang DD, Yoo GJ, Ursua RA, Villanueva C, Chesla CA, Bender MS (2017) "I don't have to explain, people understand": acceptability and cultural relevance of a mobile health lifestyle intervention for Filipinos with type 2 diabetes. Ethn Dis 27(2):143–154. https://doi.org/10.18865/ed.27.2.143

Maxwell AE, Bastani R, Warda US (2000) Demographic predictors of cancer screening among Filipino and Korean immigrants in the United States. Am J Prev Med 36(1):67–68

Maxwell AE, Bastani R, Vida P, Warda US (2002) Physical activity among older Filipino-American women. Women Health 36(1):67–79

Maxwell AE, Garcia GM, Berman BA (2007) Understanding tobacco use among Filipino American men. J Tob Res 9(7):767–776

Maxwell AE, Danao LL, Cayetano RT, Crespi CM, Bastani R (2014) Adoption of an evidence-based colorectal cancer screening promotion program by community organizations serving Filipino Americans. BMC Public Health 14(1):246. https://doi.org/10.1186/14712458-14-246

McBride M, Parreno H (1996) Filipino American families and caregiving. In: Yeo G, Gallagher-Thompson D (eds) Ethnicity and the dementias. Taylor & Francis, New York

McBride M, Mariola D, Yeo G (1995) Aging and health: Asian Pacific Islander American elders. Stanford Geriatric Education Center, Stanford, CA

Mossakowski KN (2003) Coping with perceived discrimination: does ethnic identity protect mental health? J Health Soc Behav 44(3):318–331

Munoz C, Hilgenberg C (2006) Ethnopharmacolgy: understanding how ethnicity affects drug response. J Holist Nurs Pract 20(5):227–234

Munoz C, Luckmann J (2005) Transcultural communication in nursing. Thomson Delmar Publishing Inc., New York

National Statistics Office, Philippines (2010) Index of demographic and health. http://www.census.gov.ph/data/sectordata/datandhs.html

Pacquiao DF (1993) Cultural influences in old age: an ethnographic comparison of Anglo and Filipino elders. Doctoral dissertation, Graduate School of Education, Rutgers University, New Brunswick, NJ. University Microfilms, Inc., Ann Arbor, MI (number 9320788)

Pacquiao DF (1996) Educating faculty in the concept of educational biculturalism: a comparative study of sociocultural influences in nursing students' experience in school. In: Fitzsimons VM, Kelley ML (eds) The culture of learning: access, retention, understanding and mobility of minority students in nursing. National League for Nursing Press, New York, pp 129–162

Pacquiao DF (2001) Cultural incongruities of advance directives. Bioethics Forum 17(1):27–31

Pacquiao DF (2003) Cultural competence in ethical decision-making. In: Andrews MM, Boyle JS (eds) Transcultural concepts in nursing care, 4th edn. Lippincott, Philadelphia, PA, pp 503–532

Pacquiao DF (2004) Recruitment of Philippine nurses to the US: implications for policy development. Nurs Health Policy Rev 3(2):167–180

Pasco ACY, Morse JM, Olson JK (2004) Cross-cultural relationships between nurses and Filipino Canadian patients. J Nurs Scholarsh 36(3):239–246

Philippine Statistics Authority (2019) Proportion of poor Filipinos was estimated at 16.6 percent in 2018. Philippine Statistics Authority, Republic of the Philippines. https://psa.gov.ph/system/files/Press%20Release_2018FY.pdf

Quan H, Fong A, De Coster C, Wang J, Musto R, Noseworthy TW, Richard M, Noseworthy T, Ghali W

(2006) Variation in health services utilization among ethnic populations. CMAJ 174(6):787–791

Reblando JR (2018) A comparative analysis of the Philippine nursing curriculum from other countries. Int J Adv Res 6(7):526–532. https://doi.org/10.21474/IJAR01/7393

Reeves TJ, Bennett CE (2004) We the people: Asians in the US. Census 2000 special reports. U.S. Department of Commerce, Washington, DC

Ryan C, Shaw R, Pilam M (2000) Coronary artery disease in Filipino and Filipino American patients: prevalence of risk factors and outcomes of treatment. J Invasive Cardiol 12(3):134–139

Salazar LP, Schuldermann SM, Schuldermann EH, Hunyh C (2000) The Filipino adolescents' parental socialization for academic achievement in the United States. J Adolesc Res 15(5):564–587

Spangler Z (1992) Transcultural nursing care values and caregiving practices of Philippine-American nurses. J Transcult Nurs 4(2):28–37

Tatak Pilipino (2007) Profile of the Philippines. http://www.filipinoheritage.com

Tran MT, Jeong MB, Nguyen VV, Sharp MT, Yu EP, Yu F, Tong EK, Kagawa-Singer M, Cuaresma CF, Sy AU, Tsoh JY, Gildengorin GL, Stewart SL, Nguyen TT (2018) Colorectal cancer beliefs, knowledge, and screening among Filipino, Hmong, and Korean Americans. Cancer 124(7):1552–1559. https://doi.org/10.1002/cncr.31216

Tsoh JY, Tong EK, Sy AU, Stewart SL, Gildengorin GL, Nguyen TT (2018) Knowledge of colorectal cancer screening guidelines and intention to obtain screening among nonadherent Filipino, Hmong, and Korean Americans. Cancer 124(7):1560–1567. https://doi.org/10.1002/cncr.31097

Ursua RA, Islam NS, Aguilar DE, Wyatt LC, Tandon SD, Abesamis-Mendoza N, Nur PRMQ, RagoAdia J, Ileto B, Rey MJ, Trinh-Shevrin C (2013) Predictors of hypertension among Filipino immigrants in the northeast US. J Community Health 38(5):847–855

Ursua R, Aguilar D, Wyatt L, Tandon S, Escondo K, Rey M, Trinh-Shevrin C (2014) Awareness, treatment and control of hypertension among Filipino immigrants. J Gen Intern Med 29(3):455–462. https://doi.org/10.1007/s11606-013-2629-4

Ursua RA, Aguilar DE, Wyatt LC, Trinh-Shevrin C, Gamboa L, Valdellon P, Perrella EG, Dimaporo MZ, Nur PQ, Tandon SD, Islam NS (2018) A community health worker intervention to improve blood pressure among Filipino Americans with hypertension: a randomized controlled trial. Prev Med Rep 11:42–48. https://doi.org/10.1016/j.pmedr.2018.05.002

Vargas P (2018) Dietary intake and obesity among Filipino Americans in New Jersey. J Environ Public Health 2018:6719861. https://doi.org/10.1155/2018/6719861

Vargas P, Jurado L (2015) Dietary acculturation among Filipino Americans. Int J Environ Res Public Health 13(1):ijerph13010016. https://doi.org/10.3390/ijerph13010016

Wolf DL (1997) Family secrets: transnational struggles among children of Filipino immigrants. Sociol Perspect 40(3):457–483

Wu T-Y, Bancroft J (2006) Filipino American women's perceptions and experiences with breast cancer screening. Oncol Nurs Forum 33(4):1–11

Yamane L (2002) Native-born Filipina/o Americans and labor market discrimination. Fem Econ 8(2):125–144

Yu SM, Huang ZJ, Singh GK (2004) Health status and health services utilization among US Chinese, Asian Indian, Filipino, and other Asian/Pacific Islander children. Pediatrics 113(1):101–107

Zong J, Batalova J (2018) Filipino immigrants in the United States. The Online Journal of the Migration Policy Institute. https://www.migrationpolicy.org/article/filipino-immigrants-united-states

People of German Heritage

15

Eric A. Fenkl

15.1 Introduction

Germans are reserved, formal people who appreciate a sense of order in their lives. Their love of music and celebrations has permanently influenced many of the world's cultures. The Christmas tree (*Weihnachtsbaum*) with its brightly decorated ornaments, a universal symbol of the holiday season, is a German creation. Gingerbread houses (*Lebkuchen*), Christmas carols (*Weihnachtslieder*) and cards, the "Easter hare" (*Osterhase*), hot cross buns, valentines (*Freundschaftskarten*), Groundhog Day, chain letters (*Briefe zum Himmel*), the tooth fairy, and *Kaffeeklatsch* or "gossip sessions" all have their origins in German culture.

15.2 Overview, Inhabited Localities, and Topography

15.2.1 Overview

There are 51 million Germans in the United States (U.S. Census Bureau 2018) and over 3.1 million in Canada (Statistics: Canada 2016). Ethnic groups of European origin are usually categorized as "White" on applications, in surveys, and in research studies, so there is little culturally specific information available about them. This is unfortunate, because differences in worldviews, cultural beliefs, and health-care practices among White ethnic groups hold important implications for health-care providers.

15.2.2 Heritage and Residence

The Federal Republic of Germany (*Bundesrepublik Deutschland*), comprising 16 states, boasts beautiful landscapes, high and low mountain ranges, sandy lowlands, rolling hills, lakelands, and ocean borders. Situated in the heart of Europe, Germany is a link between the East and the West and between Scandinavia and the Mediterranean. Germany has the largest economy in Europe, has the fifth largest economy in the world, and is the leading per-capita export nation in the world (CIA World Factbook 2020). With a population of over 82.9 million, it is one of the most densely populated countries in Europe. Germany is a member of the United Nations and NATO and is a founding member of the European Union (CIA World Factbook 2020).

Most of Germany is located in the temperate zone, with temperatures ranging from 27 °F in the mountains to 68 °F in the valleys of the south. Temperatures are comparable with the climate in the northwest portion of the United States. The

This chapter is an update of a chapter in the previous edition written by Jessica Steckler.

E. A. Fenkl (✉)
Florida International University, Miami, FL, USA
e-mail: efenkl@fiu.edu

© Springer Nature Switzerland AG 2021

L. D. Purnell, E. A. Fenkl (eds.), *Textbook for Transcultural Health Care: A Population Approach*,
https://doi.org/10.1007/978-3-030-51399-3_15

Upper Rhine has a mild climate; Upper Bavaria has warm Alpine winds from the south; and the Harz Mountains have cold winds, cool summers, and heavy winter snows.

15.2.3 Reasons for Migration and Associated Economic Factors

In the eighteenth century, the New World colonies from New England to the Deep South grew and flourished. Even though the colonial settlers shared an Old-World heritage, they were a diverse people. German settlers, along with other immigrants from Britain, France, Scotland, and Ireland, shared a love of family and land—a love that would eventually bond them to one another to form a nation of Americans. The earliest German immigrants to the United States settled in the colonies along the eastern seaboard, including William Penn's colony in Pennsylvania. Religious tolerance and equitable land distribution contributed to the success of these Pennsylvania settlements. Mennonites, Dunkers, Amish, and Moravians from Germany made up the new Pennsylvania communities. The area in which they settled, known as *Pennsylvania Dutch Country*, was actually mislabeled by English neighbors who thought the word *deutsch*, meaning "German," stood for "Dutch." One hundred thousand strong, these Pennsylvania Germans were the main carriers of German culture to the mid-Atlantic area (Bronner and Brown 2017).

The second wave of German immigrants arrived in the United States between 1840 and 1860. They were fleeing political persecution, starvation, and poverty in their homeland and settled on the western frontier (Weaver 1979). This group of influential Germans was less interested in taking root in the United States than in establishing a German culture. These new immigrants kept the German language in their schools, published newspapers in German, joined their own singing societies and orchestras, and married only other Germans.

The 1930s and 1940s saw a third wave of German immigration. Artists, architects, social scientists, physicists, and mathematicians came to this country to escape the Nazi Holocaust. These new arrivals were highly educated and at the height of their careers. After witnessing the horrors of the Holocaust, they had no desire to transplant Old World institutions or to establish new European-style homelands (Boorstin 1987). These third-wave immigrants became rapidly acculturated into American life and greatly enriched American culture in the fields of music, psychology, science, and mathematics. Among this prominent group were Albert Einstein and Hannah Arendt, an author and political scientist.

Germans continue to embrace the United States as their own. The desire to become American has been nurtured by the presence of American troops in Germany, and many Germans have entered the United States as spouses of military personnel. For others, business ventures and the promise of career opportunities brought them to this country. Today, about one-fourth of all American citizens can trace their ancestry to German roots. Germans are the dominant ancestral group in St. Louis, Missouri; Milwaukee, Wisconsin; Chicago, Illinois; Cincinnati, Ohio; Buffalo and New York City in New York; and Baltimore, Maryland. With an estimated size of approximately 49 million in 2016, German Americans are the largest of the self-reported ancestry groups by the US Census Bureau in its American Community Survey. German-Americans account for about one third of the total ethnic German population in the world (U.S. Census Bureau 2018). Germans have been very much a part of important events shaping U.S. socioeconomic history. They have been participants, observers, and victims in the Revolutionary War, the Civil War, the influenza epidemic, the Great Depression, World Wars I and II, the Vietnam War, the Persian Gulf War, and the current global recession.

15.2.4 Educational Status and Occupations

Germans have a deep respect for education. In Germany, credibility, social status, and level of employment are based on educational achievement. In other words, Germans are very class conscious. Germans take pride in their school system, particularly in their craftsmanship and technology. Unlike in the United States, education is free at all levels, except kindergarten, which is optional, but entrance to university education is difficult and accomplished only by passing the *Abitur* examination. Literacy rates of Germany (99 percent) and the United States (99 percent) are comparable (CIA World Factbook 2020).

In Germany, children can begin kindergarten at age 3 (Educational Aspects in the United States and Germany n.d.). This is comparable with our preschool. At age 6, they enter grade school, which includes grades 1 to 4. At grade 5, they begin one of three tracks of education: *Hauptschule*, which is special education and the most basic educational path; *Realschule*, which is general education; or *Gymnasium*, which is like U.S. college preparatory courses. German students graduate at grade 10 and can then enter into vocational education, which prepares them for a trade or for working in business, or they can continue college preparation. Those students wishing to go to the universities must pass the *Abitur* test, which is both verbal and written.

Germans who immigrated to the United States in the nineteenth century influenced American preschool and higher-educational systems. The Johns Hopkins University in Baltimore, Maryland, was founded on the model of Humboldt University in Berlin, Germany (McKinnon 1993; Janes 2004). During this same period, many American historians and political scientists attended German universities, returning with their doctoral degrees, and were instrumental in developing prototypes for American graduate education. Many of the influences of the nineteenth-century German immigrants on the educational system remain visible today.

By the mid-nineteenth century, *Turnvereins* were taking root in midwestern German American communities. These political and gymnastic organizations believed in a sound mind and body and provided opportunities to grow both physically and intellectually (Acton 1994). In this same era, schools—many of which were parochial schools—were established in which only German was spoken. German Catholics also established parochial schools in this era, but unlike the Lutherans, their ethnic identity was not tied to the church.

German immigrants were viewed as an internal threat in the US during World War II and faced turbulent times. A growing anti-immigrant sentiment leading to calls for immigration restriction intensified the political climate. Some German immigrants' desire to maintain an identity apart from the American culture was expressed through the founding of the National German American Alliance. Many German Americans changed their names, made apologies, and displayed their loyalties in an effort to attenuate suspicions, embarrassments, and persecutions.

Today, German American families continue to value education. Most German Americans have a high school education at minimum. Twenty-four percent have attained post–high school education. However, in the age group 65 and older, 43 percent have less than a high school education; no current information on educational levels of German Americans could be found. Vocational or university education is being sought more frequently by recent high school graduates attempting to prepare themselves for a highly competitive work environment and by adults who are pursuing second and third careers. By German standards, success means being employed, and education is seen as the way to achieve this success (McKinnon 1993).

The earliest German immigrants were primarily farmers. Tobacco, wheat, rice, cotton, corn, and sugar were among the most widely grown crops. Plantations grew from Virginia to the colonies in the South as a result of these prosperous ventures. Planting and harvesting crops required many workers with strong backs, and because not all Germans could pay for their passage to the New World, many worked as indentured servants. They suffered many hardships and worked long hours at the mercy of their owners. Family members were commonly separated from one another, and often children were sold to pay the debt of their parents.

In the post–Civil War era, Germans who came to the US often "chain-migrated" to the western frontier. Families and friends would leave one area to join family, friends, and neighbors in another place. These groups became farmers, miners, millers, construction workers, shopkeepers, blacksmiths, and locksmiths. Many were artists and craft workers who created pottery, leather goods, soap, candles, and musical instruments (e.g., the dulcimer). These Germans established outstanding breweries, beer gardens (*biergarten*), and pubs (*kneipen*) everywhere they settled. They also brought many trades to the US, including butchering, coppering, tailoring, and cabinetmaking. Whereas they dominated the trades, German immigrants were found less frequently in professional and management positions (Schied 1993).

In the early decades of the twentieth century, the Nazi Holocaust drove many German immigrants from their home country. Many who came in the 1930s and 1940s continued their gifted work in the US. Germans continue to establish their homes in the US. Newly arriving immigrants are highly educated and vocationally well trained. German workers are among the most skilled in the world. Germany and the US have similar industries in manufacturing, construction, and service.

15.3 Communication

15.3.1 Dominant Language and Dialects

German, the official language of the Federal Republic of Germany, is spoken in Germany; Austria; and Liechtenstein; large parts of Switzerland and South Tirol; and small parts of Belgium, France, and Luxembourg. German is the native language of over 100 million people, and many literary works have been translated into German. Within Germany, there are many dialects along with high (more formal) and low (less formal and more conversational) German. Individuals' home regions can be easily identified through their speech, and citizens from neighboring regions may have difficulty understanding one another because of the differences in regional jargon and accents.

In addition to the German language, German children learn English at grade 5, and at grade 7, they learn a third language of their choice. At grade 9, advanced English or, perhaps, a fourth language can be chosen (European Education Directory 2006).

English is the dominant language of German Americans. Germans who originally emigrated from Germany learned American English at work, in school, and through socialization. Their children grew up speaking English in public schools and German at home.

Americans and Germans have some similar patterns of speech behavior. German is a low-contextual language, with a greater emphasis on verbal than nonverbal communication, showing a high degree of social approval to people whose verbal behavior in expressing ideas and feelings is precise, explicit, straightforward, and direct. Forty-nine million people in the US claim to have German ancestors, and 1.4 million of them can speak German (U.S. Census Bureau 2018). Individuals in some German American communities mix English and German creatively when expressing humor.

15.3.2 Cultural Communication Patterns

People of German ancestry enjoy discussing topics of interest after dinner. These conversations, sometimes debates, cover a range of issues from politics, religion, food, and work experiences to life in general. Jokes, funny stories, or anecdotes about family members are interspersed within the discussion.

Germans carry on their conversations at three levels. The first, *Gespräch*, is used for casual conversation and is more informal; the second, *Besprechung*, is conversation carried on in a work setting between employees and supervisors about performance; and the third, *Diskutieren*, is the common form of social discourse used in discussions about various issues and is the most formal use of the German language. Most Americans, and others worldwide, are often ill prepared to enter the debate on philosophical and political issues that are addressed at this level and are thus placed at a disadvantage. This cultural barrier can prevent Germans from developing deeper relationships with outside groups.

Feelings among Germans and German Americans are considered private and are often difficult to share. Sharing one's feelings with others often creates a sense of vulnerability or is looked on as evidence of weakness. The act of expressing fear, concern, happiness, or sorrow allows others a view of the personal and private self, creating a sense of discomfort and uneasiness. Therefore, philosophical discussions, hopes, and dreams are shared only with family members and close friends. Emotions are intensely experienced but are not always expressed among family or friends. "Being in control" includes harnessing one's emotions and not revealing them to others.

Newer-generation German Americans, influenced by the cultural values of the US, are more overt in sharing their thoughts, ideas, and feelings with others. They have joined in the American belief that direct confrontation and open dialogue can be productive. In spite of this general pattern of acculturation, pockets of Germans in the US continue to be reserved when sharing their private affairs, thoughts, and concerns, including their health concerns, with strangers. Their reluctance for socializing may make them appear unfriendly; yet, under their stern exterior, they want to be liked.

Good manners are very important to Germans. A display of politeness and courtesy is viewed as a sign of respect. Social distance, eye contact, touch, and facial expression define boundaries. Failure to adhere to these protocols is considered rude by Germans and may alienate people who are unaware of them. The handshake, still a structured phenomenon in Germany, has been acculturated into a more casual form by German Americans and is a common method of greeting for both men and women, but the practice is to always shake hands with women first. When families and friends gather, handshaking is practiced along with pats on the arms or back and in modern day Germany, the slight embrace with faux kiss on both cheeks, as commonly used among the French, is becoming more common among Germans as well, particularly among younger people.

Practices associated with personal touch and displays of affection, such as hugging and kissing, vary among German families. In some German families, not unlike typical American families, little touching occurs between the father and the children. This relationship, however, may become more demonstrative as parents and children age. Affection between a mother and her children is more evident. In other German families, there is outward expression of love from both fathers and mothers, grandparents, and extended family members; hugs and kisses are expected and often offered as reaffirmation of love.

Whereas close friends are often extended warmth through handshakes, brief embraces, and sometimes kisses, strangers are kept at arm's length and greeted formally. Germans are careful not to touch people who are not family or close friends.

The distancing used by Germans to position themselves in relation to others is greater than the distancing used by some other cultural groups in the US. More acculturated German Americans may control their space in a manner similar to that of other Americans. In health-care situations, providers frequently enter their patients' personal space. German Americans understand the need for this intrusion and voluntarily participate in such encounters, while preserving their dignity and privacy.

Germans place a high value on their privacy. Germans may live side by side in a neighborhood and never develop a close friendship. A German neighbor would not be expected to borrow a cup of sugar from another neighbor. Germans would seldom consider dropping in on another German neighbor as this behavior is incongruent with their sense of privacy and order. Much preparation is completed to ready the house for guests. Germans use doors to protect their privacy. A closed door requires a knock and an invitation to enter regardless of whether the door is encountered in the home, business, or hospital. A closed door secures a sense of privacy and safety for Germans. Germans guard their privacy, which includes receiving phone calls at home. It is best to wait for an invitation or ask permission before contacting a new German acquaintance at home.

Germans maintain eye contact during conversation, but staring at strangers is considered rude. Even looking into a room from the outside is considered a visual intrusion; the interior of a room should not be entered without permission (Hall and Hall 1990).

Smiling is reserved for friends and family. Because smiling does not occur during introductions, Germans are often considered unfriendly. Work is considered serious business; thus, Germans smile very little at work. Dealing with illness is also considered serious business, calling for "correct responses" (i.e., reserved, direct, and unsmiling).

Several unacceptable expressions of nonverbal behavior for Germans include chewing gum in public, cleaning one's fingernails in public, talking with one's hands in the pockets, placing one's feet and legs on furniture, pointing the index finger to one's own head (an insult), and public displays of affection. Younger, more non-traditional German American youths may not adhere to these perceptions. Americans cross their fingers for luck, whereas Germans squeeze the thumb between index and middle fingers. However, allowing the thumb to protrude more than its tip length is an offensive gesture (CultureGram 1994).

15.3.3 Temporal Relationships

Germans use time to buy the future and pay for the past. Their focus on the present is to ensure the future. The past, however, is equally important, and Germans begin their discussion with background information, which always includes history. Americans generally do not understand the German people's need to lay a proper foundation for discussion. Conversely, Germans develop a deep understanding of their historical heritage through an intense analysis of past events. Friday (1989) explained this contradiction as the result of a difference in educational emphasis in German and American schools.

Germans pride themselves on their punctuality. Being on time is an obsession. People who expect to be late for appointments should call and explain. If this is not done, the German sense of order is disturbed. Work is completed by setting and meeting deadlines. "Keeping to the schedule" is extremely important. There is a sense of impatience and often intolerance in the German American who encounters a situation in which someone else is not performing on schedule. This impatience can be stirred to anger in the work setting, in the supermarket, on the highway, in the hospital, or in the health-care provider's office. In the mind of a German, who is always on time, there are rarely good excuses for tardiness, delays, or incompetence that disturb the "schedule" of events. Within this cultural continuum model, Western Europeans and North Americans attend to details in a linear, orderly manner, measuring days, hours, and seconds. Time has value for both groups, often equated with money.

15.3.4 Format for Names

Traditionally, Germans keep social relations on a formal basis. Even neighbors of long-standing acquaintance are addressed as *Herr* (Mr.), *Frau* (Mrs.), or *Fräulein* (Miss) and their last name. Those in authority, older people, or subordinates are always formally addressed. Only family members and close friends address one another by their first names. Many German Americans born in the 1930s and 1940s continue to be formal in their social and business interactions. If this consideration is not returned, or if someone presumptuously calls them by their first name, it may be considered a sign of disrespect or poor upbringing. Hall and Hall (1990) explain, "The taboo against first-naming should not be dismissed as an empty convention." In their book they describe an old custom, *Brüderschaft-trinken*, in which "two friends formalize their shift to the more intimate form of address. They hook arms and each sips from a glass. Then they shake hands and announce their first names" (p. 49).

Germans combine a person's professional title with *Herr*, *Frau*, *Fräulein*, or other titles and their last name. For example, a director of a business is addressed as *Frau* or *Fräulein Direktorin*. The title is often used without the name. A physician may be addressed simply as *Doktor*. Younger generations or more acculturated Germans may be less formal in their interactions. Because of cultural blending, health-care providers will find that German Americans vary widely in their observance of these rules of etiquette. Therefore, these health-care providers should ask their patients how they would like to be addressed. This approach lessens the possibility of the provider unintentionally offending the patient.

15.4 Family Roles and Organization

15.4.1 Head of Household and Gender Roles

Traditional German families view the father as the head of the household. In the US, the husband and wife are more likely to make decisions mutually and share household duties. Stay-at-home dads are uncommon in Germany (S. Maubach, personal communication, 2006). Often, when illness, dependence, and disability interfere and prevent family members from carrying out their roles, others assume decision-making responsibilities either temporarily or permanently.

In Germany, where emphasis is on ***Ordnung*** (order), and ***Gemeinschaft*** (community), older people are not expected to be self-reliant. Health and social programs for older people are considered part of the institutional approach of European programs. Because of the comprehensiveness of these benefits, there is less financial reliance on the family. One home may remain in the same family for generations. Often, more than one generation live under the same roof. Older family members who live with their children are included in family celebrations as well as in the daily routine of the families. As they become unable to perform their roles and duties, other family members assume their responsibilities.

Older people within German American families are sought for their advice and counsel, although the advice may not always be followed. They are admired for maintaining their level of independence and their continued contributions to society. Many live alone or with aging spouses. Helping older parents or grandparents to remain in their own home is important to German American families. By providing a helping hand with home maintenance, shopping, and finances, the family is able to safeguard and prolong a state of independence, even when living hundreds of miles away. For those who grow dependent, moving in with children or residing in a nursing home is a viable choice for German American families.

The differences in the family role for older people in Germany and in the US may be due to the far-reaching mobility of the American population that does not exist in Germany, where families generally live in close proximity. When Americans moved to the western frontier, they were required to adopt attitudes that included a degree of individualism, self-reliance, and initiative not demanded in a more geographically stable and settled society in which families had support because they were geographically close. The emphasis on these traits, as well as the concept of "America, land of unlimited opportunity," has made life in the US difficult.

The Older Americans Act, Medicare, and Medicaid legislation, which are considered residual approaches for meeting one's social needs, support the context of the German belief in self-reliance and the supportive role of the family. Such residual approaches are offered when the normal channels such as family, marketplace, and church are not sufficient for meeting needs. Strong advocacy groups such as the American Association of Retired Persons and the National Council of Senior Citizens, which have mobilized older Americans as a self-interest group, also support this idea of self-reliance (Pinquart et al. 2003).

An interesting fact is that the Germans love their dogs, and in Germany, it is acceptable to take the family dog everywhere—restaurants, visiting, and the hospital. In the US, however, animals, except for seeing-eye dogs, are restricted from most public places. Other pets in German households may be cats, rabbits, birds, hedgehogs, and, of course, horses (S. Maubach, personal communication, 2006).

15.4.2 Prescriptive, Restrictive, and Taboo Behaviors for Children and Adolescents

Prescriptive behaviors for children include using good table manners, being polite, doing what they are told, respecting their elders, sharing, paying attention in school, and doing their chores. Additional behaviors include keeping one's nose clean, eating all food that is placed on their plates, looking at a person who is talking, and sitting up straight. Prescriptive behaviors for adolescents include staying away from bad influences, obeying the rules of the home, sitting "like a lady," and wearing a robe over pajamas. Restrictive and taboo behaviors for children include talking back to adults, talking to strangers, touching another person's possessions, and getting into trouble. Restrictive and taboo behaviors for adolescents include smoking, using drugs, chewing gum in public, having guests when parents are not at home, and having run-ins with the law.

Germany has regulations about noise levels in public areas such as athletic fields where people gather to watch soccer games, tennis, and riding events. These regulations are enforced for both children and adults. On occasion, schools in highly populated areas apply similar restrictions for playground activities (German Noise Law 2010).

15.4.3 Family Goals and Priorities

In Germany, history, family, and lifelong friendships are highly valued. Concern for one's reputation is a strong value. One's family reputation is considered part of a person's identity and serves to preserve one's social position (good and bad). Family is also fundamentally important to most Germans. People often describe their unique personal relationships one has with each family member and the support that is provided through family relationships. In Germany, the family home provides a place where an individual's unique perspectives can be fully revealed, unlike in the general public where more reserved, introverted behavior is more the norm. Families are expected to help nurture a person's goals and priorities but also encourage the individual family member to become increasingly more self-reliant throughout childhood so that they are prepared to be independent as adults (Facts about Germany 2016).

15.4.4 Alternative Lifestyles

Historically, pregnancy outside of marriage resulted in disapproval; however, in recent years, child-bearing outside of marriage has become far more common, and like the US, there is decline in the number of people getting married.

As in the US, lesbian, gay, bisexual and transgender (LGBT) rights in Germany have evolved significantly over the course of the last decades. In the early part of this century, homosexuality was generally tolerated in German society and German scholars were at the forefront of investigating the concept of homosexuality and the establishment of gay rights (Beachy 2010). Although same-sex sexual activity between men was already had been illegal under Paragraph 175 by the German Empire in 1871, homosexuality flourished until the laws were extended during the Nazi regime. Nazi extensions were repealed in 1950 and same-sex sexual activity between men was decriminalized in both East and West Germany in 1968 and 1969, respectively. Same-sex marriage has been legal since 1 October 2017, after the Bundestag passed legislation giving same-sex couples full marital and adoption rights on 30 June 2017. Prior to that, registered partnerships were available to same-sex couples, having been legalized in 2001. Gender reassignment surgery is available in Germany (Trans Health Care 2020).

15.5 Workforce Issues

15.5.1 Culture in the Workplace

Germans are among the most skilled and educated workers in the world. Much of Germany's success is due to advanced technologies, and it is a leading nation in Nobel Prizes for physiology and medicine. Some of its most important contributions are in rocketry, material science, and chemical products (Solar Navigator 2006). German workers are educated to meet the needs of a highly industrialized country. The atmosphere of German business is very formal.

Several considerations must be remembered when working with Germans and some German Americans. First, it is important to be on time for work and business appointments and to complete work assignments on time. Second, business communication should remain formal: shaking hands daily, using the person's title with the last name, keeping niceties to a minimum, and avoiding the adjustment of office furniture during meetings. Employees are not addressed by their first names. Third, one should respect privacy by not entering rooms with a closed door before knocking and being invited inside. Fourth, dress, opinions, and activities should be conservative. Finally, learning to speak German is important if an employee is living in Germany and working for a German company (Hall and Hall 1990).

The current trend toward a global economy has encouraged many American companies to establish sites in Germany and many German corporations to have subsidiaries in the US as well as other places throughout the world. Many German managers are transferred to the US by their companies and easily enter and adapt to the American business climate. Others trained in the health professions, the physical sciences and education, and technologies join the ranks of practicing professionals in the US.

In the workplace, American values and beliefs often oppose German traditions. Friday (1989), in exploring the problems of transcultural adaptation for American and German managers, noted that "the management style of German and American managers within the same multinational corporation is more likely to be influenced by their nationality than by the corporation culture" (p. 436) and this is largely still true today. Although Friday's work was done outside the health-care industry, some of his findings have implications for relationships across a broad range of work settings, including health-care services. For one, German and American managers hold different perceptions of their relationship with their employer. Germans see themselves as part of the corporate family, whereas many Americans do not identify with their corporation. Germans anticipate lifelong employment with the same company, whereas Americans may move to other companies should a good opportunity arise. Another difference is that American managers expend much energy to be liked, whereas Germans prefer being credible in their positions to being liked. To satisfy their need to be liked, American managers encourage informality in the workplace, such as by addressing peers, subordinates, or superiors by their first names; by asking personal questions; and by believing in equality and making themselves at home in one another's offices. For the German manager, credentials and education confirm their credibility and lead to power.

15.5.2 Issues Related to Autonomy

Germans and German Americans expect to receive respect for their work and for their ability to make decisions about their work. They find a hovering supervisor annoying and demeaning. Balancing control and freedom in the workplace is necessary to foster productivity in German and German American workers (Hall and Hall 1990). American and German managers use different styles of assertiveness. Whereas Americans model their approach within the idea of equality or "fair play," Germans, who have no translation for "fair play," are assertive by putting other people in their place. As in all languages, nuances and jargon can frustrate the individual whose second language comes only from the textbook and who does not understand idioms and colloquial expressions. The Germans' use of two distinctive manners of communication—*gespraech*, casual talking, and *besprechung*, the workplace discussion about performance—continues into the workplace.

Reflective Exercise 15.1

Hans Bitner, a German American, has accepted a new computer technology position with a German-owned company. His job orientation will be in Germany, but he will work in the US under a German supervisor. He is very concerned about making a good impression while in Germany and during his first weeks on the job.

What advice about each of the following issues can be given to Hans to make his transition into the company as an employee easier?

1. Conversations
2. Manners in the workplace
3. Privacy
4. Public behavior

15.6 Biocultural Ecology

15.6.1 Skin Color and Other Biological Variations

Germans range from tall, blond, and blue-eyed to short, stocky, dark-haired, and brown-eyed. Over the past half-century, Germany has become increasingly more diverse due to the many foreigners coming in to Germany as guest workers initially and as people seeking asylum more recently. Because many Germans have fair complexions, skin color changes, and disease manifestations can be easily observed. Due to the northerly European climate, many Germans are sun worshipers when they have the opportunity to do so. For those with fair skin, prolonged exposure to the sun increases the risk for skin cancer.

15.6.2 Diseases and Health Conditions

Because Germany is highly industrialized, Germans suffer from many of the same life-threatening diseases that afflict groups from other highly industrialized countries. Leading causes of death for German Americans follow the patterns of the dominant American society and include heart disease, cancer, cerebrovascular disease, and accidents. Because of the poor management of industrial contaminants, people in the Eastern regions often suffer from pollution-related illnesses (Health Industry Today 2011). When assessing recent German immigrants, it is helpful for health-care providers to know where in Germany the patient resided before entering the US.

The prevalence of HIV, total (% of population ages 15–49) in Germany was reported at 0.1% in 2018, according to the World Bank collection of development indicators, and antiretroviral therapy coverage (% of people with advanced HIV infection) in Germany was reported at 80% in 2018 (Trading Economics 2020). Germany offers guidance and care to those who are infected, as well as a comprehensive prevention program for its citizens. Because prostitution has been legal in Germany since 1987, frequent health checks are required for those in this profession (WordIQ Dictionary 2010).

In 1998, research localized the genetic cause for a syndrome of symptoms for a new form of myotonic muscular dystrophy. A second study conducted in Minnesota, Texas, and Germany identified the same causative mutation (Mackle 2001). This new form of the disease, called *DM2*, appears to be most common in Americans of German descent (Mackle 2001).

Another genetic disease, hereditary hemochromatosis, is also found in German Americans. Hemochromatosis, a toxic level of iron accumulation, can cause diabetes, chronic fatigue, liver disease, impotence, and even heart attacks. The disorder is due to a mutation in the *HFE* gene located on chromosome 6. German Americans can avoid, prevent, and treat these maladies with genetic testing and early diagnosis. Hemochromatosis is treatable through the removal of iron through phlebotomy (withdrawal of blood or bloodletting). The person can expect a normal life expectancy with aggressive treatment. Diagnosis can be established through a blood test known as an *iron profile*.

Sarcoidosis, a disorder found mostly in women between the ages of 20 and 40, occurs in all races, but people of German descent are at a higher risk (Gottfried 2001). Sarcoidosis causes persistent cough or no symptoms. The cause is unknown, but doctors speculate that it involves an adverse reaction of the immune system; the diagnosis is often missed.

Dupuytren's disease, a slowly progressive disorder, is a deformity of the hand in which the fingers are contracted toward the palm. This often results in a functional disability. Dupuytren's disease is frequently found in people of German descent. Affecting mostly older males, the disease causes the synthesis of excessive amounts of collagen. The excess collagen is deposited in a ropelike fashion from the palm into the fingers, permanently fixing the fingers in a state of flexion. Although the cause is uncertain, Peyronie's disease is often found in people with Dupuytren's disease (NIH n.d.). A benign plaque forms within the erectile tissue of the penis, which causes it to bend, resulting in reduced flexibility and causing pain during erection. This can prohibit sexual intercourse. The disease occurs mostly in middle-age men and often in men who are related, suggesting that genetic factors may increase the likelihood of developing this condition. Some researchers have theorized that Peyronie's disease may be an autoimmune disorder. A surgical approach to treatment has had some success. Candidates for surgery are men with curvature so severe that it prevents sexual intercourse.

Lowenfels and Velema (1992) examined the incidence of cholelithiasis in people from Denmark, Germany, India, Italy, Norway, and England. Although the study revealed prevalence rates from each of these countries, Norway ranked first and Germany ranked second for the overall incidence of gallbladder disease. Although the study addresses populations in Germany, the results may be applicable to Germans in other parts of the world.

A cohort study of White men of Norwegian, Swedish, and German ancestry conducted between 1966 and 1986 revealed an increased risk of stomach cancer among foreign-born and first-generation German Americans living in the north-central states (Kneller et al. 1991). This study suggests an interrelationship among ethnic, geographic, and dietary factors as the cause. High concentrations of immigrants from northern Europe, which includes the high-cancer-risk countries of Germany and Scandinavia, settled in the north-central region of the US. Low educational attainment; employment in laboring and semiskilled occupations; and ingestion of salted fish (at least once a month), bacon, milk, cooked cereal, and apples increased the risk factors for the foreign-born and first-generation individuals. These findings support the theory of ethnic risk. Subjects who smoked 30 or more cigarettes per day exhibited a fivefold risk for the development of stomach cancer. In addition, those who smoked a pipe and chewed smokeless tobacco had an increased risk for stomach cancer (Kneller et al. 1991).

According to Zielenski et al. (1993), an increased incidence of cystic fibrosis (CF) is found among Hutterite German–speaking communal farmers living on the Great Plains of North America. Mutations in the Hutterite population, a genetic isolate with an average inbreeding coefficient of about 0.05, exhibit an increased prevalence of CF carriers. Maternal-child health professionals providing care to this ethnic group can assist patients by encouraging genetic counseling to ensure early diagnosis of CF in their infants.

Hemophilia, a genetic bleeding disease found in Germany and the US, can be traced from

Queen Victoria of England, who, through a gene mutation, passed hemophilia to her son and through her daughters (Kilcoyne 2004). The disease was then spread into Europe through the royal families, including the House of Hohenzollern, which consisted of kings and emperors of Prussia, Germany, and Romania. World War I led to the German Revolution, and the House of Hohenzollern abdicated, ending the monarchy. Historians believe that the source of hemophilia in the US is a woman in Plymouth, New Hampshire, most likely English. There are currently over 20,000 people in the US with hemophilia, accounting for over 75 percent of all cases of hemophilia (CDC 2011). As in the US around 1993, those with hemophilia in Germany were contaminated with the AIDS virus through the administration of blood products and anti-clotting factors. Health-care providers may want to be mindful of the German history of hemophilia and the AIDS issues while diagnosing bleeding issues in newly arrived German immigrants.

15.6.3 Variations in Drug Metabolism

Few research studies have been completed on variations in drug metabolism and interactions specific to people of German ancestry. Aggregate data on White populations report that there are no slow metabolizers of alcohol in this population (Levy 1993). One study reported that 5 percent of Germans are poor metabolizers of debrisoquine (Levy 1993), and therefore this group may need lower dosages of propranolol to control blood pressure.

15.7 High-Risk Behaviors

Germans are known for their breweries and their *Gasthäuser*, or "restaurant that serves spirits." Beer is also served at the pubs (*kneipen*). In Germany, drinking beer is a way of life. Total alcohol consumption per capita (liters of pure alcohol, projected estimates, 15+ years of age) in Germany was reported at 13.4 in 2016 (Trading Economics 2020). German youth can legally drink beer at age 16 and drive at age 18. Beer is often served with meals, whereas tap water is rarely consumed. Sparkling mineral water (*mineralwasser*) is commonly served if water is requested by a patron. Even lactating mothers are encouraged to drink malt beverages to increase breast milk production. This long-standing tradition of beer consumption is not without its abuses.

15.7.1 Health-Care Practices

Germans, whether born in Germany or in the US, share a love of nature. They enjoy the great outdoors. Fresh air and exercise are highly valued. Hiking, walking, swimming, skiing, cycling, soccer, horseback riding, and playing tennis are just a few of the activities enjoyed by people of German ancestry. Walking is a way of life. Sports are played for exercise and the pleasure of participating in group activities. Water sports are very popular and are encouraged among older people, disabled people, mothers, and small children. Because many German Americans are joiners, health club memberships appeal to German Americans.

Ruhezeit, or quiet time, is nearly sacred in Germany. However, in modern times, this time-honored tradition which occurs between 1 p.m. and 3 p.m. is seldom practical in today's fast-moving German economy. During this time, older Germans take naps, and older retired German Americans may follow this ritual as well. Neighbors and friends are expected not to create noise, telephone, or interrupt in any other manner. This quiet time is often followed by *Kaffee and Kuchen*, coffee and cake time, around 4 p.m. (The German Connection 2006).

15.8 Nutrition

15.8.1 Meaning of Food

Food is a symbol of celebration for Germans and is often equated with love. Food and food rituals are powerful identification symbols for ethnic groups. The diet of immigrants is modified by the availability of foods and their financial status. The desire to maintain ethnic food habits has prompted children and grandchildren of immigrants to retain their ethnic heritage.

15.8.2 Common Foods and Food Rituals

Traditional methods of food preparation with high-fat ingredients add to nutritional risks for many German Americans. Real cream and butter are used in German cooking. Gravies and sauces that are high in fat content, as well as fried foods, rich pastries, sausages, and boiled eggs, are only a few of the culinary favorites. Germans have traditional ways to prepare their favorite foods. Meats, turkey, chicken, pork, and fish are stewed, roasted, or marinated and are often served with gravies. Vegetables (fresh is preferred) are often served in a butter sauce. Foods are also fried in butter, bacon fat, lard, or margarine. *Bratwurst* (*currywurst*) served with curry ketchup and *pommes frites* (french-fried potatoes) with mayonnaise are found at the top of the list of fast foods in Germany.

One-pot meals such as string beans and potatoes, cabbage and potatoes, pork and sauerkraut, stews, and soups are served as family meals. Casseroles are also popular. Foods prepared with vinegar and sugar as flavorings are also favorites. Potato salad, cucumber salad, coleslaw, pickled eggs, pickled cucumbers, cauliflower, tongue, and herring are common examples of favored foods prepared with these flavorings. Sour cream, mayonnaise, and mustards are used frequently in food preparation.

The nutritional habits of some Germans may be a significant health risk factor. Food is an integral part of a German's life. Food is served at celebrations and during visits and is taken on trips. The German infatuation with food can lead to overeating, which results in obesity. Children are rewarded for good behavior with food. Those who are ill receive Jell-O, egg custards, ginger ale, or tomato soup (not creamed) to settle their stomachs. Sending food with loved ones who will be away from the family for a time is quite common: Homemade cakes, cookies, and jams are a few of the offerings.

Nothing pleases German cooks more than witnessing people with hearty appetites at the table. Generous amounts of food are prepared, and second helpings are encouraged. In choosing foods for German Americans, the health-care provider should consider cutting portion size, overcoming harmful food rituals, and reducing fat intake.

Corn, frequently served as a vegetable in North America, is not eaten in Germany, where it is considered food for farm animals. Visitors from Germany are often startled when corn is served to them, but once they taste it, they are easily converted. Many early German immigrants turned to farming to conquer starvation, raising grains (including corn), fruits, and vegetables that were popular in North America. Foods associated with special events such as weddings, holidays, and religious occasions are the last to yield to acculturation. German cooks produce their best culinary efforts for holidays. Weeks of baking and preparation often precede the actual holidays. Selection of foods for the meal, proper preservation, and artistic presentation of tasty dishes are attended with care.

Table 15.1 lists common foods in the German American diet, based on the author's experience, personal interviews, the literature, and a marketing analysis conducted at a meeting of a local DANK (*Deutsch Amerikanischer National Kongress* [German American National Congress]) for a new food chain planning an international market concept. DANK has been in existence in the US since 1959 (DANK 2020).

Table 15.1 Common foods in the German American diet

Beverages
Coffee (with sugar and cream)
Herbal teas
Kümmel (caraway seed)
Light and dark beers
Schnapps
Steinhager (juniper beverage)
White wine
Breads, noodles, and dumplings
Rolls
Knöpfle
Potato dumplings
Pretzels
Pumpernickel
Ribbles
Spätzle
Cheese
Camembert
Limburger
Desserts
Baumkuchen
(tree trunk cake)
Kranz (almond and hazelnut cake)
Lebkuchen
(honey cakes)
Lübecker marzipan
Pfannkuchen
Pfefferkuchen
(gingerbread)
Rice pudding
Springerle
(cookies)
Stollen
Strudel
Fish
Anchovy paste
Carp (*karpfen*)
Dover sole
Pickled herring
Roe
Rollmops
Smoked cisco
Fruits
Apfel (apple)
Dried apples
Dried pears
Madelkerr (fruits)
Nüsse (nuts)
Prunes
Meats and fowl
Bacon
Beef
Bratwurst
Chicken
Duck
Frankfurter
Game bird
Gänseleberwurst
(goose liver)
Goose
Knockwurst
Liver dumplings
Mettwurst
Mutton
Pork
Salami
Saage (veal)
Tongue
Veal
Venison
Vonname (smoked pork chop)
Weissbratwurster
Wild boar
Preserves
Apple butter
Crabapple jelly
Vegetables
Beets
Cabbage
Carrots
Celery root
Mushrooms
Onions
Potatoes
Sauerkraut
White asparagus
White radishes
Miscellaneous
Caraway seeds
Castor sugar (pearl sugar)
Cilantro
Honey
Juniper berries
Molasses
Paprika
Vanilla beans

Reflective Exercise 15.2

Marian Graybill is a 27-year-old single mother. She has a 6-year-old daughter. Marian has been diagnosed with hypertension and is on Lasix 20 mg daily, which she takes only when she has swelling in her feet. Her doctor has asked her to be mindful of her sodium consumption. Marian's family lives in Germany and delights in sending packages of German food favorites. Often these are envelopes of dried seasonings that can be added to fish and meat. The sodium content of the seasonings is very high. Marion loves these dishes and prepares them for her daughter and herself. It is important to Marian that her daughter be familiar with food from Germany.

1. Marian is only 27 years old and is hypertensive. Understanding Marian's need to enculturate her daughter in light of German cooking, how would she be impressed with the importance of following the doctor's request to lower her sodium intake?
2. What could be done to help ensure that she will take her Lasix on a daily basis?

15.8.3 Dietary Practices for Health Promotion

Because of apartment living in Germany, many Germans love to garden, and they bring this love to the US. Gardening provides the fresh vegetables that Germans enjoy. What is not eaten is canned, pickled, dried, or frozen for future use. Having a full larder is very important to Germans and German Americans.

Some foods are used to prevent or treat illnesses. Prune juice is given to relieve constipation. A special soup from fresh tomato juice is used to treat a migraine headache. Ginger ale or lemon-lime soda relieves indigestion and settles an upset stomach. After gastrointestinal illnesses, a recuperative diet is administered to the sick family member, beginning with sips of ginger ale over ice. If this is retained, hot tea and toast are offered. The last step is coddled eggs, a variation of scrambled eggs prepared with margarine and a little milk. If these foods are tolerated, the sick person returns to the normal diet. Garlic and onions are commonly eaten to prevent heart disease.

15.8.4 Nutritional Deficiencies and Food Limitations

The literature does not report any enzyme deficiencies or food intolerances specifically related to Germans. However, those of lower socioeconomic status may lack the financial ability to purchase foods essential for a nutritious diet.

15.9 Pregnancy and Childbearing Practices

15.9.1 Fertility Practices and Views toward Pregnancy

Large families are rare in Germany. Most couples have only two children. The German government recognizes the importance of family and provides child-rearing allowances and work leaves. German law allows employees to go on paid parental leave in addition to mandatory paid maternity leave after childbirth. The allowance is paid by the state and ranges from EUR 300.00 to up to EUR 1800.00 a month depending on the employee's prior income. Recently, an additional option dubbed "Parental Allowance Plus" was made available for parents of children born on and after 1 July 2015. Under the new rules, employees will have a right to request up to 24 months of paid parental leave (instead of 12 months) or, if both parents decide to go on parental leave, they will be entitled to 28 months of paid parental leave (Schnabel 2014).

A variety of birth control practices and interventions for improving fertility among Germans are readily available. On the one hand, the German respect for authority and love for scientific facts and data encourage the use of methods to control, as well as to enhance, fertility practices. On the other hand, the use of medication or devices might be viewed as interrupting the natural progression of things. Although abortion remains illegal in Germany, abortions are provided on request if within the gestational limit of 14 weeks, indicating that the law itself is not enforced (Center for Reproductive Rights 2020). Germans in general see fertility practices and abortion as largely a personal decision left to the mother.

Reflective Exercise 15.3

Fourteen-year-old Lydia Shultz is 2 months pregnant. She is being seen in the health-care provider's office and is accompanied by her mother and grandmother. Both her mother and grandmother occasionally weep as they wait in the office with Lydia. They tell of their disbelief that Lydia is pregnant at 14 and how embarrassed they are. They ask for names of homes for unwed mothers.

1. Discuss the impact of teenage pregnancy on this German American family.
2. Understanding the German American family culture, how can this family be helped through this life crisis?

15.9.2 Prescriptive, Restrictive, and Taboo Practices in the Childbearing Family

Germans share some of the prescriptive, restrictive, and taboo practices of other cultures concerning pregnancy. Some examples of prescriptive practices include getting plenty of exercise and increasing the quantity of food to provide for the fetus. Some restrictive practices include not stretching and not raising the arms above the head to minimize the risk of the cord wrapping around the baby's neck.

A review of the literature and personal interviews did not reveal any prescriptive, restrictive, or taboo practices related to the birthing process. Birthing rooms that allow fathers and other family members to be present are popular among German Americans. In Germany, midwives commonly deliver babies ("Birth and Midwifery in Germany," 2011). Prescriptive practices for the postpartum period include getting plenty of exercise and getting fresh air for the baby; if the mother is breastfeeding, she should eat foods that enhance the production of breast milk.

15.10 Death Rituals

15.10.1 Death Rituals and Expectations

Germans and German Americans traditionally observe a period of mourning activities after the death of a family member. German Americans usually have a family funeral director. The family may go to the funeral home together to select a coffin. Following the directions of loved ones about what should be done after their death is very important. Careful selection of the clothes to be worn by the deceased and the flowers that represent the immediate family is equally important. These selections are based on their knowledge of the deceased's way of life.

15.10.2 Responses to Death and Grief

The viewing provides an opportunity for family, friends, and acquaintances to view the body; offer their condolences; and extend their offers of assistance should the family need help in the future. Crying in public is common, but in some German American families, the display of grief may be done privately. A tradition of wearing black or dark clothing when attending a viewing or a funeral may be expected of both family and friends. Another expectation is that the bereaved family limits socialization activities for the following several months.

The traditions that surround the provision of food for the mourners have changed over the years. From the 1940s through the early 1960s, women in the neighborhood prepared the food and served it as people arrived at the home following the burial. More recently, families have become the primary providers of food and may hire caterers to prepare food or use a restaurant, as is done in Germany, where homes are too small to accommodate large groups of people.

For Germans and German Americans, death is seen as part of the life cycle, a natural conclusion to life. Individuals who embrace a set of religious beliefs may look forward to a life after death, often a better life. Death is a transition to life with God. Because illness is sometimes perceived as a punishment, the length and intensity of the dying process may be seen as a result of the quality of the life led by the person.

15.11 Spirituality

15.11.1 Dominant Religion and Use of Prayer

Martin Luther launched the Reformation in the early sixteenth century. Ninety percent of the population has some religious affiliation. Protestants and Catholics share equal portions of the population (33 percent). Other religions of German Americans include Judaism (the third largest population of Jews in Western Europe), Islam, and Buddhism (Solar Navigator 2006). Similar to the US, Germany has no state church; church and state remain separate. Religion is seen as a personal matter for German Americans, and secularism embraced.

Freedom of religion in Germany is guaranteed by article 4 of the German constitution. This states that "the freedom of religion, conscience and the freedom of confessing one's religious or philosophical beliefs are inviolable. Uninfringed religious practice is guaranteed."

Although there is no state church in Germany, churches, as independent public corporations, have a partnership relationship with the state. They can claim state grants, which in turn support schools and kindergartens. Churches can levy taxes on their membership, but the taxes are collected by the state. German churches also serve a charitable and social purpose by running nursing homes, retirement centers, hospitals, schools, training centers, and consultation and caring services. Generally speaking, Germany has an open and tolerant attitude towards people of all kinds of backgrounds and religions.

Table 15.2 reflects the formal positions or the relationships between spiritual beliefs and health practices of several Protestant religions and the Roman Catholic Church. The Jewish, Muslim, and Greek Orthodox faiths are addressed in other chapters. Health-care providers must recognize that individuals' decisions may vary from the formal position of their religious groups. Therefore, the table serves only as a guide, not as an exclusive basis for decision making in health care.

Most German religious philosophies do not divorce physical health from the actions of God. Many hold the view that God works through health-care providers as well as through the resources of medicine. Prayer is used to ask for healing, for effectiveness of treatments, for strength to deal with the symptoms of the illness, and for acceptance of the outcome of the illness. Prayers are often recited at the sickbed, with all who are present joining hands, bowing their heads, and receiving the blessing from the clergy.

Reading the Bible is also an important spiritual activity. Most German and German Americans have a family Bible, which is passed down through the generations. It serves as spiritual comfort and as a reservoir of family historical data such as the dates of births, marriages, and deaths.

15.11.2 Meaning of Life and Individual Sources of Strength

Individual sources of strength for most Germans and German Americans are their beliefs in God and in nature. Although they may not attend church on a regular basis, a German's faith is deep. Family and other loved ones are also sources of support in difficult times. Home, family, friends, work, church, and education provide meaning in life for individuals of German heritage. Family loyalty, duty, and honor to the family are strong values. Germany itself is a pragmatic country. Germans believe that truth depends very much on situation, context, and time. They show an ability to adapt traditions easily to changed conditions, a strong propensity to save and invest, thriftiness, and perseverance in achieving results.

15.11.3 Spiritual Beliefs and Health-Care Practices

Teachings of the churches joined by German people provide direction and counsel on many health-care issues. Many of these churches have taken a formal position on abortion, artificial insemination, and prolongation of life. The church prescribes when individual choice is important in deciding on accepting or refusing treatments and provides advice when seeking spiritual counsel-

Table 15.2 Positions of Roman catholic and selected protestant religions regarding various health-care practices

Health-Care	Baptist	Roman catholic	Brethren	American Lutheran	Missouri methodist	Presbyterian	Synod Lutheran	Wisconsin Lutheran	Church of Christ	Salvation army
Administration of drugs, blood, and vaccine	Acceptable	Justifiable as long as for the good of the whole	Acceptable	Acceptable	Acceptable	Acceptable	Acceptable	Acceptable	No position	Acceptable
Biopsies	Acceptable	Acceptable	Acceptable	Encouraged	Acceptable	Acceptable	Acceptable	Acceptable	No position	Acceptable
Loss of limb	Acceptable	Acceptable (principle of totality)	Acceptable	Acceptable	Acceptable	Acceptable	Has no position	Acceptable if done to save a life	No position	Acceptable
Transplants	Acceptable	Permissible	Acceptable	Encouraged	Acceptable	No position	Acceptable	Acceptable	No position	Acceptable
Prolongation of life	Discouraged in clearly terminal cases	Advocates taking into consideration benefit and burden to patient	Allowed to preserve individual's freedom and dignity	Extraordinary and heroic efforts sometimes deemed justifiable	When death inevitable, allows freedom to direct or encourage a physician to remove artificial support systems	Quality of life viewed as more important than length of life	Permitted only in extraordinary cases after careful deliberation	Disapproves of prolongation of the agony of death	No position	Acceptable if prolongation of life is desired by individual; comfort and appropriate measures should be provided
Euthanasia	Left to individual choice for no extraordinary measures	Not permissible	Decision left to doctor and family	Advocates death with dignity	Discouraged	Favors right to die	Opposed	Advocates to relieve pain	Favors death by natural means and processes	Unacceptable
Donation of body parts	Encouraged	Deemed justifiable	Acceptable	Encouraged	Encouraged	No official position	No position	Acceptable	No position	Acceptable
Autopsy	Encouraged	Permissible	Acceptable	Strongly approved	Encouraged	No position	Left to individual choice	Relatives encouraged to give permission	No position	No restriction
Disposal of body	Burial or cremation permitted	Burial favored; cremation permitted if unusual circumstances exist(e.g., infection)	Left to individual or family	Burial favored, but cremation acceptable	Burial and cremation permitted, and customary procedures for disposal of body after use for research	Burial favored, but cremation acceptable	Allowed if done with honor	Burial favored, but cremation permitted under epidemic conditions	No position	Left to individual and family

Eugenics, genetics	No position	Opposed	Advised for assessment of serious illness	Acceptable, marriage and procreation discouraged if offspring likely to inherit hereditary defects	Research and parental counseling encouraged	No position	No position	Viewed as great blessing	No position	No position
Birth control	Left to individual choice	Only naturalmeans permitted	Acceptable	Disapproved	Family planning favored	Acceptable	No position	Unacceptable, except if hereditary defects likely to occur	No position	Acceptable within marriage bond
Artificial insemination	No restrictions for mates	Viewed as illicit	Left to individual choice	Acceptable	Theological barriers	No position	No position	Left to husband and wife; adoption acceptable	No position	No position
Sterility tests	Acceptable	Permissible	Left to individual choice	Approved if necessary to ensure health of mother	Acceptable, but counsel with physician and pastor encouraged	No position	Left to individual choice	Acceptable	No position	Acceptable
Therapeutic abortion	Left to individual choice	Permitted only indirectly (e.g., removal of uterine cancer)	Left to individual choice	Approved if necessary to ensure health of mother	Acceptable, but counsel with physician and pastor encouraged	Justifiable by circumstances (e.g., rape or incest)	Permitted to save life of mother	Justifiable to save mothers life or in case of rape or incest	Decision of physician accepted	Opposed
Abortion on demand	Left to individual choice	Prohibited	Opposed	Opposed	Same as above	Deemed justifiable by circumstances (e.g., rape or incest)	No position	Opposed	No position	Opposed

ing. In their study (2005) on the role of religion and spirituality in medical patients in Germany, Büssing, Ostermann, and Matthiessen examined how German patients with cancer, multiple sclerosis and other diseases view the impact of spirituality and religiosity on their health and how they cope with illness (Büssing et al. 2005). Patients with both a religious and spiritual attitude had significantly higher values in the subscales dealing with the search for meaningful support, and the stabilizing effects of religion and spirituality than patients without such attitudes.

15.12 Health-Care Practices

15.12.1 Health-Seeking Beliefs and Behaviors

Germany has the world's oldest national social health insurance system, with origins dating back to Otto von Bismarck's social legislation, which included the *Health Insurance Bill of 1883*. A universal, multi-payer health care system paid for by a combination of statutory health insurance (*Gesetzliche Krankenversicherung*) and "*Private Krankenversicherung*" (private health insurance) ensures that most Germans receive regular medical and dental checkups, immunizations, and routine screening. Complementary and alternative therapies are all widely popular in Germany albeit not generally covered by health insurance. German health insurance does provide for health spa (*Kur*) treatments for illnesses that would benefit from such activities. These "Kur" treatments occur in Kur hotels with therapeutic certified by government agencies and provide meals and treatments that will aid in the recovery of diagnosed ailments.

15.12.2 Responsibility for Health Care

Although health care in Germany is considered the individual's own responsibility, it is also a concern of the society as a whole. Germans take health and healthcare very seriously and consider it a right vs. a privilege. The life expectancy for Germans in 2018 was 81.10 years, a 0.16% increase from 2017 in comparison the US with 78.7 years for Americans, putting the US behind other developed nations. Germany's infant mortality rate of 2.978 per 1000 infants is low as compared with the US with an infant mortality rate of 5.9 per 1000 (CIA World Factbook 2020).

Women in the family often administer remedies and treatments. In traditional families, the mother usually sees that children receive checkups, immunizations, and vitamins. German Americans use a variety of over-the-counter drugs. Many Germans also use folk and homeopathic substances as remedies. In fact, Germany is the only member state of the EU in which homeopathic remedies based on minerals or plants, and produced only in very low quantities, do not need to be registered. The use of over-the-counter drugs may stem from the belief that individuals are responsible for their own health and from the beliefs and traditions about the treatment of sickness learned within the family system. In Germany, however, over-the-counter drugs can be purchased only from a pharmacy, which increases the cost to the consumer. Therefore, over-the-counter drugs are not as accessible to Germans as they are to German Americans. Today, prescription drugs are more complex, and numerous over-the-counter medications have become more accessible to German Americans. The two used in combination may lead to dangerous drug interactions for those who practice self-medication. Thus, health-care providers need to ascertain if over-the-counter and folk remedies are being used to determine whether there are contraindications with prescription medications.

15.12.3 Folk and Traditional Practices

Among the early German immigrants, women practiced folk medicine, which often included singing and the laying on of hands. Families passed this knowledge on from mother to daughter. Common natural folk medicines included

Table 15.3 German folk remedies for various afflictions

Affliction	Remedy
Abrasions, burns	Vaseline
Boils	Black salve
Bumps and burns	Butter
Cleaning cuts and abrasions	Hydrogen peroxide
Colds	Vicks VapoRub as chest rub or placed in a vaporizer
Colds	Camphorated oil (chest rub; soft cloth covered with oil is placed over chest and neck area)
Colic in infants	Catnip and fennel (diluted in water and flavored with a little sugar)
Constipation	Castor oil
Cuts	Mercurochrome
Diaper rash	Cornstarch
Diarrhea	Paregoric in water
Earache	Warm oil
Headaches	Warm oil
Menstrual cramps	Hot tea
Muscle aches	Alcohol with wintergreen
Muscle stiffness	Hot or cold compresses
Nervousness	Spirits of ammonia in water
Sunburn	Noxzema
Teething in infants	Whiskey in water (rubbed on infant's gums)
To enhance health	Cod liver oil
Toothache	Oil of cloves
Upset stomach	Hot tea with peppermint oil

roots, herbs, soups, poultices, and medicinal agents such as camphor, peppermint, and spirits of ammonia. A list of these remedies and their uses can be found in Table 15.3.

15.12.4 Barriers to Health Care

Germany blends a private health-care delivery system with universal coverage and social solidarity. The financing is inexpensive and equitable with portable coverage. People are never uninsured in Germany, so families are not burdened with hefty health-care bills (Underwood 2009). In the US, access to care is limited for those who live in rural areas. Although efforts are being made to reduce these barriers, economic and geographic barriers to health care continue to exist for a large number of German Americans.

15.12.5 Cultural Responses to Health and Illness

When asked to describe a German's response to pain, the word most often used is "stoic." Even when Germans are experiencing pain, they may continue to carry out their family and work roles. Research reveals that older German Americans are less likely to complain, more accurate in their description of pain, and more likely to follow the physician's advice (Wright et al. 1983). Although results of studies that examine ethnicity and pain remain problematic, one significant finding does exist: Regardless of the degree of acculturation, individual expressions of pain may follow those of the more traditional members of the culture. Thus, health-care providers may not be able to identify verbal or nonverbal cues among Germans. Careful interviewing and astute observation must be used to accurately assess the level of pain experienced by Germans.

Although both Germany and the US provide care for the mentally ill, mental illness may continue to be viewed as a flaw and is perhaps not as acceptable to German Americans as it is for some other cultures. If this is accurate, members of this group may be slow to seek help because of the lack of acceptance as well as the stigma attached to needing help. German people's discomfort with expressing personal feelings to strangers may impede the counseling process and influence the counseling methods used. The German need to discuss the past without expressing personal feelings should be recognized within the counseling process.

Even though the people with mental illness have been assimilated into American culture, many may remain stigmatized in the German American culture. Since the passage of the Americans with Disabilities Act, more people are aware of the needs of the physically disabled, including acculturated German Americans. Physical disabilities caused by injury are more acceptable to German Americans than those caused by genetic problems. The latter bring feelings of guilt and a sense of responsibility.

Returning people to the highest level of health possible appeals to the German nature. The European American culture believes in helping people, including older people, to recover their health. Rehabilitation has become an integral part of patient care in both Germany and the US, and rehabilitation facilities abound in both countries. For Germans, the rapid return to their roles in society is paramount, and rehabilitation represents the transition to these roles.

Once others become aware of illness, sick individuals are excused from their responsibilities. Even through German Americans are allowed to assume the sick role, some individuals may have difficulty doing so. The stoicism of some may delay their seeking medical care and allow the problem to become more severe or chronic. This may result in the need for more complex treatments for relief of symptoms. As individuals recover, they are expected to relinquish the sick role and resume their normal responsibilities. It is important to note that it is the physician in Germany who determines whether a person can attend work. The physician determines the length of absence from work, and the employer must provide employees with their salaries.

15.12.6 Blood Transfusions and Organ Donation

German Americans identify blood transfusions, organ donation, and organ transplants as acceptable medical interventions. Many religions followed by German Americans provide guidance on each of these issues. See Table 15.2 for a more complete description of these beliefs and practices.

15.13 Health-Care Providers

15.13.1 Traditional Versus Biomedical Providers

In Germany, folk medicine and midwifery are highly revered. Midwifery in Germany is a health care profession which takes care of women in pregnancy, labor, birth and postpartum. There are circa 18,000 midwives in Germany who have passed their exam at one of 58 schools of midwifery in Germany (Ritter 2011). Currently, in Germany, medical-care regulations deem that a physician must have a midwife (*Hebamme*) present during a birth. However, a physician does not have to be present if the midwife is doing the delivery ("Birth and Midwifery in Germany," 2011). This is the opposite of the practice in the US, where a physician must be present if the birth is complicated. In Germany, alternative medicine such as acupuncture and homeopathy, is used also during childbirth to control pain. For many Germans and German Americans traditional and biomedical providers have a place in the health-care delivery model depending on the circumstances at hand.

15.13.2 Status of Health-Care Providers

Health-care providers hold a relatively high status among Germans. This admiration stems from the German love of education and respect for authority. German Americans appreciate the status symbols of money, power, and institutional affiliations held by these professionals. German families are proud to have a health-care provider in their midst, and it is common for family members to seek counsel from them. Because Germans may find asking for help difficult, they may feel more comfortable confiding in a family member.

Health-care providers' esoteric language, practices, and body of knowledge in the sciences of health care can often create barriers to forming relationships with patients. Because of their training into the culture of the health professions, health-care providers can become short-sighted and fail to meet the personal needs of patients of cultures other than their own such as Germans and German American. To deliver culturally competent health care, providers must understand their own ethnic and professional culture as well as the ethnic cultures of their patients.

References

Acton R (1994) A remarkable immigrant: the story of Hans Reimen Claussen. The Palimpsest 75(2):87–100

Beachy R (2010) The German invention of homosexuality. J Mod Hist 82(4):801–838. https://doi.org/10.1086/656077

Birth and Midwifery in Germany (2011) Midwifery today. http://www.midwiferytoday.com/international/Germany.asp

Boorstin DJ (1987) Hidden history: exploring our secret past. Harper, New York

Bronner SJ, Brown JR (2017) Pennsylvania Germans: an interpretive encyclopedia. JHU Press, Baltimore

Büssing A, Ostermann T, Matthiessen PF (2005) The role of religion and spirituality in medical patients in Germany. J Relig Health 44:321–340. https://doi.org/10.1007/s10943-005-5468-8

Center for Reproductive Rights (2020) The worlds abortion laws. https://reproductiverights.org/worldabortion laws?country=DEU

Centers for Disease Control and Prevention (CDC) (2011) Hemophilia. http://www.cdc.gov/ncbddd/hemophilia/facts.html

CIA World Factbook (2020) Germany. https://www.cia.gov/library/publications/the-world-factbook/

CultureGram (1994) Germany '95. David M. Kennedy Center for International Studies, Provo, UT

DANK (2020). http://www.dank.org/

Educational Aspects in the United States and Germany (n.d.). http://sitemaker.umich.edu/schubert.356/kindergarten

European Education Directory (2006). http://www.euro-education.net/prof/germanco.htm

Facts about Germany (2016) Diverse living arrangements. https://www.tatsachen-ueber-deutschland.de/en/categories/society/diverse-living-arrangements

Friday R (1989) Contrasts in discussion behaviors of German and American managers. Int J Intercult Relat 13(42):429–446

German Noise Law (2010). http://www.guardian.co.uk/world/2010/aug/16/germany-children-noise-law

Gottfried M (2001) Duff's a true model patient. Life and Breath Foundation. www.lifeandbreath.org/

Hall ET, Hall MR (1990) Understanding cultural differences. Intercultural Press, Yarmouth, ME

Health Industry Today (2011). http://health.einnews.com/news/germany-diseases

Janes J (2004) A spirit of reason—Festschrift for Steven Mulle. American Institute for Contemporary German Studies, Washington, D.C.

Kilcoyne RF (2004) Hemophilia, musculoskeletal complications. Medscape, 10. http://emedicine.com/radio/topic909.htm

Kneller RW, McLaughlin JK, Bjelke E, Schuman LM, Blot WJ, Wachouslder S, Gridley G, Cochien HT, Fraumeni JF (1991) A cohort study of stomach cancer in a high-risk American population. Cancer 68:672–678

Levy R (1993) Ethnic and racial differences in response to medicines: preserving individualized therapy in managed pharmaceutical programmes. Pharm Med 7:139–165

Lowenfels AB, Velema JP (1992) Estimating gallstone incidence from prevalence data. Scand J Gastroenterol 27(11):984–986

Mackle B (2001) New gene found for myotonia muscular dystrophy: unusual mutation involved. MDA News. http://www.mdaa.org/news/010803dm_mutation.html

McKinnon M (1993) In the American grain: the popularity of living history farm. J Am Cult 3:168–170

National Institutes of Health (NIH) (n.d.) Kidney and urological disease. http://kidney.niddk.nih.gov/kudiseases/a-z.asp

Pinquart M, Sörensen S, Davey A (2003) National and regional differences in preparation for future care needs: a comparison of the United States and Germany. J Cross Cult Gerontol 18:53–78

Ritter K (2011) Midwifery in Germany. https://ergobaby.com/blog/2011/07/midwifery-in-germany/. Accessed 2020

Schied FM (1993) Learning in a social context. LEPS Press, DeKalb, IL

Schnabel A (2014) Reform of the German parental allowance and parental leave act. https://blogs.dlapiper.com/employmentgermany/2014/12/01/reform-of-the-german-parental-allowance-and-parental-leave-act-2/

Solar Navigator (2006). www.solarnavigator.net

Statistics Canada (2016). http://www12.statcan.ca/census-recensement/index-eng.cfm

The German Connection (2006) Autrata family's home page—behavioral norms in German: siestas and sundays. www.seoprofiler.com/analyze/autrata.com

Trading Economics (2020) Germany—prevalence of HIV, total (% of population ages 15–49). https://tradingeconomics.com/germany/prevalence-of-hiv-total-percent-of-population-ages-15-49-wb-data.html

Trans Health Care 2020. https://www.transhealthcare.org/germany/

U.S. Census Bureau (2018) American community survey. http://www.census.gov/acs/www/

Underwood A (2009) Health care abroad: Germany. http://prescriptions.blogs.nytimes.com/2009/09/29/health-care-abroad-germany/

Weaver W (1979) Food acculturation and the first Pennsylvania-German cookbook. J Am Cult 2(3):420–429

WordIQ Dictionary (2010) Prostitution in Germany. http://www.wordiq.com/definition/Prostitution_in_Germany

Wright R, Saleebey D, Watts R, Lecca P (1983) Attitudes toward disabilities in a multicultural society. Soc Sci Med 36:616–620

Zielenski J, Fujwara TM, Markiewicz D, Paradis AJ, Anacleto AI, Richards B, Schwartz RH, Klinger K, Tsui L, Morgan K (1993) Identification of the M1101K mutation in the cystic fibrosis transmembrane conductance regulator (CFTR) gene and complete detection of cystic fibroses mutations in the Hutterite population. Am J Hum Genet 52:609–615

People of Greek Heritage

16

Irena Papadopoulos

16.1 Introduction

Focusing on people from Greece and Cyprus, this chapter provides an overview on many aspects of the Greek culture. It starts with current geographical and population data and includes immigration patterns, languages, communication patterns, and family relationships. In addition, we focus on health-related practices and behaviors in an effort to sensitize the reader on issues of particular interest to people of Greek heritage that might influence their health, responses to illness, and health related decisions.

16.2 Overview, Inhabited Localities, and Topography

16.2.1 Overview

This chapter presents two groups of people with Greek heritage. The first group refers to those people or their ancestors who emigrated from Greece. The second group originated in Cyprus. Both groups share the same history and have a common language and religion. The Greek and Greek Cypriot diaspora is of considerable size and has spread to all continents and numerous countries. The largest Greek community outside Greece is in North America; the largest Greek Cypriot community outside Greece is in Britain. Therefore, the main focus of this chapter is on the large North American Greek community, with a secondary focus on the British Cypriot community. Although geographic location and social context are important, many of the issues and principles can be applied to the broader diaspora. When the term *American* is used in this chapter, it refers to residents of both Canada and the United States.

Greece, a small country in southern Europe with a climate similar to that of southern California, covers slightly more than 50,000 square miles (131,940 sq. km) with an estimated population in 2019 of just over 10.7 million (Eurostat 2018). The metropolitan area of Athens, the capital of Greece, has a population of 3.15 million (Athens population 2019). The population is 91.6% Greek. 1.8% EU countries, 6.6%, other countries (Hellenic Statistical Authority 2011). Greece has, over the past 10 years, been the host of huge waves of refugees and migrants from middle eastern countries as well as further afar such as Afghanistan, Pakistan, and north-western African countries. In 2019,

This chapter is an update of that which appeared in a previous edition and was written by Irena Papadopoulos.
We wish to acknowledge Christina Koulougliotí's contribution in obtaining data that helped to update this chapter.

I. Papadopoulos (✉)
Research Centre for Transcultural Studies in Health, Middlesex University, London, UK
e-mail: r.papadopoulos@mdx.ac.uk

© Springer Nature Switzerland AG 2021

L. D. Purnell, E. A. Fenkl (eds.), *Textbook for Transcultural Health Care: A Population Approach*,
https://doi.org/10.1007/978-3-030-51399-3_16

more than 50,000 refugees arrived in Greece (The UN Refugee Agency 2019). This has inevitably created large population fluctuations with unavoidable strains on the socioeconomic and cultural systems of the country. Greece is a mountainous country with small patches of fertile land separated by hills, mountains, and a plethora of small and medium-sized islands. The main crops are wheat, grapes, olives, cotton, and tobacco. Geopolitical boundaries have shifted dramatically over time. Greeks struggled under 400 years of Turkish rule, which ended in 1829. At that time, the Peloponnese peninsula, central Greece, and some of the Aegean Islands were freed. Later, Thessaly, Macedonia, Crete, the Ionian Islands, Epirus, Thrace, and the Dodecanese were incorporated into Greece's boundaries. Greece joined the European Union in 1981 (Hellenic Republic, Ministry of Foreign Affairs 2018).

Cyprus, located in the most eastern part of the Mediterranean Sea, is a small mountainous island with an area of 5749 square miles (9251 sq. km). The capital is Nicosia with a population of 200,452 people (Worldometers 2019). In 2019, the total population of Cyprus was estimated to be nearly 1.2 million (United Nations DESA/ Population Division 2019). Since the entry of Cyprus into the European Union in 2004, a significant increase of economic migrants has been recorded. The number of EU nationals living in Cyprus is increasing year to year. In 2011, the top four EU countries of migration to Cyprus were Greece (31,044), United Kingdom (26,659), Romania (24,376), and Bulgaria (19,197). In addition, Cyprus regularly receives refugees and economic migrants from all over the world. Asylee applications in Cyprus increased from about 2000 in 2015 to nearly 8000 in 2018. (Asylum Information Database 2018).

Cyprus has a rich history and culture, the result of many influences over 10,000 years. Mycenean and Achaean Greeks settled in Cyprus around the fourteenth century B.C. After the Trojan War, legendary Greek heroes visited the island, where they settled and founded great cities such as Salamis, Kourion, and Paphos. The Achaean Greeks had a profound and lasting influence on the culture of Cyprus, introducing their language, religion, and customs. After the death of Christ, St. Paul travelled to Cyprus, where he was joined by St. Barnabas and St. Mark. The island was the first country to have a Christian ruler when Sergius Paulus was converted.

Cyprus gained its independence from Britain in 1960; however, the Constitution of the Republic of Cyprus proved unworkable, making a smooth implementation impossible. Following episodes of ethnic conflict between Greek and Turkish Cypriots, Cyprus was divided in 1974 following its invasion by Turkey. Almost half the population was displaced, with Greek Cypriots settling in the south and west of the island and Turkish Cypriots settling in the north and east.

Both Greek and Greek Cypriots have long and varied histories of migration and settlement in numerous countries across the world including North America, Australia, and the United Kingdom. The characteristics of the Greek and Greek Cypriot migrant communities vary considerably according to the time of immigration (with earlier immigrants being predominantly younger, rural males), the characteristics of the site of immigration (rural, island, or urban), the variant cultural characteristics (refer to Chap. 2 in this book), and the number of generations since initial immigration. Despite considerable temporal and geographic variation, several core themes are common to Greek and Greek Cypriot migrants such as emphasis on family, honor, religion, education, and Greek heritage.

The core values of ***philotimo*** (includes notions of honor, respect, pride, self-sacrifice, and hospitality) and ***endropi*** (shame) are key when considering the experience of Greeks and Greek Cypriots. Values of honor and shame are found in all societies; however, these attain immense importance among Mediterranean groups. Although *philotimo* is a characteristic of one's family, community, and nation, it most centrally implies concern for other human beings. *Philotimo* is a Greek's sense of honor and worth, derived from one's self-image, reflected image (respect), and sense of pride. *Philotimo* is enhanced through courage, strength, fulfilling family obligations, competition with other people, hospitality, and right behavior. Shame results from any conduct that is considered deviant. The system of honor and shame in the Mediterranean countries derives from the complementary opposition of the sexes, the solidarity of the family, and the relationships of hostility and competition between unrelated or unconnected families.

Reflective Exercise

Mr. Marios Stavrakis is a 49-year-old Greek who arrived in New York from Crete at the age of 21. After working very hard doing different jobs for a number of years he saved enough money with which he started a business with his best friend Mr. Soteris Ioannou, who is also his son's godfather. As the business grew, the partners spent less time with each other since each one had separate responsibilities within the company.

About a year ago Mr. Stavrakis developed signs of depression. His wife noticed that he was worried about something, was frequently anxious, and at the same time appeared to have less energy and vitality than usual. When he started neglecting the business that he so much loved and had worked so hard to make successful, his wife insisted that he see a physician. Mr. Stavrakis was prescribed antidepressants but took the medication infrequently for a while and then stopped it all together.

His condition deteriorated and he began to obsessively talk about *philotimo*. When he eventually saw a psychiatrist, he explained that he discovered that his best friend and business partner was making deals behind his back and that he was embezzling money from the company. Although he had suspected this for some time, he did not want to report his best friend to the police. At the same time, he could not deal with his anger and disappointment as he felt totally betrayed by a man whom he trusted.

1. How has the belief about the importance of *philotimo* influenced the behavior of Mr. Stavrakis?
2. Why was Mr. Stavrakis reluctant to report his friend to the police? What cultural values influenced his actions?
3. Why did he not seek medical help and why was he eventually persuaded by his wife to see a doctor?

16.2.2 Heritage and Residence

Greeks in America used to be a composite of three immigrant groups: an older group who came before or just after World War I, a second group who arrived after the relaxation of immigration laws in the mid-1960s and who constituted the main group in the Greek and Greek Cypriot American community, and the American-born children and grandchildren of these immigrants. Today, almost all the Greek migrants from the first group have died. The second group is also shrinking, while the third group is growing. The majority of those who today refer to themselves as Greek Americans belong to the third, fourth and fifth generations.

The earlier Greek immigrants congregated for the most part in the western states of Utah, Colorado, and Nevada where they worked in mines and on railroad crews; in the New England states of New Hampshire, Massachusetts, and Connecticut where they worked in shoe and textile factories; and in the large northern cities of Chicago, Detroit, Toledo, Milwaukee, Philadelphia, Buffalo, Cleveland, and New York where they worked in factories or found jobs as shoe shiners or peddlers. The greatest proportion of Greeks and Greek Cypriots in America continues to live in the Northeast and the Midwest. Most live in large urban areas such as New York, Boston, Washington, and Chicago. Whereas new immigrants still tend to gravitate toward the established Greek communities in cities, many Greeks in America have relocated to the suburbs (Moskos 2017). The Greek communities in the US and Canada are the largest Greek diasporic communities. It is estimated that there are 1.4 million people of Greek heritage living in the US (Greek Americans 2019).

Early Greek Cypriot immigrants mainly settled in large urban centers in New York, New Jersey, and Florida. In 2000, a US Census Report stated that there were 7663 Cypriots mainly in the New York and New Jersey regions. The 2011 Canadian Census reported that there were 252,960 Canadians who claimed Greek ancestry (Greek Canadians 2019). In 2019 there were 25,000 Greek Cypriots living in Canada (Government of Canada 2019).

16.2.3 Reasons for Migration and Associated Economic Factors

Significant Greek migration occurred during the late 19th and early 20th centuries. During this period, migration depleted the population of Greece by about one-fifth. Economic factors were largely responsible for this mass exodus. In the latter part of the nineteenth century, Greece suffered a major economic crisis resulting from a nearly complete failure of its major crop, currants; relatively heavy governmental taxation to sustain an army against hostilities with Turkey; and family pressure on fathers and brothers to supply a substantial dowry for unmarried women in the family. Before the 1880s, relatively few Greek immigrants entered the US. It was not until the start of the twentieth century that massive numbers of Greek immigrants arrived in America. It is estimated that between 1900 and 1920, almost 350,000 Greeks came to America, 95% of them men (Scourby 1984). They came with dreams of economic opportunity, hoping to make enough money to provide good dowries for their sisters and daughters and to be able to return to Greece with enough money to live comfortably in their villages. At the time, Greece was beleaguered by turbulent internal politics and was a difficult place for the average Greek peasant to earn a decent living.

Most Greek migrants planned to only stay in the US for a short time. As the arrival of young Greek women—potential wives—post 1920s increased, a number of men put more permanent roots in their host country. With growing communities and the establishment of small family businesses, Greek migrants began to integrate into the American society (Kitroeff 2009).

Legislation passed in 1921 and 1924 transformed America's open-door policy toward European immigrants into a closed-door policy greatly affecting the number of Greek immigrants who came into the country. While in 1921, 28,000 Greek immigrants came to America, the next year, the quota of Greeks allowed into the country was reduced to 100. This was raised to 307 in 1929, and remained at that level for three decades (Moskos 2017). Greek immigrants who had cared little about becoming American citizens saw citizenship as the only chance to bring other family members to America or to be able to return to America after visiting Greece. In addition, because fewer people were emigrating from Greece, membership in the Greek American community consisted of increasing numbers of American-born Greeks.

During most of the 1930s, the number of Greeks returning to Greece exceeded the number coming to America (Moskos 2017). Despite the economic downturn in the US, Greeks in America managed to invest a great deal of energy in their communities. Greek-language schools were started for their children, the Greek Orthodox Archdiocese centralized, and charitable organizations were established for the poor. When the Great Depression came, everyone in America was affected, including Greek immigrants. Many businesses failed, jobs were lost, and fortunes disappeared.

The Italian invasion of Greece in 1940 precipitated Greece's entry into World War II and a great outpouring of support from the Greek American community for the home country. After America entered the war in 1941, the intermingling of Greek and American interests produced a combination of American patriotism with Greek ethnic pride that underscored the great love that Greeks in America felt for both their home and their adopted countries. The immigration laws however, kept the actual number of new Greek immigrants to a minimum until the 1950s (Imai 2013).

Although the quota system was maintained, special legislation in 1953 allowed those who had been displaced by the war and those who wished to reunite with their families to enter America. In addition, countries were allowed to "borrow" on quotas for future years. As a result, approximately 70,000 Greeks entered the US between World War II and 1965. During this time, the immigration laws dating from the 1920s were liberalized. This large influx rejuvenated the Greek American community's ties to Greece and changed the composition of the Greek community from Greeks with American citizenship to Americans of Greek descent. By this time, the third generation of Greek Americans was being born. The Immigration Act of 1965 lifted the earlier restrictive quotas, allowing more Greeks to immigrate to America (History.com Editors 2019).

Whereas the U.S. Census 2000 reported that 1,153,307 people of Greek descent lived in America, in 2006, only 12,723 Greeks emigrated to the US (Alexiou 2013). The decline in Greek immigration to the US is attributed to several factors that are largely economic. Improvement of economic conditions in Greece has lessened the impetus to emigrate. More people migrated to Canada and Australia which had more lenient visa requirements than the US. Finally, with the entry of Greece into the European Union (EU) in 1981, Greeks were able to freely move within the EU, thus reducing the number of people emigrating to the US to an estimated 2000 per year (Michopoulos 2017). Immigration for Greek Cypriots is a very old phenomenon (Panayides 1988). This is exemplified by the figures from a survey published by the Ministry of Education in Cyprus and cited by the Cyprus High Commission in Britain (1986), which numbered the Cypriot population in London as 208 in 1911; 1059 in 1931; 10,208 in 1941; 41,898 in 1961; and 78,476 in 1964. The first major group of Greek Cypriots who emigrated to Britain arrived in the 1930s. Because Cyprus was a British colony, young men seeking employment made their way to Britain and primarily settled in the Camden Town and Soho areas of London but later spread to Islington, Hackney, and northward to Haringey.

The second wave of emigration occurred in 1960–1961 when 25,000 Cypriots left for Britain when Cyprus became a republic. This number was reduced to less than 2000 a year after the UK Commonwealth Immigrants Act of 1962. The last wave of emigration occurred in 1974 following the conflict between the Turkish and the Greek Cypriots when an estimated 50% of Cypriots became refugees in their own country. By 1974, an estimated 120,000 Cypriots were in Britain, of whom five out of six were of Greek origin and the remainder of Turkish origin.

In 1986, the Cyprus High Commission reported that some 200,000 Cypriot-born people and descendants of Cypriots (Greek and Turkish) were living in Britain. In 1996, the Greek Orthodox Archdiocese in Great Britain estimated that London alone was home to more than 250,000 Greek and Greek Cypriots. These figures were derived from church attendance, numbers of weddings, baptisms, and funerals performed as well as by the number of children attending the church-run and independent Greek schools. In addition to the London-based Greek Cypriot population, large communities are found in many other British cities, particularly Birmingham, Bristol, Manchester, Great Yarmouth, and Glasgow. The Greek and Greek Cypriot communities in Great Britain continue to increase, and in 2011 it was estimated that the Greek Cypriot community was in excess of 300,000. However the official number provided by the UK office for national statistics report that in 2016, 62,000 people identified themselves as Greeks, a number which includes both Greek and Greek Cypriot (Office for National Statistics 2017). The main reason for this gross underestimation is that the British census form does not include a separate 'Greek' or 'Greek Cypriot' ethnicity category. This low number reflects people who ticked the 'other' category and qualified this by adding Greek or Greek Cypriot in the explanation box.

16.2.4 Educational Status and Occupations

Most early Greek and Greek Cypriot immigrants were poor men who had limited education. However, they had a very strong work ethic, determination, and ethnic pride. Their achievements are evident in the schooling patterns of their children and by the competitive dimension of the Greek character. Greek children are expected to succeed in school. This attitude is fostered by an achievement orientation, high educational and occupational aspirations, a cohesive family unit that exhorts children to succeed, nationalistic identification with the cultural glories of ancient Greece, and private schools that teach the Greek language and culture (Marjoribanks 1994). Typically, this pattern of achievement continues into adulthood and is reflected in career success. Most third-generation Greeks in America have attended college. During the 1965 immigration, Greeks coming to America included educated professionals and students in professional fields such as engineering, medicine, surgery, and other academic areas (Moskos 2017).

A common theme (repeated so often it has become an archetype) is that of Greek parents who came from an impoverished land with no money or education. Lacking English language skills, most of the immigrants had no recourse except to accept low-paying jobs as peddlers pushing carts and shoe shiners. Greek and Greek Cypriot men disliked working for others and considered it a violation of their pride. They were industrious and frugal and eventually saved enough money to start their own businesses, such as restaurants and cigar and candy stores (Lovell-Troy 1990). In Britain, a number of Greek Cypriots established small clothing factories and some opened shops specializing in foods imported from Cyprus. Initially, they sought these opportunities to save money to return to their homeland, but the more successful they became, the more likely they were to remain in America and Britain.

In America, Greek immigrants who earned only marginal wages were more likely to return to Greece. This description represents the typical pattern in the eastern and northern parts of America. In the west, men worked on railroads and in mines and exhibited greater rates of marriage outside the Greek community because of their smaller numbers in these more-remote communities. Often, once they had settled, worked hard, and acquired some capital, they too became entrepreneurs, opening shops and small businesses and eventually acquiring American citizenship.

In the US, Greek immigrants attained middle-class status more rapidly than most of their fellow immigrants. As America grew more affluent in the 1920s, so did the Greek immigrants. During the 1950s, even more Greeks in America ascended into the middle class. American-born Greeks held mostly white-collar jobs. Professions such as engineering, medicine, pharmacy, scientific research, and teaching are favored by Greek Americans (Kunkelman 1990). Second and subsequent generations of Greeks and Greek Cypriots continue to establish their own or run family businesses, although more of them are currently entering professions such as medicine, accounting, and law.

16.3 Communication

16.3.1 Dominant Languages and Dialects

Although all Greeks, whether in Greece, Cyprus, or the diaspora, use the same form of written Greek, regional and country variations in spoken Greek exist. Diasporic Greek communities regard the retention of the Greek language as an essential part of their Greek identity; numerous efforts are continually being made to encourage second and subsequent generations to speak Greek. Papadopoulos and Papadopoulos (2000) surveyed young British-born Greeks and Greek Cypriots living in Britain to determine how they defined themselves in terms of ethnic identity. Of the 94 people who responded, 87 defined themselves as British Greek/Greek Cypriots or just Greek/Greek Cypriots. Forty-six reported that they spoke Greek fluently, 35 spoke enough to "get by" and 10 spoke "basic" Greek. Only three respondents reported not being able to speak any Greek. The spread of the Greek language is achieved by attending Greek-language schools, using Greek in the home, and regularly visiting Greece or Cyprus. Robins and Askoy (2001) argued that people of second and subsequent generations of any migrant community who are able to speak their mother tongue are more successful as they achieve greater cultural mobility. Knowledge of both Greek and English (or any other language, depending on the country of residence) enables people of Greek heritage to move through the cultural spaces both of their ancestors and of their adopted country. This is a helpful and nourishing process for both the individual and the collective (Portes and Rivas 2011).

16.3.2 Cultural Communication Patterns

Because Greeks and Greek Cypriots value warmth, expressiveness, and spontaneity, northern Europeans are often viewed as "cold" and lacking compassion. Protection of family members and maintenance of family solidarity tend to be foremost among their values. As a consequence, they are often friendly but somewhat superficial and distant with those considered "outsiders".

Greek and Greek Cypriot people tend to be expressive in both speech and gestures. They embrace family and friends when they meet, to indicate solidarity. Eye contact is generally direct, and speaking and sitting distances are usually closer than those of north-central European Americans. They gesture frequently with their hands while talking. Whereas innermost feelings such as anxiety or depression are often shielded from outsiders, anger is expressed freely, sometimes to the discomfort of those from less-expressive groups.

In health-care situations, patients often appear to be compliant in the presence of the health-care worker, but this may be only a superficial compliance, employed to ensure a smooth relationship. Greeks consider deeds to be much more important than what one says.

16.3.3 Temporal Relationships

Greeks and Greek Cypriots demonstrate a variety of temporal orientations. First, they are oriented to the past because they are highly conscious of the glories of ancient Greece. They are present oriented with regard to *philotimo*, family life, and situations involving family members. Finally, they tend to be future oriented with regard to educational and occupational achievements.

Greek Americans differentiate between "Greek time" which is used in family and social situations, and "American time," which is used in business situations. Greek time emphasizes participating in activities until they reach a natural ending point, whereas American time emphasizes punctuality.

16.3.4 Format for Names

It is customary for honorific titles to be given to community members who are older people or otherwise respected. Terms such as *Thia/Thios* (aunt/uncle), *Kyria/Kyrios* (Mrs/Mr), or *Yiayia/Papoo* (grandma/grandad) may be used. For Greeks and Greek Cypriots everywhere, having a Greek name is an important sign of their heritage. First names come either from the Bible, such as Maria and Petros (Peter), or from ancient Greek mythology and history, such as Eleni (Helen) and Alexandros (Alexander). Ideally, first daughters are named after the mother's mother, and first sons after the father's father. Following tradition, middle names are the first name of the father; thus, all children of Stavros might carry his first name as their middle name.

In health-care situations, it is not appropriate to call older women or men by their first names. The prefix "Kyria" (Mrs) or "Kyrie" (Mr.) should be used with the first name; for example, Kyria Maria or Kyrie Alexandre. A more formal mode of address is to use their surname preceded by Kyria Georgiou, Kyrios Georgiou.

16.4 Family Roles and Organization

16.4.1 Head of Household and Gender Roles

The father is considered the head of the household in Greek and Greek Cypriot families. However, the complexity of household dynamics is noted in the well-known folk phrase "the man is the head, but the wife is the neck that decides which way the head will turn". This saying acknowledges the primacy of fathers in the public sphere and the strong influence of women in the private sphere. In recent years, equality between the genders is expected and equality in decision making is normally practised.

Most important however, in consideration of gender roles are the complementary values of *philotimo* and *endropi*. These core values tend to set the pattern for the family and for the enactment of gender roles. Although in the past the educational levels of women had often matched those of their brothers, women usually did not work outside the home, particularly after they married. A woman may however, have worked in her husband's store or restaurant. Women of later generations who obtained professional degrees tended to work once their children were in school. The roles of husband and wife are characterized by mutual respect (a partnership). However, their relationship is less significant than that of the family as a unit. Fathers are responsible for providing for the family, whereas women are responsible for the management of the home and children. Traditionally, the cleanliness and order of the home was said to reflect the moral character of the woman. However, these beliefs are considered inappropriate by modern Greek people, irrespective of where they live.

16.4.2 Prescriptive, Restrictive, and Taboo Behaviors for Children and Adolescents

Children are included in most family social activities and tend not to be left with babysitters. The child is the recipient of intense affection, helpful interventions, and strong admiration. The child may be disciplined through teasing, which is believed to "toughen" children and make them highly conscious of public opinion. The family environment can be described as strongly encouraging inter-dependence and achievement. The family goals of achievement are directed toward and internalized by the children.

Reflective Exercise

Mr. Andreas Georgiou was born in the United States in 1955. His parents had left Greece in 1952 to join his father's brother, who had migrated a few years earlier. They both worked in his uncle's small restaurant until they were able to open their own in partnership with his uncle. Andreas has two younger sisters. His parents spent whatever little time they had helping at the local Greek Orthodox Church and insisted that he and his sisters attend the Saturday Greek school. Andreas remembers his father saying, "We must never forget where we come from." He also remembers how protective his parents were, particularly toward his sisters who, in his view, did not have the freedoms he had. "My parents always said that young women with sexual freedom have bad reputations and decent men do not want to marry them." Both his sisters did well at school and were able to find good jobs and good husbands. He studied art at the university and has his own printing business. Ten years ago, Andreas suffered from depression. "This started when I found out that my second child was severely disabled. I could not cope with it. We consulted numerous specialists searching for a cure. We prayed and prayed. At first, I could not speak about my son to anyone other than my closest family. I never shared my emotional turmoil with my work colleagues, and this was a major stress for me. When I eventually had to share my 'secret,' they were all very understanding."

Today, Andreas was visiting his therapist for the last time. The therapist had helped him work through his self-blame, anxiety, and sadness. He has come to love his son for who he is.

1. What cultural values drove Andreas' parents after their migration to the United States?
2. Why were his parents so protective toward their daughters?
3. What cultural values might have led Andreas to feel so devastated that he tried to hide his son's disabilities?

Greek American and British Greek Cypriot families stay intact because adolescents, particularly young women, tend to reside with their parents until they get married. Formerly, men did not marry until their sisters' *prika* (dowry) was established and they had married. Among first-generation immigrants, single men often returned to Greece or Cyprus for a bride. A *proxenistra* (matchmaker) and the families, pending the approval of the young person involved, usually arranged these marriages. Today, spouse selection is left to the young person, with parental approval.

Girls have less freedom than their brothers in dating, and it is common for them to be prohibited from dating until they are in the upper grades in high school. Adolescents in more-traditional families may experience stress as the differences in family and peer values precipitate family conflict. In the past, suppression of personal freedom by parents was reported to be a major risk factor for suicidal attempts among Greek and Greek Cypriot adolescent girls (Beratis 1990). Additional areas of high stress for Greek adolescents include family pressure to avoid shaming the honor of the family (endropi) with behaviours which are considered inappropriate, pressure for school achievement, and the lack of openness in the home to discuss with their parents matters of sexuality and sexual education.

16.4.3 Family Goals and Priorities

Greek and Greek Cypriot families tend to be very close. Within the family, members are expected to express unlimited respect, concern, and loyalty. *Symbetheri* (in-laws) are considered first-degree relatives. Family solidarity is the context in which the values of honor and shame are measured. Prestige is connected to the idea that honor is not individualistic but collective. Because a person loses honor if their kin act improperly, the honor of each family member is a matter of concern for all family members.

Older people hold positions of respect within the Greek and Greek Cypriot communities. Their stories, whether as pioneers, veterans, or hard-working businessmen, are well known throughout the community. Treatment of the ***yiayia*** (grandmother) and the ***papoo*** (grandfather) reflects the themes of closeness and respect emphasized in the family. Grandparents tend to participate fully in family activities. Families feel responsible for caring for their parents in old age, and children are expected to take in widowed parents if they are unable to live independently in their own homes. Failure to do so results in a sense of dishonor for the son and guilt for the daughter. If the older person is ill, living with the family is the first preference, followed by residential-care facilities. In recent times many older people are choosing to live alone in their own home, supported by family, friends, and health-care providers. Older Greek and Greek Cypriot widows and widowers, particularly those who speak little or no English, may experience social isolation if they do not have close contact with their children.

An important role is that of fictive kin, termed ***koumbari*** (coparents), who serve as sponsors in either (or both) of two religious ceremonies: baptism and marriage. Ideally, the baptismal sponsor also serves as the sponsor of the child's marriage. The relationship of sponsor is so important that families who are joined by this bond of fictive kinship are prohibited from intermarrying, although this is not always adhered to nowadays.

The basis of social status and prestige is family *philotimo* and cohesiveness. However, social status is also received from attributes such as wealth, educational achievement, and other achievements of its members. A family's status and integrity are validated when they support one another during times of misfortune such as poverty, illness or dishonor.

16.4.4 Alternative Lifestyles

Greek and Greek Cypriot communities tend to be relatively conservative. As a consequence, alternative lifestyles encompassing premarital sex, same-sex relationships, and to a lesser extent divorce, are considered sources of concern for family members and the community. A number of religious and community support organizations in the US and elsewhere provide support.

16.5 Workforce Issues

16.5.1 Culture in the Workplace

In the US, the high achievement orientation and work ethic have resulted in Greeks serving as a "model" ethnic group. Although incidents of discrimination and segregation, including acts of physical violence and murder directed at Greek immigrants, were common early in the twentieth century, less discrimination occurs in the workplace today. The Greeks' and Greek Cypriots' rapid, selective, acculturation has been addressed in earlier sections on migration, occupation, and education.

16.5.2 Issues Related to Autonomy

Probably no single characteristic applies so completely to members of the Greek and Greek Cypriot communities as the emphasis on self-reliance within a family context. Greeks and Greek Cypriots in North America, Britain, and Australia stress this trait. It is seen as an aversion to be told what to do, and this is given as a major reason for their pattern of establishing their own businesses as soon as possible. The desire for autonomy may also have strong links to the past history of the Greek people who had been oppressed for hundreds of years by a number of rulers. Their endeavours to gain freedom is deeply engraved on the Greek psyche which reacts strongly to anything that reminds them of their suffering and struggles against oppression in their past history.

16.6 Biocultural Ecology

16.6.1 Skin Color and Other Biological Variations

Greeks and Greek Cypriots are most commonly of medium stature, shorter than northern Europeans, but taller than other populations of southern Europe. Although some Greeks are blue eyed and blond, usually those from the northern provinces of Greece, most Greeks have dark hair and olive skin.

16.6.2 Diseases and Health Conditions

Current causes of death among Greeks and Greek Cypriots are those of developed countries and include cancer, cardiovascular and cerebrovascular diseases (Institute for Health Metrics and Evaluation 2017a, 2017b). Since the early 1970s, an increase in diabetes and heart disease has occurred in both Greece and Cyprus. Tokas (1995) reported that in Cyprus, 2000 people die each year from heart disease, whereas each year 600 Greek Cypriots are sent abroad for cardiac surgery. In a study of the health needs of Greek Cypriots living in London, Papadopoulos (1998) found that 96 percent of the respondents ate red meat; of these, 35 percent did so on most days of the week. Nearly 50 percent of the women and 36 percent of the men in the study were overweight. Allender et al. (2008) reported that cardiovascular diseases continue to be the main reason for deaths in Greece and Cyprus and that this is due to bad dietary habits, smoking, and lack of exercise. In Cyprus 37 percent of the deaths in men and 40 percent of those in women are due to cardiovascular conditions. In Greece the corresponding figures are 45 percent and 52.2 percent (Allender et al. 2008).

Two important genetic conditions, thalassaemia and glucose-6-phosphate dehydrogenase (G-6-PD) are seen in relatively high proportions among Greek populations. In the red blood cell, G-6-PD is a key enzyme in the hexose monophosphate shunt, which prevents oxidation of hemoglobin to methemoglobin. This pathway is essential to maintaining the integrity of the red blood cell and to preventing hemolysis. G-6-PD is important in the metabolism of glutation, an antioxidant agent. G-6-PD deficiency leads to haemolysis that is generally well tolerated except under specific circumstances, including exercise, infections, and the presence of oxidant drugs. The genetic locus for the deficiency is on the X chromosome, making it more common among males than females. The possibility of G-6-PD deficiency should be considered in Greek patients with unconjugated jaundice (Todd et al. 1994).

Thalassaemia is an inherited genetic disorder manifested by a slow production of or failure to synthesize hemoglobin A or B chains. Two main types are commonly known: thalassemia major (sometimes known as *Cooley's anemia*, *homozygous*, or *beta thalassemia major*) and thalassemia minor (referred to as *thalassemia trait*, or *beta thalassemia minor*). Thalassemia major is a serious condition that, if untreated, will result in death owing to very low levels of hemoglobin and the fragility of the abnormal red blood cells. Undiagnosed infants become pale and irritable, do not eat, suffer from recurrent fever, and failure to thrive. Eventually, the liver, spleen, and heart are damaged as a result of the accumulation of iron contained in the red blood cells. However, if the child is correctly diagnosed, treatment with regular blood transfusions (usually monthly) and prevention of iron overload with deferoxamine (Desferal) will provide an average life expectancy. Various other treatments such as bone marrow transplants are now available.

Conversely, most individuals with thalassemia minor are not aware of it unless they are tested for it. The UK Thalassemia Society reports that 1 in 7 Cypriots (including Turkish Cypriots) and 1 in 12 Greeks are thalassemia carriers (Todd et al. 1994; De Sanctis et al. 2017). In recent years, prenatal screening programs in Greece and Cyprus, as well as other countries such as Britain, the US, Canada, and Australia where most of the diaspora resides, have drastically reduced the number of babies being born with thalassemia major. Most Greek and Greek Cypriot women choose to have an abortion if they are found to carry an affected fetus.

16.6.3 Variations in Drug Metabolism

G-6-PD deficiency can result in a life-threatening hemolytic crisis after oxidating drugs (including primaquine, quinidine, thiazolsulfone, dapsone, furzolidone, nitrofural, naphthalene, toluidine blue, phenylhydrazine, and chloramphenicol) are taken. Even common medications such as aspirin can induce a hemolytic crisis. This threat is sufficiently severe that the World Health Organization (1994) (WHO) recommends that all hospital populations in areas with high proportions of Greeks and Greek Cypriots be screened for G-6-PD deficiency before drug therapy is instituted (Todd et al. 1994; Kaplan et al. 2015).

16.7 High-Risk Behaviors

Greeks in Greece, the United States, Canada, and Australia demonstrate lower rates of nontherapeutic drug use, alcohol misuse, and high-risk sexual behaviors than other groups in European or North American countries (Rosenthal et al. 1994). These patterns are not due to an emphasis on health promotion, but rather to a hyperawareness of the social consequences of these behaviors for the family. For example, alcohol is most often considered a food item and is consumed with meals. However, losing control by being "under the influence" engenders considerable gossip and social disgrace, focused not only on the individual but also on the family (Tripp-Reimer and Sorofman 1994; Foster et al. 2007).

Concern for the reputation and standing of the family is a prime deterrent to many high-risk behaviors. Conversely, high-risk behaviors such as obesity among both sexes and smoking among men are higher among Greeks (Wilson et al. 1993; Mammas et al. 2003). Despite the concern for the family reputation, many of the Greek Cypriots who took part in Papadopoulos' (1999) study described themselves as "risk takers" or "living dangerously." This may be due to a mixture of cultural and religious beliefs and the experiences of migration. The narratives of first-generation migrants describe the risks they had to take to survive in a foreign and often hostile land. Whereas a high level of risk taking appeared to be part of survival, it was perpetuated by the belief that "God will look after me"—that God will prevent anything untoward happening, but if anything should happen, God will heal and sustain the person. Knowing these behavioral characteristics can assist health-care providers in planning culturally sensitive interventions.

Reflective Exercise

Mr. and Mrs. Christou have a 17-year-old daughter, Helen, and a 15-year-old son, George. Mr. Christou is a second-generation British-born Greek Cypriot, while Mrs. Christou moved to England from Cyprus at the age of 18. Both children are attending the same school that is located in a predominantly middle-class area of London. The couple and the children's grandparents, who also live in London, are very proud of the children.

Helen and George are given a generous allowance each week, but lately George has been borrowing money from his sister. At first, she did not mind, but when this became more frequent she began to wonder what George did with the extra money. She also saw him asking his grandmother for money. George frequently told his parents that he was going to the snooker club with his friends after school and on Saturdays.

One day, while waiting to catch the bus home after school, Helen saw George with a group of boys smoking behind the bus shelter. That evening she confronted her brother who admitted that he occasionally used cannabis. She told him that unless he gives this up within 2 weeks, she will tell their parents. He promised to do so, but after 2 weeks he was still using cannabis. Helen informed their parents. Their first reaction was one of surprise; they could not believe that their "perfect and clever" son whom they brought up to know right from wrong, with whom they trusted and had a loving relationship, could get into "bad company" and take drugs. Following their disbelief, they had a very angry argument with their son who told them that he realized how hurt they would be if they found out. The parents told him that although they were concerned about his health, they were also concerned about the effect that his behavior might have on family and friends if they found out about his drug taking. When the parents calmed down, they told George that they still loved him and that they would do whatever they could to help him stop using cannabis. They also told him and Helen that this incident must stay within their home and they should not discuss this with anyone else.

1. What cultural value is guiding and influencing the parents' decision to keep their son's problem within the confines of the nuclear family and not seek help from their extended family and friends?
2. What cultural values underpin the parent's decision to support their son even though they are angry with him?
3. Should the grandmother be informed of the situation and told not to give George any money?

16.7.1 Health-Care Practices

Greeks and Greek Cypriots have tended to disregard standard health-promotion behaviors. Safety measures for adults, such as seat belts and helmets, are often viewed as infringements on personal freedom and are frequently ignored, particularly by the older generation. The first most common cause of deaths in Greece and the second in Cyprus for those aged 15–44 are road traffic accidents. However, relevant legislation and severe penalties are having a positive effect, and a reduction in these deaths was recently recorded (Kouta et al. 2010).

However, the gap between health-related knowledge and appropriate health action remains. Papadopoulos et al. (1998) found that Greek Cypriot people had a good knowledge of health-promotion practices. However, good knowledge did not correlate with positive health behaviors. For example, people who know what a balanced diet is, often do not eat a balanced diet. Worryingly, the new generations of Greek and Greek Cypriots, while not abandoning the traditional Mediterranean diet that is characterized by low levels of meat and fat intakes and high levels of fresh fruit and vegetables, are regularly consuming fast foods, a phenomenon that is now global. Encouragingly, Papadopoulos (1999) found that the use of screening for problems such as blood pressure, cervical cytology, and breast cancer was high among the Greek and Greek Cypriot community in the UK.

16.8 Nutrition

16.8.1 Meaning of Food

Greeks describe their culture as an "eating culture." By this they mean that food is a center piece of everyday life as well as of social and ritual events. Greek hospitality nearly always includes a ritual of food and drink.

Fasting is an integral part of the Greek Orthodox religion. General fast days are Wednesdays and Fridays; nowadays, these are observed only by some older or some strict Greek Orthodox people. During fasts, it is forbidden to eat meat, fish, and animal products such as eggs, cheese, and milk. Greek Orthodox wishing to take Holy Communion will observe at least 3 days of such fasting. However, people with health conditions and small children are exempt from fasting. Some first-generation Greeks and Greek Cypriots and some strict Greek Orthodox observe the three major fasting periods, including:

The Great Fast, Lent, for 7 weeks before Easter
The Assumption fast, from August 1 to August 14
The Christmas fast, 40 days before Christmas

Many second and subsequent generations of Greeks and Greek Cypriots usually fast only the last week before Easter, when many will also take Holy Communion.

16.8.2 Common Foods and Food Rituals

Greeks and Greek Cypriots base their diet on cereals, pulses (such as lentils, peas, and beans), vegetables, fruits, olive oil, cheese, milk, and some fish and meat. They are also relatively high consumers of sweets and snacks. In both Greece and Cyprus, consumption of pulses has decreased, and consumption of meats has increased.

For adults, dairy products are consumed in the form of yogurt or cheeses such as feta, kopanisti, kefaloteri, kasseri, and halloumi. Fats are consumed in the form of olive oil and butter. Meats include chicken, lamb, pork, and beef. Eggs, lentils, fish such as shrimp and other shellfish, and white fish are additional sources of protein.

Vegetables such as potatoes, eggplant, courgettes (zucchini), spinach, garlic, onions, peas, artichokes, cucumbers, tomatoes, asparagus, cabbage, and cauliflower are common Greek food choices. Bread choices include pita, crescent rolls, and round white bread covered in sesame seeds. Other foods include rice, tabouli, macaroni, and cracked wheat (bourgouri). Papadopoulos et al. (1998) found that the level of vegetable consumption of Greek Cypriots was almost twice that of their meat and poultry consumption. Tyrovolas et al. (2009) reported that adherence to Mediterranean diet among older Greek adults was associated with lower prevalence of obesity.

Common seasonings used by Greeks are aniseed, basil, lemon, mint, cumin, cinnamon, citron, cloves, coriander, dill, fennel, garlic, marjoram, mustard, nutmeg, oregano, parsley, rosemary, sage, sesame, thyme, vinegar, bay leaf, and honey. Fruit preferences include grapes and currants, figs, prunes, oranges, melons, watermelons, peaches, apples and apricots. Beverages such as coffee, tea, chocolate milk, and wine are common choices. Common food items are listed in Table 16.1.

Reflective Exercise

Mrs. Glitnatsis, age 82 years, has diabetes mellitus and takes insulin. She is slightly underweight and does not follow her recommended diet because as a Greek Orthodox she follows the fasting schedule of the Greek Orthodox Church. During fasting, she does not take her insulin. She is currently seeing the diabetic nurse specialist because of her uncontrolled diabetes.

1. What are the required fast days for the Orthodox Greek religion?
2. What foods are forbidden to eat during fast days?
3. If Mrs. Glitnatsis insists on fasting, what is your recommendation for food intake and insulin administration?
4. What is the Greek Orthodox position on fasting for people who are ill?

Table 16.1 Common Greek foods

Name	Description	Ingredients
Avgholemono	Soup	Chicken stock, eggs, lemon, rice
Hummus	Thick sauce for dipping bread	Chick peas (mashed), *tahini* sauce (sesame and olive oil), garlic, lemon
Maroulousalata	Salad	Lettuce, onions, cucumbers, radishes, parsley, tomatoes, feta cheese, olives, olive oil
Tsatziki	Dip	Cucumbers, yogurt, vinegar, mint, garlic, salt
Spanakopita	Cheese tarts	Spinach and feta cheese in phyllo dough pastry
Dolmathes	Stuffed grape leaves	Meat, rice, grape leaves
Keftedes	Meatballs	Ground beef/pork, grated potatoes, onions, bread, parsley, oregano, eggs, garlic
Souvlaki	Meat	Marinated pork or lamb on skewer
Moussaka	Casserole	Eggplant, potatoes, ground lamb, onions, tomato, garlic, parsley, white sauce
Pastichio	Casserole	Ground beef, macaroni, cinnamon, white sauce, cheese, parsley, tomato sauce, butter
Psiti Kota	Lemon chicken	Chicken, lemon, oil
Loukomades	Pastry	Flour, water, honey, oil, sugar, cinnamon
Kourambiedes	Wedding cookie	Flour, almonds, cloves, powdered sugar, egg, brandy
Baklava	Sweet dessert	Phyllo dough, pistachio nuts, honey, sugar, cinnamon, cloves, butter
Greek coffee	Coffee	Ground coffee, sugar

Specific foods are linked with holidays or ceremonies throughout the year. For example, several different special breads, pastries, and cakes are served at traditional ceremonies: New Year's bread, *vasilopita*; Easter pastries, *tsoureki* and *flaouna*; Christmas bread, *christosomo*; *prosfora*, a traditional bread for funerals and remembrance ceremonies, which is served with *koliva*, a mixture of boiled wheat, almonds, pomegranate seeds, sesame seeds, and raisins; and traditional, small, individual wedding cakes called *kourapiedes*.

16.8.3 Dietary Practices for Health Promotion

Although no specific classification exists of foods for health or illness, a general consensus is that people will naturally choose foods that are healthy. Therefore, an effort is made to provide ill people with the food they request. This pattern is most pronounced for pregnant women. In fact, numerous folk prescriptions exist regarding the food for pregnant women that in the past was provided to them even if they were not close family or friends.

16.8.4 Nutritional Deficiencies and Food Limitations

Although nutritional deficiencies per se are rare among Greeks and Greek Cypriots, two important enzymatic conditions merit attention. First, for people with G-6-PD deficiency, broad beans (fava beans) can induce hemolysis and an acute anemic crisis (Riepl et al. 1993). Second, Greece has one of the highest frequencies of lactase deficiency among adults in Europe (European Food Safety Authority 2010) and the prevalence of lactose maldigestion in Greek adults is about 75 percent; however, milk intolerance is rarely seen in children (Ladas and Katsiyiannaki-Latoufi 1991). Health-care providers should use this knowledge when counselling patients with these conditions.

Virtually all food items used in the traditional Greek diet are available in the US, Britain, Australia, and Canada. Even specialty items such as phyllo dough for pastries and appetizers, grape leaves in brine, and olives and halloumi can be found in specialty areas of major supermarkets. A trend has moved away from lamb to beef for many dishes among the U.S. population. The popularity of Greek food is evidenced by the success of Greek restaurants even in places where there is not a noticeable Greek community.

16.9 Pregnancy and Childbearing Practices

16.9.1 Fertility Practices and Views Toward Pregnancy

The trend for smaller families in Greece and Cyprus had been noted towards the end of the twentieth century. The birth rate in Greece has decreased dramatically over the years. From 18.4/1000 people in 1968 to 10.1 in 2007 and to 8.2 in 2017 (World data atlas 2017). This change has not diminished the high value placed on couples who have children. In large part, this decreased fertility has resulted from the desire of parents to provide adequately for their children and to have them educated so they can achieve professional status.

In North America, family size has been deliberately limited in order to adequately care for and educate children. In Britain, early immigrants had small families primarily because of housing problems. Many families lived in very cramped conditions, and many landlords did not rent to those with children. Whether one lives in Greece, Cyprus, North America, or Britain a wide variety of birth control measures such as intrauterine devices, birth control pills, and condoms, are used. The strong pro-life Greek Orthodox Church condemns birth control, while at the same time is silently accepting the reality. However, abortion is absolutely condemned as an act of murder except in certain circumstances such as when the life of the mother is in grave danger or a young woman becomes pregnant as a result of rape. In practice, a number of women, particularly those who are unmarried, have legal abortions because of the negative consequences of having a baby out of wedlock. Inflicting *endropi* on the family is believed to be more severe than the consequences of abortion. Although adoption is rare among Greeks and Greek Cypriots, it is becoming a more acceptable option for couples who cannot conceive.

Whereas the family may experience shame and dishonor if an unmarried woman becomes pregnant, also of concern is infertility in a married couple. For these couples, the reputation of the husband may be at risk and the woman is unable to achieve her highest role. Infertile couples are reported to experience mental stress, evidenced as depression for women and anxiety for men (Gourounti et al. 2012). In vitro fertilization is often not disclosed to members of the extended family to avoid gossip and stigmatization. In a group with such a strong emphasis on fertility, the attitude toward the pregnant woman is very positive and protective.

16.9.2 Prescriptive, Restrictive, and Taboo Practices in the Childbearing Family

Pregnancy is a time of great respect for women and a time when they are given special considerations. Proscriptions include not attending funerals or viewing a corpse, refraining from sinful activity as a precaution against infant deformity, and praying to specific saints who are associated with fertility, 'good pregnancy' and 'good freedom' (birth). However, most pregnancy admonitions are related to diet. Pregnant women are encouraged to eat large quantities; foods high in iron and protein are particularly important. A number of tales surround the provision of food to pregnant women. It is commonly believed that giving meat to a pregnant woman makes the dish turn out well. If a pregnant woman remarks that a food smells good, or if she has a craving for a particular food, it should be offered to her; otherwise, the child may be "marked." This is the usual explanation for birthmarks.

During childbirth in rural Greece, a midwife and female kin generally attend the woman. The pattern of the non-involvement of the fathers may still hold in immigrant generations but is being replaced with greater participation.

After delivery, the mother, termed *lehoosa/lehona*, is considered by most traditional Greeks to be ritually impure and particularly susceptible to illness for 40 days. During this time, she is admonished to stay at home and not attend church. In the past, her mother, her mother-in-law and her sisters (if she had any) would take care of most of the housework, including cooking and even caring for the baby, to allow the *lehoosa/lehona* time to rest and gain her strength. At the end of the 40 days, the mother and child attend church and receive a ritual blessing.

For breastfeeding mothers, early showering is sometimes believed to result in the infant's developing diarrhea and becoming allergic to milk. Newborns are generally breastfed, and solids are not introduced early. When relatives visit a child in the hospital, silver objects or coins may be placed in the crib for good luck. Nowadays these rules have been relaxed considerably. Greek and Greek Cypriot pregnant women and *lehoosas/lehones* are better educated and therefore more informed about all aspects of pregnancy, labour, and early motherhood. While respecting their own mothers' advice and welcoming the help that is provided, they are more independent. Childbearing within the Greek and Greek Cypriot communities has become an amalgam of tradition and science.

16.10 Death Rituals

16.10.1 Death Rituals and Expectations

Last rites (sacrament of the sick) are administered in the sacrament of Holy Communion given by a priest or, occasionally, a deacon. On the death of a community member, close relatives are notified in person, whereas other community members are notified by telephone. The wake, *klama*, is held in the family home or, more commonly in North

America today, a funeral home. All relatives and friends are expected to attend for at least a brief time. Even people with whom the deceased had considerable strife are expected to attend. The wake ends when the priest arrives and offers prayers. In Greece and Cyprus, the funeral, ***kidia***, is held the following day at the Orthodox church, with internment in a cemetery. For practical reasons, it is not always possible for the burial to occur the day after death. Furthermore, many Greek Cypriots living in Britain wish to be buried in Cyprus. After internment, family and friends gather for a meal of fish (symbolizing Christianity), wine, cheese, and olives in the family home, a church hall, or a restaurant.

On the basis of the Orthodox belief in the physical resurrection of the body, Greeks and Greek Cypriots reject cremation. The degree of adherence to this precept may vary in North America, but cremation is not practiced by Greeks and Greek Cypriots in Britain.

16.10.2 Responses to Death and Grief

After a death, pictures and mirrors may be turned over. During the wake, women may sing dirges or chant. In some regions, people practice "screaming the dead," in which they cry a lament, the *miroloyi*. This ritual may involve screaming, lamenting, and sobbing by female kin. After death, family and close relatives, who may stay at home, mourn for 40 days. Close male relatives do not shave as a mark of respect.

Black is the color of mourning dress and is often worn by family members throughout the 40-day mourning period; for widows, it may be worn longer. Formerly, a widow was expected to wear black and no jewelry or makeup for the rest of her life. Whereas black armbands are still worn in Greece and Cyprus by men, that custom is virtually non-existent in immigrant communities. Widows generally look to their family for support; widowers often remarry (Panagiotopoulos et al. 2013). These practices should be encouraged and respected by health-care providers.

A memorial service follows 40 days after burial and at 3 months, 6 months, and yearly thereafter. At the end of this service, *koliva*, boiled wheat with powdered sugar, almonds, sesame, raisins, and pomegranate seeds is served to participants. Mourning is conducted with joyful reverence. Dreams have special importance to family members after a death. Family members may even experience visitations by the deceased in their dreams. Often, these dreams provide reasons for the deceased to leave the earth, because the first 40 days following death is a transitional period when the deceased is between earth and heaven.

16.11 Spirituality

16.11.1 Dominant Religion and Use of Prayer

In Greece, 81–90% of the population belongs to the Greek Orthodox religion; 2% is Muslim, and 3% are Jewish or Catholic, 4–15% none (CIA World Factbook 2015). In the Republic of Cyprus, 89.1% of the population identifies with the Greek Orthodox religion, 2.9% are Catholic, 1.8% are Muslim, and approximately 4% percent are other (Maronite, Armenian Christian, Protestant) (CIA World Factbook 2011). Greeks and Greek Cypriots in America are affiliated with the American Achdiocese for the Greek Orthodox Church; Greeks and Greek Cypriots in Britain are affiliated with the Archdiocese of Thyateria and Great Britain. The central religious experience is the Sunday morning liturgy, which is a high church service with icons, incense, and singing or chanting by two small choirs (*the psaltes*). Services that previously lasted over 3 hours are now generally shortened to 1–2 hours.

The Greek Orthodox religion emphasizes faith rather than specific tenets. Unlike most Protestant denominations, Greek faith does not emphasize Bible reading and study. Whereas some parishioners attend church services weekly, others attend only a few times a year. Easter is considered the most important of holy day, and nearly all Greeks and Greek Cypriots in America and Britain attempt to honor the day.

Daily prayers may be offered to the saints. Women often consider faith an important factor in regaining health. Family members may make "bargains" with saints such as promises to fast, be faithful, or make church donations if the saint acts on behalf of the ill family member. They may call on an individual's namesake saint or a saint believed to have special affinity with healing or to *Panagia* (Virgin Mary). Two saints frequently invoked are Cosmas and Damian, early Christian physicians. In invoking the saint, the supplicant may say special prayers, light candles, place small gold medals at the base of the icon, or carry out some combination of these rituals. There is also a strong belief in miracles, even among second and subsequent generations of Greeks and Greek Cypriots in America and Britain.

16.11.2 Meaning of Life and Individual Sources of Strength

The world is viewed as a cosmic battleground in which the individual must be continuously vigilant and resourceful. Based on the principle of limited good, one family's gain is another's loss. Only within the bonds of the family can an individual find protection from a dangerous world. The primary set of relationships, even in an extended household, is between members of the nuclear family. From this central group, relationships radiate outward, lessening in strength to include all affinity and consanguineous relationships to the second degree (e.g., second cousins) and *koumbarous* (male) and *koumbares* (female) individuals who witness the wedding of the couple and/or are godparents to the couple's children).

Life has meaning as family roles are enacted successfully. Parents strive to provide for their children, and in turn they expect the children to achieve success in academic and occupational endeavors. Through wrong behaviors or misfortunes individuals can bring public sanction and shame (*endropi*) on the family.

The Greek concept of self consists of the interrelationship of the three values: self-respect, a sense of freedom, and the concept of the ideal person. The self emerges through relationships with other people but primarily from the family. Freedom is a central element in the self-concept. Self-reliance, that is nonreliance on people outside the family, is a virtue. For Greeks and Greek Cypriots, sources of strength are the family, the close network of extended family and friends, and the history of the glories of ancient Greece. Particularly in immigrant communities, the Greek Orthodox Church serves as a base for spirituality, language, social and political organization, and an ongoing identity with Greece and Cyprus.

16.11.3 Spiritual Beliefs and Health-Care Practices

A distinctive feature of the Greek Orthodox religion is the place it assigns to icons such as paintings of saints, the Virgin Mary, and Christ. These icons are not religious art but have sacred significance as sources of connection to the spiritual world. Icons grace the walls and ceilings of churches and cathedrals and are also found in personal altars in homes. In the homes, holy vigil candles are often kept burning.

To ensure safety and health, many begin each day by kissing the blessed icons and making the sign of the Greek cross. When a person is ill, the icon of the family saint or the Virgin Mary may be placed above the bed. Many Greeks and Greek Cypriots, regardless of where they live, also may sprinkle their homes with holy water from Epiphany Day church services to protect the members of their household from evil. They may also perform the ceremony of *kapnisma*, in which dried olive leaves, which had been kept in church during the fasting periods before Christmas and Easter, are placed in a small container called *kapnistiri* and are lighted while they say a prayer. The person who performs this ceremony (usually the older woman of the household) first goes outside to offer the *kap-*

nisma to God, then each member of the family will cross themselves and accept the *kapnisma*. The person performing the ceremony then goes around to all the rooms in the house to allow the smoke from the burning leaves to penetrate the atmosphere. It is believed that any bad spirits present will leave the house, and thus, the ceremony will protect members of the household.

16.12 Health-Care Practices

16.12.1 Health-Seeking Beliefs and Behaviors

The degree of acceptance and use of biomedicine is highly related to one's level of education and generation of immigration. For example, although fourth-generation Greek Americans in Ohio are highly traditional in many aspects of their lives such as religion, language retention, and food preferences, they have not retained many of the folk beliefs and practices concerning health care. The traditional practices discussed subsequently tend to apply to earlier immigrant generations.

Greek immigrants tend to be anxious about health, to lack trust in health professionals, and to rely on family and community for advice and remedies. Problems are seen as originating outside the individual's control and are attributed to God, the devil, spirits, and envy or malice of others. The gods may punish the nonreligious with illness (*asthenia* or *arostia*), but more often, the forces of evil are believed to cause illness. Another example of an external cause of illness is that of the ***matiasma*** (evil eye), which is often unintentionally caused through the envy of others.

To be healthy means to feel strong, joyful, and content, to be able to take care of oneself, and to be free from pain (Tripp-Reimer and Sorofman 1989). Threats to health result from a lack of balance in life, departure from family, neglect of education or work, and failure to demonstrate right behaviors such as respect toward parents, sharing with family, upholding religious precepts, and staying out too late (Tripp-Reimer and Sorofman 1989).

16.12.2 Responsibility for Health Care

Central to issues of responsibility is the orientation toward honor, independence, and distrust of people outside the family. In traditional thought, the cause of illness and misfortune falls outside the person. The family generally assumes responsibility and care for a sick member and works to control interactions with health professionals.

In the US, insurance enrollment among Greek Americans is high, which correlates with high rates of employment among this group. Even when coverage is available, they often delay seeking professional care. Greek Americans are extremely reluctant to use welfare services or other forms of governmental assistance to meet their health-care needs. Reliance on public assistance would indicate that the sick person and family are not self-reliant. Greeks in America are also reluctant to rely on Greek community organizations such as the women's Philoptochos Society (Friends of the Poor), for support.

In Britain, the National Health Service (NHS) provides free health care to all citizens. Greek Cypriots readily take their children to the family doctor and do not delay seeing the physician for their own health problems when these are severe enough to prevent them from going to work or from executing important family functions and roles. However, men in particular tend to delay seeking medical help when they can self-care for something considered non-acute or when they suspect they may be suffering from something more serious such as cancer (Papadopoulos and Lees 2004). Many first-generation Greek Cypriots speak very little or no English and if a member of their family or a friend is unable to accompany them to the doctor or hospital, an interpreter must be found. First-generation Greek and Greek Cypriots in Britain use voluntary organizations to obtain information and advice in Greek and to receive help completing various documents that enable them to receive the financial and other benefits to which they are entitled. No stigma is attached to the utilization of Greek/Greek Cypriot community organizations in

Britain because their function is not associated with the *philoptochos* movement, whose function is to help the poor.

16.12.3 Folk and Traditional Practices

Three traditional folk-healing practices are particularly notable: those related to ***matiasma*** (bad eye or evil eye), ***practika*** (herbal remedies), and ***vendouses*** (cupping). *Matiasma* results from the envy or admiration of others. Whereas the eye is able to harm a wide variety of things including inanimate objects, children, and pretty young women are particularly susceptible to attack. Common symptoms of *matiasma* include headache, chills, irritability, restlessness, and lethargy. Anecdotally, some people used to believe that in extreme cases, *matiasma* could result in death.

Greeks employ a variety of preventive mechanisms to thwart the effects of envy or evil eye, including protective charms in the form of ***phylacto***, amulets consisting of blessed wood or incense and blue "eye" beads which "reflect" the evil eye back to the person who is supposedly looking at a person with evil intentions. When the diagnosis of *matiasma* is suspected, the most common method of detection consists of placing olive oil in a glass of water. If the oil disperses, then the evil eye has been cast. Subsequent treatment consists of physical acts such as making the sign of the Greek cross over the glass of water or reciting ritual prayers. In particularly severe cases, the Orthodox priest may recite special prayers of exorcism and use incense to fumigate the afflicted person.

Practika are herbal and humoral treatments used for initial self-treatment. Chamomile, the most popular herb, is generally used in teas for gastric distress or abdominal pain, including infant colic and menstrual cramps. It is also used as an expectorant to treat colds. Liquors, such as anisette, ouzo, raki, and mastiha are used primarily for colds, sore throats, and coughs and are consumed alone or in combination with tea, lemon, honey, or sugar. Occasionally, liquors are used for treatment of *nevra* (nerves). Raki is also used as a massage fluid to rub on painful joints. Raw garlic is used as a preventive for colds, and cooked garlic is used for blood pressure and heart disease. Other *practika* include those described in Table 16.2.

Table 16.2 Practika used in Greek folk medicine

Practice	Indication
Frascomilo (sage) tea	Colds
Moroha	Stomachache
Linden tea	Stomachache
Mustard plasters	Colds
Rice (boiled)	Diarrhea
Cinnamon tea	Menstrual cramps
Tobacco	Inhibition of infection

Source: Adapted from Tripp-Reimer, T. (1981). Ethnomedical beliefs among Greek immigrants: implications for nursing interventions. *Transcultural Nursing Care*, *6*, 129–140; and Tripp-Reimer, T., & Sorofman, B. (1989). *Illness related self-care responses in four ethnic groups*. Final report no. R01Nu 1101. Submitted to National Institute of Nursing Research. Bethesda, MD: National Institutes of Health

Vendouses is a healing practice known throughout the Mediterranean area. Most frequently, it is used as a treatment for colds but other indications include high blood pressure and backache. The technique consists of lighting a swab of cotton held on a fork, then placing the swab in an inverted glass creating a vacuum in the glass that is then placed on the back of the ill person. The skin on the back is drawn into the glass. This procedure is repeated 8–12 times. An extreme alternative method for particularly serious cases, is the *koftes* (cut *vendouses*). Here, the same procedure is followed, except that a cut in the shape of a small cross is made on the skin. When the glass is placed over this cut, blood is drawn into the glass. The therapeutic rationale for using *vendouses* surrounds its counterirritant effect; the technique increases and revitalizes the circulation, draws out poisons and "cold," and prevents coagulation of blood. These practices are rarely used in the twenty-first century. Lists of herbs and recipes for Greek home remedies can be found on many Internet sites (The Greek food blog 2014).

16.12.4 Barriers to Health Care

The primary barriers to health care for Greek and Greek Cypriots in America, Australia, and Britain include a reliance on self-care in the family context and a general distrust of bureaucracies. Self-medicating behaviors are normative with herbal remedies and over-the-counter medications used widely for specific symptoms. Self-care, however, is usually undertaken with the advice of family members. Inability to speak English is an additional barrier for a number of first-generation Greek and Greek Cypriot migrants (Hurley et al. 2013).

In Greece and Cyprus, pharmacies are a common source of self-medication. In both countries, pharmacies have considerably greater authorization to advise and dispense medications. Self-treatment is still used as a first-line response to illness. When self-care fails, families often bring the ill person to a specialist rather than a primary-care physician. The rate of use of all physicians in America is lower for Greeks than for other groups of European descent; however, their infrequent use of physicians for primary care is remarkable. However, in Britain, people do not have direct access to specialist medicine unless they first consult their family doctor.

One of the main barriers to health care in Britain is racism. This is now well documented and accepted by the British government which has been trying through various policies to eliminate it. In a study that examined the primary health-care needs of Greek and Greek Cypriot women in London, Papadopoulos and Worrall (1996) found numerous cases of racist behavior by primary health-care staff, particularly toward first-generation immigrant women.

16.12.5 Cultural Responses to Health and Illness

Mental illness is accompanied by social stigma with negative consequences for the afflicted person as well as the family and relatives (Tzouvara et al. 2016). Shame (*endropi*) originates in the notion that mental illness is hereditary; afflicted people are viewed as having lifelong conditions that "pollute" the bloodline. The stigma is so wide-ranging that people labelled as mentally ill and their families may experience the loss of friends and social isolation. As a result, families place a wide variety of behaviors within the range of "normal" to delay receiving the stigmatized label (Madianos et al. 2011). Individuals with mental illness often present with somatic complaints such as dizziness and paresthesia on initial visits to health-care practitioners (Marmanidis et al. 1994). Recent immigrants tend to have higher rates of mental disorders, which perhaps result from the stress of culture change (Mavreas et al. 1990; Sands and Berry 2009).

However, a folk model for ***nevra*** (nerves) is a socially acceptable and culturally condoned medium for the expression of otherwise unacceptable emotions. *Nevra* is experienced most commonly by those in positions of least power, such as women and people living in poverty. It encompasses a wide variety of symptoms and provides a metaphor for social disorder, such as conflict between close kin or intergenerational conflict. Ideally, *nevra* are treated through medications for the relief of symptoms instead of through talk therapy.

Greek culture socially stratifies people by the nature of their disorders. Individuals with physical illnesses such as asthma, diabetes, and arthritis are most accepted, followed by people with disfiguring illnesses such as cerebral palsy. Individuals and families experiencing psychiatric illness, sexually transmitted diseases such as AIDS, and those with learning difficulties, are less accepted by the community, although close members of the family are normally very supportive. However, people in the most-stigmatized group are those with social deviance, such as addictions or delinquency.

Ponos (pain) is the cardinal symptom of ill health. It is viewed not as something to be endured, but as an evil that needs eradication. Pain mobilizes considerable family concern. The person in pain is not expected to suffer quietly or stoically in the presence of family. The family is relied on to find resources to relieve the pain, or failing that, to share in the experience of suffering. However, in the presence of outsiders, the lack of restraint in pain expression suggests lack of self-control and, therefore is considered *endropi*. Although the experience of physical pain is acknowledged publicly, emotional pain is hidden within the privacy of the family.

The key aspect of the sick role is for sick people to fully rely on the family to sustain them. Solitude is considered unpleasant and is avoided even when people are well. When an individual is ill, it is considered particularly important that he or she is not left alone. A small crowd of family and neighbours often surrounds sick people. Family and visitors provide advice regarding appropriate treatments and recall similar situations experienced by themselves or others.

When hospitalization occurs, family members expect to stay with sick people, even during examinations and therapeutic procedures. They are expected to ensure that sick people are not harmed and are receiving the best care possible. Protection of patients even includes shielding them from a serious diagnosis, such as cancer, until the family feels they are ready to learn about the diagnosis.

16.12.6 Blood Transfusions and Organ Donation

On the basis of the Christian Orthodox belief in the physical resurrection of the body, some Greeks and Greek Cypriots in the past rejected the concept of autopsy and did not readily accept organ donation. However, the Greek Orthodox Church is strongly pro-life and lately has been encouraging organ donation as an act of love. Blood transfusions are wholly acceptable and are common for people with thalassemia.

Reflective Exercise

Mrs. Fotini Papanicolaou, a 69-year-old Greek Cypriot who emigrated to England in 1962, was admitted to a surgical unit following several episodes of rectal bleeding and weight loss. She delayed going to the physician because she felt embarrassed owing to the nature of her problem. She treated herself with rice soup while also praying to God for healing. Upon admission, she was dressed in a smart black dress and was accompanied by her daughter. Mrs. Papanicolaou's husband died 2 years earlier from a heart attack. She told the nurse that although she lived on her own, her married daughter lived nearby. She also has two married sons and several grandchildren. She is very proud of her children, one of whom is a dentist, another an accountant, and her daughter is a teacher. She is particularly proud that they all did very well, despite her and her husband's having to work very hard to raise them. She said, "I have a wonderful and loyal family. I am very happy because my children and grandchildren are honorable human beings. They always follow my advice and guidance, and they all speak Greek, even my English daughter-in-law. My grandchildren's Greek is not perfect but we understand each other." The nurse noticed that Mrs. Papanicolaou carefully unpacked an icon of St. Cosmas and St. Damian and placed it with great care on her bedside table. "They will look after me," she told her while crossing herself.

In the meantime, her daughter went to the reception desk to speak to the ward manager whom she told that under no circumstances should anyone tell her mother she has cancer, if this were the diagnosis. She explained that her mother was still very sad following the death of her husband, and if she were given bad news, "this would kill her. My brothers and I want my mother to keep up her hopes that she will recover. We would like to handle any information about her condition as we know our mother very well and we know what is best for her."

1. What cultural and religious values underpin Mrs. Papanicolaou's health beliefs?
2. Is the fact that Mrs. Papanicolaou emphasizes that her children and grandchildren speak Greek important to the nurse?

3. Should nurses and other health-care providers go along with the request of Mrs. Papanicolaou's children not to tell her what is wrong with her if her diagnosis happens to be cancer?
4. How can they deal with this if the patient asks to know exactly what is wrong with her?

16.13 Health-Care Providers

16.13.1 Traditional Versus Biomedical Providers

In rural Greece, lay providers included midwives, bonesetters, and herbalists (who peddle herbs and tonics around neighbourhoods). A woman who cures, particularly the evil eye, is known as a ***magissa***, which is usually translated as "witch" but means "magician"; she may also be called *doctor*. However, in modern Greece, it is becoming rare to find people who practice traditional "medicine" (Athanasopoulou and Papadopoulos 2011). In Cyprus, too, with the exception of people who claim to exorcise the evil eye, lay providers no longer exist because modern medicine is readily available.

In the United States and Britain, traditional healers are not readily available. "Wise women" from one's own family conduct most lay healing. Occasionally, for particularly difficult cases of *matiasma*, a woman with particular gifts in diagnosis and healing may be called. The priest may also be called on for advice, blessings, exorcisms, and direct healing.

16.13.2 Status of Health-Care Providers

Many Greeks and Greek Cypriots display a general distrust of all professionals. Considerable "shopping around" for physicians and other professionals to obtain additional opinions is relatively common, even in Britain where health care is provided free to all citizens. This is particularly true if the sick person does not receive the diagnosis or the degree of sympathy judged appropriate by the family. Receiving empathy by a health professional is a major expectation for Greek patients (Karydis et al. 2001). Inconsistencies in opinions or recommendations result in further concern. In addition, the use of several physicians simultaneously may result in untoward drug interactions from multiple-drug use.

Hospitals are a particular source of mistrust for both Greeks and Greek Cypriots. When hospitalizations occur, the family may be perceived by staff as demanding or "interfering" as they enact their protective advocacy roles. Mothers may demand to sleep with their children and fear they may not receive appropriate care. There is also a fear that the sick person may be used as a subject for experimentation (Tripp-Reimer and Sorofman 1989; therefore, the provision of sensitive care is advocated for patients of Greek origin (Georgiades 2010).

References

Alexiou N (2013) Greek immigration in the United States: a historical overview. Hellenic-American oral history project: Greek Americans, Queens College, City University of New York. https://www.qc.cuny.edu/Academics/Degrees/DSS/Sociology/GreekOralHistory/Pages/Research.aspx. Accessed 24 Oct 2019

Allender S, Scarborough P, Peto V, Rayner M, Leal J, Luengo-Fernandez R, Alastair G (2008) European cardiovascular disease statistics. http://swissheartgroups.ch/%5C/uploads/media/European_cardiovascular_disease_statistics_2008.pdf. Accessed 24 Oct 2019

Asylum Information Database (2018) Cyprus country report. https://www.asylumineurope.org/reports/country/cyprus. Accessed 24 Oct 2019

Athanasopoulou M, Papadopoulos I (2011) Traditional and complementary therapies in Greece and Cyprus. In: Papadopoulos I, Kalokairinou A, Kouta C (eds) Transcultural nursing and cultural competence for health professionals. Paschalides Publications, Athens, pp 19–32; (in Greek)

Athens Population (2019). http://worldpopulationreview.com/world-cities/athens/. Accessed 25 Oct 2019

Beratis S (1990) Factors associated with adolescent suicidal attempts in Greece. Psychopathology 23:161–168

CIA World Factbook (2011) Europe: Cyprus. https://www.cia.gov/library/publications/the-world-factbook/geos/cy.html. Accessed 24 Oct 2019

CIA World Factbook (2015) Europe: Greece. https://www.cia.gov/library/publications/the-world-factbook/geos/gr.html. Accessed 24 Oct 2019

Cyprus High Commission (1986) Geographical distribution of Cypriots living in the UK. Cyprus High Commission, London

De Sanctis V, Kattamis C, Canatan D, Soliman AT, Elsedfy H, Karimi M et al (2017) β-Thalassemia distribution in the old world: an ancient disease seen from a historical standpoint. Mediterr J Haematol Infect Dis 9(1):e2017018

European Food Safety Authority (2010) Scientific Opinion on lactose thresholds in lactose intolerance and galactosaemia. EFSA panel on dietetic products, nutrition and allergies (NDA). EFSA J 8(9):1777. https://efsa.onlinelibrary.wiley.com/doi/epdf/10.2903/j.efsa.2010.1777

Eurostat (2018) Population. Tables, graphs and maps interface. https://ec.europa.eu/eurostat/tgm/table.do?tab=table&init=1&language=en&pcode=tps00001&plugin=1. Accessed 26 Oct 2019

Foster J, Papadopoulos C, Dadzie L, Jayasinghe N (2007) A review of tobacco and alcohol use literature in the native and migrant Greek community. J Subst Abus 12(5):323–335

Georgiades S (2010) Sensitive practice with Greek immigrants: a review of the evidence. Int J Cult Ment Health 3(1):52–60

Gourounti K, Lykeridou K, Vaslamatzis G (2012) Increased anxiety and depression in Greek infertile women results from feelings of marital stress and poor marital communication. Health Sci J 6(1):69

Government of Canada (2019) Library and archives Canada. Greek. Genealogy and family history. https://www.bac-lac.gc.ca/eng/discover/immigration/history-ethnic-cultural/Pages/greek.aspx. Accessed 24 Oct 2019

Greek Americans (2019). https://en.wikipedia.org/wiki/Greek_Americans. Accessed 24 Oct 2019

Greek Canadians (2019). https://en.wikipedia.org/wiki/Greek_Canadians.Accessed 24 Oct 2019

Hellenic Republic, Ministry of Foreign Affairs (2018) Greece's course in the EU. https://www.mfa.gr/en/foreign-policy/greece-in-the-eu/greeces-course-in-the-eu.html. Accessed 24 Oct 2019

Hellenic Statistical Authority (2011) Demographic characteristics/2011. http://www.statistics.gr/en/statistics/-/publication/SAM03/-. Accessed 25 Oct 2019

History.com Editors (2019) U.S. immigration since 1965. https://www.history.com/topics/immigration/us-immigration-since-1965, https://www.history.com/topics/immigration/us-immigration-since-1965. Accessed 24 Oct 2019

Hurley C, Panagiotopoulos G, Tsianikas M, Newman L, Walker R (2013) Access and acceptability of community-based services for older Greek migrants in Australia: user and provider perspectives. Health Soc Care Community 21(2):140–149

Imai, S. (2013) Immigration act of 1924. Densho encyclopedia. https://encyclopedia.densho.org/Immigration%20Act%20of%201924/. Accessed 24 Oct 2019

Institute for Health Metrics and Evaluation (2017a) Greece. http://www.healthdata.org/greece. Accessed 25 Oct 2019

Institute for Health Metrics and Evaluation (2017b) Cyprus. http://www.healthdata.org/cyprus. Accessed 25 Oct 2019

Kaplan M, Hammerman C, Bhutani VK (2015) Parental education and the WHO neonatal G-6-PD screening program: a quarter century later. J Perinatol 35(10):779–784

Karydis A, Komboli-Kodovazeniti M, Hatzigeorgiou D, Panis V (2001) Expectations and perceptions of Greek patients regarding the quality of dental health care. Int J Qual Health Care 13(5):409–416

Kitroeff A (2009) The roads of the Greeks. The history of Greek communities in the five continents. Publications Polaris, Athens

Kouta C, Papadopoulos I, Sourtzi P (2010) Public health in Cyprus and Greece. In: Scriven A, Kouta C, Papadopoulos I (eds) Health promotion for health professionals. Paschalides Publications, Athens; (in Greek)

Kunkelman GA (1990) The religion of ethnicity: belief and belonging in a Greek-American community. Garland, New York

Ladas SD, Katsiyiannaki-Latoufi E (1991) Lactose maldigestion and milk intolerance in healthy Greek school children. Am J Clin Nutr 53(3):676–680

Lovell-Troy LA (1990) The social basis of ethnic enterprise: Greeks in the pizza business. Garland, New York

Madianos MG, Zartaloudi A, Alevizopoulos G, Katostaras T (2011) Attitudes toward help-seeking and duration of untreated mental disorders in a sectorized Athens area of Greece. Community Ment Health J 47(5):583–593

Mammas IN, Bertsias GK, Linardakis M, Tzanakis NE, Labadarios DN, Kafatos AG (2003) Cigarette smoking, alcohol consumption, and serum lipid profile among medical students in Greece. Eur J Public Health 13(3):278–282

Marjoribanks K (1994) Cross-cultural comparisons of family environments of Anglo, Greek, and Italian Australians. Psychol Rep 74(1):49–50

Marmanidis H, Holme G, Hafner RJ (1994) Depression and somatic symptoms: a cross-cultural study. Aust N Z J Psychiatry 28:274–278

Mavreas V, Bebbington P, Der G (1990) Acculturation and psychiatric disorder: a study of Greek Cypriot immigrants. Psychol Med 20:941–951

Michopoulos A (2017) New Greek immigration: the case of the United States. http://seesoxdiaspora.org/publications/briefs/new-greek-immigration-the-case-of-the-united-states. Accessed 24 Oct 2019

Moskos PC (2017) Greek Americans: struggle and success. Routledge, New York

Office for National Statistics (2017) Greek residents in the UK. https://www.ons.gov.uk/aboutus/transparencyandgovernance/freedomofinformationfoi/greekresidentsintheuk. Accessed 24 Oct 2019

Panagiotopoulos G, Walker R, Luszcz M (2013) A comparison of widowhood and Well-being among older Greek and British-Australian migrant women. J Aging Stud 27(4):519–528

Panayides F (1988) Address by his excellency the high commissioner for Cyprus. In: Charalambous J, Hajifanis G, Kilonis L (eds) The Cypriot community in the UK: issues of identity. Polytechnic of North London Press, London

Papadopoulos I (1998) The health needs of the Greek Cypriot people living in Britain. In: Papadopoulos I, Tilki M, Taylor G (eds) Transcultural care: a guide for health care professionals. Quay Publications, Dinton, Wilts

Papadopoulos I (1999) The health needs of the Greek Cypriots living in London. Unpublished doctoral dissertation, University of North London, London

Papadopoulos I, Lees S (2004) Cancer and communication: similarities and differences of men with cancer from six different ethnic groups. Eur J Cancer Care 13(2):154–162

Papadopoulos I, Papadopoulos C (2000) The nature of the changing identity of second and subsequent generations of Greek and Greek Cypriot people living in North London. Parikiaki (Greek newspaper, English section), Aug 24 p. 14; Aug 31, p. 15; Sep 7, p. 15

Papadopoulos I, Worrall L (1996) All health care is good until you have a problem. An examination of the primary health care needs of the Greek and Greek Cypriot women. GGCWE (Greek and Greek Cypriot Women of Enfield, England), London

Portes A, Rivas A (2011) The adaptation of migrant children. Futur Child 21:219–246

Riepl RL, Schreiner J, Muller B, Hildemann S, Loeschke K (1993) Broad beans as a cause of acute hemolytic anemia. Dtsch Med Wochenschr 118:932–935

Robins K, Askoy A (2001) From spaces of identity to mental spaces: lessons from Turkish-Cypriot cultural experiences in Britain. J Ethn Migr Stud 27(4):685–711

Rosenthal D, Moore S, Buzwell S (1994) Homeless youths: sexual and drug-related behaviors, sexual beliefs and HIV/AIDS risk. AIDS Care 6(1):83–94

Sands EA, Berry JW (2009) Acculturation and mental health among Greek-Canadians in Toronto. Can J Commun Ment Health 12(2):117–124

Scourby A (1984) The Greek Americans. Twayne, Boston

The Greek Food Blog (2014) Herbs and spices used in Greek food. https://www.thegreekfood.com/ingredients/herbs-spices/. Accessed 24 Oct 2019

The UN Refugee Agency (2019) Operational portal refugee situations. Mediterranean situation. https://data2.unhcr.org/en/situations/mediterranean/location/5179. Accessed 24 Oct 2019

Todd P, Samaratunga IR, Pembroke A (1994) Screening of glucose-6-phosphate dehydrogenase deficiency prior to dapsone therapy. Clin Exp Dermatol 19(3):217–218

Tokas N (1995) 2000 die from heart disease. Parikiaki (Greek newspaper), p 5

Tripp-Reimer T, Sorofman B (1989) Illness related self-care responses in four ethnic groups. Final report no. R01Nu 1101. Submitted to National Institute of Nursing Research. Bethesda, MD: National Institutes of Health

Tripp-Reimer T, Sorofman B (1994) Drinks, foods, medicines and markers: Alcohol issues in four ethnic groups. Paper presented at the meeting of the American Anthropology Association, Atlanta, GA

Tyrovolas S, Bountziouka V, Papairakleous N, Zeimbekis A, Anastassiou F, Gotsis E et al (2009) Adherence to the Mediterranean diet is associated with lower prevalence of obesity among elderly people living in Mediterranean islands: the MEDIS study. Int J Food Sci Nutr 60(suppl 6):137–150

Tzouvara V, Papadopoulos C, Randhawa G (2016) Systematic review of the prevalence of mental illness stigma within the Greek culture. Int J Soc Psychiatry 62(3):292–305

United Nations DESA/Population Division (2019) World population prospects 2019. https://population.un.org/wpp/Download/Standard/Population/. Accessed 24 Oct 2019

Wilson A, Bekiaris J, Gleeson S, Papasavva C, Wise M, Hawe P (1993) The good heart, good life survey: self-reported cardiovascular disease risk factors, health knowledge, and attitudes among Greek-Australians in Sydney. Aust J Public Health 17:215–221

World Data Atlas (2017) Greece-crude birth rate. https://knoema.com/atlas/Greece/Birth-rate. Accessed 24 Oct 2019

World Health Organization (1994) World health statistical annual 1993. WHO, Geneva

Worldometers (2019) Cyprus population. https://www.worldometers.info/world-population/cyprus-population/. Accessed 24 Oct 2019

People of Guatemalan Heritage

17

Tina A. Ellis and Larry D. Purnell

17.1 Introduction

Civil war between the Ladinos and Maya during the 1960s–1990s created widespread casualties (approximately 200,000), loss of land rights, economic instability, and disruption of their normal way of life for most Guatemalans. In addition, natural disasters such as earthquakes and volcanic eruptions have caused widespread devastation. The indigenous population has been most adversely affected and, as a result, it is estimated as many as 1 million Guatemalans fled to Mexico and the US seeking safety and political asylum, requiring healthcare providers to learn about the various Guatemalan cultures to provide culturally congruent care.

This chapter is an update written by Tina Ellis and Larry Purnell in a previous edition of the textbook.

T. A. Ellis (✉)
Florida Gulf Coast University, Fort Myers, FL, USA
e-mail: tellis@fgcu.edu

L. D. Purnell
College of Health Sciences University of Delaware/ Newark, Sudlersville, MD, USA

17.2 Overview, Inhabited Localities, and Topography

17.2.1 Overview

People of Guatemalan heritage comprise a growing number of Hispanic/Latino populations in the United States (US). Little is known about their migration to other countries. While Guatemalans may share a common Spanish language with other Hispanic ethnic groups, they are, nonetheless, a unique cultural group. Health care providers must be knowledgeable regarding distinct Hispanic cultural characteristics to provide culturally competent and congruent care for patients of Guatemalan heritage.

17.2.2 Heritage and Residence

Guatemala, officially the Republic of Guatemala, is a country in Central America bordered by Mexico to the north and west, Belize and the Caribbean to the northeast, Honduras to the east, El Salvador to the southeast, and the Pacific Ocean to the south. This land, referred to as "eternal spring," has an estimated population of over 16.6 million (Central Intelligence Agency (CIA) 2018a).

The country covers 41,700 square miles, comparable to the U.S. state of Tennessee, and consists of three main regions: the cooler high-

© Springer Nature Switzerland AG 2021
L. D. Purnell, E. A. Fenkl (eds.), *Textbook for Transcultural Health Care: A Population Approach*, https://doi.org/10.1007/978-3-030-51399-3_17

lands, the tropical Pacific and Caribbean coasts, and the tropical jungle in the far north (Information Please Almanac 2019). The capital, Guatemala City, with a population of over 2 million, is located within the highland area. Guatemala is inhabited by people of numerous heritages: Mestizo (mixed Amerindian-Mayan and Spanish called *Ladino*) and European decent 60.1 percent, indigenous purely Mayan Indian decent 39.3 percent, indigenous non-Mayan (Xinca) decent 0.15 percent, and Garifuna (West and Central African, Caribbean) decent 0.5 percent. More than half of Guatemala's population resides in the rural highlands. Ladinos comprise the majority population in and surrounding Guatemala City (CIA 2019).

Guatemala was inhabited for approximately 3000 years by indigenous Amerindian (Mayan and Xinca) people who lived a simple life based on the culture, customs, and religion of their ancestors. Although the Maya excelled in architecture, weaving, pottery, and hieroglyphics, farming was the primary occupation for the majority. A few cities (*pueblos*) were established during Mayan times although many Guatemalans now live in small villages (*aldeas*).

Many of the Spaniards came to Guatemala in the fifteenth century seeking a route to the East Indies, remained, establishing a formal government, economy, and way of life based on Spanish culture, traditions, and the Catholic religion (Information Please Almanac 2019).

17.2.3 Reasons for Migration and Associated Economic Factors

As noted above, civil war between the Ladinos and Maya during the 1960s–1990s has fuelled the increased migration of Guatemalans. A Quiche woman, Rigoberta Menchu, lost her father in the civil war and became a strong advocate for indigenous respect and rights and, in 1996, received a Nobel Peace Prize (Gall and Hobby 2009). A peace accord signed in 1996 formally ended 38 years of war between the Ladinos and Maya in Guatemala and the country is now led by an elected president. Despite the many years since the civil war, Guatemala continues to be plagued by "abject" poverty with over 59 percent of the population living below the poverty level (CIA 2019), families with one member earning as little as 14 *quetzales* a day (equivalent to US $2) struggle to survive (Walsh 2006). Although, during harvest, families may earn as much as 50 *quetzales* a day; those earnings cannot be calculated into an annual salary due to the seasonal nature of agricultural work.

Guatemala is a poor country characterized by poverty, malnutrition, and infant, child, maternal mortality (CIA 2018b). Great health disparities exist between the Ladino and the indigenous Mayan population. When a Mayan child survives until their sixth birthday, it is considered a good indicator the child may reach adulthood as they are less susceptible to dying from diarrhea and respiratory diseases.

Two distinct types of migration, internal and external, occur among people of Guatemalan heritage. Within the country of Guatemala, large company-owned farms (*fincas*) along the Pacific lowlands grow coffee beans, cotton, cardamom, and sugar. During the harvest season, indigenous people from the highlands migrate to the Pacific lowlands to harvest the crops grown there. These migrants may remain in the *fincas* for months before returning home. Although in many cases only the male head of household and mature boys migrate, sometimes the entire family migrates. Another in-country migration is the relocation of indigenous people to Guatemala City. Economic necessity requires young women to leave their families to work as housekeepers for wealthy Ladino families. Their earnings are sent home to supplement the family income. In some cases, whole families relocate to Guatemala City seeking lucrative employment. Some can secure adequate income for their needs, but many cannot and must resort to living as squatters on Guatemala City streets or within the Guatemala City dump.

External migration occurs primarily to Mexico and North America. Over 57 million Hispanics reside in the US, 3.9% are estimated to be Guatemalan (US Bureau of the Census 2017). Well educated Ladino Guatemalans from Guatemala City including physicians, lawyers, and educators primarily immigrate to U.S. cities, seeking the opportunities and rewards of life in America. The exact numbers are unknown.

Less educated Guatemalans or those with no formal education often emigrate to rural areas of the United States for employment primarily in the agricultural industry. Approximately 73 percent of America's farm workers are Hispanic, including Guatemalans (National Center for Farmworker Health [NCFH] 2018). While earnings for farmworkers in the US are significantly higher than in Guatemala, over 71% of farm workers in the United States still earn incomes below the federal poverty level (NCFH 2017). These wages, although limited, provide Guatemalans with enough income to help support their families back home. When the opportunity presents itself, many Guatemalans shift to non-agricultural types of jobs such as construction and landscaping, as these generally offer more stability and higher wages.

17.2.4 Educational Status and Occupations

Prior to the Spanish conquest, all education in Guatemala was informal. The Mayan languages and cultures were passed down orally from generation to generation through parents, extended family, and community as there was no written language. Ladinos established the first public education system within the country, including higher education. This education system teaches Spanish, the official language of the country, Spanish culture, and Spanish traditions. High school and university graduates are primarily Ladinos (Shea 2001). More and more Mayan males are becoming educated as school teachers, choosing to teach in the rural areas in which they were raised, integrating Mayan history, culture, and language into the school curriculum. For the first time in the history of public education in Guatemala, Mayan children are learning to read and write their Mayan language.

Traditionally, within the Mayan culture, only male children are permitted to attend school while female children spend as much time as possible with their mothers to learn homemaking skills; although this is changing. In 2015, male literacy was estimated at 87.4% and female literacy 76.3% (CIA 2015). Female illiteracy, however, is as high as 90% in many rural communities.

Most Guatemalans are involved in occupations related to agriculture. About 2% of Guatemalans own 70% of the land profitable for farming. Rural families may own a small plot of land passed on by their ancestors; however, many of these plots were confiscated by the government during the civil war. Often, the plot is just large enough for a small adobe brick one room home with a tin roof and an area to grow corn (*maize*), beans, and a few other vegetables for family consumption. Many are without electricity or running water. With time, families with one or more members working in the US, can afford to upgrade to a cement block home with more rooms and purchase more than just the necessities.

Elementary school (*primaria*) is free and compulsory for children from 7 years of age to 13, although, there is very little enforcement of this and missing school days is common when children are needed to help with farming or household duties. There is a charge for children to attend middle and high school (*secondaria*) so many children around the age of 12 years of age begin working in order to supplement the family income.

Women in rural areas often earn money by selling woven clothing, handmade items, or small animals they have raised and sell at the local market. Weavings using a traditional back-strap loom contain colors and designs related to the specific indigenous group, village, and family by whom they are made.

17.3 Communication

17.3.1 Dominant Languages and Dialects

The major languages in Guatemala include the official language, Spanish, which is spoken by 60 percent of the population, and Amerindian languages, which are spoken by the remaining 40 percent. Ladinos speak Spanish. Officially, 21 Mayan languages are recognized, including Quiche, Cakchiquel, Kekchi, Mam, and Chuj. In addition, there are various unofficial Mayan languages (Shea 2001). Each Mayan ethnic group speaks one dialect as their primary language. If Mayans attend school in Guatemala, they learn to speak, read, and write Spanish. Two non-Mayan languages are also officially recognized: Xinca and Garifuna.

Some Mayan men have no formal education but can speak Spanish because of frequent interactions with Spanish speakers in business relationships. Mayan men may be bilingual, speaking their Mayan dialect and Spanish. This is the case on the East Coast of the US where one of the authors lives (Purnell). This rural area has many Guatemalan migrant workers in the ornamental nursery and farming industry. Most of the men speak one of the Guatemalan dialects but have learned a few Spanish words as they have migrated across the US with other migrant Hispanic workers. Health care providers need to assess the preferred language of the client, their education level, and whether the client reads and writes in the preferred language.

Blacks living along the Caribbean coast of Guatemala are Garifuna which denotes their St. Vincent island ancestry, language, and culture. They may also speak French and learn to speak Spanish in Guatemalan schools. When the Garifuna people emigrate to the US, they usually seek employment in large cities like New York, Houston, and New Orleans (Global Sherpa n.d.)

17.3.2 Cultural Communication Practices

Although topics such as income and investments are taboo among many Guatemalans, they generally like to express their inner beliefs, feelings, and emotions once they get to know and trust a person. Meaningful conversations often become loud and seem disorganized. To the outsider, the situation may seem stressful or hostile, but this intense emotion means the conversants are enjoying each other's company. Within the context of ***respeto***, respect, health care providers can encourage open communication and sharing and develop a trusting relationship with the patient by introducing themselves to the patient and each person accompanying them and asking about the patient and others before proceeding with routine health care if time and the patient's condition allows. Engaging in "small talk" is important before addressing the actual health care concern with the patient and others. Relationships are very important to Guatemalan people including the relationship with the health care provider. Therefore, taking time to develop the relationship with the Guatemalan client during the initial and subsequent encounters is expected and leads to trust and positive outcomes.

Guatemalans place great value on closeness and togetherness, including when they are in an in-patient health care facility. They frequently touch and embrace and like to see relatives and significant others. Touch between men and women, between men, and between women is acceptable, including in the health care environment. Allowing for visits and demonstration of closeness in the health care setting, whenever possible, is advisable.

Traditional Guatemalans consider sustained eye contact when speaking directly to an older person to be rude. Direct eye contact with teachers or superiors may be interpreted as disrespectful. Avoiding direct eye contact with superiors is a sign of respect. This practice may or may not be seen with second- or third-generation Guatemalan Americans. Health care providers must take cues from the patient, family, and others and avoid inaccurate perceptions of indirect eye contact. Health care providers who speak Spanish need to use the formal you (*usted*) rather than informal you (*tu*) form as it conveys both professionalism and respect.

17.3.3 Temporal Relationships

Guatemalan people tend to value the past and live in the present, being more concerned with today than the future because the future is uncertain for many. Some of life's uncertainties for Guatemalans are related to the daily challenges they face securing adequate employment, housing, food, and other necessities such as health care for themselves and their families.

For those in poverty, there is usually not enough money to save for the future and send their children to high school (*secondaria*) or for higher education. While health insurance, social security, and retirement benefits may be available for Ladinos, these benefits are not available to the average citizen in Guatemala. Most Guatemalans hope that they can work until the time of their death. When an individual is unable to work, it is customary for him or her to be taken care of in the home by the family.

Guatemala has numerous public hospitals in the cities that provide health care free of charge; however, rural Mayan Guatemalans tend to avoid them because the cost of transportation to the hospital from villagers with a vehicle is high, health care providers in the hospitals are Ladino(a) so do not understand the Mayan culture or languages, and it is thought "hospitals are places people go and die". One explanation for this way of thinking is by the time all the resources to access the hospital are secured, the person in need of care is near death. Some even die while being transported to the hospital and families would rather their loved one die at home if there is no hope of recovery.

Because work is such a priority in the life of many Guatemalans, they seek health care only when their illness or health issue has progressed to the point of preventing them from working or carrying out their duties or roles within the family. Taking time to go to the doctor means time lost from work, loss of pay, and in some cases risk for loss of the job itself.

Most Guatemalans living in rural areas do not own a watch or clock and are unable to tell time. Time is related to the natural environment, such as sunrise, sunset, and the rainy or dry seasons. Businesses in Guatemala are much more relaxed about time. Punctuality is difficult for many because of limited transportation and unexpected family needs that take priority over appointments. Written medication education for patients using icons such as sunrise, sun at midday, sunset, and the moon for night time have been found more understandable than instructions with numeral times such as 9:00 am, 12:00 pm, 5:00 pm, and 9:00 pm.

Future orientation and punctuality are highly valued in the US. Because these are in direct contrast to Guatemalan values, patient and health care provider relationships often result in conflict. Guatemalans may be late for appointments or not show up at all. Health care providers may interpret this behavior as immature or disrespectful, when the reason may have to do with a family emergency, lack of transportation, inability to get time off from work, limited finances, or fear of losing one's job. Moreover, health care providers may expect Guatemalan patients to participate in preventive health screenings and adopt behaviors to reduce their risk of long-term complications of disease. These behaviors require a future orientation that conflicts with the present orientation of many Guatemalans.

Hargrove et al. (2009) conducted a study among Mexican and Guatemalan male farmworkers in the US to identify beliefs regarding adult immunization. The researchers concluded while some men expressed lack of knowledge of immunizations other than those important in childhood and adult immunizations being good and useful, most men identified barriers as the reason they lacked recommended adult immunizations including poverty, lack of transportation, language differences, and their own health illiteracy. Removing these barriers rather than convincing people that immunizations are necessary and safe is recommended.

17.3.4 Format for Names

Guatemalans from Hispanic heritage use the Spanish format for names. At birth, a child is given a first name, for example, (Ovidio) followed by the surname of his father (Garcia), and then the surname of his mother (Salvador), resulting in Ovidio Garcia Salvador. In referring to him as Mr. (*señor*) it would be appropriate to use Señor Garcia. Men's names remain the same through their lifetime. However, when a woman named Jovita Garcia Salvador marries Francisco Vasquez Gutierrez, she then becomes Jovita Garcia de Vasquez or simply Jovita Garcia Vasquez. She also becomes Mrs. (*señora*) Vasquez. To avoid errors, including the medical record, it is important to identify the Guatemalan client by the full name. Mayan Guatemalans have a Mayan name also; however, may use their Spanish name in more formal settings. Sometimes, health care providers incorrectly assume Guatemalan couples are not married because of the format of their names. Guatemalan couples use the Spanish format when married in either the church or a civil ceremony. Couples who live together (*unidos*) may use it or choose for the woman to retain her unmarried surname.

To convey respect, health care providers should address the Guatemalan in a formal manner unless otherwise requested by the patient. Male adults are referred to as Mr. (*señor*), unmarried females are referred to as Ms. (*señorita*) and married females as Mrs. (*señora*). Highly respected persons in the community are often referred to as male (*Don*) or female (*Doña*) followed by the person's first name. For example, one respected gentleman is referred to as Don Martin, and a respected nurse is referred to as Doña Alma.

Guatemalans are customarily greeted with a handshake. In rural areas, people shake hands softly. To offer a firm handshake is perceived as aggressive behavior. In the cities, however, the handshake tends to be more firm. Guatemalans avoid direct eye contact with others, including health care providers, which is a way of demonstrating respect and should not be misinterpreted as avoidance, low self-esteem, or disinterest. They speak softly in public. Speaking loud in public is considered rude. Understanding these customs are important for positive health care provider/Guatemalan client relationships.

17.4 Family Roles and Organization

17.4.1 Head of Household and Gender Roles

Many Guatemalan families follow traditional roles for husbands, wives, and children, although this is changing for some. Traditionally, the man has been the head of household and is the primary "breadwinner" and provider for the family. He generates the family income and manages the household resources. This usually requires men to work outside of the home. Ultimate decision-making power resides with the man of the house. Families in Guatemala have been known to delay naming a newborn until the father can return from the US to make the decision on the name.

Women's roles have traditionally involved domestic work and care for family and the home. In rural areas, the wife usually rises around 5:00 a.m. or even earlier, to carry water for drinking and firewood from the nearest sources available, returns home to build a fire, and begins cooking for the family on a wood-burning stove. Once the family has eaten breakfast, the wife cleans house by sweeping the dirt floors, grinds corn to make *tortillas*, washes clothes, cares for the young children, breastfeeds as needed, sews, weaves, cares for chickens and other small animals, raises vegetables and beans, and prepares for market day (Shea 2001). Little girls begin fetching water at the age of 3–4 years and by 7 years of age are washing clothes, gathering wood, and caring for younger siblings. At age 9 years, young girls learn to weave and embroider (Glittenberg 1994).

Guatemalans place high value on the family and the extended family. They tend to be very close-knit. Gall and Hobby (2009) explain that family is the only dependable source of help in their society where church and state have limited impact on daily lives. Most families are nuclear—composed of a father, mother, and children. However, the extended family is also very important to Guatemalans and may include grandparents, aunts, uncles, and cousins. Multiple generations often reside in the same home.

A young woman's 15th birthday (*quinceñera*) is celebrated as her passage to womanhood. Coming of age for a young man is 18 years. A young man asks the young woman's father for her hand in marriage. Engagements may last several years. In the indigenous population, a young man's father may seek out a matchmaker to find a suitable bride under age 16 years. Once an arrangement is reached, the young man offers a dowry. A wedding feast celebrates the marriage. Recently married couples often live with the husband's parents.

More Guatemalan women are entering the workforce outside of the home, resulting in more egalitarian male/female roles. Husbands and wives share more of the child-rearing and household responsibilities and selecting one's own mate is becoming more common.

17.4.2 Prescriptive, Restrictive, and Taboo Behavior for Children and Adolescents

Guatemalans place a high value on the institution of family, which includes nuclear and extended family members who most often live together or very close. They are involved in each other's lives on a daily basis. Family is structured as a patrilineal system. Males are the only family members who receive an inheritance. However, the US Agency for International Development (USAID) programs for gender equality are working to correct this situation (USAID 2018).

Children are believed to be a gift from God and are highly valued in Guatemalan society. Sons are more valued than daughters (Glittenberg 1994). Children are taught to be obedient and demonstrate respect for older people. In Mayan communities, family members and other adults take an active part in raising a child. They believe it takes a village to raise a child to become a productive member of the community and to continue their culture (Menchu 1984). Values include being humble, content, and respectful of others. Working hard, avoiding arguments, and placing the needs of the family before one's own individual needs is common. In Ladino families, individualism and competition are more highly valued (Glittenberg 1994).

Disobedience among children in Guatemala may be handled with physical punishment (Menchu 1984). Health care providers in the US need to use care in assessing evidence of this to avoid misdiagnosing it as child abuse.

17.4.3 Family Goals and Priorities

The Guatemalan family demonstrates a desire to provide for the needs of each member. Parents provide for their children in hopes that they will grow up, marry, work, and have children. When family members are unable to take care of themselves, the expectation is that their family will take care of them. Life is hard in Guatemala. Families work much of their lives without luxuries such as days off from work or vacations. Most know poverty will likely prevail in their lives regardless of how hard they work. Parents feel they must prepare their children for the same hard life they have lived because very little changes from generation to generation (Menchu 1984).

Guatemalans expect to work hard and do believe their hard work will result in an increased quality of life in North America. Guatemalan families who migrate do so with the hope of a better life for themselves and their children. More opportunities are available in the US and Canada, which is why Guatemalans often risk everything, including their lives, to migrate. When families immigrate together it is often with the intention of staying in North America or Mexico, whereas external migration for men without family is based on the hope to return home one day when they have earned enough to provide for a better life in Guatemala.

17.4.4 Alternative Lifestyles

When "family" falls outside of defined norms, there is a lack of understanding and acceptance toward those involved. Religious beliefs and tradition often dictate the attitude one holds about what does and does not constitute family in Guatemala.

Catholic, Protestant, and Evangelical Christian Guatemalans do not believe in homosexuality, sexual activity among the unmarried, or infidelity. Persons involved in these activities must do so in secrecy. Even today, indigenous women dress conservatively with a woven long skirt (*corte*), blouse (*huipil*), a scarf (*tzute*), and shawl (*rebozo*) that promote modesty. A single woman is believed to be a prostitute if she is out in public alone.

Despite a prevailing "macho" attitude with a deep-rooted homophobia, some inroads have been made for gays, lesbian, and transgendered populations in Guatemala with *Lesbiradas*, an organization for lesbians and bisexual women, and an annual Gay Pride March in Guatemala City since 2000. According the Sonia Perez of the Associated Press (Perez 2019), "Guatemala has taken baby steps toward guaranteeing lesbian, gay, bisexual, transgender, queer/questioning (LGBTQ) rights such as adopting measures to identify hate crimes against members of the community which let people better express their identity". Larger cities in the US offer organizations, a support group for Latina lesbians; El Hotline of Hola Gay, an organization with information and referrals in Spanish; and Dignity, a gay Catholic support organization. Currently, more migrant farm workers in the US are identifying as LGBT (National LGBT Health Education Center 2015). Health care providers need to be sensitive toward and knowledgeable regarding the barriers to accessing care and resources in their community for Guatemalans who present with an alternative lifestyle.

17.5 Workforce Issues

17.5.1 Culture in the Workplace

During the civil war in Guatemala, residents were permitted to migrate to the US and apply for political asylum. If granted, this allowed Guatemalans to stay permanently in the US, but they were not permitted to ever return to Guatemala. Many who chose this route had no family left in their home country; their lives were at risk should they return. In 1986, the US Immigration Reform and Control Act allowed many Guatemalan agricultural workers to bypass the process involved in political asylum and simply obtain residency as farm laborers. This promoted an increase in emigration (Wellmeier 1998). The H-2A Temporary Agricultural Program, often called the H-2A visa Program (2019) is utilized. Employers who anticipate a lack of available domestic workers to bring foreign workers to the US to perform temporary or seasonal agricultural work including, but not limited to, planting, cultivating, or harvesting labor. Thus, Guatemalans migrating to the US today often find employment in the agricultural industry, including hand-harvesting crops or working on egg and chicken farms, citrus and vegetable processing plants, and ornamental plant and flower nurseries.

Wellmeier (1998) found that Guatemalans in Indiantown, Florida acquired a reputation for hard work, complained little, and paid close attention to detailed work in agriculture. However, Mayans were not often hired for picking oranges or cutting sugarcane because of their small stature.

When possible, Mayan farm workers in the US secure non-agricultural employment that offers less migration and more economic stability. Some find positions in housekeeping or maintenance in businesses or schools. Those with skills in English and Spanish may qualify for interpreter positions. If they speak a Mayan dialect as well, their services are even more highly sought by agencies that serve that particular population (Wellmeier 1998).

Guatemalan women may work in agriculture or housekeeping in the US or cook instead for single Guatemalan men working in the US without their families. This provides men with a connection to home, convenience, and the enjoyment of consuming foods prepared according to cultural traditions. Unfortunately, some non-Guatemalans believe these women are prostitutes. Prices for the meals are generally affordable.

Undocumented Guatemalans have difficulty securing a bank account or driver's license. Many carry the cash they earn with them. This habit and their small stature have resulted in Guatemalans being victims of robbery. Some Mayan agencies in the US have developed co-ops that offer simple banking services to reduce the incidence of robbery among Guatemalans (Wellmeier 1998).

Guatemalans may purchase a vehicle in order to secure employment far from home while in the US. Most have never had formal driver's training or a driver's license or driven a vehicle prior to arrival in the US. Moreover, they may be unaware of driving rules and regulations in the US owing to unfamiliarity with the English language. Driving under these circumstances places the Guatemalan and other drivers at risk for serious accidents. In addition, Guatemalans may be ticketed or arrested for driving violations owing to their lack of understanding regarding laws in the US, such as the law requiring seat belts be worn. On the Eastern Shore of Maryland, it is not unusual to see a small van loaded with migrant workers but only one of them has a driver's license. The driver is usually very vigilant to speed and other traffic laws.

Because family is highly valued among Guatemalans, needs of the family take priority over obligations in the employment setting. Guatemalans may miss work owing to an illness of a loved one, a need for transportation to an appointment, or a lack of child care. When Guatemalans living in the US learn that a loved one in Guatemala is ill or has passed away, they feel compelled to return to Guatemala for an extended period, risking loss of their job if a leave of absence is not permitted.

Because punctuality is not valued in Guatemala, Guatemalan employees in the US may arrive late for work. They may not wear a timepiece, be able to tell time, or understand the importance of punctuality in the US. This misunderstanding also occurs in health care settings. The Guatemalan patient may be late for an appointment because of circumstances related to family, work, or transportation and find the appointment has been cancelled or rescheduled when she or he arrives. To achieve a positive relationship, health care providers need to demonstrate respect for the Guatemalan patient's need for flexibility and help her or him to understand the expectations within the health care setting.

17.5.2 Issues Related to Autonomy

Guatemalans tend to respect persons in positions of authority. Those of lower socioeconomic status or with no formal education and English language skills usually acquire positions with responsibility but little authority. They prefer to get along well with others and not criticize or voice complaints when treated poorly. Moreover, the Guatemalan is likely to remain in a position equal to his or her peers rather than seek a promotion to maintain harmony with others.

17.6 Biocultural Ecology

17.6.1 Skin Color and Other Biological Variations

Most Guatemalans are a mixture of Spanish and Mayan Indian heritage. There is a small population of Black Guatemalans with ancestry from the Caribbean and Africa. This accounts for variations in skin color, facial features, hair, body structure, and other biological variations. Intermarriage among the racial groups in Guatemala has produced variations in their appearance. No one appearance is "typical" for a Guatemalan individual.

Predominantly Spanish Guatemalans have the appearance of Caucasians. They may have blonde or brown hair, fair (white) complexion, blue eyes, and be of average or taller height with a medium to large build. Guatemalans with predominantly Mayan Indian ancestry tend to have black hair, brown skin, and dark eye color and are of small stature. Black Guatemalans tend to have black hair, black skin, and dark eyes and be of average or taller height with a medium to large build.

Health care providers must be aware that cyanosis and anemia are assessed differently in dark-skinned people than in white-skinned people. Instead of being bluish in color, the skin color of a cyanotic patient of Indian or black ancestry may appear more ashen. Physical assessment of these individuals should include examination of the sclera, conjunctiva, buccal mucosa, tongue, lips, nailbeds, and palms of hands and soles of feet. In addition, jaundice is more difficult to detect among dark-skinned Guatemalans. Examination of the sclera and buccal mucosa for evidence of bilirubin is important for accurate assessment.

17.6.2 Diseases and Health Conditions

Undocumented noncitizens had similar likelihood of anxiety and depression, but lower likelihood of antidepressant use, compared with documented noncitizens. These results may reflect the resilience of an undocumented population facing multiple stressors but suggest that this group may be undertreated for depression (Sabin et al. 2013).

Health literature regarding Guatemalans is limited and most often does not differentiate Ladino, indigenous Maya, and other ethnic groups. The infant mortality rate in Guatemala is significant at 23.3 deaths per 1000 live births. Longevity for men and women has improved from the early 1990s with an estimated 69.8 years for males and 73.8 years for females (CIA World Factbook 2018a, b). Available data regarding farm workers in the US use the broad designation of Hispanic farm workers when discussing health.

The leading causes of mortality in Guatemala are influenza and pneumonia, diabetes, heart disease, and homicide (PAHO 2014). These may be directly linked to environmental factors, lifestyle, and lack of adequate health education and care. Chickering (2006) compiled data of the most frequent complaints of Guatemalans to North American health care providers during *jornadas*. These are described in order of highest to lowest frequency:

1. Musculoskeletal pain: This is not really surprising, considering the lifestyle of many Guatemalans. They are exposed to difficulties in life, including violence, malnutrition, disease, and high childbearing rates, all of which take their toll on one's health and well-being.

   Much of the work involved in the daily lives of Guatemalans is difficult. For example, young girls start fetching water, which they carry on their heads, and caring for siblings, whom they carry on their backs, at very young ages.

   Boys begin farm work with their fathers as soon as they are physically able. Farming continues much as it traditionally has with primitive tools, requiring lifting, bending, climbing steep mountains, and carrying heavy loads without the use of modern machinery.

   Adulthood brings requirements for more work, along with an increased need for physical stamina to maintain economic stability, the family, and the home. Guatemalans often believe that life is hard and pain is expected; instead of complaining about pain, they usually learn to endure the pain and continue their duties until the pain is unbearable or they physically are unable to function as desired.

   Other causes for musculoskeletal pain in Guatemalans are diseases such as malaria or dengue fever. "Whole body pain with a history of fever should be considered malaria until proven otherwise" (Chickering 2006). Guatemalans with dengue are not as likely to be seen by a health care provider because they are usually too ill to even leave their homes (Chickering 2006).
2. Abdominal pain: These complaints are divided into epigastric and no epigastric pain. Most commonly, epigastric pain is due to gastritis (*gastritis*), intestinal worms, or chronic giardia. Guatemalans may hold traditional beliefs about the symptoms caused by intestinal worms. These include thunder (*trueño*) noise in the stomach, swelling (*hinchazón*) in the stomach (especially in the afternoon following the largest meal of the day), chronic abdominal swelling (*panzudo*) in children, a ball-like mass (*bola*) in the stomach, gas mov-

ing back and forth in the upper abdomen, itchy nose, pica (especially for eating dirt, ashes, and paper), loss of appetite, and sleeping prone (*embrocado*) (Chickering 2006).

3. Cough and upper respiratory symptoms: A complaint of chronic coughing or a cough that persists can be a symptom of pulmonary tuberculosis (TB). This may be accompanied by fever, night sweats, or weight loss. Typhoid can also have symptoms of a mild, dry cough, although typhoid is not a very common disease in Guatemala (Chickering 2006).

   A factor increasing Guatemalans risk for acute upper respiratory infections is related to poverty and living conditions. Many Guatemalans of low socioeconomic status are malnourished, so their immune function is compromised. They often live in one- or two-room homes, placing them in close quarters with other family members and enabling germs to spread easily from one person to another.

   In addition to acute upper respiratory infections, chronic lung disease is seen among Guatemalans. Some areas of Guatemala have very poor air quality, exposing people to high levels of toxic inhalants. Regulations for air quality are few and poorly monitored. Another source of exposure for many Guatemalans is from cooking on wood-burning stoves. Cooking remains primitive for many, and dependence on wood for fuel is high. Exposure to carbon monoxide presents a risk in itself, but the danger becomes even higher because the cooking area is often enclosed in the home or a building adjacent to the home, with little, if any, ventilation to the outside. Chronic lung disease may result from years of one or both of these exposures.

   Health care providers may misinterpret signs and symptoms of chronic lung disease in Guatemalans as related to cigarette smoking, when, far more people are exposed to environmental toxins than smoke cigarettes.

4. Headaches: Headaches can be due to a number of factors. Malnutrition and anemia are common, and some headaches are attributed to these. In addition, cooking on the wood-burning stove causes women, especially, to be exposed to high levels of carbon monoxide, which can produce headaches.

   Although men usually carry heavy loads on their backs, women carry them on their heads, which leads some women to experience headaches. Finally, worms can cause headaches. Guatemalans themselves often relate the cause of headaches to be from anger or exposure to sun. Relationships that do not involve conflict are highly valued among Guatemalans. Unresolved issues in relationships may increase one's stress, thereby creating headaches. There is no literature to suggest sun exposure may cause headaches among Guatemalans, but perhaps the belief is related to the hot/cold paradigm. Health care providers should also assess blood pressure and neurological function in Guatemalans complaining of headaches (Chickering 2006).

5. Weakness: Weakness (*debilidad*), fatigue (*cansancio*), and dizziness (*mareos, baidos,* or *tarantamiento*) are common complaints. Anemia is the most common cause for these symptoms. Health care providers must note that weakness can also be used to describe a health problem involving one body organ. For example, weakness of the stomach (*debilidad del estomago*) can refer to a Guatemalan patient's self-report of poor nutrition, and weakness of the heart (*debilidad del corazon*) can indicate being upset emotionally (Chickering 2006).

   When Guatemalan patients complain of dizziness with or without the weakness and fatigue, dehydration should be considered. Guatemalans consume very little water on a daily basis because of the scarcity and poor quality of the accessible drinking water. Adults and children are more likely to drink coffee than water, which can lead to dehydration and gastritis.

6. Skin lesions and/or itching: The most common skin lesions are skin spots (*manchas*) known scientifically as the yeast *Malassezia furfur*. Pruritus may result from dry skin, fungi, scabies, worms, or acquired immunodeficiency disease (AIDS), known in Spanish as SIDA. In

Guatemala, human immunodeficiency virus (HIV), known in Spanish as VIH, is most often transmitted by heterosexual intercourse. Men may contract it through sexual activity with women they meet at bars; wives then contract it from their infected husbands (Chickering 2006). Guatemalan men working in the US without their significant others may be at increased risk for HIV related to engaging in sexual activity without knowledge regarding the diseases, risk factors, or safe sex practices. Prostitutes are known for frequenting migrant farm worker labor camps. Men may contract HIV in the US from a prostitute and, unknowingly, return to Guatemala spreading the virus.

Health care providers should suspect HIV/AIDS when the Guatemalan patient offers the next most common presenting complaints of anorexia and weight loss and their history includes diarrhea, cough, pruritus, painful swallowing, persistent unexplained fever, weakness, or night sweats. The health care provider should not hesitate to ask about risky sexual behaviors with which the patient may have been involved to further determine if HIV/AIDS may be the most likely diagnosis; testing/counseling should follow (Chickering 2006).

7. Diarrhea (*asientos, chorrio*): Causes for diarrhea primarily include shigella, amoebas, and giardia, although it may accompany HIV/AIDS as well (Chickering 2006). Exposure to these pathogens is high in Guatemala, especially in the rural areas. The infrastructure of the communities simply cannot provide water and sewage services necessary for promoting healthy lifestyles because these are nonexistent for most Guatemalans.

   Drinking water in Guatemala is often obtained from polluted waterways and untreated community water systems, so it is non-potable. Either bottled water must be purchased, which is cost prohibitive for most Guatemalans, or water must be boiled for at least 20 min before consumption to ensure pathogenic organisms are eradicated. Factors such as depending on wood for cooking to boil drinking water and the time required for this option lead to low levels of compliance.

   Dehydration, obviously, is the most serious side effect of diarrhea. This is especially true for children and is a leading cause of death for children under 5 years of age in Guatemala (PAHO 2014). Severe dehydration needs to be treated promptly with intravenous (IV) fluids. When IV fluids are not available to administer, oral rehydration solutions should be used. They are available in prepared packets or can be made with ingredients readily available at low cost. Whereas the pathogens mentioned are the most common causes of diarrhea in Guatemala, health care providers who treat immigrant populations in North America should consider these causes when patients have a history of traveling to and from Guatemala.

8. Eye disorders: Wind, dust, sunlight, and cooking smoke contribute to eye pain in Guatemala through conjunctiva irritation or by the stimulation of ptergia (*carnosidades*). Relief from discomfort and pain can be accomplished through the use of cool, moist compresses over the eyes. Moistening the compresses with chamomile tea (*té de manzanilla*) is ideal (Chickering 2006).

   Other painful eye disorders among Guatemalan patients include dacrocytosis and tracoma. Cataracts (even in children), congenital toxoplasmosis, toxic optic neuropathy (usually related to TB treatment), vitamin A deficiency, and actinic allergy are examples of nonpainful eye conditions common to Guatemalans. Vitamin A deficiency, however, is decreasing because sugar has been fortified with vitamin A (Chickering 2006).

   Chickering (2006) identified additional presenting complaints of Guatemalan patients as falling into categories related to menstrual/vaginal, psychiatric, urinary tract infection, pregnancy, pure fever and chills, ear and hearing, chest pain, and male genitourinary issues.

17.6.3 Variations in Drug Metabolism

Although some studies have identified differences in drug metabolism related to racial/ethnic groups, it is difficult to use these for Guatemalans owing to the mixed heritage of many. These variations, however, can influence the determination of a therapeutic dose and affect absorption, distribution, metabolism, and excretion of drugs. A few studies using one subgroup of Hispanics noted that they required lower doses of antidepressants and experienced more side effects than non-Hispanic Whites (Young et al. 2011).

17.7 High-Risk Behaviors

Although Guatemala as a country struggles to maintain traditional and faith-based values, the reality is that increasing numbers of residents seek relief from the hardships of life through abuse of substances. The prevalence of substance abuse is highest among men. Alcohol is the most readily available substance and the most widely misused. In Guatemala, it is not uncommon to see men staggering in public owing to drunkenness and unconscious on sidewalks and roads from excessive intoxication. Accidents are more prone to occur in these settings. Injuries and even deaths occur from falls, confrontations, and collisions with vehicles.

Men immigrating to the US from Guatemala for work may find themselves drinking alcohol excessively, even if they did not prior to migration. This may be due to such factors as (a) the stress of living in another country illegally; (b) being away from family, friends, and support systems; (c) fears of inadequate work and deportation; and (d) illness and being victims of violence and injury. They may work and live closely with other men they hardly know and yet develop a camaraderie based on their shared lifestyle. Their role changes from being actively involved in family to that of a long-distance economic provider, which leads to loneliness, emptiness, and unstructured time when the work day is done. Alcohol can become the antidote.

17.7.1 Health Care Practices

Guatemalans of low socioeconomic status receive little health education, have limited access to health-care, and often believe illness is punishment from God. These factors result in their poor participation in illness prevention practices. In addition, health care in rural Guatemalan areas is mostly provided by government health promoters who focus on episodic illness care rather than preventive measures such as sanitary housing, potable water, balanced diets, and family planning (Icu 2000).

Guatemalan families readily participate in immunization programs for their children, yet do not participate themselves. Adult immunizations such as tetanus, flu, and pneumonia are underutilized by Guatemalans. Moreover, women do not participate in routine screening for breast and cervical cancer, and Guatemalan men do not participate in routine screening for testicular or colon cancer.

Lack of routine health care screenings place Guatemalans at high risk for communicable diseases and cancers identified at advanced stages of disease, severely compromising positive outcomes. Health care professionals need to educate Guatemalans in a family context for disease, illness, and injury prevention; assist with low-cost referrals; and respect their cultural beliefs. When Guatemalan patients learn to view their own health and well-being as an asset to the entire family, they are more willing to participate in preventative health care measures.

17.8 Nutrition

17.8.1 Meaning of Food

Food to Guatemalans signifies physical, spiritual, and cultural wellness. Foods vary among Guatemalans based on cultural traditions and accessibility.

17.8.2 Common Foods and Food Rituals

Garifuna cuisine reflects the Caribbean coast and includes recipes from African ancestors. Common foods include sea bass, flounder, red snapper, tarpon, shrimp, seviche, coconut milk, tomato, onion, garlic, lime, and lemon (Shea 2001).

The Quiche or Kiche, Mayans believe the first humans were made from corn. A ritual occurs in spring before planting corn to bless it (Gall and Hobby 2009). They are breastfed. Corn is highly valued in the Mayan culture. Corn is the chief crop and the basis for many food products and meals. Foods are believed to bring strength, good health, and a spiritual connection to the past. The Mayan diet primarily consists of simple food and not highly spiced (Gall and Hobby 2009). Maize (corn), tamales, black beans, white rice, chicken, squash, tomatoes, carrots, chilies, beets, cauliflower, lettuce, cabbage, chard, leek, onion, and garlic. These foods are used to make tortillas, *atole* (liquid corn drink), *pinol* (chicken-flavored corn gruel), *pepi'an* (chicken stew with squash seeds, hot chilies, tomatoes, and tomatillos [small green tomato]), and *caldos* (soups made of chicken stock and vegetables) (Shea 2001). Green leafy plants are often added to soups and stews. Preferred beverages are hibiscus tea and coffee. Coffee is brewed weaker than that in North America.

Ladino foods reflect their Spanish ancestry and include maize, white rice, black or red beans, beef, chicken, pork, milk, cheese, plantains, carrots, peppers, tomatoes, squash, avocado, cilantro, chilies, onion, garlic, lemon, lime, and chocolate. Common dishes include *arroz con pollo* (chicken with rice), *chile rellenos* (peppers stuffed with pork or beef with carrot, onion, and tomato and fried in egg batter), and *tamales* (chicken or pork in a corn paste steamed in banana leaves or corn husks) (Shea 2001). Again, green leafy plants are often added to soups and stews. Foods are seasoned with toasted squash seeds ground to a powder. It is not uncommon for the poorest of the poor to consume nothing more than black beans and white rice daily.

17.8.3 Dietary Practices for Health Promotion

Guatemalans value corn because it is believed to bring good health. Corn is eaten at every meal, most often in the form of tortillas. Tortillas are made by soaking corn kernels in lime, creating dough by adding animal fat, flattening the dough, and cooking it over an open fire or wood-burning stove. Corn tortillas made in this way are whole grain whereas tortillas purchased in US grocery stores may or not be whole grain. The lime used in home-made tortillas in Guatemala is limestone rather than the fruit lime and is believed to strengthen bones. Consuming foods with limestone, however, has also been associated with dental problems and gallstones (Glittenberg 1994).

17.8.4 Nutritional Deficiencies and Food Limitations

Malnutrition is widespread in Guatemala. The diet of many Guatemalans is low in protein (Steltzer 1983), iron, and vitamin C (Chickering 2006). According to the US Food and Drug Administration (n.d.), two or more incomplete protein food sources consumed together comprise a complete protein and provide many of the essential amino acids necessary for health. Consumption of black beans together with whole grains offer a complete protein source; for example, the whole grain corn used to make tortillas. In addition, black beans and whole grain brown rice comprise a complete protein and many Guatemalans in the US find switching from white rice to whole grain brown rice acceptable.

Some people in Guatemala avoid consumption of milk, probably due to lactose intolerance. Small farms may raise cows to provide fresh cow milk for villagers; however, pasteurized milk sold in market settings is often cost prohibitive and requires refrigeration which many families do not have. Nutrients from milk then are lacking or inadequate in most diets. Whether lactose intolerance is common among Guatemalans is unknown. Research regarding this topic is unavailable.

Zinc deficiency has been found among women and children in Guatemala. Zinc is a vitamin necessary for a healthy immune system. When a pregnant mother is deficient in zinc, the fetus is adversely affected. Lactating mothers deficient in zinc pass the deficiency along to their breast feeding infant or child. This creates an increased risk of death from infectious diseases for infants and children under 5 years of age (Grossman et al. 2015).

Some families encourage their children to drink coffee with sugar when they refuse the poor-tasting, unhealthy drinking water. This practice leads to gastritis, dehydration, and dental caries. Health care providers need to provide nutrition education for clients with consideration of cultural preferences and economic status.

17.9 Pregnancy and Childbearing Practices

17.9.1 Fertility Practices and Views Toward Pregnancy

Guatemalans value life beginning from conception; a baby is a gift from God. For religious reasons, most do not believe in contraception or abortion. A Guatemalan woman may bear ten or more children in her lifetime. Of these, many will die before the age of 5 years (PAHO 2014). Women desiring contraception will not actually seek it because of lack of support from their spouse, family, and church. Access to modern birth control methods is difficult, if not impossible, for many Guatemalan women who may consider alternatives to pregnancy. When women decide to use family planning, the preferred method is Depo-Provera injection. They find this more convenient and undetectable by their spouse or family who disagree with their use of contraceptives.

In Guatemala, Mayan midwives (*comadronas*) deliver 80 percent of all children in the home. The other 20 percent are delivered by medical professionals in a hospital setting. *Comadronas* are "wise" women who are trusted in their community. They feel midwifery is a sacred "calling" by God or a saint who requires special knowledge and rituals. The Ministry of Health has permitted midwives to practice since 1935 (Lang and Elkin 1997). Many midwives do not have formal training. They may have learned the practice from their mother (*mama*) or grandmother (*abuela*) or through "dreams or visions." Prior to delivery, the midwife prays at her own home and then again at the home of the pregnant mother. Some use candles, incense, or religious icons during prayer. The delivery is followed by additional prayers. If the baby dies during delivery, the midwife and family accept it as God's will (Walsh 2006).

While health care reform is changing the practice of midwifery in Guatemala, most births in rural areas continue to be attended by lay midwives with a wealth of experiential learning yet little professional health training (Maupin 2008). Many births by midwives occur in the home, and the midwife is well known and highly respected so the idea that a Guatemalan woman in the US needs to deliver her baby in the hospital may be perceived as foreign and frightening.

Health care providers need to ask the Guatemalan woman whether she is interested in family planning instead of directing the question as "which form of birth control do you prefer?" The latter clearly arises from the health care provider's own values, which may be in direct conflict with those of the patient. Moreover, when the patient does express a desire to learn more about family planning, she may request the session include her husband. Arrangements should be made to accommodate the patient's request. Immigrant pregnant Guatemalan women need to be aware if nurse midwifery services are available in their location in their new country as they may find this preferable for perinatal health care.

17.9.2 Prescriptive, Restrictive, and Taboo Practices in the Childbearing Family

Women in rural Guatemala do not have access to laboratory tests for pregnancy so often become aware they are pregnant based on the symptom of amenorrhea. Women with irregular menstrual cycles are especially at risk for delays in awareness of pregnancy and, early care from a midwife. On the day a Guatemalan woman becomes aware she is pregnant, she and her husband share the news with respected elders of the village. The community becomes the grandparents (*abuelos*) of the unborn child. Godparents are also selected at this time (Menchu 1984).

In the seventh month of pregnancy, the woman introduces her fetus to the environment. She walks the fields and hills and goes through her daily activities, showing and telling her fetus about the life she leads. The mother tells the fetus to be honest and never abuse nature. If someone eats in front of the pregnant woman without offering her food, she believes she will have a miscarriage. Children are permitted at the delivery. The woman's husband, village leaders, and parents of the couple may be present. A single woman must not observe the birth of the baby (Menchu 1984). A midwife and a witch (*brujo*) may both attend the birth. The midwife helps with delivery, while the *brujo* prays for long life, good health, and protection from the evil eye (*mal ojo*). A breech delivery or one in which the baby's cord is around the neck is considered good luck.

Mayan women prefer to deliver at home or a maternity clinic (*casa materna*) rather than a hospital. Squatting is the birthing position they find best facilitates delivery. Following delivery, the placenta must be burned, not buried, because it is disrespectful to the earth to do so. The placenta can be burned on a log and then the ashes used for a steam bath, *temascal* (Menchu 1984). *Temascal* is a sauna (steam filled) room or building at the Guatemalan homesite using an open fire for heat and water for creating steam for bathing and believed to restore one's health using herbs for cleansing the skin or for inhalation.

Traditionally, in celebrating the birth of a baby, the villagers slaughter a sheep, for the poor a pig or chicken. The mother and baby are kept separated from others for 8 days. When the baby is born, its hands and feet are bound for 8 days. This signifies that they are meant for hard work, not for stealing, and they are not meant to have things the rest of the community does not have (Menchu 1984). Babies are carried on their mother's back. They are breastfed on demand and mothers typically continue breastfeeding until the child reaches the age of 5 years. Moreover, they may be breastfeeding a new baby while continuing to breastfeed a toddler.

Many Guatemalan women believe colostrum is harmful to their baby so will not begin breast feeding until colostrum has been expressed from the breasts and "milk" flow begins. In addition, many women are malnourished and feel they have inadequate breast milk for the health needs of their baby and prefer to supplement breast feeding with infant formula which, may in turn, lead to a reduced quantity of breast milk available for the baby. Formula is costly.

During the first eight postpartum days, friends and extended family bring food, clothing, small animals, or wood as gifts for the newborn's family. They also offer their services, like carrying water or chopping wood. The family of the newborn does nothing for these 8 days. All their needs are taken care of by others (Menchu 1984).

After 8 days, the newborn's hands and feet are untied. While the newborn is in bed with the mother, all community members visit to officially welcome him or her. The baby is told he or she is made of corn. A bag with garlic, lime, salt, and tobacco is hung around the baby's neck. A red thread is used to tie the umbilical cord to protect the baby, provide strength, and denote respect for the ancestors. If the baby is a female, the midwife pierces her ears at birth (Menchu 1984). Guatemalan parents typically sleep with their infant which US health care providers discourage as it is found to increase risk for suffocation by 50% (Moon 2019). Guatemalan mothers in the US have found the recommendation by health care providers for an infant basket, sometimes referred to as a "Moses" basket, acceptable as parents and infant can co-sleep yet suffocation risk is decreased.

17.10 Death Rituals

17.10.1 Death Rituals and Expectations

Death rituals among Guatemalans vary depending on traditional and religious beliefs.

Guatemalans grow up experiencing far more death than most Americans. They witness infants and children dying from malnutrition and disease; women experiencing perinatal death; parents and grandparents dying from injuries, illness, and violence; and loved ones dying because the health care they needed was too far away or was too expensive. The family may decide the cost for treatment of one family member is too much and decide against it because of the financial strain on the entire family.

When a person is seriously ill, she or he is usually cared for at home. Family members and the community provide assistance and support. When family must take a seriously ill family member to a hospital, extended family and the community help care for members left at home. Often, hospitals are such a great distance that the ill person dies before she or he arrives or during the hospitalization. Prayers are offered for the sick. Rituals and ceremonies are offered to prepare the body and spirit of the dying for death.

17.10.2 Responses to Death and Grief

When death occurs in Guatemala, it is customary to place the deceased in a simple wooden coffin or casket and conduct a funeral. If the family cannot afford a coffin, the community will assist in providing one. In small villages, the entire community is present for the funeral. The casket may be carried by key family members through town from the church to the cemetery while onlookers show their respect, mourn, and offer flowers. Graves are decorated with flowers on All Saint's Day in memory of the deceased. Some Guatemalans relate their illness to "punishment" or impending death to "God's will" and refuse an intervention or heroic measures prior to death to reverse the outcome. Guatemalan immigrants to the US are often unaware of technological services and advanced directives available in acute care settings. They do not hold beliefs against blood transfusions and decisions related to organ donations, and living will. Do not resuscitate options are unfamiliar to most and may be frightening to discuss with the Guatemalan patient and family. Health care providers need to demonstrate compassion in discussing these matters and respect when a decision is reached.

When death of a Guatemalan occurs in the US, the family may request repatriation because it is important for the final resting place to be the home country. Often, immediate and extended family and social service agencies pool their resources to send the body of the deceased home.

Guatemalans believe in burial; they do not practice cremation. At Indian funerals, the Mayan priest may "spin the coffin at the grave to fool the devil and point the spirit of the deceased toward heaven" (Menchu 1984). Yellow flowers are placed at the grave; yellow is the color for mourning. Food is placed at the head of the coffin for the spirit of the departed. Church bells are rung to gain favor with the gods. Ladinos mourn the dead by wearing black, but the Maya do not believe in this practice (Menchu 1984).

17.11 Spirituality

17.11.1 Dominant Religion and Use of Prayer

Approximately 65–80% of Guatemalans are Roman Catholic. Regular church attendance is not possible for many due to long distances to churches and a shortage of priests. Most priests in Guatemala are foreigners (Gall and Hobby 2009). Village leaders will often hold weekly church services; however, confessions, marriages, and baptisms may only be conducted by a priest. Priests serve a large geographic region, visiting churches on a rotating schedule. Depending on availability, the wait time for a priest may be weeks or months.

Prior to the Spanish conquest, the Maya practiced a religion based on gods associated with the sun, moon, and other natural phenomena. The causes of natural disasters were related to punishments from heavenly beings. They believed animal spirits (*nahuales*) inhabited the human spirit (Shea 2001). *Maximon* is a Mayan saint popular among Maya and Ladinos. He is believed to represent ordinary Guatemalans, has supernatural powers, and produces witches (*brujos*) and shaman (*zajorines*) (Sexton 1999).

When the Spanish brought Roman Catholicism to Guatemala, some Maya converted to Christianity. Others continue to practice their Mayan religion. Many Guatemalans combine beliefs and practices of the two. In some cases, Guatemalans integrate aspects of Catholicism into their lives while continuing to believe in the spirituality of their ancestors in private (Shea 2001).

Two practices influenced by the Spanish are *guachibal* and *cofradia*. *Guachibal* involves the practice of keeping an image of a Christian saint in the home and celebrating on the particular saint's day. *Cofradia* refers to a "religious brotherhood" that serves to maintain the "cult" of a particular saint. *Cofradias* consist of dues-paying members and elected leaders. In addition to religious activities, they often serve needy persons in the community by visiting the sick and paying for funerals (Shea 2001).

Protestantism arrived in Guatemala in the nineteenth century. Today, there are an increasing number of evangelical Christian religions in Guatemala. The Christian holidays of Holy Week, All Saint's Day, and Christmas are celebrated by Catholic and Protestant Guatemalans (Shea 2001). Some Guatemalans who continue to practice the traditional Mayan religion exclusively are finding their children and grandchildren prefer Catholicism or Protestantism. Health care providers need to determine whether the Guatemalan patient has a religious preference and demonstrate respect and sensitivity for his or her beliefs. It's important to keep in mind not all patients identifying a religious preference, believe or follow all practices associated with the religion.

17.11.2 Meaning of Life and Individual Sources of Strength

Family provides Guatemalans with meaning in their lives. Life revolves around the nuclear and extended family. Spirituality helps to explain life and the circumstances faced by Guatemalans. For example, whether Catholic, Protestant, or traditional Mayan, many believe that life's events happen for a reason. The reason may be attributed to favor from God or gods when positive experiences occur and to punishment or disfavor from God or gods when negative events occur. Some feel nothing can be done, or should be done, to change the outcome of these experiences. This belief is referred to as ***fatalism***. Illness, disease or injury may be attributed to supernatural causes.

17.11.3 Spiritual Beliefs and Health Care Practices

When illness occurs, many Guatemalans turn to their faith for strength, wisdom, and hope. Health care providers may be uncomfortable with patients who have a statue of a patron saint, picture of a saint, or candles burning while prayers are being said. The family may pray together; they need time and space to do so. They may even ask the health care professional to join prayer, which is acceptable if the provider wishes. Praying with the family promotes confidence in the relationship. The patient and family feel the compassion and respect such actions demonstrate. The health care provider may also help arrange for spiritual care services for the patient and family through their church or community and hospital resources.

17.12 Health Care Practices

17.12.1 Health-Seeking Beliefs and Behaviors

Transculturation is continually occurring in Guatemala. Characteristics between Ladinos and Indians used to vary greatly; Ladinos were Spanish speakers, wore Western-style clothing, practiced Catholicism, were better educated, worked in non-agricultural occupations, were economically better off, lived in superior housing with sanitation, had a better diet, and were healthier. Indians more often than not spoke a Mayan dialect; wore traditional woven clothing (*traje*); worked in agriculture; used primitive technology; and maintained social, political, and religious life through *cofradias* (Woods and Graves 1973). Today, the distinction between the Ladino and the Indian in Guatemala is less clear. For example, more Indian men speak Spanish, wear Western clothing, and receive an education.

Health care seeking among Guatemalans generally occurs by first seeking advice from a mother, grandmother, or other respected elder. When unsuccessful the family seeks health care from folk healers. Modern medical care may be sought as the last resort. Several medications, including antibiotics, purchased in the US only with a prescription are available without a prescription in Guatemala. The Guatemalan presents symptoms to a pharmacist, the pharmacist recommends medication and medication is purchased without an examination by a medical provider. Even injectable medications are readily available through a pharmacist. The preferred mode of treatment among Ladinos is medication administered by hypodermic injection. For example, if an infant has a cold, Ladinos believe an injection is more effective than oral medication. If someone has the flu, an intravenous (IV) infusion is preferred. Intramuscular medications are preferred to those taken orally (Kunkel 1985).

Many Guatemalans are fearful of hospitals. In Guatemala, when hospital care is necessary, patients are often seriously ill and hospitalization results in death, which perpetuates the belief that "hospitals are places where patients go to die". Most rural Guatemalans reside hours away from a hospital so even if transportation is available during an emergency, it is likely the ill person will die en route.

17.12.2 Folk and Traditional Practices

Most Guatemalans engage in folk medicine practices and use a variety of prayers, herbal teas, and poultices to treat illnesses. Many of these practices are regionally specific and vary between and among families. Lower socioeconomic groups and well-educated upper and middle socioeconomic Guatemalans practice traditional and folk medicine to some degree. Many of these practices are harmless, but some may contradict or potentiate therapeutic interventions. The providers must specifically ask their Guatemalan patients whether they are using folk medicine without appearing judgemental or the information may be withheld.

Physical or mental illness may be attributed to an imbalance between the person and the environment. Influences include emotional, spiritual, and social state, as well as physical factors such as humoral imbalance expressed as either too much hot or cold. According to this theory, many diseases are caused by a disruption in the hot-and-cold balance of the body. Thus, eating foods of the opposite variety may either cure or prevent specific hot-and-cold illnesses and conditions.

To provide culturally competent care, health care providers must be aware of the hot-and-cold theory of disease when prescribing treatment modalities and when providing health teaching. As health care providers, it is important to understand that if people of Guatemalan heritage believe in the hot-and-cold theory, it does not indicate they do not believe in or use professional Western practices. Unless a level of trust and confidence is maintained, Guatemalans following these beliefs may not express them to health care providers.

Folk practitioners are consulted for several notable conditions. These are not confined to Guatemala, but commonly in Latin America. Evil eye (*mal de ojo*) is a folk illness that occurs when one person (usually older) looks at another (usually a child) in an admiring fashion. Another example of *mal de ojo* is if a person admires something about a baby or child, such as beautiful eyes or hair. Such eye contact can be either voluntary or involuntary. Symptoms to the admired one following the encounter are numerous, ranging from fever, anorexia, and vomiting to irritability. It is believed the "spell" can be broken if the person doing the admiring touches the person admired while it is happening. Children are more susceptible to this condition than women, and women are more susceptible than men. Another method of treating the evil eye is to use an egg. The egg is passed over the child's body. Commonly a prayer is said along with this and then the egg is placed in bowl beneath a pillow and left there during the night and check to see if the white is foggy in the morning, if it is, the child was affected by the evil eye. This method also cures the evil eye at the same time.

Fright (***susto***), sometimes referred to as soul loss, is believed to be caused by experiencing one or more frightening events or stressors. As an example, susto may occur after an individual is present for or the victim of a traumatic event such as rape, murder, or an unexpected traumatic injury or death in the family. The client may present with specific physical symptoms, a cluster of vague symptoms including lack of appetite, weakness, fatigue, loss of motivation, evidence of psychiatric impairment, or inability to fulfil their roles. Individuals experiencing *susto* believe it is very serious and, if untreated, believe it will result in a deterioration in health to the point of death (Rubel et al. 1984). Whether or not the client identifies susto as the cause of their condition, health care providers benefit from answers to the following questions when a folk illness is suspected to understand the meaning and significance of ill health from the client's perspective (explanatory model):

(1) "What has happened? (2) Why has it happened? (3) Why has it happened to me? (4) Why has it happened now? (5) What would happen of nothing were done about it? (6) What are the likely effects on other people (family, friends, neighbors, employers, etc.)? (7) What should I do about it- or to whom should I turn for help (self-treatment, consultation with lay advisors, folk healers, or health professionals)?" (Helman 1990).

17.12.3 Responsibility for Health Care

Guatemalans often delay seeking health care until they are incapacitated by illness, disease, or injury. Many times, they are unaware of the dangers associated with working in agriculture in the US. They may be exposed to pesticides and dangerous equipment without proper training. Although the government has specific laws in place to protect farm workers, enforcement is limited. Companies may not tell the workers the dangers or the workers may not understand due to language differences (Calvert et al. 2009). It is important for health care providers caring for Guatemalan farm workers and those who work in hazardous occupations to be aware of the chemical exposures, laws, and resources available in order to adequately educate, provide care for, and advocate for the patient. For example, a client may present with vague symptoms such as malaise, muscle weakness, dizziness, and diaphoresis. Identifying occupational exposures the client may have had is critical as these symptoms can result from acute pesticide poisoning. These mild symptoms may progress to severe central nervous system depression, seizures, coma, and, eventually, death.

17.12.4 Barriers to Health Care

Only three out of every ten Guatemalans living below the poverty line in Guatemala seek professional medical services (PAHO 2012). According to Purnell (2013), Guatemalans experience barriers to health care, whether in Guatemala or in the US, that include:

1. **Availability:** The hours many health services are offered are inconvenient for Guatemalan patients who cannot afford to take time off work.
2. **Accessibility/reliability:** Transportation is not available to access health care services.
3. **Affordability:** Most Guatemalans do not have health insurance and are limited to health service agencies offering low-cost care.
4. **Appropriateness:** Low-cost health care offers limited services; dental, women's, or pediatric care may not be available.
5. **Accountability:** Health care providers lack education regarding Guatemalan cultures,

languages, and lifestyles required to offer culturally competent care.

6. **Adaptability:** Many health care settings do not offer one-stop shopping in which the patient may be seen for dental care and a mammogram the same day in the same location.
7. **Acceptability:** Many health care providers do not assess or adequately assess language and literacy levels necessary to offer appropriate patient education and materials to Guatemalan patients.
8. **Awareness:** Many times, Guatemalan patients are unaware of the services available in the area in which they live.
9. **Attitudes:** Sometimes health care providers convey negative attitudes toward the Guatemalan patient, making it unlikely she or he will return to that setting for care. In addition, the patient may prefer to avoid any health care setting due to the negative experience.
10. **Approachability:** Guatemalan patients may not feel welcome in the health care setting because they interpret verbal and nonverbal communication from providers as lacking in care and compassion.
11. **Alternative practices and providers:** Guatemalan patients may prefer to integrate modern medicine with traditional therapies; yet, the health care provider may be resistant to this.
12. **Additional services:** The health care setting may lack child care; the Guatemalan family that does not have extended family members in the US to assist is required to bring children and older family members with them to the visit, which is disruptive.

17.12.5 Cultural Responses to Health and Illness

Guatemalans tend to view health and illness in relation to their ability to perform duties associated with their roles. If women are functioning in their role of caring for the home and family, and men are functioning in their employment, then they feel "healthy". Aches, pains, and minor illnesses that do not prevent functioning are tolerated. When an illness prevents normal functioning required for their roles, then Guatemalans view it more seriously.

The cause of debilitating illness or disease may be viewed as punishment from God rather than lack of prevention or early detection. Sometimes, early warning signs of illness or disease are ignored in hopes they will go away on their own. When symptoms persist, fear may keep the Guatemalan patient from seeking medical attention.

When a member of a Guatemalan family needs to be cared for, others gladly comply. This occurs for reasons of advanced age, illness, and mental or physical disability, among others. Residential homes for persons with these conditions are not readily available in Guatemala and are usually cost prohibitive for families living in the US and other countries. Family members would rather care for their loved one at home if at all possible. Health care providers may demonstrate respect for the cultural values of Guatemalan patients by providing community resources that enable them to adequately care for a family member at home if they so desire.

17.12.6 Blood Transfusions and Organ Donation

Blood transfusions and organ donation are not common in Guatemala, and with many living outside of the realm of modern medicine, residents may be uneducated regarding the medical situations in which they are indicated, benefits, and risks. Some Guatemalans fear venipuncture because they believe taking blood from the body leaves it without enough to keep them strong and healthy. Catholic and Protestant Guatemalans do not hold religious beliefs prohibiting blood transfusions. Questions related to organ donation will be puzzling and elicit fear and anxiety. The Guatemalan patient may think the health care provider is asking him or her to consent to organ donation because he or she is going to die rather than understanding the context to which the question applies. The same response may occur when the Guatemalan patient is asked to complete a living will, do not resuscitate, and power of attorney for health care forms. Health care providers need to carefully assess the level of education and understanding of the Guatemalan patient regarding these issues and use an interpreter who is competent in the language and culture of the patient in order to promote successful communication.

17.13 Health Care Providers

17.13.1 Traditional Versus Biomedical Providers

Three distinct health care systems exist in Guatemala: modern medicine, Ladino folk medicine, and Indian folk medicine. Modern medicine refers to health care provided by educated physicians and nurses. Ladino folk medicine is provided by Ladino pharmacists, spiritualists, and lay healers (***curanderos***). Mayan Indians seek medical care from Mayan *shaman, bonesetters, herbalists,* and traditional midwives (***comadronas***). When Ladinos and Mayan Indians have access to modern medicine, the utilization of these services increases. However, it is estimated in Guatemala only three of every ten residents living below the poverty line seek modern medical care (PAHO 2012). In the US, Guatemalans may seek care from Mexican traditional healers.

17.13.2 Status of Health Care Providers

Guatemalans have great respect and admiration for health care providers who are viewed as authority figures with clinical expertise. Guatemalans expect their health care provider to have the appearance and manners of a professional (Purnell 2001). When this is not the case, Guatemalans lose confidence in the provider.

Guatemalans are very private and are not accustomed to discussing issues and concerns openly. It may take a while to develop the trust and rapport with the provider necessary for them to share. They fear disclosure may result in deportation or rejection. Patients also fear confidentiality will not be maintained in the health care setting.

Health care providers who are most successful in caring for Guatemalan patients practice *personalismo*, respect, and genuine compassion in their approach to care.

Moreover, Purnell (2001) found the following behaviors and comments by health care providers best conveyed respect to their Guatemalan patients:

1. Greets me when I come in
2. The way the doctor/nurse talks to me (denotes respect)
3. Asks questions about what bothers me
4. Says things to make me feel better
5. Explains things to me

Guatemalan women are usually very modest. They may refuse to discuss personal issues or receive an examination by a male health care provider. During childbirth in Guatemala, women prefer to remain fully dressed raising their skirts (*faldas*) just enough for birth of their baby. A male Guatemalan patient may refuse a female health care provider. Because Guatemalans dislike conflict, they may not actually refuse care; yet they withhold personal information due to discomfort with the health care provider. Incorporating these preferences into the encounter with Guatemalan patients enhances the development of relationships that result in effective and meaningful care.

References

Calvert GM, Karnik J, Mehler L, Beckman J, Morrissey B, Sievert J et al (2009) Acute pesticide poisoning among agricultural workers in the United States, 1998–2005. Am J Ind Med 51(12):883–898. https://doi.org/10.1002/ajim.20623

Central Intelligence Agency (2015) World factbook: Guatemala literacy. https://www.cia.gov/library/publications/the-world-factbook/geos/gt.html

Central Intelligence Agency (2018a) World factbook: Guatemala population. https://www.cia.gov/library/publications/the-world-factbook/geos/gt.html

Central Intelligence Agency (2018b) World factbook: Guatemala infant mortality rate & life expectancy at birth. https://www.cia.gov/library/publications/the-world-factbook/geos/gt.html

Central Intelligence Agency (2019) World factbook: Guatemala City population. https://www.cia.gov/library/publications/the-world-factbook/geos/gt.html

Chickering WH (2006) A guide for visiting clinicians to Guatemala: common presenting symptoms and treatment. J Transcult Nurs 17(2):190–197

Gall TL, Hobby J (eds) (2009) Worldmark encyclopedia of cultures and daily life: Vol. 2, Americas. Gale Research, Detroit, MI, pp 250–254

Glittenberg J (1994) To the mountain and back: the mysteries of Guatemalan highland family life. Waveland Press, Lake Zurich, IL

Global Sherpa (n.d.) Garifuna people, history, and culture. http://globalsherpa.org/garifunas-garifuna/

Grossman VM, Turner BS, Snyder D, Stewart R, Bowen T, Cifuentes AA, Cliff C (2015) Zinc and vitamin supplementation in an under-5 indigenous population of Guatemala: influence of lay health promoters in decreasing incidence of diarrhea. J Transcult Nurs 26(4):402–408

H-2A Temporary Agricultural Program (2019). https://www.farmers.gov/manage/h2a

Hargrove I, Anderson K, Retzlaff C, Garcia D (2009) Working with Hispanic indigenous migrant men from Mexico and Guatemala to promote immunization. Migrant Clin Network Streamline 15(2). file:///E:/Guatemala%20Chapter/ImmunizationArticle.pdf

Helman CG (1990) Culture, health and illness, 2nd edn. Wright, London, pp 95–96; p 288

Icu H (2000) Health care reform in Guatemala. http://phmovement.org/pha2000/stories/icu.html

Information Please Almanac (2019) Guatemala. http://www.infoplease.com/encyclopedia/

Kunkel P (1985) Paul Kunkel: Guatemala journal part 2. Wash Nurse 15(4):34–35

Lang JB, Elkin ED (1997) International exchange: a study of the beliefs and birthing practices of traditional midwives in Guatemala. J Nurse Midwifery 42(1): 25–31

Maupin JN (2008) Remaking the Guatemalan midwife: Health care reform and midwifery training programs in highland Guatemala. Med Anthropol 27(4): 353–382

Menchu R (1984) I, Rigoberta Menchu: an Indian woman in Guatemala, vol 13. Verso, London, p 18

Moon RY (2019) How to keep your sleeping baby safe: AAP policy explained. https://www.healthychildren.org/English/ages-stages/baby/sleep/Pages/A-Parents-Guide-to-Safe-Sleep.aspx

National Center for Farmworker Health (NCFH) (2017) A profile of migrant health: 2017. http://www.ncfh.org/fact-sheets%2D%2Dresearch.html

National Center for Farmworker Health (NCFH) (2018) Facts about farmworkers. http://www.ncfh.org/uploads/3/8/6/8/38685499/fs_demographics_2018.pdf

National LGBT Health Education Center (2015) Promoting health care access to lesbian, gay, bisexual, transgender (LGBT) farmworkers. https://www.lgbthealtheducation.org/wp-content/uploads/Promoting-Health-Care-Access-to-LGBT-Farmworkers-Final.pdf

Pan American Health Organization (PAHO) (2012) Health in the Americas, 2012 edition: Country volume. https://www.paho.org/salud-en-las-americas-2012/index.php?option=com_docman&view=download&category_slug=hia-2012-country-chapters-22&alias=132-guatemala-132&Itemid=231&lang=en

Pan American Health Organization (PAHO) (2014) Guatemala: Leading causes of death. http://www.paho.org/data/index.php/en/indicators-mortality/mnu-lcd-en.html

Perez S (2019) Aldo Dávila set to be Guatemala's first openly gay congressman. https://www.apnews.com/6ee9f7e7a64944e68531fd02a01a3ce1

Purnell L (2001) Guatemalan's practices for health promotion and the meaning of respect afforded them by health care providers. J Transcult Nurs 12(1): 40–47

Purnell L (ed) (2013) Transcultural health care: a culturally competent approach, 4th edn. F. A. Davis, Philadelphia, pp 39–40

Rubel AJ, O'Nell CW, Collado-Ardon R (1984) Susto: a folk illness. University of California Press, Berkley, CA

Sabin M, Sabin K, Kim HY, Vergara M, Varese L (2013) The mental health status of Mayan refugees after repatriation to Guatemala. https://www.scielosp.org/scielo.php?pid=S1020-49892006000300004&script=sci_abstract

Sexton JD (1999) Heart of heaven, heart of earth and other Mayan folktales. Smithsonian Institution Press, Washington, DC

Shea ME (2001) Culture and customs of Guatemala. Greenwood Press, Westport, CT

Steltzer V (1983) Health in the Guatemalan highlands. Anthropol Q 57(3):141–142

United States Agency for International Development (USAID) (2018) Guatemala: gender analysis final report. https://www.usaid.gov/sites/default/files/documents/1862/FINAL_FINAL_Gender_Anaylsis_GITA_GUATEMALA_Sept_14_2018.pdf

United States Bureau of the Census (2017) Population: Hispanics. https://www.census.gov/newsroom/facts-for-features/2017/hispanic-heritage.html

United States Food & Drug Administration (n.d.) Protein. https://www.fda.gov/nutritioneducation

Walsh LV (2006) Beliefs and rituals in traditional birth attendant practices in Guatemala. J Transcult Nurs 17(2):148–154

Wellmeier N (1998) Ritual, identity, and the Mayan diaspora. Garland, New York

Woods CM, Graves TD (1973) The process of medical change in a highland Guatemalan town. University of California, Los Angeles

Young HN, Dilworth TJ, Mott DA (2011) Disparities in pharmacists' patient education for Hispanics using antidepressants. J Am Pharm Assoc 51(3):338–408

People of Haitian Heritage

18

Jessie M. Colin

18.1 Introduction

Haiti, located on the island of Hispaniola between Cuba and Puerto Rico in the Caribbean, shares the island with the Dominican Republic. Unemployment and underemployment exist; more than two-thirds of the labor force do not have formal jobs owing to the marked decrease in assembly sector jobs. In addition, Haiti's economy suffered a severe setback when a magnitude 7.1 earthquake devastated its capital city, Port-au-Prince, in January 2010, fueling additional migration.

18.2 Overview, Inhabited Localities, and Topography

18.2.1 Overview

With a population of 1,106,777 million inhabitants, Haiti covers an area of 27,750 square kilometers (10,714 square miles), about the size of the state of Maryland in the United States (CIA World Factbook 2020). In 1492, Christopher Columbus landed on the island and named it *Hispaniola*, which means "Little Spain." Haiti, or *Ayti*, meaning "land of mountain," was given its name by the first inhabitants, the Arawak and the Caribe Indians. Before 1492, there were five well-organized kingdoms: the Magua, the Marien, the Xaragua, the Managua, and the Higuey (Dorestant 1998). Two-thirds of Haiti contains mountains, great valleys, and extensive plateaus; small plains mark the rest of the country.

The capital and largest city, Port-au-Prince, has a population of over 2,774,000.

Despite a decline in both monetary and multidimensional poverty rates since 2000, Haiti remains among the poorest and most unequal countries in Latin America. Ten years after the 2010 earthquake, poverty was still high, particularly in rural areas. Today, more than one in two Haitians is poor, living on less than $2.41 a day, and one person in four was living below the national extreme poverty line of $1.23 a day (World Bank 2020). The current exchange rate is $1 US dollar for 94 *gourd*, the Haitian monetary unit. The national poverty line is 58.6 percent and the abject poverty is 27.4 (CIA World Factbook 2020).

After the earthquake, the GDP per capita was $1200. Prior to the earthquake, two-thirds of Haitians depended on the agricultural sector, mainly small-scale subsistence farming, and are still vulnerable to damage from frequent natural disasters exacerbated by the country's wide-

This chapter is an update of a previous edition written by Jessie Colin.

J. M. Colin (✉)
Barry University, Ft. Lauderdale, FL, USA
e-mail: jcolin@barry.edu

© Springer Nature Switzerland AG 2021

L. D. Purnell, E. A. Fenkl (eds.), *Textbook for Transcultural Health Care: A Population Approach*,
https://doi.org/10.1007/978-3-030-51399-3_18

spread deforestation. Progress is evident, but much remains to be done. Extreme poverty declined from 31 to 24 percent between 2000 and 2012, and there have been some gains in access to education and sanitation, although access to basic services is generally low and is characterized by important inequalities. Urban areas have relatively fared better than rural areas, reflecting more non-agricultural employment opportunities, larger private transfers, more access to critical goods and services and narrowing inequality compared to rural areas (World Bank 2015). **According to the** CIA World Factbook (2015), **Haiti has a large wealth disparity**; it ranks fourth for income inequality. The top 20 percent of households hold 64 percent of the total wealth in the country.

Haitians suffer from lack of or unreliability of electricity. According to US AID (2020), the two key energy issues for Haiti are a broken electricity sector and a dependency on charcoal. This means that a large percentage of the population is without power; those who do have it, the power is unreliable power. The issue remains at a standstill as a quarter of the population did not have power before the 2010 earthquake and that figure remains the same to this day.

U.S. economic engagement under the Haitian Hemispheric Opportunity through Partnership Encouragement (HOPE) Act, passed in 2006, has boosted apparel exports investment by providing tariff-free access to the United States.

The infant mortality rate is high, with 54.02 deaths per 1000 live births; the average life expectancy is 62.17 years (CIA World Factbook 2020); and in 2008, only 70 percent of the urban population (50 percent in rural areas) had access to improved drinking water sources. The World Health Organization (WHO) estimated that prior to the disaster in 2010, diarrheal diseases accounted for 16 percent of deaths among children less than 3 years of age. In October 2010, an outbreak of cholera added to the devastation of the earthquake, killing an additional 3000 people and infecting approximately 130,000 more (BBC News 2011).

The Haitian population in the United States is not well documented; this may be because of the U.S. Census Bureau's inability to track the large numbers of undocumented immigrants. According to the 2010 census, over 830,000 Haitians, or 0.3 percent of the population, live in the United States. Most of them live in Florida, New York, Massachusetts, New Jersey, and Connecticut. However, some Haitian leaders and activists believe that close to 1.5 million Haitians live in the United States. An additional 122,000 live in Canada, of which 90% live in Quebec (Statistics Canada 2006). Haitians, like other ethnic groups, are very diverse. They come from urban and rural Haiti and represent all socioeconomic classes. Factors affecting Haitians' acculturation and assimilation include variant cultural characteristics (see Chap. 2).

Beyond political instability, endemic poverty and natural disasters, including a devastating 2010 earthquake, have influenced the rate of migration to the United States. Following the earthquake, the US government had added Haiti to the list of Temporary Protected Status (TPS) designation allowing them to remain in the United States. TPS expired on January 22, 2018 but was extended to January 2, 2021. More than 58,000 Haitian immigrants already in the United States prior to the 2010 earthquake have been granted TPS, which provides work authorization and relief from deportation.

18.2.2 Heritage and Residence

Before the time of Columbus, the various indigenous tribal groups intermarried. With the arrival of Europeans, and then Africans, the people of Haiti became more diverse. Today, Haitians range from light- to dark-skinned, and social identity is shaped by sharp class stratification and color consciousness.

In 1697, Haiti came under French rule. By the end of the eighteenth century, the slave population numbered 500,000. In 1791, a slave insurrection broke the chain of slavery, and on January 1, 1804, Haiti gained its independence from

France. The French plantation owners were removed and replaced by the generals of the indigenous Haitian Army, which ruled mercilessly (Louis-Juste 1995). Agricultural workers and peasants were trapped in a semifeudal system: They were exploited by landowners, terrorized by the section chiefs of police, and forced to obey laws explicitly. The coffee fields of the peasants served as the primary source of revenue for the government coffers, thereby guaranteeing all government debt payments between 1826 and 1932 (Louis-Juste 1995). These harsh conditions did not prevent the peasants from rising up against injustice and exploitation, as evidenced by the Goman uprising in 1820, the Acaau in 1880, and the peasant movement of Jean Rabel (Louis-Juste 1995).

Haitian immigrants have a sense of national pride, including a high level of self-esteem regarding their blackness, although in both public and private discourse, they may focus on color and class division—two painful wedges within Haitian society.

Haiti's independence from France in 1804 did not resolve the division among the descendants of French colonists, the African slaves, and the core of the population who were largely of African descent and culture. Many members of the upper class used the markers of **mulatto** (color), the French culture, and the French language to differentiate themselves from the lower class, who were mostly Black and **Creole** and spoke a predominantly African language.

Ti Manno, a Haitian singer who migrated to New York, used satire and irony to expose and deride the type of thinking that divides Haitians in Haiti and abroad. The following lyrics depict the turmoil and struggle that promote the division within the Haitian society (Jean-Baptiste 1985):

- *The Black Man*
- *Neg Kwens dil pa Kanmarad neg Brooklyn.*
- *Neg Potopwens dil pa anafe ak neg pwovens.*
- *Mon Che se-m nan fe yon ti pitit.*
- *M'rayi ti pitit la*
- *A fos li led.*
- *Li nwa tankou bombon siwo.*
- *Nen-l pa pwenti.*
- *Ti neg mwe ala nou pa gen chans o.*
- *La vi nou toujou red o.*
- *Nou deyo, pi red.*
- *Se neg nwe cont milat o.*
- *Nou deyo nap soufri.*
- *Nou lakay se pi red.*

Translation:

- Haitians in Queens feel superior to those who live in Brooklyn.
- Haitians in Port-au-Prince despise those who live in the provinces.
- My dear, my sister had a little baby.
- I hate this little kid.
- This baby is ugly.
- He is as dark as sugarcane syrup cake.
- His nose is not pointy.
- We Haitians, we are so unlucky.
- Life is always hard for us.
- Away from home we suffer more.
- It's black against mulatto.
- Abroad we suffer.
- At home it is even worse.

Despite independence, colonial prejudices about skin color have persisted. Internal social rivalries and the scale of Haitian mobility are tied to a European color, race, and class model. This model relates to skin pigmentation, hair texture, the shape of the nose, and the thickness of the lips. Whereas the structure of Haitian society continues to be built on a neo-colonial model, relationships based on color are extremely complex. For example, dark skin color tends to be associated with underprivileged status. Although more black-skinned people have entered the circle of the privileged, most Blacks are poor, underprivileged, and unemployed.

Haiti defines itself as a black nation. Therefore, all Haitians are members of the black race. In Haiti, the concept of color differs from the concept of race. The Haitian system has been described as one in which there are no tight racial categories but in which skin color and other phenotypic demarcations are significant variables.

In the 1940s, a Black middle class emerged in Haiti and claimed to represent the majority. The development of this class and its rhetoric served as a springboard for Francois Duvalier, a rural physician who was elected president for a 4-year term in 1957. In 1964, he became president-for-life, using the issue of black empowerment and a promise to eliminate the color and class privileges of the mulattos. By the late 1970s, a group of dark-skinned, primarily American-educated and English-speaking technocrats had attained positions of prominence and influence in the government. However, the mulatto retained social prominence and color continued to play a major role in the perception of class in Haiti.

18.2.3 Reasons for Migration and Associated Economic Factors

Haitian immigration and travel to the United States have continued for many years. Most, but not all, of those who emigrated were members of the upper class. Before 1920, Haitians traveled to North America and Europe only for educational purposes. In 1920, the United States occupied Haiti, and the first wave of Haitian migration to North America soon followed. Over the next decade, more than 40,000 Haitian peasants were forced to go to Cuba and the Dominican Republic to cut sugarcane in the *bateys* (plantations). Haitian land was taken and used for apple and banana plantations, and many acres of land throughout Haiti were controlled by the United States (Haiti: Early History to Independence 2007).

The late 1950s showed signs of weakness in Haitian agriculture. The peasants started leaving the provinces in search of work and a better life. Migrating to the capital, Port-au-Prince, they established Lasaline, today known as Cité Soleil, is the first slum of Port-au-Prince (Aristide 1995). Today, over 2 million people live in and around the capital, many in large slums. A significant turning point in Haitian migration occurred in 1964 when Duvalier declared himself president-for-life. As a result of his government, many Haitians began fleeing the island. These immigrants were primarily relatives of politicians who opposed the political philosophy of Duvalier. When Duvalier died in 1971, his son, Jean Claude (a.k.a. "Bébé Doc"), age 19, was appointed president-for-life. In addition, during this era, Haiti was suffering from economic deprivation, which motivated a major exodus of urbanites and peasants. Because many Haitians were unable to pay for their transportation, passports, and visas, some covertly emigrated to the United States in small sailboats.

From 1980 until recently, Haitian immigrants have been divided into two groups: those who have arrived in the United States legally and those who have entered through the underground. An explosion of immigration took place in 1980, in part because of a short-lived (April to October) change in U.S. immigration policy during the period of the Mariel boat lift from Cuba. The influx of Cuban refugees required that a special status be created by the State Department called "Cuban-Haitian entrant: status pending." According to Health and Rehabilitation Services, Haitian refugees were included in this status to prevent the policy from being discriminatory. This group of immigrants were referred to as **boat people**, a term associated with extreme poverty. Today, this term does not evoke as much negativism, although it continues as a reminder of a painful emigration period in Haitian history.

From the 1990s to 2010, political unrest, coups, and protests occurred. The tides of history were changing, and Jean-Bertrand Aristide was elected in the first democratically held election in many years. The democratic process did not last; in that same year, a coup d'état on Aristide and a hemisphere-wide embargo was imposed on Haiti. In 2001, Aristide was re-elected in a flawed election. In February 2004, an armed rebellion led to the departure of President Jean-Betrand Aristide; an interim government took office to organize new elections under the auspices of the United Nations Stabilization Mission in Haiti (MINUSTAH). Continued violence and technical delays prompted repeated postponements, but Haiti finally did inaugurate a democratically elected president, Réné Preval, and parliament in May of 2006. Haitian migration took on a new face when the earthquake of January 2010 occurred. Today, more than 1 million people still remain displaced—380,000 being children (Simon et al. 2011). Although thousands of Haitians remain in an immigration holding pattern since before the earthquake, 55,000 Haitians have gained family visas but continue on waiting lists because of immigrations quotas (Zissis 2010). The Dominican Republic has accepted as many as 50,000 people since the earthquake (Paravisini 2010). Since the earthquake, 2500 Haitians have been granted temporary resident visas or permits allowing them to go to Canada. In addition 3700 students and temporary workers from Haiti have been permitted to stay in that country (Power 2010). France is home to approximately 80,000 Haitians and allowed for a temporary residence by undocumented Haitians soon after the disaster in Haiti (McKenzie 2010).

Prior to the earthquake, more than two-thirds of the population, living on less than US $2,00 daily (PAHO 2011). Approximately 250,000 people lost their lives in this catastrophic event, marked as one of the worst in world history. Roughly 2.8 million people were affected, and nearly 1.5 million became homeless. After 1 year, many countries and organizations, including the Pan American Health Organization (PAHO) and the World Health Organization (WHO) launched initiatives to assist Haiti in restructuring and rebuilding their infrastructure. A Post Disaster Needs Assessment (PNDA) was initiated on February 18, 2010, by the United Nations, the World Bank (2020), the European Commission, and the inter-American Development Bank, at the request of the prime minister of Haiti. This group led other groups in assessing restructuring needs (PAHO 2011). Disease, structural instability, hunger, and an inability to reach all those outside of the city have been some of the many obstacles after the earthquake (Simon et al. 2011).

A special focus was placed on the 1.5 million people in IDP (internally displaced person) camps (PAHO 2011). In late October 2010, there was an outbreak of cholera that required specific reporting and handling (PAHO 2011). Many hospitals were totally destroyed, and many others were seriously damaged. The disposal of medical waste continues to pose an environmental risk to everyone in Haiti. Haiti has suffered a catastrophic tragedy that will take continued support from many to rebuild.

18.2.4 Educational Status and Occupations

Following Haiti's independence in 1804, the new rulers of Haiti began advocating French cultural patterns and replicating the French value system. A French model of education was informally adopted and codified in 1860, in accord with the Roman Catholic Church. This resulted in two major changes: The Catholic Church became the official church of Haiti, and Catholic missionaries became responsible for education. The accepted language for communication was now French. During this era, Creole, the language of the uneducated, was perceived as inferior. Social mobility was possible only for French-speaking Haitians. While the educated elite became acculturated into the European value system, the illiterate masses tended to perpetuate the traditional values and customs of their African heritage.

Even though Haitians value education, few are privileged enough to attain a formal education. The Haitian school system is based on the French model and offers free primary and secondary education. Public schools include those operated and controlled by religious orders as well as those under the direct jurisdiction of the Minister of Education. Children from families with financial means attend private schools. The educational model emphasizes liberal arts and humanities rather than technical and vocational studies.

The Haitian educational system continues to emphasize nineteenth-century values, which promote good manners, the classics, literature, philosophy, Latin, and Greek. It deemphasizes the physical and social sciences. The Haitian educational system is based on a two-level curriculum. In the first level, the student receives a certificate of primary education. To receive this certificate, the student must sit for a rigorous test, which includes spelling, reading comprehension, composition, Haitian history and geography, general knowledge, arithmetic, and biology. At this level, the student can speak, read, and write French at the basic level.

The next level consists of two parts: The first is reached after 6 years of secondary education. To receive this diploma, the student must pass examinations in French, English, and Spanish; Haitian literature and history; mathematics; and sciences such as physics, chemistry, biology, and botany. Students in the classical track also take Latin and Greek examinations. A student who has received the first-level certificate should be able to enter the first year of college in American schools. The second-level baccalaureate is likened to the first year of college in North America; the emphasis is on the liberal arts. Again, the student must pass an examination in all the areas covered in the first level, plus philosophy. The results of these national examinations are announced on the radio over a 2-day period or posted on a board in front of the school.

Although Haiti has several universities, they are mainly located in Port-au-Prince. Most of them are state universities. With proper credentials, anyone can enter the university system. However, since the early 1980s, only those in positions of influence have been able to benefit from the state universities. Haitian professionals mirror those of American society; they are lawyers, physicians, nurses, engineers, educators, electricians, plumbers, and construction workers.

Health care professionals in Haiti number around 5400, or about 2.8 health workers per 1000 inhabitants. More specifically, there is one physician and 1.8 nurses per 10,000 population. Nursing education in Haiti is primarily based on a diploma nursing curriculum. It is estimated there are approximately 1400 qualified nurses in Haiti; about 1000 nurses and 1500 auxiliaries are employed by the MSPP, and perhaps 400 more work in the private sector. Seventy percent of all nurses work in Port-au-Prince, where a third of the people in the country live.

The literacy rate, which means that those age 15 and over can read and write, is 61% (64.3% for males and 57.3% for females) is below the 90% average literacy rate for Latin American and Caribbean countries. The level of illiteracy continues to be a major concern in Haiti. Since 1940, the government has conducted several literacy programs.

In 1948, Haiti had its first experience with community education. This public educational system was based on the growth model of development, a UNESCO education project, which duplicated experiences in Latin America (Jean-Bernard 1983). Among Haitian immigrants, women work in hotels, hospitals, and other service industries in domestic and nursing assistant roles. Men work as laborers and factory helpers. Many more Haitians are in the workforce today than there were in the early 1980s, although data for the years 1974 and 1994 from the U.S. Immigration and Naturalization Service (2006) revealed that a disproportionate number of legal Haitians were not employed. In addition, when comparing data by specific groups, a dramatic increase in the number of Haitians in all work environments is found. Data about the work structure of undocumented people are not available because these people technically are "underground" and do not exist.

In Haiti, most major industries are owned and operated by the government. Unemployment is 66 percent (CIA World Factbook 2020). Those who are employed often work under such poor conditions that they have become unmotivated and take little pride in their work, which results in low productivity. In general, Haitians are entrepreneurial, operating their own shops, marketplaces, or schools. Among these entrepreneurs, the motivation, spirit, and pride in their work are readily apparent.

18.3 Communication

18.3.1 Dominant Language and Dialects

The two official languages in Haiti are French and Creole. Creole, a rich, expressive language, is spoken by 100 percent of the population, whereas French is spoken by 15 percent of the population. Since 1957, Creole has been the unofficially accepted language in the internal affairs of the Haitian government, but in 1987, during the Aristide presidency, it was designated in the Haitian Constitution as one of the official languages. Because Creole is the official language, it is used for internal communication within the island.

In contemporary society, the Haitian dilemma can best be understood through this dual-language system. Language is one of the vehicles used to depersonalize those of the lower classes. French is the dominant language of the educated and the elite, whereas Creole is the language of those who are suppressed, the lower classes. The emphasis on French served as a barrier to the early social dynamism that permitted Creole to develop and serve as a unifying force among the African slaves, who came from many different tribes and spoke different languages. In spite of its suppression in formal education, Creole has inspired a very rich and interesting oral literature comprising songs, proverbs, and tales. This oral literature is the most significant aspect of Haitian folklore.

Understanding the language dilemma and the literacy issues assists health-care providers in developing creative tools for educating Haitians. Some of these tools may include video programs, audiocassettes, and radio programs in Creole. Because of the masses of people who are unable to read, printed literature in Creole is not a helpful educational tool.

18.3.2 Cultural Communication Patterns

Haiti has an oral culture with a long tradition of proverbs, jokes, and stories reflecting philosophical systems. These are used to pass on knowledge, convey messages, and communicate emotions. For example, the Creole phrase *Pale franse pa di lespri pou sa* translates to "To speak French does not mean you are smart." *Crayon Bon Die pa gin gum* ("God's pencil has no eraser") conveys the concept of fatalism. Another proverb frequently used is *Sonje lapli ki leve mayi ou* ("Remember the rain that made your corn grow"), which means that one must show gratitude to those who have helped them or done good for them.

Haitians are very expressive with their emotions. By observing them, one can tell whether they are happy, sad, or angry. Haitians' communication patterns include loud, animated speech and touching in the form of handshakes and taps on the shoulder to define or reconfirm social and emotional relationships. Pain and sorrow are very obvious in facial expressions. Most Haitians are very affectionate, polite, and shy. Uneducated Haitians generally hide their lack of knowledge to non-Haitians by keeping to themselves, avoiding conflict, and, sometimes, projecting a timid air or attitude. They smile frequently and often respond in this manner when interacting with Americans or when they do not understand what is being said. Many may pretend to understand by nodding; this sign of approval is given to hide their limitations. Therefore, health-care providers must use simple and clear instructions. Because Haitians are very private, especially in health matters, it is inappropriate to share information through friends. Many may prefer to use professional interpreters who will give an accurate interpretation of their concerns. Most importantly, the interpreter should be someone with whom they have no relationship and will likely never see again.

Voice intonations convey emotions. Haitians speak loudly even in casual conversation among friends and family; the pitch is moderated in formal encounters. When the conversation is really animated, the conversants speak in close proximity and ignore territorial space, especially when emphasizing a point or an issue. Sometimes, the conversation is at such a high pitch and speed that, to an outsider, the conversation may appear disorganized or angry. Haitians love political discussions. In these instances, the conversation may appear stressful and hostile; however, to the participants, the conversation is enjoyable, motivating, and meaningful.

Traditional Haitians generally do not maintain eye contact when speaking with those in a position of authority. In the past, maintaining direct eye contact was considered rude and insolent, especially when speaking with superiors (e.g., children speaking with parents, students with teachers, or employees with supervisors). However, the influence of American education seems to be changing this trend. Most adults maintain eye contact, which means "We are on equal terms, no matter who you are. I respect you and you respect me as an equal human being." For children, however, the custom of not maintaining eye contact with superiors remains deferential. Thus, health-care providers may need to assist children in dealing with conflicting messages.

Haitians touch frequently when speaking with friends. They may touch you to make you aware that they are speaking to you. Whereas Haitian women occasionally walk hand-in-hand as an expression of their friendship, this trend is disappearing both in Haiti and in Haitian communities in North America. This behavior may be changing because of the concept of homosexuality, which is taboo within the Haitian culture.

Haitians greet one another by kissing and embracing in informal situations. In formal encounters, they shake hands and appear composed and stern. Men usually do not kiss women unless they are old friends or relatives. Children greet everyone by kissing them on the cheek. Children refer to adult friends as Uncle or Auntie out of respect, not necessarily because they are related by blood.

18.3.3 Temporal Relationships

The temporal orientation of Haitians is a balance among the past, the present, and the future. The past is important because it lays the historical foundation from which one must learn. The present is cherished and savored. The future is predetermined, and God is the only Supreme Being who can redirect it. One often hears *Bondye bon* ("God is good"), meaning if you conduct yourself conservatively and the right way, God will be there for you. The future is left up to God, who is trusted to do the right thing. In a study by Prudent et al. (2005), several of the informants voiced their belief in God's will when talking about whether or not they would survive being HIV positive and/or having AIDS.

Haitians have a fatalistic but serene view of life. Some believe that destiny or spiritual forces are in control of life events such as health and death, so they say, *Si Bondye vle* ("If God wants"). Given the belief in a predetermined path of life, one can understand this view. Haitians believe that they are the passive recipients of God's decisions. Health-care providers must be clear, honest, and open when assessing Haitian individuals' perceptions and how they perceive the forces that have an influence over life, health, and illness. Acceptance of these beliefs is an important factor in building trust and ensuring adherence.

Most Haitians do not respect clock time; flexibility with time is the norm, and punctuality is not valued. They hold to a relativistic view of time, and although they try, some find it difficult to respond to predetermined appointments. Arriving late for appointments, even medical appointments, is not considered impolite. In North America and elsewhere, Haitians may be more readily compliant with business appointments, but socially, the margin around expected time is very wide—anything or anyone can wait. It is not unusual to see an invitation to a social function listed with an invitation time an hour earlier than the actual time of the function. For example, a wedding invitation may say 6:00 p.m. when the ceremony is actually scheduled for 7:00 p.m. to ensure that all invitees are there on time. Health-care providers should be mindful of this time orientation by making reminder calls for appointments and encouraging the patient in a respectful and caring manner about the importance of timeliness. A thorough assessment of time and temporal view helps health-care providers to plan appointments so that clinic or office backlogs and disruptions are minimized.

18.3.4 Format for Names

Haitians generally have a first, middle, and last name—for example, Marie Maude Guinard.

Sometimes the first two names are hyphenated as in Marie-Maude. The family name, or *nom de famille,* is very important in middle- and upper-class society; it helps to promote and communicate tradition and prestige. However, friends call individuals by their first names. Families usually have an affectionate name or nickname for individuals. The father, mother, grandparent, or any close family member gives this affectionate name at birth.

When a woman marries, she takes on her husband's full name. For example, if Marie-Carmel Guillaume marries Charles Guy Lespinasse, she is always called Mrs. Lespinasse. In an informal setting, she might even be called Mrs. Charles. She loses her name except on paper. Her name and identity are subsumed by her husband's name. This is a reflection of Haitian society in which women are considered subservient to men. Haitian names are primarily of French origin, although many Arabic names are now heard since the migration of Arabs and Jews to Haiti in the 1920s. Haitians are formal and respectful and, as such, should be addressed by their title: Mr., Mrs., Miss, Ms., or Dr.

18.4 Family Roles and Organization

18.4.1 Head of Household and Gender Roles

Traditionally, the head of the household was the man, but in reality, most families today are matriarchal. Haitian men prefer and choose to believe that they make the decisions, but most major decisions are made by the wife and/or mother, with the man remaining a distant figure with a great deal of authority. Today, joint decisions are common. The man is generally considered the primary income provider for the family, and governance, rules, and daily decision making are considered his province. Sociopolitical and economic life centers around men. Men are expected to be sexual initiators, and the concept of ***machismo*** prevails in Haitian life. Women are expected to be faithful, honest, and respectable. Men are usually permitted freedom of social interaction, a freedom not afforded to women. The opportunities offered in North America for women to become income providers, together with their observations of different male-female interactional styles, have encouraged many Haitian women to reject their native, subservient role. This change in the marital interaction has created much stress on marital relationships and an increase in domestic violence, although domestic violence remains one of those closeted issues that are not publicly discussed.

18.4.2 Prescriptive, Restrictive, and Taboo Practices for Children and Adolescents

Children are valued among Haitians because they are key to the family's progeny, cultural beliefs, and values. Children are expected to be high achievers because *Sa ki lan men ou se li ki pa ou* ("What's in your hand is what you have"). In other words, education can never be taken away from you. Children are expected to be obedient and respectful to parents and elders, which is their key to a successful future. They are not allowed to express anger to elders. **Madichon** is a term used when children are disrespectful; it means that their future will be marred by misfortune. Another proverb used to scare and compel children to behave is *Ti moun fwonte grandi devan baron* ("An impudent or insolent child will grow under the Baron's eye [Baron Samedi is the guardian of the cemetery in the voodoo religion] and therefore won't have a long life").

Physical punishment, which is often used as a way of disciplining children, is sometimes considered child abuse by American standards. Fear of having their children taken away from them because of their methods of discipline can cause parents to withdraw or not follow through on health-care appointments if such abuse is evident (e.g., bruises or belt marks). Haitians need to be educated about American methods of discipline and laws so that they can learn new ways of disciplining their children without compromising their beliefs or violating American laws.

Many parents feel confused about how to raise their children in the United States. Their authoritarian behavior is challenged in American society, which they perceive as being too permissive. They feel powerless in understanding how to raise their children in America while still retaining Haitian traditions. The liberal American approach to child rearing poses a great dilemma for Haitian children. They find themselves living in two worlds: the American world, which allows and supports self-actualization and oneness, and the Haitian world, which promotes silence, respect, and obedience.

In the summer, Haitian parents engage their children in certain health-promotion activities such as giving them *lok* (a laxative), a mixture of bitter tea leaves, juice, sugarcane syrup, and oil. In addition, children are also given *lavman* (enemas) to ensure cleanliness. This is supposed to rid the bowel of impurities and refresh it, prevent acne, and rejuvenate the body.

Because Haitian life is centered on males, particularly firstborns, the education of boys is different from that of girls. The family is more indulgent of the behavioral deviations of boys. Boys are given more freedom and are even expected to receive outside initiation in social and sexual life. However, girls are educated toward marriage and respectability. Their relationships are closely watched. Even when they are 16 or 17 years of age, they cannot go out alone because any mishap can be a threat to the future of the girl and bring shame to her family. These beliefs increase Haitians' frustrations and challenges of rearing their children, especially girls, in America.

Health-care providers need to be aware of these various challenges and be prepared to assist children and family members to work through these cultural differences, while still conveying respect for family and cultural beliefs. Health-care providers can play a significant role by helping children and their parents to better understand American practices.

Approximately 500,000 ***restavec*** children are in Haiti (Humanium 2016). *Restavec* is translated to mean "to live with." It was started as an economically motivated action to relieve some parents of the hardship of feeding, clothing, and paying for the education of their children by loaning them out to relatives (Saint-Domingue 2011). Unfortunately, this has not proven to be true and has not met its original intent. *Restavec* children work long hours and rarely go to school. They are regularly abused. They usually eat scraps of food and sleep on the floor (Schaaf 2009). Although they are not chained or locked up, they stay to avoid severe abuse and beatings (Schaaf 2009). Sixty-five percent of the population of children are girls between 6 and 14 years old. After the earthquake, the incidence of *restavec* rose dramatically because many lost their parents or were abandoned (Schaaf 2009). Organizations like International Organization for Migration (IOM) have started an initiative to end the *restavec* system (Breyer 2018). Because of the homelessness and desperation after the quake, there has been a surge in the practice. In 2009, CNN aired a program describing the practice and posted it into a blog so people around the world would become aware of the situation (Schaaf 2009). IOM is working to stop it, along with an organization headed by a man who was a *restavec* as a child: the Jean Robert Cadet Foundation (Breyer 2018; Saint-Domingue 2011).

18.4.3 Family Goals and Priorities

The family is a strong component of the Haitian culture. The expression "Blood is thicker than water" reflects family connectedness. An important unit for decision making is the *conseil de famille,* the family council. This council is generally composed of influential members of the family, including grandparents. The family structure is authoritarian and includes linear roles and responsibilities. Any action taken by one family member has repercussions for the entire family; consequently, all members share prestige and shame.

The family system among Haitians is the center of life and includes the nuclear, consanguine, and affinal relatives, some or all of whom may live under the same roof. Families deal with all aspects of their members' lives, including counseling, education, crises, and marriage. Each family has its own traditions, which form the basis for a family's reputation and are generalized to all members of the family. The prestige of a family is very important and is based on attributes such as honesty, pride, trust, social class, and history. Even families who experience economic difficulties are well respected if they are from a *grande famille*. Wealthy families who have no historical background or tradition are referred to as *nouveaux riche* and find it difficult to marry into the more well-established *grandes familles*, even though they have money.

The family is an all-encompassing concept in the Haitian culture. By including family members in the care of loved ones, health-care providers can achieve more trusting relationships, which foster greater adherence to treatment regimens. Haitians believe that when family members are ill, there is an obligation to be there for them. If a family member is in the hospital, all family members try to visit. Many visitors may cause concern to health-care providers who are not accustomed to accommodating large numbers of visitors. Health-care providers need to be patient with them and facilitate their visits.

When grandparents are no longer able to function independently, they move in with their children. The house is always open to relatives. Elders are highly respected and are often addressed by an affectionate title such as "Aunt," "Uncle," "Grandma," or "Grandpa," even if they are not related. Their children are expected to care for and provide for them when self-care becomes a concern. The elderly are family advisers, babysitters, historians, and consultants. Migration to America poses a tremendous challenge in caring for elderly Haitians. The nursing home concept does not exist in the Haitian culture; therefore, Haitians are generally very reluctant to place their elderly family members in nursing homes.

18.4.4 Alternative Lifestyles

Homosexuality is taboo in the Haitian culture, so gay and lesbian individuals usually remain closeted. If a family member discloses that he or she is gay, everyone keeps it quiet; there is total denial. Gay and lesbian relationships are not talked about; they remain buried. There are no gay bars in Haiti, and overt homosexual conduct is not publicly displayed, although this trend seems to be changing.

Although divorce is common among Haitians, before it becomes final, family members, friends, the church, and elders try to counsel the couple. Health-care providers must approach this issue carefully and establish a trusting relationship before discussing divorce.

Single parenting, widespread in Haiti, is well accepted, and closely tied to the issue of concubinage. In Haitian society, a well-accepted practice is for men to have both a wife and a mistress, with the latter relationship referred to as *placage*. Both women bear children. The mistress raises her children alone and with minimal support from the father. These children are often known by the man's family but are not known to the wife. Haitian women in general know that their husbands are involved in extramarital relationships but pretend not to know. Health education, birth control, and safe sex are issues that should be approached with sensitivity and acceptance within cultural boundaries.

18.5 Workforce Issues

18.5.1 Culture in the Workplace

Haitians living in America have demonstrated a very strong motivation for work and a continued commitment to the entrepreneurial spirit. They can be found in every sector of the American workforce. They are hard workers, and many work two jobs to provide for their American family while sending money to Haiti for those left behind. In the first year of migration, they are generally forced to take lower-status and low-paying jobs. These jobs are used as steppingstones to better jobs until they are able to communicate in English and legalize their immigrant status. According to the U.S. Census Bureau (2010), in 2009, 71 percent of Haitians over the age of 16 years were in the civilian workforce, compared with 65 percent of the total workforce. At the same time, median earnings for Haitian males were $33,000 for men and $29,000 for Haitian women, compared with $45,000 for men and $36,000 for women in the total workforce. In addition, 20 percent of Haitians were living below the poverty line compared with the total population of 14 percent. Work is a necessity, and they conform to the rules and regulations of the workplace. Haitian immigrants have taken menial, low-paying jobs that many Americans would not accept even when unemployed. Haitians appreciate comfort, and they work to be able to afford the necessities of life. The economic survival of Haiti is closely tied to the financial support provided to family members in Haiti by Haitians who have migrated to the United States, Canada, and France.

18.5.2 Issues Related to Autonomy

In America, educated Haitians seek job opportunities in their fields. Those who have a trade try to find employment in that area. Uneducated, undocumented, and illiterate individuals experience much more difficulty in entering the job market, where employment opportunities are restricted to working in places in which there is overcrowding, poor ventilation, and high pollution, all of which place them at high risk for occupational diseases.

Immigrants from various Haitian villages and cities tend to settle in clusters with their relatives or neighbors from their areas of origin. This pattern of settlement by area of origin helps immigrants adapt to the demands of their new environment and ensure that they have someone living nearby whom they can call on in times of illness or other crises. However, when people live and work primarily in an ethnic enclave, the native culture becomes a barrier to assimilation and acculturation into the dominant society.

The educational level of health-care providers in Haiti is different from that in America. For example, medical education is not research-based, and nursing programs for the most part are at the diploma level with an apprenticeship. The only nursing baccalaureate program is the *Faculté des Sciences Infirmières de L'Université Episcopale D'Haiti* (Faculty of Nursing Science of the Episcopal University of Haiti), in Leogane on the southern coast of the island. Establishing this school and adopting this name were major accomplishments. Nursing is finally accepted on par with the medical community, as well as with the other professional schools. All other professional schools start with the words "*Faculté des Sciences*" and end with whatever the science is (e.g., medicine, law, engineering).

Haitian health-care providers who migrate to the United States have experienced a great deal of difficulty in obtaining licensure to practice. Those who learned their profession in Haiti were taught in French and the test-taking approach is different; multiple-choice examinations are a new and difficult concept for Haitians.

Haitian nurses are very skilled clinically; however, sometimes they may experience difficulty in applying theoretical knowledge to practice. This may be due in part to language barriers, socialization, and their diploma education, which focuses on tasks and skills development. Haitian professionals struggle with professional cohesiveness and collegiality. Many groups have established professional societies whose goals are to support one another, to promote professional development, and to promote collegial relationships. Some examples of these professional groups are the Haitian Nurses Association, the Haitian-American Medical Association, the Haitian Educator Association, the Haitian-American Engineers, and the Haitian-American Lawyers.

Sometimes Haitians in the workplace greet one another in their native tongue because it is easier to articulate ideas and feelings and to express support in their native language. This may be irritating to non-Haitians, who consider it rude.

18.6 Biocultural Ecology

18.6.1 Skin Color and Other Biological Variations

Different assessment techniques are required when assessing dark-skinned people for anemia and jaundice. One must examine the sclera, oral mucosa, conjunctiva, lips, nailbeds, palms of the hands, and soles of the feet when assessing for cyanosis and low blood hemoglobin levels. To assess for jaundice, one must examine the conjunctiva and oral mucosa for patches of bilirubin pigment because dark skin has natural underlying tones of red and yellow.

18.6.2 Diseases and Health Conditions

Because Haiti is a tropical island, prevalent diseases include cholera, parasitosis, and malaria. Haiti has no mosquito control, so newer immigrants should be assessed for signs of malaria such as chills, fever, fatigue, and an enlarged spleen. Other diseases of increased incidence among Haitian immigrants are hepatitis, tuberculosis, HIV/AIDS, venereal diseases, and parasitosis from inadequate potable water sources in their homeland. Actual tuberculosis rates for Haitians are misleading because, until a few years ago, Haitians living in Haiti were routinely vaccinated with *Bacille bilié de Calmette-Guérin*, thus making all subsequent skin tests positive, even though they may not actually have had the disease. Unfortunately, upon immigration, many Haitians continue to live in overcrowded areas, are malnourished, and live in very poor sanitary conditions, factors that increase their risk for infectious diseases.

Haitians are prone to diabetes and hypertension—a reflection of genetics and their diet, which is high in fat, cholesterol, and salt. Data on the prevalence of diabetes and hypertension among Haitian Americans are difficult to assess because they are categorized as black. In addition to type 1 and type 2 diabetes, there is a type 3 malnutrition-related diabetes, also known as *tropical diabetes*. The prevalence ranges from 2 to 8 percent, accounting for different parts of the island (PAHO 2001). In addition, Haitians experience a high incidence of heart disease. Cerebrovascular diseases are the third leading cause of death; other cardiopathies are in fifth place, and arterial hypertension is in 11th place. More deaths are registered among females than males. In addition to cardiovascular diseases, there is a high incidence of cancer. The National Cancer Institute statistics showed that the most frequent type of cancer treated was cervical cancer, representing 40 percent of cases. Breast cancer ranked second with 30 percent. Nasopharyngeal cancer ranked in third position with 10–15 percent of the cases (PAHO 2001). Both cancer and heart disease are related to a high-fat diet. Today, Haitians in Haiti and in the United States are very conscious of the need to limit the fat content in their diets; as a result, the Haitian diet is not as fatty as it once was.

Reflective Exercise 18.1

Marie-Sandra is a 36-year-old Haitian woman. She was para 2 gravida 2, is 18 months postpartum, and has been breastfeeding her child. She noticed a change in the color of the breast milk from the right breast. She previously had a lesion in her right breast that was initially diagnosed as an abscess and appeared to have been there for 3 months. She returned to her physician after seeing the change in the color of her breast milk. The examination revealed a mass measuring 8 × 10 cm in the superior aspect of the breast.

A biopsy confirmed carcinoma infiltrate of the right breast. Marie had a sister who died of breast cancer at age 31. Her mother died at age 51 from "some intra-abdominal cancer." The oncologist believed that it was suggestive of breast and ovarian syndrome of a mutation gene.

Marie-Sandra had chemotherapy in Haiti that made her very ill, so she went to

Cuba for continuation to complete four cycles of chemotherapy. Because definitive care was not available in Haiti, a university medical center in the United States enrolled her in a pro bono program. She came for the first time to the United States alone and that afternoon saw the surgeon, had a mammogram, and had preoperative diagnostic studies. A French interpreter was used.

The next day, Marie-Sandra had bilateral mastectomies. The left mastectomy was prophylactic because of her family history and no ability for mammography monitoring in Haiti. She remained in the United States without any family for 1 year while undergoing treatment and additional surgery. She received a 1-year course of chemotherapy, radiation therapy to the chest wall, genetic testing, Herceptin therapy, and prophylactic bilateral oophorectomies. Genetic BRCA 1 and BRCA 2 results were negative.

Marie-Sandra did well, and after 1 year, she returned to Haiti and started working again.

1. Given what you know about Marie-Sandra's history, how could she be helped to understand to change this major health event?
2. What suggestions might be provided for Marie-Sandra regarding her nutrition?
3. How might the health-care team assist with Marie-Sandra's acculturation in the United States?
4. How do traditional Haitians deal with family separation?
5. How might she be helped with being separated from her family?

Attention-deficit/hyperactivity disorder (ADHD) is a commonly diagnosed chronic mental condition in Haitian children (Prudent et al. 2005). This disease has a large genetic component (McCann et al. 2006). In the Haitian culture, there is no conceptual term for ADHD, nor is there a Creole term to describe it. Unfortunately, in the Haitian culture, the behavior displayed with this diagnosis may be interpreted as an ill-behaved or a "poorly raised" child or a psychically victimized child suffering from an "unnatural" condition. Parents may believe that this behavior can be controlled by parental discipline, or they may seek an alternative health consult such a *Hougan* or voodoo priest. Although medications are the preferred treatment for ADHD, which may be combined with psychological intervention, Haitians are fearful of psychoactive drugs because they see them as the cause of substance abuse and even possibly mental illness (Prudent et al. 2005). Therefore, assessing the parents' perceptions of the cause of the ADHD behavior and assisting them in holistic treatment are important.

18.6.3 Variations in Drug Metabolism

The literature reveals no studies on drug metabolism specific to Haitians or Haitian Americans. When Haitians are included in drug studies, it is assumed that they are included under the category of African American. Therefore, health-care providers may need to start with the literature for this broad category of ethnicity to posit and test theories of ethnic drug metabolism among Haitian Americans.

18.7 High-Risk Behaviors

Haitian refugees are one of the most at-risk populations living in the United States. Therefore, it is important for health-care providers to consider a number of factors in providing health-care services. An in-depth assessment of the person's environmental, occupational, socioeconomic, demographic, educational, and linguistic status enables the development of strategies that are culturally appropriate, adequate, and effective. As a new group of immigrants, Haitians bring to the health-care system a different set of beliefs and values about health and illness. These differences challenge health-care providers who must try to explain treatments while acknowledging, but not changing, their patients' cultural convictions. Attempts to change firmly held beliefs are counterproductive to establishing trusting health-care provider–patient relationships.

Behaviors that may be considered high risk in American society are generally viewed as recreational or unimportant among Haitians. Alcohol, for example, plays an important part in Haitian society. Drinking alcohol is culturally approved for men and is used socially when friends gather, especially on weekends. Women drink socially and in moderation. Cigarette smoking is another high-risk behavior practiced by Haitian men, whereas Haitian women have a very low rate of tobacco use. The trend toward decreasing cigarette use in America has not influenced Haitian society. Drug abuse among Haitians used to be low, but drug abuse in the adolescent population is increasing. In 1982, Haiti became the first developing country to be incorrectly blamed for the beginning of the AIDS epidemic. As a result, Haitians have had to endure the stigma associated with the belief that Haitians are "AIDS carriers." Unfortunately, HIV/AIDS has continued to spread in the Haitian community both in Haiti and the United States. Heterosexual transmission is the primary mode of HIV transmission in the Haitian community and is rapidly becoming an infection of women and children (Santana and Dancy 2000). Health-care providers need to recognize the impact the stigma has had on male–female relationships, as well as familial relationships, in the Haitian community. Health providers must broaden their scope and approaches to HIV prevention by incorporating societal, contextual, and economic factors designed to modify traditional gender roles germane to influencing beginning negotiations of safer sex practices.

High-risk behavior in the Haitian culture includes the nonuse of seat belts and helmets when driving or riding a motorcycle or bicycle. Most cars in Haiti do not have seat belts, and there are no laws regarding the use of seat belts and helmets. Haitian cities are extremely overpopulated and traffic laws are very loose, resulting in hazardous driving conditions. Everyone tries to gain the upper hand. Haitian Americans must be educated about traffic laws, seat belt use, car seats for youngsters, and the need for helmets. Health-care providers may have to use graphic videos or skits when instructing patients about these safety practices. Health-care providers may also use Haitian radio stations for educational programs when they are available. Other strategies that may be used to help promote behavioral changes are through church and community group activities. Through these avenues, health-care providers can have a significant impact on health promotion and health risk prevention among Haitian Americans.

18.7.1 Health-Care Practices

To Haitians, good health is seen as the ability to achieve internal equilibrium between ***cho*** (hot) and ***fret*** (cold) (see Sects. 18.8 and also 18.12). To become balanced, one must eat well, give attention to personal hygiene, pray, and have good spiritual habits. To promote good health, one must be strong, have good color, be plump, and be free of pain. To maintain this state, one must eat right, sleep right, keep warm, exercise, and keep clean.

Haitians who believe in voodoo (see sect. 18.11.1) and other forms of folk medicine may use several types of folk healers. These healers include a voodoo practitioner, a ***docte fey*** (leaf doctor), a ***fam saj*** (lay midwife), a ***docte zo*** (bonesetter), and a ***pikirist*** (injectionist). Depending on whether the individual believes that the illness is natural or unnatural, she or he may seek help other than Western medicine from one of these healers.

18.8 Nutrition

18.8.1 Meaning of Food

For many Haitians in lower socioeconomic groups, food means survival. However, food is relished as a cultural treasure, and Haitians generally retain their food habits and practices after emigrating. Food practices vary little from generation to generation. Most Haitians are not culinary explorers. They prefer eating at home, take pride in promoting their food for their children, and discourage fast food. When hospitalized, many would rather fast than eat non-Haitian food. Haitians do not eat yogurt, cottage cheese, or "runny" egg yolk. Haitians drink a lot of water, homemade fruit juices, and cold, fruity sodas.

18.8.2 Common Foods and Food Rituals

The typical Haitian breakfast consists of bread, butter, bananas, and coffee. Children are allowed to drink coffee, which is not as strong as that consumed by adults. Generally, the largest meal for Haitians is eaten at lunch. At lunchtime, a basic Haitian meal might include rice and beans, boiled plantains, a salad made of watercress and tomatoes, and stewed vegetables and beef or cornmeal cooked as polenta. Table 18.1 lists popular foods in the Haitian community.

Table 18.1 Popular foods in the Haitian community

Bouillon	Soup made with beef broth mixed with various green vegetables (e.g., spinach, cabbage, watercress, string beans, carrots), meat or poultry, plantain, sweet potato, and Malaga, a sweet aromatic wine
Chiquetaille	Codfish or smoked herring, unsalted, shredded finely, mixed with onions, shallots, finely chopped hot pepper, vinegar, and lime
Fritters	*Marinade:* Flour, water, eggs, parsley, onions, garlic, salt and pepper, chicken, hot pepper, and a pinch of baking soda, mixed together to pancake consistency and deep-fried *Acra:* Chopped parsley, eggs, garlic, and onion mixed with Malaga; finely shredded codfish or smoked herring and hot pepper may be added *Beignet:* Sweet ripe banana, sugar, and eggs, mixed with cinnamon, milk, margarine, flour, nutmeg, and vanilla extract
Green plantain	Boiled or fried, usually eaten with *griot*
Griot	Marinated pork cut up in small pieces and fried
Lambi	Conch meat softened and prepared in a sauce
Legume	Vegetables such as chayote and eggplant cooked with meat
Patee	Pastry dough filled with choice meat, chicken, or smoked herring
Pumpkin squash soup	Meat or poultry mixed with vegetables and pureed cooked squash and spices
Tomtom	Similar to dumplings, cooked and made into round balls and eaten with beef stew and okra

18.8.3 Dietary Practices for Health Promotion

Hot and cold, acid and nonacid, and heavy and light are the major categories of contrast when discussing food. Illness is caused when the body is exposed to an imbalance of cold (*fret*) and hot (*cho*) factors. For example, *soursop*, a large, green prickly fruit with a white pulp that is used in juice and ice cream, is considered a cold food and is avoided when a woman is menstruating. Eating white beans after childbirth is believed to induce hemorrhage. Foods that are considered heavy, such as plantain, cornmeal mush, rice, and meat, are to be eaten during the day because they provide energy. Light foods, such as hot chocolate milk, bread, and soup, are eaten for dinner because they are more easily digested. Table 18.2 presents a classification of hot and cold foods.

To treat a person by the hot-and-cold system, a potent drink or herbal medicine of the class opposite to the disease is administered. Cough medicines, for example, are considered to be in the hot category, whereas laxatives are in the cold category. Certain food prohibitions are related to particular diseases and stages of the life cycle. Teenagers, for example, are advised to avoid drinking citrus fruit juices such as lemonade to prevent the development of acne. After performing strenuous activities or any activity that causes the body to become hot, one should not eat cold food because that will create an imbalance, causing a condition called *chofret*. A woman who has just straightened her hair by using a hot comb and then opens a refrigerator may become a victim of

Table 18.2 Haitian hot and cold food classification

Very cold (−3)	Quite cold (−2)	Cool (−1)	Neutral (0)	Warm (+)	Very hot (+2)
Avocado	Banana	Tomato	Cabbage	Eggs	Rum
Cashew nuts	Grapefruit	Cane syrup	Conch	Pigeon	Nutmeg
Mango	Lime	Orange	Carrot	Soup	Garlic
Coconut	Okra	Cantaloupe	Watercress	Bouillon	Tea
Cassava	Watermelon	Chayote	Brown rice	Pork	Cornmeal mush

Source: Adapted from M.S. Laguerre (1981, pp. 194–196)

chofret. This means she may catch a cold and/or possibly develop pneumonia.

When they are sick, Haitians like to eat pumpkin soup, bouillon, and a special soup made with green vegetables, meat, plantains, dumplings, and yams. The Haitian diet is high in carbohydrates and fat. Eating "right" entails eating sufficient food to feel full and maintain a constant body weight, which is often higher than weight standards medically recommended in the United States. Men like to see "plump" women. Furthermore, weight loss is seen as one of the most important signs of illness. Additional components of what Haitians consider a healthy diet are tonics to stimulate the appetite and the use of high-calorie supplements such as *Akasan*, which is either prepared plain or made as a special drink with cream of cornmeal, evaporated milk, cinnamon, vanilla extract, sugar, and a pinch of salt.

A thorough nutritional assessment is very important to effectively promote nutritional health. Understanding food rituals assists health-care providers in designing individualized dietary plans, which can be incorporated into the diet to facilitate compliance with dietary regimens that promote a healthier lifestyle.

18.8.4 Nutritional Deficiencies and Food Limitations

Many Haitian women and children who come from rural areas have significant protein deficiencies owing to Haiti's economic deprivation. A cultural factor contributing to this problem is the uneven distribution of protein among family members. However, the problem is not one of net protein deficiency in the community but, rather, the unwise distribution of the available protein among family members. Whenever meat is served, the major portion goes to the men, under the assumption that they must be well fed to provide for the household. This same pattern exists today among Haitian immigrants. Being aware of this cultural factor enables health-care providers to prepare nutritional plans that meet patients' dietary needs.

Another major concern in this area is that of food insecurity and short intervals between births, chronic malnutrition, and anemia, which are widespread among Haitian women of childbearing age. These health inequalities result in a high prevalence of low birth weight, estimated at 15 percent; anemia, ranging from 35 to 50 percent; a body mass index under 18.5 kg/m^2, estimated at 18 percent; and a high maternal mortality rate, estimated at 456 per 100,000 live births (PAHO 2001).

18.9 Pregnancy and Childbearing Practices

18.9.1 Fertility Practices and Views Toward Pregnancy

Pregnancy and fertility practices are not readily discussed among Haitians. Most Haitians are Catholic and are unwilling to overtly engage in conversation about birth control or abortion. This does not mean that these two practices do not occur, but rather that they are just not openly discussed. Abortion is viewed as a woman's issue and is left to her and her significant other to decide. Accurate assessments and teaching related to these sensitive areas require tact and understanding. Initially, health-care providers should be cautious in assessing and gathering information related to fertility control. Pregnancy is not considered a health problem but rather a time of joy for the entire family. Pregnancy does not relieve a woman from her work. Because pregnancy is not a disease, many Haitian women do not seek prenatal care. Pregnant women are restricted from eating spices that may irritate the fetus. However, they are permitted to eat vegetables and red fruits because these are believed to improve the fetus's blood. They are encouraged to eat large quantities of food because they are eating for two. Pregnant women who experience increased salivation may rid themselves of the excess at places that may seem inappropriate. They may even carry a "spit" cup in order to rid themselves of the excess saliva. They are not embarrassed by this behavior because they feel it is perfectly normal.

Fifty percent of women living in Port-au-Prince give birth in a hospital, compared with 31 percent of births in other urban areas, and only 9 percent of births in rural areas. The leading causes of maternal deaths are obstructed labor (8.3 percent), toxemia (16.7 percent), and hemorrhage (8.3 percent). The high maternal mortality rate is mainly the result of inadequate prenatal care (PAHO 2001).

The most popular methods of contraception are the birth control pill, female sterilization, injections, and condoms (3 percent each). Among sexually active women, 13 percent use a modern method of contraception and 4 percent rely on traditional methods. Among sexually active men, 17 percent use a modern method (6 percent use condoms) and 16 percent rely on traditional methods (PAHO 2001).

18.9.2 Prescriptive, Restrictive, and Taboo Practices in the Childbearing Family

During labor, the woman may walk, squat, pace, sit, or rub her belly. Generally, Haitian women practice natural childbirth and do not ask for analgesia. Some may scream or cry and become hysterical, whereas others are stoic, only moaning and grunting. What they need is support and reassurance; for example, applying a cold compress on the woman's forehead demonstrates caring and sensitivity on the part of the health-care provider. Since migrating, some Haitian women have adopted American childbearing practices and request analgesics. Cesarean birth is feared because it is abdominal surgery. Women in higher social strata are more amenable to having cesarean deliveries. Fathers do not generally participate in the labor and delivery, believing that this is a private event best handled by women. The woman is not coached; female members of the family give assistance as needed.

The crucial period for the childbearing woman is postpartum, a time for prescription and proscription. The woman takes an active role in her own care. She dresses warmly after birth as a way to become healthy and clean. Haitians believe that the bones are "open" after birth and that a woman should stay in bed during the first 2–3 days postpartum to allow the bones to close. Wearing an abdominal binder is another way to facilitate closing the bones.

The postpartum woman also engages in a practice called *the three baths*. For the first 3 days, the mother bathes in hot water boiled with special leaves that are either bought or picked from the field. She also drinks tea boiled from these leaves. For the next 3 days, the mother bathes in water prepared with leaves that are warmed by the sun. At this point, the mother takes only water or tea warmed by the sun. Another important practice is for the mother to take a vapor bath with boiled orange leaves, a practice believed to enhance cleanliness and tighten the internal muscles. At the end of the third to fourth week, the new mother takes the third bath, which is cold. A cathartic may be administered to cleanse her intestinal tract. When the process is completed, she may drink cold water again and resume her normal activities.

In the postpartum period, Haitian women avoid white foods such as lima beans, as well as other foods, including okra, mushrooms, and tomatoes. These foods are restricted because they are believed to increase vaginal discharge. Other foods are eaten to give the new mother strength and vitality. Foods associated with this prescriptive practice are porridge, rice and red beans, plantains boiled or grated with the skins and prepared as porridge (the skin is high in iron, which is good for building the blood), carrot juice, and carrot juice mixed with red beet juice.

Breastfeeding is encouraged for up to 9 months postpartum. Breast milk can become detrimental to both mother and child if it becomes too thick or too thin. If it is too thin, it is believed that the milk has "turned," and it may cause diarrhea and headaches in the child and, possibly, postpartum depression in the mother. If milk is too "thick," it is believed to cause impetigo (*bouton*). Breastfeeding and bottle-feeding are accepted practices. If the child develops diarrhea, breastfeeding is immediately discontinued. Practices that do not put the mother or the child at risk should be supported and encouraged. Respecting the patients' cultural beliefs and practices helps to establish trust between the patient and the health-care provider and demonstrates caring. By being familiar with these health practices and beliefs, health-care providers can assist women in making culturally safe decisions related to pregnancy and plans for delivery.

Another prescriptive postpartum practice among Haitian women is to feed their infant a *lok* similar to the one administered to the older children in the summer. The laxative is administered with the initial feeding and is intended to hasten the expulsion of meconium. Because Haitians are fearful of diarrhea in children, health-care providers should stress the risks associated with *lok* and any other type of bowel-cleansing cocktails in infants and children. It is important to stress the impact of laxative use on the body system and educate the woman about the need to prevent dehydration.

18.10 Death Rituals

18.10.1 Death Rituals and Expectations

Generally, Haitians prefer to die at home rather than in the hospital. Since migrating to America, many have accepted death in a health-care facility to alleviate the heavy burden on the family during the last stage of the loved one's life. When death is imminent, the family may pray and cry uncontrollably, sometimes even hysterically. They try to meet the person's spiritual needs by bringing religious medallions, pictures of saints, or fetishes. When the person dies, all family members try, if possible, to be at the bedside and have a prayer service. If possible, and if it is not too disturbing to other patients, health-care providers should encourage this practice and involve a family member in the postmortem care.

Reflexive Exercise 18.2

Manou is a 59-year-old Haitian American woman who lives in the Midwest United States. About 6 years ago, Manou lost her only son, age 20, who died tragically after dropping out of college and joining the military. Two years after her son's death, Manou fell ill and was diagnosed with cancer of the gallbladder. Her husband, a Lutheran pastor, had moved to Florida to build a church and to serve the Haitian American population living in the area. Manou stayed in the Midwest to care for her youngest daughter, who was then finishing high school. Manou was able to function for a number of years without ever mentioning her illness to her husband or daughter. She isolated herself from her family, including her parents.

Recently, Manou fell gravely ill while she was alone in the house. Her husband, who was still in Florida at the time, had to call a family member to check on her condition. She was taken to the nearest hospital emergency room and then transferred to a nursing home. She suffered with pain on her left abdominal quadrant and had difficulty eating. Her family members rushed to care for her, although they were unaware of her condition. They made leaf teas (parsley, garlic) in the hope of alleviating her pain and epigastric discomfort.

Manou was transferred from the nursing home to another hospital for further testing. There, it was revealed to her family that she was terminally ill and needed to be admitted to a hospice care facility. The family refused and wanted to take her home to care for her.

You happen to be a nurse and a member of the family. Manou's family was in disbelief; her husband and daughter looked to you for answers and to assist them in coping with this news. They need to be prepared for her imminent death.

1. What do you need to know about the health practices that Manou had engaged in at home? Why would this be important?
2. How can you help the family come to terms with this major event that Manou kept from them?
3. Do you think Manou kept her illness a secret because of lack of trust, or was she trying to protect her family? Is this behavior typical in the Haitian community, or is this out of the ordinary?
4. How can you assist this family in their grief?
5. What can you do to assist Manou in coming to terms spirituality/religiously with her imminent death?

18.10.2 Responses to Death and Grief

Death in the Haitian community mobilizes the entire family, including the matrilineal and patrilineal extensions and affines. Death arrangements in America are similar to those in Haiti. Generally, a male relative of the deceased makes the arrangements. This person may also be more fluent in English and more accustomed to dealing with the bureaucracy. This person is also responsible for notifying all family members wherever they might be in the world, an important activity because family members' travel plans influence funeral arrangements. In addition, he is responsible for ordering the coffin, making arrangements for prayer services before the funeral, and coordinating plans for the funeral service.

The preburial activity is called *veye*, a gathering of family and friends who come to the house of the deceased to cry, tell stories about the deceased's life, and laugh. Food, tea, coffee, and rum are in abundant supply. The intent is to show support and to join the family in sharing this painful loss. Another religious ritual is called the *dernie priye*, a special prayer service consisting of seven consecutive days of prayer. Its purpose is to facilitate the passage of the soul from this world to the next. It usually takes place in the home. On the seventh day, a mass called *prise de deuil* officially begins the mourning process. After each of these prayers, a reception/celebration in memory of the deceased is held.

Haitians have a very strong belief in resurrection and paradise; thus, cremation is generally not an acceptable option (Fr. Jadotte, personal communication, 2020). Haitians are very cautious about autopsies. If foul play is suspected, they may request an autopsy to ensure that the patient is really dead. This alleviates their fear that their loved one is being *zombified*. According to this belief, this can occur when the person appears to have died of natural causes but is still alive. About 18 h after the burial, the person is stolen from his or her coffin; the lack of oxygen causes some of the brain cells to die, so the mental facilities cease to exist while the body remains alive. The zombie then responds to commands, having no free will, and is domesticated as a slave.

Reflective Exercise 18.3

Lélé, a young Haitian man, survived the earthquake in Haiti on January 12, 2010. Prior to the earthquake, he was active, full of life, and pursing his studies. He lived in one of the small towns in Haiti and was going to school at the same time. Soon after the earthquake, Lélé developed some signs and symptoms that baffled many of the health-care providers who were giving assistance to the earthquake survivors. Lélé started losing weight, his skin color changed, and he became discolored. As his condition became worse, he had difficulty swallowing. Suspecting scleroderma, Lélé was brought along with his mother to the United States for treatment.

After 6 weeks, Lélé's mother returned home because there was nothing that Western/conventional medicine could do for him. She stated that she did not want to witness Lélé's death and would rather remember him alive. After 3 months of a languishing illness, Lélé died alone and far away from his young wife, his mother, and the rest of his family. Lélé's wish was to be buried in his homeland. Given the high cost of sending the body home for burial, his wife contemplated cremation and sending his ashes home. However, Lélé had converted to the Mormon faith, which prohibits cremation. When Lélé's wife was informed of this, she sought out a spiritual leader from the Mormon faith to assist her in making such an important decision.

1. Should Lélé's mother be brought into the decision-making process regarding cremation?
2. How important is family in the Haitian culture?
3. What resources might be made available to have Lélé's body returned to Haiti?
4. If the decision was made for cremation, which is contrary to Haitian culture, how might a nurse assist his wife with the grieving process?

18.11 Spirituality

18.11.1 Dominant Religion and Use of Prayer

Patients' cultural beliefs and religion can have a great impact on their acceptance of and adherence to health care and, therefore, on the outcomes of treatment. Catholicism is the primary religion of Haiti. Since the early 1970s, however, Protestantism has gained in popularity throughout the island and has seriously challenged the Catholic Church, especially among the lower socioeconomic classes. Even though Haitians are deeply religious, their religious beliefs are combined with **voudou** (voodooism), a complex religion with its roots in Africa. Voudou, in the most simplistic sense, involves communication by trance between the believer and ancestors, saints, or animistic deities. Voudou is not considered paganism among those who practice it, even though many of the rituals resemble paganism. Participants gather to worship the *loa* or *mystere*, deities or spirits who are believed to have received their powers from God and are capable of expressing themselves through possession of a chosen believer. With their great powers, the *loa* or *mystere* can provide favors such as protection, wealth, and health to those who worship and believe in them.

18.11.2 Meaning of Life and Individual Sources of Strength

The family system among Haitians is the center of life and includes the nuclear, consanguine, and affinal relatives. They may all live under the same roof. The family deals with all aspects of a person's life, including counseling, education, crises, marriage, and death.

The best way to understand and assess the spiritual beliefs and needs of Haitian American patients is to understand their culture. This is especially important because Haitian patients may express their concerns in ways that are unique to their cultural and religious beliefs. To ensure accurate assessments of these patients, it is essential to ask questions carefully and to completely understand the answers in order to gain an understanding of patients' perceptions of health and illness as dictated by their culture and religious beliefs. By recognizing and accepting patients' beliefs, health-care providers may alleviate barriers, and patients may feel more at ease to discuss their beliefs and needs.

18.11.3 Spiritual Beliefs and Health-Care Practices

Voudou believers may often attribute their ailments or medical problems to the doings of evil spirits. In such cases, they prefer to confirm their suspicions through the *loa* before accepting natural causes as the problem, which would lead to seeking Western medical care. For Haitian patients, the belief in the power of the supernatural can have a great influence on the psychological and medical concerns of the patients.

18.12 Health-Care Practices

18.12.1 Health-Seeking Beliefs and Behaviors

For Haitians, illness is perceived as punishment, considered an assault on the body, and may have two different etiologies: natural illnesses, known as *maladi Bondye* ("disease of the Lord"), and supernatural illnesses. Natural illnesses may occur frequently, are of short duration, and are caused by environmental factors such as food, air, cold, heat, and gas. Other causes of natural illness are movement of blood within the body, disequilibrium between hot and cold, and bone displacement. Supernatural illnesses are believed to be caused by angry spirits. To placate these spirits, patients must offer feasts called *manger morts*. If individuals do not partake in these rituals, misfortunes are likely to befall them. Illnesses of supernatural origin are fundamentally a breach in rapport between the individual and her or his protector. The breach in rapport is a response from the spirit and a way of showing disapproval of the protégé's behavior. In this instance, health can be recovered if the patient takes the first step in determining the nature of the illness. This can be accomplished by eliciting the help of a *voudou* priest and following the advice given by the spirit itself. To accurately prescribe treatment options, health-care providers must be able to differentiate between these belief systems.

Physical illnesses are thought to be on a continuum beginning with "*Kom pa bon*" ("I do not feel well"). In this phase, the affected person is not confined to bed; illness is transitory, and the person should be able to return to his or her normal activities. The next phase is *Moin malad* ("I am sick"), in which the individuals stay at home and avoid activity. The third phase is *Moin malad anpil* ("I am very sick"). This means that the person is very ill and may be confined to bed. The final phase is *Moin pap refe* ("I am dying").

Haitians believe that gas (*gaz*) may provoke pain and anemia. Gas can occur in the head, where it enters through the ears; in the stomach, where it enters through the mouth; and in the shoulders, back, legs, or appendix, where it travels from the stomach. When gas is in the stomach, the patient is said to suffer *kolik*, meaning stomach pain. Gas in the head is called *van nan tet* or *van nan zorey*, which translates to "gas in one's ears," and is believed to be a cause of headaches. Gas moving from one part of the body to another produces pain. Thus, the movement of gas from the stomach to the legs produces rheumatism, to the back causes back pain, and to the shoulder causes shoulder pain. Foods that help dispel gas include tea made from garlic, cloves, and mint; plantains; and corn. To deter the entry of gas into the body, one must be careful about eating "leftovers," especially beans. Since migrating to the United States, Haitians have begun eating leftovers, which is believed to cause many of their ailments. After childbirth, women are particularly susceptible to gas, and to prevent entry of gas into the body, they tighten their waist with a belt or a piece of linen.

18.12.2 Responsibility for Health Care

Haitians engage in self-treatment and see these activities as a way of preventing disease or promoting health. Haitians try home remedies as a first resort for treating illness. They are self-diagnosticians and may use home remedies for a particular ailment, or if they know someone who had a particular illness, they may take the prescribed medicine from that person. They keep numerous topical and oral medicines on hand, which they use to treat various symptoms. For example, an individual who suspects a venereal disease may buy penicillin injections and have someone administer them without consulting a physician. In Haiti, many medications can be purchased without a prescription, a potentially dangerous practice. However, health-care providers must be very discrete in assessing, teaching, and guiding the patient toward safer health practices. Admonishing patients may cause them to withdraw and not adhere to instructions. Haitians may also lead health-care providers to believe that they are interested, when in fact they have already discredited the health-care provider. When taking the patient's history, the health-care provider should inquire if the patient has been taking medication that was prescribed for someone else. Moreover, when prescribing a potentially dangerous drug, the health-care provider should be sure to caution the patient not to give the medication to ailing friends or relatives. Even though the health-care provider may not be completely successful at stopping the practice of exchanging medications, with continued reminders, she or he may be successful later.

18.12.3 Folk and Traditional Practices

Haitians may use others' experiences with a particular illness as a barometer against which to measure their symptoms and institute treatment. If necessary, a person living in the United States may ask friends or relatives to send medications from Haiti. Such medications may consist of roots, leaves, and European-manufactured products that are more familiar to them. Therefore, it is very important to ascertain what the patient is taking at home to avoid serious complications.

Constipation, referred to as *konstipasyon*, is treated with laxatives or herbal teas. Sometimes, Haitians use enemas (*lavman*). Diarrhea is not a major concern in adults; however, it is considered very dangerous in children and sometimes interpreted as a hex on the child. Parents may try herbal medicine, may seek help from a *voudou* priest or *hougan*, or if all else fails, may consult a physician. It is very important to assess the child carefully because he or she may have been ill for quite some time.

A primary respiratory ailment is *oppression*, a term used to describe asthma. However, the term really describes a state of anxiety and hyperventilation rather than the condition. *Oppression* is considered a cold state, as are many respiratory conditions. A home remedy for *oppression* is to take a dry coconut and cut it open, fill it with half sugarcane syrup and half honey, grate one full nutmeg and add it to the syrup mix, reseal the coconut, and then bury it in the ground for a month. The coconut is reopened, the contents are stirred and mixed together, and 1 tablespoon is administered twice a day until all of the contents have been consumed. By the end of this treatment, the child is supposed to be cured of the respiratory problem.

18.12.4 Barriers to Health Care

Because orthodox medicine is often bypassed or perceived as a second choice among Haitians, the potential delay of medical care can pose an increased risk to patients. The view that physicians of conventional medicine do not understand *voudou* and, therefore, cannot cure magical illness or that an illness worsens if the bewitched person seeks a physician is enough to persuade these individuals to seek unconventional modes of therapy with which they are more comfortable. The health-care team should understand some of the basic principles and practices of folk medicine, particularly root medicine, because this can play a significant role in determining the progress of the client's health status.

Many Haitians are in low-paying jobs that do not provide health insurance, and they cannot afford to purchase it themselves. Thus, economics acts as a barrier to health promotion. In addition, for those who do not speak English well, it is difficult for them to access the health-care system, fully explain their needs, or understand prescriptions and treatments.

Reflective Exercise 18.4

Marie was raised by her grandmother since typhoid took the lives of her parents. Marie said that when her brother, Jean-Claude, contracted the disease, her grandmother used a paste-like mixture of sour oranges, the leaves of a sour orange tree, and papaya leaves and placed it on his forehead to reduce the fever. This was used for 3 days, at which time her grandmother realized that the treatment was ineffective. At that point, Marie and her grandmother took Jean-Claude via a donkey-pulled cart to the nearest clinic. The trip took 8 h. Even though the staff immediately started intravenous fluids and medication, Jean-Claude died the next day.

1. What were some of the major obstacles to treating Jean-Claude?
2. What are some of the variant cultural characteristics from the Purnell Model in this vignette?
3. What other traditional remedies do Haitians use?
4. What are traditional Haitian burial practices?

18.12.5 Cultural Responses to Health and Illness

The *root-work system* is a folk medicine that provides a framework for identifying and curing folk illnesses. When illness occurs, or when a person is not feeling well or is "disturbed," root medicine distinguishes whether the symptoms and illness are of natural or unnatural origin. An imbalance in harmony between the physical and the spiritual worlds, such as dietary or lifestyle excesses, can cause a natural illness. For example, diabetes is considered a natural illness, but most Haitians do not seek immediate medical assistance when they detect the symptoms of polyuria, excessive thirst, and weight loss. Instead, they attempt symptom management by making dietary changes on their own by drinking potions or herbal remedies. When the person finally seeks medical attention, she or he may be very sick. At this point, the health-care provider should be cautious in explaining the condition and use a culturally specific approach when explaining the medical regimen, diet, and medications.

Pain is commonly referred to as ***doule***. Many Haitians have a very low pain threshold. Their demeanor changes, they are verbal about the cause of their pain, and they sometimes moan. They are vague about the location of the pain because they believe that it is not important; they believe that the whole body is affected because disease travels. This belief makes it very difficult to accurately assess pain. Injections are the preferred method for medication administration, followed by elixirs, tablets, and capsules.

Chest pain is referred to as *doule nan ke mwen*, abdominal pain is *doule nan vent*, and stomach pain is *doule nan ke mwen* or *doule nan lestomak mwen*. Oxygen should be offered only when absolutely necessary because the use of oxygen is perceived as an indicator of the seriousness of the illness.

Nausea is expressed as *lestomak/mwen ap roule*, *M santi m anwi vomi*, *lestomak/mwen chaje*, or *ke mwen tounin*. Those who are more educated may express their discomfort as nausea. Because of modesty, they may discard vomitus immediately so as not to upset others. Specific instructions should be given regarding keeping the specimen until the practitioner has had a chance to see it.

Fatigue, physical weakness known as *febles*, is interpreted as a sign of anemia or insufficient blood. Symptoms are generally attributed to poor diet. Patients may suggest to the health-care provider that they need special care—that is, to eat well, take vitamin injections, and rest. To counteract the *febles*, the diet includes liver, pigeon meat, watercress, bouillon made of green leafy vegetables, cow's feet, and red meat.

Another condition is fright or *sezisman*. Various external and internal environmental factors are believed to cause *sezisman*, thereby disrupting the normal blood flow. *Sezisman* may occur when someone receives bad news, is involved in a frightful situation, or suffers from indignation after being treated unjustly. When this condition occurs, blood is said to move to the head, causing partial loss of vision, headache, increased blood pressure, or a stroke. To counteract this problem, the patient may sit quietly, put a cold compress on the forehead, drink bitter herbal tea, take sips of water, or drink rum mixed with black, unsweetened coffee.

Haitian Americans may strongly resist acculturation, taking pride in preserving traditional spiritual, religious, and family values. This strong hold on cultural views sometimes creates stress leading to depression. The stigma attached to mental illness is strong, and most Haitians do not readily admit to being depressed. A major factor to remember is the strong prevalence of *voudou*, which attributes depression to possession by malevolent spirits or punishment for not honoring good, protective spirits. In addition, depression can be viewed as a hex placed by a jealous or envious individual. Factors that may trigger depression are memories of family in the homeland, thoughts about spirits in Haiti, dreams about dead family members, or guilt and regrets about abandoning one's family in Haiti for the abundance in America. Health-care providers need to be sensitive to the underlying causes of problems and ascertain the need for comfort within specific religious beliefs.

In the case of an unnatural illness, the person's poor health is attributed to magical causes such as a hex, a curse, or a spell that has been cast by someone as a result of family or interpersonal disagreement. The curse takes place when the intended victim eats food containing ingredients such as snake, frog, or spider egg powder, which cause symptoms of burning skin, rashes, pruritus, nausea, vomiting, and headaches (Fishman et al. 1993). These symptoms often coincide with psychological problems manifested by violent attacks, hallucinations, delusions, or "magical possession." Because, under Western medical standards, an evil spirit would be classified as a true psychiatric problem with "culturally diverse manifestations" and not as an actual case of possession, the health-care provider is challenged in assessing and making the appropriate intervention (Fishman et al. 1993). If the health-care provider is aware of witchcraft, *voudou* practices, and the symptoms associated with them, it may prevent (1) incorrectly diagnosing an individual as mentally ill, (2) giving advice that frightens or confuses the patient into thinking an illness is unnatural in origin, or (3) initiating symptomatic treatment that does not reach the underlying stress. The role of the health-care provider is to be sensitive and understanding toward the patient who holds a belief in these traditional practices. Health-care providers should realize that hesitating to offer a specific diagnosis might be more detrimental to the patient than a negative diagnosis.

18.12.6 Blood Transfusions and Organ Donation

Most Haitians are extremely afraid of diseases associated with blood irregularities. They believe that blood is the central dynamic of body functions and pathological processes; therefore, any condition that places the body in a "blood-need" state is believed to be extremely dangerous. Patients and their families become emotional about blood transfusions. Thus, these are received with much apprehension. In addition, as in all societies, blood transfusions are feared because of the potential for HIV transmission. Health-care providers should explain the need for a blood transfusion factually and carefully clarify the procedure along with the involved risks. Health-care providers should involve patients and their families in the care as much as possible. Precautionary measures that have been taken to prevent blood contamination should also be explained.

Because Haitians hold strong religious beliefs about life after death, the body must remain intact for burial. Thus, organ donation and transplantation are not generally discussed. Since migrating to the United States, some Haitians have, with considerable distress, participated in organ transplantation. A prime concern is transference, believing that through the organ donor, the donor's personality will "shift" to the recipient and change his or her being. Health-care providers should assess Haitian patients' beliefs about organ donation and involve a religious leader to provide support and help facilitate a decision regarding organ donation or transplantation. Because some Haitians' knowledge and understanding in this area is limited, the health-care provider should be proactive by promoting health education.

18.13 Health-Care Providers

18.13.1 Traditional Versus Biomedical Providers

In general, most Haitians resort to symptom management with self-care first and then spiritual care. They commonly use traditional and Western health-care providers simultaneously (see Sects. 18.11 and 18.12.3).

18.13.2 Status of Health-Care Providers

Haitians are very respectful of physicians and nurses. Physicians are men and nurses are women. Nurses are referred to as "Miss." By incorporating culturally specific strategies in their program, health-care providers inspire confidence and trust. Haitian patients who have had limited contact with American health-care systems may have limited understanding of biomedical concepts. Health-care providers need to take the time to explain and re-explain relevant points to compensate for patients' deficient knowledge or language limitations. Health-care providers who show compassion and sensitivity toward Haitian patients achieve greater success in educating patients, families, and the community.

References

Aristide MV (1995) Economics of liberation. Roots 1(2):20–24

BBC News (2011) BBC. http://www.bbc.co.uk

Breyer SJ (2018) Using the organization of American states to end the abuse of restaveks. Columbia Human Rights Law Rev 48(1):146

CIA World Factbook (2020). https://www.cia.gov/library/publications/the-world-factbook/geos/ha.html

Dorestant N (1998) A look at Haitian history from a Haitian perspective. http://www.geocite.com

Fishman BM, Bobo L, Kosub K, Womeodu RJ (1993) Cultural issues in serving minority populations: emphasis on Mexican Americans and African Americans. Am J Med Sci 306:160–166

Haiti: Early History to Independence (2007). http://www.infoplease.com/ce6/world/A0858544.html

Humanium (2016) Humanium. http://www.ilo.org/public/french/comp/child/download/pdf/esclavage.pdf

Jean-Baptiste AR (1985) The black man: Ti Manno in public. St. Aude Records, New York

Jean-Bernard L (1983) Impossible alphabetization. Des Antilles SA, Port-au-Prince, Haiti

Laguerre MS (1981) Haitian Americans. In: Sana L (ed) Handbook of immigrant health. Springer, New York, NY, pp 194–196

Louis-Juste A (1995) Popular education and democracy. Roots 2(1):14–19

McCann BS, Scheele L, Ward N, Roy-Byrne P (2006) Childhood inattention and hyperactivity symptoms self-reported by adults with Asperger syndrome. Psychopathology 39:45–54

McKenzie AD (2010) France: time to pay back Haiti. Available from IPS Interpress Service. http://ipsnews.net

Pan American Health Organization (PAHO) (2001) Regional core health data. http://www.paho.org/English/SHA/glossary.htm

Pan American Health Organization (PAHO) (2011) Disaster in Haiti—one year later. http://www.paho.org/disasters

Paravisini L (2010) Dominican Republic fears increased migration from Haiti. http://repeatingislands.com

Power C (2010) Citizenship and immigration Canada secures fast family reunification or Haitians affected by earthquake. http://www.cic.gc.ca

Prudent N, Johnson P, Carroll J, Culpepper L (2005) Attention deficit disorder: presentation and management in the Haitian American child. Prim Care Companion J Clin Psychiatry 7(4):190–197

Saint-Domingue (2011) Stamping down on Haiti's restavek shame. http://citizenhaiti.com

Santana MA, Dancy BL (2000) The stigma of being named "AIDS carriers" on Haitian-American women. Health Care Women Int 21:161–171

Schaaf B (2009) Child slavery in Haiti: CNN covers Jean Robert Cadet Foundation. http://haitiinnovation.org/fr/about/about_haiti_innovation

Simon JJ, Kleschnitzki S, Shusterman J (2011) Children in Haiti: One year after—the long road from relief to recovery. http://unicef.org
Statistics Canada (2006). http://www40.statcan.gc.ca/z01/cs0001-eng.htm
U.S. Census Bureau (2010) The population with Haitian ancestry in the United States: 2009. http://www.census.gov/prod/2010pubs/acsbr09-18.pdf
U.S. Immigration and Naturalization Service (2006) Statistical yearbook of the immigration and naturalization service 2005. U.S. Government Printing Office, Washington, DC
World Bank (2020) The World Bank annual report 2020. http://www.worldbank.org
World Bank (2015) Investing in people to fight poverty in Haiti: Reflections for evidence-based policy making (English). World Bank Group, Washington, DC. http://documents.worldbank.org/curated/en/222901468029372321/Reflections-for-evidence-based-policy-making
Zissis C (2010) The Haitian migration debate. http://www.as-coa.org

19 People of Hindu Heritage

Monica Scaccianoce and Maria De Los Santos

19.1 Introduction

The Indus Valley civilization, one of the oldest in the world, dates back at least 5000 years. Aryan tribes from the northwest infiltrated onto the Indian subcontinent about 1500 B.C.; their merger with the earlier Dravidian inhabitants created the classical Indian culture. Arab incursions starting in the eighth century and Turkish in the twelfth were followed by those of European traders, beginning in the late fifteenth century. By the nineteenth century, Britain had assumed political control of virtually all Indian lands. Indian armed forces in the British army played a vital role in both world wars. Nonviolent resistance to British colonialism led by Mohandas Ghandi and Jawaharlal Nehru brought independence in 1947. The subcontinent was divided into the secular state of India and the smaller Muslim state of Pakistan. A third war between the two countries in 1971 resulted in East Pakistan becoming the separate nation of Bangladesh (CIA World Factbook, 2020). Despite problems related to overpopulation, environmental degradation, extensive poverty, and ethnic and religious strife, India is rising on the world stage due to rapid economic development. In January 2011, India assumed a nonpermanent seat in the UN Security Council for the 2011–2012 term (CIA World Factbook, 2020).

This chapter is an update from a previous edition written by Jaya Jambunathan.

M. Scaccianoce (✉) · M. D. L. Santos
Florida International University, Miami, FL, USA
e-mail: mscaccia@fiu.edu; delossan@fiu.edu

19.2 Overview, Inhabited Localities and Topography

19.2.1 Overview

India, located in southern Asia, has a landmass approximately one-third that of the United States. with over 1.18 billion people. The population demographics are Indo-Aryan 72%, Dravidian 25%, Mongoloid and other 3%. More than 80 percent are Hindus; 13.4 percent are Muslim; 2.3 percent are Christian; 1.9 percent are Sikh; 1.8 percent are other; and 0.1 percent are unspecified (CIA World Factbook, 2020). These divisions have historically caused tensions between different religious groups. Although different religious sectors share many common cultural beliefs and practices, they differ according to the variant cultural characteristics (see Chap. 2). Hindi and English are India's official languages, but there are also 17 regional languages that are considered official. India has several cities that have undergone place name changes, such as Bombay being renamed Mumbai. These changes were mainly done in an effort to return the city names to local dialects instead of British translations.

© Springer Nature Switzerland AG 2021
L. D. Purnell, E. A. Fenkl (eds.), *Textbook for Transcultural Health Care: A Population Approach*, https://doi.org/10.1007/978-3-030-51399-3_19

Physical characteristics influencing the history and civilization of India are the size of the country and the comparative isolation provided by the Himalayas. The country suffers from droughts, flash floods and widespread and destructive flooding from monsoonal rains, severe thunderstorms, earthquakes, deforestation, soil erosion, overgrazing, desertification, air pollution from industrial effluents and vehicle emissions, and water pollution from raw sewage and runoff of agricultural pesticides. Tap water is not potable throughout the entire country.

India's long-term challenges include widespread poverty, inadequate physical and social infrastructure, limited non-agricultural employment opportunities, insufficient access to quality basic and higher education, insufficient access to quality basic and advanced healthcare, and accommodating rural-to-urban migration. Despite these challenges, India has capitalized on its large educated English-speaking population to become a major exporter of information technology services and software workers (CIA World Factbook, 2020).

19.2.2 Inhabited Localities

According to the UN department of economic and social affairs data, India has the largest diaspora in the world since 1997 (2020). The diaspora has significant numbers in Asia, Europe, Africa, North America, South America, and Oceania creating extremely diverse profiles (Statistics of Indians Abroad, 2020).

19.2.3 Heritage and Residence

Immigrants from India come predominantly from urban areas and include all major Indian states. Earlier immigrants represented a small and transitory community of students, Indian government officials, and businesspeople, and came from a diverse linguistic, religious, regional, and caste population (*caste* is a hereditary social class, discussed later under Spirituality). Asian Indian immigrants to North America came in two waves. The first wave began in the early twentieth century and continued to the mid-1920s. Conditions such as racial discrimination and lack of access to economic advancement made it difficult for the first wave of Asian Indians to sustain themselves or their culture. More than three-quarters of the 7000 Asian Indian immigrants in this wave came from the northwest of India, primarily from Punjab, and 90 percent of the Punjabis were Sikhs. Most Punjabis worked as manual laborers, first in Canada and later on the West Coast of the United States. Other Asian Indian immigrants, who were professionals and businesspeople, were mostly Hindus and Muslims who settled in San Francisco, Los Angeles, New York City, and the Midwest. The second wave of immigration began after 1965 and still continues. Most individuals from this wave are highly educated, skilled professionals and were predominantly from the urban middle class. The colonial authority of the British Raj engrained in the Indian mentality that foreign education is better than indigenous training.

Asian Indians in the United States as reported by the American Community Survey and reported in *Little India*, currently number over 2.4 million, with the largest Indian American populations in New York, Illinois, California, in that order (Pew Research Center 2013–2015). There are also large Indian American populations in Florida, Georgia, Maryland, Michigan, Pennsylvania, Ohio, Texas, and Virginia. The New York metropolitan area, consisting of New York City and adjacent areas in the state of New York, as well as nearby areas in New Jersey, Connecticut, and Pennsylvania, are home to approximately 600,000 Indian Americans (Asian Indian Population in 2013–2015).

In relation to cultural value systems of the first-generation and second-generation Asian Indian immigrants, first-generation Asian Indians are acutely aware of readily apparent cultural differences. Their modern and traditional ideas are in conflict with Indian culture clashing, with American culture, and theory clashing with practice, inside and outside the home. The basis for interactions outside the home is the dominant culture, whereas inside the home, first-generation

Asian Indians attempt to preserve their cultural and religious heritage and abide by Indian cultural values. For second-generation Asian Indians, the conflict of being the "in-betweens" become accentuated. Like their parents, the second-generation Indians also compartmentalize their life inside and outside the home. Conflicts typically arise from the cultural clash of how second-generation Asian Indians perceive American Individualism versus Indian communitarianism, in which career decisions are based on their impact on the family's financial well-being, not the individual's.

19.2.4 Reasons for Migration and Associated Economic Factors

Asian Indians leave their country for a variety of reasons, the most important of which is to attain a higher standard of living. The reasons for an overwhelming majority were financial factors. Although Asian Indians did migrate for financial reasons, they also left for professional, educational, and social opportunities. For many Asian Indians, emigration was thought prestigious. The prospect of greater material prosperity, combined with better working conditions, enhanced the appeal of a wider range of job opportunities. Secondary reasons included opportunities for additional education as well as Indian perceptions of the United States as a land of opportunity and freedom. Immigrants include parents who come for the sake of their children and those who come on student visas and later change to permanent resident status.

India is a source, destination, and transit country for men, women, and children trafficked for the purposes of forced or bonded labor and commercial sexual exploitation. The large population of men, women, and children—numbering in the millions—in debt bondage faced involuntary servitude in brick kilns, rice mills, and embroidery factories, whereas some children endured involuntary servitude as domestic servants.

Internal trafficking of women and young girls for the purposes of commercial sexual exploitation and forced marriage also occurs; the government estimates that 90 percent of India's sex trafficking is internal. Young boys from Afghanistan, Pakistan, and Bangladesh are trafficked through India to the Gulf States for involuntary servitude as child camel jockeys. Indian men and women migrate willingly to the Persian Gulf region for work as domestic servants and low-skilled laborers, but some later find themselves in situations of involuntary servitude, including extended working hours, non-payment of wages, restrictions on their movement by withholding their passports or confinement to the home, and physical or sexual abuse. Despite the reported extent of the trafficking crisis in India, efforts are in progress to prosecute traffickers and protect trafficking victims and rescue victims of commercial sexual exploitation, forced child labor, and child armed combatants. The critical challenge overall is the lack of punishment for traffickers, effectively resulting in impunity for acts of human trafficking (CIA World Factbook, 2020).

19.2.5 Educational Status and Occupations

Most Asian Indians speak English, and many also speak another language. Because the immigration laws of 1965 granted immigrant visas only to people with certain professional and educational backgrounds, most Hindus in the United States possess high educational qualifications. However, those granted visas on the basis of marriage or relationships, such as parents, do not necessarily have the same educational backgrounds (see Table 19.1).

19.3 Communication

19.3.1 Dominant Language and Dialects

Although English enjoys associate status in India, it is the most important language for national, political, and commercial communication. **Hindi**, with 1652 dialectical variations, is the

Table 19.1 Demographic overview—custom region—India

Demographic Indicators	1995	2005	2015	2020	2025
Population					
Midyear population (in thousands)	920,585	1,090,973	1,251,696	1,326,093	1,396,046
Growth rate (percent)	1.9	1.5	1.2	1.1	1.0
Fertility					
Total fertility rate (births per woman)	3.4	2.8	2.5	2.4	2.3
Crude birth rate (per 1000 population)	28	23	20	18	17
Births (in thousands)	25,970	25,507	24,471	24,108	23,719
Mortality					
Life expectancy at birth (years)	60	65	68	70	71
Infant mortality rate (per 1000 births)	75	58	42	35	30
Under 5 mortality rate (per 1000 births)	109	81	56	47	39
Crude death rate (per 1000 population)	10	8	7	7	7
Deaths (in thousands)	8819	8695	9162	9614	10,191
Migration					
Net migration rate (per 1000 population)	-0	-0	-0	-0	-0
Net number of migrants (in thousands)	−74	−55	−50	−53	−56

From U.S. Census Bureau, International Data Base (2019)

national language and primary tongue of 41 percent of the people. Other official languages are Bengali, Telugu, Marathi, Tamil, Urdu, Gujarati, Malayalam, Kannada, Oriya, Punjabi, Assamese, Kashmiri, Sindhi, and Sanskrit. Hindustani is a popular variant of Hindi/Urdu spoken widely throughout northern India, but it is not an official language (CIA World Factbook, 2020).

Because of regional dialects in the main language, health-care providers must be simple and direct in their communication and clear in their enunciation. Communication difficulties may not be apparent in well-educated Hindus. However, with the arrival of parents and grandparents who may not speak English, it is of utmost importance for health-care providers to have an interpreter available to provide quality health care. Hindus, especially women, often speak in a soft voice, making it difficult to understand or decipher what they say. The speech is coupled with an accent, further compromising communication with individuals of other cultures.

19.3.2 Cultural Communication Patterns

Hindus have close-knit family ties. Men especially may become intense and loud when they converse with other family members. To an onlooker, it might seem disruptive, but in general, this form of communication can be construed as meaningful when it is conducted with close friends.

Women are expected to strictly follow deference customs—that is, direct eye contact is avoided with men, although men can have direct eye contact with one another. Direct eye contact with older people and authority figures may be considered a sign of disrespect. More often, men and women use head movements and hand gestures to emphasize the spoken word. Strangers are greeted with folded hands and a head bow that respects their personal territory. Touching and embracing are not acceptable for displaying affection. Even between spouses, a public display of affection such as hugging or kissing is frowned upon, being considered strictly a private matter. Despite these societal constraints regarding the outward display of affection, Hindus are extremely family oriented and nurture one another in sickness, whether at home or in the hospital setting.

19.3.3 Temporal Relationships

According to the Hindu theory of creation, time (in Sanskrit *kal*) is a manifestation of God. The past, the present, and the future coexist in God simultaneously. Hence, the Hindu concept of time is past, present, and future oriented, depending on generation, socioeconomic status, and educational level. The Hindu value on educational attainment denotes a futuristic temporality.

Because of the Indian worldview of the cyclic nature of the universe and belief in reincarnation, Indians have a relaxed attitude toward time (Jain, 1992). Due to the Hindu broad concept of time, adherence to the North American parameters of time may not be rigid. Punctuality in keeping scheduled appointments may not be considered important. Health-care providers must understand the value placed on time by Hindus and not misconstrue being late for appointments as a sign of irresponsibility or not valuing their health.

19.3.4 Format for Names

Women adhere to a specified linguistic style when talking with their husbands. The hierarchical structure of interrelationship is built into the structure of language. The woman refers to the man in the plural *Avar* and *Aap*, meaning "you" (with respect), whereas the man can use a singular "you" like *Ne*, *Aval*, or *Thum*. *Aap* means "thou" and is used for elderly family members and for strangers. Older family members are usually not addressed by name but as elder brother, sister, aunt, or uncle. A woman never addresses a man by name because the woman is not considered an equal or a superior. However, exceptions to this practice occur when the woman is older than the man.

The system of "naming" customs in India is complex and relative to the social and cultural structures. The naming customs are closely related to Hindusim, chaturvarna system of castes (see Sect. 19.11), clan, and lineage (Jayaraman, 2005).

19.4 Family Roles and Organization

19.4.1 Head of Household and Gender Roles

No institution in India is more important than the family. The family was originally patriarchal and the joint family evolved from it, the transition arising from the death of the common ancestor or the patriarch of the family. The hierarchical structure of authority in the patriarchal joint family, based on the principle of superiority of men over women, is the most important instrument of social control. The rights and duties of individuals are prescribed by the hierarchical order of power and authority. The male head of the family is legitimized and considered sacred by caste and religion that delineate relationships.

The central criteria of the Hindu joint family include (a) family property jointly owned by men and inheritable only by the male lineage (although by law it is to be shared equally among both male and female offspring), (b) the hierarchical structure of authority according to gender and age, and (c) the dependence of women and children. Central relationships in this system are based on continuation and expansion of the male lineage through inheritance and ancestor worship, related to the father–son and brother–brother relationships.

Family plays a significant role in the Indian culture. For generations, India has had a prevailing tradition of the joint family system. This is a system under which extended members of a family—parents, children, the children's spouses and their offspring, and so on—live together. Usually, the eldest male member is the head in the joint Indian family system. He makes all important decisions and rules and other family members abide by them.

Within the joint family system, the patrilineal system created a sense of worthlessness, servitude, and dependence for women characterized by a lack of freedom, as well as constraints and limitations that suppressed individual development. A submissive and acquiescent role is expected of women in the first few years of married life, with little or no participation in decision making. Strict norms govern contact and communication with the men of the family, including a woman's husband. However, in recent times, many families, especially in urban areas, have stopped abiding by the extended family system and have started living as a nuclear family.

Although a patrilineal system is not characteristic of the nuclear family, the distinctions between men and women persist. This is also true of the matrilineal system that exists in a few areas in the southwestern and northeastern regions of the country. In a matrilineal system, the lineage is counted through the woman, but power rests with the men in the family. Hence, even in a matrilineal system, constraints abound for women because of power distribution that promotes male dominance.

In the Indian household, lines of hierarchy and authority are clearly drawn, shaping structurally and psychologically complex family relationships. Ideals of conduct are aimed at creating and maintaining family harmony. All family members are socialized to accept the authority of those ranked above them in the hierarchy. In general, elders rank above juniors, and among people of similar age, males outrank females. Among adults in a joint family, a newly arrived daughter-in-law has the least authority. Males learn to command others within the household but expect to accept the direction of senior males. Ideally, even a mature adult man living in his father's household acknowledges his father's authority on both minor and major matters. Women are especially strongly socialized to accept a position subservient to males, to control their sexual impulses, and to subordinate their personal preferences to the needs of the family and kin group. Reciprocally, those in authority accept responsibility for meeting the needs of others in the family group.

Much has changed in the status and roles of women; however, most Hindu women remain subservient to their closest male relatives, a situation that is gradually changing. Hindu society is trying to redefine the role of women in the institution of family and society. Politically, Hindu women today enjoy an equal status with men and wider opportunities than their counterparts in many Western countries. Although there is ongoing discussion to provide them with new privileges and rights, including inheritance rights, much still needs to be done on the social and economic front. Women in Hindu society still suffer from gender bias and a number of other problems such as dowry, inheritance, domestic abuse, sexual exploitation, rape, and harassment.

Changing role status is further illustrated by Varghese and Jenkins (2009), who studied the variables that might be related to high cultural conflict among first- and second-generation Asian Indian immigrant women and the psychological consequences of cultural value conflict. Self-report data from 73 community-dwelling women were used to examine women's recollections of parental overprotection, their reports of cultural value conflict, and their ratings of self-esteem and depression symptoms. Varghese and Jenkins found that the results supported the hypotheses that unmarried and second-generation women would report greater maternal control and cultural value conflict than married and first-generation women. Second-generational status, high maternal control, and high cultural value conflict correlated with higher depressive symptoms, while being married, low maternal control, and low cultural conflict were related to high self-esteem.

19.4.2 Prescriptive, Restrictive, and Taboo Behaviors for Children and Adolescents

Hindu parents in general want their children to be successful and strongly encourage and emphasize scholastic achievement in fields that promise good secure employment and a high social status. Hindu parents in America and elsewhere want their children to be successful and maintain ties with their families and the Indian community. Thus, parents face a dilemma between aspiring for the American dream of success for their children and holding on to their desire to maintain Indian customs and values. Status indicators such as education, income, community and occupational leadership tend to replace ascribed social status.

The birth of a male child is considered important, and the desire for a male child rather than a female child is still prevalent. Furthermore, widowhood, especially for women, is considered a negation of marriage (Perkins et al. 2016). In America, Hindu parents may have reservations about eventual marriage partners for their children and concerns about the issues of dating, premarital sex, and freedom. Although many Hindu parents expect and accept the Westernization of their children, the question of marriage is still a concern for Hindu parents who have opinions about how their children should be married, whether "arranged" or partly arranged. Hindu parents or Indians from all religious traditions want their children to marry other Indians. Health-care providers should understand the various types of families (joint, extended, or nuclear) and should determine which individual has control within the hierarchy.

Arranged marriages at a young age are considered most desirable for women. This practice is related to the importance of virginity and restrictions placed on marriage within the same clan. For centuries, arranged marriages have been the tradition in Indian society. Even today, the vast majority of Indians have their marriages planned by their parents and other respected family members, with the consent of the bride and groom. They also demand a dowry, which has been outlawed by the Indian government, but Indian society and culture still promote and maintain it. They avoid detection by not letting authorities know about any money arrangements. Arranged matches are made after taking into account factors such as age, height, personal values and tastes, the backgrounds of their families (wealth, social standing), their castes, and the astrological compatibility of the couple's horoscopes.

In India, since marriage is thought to be for life, the divorce rate is extremely low, and arranged marriages generally have an even lower divorce rate. Divorce rates have risen significantly in recent years. There is conflict of opinion over what the phenomenon means—whether, for traditionalists, the rising number of divorces portends the breakdown of society or, for modernists, creates a healthy new empowerment for women.

Although arranged marriages are still a preferred choice among the younger generation, education and liberalization of ideas in urban areas have led to changes in selecting a marriage partner. The practice of an arranged marriage continues in the United States to minimize the stress associated with differences in caste, lifestyles, and expectations between the male and the female hierarchy.

The two major types of transfer of material wealth accompanying marriage are bride price and a dowry. The bride price is customarily prevalent among patrilineal tribes and the middle and lower castes of nontribal populations. Bride price is payment in cash and other materials to the bride's father in exchange for authority over the woman, which passes from her kin group to the bridegroom's kin group. In communities that follow this custom, a daughter is not regarded as a burden, and parents do not dread the thought of marriage. A daughter brings wealth to the family as a result of marriage.

Regional variations exist in understanding the dowry system. Dowry may be seen as the gift given to the bride and often settled prior to marriage, which may not be regarded as her property; as a gift given to the bridegroom before and at the time of marriage; or as a present to the groom's relatives. The practice of dowry in the Hindu community has a number of cultural and social sanctions. Dowry is regarded as essential to obtain a suitable match for a young woman, ensuring a high standard of life.

The increase in social and economic inequality is one of the most important inducements for a dowry and operates at all levels of society. Wealth ranges from a few hundred to thousands of rupees (Indian currency), and behind this transaction is a direct desire to improve the daughter's social status, which indirectly assists the social status of the bridegroom's family. The desire to obtain security and good status for the daughter places the bride's parents in a vulnerable position, in which they are faced with demands bearing no relation to their economic capacity. This may reduce them to a state of indebtedness.

19.4.3 Family Goals and Priorities

In the joint family structure, Hindu women are considered "outsiders" and are socialized and incorporated in such a way that the "jointness" and residence are not broken up. This means that a close relationship between the husband and the wife is disapproved because it induces favoring the nuclear family and dissolving the joint family.

The Hindu family's goals and priorities include the most important Hindu sacrament, the Vivaha (marriage), which is a religious and social institution. The marital union is a matter for the husband and wife, society, guardians, and supernatural powers that symbolize spirituality. Therefore, a marriage is regarded as indissoluble. The sacrament of marriage impresses on a person that earthly life is not to be despised; rather, it should be consciously accepted and elevated to the level of a spiritual experience.

In patrilineal societies, marriage signifies a transfer of a woman from her natal group to that of her husband. Marriage is not considered primarily an affair of the man and woman who are getting married but an event that involves the kin of both spouses. Hence, the institution of marriage is a means by which alliances are created or strengthened between two or more groups.

Family elders are held in reverence and cared for by their children when they are no longer able to care for themselves. Families believe that knowledge is transmitted through an oral tradition derived from experience, and the elderly are repositories of such knowledge.

19.4.4 Alternative Lifestyles

Religion has played a role in shaping Indian customs and traditions. While homosexuality has not been explicitly mentioned in the religious texts central to Hinduism, it has taken various positions, ranging from positive to neutral or antagonistic. Historical literary evidence indicates that homosexuality has been prevalent across the Indian subcontinent throughout history and that homosexuals were not necessarily considered inferior in any way. Whereas homosexuality probably occurs as frequently as in any group, this lifestyle may cause a social stigma, and there is a high degree of stigma associated with homosexuality in India. However, attitudes toward homosexuality have shifted slightly in recent years. In particular, there have been more depictions and discussions of homosexuality in the Indian news media. In mid-2009, the New Delhi High Court decriminalized homosexual intercourse between consenting adults, throughout India, making HIV education and surveillance very difficult. Outreach workers are often harassed and even arrested for "promoting homosexuality." It is estimated that 42 percent of all men who have sex with men are also married, fueling the increasing HIV infection rate among women of India (UNAIDS 2020).

Reflective Exercise 19.1

Revathy Srinivasan, a Hindu woman aged 25 years, has lived in the United States for 9 years, and her parents have just arrived. She wishes to marry Velayudham Mani (Vel), a man she met in graduate school, who is also a Hindu. However, her parents will not allow her to marry Vel because he is of a lower caste than she is. Besides, her parents want her to marry their friends' son Ajay, who is well established. Revathy's parents feel that both their and their friends' family backgrounds (wealth and social standing) are congruent, in addition to the astrological compatibility of both children's horoscopes. If Revathy insists on marrying Vel, they will not give Vel's family any dowry.

1. How common are arranged marriages among Hindus in India? In the United States?
2. What is a dowry?
3. What is the significance of a dowry?
4. Describe the caste system of Asian Indians.
5. Should Revathy seek counseling to help her solve her dilemma? If so, from what type of counselor should she seek help?

Gay people in India may not be completely liberated, but they are more willing than ever before to challenge curiosity, even rejection, without allowing it to damage their conscience. In recent years, many 18- to 24-year-olds have come out. They are honest with their parents. This is different from the times when parents would force a heterosexual marriage on them to "normalize" things. But in modern India the acceptance is different for male gays and lesbians because of lack of resources or educational opportunities. The Sahayatrika group of Kerala, which recently did a study on lesbian suicides, has catalyzed a reconsideration of same-sexuality. There has been an increase in support groups with separate help lines catering to Hindi-speaking, English-speaking, and transgender groups. In West Bengal, the "coming out" phenomenon is seen even in smaller districts, with networks like Manas Bangla, headquartered in Kolkata (Ali et al. 2012; Kalra, 2012; Aich et al. 2015).

A paucity of information exists regarding Hindu gay or lesbian couples in the professional literature. Health-care providers can refer lesbian, gay, or bisexual Hindu Americans to the Gay and Lesbian Vaishnava Association Inc. (2005), a resource for Hindus and other Southeast Asian groups. Single-parent, blended, or communal families are not well accepted by Hindus. In addition, two magazines, *Trikone*, the first quarterly magazine for gay South Asian men and women, and *India Currents*, a monthly arts and entertainment magazine, are targeted to Indo-Americans living in California.

19.5 Workforce Issues

19.5.1 Culture in the Workplace

With comparative ease, most Hindus have become part of the skilled workforce in America and elsewhere. Hard work, interest in saving and investment, and business acumen enable many to become financially successful. Because of their educational and professional background, it is not difficult for most to find suitable employment and improve their economic status.

Many Hindus have a singular devotion to their career, profession, or business, resulting in a personal cost evidenced in family relationships or in health status. Demonstrating hospitality is important to Hindus. A new friendship is not formally acknowledged among Hindus without the reciprocity of home visits. Thus, Americans who refer to their Indian acquaintances and colleagues as "friends" without having extended hospitality to them in their homes might confuse the Hindu immigrant's notions of friendship.

At work, Hindus upon emigration, adopt the practices and cultural habits of their new home, however, at home and at Indian gatherings, they retain many of their own cultural practices. Active participation in Indian organizations is a growing phenomenon, especially in the absence of the cultural milieu available in India. Hindu Americans believe that such participation is the only way their children can become aware of their Indian heritage. Currently in America and elsewhere, numerous regional Indian organizations are available throughout the country, resulting in a vast network of communication. Religious revivalism and social conviviality are the hallmarks of Hindu adaptation.

19.5.2 Issues Related to Autonomy

An early realization of immigrants in a new country is that they must build new relationships and find new reference groups. Hierarchies of age, gender, and caste prescribe transactions among Hindus. At work, relationships are a reproduction of the authority-dependence characteristic of family and social relationships. In seeking to establish a personal and benevolent relationship, Hindus may be seen as too eager to please, ingratiating, or docile, all antithetical to the task of assertion and independence.

Hindus speak English as well as their regional languages at home. Therefore, they rarely have any difficulty with communication in the American workforce. However, because most Hindus have learned and speak British English, those unfamiliar with British English and idioms may have difficulty understanding them and should ask for clarification.

19.6 Biocultural Ecology

19.6.1 Skin Color and Other Biological Variations

Asian Indian Hindus evidence a diversity of physical types. Asian Indians can be divided into three general groups according to the color of their skin: white in the north and the northwest, yellow in the east, and black in the south. Whites, Indids, have a light brown skin color, wavy black hair, dark or light brown eyes. They are tall or of medium height and are either dolichocephalic (i.e., long-headed) or brachycephalic (i.e., short-headed). The physical type of the Indids varies according to regions, ranging from a light to a brown skin color. The yellow races are found in the periphery of India in the areas bordering Tibet and Assam.

Black-skinned people, Melanids, are often referred to as the *Dravidians*, the population of southern India. The Melanids have dark skin (ranging from light brown to black), elongated heads, broad noses, thick lips, and black, wavy hair. They are usually less than 5 feet 6 inches tall. The most characteristic Melanids are the Tamils, a major linguistic and cultural group in South India.

Because skin color varies regionally, healthcare providers must be careful when arriving at a diagnosis that may be applicable only to white-skinned people. Pallor in brown-skinned patients may present as a yellowish-brown tinge to the skin. Pallor in dark-skinned individuals is characterized by the absence of the underlying red tones in the skin. Furthermore, jaundice may be observed in the sclera and should not be confused with the normal yellow pigmentation of the dark-skinned black patient. In addition, the oral mucosa of dark-skinned individuals may have a normal freckling or pigmentation. Cyanosis can often be difficult to determine in dark-skinned individuals. A close inspection of the nailbeds, lips, palpebral conjunctivae, and palms of the hands and soles of the feet shows evidence of cyanosis.

19.6.2 Diseases and Health Conditions

The rainy season in the tropics is associated with an increase in malaria. Asian Indians migrating from the tropical regions may be susceptible to malaria, which intensifies during the monsoon season. Filariasis is prevalent in some parts of India. Respiratory infections such as tuberculosis and pneumonia are also widely prevalent in the midlatitudes and, in the rainy season, in monsoon areas. Respiratory infections occur in the most densely populated river valleys and coastal lowlands and in dark, intensely crowded urban areas. Major infectious diseases include food- or waterborne diseases like bacterial diarrhea, hepatitis A and E, and typhoid fever; vector-borne diseases like chikungunya, dengue fever, Japanese encephalitis, and malaria; animal contact diseases like rabies; and water contact diseases like leptospirosis. Highly pathogenic H5N1 avian influenza has also been identified (CIA World Factbook, 2020).

When performing health assessments, health screenings, and physical examinations, healthcare providers must be alert to possible signs and symptoms of the risk factors associated with different diseases linked to migration from different regions of India.

The four leading chronic diseases in India, as measured by their prevalence, are, in descending order, cardiovascular diseases (CVDs), diabetes mellitus (diabetes), chronic obstructive pulmonary disease (COPD), and cancer. All four of these diseases are projected to continue to increase in prevalence in the near future given the demographic trends and lifestyle changes underway in India (WHO, 2020). Heart disease tends to develop at a very early age in Asian Indians. The major causes of cardiovascular disease are tobacco use, physical inactivity, and an unhealthy diet. India suffers disproportionately from cardiovascular disease. In 2008, Crosta reported that India would bear 60 percent of the world's heart disease burden in the near future. In addition, researchers have determined that compared to people in other developed countries, the average age of patients with heart disease is lower among Indian people, and Indians are more likely to have types of heart disease that lead to worse outcomes.

Diabetes mellitus, insulin resistance, and central obesity are also prevalent among this population, as are high serum levels of lipoprotein (Blesch et al. 1999). Diabetes is second only to CVD as a health burden in India, and, of course, the two are highly correlated and interdependent. The International Diabetes Federation (IDF) reports a projected prevalence of 70 million patients in India by the year 2025 (Sicree et al. 2006; Taylor, 2010), and the World Health Organization (WHO) estimates that India will have 80 million cases of diabetes by 2030 (Wild et al. 2004).

Hindu immigrants have a higher mortality rate than that of the local population. Rheumatic heart disease, together with high blood pressure, is a major cardiac problem. Dental caries and periodontal disease affect 90 percent of the adult population. Sickle cell disease is highly prevalent: the gene is detected in 16.48 percent of selected populations.

Breast cancer is one of the leading causes of morbidity and premature death among women in India (Azamjah, 2019). Based on studies in India, Choudhry et al. (1998) implied that immigrant women from India share the same risk as their Western counterparts.

The most prevalent forms of cancer among men are tobacco-related cancers, including lung, oral, larynx, esophagus, and pharynx. Almost 50 percent more Indian men smoke than men in the United States. Among Indian women, in addition to tobacco-related cancers, cervix, breast, and ovarian cancers are also prevalent. India currently has the highest prevalence of oral cancer cases in the world as a result of the popularity of chewing tobacco in its rural regions.

Sexually transmitted infections (STIs) in children are not uncommon in India, though systematic epidemiological studies to determine the exact prevalence are not available. STIs in children can be acquired via sexual routes or, uncommonly, via nonsexual routes such as accidental inoculation by a diseased individual. Neonatal infections are almost always acquired intrauterine or during delivery.

Sexual abuse and sex trafficking remain important problems in India. Surveys indicate that nearly half of the children are sexually abused. Most at-risk children are street-based, homeless, or those living in or near brothels. The last two decades have shown an increase in the prevalence of STIs in children, though most of the data are from the northern part of the country and from major hospitals. However, due to better availability of antenatal care to the majority of women, cases of congenital syphilis have declined consistently over the past 2–3 decades. Other bacterial STIs are also on the decline. On the other hand, viral STIs such as genital herpes and anogenital warts are increasing. This reflects trends of STIs in the adult population. Concomitant HIV infection is uncommon in children. Comprehensive sex education, stringent laws to prevent sex trafficking and child sexual abuse, and antenatal screening of all women can reduce the prevalence of STIs in children (Dhawan et al. 2010).

19.6.3 Variations in Drug Metabolism

Asians are known to require lower doses and to have side effects at lower doses than Whites for a variety of different psychotropic drugs, including lithium, antidepressants, and neuroleptics (Levy, 1993). Asians are also more sensitive to the adverse effects of alcohol, resulting in marked facial flushing, palpitations, and tachycardia.

Dietary variations may significantly alter the metabolic rate or plasma levels of medicines in Asian Indians (Levy, 1993). The metabolism of antipyrine in Asian Indians living in rural villages in India has been compared with that of Indian immigrants in England. The results indicate that drug metabolism among Indian immigrants becomes more rapid when they adopt the British lifestyle and dietary habits (Levy, 1993). Hence, because of ethnic differences in the rates of drug metabolism, more consideration should be given to individualizing treatment regimens in special population groups, such as Hindus. Health-care providers should question therapeutic regimens that do not consider racial or ethnic differences.

19.7 High-Risk Behaviors

Alcoholism and cigarette smoking among Hindus, especially among men, may cause significant health problems. Adolescents face tremendous pressure to keep up the image of "wiz kids" and to meet the expectations of their parents, which may override individual aptitudes and choices. This may create anxiety and frustration, thereby leading to failure and anger toward parents, which may predispose them to using drugs as a coping strategy. However, adequate literature or studies to substantiate these behaviors are not available.

Other high-risk behaviors include those that lead to contracting HIV. According to Cichocki (2007), the following populations are at the highest risk:

- *Sex workers*—Because of widespread poverty throughout India, women often resort to prostitution as a means of making money for their families. Others are forced into sex work due to an underground of violence and disrespect toward women. Finally, women involved in marital breakups will often begin prostituting themselves as a means of surviving financially after being left with children to feed and a household to support. In some areas of India, it is estimated that one in every two sex workers are HIV infected, many of whom are unaware of it.
- *IV drug use (IVDU)*—The recreational use of drugs often overlaps with the sex trade. While IVDU seems to be worse in the northeastern parts of India, it is common throughout the country. Many attribute the widespread problems of IVDU and HIV to government policies that do not support HIV prevention and risk reduction among IV drug users. Because IVDU is a crime and is consistently enforced and prosecuted, getting prevention messages to users is very difficult. There have been instances of prevention workers themselves being arrested while trying to help and educate IV drug users. Official estimates actually report the HIV prevalence among IV drug users to have gone down from 13 percent in 2003 to 10 percent in 2005 with a recent prevalence reduction of 50%.
- *Truck drivers*—India's economy depends a great deal on its very large trucking network across the country. While truckers help move goods and services throughout India, they also contribute a great deal to the huge HIV population and the spread of HIV from one area of India to another. Truckers will pick up sex workers along their route, engage in unprotected sex activity, and then drop off the sex worker at the trucker's next stop along the route. This has contributed to the spread of HIV from urban areas into the rural towns and villages. The most damaging fact about truckers and their use of sex workers is that they usually do not know they are infected.
- *Migrant workers*—As is the case in the United States, migrant workers in India are very transient and mobile, moving from town to town wherever the work takes them. Unfortunately, they take their risky behavior along with them, fueling the spread of HIV throughout India. While there are attempts at HIV education, the variety of languages, dialects, and cultures makes HIV education very difficult.

High-risk behaviors contribute to the spread of HIV, which has progressed from a disease found in only the highest-risk populations to one found in all segments of the Indian population, including men and women, rich and poor, urban and rural. Populations thought at one time to be low risk are now being infected, as are high-risk groups. Some groups are being infected at a higher rate than others. For instance, women are being infected by way of heterosexual transmission at an alarming rate. Women now make up about 41 percent of those living with HIV. Most are being infected by husbands or boyfriends who have multiple sexual partners, many of whom are infected with HIV and do not know it.

Reflective Exercise 19.2

Dabeet Singh, a practicing Hindu man aged 22 years, has been a student majoring in computer technology for 2 years at a nearby university. His roommate convinced him to see the nurse at the school health clinic because he has been demonstrating high-risk behaviors such as smoking two packs of cigarettes a day and drinking four 12-oz cans of beer every day. In addition, over the last 2 weeks, he has not been attending his classes but sitting in his room, acting anxious and frustrated with his friends and teachers. In a low voice, he tells the nurse that he wants to quit school and return home where people are more polite and can understand him better.

1. Give two plausible reasons for Dabeet's high-risk behaviors from a cultural context.
2. What evidence is there to show that Dabeet might be depressed?
3. What type of spiritual counseling might the nurse suggest?

19.7.1 Health-Care Practices

Patterns of health-seeking behaviors among Hindus are strongly influenced by their sociocultural networks. Customs and beliefs often affect medical-care decisions and choice of health-care services. The actions of supernatural forces and certain human excesses are considered important in causing illness, even among highly educated Hindus. Furthermore, regional variations in the intensity and strength of the belief system are significant. For example, some individuals believe that excessive consumption of sweets may cause roundworms and that too much sexual activity and worry are associated with tuberculosis. In addition, some believe that diarrhea and cholera are caused by a variety of improper eating habits. Therefore, health-care providers cannot take for granted that all Hindu immigrants have the same belief systems with the same degree of intensity. Deep-rooted beliefs about illnesses will inhibit the acceptance of scientific causes for diseases. This may result in difficulty with treatment.

19.8 Nutrition

19.8.1 Meaning of Food

Many cultures have influenced Indian food practices. Dietary habits within the Indian subcontinent are complex, regionally varied, and strongly influenced by religion. Hindus believe that food was created by the Supreme Being for the benefit of humanity; thus, growing, harvesting, preparing, and consuming food are steeped in rituals. Sacred Hindu texts contain aspects related to food, dietary habits, and recommendations.

The influence of religion is pervasive in food selection, customs, and preparation methods. Regional food habits are based on the types of cereals and fresh foods consumed. In the first category are rice and bread eaters; in the second category are vegetarians and nonvegetarians. Whereas Buddhism and Jainism turned people to vegetarianism, the influence of the Vedic religion and later the influx of outsiders made Indians nonvegetarians. In modern India, vegetarianism is firmly rooted in culture and the term *nonvegetarian* is used to describe anyone who eats meat, eggs, poultry, fish, and, sometimes, cheese. Many Brahmins in northern India consider eating meat to be religiously sanctioned. In some parts of India, Brahmins' eating fish is acceptable, whereas in other parts, eating meat of any source is sacrilegious (Kilara & Iya, 1992). Many Indians are vegetarians because of agricultural traditions and adverse economic conditions.

19.8.2 Common Foods and Food Rituals

Although India is essentially an agricultural country, food production is insufficient to adequately feed the entire population. Geographic influences favor the production of grains, rice, wheat, jowar, *bajra*, jute, oilseeds, peanuts, and mustard. The principal food crop consists of rice in the better-watered regions and wheat in the Punjab. Sesame millet, maize, and peas grow throughout the country. Sugarcane and jute are cultivated extensively. The coconut palm is a valuable resource in the southern coastal areas. Cereals supply a large percentage of the total calorie requirements. Rice, wheat, millet, barley, maize, and *ragi* make up the bulk of the diet.

A variety of "pulses" plant foods (legumes such as lentils, chickpeas or dried peas), cooked vegetables, meat, fish, eggs, and dairy products are also consumed. Heavily spiced (curry) dishes with vegetables, meat, fish, or eggs are favored, and hot pickles and condiments are common. Spice choices include garlic, ginger, turmeric, tamarind, cumin, coriander, and mustard seed. Vegetable choices include onions, tomatoes, potatoes, green leaves, okra, green beans, and root vegetables. Milk is used in coffee and tea and in preparing yogurt and buttermilk. Water is the beverage of choice with meals and as a thirst quencher.

One of the most common food items in southern India is boiled rice containing spices and vegetables; it is usually served with a lentil-based sauce, *sambar*. Other common foods in the south are *rasam* (a dilute liquid made from tomatoes, tamarind, and boiled rice served with spices) and *thayir* (yogurt with boiled rice). The traditional southern Indian vegetarian preparations of rice and lentil flour dishes are called *idli*, *dosai*, or *vadai*. Snacks are also consumed either as breakfast items or substituted in place of regular meals. Food is usually served in a *thali*, a round plate. Coffee is popular in southern India, whereas tea is the beverage of choice in the rest of the country. In the north, *chapati*, a bread made from wheat flour, is common, as is *puris*, which is similar to *chapati* except that it is deep-fried, whereas the former is baked on a round iron plate. Seasonal products such as groundnuts (peanuts), mangoes, and bananas are consumed between meals. Savory items such as deep-fried preparations of grains, vegetables, and spices are also consumed as snacks between meals. In northern India, wheat is the staple food. Other cereals are *jowar*, *bajra*, and *ragi*, which are consumed in porridges, gruels, and *rotis* (baked pancakes).

Customs and prejudices often remove certain food items from the diet, although the prohibitions vary from place to place. Thus, *bajra*, a staple food in Maratha families, is not looked on favorably in Uttar Pradesh. People from Punjab do not favor fish, whereas people from the south generally dislike the idea of meat of any kind. In Saurashtra in the south, fish, fowl, meat, and eggs are taboo practically everywhere.

Women generally serve the food and may eat separately from men. This practice may still continue in rural areas and is related to malnutrition in women. Programs have been instituted to change this practice so that females may eat at the same time as their families. (Pandey, 2017). Food preparation has strict rules. In some parts of India due to religious beliefs women are not allowed to cook or have contact with other members of the family during their menstrual periods. This practice is also changing as education about sexual reproductive practices is increasing (Garg & Anand, 2015). Brahmins are the preferred cooks because the cook must be as pure as the eater (Kilara & Iya, 1992). Health-care providers must assess for food rituals practiced by Hindus in relation to mealtimes and food selections before attempting to teach them about medication regimens.

19.8.3 Dietary Practices for Health Promotion

Foremost among the perceptions of Hindus is the belief that certain foods are "hot" and others "cold," and, therefore, should be eaten only during certain seasons and not in combination. The geographic differences in the hot and cold perceptions are dramatic; many foods considered hot in the north are considered cold in the south. Such perceptions and distinctions are based on how specific foods are believed to affect body functions. The belief is that failure to observe rules related to the hot-and-cold theory of disease results in illness. When the three basic principles or humors—*vata*, *pitta*, and *kapha*—are in the state of equilibrium, digestion and metabolism are in order, the foundations of the tissues and excretion of waste products are normal, and an individual is physically and mentally happy.

19.8.4 Nutritional Deficiencies and Food Limitations

Nutritional deficiencies are regionally patterned, indicating preferences for a certain variety of cereals. For example, beriberi is found in rice-growing areas, pellagra in maize-millet areas, and lathyrism in Central India. Thiamine deficiency is common among people who are mostly dependent on rice. Thoroughly milling rice, washing rice before cooking, and allowing the cooked rice to remain overnight before consumption the following day result in the loss of thiamine.

Commitment to the concept of the "sacred cow" has a significant impact in India's economic life and ecology, most notably by encouraging dairy farming and milk use. However, lactose intolerance affects up to one percent of infants and more than 10 percent of adults, resulting in an inability to produce the enzyme needed to digest lactose, or milk sugar. The ability or inability to digest lactose may be due to genetic differences among Asian Indians.

Cereals are still the main source of protein in India. Approximately 50–60% of rural and urban protein intake is from cereals (Minocha et al., 2019) The consumption of a single cereal, such as rice, as the bulk of a diet results in a poor intake of lysine and other essential amino acids. Pellagra, a nutritional deficiency causing skin and mental disorders and diarrhea, occurs largely where people consume mostly maize and sorghum (*jowar*). Both cereals have high leucine content and provide strong evidence for the pathogenesis of pellagra. Lathyrism is a crippling disease-causing paralysis of leg muscles that is seen mostly in adults who consume large quantities of seeds of the pulse *khesari* (*Lathyrus sativus*) over a long period of time. Thus, protein malnutrition is serious and widespread in India. Bharati et al. in 2019 reported that anaemia is a grave concern for Indian women. More than 50% suffer from anaemia due to malnutrition. Malnutrition is lower in urban areas and among families with higher levels of education. Christians have a lower rate of malnutrition as compared to Hindu women.

Goiters are common along the sub-Himalayan tracts, resulting from an iodine deficiency in food and water. Fluorosis occurs in parts of Punjab, Haryana, Andhra Pradesh, and Karnataka, resulting from drinking water with high fluoride content.

India is a diverse country that has experienced modernization in agricultural techniques and an increase in urban inhabitants. Workers may spend hours indoors and due to the increased use of sunscreen while outdoors and poor nutrition vitamin D and calcium deficiencies still exist. Fifty-two percent of the population suffers from nutritional bone disease. The disorders included osteomalacia and rickets where diets are deficient in calcium and vitamin D. Approximately 40% of the population suffered from dietary calcium deficiency during childhood (Harinarayan & Akhila, 2019). Endemic dropsy was prevalent in Northern India as a result of the use of mustard oil for cooking however due to governmental education and labelling requirements the number of cases has declined (Sonkar et al., 2019).

The high incidence of stomach cancer in the south may be due to the excessive intake of fried fatty foods and chilies and rapid food consumption. By contrast, cancer of the stomach is infrequent in people in the north who consume milk and dairy products. In several north Indian states, many people chew betel (paan), which is offered as a sign of hospitality. Paan contains arecanut, cardamom, fennel, lime (calcium hydroxide), tobacco, and other ingredients, and among users cancers of the mouth and lip are common, especially since tobacco can induce oral cancer. Beta carotene from Spirulina or other sources such as carrots can prevent such cancer (Garewal, 1995). Strachan et al. (1995) found a vegetarian diet to be an independent risk factor for tuberculosis among immigrant Asians in south London. Using a case-control method, Asian immigrants from India diagnosed with tuberculosis during the previous 10 years were compared with two Asian control groups. The results confirmed earlier findings that Hindu Asians had an increased risk for tuberculosis compared with Muslims. Religion had no independent influence after adjustment for vegetarianism. The authors concluded that decreased immunocompetence associated with a vegetarian diet might result in increased mycobacterial reactivation among Hindu Asians.

Food practices of the Hindus may remain unchanged with increasing numbers of immigrants to America. In a study of the dietary habits of 73 Asian Indians in relation to the length of residence in the United States, Raj et al. (1999) found that, in contrast to recent immigrants (less than 10 years), long-time immigrants reported eating mostly Indian foods for only dinner and weekend meals. The authors also found that regardless of the length of residence in the United States, consumption of white bread, roots, tubers, vegetable oils, legumes, and tea changed little. Self-reported data indicated that high serum cholesterol levels, increased weight, hypertension, arthritis, and diabetes were diagnosed in respondents older than age 30 years. The authors concluded that despite the small sample size, although Asian Indians in the study included many American foods in their diets, they continued to eat many traditional foods, perhaps in an effort to retain cultural identity.

All major food groups of the Hindus are generally available through Indian grocery and spice stores located in major metropolitan areas throughout the United States. The flavor, spices, and diversity of ethnic foods are making many Indian restaurants popular. Given the diversity of Hindus in America, health-care providers must individually assess dietary practices and nutritional deficiencies of their patients according to their ethnic origins and area of residence.

For the majority of Hindu Americans who are vegetarians, their protein comes from pulses, lentils, legumes, and dairy products. Studies have shown the dietary acculturation gradually occurs in Hindus who live in America (Mahadevan et al., 2014). These individuals may gradually adapt an American diet which includes high-fat dairy products, and other high-calorie and processed foods including desserts and sugary beverages. This change in diet may result in disorders such as obesity, non-insulin dependent diabetes and heart disease.

In remote and inaccessible areas where there is a lack of qualified physicians and modern health facilities, medicinal plants have traditionally been used. Certain foods and/or food additives such as *Spirulina* microalgae enhance the immune system and have antiaging properties and, in combination with turmeric and oil from the *neem* tree (an evergreen tree found in India), can ward off many infectious and noninfectious diseases, including cancer, gastrointestinal disorders, diabetes, skin troubles, dental problems, and cardiovascular disorders. Also, beta carotene (proVitamin A) as an important antioxidant has been reported to inhibit oral carcinogenesis (Sankaranarayanan et al., 1997).

19.9 Pregnancy and Childbearing Practices

19.9.1 Fertility Practices and Views Toward Pregnancy

Methods of birth control among Hindus include intrauterine devices (IUDs), condoms, and the rhythm and withdrawal methods and sterilization (Muttreja & Singh, 2018). Despite governmental family planning initiatives over half of Indian women still prefer to use the aforementioned traditional methods.

Because of their cultural orientation, Hindu women may desire education in family planning from a same-sex health-care provider, as well as assistance with delivery from female physicians, midwives, or nurse practitioners. Numerous issues surround sexuality and childbirth in Asian

Indian women. For many Indian women, intercourse experiences are painful because of lack of sex education. Young, rural Indian women lack information about sex, pregnancy and contraception and abortion. In addition, Blanchard et al. in 2015 (Blanchard et al. 2015a, b) reported that married young women bore children early by the age. In general, women are not educated about contraceptive options until after the first child. For this reason, most couples have their first child within the first year of marriage. The birth of a healthy first child reassures both families about the couple's health and that they are a "good match." Couples are commonly sent to fertility specialists if a child is not born after the first or second year of marriage. Husbands do not accompany pregnant women to physician visits, but mothers do. Only the nurse and the obstetrician (who is a female) attend to the patient at the time of delivery, although the woman's mother and other female relatives are usually nearby for assistance with her personal needs. Husbands usually come to the hospital or birth center, but they are not permitted to watch the delivery. If the husband is visiting during postpartum rounds, he will leave the room for the wife's examination.

Fisher et al. (2003) also state that pregnancy can be a frightening time for a young wife, especially if there are no female relatives available to educate her. Often the decision to have a home birth or to deliver in a hospital is decided by mothers in law and husbands (Blanchard et al., 2015a, b). Indian men might prefer to wait outside the room or stay away from the hospital during this time in contrast to American hospitals and birthing centers, where it is the norm for the new father to be present at the birth of his child. Most often in the United States, the woman and her physician discuss birth control options. In the case of the Asian Indian woman, it is important for the physician to ask if contraception should be discussed first between the physician and her husband. The physician might want to ask the husband for his permission to discuss birth control with his wife.

It is important to understand Indian cultural mores and values surrounding sexual education, sexual behavior, and the childbirth experience, as otherwise these might serve as barriers for Indian immigrants in need of health care. The lack of formal sexual education, importance of the birth of the first child, premarital contraceptive education, dominance of the husband in contraceptive decisions, and predominant role of women and lack of role for men (including the husband) in the childbirth process are all factors that can enhance the understanding of the health-care provider in providing effective care with a positive outcome.

19.9.2 Prescriptive, Restrictive, and Taboo Practices in the Childbearing Family

In the traditional East Indian culture, a family member's advice is highly valued and implemented. Grandmothers, mothers, and mothers-in-law are considered to have expert knowledge in the use of home remedies during pregnancy and the postpartum period. Many older women frequently travel to the United States to assist new mothers in antenatal and postnatal care consistent with traditional customs. For example, some women in India still believe that colostrum is unsuited for infants. They think that the milk does not "descend to the breast" until their ritual bath on the third day; as a result, newborns are fed sugar water or milk expressed from a lactating woman.

India has the highest number of child deaths in the world and almost 50% of these deaths are due to malnutrition. In a recent study only 71% of women felt that breast milk was the best food for a newborn. In addition, only 45% of women who delivered in hospitals initiated breastfeeding within one hour of delivery and only 25% of women who started breastfeeding immediately after birth. This is an area that requires more education and assessment to prevent infant death and malnutrition (Sultania et al., 2019).

Many Indian women, especially in rural areas, seek medical advice only when all available resources fail, so they may not go to a health-care provider for regular prenatal checkups. The Indian government has aimed to increase maternal health and education through various public initiatives (Blanchard et al., 2015a, b). Nonetheless, health-care providers may experience difficulty assessing the pregnant mother's sexual history because of the personal and private nature of the information and the discomfort associated with responding to a stranger about their personal lives.

Physical examinations and procedures, particularly pelvic examinations, are especially traumatic to Hindu American women who may not have experienced or heard about these examinations in the past. It is important to explain the procedures, provide privacy, and assign a female health-care provider to decrease the stress and perceived discomfort associated with a pelvic examination. Most Hindu women are not accustomed to being cared for by male health-care providers.

Childbirth is a social and religious event in the Hindu culture. Pregnancy rituals to protect the pregnant mother and the unborn child from evil spirits are performed during specific months of pregnancy. Pregnancy rites are performed in the woman's house during the fifth month of pregnancy. Another ritual is performed in the husband's house during the eighth month of pregnancy. A *bangle*, meaning to surround, must be worn by all auspicious women (barring widows, who are not considered auspicious) especially during pregnancy, when women are considered susceptible to the influence of evil spirits. Bangles act as a sort of "ring-pass-nots" and are believed to create barriers that prevent evil spirits from approaching pregnant women (Gatrad et al. 2004).

Dietary restrictions also exist during pregnancy. There are a diversity of practices related to foods that can be consumed or avoided during pregnancy, and the perceptions of foods as "hot" and "cold." The general belief is that hot foods are harmful and cold foods are beneficial during a pregnancy. Since pregnancy is thought to produce a state of "hotness," it is desirable to balance it out by eating cold foods (Chakrabarti & Chakrabarti, 2019). During early pregnancy, cold foods are recommended to avoid miscarriage, while hot foods are recommended during the last stages of pregnancy to aid in the delivery (Lakshmi, 2018).

Based on the hot-and-cold theory of disease, certain hot foods like eggs, jaggery (traditional, unrefined, whole-cane sugar), coconut, groundnut, maize, mango, papaya, fruit, and meat are avoided during pregnancy because of a fear of abortion caused by heating the body or inducing uterine hemorrhage. Pregnancy is a time of increased body heat, so cold foods such as milk, yogurt, and fruits are considered good. (Higginbottom et al., 2016). Whereas increased heat is deemed natural during pregnancy, overheating is considered dangerous. Minor swelling of the hands and feet is seen as increased heat and is not of much concern. However, a burning sensation during urination, scanty urine, and a white vaginal discharge are considered serious signs of significant overheating. Mukhopadhyay and Sarkar (2009) surveyed 199 women in Sikkim in northeast India about pregnancy-related food beliefs. The authors found that women of social, literacy, or economic standing were more likely to eat special foods. Women with fewer children were more likely to follow dietary practices, in contrast with mothers who had several children. Pregnant women tended to increase their intake of foods such as milk and green vegetables, while decreasing their fruit intake.

Beliefs surrounding what facilitates a good pregnancy and associated outcome, as well as negative sanctions, are often held by immigrating women from India. Most Indian women have fatalistic views about life, including pregnancy. The practice of eating less or "eating down" during pregnancy is common, as it is believed that excessive eating results in large newborns and difficult deliveries (Choudry, 1997). Also, the consumption of high-protein foods, including milk, are avoided because they result in an exaggerated growth of the baby that may lead to a difficult delivery.

In addition to concern about the size of the baby, other factors that influence dietary practices of pregnant women include bodily movement, constitution, and morning sickness. These factors influence both the quality and the quantity of foods consumed. Morning sickness is caused by an increase in *pitta* or bodily heat. *Pitta*—an ayurvedic (*ayur*, longevity; *veda*, science; ***Ayurveda*** is the traditional system of medicine in India)—means "bile" and is a symptom complex associated with dizziness, nausea, yellow body excretions, a bitter taste in the mouth, and overheating of the body.

Anemia caused by iron deficiency is one of the nutritional disorders affecting women of childbearing age. This condition may be aggravated because of the practice of reducing the consumption of leafy vegetables to avoid producing a dark-skinned baby (Higginbottom et al., 2016).

Other beliefs during pregnancy, such as physical activity like fetching water and carrying heavy loads until labor begins, continue in working-class class and farm women, in contrast to wealthy women who are coddled by their families. Although some beliefs can be rationalized, others may seem to have a lack of explanation. Profuse bleeding prior to delivery is seen as a good sign because it will purify the uterus and produce a clean child (Matsuyama & Moji, 2008) bleeding during the fifth month is a sign that the baby is a male (Choudry, 1997).

There is no taboo against the father being in the delivery room, but men are usually not present during birthing. The men do not stay in the delivery rooms and hold their wives' hands during delivery. Instead, they tend to wait outside the delivery room and allow female relatives to support the pregnant mother during labor and delivery. Hence, in relation to the role of men during childbirth, the role of the husband is minimal. Traditional families are reluctant to accept the changes such as shared responsibility and joint decisions. However, these changes are embraced by affluent, educated, and urban families. An awareness of these cultural practices will enable nurses to better understand husbands' reluctance to be present in the delivery room. Miller and Goodin (1995) stated that for Hindu American women, one important factor to achieve a balance in health and wellness is self-control of strong feelings. These women often manifest this belief by suppressing their feelings and emotions during labor and delivery. Nurses can assist in meeting new mothers' needs by closely observing their nonverbal communication such as a change in body posture, restlessness, and facial expressions.

In the Hindu culture, the birth of a son is considered a blessing, not only because the son can carry the family name but also because he can take care of the parents in their old age. Son preference is in the interest of lineage which is passed down through males (Bandyopadhyay, 2003). Furthermore, a son is also required for the performance of many sacred rituals. In contrast, the birth of a daughter is cause for worry and concern because of the traditions associated with dowry, a ritual that can impoverish the lives of those who are less affluent. Laws have been enacted to outlaw dowries however the practice still continues due to deep rooted cultural norms (Banerjee, 2014).

In India a male child is regarded to be superior to a female child, but this trend is changing. In a 2010 study, Kansal et al. surveyed 203 pregnant women in India. Two thirds of these women did not show any gender preference and 23% reported that they preferred a male child.

Bandyopadhyay in 2015 noted some women in India still consume herbal medicines that are thought to be sex selective drugs. In addition, they engage in practices such as consuming a special diet before and after conception, and reciting chants in hopes of having a male offspring. The preference for a male child exists even among immigrating women from India.

After the birth of a child, both the mother and the baby undergo purification rites. The postpartum mother and baby are considered to be impure and are confined to a room. This period of necessitated and mandatory confinement assists in bonding between the mother and the newborn, with the mother given adequate rest and time to tend to the baby's needs. A ritual bath and religious ceremony are performed by the priest to purify the mother and to end the mandated confinement (Hrodrigues, 2008). The baby is officially named on the 11th day during the "cradle ceremony or naming ceremony" and several rituals are performed to protect the baby from evil spirits and to ensure longevity (Gatrad et al. 2004).

During the postpartum period, hot foods such as chicken drumsticks, dried fish, and greens are considered good for lactation and to help the body heal and shed "bad blood" ie cleanse the body (Higginbottom et al., 2016). During this time cold foods are believed to produce back pain in the mother and diarrhea and indigestion in the infant. Cold foods such as buttermilk and curds, gourds, squach, tomatoes, and potatoes are restricted because they produce gas. Such abstentions are practiced primarily for the baby's health because harmful influences might be transmitted through the mother's breast milk. Sources of protein such as eggs, curds, and meat are avoided because they might adversely affect the baby.

There are strict taboos relating to the postnatal diet. The mother's diet the first few days may be restricted to liquids, rice, gruel, and bread. Special foods are eaten during the confinement period which may last for 3 weeks for upper caste women or about 10 days for lower caste women. Diets may vary according to socioeconomic status with rural women often receiving and inadequate diet. (Bandyopadhyay, 2009) Thus, for teaching to be effective, health-care providers must obtain dietary preferences and practices from the family before planning nutritional counseling.

During the postpartum period, the mother remains in a warm room and often keeps the windows closed to protect herself against cold drafts. Exposure to air conditioners and fans, even in warm weather, may be considered dangerous. Nurses can help the new mothers wear warm clothing and provide additional blankets to keep them warm. In summary, in order to provide culturally congruent care, the health-care provider must determine the belief systems regarding prescriptive, restrictive, and taboo practices in the Hindu American childbearing family.

In the United States, childbirth is viewed within the context of the nuclear family. The role and extent of involvement of grandparents and other immediate family members are decided by the new parents. This is in contrast with the role of the family in the Indian culture, which extends beyond immediate family relatives because it is considered a part of social order providing emotional and social support during a time of need.

The postpartum period has its own taboos in that the mother's movements are constrained within the house. This confinement period is usually 40 days, during which the mother is assisted in her personal care, fed a special nourishing diet, and receives body massages. It is believed that pregnancy produces a state of hotness, with delivery disturbing the balance achieved during pregnancy, and weakens the woman. In order to regain the balance, milk, ghee (clarified, unsalted butter without any solid milk particles), nuts, and jaggery are included in the diet of the new mother. Dried ginger is also eaten, since it is believed to help control postpartum bleeding and acts as a uterine cleansing agent. The newborn may be cared for by the local woman or midwife, who provides massages for both mother and baby. Cold baths or showers are generally avoided. (Bandyopadhyay, 2009).

Breastfeeding is practiced by the majority of mothers in India. However there is a significant need for more education about this practice. In the aformentioned study by Sultania et al. (2019) less than half of Indian mothers initiated breastfeeding within one hour of delivery. Most mothers (82%) fed colostrum. Lower levels of education in the women was associated with exclusive breastfeeding without initiating additional feeds even after 6 months. Solid food is introduced when the infant begins to reach out for food from the mother's plate. The introduction of solid food is celebrated with rituals and religious ceremonies and is called *annaprasan* (Gatrad et al. 2004).

Bhattacharyya et al. (2016) conducted a study on family, community, and provider practices during labor and childbirth—factors likely to influence newborn health outcomes. Data were collected through interviews of 500 women in Jharkhand state. The authors found that there were still strongly held beliefs in favor of home-based childbirth (210 delivered at facility and 290 delivered at home). This is despite the fact that India had adopted a cash transfer to be given to mothers who deliver in facilities. Birth in India is considered a natural process and the services of traditional birth attendants are widely utilized. Utilization of health care facilities is an index of wealth and status. The educational level of the mother also was one of the factors that led to facility deliveries.

During home childbirth, a team of birth attendants or an older female relative often the mother in law make decisions and perform key functions. The government now requires that a skilled health attendant be present at the delivery (Contractor et al., 2018). Also, to hasten home delivery, providers were commonly invited to administer oxytocin injections, whereas health staff did the same during facility deliveries. The practice of applying forceful fundal pressure was universal in both situations. In rural areas women still go through child birth with little interruption of regular activities. The process of delivery is considered impure so the delivery takes place outside the house. The placenta (impurity) is buried in a pit. This ritual is completed to protect the baby.

Implications for nursing care abound in how nurses' approach and care for the childbearing women emigrating from India. Nurses should avoid stereotyping Asian Indian women because there are regional and cultural variations, and a lack of understanding of the variations might lead to misinterpretation of behaviors. Nurses can provide culture-specific perinatal education and care by understanding beliefs and practices related to pregnancy, since many immigrant families may want to preserve their tradition and values.

19.10 Death Rituals

19.10.1 Death Rituals and Expectations

Death is seen as a family and communal affair. Family members conduct prayers with the dying person to assist their transition to the next life. In India the body can be disposed of by cremation, burial or leaving the body in the Ganges river. An open box or casket is used for cremation and this usually occurs within 24 hours of death. The eldest son is a pall barer and he and other and male relatives carry the body to the cremation site. In the United States a hearse carries the body and is accompanied by family members (Gupta, 2011).

The death rite is called ***antyesti***, or last rites. The basic purpose is to purify the deceased and console the bereaved. A tenet of Hinduism is that the soul survives the death. Therefore, performing a ritual bath, sprinkling holy river water over the body, covering the body with new clothes, daubing parts of the body with ghee, and chanting Vedic utterances purify and strengthen the deceased for the postmortem journey (Dasa, 2003). The priest pours water into the mouth of the deceased and blesses the body by tying a thread around the neck or wrist. The priest may anoint with water from the holy Ganges River or put the sacred leaf from the *Tulsi* plant in the mouth. At the yearly death anniversary (according to the lunar calendar), *Shradda* ceremonies (usually associated with funeral and postfuneral activities) are held in the home with the offering of *pinda* or balls of cooked rice (pindadana) to one's ancestors (Manguso, 2017). Although individual Hindu community rites vary and can be simple or exceedingly complex, the basic rationale is the same.

19.10.2 Responses to Death and Grief

Hindu families share sacred moments and celebrate important events as a unit, and deaths are considered family events. Women may respond to the death of a loved one with loud wailing, moaning, and beating their chests in front of the corpse, attesting to their inability to bear the thought of being left behind to handle situations by themselves. This is significant for women because widowhood is considered inauspicious (Gupta, 2011).

Mourning is a family as well as a social and communal affair. During the 12 days of mourning, female mourners visit female members of the bereaved family at a fixed hour every afternoon. With progression of days, wailing becomes less intense. The functional value of these practices may indicate the provision of intense security and comfort for bereaved people. At a psychological level, it provides catharsis for the entire family and may assist with speeding up the process of recovery from the loss of a loved one and making positive adjustments. Hence, healthcare providers need to offer support and understanding of the Hindu culture with respect for death and grief beliefs.

19.11 Spirituality

19.11.1 Dominant Religion and Use of Prayer

The cultural heritage of India is found primarily in philosophy and religion. Sources of philosophical ideas and religious beliefs lie in the *Vedas* and *Upanishads*, repositories of Hindu culture. They explain the two great objects of human life: duty and liberation. The relationship between religion and social structure is intricate. Religion provides the legitimacy and ideology for social and economic practices, whereas social structures produce particular religious beliefs. Two concepts are primary in the Hindu belief system: *karma* (all human actions lead to consequences (as you sow, so shall you reap) and *dharma* (righteousness action). Dharma forms the basis of karma, and the principles of dharma come from the karma theory (Rao, 2010). The doctrine of karma, dharma, reincarnation, the concept of the four ends or stages of life, and the caste system are conducive to maintaining these beliefs.

In Hindu philosophy, the external world is seen as being illusory, called *maya*. Since the world of the senses, the empirical world, is constantly changing, it is seen as an inconstant, illusory world. The ultimate purpose of human existence is to attain *moksha* (Avci, 2019).

Hinduism represents a set of beliefs and a definite social organization. Hinduism connotes the belief in the authority of Vedas and other sacred writings of the ancient sages, the immortality of the soul and belief in a future life, the existence of a Supreme God, the theory of karma and rebirth, the worship of ancestors, a social organization represented by the four castes, the theory of the four stages of life, and the theory of the four *Purusarthas*, or ends of human endeavor.

The social structure of Hinduism revolves around two fundamental institutions: the caste and the joint family (explained earlier in this chapter) relate to everything connected with the Hindu people outside their religion. The Orthodox Hindu view is that society has been divinely ordained on the basis of the four castes: Brahmanas, Kshatriyas, Vaisyas, and Sudras. The fourfold caste system— *Chaturvarna*—is a theoretical division of society in which tribes, clans, and family groups are affiliated. Yet, the theory of society based on caste still governs Hindu life (Saxena et al., 2015). All of the innumerable subcasts claim to belong to one of the four castes. The essential principles of *Chaturvarna* are unchangeable inequality based on birth, the gradation of professions and their inequality, and restrictions on marriage outside one's own group. Although religion does not bestow the caste system with a religious sanction, the great Hindu legal codes are based on the caste system.

In America and elsewhere, individual worship may take different forms within the Hindu religious tradition. Popular Hindu forms of worship require no special arrangements and can be carried out in private. A household shrine is an aid rather than a requirement for worship. Shrines may be set up in the living room or the dining room but are most often located in a back room or a closet. The shrine typically contains representations or symbols of one or more deities. This practice is still continued by some families in the United States but may not be thought of as a requirement anymore (Andrews, 2018).

Almost all family and group religious observances take place on the weekend to fit the American work schedule, even though the lunar liturgical calendar could fall within the normal workweek. Indian worship includes praying, singing hymns, reciting scripture, and repeating the names of deities. For some Hindus, worship is the identification with, or merging of, the inner self with the ultimate reality, Brahma. Temples serve as important support institutions for the practice of the Hindu religion. The installation of a Hindu temple and the invocation of God into its central image make God present in that place, and the land becomes holy. The first Hindu temple constructed in America was in Pittsburgh, Pennsylvania, in 1976; it was modeled after the most popular Hindu temple in India: the Sri Venkateswara temple at Tirupathi.

19.11.2 Meaning of Life and Individual Sources of Strength

One of the main concepts that form the basis of the Hindu attitude toward life and daily conduct is the *Purusartha*, the four ends of humanity. The first of these is characterized by righteousness, duty, and virtue. Other activities through which a person seeks to gain something for self or pursue pleasures are material gain and love or pleasure. Finally, the renunciation of all these activities is to devote oneself to religious or spiritual activities for liberation from a worldly life.

Karma stresses the individual's responsibility for one's actions and is interpreted in terms of past life. One's present condition is seen as a result of one's actions in a past life or lives. Hence, the doctrine of karma by itself enunciates only the principle of an individual's moral responsibility for his or her own deeds. Actions lead to certain consequences, and an individual needs to be aware of this when taking an action. The doctrine of karma has persisted in India from the Vedic times of about 1000 B.C. and is a vital concept that permeates the lives and thoughts of the rich and poor (Whitman, 2007).

To Hindus, religion and family are considered primary sources of strength. Dharma places a high priority on the family. Family is considered to be a critical stage in the path of action, which leads to ultimate spiritual liberation (Avci, 2019). A number of rituals and spiritual practices are connected with the family because it is through the families that Hindus fulfil many religious obligations. Common life-cycle rituals of Hindus in the United States include prenatal rituals, birth and childhood naming ceremonies, marriage, and cremation within 24 hours after death. All these involve the extended family whenever possible.

Hinduism is concerned with questions regarding ultimate reality and the individual's relationship with it. Spiritual support gives hope to life, ensures courage to face the consequences of illness, and directs the thinking of the person in a positive direction. Because of the strength of the kinship organization and a sense of kinship obligation, the individual seeks solace and strength in such an organization.

19.11.3 Spiritual Beliefs and Health-Care Practices

Hindus believe that all illnesses attack an individual through the mind, body, and soul. The body is the objective manifestation of a subjective mind and consciousness. Spiritual beliefs act mainly as diversional therapies during illness. Suffering of any kind produces hope, which is essential to life. Spiritual support gives hope and helps control emotions and behavior. The Ayurvedic view of health emphasizes social, environmental, and spiritual contexts. The key concept is harmony within the organism and within the system of which the organism is a part. In the Ayurvedic philosophy, people, health, and the universe are said to be related, and when these relationships are in imbalance, health problems can result. Herbs, metals such as copper or zinc, massages, and other techniques are used to clean the body and restore balance. The goals of Ayurvedic practice include health promotion, disease prevention and measures to promote heathy aging (Basisht, 2014).

Complementary and alternative medicine has been used in India for thousands of years. Roy, Gupta and Ghosh (2015) studied the use and perception of CAM in a tertiary care hospital in India. Subjects consisted of physicians and patients. Fifty-eight percent of the physicians used CAM while only 28% of the patients admitted to using it. Patients reported that they felt it was safer and less costly and worked better than other medications.

The concept of palliative care in India is in a relatively early stage of development. The first hospice was established in 1986 and today there are over 900 facilities that provide palliative care. A national policy supporting palliative care was enacted in 2008 however it lacked the funding needed for implementation. Rajagopal (2015) reported that only one percent of the population has access to these services.

The use of opioids for pain management is also limited due to previous strict regulations that prohibited the use of these substances. The government has enacted legislation that supports use of opioids in palliative care however pain management is still in its' infancy. Most Indians with terminal illnesses are still cared for and prefer to die in their homes.

Limited studies have been published on how Asian Indian immigrants view hospice services as they may face or are currently facing end-of-life care decisions. Doorenbos (2003) examined whether absence of information about hospice, lack of financial resources, and cultural differences explain the lack of hospice service use in Asian Indian immigrant populations. Results indicated that in a sample of 43 first-generation Asian Indian immigrants, only 12 percent knew what a hospice program was, 22 percent had a little knowledge of hospice, and 22 percent had no knowledge of hospice. The results also indicated that some hospice staff misunderstood Asian Indian death and dying rituals. There was no indication from the participants that financial resources were a barrier to hospice use. Most of the respondents (86 percent) indicated their preference to die at home. However, only 11 percent were aware that the individual's home is the primary site for hospice care. The results demonstrate that although hospice would be the appropriate end-of-life care for this population, the main barrier was knowledge related to the site of hospice care. The majority also rated death and dying beliefs and rituals as important to them.

Gupta (2011) explored Asian Indian American Hindu (AIAH) cultural views related to death and dying through three focus group (senior citizens, middle-aged adults, and young adults) interviews, using both open-ended and semi structured questions. Focus group discussions were related to meaning attributed to death and pre- and post-death practices. Results indicated that while all three generations believed in the afterlife and karmic philosophy, they exhibited differences in the degree to which Hindu traditions surrounding death and bereavement were influenced by living in the United States.

Health-care providers should assess the extent to which religion, beliefs and values, or socioeconomic status is a part of the individual's life, as these are related to the individual's perception of health and illness and daily practices. Also, assessing spiritual life is essential for identifying resources and solutions for therapy. In the Doorenbos (2003) study, completion of a cultural assessment at the time of admission into hospice care would assist hospice staff in identifying the beliefs that Asian Indian immigrants considered important.

19.12 Health-Care Practices

19.12.1 Health-Seeking Beliefs and Behaviors

In Indian culture, rituals are closely connected with religious beliefs about the relationships of human beings with supernatural forces. To maintain harmony between the self and the supernatural world, the belief that one can do little to restore health by oneself provides a basis for ceremonies and rituals. Worshipping goddesses, pilgrimages to holy places, and pouring water at the roots of sacred trees are believed to have medicinal effects for healing the sick person.

Asian Indians experience mental distress biomedically and assign it to ill-defined medical conditions, a phenomenon called somatization. Instead of admitting that they feel sad or depressed, Asian Indian women may say that they are experiencing weakness. Since most individuals who present with somatic complaints report some psychological distress on closer scrutiny, the health-care provider should probe about mental distress when presented with ill-defined somatic complaints.

19.12.2 Responsibility for Health Care

In general, Hindus are responsible for their own health care, but they mobilize personal, social, and religious resources in the face of a crisis. The resolution ranges from a denial of discomfort to acceptance of limitations of somatic or other psychological symptoms. Medical beliefs are a blend of modern and traditional theories and practices. In *Ayurveda*, the primary emphasis is on the prevention of illnesses. Individuals have to be aware of their own health needs. One of the principles of *Ayurveda* includes the art of living and proper health care, advocating that one's health is a personal responsibility. In *Ayurvedic* theory, the key to health is an orderly daily life in which personal hygiene, diet, work, and sleep and rest patterns are regulated. Depending on an individual's constitution, a daily routine has to be established and changed according to the season. Individuals must have information and awareness about living well. Hence, it is important to include prevention, health education, and health-care services.

A common health-care problem among Hindus is self-medication. Pharmacies in India generally allow the purchase of medications such as antibiotics without a prescription. Thus, Hindus migrating to America are accustomed to self-medicating and may bring medications with them or obtain them through relatives and friends. Self-treatment is also more likely if the symptoms are stigmatizing, such as psychiatric or STI symptoms. Use of CAM is widely prevalent among Hindus. Many feel that modern medicine may be good only for acute conditions, while the traditional systems of medicine are better and more effective for chronic conditions (Gupta, 2010). The use of CAM depends on the severity of the illness. Individuals will use CAM for minor illnesses before taking allopathic medications, while they use allopathic medications for severe conditions as the first choice. Often patients do not tell physicians they are using CAM for fear of offending the physician (Gupta, 2010).

19.12.3 Folk and Traditional Practices

Numerous practices taken from the hot-and-cold theory of disease causation and folk practices are related to illnesses. Traditional healers, *nattu-vaidhyars*, use *Ayurvedic*, *Siddha*, and *Unani* medical systems. These systems are all based on the Tridosha theory. The *Ayurvedic* system uses herbs and roots; the *Siddha* system, practiced mainly in the southern part of India, uses medicines; and the *Unani* system, similar to the *Siddha*, is practiced by Muslims (Pal, 1991).

According to the *Tridosha* theory, the body is made up of modifications of the five elements: air, space, fire, water, and earth. These modifications are formed from food and must be maintained within proper proportions for health. A balance among three elements or humors—phlegm or mucus, bile or gall, and wind—corresponds to three different types of food required by the body. The following are some types of foods and the allopathic equivalents of diseases associated with them:

1. **Heat-producing foods:** *Brinjals* (Indian eggplant), dried fish, green chilies, raw rice, and eggs. *Pittham* foods include cluster beans, groundnuts, almonds, millet, oil, and runner beans. Allopathic equivalents of heat-producing diseases include diarrhea, dysentery, abdominal pain, and scabies. Allopathic equivalents of *pittham* diseases include vomiting, jaundice, and anemia.
2. **Cooling foods:** Tomatoes, pumpkin, gourds, greens, oranges, sweet limes, carrots, radishes, barley, and buttermilk. Cold, headache, chill, fever, malaria, and typhoid are allopathic equivalents of cool diseases.
3. **Gas-producing foods:** Root vegetables like potato, sweet potato, and elephant yam; plantain; and chicken drumsticks. Joint pains, paralysis, stroke, and polio are disorders related to gas-producing foods.

Heating and cooling effects are produced in the body and thus are not related to the temperature or spiciness of foods. An imbalance leads to disease. If too much heat is in the body from consuming heat-producing foods, then cold foods need to be eaten to restore balance.

Blood is considered one of the seven *dhatus* (body tissues) in the *Tridosha* theory of *Ayurveda*. The strength or weakness of the *dhatus* depends on the "richness or poverty of the blood." Blood is equated with life and is preserved with great care. Special foods like beet root (red foods) are required for good blood, whereas "no blood" is the concept nearest to that of malnutrition.

Thus, in terms of health practices, cultural patterns in India are regionally specific. Health-care providers must be extremely careful in their assessments and not stereotype health-care practices. In addition, providers must also be aware of practices related to the hot-and-cold theory of disease causation and treatment.

19.12.4 Barriers to Health Care

Dependency and reliance on family and friends may be considered a barrier among Hindus. In addition, the practice of self-medicating behaviors may mask disease symptoms until the health condition is at a more advanced stage, making treatment regimens more complex.

In relation to mental health, the stigma associated with mental illness can serve as a barrier to seeking professional treatment. Families may attempt to deny the illness because they may not want outsiders to know about the family member's mental illness (Kumar & Nevid, 2010).

Barriers to seeking preventive services for terminal illnesses like cancer are low awareness of cancer risk and methods of early detection and stigma. Cancer may still be considered a terminal illness by many Indians and there is still a misconception that the individual may not be able to work with this illness. (Sahoo et al., 2019). The Indian government has launched public health programs aimed at education and primary prevention to attempt to address these concerns.

A barrier to mammography and other screening procedures to detect reproductive organ cancer is fear of taking off clothes or modesty. Hindu women also do not like to discuss genitourinary symptoms or undress in front of others, nor do they want to see a physician during their menstrual periods because it is considered "dirty." However, this might be changing as Hindu women are increasingly more educated and are joining the workforce (Gupta, 2010).

19.12.5 Cultural Responses to Health and Illness

Many Hindus have a fatalistic attitude about illness causation. Because of their religious beliefs of *karma*, they attempt to be stoic and may not exhibit symptoms of pain. Furthermore, pain is attributed to God's will, the wrath of God, or a punishment from God and is to be borne with courage. As a result, health-care providers may need to rely more on the nonverbal aspects of pain when assessing Hindu patients.

Hindus view pain and suffering as a part or life. They believe that suffering is a consequence of actions in a current or previous life. Suffering is thought of as temporary and methods of detaching from the pain such as mediation and yoga are often practiced as pain control measures (Whitman, 2007).

Many Hindus are steadfast in their fatalistic spiritual belief. An individual's *dharma* and *karma* mold one's destiny and worldview. This may be a reason for Hindus' underutilization of psychological or counselling services as options for coping. Health-care providers must assess individual attitudes and comfort levels when counselling Asian Indian patients.

Because of the stigma attached to seeking professional psychiatric help, many Hindus do not access the health-care system for mental health problems. Instead, family and friends seem to be the best help, and there is a general belief that time is the best healer. Physical and mental illnesses are considered God's will, past *karma*, and are associated with a fatalistic attitude. The sick role is assumed without any feeling of guilt or ineptness in doing one's tasks. Because of strong family and kinship ties, the sick role is well accepted. The individual is cared for and relieved of responsibilities for that time. Because of strong family ties and joint and extended families, Hindus are not likely to use long-term-care facilities.

Psychological distress may be demonstrated through somatization, which is common, especially in women. The symptoms may be expressed as headaches, a burning sensation in the soles of the feet or the forehead, and a tingling pain in the lower extremities. Also, the belief in *Ayurveda* in the interrelatedness of mind, body, and spirit may make those with mental health symptoms delay treatment, since they seek spiritual, mind, and body treatments before seeking professional mental health services. Because family is important and family members may accompany the patient to the health-care provider's office, it may pose a problem for the American health-care system's emphasis on autonomy and privacy.

Janardhana et al. (2015) studied 200 families who cared for individuals with severe mental illness. This study reported that women provided most of the care for the client with the mental disorder and that they did accompany the client when they left the house. A noteworthy finding was that men were also accepting the role of caregivers. It is common for Indians with mental disorders to be cared for at home.

Lack of psychiatric providers and psychotropic medications and stigmatization are common problems in India. Zieger et al. (2017) estimated that only 10% of the patients who need care are able to be treated by a provider. Other individuals seek help from medical practitioners, faith healers and primary care facilities. Patients in India may still have negative attitudes toward psychiatric care and this then leads to noncompliance with medications. These beliefs may extend to Indians in the United States and elsewhere and need to be explored if they present with a mental disorder.

India is the second most populous country in the world. The adolescent population of India is the highest in the world. Various studies report that from 12 to 29% adolescents suffer from mental illness. The rate of mental disorders in adolescents is higher in urban areas. The disorders with the highest prevalence rates include: depressive disorders, intellectual disabilities, psychotic disorders and anxiety disorders. Suicide, alcohol and tobacco use are also areas of concern in this population. India does not have Child and Adolescent Mental Health policies and it is estimated that less than one percent of children and adolescents suffering from mental disorders receive services (Hossain & Purohit, 2019).

Families tend to be protective of an ill member. They may not want to disclose the gravity of an illness to the patient or discuss impending disability or death for fear of the patient's vulnerability and loss of hope, resulting in death. The conflict between medical ethics and patients' values may pose a problem for health-care providers, who need to be cognizant of the importance of the family members' wishes and values regarding the care of their loved ones.

19.12.6 Blood Transfusions and Organ Donation

India's first successful corneal transplant and kidney transplants occurred in the 1960s. The first successful cardiac transplant was in 1994. Despite this late start in organ transplantation India has made strides in the area of corneal donations (Sachdeva, 2017). India enacted legislation in 1994 and 2011 to promote organ donation and is providing public education. Nonetheless India organ donation rate is extremely low. This low rate is due in part to lack of transplant facilities, superstition and lack of knowledge by medical providers and the public about key issues such as the definition of brain death (Srivastava & Mani, 2018). In addition, the cost of organ transplantation is often paid by families.

Religious issues and medical-legal concerns are not a major hurdle for organ donation. There is no Hindu policy exists that prevents receiving blood or blood products. Donating and receiving organs are both acceptable.

19.13 Health-Care Providers

19.13.1 Traditional Versus Biomedical Providers

Although Hindus in general have a favorable attitude toward American physicians and the quality of medical care received in the United States, relatives and friends are consulted first rather than a health-care provider. Kinship and friendship ties remain strong, even in medical matters.

Because any open display of affection is taboo, Hindu women are especially modest. Women generally seek female health-care providers for gynecologic examinations. Health-care providers need to respect their modesty by providing adequate privacy and assigning same-gender caregivers whenever possible.

In the area of mental health, traditional healers, such as *Vaids*, practice an empirical system of indigenous medicine; *mantarwadis* cure through astrology and charms, and *patris* act as mediums for spirits and demons. Health-care providers must specifically ask if their Hindu patients are using these folk practitioners and what treatments have been prescribed.

Reflective Exercise 19.3

Harini Chaturvedi, a Hindu woman aged 60 years, recently moved to the United States from India to assist her son, his wife, and their four children (two sons, ages 13 and 11 years, and two daughters, ages 6 and 2 years) with child rearing while the parents work. One day, Ms. Chaturvedi complains of lower abdominal pain. Her son takes her to the emergency room. The male nurse who is on duty explains that he needs to assess her in order to obtain data about her pain. Ms. Chaturvedi does not maintain eye contact with the male nurse, and she is reluctant to let him examine her. She states she would prefer to talk with a female nurse. She also states that she does not want any pain medications.

1. Why would Mrs. Chaturvedi not maintain eye contact with the male nurse?
2. Why is Mrs. Chaturvedi reluctant to let a male nurse examine her?
3. Identify from a cultural and religious standpoint reasons for not wanting pain medications.

19.13.2 Status of Health-Care Providers

The Indian patient's view of the physician used to be that of omnipotence and paternalism. Indian patients tended to be subservient and often would not openly question physicians' behaviors or treatments. The physician was also viewed as an older person who is protective, authoritative, and someone who deserved a high level of respect. An article by Paul and Bhatia (2015) reported that these views are changing due to consumer access to information via the internet, corruption, inequality in health care and the desire for patients to be involved in their care. Assessment of each individual's view of medical providers should be completed to provide patient centered, culturally appropriate care.

References

Aich P, Andersen KL, Banerjee SK, Rawat A, Upadhyay B, Warvadekar J (2015) How prepared are young, rural women in India to address their sexual and reproductive health needs? A cross-sectional assessment of youth in Jharkhand. Reprod Health 12(97):1–10. https://doi.org/10.1186/s12978-015-0086-8

Ali NA, Burgard SA, Hogue DE, Lori JR, Story WT, Taleb F (2012) Husband's involvement in delivery care utilization in rural Bangledesh: a qualitative study. BMC Pregnancy Childbirth 12:28. https://doi.org/10.1186/1471-2393-12-28

Andrews A (2018) Making homes in the American world: Bengali Hindu Women's transformations to home shrine care traditions in the U.S. Nidan 3(1):1–15

Avci E (2019) A comparative analysis on the perspective of Sunni theology and Hindu tradition regarding euthanasia: the impact of belief in resurrection and reincarnation. J Relig Health 58(5):1770–1791. https://doi.org/10.1007/s10943-019-00836-4

Azamjah N, Soltan-Zadeh Y, Zayeri F (2019) Global trend of breast cancer mortality rate: a 25-year study. Asian Pac J Cancer Prev: APJCP 2015–2020 20(7). https://doi.org/10.31557/APJCP.2019.20.7.2015

Bandyopadhyay M (2003) Missing girls and son preference in rural India: looking beyond popular myth. Health Care Women Int 24:910–926

Bandyopadhyay M (2009) Impact of ritual pollution on lactation and breast feeding in rural West Bengal India. Int Breastfeed J 4:2. http://www.internationalbreastfeedingjournal.com/content/4/1/2

Banerjee P (2014) Dowry in 21st-century India: the sociocultural face of exploitation. Trauma Violence Abuse 15:34–40. https://doi.org/10.1177/1524838013496334

Basisht G (2014) Exploring insights toward definition and laws of health in Ayurveda: global health perspective. AYU 35(4):351–355. https://doi.org/10.4103/0974-8520.158975

Bharati P, Bharati S, Pal M, Sen S (2019) Malnutrition and anaemia among adult women in India. J Biosoc Sci 51(5):658–668. https://doi.org/10.1017/S002193201800041X

Bhattacharyya S, Avan B, Srivastava A, Roy R, R. (2016) Factors influencing women's preference for health facility deliveries in Jharkhand state, India: a cross sectional analysis. BMC Pregnancy Childbirth 16(50):50. https://doi.org/10.1186/s12884-016-0839-6

Blanchard AK, Bruce SG, Jayanna K, Gurav K, Mohan HL, Avery L, Ramesh BM (2015a) An exploration of decision-making processes on infant delivery site from the perspective of pregnant women, new mothers, and their families in northern Karnataka, India. Matern Child Health J 19(9):2074–2080. https://doi.org/10.1007/s10995-015-1720-3

Blanchard KA, Avery L, Bruce SG, Frederick J, Gurav K, Jayanna K, Mohan HL, Moses S, Ramesh BM (2015b) An exploration of decision-making processes on infant delivery site from the perspective of pregnant women, new mothers, and their families in northern Karnataka, India. Matern Child Health 16(1):2074–2080. https://doi.org/10.1007/s10995-015-1720-3

Blesch KS, Davis F, Kamath S (1999) A comparison of breast and colon cancer incidence rates among native Asian Indians, U.S. immigrant Asian Indians, and whites. J Am Diet Assoc 99(10):1275–1277

Chakrabarti A, Chakrabarti S (2019) Food taboos in pregnancy and early lactation among women living in a rural area of West Bengal. J Family Med Prim Care 8(1):86–90. https://doi.org/10.4103/jfmpc.jfmpc_53_17

Choudhry UK, Srivastava R, Fitch M (1998) Breast cancer detection practices of south Asian women: knowledge, attitudes, and beliefs. Oncol Nurs Forum 25(10):1693–1701

Choudry UK (1997) Traditional practices of women from India: pregnancy, childbirth and newborn care. JOGNN 26(5):533–539. https://doi.org/10.1111/j.1552-6909.1997.tb02156.x

CIA World Factbook (2020) India. https://www.cia.gov/library/publications/the-world-factbook/geos/in.html

Cichocki M (2007) HIV around the world—India. http://aids.about.com/od/clinicaltrials/a/india.htm

Contractor S, Das A, Dasgupta J, Van Belle S (2018) Beyond the template: the needs of tribal women and their experiences with maternity services in Odisha, India. Int J Equity Health 17:134. https://doi.org/10.1186/s12939-018-0850-9

Dasa S (2003) Hindu funeral rites and ancestor worship. Sanskrit Religions Institute. Sanskrit.org

Dhawan J, Gupta S, Kumar B (2010) Sexually transmitted diseases in children in India. Indian J Dermatol Venereol Leprol 76:489–493

Doorenbos A (2003) Hospice access for Asian Indian immigrants. J Hosp Palliat Nurs 5(1):27–33

Duvvurry VK (1991) Play, symbolism, and ritual. Peter Lang, New York

Fisher JA, Bowman M, Thomas T (2003) Issues for south Asian Indian patients surrounding sexuality, fertility, and childbirth in the U.S. health-care system. J Am Board Fam Pract 16(2):180–181

Garewal H (1995) Antioxidants in oral cancer prevention. Am J Clin Nutr 62:1410S–1416S

Garg S, Anand T (2015) Menstruation related myths in India: strategies for combating it. J Family Med Prim Care 4(2):184–187. https://doi.org/10.4103/2249-4863.154627

Gatrad AR, Ray M, Sheikh A (2004) Hindu birth customs. Arch Dis Child 89:1094–1097

Gay and Lesbian Vaishnava Association Inc (2005). http://www.galva108.org/faq.html

Gupta VB (2010) Impact of culture on healthcare seeking behavior of Asian Indians. (2010). J Cult Divers 17(1):13–19

Gupta R (2011) Death beliefs and practices from an Asian Indian American Hindu perspective. Death Stud 35(3):244–266. https://doi.org/10.1080/07481187.2010.518420

Harinarayan CH, Akhila H (2019) Modern India and the tale of twin nutrient deficiency-calcium and vitamin D-nutrition trend data 50 years-retrospect, introspect, and prospect. Front Endocrinol 10. https://doi.org/10.3389/fendo.2019.00493

Higginbottom GMA, Davey C, Osswald B, Shankar J, Vallianatos H (2016) Understanding south Asian immigrants' women's food choices in the perinatal period. Int J Women's Health Wellness 2(1):1–7. https://doi.org/10.23937/2474-1353/1510013

Hossain M, Purohit N (2019) Improving child and adolescent mental health in India: status, services, policies, and way forward. Indian J Psychiatry 61(4):415–419. https://doi.org/10.4103/psychiatry.IndianJPsychiatry_217_18

Hrodrigues (2008) Mahavidya. Scholarly resources for the study of Hinduism. http://www.mahavidya.ca/2008/04/15/birth-rituals-in-hinduism/

Jain NC (1992) Teaching about culture and communicative life in India. Paper presented at the annual convention of the western states communication association, Boise, ID

Janardhana N, Naidu DM, Raghunandan S, Saraswathi L, Seshan V (2015) Care giving of people with severe mental illness: an Indian experience. Indian J Psychol Med 37(2):184–194. https://doi.org/10.4103/0253-7176.155619

Jayaraman R (2005) Personal identity in a globalized world: cultural roots of Hindu personal names and surnames. J Pop Cult 38(3):467–490

Kalra G (2012) A psychiatrist's role in "coming out" process: context and controversies post-377. Indian J psychiatry 54(1):69–72. https://doi.org/10.4103/0019-5545.94652

Kilara A, Iya KK (1992) Food and dietary habits of the Hindu. Food Technol 46:94

Kumar A, Nevid JS (2010) Acculturation, enculturation, and perception of mental disorders in Asian Indian immigrants. Cult Divers Ethn Minor Psychol 16(2):274–283

Lakshmi E (2018) Traditional foods of Indian origin in pregnancy. MOJ Food Process Technol 6(3):280–281. https://doi.org/10.15406/mojfpt.2018.06.00176

Levy RA (1993) Ethnic and racial differences in response to medicines: preserving individualized therapy in managed pharmaceutical programmes. Pharm Med 7:139–165

Mahadevan M, Blair D, Raines ER (2014) Changing food habits in a south Indian Hindu Brahmin community: a case of transitioning gender roles and family dynamics. Ecol Food Nutr 53(6):596–617. https://doi.org/10.1080/03670244.2014.891993

Manguso E (2017) The anniversary death ritual of Hinduism

Matsuyama A, Moji K (2008) Perception of bleeding as a danger sign during pregnancy, delivery, and the postpartum period in rural Nepal. Qual Health Res 18(2):196–208. https://doi.org/10.1177/1049732307312390

Miller S, Goodin J (1995) East Indian Hindu Americans. Transcultural nursing: assessment and intervention. St. Louis: CV Mosby.

Minocha S, Makkar S, Swaminathan S, Thomas T, Webb P, Kurpad AV (2019) Supply and demand of high quality protein foods in India: trends and opportunities. Glob Food Sec 23:139–148. https://doi.org/10.1016/j.gfs.2019.05.004

Mukhopadhyay S, Sarkar A (2009) Pregnancy related food habits among women of rural Sikkim, India. Public Health Nutr 12:2317–2322. https://doi.org/10.1017/s1368980009005576

Muttreja P, Singh S (2018) Family planning in India: the way forward. Indian J Med Res 148(Suppl):S1–S9. https://doi.org/10.4103/ijmr.IJMR_2067_17

Noble AG, Dutt AK (1982) India: cultural patterns and processes. Westview Press, Boulder, CO

Pal M (1991) The Tridosha theory. Anc Sci Life 10(3):144–155

Pandey G (2017) The Indian women eating with their families for the first time. BBC News, Delhi. https://www.bbc.com/news/world-asia-india-41148492

Paul P, Bhatia V (2015) Doctor patient relationship: changing scenario in India. Asian J Med Sci 7(4):1–5. https://doi.org/10.3126/ajms.v714.13929

Perkins JM, Lee H, James KS et al (2016) Marital status, widowhood duration, gender and health outcomes: a cross-sectional study among older adults in India. BMC Public Health 16:1032. https://doi.org/10.1186/s12889-016-3682-9

Raj S, Ganganna P, Bowering J (1999) Dietary habits of Asian Indians in relation to length of residence in the United States. J Am Diet Assoc 99(9):1106–1108

Rajagopal MR (2015) The current status of palliative care in India. Cancer Control 22:57–62

Rao RN (2010) Talking to the dying: Hindu views, Hindu ways. Int J Sociol Fam 36(1):65–76

Roy V, Gupta M, Ghosh RK (2015) Perception, attitude and usage of complementary and alternative medicine among doctors and patients in a tertiary care hospital in India. Indian J Pharmacol 47(2):137–142. https://doi.org/10.4103/0253-7613.153418

Sachdeva S (2017) Organ donation in India: scarcity in abundance. Indian J Public Health 61:299–230. https://doi.org/10.4103/ijph.IJPH_230_16

Sadler GR, Dhanjal SK, Shah RB, Ko C, Anghel M, Harshburger R (2001) Asian Indian women: knowledge, attitudes and behaviors toward breast cancer early detection. Public Health Nurs 18(5):357–363

Sahoo SS, Sahu DP, Verma M, Parija PP, Panda UK (2019) Cancer and stigma: present situation and challenges in India. Oncol J India 3:51–53. https://doi.org/10.4103/oji.oji_51_19

Sankaranarayanan R, Mathew B, Varghese SPR, Menon V, Jaydeep A, Nair MK, Mathews C, Mahalingam TR, Balaram P, Nair PP (1997) Chemoprevention of oral leukoplakia with vitamin a and betacartoene: an assessment. Oral Oncol 33:231–236

Saxena A, Saini JS, Gupta Y, Parasuram A, Iyengar SRS (2015) Social network analysis of the caste-based reservation system in India. arXiv preprint arXiv:1512.03184

Shetty PS (2002) Nutrition transition in India. Public Health Nutr 5(1):175–182

Sicree R, Shaw J, Zimmet P (2006) Diabetes and impaired glucose tolerance. In: Gan D (ed) Diabetes atlas. International Diabetes Federation, 3rd edn. International Diabetes Federation, Belgium, pp 15–103

Sonkar SK, Kumar S, Alam M, Atam V (2019) A case series of ten patients of epidemic dropsy June 2018 in a tertiary care Centre. J Curr Med Res Opin 2(12):375–377. https://doi.org/10.15520/jcmro.v2i12.240

Srivastava A, Mani A (2018) Deceased organ donation and transplantation in India: promises and challenges. Neurol India 66(2):316–322. https://doi.org/10.4103/0028-3886.227259

Statistics of Indians Abroad. (2020). https://www.nriol.com/indiandiaspora/statistics-indians-abroad.asp

Strachan DP, Powell KJ, Thaker A, Millard FJC, Maxwell JD (1995) Vegetarian diet as a risk factor fortuberculosis in immigrant South London Asians. Thorax 50:175–180

Sultania P, Agrawal N, Charles R, Dharel D, Dudani R, Rani A (2019) Breastfeeding knowledge and behavior among women visiting a tertiary care center in India: a cross-sectional survey. Ann Global health 85(1):65. https://doi.org/10.5334/aogh.2093

Taylor WD (2010) The burden of non-communicable diseases in India. The Cameron Institute, Hamilton, ON

UNAIDS (2020) https://www.unaids.org/en/regionscountries/countries/india, retrieved

U.S. Census Bureau (2019) International data base. http://www.census.gov/ipc/www/idb/worldpopinfo.php

Varghese A, Jenkins SR (2009) Parental overprotection, cultural value conflict, and psychological adaptation among Asian Indian women in America. Sex Roles 61:235–251

Vasudev S, Radhakrishnan GL, Ravindran N, Dangor K (2004) The gay spirit. India Today 46:235–251

Whitman SM (2007) Pain and suffering as viewed by the Hindu religion. J Pain 8(8):607–613. https://doi.org/10.1016/j.jpain.2007.02.430

Wild S, Roglic G, Green A, Sicree R, King H (2004) Global prevalence of diabetes: estimates for the year 2000 and projections for 2030. Diabetes Care 2:1047–1053

WHO (2020). https://www.who.int/chp/chronic_disease_report/media/INDIA.pdf

Zieger A, Mungee A, Schomers G, Ta TMT, Weyers A, Boge K, Detting M, Bajbouj M, Von Lersner U, Angermeyer MC, Tandon A, Han E (2017) Attitudes toward psychiatrists and psychiatric medication: a survey from five metropolitan cities in India. Indian J Psychiatry 59(3):341–346. https://doi.org/10.4103/psychiatry.indianjpsychiatry_190_17

20 People of Iranian Heritage

Homerya Haifizi and Melinda Steis

20.1 Introduction

Iran is a geographically and ethnically diverse, non-Arab, Shite Muslim country. Iranians, also known as Persians, speak Farsi and are of Indo-European origin, originally the Aryans of India. In 1935, the country's name was changed from Persia to *Iran* (from the word *aryana*) to present an image of progress and to unify the many ethnicities, tribes, and social classes. The Persian Empire, founded by Cyrus the Great in 559 BC, covered an area from the Hindu Kush (now in Afghanistan) to Egypt. Iran's 1979 Revolution generated a steady wave of immigration to the North America, Europe, and Australia.

20.2 Overview, Inhabited Localities, and Topography

20.2.1 Overview

Prior to the 1979 Revolution, the main reason for immigration was educational advancement. The 2010 U.S. Census report described the approximately 289,465 immigrants as a highly educated community with a high social status. Today, there are an estimated 1 million Iranians living in the US (Organization of Iranian American Communities n.d.) and close to 5 million outside of Iran. The Los Angeles area in California has the largest concentration of Iranians outside of Iran. The population of Iran continues to grow and is currently at more than 84.9 million (Central Intelligence Agency 2020). Immigrants of the pre 1979 Revolution are demographically more cohesive and exited by choice. For this population, a primary source of stress was distance from family and friends (Jalali 1996). Post 1979 immigrants were less cohesive in nature and their departure was more forced than voluntary. Some of these families experienced multiple losses when leaving Iran under duress. They lost financial assets and status. Many experienced a profound degree of hardship, such as fleeing Iran by relying on smugglers and other high-risk means only to seek refugee status (Koser 1997).

In recent years, geopolitical issues have generated deep concerns for all Iranian immigrants regardless of when and why they immigrated. Among immigrants, a deep generational gap within the family unit and with the larger population frequently occurs. First-generation Iranian-born immigrants often live between two worlds. Their age and reason for immigration are mitigating factors. The generational gap widens as each subgroup adopts the new culture and garners new

This is a revised chapter in the previous edition written my Homerya Hafizi.

H. Haifizi (✉) · M. Steis
Veterens Administration, Merritt Island, Viera, FL, USA
e-mail: Homerya.Haifiz1@va.gov; Melinda.steis@va.gov

© Springer Nature Switzerland AG 2021

L. D. Purnell, E. A. Fenkl (eds.), *Textbook for Transcultural Health Care: A Population Approach*,
https://doi.org/10.1007/978-3-030-51399-3_20

manners of self-expression. As stated, Iranians experience discrimination and stigmatization because of the continued political and governmental conflicts between Iran's government and the many Western nations, particularly the US (Public Affairs Alliance of Iranian Americans 2020).

Iran's steady presence in the international news and media is mostly unfavorable and bares a constant reminder in the immigrants' psyche. Years of economic sanctions have weighed heavily on the people of Iran as goods, services, and trades are vastly restricted. It is understood that the quality of an immigrant's life and the process of acculturation are directly related to the actual and perceived acceptance of the host community and/or country. Iranian immigrants have endured these blemished optics for over 40 years (Lipson 1992; Martin 2012).

Iranians are mostly secular and nationalistic in their beliefs unlike Sunni Arab nations who hold a more common Islamic identity (Sayyedi 2004). Iranians are proud of their heritage, which includes ancient empires, the Zoroastrian religion, and some of the world's greatest poets and leaders in philosophy, astronomy, and medicine.

Immigration of Iranians occurred over several phases and across the past five decades. The first-generation immigrants are now senior citizens in these host countries, making the Iranian community at least three generations deep. As age conversely affects acculturation, family dynamic is impacted because of the widening intergenerational gaps. Evidence suggests that, in general, Iranian women acculturate at a faster rate than men and begin to undermine the patriarchal cultural norms (Darvishpour 2002). Older immigrants often feel isolated because of limited language proficiency, dependence on their children, and a perpetual sense of their temporary status that makes permanent commitments to the host nation less likely. They are concerned about their children "over assimilating" and have greater mental health needs, although poorly self-reported. They don't take part in wellness programs and screenings because "they are too old". It is only with age-associated illnesses that older immigrants become consumers of the healthcare system of the host country (Martin 2012).

20.2.2 Heritage and Residence

With the fall of the Persian Empire to the Greeks, followed by many years of occupation by the Arabs, Mongols, and Turks, over time Iranians learned to simultaneously assimilate and preserve their national identity (internalized self-repression) (Lipson 1992). Iran is a collective of many ethnic groups, languages, traditions, cultures and beliefs, and has been ruled by a series of autocratic dynasties that ended with the Pahlavi Dynasty in 1979.

Eid Norouz is a significant, non-denominational, non-religious culturally unifying event. The practice of visiting family and friends during *Eid Norouz* is an important expression of comradery and care (Omeri 1997). The Iranian diaspora continues this observation which coincides with the Spring Equinox in the Northern Hemisphere and the first day of the new calendar year in Iran. Unlike other Sunni Muslim nations, Iran uses the solar not the lunar calendar.

A central tenet of the Iranian social life and personal development is the boundary between inside/private (***baten***) and outside/public (***zaher***). The concepts of "Inside" and "Outside" operate on a continuum and define both the person and the family. Honor and social shame play powerful roles in the societal tone.

20.2.3 Educational Status and Occupations

Iran does not have a formal caste system. Affluence, class, and social status are gained through education, if not inherited. Iranian immigrants strive to maintain a social façade (to save face/*zaher*) because family judgment and social shame are weighed heavily in this community. These issues rarely are mentioned in studies addressing physical and mental health needs of Iranian immigrants in the US (Sayyedi 2004).

Many former white-color middle-aged immigrants were unable to find comparable work resulting in underemployment or self-employment. For example, in Los Angeles, 61% of Iranian heads of household claimed to be self-employed in 1987 and 1988 (Dallalfar 1994) while 10% were employed in blue-collar jobs (Bozorgmehr et al. 1993). Health-care providers should not assume education and social class from reported occupation alone.

20.3 Communication

20.3.1 Dominant Languages and Dialects

Farsi is the national language but nearly half the country's population also speak different languages and dialects, such as Turkish, Kurdish, Armenian, or Baluchi. Most newcomers are not familiar with the host country's vernacular or slang. Unspoken attributes directed at individuals with an accent, where having an accent is paralleled with diminished knowledge or proficiency, is a continued source of stress.

20.3.2 Cultural Communication Patterns

The social communication style is underlined by history and social norms. Iranians cautiously interact with outsiders (non-family and friend). Expressed thoughts and emotions are filtered (*ta'arof*) as a defensive behavior. This filtering includes a distinct form of speech and its underlying dynamics set boundaries (Sayyedi 2004). The continuum of *baten* and *zaher* is continuously operating in the backdrop. This buffering zone leaves very little room for spontaneity. In transactional relationships, such as interactions with health-care professionals, this cautious approach hinders absolute information sharing particularly in gender-based illness where individual or social shame is center stage.

A family spokesperson may hold multiple roles, e.g., interpreter, health-care surrogate, and/or financial advisor. Although commonly rejected, health-care providers should offer a facility sponsored interpretation service to declutter roles. Children and grandchildren are used as interpreters creating another layer of complexity. Respected friends and friends of the family are used as subject matter experts. Immediate and extended family members are very present so if hospitalized, health-care providers should proactively establish a plan for professional care that considers time, privacy, and rest (Omeri 1997).

Personal distance is close, like other Mediterranean cultures. Iranians maintain intense eye contact between intimates and equals of the same gender. This behavior may be observed less among traditional Iranians. Conversations are expressive in tone and body language.

20.3.3 Temporal Relationships

Time orientation is a balance between present and future making Iranians open to health promotion and education, especially when culturally sensitive and appropriately are designed (Jafari et al. 2010). Vahabi (2011) reported that direct translation of generic breast cancer material without considering women's historical, political, and cultural experiences is insufficient. This population of women faced chemical warfare during the 8-year Iran-Iraq conflict, dealt with air and water pollution, and a sociopolitical environment limited their access to accurate breast cancer information and screening. The fatalistic concept of ***Kizmit*** (fate) however, also runs deep hindering the acceptance of health risk assessment and risk reduction. Acknowledged as a highly educated immigrant population with higher status positions and high income, paradoxically Iranians are under-utilizers of the health-care system (Emami et al. 2001; Ghaffarian 1998).

In business, Iranians portray a strong work ethic. Although social time is very flexible, they are highly time conscious and intensely competitive at work and school.

20.3.4 Format for Names

Iranians refrain from calling older people and those in higher status by their first names. Typically, a younger person initiates the greeting. Traditionally women keep their maiden name after marriage but are socially referred to by their husband's surname. This duality is important for accuracy of records and equally as important when identifying a health-care proxy.

20.4 Family Roles and Organization

Consistent with other collectivistic cultures, Iranian families are patriarchal in nature. Open display of conflicts are avoided, parents are highly respected, and communication is rule based to ensure group harmony (Daneshpour 1998).

20.4.1 Head of Household and Gender Roles

Today's families have fewer children and the authority figure may be a working female adult. Most siblings have deep and lively relationships. Iranian Americans concurrently adopt the culture of the host country and maintain their original culture (Ghaffarian 1998). They have more freedom in the new country to select their life partners; families prefer to behold the final approval and some actively take part in matchmaking. Husbands are often a few years older than their wives.

20.4.2 Prescriptive, Restrictive, and Taboo Behaviors for Children and Adolescents

Iranian families are child oriented and protective. Taboo behaviors for Iranian teens resemble their counterparts in the host countries such as smoking, drugs, alcohol and influence of bad peers. The fear of shaming the family and losing face in public is a strong deterrent. There is a double standard for daughters and sons. Behavioral norms are more lax for sons. Shoes are often removed at the door. Older immigrants dress conservatively. Devout Muslim women conceal their hair with head scarves (*hejab*). In Iran, *hejab* is compulsory.

20.4.3 Family Goals and Priorities

The family unit is an important institution. Residing in close-proximity mitigates a sense of isolation, maintains strong intergenerational ties, and creates a safety network. Finding a healthy balance within the family unit is very stressful, boundaries are not clear and family-based roles are emphasized over individuality.

Parenting values and behaviors vary dramatically across the immigrant families. Parents are conscientious financial supporters of their children's education, business, and even marriage. Their children's academic and career achievement is considered that of the collective family unit. As seen in other collectivistic cultures, parents expect their children's commitment. Some immigrant parents are challenging these traditional views the longer they live outside of Iran and acculturate into the host nation (Sayyedi 2004).

Age is a respected attribute particularly in the family structure; therefore, care of elderly parents is a duty and a social expectation. The elderly are expected to be cared for at home using institutional placement as a last resort. Iranian caregivers are especially challenged with many competing priorities including work and family responsibilities, parental expectations, caregiving duties and finances, cultural norms of elder care (all eyes are on them), parents with limited language proficiency, and their reliance on their children for coordination of all healthcare needs.

With the rise in an older culturally, ethnically, and linguistically diverse population of immigrants addressing age-related conditions in a culturally proficient manner is an emerging concern (Antelius and Kiwi 2015). Dementia is especially challenging. Immigrants with dementia, whether proficient in the host language or not, mix languages, confuse them, or revert back to their native tongue. As the disease progresses and they begin to mentally "live in their past," they also prefer to physically live in a culturally relevant environment. In response to this growing issue, some highly populated enclaves, e.g., southern California and Sweden, have established adult day-care centers and culturally profiled facilities (Antelius and Kiwi 2015; Emami et al. 2001). Those not residing in immigrant enclaves are at a disadvantage with extremely limited ability to routinely integrate into an ethnically similar network of like-minded individuals (Momeni et al. 2011).

20.4.4 Alternative Lifestyles

In contrast to the older generation, younger Iranian immigrants are increasingly tolerant of alternative lifestyles and pre-marital cohabitation. Divorce is less stigmatized; even recent immigrants hold similar beliefs. Root causes might be realities of Iran where the rate of divorce has tripled (1 in 7) and marriages are failing earlier, in the first 5 years, based on a 2010 Iranian government report. This trend reflects the resilience of Iranian women and their continued efforts for equality under the Islamic regime and its male-centered legal system (Yong 2010).

Rezaian (1989) found that intraculturally married Iranians reported greater marital satisfaction and a mutually driven cultural mores that advocate for family stability and role-identity (parent, spouse). Out of wedlock pregnancy and homosexuality remain in the conservative sphere so the immigrant Iranian LGBTQ community has to balance openness with familial values. In Iran, homosexuality is considered sacrilegious and according to the Iranian Constitution, a crime punishable by death (Arlanson 2010).

Reflective Exercise #1

1. What is the national language of Iranians? What other languages are spoken by Iranians?
2. What is *zerangi* and how is it displayed?
3. How should traditional Iranian men and women be addressed?
4. Describe Iranian family hierarchy and decision-making practices.
5. What are the primary health conditions among Iranians?
6. Explain *garm* and *sard*? What might you do if you are unfamiliar with this category of foods?
7. How would you vary meals and medication administration during Ramadan?
8. What is *narahati*, what is its significance, and what are its symptoms?

20.5 Workforce Issues

20.5.1 Culture in the Workplace

Lack of legal residence and professional degrees that are not recognized in the host country hinder procurement of comparable employment. Underemployment concomitant with continual bitterness can manifest itself as outward anger or somatization, both with serious outcomes to personal health and familial relations.

Prejudice is less evident in multicultural and metropolitan cities. The general lack of understanding that shrouds the Middle Eastern countries adds fuel. More acculturated immigrant professionals respond flexibly in the workplace, efficiency and efficacy are means to contest prejudice.

20.6 Biocultural Ecology

20.6.1 Skin Color and Other Biological Variations

Iranians are white Indo-Europeans. Their skin tones and facial features resemble those of other Mediterranean and Southern European groups.

20.6.2 Diseases and Health Conditions

The prevalence of multiple sclerosis (MS) is on the rise in the Middle East. The prevalence rate in Iran is medium to high (Etemadifar et al. 2019). Nasr et al. (2016) reported that Iran has the highest rate of MS in the Middle East and Asia. It is reasonable to consider a genetic disposition. However, studies also indicate that migration to a low-risk area reduces the risk of MS incidence suggesting the impact of environmental factors (Guimond et al. 2014).

Similar to US, mental illness still carries a stigma but the Iranian community is becoming more accepting. Seeking professional help from a psychiatrist or psychologist may be referred to as seeking a "counselor" for child rearing or marital issues. Focus is less placed on the mental illness and more on relationship and personal development, all to save public face (*zaher*) and to deter family shame. In another study, Bagheri (1992) found that mental illness is likely to be called a "neurological disorder" or ***narahati-e-asa'b.***

Iranians consider psychopharmacological treatment to be most effective for a somatic illness. Two studies suggested that self-reported mental health in Iranian men were associated with income (under or unemployment), smoking, and social capital (a sense of connection with one's roots and culture, satisfaction with social life, group belonging) (Jafari et al. 2010; Momeni et al. 2011). The rate by which depressive disorders were reported merited attention in both men and women.

Iranian immigrants experience numerous stressors related to resettlement in a foreign culture. As measured by the Health Opinion Survey, 44 percent of Lipson's (1992) newer immigrant interviewees experienced medium or high stress compared with 14% of the long-term-resident group. With reference to mood, about 35% of the informants answered yes when asked if they considered themselves to be "nervous," and about the same percentage stated that they did not have "peace of mind." The reasons were adjusting to their new life in the US, missing family members, and having concerns about relatives left behind. Despite these problems, most Iranian immigrants had no plans to seek counseling or treatment, preferring to rely on family support. In recent years, psychotherapy and counseling have become acceptable treatment modalities, particularly in dealing with children (Sayyedi 2004).

Cardiovascular disease, hypertension, and diabetes are on the rise in Iran and all conditions respond adversely to stressors of an immigrant life. Sharifi and Shah (2019) identified cardiovascular risk factors and events in the Iranian immigrants to be lower than their Canadian and other Middle Eastern counterparts. Ischemic heart disease is on the rise secondary to the stress of living under economic and social constraints. Estimates from 2010 are that 18% of adults are affected by this condition, down from 22% in 2002 (Azizi et al. 2002).

The most common health problems in Iran are linked to underdevelopment, economic downturn, mental stress, and scarcity of resources. Examples of common health conditions are malnutrition, hepatitis A and B, rising rates of tuberculosis and syphilis, hypertension, and genetic blood dyscrasias. Twenty percent of all Iranians residing in Iran suffer from an episode of mental illness requiring specialized medical intervention (Mehr News Agency 2008). In 2010, the reported prevalence of diabetes in Iran was 8%, with an annual increase of 1%. Thalassemia is prevalent in the northern and eastern provinces of Iran. Mediterranean glucose-6-phosphate dehydrogenase (G-6-PD) deficiency is also common among people of Iranian heritage and can precipitate a hemolytic crisis when fava beans are eaten. This deficiency also affects drug metabolism.

McDonald et al. (2017) determined a healthy immigrant effect on cancer diagnosis, finding significantly lower cancer diagnosis in immigrants compared to non-immigrant Canadians, especially for foreign born individuals originating from developing countries. Endeshaw et al. (2019) found an elevated risk of liver cancer mortality in foreign-born individuals in comparison to the comparable US population. With the burden of liver cancer in the US on a steady rise, immunization with Hepatitis B vaccine is recommended. Gastric cancer claims a higher mortality rate in foreign-born males, reducing the prevalence of *H. Pylori* in first generation immigrants of high burden countries is strongly recommended (Hallowell et al. 2019).

Family's response to addiction range from complete support to disownment. Addiction carries a burden of shame. Prevalence of substance abuse in this population is related to low levels of acculturation, a perception of prejudice, a sense of helplessness and loneliness, and poor coping skills. Iranian men may demonstrate their "masculinity" by claiming to "hold" their liquor well.

20.6.3 Health-Care Practices

Changes in environment and personal attitudes affects immigrants' physical activity, food habits, and body image (Delavari et al. 2013). Iranian immigrants are accustomed to walking and using mass transit for city mobility. Recent immigrants may miss the collegiality that is facilitated by these activities and feel limited and "trapped". Conversely, less air pollution and more parks and open spaces have led to an increase in physical activity not connected to activities of daily chores (walking to the bodega or using mass transit). The sense of body image shifted from being thin to being fit. Delavari et al. (2013) reported that Iranian immigrants in Victoria, Australia had rates of obesity similar to high-income countries. The study group was identified as well educated and often emigrated for personal freedom not material deprivation.

Humoral medicine, agricultural availability, and Islamic dietary rules play a significant role in health-care practices and food intake. Refer to Healthcare Practices section for more information.

20.7 Nutrition

20.7.1 Meaning of Food and Food Rituals

Food is a symbol of hospitality and kinship. There are strict religious and regional dietary norms and humoral medicine has greatly impacted dietary beliefs. Immigrants believe their native food is more nutritious and with minor modifications it can be even more advantageous than subscribing to the host countries food habits.

Iranian food is flavorful.Health conscious immigrants have modified recipes and prep time. A pleasing mixture of colors and ingredients comprise the balance of ***garm*** (hot) and ***sard*** (cold). The key to humoral theory is balance and moderation. Too much of any one category leads to being "overheated-sweating, itching, hives" or "chilled-dizziness, nausea, vomiting." Susceptibility is believed to be gender dependent. Women are more susceptible to *sardie*. Fresh ingredients are preferred although cost and availability can be limiting factors.

Consumption of fast food is less common among older immigrants because of nutritional value, budgetary constraints, and taste preferences. Tea is the hot beverage of choice; it is consumed mostly with cubed sugar, dates, and pastries. Long grained white rice, sheet bread (wheat and white), and potatoes are the main carbohydrates. Corn is used but less favored. Beans, legumes, herbs, and vegetables are liberally included in meals. Dairy products are dietary staples, particularly eggs, milk, yogurt, feta cheese, and two dairy by-products, *doog* (yogurt based) and *kashk* (milk based). Meat based protein choices are beef, lamb, poultry, and fish. Wild meat is rarely consumed; some are considered *haram* (prohibited in Islam). Shellfish is regionally favored. Fresh fruit and vegetables are staples. Similar to Judaism, Islam has a strict set of dietary laws; ***halal*** means permissible and prescribed, ***haram*** means forbidden. Pork and some shellfish are *haram*. Compliance with proscriptive food items is seen less frequently among younger generations. Medical conditions that require strict dietary guidelines or when food-drug interactions are a concern, can be managed via adjustments to native food practices without significant deviations.

20.8 Pregnancy and Childbearing Practices

20.8.1 Fertility Practices and Views Toward Pregnancy

Vasectomies are less commonly adopted for birth control. Concepts of hot and cold are applied during pregnancy and postpartum. Menstruating women may refrain from participating in religious activities, vigorous physical activity, or intimacy. Historically, infertility was attributed to the woman. In the past, Iran's Health Ministry launched a nationwide campaign and introduced contraceptives—pills, condoms, IUDs, implants, tubal ligations, and vasectomies. However, the campaign has had problems and it changes with leadership. It is unknown what their use is in the US and elsewhere.

20.8.2 Prescriptive, Restrictive, and Taboo Practices in the Childbearing Family

Access to material care is confounded by the factors discussed earlier. Insurance and logistics of a complex healthcare system, language, and finances are elements to consider. In the more traditional families, the spouse may not be present at birth. The choice of location for delivery is a medical decision, not cultural. The postpartum period can be extensive, up to 30–40 days. Some families may strictly limit visitations to ensure the infant's safety and the mothers' recovery. The more-acculturated families take advantage of mother and child education classes, group support, books, and social media to prepare for delivery and parenting.

20.9 Death Rituals

20.9.1 Death Rituals and Expectations

Outpouring of support is followed by a memorial wake after death. Devoted Muslims will use services of a mosque and an Islamically profiled mortuary service for prescriptive preparation of the body and burial ceremonies. In the absence of these services, families work with local mortuaries to observe as much of the rituals as possible. In Iran, embalming is not practiced, neither is cremation. Bodies are shrouded for burial; coffins are not used. In the immigrant population, cremation is practiced and burial practices are more in line with the host country. The belief in a rise to heaven or a fall into hell are also shared in this population. There are no religious or cultural prohibitions to autopsy.

20.9.2 Responses to Death and Grief

Loss of a loved one is met with outward and expressive grieving. Death is a journey, from mortal life to spiritual existence and unification with God. The practice of donating to a charity to help those in need (in the memory of the deceased) is customary. Black is the color of mourning. Gravesite is frequently visited and gravestones mark graves.

Reflective Exercise #2

A 59-year-old Iranian male collapses at work after he tells a co-worker that he is having a panic attack. Emergency response arrives in 4 min and establishes airway via intubation and works on the individual for 15 min to reverse pulseless electrical activity (PEA). At the hospital the individual undergoes emergent catheterization to address a 90% occlusion in the left anterior descending artery. In his employment records, the patient had listed a co-worker as his next of kin.

Approximately 2 h after the event onset, through a series of calls, his children and former spouse are located. The children are approached as his immediate kin to establish their roles as health-care proxies. They informed hospital staff that their parents were divorced. They also learn that his condition is gravely critical and a DNR (do not resuscitate) status had to be determined. Both children, due to their self-reported lack of knowledge in the US health-care system, relegated the responsibility to their mother (the divorced spouse). The spouse was able to produce a power of attorney as well as an explicitly detailed living will that was signed a few years prior by the patient and in the presence of an attorney.

The patient had clear instruction for withdrawal of all life support, including but not limited to nourishment and hydration. The patient had an aunt and several cousins in the US. His siblings lived in Iran, as did his mother who had taken ill with a stroke and dementia some years earlier. Several days into his hospitalization he was diagnosed with brain damage due to a long period of pulseless electrical activity in the field. The patient was an organ donor. He had lived in the US for 27 years.

1. How would you address the family dynamics in this scenario?
2. Who has the right to make decisions for this individual?
3. How customary is it to see end of life planning in Iranian immigrants and how would you address this issue in the absence of one?
4. Is a DNR and organ donation concepts well accepted in the Iranian culture?

20.10 Spirituality

20.10.1 Dominant Religion and Use of Prayer

Iranian immigrants are more secular, although some may observe the month of *Ramadan* and take part in daily prayers. *Ramadan* is associated with fasting from sunrise to sunset. Certain individuals are exempt from fasting: for example the ill, the elderly, and children. The intent of *Ramadan* is to cleanse the body, soul, mind, and thought.

20.10.2 Meaning of Life and Individual Sources of Strength

Family, friendship, and social support are sources of strength and comfort, particularly in times of illness or crisis (Omeri 1997). Iranians are highly affiliative and thrive on social relationships. Given the importance of such contact, health-care providers may need to adjust visiting policies.

20.10.3 Spiritual Beliefs and Health-Care Practices

Tagdir means God has power over one's fate. The belief is more characteristic of older immigrants. Hafizi (1990) illustrated the integration of religion and health. In the words of a highly educated and devout Muslim man: *To ask me what health means is to ask me how I see myself in relation to God, my family, the society, and my material body. Man is the embodiment of a celestial being. Death is "graduation" to a higher level. I believe in God and His plan for my future. Being sick is not having a cold, it is not having the vision and the interest to deal with the cold.* Spirituality and religious affiliations decrease rates of advance planning in end of life matters (Rahemi 2019).

20.11 Health-Care Practices

20.11.1 Health-Seeking Beliefs and Behaviors

Health beliefs are a culmination of Galenic medicine (humoral), Islamic practices, and elements of biomedicine. The distinctive balance of four humors *(mezaj)* result in the individual's unique temperament *(tabi'at)*. At symptom's onset, oftentimes the first question is "Did you eat something that did not agree with you (your *mezaj*)?" Climate and weather are believed to significantly affect health. Fright or the startling effect of bad news is believed to lift hope and accompany physical symptoms such as chill and fever. The idea of an evil eye/spirit circulates in the traditionally inclined families attributing illness to an outside person or force, giving meaning to an occurrence of puzzling origin.

Galenic medicine and Islamic dietary rules are embedded in daily practices. They are preventative in nature while modern medicine is sought for its curative qualities. ***Narahati*** is a general term used to express a wide range of symptoms. Somatization is not uncommon as a form of communicating emotional distress in a more culturally sanctioned approach. The "personal self" (***baten***) presented as the "somatic self" (***zaher***).

Hafizi's (1990) research found that Iranians' concepts of health represented two of four domains: the clinical view (health as absence of disease) and the adaptive view (health as the ability to cope successfully). Health is a lifestyle marked by demands and adaptations (Hafizi 1990). Similar health concepts were found among older Iranian people in Sweden (Emami et al. 2000).

20.11.2 Responsibility for Health Care

Care-seeking behaviors are influenced by cultural values (Rahemi 2019). Iranians seek treatment relatively soon after the onset of symptoms if a combination of advice from knowledgeable acquaintances and family and home remedies fail.

Iranians value a relationship-based approach in health care (Martin 2009). They criticize Western medicine as time constrained and transactional. These beliefs create a loosely build trust with the host health-care system. Dastjerdi (2012) reported that Iranians residing in Canada frequently organize their care from Iran. Some antibiotics, codeine-based analgesics, mood-altering drugs in the benzodiazepine family, and intramuscular vitamins are available as OTC in Iran. Self-adjustment of medications may occur if in financial hardship or symptoms have not resolved.

20.11.3 Folk and Traditional Practices

Herbal remedies are used in a complementary manner to prevent illness, to maintain health, and to manage symptoms. For example, Starflower, for management of the common cold, relief of anxiety, or cleansing of the urinary tract; or quince seeds soaked in hot water for a sore throat.

20.11.4 Barriers to Health Care

Lack of language proficiency in host countries is a major predictor of underutilization of the health-care system and poor health (Alizadeh-Khoei et al. 2011). Additional barriers are insufficient finances, lack of insurance, immigration status, and lack of transportation and trusted interpreter.

20.11.5 Cultural Responses to Health and Illness

Iranians are expressive about their pain and suffering. The grandmother of a young woman with a brain tumor consoled herself and the patient when declaring "Suffering in this world assures her a place in heaven."

Understanding spirituality, attitudes about end of life planning, and trust of the health-care system become pivotal points for health-care professionals to interpret. Also important to understand is that as immigrants age and acculturate into their host countries' practices, their own perceptions of health and illness may also flux (Emami et al. 2001). In a 2019 study by Rahemi, researchers reported that nearly half (47.4%) of the sample's more acculturated and educated older Iranians had a written or oral end of life plan. An indication of a shift from their native beliefs of tagdir/kizmat/Will of God to an embrace of autonomy in the dying process. Cultural norms and values weigh heavily in this stage; home is preferred over hospital and long-term placement for chronic disabling conditions is less favored. Palliative and hospice care are less likely sought and somewhat misunderstood.

20.11.6 Blood Transfusions and Organ Donation

Blood transfusions, organ donations, and organ transplants are widely accepted among Iranians. In Iran, organ donations have become a business transaction—if a kidney is needed, it can be purchased.

Reflective Exercise #3

The members of an Iranian family immigrated to the US over a 45-year period. Initially, two siblings came to United States for education, obtained US residency, and never returned. The parents joined them to take care of their grandchildren. After 30 years, the third sibling immigrated because he felt a complete sense of hopelessness and helplessness under Iran's Islamic Republic; not for himself but for his teenage children. Neither of the parents became fluent in English, although they attended several years of English classes for adults. To them their stay was temporary because the Islamic Republic was temporary. The parents were able to manage conversational English but for business and health care they relied on their children. The father of the family had advanced cardiac disease and after several hospitalizations he opted for no further treatment and requested to die at home. The mother of the family was showing signs of dementia but this went unnoticed by the siblings. They noticed their mother being slightly agitated, apprehensive, and suspicious but they never attributed it to dementia, rather to aging. She would ask why they were not truthful and open with her, even though they were. Family members lived in close proximity, they shared parental duties, and lived in a state not populated by many Iranians.

1. How might this family respond to palliative care or hospice? Do you think it would be positive?
2. In the absence of an Iranian enclave, what options do the siblings have to care of their elderly parents?

20.12 Health-Care Providers

20.12.1 Traditional Versus Biomedical Providers

Health care is a widespread concern. The purity of the biomedical model is less appealing. Iranian immigrants expect to receive a definitive diagnosis from an authoritative source with a clear road map for treatment including prescriptions and therapies. They may not ask too many questions or inquire about different modalities thinking that the provider knows best.

20.12.2 Status of Health-Care Providers

Religious and folk providers are not sought by most Iranian immigrants. The most respected health-care provider is an educated and experienced physician. Nursing care in the US is found to be far more interactive, communicative, and people oriented. Allied health professionals and their modalities of care are accepted as necessary and imperative to health.

References

Alizadeh-Khoei M, Mathews RM, Hossain SZ (2011) The role of acculturation in health status and utilizatino of health services among the Iranian elderly in metropolitan Sydney. J Cross Cult Gerontol 26:397–405

Antelius E, Kiwi M (2015) Frankly, none of us know what dementia is: dementia caregiving among Iranian immigrants living in Sweden. Care Manag J 16(2):79–94

Arlanson J (2010) Mohammad and the Homosexual. http://www.answering-islam.org/Authors/Arlandson/homosexual.htm

Azizi F, Ghanbarian A, Madjid M, Rahmani M (2002) Distribution of blood pressure and prevalence of hypertension in Tehran adult population: Tehran lipid and glucose study (TLGS), 1999-2000. J Hum Hypertens 16(5):305–312

Bozorgmehr M, Sabagh G, Der-Martirosian C (1993) Beyond nationality: Religio-ethnic diversity. In: Kelly

R (ed) Irangeles: Iranians in Los Angeles. University of California Press, Berkeley, pp 59–79
Central Intelligence Agency (2020) The world fact book. https://www.cia.gov/library/publications/the-world-factbook/geos/ir.html. Accessed 8 Feb 2020
Dallalfar A (1994) Iranian women as immigrant entrepreneurs. Gend Soc 8(4):541–561
Daneshpour M (1998) Muslim families and family therapy. J Marital Fam Ther 24(3):355–390
Darvishpour M (2002) Immigrant women challenge the role of men: how the changing power relationship within Iranian families in Sweden intensifies family conflicts after immigration. J Comp Fam Stud 33(2):270–296
Dastjerdi M (2012) The case of Iranian immigrants in the greater Toronto area: a qualitative study. Int J Equity Health 11:9
Delavari M, Farrelly A, Andre MD, Swinburn B (2013) Experiences of migration and the determinants of obesity among recent Iranian immigrants in Victoria, Australia. Ethn Health 18(1):66–82
Emami A, Torres S, Lipson J, Ekman S-L (2000) An ethnographic study of a day care center for Iranian immigrant seniors. West J Nurs Res 22:169–171
Emami A, Benner P, Ekman S-L (2001) A sociocultural health model for late-in-life immigrants. J Transcult Nurs 12(1):15–24
Endeshaw M, Hallowell BD, Razzaghi H, Senkomago V, McKenna MT, Saraiya M (2019) Trends in liver cancer mortality in the United States: dual burden among foreign- and US-born persons. Cancer 125:726–734
Etemadifar M, Ghourchian S, Sabeti F, Akbari M, Etemadifar F, Salari M (2019) The higher prevalence of multiple sclerosis among Iranian Georgians; new clues to the role of genetic factors. Rev Neurol 175(10):625–630
Ghaffarian S (1998) The acculturation of Iranian immigrants in the United States and the implications for mental health. J Soc Psychol 138(5):645–654
Guimond C, Lee JD, Ramagopalan SV, Dyment DA, Hanwell H, Giovannoni G et al (2014) Multiple sclrosis in the Iranian immigrant population of BC, Cnada: prevalence and risk factors. Mult Scler J 20(9):1182–1188
Hafizi H (1990) Health and wellness: an Iranian outlook. Unpublished Master's Thesis, University of California
Hallowell BD, Endeshaw M, Senkomago V, Razzaghi H, McKenna MT, Saraiya M (2019) Gastric cancer mortality rates among US and foreign-born persons: United States 2005-2014. Gastric Cancer 22:1081–1085
Jafari S, Baharlou S, Mathias R (2010) Knowledge of determinants of mental health among Iranian immigrants of BC, Canada: "a qualitative study". J Immigr Minor Health 12:100–106
Jalali B (1996) Iranian families. In: McDGoldrick M, Giorano J, Pearce JK (eds) Ethnicity & family therapy, 2nd edn. Guilford Press, New York, pp 347–363
Koser K (1997) Social networks and the asylum cycle: the case of Iranians in the Netherlands. Int Migr Rev 31(4):591–612
Lipson J (1992) Iranian immigrants: health and adjustment. West J Nurs Res 14:10–29
Martin SS (2009) Healthcare-seeking behaviors of older Iranian immigrants: health perceptions and definitions. J Evid Based Soc Work 6(1):58–78
Martin SS (2012) Exploring discrimination in American health care system: perception/experiences of older Iranian immigrants. J Cross Cult Gerontol 27:291–304
McDonald JT, Farnworth M, Liu Z (2017) Cancer and the healthy immigrant effect: a statistical analysis of cancer diagnosis using a linked census-cancer registry administrative database. BMC Public Health 17:296
Mehr News Agency (2008) 20% of Iranians depressed: roll with the punches in life, from payvand.com/news/08/aug/1323.html
Momeni P, Wettergren L, Tessma M, Maddah S, Emami A (2011) Factors of importance for self-reported mental health and depressive symptoms among ages 60-75 in urban Iran and Sweden. Scand J Caring Sci 25:696–705
Nasr Z, Majed M, Rostami A, Sahraian MA, Minagar A, Amini A et al (2016) Prevalence of multiple sclerosis in Iranian emigrants: review of the evidence. Neurol Sci 37:1759–1763
Omeri A (1997) Culture care of Iranian immigrants in New South Wales, Australia: sharing transcultural nursing knowledge. J Transcult Nurs 8(2):5–17
Organization of Iranian American Communities (n.d.) Iranian community USA. https://oiac.org/iranian-community-usa/. Accessed 9 Feb 2020
Public Affairs Alliance of Iranian Americans (2020). https://paaia.org/. Accessed 8 Feb 2020
Rahemi Z (2019) Planning ahead for end-of-life healthcare among Iranian-American older adults: attitudes and communication of healthcare wishes. J Cross Cult Gerontol 34:187–199
Rezaian F (1989) A study of intra- and inter-cultural marriages between Iranians and Americans. Unpublished Doctoral Dissertation. California Institute of Integral Studies
Sayyedi M (2004) Psychotherapy with Iranian-Americans: the quintessential implementatino of multiculturism. California Psychologist 37:12–13
Sharifi F, Shah B (2019) Cardiovascular risk factors and events in Iranian immigrants versus other immigrants from the Middle East. J Immigr Minor Health 21:788–792
Vahabi M (2011) Knowledge of breast cancer and screening practices among Iranian immigrant women in Toronto. J Community Health 36:265–273
Yong W (2010) Iran's divorce rate stirs fear of society in crisis. New York Times. http://www.nytimes.com/2010/12/07/world/middleeast/07divorce.html

People of Italian Heritage

21

Alessandro Stievano, Gennaro Rocco, Franklin A. Shaffer, and Rosario Caruso

21.1 Introduction

This chapter provides the most updated issues regarding the various aspects of Italian culture. This chapter gives an interesting perspective of the various elements that compose the Purnell Model for Cultural Competence because it blends views regarding first, second, and third generation of Italian-Americans and the real social life about the same issues in Italy. It is a meaningful perspective, in our opinion, because it allows to have a big picture of Italian culture not only in determined settings (US settings) but also on a wider scale.

This chapter is an update from a previous edition written by Sandra Hillman.

A. Stievano (✉)
University of Rome Tor Vergata, Rome, Italy

G. Rocco
Centre of Excellence for Nursing Scholarship, Opi Rome, Rome, Italy

F. A. Shaffer
CGFNS International, Inc., PA, USA

R. Caruso
S. Donato's Group, Milan, Italy

21.2 Overview, Inhabited Localities and Topography

21.2.1 Overview

Italy is a European republic, being a peninsula delimited by the Alps and surrounded by several islands, such as Sicily, Sardinia, and other minor islands. Geographically, Italy is located in South-Central Europe. The country covers a total area of 301,340 km^2 (116,350 sq mi) and shares land borders with France, Switzerland, Austria, and Slovenia. Vatican City and San Marino are two enclaved microstates located in the Italian peninsula.

Italy is divided into 20 regions and 107 provinces (Cei et al. 2018) with a population of over 60 million. In the Northern parts of the country, the Alps separate Italy from France, Switzerland, Austria, and Slovenia. On the West, the borders are given by the Ligurian and Tyrrhenian seas, while the Adriatic Sea delimits the Eastern border; in the South is the Ionian Sea. All these seas are part of the Mediterranean Sea. Overall, the climate of Italy varies significantly according to its diverse topography given by both mountains and plains. Specifically, Northern Italy consists of a great plain crossed by the river Po, the longest of the country that originates from the Cottian Alps in the Piedmont Region. Central Italy contains the Apennine Mountains, whereas Mezzogiorno in the South consists of lower mountains and stony hills with large areas with scarce vegetation result-

© Springer Nature Switzerland AG 2021
L. D. Purnell, E. A. Fenkl (eds.), *Textbook for Transcultural Health Care: A Population Approach*,
https://doi.org/10.1007/978-3-030-51399-3_21

ing in soil erosion and, in turn, deterioration of the soil into clay (Cei et al. 2018).

21.2.2 Heritage and Residence

In the distant past, Italy was the mother country of the Romans and the metropolis of the Roman Empire. The Roman Empire dominated Western Europe and the Mediterranean for many centuries, making immeasurable contributions to humanity (Hay 2019). For instance, Romans led to the development of Western philosophy, science, and art that remained central during the Middle Ages and the Renaissance (Trapp 2017). After the fall of Rome in AD 476, Italy remained fragmented in numerous city-states and regional polities until the Italian unification led to the establishment of an Italian nation-state (Hay 2019). The new Kingdom of Italy was established in 1861, quickly modernized and built a colonial empire, colonizing parts of Africa (Libya) and countries along the Mediterranean (Alfani and Ryckbosch 2016). However, the Southern regions of the young nation remained rural and poor, characterized by a great emigration to different countries as Argentina, Brazil, and the United States of America (USA). This great emigration occurred between 1880 and 1914 and brought more than four million Italians to the USA (Klein 1983).

Italy in World War II was on the side of Germany and Japan, and ended in military defeat and an Italian Civil War (Pilat 2009). Following the liberation of Italy, the country abolished the monarchy with a referendum, reinstating democracy and was characterized by an "economic miracle" in the period between 1958 and 1964. Italy co-founded the European Union, North Atlantic Treaty Organization (NATO), and the Group of Six (later G7 and G20). Overall, Italy is a country rich in history, famous for the heritage and marvels of ancient Rome, such as the Coliseum, the Spanish Steps and St. Peter's Square; the Leaning Tower of Pisa; the canals and Piazza San Marco in Venice; the Duomo in Milan; the ruins of Pompeii; and the Arena in Verona (Giambona and Grassini 2020).

21.2.3 Reasons for Migration and Associated Economic Factors

Nearly five centuries before modern Italy became a nation, Italians began migrating to the Americas in large numbers. Early Italians, under the sponsorship of French, English, Portuguese, and Spanish governments, sought adventures as explorers, warriors, sailors, soldiers, and missionaries. The 12,000 or more Italians who came to America between the founding of the American Republic in 1783 and the establishment of modern Italy as a nation in 1871 were scattered throughout North America, with large concentrations in the Northeast and the lower Mississippi Valley. Early Italian immigrants came from an agricultural background and differed in several respects from later immigrants who began arriving at the close of the nineteenth century. Many of the later immigrants were political refugees who had a variety of skills and occupations, such as tradesmen, artists, musicians, and teachers. These Italian immigrants exercised a civilizing influence on a society largely dependent on Europe for cultural guidance. Between 1810 and 1820, when immigration statistics were first compiled, 439 Italian immigrants came to the United States. By 1860, only 14,000 Italians immigrants had been recorded, the majority of whom came from Northern Italy. After 1880, Italian men came massively from the South of Italy, the poorest part of the country even nowadays. By 1901, Southern Italians composed 83% of the Italian immigrants in the United States. Of all Europeans who immigrated during those years, Italians had the smallest proportion of women and children (Mangione and Morreale 1993).

Over 3.8 million Italians immigrated between 1899 and 1924, with the peak of migration between 1901 and 1910. Many in this group were women and children. However, 2.1 million Italians returned to their homeland during those same years, with only 1.7 million immigrants remaining in the USA. During the peak years of immigration, 97% of the Italians entered the United States through New York City (Ellis Island), giving it the largest Italian population of any city in the country (Mangione and Morreale 1993).

During the '60s, Italian immigration to the USA stabilized around 25,000 people per year. Between 1974 and 1986, a steady decline occurred in Italian immigration, with less than 5000 new immigrants entering the United States annually (Mangione and Morreale 1993). The great Italian migration to North America that began in the nineteenth century ended. The current Italian American population of almost 18 million ranks fifth in the United States, preceded by German, Irish, English, and African Americans. The Italian culture with its values and cuisine has had an impact on the cultural landscape of North America. It would be a mistake to conclude that Italian culture in America is disappearing. Beyond pasta and pizza, Italian immigrants and their children have made and continue to make an impact on the arts, architecture, and commerce.

21.2.4 Educational Status and Occupations

In Italy, tertiary educational attainment is increasing for younger generations, even though it remains relatively low in the population compared to other advanced countries. The share of tertiary-educated 25–64 year-olds was 19% in 2018, compared to 28% among 25–34 year-olds.

In Italy, the employment rate for adults with a tertiary qualification in fields as science, technology, engineering and mathematics (STEM) is relatively close to the Organization for Economic Co-operation and Development (OECD) average: this is the case for information and communication technologies (87%), engineering, manufacturing, and construction (85%). The share of adults with a tertiary education in engineering, manufacturing, and construction is comparatively low (15%), but it is slightly higher among recent graduates (17%). The employment rate for adults with a tertiary education in arts and humanities, social sciences, journalism, and information is relatively low (77%), although these fields remain relatively popular. Italy has the largest share of teachers over the age of 50 across OECD countries (59%) and will have to renew half of its teaching workforce in the next decade or so. However, Italy has the lowest share of teachers in the population aged 25–34 year-olds across OECD countries. Technical and vocational education and training is an alternative pathway to enter the labor market: young adults (25–34 year-olds) with upper secondary or post-secondary non-tertiary vocational attainment have similar employment prospects to those with a tertiary qualification.

Italians living in the US seek to do something that can demonstrate their success to their families. Because of their general distaste for abstract values, ambivalent attitude toward formal schooling, and desire to remain close to family, a disproportionate number of second- and third-generation Italian Americans are employed in blue-collar jobs and, in general, have a better schooling than their fathers.

21.3 Communication

21.3.1 Dominant Language and Dialects

The official language of Italy is Italian since the unification of the Country in 1860. Grammatically correct Italian is musical and romantic because vowels predominate over consonants, expressing the many subtleties of thoughts and feelings in a delicate manner. In many Italian households, discussions can become quite passionate, with voice volumes raised and many people speaking at once. However, courtesy is a quality that is very much appreciated in Italy.

There are specific etiquettes and protocols for individual social and business situations:

Italians are well-known for nonverbal communication. They communicate with gestures as well as with words. Nonverbal communication is conveyed via different means. Eye contact for an Italian means respect and understanding. Facial expressions show how a person feels and often what they are thinking. Italians usually smile and have relaxed facial expressions when they are calm and when they wish to make people feel at ease. Spatial distancing is important in social interactions, usually an arm's length from the person to whom they are speaking. This is especially true if Italians are speaking to someone they do not know or do not know well.

Discussions are frequent among Italians when they express different opinions. If you are talking to someone and your opinion is different from theirs, they will probably start a discussion with you to prove that they are right. This is normal and does not mean the persons are having an argument. In Italy many gestures have very specific meanings that can be learned after prolonged contacts with Italians. For example, a slowly raised chin means "I don't know."

As mentioned before, the official language of Italy is Italian, a Romance language derived from Latin. However, all socioeconomic groups in the 20 regions of Italy speak different dialects (Lepschy and Lepschy 2013). The dialects of Northern Italy contain numerous German words. Spanish, French, and German languages influenced Italy during the Neapolitan period. For example, Piedmontese is strongly affected by the French and Spanish languages, whereas the dialects of Sicily have been strongly influenced by French, Spanish, Greek, Albanian, and Arabic languages. Accordingly, many dialects are sometimes incomprehensible for Italians coming from different regions (Lepschy and Lepschy 2013).

21.3.2 Cultural Communication Patterns

Italians are not afraid to express their feelings (Shuter 1977). For example, the "typical" kiss is Eastern European style, with a kiss on each cheek. They frequently touch and embrace family and friends. Touching between men and women, between men, and between women is frequently seen during verbal communication. Whereas nonverbal methods of communicating are common to all societies, Italians have elaborated, this art especially in the Southern Regions of the country. Gestures convey a range of feelings from eloquence to intense anger (Kendon 1995). These messages, however, are conveyed in an economical, subtle, flowing, and almost-imperceptible manner.

21.3.3 Temporality

Broadly speaking, punctuality is not the utmost priority for Italians. Some delays are common in many working environments and settings, even if adherence to clock time in important situations is very well appreciated. These small delays should not be considered as a sign of lack of respect, but rather as a cultural habit, especially in the Southern parts of the country where people also have different customs compared with the Northern parts. For example, dining at around 20.30–21.00, especially during summertime is absolutely normal but is unusual in the North.

As a general guideline, work plans are often not taken too strictly so that some flexibility can be built around them. For example, a conference that is supposed to start at 9:00 a.m., may actually begin 30 min later without too much embarrassment by the organizers.

21.3.4 Format for Names

The most common names for males are Marco, Alessandro, Giuseppe, Flavio, Luca, Giovanni, Roberto, Andrea, Stefano, Angelo, Francesco, Mario, Luigi. For females, the most names are Anna, Maria, Sara, Laura, Aurora, Valentina, Giulia, Rosa, Gianna, Giuseppina, Angela, Giovanna, Sofia, Stella. The following are the most common surnames in Italy: Rossi, Russo, Ferrari, Esposito, Bianchi, Romano, Colombo, Ricci, Marino, Greco, Bruno, Gallo, Conti, and De Luca. Nicknames are used but without a particular frequency.

Reflective Exercise #1 (communication)
Armando, an Italian high school senior, is expected after he graduates to work in his father's *bakery* where he has worked after school since he was 14. His father told him that upon graduation, he could earn a real salary in the business and soon start doing some of the buying.

His father is very proud to be the owner of a store. However, Armando is not interested in doing a career managing a bakery and does not know how to tell to his father. Armando actually wishes to became a nurse.

The problem is that he could only imagine how disappointed his father would be. His father considers university as impractical. Furthermore, the sister of Armando, Arianna, has a real head for business in managing bake-house, but his father does not even see it. She could run the store; however, if his father has his way that will never happen.

Armando feels stuck. His dad thinks he should go to work, when he would like to apply for starting a bachelor in nursing science.

Considering the story here above, please think about the following questions, imagining you as a friend of Armando that has already started the bachelor in nursing:

1. How would you counsel Giorgio Armando?
2. What do you see as the important issues for him?
3. What do you need to know about the Italian culture and history to be useful in giving advises to Armando?

21.4 Family Roles and Organization

The family (*La famiglia*) is one the most important aspect of an Italian's life (Bertocchi and Bozzano 2019). However, in the twenty-first century Italian families have become smaller in size. In the past decades, the fertility rate declined. The economic crisis severely hit the country in the second decade of the twenty-first century (Goldstein et al. 2003). In the last 40 years, the structure of families has been significantly influenced by demographic, economic, and professional changes, determining a transition from a patriarchal to a nuclear family model with a higher number of single-parent families, single-person households, childless couples, and same-sex couples (Gauthier 2007). More than one third of the families are currently comprise single-person households; the total fertility rate is the lowest in Europe (Cooke 2009). Overall, the fast-economic pace in the twenty-first century has changed many dynamics within Italian families; for example, people are often away during the week due to fast-paced working rhythms. Furthermore, major demographic changes in Italian families are given by a decreased number of marriages, delays in getting married with a high number of civil ceremonies, and by a reduced birth rate (Luciano et al. 2012). Under an economic lens, Italy is becoming one of the European countries with the lowest growth rate, and with an increasing number of births out of wedlock, an increased marital instability, and with a constantly growing number of legal separations (Luciano et al. 2012).

21.4.1 Head of Household and Gender Roles

In Italy, family is characterized by strong ties and is based on mutual aid of all its members. The role of head of the family is still important and usually is a male and the breadwinner for the entire family. Nowadays, in Italy the most common reasons people decide to marry later or not to marry at all are education (as per the expenses linked to attending university), lack of economic independence, and lack of a steady job. In Italy, those who attend university are not, on average, economically independent, so with no third party. Moreover, nowadays Italians have difficulty finding a steady job and they do not want to start a family without having economic solidity (Tomassini et al. 2003).

As described by the Italian National Institute of Statistics (ISTAT), Italian women usually have their first child at 30.8 years old; accordingly, Italian motherhood generally starts later than other European countries where women generally have their first child between 26 and

30 years of age (Blangiardo and Rimoldi 2014). Moreover, there is a substantial increase in the number of families made up of a single parent; two millions roughly, 83.6% of whom are run by women (Blangiardo and Rimoldi 2014).

It is also significant to highlight is a deep respect for elderly family members in the Italian culture. Senior family members are usually taking care of their children and grandchildren. Their care comes with the expectation that their children will support and assist them throughout old age later in life. Residential care is avoided unless the family has no other option.

21.4.2 Prescriptive, Restrictive and Taboo Behaviors for Children and Adolescents

Italian American children who grow up in a home in which the parents speak an Italian dialect are able to more readily absorb their culture. Although this creates closer family ties, it can produce a conflicting sense of identity when these children speak another language outside the home. Italian American children are taught to have good manners and respect for their elders. Both male and female children are encouraged to be independent and are expected to contribute to the family's support as soon as they are old enough to work. This work ethic continues in second- and third-generation families in the US.

In recent decades, social research on youth in Italy has explored a range of issues through different interpretative and methodological approaches. Collective identities and forms of identification among youth are shaped more and more frequently through the sharing of social and social digital practices. With reference to four main fields (sport, music, politics, religion) and focusing on youth cultures, it is easy to identify transversal processes through which young people today elaborate and adopt social more digitalized practices, create new interactions, and develop innovative signification processes.

21.4.3 Family Goals and Priorities

Italian American families maintain close relationships. Love and warmth, security, and the expression of emotions are the most common characteristics of the Italian family. Daughters have close ties with both parents, particularly as they approach old age. Among first-generation Italians, the welfare of the family was considered the primary responsibility of each of its members. Although many second- and third-generation Italian Americans no longer live in an immediate Italian enclave, they return home frequently to maintain family, community, and ethnic ties.

While parents are alive, their home is most often the focus of kinship gatherings. Sons and daughters visit frequently during the week and after church on Sundays to share a large meal at the parents' house. Frequent contact with parents generally means contact with siblings and, often, aunts and uncles. Italian Americans are almost twice as likely as other ethnic groups to see a parent daily or at least several times a week. If personal contact is not possible, frequent telephone contacts are made, sometimes several times a day. Love, respect, self-sacrifice, and mutual responsibility essentially summarize the diffuse sentiments most respondents express in regard to their parents. The status of Italian Americans in old age is one of continuity. Although they are more dependent in terms of their social and psychological needs and more disengaged from work and roles connected with formal and informal associations, they are firmly entrenched in the extended family system. Even though they may have major worries about maintaining traditional family values, changes have not resulted in their exclusion from the lives of their children. Continuity in parental roles is an important factor in maintaining the high status of older people. For older women, motherhood and domestic roles change in intensity but do not lose their centrality. As grandmothers, their nurturing functions continue. For men, the loss of the work role does not noticeably affect their central role in the family. Instead, the absence of the work role permits greater family involvement.

To many older Italians, the ideal living situation is to maintain one's own home near one's children because many believe they do not feel the same when they are not in their own homes. Parents receive respect, gratitude, and love in return for their many sacrifices.

21.4.4 Alternative Lifestyles

Despite values clearly defined around family obligations, Italians and Italian Americans generally do not reject another family member because of an infraction or alternative lifestyle. They may complain and argue with the deviant member, hoping for a change, but if their complaining and arguing fail to bring about the desired change, they still accept the individual and live with the consequences.

Gay and lesbian lifestyles, exist as they do in all cultures, but they are not readily accepted. Gender reassignment in Italy is rare, but it does occur. Actual statistics are not complete.

Reflective Exercise #2

Francesco, age 43 years, is the only child of Marco and Anna. He is a teacher and has lived with his parents in the family home since birth. Over the past few years he has become increasingly more aware of the fact that his parents are aging and that if he remains single he will have to face their loss alone. He has never been married or been seriously involved with a woman until recently when he met Alessia online who lives a great distance from his hometown. Francesco and Alessia have been seeing each other for the past 2 years and have become very connected.

Francesco's parents, especially his father, are not happy with his involvement with a woman such a distance away and have spoken very clearly against the relationship. As an only son Francesco is extremely dedicated to his parents and for whom he has great respect. Even though he is in his 40s, he is torn between continuing the relationship with Alessia or honoring his parents and looking for another woman closer to home. He is quite confused about the right choice to make and has become anxious and is now smoking a pack of cigarettes a day.

Given the role of authority and decision making that Italian parents play into old age over their children, how might you advise Francesco with his difficultly choosing to continue his long-distance relationship or finding someone closer to home who meets his parents' expectations.

1. What do you see as the important issues for Francesco?
2. What counseling approach would you use?
3. What do you need to know about the Italian culture and history to work effectively with Francesco?
4. How would you counsel him on the relationship between the choice he makes and its effects on his well-being and physical health?

21.5 Workforce Issues

In the last 20 years, the Italian economy underwent three main economic recessions (OECD 2020). Currently, the Italian labor market suffers a sizable negative shock from the economic crisis experienced since 2010 and has subsequently experienced a very moderate recovery from 2017. Despite some improvements, unemployment remains higher than pre-crisis levels, especially for the youngest workers and in the Southern parts of the Country. In this context, female participation has been slowly increasing. Economic heterogeneity between Italian regions is still high, with the South unable to catch up with the economic rates of the Northern parts. Productivity rate is stable at relatively low levels compared to other European countries. Finally, undeclared employment is high, especially in the South.

21.5.1 Culture in the Workplace

Italians strongly believe in the work ethic; for example, they rarely miss work commitments owing to a cold, headache, or minor illnesses. if completing their work requires staying later, they do so. Although the family is of utmost importance to Italians, work takes priority over family unless serious family situations arise.

21.5.2 Issues Related to Autonomy

Young Italians seek autonomy and independence through work. Unfortunately, many of them stay at home with parents for years into their adulthood as a result of the changed economic climate (Blangiardo and Rimoldi 2014). Indeed, Italians leave their parents' home at one of the highest ages in Europe. However, when children move away, family ties are still very strong.

Italian immigrants in the US and their descendants regard work as moral training for the young. Among Italian Americans, work is viewed as a matter of pride, demonstrating that one has become a man or a woman and is a fully functioning member of the family. So strong is this ethic that it governs behavior apart from monetary gain derived.

Even though Italian Americans have the utmost respect for their employer, they are emotional and passionate people. When a confrontation arises, Italians are likely to get involved. For example, the first Italian immigrants working in New York City defended themselves against deplorable working conditions by banding together to form one of the city's largest unions (Mangione and Morreale 1993).

Second- and third-generation immigrants, having a command of the English language, are more apt to seek positions of authority, take responsibility, and become managers or business proprietors. Italians born and educated in the United States usually have little difficulty communicating with others in the workforce. Newer immigrants and those with limited English language skills have the most difficulty assimilating into the workforce, a problem that is common to all non-English-speaking immigrants.

21.6 Biocultural Ecology

21.6.1 Skin Color and Other Biological Variations

Because of Italy's proximity to Switzerland, Austria, and Germany in the North and to North Africa in the South, Italians as a group have different physical characteristics. Those from a predominantly northern background have lighter skin, lighter hair, and sometimes blue eyes, whereas those from the South of Rome, particularly from Sicily, have dark, often curly hair, dark eyes, and olive-colored skin. Accordingly, health-care providers should be aware of skin variations among Italians, especially when assessing for anemia, cyanosis, and lowered oxygenation levels in those who are darker skinned. In dark-skinned patients, the skin turns ashen instead of blue in the presence of cyanosis and decreased hemoglobin levels. To observe for these conditions, the practitioner has to examine the sclera, conjunctiva, buccal mucosa, tongue, lips, nail-beds, and palms and soles of the feet. To assess for jaundice, the practitioner needs to look at the conjunctiva and in the buccal mucosa for patches of bilirubin pigment.

21.6.2 Diseases and Heath Conditions

Italians have some notable genetic diseases such as familial Mediterranean fever, Mediterranean-type glucose-6-phosphate dehydrogenase (G-6-PD) deficiency, and beta-thalassemia (Caprari et al. 2001). Mediterranean-type

G-6-PD deficiency is an inherited, X-linked, recessive disorder most fully expressed in homozygous males, with a carrier state found in heterozygous females. Red blood cell damage begins after intense or prolonged administration of sulfonamides, antimalarial agents, salicylates, or naphthaquinolones; after ingestion of fava beans; or in the presence of hypoxemia or acidosis. Supportive therapy includes withdrawing the causative agent and administering blood transfusions and oral iron therapy, which usually results in spontaneous recovery (Caprari et al. 2001).

Beta-thalassemia, of which there are two types, is prevalent among Greeks, Italians, Sephardic Jews, and Arabs. Both are caused by genetic defects affecting the synthesis of the hemoglobin A or B chain. Beta-chain production is depressed moderately in the heterozygous form, beta-thalassemia minor, and severely depressed in the homozygous form, thalassemia major, which is also called *Cooley's anemia*, after the American physician who described it. Beta-thalassemia minor causes mild to moderate anemia, splenomegaly, bronze coloring of the skin, and hyperplasia of the bone marrow. Affected persons are usually asymptomatic. Persons with beta-thalassemia major may experience severe anemia; death owing to high-output cardiac failure can occur in early childhood if this condition is left untreated. No cure exists, but palliative therapy includes repeated transfusions of packed red blood cells (Angelucci et al. 2016).

21.6.3 Variations Is Drug Metabolism

The medical literature does not report any variations in drug metabolism or interactions specific to Italians or Italian Americans in general. However, health conditions such as Mediterranean-type G-6-PD deficiency and thalassemia have a profound effect on drug metabolism. Because conditions such as hypoxemia and acidosis, ingestion of fava beans, and the administration of sulfonamides, antimalarial agents, salicylates, and naphthaquinolones can exacerbate these conditions, health-care professionals must take extra precaution when prescribing these drug therapies for Italian Americans.

21.7 High-Risk Behaviors

Italy is the second European country for life expectancy at birth; two thirds of Italians claim to be in good health. No particular high-risk behaviors for Italians have been identified. However, some areas of healthcare could indicate some risky behaviors, such as the significant rates of dental caries and high rates of smokers among Italians. Lower dental care could be associated with lower access to dental care services observed during the last decades, as a result of the economic crisis. Overall alcohol consumption per adult also fell and is below the European average. The proportion of adults reporting regular heavy alcohol consumption is also much lower than in most other European countries. On the other hand, in contrast to low levels in adults, overweight or obesity problems among children have grown and are now above the European average.

21.8 Nutrition

21.8.1 Meaning of Food

To Italians food is a symbol of life, particularly family life (Chytkova 2011). Respect for foodstuff is upheld among all social classes; Italians are very proud of the high quality of their products. The ceremony of eating is usually honored on the weekends when families gather together. Food conveys a link between family members as it represents the tradition. In a symbolic sense, meals are a communion of the society, and food is highly respected because it is a tangible means of that communion (Guarrera et al. 2006).

21.8.2 Common Foods and Food Rituals

The traditional diets of countries around the Mediterranean Sea slightly differ, so there are different versions of the Mediterranean diet. The Italian Mediterranean diet—rich in vegetables, pasta, fruit, fish, and cheese—varies according to the region of Italy (De Lorenzo et al. 2001). Northern Italian foods are rich in cream and cheese, resulting in a potentially high intake of fat (Capurso and Vendemiale 2017). Southern Italian foods are prepared according to the Mediterranean diet (Renna et al. 2015). If someone has a chronic condition like heart disease or high blood pressure, this type of diet is often promoted to decrease the risk of heart and circulatory diseases but also depression, and dementia (Laidlaw et al. 2014).

The staples of the Italian diet are all types of pasta (i.e. fusilli, maccaroni, spaghetti, gnocchi, lasagne, bucatini, pappardelle, fettuccine, tortellini, penne, rigatoni, conchiglie, etc.). Vegetables, fresh fruit, garbanzos beans, and fruit are also common. Popular Italian foods include lentils, tomatoes, eggplant, ham (prosciutto), olive oil, wine, ice cream (gelato), pastries such as cannoli and biscotti, and different cheeses coming from various Regions, such as provolone, ricotta, pecorino romano, taleggio, and so on. Other common dishes include pizza and different kind of meats (see Table 21.1). Italian wine is part of almost every meal. Inevitable at the end of every major food gathering is the espresso coffee usually consumed one single shot. Breakfast has never been an important meal as in other traditions.

As per the third millennium lifestyles, the working routine dictates more the eating rhythms and dinner has become the main meeting of the nuclear family. Time and circumstances permitting in the weekends, the major meal of the week is the most leisurely and is shared with the largest gathering of '*la famiglia*'. Overall, Italians love ceremonies and feasts and they are celebrated in the main social

Table 21.1 Some popular Italian dishes

Common Name	Description	Ingredients
Acquacotta	Acquacotta—an Italian soup that was originally a peasant food	Historically, its primary ingredients were water, stale bread, onion, tomato and olive oil, along with various vegetables and leftover foods that may have been available
Minestrone	Soup with greens	Escarole and beans with garlic and other herbs
Melanzane alla parmigiana	Eggplant parmesan	Eggplant, tomato sauce, bread crumbs, parmesan cheese, and mozzarella cheese
Pasta con pesto	Sauce served over spaghetti	Sauce of basil, nuts, olive oil, and garlic
Pasta e fagioli	Fresh-made pasta and beans	Shell-shaped pasta, kidney beans, and tomato sauce
Calzone	Folded over dough	Usually filled with mozzarella, tomato and other ingredients
Pizza Margherita	Stretched, flattened dough	Tomato and buffalo mozzarella
Lasagne	Lasagne are a type of wide, flat pasta	Lasagne, or the singular lasagna, commonly refers to an Italian cuisine dish made with stacked layers of flat pasta alternated with sauces and ingredients such as minced meat, tomato sauce, cheese, and seasonings and spices such as garlic, oregano and basil
Spaghetti aglio e olio	Spaghetti with olive oil	Spaghetti, olive oil, garlic, and red pepper
Tortellini	Little rounds of pasta in white or red sauce	Pasta dough stuffed with meat and cheese
Scaloppine	Scaloppine is a type of Italian dish that comes in many forms. It consists of thinly sliced meat, most often beef, veal, or chicken	Scaloppine is dredged in wheat flour and sautéed in a variety of sauces

occasions such as marriages, baptisms, communions, confirmations, and so on. Broadly speaking, Italians love dancing, eating, drinking, and firework festivities associated with Italian feasts.

21.8.3 Dietary Practices for Health Promotion

The Harvard School of Public Health, Oldways Preservation and Exchange Trust, and the European Office of the World Health Organization introduced the Mediterranean Diet Pyramid as a guide to help familiarize people with the most common foods of the region (Oldways Trust 2018). More of an eating pattern than a strictly regimented diet plan, the pyramid emphasize certain foods based on the dietary traditions of Greece and Southern Italy during the mid-twentieth century (Sotos-Prieto et al. 2017).

21.9 Pregnancy and Childbearing Practices

21.9.1 Fertility Practices and Views Towards Pregnancy

Most first-generation Italian Americans did not practice birth control and rarely discussed matters related to sex. Second-generation Italian women often introduced birth control years after the marriage. Sex was rarely discussed in the family, premarital sex was restricted, and adultery was a strict taboo. However, many third-generation Italian Americans use birth control from the beginning of the marriage, and sex is commonly discussed in the family. External restrictions on premarital sex have weakened.

The proportion of Italian Americans women aged 35–44 years with five or more children is second lowest in comparison with women of other ethnic groups in the same age range. A strong sense of modesty and embarrassment among Italian Americans results in the avoidance of discussions related to sex and menstruation, hindering early diagnosis and primary prevention interventions.

21.9.2 Prescriptive, Restrictive, and Taboo Practices in the Childbearing Family

Within Italian families, educational discussions about birth-control were not common in the 50s of the last century for cultural reasons. However, new generations are better educated about birth-control. Despite a number of myths and traditions that were present among Italians until few decades ago, currently no particular beliefs are involved in the management of pregnancy. Women are generally assisted by a gynecologist during pregnancy and tend to follow the medical and midwifery indications. Overall, pregnant women in Italy have access to good health care services. The delivery is generally organized at the hospital level with supervision of midwives.

Many traditional Italians fear hospital care in North America, except in the case of childbirth. Although some women still prefer the family physician or a midwife and home confinement, many Italian women now deliver their babies in the hospitals.

21.10 Death Rituals

21.10.1 Death Rituals and Expectations

In the Italian family, death is a great social loss and brings an immediate response from the community (Di Mola and Crisci 2001). Sending food and flowers (chrysanthemums) and congregating at the home of the deceased are expected. Italian death rituals can be demonstrative. The funeral procession to the cemetery is a symbol of family status especially in villages. There is great pride in the size of the event, determined by the number of cars in the procession. Although there is a tendency today to decrease the elaborateness of the funeral, it remains very much a family and community event. Its ritual recognition deemphasizes death. Within the context of fatalism in Catholicism, many Italians view death as "God's will"; thus, a fatal diagnosis may not be discussed with the ill family member.

21.10.2 Responses to Death and Grief

Emotional outpourings might be profuse, and the activities around a funeral provide a means of closeness for the whole community (Seymour 2012). Women may mourn dramatically especially in the Southern parts of Italy for the whole family. Family members get up constantly to touch and talk to the deceased loved one. The real time of sorrow comes at the end of the ceremony when the priest and nonfamily congregation say good-bye to the deceased. At this time, the family is left alone for a time with their loved one.

21.11 Spirituality

21.11.1 Dominant Religion and Use of Prayer

In Italy, religion and spirituality are very present and often deeply intertwined. Rome is the center of the Roman Catholic Church, the largest Christian denomination in the world. The predominant religion of Italians is Roman Catholicism. The Pope, the leader of Roman Catholicism is inside Vatican City based in Rome.

21.11.2 Meaning of Life and Individual Sources of Strength

Spirituality has always been present in the history of nursing and continues to be a topic of nursing interest, as shown in the literature (Mcsherry and Jamieson 2013). The possible range of interpretations of spirituality has branched out in recent years, influenced by religious cultures and by historical time, by societies that are increasingly pluralistic and also depending on the perspective of the person using the construct (Mcsherry and Jamieson 2013).

21.11.3 Spiritual Beliefs and Health-Care Practices

The center of Roman Catholic worship is the celebration of Mass, the Eucharist, which is the commemoration of Christ's sacrificial death and of His Resurrection. Other sacraments are baptism, confirmation, confession, matrimony, ordination, and anointing of the sick that are sometimes practiced also by people who are not very interested in religion or are not practicing it very much. Most Catholic Italians have broad beliefs in Catholics. Most Italians pray to the Virgin Mary, the Madonna, and a number of saints. Prayers and having faith in God and saints can help the majority of Italians through illnesses and hardships in life. In times of illness, the health-care providers may need to help patients obtain the basic rites of the sacrament of the sick, which includes anointing, communion, and if possible, a blessing by the priest.

21.12 Health-Care Practices

21.12.1 Responsibility for Health Care

The family, the most dominant influence on the individual, is viewed as the most credible source of health-care practices. Italians believe that the most significant moments of life should take place under their own roofs. Research completed in Italian neighborhoods in Baltimore suggested that the extended family is the front-line resource for intensive advice on emotional problems. Mental-health specialists are frequently perceived as inappropriate agents for meeting problems that are beyond the expertise of the family and local community. Most second- and third-generation Italians take responsibility for their own health care and engage in health promotion more than those of the first generation. Most also have health insurance coverage. From the family perspective, the mother assumes responsibility for the health of the children.

21.12.2 Folk and Traditional Practices

The powers of the occult are not limited to saints. For traditional Italian Americans, certain humans are believed to have immediate and potent access to magical powers. These are the *maghi* (male witch) and the *maghe* (female witch) who are granted various degrees of black-magic power at birth. These traditional beliefs may be practiced by first-generation and newer Italian immigrants but hold little value for second- and third-generation Italian Americans. Health-care providers should accept and incorporate these practices into their treatment plans, along with providing written instructions for biomedical treatments.

21.12.3 Responsibility for Health Care

The family, the most dominant influence on the individual, is viewed as an important advisor for health-care practices. Research completed in Italian neighborhoods in Baltimore suggested that the extended family is the front-line resource for intensive advice on emotional problems. Most second- and third-generation Italians take responsibility for their own health care and engage in health promotion more than those of the first generation. Most also have health insurance coverage. From the family perspective, the mother assumes responsibility for the health of the children.

21.12.4 Cultural Responses to Health Care

Both age and gender mediate ethnic differences in the expression of pain for Italian Americans. Older Italian Americans, especially women, are more likely to report pain experiences, express symptoms to the fullest extent, and expect immediate treatment. Italians tend to be more verbally expressive with chronic pain than do some other ethnic groups. Neill's (1993) study on pain in acute myocardial infarction indicates that patients of Jewish and Italian ancestry exhibit more-expressive pain behaviors than patients of Irish or English ancestry.

Vincente's (1993) study examined Italian Americans' attitudes and beliefs about mental-health services and mental-health workers. The results suggested that Italian American professionals have the highest level of satisfaction with their work and show the most tolerance for deviant behavior in the community. Because Italian Americans tend to report more symptoms and report them more dramatically, physicians tend to diagnose emotional problems in Italian patients more frequently than in other ethnic groups.

A person with a physical or mental disability is absolutely not stigmatized in the Italian culture because the condition is believed to be due to God's will. Illness can be caused by suppressing emotions and stress from fear, guilt, and anxiety. If a person does not vent these feelings, the person may burst.

Families may be ashamed to let neighbors know of an incident that may impair the social status of a family member. When a family member is sick, other women in the family take over and assist the sick person until they are well.

21.12.5 Blood Transfusions and Organ Donation

Judicious use of medications and blood transfusions is permissible and morally acceptable as long as the benefits outweigh the risks to the individual; thus, Italian Americans have little objection to accepting a blood transfusion when needed. First-generation immigrants are, in general, not prone to donating their organs. Second- and third-generation Italian Americans may also reflect this perspective. Organ donation is morally permissible when the benefits to the recipient are proportionate to the loss of the organ to the donor and when the organ does not deprive the donor of life or the functional integrity of the body. Otherwise, organ transplant is an individual decision.

21.13 Healthcare Providers

21.13.1 Traditional Versus Biomedical Providers

The World Health Organization through the official set up of the 'Year of the Nurse and the Midwife' in 2020 has set out an agenda to realize an unprecedented shift in the delivery of health care (Peate 2020; Waldrop 2020). This transformation is motivated by the need to provide quality, affordable, and sustainable health care in ageing societies worldwide.

Many nursing associations are planning hundreds of events to mark 2020, which will also see celebrations of the 200th anniversary of the birth of Florence Nightingale, and the publication of the first WHO State of the World's Nursing Report. Nurses and midwives are key to the achievement of WHO's goal of Universal Health Coverage because they play a critical role in health promotion, disease prevention and the delivery of care in all settings (Peate 2020).

Undergirding these paradigm shifts are efforts to understand the nature of professional culture and cultural change in different health care systems. "Nurses and midwives play a vital role in providing health services. These are the people who devote their lives to caring for mothers and children; giving lifesaving immunizations and health advice; looking after older people and generally meeting everyday essential health needs. They are often, the first and only point of care in their communities. The world needs nine million more nurses and midwives if it is to achieve Universal Health Coverage by 2030. That's why the World Health Assembly has designated 2020 the International Year of the Nurse and the Midwife" (Waldrop 2020).

Professional cultural evolutions or reformations of different health systems is key to achieve sustainability and quality care in societies strongly affected by epidemiological transitions. In this ongoing process, in Italy, nursing has been evolving since the beginning of the nineties (1992) and has resulted in major changes in all health-care allied professions. The Decree of the Ministry of Health 739/1994 (called Professional Profile) provided the first recognition of professional autonomy for nurses in Italy (Destrebecq et al. 2009; Stievano et al. 2012). Other laws followed that officially ratified nursing professional autonomy and established key competencies to be enacted (Sala and Manara 1999). A primary outcome of these efforts was critical academic nursing changes: a 3-year university degree in 1996 was acknowledged as the unique pathway for nurses and allied health professionals to enter their professions (e.g., physiotherapists, speech therapists, occupational therapists, radiologic technologists, optometrists). For example, the enactment of Law 42 on 26 February 1999 'Disposition concerning health professions' was a milestone in the history of the Italian nursing and allied health professions, ratifying autonomy in their full scope of practice.

The educational changes began in 1991, when the hospital-based nursing diploma was nearly abandoned. The following steps were first the continuation of the nursing diploma and a university diploma as a double pathway between 1992 and 1996, and then a university-based diploma from 1996 to 2001. From 2001, a nursing degree at bachelor level and also for the above-mentioned health professionals exist as the single pathway to entry different healthcare professions. Despite this recent educational evolution, nursing in Italy achieved important goals in terms of a self-regulated profession by legislative norms since 1954.

The National Federation of OPI Colleges (former IPASVI) is a nonprofit body by public law established firstly in 1954 to represent the nursing profession on a national basis with the name of IPASVI which represented nurses, child-care workers and health visitors. OPI coordinates Italy's Provincial Colleges, which hold the registry of the nurses. It addresses two primary aims: the first is "protection of the citizen" which establishes the right of the individual, as sanctioned by the Italian Constitution, to receive health care services by qualified personnel who hold a specific fitness-to-practice title. The second aim sustains "professionalism" by (a) supporting equitable treatment, (b) ensuring that the ethical code is respected, (c) endorsing the professional cultural growth of registered nurses, and (d) institutionalizing good professional practice. OPI regulations were created to ensure maintenance of practice, competencies and compliance with their stated ethical code. In recent years, the Law of 11 January 2018 (Law Lorenzin), which came into force on 15 February 2018, established the National Federation of Orders of Nursing Professions (FNOPI) that replaced the IPASVI National Board of Colleges.

At the provincial/interprovincial level, operate the nursing professions orders (OPI), which only include nurses and pediatric nurses. This was an important transition insofar as the IPASVI National Board of Colleges was a simple auxiliary body of the Italian State while the National Federation of Orders of Nursing Professions (FNOPI) operates as a subsidiary body of the State and can therefore carry out administrative tasks in place and on behalf of the State.

Currently, managing professional change is viewed as an essential component of health care system reforms in Italy and in several countries (Ayala et al. 2014). Patient safety and the adoption of evidence-based practice have triggered changes that have transformed the "culture" of how and by whom health care is delivered (Manojlovich et al. 2008). These cultural shifts are underscored by reforms in health disciplines in Europe and are particularly evident in nursing within recent professional educational directives (Lahtinen et al. 2014).

References

Alfani G, Ryckbosch W (2016) Growing apart in early modern Europe? A comparison of inequality trends in Italy and the Low Countries, 1500–1800. Explor Econ Hist 62:143–153. https://doi.org/10.1016/j.eeh.2016.07.003

Angelucci E, Burrows N, Losi S, Bartiromo C, Hu XH (2016) Beta-thalassemia (BT) prevalence and treatment patterns in Italy: a survey of treating physicians. Blood 128(22):3533–3533. https://doi.org/10.1182/blood.v128.22.3533.3533

Ayala RA, Fealy GM, Vanderstraeten R, Bracke P (2014) Academisation of nursing: an ethnography of social transformations in Chile. Int J Nurs Stud 51(4):603–611. https://doi.org/10.1016/j.ijnurstu.2013.08.010

Bertocchi G, Bozzano M (2019) Origins and implications of family structure across Italian provinces in historical perspective. In: Studies in economic history. Springer, Berlin, pp 121–147. https://doi.org/10.1007/978-3-319-99480-2_6

Blangiardo GC, Rimoldi S (2014) Portraits of the Italian family: past, present and future. In: Journal of comparative family studies, vol 45. University of Toronto Press, Toronto, pp 201–219. https://doi.org/10.2307/24339606

Caprari P, Caforio MP, Cianciulli P, Maffi D, Pasquino MT, Tarzia A, Amadori S, Salvati AM (2001) 6-phosphogluconate dehydrogenase deficiency in an Italian family. Ann Hematol 80(1):41–44. https://doi.org/10.1007/s002770000233

Capurso C, Vendemiale G (2017) The Mediterranean diet reduces the risk and mortality of the prostate Cancer: a narrative review. Front Nutr 4:38. https://doi.org/10.3389/fnut.2017.00038

Cei L, Stefani G, Defrancesco E, Lombardi GV (2018) Geographical indications: a first assessment of the impact on rural development in Italian NUTS3 regions. Land Use Policy 75:620–630. https://doi.org/10.1016/j.landusepol.2018.01.023

Chytkova Z (2011) Consumer acculturation, gender, and food: Romanian women in Italy between tradition and modernity. Consum Mark Cult 14(3):267–291. https://doi.org/10.1080/10253866.2011.574827

Cooke LP (2009) Gender equity and fertility in Italy and Spain. J Soc Policy 38(1):123–140. https://doi.org/10.1017/S0047279408002584

De Lorenzo A, Alberti A, Andreoli A, Iacopino L, Serrano P, Perriello G (2001) Food habits in a southern Italian town (Nicotera) in 1960 and 1996: still a reference Italian Mediterranean diet? Diabetes Nutr Metab 14(3):121–125

Destrebecq A, Lusignani M, Terzoni S (2009) A national survey on the activities performed by nurses and aids in Italian outpatients' services. Cah Sociol Demogr Med 49(2):137–166. http://www.ncbi.nlm.nih.gov/pubmed/19634613

Di Mola G, Crisci MT (2001) Attitudes towards death and dying in a representative sample of the Italian population. Palliat Med 15(5):372–378. https://doi.org/10.1191/026921601680419410

Gauthier AH (2007) The impact of family policies on fertility in industrialized countries: a review of the literature. Popul Res Policy Rev 26(3):323–346. https://doi.org/10.1007/s11113-007-9033-x

Giambona F, Grassini L (2020) Tourism attractiveness in Italy: regional empirical evidence using a pairwise comparisons modelling approach. Int J Tour Res 22(1):26–41. https://doi.org/10.1002/jtr.2316

Goldstein J, Lutz W, Testa MR (2003) The emergence of sub-replacement family size ideals in Europe. Popul Res Policy Rev 22(5–6):479–496. https://doi.org/10.1023/b:popu.0000020962.80895.4a

Guarrera PM, Salerno G, Caneva G (2006) Food, flavouring and feed plant traditions in the Tyrrhenian sector of Basilicata, Italy. J Ethnobiol Ethnomed 2(1):37. https://doi.org/10.1186/1746-4269-2-37

Hay D (2019) From Roman empire to renaissance Europe. Routledge, London. https://doi.org/10.4324/9780429059919

Kendon A (1995) Gestures as illocutionary and discourse structure markers in southern Italian conversation. J Pragmat 23(3):247–279. http://www.cogsci.ucsd.edu/~nunez/COGS160/Kendon_1995_PS.pdf

Klein HS (1983) The integration of Italian immigrants into the United States and Argentina: a comparative analysis. Am Hist Rev 88(2):306. https://doi.org/10.2307/1865404

Lahtinen P, Leino-Kilpi H, Salminen L (2014) Nursing education in the European higher education area—variations in implementation. Nurse Edu Today 34(6):1040–1047. https://doi.org/10.1016/j.nedt.2013.09.011. Churchill Livingstone

Laidlaw M, Cockerline CA, Rowe WJ (2014) A randomized clinical trial to determine the efficacy of manufacturers' recommended doses of omega-3 fatty acids from different sources in facilitating cardiovascular disease risk reduction. Lipids Health Dis 13(1). https://doi.org/10.1186/1476-511X-13-99

Lepschy A, Lepschy G (2013) The Italian language today, 2nd edn. Routledge, London. https://doi.org/10.4324/9781315003214

Luciano M, Sampogna G, del Vecchio V, Giacco D, Mulè A, de Rosa C, Fiorillo A, Maj M (2012) The family in Italy: cultural changes and implications for treatment. Int Rev Psychiatry 24(2):149–156. https://doi.org/10.3109/09540261.2012.656306

Mangione J, Morreale B (1993) La storia : five centuries of the Italian American experience, 1st edn. Harper Collins, New York

Manojlovich M, Barnsteiner J, Bolton LB, Disch J, Saint S (2008) Nursing practice and work environment issues in the 21st century. Nurs Res 57(Suppl 1):S11–S14. https://doi.org/10.1097/01.NNR.0000280648.91438.fe

Mcsherry W, Jamieson S (2013) The qualitative findings from an online survey investigating nurses' perceptions of spirituality and spiritual care. J Clin Nurs 22(21–22):3170–3182. https://doi.org/10.1111/jocn.12411

OECD (2020) OECD based recession indicators for Italy from the period following the peak through the trough. https://fred.stlouisfed.org/series/ITARECD

Oldways Trust (2018) Mediterranean diet | oldways. https://oldwayspt.org/traditional-diets/mediterranean-diet

Peate I (2020) The 2020 year of the nurse and midwife and nightingale bicentenary celebrations. Pract Nurs 31(1):44–45. https://doi.org/10.12968/pnur.2020.31.1.44

Pilat SZ (2009) Reconstructing Italy: the Ina-casa neighborhoods of the postwar era. Ashgate Publishing, Farnham

Renna M, Rinaldi VA, Gonnella M (2015) The Mediterranean Diet between traditional foods and human health: the culinary example of Puglia (Southern Italy). Int J Gastron Food Sci 2(2):63–71. https://doi.org/10.1016/j.ijgfs.2014.12.001. AZTI-Tecnalia

Sala R, Manara D (1999) The regulation of autonomy in nursing: the Italian situation. Nurs Ethics 6(6):451–467. https://doi.org/10.1177/096973309900600602

Seymour M (2012) Emotional arenas: from provincial circus to national courtroom in late nineteenth-century Italy. Rethink Hist 16(2):177–197. https://doi.org/10.1080/13642529.2012.681190. Routledge

Shuter R (1977) A field study of nonverbal communication in Germany, Italy, and the United States. Commun Monogr 44(4):298–305. https://doi.org/10.1080/03637757709390141

Sotos-Prieto M, Bhupathiraju SN, Mattei J, Fung TT, Li Y, Pan A, Willett WC, Rimm EB, Hu FB (2017) Association of changes in diet quality with total and cause-specific mortality. N Engl J Med 377(2):143–153. https://doi.org/10.1056/NEJMoa1613502

Stievano A, De Marinis MG, Russo MT, Rocco G, Alvaro R (2012) Professional dignity in nursing in clinical and community workplaces. Nurs Ethics 19(3):341–356. https://doi.org/10.1177/0969733011414966

Tomassini C, Wolf DA, Rosina A (2003) Parental housing assistance and parent-child proximity in Italy. J Marriage Fam 65(3):700–715. https://doi.org/10.1111/j.1741-3737.2003.00700.x

Trapp M (2017) Philosophy in the Roman empire. Routledge, London. https://doi.org/10.4324/9781315246895

Waldrop J (2020) 2020 is the "Year of the Nurse and Midwife.". J Nurse Pract 16(1):A7–A8. https://doi.org/10.1016/j.nurpra.2019.11.006. Elsevier Inc

22 People of Jewish Heritage

Janice Selekman and Polly Zavadivker

22.1 Introduction

This chapter aims to provide an overview of Jewish American heritage as a whole, but will focus on the specific needs of religiously observant people and families (referred throughout the chapter as Orthodox Jews). Readers should bear in mind that while all Orthodox Jews observe the same religious laws, there are many distinct sects within these communities, each of which follows its own customs, spiritual leaders, social norms, and political priorities. Therefore, an individual assessment is crucial to prevent stereotyping.

22.2 Overview, Inhabited Localities, and Topography

22.2.1 Overview

The Jewish people, or Jews, are a group who share a common religious, cultural, and ethnic background. Jewish identity encompasses more than religion; it also refers to belonging to the Jewish people, a group composed of communities that can be found throughout the world. Jews tend to identify with the idea of a shared ancestry, upbringing, and history. Jewish communities developed distinct cultures over time that were shaped by where they lived. Today one will hear the terms *Hebrew*, *Israelite*, and *Jew* used interchangeably when discussing the Hebrew Bible and origins of Jewish history.

Jews consider Abraham to be the founder of their tradition, and thus the first Jew. In the Torah, the primary Jewish scripture, God gives Abraham's grandson Jacob a new name, Israel, which in Hebrew means "struggled with God" (Genesis, 32: 22–32). Israel's 12 sons and their descendants became known collectively as the "children of Israel." The term *Jew* is derived from the name Judah, one of Jacob's sons.

Hebrew is the language of Jewish scripture and prayer, and is also one of the official languages of the modern state of Israel. The modern Hebrew language was developed in the late nineteenth century on the basis of ancient Hebrew grammar and vocabulary. It is the spoken language of Israelis, and is used for prayer and Torah study by Jews wherever they live. Thus one would refer to the people as Jews or Jewish people. Their religious faith is Judaism, their religious language is Hebrew, and the modern Jewish nation-state is Israel. Nearly half of the global Jewish population today lives outside of Israel. The global dispersion of Jews outside of Israel is known as the **Diaspora**.

This chapter in an update written by Janice Selekman in a previous edition of the book.

J. Selekman (✉) · P. Zavadivker
University of Delaware, Newark, DE, USA
e-mail: selekman@udel.edu; pollyz@udel.edu

© Springer Nature Switzerland AG 2021
L. D. Purnell, E. A. Fenkl (eds.), *Textbook for Transcultural Health Care: A Population Approach*,
https://doi.org/10.1007/978-3-030-51399-3_22

The religion of **Judaism** is practiced in a variety of denominations (also referred to as movements, streams or branches). In the United States, the Jewish population affiliates with these denominations as follows: **Reform** (36%); **Conservative** (18%), **Orthodox** (10%), and Reconstructionist and Jewish Renewal (6%). Another 30% of American Jews identify as Jewish without claiming any specific denominational affiliation. Nearly one in five Jews in the United States describe themselves as **Secular** or cultural Jews; that is, they claim to have no religion but nonetheless identify as Jewish (Pew Research Center 2013).

The Reform movement maintains that traditional Jewish law (in Hebrew, ***halakhah***) should be practiced in a way that enables one to participate in the surrounding community and to incorporate progressive principles of gender equality and social justice into religious life. Reform Jews do not usually follow daily religious practices but often still eat traditional Jewish foods, observe holidays such as Yom Kippur and Passover, and participate in religious rites of passage such as circumcision and bar/bat mitzvah.

Conservative Jews have also made progressive reforms to Jewish law, while attempting to preserve traditions such as weekly Sabbath observance and *kashrut* (eating and preparing food according to kosher laws). Conservative Jews emphasize the importance of modern Jewish education for children and tend to oppose intermarriage between Jews and non-Jews. Orthodox Jews attempt to strictly adhere to *halakhah* in all aspects of life, including daily prayer, kashrut, Sabbath and holiday observance, laws of family purity, and modest dress. There are also ultra-Orthodox Jewish communities, whose members tend to live in close-knit neighborhoods and can be found in most major cities in the Northeast United States. Ultra-Orthodox Jews include several different groups, including Hasidic Jews, and others. Their fervent devotion to religion is often combined with a conscious rejection of popular culture and secular society.

The world Jewish population does not follow any single authority figure (there is no Jewish "Pope," for example); however, many Orthodox and ultra-Orthodox individuals will choose to consult their own rabbis or other legal authorities before making life and health-related decisions.

According to Jewish law, a child born to a Jewish mother is Jewish. Because the rate of intermarriages between Jews and non-Jews has rapidly increased in the United States and Europe since the mid-twentieth century, a debate over Jewish identity and patrilineal descent has ensued. The Reform movement recognizes the child of a Jewish father and non-Jewish mother as Jewish; the Conservative and Orthodox movements do not (Robinson 2016). Although adherents of Judaism are forbidden to actively proselytize to other religions, all denominations welcome converts as full members of their communities. Clergy in each respective movement offer pre-conversion classes for adults and perform conversions.

22.2.2 History and Residence Patterns

The history of Jewish Americans (also known as American Jews) began in 1624 when a group of 23 Jews arrived at the Dutch-ruled port city of New Amsterdam (now New York) after fleeing the Portuguese Inquisition in Brazil. Additional immigrants from Europe brought the number of Jews to between 1000 and 2500 by the time of the American Revolution. Jews fought on both sides of the Revolutionary War. Among the most well known of the colonial sympathizers was Haim Solomon, a Philadelphia banker who raised significant funds for George Washington's army. Their numbers reached approximately 250,000 by mid-nineteenth century. By the time of World War I, Jewish immigrants from Russia and Poland had brought the number of Jews in the United States and Canada to over two million.

As of 2018, the Jewish population of the US was estimated at roughly 5.7 million, accounting for just over 2% of the total U.S. population (DellaPergola 2018); this is 40% of the world's Jewish population (Templer and Tangen 2014). In comparison, there are about 6.7 million Jews in the state of Israel, where they make up 74% of the country's total population (DellaPergola). The American Jewish population is concentrated heavily in certain geographical regions, with 43% of Jews living in the Northeast United States; 23% in the South; 23% in the West; and 11% in the Midwest. States where Jews comprise over 3% of the total population include New York (9.1%), New Jersey (6.1%), Massachusetts (4.2%), Maryland (3.9%), Connecticut (3.3%), Florida (3%), and California (3%) (Jewish Virtual Library 2019). The great majority of American Jews (96%) live in urban or suburban areas, while just 4% reside in rural regions (Pew Research Center 2013). These residential patterns reflect the important role that centralized institutions such as synagogues, schools and community centers have played in the formation of Jewish community life.

22.2.3 Reasons for Migration and Associated Economic Factors

Jews in Europe began to emigrate in large numbers starting in the mid-1800s, compelled both by economic opportunities and the experience of religious persecution. The greatest influx of immigrants to North America occurred between 1880 and 1920. Many of these immigrants came from Russia and Eastern Europe. Once in America, many of these immigrants rapidly acculturated to American society in order to become socially mobile and provide new educational and economic opportunities for their children.

A majority of Jewish families in the US today are descendants of these Eastern European and Russian immigrants. They are referred to as **Ashkenazi** Jews. Ashkenazi Jews make up roughly 65–75% of the world's Jewish population (DellaPergola 2018). This is an important factor as it relates to genetic diseases prevalent among the Jewish population (see the section later in this chapter). Many American Jews of Ashkenazi descent have stories of how some members of their families immigrated to America; some had relatives who were killed in a wave of **pogroms** (anti-Jewish riots) in Russia, or were among the nearly six million people killed during the Holocaust, carried out by Germany during World War II.

Jews who trace their ancestry to Spain, Portugal, the Mediterranean area, and North Africa, are referred to as **Sephardic** Jews. Many Sephardic religious customs and cultural forms differ from those of Ashkenazi Jews. Jews who trace their family roots to Middle Eastern countries, including Iraq, Yemen, and Iran are referred to as **Mizrahi**, or Eastern Jews. Nearly all Mizrahi Jews left their countries of origin following the establishment of the state of Israel in 1948, and have formed strong communities in Israel and elsewhere in the Diaspora.

The fall of the Soviet Union in 1991 prompted a wave of Jewish migration from Russian-speaking countries, primarily to Israel, the United States, and Germany. Because the practice of religion was illegal in the Soviet Union for more than 70 years, these Jews grew up without the capacity to study or observe Judaism. The population of Jews from the former Soviet Union in the US is about 700,000, or nearly 10% of the American Jewish community (Pew Research Center 2013).

There is also a small minority of black Jews, known as **Falasha** Jews. This community originated in Ethiopia and participated in a mass exodus in 1984 to Israel; subsequently, a small number continued on to the US.

22.2.4 Educational Status and Occupations

Throughout their history, Jews have placed a major emphasis on literacy, education, and social welfare. Lifelong learning is one of the most respected values and priorities in Jewish cultures throughout the world. The Jewish people are often called the "People of the Book," a reference to a strong tradition of scriptural study and interpretation (Robinson 2016). Historically, Jews prioritized the teaching of Torah to boys from a young age, regardless of social status or ability. In the past century, all Jewish denominations have emphasized the need for girls and boys alike to learn Torah as a lifelong commitment. Whereas this emphasis on learning traditionally referred only to sacred texts, during the modern period it came to include secular learning as well.

Evidence suggests that Jews place a high value on formal education and the acquisition of advanced degrees. A high percentage of American Jews have obtained university and postgraduate education; 59% of Jewish Americans hold degrees from 4-year universities, compared to 27% of the total population and 31% hold an advanced degree, compared to 11% in the general population (Pew Research Center 2013).

There are also many options for Jewish education. Most synagogues offer religious school classes 1 or 2 days a week for children up through approximately tenth grade. Many larger Jewish communities have private day schools where children are taught Hebrew and Judaism in addition to a secular curriculum. All denominations offer Torah study classes for adults, as well.

Jews hold occupations in all professional sectors of the economy, including medicine, finance, media, law, academia, and business. Professions that involve social action, volunteerism, and public welfare are also common as vocations or avocations. Healthcare professions, social work, teaching, and the legal profession have been popular occupational pursuits. The term ***tzedakah*** (justice) is used to indicate charity or voluntary giving, a central commandment in Judaism. The concept and practice of tzedakah among Jews may have played a role in influencing choice of professions among Jews.

This group is also well represented in the fields of medicine, dentistry, law, sociology and finance. The participation of Jews in these professional sectors developed as the result of historical factors over a period of many centuries. Some scholars believe that Jews entered these professions in large numbers because they were excluded from many other occupations, beginning in medieval Europe, as a result of theological and cultural differences between Jews and Christians (Templer and Tangen 2014).

American Jews have made important advancements in the above fields during the twentieth century. A notable example is the first appointment of Jews as Supreme Court justices, starting with Louis Brandeis in 1916, and Ruth Bader Ginsburg and Elena Kagan in more recent decades. Many people of Jewish descent have been recognized for ground breaking scientific and social research, having won 22.5% of all Nobel Prizes awarded between 1901 and 2019, especially in the fields of economics, physics, chemistry, physiology and medicine (JINFO. ORG 2019). Eight percent of Nobel Peace Prize winners have been of Jewish descent (Templer and Tangen 2014).

Jews in the United States have also made significant contributions to fine arts and popular culture. For example, they have won 52% of Pulitzer Prizes for non-fiction (Templer and Tangen 2014). Well-known American composers with Jewish ancestry include George Gershwin, Aaron Copeland, Leonard Bernstein, Irving Berlin, Richard Rogers, and Stephen Sondheim; the American theater counts Arthur Miller as one of its most celebrated playwrights, along with Woody Allen, Mel Brooks, Oscar Hammerstein II, Alan Lerner, and Neil Simon.

22.3 Communication

22.3.1 Dominant Language and Dialects

English is the primary language of Jewish Americans. Although Hebrew is a spoken language in Israel and is used for prayers and public reading of the Torah, it is generally not used in conversation among American Jews. Many older Ashkenazi Jews who immigrated to the US early in the twentieth century or are first-generation Americans speak **Yiddish**, a hybrid language with Hebrew and German components. Yiddish is also a spoken language in many Hasidic and other ultra-Orthodox communities, and may be the primary language spoken at home in some of them, particularly in the New York metropolitan area (Robinson 2016).

Many Yiddish terms have worked their way into the English language, including *kvetch* (to complain); *chutzpah* (boldness or audacity); *bagel* (a boiled roll with a hole in the middle); *challah* (a rich, braided white bread); *knish* (a dumpling with filling); *mentsch* or *mensh* (a respected person with dignity); *shlep* (to drag or carry); *kosher* (technically applying to food preparation, but idiomatically meaning legal or appropriate; and *oy* or *oy vey* (oh my).

Common Hebrew expressions include "*l'chaim*" (to life), which is said after blessing wine or giving a toast; "*shalom alechem*" (peace be with you), a traditional salutation; "*mazel tov*" (congratulations); and "*shabbat shalom*" (a good and peaceful Sabbath), which is said from Friday evening at sundown until Saturday at sunset.

22.3.2 Cultural Communication Patterns

No religious ban or cultural norms prevent Jews from openly expressing their feelings. Among the majority of American Jews, communication practices are more related to their American upbringing than to their religious practices. Some Orthodox Jews may or may not choose to share with medical professionals if their decisions are related to the observance of Jewish laws, for example, in regard to nutrition or reproductive health.

Dark humor and complaints are frequently used as coping mechanisms and as a way to communicate with others. This might be especially true among older populations, and people born outside of the United States. However, jokes are considered to be insensitive when they reinforce mainstream stereotypes about Jews, such as implying that Jews are cheap or pampered (e.g., American Jewish princesses). Any jokes that refer to the Holocaust are also inappropriate.

Modesty is a primary value in Orthodox Judaism. It is seen in the style of dress and in all forms of behavior. Modesty is meant to express the principle of humility. The idea is to not "show off" or try to impress others.

Orthodox Jewish men and women observe a religious law known as "guarding touch." They will not touch people of the opposite sex other than their spouses, parents, or children. Their failure to shake one's hand should not be interpreted as a sign of rudeness. Some Orthodox men and women will tend not to engage in casual conversation with the opposite sex and some may choose not to make eye contact. Jews who are not Orthodox tend to be more informal and will use touch and short spatial distance when communicating. Healthcare providers should not touch Orthodox men and women unless they are providing direct care. Hands-on "therapeutic touch," as in holding a patient's hand to give comfort, may not be appropriate with these patients, but it is appropriate to ask if they want to hold your hand.

22.3.3 Temporal Relationships

Jewish tradition has developed over more than 2000 years and places a strong emphasis on remembering its long past. This principle originated from the scripture itself. The commands "to keep and remember" are stated dozens of times throughout the Hebrew Bible. In addition, each major holiday includes prayers and rituals meant to commemorate events of the Jewish people's ancient past, such as the Exodus from Egypt, which is remembered on Passover. However, Jewish tradition also underscores the importance of planning and preparing for a better future. The liturgy for the holiday of Yom Kippur, for example, includes prayers asking God to write and seal the people in the "book of life," a metaphor for the good life one hopes to live in the coming year. Similarly, stories of the Holocaust and other Jewish historical tragedies are often told with the explicit message that one should "never forget," lest the tragic history be repeated. Yet many Holocaust stories are told in order to relate an uplifting message, namely the idea that Jews have survived against great odds and must continue to do the same in the future. Therefore, the orientation to time within Jewish tradition is simultaneously to the past, present, and future.

The Jewish calendar is based on both a lunar and a solar year. Each month begins with the appearance of the new moon and lasts either 29 or 30 days. The Jewish festivals and holidays are based on lunar phases, whereas the seasons are based on the solar year, which is 11 days longer than the lunar year. Therefore, an extra (thirteenth) month is periodically added to the Jewish calendar during leap years, usually at the end of winter, 7 times every 19 years. All Jewish holidays, as well as the weekly Sabbath, begin when the sun sets and continue until sunset of the following day. The start of a "day" in Judaism is thus at sunset, not sunrise. The basis for this practice is the line in Genesis 1:5 "And there was evening and there was morning."

22.3.4 Format for Names

For secular use, the Jewish format for names follows the Western tradition. The given name comes first, followed by the family surname. Only the given name is used with friends and in informal situations. In more formal situations, the surname is preceded by the appropriate title of Mr., Miss, Ms., Mrs., Dr., and so on.

It is an Ashkenazi custom that babies may be named after someone who has died, in order to keep their memory alive. It is a custom among Sephardi Jewish families to name a child after a living person to honor him or her. In Orthodox circles, children are not referred to by their names until after the *bris* or *brit milah* (circumcision ritual), performed on the 8th day after birth. Parents may choose to give their children names from the Hebrew Bible (popular names for girls are Sarah, Rachel, Rebecca, and Leah; for boys, they include Noah, Jacob, David, Samuel, and others). Some parents opt for one of several popular names in modern Hebrew, such as Ilan or Ilana (meaning tree), Aviva (spring), or others. Some parents also choose a non-Hebrew, English name for their child, in addition to a corresponding Hebrew name (i.e., Robert in English, Reuven in Hebrew; Melissa in English, Malka in Hebrew). A person's Hebrew name will be used if he or she is mentioned in prayers of healing, called to read from the Torah, or in life cycle events, such as a wedding or funeral. Traditionally the person's given name is followed by their father's or both parents' Hebrew names. An example would be Miriam bat Yitzhak (Miriam, daughter of Isaac), or Moshe ben Shimon v Hava (Moses, son of Simon and Eve).

22.4 Family Roles and Organization

22.4.1 Head of Household and Gender Roles

The family is the core unit of Jewish society. Whereas men have traditionally been considered the breadwinners for the household and women recognized for running the home and being responsible for the children, in recent times there has been great flexibility regarding gender roles among all varieties of Jews. As most Jews have integrated into society, they have also adopted the dominant culture's gender norms, and today there are few differences between Jewish and non-Jewish families in this regard. In most Jewish families, both parents share the responsibilities for supporting the home and raising the children. However, in families that observe Jewish rituals, it is still common to find the mother lighting the Sabbath candles and the father blessing the wine (Bernstein and Fishman 2015).

Some commandments in Judaism are gender-specific. Orthodox Jews practice gender-specific laws, while the Reform and Conservative movements have altered or discarded some of them. For example, Jewish laws state that men are required to recite prayers in public quorums three times a day, and to regularly study and hear the Torah read aloud, whereas women are exempt from those commandments. On the other hand, Jewish law requires married women to immerse themselves in a ritual bath (*mikveh*) once a month following menstruation, before they can resume conjugal relations with their spouses. In addition, women customarily bake braided egg bread (*challah*) and light candles for the Sabbath. According to Jewish law, the father has the legal obligations to teach his children content on Jewish law, ethics and morals, how to swim, and for sons, how to practice a trade. He must also provide his daughters with financial means for marriage (akin to a dowry) (Cohn-Sherbok 2010).

The *ketubah*—marriage contract—states the obligations of spouses to one another; traditionally, Jewish husbands are obligated to financially support their wives, including in the event that the marriage is dissolved. Divorce laws in traditional Judaism are also gender-specific. Married couples who wish to obtain a divorce must go before a rabbinic court to receive a *get*, or divorce contract. Jewish law states that a *get* cannot be granted without the husband's consent. All Jewish denominations, including Reform, Conservative and Orthodox, have to varying degrees adopted progressive positions on gender-specific laws and traditions, especially by encouraging more Jewish education for girls and expanding opportunities for women as spiritual leaders.

Although traditional Jewish law is clearly patriarchal, Jewish women have historically been active in secular social movements to demand and protect all human rights, especially those of women. In the United States, Jewish women played prominent roles in the suffrage movement of the early twentieth century and women's movements for reproductive and employment rights in the 1960s and 1970s. Women are now expected to achieve an optimal level of education and to seek gainful employment if they so desire.

22.4.2 Prescriptive, Restrictive, and Taboo Behaviors for Children and Adolescents

Children are highly valued in Jewish tradition. Jewish teachings exhort parents to treat children with respect and provide them with love. Jewish law also obligates parents to provide their children with an education that helps them gain employment and transmits their Jewish heritage. Children are welcomed and incorporated into most holiday celebrations and services. One important ritual of Passover, for example, requires that children ask four questions about the purpose and meaning of the *seder* (order of stories, blessings and foods recited at the Passover meal).

The fifth of the Ten Commandments is to "honor your mother and father" (Exodus 20:12). This is interpreted in Jewish tradition to mean that children must treat their parents with respect at all times. In practice, this means, for example, that they must refrain from embarrassing their parent, or provide them with dignified care in old age. The commandment to honor one's parents continues even after death. One is supposed to recite the memorial prayer (*kaddish*), and in conversation, to praise their memory with the words "may his (or her) memory be a blessing."

In Judaism, the age of adulthood is 13 years and 1 day for a boy, and 12 years and 1 day for a girl (Robinson 2016). Children are deemed to be capable at this age of differentiating right from wrong and having the capacity to commit themselves to performing the commandments. Recognition of religious adulthood and assumption of its responsibilities occur during a religious ceremony called a *bar* or *bat mitzvah* (literally, son or daughter of the commandment, respectively); during this ceremony, the child is called to read a portion from the Torah and takes part in leading services for the first time in a public gathering (Robinson 2016). The *bat mitzvah* ceremony for girls has increased in prevalence among Reform and Conservative Jews in the twentieth century. In America, this rite of passage is usually accompanied by a family celebration. However, because sons and daughters are still teenagers living at home, it is recognized that they are still the responsibility of their parents. In Orthodox communities, boys who have reached their *bar mitzvah* are now responsible to perform the rituals of daily prayer, Torah study and all other commandments.

22.4.3 Family Goals and Priorities

A traditional expression said after the birth of a baby is that the parents should be blessed to raise and prepare their child for a life of "Torah, marriage, and good deeds." Similarly, on the occasion of a wedding, a typical blessing for couples is that they should “build a home on the foundations of Torah and the commandments.” The goals and priorities of Jewish families focus on the ideal of raising children who will have knowledge of the Jewish traditions, as well as connections with rituals and community. The principle is to prepare children to grow into adults who will marry and raise Jewish children of their own. The emphasis on childbearing relates to God’s command to Adam and Eve, to “be fruitful and multiply” (Genesis 1:28).

Marriage is considered to be an ideal state for adults; it is considered a sacred bond between adults and a means of personal fulfilment (Lieber 2012). The Torah states that “it is not good for a person to be alone” (Genesis 2:18). This verse has been interpreted to mean that the purpose of marriage is to provide companionship and allow individuals to care for another person. Marriages must be monogamous. Rabbinic laws permit marriage between cousins, although this is rarely practiced among Jews today. Jewish law forbids intermarriage between Jews and non-Jews, although opinions about this vary considerably among the denominations in light of the increase in interfaith unions between Jews and non-Jews in the past century.

Sexuality is a right of both men and women. The sexual instinct is considered to be neither sinful nor shameful, but restraint and modesty are expected. In addition to procreation requirements, conjugal rights for women exist. Jewish husbands are obligated to provide their wives with non-procreative intercourse if they desire it (Robinson 2016). Sexual intercourse is viewed as a pure and holy act when performed mutually within the relationship of marriage. With some exceptions, a husband’s refusal to have sex with his wife can be grounds for a divorce. However, the act of sex, if not performed with sobriety and modesty and the wife’s willingness, is considered impermissible by Jewish tradition. Premarital sex is not condoned.

Orthodox Jewish women who observe the *halakhah* of “ritual purity” will physically separate themselves from their husbands during their menstrual periods and for up to 7 so-called “clean” days after the end of menstruation (Robinson 2016). Sexual relations and touching are forbidden during these days, until the woman has immersed in a *mikveh*, a ritual bath. Sexual contact for this group may, therefore, occur only during 2 weeks out of each month.

Reform Judaism supports the need for sex education. In 2011 the Union for Reform Judaism developed a curriculum for students in synagogue classes called “Sacred Choices: Adolescent Relationships and Sexual Ethics” (Winer 2011). This sexual ethics curriculum was designed for middle school and high school students to teach Jewish teens that their bodies are gifts from God and that Judaism provides relevant guidance on how to use and care for that gift and deal with the consequences of their choices regarding sexuality. In general, the Jewish community supports providing youth with accurate information so they can make informed choices.

Whereas it is recognized that one’s later years are a time of physical decline, older people are to be venerated, especially for the wisdom they have to share. Old age is a state of mind rather than a chronological age; one may continue to “give” to society in a variety of ways other than employment. In addition, one may never “retire” from practicing the commandments.

Honoring one’s parents is a lifelong endeavor and includes maintaining their dignity by feeding, clothing, and sheltering them, even if they suffer from senility. Respect for older people is essential even when their actions are irrational. The care of an older family member is the responsibility of the family; when the family is unable to provide care owing to physical, psychological, or financial reasons, the responsibility falls to the community. This role has always been a hallmark of Jewish communal life.

22.4.4 Alternative Lifestyles

Views on homosexuality vary with each branch of Judaism. Orthodox Jews are largely non-accepting of same sex-unions, based on scripture (Lev. 18:22). The Torah, especially as interpreted by the Orthodox, prohibits homosexual intercourse for men; it says nothing specifically about

Table 22.1 Jewish holidays: 2020–2026

Holiday	2020–2021 (5781)[a]	2021–2022 (5782)[a]	2022–2023 (5783)[a]	2023–2024 (5784)[a]	2024–2025 (5785)[a]	2025–2026 (5786)[a]
Rosh Hashanah	9/19–9/20	9/7–9/8	9/26–9/27	9/16–9/17	10/3–10/4	9/23–9/24
Yom Kippur	9/28	9/16	10/5	9/25	10/12	10/2
Sukkot[b]	10/3–10/10	9/21–9/28	10/10–10/17	9/30–10/7	10/17–10/24	10/7–10/14
Chanukah[c]	12/1–12/18	11/29–12/6	12/19–12/26	12/8–12/15	12/26–1/2	12/15–12/22
Purim[c]	2/26	3/17	3/7	3/24	3/14	3/3
Passover[b]	3/28-4/4	4/16–4/23	4/6–4/13	4/23–4/30	4/13–4/20	4/2–4/9
Shavuot	5/17–5/18	5/5–5/6	5/26–5/27	6/12–6/13	6/2–6/3	5/22–5/23

Note: Jewish holidays always begin at sundown the evening before the date noted on this calendar; holidays end at sundown on the date shown

[a]Year on the Jewish calendar

[b]Non-work days include the first 2 and last 2 days for Sukkot and Passover

[c]Work is permitted during these holidays

sex between women. Some of the objections to gay and lesbian lifestyles are socially and culturally motivated, and include the inability of these unions to fulfil the commandment of procreation (to "be fruitful and multiply") and the possibility that acting on the recognition of one's homosexuality could ruin an existing heterosexual marriage. The Conservative movement held this official position until as recently as 2006, but now allows clergy to perform same-sex marriage ceremonies and welcomes gay and lesbian congregants. The Reform and Conservative movements can also ordain gay and lesbian clergy, and support full legal and social equality for homosexuals (Krasner 2015).

Approximately 5.1–7% of American Jews self-identify as lesbian, gay, or bisexual (Krasner 2015). Of these, 31–34.8% were married or living with long-term partners and 9–14% were raising their own children. The American Jewish community is generally more accepting of homosexuality and same-sex marriages than the general population (Krasner 2015).

22.5 Workforce Issues

22.5.1 Culture in the Workplace

Specific workforce issues may occur, especially for those who are observant of the Sabbath and the High Holy days. Jews who observe the Sabbath must have Friday evenings prior to sunset and Saturdays off, as work of any kind is forbidden during that time. They may work on Sundays. Supervisors must be sensitive to the needs of Jewish staff and recognize the holiness of the Sabbath to many. Jewish staff should be allowed to request time off for the major Jewish holidays (Table 22.1). Since holidays begin the evening before, they must have time off during the evening shift before *and* the following day. Staff should not be penalized by having to use this time off as unpaid holidays or vacation time, but they should have the option to exchange for the Christmas and Easter holidays, time that is usually afforded to non-Jewish staff. The best approach is to have policies that allow one's employees to request time off to follow Jewish law without penalty.

Jewish healthcare providers are fully acculturated into the American workforce. Judaism's beliefs are congruent with the values that American society places on the individual and family. As English is the primary language for Jewish Americans, no language barriers to communicating in the workplace exist.

22.5.2 Issues Related to Autonomy

Jewish nurses have begun to speak out on their needs in the workplace. With the inclusion of the American Nurses Association (ANA) Standard of Practice #8 related to culturally congruent practice that incorporates cultural competence and cultural sensitivity, many are now addressing this long-ignored area (American Nurses Association 2015). In 1990, a National Nurses Council (now called the Nurses Allied Health Professions Council) was established through Hadassah, the Zionist women's organization (Benson 1994). This group promotes solidarity and empowerment to enhance sensitivity toward Jewish cultural needs within the healthcare community. In 2008 the Orthodox Jewish Nurses Association was established to provide a forum to discuss professional issues related to Orthodox Jewish nurses and to serve the needs of its members. Healthcare professionals in other countries might find this helpful.

Jewish nurses have made significant contributions in nursing education, research, and leadership. Examples include Lillian Wald, who established the Henry Street Settlement in New York and began the specialty of School Nursing; Josephine Goldmark, whose report on the poor state of nursing programs and nursing students initiated significant curricular changes; Mathilda Scheuer, President of the American Nurses Association; and Claire Fagin, who served as both Dean of the School of Nursing and President of the University of Pennsylvania (Mayer 2009).

Ways in which the professional nursing community can demonstrate its sensitivity to Jewish nurses can include the following: scheduling major nursing conferences or professional meetings so as not to occur during the High Holy Days in the fall or during Passover in the spring; avoiding non-kosher foods such as pork and shellfish during catered affairs, or alternatively, providing kosher alternatives; and eliminating prayers before conference meals that invoke the name of Jesus.

22.6 Biocultural Ecology

22.6.1 Skin Color and Other Biological Variations

Jews share no physical characteristics that distinguish them from non-Jews. Ashkenazi Jews are usually White, and exhibit a range of complexions and hair colors. Some groups of Sephardic and Mizrahi Jews may have dark skin tones and hair coloring, similar to the populations around the Mediterranean and Middle East, or other places to where they trace their origins. There are also Jewish groups throughout Africa who are Black, most notably the Jews originally from Ethiopia, known as *Falasha* Jews. In the United States, there are African-American and African-Israeli groups who identify as Jewish.

22.6.2 Diseases and Health Conditions

Because Jews reside throughout the United States, no specific risk factors are based on topography. Genetic risk factors vary based on whether the family immigrated from Ashkenazi (Eastern or Central Europe [Russia, Poland, Germany, Ukraine, and Lithuania]), Sephardic (Northern Africa or Spain), or Mizrahi (Middle East) areas. There is a greater incidence of some genetic disorders among individuals of Jewish descent, especially those who are Ashkenazi. It is estimated that 1 in 4 Ashkenazi Jews carries one of these mutations (Victor Center for Jewish Genetic Diseases 2019). Most of these disorders are autosomal recessive, meaning that both parents must carry the affected gene. Although the best known is Tay-Sachs disease, Gaucher disease—Type 1, Canavan disease, familial dysautonomia, torsion dystonia, Niemann-Pick disease—Type A, Bloom syndrome, Fanconi anemia—Type C, and mucolipidosis IV are other prominent conditions seen in this population (Victor Center for Jewish Genetic Diseases 2019).

Nineteen diseases are now available for genetic screening for people of Ashkenazi Jewish descent. This saliva-based panel can identify those at risk (Grinzaid et al. 2015). Seven of the conditions have carrier frequencies higher than 1 in 100 and nine more have frequencies between 1:100 and 1:200 (Shi et al. 2017).

Gaucher disease (pronounced Go-shay) is the most common genetic disease affecting Ashkenazi Jews, with 1 in 10–14 carrying the gene (National Gaucher Foundation 2019; Victor

Center for Jewish Genetic Diseases 2019). Gaucher disease is a lipid-storage disorder. This inborn error of metabolism results in lack of an enzyme that normally breaks down glucocerebroside, a lipid by-product of erythrocytes. The glucocerebroside accumulates in the body via fat-laden Gaucher cells that build up in the spleen, liver and bone marrow, resulting in weakening and fracturing of the bones due to infarctions, anemia, and platelet deficiencies. The spleen becomes painfully enlarged. There are 400 different genetic mutations of the disease; 4 of them account for 95% of cases in Ashkenazi Jews. Treatment for Gaucher disease-Type 1 consists of either enzyme replacement therapy or substrate reduction therapy (National Gaucher Foundation 2019).

The gene for Tay-Sachs disease is carried by approximately 1 in 25 to 1:30 Ashkenazi Jews. This autosomal recessive condition is a lysosomal sphingolipid storage disorder caused by an absence of hexosaminidase A, resulting in an accumulation of a lipid called *GM2 ganglioside* that causes progressive damage to the neural cells. The onset of intellectual and developmental delay begins between 3 and 6 months of age, with progressive deterioration, increasing seizure activity, blindness, hearing loss, and death by approximately age 5 (Harvard Health Publishing 2018). The incidence of Tay-Sachs disease has decreased significantly since the early 1980s, due to greater ease of testing for carriers and the fetus during pregnancy, as well as a concerted effort in the Jewish American community to provide testing.

Canavan's disease is a rare, fatal, degenerative brain disease similar to Tay-Sachs disease, but caused by a defective gene that impairs the formation of myelin in the brain. Approximately 1 in 40 Ashkenazi Jews carry the gene. The resulting symptoms begin in mid-infancy and include developmental delay, loss of vision, and a loss of reflexes resulting in death by the age of 10 years (Victor Center for Jewish Genetic Diseases 2019).

Familial dysautonomia, or hereditary sensory and autonomic neuropathy, type III, is also an autosomal recessive genetic disease, with the gene located on chromosome 9q31. It causes dysfunction of the autonomic and peripheral sensory nervous systems. Affected children have decreased myelinated fibers on nerves that lead to afferent impulses but maintain a normal intelligence. Symptoms include significant temperature and blood pressure fluctuations; a decrease in the number of taste buds; altered pain sensation; increased salivation and sweating; abnormal sucking or swallowing difficulties and vomiting, resulting in failure to thrive; and decreased tears, resulting in increased risk of corneal ulceration. With aging, scoliosis, altered pain sensation, kidney insufficiency and unsteady walking may occur (National Organization for Rare Disorders 2019). One in 30 Ashkenazi Jews is a carrier (Victor Center for Jewish Genetic Diseases 2019).

Other conditions that have a higher incidence among Ashkenazi Jews than the general public include the following:

- Niemann-Pick disease type A is an autosomal recessive severe neurodegenerative disorder that starts at 6 months of age. It involves an abnormal storage of sphingomyelin and cholesterol in organs. It is caused by an enzyme deficiency and leads to central nervous system degeneration, including a loss of mental abilities and movement. There is also lung involvement leading to respiratory failure. Those with type A usually do not survive past early childhood. The gene for Type A is carried by 1 in 90 Ashkenazi Jews.
- Bloom syndrome, a rare genetic condition that results in abnormal breakage of chromosomes, results in respiratory and gastrointestinal infections, erythema, telangiectasia, sun sensitivity, and dwarfism. Whereas the intelligence of those affected is usually normal, they face an increased risk of cancer, infertility and diabetes (Bloom's Syndrome Association 2019). The average age of death is 27; the gene is carried by 1 in 100 Ashkenazi Jews.
- Fanconi anemia—Type C is a type of aplastic anemia with symptoms of decreased red blood cells, white blood cells and platelets; those affected are also at an increased risk of cancer,

especially leukemia. Many die by early adulthood. One in 89 Ashkenazi Jews are carriers (St. Jude Children's Research Hospital 2019; Victor Center for Jewish Genetic Diseases 2019).

- Mucolipidosis IV is found in 1 of 100 Ashkenazi Jews. This lipid-storage disease results in central nervous system deterioration during the first year with motor and intellectual delay, as well as various eye disorders. The prognosis varies (Victor Center for Jewish Genetic Diseases 2019).
- Torsion dystonia, an autosomal dominant condition, is carried by 1 in 1000 to 1 in 3000 Ashkenazi Jews in the United States. The disease is first seen around age 11 years and leads to rapid progression in loss of motor control and twisting painful spasms of the limbs, resulting in contractures. Affected individuals lead a full life and have a normal intelligence.

A number of genetic conditions have also been seen at higher frequencies among Sephardic Jews. These conditions include alpha and beta thalassemia, ataxia telangiectasia, corticosterone methyloxidase Type 2 Deficiency, Costeff Optical Atrophy, cystic fibrosis, Familial Mediterranean Fever, Glycogen Storage Disease Type 3, G6PD Deficiency, and Hereditary Inclusion Body Myopathy (Forward 2014).

Both the American College of Medical Genetics and the American Congress of Obstetricians and Gynecologists recommend carrier screening for Jewish adults of child-bearing age (Grinzaid et al. 2015). The Reform movement supports a couple's right to make the decision as to whether or not to have the testing done. Testing is allowed because the knowledge is available, and may contribute to the couple's emotional and psychological well-being.

Orthodox rabbis often do not support genetic testing because it might cause couples to refrain from marrying or having children, thus preventing them from fulfilling the *mitzvah* of procreation. However, a service called Dor Yeshorim (2019) carries out anonymous genetic testing before marriage and, although they will not identify one's carrier status, they will inform couples if they are compatible or not. The goal of this organization is to eliminate the occurrence of debilitating genetic disease in Jewish families.

Some Orthodox rabbis allow the practice of pre-implantation screening of in vitro fertilized zygotes if both husband and wife are known carriers of Tay-Sachs and then to only use the healthy ones for implantation. "The discarding of the affected zygotes is not considered as abortion, since the status of a fetus or a potential life in Judaism applies only to a fetus implanted and growing in the mother's womb" (Rosner n.d.).

Other conditions with increased incidence in the Jewish population include inflammatory bowel disease (ulcerative colitis and Crohn's disease), which is seen more often in Ashkenazi Jews than in other ethnic groups (Santos et al. 2018), as is colorectal cancer and pancreatic cancer; the BRCA1 and BRCA2 genes that cause breast and ovarian cancer are found in 1 out of every 100 Jewish women of Ashkenazi background (Johns Hopkins 2018). There is also an increased incidence of lactose intolerance in this population (Glausiusz 2015).

22.6.3 Variations in Drug Metabolism

The pharmacogenetic variability is different for Ashkenazi Jews compared to other groups, especially for metabolism of medication classes that are affected by the enzymes CYP2A6, CYP2C9, NAT2, and VKORC1 (Zhou and Lauschke 2018). The genes that regulate these enzymes have implications for dosing a number of antidepressants, benzodiazepines, mephenytoin, proton pump inhibitors, and antiplatelet drugs (Yang et al. 2014). One of the few drugs found to have a higher rate of side effects in people of Ashkenazic ancestry is clozapine, used to treat schizophrenia. Twenty percent of Jewish patients taking this drug developed agranulocytosis, compared with about 1% of non-Jewish patients. A specific genetic haplotype has been identified to account for this finding (Schatzberg and Nemeroff 2009). Thus, healthcare providers must order testing for agranulocytosis when Jewish patients are prescribed clozapine.

22.7 High-Risk Behaviors

According to Jewish law, individuals may not intentionally damage their bodies or place themselves in danger. The basic philosophy is that the body must be protected from harm. The body is viewed as belonging to God; therefore, it must be returned intact when death occurs. Consequently, any substance or act that harms the body is not allowed. This includes suicide or taking non-prescription or illegal medications. The Torah prohibits the act of making marks on the body, including tattoos, branding or piercings other than the earlobes (My Jewish Learning 2019). However, having tattoos would not result in exclusion from participating in any religious activities.

The Hebrew Bible includes several cautionary tales about the dangers of alcohol. However, wine is an essential part of religious holidays and festive occasions and is a traditional symbol of joy. The Bible speaks of the undesirable effects of wine on the person, as well as its positive use as a medicine. Consequently, wine is appropriate and acceptable as long as it is used in moderation. Substance and alcohol abuse exist in the Jewish community. Jewish Alcoholics, Chemically Dependent Persons and Significant Others (JACS) is a mutual support network with groups throughout North America, Israel and other countries that seek to encourage and assist Jews and their families to explore recovery in a nurturing Jewish environment. There are also Jewish-oriented residential treatment centers in New York, California, and Florida (Chabad.org 2019a, b). A pilot study conducted in Canada found that over 20% of Jewish respondents had a family history of alcohol or drug abuse (Baruch et al. 2015).

22.7.1 Healthcare Practices

Jewish Americans tend to be health conscious. Taking care of one's body is a *mitzvah* and there are many Jewish laws that address the care of the body (Barilan 2014, p. 44). Given the higher rates of advanced education among Jews, they are more likely to practice preventive health care, with routine physical, dental, and vision screening. This is also a well-immunized population. Although the older generation is still more likely to defer to medical authority, Jewish adults tend to want to participate in healthcare decision-making.

22.8 Nutrition

22.8.1 Meaning of Food

Eating is important to Jews on many levels. Besides satisfying hunger and sustaining life, it also teaches discipline and reverence for life. For those who follow the dietary laws (*kashrut*), a tremendous amount of attention is given to the slaughter of animals and the preparation and consumption of all foods. In addition, the family dinner table is often the site of religious holiday celebrations and services, especially the Sabbath, the *seder* on Passover, Rosh Hashanah (Jewish New Year), and breaking the fast for Yom Kippur (Day of Atonement). Jewish dietary practices serve as a unifying factor in ethnic identity; for many, they also represent a spiritually refining act of self-discipline.

22.8.2 Common Foods and Food Rituals

One food that is often identified as "Jewish" is chicken soup. This has been colloquially referred to as "Jewish penicillin," and is often served with *knaidlach or matzah balls* (dumplings made of matzah meal). Although it has no intrinsic meaning or religious value, it is a customary food course in many religious homes, especially on Friday evenings as part of the Sabbath meal, and also during times of illness or recovery. It is frequently associated with feelings of care and nurturing.

Common Ashkenazi foods include challah (a thick, braided white bread), kugel (noodle pudding, either sweet or savory), blintzes (crepes filled with a sweet cottage cheese), chopped liver (served cold, similar to liverwurst), gefilte fish (ground karp molded into oblong balls, steamed, then served cold with horseradish), hamentashen (a triangular pastry with different types of filling), and Nova Scotia or "belly" lox (cold smoked salmon) served with cream cheese and salad vegetables on a bagel. Slow-cooked beef brisket is often the entrée at the Rosh Hashanah (New Year) dinner. The Sephardic and Mizrahi communities have their own distinct culinary traditions.

The laws regarding food are found in the Torah, especially in the books of Leviticus and Deuteronomy. They are commonly referred to as the laws of **kashrut**, or the laws that dictate which foods are permissible under religious law. The term **kosher** means fit or proper to eat as it relates to the Jewish dietary laws (Orthodox Union 2019); it is not a brand or form of cooking. Whereas some believe that the mandatory statutes were developed and implemented for health reasons, religious scholars dispute this view, claiming that the only reason one needs to follow the laws is because they are mandatory commandments of God. Therefore, the laws are followed as a personal attachment to the religion, community and family customs; and also, the belief that God has mandated them. The idea that kosher laws promote health is only a secondary gain. Issues related to *kashrut* may be a significant part of an inpatient hospital stay, making it helpful to know what is and is not acceptable.

Foods are divided into two categories: those considered kosher (permitted or clean) and those considered ***treyf*** (forbidden or unclean). A permitted animal may become *treyf*, or forbidden, if it is not slaughtered, cooked, or served properly. Kosher slaughter of animals is done in a way that prevents undue cruelty to the animal and ensures the animal's health for the consumer (Lieber 2012). The jugular vein, carotid arteries, and vagus nerve must be severed in a single quick stroke with a sharp, smooth knife, causing the animal to die instantly. No sawing motion or second stroke is permitted. This also allows the maximal amount of blood to leave the body. Care must be taken that all blood is drained from the animal before it is eaten. The consumption of blood is prohibited. An animal that dies from old age or disease may not be eaten, nor may it be eaten if it meets a violent death or is killed by another animal. In addition, flesh that is cut from a live creature may not be eaten (Robinson 2016).

Milk and meat may not be mixed together in cooking, serving, or eating, based on the Torah prohibition, "You must not boil a calf in its mother's milk" (Deut. 14:21). To avoid mixing foods, utensils and plates used to serve them are separated. Contemporary religious Jews who follow the dietary laws will often have two sets of dishes, pots, and utensils: one set for milk products (*milchig* in Yiddish) and the other for meat (*fleishig*). Some homes will have different sets of dish towels and even different sinks. Because glass is not absorbent, it can be used for either meat or milk products, although religious households still usually have two sets. Therefore, foods such as cheeseburgers, lasagna made with meat, or grated cheese on meatballs are unacceptable for those who keep kosher. Milk cannot be used in coffee if it is served with a meat meal. Nondairy creamers can be used instead, as long as they do not contain sodium caseinate, which is derived from milk. Those who observe ritual separation between meat and milk will also wait for a period of time after consuming meat before they eat milk products. Some will wait 3 h after eating meat, and others up to 6 h, with the premise that food takes that long to digest from the stomach, and to allow any residue on the palate and trapped in the teeth to dissolve (Robinson 2016).

A number of foods are considered ***pareve*** (neutral) and may be used with either dairy or meat dishes. These include fish, eggs, anything grown in the soil (vegetables, fruits, coffee, sugar, and spices), and chemically produced goods (Lieber 2012). Vegetables and fruits must be washed carefully to ensure that they are free of insects. A "U" with a circle around it is the seal of the Union of Orthodox Jewish Congregations and is used on food products to indicate that they are kosher. A circled "K" and other symbols may also be found on packaging to indicate that a product is kosher. There are several other kosher symbols one might see on packaging.

When working in a Jewish person's home, the healthcare provider should not bring food into the house without knowing whether or not the patient adheres to kosher standards. If the patient keeps a kosher home, do not use any cooking items, dishes, or silverware without knowing which are used for meat and which for dairy products. Healthcare providers must fully understand the dietary laws so they do not offend the patient, can advocate for kosher meals if they are requested during hospital stays, and can plan medication times accordingly. It may also be important to know that Orthodox Jews refrain from cooking food on the Sabbath and other holidays; instead, any foods that need to be cooked with heat (baking, boiling, frying, etc.) must be prepared before the holiday begins.

Mammals are considered kosher if they have split (cloven) hooves and chew their cud, and also meet the other requirements for kosher slaughter and consumption. These animals include buffalo, cattle, goat, deer, and sheep. Pigs and rabbits are examples of animals that do not meet these criteria. Although liberal Jews decide for themselves which dietary laws they will follow, many still avoid pork and pork products out of a sense of obligation to community or family. Serving pork or other unkosher foods to a Jewish patient or colleague, unless specifically requested, is insensitive.

Birds of prey are considered "unclean" and unacceptable because they grab their food with their claws. Acceptable poultry are chicken (one of the most frequently consumed forms of protein), turkey, goose and duck. Fish, which is considered *pareve*, can be eaten if it has both fins and scales. Nothing that crawls on its belly is allowed, including clams, crabs, shrimp, lobsters, and other shellfish; other seafood that does not have fins or scales and is therefore considered nonkosher includes squid, octopus, scallops, mussels, and calamari (Robinson 2016).

Kosher meat must be prepared for cooking by soaking and salting to drain all the blood from the flesh. It is usually purchased with the required preparations already having been completed from kosher butchers and some grocery stores. As increased residual salt may result from the salting method, patients with sodium restrictions may need counselling to assist them in making dietary adjustments. Broiling is an acceptable cooking method, especially for liver, because it drains the blood. Care must be taken in serving cheese or other dairy products to ensure that no animal substances are served at the same time. Breads and cakes made with lard are *treyf*, and breads made with milk or milk by-products (e.g., casein) cannot be served with meat meals. Eggs from nonkosher birds, milk from nonkosher animals, and oil from nonkosher fish are not permitted. Butter substitutes (commonly margarine) are used with meat meals. Honey is allowed, even though bees are not kosher for consumption, and is considered pareve.

It should be noted that some medications have a beef or pork (porcine) base, such as heparin. Gelatin is often used in making the capsule shells for many medications and is also used as one of the stabilizers in some medications such as vaccines. "Gelatin is a partially hydrolysed collagen which is usually bovine (beef) or porcine (pig) in origin" (Queensland Health Guideline 2019).

Reflective Exercise 22.1

Mr. Orr, an 80-year-old Jewish patient, is a first-generation American whose parents emigrated from Eastern Europe; he has significant heart disease. He is being cared for in his home by visiting nurses. His wife of 55 years has some physical limitations but is self-sufficient to maintain her home. The nurse may want to teach his wife how to make meals that will meet his health needs and enters the kitchen to obtain a measuring cup. Concerned because she has a kosher kitchen, Mrs. Orr starts yelling at the visiting nurse.

1. What needs to be known in advance before entering Mrs. Orr's kitchen?
2. What questions should be asked during the history taking of this patient about his cultural practices related to food and keeping kosher?
3. What specific sources of protein would be appropriate for this patient?

4. Mr. Orr loves to have a cup of hot tea with cream. What is the best timing for this treat related to his meals?
5. Mr. Orr does not want to take some of his prescribed medications because he fears they are made from porcine or non-kosher beef. How might you resolve this issue for him?

There are many variations of how Jews observe the laws of kashrut. While many Jews may keep a kosher home, they may eat non-kosher foods outside the home, allowing them to eat in restaurants and the homes of friends. Others do not follow any of the laws. There are some who connect the laws of kashrut with the precepts of environmentalism, such as avoiding pesticides used in the growing of foods (Lieber 2012).

Kosher meals are available in many hospitals or can be obtained from frozen food suppliers. They are usually provided in sealed packages and should be served with disposable plates and utensils. Healthcare providers should not open the packages or transfer the foodstuffs to another serving dish. Frozen kosher meals are available on a commercial basis. Help may be needed for a patient to choose from a facility's menu options. No milk or yogurt should be placed on a tray with meat, and butter cannot be served with bread. Even salad dressing needs to be made without dairy ingredients if it is served with meat. If healthcare providers have difficulty locating a supplier, they should contact a local synagogue or rabbi. Determining a patient's dietary preferences and practices regarding dietary laws should be done during the admission assessment. It must be stressed that most Jews choose if and to what degree they wish to follow the laws of kashrut.

Before eating, some Jews will recite a prayer over a piece of bread, often in Hebrew (referred to as *Hamotzi*). Orthodox Jews will ritually wash their hands before eating bread, and also recite prayers after eating.

22.8.3 Dietary Practices for Health Promotion

While it has been claimed that many Jewish dietary practices provide a secondary gain of preventing disease, the primary purpose of kashrut observance is to keep a commandment. Many Jews interpret the dietary laws as a means to elevate the act of eating to a spiritual level, a belief that may also apply to the practice of washing one's hands and praying before and after eating.

22.8.4 Nutritional Deficiencies and Food Limitations

No nutritional deficiencies are common to individuals of Jewish descent. As with any ethnic group, nutritional deficiencies may occur in individuals of a lower socioeconomic status, because of the expense or availability of certain foods. It should be noted that kosher meat is usually more expensive than non-kosher meat because of the production methods involved.

In addition to the dietary laws discussed previously, other dietary laws are followed at specified times. For example, during the week of Passover, no bread or products with yeast may be consumed, including pasta, pastries, and beer. *Matzah* (unleavened bread) is eaten instead. Any product that is fermented or can cause fermentation may not be eaten (Lieber 2012). Rather than attend synagogue, the family conducts the festival meal (*seder*) around the dinner table during the first two nights of the holiday, and incorporates dinner into a service that allows all participants to retell the story of Moses and the Exodus from Egypt.

The Jewish calendar has a number of fast days. The most widely observed is the holiest day of the year, Yom Kippur. On this Day of Atonement, Jews abstain from food and drink as they pray to God to forgive any sins they have committed during the past year. They eat an early dinner on the evening before the holiday begins and then fast until after sunset the following day. Individuals who are ill, older people, young children, pregnant or lactating mothers, the physically incapacitated, and those whose health would be "injured by fasting are forbidden to do so," such as those with diabetes (Robinson 2016, p. 100); those affected may need to be reminded of this exception to Jewish law. Maintaining an ill person's health supersedes the act of fasting. If concerns arise, a consultation with the patient's rabbi may be necessary.

22.9 Pregnancy and Childbearing Practices

22.9.1 Fertility Practices and Views Toward Pregnancy

God's first commandment to humanity is to "be fruitful and multiply" (Genesis 1:28). Having children is considered to be a gift and responsibility in Judaism, not a right. Some ultra-Orthodox communities place a cultural value on the importance of having male children because they will be able to say ***kaddish*** (the prayer for the dead) for their parents after their passing. In other branches of Judaism, both sexes may recite the *kaddish*. Traditionally, families have been encouraged to have *at least* two children; however, it is ultimately more important that parents decide how many children to have based on the best interests of their future children (Dubov 2019). While the traditional ideal was to have one male and one female, it is recognized that these outcomes are beyond the parents' control (Barilan 2014).

Couples who are unable to conceive should try all possible means to have children. This includes infertility counseling and interventions such as egg and sperm donation. In Orthodox communities, *artificial insemination* is usually allowed if the sperm and egg are taken from the married Jewish partners (Loike 2016). The reason for this rule is that if the egg and sperm are taken from individuals other than the married couple, there is a risk that that rabbinic authorities might consider the offspring to be the product of incest, for example, if it is suspected that the egg or sperm was donated by one of the parent's siblings. Others hold the opinion that the use of donor eggs or sperm may be seen as an act of adultery, although some argue that it is not considered an act of adultery if no sexual intercourse has occurred.

Orthodox Jews have differing opinions as to whether the woman may receive donor sperm from a man who is other than her husband (Margalit 2017). There is even greater debate about who is to be defined as a child's legal mother according to Jewish law in cases where egg donation or surrogates are used. According to Jewish law, both the gestational and the genetic mothers could be considered the child's legal mothers (Laufer-Ukeles 2016; Margalit 2017). However, surrogacy, along with other forms of assisted reproductive technology, do enable couples to fulfil the commandment to have children. Because it is believed that God created an incomplete world, "human beings are encouraged to partner with God to improve and perfect the world" (Loike 2016, p. 226). Therefore, cloning is allowed "when there are medical benefits that justify a new technology" (Loike, p. 227).

When all natural attempts have been made, adoption may be pursued, although adoption is not mentioned in Jewish law because children are not regarded as "property;" that is, they cannot be "owned" or transferred from one set of parents to another (Barilan 2014). To some Orthodox Jews, adoption does not fulfil the commandment to "be fruitful and multiply" (Laufer-Ukeles 2016). However, the Talmud states, "He who raises someone else's child is regarded as if he had actually brought him into the world physically" (Sanhedrin 19b, as quoted in Robinson 2016, p. 145). Most Jews accept and encourage the practice of adoption; American Jews adopt at a rate that is double that of non-Jews, with 5% of Jewish homes raising an adopted child (Sartori 2016). Because some of these children will have been adopted from non-Jewish parents, the genetic factors discussed earlier in this chapter would not apply to them.

During the late twentieth century, the average family size among American Jews decreased, while the rate of marriage between Jews and non-Jews increased. This has resulted in a decreased rate of births among the Jewish population in the twenty-first century. While the average Jewish family in the United States has fewer than 2 children (1.9), in Israel, the average Jewish family in 2017 had 3.72 children (Bassok 2017; Sheskin 2019). Some believe that the historical tragedy of the Holocaust, in which one-third of the global Jewish population was killed, places a moral obligation on Jews today to bring one more child into the world than they would have otherwise.

For most Jews, the use of *birth control* is up to the individual, regardless of age or the situation. Judaism insists that parents must keep children's best interests in mind, and only bring children into the world if they can meet their needs. Therefore, the use of temporary birth control may be acceptable. "Many modern Jews feel that the benefits of contraception, be they female health, family stability, or disease prevention, uphold the commandment in Judaism to 'choose life' much more strongly than they violate the commandment to 'be fruitful and multiply'" (Jewish Views on Contraception 2019).

Some Orthodox Jews may hold a negative view of birth control, and interpret the prevention of pregnancy as a way to defer the commandment to "be fruitful and multiply." Also, while Jewish law does not prohibit the use of birth control, the Torah does prohibit "the wasting of seed" (Genesis 38:9). Most Orthodox Jews do not have a favourable view of contraception unless a pregnancy has the potential to place the mother's life or health (including emotional wellbeing) in jeopardy. However, some individuals may request permission from their rabbis to use birth control. Condom use is supported in this population only if unprotected sexual intercourse would pose a medical risk to either spouse (Jewish Views on Contraception 2019), as it is seen as a "wasting of seed." Orthodox Jews generally oppose the practice of sterilization, unless the life of the mother is in danger. They also forbid vasectomies based on the biblical prohibition against interfering with male reproductive physiology (Jewish Views on Contraception 2019).

Because the primary focus of Jewish tradition is to uphold the sanctity of life, it is important to identify *when life begins*. The fetus is not considered to be a living soul or person until it has been born. Birth occurs when the baby's head or "greater part" of the body emerges from the womb (Robinson 2016). Until that time, the fetus is considered to be part of the mother's body and has no independent identity. The unborn is not defined as a person and has no independent life.

The mother's health is of paramount concern and takes precedence over the unborn in all cases. Jewish law states that if a fetus endangers the mother's physical or mental health, it is defined as an "aggressor" which must be aborted. This is observed in all branches of Judaism (Robinson 2016). Therefore, a federal government ban on *abortion* could be considered a violation of religious freedom. Orthodox Jews do not allow abortions for any reason other than to save the mother's life, as the fetus is considered to be part of the mother's body, and it is forbidden to harm one's body. Other branches of Judaism permit abortion if the mother desires it, or in cases of incest or rape. Reform Jews believe that a woman maintains control over her own body, and it is up to her whether to abort a fetus. Although no connotation of sin is attached to abortion, the decision is not to be made without serious deliberation. Most Jews favor a woman's right to choose.

22.9.2 Prescriptive, Restrictive, and Taboo Practices in the Childbearing Family

While pregnancy is an exciting time for all parents, it is not a Jewish custom to hold baby showers or purchase items for the unborn child. However, it is acceptable to hold a baby shower if the mother-to-be so desires. Baby names are usually not discussed in public until after the child is born, although most parents will choose the child's name before birth. This is based on a cultural belief that one should not prepare for children before they exist in the world, as well as being a folk belief that drawing attention to the pregnancy may result in bad luck for the child-to-be.

During labor, Orthodox men and women will not touch each other due to the bleeding that occurs during labor (see "Family Goals and Priorities" above). Some Orthodox men may choose not to attend the delivery because they observe the Jewish law that prohibits a husband to view his wife's genitals (Robinson 2016). These behaviors should never be interpreted as insensitivity on the part of the husband. Healthcare professionals should initiate the following interventions during the delivery of a child in an Orthodox family: the mother should be given hospital gowns that cover her in the front and back to the greatest extent possible; she may prefer to wear a surgical cap so that her hair remains covered; the father should be given the opportunity to leave during procedures and the birth; and, if he chooses to stay, the mother can be draped so that the husband may sit by his wife without viewing her perineum, including by way of mirrors.

Because he is not permitted to touch his wife, he may offer only verbal support. The female nurse may need to provide all of the physical care. Pain medication during delivery is acceptable. Healthcare providers should be aware that if an Orthodox family stays in the hospital during Shabbat (from Friday night to Saturday night) or other Jewish holidays, during the holiday they will not perform tasks such as writing, turning on and off lights, pushing call buttons from their beds or for elevators, using phones or computers, or others that require the use of electricity.

There are several Jewish folk practices connected to childbirth and newborn babies. Some have the custom of hanging amulets, a red ribbon or piece of yarn from the baby's crib, from the belief that they ward off evil spirits and keep the baby safe. Some Orthodox women may choose to read or place special psalms near the bed while they are in labor. Some families will choose not to announce or even discuss the baby's name until after the male's circumcision or the girl's baby-naming ceremony in the synagogue. Children are often named after a relative to honor that person. Many Ashkenazi Jewish families only name children after deceased relatives, whereas there are Sephardic and Mizrahi traditions for naming after living elders. The baby may be given both a conventional name for the birth records and a Hebrew name used for religious ceremonies (Robinson 2016).

For male infants, *circumcision*, which is both a medical procedure removing the foreskin and a religious rite, is performed. A *brit milah* (more commonly called a *bris*) is a ritual that originates with God's command to Abraham to circumcise his son Isaac as a token of an "everlasting covenant" between God and all of his future descendants (Genesis 17:13). The procedure itself and the accompanying ceremony are performed on the eighth day of life by a person called a **mohel**, an individual trained in the circumcision procedure, asepsis, and the religious ceremony. Although a rabbi is not necessary, it is also possible to have the procedure done by a physician with a rabbi present to say the blessings. Jewish parents who do not observe Jewish laws often still opt for medical circumcision. This preference illustrates how the power of this ritual has endured over thousands of years, even among the secular or religiously unaffiliated.

Attending a *brit milah* is the only mitzvah for which religious Jews must violate the Sabbath so that the *bris* can be completed at the proper time (Robinson 2016). The naming of the newborn son occurs during the *bris* ceremony (girls are named in the synagogue or in the family's home). The *brit milah* is a family festivity, and many relatives and friends are invited. In most cases today, the ceremony is performed in the home or synagogue; however, if the child is still in the hospital, it is important for the hospital to provide a room for a small private party to celebrate. Some in the medical community debate whether circumcision should be practiced, and some label it as "barbaric"; however, it is insensitive and inappropriate to share these opinions with a family desiring a ritual circumcision.

A circumcision may be delayed for medical reasons, including unstable condition owing to prematurity, life-threatening concerns during the early weeks after birth, bleeding problems, or a defect of the penis, which may require later surgery (Robinson 2016). At birth, a child is free of all sin; failure to circumcise carries no eternal consequences for the child should the child die. Although there is no rule against designating godparents for a newborn, it is considered a local and familial preference, rather than a widely shared custom.

22.10 Death Rituals

22.10.1 Death Rituals and Expectations

Death is an expected part of the life cycle for all people. It is a Jewish value that each day is to be appreciated and lived as fully as possible, as expressed by the prayer that asks God to "teach us to number our days so that we may gain a heart of wisdom" (Psalm 90:12). Some Jews start each day with a prayer of gratitude for God having "returned" the soul of life to them. The goal is to appreciate things and people while one still has them. There are different Jewish beliefs about the afterlife. Some views hold that the soul continues to flourish, although many dispute this interpretation because it is not mentioned in the Torah. Most Jews do not dwell on ideas about life after death; rather, their focus is on how to conduct one's life in the present.

Death is defined as the cessation of brain activity and cessation of respiration without support, although there continues to be debate regarding this definition (Bleich and Jacobson 2015). It is believed that while the body disintegrates, the soul or spirit returns to God (Bar-Levav 2014).

The act of *organ donation* is also debated. Some consider it to be an act that saves a life, and thus a mitzvah or good deed (Barilan 2014). Jewish law prohibits organs from being taken from someone who is still alive, which is a medical issue. However, Orthodox authorities increasingly support organ donation out of the principle that saving a life supersedes Jewish law.

Active *euthanasia*, also sometimes referred to as physician-assisted suicide, in which something is given or done that results in the patient's death, is forbidden for Orthodox Jews (Robinson 2016). One of the Ten Commandments is "Thou shalt not kill," and euthanasia is considered murder. A dying person is considered a living person in all respects. Sufficient pain control should be provided, even if it decreases the person's level of consciousness. Withholding food from a patient in order to speed up the dying process is considered active euthanasia and is forbidden.

Passive euthanasia may be allowed, depending on how this act is interpreted. When treatment can no longer cure, reverse or control the condition that will ultimately lead to death, it is no longer regarded as medicine and therefore may be withdrawn (Robinson 2016). Nothing may be used or initiated that prevents a person from dying naturally or prolongs the dying process. Therefore, anything that artificially prevents death (e.g., cardiopulmonary resuscitation or use of ventilators) may possibly be withheld, depending on the wishes of the patient and his or her religious views. Regardless of the patient's decisions, pain control must be maintained.

Taking one's own life is prohibited and is viewed as a criminal and immoral act that violates the prohibition to harm any human being, including oneself. Orthodox Jews believe that *suicide* removes all possibility of repentance. Some believe that Jews who commit suicide should not be afforded full burial honors. They are buried on the periphery of the Jewish cemetery, and mourning rites may not be observed, unless the individual was mentally incompetent. However, some religious communities will reason that the deceased person didn't truly intend to kill him or herself, and therefore did so as an accident, and should be buried according to tradition. The more liberal view, however, is to emphasize the needs of the deceased person's living relatives, and all burial and mourning activities proceed according to the usual traditional rites and wishes of the family. Children are never considered to have intentionally killed themselves and are afforded all burial rights.

It is a crucially important Jewish value to show respect for the dead (in Hebrew, *kavod haMet*). The dying person should not be left alone (Bar-Levav 2014). It is considered a good deed to stay with a dying person, unless the visitor is physically ill or incapable of handling their emotions. Judaism does not have any ceremony similar to the Catholic sacrament of the sick. Any Jew may ask for God's forgiveness for her or his sins; no confessor is needed. However, it is not commonly known that Jews also have a personal confessional prayer called ***Viddui*** (Hebrew: confession) in which a person admits to their sins before God, and which is recited when death is imminent. It may be said by the dying person or by somebody for her or him. Some Jews also feel solace in saying the *Shema* ("Hear, O Israel"), in Hebrew or English. This prayer confirms the Jewish belief in one God.

Reflective Exercise 22.2

Samuel is an older Jewish adult who is in the final stage of terminal cancer. His nurse has developed a close relationship with him, but has never discussed religion. Samuel states that he is not afraid of dying. The nurse asks him if he wants a clergy in order to make confession; he says no and appears annoyed. Samuel decides he wants to end all treatments, but the nurse tries to encourage him to continue the treatments in the hope that his cancer will enter remission. Then the nurse tells Samuel's wife, "God doesn't give you anything that you can't handle" and "This is your cross to bear." The nurse is perplexed as to why her well-intentioned comfort modalities are not effective with either the patient or his wife.

1. What does the nurse need to understand about a Jew's relationship with God?
2. Why are comments such as one's 'cross to bear' or the suggestion of confession inappropriate to use with a Jewish patient?
3. What does the nurse need to understand about the Jewish view of death?
4. Given personal preferences and cultural beliefs, what interventions might be most helpful to the family in the final stage of life?

At the time of death, the nearest relative can gently close the deceased person's eyes and mouth, and cover their face with a sheet to maintain the person's dignity. The body is treated with respect and revered for the function it once fulfilled. Healthcare providers need to ask the closest relative of the deceased specifically about the practices to follow after death. Healthcare providers who have acquired some familiarity with Jewish practices associated with death can be very helpful with their patients and families. This is considered to be one of the most important *mitzvot* (good deeds) in Judaism. Their informed presence will continue to benefit all involved, as the long process of integrating loss into their lives continues.

According to traditional burial practices, the dead body should not be left alone from the time of death until the funeral, so as not to leave it defenseless (Lieber 2012). Jews who opt to have traditional burial rites for the deceased may arrange for the body to be transferred from the hospital to the care of a Jewish burial society (in Hebrew, *khevra kadisha*). Prior to burial, a specially trained person in the burial society will perform a Jewish ritual known as *taharah*, or purification, in which they gently wash and wrap the body in a linen shroud. Many Jewish communities sponsor burial societies whose members are trained to perform all of the traditional burial rites. Healthcare providers can help the deceased person's family to locate a burial society by consulting with a local synagogue or rabbi. At the funeral, the immediate family may tear a piece of their clothing or wear on their lapel a small black ribbon that is torn (Robinson 2016).

Autopsy is usually not permitted among religious Jews because it results in desecration of the body, and it is important that the body be interred whole. Allowing an autopsy might also delay the burial, something that is not recommended. Conversely, autopsy is allowed if its results would save the life of another patient (Beal and Collins 2019). Many branches of Judaism currently allow an autopsy if it is required by law; the deceased person has willed it; or it saves the life of another, especially an offspring. The body must be treated with respect during the autopsy.

Any attempt to hasten or retard decomposition of the body is discouraged. Cremation is prohibited because it unnaturally speeds the disposal of the dead body. Embalming is prohibited because it slows the process of returning the body to the earth. However, in circumstances in which the funeral must be delayed, some embalming may be approved. Cosmetic restoration for the funeral is discouraged.

Jewish *funerals* and burials follow certain practices; they usually occur within 24–48 h after the death. The funeral service is directed at honoring the departed by speaking only well of him or her. The practice of having flowers at the funeral and cemetery was a Christian custom used to offset the odor of decaying bodies; therefore, some Jews choose not to have flowers. In lieu of flowers, a donation to a charity in the name of the deceased may be a more meaningful tribute.

Simplicity is characteristic of the Jewish burial. The casket is often a simple pine box with no ornamentation. The body may be wrapped only in a shroud to ensure that the body and casket decay at the same rate. A wake or viewing is *not* part of a Jewish funeral. The prayer said for the dead, the *kaddish*, is usually not said alone, but is recited in the company of others. Among Orthodox Jews, this prayer requires a quorum of at least 10 Jewish men (a *minyan*, in Hebrew); other branches of Judaism allow women to be counted as part of the quorum. The prayer says nothing about death, but rather, praises God and reaffirms one's own faith. During the funeral ceremony, mourners are all invited to shovel a bit of dirt onto the casket. This is considered to be the last loving act one can perform for the deceased in this world.

A funeral according to *halakhah* (Jewish law) emphasizes the finality of death. It is believed that the funeral is the moment at which the soul departs the body. After the funeral, the focus of concern shifts from the deceased to the mourners. Mourners are welcomed at a family home. A customary greeting is to tell the mourner, "May you be comforted among the mourners of Zion and Jerusalem." It is a custom to ritually wash one's hands after leaving a cemetery, as a symbol of cleansing the impurities associated with contact with the dead. It is also a custom that water is poured over the hands and into a drain or onto the ground outside of the house, based on a belief that the tragedy of death will not be passed along in reused water. At the home, a meal is served to all the guests. This "meal of condolence" or "meal of consolation" is traditionally provided by the neighbors and friends; it frequently includes hard-boiled eggs to remind all of the continuing cycle of life.

Shiva (Hebrew for "seven") is the 7-day period that begins with the burial. *Shiva* helps the surviving individuals face the actuality of their loved one's death. During this period when the mourners are "sitting *shiva*," they do not work. When healthcare providers are the ones experiencing the loss, it is important for supervisors to understand these Jewish mourning customs. In some homes, mirrors are covered, with the intent to decrease the focus on one's personal vanity; no activity is permitted to divert attention from thinking about the deceased, and evening and morning prayer services may be conducted in the closest relative's home. Condolence calls, cards or letters, and the giving of consolation are appropriate during this time (Bar-Levav 2014).

After *shiva*, the mourning period varies based on who has died. Mourning for a relative lasts 30 days, and for a parent, 1 year. Judaism does not support prolonged mourning. A tombstone is erected within 1 year of the death, at which time a graveside service is held. This is called an *unveiling*. According to the Jewish calendar, the anniversary of the death is called *yahrzeit*, and at this time, special 24-h candles are lit and the *kaddish* prayer is said.

Mourning is not required for a fetus that is miscarried or stillborn. This may also be true for any premature infant who dies within 30 days of birth. However, most parents do mourn for both fetal loss and premature death and should be allowed to follow the mourning practices described earlier to whatever degree they choose (Goldstein 2019). Although some believe that the baby should be named, not all of the traditional burial customs are followed.

Orthodox Jews follow the practice of burying any limbs that have been amputated before death in the person's future gravesite. Because the limb was part of the person's body, it is to be re-buried with the person when they die (Chabad.org 2019b). No mourning rites are required. In the case of an amputation, the healthcare provider may need to assist with arrangements for burial of the body part.

One noticeable difference between Jewish and non-Jewish cemeteries is that Jews will often leave small stones on the top of the gravestone. While the original reason for this practice is unknown, it is now used as a token to indicate that the grave has been visited and the person has not been forgotten.

22.10.2 Responses to Death and Grief

The period following a death has discrete segments. These distinct segments are intended to assist mourners in their gradual adjustment to the loss. The period of time between the death and the burial is short, and provides time for the emotional reaction to the death. The burial may be delayed only if required by law, if relatives must travel great distances, or if it is the Sabbath or another holiday during which work is not permitted. Mourners are absolved from praying during this time. Crying, anger, and talking about the deceased person's life are acceptable. During the 7 days of *shiva*, the mourner sets the tone and initiates conversation. These discrete periods of mourning are meant to limit the formal mourning process to not more than 30 days for a relative and 1 year for parents. Mourning in this discussion only relates to the laws of mourning and not the act of grieving the loss of the individual, which of course will vary and is likely to continue well beyond the formal mourning period.

22.11 Spirituality

22.11.1 Dominant Religion and Use of Prayer

Judaism is over 4000 years old, and is the oldest of the three monotheistic religions, with elements that Christianity and Islam later adopted. Its scripture is known as the Hebrew Bible, or *Tanakh*, in Hebrew, and includes three sections. The first section, the Torah, includes five books, and is sometimes called the Five Books of Moses, based on the traditional view that God revealed the Torah to Moses, who then wrote it down. Its five books include Genesis, Exodus, Leviticus, Numbers, and Deuteronomy. The Torah contains stories of Creation, the Jewish people's founding patriarchs and matriarchs, the exodus of the ancient Israelites from slavery in Egypt, and their wanderings toward the land of Canaan that God had promised to Abraham and his descendants. The Torah also contains all of the laws from which Judaism would be developed. The second section of the Hebrew Bible is called "Prophets" and consists of chronicles of the prophets Jeremiah, Isaiah, Deborah, and several others, as well as royal figures such as the kings Solomon and David. The third section, the "Writings," includes the books of Psalms, Lamentations, Ruth, Esther, and others. The Hebrew Bible is known to Christians as the *Old Testament*. Jews do not regard the Christian New Testament as sacred.

The Jewish people have survived over a long history, also enduring some periods of oppression. They have been expelled from different localities and countries at many different times, and in some places were forbidden to practice their religion. They have also faced both official and unofficial kinds of discrimination, including in the US. Jews have been limited to certain occupations, housing sectors, and admission to college. In extreme cases anti-Jewish hatred became violent, involving mass killing and extermination. Many Jews have immediate family members who were killed in the Holocaust during the Second World War in Europe. Yet, despite periods of persecution and crisis, Judaism and Jewish people have nonetheless survived and even flourished.

Jewish theology is based on a belief in one God as the Creator of the universe. The watchword of the faith is found in Deuteronomy (6:4): "Hear O Israel, the Lord is our God, the Lord is One." God is considered to be all-powerful and ineffable, and is never depicted in human form. Making and praying to statues or graven images are forbidden by the second commandment, which forbids the worship of idols.

The spiritual leader in Judaism is the **rabbi** (teacher). He (or she, in liberal branches) is regarded as an interpreter and teacher of Jewish law and ethics. Rabbis are not considered to be any closer to God than any other Jews. All Jews pray directly to God. They do not need the rabbi to intercede, to hear confession, or to grant atonement.

For Orthodox Jews, daily prayer is the mainstay of their spiritual discipline; it is an obligation (Lieber 2012). These prayers, which can be done in a public forum or privately, are required at specific times throughout the day. Prayers are usually personal and sometimes said collectively on behalf of all those present.

The following are some of the major principles that guide Judaic bioethics:

- Humanity's purpose on earth is to live a righteous life according to certain God-given commandments. These include kindness to the needy, benevolence, and compassion for the suffering.
- Life possesses enormous intrinsic value, and its preservation is of great moral significance.
- All human lives are equal.
- Our lives are not our own exclusive private possessions.
- Justice, truth and peace are essential (Jewish Ethics 2019).

The Torah that is used in synagogues for public readings is called a Sefer Torah. It is handwritten in Hebrew on scrolls made of animal skin parchment, which are bound to two wooden dowels. These scrolls are usually covered in fine ornamented fabrics and kept in the "Holy Ark" within each synagogue under an "eternal light." The Torah is the foundational text that gives laws and provides guidance on human life. Portions of the Torah are read each week from the Sefer Torah during Sabbath services. When the Torah scrolls are removed from the ark, the congregants are expected to show reverence by standing and refraining from idle talk.

There are 613 commandments derived from the Torah (also called *mitzvot*); these and the subsequent oral law derived from the biblical statutes determine Jewish law, or *halakhah*. These commandments ask for a commitment in behavior and also address ethical concerns. Because they are derived from the Torah, it is believed that the commandments reflect the will of God, and religious Jews feel it is their duty to carry them out in order to fulfil their covenant with God. This makes Judaism not only a religion but also a way of life.

The current practice of Judaism spans a wide spectrum. There is only one religion, but there are multiple branches or denominations of Judaism. The Orthodox are the most traditionally observant. They maintain the most conservative interpretation of Jewish laws while also living as members of society. They observe Jewish holidays throughout the year, as well as the weekly Sabbath, often by attending the synagogue on Friday evening and Saturday morning.

In most Orthodox synagogues, the services are conducted primarily in Hebrew, and men and women sit separately, in some cases with some type of divider between them. During the Sabbath and holidays, they will abstain from many different kinds of work. These include spending money, cooking, driving, writing, and using phones, computers, lights, elevators, and other forms of technology. Orthodox Jews also observe the Jewish dietary laws. In addition, they often wear clothing that identifies them as observant. Men wear head coverings at all times in reverence to God, including hats, or a ***yarmulke*** or *kippah*. They will also wear a special garment with long strings, or *tzitzit*, under their shirts, as commanded in Deuteronomy, as a reminder of the laws of the Torah. Orthodox women usually cover their hair with hats, scarves or wigs, and wear modest clothing, including long sleeves, and skirts or dresses.

Orthodox Jewish men and some Conservative men and women pray with *tefillin*, or phylacteries, during morning prayer services. These are two small black boxes connected to long leather straps that contain parchments with biblical passages. These straps are wrapped around the arms and forehead as reminders of the laws of the Torah. The *tallis* (or *tallit*) is a rectangular prayer shawl with fringes. This is also used only during prayer but is frequently used by both Conservative and Orthodox Jews.

A *mezuzah* is a small elongated container that is posted on door frames of the entry door to the house and sometimes on doors of rooms inside as well. It contains a small, rolled parchment inscribed with the verses that begin "Hear O Israel" (*Shema, Yisrael*) from Deuteronomy 6:4–9. This passage states God's commandment that Jews must post the words of the *Shema* on the doorways of their house. The mezuzah is often marked with a Hebrew letter on the outside that denotes a symbol of God. Some interpret the mezuzah as a sign of God's protection, presence, and commandments, and the Jew's duties to perform those commandments; some people have the custom to reverently touch a mezuzah when entering a doorway. Some individuals also wear a small *mezuzah* as a necklace. Other prominent religious symbols include the *Star of David*, a six-pointed star that has been a symbol of the Jewish community since the 1350s, and the *menorah* (candelabrum originally lit in ancient temples).

The Conservative branch of Judaism is not quite as strict in its interpretations of Jewish law. Whereas Conservative Jews observe most of the *halakhah*, they also interpret the laws in ways that are more compatible with contemporary life. According to DeLange (2010), "The aim of the founders [of Conservative Judaism] was to embrace the liberalism and pluralism of American Reform [Judaism] while safeguarding traditional practice" (p. 13). Many drive to the synagogue on the Sabbath, for example, and men and women sit together during prayer. Many Conservative Jews keep a kosher home, but they may or may not follow all of the dietary laws outside the home. Women can be ordained as rabbis and are counted in a *minyan*, the minimum number of 10 required for communal prayer. Whereas a yarmulke is customarily worn in the synagogue, it is optional outside of that environment.

The liberal movement is called *Reform* Judaism. Reform Jews claim that the laws of the Torah should be practiced in a way that is relevant and meaningful to the time and place in which one lives. Reform Jews also place an emphasis on practicing the moral laws of the Torah. They practice fewer rituals, although they frequently have a *mezuzah* for their homes, celebrate the holidays, and have a strong ethnic identity. They consider education and ethics to be of paramount importance in one's personal life and often integrate Jewish religious values with American social and political norms. Reform Jews are known for engaging in progressive social movements on behalf of causes such as civil rights, the environment, and the state of Israel. They may or may not follow the Jewish dietary laws, but they may have specific foods (e.g., pork) that they abstain from eating. Men and women share full equality in all religious practices.

Of all the types of Jews, the so-called ultra-Orthodox are perhaps the most visually recognizable. They usually live, work, and study within close-knit neighborhoods, often in walking distance to synagogues, schools, and businesses. The men are visually identifiable by their full beards, side locks of hair around the ears (*payes*), black hats (or on special occasions, fur hats known as *streimels*) and dark clothing (often suits). Women wear modest clothing. Married women will cover their hair with wigs, scarves and sometimes a hat as well.

A newer denomination, Reconstructionism, is a product of the twentieth century American Jewish experience. It views Judaism as an evolving religion of the Jewish people and seeks to adapt Jewish beliefs and practices to the needs of the contemporary world. It views Judaism as a civilization and a culture, with its own distinct customs, common history, and languages (Robinson 2016). It attempts to bridge the Conservative principles of religious observance with progressive ideological imperatives of Reform Judaism.

The terms "secular" or "cultural" Jew refer to individuals who identify as Jewish but claim to have no religion. It is estimated that one in five American Jews identify as secular (Pew Research Center 2013). Secular Jews often participate in the cultural components of the religion, including holidays, but may or may not believe in God, or belong to a synagogue. They have not rejected their Jewish identity or their attachment to the Jewish people; rather, they will engage with aspects of secular Jewish culture (including Israeli culture) such as philosophy, literature, music, film, fine arts, and more. They often remain very supportive of Jewish causes without necessarily holding a belief in God as a supernatural being. Many Jews, however, do not indicate any affiliation.

The Jewish house of prayer is called a **synagogue**, **temple**, or **shul** (in Yiddish). It is never referred to as a "church." Jews may pray alone, or communally when 10 Jews over the age of 13 who have had their *bar mitzvah* are gathered together for prayer. This group is called a *minyan*. Traditionally Jews pray three times a day: morning, late afternoon, and evening. They wash their hands and say a prayer on awakening in the morning and before meals.

Religious patients in hospitals may want their prayer items (*yarmulke* or *kippah*, *tallit, tzitzit, tefillin*) and may request a minyan. Hospital policies regarding the number of visitors in the sick person's room may have to be altered in such instances.

One of the most common religious practices related to patients involves "visiting the sick" (*bikkur cholim*). This commandment is one of the social obligations of Judaism and ensures that Jews look after the physical, emotional, psychological, and social well-being of others in their community, and provides hope as well as companionship (Robinson 2016). Moreover, one must consider the patient's welfare and not stay too long, tire the patient, or come only to satisfy one's own needs.

22.11.2 Meaning of Life and Individual Sources of Strength

The preservation of life is one of Judaism's greatest priorities (Orthodox Union 2018). Even the laws that govern the Sabbath and other holidays may be broken if doing so can help to save a life, including one's own. Healthcare providers should do everything in their power to save a life. Each individual is considered to have been created in God's image, and the individuality of the human experience is one of the precepts of the faith. Good health is considered an asset.

22.11.3 Spiritual Beliefs and Healthcare Practices

The fourth of the Ten Commandments is to "remember the Sabbath day and keep it holy." The Sabbath begins 18 min before sunset on Friday. Lighting candles, saying prayers over challah and wine, and participating in a festive Sabbath meal usher in this holy day every week. It ends 42 min after sunset (or when three stars can be seen in the sky) on Saturday, with a service called *Havdalah* (Hebrew for separation). Many Jews observe the Sabbath as a release from weekday concerns and pressures. During this time, religious Jews engage in congregational prayer, study and socializing. They do no manner of work, including answering the telephone, operating any electrical appliances, handling money, driving, or operating a call bell from a hospital bed.

If an Orthodox patient's condition is not life-threatening, medical and surgical procedures should not be performed on the Sabbath or on holy days. Although synagogue attendance is an obligation for Orthodox men on the Sabbath and holidays, extenuating circumstances such as illness or bad weather are legitimate reasons for not attending the services. Although the Sabbath is holy, and holidays that require fasting are part of Jewish law, matters involving human life take precedence over them. Therefore, a gravely ill person and the work of those who need to save her or him are exempted from following the commandments regarding the Sabbath and fast days. This even includes eating non-kosher food if there is the slightest chance that human life will be saved.

In addition to the Sabbath, a number of Jewish holidays are celebrated with special traditions. Rosh Hashanah (Jewish New Year) and Yom Kippur (Day of Atonement) are called the *High Holy Days* or *Days of Awe*, and usually occur in September or early October. They mark a 10-day period of self-examination and repentance. This is a time when Jews apologize for wrongs they have committed knowingly or unknowingly against others during the previous year. According to tradition, during these 10 days, each person stands before God, and their fate for the coming year is determined. Thus, the greeting during this time is "May you be written and sealed into the book of life for a good year." Yom Kippur is the most solemn of the Jewish holidays.

On Rosh Hashanah it is customary to eat apples and honey to symbolize the wish for a sweet year, and on Yom Kippur, one is commanded to fast for a day as a form of affliction, with the goal to atone and purify oneself. As noted before, fasting for Yom Kippur may be broken for reasons of critical illness, for labor and delivery, or for children under the age of 12. The liturgy for Rosh Hashanah and Yom Kippur includes the blowing of the *shofar* (a ram's horn), which is thought to represent a reminder for individuals to repent or atone for their sins before God.

Other major holidays include Passover, also called the Feast of the Unleavened Bread, which lasts 8 days and celebrates the Exodus from Egypt and freedom from slavery. Sukkot, a festival of the harvest in which individuals may sleep or eat in temporary huts built outside their homes for a week, concludes with Simkhat Torah (rejoicing the Torah). Shavuot takes place exactly 7 weeks after the end of Passover and celebrates God's revelation of the Torah to the wandering Israelites at Mt. Sinai. Other holidays include Chanukah, an 8-day holiday during the winter; and Purim, usually in the late winter or early spring. Both of these holidays celebrate the survival of Judaism and Jewish people in ancient times. Table 22.1 provides a list of Jewish holidays for the years 2020 through 2026.

Reflective Exercise 22.3

Emma is 8 years old and hospitalized for an acute infection requiring a central line and antibiotics. It is December 24, and the hospital is decorated for Christmas. Multiple organizations have paraded through the unit with someone dressed as Santa Claus handing out presents. Emma thinks it is "neat" to get all these presents. However, when the staff ask her if she is excited that Santa Claus is there, she says "no." When they ask if she misses putting up a Christmas tree, she also says "no." The nurse is even more perplexed when she finds out that there are no decorations in the home at all, including a wreath on the door. The nurse asks if she wants to hang up a stocking, and, being Jewish, Emma does not know what they are talking about. When Emma's mother hears the nurse telling her that this is the night to celebrate Jesus's birth, she becomes angry. When the mother explains that they do not believe in Jesus, the nurse responds, "I feel sorry for you."

1. What discussion could have been held with Emma's parents in the days before Christmas?
2. How might Emma still enjoy the "presents" without attaching it to religion?
3. It is disrespectful of the nurse to say that she feels sorry for Emma's family. How might the nurse have handled the situation more professionally?
4. It is inappropriate to compare Christmas with Chanukah. How else might the nurse have recognized the distinct features of Emma's Jewish heritage?

22.12 HealthCare Practices

22.12.1 Health-Seeking Beliefs and Behaviors

According to Jewish law, all people have a duty to keep themselves in good health. This encompasses physical and mental well-being and includes not only the need to seek early treatment for illness but also prevention of illness. Medical treatments are endorsed as a way to "cultivate and preserve the body" (Barilan 2014, p. 43). It is hoped that having a healthy body can assist one with the pursuit of their spiritual goals. Measures that prevent disease or illness are promoted as long as the benefits outweigh the risks. Among ultra-Orthodox Jews, there are some who oppose mass screening to identify hidden asymptomatic conditions (Barilan).

Judaism teaches its members to "choose life" (Deuteronomy 30:19) and places the highest value on preserving human life (Orthodox Union 2018). To refuse lifesaving medical treatment is seen by some as committing suicide, as one is choosing death over life. All denominations within Judaism recognize that religious requirements may be laid aside if a life is at stake or if an individual has a life-threatening illness. However, once it is clear that an individual is dying and that medical treatment is no longer working, individuals may choose not to interfere with death. Hospice care is fully consonant with Jewish beliefs.

In Orthodox denominations of Judaism, taking medication on the Sabbath or other holidays that is not necessary to preserve life may be viewed as "work" (i.e., an action performed with the intention of bringing about a change in existing conditions) and is unacceptable. As a possible result of this belief, some people with conditions such as asthma may not recognize the severity of their condition; they may also be unaware of Jewish laws that *do* allow them to take their necessary medications. These patients need to be taught about the potential life-threatening consequences of their condition as well as the exceptions to Jewish law that permit them to take their medications. In the Jewish faith, all individuals have value regardless of their condition. This applies to all individuals, including those with developmental disabilities, AIDS or any other conditions that often have social stigmas or taboos attached to them.

Judaism promotes the goals of preventing disease, restoring health, and prolonging life. Therefore therapeutic genetic engineering via gene therapy is permitted if it prevents disease and disability in future generations. Because genomics is the interaction between one's genes and the environment, improving the environment in order to improve healthy lives is also valued.

22.12.2 Responsibility for Health Care

Although it is the responsibility of healthcare providers to heal, individuals must seek the services of the healthcare provider to ensure a healthy body. Once individuals have the knowledge necessary to impact on their healing, it is their obligation to do so unless the treatments are overly experimental (My Jewish Learning n.d.). Deuteronomy 4:15 states to take very good care of oneself. This includes the discovery of new medications and treatments to eliminate or modify disease and suffering; therefore, humanity can help God to heal and cure. Jews also believe that God gives humans freedom of choice.

Because the preservation of life is paramount, all ritual commandments are waived when danger to life exists. Physical and mental illnesses are legitimate reasons for not fulfilling some of the commandments. Because adult Jews tend to have more advanced education than that of the general population, they may be familiar with and interested in trying the newest available treatments. This could have both positive and negative consequences. The literature reveals no studies regarding Jews' self-medicating practices.

22.12.3 Folk and Traditional Practices

Jewish folk practices are historically and biblically based. Jews have adopted and adapted to customs from the cultures and countries in which they have lived during the centuries of the *Diaspora*. Specific practices are explained in the sections of this chapter on Nutrition and Spiritual Beliefs and Healthcare Practices.

22.12.4 Barriers to Health Care

There are no major barriers to health care that are specific to Jews in contemporary America, apart from the universal problems of lack of health insurance or being underinsured. The Jewish community endeavors to help those in need, including new immigrants, and works to assist fellow Jews in becoming self-sufficient. Community organizations, especially the Jewish Federation in each state, include programs to help individuals in need; these agencies are ubiquitous today wherever Jews live in the United States.

22.12.5 Cultural Responses to Health and Illness

The verbalization of pain is acceptable and common. Individuals want to know the reason for the pain, which they consider just as important as obtaining relief from it. The sick role for Jews is highly individualized and may vary among individuals according to the severity of symptoms. As the family is central to Jewish life, family members share the emphasis on maintaining health and assisting with individual responsibilities during times of illness.

Jews are well represented among physicians, psychoanalysts, psychiatrists, and psychologists. The maintenance of one's mental health is considered just as important as the maintenance of one's physical health. This designation includes psychiatric conditions.

22.12.6 Immunizations

There is nothing in Jewish law that prohibits immunizations. To the contrary, the Jewish population is generally very supportive of immunizations (Turner 2017) and the prevention of disease. In November, 2018, the nation's largest Orthodox Jewish organization, the Orthodox Union (2018), along with the Rabbinical Council of America, came out supporting the administration of all vaccines for healthy children according to the timetable recommended by their pediatrician. During the measles outbreak in 2019, there was a focus on the epicenter of the outbreak among communities of ultra-Orthodox Jews in New York; these groups were receiving a significant amount of mis-information from an anonymous anti-vaccine group (Bursztynsky 2019). The Orthodox Union, in contrast, indicated that the measles outbreak could have very serious consequences.

Reflective Exercise 22.4

A young family brings their toddler to your clinic for routine well-child care. It is time for the child's MMR vaccine (measles, mumps, rubella). The mother refuses, stating it is against her religion (Jewish).

1. What questions do you want to ask?
2. What information do you want to provide?

22.12.7 Blood Transfusions and Organ Donation

Jewish law views organ transplants from four perspectives: the recipient, the living donor, the cadaver donor, and the dying donor. Because life is sacred, if the recipient's life can be prolonged without considerable risk, then transplant is favorably viewed. For a living donor to be approved, the risk to the life of the donor must be considered. One is not obligated to donate a body part unless the risk is small. Examples include kidney and bone marrow donations (Barilan 2014; Kochen 2014). If there is more than a 50% chance of either the patient or the donor dying, the organ donation is not permitted. The action of donating an organ to save another is considered a great *mitzvah* (a good deed). According to ancient scripture, saving the life of another person is a great act that one should perform if they can (Kochen 2014).

Conservative and Reform Judaism approve of using the flat EEG as the determination of death so that organs, such as the heart, can be viable for transplant. Burial may be delayed if organ harvesting is the cause of the delay. However, among other groups, this definition of death remains controversial (see discussion in earlier section "Death Rituals"). Healthcare providers may need to assist Jewish patients to obtain a rabbi's opinion when they are making a decision regarding organ donation or transplant.

The use of a cadaver for transplant is generally approved if it is to save a life. No one may derive economic benefit from the corpse. Although desecration of the dead body is considered purposeless mutilation, this does not apply to the removal of organs for transplant. Use of skin for burns is also acceptable.

22.13 Health-Care Providers

The ancient Israelites are credited with promoting hygiene and sanitation practices and basic principles for public health care. More recently, Jewish values have played a role in the development of modern public health programs. Lillian Wald, a well-known American Jewish nurse, was inspired by the religious principles of visiting the sick and disease prevention to develop the Henry Street Settlement in New York City, which became a prototype of public health nursing for those in need and initiated the idea of school nursing. Jewish physicians have made significant contributions, including the development of immunizations (Baruch Blumberg [hepatitis B]; Jonas Salk and Albert Sabin [polio]); as well as psychotherapy (Aaron Beck [cognitive therapy] and Sigmund Freud [psychoanalysis]).

22.13.1 Status of Health-Care Providers

Physicians are held in high regard. Whereas physicians must do everything in their power to preserve life, they are prohibited from initiating measures that prolong the act of dying. Once standard therapy has failed, or if additional treatments are unavailable, the physician's role changes from that of curer to providing supportive care, such as food and water, good nursing care, and optimal psychosocial support.

Reflective Exercise 22.5

Your nurse colleague is Jewish. What questions do you want to ask to be culturally congruent in your working relationship?

You are working on the scheduling for the next 6 weeks and she has requested to be off for Rosh Hashanah and Yom Kippur. What further information do you need from her to complete your schedule?

You are planning a nursing conference for the nurses in your town. As you plan the food options, what needs to be considered without knowing the cultural backgrounds of those who will be participating? What do you need to consider when you select a date for the conference? What needs to be said to the person who wants to say 'Grace' before the lunch meal is served?

Acknowledgements The authors would like to thank Rabbi Nick Renner for his thorough and thoughtful review of the chapter content.

References

American Nurses Association (2015) Nursing: scope and standards of practice, 3rd edn. American Nurses Association, Silver Spring MD

Barilan Y (2014) Jewish bioethics. Cambridge University Press, New York

Bar-Levav A (2014) Jewish attitudes toward death: a society between time, space and texts. In: Reif S, Lehnardt A, Bar-Levav A (eds) Death in Jewish life. Walter de Gruyter GmbH, Berlin/Boston

Baruch M, Benarroch A, Rockman GE (2015) Alcohol and substance use in the Jewish community: a pilot study. J Addict 2015:763930–763934. https://doi.org/10.1155/2015/763930

Bassok M (2017) Average Israeli family has 3.72 people; most common name is Cohen. Haaretz. https://www.haaretz.com/israel-news/.premium-average-israeli-family-has-3-72-people-most-common-name-is-cohen-1.5440222

Beal S, Collins K (2019) Religions and the autopsy. https://emedicine.medscape.com/article/1705993-overview#a3

Benson E (1994) Jewish nurses: a multicultural perspective. J N Y State Nurses Assoc 25(2):8–10

Bernstein R, Fishman S (2015) Judaism as the "third shift". In: Fishman S (ed) Love, marriage, and Jewish families. Brandeis University Press, Waltham MA, pp 196–217

Bleich JD, Jacobson A (2015) Jewish law and contemporary issues. Cambridge University Press, New York

Bloom's Syndrome Association (2019) About bloom's syndrome. https://www.bloomssyndromeassociation.org/page/aboutbloomssyndrome

Bursztynsky J (2019) Jewish nurses debunk anti-vaxxer misinformation as measles spreads in NYC ultra-orthodox community. Health Sci. https://finance.yahoo.com/news/jewish-nurses-debunk-anti-vaxxer-143056182.html

Chabad.org (2019a) Jewish recovery. https://www.chabad.org/library/article_cdo/aid/714274/jewish/Jewish-Recovery.htm

Chabad.org (2019b) The internment in judaism. https://www.chabad.org/library/article_cdo/aid/281565/jewish/The-Interment-in-Judaism.htm

Cohn-Sherbok D (2010) Judaism today. Continuum International, London

DeLange N (2010) An introduction to Judaism. Cambridge University Press, Cambridge

DellaPergola S (2018) World Jewish population, 2018. In: Dashefsky A, Sheskin I (eds) The American Jewish year book, 2018. Springer, Dordrecht, pp 361–452

Dor Yeshorim (2019) Our mission. https://doryeshorim.org/our-mission/

Dubov M (2019) Why do observant Jews have so many kids? https://www.chabad.org/library/article_cdo/aid/4372320/jewish/Why-Do-Observant-Jews-Have-So-Many-Kids.htm

Forward (2014) All about genetic diseases that strike Sephardic Jews. https://forward.com/culture/203321/all-about-genetic-diseases-that-strike-sephardic-j/

Glausiusz J (2015) Jewish genetics: 75% of Jews are lactose intolerant and 11 other facts. Haaretz. https://www.haaretz.com/science-and-health/.premium-12-facts-about-jewish-genetics-1.5375744

Goldstein Y (2019) The Laws & customs of mourning, vol 1. Lachamei Hagetaot, Safed

Grinzaid K, Page P, Denton J, Ginsberg J (2015) Creation of a national, at-home model for Ashkenazi Jewish carrier screening. J Genet Couns 24:381–387

Harvard Health Publishing (2018) Tay-Sachs disease. Harvard Medical School, Harvard. https://www.health.harvard.edu/a_to_z/tay-sachs-disease-a-to-z

Jewish Ethics (2019) Wikipedia. https://en.wikipedia.org/wiki/Jewish_ethics

Jewish Views on Contraception (2019). https://en.wikipedia.org/wiki/Jewish_views_on_contraception

Jewish Virtual Library (2019) Jewish population of the United States by state (1899—Present). https://www.jewishvirtuallibrary.org/jewish-population-in-the-united-states-by-state

JINFO.ORG (2019) Jewish nobel prize winners. http://www.jinfo.org/Nobel_Prizes.html

Johns Hopkins (2018) Specific to individuals of Ashkenazi Jewish ancestry. The Sol Goldman Pancreatic Cancer Research Center, Baltimore MD. https://pathology.jhu.edu/pancreas/PartAJA.php?area=pa

Kochen M (2014) Organ donation and the divine lien in talmudic law. Cambridge University Press, New York

Krasner J (2015) We all still have to potty train: same-sex couple families and the American Jewish Community. In: Fishman S (ed) Love, marriage, and Jewish families. Brandeis University Press, Waltham MA, pp 73–107

Laufer-Ukeles P (2016) Multiplying motherhood: gestational surrogate motherhood and Jewish law. In: Greenspoon L (ed) Mishpachah: the Jewish family in tradition and in transition. Purdue University Press, West Lafayette IN, pp 235–268

Lieber A (2012) The essential guide to Jewish prayer and practices. Penguin, New York

Loike J (2016) The Jewish perspective in creating human embryos using cloning technologies. In: Greenspoon L (ed) Mishpachah: the Jewish family in tradition and in transition. Purdue University Press, West Lafayette IN, pp 221–233

Margalit Y (2017) The Jewish family: between family law and contract law. Cambridge University Press, Cambridge

Mayer S (2009) Nursing in the United States. The encyclopedia of Jewish women. https://jwa.org/encyclopedia/article/nursing-in-united-states

My Jewish Learning (2019) The tattoo taboo in Judaism. https://www.myjewishlearning.com/article/the-tattoo-taboo-in-judaism/

My Jewish Learning (n.d.) Jewish health & healing practices. https://www.myjewishlearning.com/article/jewish-health-healing-practices/

National Gaucher Foundation (2019). https://www.gaucherdisease.org/about-gaucher-disease/what is/

National Organization for Rare Disorders (2019) Dysautonomia, familial. https://rarediseases.org/rare-diseases/dysautonomia-familial/

Orthodox Union (2018) Statement on vaccinations from the OU and rabbinical council of America. https://www.ou.org/news/statement-vaccinations-ou-rabbinical-council-america/

Orthodox Union (2019) Kosher food: the kosher primer. https://oukosher.org/the-kosher-primer/

Pew Research Center (2013) A portrait of Jewish Americans. https://www.pewforum.org/2013/10/01/jewish-american-beliefs-attitudes-culture-survey/

Queensland Health Guideline (2019) Medicines/pharmaceuticals of animal origin. Document number: QH-GDL-954:2013. https://www.health.qld.gov.au/__data/assets/pdf_file/0024/147507/qh-gdl-954.pdf

Robinson G (2016) Essential judaism: a complete guide to beliefs, customs, and rituals. Atria, New York

Rosner F (n.d.) Jewish medical ethics: genetic screening & genetic therapy. Jewish Virtual Library. https://www.jewishvirtuallibrary.org/genetic-screening-and-genetic-therapy-in-judaism

Santos M, Gomes C, Torres J (2018) Familial and ethnic risk in inflammatory bowel disease. Ann Gastroenterol 31(1):14–23

Sartori J (2016) Modern families: multifaceted identities in the Jewish adoptive family. In: Greenspoon L (ed) Mishpachah: the Jewish family in tradition and in transition. Purdue University Press, West Lafayette IN, pp 197–218

Schatzberg A, Nemeroff C (2009) Textbook of psychopharmacology. The American Psychiatric Publishing, Arlington VA

Sheskin I (2019) Jews in America: trends in American Jewish demography. https://www.jewishvirtuallibrary.org/trends-in-american-jewish-demography
Shi L, Webb B, Birch A, Elkhoury L, McCarthy J, Cai X, Oishi K, Mehta L, Diaz G, Edelmann L, Kornreich R (2017) Comprehensive population screening in the Ashkenazi Jewish population for recurrent disease-causing variants. Clin Gen 91(4):599–604. https://doi.org/10.1111/cge.12834
St. Jude Children's Research Hospital (2019) Fanconi anemia. https://www.stjude.org/disease/fanconi-anemia.html
Templer D, Tangen K (2014) Jewish population percentage in the U.S. states: an index of opportunity. Comprehen Psychol 3:8. https://doi.org/10.2466/17.49.CP.3.8
Turner A (2017) Jewish decisions about childhood vaccinations: the unification of medicine and religion. Paediatr Health 5:1. http://www.hoajonline.com/journals/pdf/2052-935X-5-1.pdf
Victor Center for Jewish Genetic Diseases (2019) Judaism: ashkenazi Jewish genetic diseases. https://www.jewishvirtuallibrary.org/ashkenazi-jewish-genetic-diseases
Winer L (2011) Sacred choices: adolescent relationships and sexual ethics. Am J Sex Educ 6(1):20–31
Yang Y, Peter I, Scott S (2014) Pharmacogenetics in Jewish populations. Drug Metabol Drug Interact 29(4):221–233. https://doi.org/10.1515/dmdi-2013-0069
Zhou Y, Lauschke V (2018) Comprehensive overview of the pharmacogenetics diversity in Ashkenazi Jews. J Med Genet 55(9):617–627. https://doi.org/10.1136/medgenet-2018-105429

23 People of Korean Heritage

Eun-Ok Im

23.1 Introduction

This chapter describes background information, health-related beliefs and values, and health related behaviors of people of Korean heritage. The chapter begins with an overview on geographical characteristics, history, culture, and populations on the Korean peninsula. Typical communication styles and relationships, traditional and typical family roles and organization, workforce issues, and biocultural ecology are presented to provide background information of Korean populations. More details are presented on high-risk health behaviors, health-care practices, typical meals and dietary practices, pregnancy and childbearing practices, death rituals, spirituality, barriers to health care, and traditional versus biomedical health-care providers.

This chapter is an update written by Eun-Ok Im in a previous edition of the book.

E.-O. Im (✉)
Emory University, Atlanta, GA, USA
e-mail: eun.ok.im@emory.edu

23.2 Overview, Inhabited Localities, and Topography

23.2.1 Overview

This chapter focuses on the commonalities among people of Korean heritage with historical reference to the mother country, South Korea. The word *Korea* limitedly refers to the Republic of Korea. Because some information may not be pertinent to every Korean, this chapter serves as a guide for health-care providers rather than as a mandate of facts. Differences in beliefs and practices among Koreans in Korea, the United States, and other countries vary according to variant cultural characteristics as presented in Chap. 2. An understanding of Korean culture and history gives health providers the insight needed to perform culturally congruent assessments, plan effective care and follow-up, and work effectively with Koreans in the workforce.

South Korea is a peninsula separated by North Korea to the north at the 38th parallel and surrounded by the former Soviet Union to the northeast, the Yellow Sea to the west, and the East Sea to the east. South Korea has a landmass of 98,480 sq. km (38,031 sq. mi.), and a population of 51.63 million in 2018 (Statistics Korea 2019). The population in the Seoul Capital Areas including Seoul itself, Incheon, and Gyeonggi is about 25.71 million, which means that about 49.8% of the total population of

© Springer Nature Switzerland AG 2021

L. D. Purnell, E. A. Fenkl (eds.), *Textbook for Transcultural Health Care: A Population Approach*,
https://doi.org/10.1007/978-3-030-51399-3_23

South Korea are residing in the Seoul areas (Statistics Korea 2019). The proportion of foreign nationals in South Korea is up to 1.65 million people in 2018 (3.2% of the total population in South Korea), which indicates a significant increase (11.6%) of foreign nationals from 2017 (Statistics Korea 2019).

A new international state-of-the-art airport is located in Incheon, 60 km from the center of Seoul. The airport has recently been extended to include its second terminal due to a drastically increasing number of tourists. Other large cities are Busan (Pusan) and Daegu (Taegu). Planes, trains, and buses link all South Korean major cities, making travel easy and efficient. With the recent increase in the number of automobiles and the construction of highways, motorways are becoming more congested. Major industries in South Korea are electronics, automobiles, telecommunications, shipbuilding, chemicals and steel (Bajpai 2015). South Korea is now well known as riding on the "*hallyu* movement" or the "Korean wave," which is the globalization of Korean dramas throughout Singapore, Malaysia, Japan, China, and the US. Since the 1990s, the entertainment industry of South Korea has grown explosively, producing Asia-wide successes in music, television, and film (Korean Culture and Information Service 2019).

The continental and monsoon climate of Korea is fairly consistent throughout the peninsula, except during the winter months. North Korea has cold, snowy winters, with an average temperature in January of 17 °F. South Korea is milder, with an average January temperature of 23 °F. During the summer months, the monsoon winds create an average temperature of 80 °F, with high humidity throughout the peninsula. August is the hottest month of the year, when temperatures reach over 100 °F in many areas. Precipitation occurs mostly during the summer months and is heavier in the south. The peninsula is mountainous; only 20% of the terrain is located in lowlands. Such topography encourages the development of concentrated living areas. Most cities and residential areas are located along the coastal plains and the inland valleys opening to the west coast.

23.2.2 Heritage and Residence

Korea is one of the two oldest continuous civilizations in the world, second only to China. Koreans trace their heritage to 2333 B.C. In the first century A.D., tribes from central and northern Asia banded together to form this "Hermit Kingdom," spattering the countryside with palaces, pagodas, and gardens. Over the ensuing centuries, Mongols, Japanese, and Chinese invaded the Korean peninsula. Japan forcibly annexed Korea in the early twentieth century, ruling it harshly and leaving ill will that persists to this day. As a result of the Potsdam Conference after World War II, the United States took over the occupation of South Korea, with the USSR occupying North Korea. By 1948, Korea's new government was recognized by the United Nations, only to be followed by the North Korean Communist forces invading South Korea in 1950. The result was the Korean War, which lasted until 1953 and caused mass devastation, from which the country has made a remarkable recovery. Open aggression between North and South Korea again occurred in 1998 and 1999. In 2000, the two Koreas signed a vague, yet hopeful, agreement that the two countries would be reunited. Yet, the tension between South Korea and North Korea has fluctuated during the past decades. In 2010, North Korea's resumption of its nuclear weapons program has set its neighbors and much of the rest of the world on edge (CNN 2010). Then, in 2018, there was a reconciliation mood between South Korea and North Korea, but the situation became vague due to political changes in the involved countries that included South Korea, North Korea, the US, China, Japan, and Russia (CNN 2019).

In 1988, Seoul hosted the Olympic Games, elections were held, and relations were re-established with China and the Soviet Union. Intermittent corruption among political officials has continued to surface, threatening internal relationships and the economy. In 1997, South Korea's economy tumbled dramatically, resulting in economic and democratic reforms. With unwavering persistence, Koreans have rebuilt their major world economy, and Korea is now the fourth largest economy in Asia and the 11th largest in the world (Global Edge 2019). The United States continues to maintain a strong military presence throughout South Korea.

23.2.3 Reasons for Migration and Associated Economic Factors

Koreans are one of the most rapidly increasing immigrant groups in the US (Pew Research Institute 2019). The first major immigration from Korea to the US occurred between 1903 and 1905 when the Korean government prohibited further emigration; about 10,000 Koreans had entered Hawaii, and 1000 reached the U.S. mainland. The U.S. Immigration Act of 1924 practically closed the door to Japanese and Koreans. During the civil rights movements of the 1950s and 1960s, new immigration laws repealed the earlier limitations on Asian immigration. Koreans continue to immigrate to the US to pursue the American dream, to increase socioeconomic opportunities, and to attend colleges and universities. The 2010 U.S. Census Bureau survey indicated that about 1.8 million Koreans were residing in the US (U.S. Census Bureau 2019). About 39% of Korean Americans were U.S. born, and 62% were foreign born (Pew Research Institute 2019). Among the foreign born, about 60% were U.S. citizens. About 53% had lived in the US for more than 20 years, and 14% had lived in the US for less than 5 years (Pew Research Institute 2019). The median annual household income of Korean Americans was $60,000; that of the U.S. born Korean Americans was $68,000 and that of the foreign-born Korean Americans was $57,000 (Pew Research Institute 2019).

Most of the population pursues higher education, and South Korea has more citizens with PhDs per capita than any other country in the world. Owing to Confucian cultural influence, education is emphasized as a virtue of human beings (all human beings should be educated) and is highly valued in the Korean culture (Im 2002). Due to this cultural heritage, South Korea is ranked as one of the top OECD countries that have the highest performances in reading literacy, mathematics and sciences (the ninth in the world; OECD 2015). Also, South Korea is one of the top countries with the highest-educated labor forces (The Economist 2019).

Before the late nineteenth century, education was primarily for those who could afford it. State schools educated the youth from the *yangban* (upper class), focusing on Chinese classics in the belief that these contained the tools of Confucian morality and philosophy that also apply in politics. In the late 1800s, the state schools were opened to all citizens. Early Christian missionary work introduced the Western style of modern education to Korea. Initially, many Koreans were skeptical of the radical curriculum and instruction for females, but the popularity of this style grew rapidly.

After the takeover of Korea by the Japanese in 1910, two types of schools emerged—one for Japanese and the other for Koreans. The Korean schools focused on vocational training, which prepared Koreans for only lower-level positions. Japanese colonial education was designed to keep Koreans subordinate to Japanese in all ways (Sorensen 1994). In 1949, South Korea allowed for the implementation of an educational system similar to that of the US. This 6–3–3–4 ladder (6 years in elementary school, 3 years in junior high, 3 years in high school, and 4 years in college) continues today in contemporary South Korea.

In the US, many Koreans own their own small businesses, which vary from mom-and-pop stores and gas stations to grocery stores and real estate agencies to retail shops. Their reputation for hard work, independence, and self-motivation has given them the label "the model minority." However, this has caused a backlash in some communities, such as Washington, DC, where they have been compared with other minority groups. The message has become "If the Koreans can do it, why not other groups?" The turmoil and riots that took place in Los Angeles in April 1992 between the African American community and the Korean American merchants are examples of conflicts that arise from such labeling.

Many Korean small businesses are located in African American neighborhoods because of low capital investment requirements and limited resources of the owners. Korean merchants begin dealing in inexpensive consumer goods as a practical way to start a business in a capitalistic society. Koreans often assist each other in establishing businesses by pooling their money and taking turns with rotating credit associations to provide each family with the opportunity for financial success.

23.3 Communication

23.3.1 Dominant Language and Dialects

The dominant language in Korea is *Korean*, or *han'gul*, which originated in the fifteenth century with King Se Jong, and is believed to be the first phonetic alphabet in East Asia. Several dialects exist in the Korean language, called *saturi* in Korean (Hallen and Lee 2019). The Korean standard language in South Korea is based on the dialect of the area around *Seoul*, and the Korean standard language in North Korea is based on the dialect of the area around *Pyongyang*. All the dialects of Koreans are similar to one another except that of Jeju Island (Hallen and Lee 2019). The most notable difference among dialects is the accent. For example, the Korean standard language has a very flat intonation, while the Gyeongsang dialect has a very strong accent and intonation (Hallen and Lee 2019).

The Korean language has four levels of speech that are determined based on the degree of intimacy between speakers. These varying levels reflect inequalities in social status based on gender, age, and social positions. Use of an inappropriate sociolinguistic level of speech is unacceptable and is normally interpreted as intended formality to, disrespect for, or contempt to a social superior.

Chinese and Japanese have influenced the Korean language. Before the Japanese occupation in Korea, highly educated Koreans used Chinese characters, and Chinese characters were taught in Korean traditional schools. Then, during the Japanese occupation in the early twentieth century, the Japanese forbade public use of the Korean language, requiring the use of the Japanese written and spoken language, which introduced some Japanese terms and words into contemporary Korean language.

Most Koreans in the US can speak, read, write, and understand English to some degree. However, some Americans may have difficulty understanding the English spoken by Koreans, especially those who learned English from Koreans who spoke with their native intonations and pronunciations.

23.3.2 Cultural Communication Patterns

Sharing thoughts, feelings, and ideas is very much based on age, gender, and status in Korean society. Traditionally, the Korean community values the group over the individual, men over women, and age over youth. Those holding the dominant position are the decision makers who share thoughts and ideas on issues.

Koreans prefer indirect communication because they perceive direct communication as an indication of intention or opinions as rude. Moreover, Koreans may agree with the health-care provider in order to avoid conflict or hurting someone's feelings, even if something is impossible (Im 2002). Thus, it is important to be aware of, and sensitive to, communication patterns when working with these families and remember those growing up in the US may adopt the dominant American communication style.

Traditional Koreans tend to avoid direct eye contact, especially with older people, perceived authorities (e.g., health-care providers), and strangers. Avoiding direct eye contact with older people and perceived authorities indicates respect, and women's avoiding direct eye contact with men shows modesty. Younger generations of Koreans educated in the United States may adopt the dominant communication style of eye contact. Koreans are usually comfortable with silence owing to the Confucian teaching "Silence is golden." Silence was traditionally emphasized as a virtue of educated people. Even among Korean Americans, people who are silent, especially men, are viewed as humble and well educated. However, the social fabric and cultural norms of Koreans are changing as they interact with Western societies and culture. Younger generations of Koreans, even in South Korea, are noted as being very sociable and kind to visitors (Park 2011).

Close personal space (less than a foot) is shared with family members and close friends, but it is inappropriate for strangers to step into "intimate space" unless needed for health care (Im 2002). Visitors from America may be uncomfortable with Koreans' spatial distancing in public spaces. Koreans stand close to one another and do not excuse themselves if they bump into someone on the street. This may be due to the high population density in the metropolitan areas of South Korea and Koreans' cultural attitudes toward strangers (e.g., they usually do not speak with strangers). Among family members and close friends, touching, friendly pushing, and hugging are accepted. However, among strangers, touching is considered disrespectful unless needed for care. Also, touching among friends and social equals of the same sex is common and does not carry a homosexual connotation as it might in Western societies. However, more social etiquette rules apply when it comes to touching older family members or those of higher social status. Hugging and kissing recently have become common among parents and young children, as well as among young children and aunts or uncles.

Feelings are infrequently communicated in facial expressions. Smiling a lot shows a lack of intellect and disrespect. One would not smile at a stranger on the street or joke around during a serious conversation. Joking and amusement have their designated times. In South Korea, men frequent bars after work and may express their sense of humor in this setting. Men and women alike appreciate and encourage jokes and laughter in appropriate settings. Koreans generally do not express their emotions directly or in public; expressing emotions in front of others, including family members, is regarded as shameful, especially among men (Im 2002). An old Korean belief related to men's emotions is that men should cry only three times in their lives: when they are born, when their parents die, and when their country perishes (Im 2002). Given these cultural communication patterns, health-care providers should not interpret these nonverbal behaviors as meaning that Korean patients are not interested in, do not care about, or emotionally insensitive to the information presented during health teaching and health promotion interventions.

23.3.3 Temporal Relationships

Traditional Koreans are past oriented. Much attention is paid to the ancestry of a family. Yearly, during the Harvest Moon in Korea, *chusok* (respect) is paid to ancestors by bringing fresh fruits from the autumn harvest, dry fish, and rice wine to gravesites. However, the younger and more educated generation is more futuristic and achievement oriented.

In Korea, Korean traditional shamans (called *modang* and/or *jumjangi*) are visited to determine the best home to purchase, the best date for having a wedding, and the best time to start a new business. The busiest time of the year for the shaman is just before the Chinese New Year. Koreans are eager to know their fortune for the coming year. Many believe that misfortunes occur because ancestors are unhappy. During these times, families show respect to ancestors by more frequent visits to their gravesites in the hope of appeasing the spirits. Shamans, who may be used by Koreans of all socioeconomic levels, are also used in Korea to eliminate bad sprits from their homes and new places of business. The Korean concept of time depends on the circumstances. Koreans embrace the Western respect for time for important appointments, transportation connections, and working hours, all of which are recognized as situations in which punctuality is necessary. Yet, socially, Korean Americans arrive at parties and visit family and friends up to 30 min later (and sometimes 1–2 h later; they call it "Korean time") than the agreed-upon time. This is socially acceptable when the person or family is waiting at home. If the social meeting is being held in a public setting, a half-hour time span for arrival at the meeting place can be expected. However, again the social fabric and cultural norms of Koreans have drastically changed during the past decade, and younger generations of Koreans would have different time perception and management compared with older generations.

Reflective Exercise 23.1

Lisa, a school nurse, is examining Mina Lee, who was referred by her teacher because of her reluctance to speak in class. Mina Lee is a first-grader whose family has recently immigrated from South Korea. She is the second child of her family, and she is living with her mother, father, and brother who do not speak English fluently.

1. What should the nurse to do assess Mina's reluctance to speak in class?
2. Could language be a possible barrier?
3. What cultural barriers besides language might explain Mina's silence in class?
4. What are the traditional Korean cultural attitudes and values related to silence and teacher–student relationships?
5. What are some implications for nursing practices for children who recently immigrated from Korean culture?

23.3.4 Format for Names

The number of surnames in Korea is limited, with the most common ones being Kim, Lee, Park, Rhee or Yi, Choi or Choe, and Chung or Jung. Korean names usually contain two Chinese characters (recently, just one Chinese character or pure Korean names are frequently used though), one of which describes the generation and the other the person's given name. The surname comes first; however, because this may be confusing to many Americans, most Korean Americans follow the Western tradition of using the given name first followed by the surname. Adults are not addressed by their given names unless they are on friendly terms; individuals should be addressed by their surname with the title Mr., Mrs., Ms., Dr., or Minister.

Given the diversity and acculturation of Koreans, health-care providers need to determine the Korean patient's language ability, comfort level with silence, and spatial-distancing characteristics. In addition, Koreans should be addressed formally until they indicate otherwise.

23.4 Family Roles and Organization

23.4.1 Head of Household and Gender Roles

Fundamental ideas about morality and the proper ordering of human relationships among Koreans are closely associated with kinship values derived mainly from Confucian concepts of filial piety, ancestor worship, funerary rites, position of women, the institution of marriage, kinship groups, social status and rank, and respect for scholars and political officials. Although constitutional law in South Korea declares equality for all citizens, not all aspects of society have accepted this. Korean culture is largely based on patriarchal and Confucian norms that subordinate women (Im 2002). In Confucian traditional Korean families, the father was always the head of the family; he had power to control the family, and the family had to obey any order from the father. Wives did not share household tasks with their husbands, so they tended to be physically overloaded and psychologically distressed. Wives' exploitation was hidden under Confucian norms that praised women who sacrificed themselves for their families and nation (Im and Meleis 2001). Also, the wife was confined to the home and bore the major responsibility for household tasks; the husband was the breadwinner. Nowadays, this typical family structure has been changed, especially among younger generations. Many women and men remain single until their 30s and 40s. Even when they marry, they sometimes do not have a child; they choose to enjoy their lives rather than raise the next generation.

Among Korean immigrants in the US and other individualistic cultures, women hold the family together and play a vital role in building an economic base for the family and community, often sacrificing themselves in the immigration process. The Korean immigrant woman may have started as a cleaning woman or seamstress, then worked at a fast-food restaurant, and then in a small shop owned with her husband. However, the women's financial contributions to the family usually do not change the gender roles; their husbands still occupy central stage, exercise the authority, and make the major family decisions (Im and Meleis 2001).

23.4.2 Prescriptive, Restrictive, and Taboo Behaviors for Children and Adolescents

In contrast to the Western culture in which mothering is individually fashioned and relies on the expertise of health-care providers, in the highly ritualistic Korean culture, mothering is molded by societal rules, and information is less frequently sought from health-care providers. In this context, mothers tend to view infants as passive and dependent, and they seek guidance from folklore and the extended family (Choi 1995). In Korea, children over the age of 5 years are expected to be well behaved because the whole family is disgraced if a child acts in an embarrassing manner. Most children are not encouraged to state their opinions. Parents usually make the decisions.

Korean families have high standards and expectations for their children, and "giving a whip to a beloved child" is the basis for discipline of children (Im 2002). Thus, the pressure of high performance in school and entering a highly ranked university is prevalent among Korean children and adolescents (Im 2002). Usually, Koreans are not happy with very masculine girls or very feminine boys (Im 2002).

"Teaching to the test" is also common in Korea, but the role of teachers is also to encourage self-study. The future of Korean students is determined by their teachers' recommendations, and this pressure can be extremely intense for students who are not doing well. The teaching style is one in which students listen and learn what is being taught. Regardless of private doubts, a student rarely questions a teacher's authority. Korean children in America must be taught the teaching style in American schools, in which questioning is positive and is valued as class participation. Even if Korean American students understand the style of teaching, it can be difficult to know the appropriate timing for asking questions.

The pressure of doing well in school and attending a university of high-quality leaves Korean adolescents little room for social interactions. Activities that interfere with one's education are considered taboo for adolescents. In South Korea, students frequently attend study groups after school or special tutoring sessions paid for by their families in preparation for examinations to enter a university. Short coffee breaks or snacks at local coffee shops or noodle houses are permissible, but then it is "back to the books."

Dating is now common among high school students in South Korea. Even an elementary student would say that she or he has a boyfriend or a girlfriend. Yet, adolescent girls are usually not allowed to spend the night at their friends' houses, virginity is emphasized, and sexual activities and pregnancy at puberty stigmatize the family across social classes. Although talking about sexuality, contraception, or pregnancy in public is taboo, close girlfriends or boyfriends exchange information on these topics or get their information from women's magazines. Neither the school system nor the family assumes responsibility for sex education. In recent years, girls in elementary school are given educational classes regarding their menstrual cycle and sexual relations.

Once young adults have entered a university, they receive their freedom and are then permitted to make their own decisions about personal and study time. Group outings are common for meeting the opposite sex. Dating may occur from these group meetings and consists of movies, dinner, and walks in the park.

Issues arise between the first-generation Korean immigrant parents and the later-generation children in relation to conflicting values and communication. With rapid acculturation, the latter generations often take on the values of the dominant society or culture. Thus, parents who are of the first generation in most cases are challenged when their children do not accept traditional values and ideals that they may still hold dear. The different cultures between the first-generation parents and later-generation children are sometimes the cause of domestic violence. Most of the first generation of Korean immigrants were educated in Korea, and they have a strong stereotype of Korean patriarchal culture. However, because the generation is educated in the US (some of them never visited Korea), most second-generation individuals feel a spirit of insubordination and often quarrel. For some, physical abuse might be involved if they do not follow orders (Kim et al. 2006; Kim and Chung 2003; Park 2001; Yick and Oomen-Early 2008).

23.4.3 Family Goals and Priorities

In a Korean traditional family, family members had specific rights and duties within their family. For example, the first son inherited all the properties of his parents and had the duty of caring for his elderly parents until they died. A family member replaced the roles of another family member who died. Thus, if the first son died, the second son was in charge of all the duties of the first son. However, this traditional family system is dissolving among both Koreans and Koreans elsewhere. Usually in Korean Americans, both parents work to provide every opportunity possible for their family. As each family member learns to adjust to the changing roles in the new country, conflict can result. Children adapt most easily to the new culture and may even take on the dominant culture's values; this may cause family conflicts.

Lee and Lee (1990) studied the adjustment of Korean immigrant families in the US in relation to roles, values, and living conditions between husbands and wives and between parents and children. The findings showed a transition from an independent family structure in which the woman had little knowledge of the man's activities outside the home, to a joint family structure. Many activities were carried out together with an interchange of roles at home. Conflict centered on undefined role expectations. In Korea, the roles of men and women were very clear. However, upon immigrating to the US and elsewhere, men and women were faced with conflicting roles in the new culture and had to struggle to redefine them. Other conflict areas were the couple's ability to speak English or the new language, the woman's inability to drive, the degree of acculturation, the limited social contact, and the stressors of living in a new culture. In recent years, the gender ideology gets weaker among Korean immigrant families, but it still continues (Kim 2017).

In Korea, education is a family priority. The outcome of having a highly educated child was a secure old age for the parents. Because of the dependent relationship between parents and their children, parents were more willing to make drastic sacrifices for the advancement of their children's education. Today, status could be achieved rather than inherited in Korea, and education in Korea has been a major determinant of status, independent of its contribution to economic success. However, with recent drastic economic growths in South Korea and subsequent large disparities between the rich and the poor (e.g., educational opportunities), it becomes difficult for younger generations to move to a higher social status (An and Bosworth 2013).

Traditionally in Korea, parents expected their children to care for them in old age. *Hyo* (filial piety), which is the obligation to respect and obey parents, care for them in old age, give them a good funeral, and worship them after death were a core values of Korean ethics. The obligation to care for one's parents is written into civil code in Korea. The burden was on the eldest son, who was obliged to reside with his parents and carry on the family line. Such an arrangement made the generations dependent on each other. The son felt obligated to care for his parents because of the sacrifices they made for him. Similarly, he made the same sacrifices for his children and expected them to provide for him and his wife in their old age. Many of these traditions have changed. Some of the eldest children emigrated, leaving the responsibility for their parents to the siblings who remained in Korea.

Some older Koreans were brought to the US without their friends and with minimal or no English skills. They often felt obligated to assist the family in any way possible by preparing meals or taking care of the children when the parents were not home. Decision making for older people was hampered in their new culture. Korean older people were frequently consulted on important family matters as a sign of respect for their life experiences. Older people's roles as decision makers in the US and elsewhere have shifted with the younger generation of Korean wanting the final decision-making authority in their young families.

Traditionally, Koreans give great respect to their elders. Old age begins when one reaches the age of 60 years, with an impressive celebration prepared for the occasion. The historical significance of this celebration is related to the Chinese lunar calendar. The lunar calendar has 60 cycles, each with a different name. At the age of 60, the person is starting the calendar cycle over again. This is called ***hwangap***. This celebration was more significant in the past when life expectancy in Korea was much lower than it is today. Despite a change in the direct role of older people in their families, older Koreans are socially well respected in Korea. In public, an older woman is called *Halmoni* (grandmother), and those who are not blood relatives call an older man *Harabuji* (grandfather). Older people are offered seats on buses out of respect and honor. However, recent changes in Korean culture have made this tradition change as well. Sometimes quarrels between older people and young people, usually about respect, are reported in the daily news in South Korea.

Traditionally, the extended Korean family played an important role in supporting its members throughout the life span. With the breakup of the extended family, Koreans support one another through secondary organizations such as the church. The church assists new immigrants with the transition to life in the new country. The church is a resource for information about child care, language classes, and social activities (Choi 2015). Koreans without family support may seek other Koreans who live in the area. With Koreans dispersed throughout the country, this task can be difficult.

Whereas some Koreans inherit social status, many have the ability to change their status through their education and professions. As mentioned above, moving to a higher social status in South Korea gets difficult in recent years with higher socioeconomic disparities between the rich and the poor (An and Bosworth 2013). Traditional Korean culture espouses respect not only for older people but also for those of valued professions. In modern Korea, professors, bureaucrats, business executives, physicians, and attorneys receive a high level of respect. Historically, those with the highest education were handsomely paid. Even though the salary differences between university professors and other professions have narrowed significantly in recent years in South Korea, the status of the intellectual remains high. Similarly, the bureaucratic officer has a high social status, wielding much respect and influence.

23.4.4 Alternative Lifestyles

Alternative lifestyles have been usually frowned upon in Korean culture although the culture has recently changed drastically. In traditional Korean culture, women who divorced suffered social stigma, the degree of which depended on the situation. However, recent changes in the Family Law in South Korea now permit women to head a household, recognize a wife's right to a portion of the couple's property, and allow a woman to maintain greater contact with her children after a divorce. South Korea now has one of the highest divorce rates in the world (Statistics

Korea 2019). The number of divorces was 108,700 in 2018, which is a 2.5% increase from 2017 (Statistics Korea 2019). The stigma attached to divorce becomes less important than before because many people, even older couples, are having divorces in recent years.

Yet, there is little governmental or private assistance for divorced women in South Korea. Mixed marriages, between a Korean and a non-Korean, are highly disregarded by some, and the Korean government makes it very difficult for these marriages to occur. Korean women who have married American servicemen are often the objects of Korean jokes and are ridiculed by some.

Living together before marriage is not customary in Korea. If pregnancy occurs outside marriage, it may be taken care of quietly and without family and friends being aware of the situation. However, with recent changes in Korean culture, some celebrities began to announce their pregnancies before marriage, and a pregnancy right before marriage is not looked upon as harshly as in the past. In the United States, pregnancy outside of marriage may not carry such a great stigma among the more acculturated.

As in other Asian cultures, homosexuality has not been accepted in Korean culture (Kimmel and Yi 2004). Also, Korean's understanding and knowledge of homosexuality are ambiguous and limited (Kim and Hahn 2006). Koreans believe that homosexuality is an abnormal and impure modern phenomenon. Despite the recent coming out of several Korean homosexual entertainers in South Korea, those who have relations with a person of the same sex still remain "in the closet." Personal disclosure to friends and family could negatively influence "the face" of families and may lead to marginalization.

23.5 Workforce Issues

23.5.1 Culture in the Workplace

Korean immigrants come from a culture that places a high value on education. Many Korean immigrants are college educated and held white-collar jobs in Korea. Moreover, it is difficult for Korean immigrants to obtain work in the US and elsewhere commensurate with their experience because of language difficulties, restricted access to corporate America, and unfamiliarity with the new culture (Im and Meleis 2001). The skills and work experiences they had in Korea are often not accepted by businesses, forcing them to take jobs in which they may be over skilled while they save money to start their own businesses. Korean women frequently need to find jobs to assist the family financially, which may cause role conflicts between more traditional husbands and wives.

Koreans have a strong work ethic. They work long hours each week for the advancement of family opportunities. Family is the priority for Koreans, but on the surface this may not always be apparent when long hours are devoted to work. The goal is to save money for education and other opportunities, so the family can provide for their children in the future.

The number of Korean medical personnel working in the American health-care system is unknown. Significant numbers of Korean nurses and physicians are practicing in the US and Canada; many have received part or all of their education in the US. Yi and Jezewski's study (Yi and Jezewski 2000) of 12 Korean nurses' adjustment to hospitals in the US identified five phases of adjustment. The first three phases—relieving psychological stress, overcoming language barriers, and accepting American nursing practices—take 2–3 years. The remaining two phases—adopting the styles of American problem-solving strategies and adopting the styles of American interpersonal relationships—take an additional 5–10 years. Accordingly, orientation programs need to address language skills, practice differences, and communication and interpersonal relationships to help Koreans adjust to the workforce. These same phases may occur with other Korean health-care providers. Since this study was conducted about 20 years ago, there have been virtually no studies on adaptation experience of Korean nurses and/or physicians in the US that were identified in multiple database searches.

23.5.2 Issues Related to Autonomy

Those in supervisory positions need to recognize the roles and relationships that exist between Koreans and their employers. A supervisor is treated with much respect in work and in social settings. Informalities and small talk may be difficult for Korean immigrants. For an employee to refuse an employer's request is unacceptable, even if the employee does not want or feel qualified to complete the request. Supervisors should make an effort to promote open conversation and the expression of ideas among Koreans. Asking Korean employees to demonstrate procedures is better than asking them whether they know how to perform them. Those who have adjusted to the American business style may be more assertive in their positions, but an understanding of this work role gives supervisors the tools to more readily use Koreans' skills and knowledge.

As with any new language, it is often difficult to understand slang and colloquial language. Employers and other employees should be clear in their communication style and be understanding of miscommunications. Ethnic biases are often directed at Koreans who speak English with an accent. Employers' and co-workers' preconceived notions of immigrants can also be a deterrent to Koreans in the workforce.

23.6 Biocultural Ecology

23.6.1 Skin Color and Other Biological Variations

Koreans are an ethnically homogeneous Mongoloid people who have shared a common history, language, and culture since the seventh century A.D. when the peninsula was first united. Common physical characteristics include dark hair and dark eyes, with variations in skin color and degree of hair darkness. Skin color ranges from fair to light brown. Epicanthal skin folds create the distinctive appearance of Asian eyes. With the popularity of drastic plastic surgeries in recent years, Koreans' typical facial characteristics might not be easily found in some cases.

23.6.2 Diseases and Health Conditions

Schistosomiasis and other parasitic diseases were endemic to certain regions of Korea in the past, but a recent study reported that these are mostly disappearing at present through systematic and sustainable control and management activities (Bahk et al. 2018). Yet, health-care providers still need to consider parasite screening with Korean immigrants since they frequently enjoy sushi or sashimi (raw fishes). In 1970s, South Korea manufactured and used asbestos-containing products and had not taken the precautions necessary to adequately protect employees and meet international standards. Subsequently a number of asbestos-related events occurred in South Korea, and they made the general public concerned about asbestos. Although the events happened several decades ago, asbestosis is still a concern among Koreans because of the amount of asbestos used and the long latency period. Thus, Korean immigrants may still need to be assessed for asbestos-related health problems (Kim 2009).

The high prevalence of stomach and liver cancer, tuberculosis, hepatitis, and hypertension in South Korea predisposes recent immigrants to these conditions. High rates of hypertension lead to an increase in cardiovascular accidents and renal failure. The high incidence of stomach cancer is associated with environmental risks, such as diet and infection (*Helicobacter pylori*), and in some cases, genetic predisposition (Kim 2003). As with other Asians, a high occurrence of lactose intolerance exists among people of Korean ancestry. Dental hygiene and preventive dentistry have recently been emphasized in health promotion in South Korea. Because of the high incidence of gum disease and oral problems, however, these conditions deserve attention.

23.6.3 Variations in Drug Metabolism

Growing research in the field of pharmacogenetics has found variations in drug metabolism among ethnic groups. Studies suggest that Asian populations require lower dosages of psychotropic drugs (Han and Pae 2013). Other studies have shown variations in drug metabolism and interaction with propranolol, isoniazid, and diazepam among Asians in comparison with those of European Americans and other ethnic groups (Pinto and Dolan 2011). Although these studies primarily focus on people of Chinese and Japanese heritage, health-care providers should be aware and attentive to the possibility of drug metabolism variations among Korean Americans.

Reflective Exercise 23.2

Laurie, a nurse working at a local clinic, is assessing Young Kim, who is a high school senior entering college in the coming fall. Born in South Korea, he was 4 years old when his family emigrated from South Korea. His mother accompanied him to the clinic to get his immunization records cleared for his entrance to college. All immunization records were adequately documented in his medical records, but there was an issue related to his tuberculosis (TB) immunization because his TB skin test was positive. His mother claimed that he received the Bacille Calmette Guerin (BCG) vaccination.

1. What might explain Young's positive TB skin test?
2. Most Korean immigrants have BCG immunization records and subsequent positive TB skin tests. What are the pros and cons of BCG vaccination, subsequent positive TB skin tests, x-rays for verifying TB status, and taking preventive medications?
3. Besides tuberculosis, what are some other prevalent infectious diseases among Koreans?

23.7 High-Risk Behaviors

Because Koreans place great emphasis on education, many subject their children to intense pressure to do well in school. A national survey conducted among 80,000 middle and high school students in South Korea demonstrated such pressures: 1 out of 20 Korean youths attempted suicide, and a major reason was their lack of success in school (The Hankyoreh 2007). Similar pressures have been seen in the US, where suicide has occurred in Korean high school and college students because of intense pressure to do well in school. South Korea has the highest suicide mortality rate among the 35 OECD nations from 2003, and suicide is the top reason for death among adolescents in South Korea (Kwak and Ickovics 2019).

In 2004, South Korea had a high incidence of alcohol consumption, up from 7.0 L in 1980 to 8.1 L per adult per capita, which was similar to that of the US and Ireland at 7.8 L per adult per capita (World Health Organization 2004). Yet, the alcohol consumption became the OECD average (9.1 L per capita) in 2012 (OECD 2019b). Korean business transactions commonly occur after the decision makers have had several drinks. Koreans believe that people let their masks down when they drink and that they truly get to know someone after they have had a few drinks. Socioeconomic changes in Korea have resulted in differences in alcohol-related social and health problems, with a change from drinking mild fermented beverages with meals to drinking distilled liquors without meals.

In Korea, women drink far less than men. Sons' drinking patterns are similar to their fathers' patterns. A substantial generational difference exists among females, with daughters abstaining from alcohol less frequently than their mothers and drinking more, and more often, than their mothers (Park and Kim 2019). In the US and South Korea, drinking and vehicular accidents among Koreans and Korean Americans are a cause for concern.

In 2001, about 39% of Korean American men and 6% of Korean American women were smokers (Lew et al. 2001). However, the smoking rate among Koreans has steadily reduced from 47.8% in 2008 to 40.3% in 2015 in South Korea (Nam 2018). In Korea, a few women do smoke, and for those who do, smoking in public, such as on the street, is considered taboo.

23.7.1 Health-Care Practices

Seat belts are infrequently worn in South Korea, although seat belts are now mandatory (The Korea Bizwire 2018). Korean Americans understand the legal mandates in the US and comply with seat belt and child-restraint laws.

Hobbies such as hiking and golf are enjoyed in South Korea. Korean Americans do not identify hiking as a frequent pastime, either because of environmental constraints or because of living situations. Golf remains a significant activity among those Korean Americans who are financially able to play the sport.

23.8 Nutrition

23.8.1 Meaning of Food

Food takes on a significant meaning when one has been without food. Many Koreans over the age of 60 who lived during the Korean War experienced a time when their next meal was not guaranteed. Because of a devastated economy and agricultural base, barley and *kimchee*, a spicy pickled cabbage, were dietary staples during the war. Koreans are taught to respect and not waste food.

23.8.2 Common Foods and Food Rituals

Korean food is flavorful and spicy. Rice is served with 3–5 (and sometimes up to 20) small side dishes of mostly vegetables and some fish and meats. Seasonings in Korean cooking include red and black pepper, garlic, green onion, ginger, soy sauce, and sesame seed oil. The traditional Korean diet includes steamed rice; hot soup; *kimchee*; and side dishes of fish, meat, or vegetables served in some variation for breakfast, lunch, and dinner. Breakfast is traditionally considered the most important meal.

Kimchee is made from a variety of vegetables but is primarily made from a Chinese, or Napa, cabbage. Spices and herbs are added to the previously salted cabbage, which is allowed to ferment over time and is served with every meal in a variety of forms. Some common Korean American dishes include the following:

- *Bibimbap* is a combination of rice, finely chopped mixed vegetables, and a fried egg served in a hot pottery bowl. Hot pepper paste is usually added.
- *Bulgolgi* is thinly sliced pieces of beef marinated in soy sauce, sesame oil, green onions, garlic, and sugar, which is then barbecued.
- *Jopchae* are clear noodles mixed with lightly stir-fried vegetables and meats.

Rice is usually served in individual bowls, set to the left of the diner. Soup is served in another bowl, placed to the right of the rice. Chopsticks and large soupspoons are used at all meals. Koreans may use forks and knives, depending on their degree of assimilation into American culture. Meals are frequently eaten in silence, using this opportunity to enjoy the food. When Koreans migrate to the US, they increase their consumption of beef, dairy products, coffee, soda, and bread, as well as decrease their intake of fish, rice, and other grains. However, incorporating a larger quantity of Western foods does not make a less healthy diet. They consume diets consistent with their traditional Korean food patterns, with 60% of calories coming from carbohydrates and 16% of calories from fat (Kim et al. 2000). Yet, with economic and cultural changes, Korean food patterns have also changed during the past several decades. To increase compliance with dietary prescriptions, health teaching should be geared to the unique Korean food choices and practices.

Understanding the ritual offering of food and drink to guests is important. Koreans offer a guest a drink on first arriving at their home. The guest declines courteously. The host offers the drink again and the guest again declines. This ritual can occur 3–5 times before the guest accepts the offer. This interaction is done out of respect for the hosts and their generosity to share with their guest and to express an unwillingness to impose on the hosts. Accepting an offer when first asked is considered rude and selfish.

23.8.3 Dietary Practices for Health Promotion

Most dietary practices for health promotion apply to pregnancy, discussed later in this chapter. Someone suffering from the common cold is served soup made from bean sprouts. Dried anchovies, garlic, and other hot spices are added to the hot soup, which assists in clearing a congested nose.

23.8.4 Nutritional Deficiencies and Food Limitations

Lee et al. (2009b) conducted an in-depth assessment of the nutritional status of 202 Korean elderly in a metropolitan city on the East Coast and reported that the Korean elderly consumed more than two regular meals in a day that were considered part of a Korean food pattern. The average consumption of nutrients was generally lower than in Americans reported in the National Health and Nutrition Examination Survey III, except carbohydrates, vegetable protein, and sodium intake. The researchers noted inadequate intake of calcium, dietary fiber, and folate, and suggested that health-care providers consider ways to lower sodium intake and increase fruit and vegetable consumption.

A study by Park et al. (2005) indicated that the proportion of overweight or obesity was 31.4% in U.S.-born Korean women and 9.4% in Korean-born Korean women. They also reported that U.S.-born Korean women had higher intakes of total fat and fat as a percentage of energy and lower intakes of sodium, vitamin C, beta-carotene, and carbohydrate as a percentage of energy than Korean-born women. In addition, Cho and Juon (2006) reported that of 492 Korean American respondents, 38% were overweight and 8% were obese according to the World Health Organization for Asian populations. These findings suggest that acculturation of Korean immigrants affects dietary intakes in ways that may alter their risks of several chronic diseases. These are also consistent with the findings on obesity among Asian Americans (Gong et al. 2019).

Korean Americans, as with most other Asians, are at a high risk for lactose intolerance. Thus, milk and other dairy products are not part of the traditional Korean diet, emphasizing the need to assess them for calcium deficiencies.

Korean Americans living in or near large metropolitan cities have access to Korean markets and restaurants. When no Korean stores are available, Chinese or Japanese markets may contain some of the foods Koreans enjoy. When no Asian markets are available, the American grocery store suffices.

23.9 Pregnancy and Childbearing Practices

23.9.1 Fertility Practices and Views Toward Pregnancy

To curtail population growth in Korea, the government promoted the concept of two children per household. The government supported the use of contraception when a 10-year family planning program was adopted in the early 1960s, resulting in a mass public education program on contraception. When contraceptive devices became easily available in Korea, fertility control spread widely among married women. Contraceptive devices are covered by the present national health insurance of Korea. Recently, South Korea's fertility rate fell to a new record low in 2019 as more women engaged in economic activities and got married at older ages (CNN 2019). In 2018, the total fertility rate (the mean number of children a woman has in her lifetime) has decreased to 0.98 from 1.05 in 2018.

The average number of babies per woman of childbearing age in the US was 1.72 in 2018 (CNN 2019).

Induced abortion only with legally acceptable rationales is allowed in South Korea; yet there is an unspoken acceptance of the practice. The legally acceptable reasons for induced abortion include genetic defects, communicable diseases, pregnancy due to rape, pregnancy by family members or close relatives, and pregnancy that threatens the mother's health.

Pritham and Sammons (1993) investigated Korean women's attitudes toward pregnancy and prenatal care with regard to their beliefs and interactions with health-care providers from the US. The survey was conducted with 40 unemployed Korean women between the ages of 18 and 35 at an American military medical-care facility in a major metropolitan area of Korea. Attitudes toward childbearing practices and relationships with health-care providers were elicited. The results indicated that these women were happy about their pregnancies. Only one-third of the respondents agreed with the traditional preference for a male child. About 40% of the women reinforced strong food taboos and restrictions and acknowledged the need to avoid certain foods during pregnancy. Twenty percent disagreed with the use of prenatal vitamins, and 25% indicated needing only a 10- to 15-pound weight gain in pregnancy. The women generally had sound health habits in relation to physical activity and recognized the harm of smoking while pregnant. The study sample was homogeneous and small, limiting the ability to generalize about the findings. Since this study, few studies have been conducted on Korean Americans' attitudes toward pregnancy and prenatal care.

Pregnancy in the Korean culture is traditionally a highly protected time for women. Both the pregnancy and the postpartum period have been ritualized by the culture. A pregnancy begins with the ***tae-mong***, a dream of the conception of pregnancy. Once a woman is pregnant, she starts practicing ***tae-kyo***, which literally means "fetus education." The objective of *tae-kyo* is to promote the health and well-being of the fetus and the mother by having the mother focus on art and beautiful objects. Some beliefs include the following:

- If the pregnant woman handles unclean objects or kills a living creature, a difficult birth can ensue (Howard and Barbiglia 1997).
- Some women wear tight abdominal binders beginning at 20 weeks gestation or work physically hard toward the end of the pregnancy to increase the chances of having a small baby (Howard and Barbiglia 1997).
- In addition, expectant mothers should avoid duck, chicken, fish with scales, squid, or crab because eating these foods may affect the child's appearance. For example, eating duck may cause the baby to be born with webbed feet (Howard and Barbiglia 1997).

Reflective Exercise 23.3

Alex, a nurse working at a prenatal clinic in a hospital, is assessing Sook Park, who is 12 weeks pregnant. Sook was raised and educated in South Korea; recently married Robert Kim, a Korean American; and moved to the United States early in her pregnancy. Alex found that Sook lost 5 pounds since her last visit.

1. From a cultural standpoint, what might explain Sook's recent loss of weight?
2. Identify Koreans' cultural beliefs, attitudes, and practices related to foods during pregnancy.
3. What Korean cultural beliefs, attitudes, and practices related to foods during pregnancy might explain Sook's weight loss?
4. What are some immigration and acculturation issues that might influence Sook's nutrition?

23.9.2 Prescriptive, Restrictive, and Taboo Practices in the Childbearing Family

Few recent studies have been conducted on this topic, so the content in this section is mainly based on the literature from 1990s. Birthing practices among both Koreans and Korean Americans are highly influenced by Western methods. Women commonly labor and deliver in the supine position. After the delivery, women are traditionally served seaweed soup, a rich source of iron, which is believed to facilitate lactation and to promote healing of the mother. Bed rest is encouraged after pregnancy for 7–90 days. Women are also encouraged to keep warm by avoiding showers, baths, and cold fluids or foods.

The postpartum period is seen as the time when women undergo profound physiological, psychological, and sociological changes; this period is known as the *Sanhujori* belief system. In this dynamic process, postpartum women should care for their bodies by augmenting heat and avoiding cold, resting without working, eating well, protecting the body from harmful strains, and keeping clean (Howard and Barbiglia 1997). In Western society in which they may lack extended family members from whom to seek assistance, Korean women may be faced with a cultural dilemma.

Park and Peterson (1991) studied Korean American women's health beliefs, practices, and experiences in relation to childbirth. Using structured questions, they interviewed in Korean a non-random sample of 20 female volunteers. Those interviewed subscribed to a holistic view, which emphasized both emotional and physical health. Only one-half of the women interviewed rated themselves healthy. The authors related this to the stresses of immigration and pregnancy. Preventive practices were not found among members of this group. Only one woman regularly received Pap smears and did breast self-examinations. A common finding was that most women participated in a significant rest period during puerperium. Those who did not rest lacked help for the home. All the women ate brown seaweed soup and steamed rice for about 20 days after childbirth to cleanse the blood and to assist in milk production. Because pregnancy is a hot condition and heat is lost during labor and delivery, some women avoided cold foods and water after childbirth to prevent chronic illnesses such as arthritis. The baby should be wrapped in warm blankets to prevent harm from cold winds. Herbal medicines are also used during puerperium to promote healing and health (Howard and Barbiglia 1997).

Health-care providers can improve the health of Korean women by providing factual information about Pap smears and teaching breast self-examination. Pregnant Korean women should be asked about their use of herbal medicine during pregnancy so that harmless practices can be incorporated into biomedical care. Recommendations for improving postpartum care among Korean American women include (a) developing an assessment tool that health-care providers can use to identify traditional beliefs early in a pregnancy, (b) developing a bilingual dictionary of common foods, (c) developing pamphlets with medical terms used in the U.S. health-care system, and (d) providing time for practicing English skills (Park and Peterson 1991).

23.10 Death Rituals

23.10.1 Death Rituals and Expectations

Traditionally in Korea, it was important for Koreans to die at home. Bringing a dead body home if the person died in the hospital was considered bad luck. Consequently, viewing of the deceased occurred at home if the individual died at home and at the hospital if the person died at the hospital. Several days or more were set aside for the viewing, depending on the status of the deceased. The eldest son was expected to sit by the body of the parent during the viewing (Chu 2017). Friends and relatives paid their respects by bowing to a photograph of the deceased placed in the same room in which the body rested. The guests were then offered the favorite foods of the deceased. Today, most Korean Americans are not accustomed to viewing the body of the deceased. More commonly, relatives and friends come to pay their respects by viewing photographs of the deceased.

Although Korean Americans view life support relatively positively, the attitudes depend on types of interventions and groups of people (Yun et al. 2018). In a recent study, Yun et al. (2018) investigated the attitudes of four different groups of Koreans (patients with cancer, family caregivers, physicians and general population) towards five types of end-of-life interventions (pain control, withdrawal of life-sustaining treatment, passive euthanasia, active euthanasia and physician-assisted suicide). All the groups positively viewed pain control and withdrawal of life-sustaining treatment. However, the groups had different attitudes towards other remaining interventions. About 99% of physicians positively viewed passive euthanasia compared with other three groups, but lower percent of all four groups positively viewed active euthanasia or physician-assisted suicide.

An ancestral burial ceremony follows, with the body being placed in the ground facing south or north. Both the place and the position of the deceased are important for the future fortune of the living relatives. Koreans believe that if the spirit is content, good fortune will be awarded to the family. Unlike Western graves, a mound of dirt covers the gravesite of the deceased in Korea.

Cremation is an individual and family choice and is practiced more commonly in Korea for those who have no family or die at a young age in the past decades. For example, when unmarried people die without any children to perform ancestral ceremonies, they are often cremated and their ashes scattered over a body of water. However, according to the news report in 2018, over 80% chose cremation due to high costs of burials and related environmental concerns (The Straits Times 2018).

Rice wine is traditionally sprinkled around the grave. Korean families bow two to four times in respect at the gravesite, and then the men, in descending order from the eldest to the youngest, drink rice wine. Some Korean Americans dedicate a corner of their home to honor their ancestors because they cannot go to the gravesite.

Circumstances in which "do not resuscitate" orders are an issue need to be addressed cautiously. Families trust physicians and may not question other options. Because death and dying are fairly well accepted in the Korean culture, prolonging life may not be highly regarded in the face of modern technology. Korean hospitals focus on acute care. Families are expected to stay with family members to assist in feeding and personal care around the clock. Thus, many Korean may expect to care for their hospitalized family members in health-care facilities.

23.10.2 Responses to Death and Grief

Mourning rituals, with crying and open displays of grief, are commonly practiced and socially accepted at funerals and they signify the utmost respect for the dead. The eldest son or male family member who sits by the deceased sometimes holds a cane and makes a moaning noise to display his grief. The cane is a symbol of needing support. Health-care providers may need to provide a private setting for Koreans to be able to grieve in culturally congruent ways.

23.11 Spirituality

23.11.1 Dominant Religion and Use of Prayer

Confucianism was the official religion of Korea from the fourteenth to the twentieth centuries. Buddhism, Confucianism, Christianity, shamanism, and ***Chondo-Kyo*** are practiced in Korea today. *Chondo-Kyo* (religion of the Heavenly Way) is a nationalistic religion founded in the nineteenth century that combines Confucianism, Buddhism, and Daoism. Among Koreans in South Korea, those without a religion are 56%. Among those with a religion, Protestants are 45%, Buddhists are 35%, Catholics are 18%, and others are 2% (Korea.net 2019). In the US, the church acts as a powerful social support group for Korean immigrants (Im and Yang 2006). Jo et al. (2010) even suggest that Korean churches have a high potential to serve an important role in the health of Koreans.

Koreans might not pray in the same fashion as Westerners, but for many people, the spirits demand homage. Korean churches often have prayer meetings several times a week, some with early-morning prayers. Buddhist temples have spirit rooms attached to them. Although Buddhists believe the spirit enters a new life, the beliefs of the shamans are so strong that the Buddhist church incorporated an area of their church for those who believe that ancestral spirits need honoring and homage. With such a variety of spiritual beliefs, caregivers must assess each Korean patient individually for religious beliefs and prayer practices.

23.11.2 Meaning of Life and Individual Sources of Strength

Family and education are central themes that give meaning to life for Korean Americans. The nuclear and extended families are primary sources of strength for Koreans in their daily lives. These concepts were covered earlier under Family Roles and Organization and Educational Status and Occupations.

23.11.3 Spiritual Beliefs and Health-Care Practices

Shamanism is a powerful belief in natural spirits. All parts of nature contain spirits: rivers, animals, and even inanimate objects. The many religions of Koreans create numerous ideologies about what happens with the spirits of the deceased. Christians believe the spirit goes to heaven; Buddhists believe the spirit starts a new life as a person or an animal; and shamanists believe the spirit stays with the family to watch over them and guide their actions and fortunes. Such a variety of faith systems provide a great diversity in beliefs of the Korean people. Given this diversity of spiritual beliefs among Koreans, each patient needs an individual assessment with regard to spiritual and health-care practices.

23.12 Health-Care Practices

23.12.1 Health-Seeking Beliefs and Behaviors

Beliefs that influence health-care practices include religious beliefs (see Dominant Religion and Use of Prayer) and dietary practices (see Nutrition). Health-care providers need to be aware that the theme dominating these beliefs is a holistic approach, which emphasizes both emotional and physical health.

Health-care practices among Koreans in America have primarily focused on curative rather than preventive measures. In recent years, health promotion in Korea became a steady public-health focus. In Korea, education on dental hygiene, sanitation, environmental issues, and other preventive health measures is being encouraged. Visits to the physician for an annual physical examination, Pap smears, and mammograms are becoming common with economic growths. Among Koreans, traditional patterns of health promotion include harmony with nature and the universe, activity and rest, diet, sexual life, covetousness, temperament, and apprehension (Lee 1993).

23.12.2 Responsibility for Health Care

One American study reported that only 13.5% of Korean American men and 11.3% of Korean American women had a digital rectal examination (DRE) for occult blood. Regression analysis indicated that gender, education, knowledge of the warning signs of cancer, and length of residence in the US were significantly related to having undergone DRE. The researchers determined that this group of Koreans did not see health-care providers or health brochures as valuable sources of information; to target this group, efforts should be coordinated with church and community leaders (Jo et al. 2010).

Because of women's modesty during physical examinations and their preferences that women perform intimate examinations, many Korean women defer having Pap tests or breast cancer screening tests. A study among Korean immigrant women reported that 78% of the participants had gotten a mammogram at some point in their lives and that 38.6% had gotten one in the previous year (Lee et al. 2006). Modesty and limited knowledge about breast self-examination and causes of breast cancer were pointed out as major reasons for low rates of mammography among Koreans (Han et al. 2000; Lee et al. 2009a).

In the past, cervical cancer was the number one female cancer diagnosed among Korean women, and the reluctance for undergoing Pap tests was suspected as one of the major reasons (Lee 2000). Yet, cervical cancer became the seventh most frequently diagnosed cancer and the third leading cause of cancer-related deaths among Korean women in 2018 (Kweon 2018); however the reluctance for having pap tests still remains.

Recent Korean immigrants come from a country in which universal health insurance was implemented in the late 1980s. A government mandate established employer-based health insurance for medium and large firms. Regional health insurance systems, subsidized by the government, were later established for small firms, farmers, and the self-employed. In a recent study (Li et al. 2016), the odds of having a usual source of care were 2.33 times among Korean Americans with health insurance (95%CI = 1.71–3.19) compared with those without health insurance.

The use and availability of over-the-counter medications vary tremendously between the US and Korea. Many prescription drugs in the US such as antibiotics, anti-inflammatory and cardiac medications, and certain pain control medications could be purchased over-the-counter in Korea at any *yak bang* (pharmacy) in the past. Yet, with health-care policy changes, they are not available as over-the-counter medications in these days. However, under the influences of previous practices before the health-care policy changes, older people in Korea may illegally perform home infusions of dextrose and water or albumin when feeling "tired" or "fatigued."

Self-medication with herbal remedies is also practiced. Ginseng root is used for anything from a remedy for the common cold to an aphrodisiac. Seaweed soup is also used as a medicine. Chinese traditional herbs are used to control the degree of "wind" that may be in the body. Other herbal medications are taken for preventive or restorative purposes. Accordingly, health-care providers should query their patients about their use of traditional Korean medicine and must be aware that herbal medicine may be used in conjunction with Western biomedicine.

23.12.3 Folk and Traditional Practices

Hanyak, traditional herbal medicine used for creating harmony between oneself and the larger cosmology, is a healing method for the body and soul. ***Hanbang***, the traditional Korean medical-care system, works on the principle of a disturbed state of ***ki***, cosmological vital energy. Symptoms are often interpreted in terms of a psychological base. Treatments include acupuncture, acumassage, acupressure, herbal medicines, and moxibustion therapy. The therapeutic relationship between ***hanui*** (oriental medicine doctors) and their patients is genuine, spontaneous, and harmonious. Patients who use both Western and traditional Korean practitioners may experience conflicts because of the lack of cooperation between *hanui* and biomedical health-care providers. Even Korean Americans are known to use both *hanui* and biomedical health-care providers.

Shamans are used in healing rituals to ward off restless spirits. Shamans originated with the religious belief of shamanism, the belief that all things possess spirits. A shaman, ***mundang***, is usually a woman who has special abilities for communicating with spirits. The shaman is used to treat illnesses after other means of treatment are exhausted. The shaman performs a ***kut***, a shamanistic ceremony to eliminate the evil spirits causing the illness. Such a ceremony may take place when a young person dies to prevent his or her spirit from staying tied to the earth. Others believe a shaman can eliminate evil spirits that may be causing difficulty with financial transactions. Although shamans have been around for many years, Koreans consider them part of the lowest class. Health-care providers need to determine whether Koreans in America and elsewhere are using folk therapies and should include nonharmful practices with biomedical therapies and prescriptions.

Reflective Exercise 23.4

Maria, a nurse working in an in-patient oncology unit, is assessing Jong Kim, age 72 years, who was recently diagnosed with lung cancer. Jong Kim emigrated from South Korea about 40 years ago and has smoked since he was in his early 20s. Although Jong does not request pain medication, his facial expressions show that he is obviously in significant pain.

1. What should the nurse do to adequately assess Jong's pain level?
2. What are some Korean cultural beliefs, attitudes, and practices related to cancer?
3. What Korean cultural beliefs, attitudes, and practices related to cancer might influence Jong's pain management?
4. What are some Korean cultural beliefs, attitudes, and practices related to pain and pain medication?

23.12.4 Barriers to Health Care

Because many Koreans use various options for healing, Western medical practices may be used in conjunction with acupressure, acupuncture, and herbal medicine. Barriers for in America may result from the expense of non-Western therapies because many insurance companies do not cover alternative therapies.

As for many other American residents, the lack of insurance creates barriers to health care. Paying for health care out-of-pocket is expensive and not feasible for many Korean American families. Language, modesty, cultural attitudes toward certain illnesses, and communication problems also serve as impediments for access to health care.

23.12.5 Cultural Responses to Health and Illness

Perceptions of pain vary widely among Koreans. Some Koreans are stoic and are slow to express emotional distress from pain. Others are expressive and discuss their smallest discomforts. Family and friends are useful resources for learning some of the historical coping mechanisms of sick individuals. Nonverbal cues and facial expressions must be monitored for those who are stoic rather than expressive. Pain assessments should be conducted regularly, and education may be necessary for stoic individuals.

Mental illness is stigmatized in the Korean culture. Bernstein (2007) conducted a study of Korean American women and their reluctance to use mental health providers in the US. Her study concluded that most of the participants acknowledged the need for mental health services but did not seek professional help and coped with the stressors of immigrant life by endurance, patience, and religion. ***Hwa-byung*** is a cultural syndrome among Koreans that results from the suppression of anger or other emotions (Lee et al. 2014). *Hwa* means "fire and anger," and *byung* means "illness." The emotions that are suppressed could be sadness, depression, worry, anger, fright, and fear, and these emotions are usually related to conflicts with close relatives or family, such as sons and daughters or significant others. These could be expressed as physical complaints, ranging from headaches and poor appetites to insomnia and lack of energy. The symptoms are usually chronic in nature, and a variety of remedies are used to alleviate the symptoms. *Hwa-byung* is difficult to cure and the symptoms are viewed as inevitable.

Historically, the area of special education has not been well studied or researched in Korea. Families who have children with mental or physical disabilities often question what they have done wrong to make their ancestors angry. Families feel stigmatized for such a misfortune and cannot accept their children's disfigurement or low intellect. Korea lacks social support to assist families in caring for children with mental or physical disabilities. Some families abandon these children in their desperate need for support with long-term care and expenses. Other children are kept from the public eye in the hope of saving the family from stigmatization.

According to the National Disability Survey, there exist over 2.6 million people with disabilities in South Korea (Ock et al. 2016). About 50% of these people with disabilities have physical disabilities. In the past few decades, policies and services related to the disabled have improved. For instance, regional rehabilitation centres for people with disabilities have been established, and rehabilitation programs have been implemented in communities in South Korea.

It was estimated in 2007 that the number of students with a disability in South Korea was 77,452 and that 65,940 of them attended either a specialized or regular school, in either special or normal education. In other words, 11,512 students with a disability were not receiving any education at all. These statistics may reflect negative attitudes toward people with a disability, which may influence the idea of mainstreaming students with a mild disability in South Korea. Koreans in America and elsewhere may hold these same views regarding people with mental and physical disabilities and need special support in obtaining assistance.

In Korea, once hospitalized people are physically stable, they are discharged to their homes to be with the family. Bowel training and physical therapy activities are not the responsibility of the hospital. The families must care for family members at home. Long-term care for chronic problems or for rehabilitation is rare in Korea. Thus, Korean are familiar with the concept of family home care. Depending on their adaptation to the dominant health-care system and families' contact with health-care providers in there new country, some Korean adjust their ideologies on the sick role.

23.12.6 Blood Transfusions and Organ Donation

No beliefs held by Koreans prevent the acceptance of blood transfusions. Organ donation and organ transplantation are rare, reflecting traditional attitudes toward integrity and purity. These issues need to be approached sensitively with Koreans because they may be influenced by the individual's religious beliefs.

23.13 Health-Care Providers

23.13.1 Traditional Versus Biomedical Providers

In general, no taboos exist that prevent health-care providers from delivering care to the opposite gender. Female physicians are definitely preferred for maternity care and female problems because women feel more comfortable discussing gynecological and obstetric issues with female physicians. However, more traditional Koreans frequently prefer health-care providers who speak Korean, are older, and are of the same gender, although many will seek health care from others who do not meet these requirements if their preferred care provider is not available.

The area of social work is relatively new in South Korea. The hospitals previously had no positions for such a role in South Korea. Yet, during the past decades, an increasing number of educational programs for social workers have been in place and much development has been made in the area of social support. However, because these roles are still new to many Koreans, health-care providers may need to encourage Koreans to use these services.

23.13.2 Status of Health-Care Providers

Because traditional Korean culture accords high respect to men, older people, and physicians, the ideal physician is an older man with grey hair. This shows that he has experience and wisdom and is able to make the best decisions. With such a high status in Korea, physicians expect respect from all other health-care providers. Usually, nurses are expected to carry out physicians' orders explicitly. This is not to say that the nurse cannot question orders, but great time and effort are spent consulting other nurses before questioning physicians in the most respectful way. However, as nurses are becoming more educated in Korea, they are becoming more assertive and more closely mirror Western practice patterns.

Currently, the average number of nurses per 1000 people in South Korea has increased to 19.2, which is higher than the average number of the Organization for Economic Co-operation and Development (OECD) member countries (Hong and Cho 2017). Also, with an emphasis on increasing the educational level of nurses, they too are gaining stature and respect in Korean culture. Baccalaureate, masters, and doctoral programs are available for nurses in Korea. In addition, in South Korea, the Advanced Practice Nurse (APN) system has been established in 2003, and legal regulations on the scope of practice of APN will be activated in 2020 (Jung 2018).

References

An CB, Bosworth B (2013) Income inequality in Korea: an analysis of trends, causes, and answers, vol 354, 1st edn. Harvard University Asia Center, Cambridge MA. https://doi.org/10.2307/j.ctt1x07vsm

Bahk YY, Shin EH, Cho SH, Ju JW, Chai JY, Kim TS (2018) Prevention and control strategies for parasitic infections in the Korea Centers for Disease Control and Prevention. Korean J Parasitol 56(5):401–408. https://doi.org/10.3347/kjp.2018.56.5.401

Bajpai P (2015) Emerging markets: analyzing South Korea's GDP. Investopedia. https://www.investopedia.com/articles/investing/091115/emerging-markets-analyzing-south-koreas-gdp.asp

Bernstein KS (2007) Mental health issues among urban Korean American immigrants. J Transcult Nurs 18(2):175–180

Cho J, Juon HS (2006) Assessing overweight and obesity risk among Korean Americans in California using World Health Organization body mass index criteria for Asians. Preven Chronic Dis 3(3):A79

Choi E (1995) A contrast of mothering behaviors in women from Korea and the United States. J Obstet Gynecol Neonatal Nurs 24(4):363–369

Choi HA (2015) A postcolonial self: Korean immigrant theology and church. SUNY Press, Albany NY

Chu M (2017) Korea's funeral culture. Korea Biomed Rev. http://www.koreabiomed.com/news/articleView.html?idxno=185

CNN (2010) Complete coverage on North Korea. http://topics.cnn.com/topics/north_korea

CNN (2019) North Korea news. https://www.cnn.com/specials/asia/north-korea

Global Edge (2019) South Korea: introduction. https://globaledge.msu.edu/countries/south-korea

Gong S, Wang L, Li Y, Alamian A (2019) The influence of immigrant generation on obesity among Asian Americans in California from 2013 to 2014. PLoS One 14(2):e0212740. https://doi.org/10.1371/journal.pone.0212740

Hallen C, Lee M (2019) The history of Korean language, the overview. http://linguistics.byu.edu/classes/Ling450ch/reports/Korean3.html

Han C, Pae CU (2013) Do we need to consider ethno-cultural variation in the use of atypical antipsychotics for Asian patients with major depressive disorder? CNS Drugs 27(1):47–51. https://doi.org/10.1007/s40263-012-0033-y

Han Y, Williams R, Harrison R (2000) Breast cancer screening knowledge, attitudes, and practices among Korean-American women. Oncol Nurs Forum 27(10):1589–1591

Hong KJ, Cho SH (2017) Comparison of nursing workforce supply and employment in South Korea and other OECD countries. Perspect Nurs Sci 14(2):55–63

Howard J, Barbiglia V (1997) Caring for childbearing Korean women. J Obstet Gynecol Neonatal Nurs 26(6):665–671

Im EO (2002) Korean culture. In: Hill PSS, Lipson J, Meleis AI (eds) Caring for women cross-culturally: a portable guide. F.A. Davis Company, Philadelphia

Im EO, Meleis AI (2001) Women's work and symptoms during menopausal transition: Korean immigrant women. Women Health 33(1/2):83–103

Im EO, Yang K (2006) Theories on immigrant women's health. Health Care Women Int 27(8):666–681

Jo AM, Maxwell AE, Yang B, Bastani R (2010) Conducting health research in Korean American churches: perspectives from church leaders. J Community Health 35(2):156–164

Jung YS (2018) Voices of nurses should be included in the activation of advanced practice nurse system. http://www.bosa.co.kr/news/articleView.html?idxno=2094335

Kim KE (2003) Gastric cancer in Korean Americans: risks and reductions. Korean Am Stud Bull 13(1/2):84–90

Kim HR (2009) Overview of asbestos issues in Korea. J Korean Med Sci 24(3):363–367. https://doi.org/10.3346/jkms.2009.24.3.363

Kim C (2017) The impact of perceived childhood victimization and patriarchal gender ideology on intimate partner violence (IPV) victimization among Korean immigrant women in the USA. Child

Abuse Negl 70:82–91. https://doi.org/10.1016/j.chiabu.2017.05.010

Kim H, Chung RH (2003) Relationship of recalled parenting style to self-perception in Korean American college students. J Genet Psychol 164(4):481–492

Kim YG, Hahn SJ (2006) Homosexuality in ancient and modern Korea. Cult Health Sex 8(1):59–65

Kim KK, Yu ES, Chen EH, Cross N, Kim J (2000) Nutritional status of Korean Americans: implications for cancer risk. Oncol Nurs Forum 27(10):1573–1583

Kim E, Cain K, McCubbin M (2006) Maternal and paternal parenting, acculturation, and young adolescents' psychological adjustment in Korean American families. J Child Adolesc Psychiatr Nurs 19(3):112–129

Kimmel DC, Yi H (2004) Characteristics of gay, lesbian, and bisexual Asians, Asian Americans, and immigrants from Asia to the U.S.A. J Homosex 47(2):143–172

Korea.net (2019) Religion. http://www.korea.net/AboutKorea/Korean-Life/Religion

Korean Culture and Information Service (2019) Hallyu (Korean wave). KoreaNet (the Official Website of the Republic of Korea). http://www.korea.net/AboutKorea/Culture-and-the-Arts/Hallyu

Kwak CW, Ickovics JR (2019) Adolescent suicide in South Korea: risk factors and proposed multi-dimensional solution. Asian J Psychiatr 43:150–153. https://doi.org/10.1016/j.ajp.2019.05.027

Kweon SS (2018) Updates on cancer epidemiology in Korea, 2018. Chonnam Med J 54(2):90–100. https://doi.org/10.4068/cmj.2018.54.2.90

Lee Y (1993) Health promotion: patterns of traditional health promotion in Korea. Kanhohak Tamgu 2(2):21–36

Lee M (2000) Knowledge, barriers, and motivators related to cervical cancer screening among Korean-American women. A focus group approach. Cancer Nurs 23(3):168–175

Lee DC, Lee EH (1990) Korean immigrant families in America: role and value conflicts. In: Kim HC, Lee EH (eds) Koreans in America: dreams and realities. Institute of Korean Studies, Seoul, pp 165–177

Lee EE, Fogg LF, Sadler GR (2006) Factors of breast cancer screening among Korean immigrants in the United States. J Immigr Minor Health 8(3):223–233

Lee H, Kim J, Han HR (2009a) Do cultural factors predict mammography behaviour among Korean immigrants in the USA? J Adv Nurs 65(12):2574–2584

Lee YH, Lee J, Kim MT, Han HR (2009b) In-depth assessment of the nutritional status of Korean American elderly. Geriatr Nurs 30(5):304–311

Lee J, Wachholtz A, Choi KH (2014) A review of the Korean cultural syndrome Hwa-Byung: Suggestions for theory and intervention. Asia T'aep'yongyang Sangdam Yon'gu 4(1):49

Lew R, Moskowitz JM, Wismer BA, Min K, Kang SH, Chen AM, Tager IB (2001) Correlates of cigarette smoking among Korean American adults in Alameda County, California. Asian Am Pac Isl J Health 9(1):49–60

Li J, Maxwell AE, Glenn BA, Herrmann AK, Chang LC, Crespi CM, Bastani R (2016) Healthcare access and utilization among Korean Americans: the mediating role of English use and proficiency. Int J Soc Sci Res 4(1):83–97. https://doi.org/10.5296/ijssr.v4i1.8678

Nam K (2018) Men's smoking rate in S. Korea down in 2017. http://www.koreaherald.com/view.php?ud=20180401000253

Ock M, Ahn J, Yoon SJ, Jo MW (2016) Estimation of disability weights in the general population of South Korea using a paired comparison. PLoS One 11:9. https://doi.org/10.1371/journal.pone.0162478

OECD (2015). Korea Student Performance. Retrieved from https://gpseducation.oecd.org/CountryProfile?primaryCountry=KOR&treshold=10&topic=PI

OECD (2019b) Tackling harmful alcohol use country note—Korea. https://www.oecd.org/korea/Tackling-Harmful-Alcohol-Use-Korea-en.pdf

Park MS (2001) The factors of child physical abuse in Korean immigrant families. Child Abuse Negl 25(7):945–958

Park SY (2011) Shinsedae: conservative attitudes of a 'new generation' in South Korea and the impact on the Korean presidential election. East-West Center. https://www.eastwestcenter.org/news-center/east-west-wire/shinsedae-conservative-attitudes-of-a-new-generation-in-south-korea-and-the-impact-on-the-korean-pres

Park E, Kim YS (2019) Gender differences in harmful use of alcohol among Korean adults. Osong Public Health Res Perspect 10(4):205–214. https://doi.org/10.24171/j.phrp.2019.10.4.02

Park KY, Peterson LM (1991) Beliefs, practices, and experiences of Korean women in relation to childbirth. Health Care Women Int 12(2):261–269

Park SY, Murphy SP, Sharma S, Kolonel LN (2005) Dietary intakes and health-related behaviours of Korean American women born in the USA and Korea: the multiethnic cohort study. Public Health Nutr 8(7):904–911

Pew Research Institute (2019) Koreans in the U.S. Fact Sheet. https://www.pewsocialtrends.org/fact-sheet/asian-americans-koreans-in-the-u-s/

Pinto N, Dolan ME (2011) Clinically relevant genetic variations in drug metabolizing enzymes. Curr Drug Metab 12(5):487–497

Pritham UA, Sammons LN (1993) Korean women's attitudes toward pregnancy and prenatal care. Health Care Women Int 14:145–153

Sorensen CW (1994) Success and education in South Korea. Comp Educ Rev 38(1):10–35

Statistics Korea (2019) Population census. http://kostat.go.kr/portal/eng/pressReleases/8/7/index.board

The Economist (2019) What the world can learn from the latest PISA test results. https://www.economist.com/international/2016/12/10/what-the-world-can-learn-from-the-latest-pisa-test-results?spc=scode&spv=xm&ah=9d7f7ab945510a56fa6d37c30b6f1709

The Hankyoreh (2007) 1 out of 20 Korean youths attempts suicide: study. http://english.hani.co.kr/arti/english_edition/e_national/199975.html

The Korea Bizwire (2018) South Korea passes much-needed universal seatbelt law. http://koreabizwire.com/south-korea-passes-much-needed-universal-seat-belt-law/116142

The Straits Times (2018) More than 80 percent of South Koreans choose cremation as views on death change. https://www.straitstimes.com/asia/east-asia/more-than-80-per-cent-of-south-koreans-choose-cremation-as-views-on-death-change

U.S. Census Bureau (2019) Asian/Pacific American Heritage month: may 2015. https://www.census.gov/newsroom/facts-for-features/2015/cb15-ff07.html

World Health Organization (2004) Alcohol consumption among adults. http://www.nationsencyclopedia.com/WorldStats/WHO-alcohol-consumption-adults.html

Yi M, Jezewski M (2000) Korean nurses' adjustment to hospitals in the United States of America. J Adv Nurs 32(3):721–729

Yick AG, Oomen-Early J (2008) A 16-year examination of domestic violence among Asians and Asian Americans in the empirical knowledge base: a content analysis. J Interpers Violence 23(8):1075–1094. https://doi.org/10.1177/0886260507313973

Yun YH, Kim KN, Sim JA, Yoo SH, Kim M, Kim YA, Kang BD et al (2018) Comparison of attitudes towards five end-of-life care interventions (active pain control, withdrawal of futile life-sustaining treatment, passive euthanasia, active euthanasia and physician-assisted suicide): a multicentred cross-sectional survey of Korean patients with cancer, their family caregivers, physicians and the general Korean population. BMJ Open 8:9. https://doi.org/10.1136/bmjopen-2017-020519

People of Mexican Heritage

24

Rick Zoucha and Anelise Zamarripa-Zoucha

24.1 Introduction

People of Mexican heritage are a very diverse group geographically, historically, and culturally and are difficult to adequately describe. Although no specific set of characteristics can fully describe people of Mexican heritage, some commonalities distinguish them as an ethnic group, with many regional variations that reflect subcultures in Mexico and in the US. A common term used to describe Spanish-speaking populations in the US, including people of Mexican heritage, is **Hispanic**. However, the term can be misleading and can encompass many different people clustered together owing to a common heritage and lineage from Spain. Many Hispanic people prefer to be identified by descriptors more specific to their cultural heritage, such as Mexican, Mexican American, Latin American, Spanish American, Chicano, Latino, or Latinx. The term Latinx is becoming more common and preferred by members of the Latin LGBTQIA community. It is seen as a gender-neutral, non-binary word for people of Latin American decent (Merriam-Webster 2020). Therefore, when referring to Mexican Americans it may be important to use their preferred term instead of Hispanic or Latino. As a broad ethnic group, people of Mexican heritage historically refer to themselves as *la raza*, which means "the race." The full term "La Raza Cosmica" was first introduced by Mexican scholar Jose Vasconcelos, and means the "cosmic people". This was coined to not reflect purity but the mixture inherent in Hispanic or Latinx people. It was meant to be inclusive reflecting the common heritage and destiny Hispanics share with all other people of the world (Lopez 2016; Simon 2018).

24.2 Overview, Inhabited Localities, and Topography

24.2.1 Overview

Mexico, with an estimated population of 125,959,205 (CIA World Factbook 2019), is inhabited by Indian Spanish (mestizo), Indian (Native American), Caucasian (Central Intelligence Agency [CIA] 2019). The blend of Spanish white, Middle Eastern, African heritage and Native American (Aztec, Mayan, Inca and others from Central and South America) can be traced to descendants of Spanish and other European whites (Schmal and Madrer 2011). Lesser-known ethnic groups whose members

This chapter is an update written by Rick Zoucha in a previous edition of the book.

R. Zoucha (✉)
Duquesne University, Pittsburgh, PA, USA

A. Zamarripa-Zoucha
UPMC Presbyterian Hospital, Pittsburgh, PA, USA
e-mail: zoucha@duq.edu

© Springer Nature Switzerland AG 2021

L. D. Purnell, E. A. Fenkl (eds.), *Textbook for Transcultural Health Care: A Population Approach*,
https://doi.org/10.1007/978-3-030-51399-3_24

have historically immigrated to Mexico include Chinese and Jewish populations. Some individuals can trace their heritage to North American Indian tribes in the southwestern part of the US (Schmal and Madrer 2011).

24.2.2 Heritage and Residence

Mexico City, one of the largest cities in the world, has a population of over 22 million. Mexico is undergoing rapid changes in business and healthcare practices. Undoubtedly, these changes have accelerated and will continue to accelerate since the recent passage of a new agreement signed in 2020 called the (USMCA) United States-Mexico-Canada-Agreement strengthening the former NAFTA agreement (USTR 2020). Unfortunately, even though the United States-Mexico-Canada-Agreement may enhance economic strength, it probably will not result in people moving across the border to seek employment and educational opportunities in the US due to political disagreement on both sides of the border.

Historically, for generations, people of Mexican heritage lived on the land that is now known as the southwestern US, long before the first White settlers came to the territory. By 1853, approximately 80,000 Spanish-speaking settlers of Mexican heritage lived in the area lost by Mexico during the Texas Rebellion, the Mexican War, and the Gadsden Purchase. After the northern part of Mexico was annexed to the US, the settlers were not officially considered immigrants but were often viewed as foreigners by incoming White Americans. By 1900, Mexican Americans numbered approximately 200,000. However, during the "Great Migration" between 1900 and 1930, an additional one million Mexicans entered the US. This may have been the greatest immigration of people in the history of humanity (American Social History Productions 2020).

Hispanics, the fastest-growing ethnic population in the US, includes over 59 million people (U. S. Census Bureau 2018a, b, c). Sixty-seven percent of all Hispanics are of Mexican heritage (U. S. Census Bureau 2018a, b, c). Mexican Americans reside predominantly in California, Texas, Florida, New York, Arizona, Illinois, and New Jersey. However, the major concentration of Mexican Americans, totalling over 19 million, is found in the southern and western portions of the US (U.S. Census Bureau 2010). Ninety percent of Mexican Americans live in urban areas such as San Diego, Los Angeles, New York City, Chicago, and Houston, whereas less than 10% reside in rural areas.

24.2.3 Reasons for Migration and Associated Economic Factors

Historically, many Mexicans left Mexico during the Mexican Revolution to seek political, religious, and economic freedoms (Steinhauer 2015). Following the Mexican Revolution, strict limits were placed on the Catholic Church, and, until recently, clerics were not allowed to wear their church garb in public. For many, this restricted the expression of faith and was a minor factor in their immigration north to the US (Steinhauer 2015; Beezley 2019). Since the "Great Migration," the limited employment opportunities in Mexico, especially in rural areas, have encouraged Mexicans to migrate to the US as sojourners or immigrants or with undocumented status; the latter are often derogatorily referred to as *wetbacks* (***majodos***) by the white and Mexican American populations.

Of the undocumented immigrants in the US, an estimated 4.9 million are from Mexico (Gonzales-Barrera and Krogstad 2019). Before the Immigration Reform and Control Act of 1986, hundreds of thousands of Mexicans crossed the border, found jobs, and settled in the US. Although the numbers have decreased since 1986, border towns in Texas and California still experience large influxes of Mexicans seeking improved employment and educational opportunities. The tide of illegal immigration to the US has decreased from 2008, at which time it was over seven million. In 2010, that number decreased to 6,640,000 (Department of Homeland Security 2011). Illegal immigration and what can or should be done to control it, especially in border states with Mexico, continue to be hotly debated issues. Annually, many migrants die trying to illegally emigrate. Solutions to U.S. citizens' concerns are not forthcoming in the near future. Even though the economy of Mexico has grown, the buying power of the peso has decreased and inflation rates have increased faster than wages; thus, 41.9% of the population continues to live in poverty (World Bank Group 2019). Recent Mexican immigrants are more likely to live in poverty, are more pessimistic about their future, and are less educated than previous immigrants. Many Mexicans are among the very poor, with little hope of improving their economic status. However, between the years of 2015 and 2018, the poverty rate for Hispanics decreased from 21 to 17% (U.S. Census Bureau 2019).

24.2.4 Educational Status and Occupations

Many second- and third-generation Mexican Americans have significant job skills and education. By contrast, many, especially newer immigrants from rural areas, have poor educational backgrounds and may place little value on education because it is not needed to obtain jobs in Mexico. Once in the US, they initially find work similar to that which they did in their native land, including farming, ranching, mining, oil production, construction, landscaping, and domestic jobs in homes, restaurants, and hotels and motels. Economic and educational opportunities in the US are attainable, which allows immigrants to pursue the great American dream of a perceived better life (Radford 2019). Many Mexicans and Mexican Americans work as seasonal migrant workers and may relocate several times each year as they "follow the sun." Sometimes their unwillingness or inability to learn English is related to their intent to return to Mexico; however, this may hinder their ability to obtain better-paying jobs. In the current political environment in the US there may be a decrease in immigrants and migrants seeking opportunities in the US.

The mean educational level in Mexico is 5 years. Until 1992, Mexican children were required to attend school through the sixth grade, but since the Mexican School Reform Act of 1992, a ninth-grade education is required. However, great strides have been made in educational standards in Mexico, which now reports a 95% literacy rate among its population (The World Bank 2019). A common practice among parents in poor rural villages is to educate their children in what they need to know. This group often finds immigration to the US to be their most attractive option. For many Mexicans, high school and university education are neither available nor attainable.

Hispanics are the most undereducated ethnic group in the US, with 72% aged 25 years or older having a high school education, compared with 94% for non-Hispanic Whites. However, the number of Hispanics who completed 4 years of college has increased to 16% of the total Hispanic population, up from 6.4% in 2000 (U. S. Census Bureau 2018a, b, c). As second- and third-generation Mexican Americans acculturate and improve their socioeconomic status, these percentages are likely to increase the same as they have for immigrants from European countries in previous centuries.

24.3 Communication

24.3.1 Dominant Language and Dialects

Mexico is one of the largest Spanish-speaking countries in the world, with over 80 million speaking the language. The dominant language of Mexicans and Mexican Americans is Spanish. However, Mexico has 68 indigenous languages and more than 500 different dialects (Yucutan Times 2018). Knowing the region from which a Mexican American originates may help to identify the language or dialect the individual speaks. For example, major indigenous languages besides Spanish include Nahuatl and Otami, spoken in central Mexico; Mayan, in the Yucatan peninsula; Maya-Quiche, in the state of Chiapas; Zapotec and Mixtec, in the valley of Oaxaca; Tarascan, in the state of Michoacan; and Totonaco, in the state of Veracruz. Many of the Spanish dialects spoken by Mexican Americans have similar word meanings, but the dialects spoken by other groups may not. Because of the rural isolationist nature of many ethnic groups and the influence of native Indian languages, the dialects are so diverse in selected regions that it may be difficult to understand the language, regardless of the degree of fluency in Spanish.

Radio and television programs broadcasting in Spanish in both the US and Mexico have helped to standardize Spanish. For the most part, public broadcast communication is primarily derived from Castilian Spanish. This standardization reduces the difficulties experienced by subcultures with multiple dialects. When speaking in a non-native language, health-care providers must select words that have relatively pure meanings in the language and avoid the use of regional slang.

Contextual speech patterns among Mexican Americans may include a high-pitched, loud voice and a rate that seems extremely fast to the untrained ear. The language uses **apocopation**, which accounts for this rapid speech pattern. An apocopation occurs when one word ends with a vowel and the next word begins with a vowel. This creates a tendency to drop the vowel ending of the first word and results in an abbreviated, rapid-sounding form. For example, in the Spanish phrase for "How are you?" *¿Cómo está usted?* may become *¿Comestusted?*. The last word, *usted*, is frequently dropped. Some may find this fast speech difficult to understand. However, if one asks the individual to enunciate slowly, the effect of the apocopation or truncation is less pronounced.

To help bridge potential communication gaps, health-care providers need to watch the patient for cues, paraphrase words with multiple meanings, use simple sentences, repeat phrases for clarity, avoid the use of regional idiomatic phrases and expressions, and ask the patient to repeat instructions to ensure accuracy. Approaching the Mexican American patient with respect and ***personalismo*** (behaving like a friend) and directing questions to the dominant member of a group (usually the man) may help to facilitate more open communication. Zoucha and Zamarripa (2013) found that becoming personal with the patient or family is essential to building confidence and promoting health. The concept of *personalismo* may be difficult for some health-care providers because they are socialized to form rigid boundaries between the caregiver and the patient and family.

24.3.2 Cultural Communication Patterns

Whereas some topics, such as income, salary, or investments are taboo, Mexican Americans generally like to express their inner beliefs, feelings, and emotions once they get to know and trust a person. Meaningful conversations are important, often become loud, and seem disorganized. To the outsider, the situation may seem stressful or hostile, but this intense emotion means the conversants are having a good time and enjoying one another's company. Within the context of *personalismo* and ***respeto***—respect—health-care providers can encourage open communication and sharing and develop the patient's sense of trust by inquiring about family members before proceeding with the usual business. It is important for health-care providers to engage in "small talk" before addressing the actual health-care concern with the patient and family (Zoucha and Zamarripa 2013).

Mexican Americans place great value on closeness and togetherness, including when they are in an in-patient facility. They frequently touch and embrace and like to see relatives and significant others. Touch between men and women, between men, and between women is acceptable. To demonstrate respect, compassion, and understanding, health-care providers should greet the Mexican American patient with a handshake. Once rapport is established, providers may further demonstrate approval and respect through backslapping, smiling, and affirmatively nodding the head. Given the diversity of dialects and the nuances of language, culturally congruent use of humor is difficult to accomplish and, therefore, should be avoided unless health-care providers are absolutely sure there is no chance of misinterpretation. Otherwise, inappropriate humor may jeopardize the therapeutic relationship and opportunities for health teaching and health promotion.

Mexican Americans consider sustained eye contact when speaking directly to an older person to be rude. Direct eye contact with teachers or superiors may be interpreted as insolence. Avoiding direct eye contact with superiors is a sign of respect. This practice may or may not be seen with second- or third-generation Mexican Americans. Health-care providers must take cues from the patient and family.

24.3.3 Temporal Relationships

Many Mexican Americans, especially those from lower socioeconomic groups, are necessarily present oriented. Many individuals do not consider it important or have the income to plan ahead financially. The trend is to live in the "more important" here and now, because *mañana* (tomorrow) cannot be predicted. With this emphasis on living in the present, preventive health care and immunizations may not be a priority. *Mañana* may or may not really mean tomorrow; it often means "not today" or "later."

Some Mexicans and Mexican Americans perceive time as relative rather than categorically imperative. Deadlines and commitments are flexible, not firm. Punctuality is generally relaxed, especially in social situations. This concept of time is innate in the Spanish language. For example, one cannot be late for an appointment; one can only arrive late. In addition, immigrants from rural environments where adhering to a strict schedule is not important may not own a clock or even be able to tell time.

Because of their more relaxed concept of time, Mexican Americans may arrive late for appointments, although the current trend is toward greater punctuality. Health-care facilities that use an appointment system for patients may need to make special provisions to see patients whenever they arrive. Health-care providers must carefully listen for clues when discussing appointments. Disagreeing with health-care providers who set the appointment may be viewed as rude or impolite. Therefore, some Mexican Americans may not tell you directly that they cannot make the appointment. In the context of the discussion, they may say something like "My husband goes to work at 8:00 a.m., and the children are off to school, and then I have to do the dishes…." The health-care provider should ask, "Is 8:30 a.m. on Thursday okay for you?" The person might say yes, but the health-care provider must still intently listen to the conversation and then possibly negotiate a new time for the appointment. In the conversation, the patient may just give clues that he or she will not arrive on time, because it is important to save face and avoid being rude by saying that outright.

It is important to note that cultural behaviours and interactions for Mexican Americans are changing due to globalization, acculturation, and personal preference. It is always important to be aware of and be open to changes in cultural norms from individuals, families, and communities and to avoid stereotyping. This is critical in understanding the individual and family in the context of the evolutionary nature of culture.

24.3.4 Format for Names

Names in most Spanish-speaking populations seem complex to those unfamiliar with the culture. A typical name is La Señorita Olga Gaborra de Rodriguez. Gaborra is the name of Olga's father, and Rodriguez is her mother's surname. When she marries a man with the surname Guiterrez, she becomes La Señora (denotes a married woman) Olga Guiterrez de Gaborra y Rodriguez. However, this full name is rarely used except on formal documents and for recording the name in the family Bible. Out of respect, most Mexican Americans are more formal when addressing non-family members. Thus, the best way to address Olga is not by her first name but rather as Señora Guiterrez. Titles such as *Don* and *Doña* for older respected members of the community and family should remain—not all members are respected so not all would have the title *Don* or *Doña*. If using English while communicating with people older than the nurse or health-care provider, use titles such as Mr., Ms., Miss, or Mrs., as a sign of respect.

Health-care providers must understand the role of older people when providing care to people of Mexican heritage. To develop confidence and *personalismo*, an element of formality must exist between health-care providers and older people. Becoming overly familiar by using physical touch or addressing them by first names may not be appreciated early in a relationship (Juckett 2013). As the health-care provider develops confidence in the relationship, becoming familiar may be less of a concern. However, using the first name of an older patient may never be appropriate (Zoucha and Zamarripa 2013).

24.4 Family Roles and Organization

24.4.1 Head of Household and Gender Roles

The typical family dominance pattern in traditional Mexican American families is patriarchal, with evidence of change toward a more egalitarian pattern in recent years (Greene 2017). Change to a more egalitarian decision-making pattern is primarily identified with more educated and higher socioeconomic families. **Machismo** in the Mexican culture historically has seen men as having strength, valor, and self-confidence, which the view and behavior has been changing for many. The female historically takes responsibility for decisions within the home and for maintaining the family's health. However, this practice for women has been evolving and more men engage in home and external responsibilities. *Machismo* historically assisted in sustaining and maintaining health and well-being not only for the man but also with implications for the family (Portland State University 2020).

24.4.2 Prescriptive, Restrictive, and Taboo Behaviors for Children and Adolescents

Children are highly valued because they ensure the continuation of the family and cultural values (Zoucha and Zamarripa 2013). They are closely protected and not encouraged to leave home. Even *compadres* (godparents) are included in the care of the young. Each child must have godparents in case something interferes with the parents' ability to fulfill their child-rearing responsibilities. Children are taught at an early age to respect parents and older family members, especially grandparents. Physical punishment may be used as a way of maintaining discipline and is sometimes considered child abuse in the US. Using children as interpreters in the health-care setting is discouraged owing to the restrictive nature of discussing gender-specific health assessments. This practice has been discouraged in most health care institutions in the US.

24.4.3 Family Goals and Priorities

The concept of *familism* is an all-encompassing value among Mexicans, for whom the traditional family is still the foundation of society. Family takes precedence over work and all other aspects of life. In many Mexican families, it is often said "God first, then family." The dominant Western health-care culture stresses including both the patient and the family in the plan of care. Mexicans are strong proponents of this family care concept, which includes the extended family. By including all family members, health-care providers can build greater trust and confidence and, in turn, increase adherence to health-care regimens and prescriptions (Sobel and Sawin 2014).

Reflective Exercise 24.1

Mrs. Uribe is a 64-year-old Mexican American woman recently diagnosed with blockages in her coronary vessels, and as a result, she will undergo a coronary artery bypass graft. Mrs. Uribe is recently widowed and is grieving for her husband of 39 years. She has 7 children (3 sons aged 37, 33, and 30; 4 daughters aged 38, 35, 32, and 28) and 18 grandchildren. The youngest son lives at home with his mother along with his wife and 2 children. The other children live within ten blocks. Mrs. Uribe spends a lot of time helping to care for the grandchildren while her children work. The five youngest members of the family were born in the US, and the rest of the family was born in Guanajunato, Mexico. Mrs. Uribe has never officially worked outside of the home and receives survivor benefits from her husband's pension. The only job she has ever done is house cleaning and other domestic help for her husband's previous work acquaintances. Mrs. Uribe has one living brother who lives 8 miles away and a sister who died of heart disease 5 years ago.

The Uribe family members are of the Pentocostal faith. Mrs. Uribe is a very devout and attends services several times a week at the church three blocks away. The children attend Church services with the family on occasional Sundays. Mrs. Uribe prays and reads the Bible so that God will take care of her and her family. Mrs. Uribe is considered a good cook and prepares dinner every evening for one of her sons and his family. The daughter-in-law helps cook the meals even after a full day of work. Mrs. Uribe and her family live in a three-bedroom wood frame house. The home is located in a predominant Mexican American neighborhood 2 miles from the Mexican border in Reynosa, Texas.

Mrs. Uribe does not have any work experience and is grateful that her husband left a small but substantial life insurance policy. Mrs. Uribe receives help with shopping and rides to the doctor from her youngest daughter and many *comadres*. One of her *comadres* is a ***curandera*** who has been offering Mrs. Uribe herbs and teas to help healing. Mrs. Uribe enjoys making tamales and *menudo* in her kitchen along with her family and *comadres*. All of the Uribe children, *comadres* as well as Church members have committed to help Mrs. Uribe during and after her surgery.

1. When the home health nurse comes to assess Mrs. Uribe's incision and teaches about wound care, who should be included in the teaching and why?
2. Explain the importance of *familism* to the Uribe family.
3. Mrs. Uribe has been offered herbal tea by the *curandera* while the home nurse is making a visit. Should the nurse intervene to stop this practice? Please provide rationale for your answer.
4. The nurse is making a visit when the family is praying and reading the Bible together for the health of Mrs. Uribe. The nurse is invited to join. What should the nurse do in this situation?

Single, divorced, and never-married male and female children may live with their parents or extended families, regardless of economics. Extended kinship is common through ***padrinos***: godparents who may be close friends and are considered the same as family (Zoucha and Zamarripa 2013). Thus, the words *brother, sister, aunt, cousin* and *uncle* do not necessarily mean that they are related by blood.

When grandparents and older parents are unable to live on their own, they may move in with their children. The extended family and kinship structure may promote an obligation to visit sick friends and relatives and may result in large numbers of visitors to hospitalized family members and friends. This practice may necessitate that health-care providers relax strict visiting policies in health-care facilities. In the alternative, ask a family spokes person to assist in regulating visitors.

Social status is highly valued among Mexican Americans, and a person who holds an academic degree or position with an impressive title commands great respect and admiration from family, friends, and the community. Good manners, a family, and family lineage, as indicated by extensive family names, also confer high status for Mexicans.

24.4.4 Alternative Lifestyles

Sixteen percent of Mexican families in the US live in poverty, and many are headed by a single female parent. This percentage is lower than that for other culturally diverse groups in the US (U. S. Census Bureau 2018a, b, c).

Reflective Exercise 24.2

Mr. Perez is an 80-year-old Mexican American man who was recently diagnosed with bladder and prostate cancer. Mr. Perez has been married for 60 years and has 8 adult children (3 daughters aged 57, 51, and 40; 5 sons aged 55, 53, 44, 43, and 42), 19 grandchildren, and 4 great-grandchildren. Mr. Perez's youngest son and his family live with Mr. and Mrs. Perez. The other Perez children, except the second-oldest daughter, live within 3–10 miles from their parents. The second-oldest daughter is a teacher and lives out of state. All members of the family except for Mr. Perez were born in the US. He was born in Durango, Mexico, and immigrated to the US at the age of 16 in order to work and send money to the family in Mexico. Mr. Perez has returned to Mexico throughout the years to visit and has lived in California ever since. Mr. Perez is retired from work as a carpet layer. Mr. Perez has one living older brother who lives within 15 miles. All members of the family speak Spanish and English fluently.

Most of the Perez family are Catholic, as evidenced by the religious items hanging on the wall and prayer books and rosary on the coffee table. Statues of St. Jude and Our Lady of Guadalupe are on the living room table. Mr. and Mrs. Perez have made many mandas (bequests or promises) to pray for the health of the family, including one to thank God for the healthy birth of all the children, especially after the doctor had discouraged them from having any children after the complicated birth of their first child. The family attends Mass together every Sunday morning and then meets for a traditional Mexican breakfast at a local restaurant frequented by many of their church's other parishioner families. Mr. Perez believes his health and the health of his family are in the hands of God.

The Perez family lives in a four-bedroom ranch home they bought 40 years ago. The home is located in a predominantly Mexican American neighborhood located in the El Norte section of town. Mr. and Mrs. Perez have been active in the church and neighborhood community until recently when Mr. Perez had been experiencing abdominal pain and difficulty urinating. The Perez home is usually occupied by many people at any given time and has always been the gathering place for the family.

During his years of employment, Mr. Perez was the major provider for the family and now receives Social Security checks and a pension. Mrs. Perez is also retired and receives a pension for work period as a nurse's aide. Mr. and Mrs. Perez count on their nursing student granddaughter to guide them and advise on their health care. Mr. Perez visits a *curandero* for medicinal folk remedies. Mrs. Perez is the provider of spiritual, physical, and emotional care for the family. In addition, their nursing student granddaughter is always present during any major surgeries or procedures. Mrs. Perez, the eldest daughter, and her granddaughter (the nursing student) will be caring for Mr. Perez during his procedure for a

TURBT (transurethreal resection of bladder tumor) as well as radiation therapy.

1. Explain the significance of family and kinship for the Rodriguez family.
2. Describe the importance of religion and God for the Rodriguez family.
3. Identify two stereotypes about Mexican Americans that were dispelled in this case with the Rodriguez family.
4. What is the role of Mrs. Rodriguez in this family?

Although homosexual behavior occurs in every society, The Williams Project reported that five states—California, Texas, Florida, New York, and Arizona have the highest number of same-sex Latinx couples, totalling 146,100, living together in the US (Kastanis and Gates 2013). Newspapers from Houston, Texas; Washington, DC; and Chicago, Illinois, report on the efforts of Hispanic lesbian and gay organizations in the areas of HIV and AIDS (*La SIDA* in Spanish) and life partner benefits. In Mexico, homosexuality is not a crime, but antihate groups raised serious concerns about killings of homosexual men, causing many to remain closeted (RefWorld 2020). In Mexico, *machismo* plays a large part in the phobic attitudes toward gay behavior. Larger cities in the US may have *Ellas*, a support group for Latina lesbians; El Hotline of Hola Gay, which provides referrals and information in Spanish; or Dignity, for gay Catholics. Health-care providers who wish to refer gay and lesbian patients to a support group may use such agencies.

24.5 Workforce Issues

24.5.1 Culture in the Workplace

In the US, Hispanics are the most underrepresented culturally diverse group in the health-care workforce. Although more than 18% of the American population is of Hispanic origin, only 10.2% of registered nurses are from Hispanic heritage (U.S. Department of Health and Human Services Health Resources and Services Administration 2018). Cultural differences that influence workforce issues include values regarding family, pedagogical approach to education, emotional sensitivity, views toward status, aesthetics, ethics, balance of work and leisure, attitudes toward direction and delegation, sense of control, views about competition, and time.

Because family is the first priority for most Mexicans, activities that involve family members usually take priority over work issues. Putting up a tough business front may be seen as a weakness in the Mexican culture. Because of this separation of work from emotions in American culture, most Mexican Americans tend to shun confrontation for fear of losing face. Many are very sensitive to differences of opinion, which are perceived as disrupting harmony in the workplace. People of Mexican heritage may find it important to keep peace in relationships in the workplace.

For many Mexicans, truth is tempered by diplomacy and tact. When a service is promised for tomorrow, even when they know the service will not be completed tomorrow, it is promised to please, not to deceive. These conflicting perspectives about truth can complicate treatment regimens and commitment to the completion of work assignments. Intentions must be clarified and, at times, altered to meet the needs of the changing and multicultural workforce.

For many Mexicans, work is viewed as a necessity for survival and may not be highly valued in itself, whereas money is for enjoying life. Most Mexican Americans place a higher value on other life activities. Material objects are usually necessities and not ends in themselves. The concept of responsibility is based on values related to attending to the immediate needs of family and friends rather than on the work ethic.

Many Mexicans believe that time is relative and elastic, with flexible deadlines, rather than stressing punctuality and timeliness. In Mexico, shop hours may be posted but not rigidly respected. A business that is supposed to open at 8:00 a.m. opens when the owner arrives; a posted time of 8:00 a.m. may mean the business will open at 8:30 a.m., later, or not at all. The same attitude toward time is evidenced in reporting to work and in keeping social engagements and medical appointments. If people believe that an exact time is truly important, such as the time an airplane leaves, then they may keep to a schedule. The real challenge for employers is to stress the importance and necessity of work schedules and punctuality in the American workforce.

24.5.2 Issues Related to Autonomy

Many Mexican Americans respond to direction and delegation differently from most European Americans. Many newer immigrants are used to having traditional autocratic managers who assign tasks but not authority, although this practice is beginning to change with more American-managed companies relocating to Mexico. Mexicans who were born and educated in the US usually have no difficulty communicating with others in the workplace. When better-educated Mexican immigrants arrive in the US, they usually speak some English. Newer immigrants from lower socioeconomic groups have the most difficulty acculturating in the workplace and may have greater difficulty with the English language.

24.6 Biocultural Ecology

24.6.1 Skin Color and Other Biological Variations

Because Mexican Americans draw their heritage from Spanish and French peoples and various North American and Central American Indian tribes and Africans, few physical characteristics give this group a distinct identity. Some individuals with a predominant Spanish background might have light-colored skin, blond hair, and blue eyes, whereas people from indigenous Indian backgrounds may have black hair, dark eyes, and cinnamon-colored skin. Intermarriages among these groups have created a diverse gene pool and have not produced a typical-appearing Mexican.

Cyanosis and decreased hemoglobin levels are more difficult to detect in dark-skinned people, whose skin appears ashen instead of the bluish color seen in light-skinned people. To observe for these conditions in dark-skinned Mexicans, the health-care provider must examine the sclera, conjunctiva, buccal mucosa, tongue, lips, nailbeds, palms of the hands, and soles of the feet. Jaundice, likewise, is more difficult to detect in darker-skinned people. Thus, the health-care provider needs to observe the conjunctiva and the buccal mucosa for patches of bilirubin pigment in dark-skinned Mexicans.

24.6.2 Diseases and Health Conditions

Common health problems most consistently documented in the literature for both people from Mexico and Mexican Americans are difficulty in assessing and utilizing health care are: heart disease, cancer, unintentional injuries, stroke, diabetes, asthma, COPD, HIV/AIDS, obesity, suicide, and liver disease (US Department of Health and Human Services Office of Minority Health 2019). In Mexican American migrant-worker populations, infectious, communicable, and parasitic diseases continue to be major health risks. Substandard housing conditions and employment in low-paying jobs have perpetuated higher rates of tuberculosis in Mexican Americans.

Newer Mexican immigrants from coastal lowland swamp areas and from some mountainous areas where mosquitoes are more prevalent may also have a higher incidence of malaria. Health-care providers must take these topographic factors into consideration when performing health screening for symptoms of anemia, lassitude, failure to thrive, and weight loss among Mexican immigrants.

Cardiovascular disease is the leading cause of death and disability in diverse populations, including Mexican Americans (Heron 2019). However, current research shows that despite the adverse cardiovascular risk profile, including the incidence of obesity, diabetes, and untreated hypertension, Mexican Americans have a lower rate of coronary heart disease mortality than non-White Hispanics (Balfour et al. 2016). Cardiovascular risk factors are influenced by behavioral, cultural, and social factors. Mexican Americans have the highest prevalence of no leisure time physical activity (Arrendondo et al. 2016). In addition, poor health, low social support, lack of educational and occupational opportunities, low access to health care, and discrimination contribute to the risk factors associated with cardiovascular disease (Balfour et al. 2016).

Mexican Americans have 5 times the rate of diabetes mellitus, with an increased incidence of related complications, as that in European American cohort groups. In addition, health-care providers working with Mexican immigrants and Mexican Americans should offer screening and teach patients preventive measures regarding pesticides and communicable and infectious diseases because many work with chemicals and may live in crowded housing conditions.

24.6.3 Variations in Drug Metabolism

Because of the mixed heritage of many Mexican Americans, it may be more difficult to determine a therapeutic dose of selected drugs. Several studies report differences in absorption, distribution, metabolism, and excretion of drugs, including alcohol, in some Hispanic populations. The mixed heritage of Mexican Americans makes it more difficult to generalize drug metabolism. Few studies include only one subgroup of Hispanics; therefore, health-care providers need to consider some notable differences when prescribing medications. Hispanics require lower doses of antidepressants and experience greater side effects than non-Hispanic Whites.

24.7 High-Risk Behaviors

Alcohol plays an important part in the Mexican culture. Many of this group's colorful lifestyle celebrations include alcohol consumption. Men overall drink in greater proportion than women, but this trend is changing due to acculturation. Mexican American women are consuming more alcohol than their mothers or grandmothers (National Institute on Alcohol Abuse and Alcoholism 2013).

Because of these drinking patterns, alcoholism represents a crucial health problem for many Mexicans. More acculturated Hispanics consume more alcoholic beverages than non-Hispanic Whites, possibly expecting alcohol to make them more socially acceptable and extroverted. Low acculturation and distorted self-image problems have special implications for nursing and health care.

Marijuana is the number-two drug used by Mexican Americans because it is readily available in their native land and easily accessible from people who work in farming and ranching occupations. Some adults who can afford drugs use cocaine and heroin, and the younger population uses inhalants (McCance-Katz 2020).

The trend toward decreasing cigarette smoking in the US is extending to the Mexican American culture, in which cigarette smoking rates have steadily declined for both men and women between 2015 and 2018 (Creamer et al. 2019). However, the reported decrease in cigarette smoking rates for Mexican American men and women should not promote a sense of complacency for nurses and health-care providers.

24.7.1 Health-Care Practices

Responsibility for health promotion and safety may be a major threat for those of Mexican heritage accustomed to depending on the family unit and traditional means of providing health care. Continuing disparities in health and health-seeking behaviors have been reported in several studies. Lower socioeconomic conditions and acculturation are responsible for Latina women being overweight, exhibiting hypertension, experiencing high cholesterol levels, and having increased smoking behaviors (Paz and Massey 2016). Latino men are less likely to have cancer screening or physical examinations than their non-Latino white counterparts. High-risk health behaviors such as drinking and driving, cigarette smoking, sedentary lifestyle, and non-use of seat belts increase with fewer years of educational attainment. Through educational programs and enforcement of state laws, more Mexicans are beginning to use seat belts.

24.8 Nutrition

24.8.1 Meaning of Food

As in many other ethnic groups, Mexicans and Mexican Americans celebrate with food. Mexican foods are rich in color, flavor, texture, and spiciness. Any occasion—births, birthdays, Sundays, religious holidays, official and unofficial holidays, and anniversaries of deaths—is seen as a time to celebrate with food and enjoy the companionship of family and friends. Because food is a primary form of socialization in the Mexican culture, Mexican Americans may have difficulty adhering to a prescribed diet for illnesses such as diabetes mellitus and cardiovascular disease. Health-care providers must seek creative alternatives and negotiate types of foods consumed with individuals and families in relation to these concerns.

24.8.2 Common Foods and Food Rituals

The Mexican American diet is extremely varied and may depend on the individual's region of origin in Mexico. Thus, one needs to ask the individual specifically about his or her dietary habits. The traditional staples of the Mexican American diet are rice (*arroz*), beans, and tortillas, which are made from corn (*maíz*) treated with calcium carbonate. However, in many parts of the US, only flour tortillas are available. Even though the diet is low in calcium derived from milk and milk products, tortillas treated with calcium carbonate provide essential dietary calcium. Popular Mexican American foods are eggs (*huevos*), pork (*puerco*), chicken (*pollo*), sausage (*chorizo*), lard (*monteca*), mint (*menta*), chili peppers (*chile*), onions (*cebollas*), tomatoes (*tomates*), squash (*calabaza*), canned fruit (*fruta de lata*), mint tea (*hierbabuena*), chamomile tea (*té de camomile* or *manzanilla*), carbonated beverages (*bebidas de gaseosa*), beer (*cerveza*), cola-flavored soft drinks, sweetened packaged drink mixes (*agua fresa*) that are high in sugar (*azucar*), sweetened breakfast cereals (*cereales de desayuno*), potatoes (*papas*), bread (*pan*), corn (*maíz*), gelatin (*gelatina*), custard (*flan*), and other sweets (*dulces*). Other common dishes include chili, enchiladas, tamales, tostadas, chicken mole, arroz con pollo, refried beans, tacos, tripe soup (*Menudo*), and other soups (*caldos*). *Caldos* are varied in nature and may include chicken, beef, and pork with vegetables.

Mealtimes vary among different subgroups of Mexican Americans. Whereas many individuals adopt North American schedules and eating habits, many continue their traditional practices, especially those in rural settings and migrant-worker camps. For these groups, breakfast is usually fruit, perhaps cheese, or bread alone or in some combination. A snack may be taken in mid-morning before the main meal of the day, which is eaten from 2 to 3 p.m. and, in rural areas especially, may last for 2 h or more. Mealtime is an occasion for socialization and keeping family members informed about one another. The evening meal is usually late and is taken between 9 and 9:30 p.m. Health-care providers must consider Mexican Americans' actual mealtimes when teaching patients about medication and dietary regimens related to diabetes mellitus and other illnesses.

24.8.3 Dietary Practices for Health Promotion

A dominant health-care practice for Mexicans and many Mexican Americans is the hot-and-cold theory of food selection. This theory is a major aspect of health promotion and illness, disease prevention, and treatment. According to this theory, illness or trauma may require adjustments in the hot-and-cold balance of foods to restore body equilibrium. The hot-and-cold theory of foods is described under Health-Care Practices, later in this chapter.

24.8.4 Nutritional Deficiencies and Food Limitations

In lower socioeconomic groups, wide-scale vitamin A deficiency and iron deficiency anemia exist (Access to Nutrition Index 2018). Some Mexican and Mexican Americans have lactose intolerance, which may cause problems for schools and health-care organizations that provide milk in the diet because of its high calcium content.

Because major Mexican foods and their ingredients are available throughout the US, traditional food practices may not change much when Mexicans immigrate. Of course, Mexican foods are extremely popular throughout the US and are eaten by many Americans because of the strong flavors, spiciness, and color. Table 24.1 lists the Mexican names of popular foods, their description, and ingredients. Individual adaptations to these preparations commonly occur.

24.9 Pregnancy and Childbearing Practices

24.9.1 Fertility Practices and Views toward Pregnancy

Mexican American birth rates 886,210, or 14.8%, live births in 2018 (Martin et al. 2019). Multiple births are common, especially in the economically disadvantaged groups. The optimal childbearing age for Mexican women is between 19 and 24 years. Fertility practices of Mexican Americans are connected with their predominantly Catholic religious beliefs and their tendency to be modest. Some women practice the belief that prolonged infant breastfeeding is a method of birth control. Abortion in many communities is considered morally wrong and is practiced (theoretically) only in extreme circumstances to keep the mother's life intact. However, legal and illegal abortions are common in some parts of Mexico and the US. Despite the strong influence of the Catholic Church over fertility practices; being Catholic does not prevent some Mexican American women from using contraceptives, sterilization, or abortion for unwanted pregnancies.

Table 24.1 Mexican foods

Common name	Description	Ingredients
Arroz con pollo	Chicken with rice	Chicken baked, boiled, or fried and served over boiled or fried rice
Chili	Chili	Same as the US but tends to be more spicy
Chili con carne	Chili with meat	Chili with beef or pork
Chili con salsa	Chili with sauce	Chili with a sauce that contains no meat
Dulces	Sweets	Candy and desserts usually high in sugar, lard, and eggs
Enchiladas	Enchiladas	Tortilla rolled and stuffed with meat or cheese and a spicy sauce
Papas fritas	Fried potatoes	Potatoes usually fried in lard
Flan	Flan	Popular dessert made of egg custard; may be filled with fruit or cheese
Gelatina	Gelatin	Popular dessert made with sugar, eggs, and jelly
Pollo con molé	Chicken molé	Chicken with a sauce made of hot spices, chocolate, and chili
Salchica or *chorizo*	Sausage	Sausage almost always made with pork and spices
Tacos	Tacos	Tortilla folded around meat or cheese
Tamales	Tamales	Fried or boiled chopped meat, peppers, cornmeal, and hot spices
Tortilla (flour or corn)	Tortilla	A thin unleavened bread made with cornmeal and treated with lime (calcium carbonate) or flour and lard
Tostadas	Tostadas	Fried corn tortilla with refried beans and lettuce and tomatoes

Diaphragms, foams, and creams are not commonly used for birth control practice, mostly because they are not approved by Catholic doctrine. Birth control pills may be considered unacceptable because they are an artificial means of birth control. Physicians' offices and clinics that see large numbers of migrant workers on the Delmarva Peninsula on the U.S. East Coast report that many younger female patients are using Norplant (levonorgestrel; a long-term contraceptive system) for birth control. Family planning is one area in which health-care providers can help the family to identify more realistic outcomes consistent with current economic resources and family goals.

Foreign-born Mexicans are less likely to give birth to low-birth-weight babies than US-born Mexican women, even though US-born mothers are usually of higher socioeconomic status and receive more prenatal care. Research suggests that better nutritional intake and lower prevalence of smoking and alcohol use are some reasons for these protective outcomes (Mydam et al. 2019).

Because pregnancy among Mexican Americans is viewed as natural and desirable, many women do not seek prenatal evaluations. In addition, because prenatal care is not available to every woman in Mexico, some women do not know about the need for prenatal care. With the extended family network and the woman's role of maintaining the health status of family members, many pregnant women seek family advice before seeking medical care. Thus, *familism* may deter and hinder early prenatal checkups. To encourage prenatal checkups, health-care providers can encourage female relatives and husbands to accompany the pregnant woman for health screening and incorporate advice from family members into health teaching and preventive care services. Using videos with Spanish-speaking Mexican Americans is one culturally effective way for incorporating health education, especially for those patients who have a limited understanding of English. In addition, incorporating cultural brokers known to the Mexican American family may help to empower patients and reduce conflict for Mexicans and Mexican Americans.

24.9.2 Prescriptive, Restrictive, and Taboo Practices in the Childbearing Family

Beliefs related to the hot-and-cold theory of disease prevention and health maintenance influence conception, pregnancy, and postpartum rituals. For instance, if part of the belief, during pregnancy a woman is more likely to favor hot foods, which are believed to provide warmth for the fetus and enable the baby to be born into a warm and loving environment (Juckett 2013). Cold foods and environments are preferred during the menstrual cycle and in the immediate post delivery period. Many pregnant women sleep on their backs to protect the infant from harm, keep the vaginal canal well lubricated by having frequent intercourse to facilitate an easier birth, and keep active to ensure a smaller baby and to prevent a decrease in the amount of amniotic fluid (Lynch et al. 2012). An important activity restriction is that pregnant women should not walk in the moonlight because it might cause a birth deformity. To prevent birth deformities, pregnant women may wear a safety pin, metal key, or some other metal object on their abdomen (Alcaniz 2018). Other beliefs include avoiding cold air, not reaching over the head in order to prevent the baby's cord from wrapping around its neck, and avoiding lunar eclipses because they may result in deformities.

In more traditional Mexican families, the father is not included in the delivery experience and should not see the mother or baby until after both have been cleaned and dressed. This practice is based on the fear that harm may come to the mother, baby, or both. Integrating men into the birthing of a child is a process that requires changing social habits in relation to cultural aspects of life and gender roles. For many, the presence of men during delivery is considered an uninvited intrusion into the Mexican culture. Among less traditional and more acculturated Mexican Americans, men participate in prenatal classes and assist in the delivery room. However, based on personal experiences, men who provide support during delivery may receive friendly gibing from their male counterparts for taking the role of the wife's mother (personal communication, Larry Purnell, 2020). In any event, health-care providers must respect Mexicans' decision to not have men in the delivery room.

During labor, traditional Mexican women may be quite vocal and are taught to avoid breathing air in through the mouth because it can cause the uterus to rise up. Immediately after birth, they may place their legs together to prevent air from entering the womb (Dillon 2016). Health-care providers can help the Mexican pregnant woman have a better delivery by encouraging attendance at prenatal classes.

The postpartum preference for a warm environment may restrict postpartum women from bathing or washing their hair for up to 40 days. Although postpartum women may not take showers or sit in a bathtub, this does not mean that they do not bathe. They take "sitz" baths, wash their hair with a washcloth, and take sponge baths. Other postpartum practices include wearing a heavy cotton abdominal binder, cord, or girdle to prevent air from entering the uterus; covering one's ears, head, shoulders, and feet to prevent blindness, mastitis, frigidity, or sterility; and avoiding acidic foods to protect the baby from harm (Dillon 2016).

When the baby is born, special attention is given to the umbilicus; the mother may place a belt around the umbilicus (*ombliguero*) to prevent the navel from popping out when the child cries. Cutting the baby's nails in the first 3 months is thought to cause blindness and deafness.

Health-care providers need to make special provisions to provide culturally congruent health teaching for lactating women who work with or are exposed to pesticides, such as dichlorodiphenyldichlorothene (DDE), the most stable derivative from the pesticide DDT. High DDE levels among lactating women have a direct correlation with a decrease in lactation and increase in breast cancer, especially in women who have had more than one pregnancy and previous lactation (Gray et al. 2017). Education level and degree of acculturation are key issues when developing health education and interventions for risk reduction.

24.10 Death Rituals

24.10.1 Death Rituals and Expectations

Mexicans often have a stoic acceptance of the way things are and view death as a natural part of life and the will of God (Martin and Barkley Jr. 2016). Death practices are primarily an adaptation of their religion. Family members may arrive in large numbers at the hospital or home in times of illness or an approaching death. In more traditional families, family members may take turns sitting vigil over the sick or dying person. Autopsy is acceptable as long as the body is treated with respect. Burial is the common practice; cremation is an individual choice.

24.10.2 Responses to Death and Grief

When a person dies, the word travels rapidly, and family and friends travel from long distances to get to the funeral. They may gather for a **velorio**, a festive watch over the body of the deceased person before burial. Some Mexican Americans bury the body within 24 h, which is required by law in Mexico.

More traditional grieving families may engage in the protection of the dying and bereaved, such as small children who have difficulty dealing with the death (Andrews and Boyle 2016). Mexican Americans encourage expressions of feeling during the grieving process. In these cases, health-care providers can assist the person by providing support and privacy during the bereavement.

24.11 Spirituality

24.11.1 Dominant Religion and Use of Prayer

The predominant religion of most Mexicans and Mexican Americans is Catholicism. The major religions in Mexico are Roman Catholic with

76.5%; Protestant with 5.2%; Pentecostal with 1.4%; unspecified with 13.1%; Jehovah's Witness, 1.1%; other 0.3%; and 3.1% identified with no religion. Since the mid-1980s, other religious groups such as Mormons, Jehovah's Witnesses, Seventh Day Adventists, Presbyterians, and Baptists have been gaining in popularity in Mexico (CIA World Factbook 2019). Although many Mexicans and Mexican Americans may not appear to be practicing their faith on a daily basis, they may still consider themselves devout Catholics, and their religion has a major influence on health-care practices and beliefs. For many, Catholic religious practices are influenced by indigenous Indian practices.

Newer immigrant Mexican Americans may continue their traditional practice of having two marriage ceremonies, especially in lower socioeconomic groups. A civil ceremony is performed whenever two people decide to make a union. When the family gets enough money for a religious ceremony, they schedule an elaborate celebration within the church. Common practice, especially in rural Mexican villages and some rural villages in the southwestern US, is to post a handwritten sign on the local church announcing the marriage, with an invitation for all to attend.

Frequency of prayer is highly individualized for most Mexican Americans. Even though some do not attend church on a regular basis, they may have an altar in their homes and say prayers several times each day, a practice more common among rural isolationists.

24.11.2 Meaning of Life and Individual Sources of Strength

The family is foremost to most Mexicans, and individuals get strength from family ties and relationships. Individuals may speak in terms of a person's soul or spirit (*alma* or *espiritu*) when they refer to one's inner qualities. These inner qualities represent the person's dignity and must be protected at all costs in times of both wellness and illness. In addition, Mexicans derive great pride and strength from their nationality, which embraces a long and rich history of traditions.

Leisure is considered essential for a full life, and work is a necessity to make money for enjoying life. Mexican Americans pride themselves on good manners, etiquette, and grooming as signs of respect. Because the overall outlook for many Mexicans is one of fatalism, pride may be taken in stoic acceptance of life's adversities.

24.11.3 Spiritual Beliefs and Health-Care Practices

Most Mexicans enjoy talking about their soul or spirit, especially in times of illness, whereas many health-care providers may feel uncomfortable talking about spirituality. This tendency may communicate to Mexicans that the health-care provider has suspect intentions, is insensitive, and is not really interested in them as individuals. It may be common for a person needing care in the home or hospital to have a statue of a patron saint or a candle with a picture of the saint. Rosaries may be present, and at times, the family may pray as a group. Depending on the confidence maintained with the family and client, a health-care provider may be asked to join in the prayer. If time permits, it is very appropriate to pray with the family, even if only for a few minutes. This action promotes confidence in the relationship and can have a positive impact on the health and well-being of the patient and family (Zoucha and Zamarripa 2013).

24.12 Health-Care Practices

24.12.1 Health-Seeking Beliefs and Behaviors

The family is the most credible source of health information and the most significant impediment to positive health-seeking behavior. Mexican Americans' fatalistic worldview and external locus of control are closely tied to health-seeking behaviors. Because expressions of negative feelings are considered impolite, Mexicans may be reluctant to complain about health problems or to place blame on the individual for poor health. If a person becomes seriously ill, that is just the way things are; all events are acts of God (Zoucha and Zamarripa 2013). This belief system may impair the dominant view of communications and hinder health teaching, health promotion, and disease-prevention practices. Therefore, it is imperative for health-care providers to plan health-promoting activities and teaching that are consistent with this belief but encourage health. For instance, if a person believes that the illness is due to a punishment from God, it may be possible to ask to be forgiven by God, thereby restoring health. This may be an opportune time to call a priest or minister for official recognition of forgiveness.

24.12.2 Responsibility for Health Care

To many Mexicans, good health may mean the ability to keep working and have a general feeling of well-being (Zoucha and Zamarripa 2013). Illness may occur when the person can no longer work or take care of the family. Therefore, many Mexicans may not seek health care until they are incapacitated and unable to go about the activities of daily living. Unfortunately, many people of Mexican heritage may not know and understand the occupational dangers inherent in their daily work. Migrant workers are often unaware of the dangers of pesticides and the potentially dangerous agricultural machinery. Health-care providers must serve as advocates for these people regarding occupational safety. Often, the companies do not tell the workers of the dangers of the work, or the workers may not understand owing to the inability of the company officials to speak the language of the workers.

The use of over-the-counter medicine may pose a significant health problem related to self-care for many Mexican Americans. In part, this is a carryover from Mexico's practice of allowing over-the-counter purchases of antibiotics, intramuscular injections, intravenous fluids, birth control pills, and other medications that require a prescription in the US. Often, Mexican immigrants bring these medications across the border and share them with friends. In addition, friends and relatives in Mexico send drugs through the mail. To protect patients from contradictory or potentiating effects of prescribed treatments, health-care providers need to ask patients about prescription and non-prescription medications they may be taking.

24.12.3 Folk and Traditional Practices

Mexican Americans engage in folk medicine practices and use a variety of prayers, herbal teas, and poultices to treat illnesses. Many of these practices are regionally specific and vary between and among families. The Mexican *Ministerio de Salud Publica y Asistencia Social* (Ministry of Public Health and Social Assistance) publishes an extensive manual on herbal medicines that are readily available in Mexico. Lower socioeconomic groups and well-educated upper- and middle-socioeconomic Mexicans to some degree practice traditional and folk medicine. Many of these practices are harmless, but some may contradict or potentiate therapeutic interventions. Thus, as with the use of other prescription and non-prescription drugs discussed earlier, it is essential for health-care providers to be aware of these practices and to take them into consideration when providing treatments (Rivera et al. 2003). The provider must ask the Mexican American patient specifically whether she or he is using folk medicine.

To provide culturally competent care, health-care providers must be aware of the hot-and-cold theory of disease when prescribing treatment modalities and when providing health teaching. According to this theory, many diseases are caused by a disruption in the hot-and-cold balance of the body. Thus, eating foods of the opposite variety may either cure or prevent specific hot-and-cold illnesses and conditions. Physical or mental illness may be attributed to an imbalance between the person and the environment. Influences include emotional, spiritual, and social state, as well as physical factors such as humoral imbalance expressed as either too much hot or cold. As health-care providers, it is important to understand that if people of Mexican heritage believe in the hot-and-cold theory, it means that they do not believe or use professional Western practices (Spector 2017). Unless a level of trust and confidence is maintained, Mexicans who follow these beliefs may not express them to health-care providers (Zoucha and Zamarripa-Zoucha 2016).

Hot and cold are viewed as specific properties of various substances and conditions, and sometimes opinions differ about what is hot and what is cold in the Mexican community. In general, cold diseases or conditions are characterized by vasoconstriction and a lower metabolic rate. Cold diseases or conditions include menstrual cramps, *frio de la matriz*, rhinitis (*coryza*), pneumonia, *empacho*, cancer, malaria, earaches, arthritis, pneumonia and other pulmonary conditions, headaches, and musculoskeletal conditions and colic. Common hot foods used to treat cold diseases and conditions include cheeses, liquor, beef, pork, spicy foods, eggs, grains other than barley, vitamins, tobacco, and onions (Neff 2011).

Hot diseases and conditions may be characterized by vasodilation and a higher metabolic rate. Pregnancy, hypertension, diabetes, acid indigestion, *susto*, *mal de ojo* (bad eye or evil eye), *bilis* (imbalance of bile, which runs into the bloodstream), infection, diarrhea, sore throats, stomach ulcers, liver conditions, kidney problems, and fever may be examples of hot conditions. Common cold foods used to treat hot diseases and conditions include fresh fruits and vegetables, dairy products (even though fresh fruits and dairy products may cause diarrhea), barley water, fish, chicken, goat meat, and dried fruits (Neff 2011).

Folk practitioners are consulted for several notable conditions. *Mal de ojo* is a folk illness that occurs when one person (usually older) looks at another (usually a child) in an admiring fashion. Another example of *mal de ojo* is if a person admires something about a baby or child, such as beautiful eyes or hair. Such eye contact can be either voluntary or involuntary. Symptoms are numerous, ranging from fever, anorexia, and vomiting to irritability. The spell can be broken if the person doing the admiring touches the person admired while it is happening. Children are more susceptible to this condition than women, and women are more susceptible than men. To prevent *mal de ojo*, the child wears a bracelet with a seed (*ojo de venado*) or a bag of seeds pinned to the clothes (Kemp 2006).

Another childhood condition often treated by folk practitioners is ***caida de la mollera*** (fallen fontanel). The condition has numerous causes, which may include removing the nursing infant too harshly from the nipple or handling an infant too roughly. Symptoms range from irritability to failure to thrive. To cure the condition, the child is held upside down by the legs.

Susto (magical fright or soul loss) is associated with epilepsy, tuberculosis, and other infectious diseases and is caused by the loss of spirit from the body. The illness is also believed to be caused by a fright or by the soul being frightened out of the person. This culture-bound disorder may be psychological, physical, or physiological in nature. Symptoms may include anxiety, depression, loss of appetite, excessive sleep, bad dreams, feelings of sadness, and lack of motivation. Treatment sometimes includes elaborate ceremonies at a crossroads with herbs and holy water to return the spirit to the body (R. Zamarripa, personal communication, 2020).

Empacho (blocked intestines) may result from an incorrect balance of hot and cold foods, causing a lump of food to stick in the gastrointestinal tract. To make the diagnosis, the healer may place a fresh egg on the abdomen. If the egg appears to stick to a particular area, this confirms the diagnosis. Older women usually treat the condition in children by massaging their stomach and back to dislodge the food bolus and to promote its continued passage through the body.

Health-care providers are cautioned against diagnosing psychiatric illnesses too readily in the Mexican population. The syndromes ***mal ojo*** and *susto* are culture-bound and are potential sources of diagnostic bias. The potential culture-bound mental illness must be understood in the context of the culture and the unique symptoms that accompany each illness.

24.12.4 Barriers to Health Care

Thirty-five percent of Mexican Americans, compared with 17% of the U.S. population in general, do not have health insurance (CDC 2011). A number of factors may account for this high percentage of uninsured individuals. First, many Mexican Americans constitute the working poor and are unable to purchase insurance. Second, many are migratory and do not qualify for Medicaid. Third, many have an undocumented status and are afraid to apply for health insurance. Fourth, even though insurance is available in their native homeland, it is very expensive and not part of the culture.

Whereas wealthier Mexican Americans have little difficulty accessing health care in the US, lower socioeconomic groups may experience significant barriers, including inadequate financial resources, lack of insurance and transportation, limited knowledge regarding available services, language difficulties, and the cultures of health-care organizations. Like many other immigrant groups who lack a primary provider, Mexican Americans may use emergency rooms for minor illnesses. Health-care providers have the opportunity to improve the care of Mexican Americans by explaining the health-care system, incorporating a primary-care provider whenever possible, using an interpreter of the same gender, securing a cultural broker, and assisting patients in locating culturally specific mental health programs (Zoucha and Zamarripa 2013).

24.12.5 Cultural Responses to Health and Illness

Good health to many Mexican Americans is to be free of pain, able to work, and spend time with the family. In addition, good health is a gift from God and from living a good life (Zoucha 2011).

Mexicans and Mexican Americans tend to perceive pain as a necessary part of life, and enduring the pain is often viewed as a sign of strength. Men commonly tolerate pain until it becomes extreme (Hollingshead et al. 2016). Often, pain is viewed as the will of God and is tolerated as long as the person can work and care for the family. These attitudes toward pain delay seeking treatment; many hope that the pain will simply go away. Research has shown that many Mexican Americans experience more pain than other ethnic groups but that they report the occurrence of pain less frequently and endure pain longer (Hollingshead et al. 2016). Six themes have emerged that describe culturally specific attributes of Mexican Americans experiencing pain:

1. Mexicans accept and anticipate pain as a necessary part of life.
2. They are obligated to endure pain in the performance of duties.
3. The ability to endure pain and to suffer stoically is valued.
4. The type and amount of pain a person experiences are divinely predetermined.
5. Pain and suffering are a consequence of immoral behavior.
6. Methods to alleviate pain are directed toward maintaining balance within the person and the surrounding environment (Torres et al. 2017).

By using these themes, health-care providers can evaluate Mexicans experiencing pain within their cultural framework and provide culturally specific interventions.

Because long-term-care facilities in Mexico are rare and tend to be crowded, understaffed, and expensive, many Mexican Americans may not consider long-term care as a viable option for a family member. In addition, because of the importance of extended family, Mexican Americans may prefer to care for their family members with mental illness, physical handicaps, and extended physical illnesses at home. In Mexican American culture, someone with a mental illness is not looked on with scorn or blamed for his or her condition because mental illness, like physical illness, is viewed as God's will. It is common to accept those with mental illness and care for them in the context of the family until the illness is so bad that they cannot be managed in the home (Zoucha and Zamarripa 2013).

Mexicans can readily enter the sick role without personal feelings of inadequacy or blame. A person can enter the sick role with any acceptable excuse and be relieved of life's responsibilities. Other family members willingly take over the sick person's obligations during his or her time of illness.

24.12.6 Blood Transfusions and Organ Donation

Extraordinary means to preserve life are frowned on in the Mexican and Mexican American cultures, and ordinary means are commonly used to preserve life. Extraordinary means are defined and determined by the individual, taking into account such factors as finances, education, and availability of services.

Blood transfusions are acceptable if the individual and the family agree that the transfusion is necessary. Organ donation, although not deemed morally wrong, is not a common practice and is usually restricted to cadaver donations, because donating an organ while the person is still alive means that the body is not whole. Acceptance of organ transplant as a treatment option is seen primarily among more educated people. One reason that organ transplant is unacceptable to some groups is the belief that *mal aire* (bad air) enters the body if it is left open too long during surgery and increases the potential for the development of cancer.

24.13 Health-Care Providers

24.13.1 Traditional Versus Biomedical Providers

Educated physicians and nurses are often seen as outsiders, especially among newer immigrants. However, health-care providers are viewed as knowledgeable and respected because of their education (Zoucha and Zamarripa 2013). To overcome this initial awkwardness, health-care providers should attempt to get to know the patient on a more personal level and gain confidence before initiating treatment regimens. Engaging in small talk unrelated to the health-care encounter before obtaining a health history or providing health education is advised. Health-care providers must respect this cultural practice to achieve an optimal outcome from the encounter.

Folk practitioners, who are usually well known by the family, are usually consulted before and during biomedical treatment. Numerous illnesses and conditions are caused by witchcraft. Specific rituals are carried out to eliminate the evils from the body. Lower socioeconomic and newer immigrants are more likely to use folk practitioners, but well-educated upper- and middle-class people also visit folk practitioners and ***brujas*** (witches) on a regular basis Zoucha and Zamarripa 2013). Although often no contradictions or contraindications to folk remedies exist, health-care providers must always consider patients' use of these practitioners to prevent conflicting treatment regimens.

Even though the Catholic Church preaches against some types of folk practitioners, they are common and meet yearly for several days in Catemaco, Veracruz. Folk practitioners include the ***curandero***, who may receive her or his talents from God or serve an apprenticeship with an established practitioner. The *curandero* has great respect from the community, accepts no monetary payment (but may accept gifts), is usually a member of the extended family, and treats many traditional illnesses. A *curandero* does not usually treat illnesses caused by witchcraft.

The ***yerbero*** (also spelled ***jerbero***) is a folk healer with specialized training in growing herbs, teas, and roots and who prescribes these remedies for prevention and cure of illnesses. A *yerbero* may suggest that the person go to a ***botanica*** (herb shop) for specific herbs. In addition, these folk practitioners frequently prescribe the use of laxatives.

A ***sobador*** subscribes to treatment methods similar to those of a Western chiropractor. The *sobador* treats illnesses, primarily affecting the joints and musculoskeletal system, with massage and manipulation.

Even though Mexicans like closeness and touch within the context of family, most tend to be modest in other settings. Women are not supposed to expose their bodies to men or even to other women. Female patients may experience embarrassment when it is necessary to touch their genitals or may refuse to have pelvic examinations as a routine part of a health assessment. Men may have strong feelings about modesty as well, especially in front of women, and may be reluctant to disrobe completely for an examination. Mexican Americans often desire that members of the same gender provide intimate care (C. Zamarripa, personal communication, 2020). Health-care providers must keep in mind patients' need for modesty when disrobing or being examined. Thus, only the body part being examined should be exposed, and direct care should be provided in private. Whenever possible, a same-gender caregiver should be assigned to Mexican Americans.

24.13.2 Status of Health-Care Providers

Mexican American patients have great respect for health-care providers because of their training and experience. They expect health-care providers to project a professional image and be well groomed and dressed in attire that reflects their professional status (Zoucha and Zamarripa 2013). Whereas they have great respect for health-care providers, some Mexican Americans may distrust them out of fear that they will disclose their undocumented status. Health-care providers who incorporate folk practitioners, the concept of *personalismo*, and respect into their approaches to care of Mexican American patients will gain their patients' confidence and be able to obtain more thorough assessments.

Reflective Exercise 24.3

Vicente Leon is a 25-year-old Mexican man who was recently diagnosed with a right radial bone fracture after a work-related injury. An emergency room physician has recommended surgery and physical therapy. Vicente is unmarried and is a recent immigrant from Mexico City, Mexico. He is an undocumented worker and has been working for a construction company doing roofing and bricklaying. Vicente's family resides in Mexico. His parents, maternal grandparents, five sisters, and two brothers live in a small two-bedroom stone home in the *Colonia* region of Mexico City. Vicente is the oldest of the children and has come to the US to work and send money back to the family. Vicente's dad is being treated for colon cancer and needs money to pay for health care. Vicente is also trying to earn enough money to bring his dad to the US for further cancer treatment. Vicente speaks mainly Spanish, with limited ability in English.

Vicente is a devout Catholic who attends Mass weekly and prays the rosary to *La Virgen de Guadalupe* daily. Vicente often blesses himself with holy water he brought from San Juan de Los Lagos. Vicente believes that God will heal him and that his health is in the hands of God.

Vicente is sharing the rent on a two-bedroom apartment with three other migrant workers from Mexico. The apartment is located 15 miles from his job where new homes are being built outside the city. Vicente usually takes two buses to work. One of the migrant workers has an uncle who helped secure the jobs for them. Vicente and his co-workers cook and eat dinner together most evenings and enjoy drinking *cervezas* (beer) on the weekends.

Vicente has saved money from working over the past 18 months but is worried about health-care coverage. He usually goes to a local clinic for his health-care needs. His friends suggested that he should visit a *bruja* because he might have had a spell cast upon him. He and his friends believe that the *bruja* can rid him of the spell and heal him. Vicente's friends are able to help take care of him on weekends only because of their weekday 12-h work schedules. Vicente has an uncle from

Mexico who is trying to get money together for a trip up to help Vicente as he recovers. Vicente will require home physical therapy and nursing care after his surgery. He will also be unable to work for 6–8 weeks.

1. The home-care case manager, a registered nurse, is sending a physical therapist to the home. What should the nurse consider?
2. What does the nurse need to know about what part of Mexico Vicente and his family are from?
3. Identify potential communication needs of Vicente, his friends, and his visiting family.
4. Vicente is concerned about letting his boss down because of his illness. Why is Vicente concerned about this with his boss?

Health-care providers can demonstrate respect for Mexican American patients by greeting the patient with a handshake, touching the person, or holding the person's hand, all of which help to build trust in the therapeutic relationship. Providing information and involving the family in decisions regarding health; listening to the individual's concerns; and treating the individual with *personalismo*, which stresses warmth and personal relationships, also foster trust.

References

Access to Nutrition Index (2018) Global index 2018. https://www.accesstonutrition.org/sites/gl18.atnindex.org/files/resources/atni_report_global_index_2018.pdf

Alcaniz L (2018) Traditional Hispanic beliefs and myths about pregnancy. https://www.babycenter.com/0_traditional-hispanic-beliefs-and-myths-about-pregnancy_3653769.bc

American Social History Productions/Center for Media and Learning (2020) Early twentieth century mexican immigration to the U.S. https://herb.ashp.cuny.edu/exhibits/show/mexican-immigration

Andrews M, Boyle J (2016) Transcultural concepts in nursing care, 7th edn. Wolters Kluwer, Philadelphia

Arrendondo EM, Sotres-Alvarez D, Stoutenberg M, Davis SM, Crespo NC, Carnethon MR, Astaneda SF, Isasi CM, Esponiza RA, Daviglus ML, Perez LG, Evenson KR (2016) Physical activity levels in the U. S Latino/Hispanic adults: results from the Hispanic community health study/study of Latinos. Am J Prevent Med 50(4):500–508. https://doi.org/10.1016/j.amepre.2015.08.029

Balfour PC, Ruiz JM, Talavera GA, Allison MA, Rodriguez CJ (2016) Cardiovascular disease in Hispanics/Latinos in the United States. J Latina/o Psychol 4(2):98–113. https://doi.org/10.1037/lat0000056

Beezley W (2019) The Oxford encyclopedia of Mexican history and culture. Oxford University Press Inc., Oxford

Centers for Disease Control and Prevention (CDC) (2011) Health of Mexican American population. http://www.cdc.gov/nchs/fastats/mexican_health.htm

Central Intelligence Agency (2019) The world factbook: North America: Mexico. https://www.cia.gov/library/publications/the-world-factbook/geos/mx.html

Creamer MR, Wang TW, Babb S, Cullen KA, Day H, Willis G, Jamal A, Neff L (2019) Tobacco product use and cessation indicators among adults – United States, 2018. Morb Mortal Wkly Rep 68(45):1013–1019. https://www.cdc.gov/mmwr/volumes/68/wr/mm6845a2.htm?s_cid=mm6845a2_w#T1_down

Department of Homeland Security (2011) Estimates of the unauthorized immigrant population residing in the United States. http://www.dhs.gov/xlibrary/assets/statistics/publications/ois_ill_pe_2010.pdf

Dillon PM (2016) Nursing health assessment: the foundation of clinical practice, 3rd edn. F. A. Davis Company, Philadelphia

Gonzales-Barrera A, Krogstad JM (2019) What we know about illegal immigration from Mexico. https://www.pewresearch.org/fact-tank/2019/06/28/what-we-know-about-illegal-immigration-from-mexico/

Gray JM, Rasanayagam S, Engel C, Rizzo J (2017) State of the evidence 2017: an update on the connection between breast cancer and the environment. Environ Health 16(94):94. https://doi.org/10.1186/s12940-017-0287-4

Greene A (2017) Mexican immigration to the United States and its effect on Mexican culture and the family structure. https://info.umkc.edu/latinxkc/essays/spring-2017/mexican-immigration-family/

Heron, M. (2019). Deaths: leading causes for 2017. Natl Vital Stat Rep 68(6). https://www.cdc.gov/nchs/data/nvsr/nvsr68/nvsr68_06-508.pdf

Hollingshead NA, Ashburn-Nardo L, Stewart JC, Hirsh AT (2016) The pain experience of Hispanic Americans: A critical literature review and conceptual model. J Pain 17(5):513–528. https://doi.org/10.1016/j.jpain.2015.10.022

Juckett G (2013) Caring for Latino patients. Am Fam Physician 87(1):48–54. https://www.aafp.org/afp/2013/0101/p48.html#sec-2

Kastanis A, Gates GJ (2013) LGBT Latino/a individuals and Latino/a same-sex couples. https://

williamsinstitute.law.ucla.edu/research/census-lgbt-demographics-studies/lgbt-latino-oct-2013/
Kemp C (2006) Hispanic health beliefs and practices: Mexican and Mexican-Americans(clinical notes). http://www.nursingworld.org/MainMenuCategories/ANAMarketplace/ANAPeriodicals/OJIN/TableofContents/Volume112006/No3Sept06/ArticlePreviousTopics/CulturallyCompetentNursingCare.html
Lopez CG (2016) No, conservative media, that's not what "La Raza" means in Spanish. MediaMaters for America. https://www.mediamatters.org/sean-hannity/no-conservative-media-thats-not-what-la-raza-means-spanish
Lynch KE, Landsbaugh JR, Whitcomb BW, Pekow P, Markenson G, Chasan-Taber L (2012) Physical activity pregnant Hispanic women. Am J Prevent Med 43(4):434–439. https://doi.org/10.1016/j.amepre.2012.06.020
Martin EM, Barkley TW Jr (2016) Improving cultural competence in end-of-life pain management. Nursing 46(1):32–41. www.nursing2016.com
Martin JA, Hamilton BE, Osterman MJK, Driscoll AK (2019) Births: final data for 2018. Natl Vital Stat Rep 68(13):1–46
McCance-Katz EF (2020) 2018 National survey on drug use and health: Hispanics, Latino, or Spanish origin of descent. http://www.samhsa.gov/data/sites/default/files/reports/rpt23249/4_Hispanic_2020_01_14.pdf
Merriam-Webster (2020) Latinx. In The Merriam-Webster.com dictionary. https://www.merriam-webster.com/dictionary/Latinx
Mydam J, David RJ, Rankin KM, Collins JW (2019) Low birth weight among infants born to black Latina women in the United States. Mater Child Health J 23(4):538–546. https://doi.org/10.1007/s10995-018-2669-9
National Institute on Alcohol Abuse and Alcoholism (2013) Alcohol and the hispanic community. https://www.niaaa.nih.gov/sites/default/files/hispanicFact.pdf
Neff N (2011) Folk medicine in Hispanics in the southwestern United States. http://www.rice.edu/projects/HispanicHealth/Courses/mod7/mod7.html
Paz K, Massey KP (2016) Health disparity among Latina women: comparison with non-Latina women. Clin Med Insights Womens Health 9(Suppl 1):CMWH.S38488. https://doi.org/10.4137/CMWH.S38488
Portland State University (2020) Gender roles. https://www.pdx.edu/multicultural-topics-communication-sciences-disorders/gender-roles
Radford J (2019) Key findings about U.S. immigrants. https://www.pewresearch.org/fact-tank/2019/06/17/key-findings-about-u-s-immigrants/
Refworld (2020) Update: Treatment of homosexuals in Mexico. https://www.refworld.org/docid/3ae6a63e0.html
Rivera JO, Anaya JP, Meza A (2003) Herbal product use in Mexican-Americans. Am J Health Syst Pharm 60(12):1281–1282
Schmal J, Madrer J (2011) Ethnic diversity in Mexico [electronic version]. Mexico Connect. http://www.mexconnect.com/mex_/feature/ethnic/ethnicindex.html#imm
Simon Y (2018) Hispanic vs. Latino vs. Latinx: a brief history of how these words originated. https://remezcla.com/features/culture/latino-vs-hispanic-vs-latinx-how-these-words-originated/
Sobel LL, Sawin EM (2014) Guiding the process of culturally competent care with Hispanic patients: a grounded theory study. J Transcult Nurs 27(3):226–232
Spector R (2017) Cultural diversity in health and illness, 9th edn. Pearson, Upper Saddle River NJ
Steinhauer J (2015) The history of Mexican immigration to the U.S. in the early 20th century. https://blogs.loc.gov/kluge/2015/03/the-history-of-mexican-immigration-to-the-u-s-in-the-early-20th-century/
The World Bank (2019) Literacy rate, adult total (% of people ages 15 and above)—Mexico. https://data.worldbank.org/indicator/SE.ADT.LITR.ZS?locations=MX
Torres CA, Thorn BE, Kapoor S, DeMonte C (2017) An examination of cultural values and pain management in foreign-born Spanish-speaking Hispanics seeking care at a federally qualified health center. Pain Med 18(11):2058–2069. https://doi.org/10.1093/pm/pnw315
U. S. Census Bureau (2018a) POV-02. People in families by family structure, age, and sex, iterated by income-to-poverty ratio and race. https://www.census.gov/data/tables/time-series/demo/income-poverty/cps-pov/pov-02.html#par_textimage_10
U. S. Census Bureau (2018b) The Hispanic population of the United States: 2018. Table 1. Population by sex, age, Hispanic origin, and Race: 2018. https://www.census.gov/content/census/en/data/tables/2018/demo/hispanic-origin/2018-cps.html
U. S. Census Bureau (2018c) The Hispanic population of the United States: 2018. Table 2. Population by sex, age, Hispanic origin type: 2018. https://www.census.gov/content/census/en/data/tables/2018/demo/hispanic-origin/2018-cps.html
U. S. Census Bureau (2019) POV-01. Age and sex of all people, family members, and unrelated individuals iterated by income-to-poverty ratio and race. https://www.census.gov/data/tables/time-series/demo/income-poverty/cps-pov/pov-01.2018.html
U. S. Department of Health and Human Services (2018) 2018 National sample survey of registered nurses. https://data.hrsa.gov/DataDownload/NSSRN/GeneralPUF18/2018_NSSRN_Summary_Report-508.pdf
U. S. Department of Health and Human Services, Office of Minority Health (2019) Profile: Hispanic/Latino Americans. https://minorityhealth.hhs.gov/omh/browse.aspx?lvl=3&lvlid=64
U. S. Census Bureau (2010) The Hispanic population 2010. https://www.census.gov/prod/cen2010/briefs/c2010br-04.pdf
USTR (2020). https://ustr.gov/usmca
World Bank Group (2019). https://data.worldbank.org/indicator/SI.POV.NAHC?locations=MX

Yucutan Times (2018) Mexico's indigenous languages in danger of disappearing. https://www.theyucatantimes.com/2018/10/mexicos-indigenous-languages-in-danger-of-disappearing/

Zoucha R (2011) Understanding the meaning of confianza in the context of health for Mexicans in an urban community: a focused ethnography. Duquesne University School of Nursing, Pittsburgh

Zoucha R, Zamarripa C (2013) People of Mexican heritage. In: Purnell LD, Betty J (eds) Paulanka, 4th edn. FA Davis, Philadelphia

Zoucha R, Zamarripa-Zoucha A (2016) Health and health care in Mexico. In: Holtz C (ed) GlobalHealth care: issues and policies, 3rd edn. Jones and Bartlett, Burlington MA

25 People of Puerto Rican Heritage

Arturo Gonzalez and Mariceli Comellas Quinones

25.1 Introduction

People of Mexican heritage are a very diverse group geographically, historically, and culturally and are difficult to adequately describe. Although no specific set of characteristics can fully describe people of Mexican heritage, some commonalities distinguish them as an ethnic group, with many regional variations that reflect subcultures in Mexico and in the US. A common term used to describe Spanish-speaking populations in the US, including people of Mexican heritage, is Hispanic. However, the name can be misleading and encompass many different people clustered together due to a common culture and lineage from Spain. Many Hispanic people prefer to be identified by descriptors more specific to their cultural heritage, such as Mexican, Mexican American, Latin American, Spanish American, Chicano, Latino, or Latinx.

The current chapter is a revision previously written by Larry Purnell in the previous edition of the book.

The authors would like to thank Josue Toro Navarro for contributing the reflective exercises in this chapter.

A. Gonzalez · M. C. Quinones (✉)
Florida International University, Miami, FL, USA
e-mail: artgonza@fiu.edu; mcomella@fiu.edu

25.2 Overview Inhabited Localities, and Topography

25.2.1 Overview

In November 1493, Christopher Columbus sailed into a wide and deep harbor, one of the finest in the Caribbean. Claiming the island for Spain's burgeoning empire, he named it *Puerto Rico de San Juan Bautista,* the rich port of St. John the Baptist. The island of Puerto Rico, easternmost of the Greater Antilles, lies opposite Hispaniola along the Mona Passage, an important shipping lane to and from the Panama Canal (CIA World Factbook 2019a). Puerto Rico was originally inhabited by the Taíno peoples, who were largely exterminated and replaced by African slave labor during 400 years of Spanish rule (CIA World Factbook 2019a). Puerto Rico was claimed by the United States during the Spanish American War in 1898 and has been part of the U.S. ever since (CIA World Factbook 2019a). Puerto Rico's numerous rivers flow out of the high central mountains down to the fertile northern coastal plain, providing rich agricultural land in a tropical climate that is perfect for growing coffee, sugarcane, pineapples and plantains (CIA World Factbook 2019a).

Puerto Ricans remain one of the largest Hispanic cultural subgroups in the United States. Demographically, Puerto Ricans identify as white (75.8%), black/African American

© Springer Nature Switzerland AG 2021

L. D. Purnell, E. A. Fenkl (eds.), *Textbook for Transcultural Health Care: A Population Approach*,
https://doi.org/10.1007/978-3-030-51399-3_25

(12.4%), other (8.5%), and of mixed heritage (3.3%) (CIA World Factbook 2019a). Recent estimates show that more than 5.4 million are living in the continental United States, compared to 3.2 million still living in Puerto Rico (Wang and Rayer 2018). This number represents an increase of about 1.2 million residents in the United States since 2000 and a decline in the population of Puerto Rico by about 8% (Wang and Rayer 2018). In 2016, the largest concentrations of Puerto Ricans in the United States were found in Florida, New York, Pennsylvania, New Jersey, and Massachusetts, comprising 62% of the total Puerto Rican population (Wang and Rayer 2018). Puerto Ricans have a unique pride in their country, culture, and music. They self-identify as *Puertorriqueños*, Boricuans (the Taíno Indian word for Puerto Rican) or Niuyoricans, for those born in New York.

In 2017, the estimated mean annual income for Puerto Ricans was $39,4000 (CIA World Factbook 2019a). This can be compared to the overall U.S. mean income for the same year, which was $59,800 (CIA World Factbook 2019b). Approximately 15% of Puerto Rican families in the continental United States are living below the poverty line, which is significantly lower than the 46% still on the island living in poverty (Kaiser Family Foundation 2017).

25.2.2 Heritage and Residence

Puerto Ricans were granted United States citizenship in 1917 with the passage of the Jones Act (Rivera 2018). In 1947, Puerto Rico became the first U.S. territory to elect its governor, and in 1952 the island became a Commonwealth (Rivera 2018). However, this political status as a Commonwealth remains a sensitive topic for many Puerto Ricans. From the *jíbaros* (peasants) to the educated political leaders, a perception exists that the *Americanos* (European Americans), including their culture and politics, remains a potential threat to the Puerto Rican culture, language, and political future.

25.2.3 Reasons for Immigration and Associated Economic Factors

Puerto Ricans have been migrating to the United States for decades to seek employment, education, and better quality of life. In holding U.S. citizenship, Puerto Ricans have the unique ability among Hispanic groups to move back and forth between the mainland and the island without needing to take immigration law into account (Rosario Colón et al. 2019). For much of the twentieth century, Puerto Rican migration was driven by a need for manual labor in the continental United States. This back and forth migration pattern led to significant cultural exchange (Rosario Colón et al. 2019). This pattern of circular migration became known as *Va y Ven* (go and come), with individuals or families caught in reverse cycles of living in the United States for a few years or months and then returning to Puerto Rico.

In recent decades, many physicians, lawyers, and other professionals have left Puerto Rico in order to enhance their educational status, social mobility, and employment opportunities. Many Puerto Ricans select areas where they can seek to preserve their cultural, social, and familial wealth. Since 2000, many have chosen Florida as their first destination, with migration to that state increasing from less than half a million annually to over one million in 2016 (Meléndez et al. 2017). However, an even greater push factor for Puerto Rican migration came following Hurricane Maria, sparking what has been called a mass exodus from the island (Meléndez et al. 2017). It remains to be seen if migration patterns since Hurricane Maria will have a permanent effect on this pattern.

25.2.4 Educational Status and Occupations

Traditionally, education is highly respected among Puerto Ricans. Children are praised for their educational achievements and are encouraged to obtain an education in order to improve their future opportunities. Although Puerto Rico has a compulsory education system similar to the United States, many children migrating to the mainland often find themselves placed in lower grades because of language barriers. Unfortunately, as well, hundreds of school buildings have been abandoned since Hurricanes Maria and Irma in 2017, leaving both student enrollment and teacher employment levels significantly lower across the island (Katz 2019).

The literacy level in Puerto Rico is 93.3%, with roughly similar levels for men and women (CIA World Factbook 2019a). There are numerous well-developed and sophisticated public and private universities in Puerto Rico, including three U.S accredited medical schools. Despite the value placed on education, many Puerto Ricans have faced high dropout rates. Between 2013 and 2017, the high school graduation rate was 74.4%, and the college graduation rate 24.9% (U.S. Census Bureau 2019). Despite these numbers, many parents make significant financial sacrifices in order to obtain for their children a private education, most often Catholic schools. Private schools are often referred to as *colegios* (colleges), often creating confusion with the American English translation of undergraduate institutions and the Central American term for college education. Instead of "college," the term *universidad* (university) is most commonly used to refer to 4-year college institutions in Puerto Rico, where bachelor's and master's degrees equate to those on the mainland.

Many Puerto Ricans who migrated before the 1970s often had little more than a fifth-grade education. Most of these individuals were farmers who had worked on the rice, sugar cane, and coffee plantations, or in the garment and manufacturing industries concentrated in the northeastern and midwestern cities. Yet beginning in the mid-1970s, this pattern began to change as more educated Puerto Ricans began to migrate. Puerto Rico's economy then stalled beginning in 2006, resulting in harsh austerity measures (Katz 2019). The devasting 2017 hurricanes only increased this economic pain, with high unemployment rates leading to increases in alcoholism, drug abuse, street crime, family disruption, and conflict. The advances made in the educational status of Puerto Ricans is threatened by continuing high unemployment levels across the island. In 2011, following the Great Recession of 2009, unemployment reached 16% before dropping to 11.5% in late 2017 (CIA World Factbook 2019a). However, by September 2019, the official unemployment rate in Puerto Rico was 7.6%, albeit in a smaller labor force pool (U.S. Bureau of Labor Statistics [BLS] 2019).

25.3 Communication

25.3.1 Dominant Language and Dialects

Four centuries of Spanish colonization led to that language's prevalence across the island. However, Puerto Ricans were forced to learn English following the U.S. takeover in 1898. Many Puerto Ricans at the time were unable to read or write Spanish, which caused significant difficulty. Bilingual sensitivity often revolved around the belief that English would eventually replace Spanish, negating significant aspects of Puerto Rican culture, traditions, and practices. In 1902, for example, passage of the Language Law established that both English and Spanish would officially be interchangeable in government departments, island courts, and public offices (Rivera and López 2018).

Puerto Rico was one of the first areas in all of Latin America to conduct bilingual education in English and Spanish for children beginning in kindergarten. That Puerto Rico has two official languages has long been a sensitive issue for many residents and language has often been a political issue in Puerto Rico, disputed and debated in government since at least the 1980s. Spanish is spoken in the home, at schools, in business, and in the media. Many Puerto Ricans frequently use the phrase "***!Ay bendito!***" to express astonishment, surprise, lament, or pain. However, people from the more metropolitan cities are more likely to read, write, and speak English to varying degrees.

A strong continuing Spanish language use, however, has often disadvantaged students at public schools, where English instruction occurs an hour a day, compared to children from elite families who can attend private schools with greater English instruction (Carroll 2016). Given that Spanish remains the overwhelming language of instruction across Puerto Rico, English has become a symbol of upward mobility (Carroll 2016). Indeed, the most sizeable group of Puerto Ricans fluent in English are those who have returned to the island from the mainland (Carroll 2016).

Puerto Ricans speak a radical dialect of Spanish, one that modifies final consonants and often produces vibrant trill sounds (Fabiano-Smith et al. 2014). However, contextual differences in language remain, mainly in pronunciation between monolingual or bilingual residents, indicating that English proficiency has some impact on dialect usage (Fabiano-Smith et al. 2014). Many of these rural dwellers substitute the sound of *e* for *i* and often drop the last letters of some words. For example, *después* (after) may be pronounced as *dispu*, and *para donde vas* (where are you going?) may be pronounced as *pa'onde vas*. In addition, most Puerto Ricans exchange the letter *r* for the letter *l*; for example, *animar* (encouragement) may be pronounced as *animal*, sounding like "animal." Some use a rolling *r*, a pharyngeal pronunciation that uses double *r*, such as *arroz* (rice) and *perro* (dog). Many Puerto Ricans speak with a melodic, high-pitched, fast rhythm that may leave non–Puerto Rican healthcare providers confused. This pitch and these inflections are maintained when speaking English. Because some Puerto Ricans feel uncomfortable, or even insulted, if people comment on their accent, the healthcare provider should avoid making comments about accent, use caution when interpreting voice pitch, and seek clarification when in doubt about the content and nature of a conversation that may seem confrontational.

25.3.2 Cultural Communication Patterns

Puerto Ricans are known for their generous hospitality and the value place on interpersonal interaction such as ***simpatía*** (sympathy), a cultural script in which an individual is perceived as being likeable, attractive and fun-loving. Puerto Ricans enjoy conversing with friends and sharing information about their families, heritage, thoughts, and feelings. They often expect the healthcare provider to exchange personal information when beginning a professional relationship. The healthcare provider may wish to set boundaries with discretion and ***personalismo*** (personalism) emphasizing personal rather than impersonal and bureaucratic relationships.

Most Puerto Ricans will readily express their physical ailments and discomforts to healthcare providers, with the exception of discussing taboo subjects such as sexuality. Some studies suggest that discussion of sex-related topics remains taboo in Puerto Rico, especially within families (Colón-López et al. 2017). Yet if ***confianza*** (trust) can be established, healthcare providers can then establish open communication channels with individuals and their families.

Spatial distancing among Puerto Ricans varies with age, gender, generation, and acculturation. Personal space is often a significant issue for some older women, especially those from more rural areas, who often prefer to maintain a greater distance from men (Acello and Hegner 2016). Younger women, and those Puerto Ricans born on the mainland may be less self-conscious regarding this issue (Acello and Hegner 2016). Healthcare providers must, therefore, carefully assess each individual's perception of distance and space.

Puerto Ricans are very expressive, using many body movements to convey their messages. During conversations, hand, leg, head, and body gestures are commonly used to augment messages being expressed by words. Puerto Ricans will also express their feelings and emotions through touch and are ***cariñosos*** (loving and caring) in both verbal and nonverbal ways. Greeting Puerto Ricans with a friendly handshake is always acceptable. Once trust is established, a patient may greet a healthcare provider with a friendly hug. During conversations with close companions and family, they are likely to touch with love and affection, including a gentle hand stroke on the shoulder. Puerto Rican women greet one another with a strong familiar hug, and if among family or close friends, a kiss is included. Men may greet other men with a strong right handshake and a left hand stroking the greeter's shoulder.

Nonverbal communication plays a vital role in acquiring informed consent for healthcare, research procedures, when providing health education, and discharge planning. Nonverbal communications among Puerto Ricans may include an affirmative nod with an "aha" response, but this does not necessarily mean agreement or understanding related to the conversation. Using a respectful and friendly approach, healthcare providers should seek clarification of the information provided, ask for language preference in verbal and written information, and allow time to exchange of information with questions and answers when critical decisions need to be made. Puerto Rican patients may prefer to read or share sensitive information, options, and decisions with close family members, and some may seek to obtain verbal approval from extended family or community members who are knowledgeable in health matters. When consent is needed from a female patient, the healthcare provider may want to ask if verbal approval or consent from the partner should be obtained first.

Traditional cultural norms discourage an overt sexual-being image for women, but with family assimilation to the mainland, many of these traditional values have disappeared, especially among younger Puerto Ricans. When topics such as sex, sexual orientation, sexually transmitted infections (STIs), or other infectious diseases are discussed, an environment built on *confianza* and *personalismo* must be established if these sensitive issues are to be effectively addressed with patients. Voice volume, tone, the degree of eye contact, spatial distancing, and time are all variables that can have an impact on discussions of sensitive topics with some Puerto Ricans.

The meaning and cultural value placed on direct eye contact has changed over time. Among younger Puerto Ricans and those born on the mainland, eye contact is maintained and is often encouraged among those who believe in a non-submissive and assertive portrayal. Young Puerto Rican women may take offense to verbal and nonverbal communications that portray women as nonassertive and passive. However, among more traditional Puerto Ricans born and raised in rural areas of Puerto Rico, limited eye contact is preferred as a sign of respect, especially with older people, who are seen as figures of respect and great wisdom.

There are also important factors related to communication and culture that should be considered when working with Puerto Ricans, especially those who migrated to the U.S. following Hurricane Maria. Many migrants who have undergone post-disaster events are more likely to have decreased language proficiency and are at a higher risk of severe mental distress (Carl et al. 2019). Language barriers are often identified among Puerto Ricans who have moved to the mainland. For this reason, healthcare providers should consider working with professional translators while seeing a patient– the use of family members or friends is ill-advised within such settings, given the potential for bias and miscommunication when communicating sensitive information during a clinical visit.

25.3.3 Temporal Relationships

Puerto Ricans are present-oriented, holding a realistic and serene view of life. This is an attitude that can often frustrate healthcare providers and those in the business world, as those unaware of this cultural nuance may misinterpret this view as fatalistic. Healthcare providers should respect this point of view and assist patients in identifying options, choices, and opportunities that empower individuals towards changing high-risk behaviors.

Many Puerto Ricans also have a relativistic view of time, which often translates into not arriving on time for appointments. This flexible orientation towards time, when coupled with a relaxed attitude, may extend into healthcare appointments, which can interfere with the ability to provide health services in a time-limited, cost-containment environment. Therefore, during initial interviews, healthcare providers may want to carefully outline the expectations for arriving on-time for appointments and the time limitations that exist within a daily schedule.

Reflective Exercise 25.1

Paco and Estrellita Lopez bring their 3-year-old son to the clinic. They both speak some English and the nurse speaks no Spanish. The nurse looks at Estrellita holding Pacquito, who is crying, and asks how she can help them. Estrellita looks at her husband and speaks to him in Spanish. Paco tells the nurse that Pacquito ate well until 2 days ago, but now cries while eating. The nurse continues to look at Estrellita and asks if the child has any other health problems.

Estrellita looks at her husband while speaking to him in Spanish, after which the husband tells the nurse that the boy was born with a heart problem. Again, the nurse asks Estrellita if she knows what type of heart problem it is.

Paco looks at his wife and speaks in Spanish to her, then turns to the nurse and says that they can go to La Mirada Clinic to get help for their son.

1. How is the concept of *machismo* displayed in this scenario?
2. How is the concept of *marianismo* displayed in this scenario?
3. Why did the parents leave?

25.3.4 Format for Names

Showing respect (*respeto*) for adults, parents, and older individuals is highly valued among Puerto Ricans. This respect is reflected in the ways that children are expected to speak to and refer to adults and older individuals. Rather than the terms *Señora* (Mrs.) and *Señor* (Mr.), Puerto Ricans often use the more honorific terms *Doña* and *Don* for most adults. Aunts and uncles have their name preceded by *tití* or *tío* (auntie/uncle) and *madrina* or *padrino* (godmother or godfather). These prefixes are symbols of respect and position in the family. Healthcare providers should maintain levels of formal respect using *Doña* and *Don* or at least *Señora* and *Señor,* and avoid calling Puerto Rican patients by their first name, as well as avoiding terms of familiarity such as "honey" or "sweetheart."

Similar to other Hispanic groups, Puerto Ricans have a complex system for addressing others, specifically women. Single women may prefer to use their father's and mother's surnames, in that order. For example, a single woman may style her name as follows: Sonia López Mendoza, with López being her father's surname and Mendoza her mother's. When she is married, the husband's last name, Pérez, is added with the word *de* to reflect that she is married. This woman's married name would be Sonia López de Pérez; the mother's surname is eliminated. In both business and healthcare settings, Señora López de Pérez is the correct formal title to use when initiating conversation or building a relationship. Younger or more acculturated women may simply change their last names to that of their husband. The importance and respect given to these prescriptive name formalities are perpetuated when friendly verbal and nonverbal gestures accompany the greeting.

25.4 Family Roles and Organization

25.4.1 Head of Household and Gender Roles

Traditional Puerto Rican society can be characterized as overwhelmingly patriarchal. Puerto Rican men are often socialized in a culture of *machismo,* which emphasized qualities of bravery, strength, virility, aggression, and autonomy (Torres 1998). While *machismo* is often interpreted along the lines of aggressive and irresponsible behaviors, some aspects of this enculturation are often interpreted more along the lines of self-confidence, protectiveness towards women, children, and the family, pride, dignity, and honor

(Torres 1998). Unfortunately, *machismo* culture also supports certain negative behaviors, including alcoholism, infidelity, lack of parenting involvement, and abandonment of partners (Torres 1998). The degree to which certain behaviors are demonstrated will depend on the individual.

Socioeconomic changes over the past few decades have given rise to changes in the position and role of women in Puerto Rican society, despite the traditional values that seek to continue to define women solely in terms of their reproductive roles. These gender-role expectations are often strikingly different among affluent and more acculturated families, but more traditional families may continue to have expectations that women must be lenient, submissive, and always willing to please men. Men, as a result, demand respect and obedience from women in the family. Nevertheless, women have always played a central role in the family and the community. Puerto Rican families are moving towards more egalitarian relationships, especially in Puerto Rico where women continue to make significant contributions through their increasing participation in politics and other traditional male-oriented roles. Many of these changes are the result of both acculturation on the mainland and the increase in Puerto Rican women in the workforce.

In the twenty-first century, it can be argued that a new identity is emerging in Puerto Rico, with many leading feminists calling for changes in traditional family structures, values, power, and authority. The realities of modern life mean that more Puerto Rican families are sharing the economic and social responsibilities of running the household, but the entrenched cultural ideas of *machismo* often continue to be a source of confrontation. In fact, intimate partner violence (IPV) is often six times higher in Puerto Rico than many places in the continental United States (Villafañe-Santiago et al. 2019). Puerto Rican women continue to negotiate for power in order to equalize the dynamics that exist within sexual relationships, especially when confronting the cultural assumption by many men, who continue to believe that women must be submissive and obedient to their partners. Therefore, when assessing health risks and relationships, healthcare providers must consider these issues and assess family patterns of relating in order to identify appropriate interventions.

25.4.2 Prescriptive, Restrictive, and Taboo Behaviors for Children and Adolescents

Children are the center of Puerto Rican family life. Throughout their childhood, Puerto Ricans are socialized to have respect for adults, especially the elderly. Great significance is given to the concept of **familism**, the idea that the needs of the entire family often outweigh those of individual members, and any behaviors that depart from this idea are discouraged, as they may be perceived as bringing disgrace to the family. Traditional families who expect their children not to contradict, argue, or disagree with parents may experience difficulty when adolescents exposed to the more Americanized aspects of culture seek their independence, struggling between traditional and contemporary values. Often, the conflict between these cultural expectations can have negative effects on Puerto Rican adolescents. For example, Puerto Rican girls often report greater parental control over their relationships with the opposite sex, often dating later and keeping boyfriends a secret out of a fear of being forced to end the relationship (Villalobos Solís et al. 2017). Adolescents become caught between the desire for greater autonomy and feeling obligated to help their families and respect their parents, as many familial obligations, such as taking care of a family member, can preclude more autonomous behaviors, such as spending time with peers (Villalobos Solís et al. 2017). Mental healthcare providers addressing family conflict must work within the context of the family to resolve adolescents' mental health issues rather than using individual approaches.

Puerto Rican culture also has several prescriptive cultural values regarding health and weight which can impact health. Many families believe that a healthy child is one who is *gor-*

dita or *llenito* (diminutive for fat or overweight) and has red cheeks. Massara's (1989) early work on weight, body image perceptions, and health argued that an oversized body image may be perceived as a mirror of physical and financial wealth, even among adult women. Young mothers are often encouraged to add cereal, eggs, and *viandas* to their infant's milk bottles (see Nutrition). Nurses are in an excellent position to educate mothers about these practices and the health risks for children who are overweight.

Socialization patterns for adolescent boys and girls are quite different. Males are socialized to become macho, powerful, and strong with a healthy sexual appetite. These behaviors often encourage dominance over women, the development of values concentrated on obtaining social privileges, and the pursuit of high-paying careers that will provide them with a financial advantage over others. Girls, on the other hand, are often socialized to focus on home economics, family dynamics, and motherhood. Schooling for girls is meant to produce educated housewives, not necessarily educated professional women. Because motherhood provides a powerful social status, the value placed on it may be a precursor to teen pregnancy rates among Puerto Rican girls, as motherhood provides a source of power, support, and cultural recognition (Orshan 1996).

Some families continue to abide by cultural prescriptions that encourage, for boys, the initiation of sexual behaviors before marriage, extramarital activity, and control over their sexual relationships. Girls, in contrast, are traditionally socialized to be modest, sexually ingenious, respectful, and subservient to men, all part of their cultural script under *marianismo* (Orshan 1996). *Marianismo* is built upon the idea that Puerto Rican women model their lives on the example of the Virgin Mary, requiring them to be chaste, pure, willing to sacrifice for the family, and submissive to their husbands (Ertl et al. 2019). The ideal Puerto Rican woman, according to *marianismo,* is a loving mother committed first and foremost to her family, expected to remain quiet and polite, and many household tasks, such as cooking, cleaning, and taking care of younger siblings, falls on young girls still living at home (Ertl et al. 2019). Discussions about sexuality are considered taboo for many families, who use the term *tener relaciones* (to have relations) rather than the word *sex*. Despite this, mothers remain the primary individual responsible for the sexual education of their daughters (Colón-López et al. 2017). Modesty is highly valued, and issues such as menstruation, birth control, impotence, STIs, and infertility are rarely discussed.

Less-educated families and those from rural areas may have great difficulty educating young women about sexuality and reproductive issues. Additionally, the culture of women that exists in Puerto Rican families often delay communication regarding sex-related topics, avoid the topic altogether, or focus on abstinence as the only option outside of marriage (Colón-López et al. 2017). Many Puerto Rican adolescents are therefore dependent upon educational institutions in order to learn about mensuration or their reproductive systems. However, the ubiquity of Internet access means that many adolescents will be exposed to such information through the media and peers. Yet cultural respect of the role of healthcare providers as educators places them in an excellent position to educate the family about sexuality issues. This respect gains them entrance into a familiar and trusted family environment that must be valued for its cultural traditions and practices.

Many families expect their children to remain at home until they either get married or obtain their college degree. Families want to be able to care for their children and provide theme with both emotional and financial support to the extent that it is feasible. Children, therefore, are expected to follow the family traditions and rules. Most fathers expect to be consulted, but they see themselves mainly as financial providers. Puerto Rican families are often very rigorous with their children's discipline. Puerto Rican mothers tend to be very protective of their children and may use physical punishment. Many Puerto Rican mothers use threats of punishment, guilt, and discipline, which can create stress and difficulties for adolescents as they struggle with the more permissive cultural patterns of the United States, such as dating. Healthcare providers should assess families for these patterns and provide counseling that promotes stability. The cultural definition of physical abuse is challenging, and healthcare providers must assess each situation before determining child abuse.

25.4.3 Family Goals and Priorities

Family roles and priorities among Puerto Ricans are based on the concept of *familism*. Puerto Ricans value the unity of the family. *La familia* is the nucleus of the community and the society. The family structure may be nuclear or extended. Family members include grandparents, great-grandparents, married children, aunts, uncles, cousins, and even divorced children with their children. Two families may live in the same household. Indeed, the influence of *machismo* can have positive effects on the family, as many men view the nuclear family as the superior family structure (Mogro-Wilson et al. 2016). Fathers will even extend this paternal commitment to nieces and nephews under the expectation that the extended family is required to assist and support parenting roles and responsibilities (Mogro-Wilson et al. 2016).

After marriage, children will live away from their parents but are expected to maintain close ties with their families, especially the women. Most Puerto Rican families want a daughter because daughters are the traditional caretakers when parents reach advanced ages. In addition, women are seen as those who will continue family traditions. Male children, who are usually more independent, are valued because they continue the family name.

Extended family members are active in the care of all children, providing support and encouraging the maintenance of cultural and religious traditions. How children behave is a public statement regarding their parents and family, and that maintaining respect in the community is a means of demonstrating respect to their parents (Mogro-Wilson et al. 2016). Grandparents are given an active role in rearing grandchildren, supporting the family, babysitting, teaching traditions, disciplining, and enforcing educational activities. Women, as they age, become more widely regarded for their wisdom, providing them with a greater status in society. Often, older women have a covert power over spouses, children, and the family. Healthcare providers can use this older generation of Puerto Ricans to introduce health promotion and disease-prevention education within their families.

At times, dependent older people are expected to live with their children and be cared for emotionally and financially. Informal and formal support systems are considered critical factors in promoting the health of older Puerto Ricans, particularly older women. All members of the family provide support for financial and manpower efforts needed to keep older people at home. Those who have higher financial liquidity may take financial responsibility in exchange for the manpower and physical efforts of those who cannot provide financially. Placements in nursing homes and extended-care facilities may be seen as inconsiderate to older people, and family members who must use these organizations may feel guilty and experience depression and distress. Thus, healthcare providers must be sensitive to these issues by exploring alternatives for care and providing information to all family members involved in this decision-making process. Discharge planning, hospice care, and other situations can be addressed in a "conference-style" approach to develop strategies for providing emotional support and assistance to family members.

Friends, neighbors, and close and distant family members are expected to visit a person during times of illness, support the family, and take an active role in family decisions and activities. A family member is expected to remain at the bedside of the sick person. Healthcare providers should ask for the name of the family spokesperson and document this in the patient's chart. Nurses may need to set boundaries with patients' families about visitation, personal space, and privacy matters.

25.4.4 Alternative Lifestyles

Since the early 1980s, Puerto Rican families have experienced an increased incidence of pregnancy among teenagers and unmarried women. This trend is believed to be the result of the increased number of women in the labor force, high divorce rates, poverty, and the increased number of households headed by women. The most recent data on teen pregnancy rates in Puerto Rico is 29.6/1000 females aged 15–19, which can be compared to the overall rate of 20.3 in the rest of the United States (U.S. Department of Health and Human Services [HHS] 2019). For healthcare services to be effective in identifying appropriate interventions, healthcare providers must assess social-support factors and the socioeconomic status of individuals.

Homosexuality continues to be a taboo topic that carries a great stigma among Puerto Ricans. Same-sex behavior is often undisclosed to avoid family rejection and preserve family links and support. The culture of *machismo* and *marianismo* lead to a significant level of intolerance for homosexuality and gender non-normativity in Puerto Rico (Ramos-Pibernus et al. 2016). LGBTQ Puerto Ricans continue to face discrimination across the island, with as many as 25% of individuals experiencing some form of discrimination in school, at work, and when receiving government services (Rodríguez-Díaz et al. 2016). Given that there are significant disparities in the sexual health outcomes for LGBTQ populations, overcoming stigma and family rejection may be dependent upon family communication (Smollin et al. 2018). In many Puerto Rican families, the sexual identity of LGBTQ youth is often ignored or believed to be non-normative (Smollin et al. 2018). Communication regarding sexuality, even for parents of heterosexual children, is often vague and indirect (Smollin et al. 2018).

25.5 Workforce Issues

25.5.1 Culture in the Workplace

In general, Puerto Rican men and women readily assimilate into the U.S. work environment, which is similar to their native work environment in Puerto Rico. Nurses are among the latest group of Puerto Rican professionals who have come to the United States seeking better employment and educational opportunities. They often seek employment at local hospitals, federal health facilities such as the Army, Navy, Air Force, and the Veterans Administration. This trend has expanded since Hurricane Maria, as numerous healthcare professionals leave the island in search of employment on the mainland (Rege 2018). Criticism among Puerto Ricans regarding the federal government's response, or lack thereof, to Hurricane Maria, in providing basic services, including electricity, communication, food and water, medical and healthcare services contributed to recent increased migration (Rege 2018). As with any such disaster, uncertainty regarding job and property loss, and extended school closures affected daily life.

Despite stereotypical views of Puerto Ricans as people who do not work and depend solely on the U.S. welfare system, most Puerto Ricans are hardworking, like to be competitive, and often make extended efforts to please their employers. Many Puerto Ricans in the labor force place a high value on their occupations, positions, and businesses. They strive for high performance even in the face of oppression; they offer little resistance and maintain the ability to be happy even when confronting oppressive situations.

Several cultural differences among Puerto Ricans—such as education; the value placed on honesty, integrity, personal relationships; and relativistic views of time—may have an influence in the workplace. The educational system in Puerto Rico emphasizes theoretical and practical content as well as neatness. Consequently, whereas most migrant Puerto Ricans are task oriented and meticulous about the presentation of their work, some have a relativistic view of time and may not value regular attendance and punctuality in the workforce. Most Puerto Ricans are cheerful, have a positive attitude, and value personal relationships at work. Work is perceived as a place for social and cultural interactions, which may include listening to background music while performing job activities. This practice can lead to loud, cheerful, and noisy conversations that may require the employer's attention.

For many women, family responsibilities, pregnancy, and the health of their children and other family members take priority over work. For others, access to the welfare system becomes more convenient than the pride of having a secure job. In Puerto Rico, women are given a lengthy maternity leave because of the emphasis and value placed on the well-being of working women and their infants. According to a 1942 law, Puerto Rican women are entitled to 8 weeks of maternity leave, 4 weeks of prenatal and 4 weeks of postnatal leave (Hopgood-Jovet 2018). In the U.S. labor force, many working Puerto Rican women resent the limited maternity leave supported by the American culture.

Employers may need to negotiate more flexible work responsibilities among Puerto Ricans during religious holiday celebrations such as Easter and Christmas. In Puerto Rico, schools are closed, and the community celebrates a spiritual and religious recess from day-to-day activities and work responsibilities. The great solemnity and religious commitment among all religious groups bring Puerto Rican families to a societal halt for almost 6 weeks annually. Schools recess from early December to the middle of January, waiting for the Epiphany, *Dia de los Tres Reyes Magos* ("Day of the Three Wise Men") on January 6, and the *Octavitas*, a post-Epiphany traditional musical and cultural celebration that extends the Christmas celebration eight more days. Many Puerto Ricans on the mainland wish to use vacation and unpaid leave to spend time with their families in Puerto Rico. Traditional music, food, and folk activities during these celebrations are used to uphold ethnic pride. Holiday seasons may challenge employers, who need to manage absenteeism, increased consumption of alcohol, requests for vacation, leave without pay, and decreased productivity.

25.5.2 Issues Related to Autonomy

Puerto Rican families have traditionally socialized men into aggressive, domineering, and outspoken roles. Thus, many men display confidence at work and assume leadership positions with autonomy. However, more recent male immigrants who are less educated and have language difficulties may be reluctant to assume leadership roles, may be shy and not as outspoken, and may hesitate to challenge authority and workplace norms. Changing the conduct of these recent male immigrants in the workforce is related to the passivity and docile behaviors learned in the U.S. and Puerto Rican educational systems. These immigrants are more likely to conform to the behavioral norms of the workplace and avoid personal conflict or confrontations in an effort to maintain positive relationships.

Women from rural areas and traditional families are more likely to come from a submissive and noncompetitive environment. Thus, they may be perceived as less determined, less confident, and less outspoken than other American women in managerial and supervisory capacities. However, the decline of male employment since 2000 has led to an increased need for economic contributions to the family on the part of women, which has helped to breakdown the traditional view of the male as the sole breadwinner, with many women expressing the idea that helping out in such a manner is part of their obligation to the family (Warren 2010). Growing levels of employment for Puerto Rican women have largely reshaped attitudes regarding women's roles in society, greater control over their fertility, and significant rejection of the more open and repressive aspects of *machismo* that often leads to control, infidelity, or partner violence (Warren 2010).

Some women may still find themselves in conflict with traditional values when in a competitive, assertive work environment. Their ability to succeed in the workforce may depend on their employers' support of assertiveness and on-the-job training. In addition, women who wish to climb the career ladder may benefit from an environment that provides information, promotes confidence, fosters positive interpersonal relationships, and teaches strategies for resolving conflict. Indeed, many Puerto Rican women value the social relationships that exist in the workplace, as these environments are less isolating that traditional homemaker roles (Warren 2010).

Although most Puerto Ricans are bilingual, some may speak broken English, street English, or Puerto Rican **Spanglish** such as "I must pay *billes* [bills] and find *dinero* [money].” Younger and urban Puerto Ricans are usually more fluent in English, a skill that facilitates integration into the labor market. Older adults and people who come from a rural background may have less education, lower literacy levels, decreased English proficiency, and increased difficulty assimilating into the labor force.

25.6 Biocultural Ecology

25.6.1 Skin Color and Biological Variations

Information regarding the heritage of Puerto Ricans indicates that the ancestry of the people can be traced back to the original inhabitants of the island, the Taino or native Indians (Toro-Babrador et al. 2003). Unfortunately, much of the Taino were killed in the 1600s due to Spanish invasion of the island and the introduction of African slavery (Toro-Babrador et al. 2003). While many assume that Taino bloodlines were primarily destroyed during this time period, a genetic study of heritage among Puerto Ricans indicates that among this group, 61% had Taino DNA followed by 27% African, and 12% Caucasian (Toro-Babrador et al. 2003). This mixture of Native Indian, African, and Caucasian heritage have resulted in specific biological traits including: dark skin; thick, kinky hair; and a wide, flat nose. However, some members of this population have light skin with straight auburn hair and hazel or dark brown eyes. Biological variations in skin color may require healthcare providers to vary their techniques when assessing individual Puerto Ricans for anemia or jaundice.

Unfortunately, limited information is available regarding how biocultural variations among Puerto Ricans impact disease epidemiology. Research does correlate higher incidences of hypertension and diabetes mellitus in the Puerto Rican population with Indian and African heritage (Garcia and Ailshire 2019). Additionally, current research suggests that biocultural variations significantly contribute to the development of colorectal cancer in the Puerto Rican population (Perez-Mayoral et al. 2019). Colorectal cancer is the primary cause of cancer death among Puerto Ricans with this group having a higher mortality from this disease than U.S.-born Hispanics (Perez-Mayoral et al. 2019). The dearth of information regarding biocultural variants for this cultural group require healthcare providers to assess each person as a unique individual with awareness that standards developed for the dominant American population may not necessarily apply to the Puerto Rican patient.

25.6.2 Disease and Health Conditions

Data from the Institute for Health Metrics and Evaluation (2018) indicate that the leading causes of death for Puerto Ricans include: heart disease, Alzheimer’s disease, diabetes, stroke, and lower respiratory infections. Interestingly, overall cancer rates for Puerto Ricans are lower (314.4 new cases per 100,000 population) when compared with the U.S. (454.8 new cases per 100,000 population) (Puerto Rico Department of Health 2016). However, prostate and colorectal cancer rates for Puerto Ricans are noted to be higher, with age-adjusted mortality rates from these diseases reported to be much lower than those for individuals living in the United States (Puerto Rico Department of Health 2016). Although the overall cancer mortality rate among Puerto Ricans is lower than that for other groups, healthcare providers should continue to educate Puerto Rican families about cancer prevention. Smoked, pickled, and spiced foods should be discouraged, whereas traditional meals, fruits, and vegetables should be encouraged.

HIV rates for Puerto Ricans are noted to be higher than those for the United States: 18.4 new cases per 100,000 population in 2014 compared with 16.7 new cases per 100,000 population in the U.S. (Puerto Rico Department of Health 2016). The primary mode of transmission for the virus in Puerto Rico is through injection drug use (42%) followed by male-to-male sexual contact (27%) (AIDS United 2019). Emerging threats to health include an increase in natural disasters, injuries, and mental health issues. In most areas of the island, mental health services are unavailable, limiting patient access to this type of care (Puerto Rico Department of Health 2016). Life expectancy of Puerto Ricans at birth is 81 years of age for the population (CIA World Factbook 2019a). Breaking this down by gender, men have a significantly shorter life expectancy than women: 77.7 and 84.5 years, respectively (CIA World Factbook 2019a).

Puerto Rican women living in the United States have a high incidence of being overweight (Schneiderman et al. 2014). The prevalence of overweight/obesity in this population increases with age and is more common in women from lower socioeconomic levels (Schneiderman et al. 2014). Increased rates of obesity are associated with a higher incidence of diabetes, which is currently the third leading cause of death in Puerto Rico (Institute for Health Metrics and Evaluation 2018). Even though rates of obesity and diabetes are higher for women, data also indicates that there are some geographical variants in the disease as well. In particular, men and women from urban areas are more likely than those from rural areas to become obese and develop diabetes (Aschner et al. 2014). Consequently, healthcare providers need to develop interventions that are appropriate to gender, age, and socioeconomic status, while taking into consideration the patient's urban or rural living environment.

Vector-borne illnesses including dengue fever and Zika are also of concern in Puerto Rico. Current evidence indicates that epidemics of dengue fever occurred in Puerto Rico in 2007, 2010, and most recently in 2013 (Noyd and Sharp 2015). Dengue fever is typically transmitted by the *Aedes aegypti* mosquito (Noyd and Sharp 2015). Zika is also an emerging mosquito-borne disease impacting Puerto Rico. On December 31, 2015, the Puerto Rician Department of Health reported its first case of the disease (Thomas et al. 2016). By the end of January 2014, a total of 30 cases of the disease were reported primarily in the eastern part of the island and San Juan (Thomas et al. 2016). The *Aedes aegypti* mosquito is also responsible for the transmission of this disease and epidemiologists argue that due to the presence of the mosquito on the island, it is anticipated that the number of cases of Zika will increase in the coming years (Thomas et al. 2016). Because of the tropical climate of the island outbreaks of dengue fever and Zika can occur year round, making it imperative for healthcare providers to advise Puerto Rican patients and families traveling to the island to avoid endemic areas and to use mosquito repellant and protective clothing at all times.

25.6.3 Variations in Drug Metabolism

Evidence regarding variations in drug metabolism for Puerto Ricans is scant. One study located on the topic indicates that there are some genetic polymorphisms in genes that encode for CYP450 including CYP2C9, CYP2C19, and CYP2D6 (Claudio-Campos et al. 2015). However, the exact impact of these genetic differences on drug metabolism have not been fully identified or explored (Claudio-Campos et al. 2015). Because of the African heritage of many Puerto Ricans, drug absorption, metabolism, and excretion differences experienced by African Americans and American Indians may hold true for black Puerto Ricans. Given that some Puerto Ricans are short in stature and have higher subscapular and triceps skin folds, long trunks, and short legs, therapeutic dosages calculated for the European American patients may not be appropriate for Puerto Ricans. Healthcare providers must be aware that pharmaceutical studies conducted with European Americans may not yield the same results with Puerto Ricans. Consequently, individual patient assessment with accurate documentation of findings is needed to provide care for this population group.

25.7 High-Risk Behaviors

Puerto Ricans are at high risk for illnesses caused by health behaviors including: alcohol consumption, tobacco use, illicit drug use, physical inactivity, poor dietary choices, failure to practice safe sex, and underutilization of preventative health services. Alcohol consumption among Puerto Ricans has been identified as a significant health issue that is often overlooked when assessing the patient. In a recent study of Puerto Rican alcohol use habits a survey conducted demonstrated that 31% of men and 27% of women report being moderate drinkers while 21% of men and 8% of women report being heavy drinkers (Andrews-Chavez et al. 2015). Of those surveyed an additional 35% reported drinking alcohol while taking prescribed medications with alcohol contraindications (Andrews-Chavez et al. 2015). Alcohol consumption among Puerto Ricans is attributed, largely, to acculturation into the mainstream U.S. culture and to psychosocial factors (Alvarez et al. 2017).

There is some evidence demonstrating that greater negative attitudes towards drinking, low family cohesion, and religious practices may be related to a greater risk of alcohol use disorder (Caetano et al. 2018a). Alcohol use patterns, combined with acculturation and socioeconomic status differences between men and women, and men appear to have a higher lifetime prevalence risk for alcohol use when compared to women (Castañeda et al. 2019). Alcohol abuse can be behaviorally modifiable for many individuals. Thus, healthcare providers can be more aware of the patient's age and sex (i.e. men compared to women) as potential indicators of alcohol abuse and conduct the necessary assessments during a clinical visit. The Alcohol Use Disorders Identification Test (AUDIT) would be a quick assessment tool for alcohol consumption, drinking behaviors, and alcohol-related problems. The AUDIT is a 10-item screening tool initially developed by the World Health Organization and validated for at-risk drinking in patients visiting primary care of different racial/ethnic backgrounds (Volk et al. 1997).

Data provided by the Centers for Disease Control and Prevention ([CDC] 2018) further demonstrates that Hispanic or Latino groups living in the United States typically have a lower rate of cigarette smoking when compared with other ethnic groups. However, data indicate that the rates of smoking are different for Latino subpopulations (CDC 2018). In particular, Puerto Ricans have been noted to have the highest smoking rates (28.5%) compared with Cubans (19.8%), and Mexicans (19.1%) (CDC 2018). Even though the smoking rate for Puerto Ricans is high, current data indicate that 67.4% of Puerto Ricans report a desire to quit and Hispanics overall typically have smoking cessation rates that are higher than Whites (CDC 2018). The principle challenge for smoking cessation among Puerto Ricans is that many lack access of healthcare services making it less likely that they will have access to counseling and supports to kick that habit (CDC 2018).

Island-wide estimates of illicit drug use in Puerto Rico have not been measured since the 1990s (Caetano et al. 2018b). Although definitive population-based data regarding this health issue are not currently available, a recent study completed by Caetano et al. (2018a, b) provide some insight into the scope and impact of this problem for Puerto Ricans living in San Juan. Data captured by these authors demonstrate that 16.5% of individuals living in San Juan reported illicit drug use. The prevalence among men was higher at 20.7% compared with women at 12.9% (Caetano et al. 2018a, b). Preferred drugs included marijuana and opioids (Caetano et al. 2018a, b). Male gender, lower socioeconomic status, and younger age (18–29) were associated with a higher incidence of illicit drug use (Caetano et al. 2018a, b). Issues related to acculturation, peer factors, individual, family, parental, and gender-role issues are the most important risk factors in need of early health provider interventions to decrease susceptibility to drug addiction and delinquency. Healthcare providers should develop programs that promote early interventions for the use of illicit drugs. Interventions should focus on individual psychological differences, gender issues, and other contributing factors.

Because many Puerto Ricans support machismo and submission of women, these roles foster high-risk behaviors that impede the prevention and increase the transmission of HIV (Hernandez et al. 2012). In traditional Puerto Rican culture, most men are given free will over sexual practices, including the approval and initiation of sex before marriage and extramarital affairs with other women. Some men may perceive that sexual intercourse with men is a sign of virility and sexual power rather than homosexual behavior. Puerto Rican women are often found in a paradoxical position, because they have to deal with cultural beliefs and health-protective practices. Knowledge about HIV, beliefs about health and illness, and beliefs and practices related to condom use are common concerns encountered by health-care providers in the prevention and transmission of HIV.

Lack of condom use is perhaps one of the most significant risk behaviors that need immediate attention and intervention from health-care providers. Issues such as embarrassment, cost, gender or power struggles, and abuse are among the barriers encountered by Puerto Rican women. Some men fear that if they use condoms, they portray a less macho image, have decreased sexual satisfaction, or indicate that they have an STI or HIV. In addition, the Catholic Church's opposition to the use of condoms, lower educational levels, lower socioeconomic status, and acculturation are significant variables related to the high rates of AIDS and HIV among Puerto Ricans and other Hispanics. Health-care providers must be aware of these barriers, assess individual perceptions of high-risk behaviors, and intervene with programs designed to meet the particular needs of clients who are at high risk for HIV infection or other STIs.

25.8 Nutrition

25.8.1 Meaning of Food

Puerto Ricans celebrate, mourn, and socialize around food. Food is used to honor and recognize visitors, friends, family members, and health-care providers; as an escape from everyday pressures, problems, and challenges; and prevent and treat illnesses. Puerto Rican patients may bring homemade goods to health-care providers as an expression of appreciation, respect, and gratitude for services rendered. Refusing these offerings may be interpreted as a personal rejection.

Some Puerto Ricans believe that being overweight is a sign of health and wealth. Some eat to excess, believing that if they eat more, their health will be better, whereas others pay no attention to weight control or dietary practices. Many Puerto Ricans perceive that European Americans are more preoccupied with how they look than how healthy they are. Efforts by American health-care providers directed at weight control may be seen as the Americans' excessive preoccupation with a thin body image.

25.8.2 Common Foods and Food Rituals

Traditional Puerto Rican families emphasize having a complete breakfast that begins with a cup of strong coffee or *café con leche* (coffee with milk). Some drink strong coffees such as espresso with lots of sugar; others boil fresh milk (or use condensed milk) and then add the coffee. Many families introduce children to coffee as early as 5 or 6 years of age. A traditional Puerto Rican breakfast includes hot cereal such as oatmeal; cornmeal; or rice and wheat cereal cooked with vanilla, cinnamon, sugar, salt, and milk. Although less common, traditional Puerto Ricans may eat corn pancakes or fritters for breakfast.

Lunch is served by noon, followed by dinner at around 5 or 6 p.m. A cup of espresso-like coffee is also enjoyed at 10:00 a.m. and 3:00 p.m. Rice and stew *habichuelas* (beans) are the main dishes among Puerto Rican families. Rice may be served plain or cooked and served with as many as 12 side dishes. Rice cooked with vegetables or meat is considered a complete meal. *Arroz guisado* (rice stew) is seasoned with *sofrito*, a blend of spices such as cilantro, *recao* (a type of cilantro), onions, green peppers, and other non-spicy ingredients. Rice is cooked with chicken, pork, sausages, codfish, calamari, or shrimp, corn, several types of beans, and *gandules* (green pigeon peas), a Puerto Rican bean rich in iron and protein. Rice with *gandules* is a traditional Christmas holiday dish that is accompanied by *pernil asado* (roasted pork) and *pasteles*, made with root vegetables, green plantain, bananas, or condiments and then filled with meat and wrapped with plantain leaves. Fritters are also common foods.

Puerto Ricans eat a great variety of pastas, breads, crackers, vegetables, and fruits. *Tostones*, fried green or ripe plantains, are a favorite side dish served with almost every meal. Puerto Rican families eat a variety of roots called *viandas*, vegetables rich in vitamins and starch. The most common *viandas* are celery roots, sweet potatoes, dasheens, yams, breadfruit, breadnut, green and ripe plantains, green bananas, tanniers, cassava, and chayote squash or christophines. A list of common Puerto Rican meals is presented in Table 25.1. Because Puerto Rican meals are flavorful, patients in the health-care setting may find more traditional American meals to be flavorless and unattractive. However, more acculturated Puerto Ricans are changing their traditional food practices and often follow mainland U.S. dietary practices. Health-care providers who work with traditional Puerto Rican patients should become familiar with these foods and their nutritional content to assist families with dietary practices that integrate their traditional or preferred food selections.

Table 25.1 Common puerto rican meals

Puerto rican meal	English translation
Alcapurrias	Green plantain fritters filled with meat or crab
Arepas de maíz y queso	Cornmeal and cheese fritters
Arroz con pollo	Rice with chicken
Arroz con gandules	Rice with pigeon peas
Arroz blanco (con aceite)	Plain rice (with oil)
Arroz guisado básico	Plain stewed rice
Asopao de pollo	Soupy rice with chicken
Bacalaitos	Codfish fritters
Bocadillo	Grilled sandwich
Mondongo	Tripe stew
Paella de mariscos	Seafood paella
Pastelillos de carne, queso, o pasta de guayaba	Turnovers filled with meat, cheese, or guava paste
Pollo en fricase con papas	Stewed chicken with potatoes
Relleno de papa	Potato ball filled with meat
Sancocho	*Viandas* and meat stew
Sofrito	Condiment
Surullo de queso	Cornmeal fritters filled with cheese

25.8.3 Dietary Practices for Health Promotion

Many Puerto Ricans ascribe to the hot-cold classifications of foods for nutritional balance and dietary practices during menstruation, pregnancy, the postpartum period, infant feeding, lactation, and aging. Some of the hot-cold classifications are presented in Table 25.2. Health-care providers should become familiar with these food practices when planning culturally congruent dietary alternatives.

Understanding that iron is considered a "hot" food that is not usually taken during pregnancy can assist health-care providers to negotiate approval and educate Puerto Rican women about the importance of maintaining adherence to daily iron recommendations, even during pregnancy and lactation. An additional summary of Puerto Rican cultural food habits, reasons for practices, and recommendations for health-care providers during such developmental stages is presented in Table 25.3.

The use of herbal supplements and complementary and alternative medicine are also important for women in treating symptoms associated with menopause. While many women will use hormone replacement therapy or HRT, many also chose to supplement this intervention with a combination of herbal remedies including: black cohosh, evening primrose, St. John's wort, gingko, ginseng, valerian root, sarsaparilla, chamomile, red clover, and passion flower (Green et al. 2017). Additionally, Puerto Rican women will be more likely to use complementary and alternative medicine including: relaxation, massage, acupuncture, guided imagery, chelation, biofeedback, and therapeutic touch in addition to or as an alternative to HRT (Green et al. 2017). Healthcare providers should understand and be able to discuss the safety and efficacy of the most frequently used alternatives. The use of HRT alternatives should be included in routine health assessment among women in this stage.

Table 25.2 Puerto rican hot-cold classification of selected foods, medications, herbs, and health-illness status

Hot-cold classification	Health-illnesses status	Western medications	Traditional herbs	Foods
Hot	Gastrointestinal illnesses (constipation, diarrhea, Crohn's colitis, ulcer, bleeding)	Syrups	Teas	Cocoa products
	Gynecological issues (pregnancy, menopause)	Dark-colored pills	Cinnamon	Alcoholic beverages
	Skin disorders (rashes, acne)	Aspirin	Dark-leaf teas	Caffeine products
	Neurological disorders (headache)	Anti-inflammatory agents	Teas	Hot cereals (wheat, corn)
	Heart disease	Prednisone		Salt
	Urological illnesses	Antihypertensives		Spices and condiments
		Castor oil		Beans
		Cinnamon		Nuts and seeds
		Vitamins (iron)		
		Antibiotics		
Cold	Osteomuscular illnesses (arthritis, rheumatoid arthritis, multiple sclerosis)	Diuretics	Orange lemon chamomile	Rice
	Menstruation	Bicarbonate of soda	Linden	Rice and barley water
	Respiratory illnesses	Antacids	Mint	Milk
		Milk of magnesia	Anise	Sugar and sugar products
				Root vegetables
				Avocado
				Fruits
				Vegetables
				White meat
				Honey
				Onions

An infant is believed to be healthy if it is *gordito* (a little fat) and has red cheeks. Consequently, many mothers add ground root vegetables, eggs, hot cereals, rice, canned baby foods, and fruits and vegetables to the infant's bottle at an early age. Traditionally, when children are introduced to soft foods and vegetables, parents boil and grind root vegetables for the infant. For some, these dietary practices have changed with the availability of canned baby food. Many mothers tend to feed whole cow's milk or canned milk (Carnation) earlier than recommended in Western practice, believing that canned milk produces healthier babies. Health-care providers must educate families regarding the nutritional content of canned milk versus fresh milk, breastfeeding, and formula.

For older Puerto Ricans, a good diet includes meats, traditional meals, and vitamin supplements. Beverages such as fresh-squeezed orange juice, grape juice, and *ponches* (punches) are used as additional nutritional support, particularly for those who are immunosuppressed or chronically or terminally ill. If the older individual is believed to have low blood pressure and is weak or tired, a small daily portion of brandy may be added to black coffee to enhance the work of "an old heart." If the health-care provider criticizes these practices, it may deter the client from seeking follow-up care and decrease trust and confidence in health-care providers. Health-care providers must inquire about these practices and should incorporate harmless or nonconflicting practices into the diet.

Table 25.3 Puerto rican cultural nutrition and health beliefs and practices during particular stages

Behavioral period	Dietary and health practices	Cultural justification	Recommendation for health-care professionals
Menstruation	*Food taboos*: Avoid spices, cold beverages, acid-citric fruits and substances, chocolate, and coffee	May induce cramps, hemorrhage clots, and physical imbalance. May produce acne during menstruation	Assess individual beliefs and acknowledge them
	Foods encouraged: Plenty of hot fluids, such as cinnamon tea, milk with cinnamon and sugar. Teas such as chamomile, anise seed, linden tea, mint leaves	Fluids encourage body cleaning of impurities. Hot beverages encourage circulation and reduce abdominal colic, cramps, and pain. Teas are soothing to all body systems	Incorporate traditional beliefs in treatments as required in nonsteroidal anti-inflammatories for dysmenorrhea
	Health practices: Avoid exercise and practice good hygiene. Do not walk barefoot. Avoid wind and rain. Stay as warm as possible	Exercise may increase pain and bleeding. Good hygiene is important for health. Walking barefoot during menstruation may cause rheumatoid arthritis and other inflammatory diseases. Warm temperatures promote circulation and the health of the reproductive system as well as prevent cramps	Encourage passive exercise. Provide information about the role of exercise in the reduction of menstrual discomfort. Support other practices
Pregnancy	*Food taboos*: Hot food, sauces, condiments, chocolate products, coffee, beans, pork, fritters, oily foods, and citrus products	May cause excess flatus, acid indigestion, bulging, and constipation. Chocolate and coffee may cause darker skin in fetus. Some believe citrus products may be abortive	Encourage healthy food habits. Provide information about chocolate and coffee myths encourage fruits
	Foods encouraged: Milk, beef, chicken, vegetables, fruits, *ponches*	Considered healthy and nutritious. Increases hemoglobin, strengthens and promotes good labor	Discourage the use of raw eggs in beverages because of possibility of *Salmonella* poisoning
	Health practices: Rest and get plenty of sleep. Eat plenty of food. Follow diet cautiously. Many avoid sexual intercourse early in pregnancy. Practice good hygiene and take warm showers	Enhances health and prevents problems during birth. Sex may cause problems with baby or preterm labor	Encourage use of food recommended for pregnancy. Provide information about sexual activity. Encourage a balanced plan of exercise with emphasis on weight control and health of the baby
Lactation	*Food taboos*: Avoid beans, cabbages, lettuce, seeds, nuts, pork, chocolate, coffee, and hot food items at all times	These foods cause stomach illnesses for infant and mother, including baby colic, diarrhea, and flatus	Include a dietary plan that is balanced with substitute food items. Clarify any myths about infant diarrhea, colic, and flatus
	Foods encouraged: Milk, water, *ponches*, chicken soup, chicken beef, pastas, hot cereals	Improve health and increase hemoglobin and essential vitamins protect mother and infant from illnesses. Fluids and *ponches* increase milk supply. Red meats reduce cravings	As above with raw eggs
.	*Health practices*: Avoid cold temperatures and wind. A few may avoid showering for several days during the *cuarentena* after birth. Great attention is paid to health of the mother	Cold temperatures and winds are believed to cause stroke and facial paralysis in a new mother showering may cause respiratory diseases. Mother is believed to be at risk and fragile	Provide information about reasons for stroke and facial paralysis. Provide time to ask questions and reduce anxiety during winter season deliveries

Table 25.3 (continued)

Behavioral period	Dietary and health practices	Cultural justification	Recommendation for health-care professionals
Infant feeding	*Food taboos*: Beans, too much rice, and uncooked vegetables	Believed to cause stomach colic, flatus, and distended abdomen. Too much rice causes constipation	Provide information about appropriate dietary patterns for infant
	Foods encouraged: Hot cereals, *ponches*, chicken broth or *caldos*. Fresh fruits, cooked vegetables, *viandas*. Raw eggs, cereals, baby foods in milk bottle. Fresh fruit juices. Mint, chamomile, and anise tea. Sugar and honey used for hiccups	Believed to be nutritious and healthy and to decrease hunger *Caldos* are fortifying and prevent illness. Cooked vegetables are healthy and prevent constipation. Bottle food fills the baby. Fresh juices and fruits refresh the stomach. Teas help baby sleep and cure flatus. Sugar and honey have curing properties	Instruct about infant diet and timely introduction of food items to diet. Explain consequences of excessive weight in infants. Discourage food in bottle to prevent choking
	Health practices keep baby warm while feeding	Warm babies eat, chew, and digest food better, and choking is decreased	Discourage raw eggs because of the risk of *Salmonella* and egg allergies and the use of honey because of the risk of botulism. Teas are harmless and provide additional fluid when used in moderation without sugar
			Provide information about babies and choking

During illness, Puerto Ricans pay close attention to dietary practices. Chicken soups and *caldos* (broth) are used as a hot meal to provide essential nutrients. A mixture of equal amounts of honey, lemon, and rum is used as an expectorant and antitussive. A malt drink, *malta* (grape juice), or milk is often added to an egg yolk mixed with plenty of sugar to increase the hemoglobin level and provide strength. Ulcers, acid indigestion, and stomach illnesses are treated with warm milk, with or without sugar. Herbal teas are used to treat illnesses and to promote health. Most herbal teas do not interfere with medical prescriptions. Incorporating their use with traditional Western medicine may enhance adherence to treatment.

Although the use of dietary and herbal supplements is considered relatively safe for healthy adults, some herbal supplements, including over the counter dietary supplements may not be approved or regulated by the U.S. Food and Drug Administration (FDA). Therefore, it is vital for clinicians to become aware of the herbal supplements use in individuals who have a health condition and/or are using prescription medications. Some of these dietary supplements may interact with some medications or may even have unclear doses for individuals to safely follow. It is important to note the use of these supplements in older adults as they may have an increased health risk which has been associated with possible interactions of supplements with some prescription medications (Olivera and Palacios 2012).

25.8.4 Nutritional Deficiencies and Food Limitations

Most Puerto Ricans moving to the mainland locate in areas with Puerto Rican or Hispanic communities and where preferred foods are readily available. Traditional cooking and food practices do not necessarily change. Instead, European American foods are quickly integrated into the dietary practices, thereby increasing food diversity. Fresh fruits and juices are consumed in large quantities.

Few studies have shown significant data about nutritional deficiencies among Puerto Ricans. Studies that include small samples of Puerto Ricans show that Puerto Rican children have nutritional statuses similar to those of Mexican

American and African American children in terms of malnutrition, obesity, and short stature. Low-income Puerto Rican children and adolescents have been found to have anemia and tooth decay related to consumption of less than the recommended daily allowances of iron, folacin, thiamine, niacin, and vitamin C. Menstruation is viewed as a time when women must care for themselves and adhere to certain dietary practices to promote health. From the onset of menstruation, young girls are encouraged to avoid foods believed to produce flatus, abdominal cramps, and colic. Hot drinks are encouraged to increase circulation and promote the elimination of metabolic waste.

25.9 Pregnancy and Childbearing Practices

25.9.1 Fertility Practices and Views Toward Pregnancy

Marital status, knowledge, attitudes, beliefs about the reproductive system, the role of motherhood, sexuality, and contraceptive use are factors that need to be considered when assessing and implementing culturally congruent maternal–infant interventions and educational programs. The percent of Puerto Rican women who receive prenatal care, beginning in the first trimester, has increased from 74% in 2009 to 83% in 2017 (HHS 2018). Health-care providers should be aware that social support has been found to be one of the most significant factors related to perinatal outcomes among Puerto Rican women. Among others, social support has been found to have significant implications for stress levels, health behaviors, and infant health (Landale and Oropesa 2001).

Teenage pregnancy has been linked with *marianismo* (traditional female gender roles within the culture), mother's past communication about traditional gender role norms, and the mother's positive outcome expectations for communication with the child (Torres et al. 2016). Perceptions of the physical condition in the neighborhood have also been found to be an important factor linked with teenage pregnancy (Torres et al. 2016). Data indicated that in Puerto Rico the infant mortality rate decreased 13.2–7.6% from 2016 to 2017 respectively (HHS 2018). In 2016, Puerto Rican female teen birth rates for 15–19-year-olds had a 13% decrease from 2015 (HHS 2018). Likewise, teen birth rates decreased by approximately 19% from 2016 to 2017 (HHS 2018).

The most popular fertility control methods used by Puerto Rican women in the past were tubal ligation, called *La Operación* (the surgery) (23%), oral contraceptives (8.7%), hysterectomies (3.5%) and oophorectomies (3.2%) (Stroup-Benham and Treviño 1991). Puerto Rican health outcomes, the behavior and practices of women regarding fertility, and views towards pregnancy are under studied, which further limits our ability to develop evidence-based health care programs to support this population. Unfortunately, there is limited evidence related to the beliefs and traditions held by Puerto Ricans towards pregnancy.

Puerto Rican mothers hold both traditional and progressive beliefs given the symbiotic relationship between Puerto Rico and the U.S. (Hammer et al. 2007). Mothers may also integrate aspects of both the Puerto Rican culture and the U.S. mainstream culture into their views concerning childbearing and education (Hammer et al. 2007). Providing culturally sensitive information to new mothers about their own health as well as the baby, prenatally, is important as part of routine prenatal care as has been associated with improved health outcomes for both the child and the mother (Green 2018). Health care providers should seek to understand Puerto Rican traditional beliefs and behaviors among mothers and their extended family members, as well as providing culturally sensitive services. Healthcare providers should encourage prenatal care, which can help determine the mother's and child's health and determine the sex of the unborn child. Prenatal care may have positive long-term health outcomes on mothers' health behaviors, and their children's health and could represent an opportunity for continued follow up (i.e. prenatal and postpartum) care during and after labor.

25.9.2 Prescriptive, Restrictive, and Taboo Practices in the Childbearing Family

Hygiene is highly valued during pregnancy, labor, and the postpartum period. Pregnancy is a time of indulgence for some Puerto Rican women. Favors and wishes are granted to women for their well-being and the health of their babies. Men are socialized to be tolerant, understanding, and patient regarding pregnant women and their preferences. Pregnant Puerto Rican women are encouraged to rest and consume large quantities of food, yet carefully watch what they eat. Prescriptive dietary recommendations may differ between Puerto Rico and the mainland (Tovar et al. 2010). Common advice from family members is often influenced by the thought of being really healthy and to gain weight and keep eating (Tovar et al. 2010).

Many young Puerto Rican families prefer to attend birthing classes. Some expect women to "get fat" and place little emphasis on weight control. Strenuous physical activity and exercise are discouraged, and lifting heavy objects is prohibited. Women are strongly discouraged from consuming aspirin, Alka-Seltzer, and malt beverages because these substances are believed to cause abortion.

Reflective Exercise 25.2

Rosa Medina, age 33, is 3 days postpartum. She has brought 3-day-old Juanita to the maternal-child clinic because the baby has been crying continuously.

The nurse greets Mrs. Medina and asks how she and her family are getting along with Juanita. Mrs. Medina says that things are "mostly okay" and that her mother and sister help take care of the other children, while her husband was able to keep his job after a lot of people at the market were fired.

The nurse asks if the problem is with Mrs. Medina or her daughter Juanita, and Mrs. Medina says that Juanita has been crying all day, but nothing seems to help. They don't have any chamomile tea to offer the child either.

Nurse: When did she first start crying?

Mrs. Medina: Yesterday morning, right after I gave her the bottle.

Nurse: Was she taking her bottle okay up to that time?

Mrs. Medina: I am not sure. I have been kind of tired since the birth, and my mother has been feeding her most of the time. My mother says that she does not eat as much as she should. My mother never thinks babies eat enough.

Nurse: Can you help me examine her belly?

Mrs. Medina: Yes, but she cries every time I touch her belly. I think she has *empacho*.

Nurse: I notice you have this cloth wrapped around her belly. Can I take it off?

Mrs. Medina: Oh, of course.

Nurse: You have a coin over Juanita's belly. What purpose does it serve?

Mrs. Medina: To keep the bad spirits away and help the cord heal.

Nurse: And the cloth holds the coin in place?

Mrs. Medina: The cloth keeps her belly button from sticking out when she cries.

Nurse: Did you do this with your other children?

Mrs. Medina: Yes. Everyone in our family does this, and we all have flat belly buttons.

Nurse: You know it really is not necessary to put the cloth on her.

1. How is *familism* displayed in this scenario?
2. What is the culture-bound syndrome *empacho*? What is the equivalent Western concept of *empacho*?
3. Should the nurse actively discourage Mrs. Medina from placing the coin on the infant's umbilicus? Why? Why not?
4. Should the nurse actively discourage Mrs. Medina from using the abdominal cloth? Why? Why not?

Many women refrain from *tener relaciones* (sexual intercourse) after the first trimester to avoid hurting the fetus or causing preterm labor. Some men view this time as an opportunity for extramarital sexual affairs. Health-care providers should inquire in a nonconfrontational manner about this possibility and educate men regarding the dangers of STIs and HIV.

Women prefer the bed position for labor, wish to have their bodies covered, and prefer a limited number of internal examinations. They welcome their husbands, mothers, or sisters to assist during labor. Men are expected to be supportive. During labor, women may be loud and verbally expressive, a culturally accepted and an encouraged method for coping with pain and discomfort. Pain medications are welcomed. Health-care providers should respect these wishes and explain the necessity of invasive interventions during labor. Most women oppose having a cesarean section because it indicates a "weak woman." The health-care provider should discuss the possibility of a cesarean section early in the pregnancy.

Postpartum women receive care from their family and friends. Traditionally, their first postpartum meal should be homemade chicken soup to provide energy and strength. Women are encouraged to avoid exposure to wind and cold temperatures, not to lift heavy objects, and not to do housework for 40 days after delivery (the *cuarentena*). Some traditional women do not wash their hair during this time. Because the mother is believed to be susceptible to emotional and physical distress during the postpartum period, family members try not to contribute to stress or to give bad news to the new mother. Fathers may be reluctant to tell the new mother about a problem with the newborn. However, most Puerto Rican women want to be told immediately about any problems. This is a critical issue during the postpartum period, especially with premature babies, given the belief that a healthy baby is a "symbol of father's virility and a time for the woman to demonstrate her fertility, strength, and success during and after birth" (Crouch-Ruiz 1996).

Transition to motherhood often includes physical recovery from childbirth and mental adjustment to life-change (Fink 2013). Family influence, dilemmas of cultural practices, and cultural conflict are factors that may influence the way moms-to-be seek health services. Often, research-based information from health professionals differs from culturally influenced advice of trusted friends and family, which may force new mothers to decide which advice to follow and dividing their loyalties. There have been limited studies on post-partum care focusing on Puerto Rican women. Encouraging self-efficacy for decision making empowers Puerto Rican new mothers to care for themselves and their infants. Increasing cultural competence may be invaluable in improving health outcomes and satisfaction measures for Puerto Ricans (Fink 2013).

Some mothers might ask to talk to the pediatrician, rather than to a nurse, about infant problems. Because of the value placed on family and children, women who need to return to work early may experience great distress when they do not follow some of these cultural values or norms. Health-care providers should assess for individual perceptions and dissatisfaction with the working role and birth recuperation. Mothers who breastfeed are encouraged to drink lots of fluids such as milk and chicken soup, and if they are feeling weak or tired, to drink *ponches*, beverages consisting of milk or fresh juices mixed with a raw egg yolk and sugar. Hot foods such as chocolate, beans, lentils, and coffee are discouraged because they are believed to cause stomach irritability and flatus for the mother and colic for the infant.

Data provided by the Puerto Rico Breastfeeding Promotion Coalition (2016) indicates that in 2015, 85.9% of women breastfeed their infants for at least 1 month. These rates fall to 26.5 and 29.8% at 6 and 12 months respectively (Breastfeeding Promotion Coalition 2016). Traditional Puerto Rican mothers and those from rural areas may prefer to breastfeed their babies for the first year. Mothers who work outside the home may select breastfeeding or formula or both. However, with the introduction of formula through U.S. food stamp programs, two generations of Puerto Rican mothers have been inclined to relinquish breastfeeding and adopt formula as the primary source of infant nutrition. Because some Puerto Rican women believe that breastfeeding increases their weight, disfigures the breast, and makes them less sexually attractive, they undervalue the benefits of breastfeeding. Health-care providers need to provide information about these beliefs and educate women about breastfeeding myths and misconceptions. Because maternal grandmothers have a great influence on practices related to breastfeeding, they should be included along with significant others in educational programs that encourage this practice.

25.10 Death Rituals

25.10.1 Death Rituals and Expectations

Death is perceived as a time of crisis in Puerto Rican families. The body is considered sacred by many individuals and is guarded with great respect. Death rituals are shaped by religious beliefs and practices, and family members are careful to complete the death rituals. News about the deceased should be given first to the head of the family, usually the oldest daughter or son. Because of cultural, physical, and emotional responses to grief, health-care providers should use a private room to communicate such news and have a clergy or minister present when the news is disclosed. Family privacy at this time is highly valued.

Providers should allow time for family to view, touch, and stay with the body before it is removed. Traditionally, some Puerto Rican families keep the body in their home before burial. Cultural traditions and financial limitations influence this decision. Consequently, some older adults may wish to follow these death rituals. For some, funeral homes are viewed as impersonal, financially unnecessary, and detrimental to the mourning process because they detract from family intimacy.

Although the family may prefer to have all death rituals finished within a reasonable timeframe, it is important to extend burial rituals until all close family members can be present. The head of the family is expected to coordinate the arrival of family members, which usually creates a delay in death rituals and burial time and an emotional burden and stress on family members. Health-care providers should ensure that members of the family are provided with support, resources, and information regarding differences in U.S. legal requirements. These requirements are often confusing and are considered insensitive, particularly with a stillbirth or when an autopsy is necessary. Authorization from several family members might be essential. Because of the spiritual and religious importance of burial traditions and rituals during these events, cremation is rarely practiced among Puerto Ricans. Among Catholics, the head of the family or other close family member is expected to organize the religious ceremonies, such as the praying of the rosary, the wake (***velorio***), and the novenas, the 9 days of rosary following the death of the family member. Family may meet at the deceased's home for several days, sometimes weeks, to support the family and talk about the deceased. Food is served throughout the day as a symbol of gratitude for those who come to pay their respects.

25.10.2 Responses to Death and Grief

It is culturally acceptable for the family of the deceased to freely express themselves through loud crying and verbal expressions of grief. Some may talk in a thunderous way to God. Others may express their grief through a sensitive but continuous crying or sobbing. Some believe that not expressing their feelings could mean a lack of love and respect for the deceased. Similar to the reaction to other crisis events, some may develop psychosomatic symptoms, and others may experience nausea, vomiting, or fainting spells as a result of a nervousness attack—*ataque de nervios*. Health-care providers should be nonjudgmental with mourners' psychosomatic or other expressions of grief by providing a private environment and helping to minimize interruptions during that period.

Special prayers and readings are believed to be necessary at the moment of death, and families expect to be present to recite these prayers.

25.11 Spirituality

25.11.1 Dominant Religion and Use of Prayer

Religious beliefs among Puerto Ricans influence their approach to health and illness. Most Puerto Ricans are Catholic (85%), and the remainder have Protestant Evangelical religious affiliations (CIA World Factbook 2019a). A few practice *espiritismo*, a blend of Native Indian, African, and Catholic beliefs that deal with rituals related to spiritual communications with spirits and evil forces. ***Espiritistas***, individuals capable of communicating with spirits, may be consulted to promote spiritual wellness and treat mental illnesses.

Upon immigration, many Puerto Ricans may feel out of place and need support resources. Many join Evangelical churches because these offer a more personal spiritual approach. These religious groups provide social support and promote harmony and spiritual–physical well-being. Health-care providers should reinforce these spiritual practices, while incorporating prescribed medications, health activities, and the prevention of risk behaviors. *Espiritistas* treat patients with mental health conditions and are often consulted to determine folk remedies compatible with Western medical treatments. Health-care providers should be aware that the elderly, those who have limited access to health care, and those who are dissatisfied with or distrust the Western medical system commonly use spiritual healers.

Among Catholics, candles, rosary beads, or a special patron or figurine might accompany the patient to the health-care facility and be used during prayer rituals. To provide timely and appropriate interventions to Catholic families, health-care providers should inquire about the family's wishes regarding the Sacrament of Anointing of the Sick if the patient is gravely ill.

25.11.2 Meaning of Life and Individual Sources of Strength

Puerto Ricans consider life sacred, something that individuals should preserve. Many see the quality of life as a harmonious balance among the mind, the body, and the spirit. Spirituality helps Puerto Ricans gain strength to deal with illness, death, and grief and ultimately promotes well-being. Most Puerto Ricans are very religious, and when confronted with situations related to health, illness, work, death, or the prognosis of a terminal illness, they maintain their trust in spiritual forces. Spiritual forces assist in controlling and managing social and economic constraints. Their own personal actions are perceived as inconsequential or trivial without the trust and confidence in God's will, *Si Dios quiere* (if God wants). Rather than a fatalistic approach to life during illness, death, or health promotion, Puerto Ricans use coping mechanisms such as religious practices that are instrumental in providing control in their lives. For example, the role of religion in the lives of Puerto Ricans with chronic illnesses or with disabled children has been described as a critical source of support and a mechanism that allows for appropriate interpretation of health and illness (Desai et al. 2016). God, who is their highest source of strength, guides life. For some, scripture readings, praise, and prayer bring inner spiritual power to the soul, *el alma*.

25.11.3 Spiritual Beliefs and Health-Care Practices

Spiritual practices influenced by religious groups have a great impact on the health status of many Puerto Ricans because churches have had a great influence on the health of individuals by discouraging high-risk behaviors and promoting health.

Through prayer, church attendance, and worship, many Puerto Ricans discover spiritual courage and inner strength to avoid high-risk behaviors such as smoking and substance abuse. Clergy and ministers are a resource for spiritual wisdom and help with a host of spiritual needs.

Although amulets have lost their popularity, some Puerto Ricans still use them. In some instances, when a child is born, the family will place a *manita de azabache* (small black fist) charm pendant with the infant. Often these are placed around the infant's wrist for protection against *mal de ojo*. The term *mal de ojo* ("evil eye"), is often related to children who fail to thrive and is described as a hex conveyed by an envious glance (Juckett 2013). Among Puerto Ricans, the belief in *mal de ojo* is well-known and symbolizes the diagnosis of childhood symptoms appearing with sudden onset (Harwood 1981; Weller 2015). An *azabache* or a rabbit's foot might be used for good luck, to drive away bad spirits, and to protect a child's health. Traditionally, rosary beads and patron saint figures may be placed at the head or side of the bed or on the patient to protect him or her from outside evil sources. Health-care providers should ask permission before removing, cleaning, or moving these objects. A benediction may be requested before removing amulets or religious objects or providing spiritual support. These objects are often used as a means of dealing with a crisis or as an expression of hope. The health-care provider should assess individual and family religious preferences and support spiritual resources according to the patient's or family's request.

25.12 Health-Care Practices

25.12.1 Health-Seeking Beliefs and Behaviors

Most Puerto Ricans have a curative view of health. They tend to underuse health-promotion and preventive services such as regular dental or physical examinations and Pap smears (Colon and Sanchez-Cesareo 2016). Many use emergency health-care services rather than preventive health-care services for acute problems. Acculturation, age, access to health care, education, and income influence health-seeking beliefs and behaviors. Health-care providers must develop mechanisms to integrate individual, family, and community resources to encourage a focus on health promotion and enhance early health screening and disease prevention. In particular, a great deal of attention must be provided to improve interpersonal processes of care among providers and Puerto Rican patients. Barriers to care for this population include: lack of insurance, language, cultural beliefs, mistrust of Western healthcare providers, and illiteracy (Juckett 2013).

Good hygiene is a basic concept for health promotion among Puerto Ricans. Daily showers are essential for good health and personal appearance. Exceptions may be made during illnesses such as colds, flu, or viral infections. After surgery, some prefer to bathe using a basin of water instead of taking a shower or tub bath. Most prefer to shower and wash their hair daily; however, some women may avoid doing these activities during menstruation. During hospitalization, some refrain from having a bowel movement if they have to use a bedside commode or bedpan. Healthcare workers are in a unique position to respectfully explore those beliefs and practices and to provide a private, nonintrusive environment for the patient.

25.12.2 Responsibility for Health Care

Most Puerto Ricans believe in family care rather than self-care. Women are seen as the main caregivers and promoters of family health and are the source of spiritual and physical strength. Health-care providers should incorporate the participation of the family in the care of the ill.

Natural herbs, teas, and over-the-counter medications are often used as initial interventions for symptoms of illness. Many consult family and friends before consulting a health-care provider. Moreover, pharmacists play a vital role in symptom management. Although Puerto Rico is subject to U.S. Food and Drug Administration regulations and practices, many Puerto Ricans are able to obtain controlled prescriptions from their local pharmacist in Puerto Rico. When they are on the mainland, they try to obtain the same kind of services from local pharmacists, creating distress and frustration for both the patient and the pharmacist.

Over-the-counter medications and folk remedies are often used by Puerto Ricans to treat mental health symptoms, acute illnesses, and chronic diseases. Health-care providers should inquire about those practices and encourage patients to bring their medications to every visit. Engaging in a friendly conversation encourages patients to reveal their use of folk treatments, over-the-counter medications, and concurrent use of folk healers. Since the early 1980s, Puerto Ricans have become accustomed to the use of extended-care facilities and nursing homes. However, they prefer to keep chronically or terminally ill family members at home.

25.12.3 Folk and Traditional Practices

Espiritismo and **Santería** are magico-religious and folk-healing practices to ward off evil spirits or obtain spiritual protection which are used by some Puerto Ricans in addition to mainstream healthcare services (Desai et al. 2016; Lynch and Hanson 2011; Spector 2013). *Espiritistas* solve problems by communicating with spirits. The *Santería* focuses on health promotion and personal growth and development. Clients who use these folk practices visit ***bótanicas*** (folk religious stores) and use natural herbs, aromatic incenses, special bathing herbs, prayer books, prayers, and figurines for treating illness and promoting good health. Providers must examine their own views about traditional practices and healers and refrain from making prejudicial comments that may inhibit collaboration with folk healers. Many Puerto Ricans practice the use of a broad range of home and botanical remedies believed to be effective for treating ailments. Many of these home remedies practices are based on learned concepts transmitted through families and popular culture. Health beliefs may also originate from religious spiritual practices, either traditional or folk.

Puerto Ricans may use folk practices for instances of shortness of breath, nausea, and vomiting. Asphyxia or shortness of breath is believed to be caused by lack of air in the body. Fanning the face or blowing into the patient's face is believed to provide oxygen and relieve dyspnea. Some may use tea from an alligator's tail, snails, or *savila* (plant leaves) for illnesses such as asthma and congestive heart failure.

Nausea and vomiting may be embarrassing and cause alarm to many Puerto Rican patients. Many believe that smelling or rubbing isopropyl alcohol (*alcolado*) may help alleviate these symptoms. Some place a damp cloth on the forehead to refresh the "hot" inside the body and relieve nausea. Some put the head between the legs to stop vomiting. Mint, orange, or lemon tree leaves are boiled and used as tea to relieve nausea and vomiting. Rectal suppositories are believed to induce diarrhea. Health-care providers should provide clear information about suppositories and the etiologic cause of symptoms.

Healthcare providers must be aware that Puerto Rican folklore may hinder the ability of individuals with serious health concerns not to seek medical advice or delay seeking until it may be too late. Thus, it is key to be able to identify common diseases and conditions for which preventive care or treatment might be complex since there are individuals and families may have created belief system around the disease, one that is often false and without scientific evidence. Folk beliefs exist without scientific evidence and may lead to situations where individuals ignoring important health symptoms during a serious illness. These beliefs are often transmitted from generation to generation and can not only affect the individual but also family members who often share similar customs, values, and exist within the same environment. Sensitivity is paramount but a clinician-patient discussion of the importance of health-related beliefs and management practices should be necessary. This is an important key factor in overcoming some cultural beliefs which may preclude patients from seeking medical treatment to potentially avoid disease-related complications.

25.12.4 Barriers to Health Care

The medically indigent in Puerto Rico receive free health-care services through the Department of Health or other government led service programs. On the mainland, accessing health-care services is a complex issue for many Puerto Ricans. This problem has been exacerbated by Hurricane Maria, which resulted in a mass exodus of medical providers from the country (Roman 2017). Before the hurricane, the island was plagued with deteriorating healthcare infrastructure as well as a fragmented system that was inadequate to meet basic population health needs (Roman 2017). In the months following Hurricane Maria, the loss of qualified medical personnel not only reduced the number of facilities that could remain open to deliver care, but also the event resulted in a decline in the number of medical training programs available to those seeking entrance into the medical profession (Roman 2017). Despite these challenges, information from the Kaiser Family Foundation (2017) does indicate that the number of uninsured in Puerto Rico is 6%. This figure is lower than that for the United States: 9% (Kaiser Family Foundation 2017).

25.12.5 Cultural Responses to Health and Illness

Typically, when a family member is ill, other family members and friends become a source of support and care. Puerto Ricans may be loud and outspoken in expressing pain. Health-care providers should not censure this expression of pain or judge it as an exaggeration. This expressive behavior is a socially learned mechanism to cope with pain. *Ay!* is a common verbal moaning expression for pain (*dolor*). Because rural older individuals might have difficulty interpreting and quantifying pain, the use of numerical pain-identifying scales may be inappropriate. Most people prefer oral or intravenous medications for pain relief rather than intramuscular injections or rectal medications. In addition, herbal teas, heat, and prayer are used to manage pain.

Because mental illness carries a stigma, obtaining information or talking about mental illness with Puerto Rican families may be difficult. Some might not disclose the presence or history of mental illnesses, even in a trusting environment. In addition, Puerto Ricans may have a different cultural perception about the etiology, meaning, and treatment of mental illnesses. A mental illness may result from a terrible experience, a crisis, or the action of evil forces or spirits. Some perceive that symptoms of mental illness result from *nervios* (nerves), having done something wrong, or breaking God's commandments. When someone is anxious or overcome with emotions or problems, she or he is just *nervioso*. Similarly, someone who is experiencing despair, anorexia, bulimia, melancholy, anxiety, or lack of sleep may be *nervioso*(*a*), or suffering from *ataque de nervios* rather than being clinically depressed, manic-depressive, or mentally ill. These conditions may be used to camouflage mental illness. Given the high incidence of depression among the Puerto Rican population, this is a critical mental health issue for Puerto Ricans (Canino et al. 2019). Providers must acknowledge the confidentiality of information when obtaining a history. If trust is developed, health-care providers may get a more accurate response to their questions.

Health-care providers must become familiar with the vocabulary used to describe signs and symptoms of mental illnesses among this group. Families must be provided with clear and relevant information about the diagnosis, treatment, and etiology of mental illnesses to enhance adherence to treatment and follow-up care. In addition, health-care providers should be aware of traditional healing practices and be sensitive to mental health services for Puerto Rican families. Community-based settings such as churches, schools, and childcare centers are excellent environments for promoting physical and mental health among Puerto Ricans.

Genetic or physical defects among Puerto Ricans may be seen as a result of heredity, suffering, or lack of care during pregnancy. Less-educated individuals may place guilt and blame on the mother or father. Caregivers must provide information about the causes of genetic defects and reduce stress and guilt for parents. For decades, Puerto Rican families cared for impaired family members in a covert environment, away from the eyes of the community. At present, families are more open about these family members and care for the physically and mentally challenged at home, which is preferred over acute- or long-term-care facilities. The role of familism is of particular importance for Puerto Ricans because they provide caregiving in an interdependent network of extended family members who provide social support, solidarity, *cariño*, and resources for the family.

Sociocultural differences exist in parental beliefs and attitudes when caring for children with disabilities. Puerto Rican parents develop a sense of interdependence and overprotection that is expressed through extreme nurturing behaviors and positive caring behaviors. As a result, conventional test scores for family functioning may not be appropriate to interpret child development and family adaptation. Health-care providers should be aware of these behaviors differences and act with caution when interpreting these results (Gannotti et al. 2001). Caregivers' stress should be a key component of the health assessment of these families. Health-care providers should supply information about community resources, support groups, and culturally appropriate mental health services.

Reflective Exercise 25.3

Mrs. Martínez, a 37-year-old beauty salon worker from Puerto Rico, has been visiting her brother, who is a student at Central University. The following is the conversation that she has with a nurse at a clinic in her brother's town:

Nurse: Good afternoon. Why have you come to the clinic?

Mrs. Martínez: Shouldn't I be here?

Nurse: What is wrong with you that you came here?

Mrs. Martínez: I think that I am having a miscarriage.

Nurse: What is your name?

Mrs. Martínez: Lucero Martínez de Estrada y Rodríguez.

Nurse: That is a very long name.

Mrs. Martínez: Yes, I am from Puerto Rico, and we have longer names than you do on the mainland.

Nurse: What makes you think you are having a miscarriage?

Mrs. Martínez: I am 3 months pregnant, and I started bleeding this afternoon.

Nurse: How much bleeding are you having?

Mrs. Martínez: Oh, a lot.

Nurse: Well, do you have insurance?

Mrs. Martínez: Yes, I have insurance through my work.

Nurse: I do not know if we take insurance from Puerto Rico.

Mrs. Martínez: I can pay if I need to.

Nurse: Are you married?

Mrs. Martínez: Oh, yes, I am married.

Nurse: Where is your husband?

Mrs. Martínez: Back in Puerto Rico taking care of our daughter. I am here visiting my brother.

Nurse: Are you taking any medicine?

Mrs. Martínez: (Takes a bottle out of her purse.) Just these that I got from the pharmacist.

Nurse: I do not recognize these pills. They are in Spanish. You know you should not be taking any medicine unless it is prescribed by a doctor.

Mrs. Martínez: He is a pharmacist. And he prescribed these pills.

Nurse: Since this is not a big emergency, it might be a long wait to see the doctor.

Mrs. Martínez: I'll call my brother and tell him where I am.

1. Did the nurse approach Mrs. Martinez with *simpatia*?
2. In what ways did the nurse display *respeto*?
3. What evidence in this scenario can be construed as the nurse not displaying *respeto*?
4. What might the nurse have done to determine the English equivalent of the medicine prescribed by the pharmacist?
5. How common is it for pharmacists to prescribe medications in Puerto Rico?
6. How common is it for pharmacists to prescribe medications in the United States?
7. Does the mainland United States accept insurance from Puerto Rico?

25.12.6 Blood Transfusions and Organ Donation

For many Puerto Ricans, organ donation is seen as an act of good will and a gift of life. However, autopsy may be seen as a violation of the body. When discussions regarding autopsies and organ donations are necessary, the health-care provider must proceed with patience and provide precise and simple information. A clergy or minister may be helpful and may be expected to be present at the time of death. Although no proscriptions exist against blood donation and blood transfusion, many Puerto Ricans may be reluctant to engage in these procedures for fear of contracting HIV. Health-care providers need to carefully explore these beliefs and dispel myths.

25.13 Health-Care Providers

25.13.1 Traditional Versus Biomedical Health-Care Providers

Many Puerto Ricans use traditional and folk healers such as *espiritistas* and *Santeros* along with Western health-care providers. Some *espiritismo* practices are used to deal with the power of good and evil spirits in the physical and emotional development of the individual. *Santeros*, individuals prepared to practice *santería*, are consulted in matters related to the belief of object intrusion, diseases caused by evil spirits, the loss of the soul, the insertion of a spirit, or the anger of God.

Modesty is a highly valued quality. An intimate and unobtrusive environment is preferred for disclosing health-related concerns. Individuals expect a respectful environment; a soft tone of voice; and time to be heard, explain concerns, and ask questions when discussing health matters. Rooms without doors are considered disrespectful and conspicuous, particularly if the visit requires the removal of clothing. Some Puerto Ricans may have a gender or age bias against health-care providers. Men prefer male physicians for care and may feel embarrassed and uncomfortable with a female physician. A few individuals discount the academic and intellectual competencies of female physicians and may distrust their judgment and treatment. Some Puerto Rican women feel uncomfortable with a male physician, whereas a few prefer a male doctor. Elderly Puerto Ricans may prefer older health-care providers because they are seen as wise and mature in matters related to health, life experiences, and the use of folk practices and remedies. To build the patient's confidence, younger and female health-care providers must demonstrate an overall concern for the patient and develop respect and understanding by acknowledging and incorporating traditional healing practices into treatment regimens.

25.13.2 Status of Health-Care Providers

Puerto Ricans hold health-care providers in high regard because they are seen as wise authority figures. Distrust may develop if the health-care provider lacks respect for issues related to traditional health practices, ignores personalism in the relationship, does not use advanced technological assessment tools, and has a physical or personal image that differs from the traditional "well-groomed, white-attire" image. Overall, however, Puerto Ricans are well-educated health consumers and expect high-quality care blended with traditional practices and reliable technological approaches.

References

Acello B, Hegner BR (2016) Nursing assistant: a nursing process approach, 11th edn. Cengage Learning, Boston MA

AIDS United (2019) Puerto Rico. https://www.aidsunited.org/Programs-0024-Grantmaking/Puerto-Rico.aspx

Alvarez MJ, Frietze G, Ramos C, Field C, Zarate MA (2017) A quantitative analysis of Latino acculturation and alcohol use: myth versus reality. Alcohol Clin Exp Res 41(7):1246–1256. https://doi.org/10.1111/acer.13420

Andrews-Chavez JY, Lee CS, Houser RF, Falcon LM, Tucker KL (2015) Factors associated with alcohol consumption patterns in a Puerto Rican urban cohort. Public Health Nutr 18(3):464–473. https://doi.org/10.1017/S1368980014000433

Aschner P, Aguilar-Salinas C, Aguirre L, Franco L, Gagliardino JJ, Lapertosa SG et al (2014) Diabetes in South and Central America: an update. Diabetes Res Clin Pract 103:238–243. https://doi.org/10.1016/j.diabres.2013.11.010

Caetano R, Gruenewald P, Vaeth PAC, Canino G (2018a) DSM-5 alcohol use disorder severity in Puerto Rico: prevalence, criteria profile, and correlates. Alcohol Clin Exp Res 42(2):378–386. https://doi.org/10.1111/acer.13572

Caetano R, Vaeth PAC, Canino G (2018b) Illegal drug use and its correlates in San Juan, Puerto Rico. Drug Alcohol Depend 185:356–359. https://doi.org/10.1016/j.drugalcdep.2017.12.029

Canino G, Shrout PE, NeMoyer A, Vila D, Santiago KM, Garcia P et al (2019) A comparison of the prevalence of psychiatric disorders in Puerto Rico with the United States and the Puerto Rican population of the United States. Soc Psychiatry Psychiatr Epidemiol 54(3):369–378. https://doi.org/10.1007/s00127-019-01653-6

Carl Y, Frias RL, Kurtevski S, Gonzalez Copo T, Mustafa AR, Font CM et al (2019) The correlation of English language proficiency and indices of stress and anxiety in migrants from Puerto Rico after Hurricane Maria: a preliminary study. Disaster Med Public Health Prep 21:1–5. https://doi.org/10.1017/dmp.2019.22

Carroll KS (2016) Understanding perceptions of language threat: the case of Puerto Rico. Caribb Stud 44(1/2):167–186. https://doi.org/10.1353/crb.2016.0006

Castañeda SF, Garcia ML, Lopez-Gurrola M, Stoutenberg M, Emory K, Daviglus ML et al (2019) Alcohol use, acculturation and socioeconomic status among Hispanic/Latino men and women: the Hispanic community health study/study of Latinos. PLoS One 14(4):e0214906. https://doi.org/10.1371/journal.pone.0214906

Centers for Disease Control and Prevention (2018) Hispanics/Latinos and tobacco use. https://www.cdc.gov/tobacco/disparities/hispanics-latinos/index.htm

CIA World Factbook (2019a) Puerto Rico. https://www.cia.gov/library/publications/the-world-factbook/geos/rq.html

CIA World Factbook (2019b) United States. https://www.cia.gov/library/publications/the-world-factbook/geos/us.html

Claudio-Campos K, Orengo-Mercado C, Renta JY, Peguero M, Garcia R, Hernandez G et al (2015) Pharmacogenetics of healthy volunteers in Puerto Rico. Drug Metab Pers Ther 30(4):239–249. https://doi.org/10.1515/dmpt-2015-0021

Colon HM, Sanchez-Cesareo M (2016) Disparities in health care in Puerto Rico compared with the United States. JAMA Intern Med 176(6):794–795. https://doi.org/10.1001/jamainternmed.2016.1144

Colón-López V, Fernández-Espada N, Vélez C, Gonzalez VJ, Diaz-Toro EC, Calo WA, Savas LS, Pattatucci A, Fernández ME (2017) Communication about sex and HPV among Puerto Rican mothers and daughters. Ethn Health 22(4):348–360. https://doi.org/10.1080/13557858.2016.1246938

Crouch-Ruiz E (1996) The birth of a premature infant in a Puerto Rican family. In: Torres S (ed) Hispanic voices: Hispanic health educators speak out. National League for Nursing, New York NY, pp 26–28

Desai PP, Rivera AT, Backes EM (2016) Latino caregiver coping with children's chronic health conditions: an integrative literature review. J Pediatr Health Care 30(2):108–120. https://doi.org/10.1016/j.pedhc.2015.06.001

Ertl MM, Rentería R, Dillon FR, Babino R, De La Rosa M, Brenner RE (2019) Longitudinal associations between marianismo beliefs and acculturation stress among Latina immigrants during initial years in the United States. J Couns Psychol 66(6):665–667. https://doi.org/10.1037/cou0000361

Fabiano-Smith L, Shuriff R, Barlow RA, Goldstein BA (2014) Dialect density in bilingual Puerto Rican Spanish-speaking children. Linguist Approaches Biling 4(1):34–60. https://doi.org/10.1075/lab.4.1.02fab

Fink AM (2013) Culturally competent postpartum care for Puerto Rican women: results of a qualitative study. J Obstet Gynecol Neonatal Nurs 42(S1):S84–S85. https://doi.org/10.1111/1552-6909.12175

Gannotti ME, Handwerker WP, Groce NE, Cruz C (2001) Sociocultural influences on disability status in Puerto Rican children. Phys Ther 81(9):1512–1523

Garcia C, Ailshire JA (2019) Biological risk profiles among Latino subgroups in the health and retirement study. Innov Aging 3(2):1–14. https://doi.org/10.1093/geroni/igz017

Green TL (2018) Unpacking racial/ethnic disparities in prenatal care use: the role of individual-, household-, and area-level characteristics. J Women's

Health 27(9):1124–1134. https://doi.org/10.1089/jwh.2017.6807

Green RR, Santoro N, Allshouse AA, Neal-Perry G, Derby C (2017) Prevalence of complementary and alternative medicine and herbal remedy use in Hispanic and non-Hispanic white women: results from the study of women's health across the nation. J Altern Complement Med 23(10):805–811. https://doi.org/10.1089/acm.2017.0080

Hammer CS, Rodriguez BL, Lawrence FR, Miccio AW (2007) Puerto Rican mothers' beliefs and home literacy practices. Lang Speech Hear Serv Sch 38(3):216–224. https://doi.org/10.1044/0161-1461(2007/023)

Harwood A (1981) Mainland Puerto Ricans. In: Harwood A (ed) Ethnicity and medical care. Harvard University Press, Cambridge MA, pp 397–481

Hernandez AM, Zule WA, Karg RS, Browne FA, Wechsberg WM (2012) Factors that influence HIV risk among Hispanic female immigrants and their implications for HIV prevention interventions. Int J Family Med 2012:1–11. https://doi.org/10.1155/2012/876381

Hopgood-Jovet A (ed) (2018) An overview of Puerto Rico employment law. The Society for Human Resource Management, Alexandria VA. https://www.shrm.org/resourcesandtools/legal-and-compliance/state-and-local-updates/pages/overview-puerto-rico-law.aspx

Institute for Health Metrics and Evaluation (2018) Puerto Rico. http://www.healthdata.org/puerto-rico

Juckett G (2013) Caring for Latino patients. Am Fam Physician 87(1):48–54. https://www.aafp.org/afp/2013/0101/p48.html

Kaiser Family Foundation (2017) Puerto Rico: fast facts. https://www.kff.org/disparities-policy/fact-sheet/puerto-rico-fast-facts/

Katz JM (2019) The disappearing schools of Puerto Rico. The New York Times Magazine. https://www.nytimes.com/interactive/2019/09/12/magazine/puerto-rico-schools-hurricane-maria.html

Landale NS, Oropesa RS (2001) Migration, social support and perinatal health: an origin-destination analysis of Puerto Rican women. J Health Soc Behav 42(2):166–183

Lynch EWE, Hanson MJE (2011) Developing cross-cultural competence: a guide for working with children and their families, 4th edn. Paul H. Brookes Publishing, Baltimore MD

Massara EB (1989) !Que gordita!: a study of overweight among Puerto Rican women. AMS Press, New York

Meléndez E, Hinojosa J, Roman N (2017) Post-hurricane Maria exodus from Puerto Rico and school enrollment in Florida. Center for Puerto Rican Studies, New York. https://centropr.hunter.cuny.edu/sites/default/files/CentroReport-RB2017-02-POST-MARIA-FL-PR-EXODUS%20%281%29.pdf

Mogro-Wilson C, Rojas R, Haynes J (2016) A cultural understanding of the parenting practices of Puerto Rican fathers. Soc Work Res 40(4):237–248. https://doi.org/10.1093/swr/svw019

Noyd DH, Sharp TM (2015) Recent advances in dengue: Relevance to Puerto Rico. P R Health Sci J 34(2):65–70. https://www.ncbi.nlm.nih.gov/pmc/articles/PMC4587385/

Olivera EJ, Palacios C (2012) Use of supplements in Puerto Rican older adults residing in an elderly project. P R Health Sci J 31(4):213–219

Orshan SA (1996) Acculturation, perceived social support, and self-esteem in primigravida Puerto Rican teenagers. West J Nurs Res 18:460–473

Perez-Mayoral J, Soto-Saldado M, Shah E, Kittles R, Stern MC, Olivera MI et al (2019) Association of genetic ancestry with colorectal tumor location in Puerto Rican Latinos. Hum Genomics 13(12):1–11. https://doi.org/10.1186/s40246-019-0196-4

Puerto Rico Breastfeeding Promotion Coalition (2016) Puerto Rico breastfeeding report. http://www.usbreastfeeding.org/d/do/2890

Puerto Rico Department of Health (2016) 2015 Puerto Rico primary care needs assessment. http://www.salud.gov.pr/Estadisticas-Registros-y-Publicaciones/Publicaciones/2015%20Puerto%20Rico%20Primary%20Care%20Needs%20Assesment.pdf

Ramos-Pibernus AG, Rodríguez-Madera SL, Padilla M, Varas-Díaz N, Vargas Molina R (2016) Intersections and evolutions of 'butch-trans' categories in Puerto Rico: needs and barriers of an invisible population. Glob Public Health 11(7/8):966–980. https://doi.org/10.1080/17441692.2016.1180703

Rege A (2018) US hospitals increasingly recruiting nurses from hurricane-ravaged Puerto Rico: 5 things to know. Becker's Hospital Reviewe, Chicago IL. https://www.beckershospitalreview.com/human-resources/us-hospitals-increasingly-recruiting-nurses-from-hurricane-ravaged-puerto-rico-5-things-to-know.html

Rivera MOL (2018) Hard to sea: Puerto Rico's future under the Jones act. Loyola Marit Law J 17(1):65–137

Rivera MG, López LAO (2018) El Español y el Inglés en Puerto Rico: Una polémica de más de un siglo. Cent J 30(1):106–131

Rodríguez-Díaz CE, Jovet-Toldeo GG, Vélez-Vega CM, Ortiz-Sanchez EJ, Santiago-Rodríguez EI, Vargas Molina RL et al (2016) Discrimination and health among lesbian, gay, bisexual and trans people in Puerto Rico. P R Health Sci J 35(3):154–159

Roman J (2017) Hurricane Maria: a preventable humanitarian and health care crisis unveiling the Puerto Rican dilemma. Ann Am Thorac Soc 15(3):293–295. https://doi.org/10.1513/AnnalsATS.201710-792OI

Rosario Colón J, Domenech Rodríguez MM, Galliher RV (2019) Parenting styles and child outcomes in Puerto Rican families. Puerto Rican J Psychol 30(1):12–28

Schneiderman N, Llabre M, Cowie CC, Barnhart J, Carethon M, Gallo LC et al (2014) Prevalence of diabetes among Hispanics/Latinos from diverse backgrounds: the Hispanic community health study/study of Latinos (HCHS/SOL). Diabetes Care 37:2233–2239. https://doi.org/10.2337/dc13-2939

Smollin L, Garcia Valles JC, Torres MI, Granberry PJ (2018) Puerto Rican mothers' conversations about sexual health with non-heterosexual youth. Centro J 30(2):406–428

Spector RE (2013) Cultural diversity in health and illness. Pearson, New York NY

Stroup-Benham CA, Treviño FM (1991) Reproductive characteristics of Mexican-American, Puerto Rican and Cuban-American women. J Am Med Assoc 265:222–226

Thomas DL, Sharp TM, Torres J, Armstrong PA, Munoz-Jordan J, Ryff KR et al (2016) Local transmission of Zika virus—Puerto Rico, November 23, 2015-January 28, 2016. Mortal Morbid Wkly Rep 65(6):154–158. https://www.cdc.gov/mmwr/volumes/65/wr/mm6506e2.htm

Toro-Babrador G, Weaver OR, Martinez-Cruzado JC (2003) Mitochondrial DNA analysis in Aruba: strang maternal ancestry of closely related Amerindians and implications for the peopling of Northwestern Venezuela. Caribb J Sci 39(1):11–22

Torres JB (1998) Masculinity and gender roles among Puerto Rican men: machismo on the U.S. mainland. Am J Orthopsychiatry 68(1):16–26

Torres MI, Granberry P, Person S, Allison J, Rosal M, Rustan S (2016) Influential factors of Puerto Rican mother–child communication about sexual health topics. Matern Child Health J 20(11):2280–2290. https://doi.org/10.1007/s10995-016-2041-x

Tovar A, Chasan-Taber L, Bermudez OI, Hyatt RR, Must A (2010) Knowledge, attitudes, and beliefs regarding weight gain during pregnancy among Hispanic women. Matern Child Health J 14(6):938–949. https://doi.org/10.1007/s10995-009-0524-8

U.S. Department of Health and Human Services (2018) Puerto Rico: state snapshot. https://grants6.tvisdata.hrsa.gov/uploadedfiles/StateSubmittedFiles/2019/stateSnapshots/PR_StateSnapshot.pdf

U.S. Department of Health and Human Services (2019) Puerto Rico adolescent reproductive health facts. https://www.hhs.gov/ash/oah/facts-and-stats/national-and-state-data-sheets/adolescent-reproductive-health/puerto-rico/index.html

United States Bureau of Labor Statistics (2019) Local area unemployment statistics: Puerto Rico. https://data.bls.gov/timeseries/LASST720000000000003?amp%253bdata_tool=XGtable&output_view=data&include_graphs=true

United States Census Bureau (2019) Quick facts: Puerto Rico. https://www.census.gov/quickfacts/PR

Villafañe-Santiago A, Serro Taylor J, Jiménez-Chafey MI, Irizarry-Robles CY (2019) Parenting styles and child outcomes in Puerto Rican families. Puerto Rican J Psychol 30(1):70–81

Villalobos Solís M, Smetana JG, Tasopoulos-Chan M (2017) Evaluations of conflicts between Latino values and autonomy desires among Puerto Rican adolescents. Child Dev 88(5):1581–1597. https://doi.org/10.1111/cdev.12687

Volk RJ, Steinbauer JR, Cantor SB, Holzer CE (1997) The alcohol use disorders identification test (AUDIT) as a screen for at-risk drinking in primary care patients of different racial/ethnic backgrounds. Addiction 92(2):197–206. https://doi.org/10.1111/j.1360-0443.1997.tb03652.x

Weller SC, Baer RD, de Alba Garcia JG, Glazer M, Trotter R, Salcedo Rocha AL et al (2015) Variation and persistence in latin american beliefs about evil eye. Cross-Cultural Res 49(2):174–203. https://doi.org/10.1177/1069397114539268

Wang Y, Rayer S (2018) Growth of the Puerto Rican population in Florida and on the U.S. mainland. Bureau of Economic and Business Research, Cambridge MA. https://www.bebr.ufl.edu/population/website-article/growth-puerto-rican-population-florida-and-us-mainland

Warren AC (2010) Women's employment and changing gender relations in Puerto Rico. Caribb Stud 38(2):59–91. https://doi.org/10.1353/crb.2010.0058

26 People of Russian Heritage

Tatayana Maltseva

26.1 Introduction

This chapter focuses on the cultural beliefs, values and practices of people of Russian heritage. An extensive Russian history is included so that the reader better understands the complexity of the Russian revolution, challenges of the Soviet era, and the current formation of independent countries. Even though, there are many ethnic groups who have emigrated from Russia and other Russian independent countries, the emphasis is given to the Russian population living in the United States and other countries. Common illnesses, cultural preferences, and health care traditions described in this chapter will guide healthcare professionals in providing culturally sensitive and congruent care to people of Russian heritage.

This chapter is a revision previously written by Karen Aroian, Galina, Khatutsky, and Alexandra Dashevskaya in the previous edition of the book.

T. Maltseva (✉)
Florida International University, Miami, FL, USA
e-mail: bogopolt@fiu.edu

26.2 Overview, Heritage, and Topography

The Russian Federation, which is geographically the largest country in the world, is almost twice the size of the United States of America (US) (CIA World Factbook 2019a, b). In the early 1990s, the Soviet Union collapsed, forming the Russian Federation. Each former Soviet Union Republic became its own independent country. Currently, the Russian Federation consists of 21ethnic republics, several regions, and multiple provinces. There are 11 time zones in Russia. The climate variations are as vast as the nation itself, with temperatures ranging from approximately +94 °F (35 C) to −49 °F (−45 C) depending on the region and season. This giant nation is not only famous for encompassing over two million square miles of the arctic circle; it also contains the Kamchatka Peninsula, an area that is famous for earthquake epicenters and active volcanos (Encyclopedia Britannica 2019).

There are approximately 200 national and ethnic groups living in Russia. Approximately 77.7% of people residing in Russia are of Russian ethnicity, Tartars are 3.7%, Ukrainians are 1.4%, and 17% are from other ethnic groups. The population of Russia is approximately 142 million as of 2018 July census data (CIA World Factbook 2019a, b). The capital of the Russian Federation is Moscow with an estimated 12 million people. The second largest city of the

© Springer Nature Switzerland AG 2021

L. D. Purnell, E. A. Fenkl (eds.), *Textbook for Transcultural Health Care: A Population Approach*,
https://doi.org/10.1007/978-3-030-51399-3_26

Russian Federation is St. Petersburg with close to five million people (World Atlas 2017). Moscow and St. Petersburg are the Russian Federation's largest tourist destinations because of the Romanov dynasty and preserved historical architecture as exemplified by the tsarist-era palaces around St Petersburg, Peterhof Palace in St. Petersburg, the revolutionary Red Square with the Kremlin in Moscow, the heart of Russian Orthodoxy, and the Saint Basil's Cathedral in Moscow. Russian people take pride in their heritage and culture; the nation is famous for its plethora of museums, cultural exhibits, theatrical productions, opera, and ballet shows. Russian cultural heritage boasts an abundance of globally influential artists and musicians such as Chagal, Kandinsky, Tchaikovsky, Rachmaninov, Rimsky-Korsakov, and others.

The population growth rate is declining at −0.11% and a birth rate of 10.7 births per 1000 population, places Russia at 184 in ranking compared to the other countries in the world (CIA World Factbook 2019a, b). The average life expectancy has increased to 65 for Russian men and 77 years for Russian women, in contrast to the Soviet era. A low fertility rate (1.6 per women of reproductive age) contributes to the Russian population decline (CIA World Factbook 2019a, b; Library of Congress 2019a, b). While major cities have high population density, 29% of Russians live in remote rural areas where poverty is rampant and access to basic health care is limited (CIA World Factbook 2019a, b; Library of Congress 2019a, b).

Russian history is replete with cycles of war, civil unrests, and complete collapses of the economy. For example, the first structured state of modern Russia, Ukraine, and Belarus called, the Kievan Rus, was founded in 862. Mongols invaded Kievan Rus in 1237 through 1240 and destroyed major cities—Moscow, Chernigov, Pereyaslavl, and Kiev (Martin 2004a, b). Ivan III the Great freed Russia from Mongols and united several Russian regions (Krom 2004). His grandson Ivan IV called "*Grozny*" in translation Ivan *the Terrible* became the czar of all Russia from 1547 to 1584. During his reign he transformed Russia into multi-ethnic empire and established the strong relationship with the Orthodox church through abundant donations (Bogatyrev 2004). Michael Romanov was coronated as a tsar of Russia in 1613. Since then, the Romanov dynasty ruled Russia for three centuries (Martin 2004a, b). Russia established world power during the time of Peter I, called Peter the Great at the end of 1600 and early 1700 centuries. He built the new Capital of Russia—St. Petersburg, modernized the military, founded the Russian navy, and established diplomatic relationships with European countries (Hughes 2004). Alexander II, promoted self-government, started universal military service and strengthened the Russian borders. During his reign, the Russian territories of Alaska and Aleutian Islands were sold to the United States in 1867 (Granville 2004).

The rise of the Soviet Union started with the Bolshevik revolution on November 7th, 1917. Vladimir Lenin took power, eliminated czarist regime and became the first leader of the communist party in the Soviet Union (Williams 2004). The Bolshevik revolution occurred because the Russian people were looking toward the advances in the western Europe which included the introduction of a more democratic government, minimization of the role of royals in the government, recognition of the working class, and industrial modernization. Nicholas II failed to recognize what the revolutionaries wanted and so the country revolted.

Josef Stalin, who was one of Lenin's associates, came to power after Lenin's death in 1924 and was the long working leader of the Soviet Union. He became a dictator, whose priority was to turn Russia into an industrial, agricultural, and military superpower. He upheld communist ideals and killed millions of people who opposed his regime. There were 15 ethnically and culturally diverse republics in the Soviet Union during that time, with the largest being the Republic of Russia. Germany attacked the Soviet Union in June 22, 1941. World War II lasted 4 years and ended with a Russian army victory in 1945 (Gill 2004). After the war, the Soviet Union rebuilt the cities, strengthened the economy, and developed the first space exploration program. With the election of Mikhail

Gorbachov in 1985, Soviet government chose to depart from the autocratic rule of the communist party and underwent reorganization or *perestroika* (reform) and *glasnost* (transparency) (Brown 2004).

The Chernobyl nuclear plant explosion occurred near the city of Pripyat, Ukraine on April 26, 1986. Huge amounts of radioactive materials released into the atmosphere and were deposited over Ukraine, Byelorussia, some parts of the Russian Federation, and other parts of Europe. This nuclear explosion resulted in thousands of deaths and closures of several cities and villages in Ukraine.

The Soviet Union collapsed on August 24, 1991 as a result of an unsuccessful Communist Party coup and growing separatist movements within Soviet Republics. Russia converted to an independent country and other Soviet republics formed a Commonwealth of Independent States. The ending of the Soviet Union affected the adoption of a new constitution in 1993 and the establishment of the executive, legislative, and judiciary brunches of new government. Intense democratic reform and the development of a market economy occurred after the dissolution of the Soviet Union. However, corruption and bribery are still widespread in Russia as well as former Soviet republics and many democratic reforms never materialized.

Despite the current market economy, one–fifth of the population in Russia live in poverty. According to the Russian Federation State statistics, the poverty rate was 13.2% in 2017. The unemployment rate was 5.2% in 2017 (CIA World Factbook 2019a, b). Crime is also prevalent in Russia. Many rural areas of Russia have poor police presence. Working as a police officer is not considered a reputable occupation in Russia.

Russia is abundant in natural resources, which include rich accumulations of oil, natural gas, coal, wood, aluminium, platinum, cooper, nickel, gold, and diamonds. Russia boasts one of the highest levels of forestation in the world; however, air, water, and land pollution remain high (Energy Information Administration [EIA] 2017; Library of Congress 2019a, b).

26.2.1 Heritage and Residence

Russia is a multinational state with around 185 ethnics groups and minorities. Russian is the official state language. There are more than 100 other ethnic languages spoken in the country. Russians comprise the largest group, about 77% of the population. Other large ethnic groups in Russia include Ukrainians, Tartars, and Bashkirs. According to the 2012 Russian Censes, 86% of Russians live in the Russian Federation and 14% live in the other states, such as Ukraine and Kazakhstan. The majority of Russians identify themselves with Orthodox Christianity (World Atlas 2012). After the collapse of the Soviet Union, many ethnic Russians left former Soviet Republics in fear of retaliation and settled in the Russian Federation to start a new life. The same happened for other minority groups: non-Russian population of ethnic minorities left Russia and migrated to their native republics during 1991–2000.

Global immigration of people from the former Soviet Union to the United States, Canada, Israel, Australia, and Germany began in 1986 at the start of *perestroika*. Over 1.5 million Jews left Russia since then emigrating to other countries (Tartakovsky et al. 2017). There are 2,608,412 people of Russian descent living in the US according to the 2017 American Community Survey of United Census Bureau. In 1990–1997 alone, there were 433,427 Russians who obtained lawful permanent resident status in the US (U.S. Department of Homeland Security 2017). Russian immigration to the US has been on decline since then; 2017 statistical data indicated that only 8841 Russians obtained permanent resident status in the US (U.S. Department of Homeland Security 2017). The US serves as the main international home for Russian orphans with an estimated 58,000 Russian children adopted from 1995 to 2011 by American citizens. Adoption of Russian children by Americans ceased in 2012 when the Russian government championed a bill that prohibited the adoption of Russian children by American citizens as a retaliatory move for investigations into human right abuses by the US in Russia (DeBose and DeAngelo 2015).

The majority of Russian immigrants in the US live in urban areas and primarily settle in the areas where large Russian communities already have a presence, such as New York, Boston, Chicago, Philadelphia, Atlanta, Houston, Detroit, Baltimore, Los Angeles, San Francisco, San Diego, Portland, and Miami. Almost half, 350,000, who came between 1970 and 1990 reside in New York, with most living in Brooklyn (Isurin 2011). New York appealed to Russians due to abundant work opportunities and economic advances. Houston has an estimated population of 30,000 Russian immigrants (Isurin 2011). Many older Russian immigrants prefer to relocate to Miami, Florida due to the pleasant climate during the winter. Florida became a one of the top six states where Russian immigrants reside (Bell 2015). Russian immigrants reside also in Canada, primarily in Montreal, Toronto, and Vancouver. Jurcik et al. (2013) reported that Citizenship and Immigration Canada (2010) documented that over 7600 new permanent residents arrived in Canada in 2009 from the different republics of the former Soviet Union. Although a large population of Russian immigrants arrived in North America and other countries, research is very limited on the ethnic composition of this group (Jurcik et al. 2013). Russian immigrants are generally highly educated, willing to leave their native country due to economic and political reasons, and greatly value social support and family systems (Yakobov et al. 2019).

Immigration from Russia and other former Soviet Republics to the United States was a complicated process. In particular, the most difficulties occurred during the Soviet era from 1970 to 1980 when people had to wait more than a year to get government permission to leave while relinquishing their Soviet passport and the citizenship (Isurin 2011). Many immigrants had to leave their apartments, houses, possessions, and valuables and were only allowed by the government to take minimal luggage with them. In the literature, the definition of Russian immigrants very broadly encompassing all immigrants who speak the Russian language. Russian immigrants in foreign countries are referred to as Russians regardless of their ethnicity status or religious practices.

Coming from a collectivistic society, Russians who immigrated to the US during communism had a challenging time to adapt to the new country, learn a new language, take care of the family, and balance their lives. Some of them developed nostalgia about their motherland. Russian immigrants who left their native country after the collapse of the Soviet Union or who arrived at the younger age were more adjustable to the new culture and learned the language quicker because they were already familiar with technological advances and Western economy while living in Russia or other regions.

26.2.2 Economic Factors and Migration

The Russian population in the US represents a minority group that has been less studied compared with East Asian and European American groups (Jurcik et al. 2015). More than 3.3 million people emigrated from Russia to the US during the first wave of migration from 1880 until 1920 (Chiswick and Larsen 2015). Christian Orthodox Russians, Russian Jewish immigrants, Russians, Ukrainians, and Byelorussians left the country with the purpose to improve their economic lives, escape persecution, and find religious freedom. Many people from Orthodox clergy, demobilized army soldiers, intellectuals, artists, writers, and other individuals from different ethnic groups who opposed the communist rule and feared for their lives left the country after World War I in 1914 and the Bolshevik Revolution in 1917.

A huge exodus of Jewish immigrants from Russia, Byelorussia, Ukraine, and close republics took place between 1920 and 1922 in fear of *pogroms*—sudden Bolshevik and military raids during which they assaulted Jewish individuals and destroyed their businesses and properties. In the late 1920s, immigration almost stopped until early 1970. Soviet people had no freedom to travel outside of their country. If a person was required to go abroad, he or she would need special permission from the Soviet government and undergo extensive questioning; most of the time the permission to travel as a tourist was denied. Ordinary Russian people could not even contemplate immigration to another country (Isurin 2011).

The second wave of immigration is not very well defined. Many people displaced in European Countries after the World War II in 1945 and some of them were prisoners of war in concentration camps and could not or did not return to Russia after the war ended (Isurin 2011). The third wave of Russian immigration began in the early 1970s, when the Russian government allowed people from some minority groups to emigrate. A formal agreement was reached between the Soviet Union and the US; it primarily allowed people from Jewish ethnic minorities to apply for refugee status in the US and Israel and permanently leave Russia and other Soviet Republics. More than 600,000 Russians, primarily of Jewish heritage, were granted refugee status which allowed them to emigrate (Chiswick and Larsen 2015; Isurin 2011). During the Soviet time, Jews could not practice their religion openly, were denied admission to universities, job promotions, places to live, and employment. Parents instructed their children not to talk openly to anyone about their ethnic background, fearing the children would be bullied or, in worst cases, physically harmed. Thus, many Jewish families left the former Soviet Union to seek better opportunities for their children during the third wave of immigration. Immigration during the third wave was very difficult, as people could not go directly to the US. Russian immigrants had to first stay in Austria for approximately a month and then in Italy for up to 6 months to wait for a *garant*—a letter from a sponsor in the US, Canada, or Australia. Once Jewish immigrants reached Italy, they could choose to immigrate to Canada or Australia even though they had to wait up to a year for the *garant*. Immigrants who desired to live in Israel did not have to wait for the *garant* and were given permission to enter Israel after staying 1 month in Austria.

The fourth and the last wave of immigration occurred after the collapse of the Soviet Union in 1991. This time, people were provided more freedom to immigrate and directly enter their country of destination. During this time, the reason for immigration was primarily an economic one. People wanted to leave the county to get better jobs and salaries, study in Western universities, or simply to reunite with their families. Many scientists, doctors, and researchers left Russia and obtained work permission visas in the US and European countries. There were more than 450,000 people who immigrated from Russia between 1991 and 2000 to the US (Chiswick and Larsen 2015).

Currently, Russia is considered an independent democratic country which follows the market economy. Therefore, people, who want to permanently immigrate to the US cannot justify a reason for their immigration easily to obtain a refugee status. Also, current US immigration laws make it difficult to obtain an entrance visa without proper justification, slowing Russian immigration for the past 5–10 years. According to the U.S. Department of Homeland Security, there were 72,924 immigrants from Russia who obtained permanent US resident visa between 2010 and 2017 versus 433,427 from 1990 to 1999 (U.S. Department of Homeland Security 2017).

26.2.3 Education and Professions

Russian immigrants are generally well educated. Having a good education and profession are integral parts of the Russian culture. Russians value education and see it as the highest measure of success in life. Typically, Russians go to great lengths to achieve the highest possible degrees in education. People in the former Soviet Union were more respected if they were medical doctors or college professors because these jobs can only be attained through advanced education. A very common practice is for immigrants to sacrifice everything in order to obtain the highest education: e.g. work extra hours or maintain several jobs to support their children while they are studying.

The literacy rate in Russia is 99.7% which is one of the top literacy rates in the world where males and females are equally literate (CIA World fact book 2019a, b). The median age of Russian immigrants is 32 years compared to 38 for all foreign-born people settling in the US

(Ameredia Integrated Cultural Market 2019). Russian immigrants over 65 years of age comprise more than 20% of immigrants who arrived during the third wave of immigration. A majority of single men and women over the age of 60 years who resettled from the former Soviet Union in the United States from 1990 to 2000 were widowed (Fitzpatrick and Freed 2000).

Based on the demographic characteristics, 30% of Russian immigrants are skilled workers and 23% have obtained a post-graduate degree. Computers science, academia, and health care are preferred professional fields. Male immigrants choose to work as computer engineers while females are likely to be school teachers, physicians, and registered nurses which clearly delineate gender preferences in selecting professions (Michalikova 2018). The majority of Russian immigrants are married—64% and have children. Russian-Americans prefer to own a home and have smaller family size, usually 1–2 children per family compared to 2–3 children per family of other US foreign born populations. Over 80% of the Russian Americans have graduated from high school versus 67% for other US foreign born populations. For higher education, 53% of Russian Americans graduated from a College or University with a bachelor's degree. They favor higher education, high-paying employment, and leadership positions. Russians are also avid readers (Ameredia Integrated Cultural Market 2019). In the US, 67.5% of Russian are employed in professional, managerial, technical, and sales occupations compared with 45.6% of all foreign-born people living in the US (Ameredia Integrated Cultural Market 2019).

Education in Russia is very rigorous compared with the US educational system. On the very first school day, children are held accountable for completing the classwork and homework on time with maximum effort. Interestingly, foreign languages were not compulsory in the past. Therefore, many Russian immigrants who came between 1970 and 2000 experienced English language barriers when they arrived. The language barrier was even tougher for older people because they needed to depend on their children for interpretation and assistance with health-care services and medical treatments (Fitzpatrick and Freed 2000). More recently, the Russian education system has implemented several changes so that young graduates can easily join the global economy. These changes included mandatory studies of foreign languages at different levels of education. Moreover, many corporate organizations in Russia now require employees to have English language proficiency and even will only hire applicants who are fluent in English. The younger population in Russia have access to the Internet, use the same electronic devices as their peers in other countries, and have access to the world wide social media. Therefore, they are better able to communicate in English.

Older Russian immigrants receive assistance from the government, such as subsidized housing, food stamps, supplemental social security income, and federal medical coverage. They are educated individuals; however, due to their language barriers, their age and physical illnesses, they were unable to be part of the workforce in the US. Another important thing to keep in mind about people of Russian heritage is that many of them could not work in the US according to their occupation or specialty they attained in Russia. Professional fields in the US required licensing and proper credentials in order to be qualified for employment. Therefore, many professionals who migrated during the third wave had to join the service work force in order to support their families and/ or get requalified for another occupation. Many Russian immigrants had to give up the occupations for which they were trained and assimilated to their new lives (Chiswick and Larsen 2015). Older Russian immigrants, however, experienced true loss while learning how to live in their new country; they grieved having to sacrifice their motherland, their friends, and relatives; their inability to communicate in the host language; inability to work; and a decline in social status. Friendship is highly valued in the collectivistic society and the loss of communication with the old friends from the native country is recognized as one of the more serious concerns of immigrants as it can lead to the development of depressive symptoms (Fitzpatrick and Freed 2000; Isurin 2011).

26.3 Communication

26.3.1 Primary Language and Dialects

The Russian language is a deep, expressive, and phonetic. It was derived from the Cyrillic alphabet which was adapted from Greek and Hebrew to give Slavic populations a written language in the ninth century. Ukrainian and Byelorussian are very similar to Russian. Word order is very flexible and not as strict as it is in English and other languages. The Russian language still uses the Cyrillic alphabet in written communication. The alphabet consists of 33 letters: 10 vowels, 21 consonants, and 2 signs—hard sign ъ and soft sign ь. The language does not include articles. The United Nation established Russian as one of the six official languages. Russian was an approved language for all republics under the Soviet time and is currently the official language in the Russian Federation. Students are required to take courses in Russian and literature to master language proficiency in reading, writing, and communication.

According to the U.S. Census Bureau (2017a, b), 2,476,981 Russian Americans over age 5 have the ability to speak English at home. Out of these, 78.5% speak English only, 21.5% speak a language other than English, and 8% could not speak English at all. English speaking ability of Russian immigrants in the US is based on the 2012 U.S. Bureau data indicated that 39.6% of immigrants spoke Russian at home and 20.7% spoke English only at home. Data comparison between these 2 years indicated significant improvement in English language communication at home among immigrants. Russian speaking immigrants can write and read in English better than verbally communicating. Numerous Russian speaking services are available in many metropolitan areas for people who do not speak English. Russian communities also became popular places for living, particularly among older Russians where they have opportunities to socialize with their peers, seek medical treatment from the Russian speaking medical providers, eat gourmet food at Russian cafes and restaurants, and purchase Russian foods from authentic Russian food stores. Metropolitan cities with a large concentration of Russian immigrants have their own television and radio stations, journals, and newspapers written in the Russian.

26.3.2 Cultural Communication and Relationships

Russians value family, friendships, and traditions. They are very loyal to preservation of their culture and very proud of their cultural roots. They enjoy get-togethers with the family over family dinners and events to have valuable discussions about social, cultural, and political issues. Personal advice and word of mouth recommendations from family and friends are very powerful among Russians and greatly influence their decision-making processes about health care treatment and life style changes. Russians from older generation, in particular, prefer to rely on help from family members, relatives, and friends rather than public services. Russians are willing to see health-care providers for their health-care needs but prefer to make appointments with the Russian-speaking specialists.

Most Russians are very well-educated. They have great awareness of their values and communicate freely about their beliefs and customs. Being respectful to others is an integral part of the culture. Russians maintain eye contact, value people who obtain higher education and ranks, and are very respectful of authorities and older people. They may speak loudly to get their point across or might simply nod their head in agreement. It's important to point out that Russian-speaking people are often likely to just nod their head, even if they don't understand English and will ask for an interpreter. Handshake is very common in Russian culture among males. This is a symbol of respect, agreement, and trust. Home is considered a private place. All members of the household are expected to follow the household rules. Shoes are removed prior to entering the house and people and guests are required to change to the household slippers.

Russians maintain formal behavior in public prefer not to invade the personal space of others. Kissing and hugging with strangers on a first encounter are considered inappropriate. Intimate personal space is reserved only for the familial environment. Kissing on a cheek in public is kept for very close friends. Casual gestures in public places, such as placing feet on the desk, reclining, keeping hands in the pockets or behind the head or across the chest, are considered disrespectful and even insulting. Pointing the index finger or making a tight fist are considered a sign of anger and are viewed as insulting. Russians are very polite in communicating in the public places; they reserve emotional feelings and soften conflicts. They prefer not to bring bad news abruptly to family members and may withhold the information or soften the truth.

Russians are very responsible, dedicated, and caring people toward others, especially those who need help. They value time and oblige to the rule of work and society. They are empathetic and might be inclined to sacrifice much to help others. They are very resilient, emotionally stable, and strong due to the long history of suffering and challenges. Russians reciprocate the same level of recognition and respect they give to others. Russians think of themselves as being "strong" and "powerful" in any situation as they can overcome any life challenges. Verbalizing multiple complaints is considered a weakness.

26.3.3 Time

Russian immigrants value their time and other people's schedules and are very punctual with health-care providers. If they cannot make the appointment, they will call to reschedule. Not showing up for the appointment or to a family event without prior notification is considered disrespectful. People who live in Russia mostly operate in the present time. They focus on what they have today and might not even think about the future. Russian immigrants think about their present and future; in particular they focus on the future of their children and support them to get a better education. They are family oriented and prefer to spend free time with family and close friends. They get together for the family events, cook meals, and socialize within the family and circle of friends. It is customary in Russian culture to stay late in family gatherings.

26.3.4 Greetings

Russians value addressing other people properly. Adults are addressed by the full first name and their patronymic name. The patronymic name represents the first name of their father. The ending of the patronymic name depends on the gender of the individual. For example, a male adult is called at work and other public places by the following format—Alexander Ivanovich (Alexander, son of Ivan). The female adult is called by others as Elena Vladimirovna (Elena, daughter of Vladimir). Russians appreciate formality in greetings and, therefore, use titles such as Doctor, Mister, Miss, and Professor. Children are expected to call adult relatives by addressing them as mother, father, aunt, uncle, grandfather, grandmother. Even during the dating process, people address each other by the full first name and the patronymic name until a more close relationship has been established.

26.3.5 Family Roles and Organization

26.3.5.1 Head of Household and Gender Roles

Traditional gender roles in Russia consist of earning responsibility as the ultimate duty of fathers and parenthood as the exclusive responsibility of mothers (Lipazova 2017). This contributes to inequality of gender roles. The mother has a priority in the child's life and upbringing while the father's participation in child care may be considered optional. Even with a current socioeconomic change in Russia, traditional gender roles and practices still exist in 65% of families with dual earners, but household work and childcare are primarily the responsibility of females (Lipazova 2016).

During Soviet times, mothers were granted 1 year of paid maternity leave, had a choice to register children in the preschool and aftercare, and could even send children to sleepaway summer camps funded by the government. Fathers had the right to take parental leave but it was t not always beneficial because they served as the primary breadwinner (Federal Law 1992, 2015a, b).

Children and young adults typically seek advice from their parents and grandparents for major decisions in matters such as selecting a profession and higher education. The role reversal might be evident in immigrants as younger generations help parents and grandparents in a decision-making due to the inability of older members of the family to speak and understand English. Many young adults take their grandparents for the health-care appointments to help them be more comfortable in communicating with English speaking health-care providers. However, older Russian immigrants might feel embarrassed by unintentional family role reversal practices which can be necessitated by living in the US and other countries.

Middle aged women have multiple roles, simultaneously managing work, home, and childcare responsibilities. Younger female immigrants are very engaged in getting a good education and career and focus on negotiating family gender role equality. Multicultural or mixed marriages are becoming common for the younger generation of Russian immigrants.

Social support plays a significant part in family well-being. Russians value concrete and practical support when encountering problems. Such matter-of-fact support encourages stronger communities, allowing immigrants to take full advantage of resources (Jurcik et al. 2015).

26.3.5.2 Behaviors for Children and Adolescents

Russian children are raised to be respectful and do as they told by their parents and grandparents. Children are expected to achieve good grades in school, complete their homework on a daily basis, and graduate with honors. Societal rules exceed the individual's needs, Russians do not typically find excuses for not getting work or school work done. Children are required to care for the sick family members and help family by performing daily chores. Grandparents usually help by watching children, particularly, when both parents are working.

Older people are modest and do not usually show affection in the public. Topics about intimate relationships, including sexuality, are not usually discussed in the family. Sex education and contraception are not public topics of discussion. If a Russian teenager gets pregnant, it is considered a loss of trust and shame for the family. This pregnancy would most likely result in a clinically induced abortion. A sexual relationship outside of marriage is not supported (Lowental 2004). Immigrants try to maintain strong parental influences on their children and try to stay committed to their Russian roots. They prefer to send their children to a Russian-speaking day care, enrol them in sports or dance classes conducted by Russian coaches and performers.

More educated immigrants from larger urban cities are more positive toward raising their children in congruency with norms of the host county compared to immigrants with less education and those from rural areas of Russia who are more inclined to continue old country norms (Remennick 2015). Immigration to the new country can affect children in both positive and negative ways. Fishman and Mesch (2005) reported that adolescents from the former Soviet Union have a higher school dropout rate and lower rate of curriculum matriculation compared with their native peers. Nevertheless, many children of Russian immigrants excel in school, get accepted to colleges and universities, and become professionals (Haim 2014; Lerner et al. 2007; Eisikovits 2008).

26.3.5.3 Family Goals and Priorities

Collectivism, an integral part of Russian culture has been rooted in Russia for centuries. Family and community requirements take priority over individual needs. Their requests need to be met first. Russians rely on help from close and extended family members, consult each other, and give emotional support to each other during the difficult times and in the time of crises. Family and friends provide love and empathy to those who needs help. They prefer not to follow individualistic tendencies as seen in individualistic cultures.

Trust among family, friends, neighbours, and co-workers is greatly valued. Russians value friendship and a total commitment to each other; friendships are for life. They may establish stronger moral relationships with friends of the same sex versus those with their spouses. True Russian friendships include unconditional helping each other in difficult times and with the daily routines such as babysitting, cooking, borrowing and lending money, and offering a free space to live (Fitzpatrick and Freed 2000).

The younger generation is expected to help the family and are required to perform their household chores. Russians consider the chores to be gender specific and prefer patriarchal household leadership. Females are usually assigned to cooking, cleaning, and laundry; males perform more physical labor tasks such as fixing the house, repairing old furniture, and performing electrical work. Males and females do grocery shopping equally. Occupations also have gender preferences. For example, surgeons, judges, and engineers are typically male while nurses, school teachers, and cooks are female. Even though a prestigious job and education are important, finding a good husband is often more important for women.

Intimate partner violence remains high in Russia. Russian Americans tend not to report domestic violence due to the long history of disbelief in authority figures. Alcohol misuse and poverty often contribute to domestic violence, particularly in rural Russia. Females prefer not to report rape due to the fear of retaliation and are sometimes even encouraged by the family members to stay quiet and forget about the rape. This cultural behavior of not reporting domestic violence or rape may continue after immigration.

26.3.5.4 Alternative Lifestyles

Russians are expected to get married by the age of 25. The ability of a woman to conceive and carry the child to term is highly valued. Spouses are expected to have a child within 1 year of marriage. Child care responsibilities are the primary responsibility of the female. Males are typically considered to be the bread-winner for the family. Grandmothers often help to take care of the grandchildren in order for the mother to return to the workforce or continue her education. Divorce rates are very high in Russia, particularly in families who could not have children or are of low socioeconomic status (Shadrina 2018). Russians immigrants may experience high divorce rates as well due to the hardship of emigration.

Mothers with alcohol misuse often lose custody of their children during a divorce. In Russia, civil law favors mothers during a divorce but custody agreements are decided based on financial and parental merits. What is commonplace during the divorce in Russia is for either parent to give up parental involvement in the future rearing of the children. This, however, does not excuse them from financial child support. Divorce does not affect the social status of Russian immigrants with the exception of the Russian Religious Orthodox community who do not support dissolution of the family.

Homosexuality is now legal in Russia with a revision of the Russian penal code in 1997. However, the stigmatism of lesbians and gays is still prevalent, as homosexuality was considered a crime until 1997. Same-sex marriages are not supported in Russia. Even now, many hide homosexuality from their families as they are afraid of being rejected by their family. Russian immigrants may not disclose their sexual orientation to even health-care providers for fear of losing services, rejection from others, and loss of social support. Adoption of children from orphanages in Russia is not encouraged by same-sex couples (Gulevich et al. 2016).

26.4 Workforce Issues

26.4.1 Acculturation

Russians are very dedicated to their work duties and assignments and will mostly make sure that job responsibilities are very well-performed (Williams 2019). They are punctual with meetings and expect the same from their colleagues. They value authority figures and follow the chain of command. Russians prefer face to face communication rather than electronic communications, such as emails, business skype, or e-messenger.

Eye contact is not usually maintained in the Russian culture during meetings compared with what is expected in individualistic cultures. Moreover, they might perceive direct eye contact as an invasion of private space and feel uncomfortable during the conversation. The value of the spoken word is very important (Williams 2019). Russian are usually very polite, empathetic, respectful, and diligent with authority figures. They are good decision makers and critical thinkers. They will do everything possible to avoid conflicts in relationships, particularly, if it is related to the workplace or school environment (Lvina 2015).

Most Russians prefer an informative open exchange with a clear action plan instead of suggestive discourse and innuendos. They prefer to collaborate on projects and tasks rather than work individually. They are usually concise in their communications, state their opinions more freely, and may often interrupt a speaking person. Brisk communication is socially acceptable and is a preferred method in many instances. It is interesting to note that Russians may grin and smile when they truthfully intend to do so and say what they denote directly to a person. As a result, the interpersonal interaction could be viewed by Western cultures as less expressive (Larina 2015).

26.5 Biocultural Ecology

26.5.1 Skin Color, Diseases, and Health Conditions

Russians are Caucasian. High rates of obesity among Russians have been documented in literature (Hu et al. 2017). Stress has been associated with heart rate deregulations among Russian men and women living in Moscow (Glei et al. 2013). Cardiovascular diseases and mortality related to them are common among ethnic Russians (Davletov et al. 2016). Cardiometabolic risk factors, particularly hypertension, are high among Russian immigrants living in the US and other countries (Commodore-Mensah et al. 2016). Mortality from cardiovascular diseases is highest in Russia compared to other countries in Europe (Маколкин and Зябрев 2006 in Russian language).

With a change in the economy in Russia and transition to the capitalist system, increased levels of stress correlated with high levels of alcohol consumption and, consequently, produced a rise in cardiovascular illnesses (Karabchuk et al. 2017). Russian immigrants suffer most commonly from cardiovascular diseases, such as hypertension, hypercholesteremia, and cardiac arrhythmias. The common disease of endocrine system is diabetes and related complications—peripheral neuropathy, renal function alteration, and sensory deprivation (Hosler et al. 2004; Simolka and Schnepp 2017).

Russian immigrants and ethnic groups who came to the US from Byelorussia, Ukraine, and Western parts of Russian affected by the Chernobyl nuclear disaster have an increase in thyroid disorders, thyroid cancers, bronchial asthma, post-traumatic stress disorder, depression, increased risks of developing respiratory disorders, hypertension, and ischemic heart disease (Cwikel et al. 1997a, b; Cardis et al. 2005; Hatch et al. 2015; Slusky et al. 2017). Almost half a million people had to be evacuated from the Chernobyl area (World Health Organization 2007). In addition, many people developed long-term mental health issues related to the trauma of rapid relocation and loss of housing and social support systems (WHO 2007). Radioactive contamination created an 18 mile radius around Chernobyl determined to be unliveable for 150 years.

Recent National Health Interview Survey data indicated that overweight, obesity, and hypertension are prevalent among Russian immigrants in the US. Male immigrants have higher rates of hypertension compared with male immigrants from European countries (Commodore-

Mensah et al. 2018). Older Russian immigrants in the US, and probably elsewhere, experience chronic illnesses and emotional problems. Depressive symptoms are common among older Russian immigrants (Fitzpatrick and Freed 2000). Simolka and Schnepp (2017) reported that in New York State, which has the largest population of Russian immigrants in the US, the prevalence of diabetes over age 40 years was 16.9% compared to 6.5% of the Russian immigrants living in Germany.

Cognitive decline is also prevalent among older Russian immigrants. The resistance to change, decline in social status, inability to see relatives and friends left in the native country, and English language barriers contribute to instability and insecurity. Many prefer to live in retirement communities where many Russian immigrants reside in order to be able to socialize with their peers, overcome English language barriers, and to avoid being dependent on their children and grandchildren for health-care needs. Russian immigrants who are not able to communicate have to depend on somebody to interpret for them, may lack family support; therefore, feeling abandoned in the new culture. Such feelings can often lead to psychological distress and depression (Fitzpatrick and Freed 2000). Russian immigrants may sometimes present with vague somatic complaints such as neurological and gastrointestinal issues to health care providers. Somatic complains may be attributed to the underlying mental issue disorders and stigmatization of seeking treatment for mental illness.

26.5.2 Variations in Drug Metabolism

Proteins of cytochrome P450 enzyme produced by the liver perform an important role in pharmacodynamics of drug metabolism. Mirzaev et al. (2017) found that Russian ethnic groups from Dagestan region have a high level of cytochrome p450 isoenzyeme which predisposes them to a decreased metabolism of hypoglycemic agents, anticoagulants, and non-steroid-anti-inflammatory drugs. Antiplatelet agents' absorption and metabolism are reported in Russians to be similar to other Caucasian groups. Russians with hyperlipidemia are more predisposed to the development of statin-induced myopathy as an adverse effect of hypolipidemic agents compared to other ethnic groups. Genetic variations of the gene responsible for the transport of protein may affect the pharmacokinetics of the statins in the Russian population (Mirzaev et al. 2017).

Molecular genetic risk factors for alcohol dependence are associated with a polymorphism in GABRA2 gene in Russian population. In this population, linkage disequilibrium was observed across GABRA 2 gene as reported by Lappalainen et al. (2005). Variation in this gene predispose Russians to alcohol dependence. The effect of GABRA 2 gene mutation is more common in Russian men than women. This might contribute to the high alcohol consumption in Russian males (Lappalainen et al. 2005). A combination of genetic predisposition, extreme weather conditions, particularly in the Northern and far Eastern parts of Russia, and socially acceptable behavior of consuming many alcoholic beverages can attribute to the high rate of alcohol dependence among Russians.

26.6 High-Risk Behaviors

For centuries, Russia has been known for heavy alcohol consumption. In the early 1990s, the alcohol-related mortality rate increased due to disregulation of alcohol purchasing and production laws in Russia. Kenan et al. (2015) reported that alcohol account for 10–14% mortality in Russia at the present time, despite the wide effort by the government to restrict alcohol consumption. Alcohol misuse, vodka in particular, remains a major cause of male mortality in Russia. Combination of smoking and consumption of three or more bottles of vodka per week contributed to premature deaths in different age groups of males (Zaridze et al. 2014). Consuming copious alcoholic beverages is an integral part of Russian family celebrations, suggesting that heavy drinking is reinforced by societal norms. Alcohol is also perceived as a primary stress reliever for males but not females who have other coping skills for dealing with stressful situations (Kenan et al. 2015). Interestingly, there is an observable reduction in alcohol consumption when Russian males get married and become fathers. Performing a *breadwinner* role brings new responsibilities and results in less time and less available income to be spent on alcohol. Russian immigrants in the US have lower rates of alcohol misuse compared to those living in Russia. Lower rates of alcohol misuse are also evident in Russian Orthodox and Russian Jewish religious groups.

Smoking is embedded in the Russian culture. Russia has a high rate of male smokers with the average usage of 2786 cigarettes per adult per year (Kharlamov 2017). E-cigarettes are gaining popularity among younger generations where 13% of them are between 15 and 24 years old (Kharlamov 2017). The Russian government is working to ban the use of vaping, tobacco, and electronic cigarettes in public places and to limit the sale of tobacco and vaping items to minors (Kharlamov 2017). In fact, the Russian government has already implemented multiple policies over the past 10 years that have contributed to a 27% decline in mortality rate due to cardiovascular complications related to excessive alcohol consumption, smoking, unhealthy diet, low physical inactivity, and psychological stress (Kharlamov 2017).

Russian immigrants are less likely to smoke compared to Russians living in their native country. Smoking is not prevalent among Russian Orthodox and other Russian ethnic minority groups. However, the younger generation of immigrants may have increased usage of vaping and electronic cigarettes due to exposure to peer pressure, social media, their desire to fit in with American youths, and easy access of these smoking devices.

Opioid and other drug usage, as well as hepatitis C virus exposure are present among young adult Russian immigrants in the US. However, this population has not been studied extensively or documented in the literature. Drug usage was not congruent with Soviet social ideology and was not publicly addressed. People who used substances were often treated like criminals. In the 1970s, the Ministry of Health established special substance treatment centers, called *narcologicheskiy dispanseries.* The Russian government's current policy focuses more on drug law enforcement measures rather than on addiction prevention and treatment (Rivkin-Fish 2017). Guariano et al. (2015) reported in their study that young adults from the former Soviet Union living in New York City have a high rate of lifetime heroin use and injection drug use, despite evidence of protective factors such as a high level of education, close family relationship, stable household environment, and financial and moral support from their parents. Injection risk pattern and non-medicinal opioids usage among young Russian immigrants living in New York contribute to the recent opioid epidemic.

Polypharmacy and using excessive herbal remedies present as high-risk behaviours among immigrants. Russian pharmacies called *aptekas* are present in communities where most Russian immigrants live. These pharmacies sell copious amounts of medications and herbal remedies from Russia. It is well known that many herbal remedies have contraindications when combined with prescription medications. Another high-risk behavior among many is to share medications with their family and friends and even self-medicate themselves without seeking help from a health-care provider.

Russians may be hesitant to immunize their children. Due to lack of disposable needles in many areas of Russia and high-risk of exposure to HIV as well as lack of knowledge about available vaccines in the US, many Russian parents choose to not immunize their children unless they receive adequate education and reassurance from the health-care provider about safe immunization practices. According to the Department of Health (2012) the Russian speaking population has the lowest record of childhood immunizations in Washington State. Language barriers, dissatisfaction among parents in doctor-patient relationships in making rushed decisions about the immunizations for their children, and influence of Russian social networks and Russian language media contribute to Russian parents not immunizing their children (Department of Health 2012).

Russian adolescents are prone to developing high-risk behaviors such as excessive alcohol drinking, substance abuse, and dangerous sexual behaviors as in many other cultures. Pokhrel et al. (2018) reported that increased substance use and risky sexual behavior are evident among Russian ethnic groups. Sexual education is still lacking in Russian society (Vishnevsky et al. 2017). Acculturation to the American culture and lack of traditional family roles may predispose young Russian females to engage in early unprotected sexual activities. Higher parental education and involvement in adolescent lives can serve as protective factors for prevention of substances use and risky sexual behaviour.

26.7 Nutrition

26.7.1 Meaning of Food

People of Russian heritage, and particularly the older generation, experienced frequent food shortages in Russia due to war or poverty. Therefore, serving an abundance of food items has evolved into a custom in Russian celebrations and family gatherings. During social and family events, Russians to greet their guests with a variety of different meals to show love and appreciation of the food choices and meals they have prepared for the event. It might take several days to cook for the event. Visitors are expected to show respect to the hosts by actively participating in meal consumption and praising the quality of food. In the US, many limit their meal gathering events due to their busy working schedule and lack of time required for extensive preparation required for several Russian dishes.

26.7.2 Common Meals and Food Rituals

Nutritional concerns such as high amounts of salt, fat, and carbohydrate intake among Russians can predispose to the development of multiple chronic diseases. Russian immigrants exhibit prevalence of obesity, high blood pressure, diabetes, and sedentary life styles. Poor dietary choices combined with economic suffering, cultural, and language difficulties, predispose Russian immigrants to less positive health outcomes. Dietary choices and food access depend on economic stability of the family. Older immigrants prefer to eat Russian food and home-cooked traditional Russian meals. Typical diets contain vast quantities of saturated fats, sugars, and salt and inadequate consumption of fruits and vegetables.

Russian do not practice trimming fat from meat or removing skin from chicken. They prefer to drink whole milk, and eat fatty foods such as sour cream and processed cheeses. They also fry meat and potatoes using large amounts of butter or margarine (Hosler et al. 2004). This type of diet predisposes chronic diseases such as obesity, hypertension, diabetes, and hyperlipidemia (Lunze et al. 2015). Typically, Russian immigrants consume three meals a day, have tea and baked goods rituals throughout the day, and prefer to drink water or juices without ice. Russian grocery stores called *gastronome* and Russian restaurants are popular in Russian communities in the US. However, the younger generation frequently prefers to visit American venues and eat American food.

26.7.3 Dietary Practices for Health Promotion

Russian immigrants may turn to homeopathic medicine when they are ill. In addition, they consume hot tea with honey and hot milk, chicken soup, beet soup called *borscht,* yogurt, soft cheese, and oatmeal.

26.7.4 Nutritional Deficiencies and Food Limitations

Russian immigrants from Jewish heritage who practice kashrut or kosher diet will totally restrict consumption of dairy and meat products to different days. Meat and dairy products are always cooked separately. Kosher kitchen contains separate dishes, cooking places, utensils, and cookware for the preparation of meat and dairy products. Pareve food, which is neither meat nor dairy, can be eaten on any day. Some examples of pareve food include fruits, grains, eggs, and vegetables. The Torah does not allow consumption of non-kosher food, such as pork; meat must be prepared by following strict kosher guidelines (Kosher Certification 2019).

26.8 Pregnancy and Childbearing Practices

26.8.1 Fertility Practices and Cultural Beliefs

Marriage and childbearing are embedded into the Russian culture. Females and males form families in their early twenties. Successful childbearing is considered a to be the complete fulfilment of parental roles. Infertility is not commonly perceived as a health issue; it is often viewed as punishment for some female fault. The family might end up in divorce because of inability to conceive or bring a fetus to term.

Abortions were not supported and were banned in the former Soviet Union. Family planning was conducted poorly. By the 1990s, the country had an alarming rate of abortions, shortage of contraceptive devices, and lack of sexual education. About four million abortions are performed annually in present Russia. Abortions accounted for one third of the cases of maternal mortality (Vishnevsky et al. 2017). Introduction of the Family Planning Program by the Russian government with partnership from United States Agency for International Development (USAID) and the United Nations Popular Fund (UNFPA) contributed to regulation of childbirth, a decrease in abortions, and promotion of contraception and safe sex practices.

Effective family planning became more evident in post-Soviet Union times with a steady increase in fertility and the number of births (Vishnevsky et al. 2017). The Russian Ministry of Health established counselling programs for women to educate them about the danger of abortions, recommended health-care providers to obtain informed consent protocols from females seeking non-medical abortions, and established policies for fetus ultrasounds to be performed before termination of pregnancy (Rivkin-Fish 2017).

Currently, the most popular contraceptive methods among Russian families are condoms (37% of contraceptive users) followed by intrauterine devices (IUD) (14.2%), and birth control pills (13.2%) (Vishnevsky et al. 2017). The preferred method of contraception depends on the age group. Younger females prefer oral contraception whereas IUDs are more common among married middle-aged females (Vishnevsky et al. 2017). Russian males do not usually consider vasectomies as a method of contraception because they believe the procedure may decrease their self-esteem and their masculinity. They will use condoms as a method of contraception; however, some Russian males may prefer not to use it to experience sexual pleasures.

The incidence of HIV/AIDS and other sexually transmitted diseases remains high in Russia. Females often make their own decisions about the contraception without consulting their spouse. Russian female immigrants will often research information about birth control pills or injections before they consent to take them. The reluctance to take birth control medication is because fertility is considered a strength in the Russian culture. Furthermore, Russian immigrants are aware that hormonal contraceptives carry numerous adverse effects and often prefer not to take these medications for that reason. Some Russians also have poor prior experience with birth control medications while living in Russia.

Young Russian females learn to take care of themselves during menstruation to preserve their sense of well-being. They choose not to participate in strenuous physical activities during that time. During the former Soviet Union, feminine products were not available and many females had to make their own hygienic pads from fabric material and cotton. Currently, there is a great variety of feminine products in Russia and many of them are imported from European countries and China.

26.8.2 Childbearing Family

During pregnancy, females engage in light activities, excluding themselves from heavy work and resting more. Pregnancy is considered a big achievement and is praised in the Russian culture. Breastfeeding is encouraged as Russians believe in the health benefits of breast milk. Some breastfeed their children up to a year. Females also receive an extensive paid maternity leave from the government. Many Russians might not purchase baby clothes and other supplies during pregnancy due to a common belief that doing so may attract the evil eye before the baby's arrival. Children are dressed according to the colors pertinent to the gender—blue and black colours for the boys and pink and red for the girls. Pregnant females and mothers are respected in the Russian culture. Jewish people of Russian descent practise *briz* or circumcision of their male infants.

26.9 Death Rituals

26.9.1 Death Rituals and Expectations

The death rate in Russia was 13.4 deaths per 1000 in 2018, which is nearly twice the death rate for 2018 in the US (Center for Disease Control and Prevention 2017; CIA World Factbook 2019a, b). Preference for an open or closed casket depend son the family's wishes and religious practices. Jewish immigrants from Russia do not open the casket and practice *Shiva,* the obligatory 7 days of mourning. They do have a schedule of mourning prayers or *kaddish* (prayers for the dead) according to Torah guidelines. Burial in the Jewish tradition occurs within 24 h after the death except for the High Holidays and the Sabbath. In some instances, the burial is postponed up to several days to allow family from other countries to arrive for the funeral. Cremation is prohibited in Judaism, even though American Jews often choose cremation after death. Russian Orthodox immigrants do not practice cremation either.

Orthodox believe that the deceased soul is present for 7 days and place a glass with alcohol and piece of bread on top of the glass on the deceased's grave and in the house next to a portrait to show that he or she is still part of the family and being cared for. People wear black clothes, remove jewellery and do not apply make-up during funerals.

During anticipatory grief, family and friends get together to say goodbye to the terminally ill individual. They will hold hands, read aloud favorite books and poems to the ill family member and comfort him or her with warm clothes, blankets, pillows, and food. Elderly Russians might feel anxious about upcoming funeral expenses as many of them cannot afford it; therefore, they rely on the family and friends to cover funeral costs. Russians prefer to die at home, in their own beds, and not in the hospital where they feel a lack of dignity (Yurevich 2018). Russian immigrants do not like to participate in future end-of life funeral and burial pre-planning activities.

Russians will protect terminally ill family member from knowing the actual diagnosis and prognosis of the disease. They do not want to put extra psychological stress on a dying loved one. Physicians in Russia practice the same tradition and will not disclose a poor prognosis of a disease and will often temper the terminal diagnosis when talking to their patients. Discussing death is a not a common practice in Russian health care and is considered to be damaging to the person's mental well-being. Family members also do not want the ill person to lose hope and fall into depression. Therefore, Russian immigrants might privately talk to an American health care provider, asking them to not disclose a poor prognosis. A special approach is required for physicians and other health-care providers to speak with the family first prior to disclosing information about life-threatening diseases to the patient (Eckemoff et al. 2018). It is common in Russian culture for family to seek consultation from multiple doctors to make sure that everything is done to find a cure for the loved one. Family will often try to make the ill individual happier and focus on life rather than on the preparation for death.

Russian families are reluctant to discuss advance directive, do not resuscitate (DNR) procedures, living wills, and durable power of attorney documents. They and their ill family member will often not sign these forms. The family will often not agree to disconnect life support. They believe that everything should be done from a medical care perspective to save the life of the loved one.

Leaving major health-care decisions about end-of-life- care to the family may be perceived by Russian immigrants as a sign of incompetence of the health care provider (Newhouse 2013). Talking about death is believed to be bad luck because end-of-life-care is a very sensitive subject and should be introduced with caution (Newhouse 2013). Elderly Russian immigrants consider pain and suffering a private matter and will share it only with close family and friends and will often choose not disclose their level of pain and suffering to health-care providers. Russian immigrants might, during the health care visit, act like nothing is wrong with them. Clinicians should focus on nonverbal communication cues and observe signs and symptoms carefully. Russians prefer to keep their feelings hidden and remain stoic and strong so that other people may not view them as weak. Home health care and pain management services were hardly available in the former Soviet Union. Palliative care education is beneficial for Russian-speaking immigrants to make them informed about the services available for their dying family members and further amelioration of caregiver burden (Newhouse 2013).

26.9.2 Bereavement

The death of a loved one is considered a tragedy in the Russian culture. Family and friends gather together to support each other during this difficult time. A custom is to mourn and cry before, during, and after a funeral. Caskets are decorated with wreaths and flowers during funeral ceremonies. Russian Orthodox families mourn the death of a loved one for at least 40 days as they believe that the soul is still present on earth for the entirety of this time frame. They wear black clothes, cover all mirrors in the house, and place a black ribbon on the portrait of the deceased individual. The living spouse must wait for at least a year before starting to date or remarrying. The Russian family comforts each other with food and drinks during the time of mourning.

Family and friends get together after the burial to eat, drink and give respect to the deceased individual. It is required in the Russian culture to be present at the funeral and after the funeral for food and drinks with a family. In keeping with the collectivistic nature of the Russian culture, the entire community, particularly in the small villages, participates in the burial procession to show respect to the family who lost the loved one. The Russian Orthodox religion may offer a special prayer vigil, psalm reading, or religious hymn reading over the diseased body. This ritual is called *panikhida*.

26.10 Religion and Spirituality

26.10.1 Religious Practices and Use of Prayer

There are approximately 179,500 Russian Jews residing in Russia, currently, the world's seventh largest home to Jewish diaspora (World Jewish Congress 2016a, b). The Jewish population of Russia dramatically decreased in the early 1980s and 1990s because changes in local political climate finally allowed Jews to emigrate to Israel and the United States to avoid discrimination. In terms of religion, Russian Orthodoxy represents about 20% of the population, Muslims are 15, and 2% are from other Christian religious affiliations. Others do not have a specific religion (CIA World Factbook 2019a, b).

Russian Orthodox Christianity became the official religion of Russia in 2005 and it remains very influential with the Russian government (Library of Congress 2019a, b). Officially, however, the Russian Federation recognizes Orthodox Christianity, Buddhism, Islam, and

Judaism as established religions. Interestingly, Russians either do not practice any religion or call themselves atheists, which resulted from over 70 years of religious repression under the Soviet communist party rule. In the former Soviet Union, religion was not allowed and if people were caught worshipping, they could risk being punished by the government and losing everything. For example, people practicing Judaism in the former Soviet Union had to hide *matzo* in the house during the holiday of Passover when non-Jews visited them, and cover the windows that no one could see Shabbat candles.

Governmental policies related to religious affiliations changed with the collapse of the Soviet Union. People can now openly act according to their religious affiliations. Approximately 80–82% of the population in Russia now identify as Russian Orthodox (Ershov and Fursov 2019). The Russian Orthodox church and the Patriarchate has become a significant entity in the current government and in the lives of the Russian people. There are currently 179,500 Jews living in the Russian Federation which is the seventh largest Jewish population in the world. The Russian Federation has declared equality of rights for people of ethnic minority groups to practice religion freely (World Jewish Congress 2016a, b).

26.10.2 Meaning of Life and Individual Sources of Strength

Russian immigrants prefer to live a private life. They socialize mostly within the Russian families, their friends, and the Russian community. Religion is not usually practiced by many Russian immigrants as they focus primarily on work and family obligations. Family and close friends have more meaning in the Russian culture rather than their religious roots; therefore, Russian immigrants will rely on emotional support from family and friends first and most likely will not seek spiritual guidance from a priest, rabbi, or other spiritual leader.

26.10.3 Spiritual Beliefs and Health-Care Practices

Currently, the Russian Orthodox Church actively contributes to the medical and charitable activities in the Russian Federation. The Russian Orthodox Church provides medical and social services to people with terminal illnesses, human immunodeficiency virus (HIV), mental illnesses, and substance misuse. The Russian Orthodox Church, under the auspices of the Moscow *Patriarchate*, opened religious hospitals, clinics, and rehabilitation centers to help people who require medical treatments and have no means to pay for their medical costs. Many patients in these facilities are elderly and disabled. The average length of stay in a hospital operated by the Russian Orthodox Church is 31 days. Patients' satisfaction with care and treatment is reported to be 98% (Ershov and Fursov 2019). People who practice any religion and have major illnesses consider a prayer as an important part of the healing process. It is a accustom for the Russian Orthodox to attend the church services and light a candle while praying for the health and respect for sick or deceased loved ones.

26.11 Health-Care Practices

26.11.1 Focus of Health Care

Genetic predispositions, economic stress, family problems, increased alcohol intake, poor dietary choices, and a cold climate in Russia contribute to the development of multiple health problems. Russian immigrants expect health-care providers to holistically diagnose their illnesses and provide all possible choices for treatment. A common practice in the former Soviet Union was to admit people to the hospital for a holistic medical evaluation and treatment. If a disease was found during the hospitalization, the group of doctors, called a *consilium,* gathered together to review a case and provide a holistic medical treatment plan. Russian immigrants might consider Western medicine fragmented because it focuses primarily on the diagnostic and blood test results for prescription of medications and not holistic care.

Good health is important to Russian immigrants and so they are usually compliant with health-care appointments and meet the requirements of treatment. Russians prefer to have multiple inputs from different practitioners to get the best care. Russians often mix together prescribed medication regimens from multiple health-care resources and even supplement it with homeopathic and alternative medicines. Many health-care providers are often uninformed that Russian immigrants might seek multiple treatment plans.

Russians consider mental illness as shameful to the family and to the public. Such mistaken beliefs contributed to the Soviet ideology that every person had to fit in an ideal society as a comrade and be able to follow socialistic standards. In the Soviet era, people with mental illness or with learning disabilities who could not follow the society rules were considered unfit, were institutionalized in psychiatric facilities, and completely removed from their families and communities for life. Their families were encouraged not to visit them in psychiatric institutions and to forget about them (Petrea and Haggenburg 2014). As a result, Russian-Americans might not directly seek psychiatric treatment, ask for psychotropic medications, and may provide misleading data with vague psychosomatic complaints to health-care providers related to personal and family history of mental illness. Health-care providers should be alert that suicide rates are high among males and indigent people living in the northern and eastern parts of Russia (Norheim et al. 2016). Russian immigrants often migrate together as a unit of the entire immediate and extended family. They have strong social support from their friends and families which they believe promotes their resilience against depression (Mirsky 2009; Jurcik et al. 2015).

Russian-Americans prefer to use alternative medicine first. They often search Russian health journals to find information about their disorders and treatment. It is very common for Russian immigrants to subscribe to the Russian Health magazine, called *Zdorovie,* so they may find cures to their illnesses. They also rely on the Internet for health-related information and use a search engine *Rulist.com* (Russian Yellow Pages) that provides information about schools, businesses, services, health, and wellness.

Overweight and obesity may not be considered health concerns in Russian culture. Russian-Americans are more accepting of obesity and weight gain as these factors are common in Russian culture due to high caloric intake and lack of exercise. Currently, half of the population in Russia has increased body mass index; 1 in 4 women and 1 in 10 men are obese (The Organization for Economic Co-operation and Development (OECD) 2019). Many Russians choose to consume more food to deal with life stressors as a coping strategy. Peltzer and Pengpids (2014) identified that among obese university students, Russians underestimate their obesity factors. Being overweight is interpreted as normal in Russian culture.

26.11.2 Health-Seeking Beliefs and Behaviors

Russian immigrants are very well-educated and therefore, choose to be very active in their health-care needs. They often research the qualification of the health-care provider, seek recommendations for health-care providers from their friends and relatives, and examine the background of the specialist before making an appointment. Health-care barriers may exist in situations where Russian immigrants have a language barrier and/or are not able to communicate with an English-speaking provider, do not have insurance coverage or have high co-payments/deductibles, or lack transportation. Many immigrants may choose to rely on the emergency room services because of lack of medical insurance or time to visit a health-care provider in the office. Only the older generation of the Russian immigrants will actively seek health-care services and keep doctor's appointments because they are not working and have medical coverage.

Idehen et al. (2017) identified that Russian females are active in seeking cervical cancer screening. The age and the number of years living in a non-native country determine the compliance rates with the cervical cancer screening. This is because younger immigrants are exposed to health promotion and disease prevention programs. Moreover, knowing and comprehending the official language of a host country increases the rate of health-care screening participation. More Russian females visited a doctor within the past year and had physical exams during the last 5 years compared to Russian males in the Russian immigrant population (Tiittala et al. 2018).

Interestingly, research illustrates that Russian immigrants have greater rates of missed Hepatitis B and C or syphilis diagnoses compared to other immigrants. This may arise from factors related to poor self-care habits, living alone, smoking, and previous blood-borne pathological infections (Tiittala et al. 2018). Russian elderly immigrants are less likely to follow up with dental care if they have to pay significant out of pocket expenses. This may be related to health and dental services having been fully covered by the government in Russia (Wu et al. 2005). Russian residents in Boston, Massachusetts; Brooklyn, New York; and Sunny Isles; Miami, Florida have easy access to Russian-speaking health-care providers who opened their practices in these communities.

Lack of health-care insurance, high cost of medications, and linguistic and cultural barriers may predispose Russian immigrants to be dissatisfied with health care. Many were exposed to socialized universal health care in Russia where they experienced free home care visits from their doctor, low cost of medications, and extensive hospitalizations during an illness. In Russia, if a child up to the 3 years of age requires hospitalization, the child will be admitted together with their mother to prevent separation anxiety. Russian immigrants who have strong social support from their friends and families experience greater satisfaction and better acculturation in the US. General satisfaction and social support are important factors for Russian-speaking individuals (Brailovskaya et al. 2017).

Russian immigrants utilize the services of both male and female physicians. People of Russian heritage value professionalism, competence of a physician, cultural sensitivity, and ability to approach the treatment from a holistic perspective. Older immigrants prefer to have Russian-speaking physicians.

26.11.3 Health Care Responsibility

Family members are consulted on the major health-care decisions before Russian clients sign a consent for surgery or other procedures. Immediate family members will most likely be present during a health-care appointment to provide emotional support and advice to the person seeking health-care (Fitzpatrick and Freed 2000). Russian immigrants take pride and responsibility for their health care needs. They are very punctual with their health-care appointments. Most will seek alternative treatments and use homeopathic medicines in addition to prescribed medications. Russians respect health-care providers and listen to their recommendations. However, if a Russian-speaking individual does not understand English thoroughly, he or she might not seek clarification and might misinterpret the treatment plan. It is important for health-care providers to assess the language fluency of Russian-speaking patients and test their understanding of the diagnosis and treatment. An interpreter is required for accuracy of information. They need to give directions using simple terminology and avoid medical jargons. Older Russian immigrants will not be proactive in scheduling preventive or diagnostic screening unless a health care provider orders them to complete it.

26.11.4 Folk and Traditional Practices

Russian immigrants rely extensively on the homeopathic medicine in addition to the Western medicine. Russians like to use herbal teas, hot broths, honey, lemon, mineral water, and hot milk as remedies. The traditional culture is to use different types of oils and ointments, baths, herbs, massage, and sauna to improve health. Russians use a cupping technique at home for family members who have a cold or flu or respiratory issues. During this procedure, special glass cups are heated with a flame and placed individually on the person's back for 15–20 min. Physicians and nurses in Russia learn how to perform the cupping procedure as part of their training. When a child is sick with a chicken pox, Russians use green homeopathic tincture, called *Zelenka* to control lesion rupture.

26.11.5 Barriers to Health Care

Russians anticipate their health-care providers to be culturally sensitive, maintain a non-judgmental attitude, and be professional and respectful. Russian immigrants will establish trust with a health-care provider who is knowledgeable about homeopathic medicine in the Russian culture. Russians family members are usually very involved in the medical care and may shield a terminal diagnosis and treatment. Being diagnosed with any form of cancer in the Russian culture is considered a "death sentence". Mental illness is believed to be a stigma and is not widely disclosed. Russian immigrants and their families will not reveal to anyone any history of psychiatric illness or any psychiatric admissions knowing that these disclosures were punishable in Soviet Russia by the government or looked down upon by peers. Straiton et al. (2016) reported that Russian females are less likely to buy psychotropic medications compared to other immigrant groups. Therefore, health-care providers have to ascertain mental health history with caution as well as with a great level of confidentiality when treating Russian patients. They should also educate them about the importance of compliance in treatment.

Russian immigrants prefer direct access to specialists of their choice and to medical services. They might not be familiar with the idea of managed care and case management. Any additional steps like referrals to specialists and waiting for insurance authorization are considered unnecessary steps and may adversely affect their health by wasting time in getting valuable health services. Russian immigrants may find the American health-care system to be complicated. They might not be accustomed to such ideas as preventive medicine, medical malpractice, referrals, insurance authorizations, and diagnosis related group fees for payment.

Affordability. Many older Russian immigrants are enrolled in the Medicare and Medicaid programs. Many need help navigating the health-care system, understanding its complexities, meeting deductibles, coverage for services, and co-payments. They believe that health care should be distributed equally and at no cost. Health care was free in the former Soviet Union. The health-care system went through a transformation with the collapse of the Soviet Union. There are private and public hospitals and clinics in Russia now. Patients have to pay fees to for health care and hospitalizations. Hospitals and doctors in private health-care facilities charge for their services. Footman et al. (2014) reported in their study that from 29 to 72% people living is Russia, Armenia, Belarus, Georgia, Kazakhstan, Kyrgyzstan, Moldova, and Ukraine did not take their prescribed medications between 2001 and 2010 due to lack of affordability or availability of medications. Recent immigrants from Russia have a difficult time understanding medical insurance and fee for service as these practices are also new in the Russian Federation.

Many Russian immigrants in the US carry health care insurance either through their work or by individual purchase. Older Russian immigrants and people with disabilities are eligible to enrol in federal medical insurances such as Medicare and Medicaid. Yarova et al. (2013) reported that more than 60% of participants from the former Soviet Union did not have medical insurance due to the high cost or high insurance deductibles. Some immigrants even travel to their native country to obtain dental procedures or undergo medical examinations and treatment. Russian female immigrants are as not as vigilant about their health compared to Americans from similar socioeconomic backgrounds. Russian females, ages 50 and above, considered it selfish to take care of one's own health first. Taking care of the family and loved ones' health is viewed as more of a priority. The most common reason for

Russian female immigrants ages 50–59 not to utilize health-care services in the US is usually due to mistrust of U.S. doctors. For elderly immigrants, the biggest concern is access to health care related to affordability and cost (Yarova et al. 2013).

Language Proficiency. Older generations of Russian immigrants prefer to socialize and live in close proximity to people from their own country. Many older Russian immigrants subscribe to the Russian language-streaming TV channels, have Russian journals and newspapers delivered to their houses from either New York, Moscow, or Saint Petersburg. Older Russian immigrants prefer to seek care from Russian-speaking health-care providers as older Russian immigrants feel comfortable with health-care providers who speaks their native language. Children and grandchildren of older generation of immigrants often take them to the health-care appointments, serve as interpreters, and help them to navigate the health-care system in the US.

There are many Russian-speaking health-care providers in the US, depending on the geographical area. For example, many Russians who settled in New York, Boston, San-Francisco, and Miami have easy access to the Russian-speaking health-care providers and dentists. Some adults living facilities deliver services and care in the Russian language and employ Russian-speaking nurses, certified nursing assistants, case managers, therapists, and social workers. The Russian American Medical Association (RAMA), a non-for-profit organization, was established in 2002 to unite Russians-speaking medical providers in the US and help patients and colleagues in the Russian Federation with educational advancement and humanitarian missions (Russian American Medical Association (RAMA) 2019). A website is available for Russian immigrants to find a Russian-speaking physician by selecting the preferred language link and convenient office practice location (zocdoc.com). There are also Russian yellow pages for US and Canada (Rulist.com).

Accessibility. The Pan American Health organization reported that 9% or 12.5 million people in the workforce are employed in health care. There are more than 1000,000 physicians in the US and approximately 2,687,000 nurses (Pan American Health Organization (PAHO) 2014). Roughly speaking, there is 1 physician and nearly 3 nurses for every 300 people living in the US. In contrast, the World Health Organization 2011 reported that there are 50.1 physicians for every 10,000, or 1 for every 200, people in the Russian Federation and 73.6 nurses for 10,000 people, or 1 for every 140. This demonstrates an overall increase in health-care personnel since the failure of the Soviet Union; however, some specialties, such as obstetricians, surgeons, and radiologists are still in great need in Russia (World Health Organization (WHO) 2011). Despite an increase in health-care professionals in Russia, health care in Russia still requires more enhancement of health-care facilities and improvement in treatment quality (Muth 2017). Though the American health-care system has great availability of physicians and other health-care professionals, Russian immigrants might have difficulties in accessing health care upon emigrating to other countries.

Russian immigrants are dissatisfied with a long-time frame necessary to schedule appointments, long wait lists to see specialists, and lack of transportation to reach the health-care facilities. Moreover, absence of medical insurance or denial of insurance by certain providers contribute to the experience of the inability to access the health-care system in the US. It is a custom for Russian immigrants to compare Russian and American health-care systems. While living in the former Soviet Union, immigrants were used to having their physician visit them at home for medical care, extended time off from work due to an illness, availability of walk-in outpatient centers, and easy access to public transportation. Many outpatient centers in Russia are conveniently located and easily accessible by public transport.

26.11.6 Cultural Responses to Health and Illness

Russian immigrants are used to very strong community support and they are very well educated. They often challenge American health-care providers to find all possible medical treatment plans for their diseases. Health-care promotion and disease prevention were not extensively practiced in the former Soviet Union; therefore, older Russian immigrants have difficulties coping with a disease and do not commonly understand the meaning of a healthy life-style. It is very hard for many to change life-long established poor health-care habits. Some may even claim that English language barriers, immigration difficulties, and hardship living in the US and elsewhere justify their unhealthy life style. They will seek services from the multiple health-care providers, and will not disclose other possible treatments to the health-care providers. Russian immigrants prefer to stay longer in the hospitals as it was part of the Russian culture to be in the hospital until they completely recovered from the illness. In some cases, patients remained in the hospitals for months in Russia. Therefore, Russian immigrants might feel dissatisfied with foreign health-care systems due to lack of holistic approaches, and short hospitalizations.

26.11.7 Blood Transfusion and Organ Donation

People of Russian heritage may not willingly participate in blood transfusions and organ donation. This belief is contributing to the fact related to deficit of disposable needles in former Soviet Union and fear of contamination and exposure to HIV.

Reflective Exercise

Aleksandr Popov, age 77 years, emigrated to his new country after perestroika in 1992. He lives alone in a small apartment in a primarily low-income Russian community. In Russia, he was a respected teacher in an elementary school. Besides Russian, he speaks Turkic and Uralic. His English is very limited and relies on others in his community for English translation and interpretation. He reads Russian newspaper, watches Russian television stations, listens to Russian radio stations, and communicates via the internet with friends and relatives in Russia on a regular basis. He has no living relatives in his new country.

His health status is decreasing and has been diagnosed with alcohol misuse, cardiomyopathy, diabetes, and arthritis that affects his ability to keep appointments. He is reluctant to ask friends and neighbors for transportation and takes taxicabs to his appointments when he has the money to do so. Because of limited income, he frequently takes his medications only when he "feels bad". In addition, he admits to not understanding the instructions for his medications and dietary recommendation. One of his younger neighbors tells him that he should see a psychiatrist because he appears depressed.

1. What resources are available in your community to help Mr. Popov with interpretation and translation?
2. How might you help Mr. Popov obtain necessary financial resources to afford his medication.
3. What other transportation possibilities are in your neighborhood?
4. Food preparation appears to be a concern. What might you do to help him afford nutritious food?
5. Are nutritionists available to assist him with dietary preparation instructions?

26.12 Health-Care Providers

26.12.1 Traditional Versus Holistic Medical Care

Many Russian immigrants who arrived in the US in 1980–2010 were physicians. Some younger Russian physicians passed the Medical Board Examination for the Foreign Medical Graduates in the US, completed residency, and established their own practices, particularly in Brooklyn, NY, as many Russian immigrants resettled there at that time. Although some older Russian physicians did not pursue Medical Board Examination in the US, they found employment as ancillary personnel in health-care facilities. Russian immigrants might ask former Russian physicians informally for their opinions about health-related topics. Older Russian Americans often prefer to seek treatment from Russian speaking physicians as they feel more comfortable with a health-care practitioner who is familiar with their culture and speaks Russian language.

Health care in Russia focuses on the holistic and homeopathic medicines in addition to more traditional approaches. Russian immigrants, and particularly the older generation are dissatisfied with a lack of a holistic approach in the American health-care system and prefer to rely on the holistic medicine first and seek treatment from Russian-speaking health care providers.

26.12.2 Status of Health-Care Providers

Medical education in Russia depends on the selected speciality, and entering medical school or a nursing program is possible right after graduation from high school. A Bachelor degree in nursing is awarded after 4 years in a nursing program (Ministry of Science and Higher Education of Russian Federation 2019). An important aspect of medical and nursing training was learning a holistic approach to treatment of all patients. The training for nurses in the former Soviet Union started either after the eighth grade for 3 years or after the tenth grade for 2 years. This program was very much hands on and focused on the performance of nursing skills correctly and following physicians' orders. It was a common practice for nurses to have a work schedule which required them to be at work for 24 h at the hospital and then off for 72 h. A positive characteristic of health professionals trained in the former Soviet Union was that they promoted the Russian emphasis on holism and holistic health care.

The current nursing education system is still highly regulated by the Federal Educational board. Nursing education became mandatory for 3 years completion after the 11th grade in 1991. Nursing classes were mainly taught by physicians and there was no focus in the curriculum on health promotion and disease prevention. In 1994, the 4-year program was approved for nurses who wanted to pursue a higher education (Ivanov and Papanpegara 2003). Nurses and doctors trained in Russia who immigrated the US have to comply with the American credentialing system, take the required examinations for foreign medical graduates, and complete residency for medical doctors if they want to practice medicine. People of Russian heritage respect the knowledge and expertise of health-care providers. Once they establish trust with health-care practitioners, they will view them as responsible for their health care.

Case Study

Zinaida Ivanovna is a 71-years-old widow who immigrated with her husband and daughter as political refugees to New York from Saratov, Russia when she was 40 years old along. Zinaida Ivanovna said that she immigrated for her daughter's sake. They settled in an apartment building in Brooklyn, New York. Her husband found a job as a taxi driver and she worked as a cook at a Russian restaurant until her retirement. Her husband died of a heart attack in 2001 and her daughter died several years ago from breast cancer.

Sometimes, Zinaida wakes up in the middle of the night and starts looking for her daughter, not realizing that she passed away several years ago. Now everything seems so hard for Zinaida as she cannot manage her daily life. Zinaida suffers from high blood pressure, diabetes, and arthritis. She has no family support and relies on her Russian neighbor, Katya, for help with groceries, transportation, and doctors' visits. Zinaida is not able to communicate in English and depended on her daughter for interpretation when she was alive. Zinaida only schedules appointments with the Russian-speaking health-care providers in Brooklyn and feels comfortable with their services and treatment because they speak her language. Zinaida does not understand the U.S. medical system and gets easily frustrated when she has to wait several months for her appointment to see a specialist.

Being a cook for her entire life, Zinaida, prefers to prepare homemade traditional Russian dishes for herself and her neighbor Katya. Zinaida eats sandwiches with herring for breakfast, borscht with fried potatoes, cutlets for lunch, and pirogues or meat dumplings for dinner. The Russian food makes her comfortable. At times, Zinaida skips her medications for blood pressure explaining that when she feels good, the medications are not helping her. She prefers to visit *apteka*, located within a walking distance from her apartment in Brooklyn to buy homeopathic medicine imported from Russia to help her with arthritis and blood pressure.

1. What cultural traits does Zinaida exhibit in her values, beliefs, and practices?
2. What educational material should be incorporated for Zinaida's condition. Does she need a special diet?
3. What type of assessment, planning, implementation, and evaluation should be done by a nurse in caring for Zinaida?

References

Ameredia Integrated Cultural Market (2019) Russian American demographics. http://www.ameredia.com/resources/demographics/russian.html

Bell L (2015) Is the Russian presence in Sunny Isles waning? Miami Herald. https://www.miamiherald.com/news/business/biz-monday/article5702085.html

Bogatyrev S (2004) Ivan IV. In: Millar JR (ed) Encyclopedia of Russian history, vol. 2. Macmillan Reference USA, pp 689–691. Gale Ebooks, https://link.gale.com/apps/doc/CX3404100601/GVRL?u=miam11506&sid=GVRL&xid=e4779736. Accessed 26 Oct 2019

Brailovskaya J, Schonfeld P, Kochetkov Y, Margraf J (2017) What does migration mean to US? USA and Russia: relationship between migration, resilience, social support, happiness, life satisfaction, depression, anxiety, and stress. Curr Psychol 38:421–431. https://doi.org/10.1007/s12144-017-9627-3

Brown A (2004) Gorbachev, Mikhail Sergeyevich. In: Millar JR (ed) Encyclopedia of Russian history, vol.2. Macmillan Reference USA, pp 577–583. Gale Ebooks. https://link.gale.com/apps/doc/CX3404100508/GVRL?u=miam11506&sid=GVRL&xid=344f3537. Accessed 26 Oct 2019

Cardis E et al (2005) Risk of thyroid cancer after exposure to 131I in childhood. J Nat Cancer Inst 97:724–732. https://doi.org/10.1093/jnci/dji129

Center for Disease Control and Prevention (2017) Mortality in the United States. https://www.cdc.gov/nchs/products/databriefs/db328.htm

Chiswick B, Larsen N (2015) Russian Jewish immigrants in the United States: the adjustment of their English language proficiency and earnings in the American community survey. Contemp Jew 35:191–209. https://doi.org/10.1007/s12397-015-9137-2

CIA World Factbook (2019a) Russia (2015 literacy data). https://www.cia.gov/library/publications/resources/the-world-factbook/geos/rs.html

CIA World Factbook (2019b) Russia. https://www.cia.gov/library/publications/the-world-factbook/geos/rs.html

Commodore-Mensah Y, Ukonu N, Obisesan O, Aboagye JK, Agyemang C, Reilly CM, Dunbar SB, Okosun IS (2016) Length of residence in the United States is associated with a higher prevalence of cardiometabolic risk factors in immigrants: a contemporary analysis of the National Health Interview Survey. J Am Heart Assoc 5:1–10. https://doi.org/10.1161/JAHA.116.004059

Commodore-Mensah Y, Selvin E, Aboagye J, Turkson-Ocran R, Li X, Dennison Himmelfarb C, Ahima RS, Cooper LA (2018) Hypertension, overweight/obesity, and diabetes among immigrants in the United States: an analysis of the 2010–2016 National Health Interview Survey. BMC Public Health 18:1–10. https://doi.org/10.1186/s12889-018-5683-3

Cwikel JG, Abdelgani A, Goldsmith JR, Quastel M, Yevelson II (1997a) Two year follow up study of stress related disorders among immigrants to Israel from the Chernobyl area. Environ Health Perspect 105(Suppl 6):1545–1550

Cwikel JG, Goldsmith JR, Kordysh E, Quastel M, Abdelgani A (1997b) Blood pressure among immigrants to Israel from areas affected by Chernobyl disaster. Public Health Rev 25(3–4):317–335

Davletov K, McKee M, Berkinbayev S, Battakova Z, Zhussupov B, Amirov B, Junusbekova G, Rechel B (2016) Ethnic differences in all-cause mortality rates in Kazakstan. J Public Health 133:57–62. https://doi.org/10.1016/j.puhe.2015.11.026

DeBose C, DeAngelo E (2015) The new cold war: Russia's ban on adoption by U.S. citizens. J Am Acad Matrimonial Law 28:51–76

Department of Health (2012) Study of childhood immunization in Washington State Russian-speaking populations. https://www.doh.wa.gov/Portals/1/Documents/Pubs/348-354-RussianFocusGroup.pdf

Eckemoff EH, Sudha S, Wan D (2018) End of life care for older Russian immigrants - perspectives of Russian immigrants and hospice staff. J Cross

Cult Gerontol 33:229–245. https://doi.org/10.1007/s10823-018-9353-9

Eisikovits RA (2008) Immigrant youth who excel: globalization's uncelebrated heroes. Information Age Publishing, Charlotte

Encyclopedia Britannica (2019) Kamchatka Peninsula. https://www.britannica.com/place/Kamchatka-Peninsula

Energy Information Administration [EIA] (2017) Country Analysis brief: Russia. https://www.eia.gov/beta/international/analysis.php?iso=RUS

Ershov B, Fursov V (2019) Medical assistance and orthodox traditions in Russia. Dataset 11. https://doi.org/10.17916/P64S3R

Federal Law (1992) State protection of family, mothers, fathers, and children. [in Russian language]. http://pravo.gov.ru/

Federal Law (2015a) State allowances for citizens having children. [in Russian language]. http://pravo.gov.ru/

Federal Law (2015b) Limit size base for accruing insurance payments to the social insurance fund of Russian Federation and pension fund of Russian Federation from January 1, 2016 [in Russian language]. http://pravo.gov.ru/

Fishman G, Mesch G (2005) Acculturation and delinquency among adolescent immigrants from the FSU. J Confl Violence Res 7(2):14–40

Fitzpatrick T, Freed AO (2000) Older Russian immigrants to the United States: their utilization of health services. Int Soc Work 43:305–323. https://doi.org/10.1177/002087280004300304

Footman K, Richardson E, Roberts B, Alimbekova G, Pachulia M, Rotman D, Gasparishvili A, Mckee M (2014) Foregoing medicines in the former Soviet Union: changes between 2001 and 2010. Health Policy 118:184–192. https://doi.org/10.1016/j.healthpol.2014.09.005

Gill G (2004) Stalin, Josef Vissarionovich. In: Millar JR (ed) Encyclopedia of Russian history, vol. 4. Macmillan Reference USA, pp 1455–1459. Gale Ebooks. https://link.gale.com/apps/doc/CX3404101287/GVRL?u=miam11506&sid=GVRL&xid=d5626f7b. Accessed 26 Oct 2019

Glei DA, Goldman N, Shkolnikov VM, Jdanov D, Shkolnikova M, Vaupel JW, Weinstein M (2013) Perceived stress and biological risk: is the link stronger in Russians than in Taiwanese and Americans? Stress 16:411–420. https://doi.org/10.3109/10253890.2013.789015

Granville J (2004) Alaska. In: Millar JR (ed) Encyclopedia of Russian history, vol. 1. Macmillan Reference USA, pp 26–28. Gale Ebooks. https://link.gale.com/apps/doc/CX3404100040/GVRL?u=miam11506&sid=GVRL&xid=40c56387. Accessed 26 Oct 2019

Guariano H, Marsch L, Deren S, Straussner S, Teper A (2015) Opioid use trajectories, injection drug use and HCV risk among young adult immigrants from the former Soviet Union living in New York City. J Addict Disord 34:162–177. https://doi.org/10.1080/10550887.2015.1059711

Gulevich OA, Osin EN, Isaenko NA, Brainis L (2016) Attitudes to homosexuals in Russia: content, structure, and predictions. J High Sch Econ 13:79–110. [In Russian translated to English]. https://cyberleninka.ru/article/v/attitudes-to-homosexuals-in-russia-content-structure-and-predictors

Haim O (2014) Factors predicting academic success in second and third language among Russian-speaking immigrant students studying in Israeli schools. Int J Multiling 11:41–61. https://doi.org/10.1080/14790718.2013.829069

Hatch M, Ostroumova E, Brenner A, Federenko Z, Gorokh Y, Zvinchuk O, Shpak V, Tereschenko V, Tronko M, Mabuchi K (2015) Non-thyroid cancer in Northern Ukraine in the post-Chernobyl period: short report. Cancer Epidemiol 39:279–283. https://doi.org/10.1016/j.canep.2015.02.002

Hosler AS, Melnik T, Spence M (2004) Diabetes and its related risk factors among Russian-speaking immigrants in New York State. Ethn Dis 14:372–377

Hu Y, Malyutina S, Pikhart H et al (2017) The relationship between body mass index and 10-year trajectories of physical functioning in middle-aged and older Russians: prospective results of the Russian HAPIEE study. J Nutr Healthy Aging 21:381–388. https://doi.org/10.1007/s12603-016-0769-z

Hughes L (2004) Peter I. In: Millar JR (ed) Encyclopedia of Russian history, vol. 3. Macmillan Reference USA, pp 1168–1173. Gale Ebooks, https://link.gale.com/apps/doc/CX3404101008/GVRL?u=miam11506&sid=GVRL&xid=3264b21e. Accessed 26 Oct 2019

Idehen EE, Korhonen T, Castaneda A, Juntunen T, Kangasniemi M, Pietila A-M, Koponen P (2017) Factors associated with cervical cancer screening participation among immigrants of Russian, Somali, and Kurdish origin: a population-based study in Finland. BMC Womens Health 17:1–10. https://doi.org/10.1186/s12905-017-0375-1

Isurin L (2011) Russian diaspora: culture, identity and language change. Walter de Gruyter, New York

Ivanov LL, Papanpegara G (2003) Public health nursing education in Russia. J Nurs Educ 42(7):292–295

Jurcik T, Chentsova-Dutton YE, Solopieieva-Jurcikova I, Ryder (2013) Russians in treatment: the evidence-base supporting cultural adaptations. J Clin Psychol 69(7):774–791

Jurcik T, Yakobov E, Solopieva-Jurcikova L, Ahmed R, Sunohara M, Ryder A (2015) Unraveling ethnic density effects, acculturation, and adjustment: the case of Russian-speaking immigrants from the former Soviet Union. J Community Psychol 43:628–648. https://doi.org/10.1002/jcop.21708

Karabchuk T, Selezneva E, Kumo K (2017) Demography of Russia from past to the present. Palgrave Macmillan, London

Kenan K, Saburova L, Bobrova N, Elbourne D, Ashwin S, Leon D (2015) Social factors influencing Russian male alcohol use over the life course: a qualitative study investigating age based social norms, masculinity, and workplace context. PLoS One 10:1–15. https://doi.org/10.1371/journal.pone.0142993

Kharlamov A (2017) Cardiovascular burden and percutaneous interventions in Russian Federation: systematic epidemiological update. Cardiovasc Diagn Ther 7(1):60–84

Kosher Certification (2019) Meat, dairy, and pareve. http://www.ok.org/companies/what-is-kosher/meat-dairy-pareve-setting-boundaries/

Krom MM (2004) Ivan III. In: Millar JR (ed) Encyclopedia of Russian history, vol. 2. Macmillan Reference USA, pp 687–689. Gale Ebooks, https://link.gale.com/apps/doc/CX3404100600/GVRL?u=miam11506&sid=GVRL&xid=ac5320fc. Accessed 26 Oct 2019

Lappalainen J, Krupitsky E, Remizov M, Pchelina S, Taraskina A, Zvartau E, Somberg L, Covault J, Kranzier HR, Krystal JH, Gelernter J (2005) Association between alcoholism and y-Amino Butyric acid α 2 receptor subtype in a Russian population. Alcohol Clin Exp Res 29(4):493–498

Larina T (2015) Culture specific communicative styles as a framework for interpreting Lingustic and cultural idiosyncrasies. Int Rev Pragmat 7:2. https://doi.org/10.1163/18773109-00702003

Lerner J, Rapoport T, Lomsky-Feder E (2007) The 'etnic script' in action: the re-grounding of Russian-Jewish immigrants in Israel. Ethos 35(2):168–195

Library of Congress (2019a) Country profile: Russia. Library of Congress—Federal Research Division, Library of Congress call number: DK510.23. R883 1998. http://memory.loc.gov/frd/cs/rutoc.html

Library of Congress (2019b) Country profile: Russia. Library of Congress—Federal Research Division, Library of Congress call number: DK510.23. R883 1998. Retrieved from https://www.loc.gov/resource/frdcstdy.russiacountrystu00curt/?sp=27

Lipazova A (2016) Fatherhood models in the middle class of contemporary Russia. Russian Sociol Rev 15(4):202–214

Lipazova A (2017) Fatherhood in the Russian provinces: a theoretical and empirical analysis. J Soc Policy Stud 15:629–642. https://doi.org/10.17323/727-0634-2017-15-4-629-642. [Translated to English]

Lowental L (2004) Living through the soviet system. Routledge, New York

Lunze K, Yurasova E, Idrisov B, Gnatienko N, Migliorini L (2015) Food security and nutrition in the Russian Federation–a health policy analysis. Glob Health Action 8(1):27537. https://doi.org/10.3402/gha.v8.27537

Lvina E (2015) The role of cross-cultural communication competence: effective transformational leadership across cultures. JIP Int Multidiscipl J 3(1):1–16

Martin J (2004a) Kievan Rus. In: Millar JR (ed) Encyclopedia of Russian history, vol. 2. Macmillan Reference USA, pp 750–755. Gale Ebooks. https://link.gale.com/apps/doc/CX3404100656/GVRL?u=miam11506&sid=GVRL&xid=4ce39e23. Accessed 26 Oct 2019

Martin RE (2004b) Romanov, Mikhail Fyodorovich. In: Millar JR (ed) Encyclopedia of Russian history, vol. 3, Macmillan Reference USA, vol. 2. pp. 1299-1302. Gale Ebooks. https://link.gale.com/apps/doc/CX3404101137/GVRL?u=miam11506&sid=GVRL&xid=cf42fdb2. Accessed 26 Oct 2019

Michalikova N (2018) Segmented socioeconomic adaptation of New Eastern European professionals in the United States. Comp Migr Stud 6:1–27. https://doi.org/10.1186/s40878-018-0077-3

Ministry of Science and Higher Education of Russian Federation (2019) Medical education in Russia. https://studyinrussia.ru/en/actual/articles/medical-education-in-russia/

Mirsky J (2009) Mental health implications on migration: a review of mental health community studies on Russian-speaking immigrants in Israel. Soc Psychiatry Psychiatr Epidemiol 44:179–187. https://doi.org/10.1007/s00127-008-0430-1

Mirzaev KB, Sychev DA, Ryzhikova K, Konova OD, Mammaev SN, Gafurov DN, Shuev GN, Grishina EA, Sozaeva ZA (2017) Genetic polymorthism of cytochrome P450 enzyemes and transport proteins in Russian population and three ethnic groups of Dagestan. Genet Test Mol Biomarkers 21:747–753. https://doi.org/10.1089/gtmb.2017.003

Muth S (2017) Russian as a commodity: medical tourism and the healthcare industry in post-soviet Lithuania. Int J Biling Educ Biling 20:404–416. https://doi.org/10.1080/13670050.2015.1115002

Newhouse (2013) Working with Russian-Jewish immigrants in end-of-life care settings. J Soc Work End Life Palliat Care 9:331–342. https://doi.org/10.1080/155242256.2013.846884

Norheim AB, Grimholt TK, Loskutova E, Ekeberg O (2016) Attitudes toward suicidal behaviour among professionals at mental health outpatient clinics in Stavropol, Russia and Oslo, Norway. BMC Psychiatry 16:1–12. https://doi.org/10.1186/s12888-016-0976-5

Pan American Health Organization (PAHO) (2014) United States of America country profile. https://www.paho.org/salud-en-las-americas-2017/?page_id=165

Peltzer K, Pengpids S (2014) Underestimation of weight and its associated factors in overweight and obese university students from 21 low, middle and emerging economy countries. Obes Res Clin Pract 9:234–242. https://doi.org/10.1016/j.orcp.2014.08.004

Petrea I, Haggenburg M (2014) Mental health care. In: Rechel B, Richardson E, McKee M (eds) Trends in health systems in the former Soviet countries. European observatory on health system and policies. World Health Organization, pp 159–172. https://www.ncbi.nlm.nih.gov/books/NBK458305/pdf/Bookshelf_NBK458305.pdf

Pokhrel P, Bennett BL, Regmi S, Idrisov B, Galimov A, Akhmadeeva L, Sussman S (2018) Individualism-collectivism, social self-control and adolescent substance use and risky sexual behavior. Sub Use Misuse 53(7):1057–1067. https://doi.org/10.1080/10826084.2017.1392983

Remennick L (2015) We do not own our children: transformation of parental attitudes and practices in two generations of Russian Israelis. Int Migrat Integrat 16(2):355–376

Rivkin-Fish M (2017) Legacies of 1917 in contemporary Russian public health: addiction, HIV, and abortion. AJPH 107:1731–1735. https://doi.org/10.2105/AJPH.2017.304064

Russian American Medical Association (RAMA) (2019) About RAMA. http://www.russiandoctors.org/en/

Shadrina E (2018) Demography of Russia: from the past to the present. Popul Stud 72:279–282. https://doi.org/10.1080/00324728.2018.1441218

Simolka S, Schnepp W (2017) Subjective perspectives of diabetes mellitus among immigrants in the former Soviet Union. Central Eur J Nurs Midwifery 8:596–608. https://doi.org/10.15452/CEJNM.2017.08.0007

Slusky DA, Cwikel J, Quastel MR (2017) Chronic diseases and mortality among immigrants to Israel from areas contaminated by Chernobyl disaster: a follow-up study. Int J Public Health 62:463–469. https://doi.org/10.1007/s0038-017-0941-1

Straiton ML, Powell K, Reneflot A, Diaz E (2016) Managing mental health problems among immigrant women attending primary health care services. Health Care Women Int 37(1):118–139. https://doi.org/10.1080/07399332.2015.1077844

Tartakovsky E, Patrakov E, Nikulina M (2017) The emigration intentions of Russian Jews: the role of socio-demographic variables, social networks, and satisfaction with life. East Eur Jew Aff 47(2–3):242–254. https://doi.org/10.1080/13501674.2017.1396175

The Organization for Economic Co-operation and Development (OECD) (2019) Third lancet series on chronic diseases: Russian federation—key facts. http://www.oecd.org/els/health-systems/thirdlancet-seriesonchronicdiseasesrussianfederation-keyfacts.htm

Tiittala P, Ristola, Liitsola K, Ollgren J, Koponen P, Surcel H-M, Hiltunen-Back E, Davidkin I, Kivelä P (2018) Missed hepatitis b/c or syphilis diagnosis among Kurdish, Russian, and Somali origin migrants in Finland: linking a population-based survey to the national infectious disease register. BMC Infect Dis 18:1–11. https://doi.org/10.1186/s12879-018-3041-9

U.S. Census Bureau (2017a) Selected social characteristics in the United States: 2017 American community survey. https://www.factfinder.census.gov/faces/tableservices/jsf/pages/productview.xhtml?pid=ACS_17_1YR_DP02&prodType=table

U.S. Census Bureau (2017b) American fact finder selected population profile in the United States—Russian. https://factfinder.census.gov/faces/tableservices/jsf/pages/productview.xhtml?pid=ACS_17_1YR_S0201&prodType=table

U.S. Department of Homeland Security (2017) Yearbook of Immigration Statistics: 2017.Table 2. Persons obtaining lawful permanent resident status by region and selected country of last residence: fiscal year 1820 to 2017. https://www.dhs.gov/sites/default/files/publications/yearbook_immigration_statistics_2017_0.pdf

Vishnevsky A, Denisov B, Sakevich V (2017) Contraceptive revolution in Russia. Demograph Rev 4(1):6–34. https://doi.org/10.17323/demreview.V4I1.6986

Williams RC (2004) Bolshevism. In: Millar JR (ed) Encyclopedia of Russian history, vol. 1. Macmillan Reference USA, pp 159–161. Gale Ebooks. https://link.gale.com/apps/doc/CX3404100161/GVRL?u=miam11506&sid=GVRL&xid=8a0c1658. Accessed 26 Oct 2019

Williams E (2019) What is the difference between American & Russian cultures in business? Small Business—Chron.com. http://smallbusiness.chron.com/difference-between-americans-russians-cultures-business-56041.html

World Atlas (2012) Largest ethnic groups in Russia. https://www.worldatlas.com/articles/largest-ethnic-groups-in-russia.html

World Atlas (2017) Russia. https://www.worldatlas.com/webimage/countrys/asia/ru.htm

World Health Organization (2007) Global public health security in the 21 century, pp 29–31. https://www.who.int/whr/2007/whr07_en.pdf

World Health Organization (WHO) (2011) Russian federation health care system. http://www.euro.who.int/__data/assets/pdf_file/0005/186089/E96158-Rus.pdf?ua=1

World Jewish Congress (2016a) Russian federation. https://www.worldjewishcongress.org/en/about/communities/RU

World Jewish Congress (2016b) Federation of Jewish organizations and communities of Russia. https://www.worldjewishcongress.org/en/about/communities/RU

Wu B, Tran TV, Khatutsky G (2005) Comparison of dental care utilization among Chinese- and Russian speaking immigrant elders. J Public Health Dentist 65(2):97–103

Yakobov E, Jurcik T, Solopieieva-Jurcikova I, Ryder AG (2019) Expectations and acculturation: further unpacking of adjustment mechanisms within the Russian-speaking community in Montreal. Int J Intercult Relat 68:67–76. https://doi.org/10.1016/j.ijintrel.2018.11.001

Yarova LA, Krassen Covan E, Fugate-Whitlock E (2013) Effect of acculturation and health beliefs on utilization of health care services by elderly women who immigrated to the USA from the former Soviet Union. Health Care Women Int 34:1097–1115. https://doi.org/10.1080/07399332.2013.807259

Yurevich AV (2018) Attitudes to death as a scientific problem. Her Russ Acad Sci 88(1):75–80

Zaridze D, Lewington S, Boroda A, Scelo G, Karpov R, Lazarev A, Konobeevskaya I, Igitov V, Terechova T, Bofetta P, Sherliker P, Kong X, Whitlock G, Boreham J, Brennan P, Peto R (2014) Alcohol and mortality in Russia: prospective observational study of 151,000 adults. Lancet 383:1465–1463. https://doi.org/10.1016/S0140-6736(13)62247-3

Маколкин ВИ, Зябрев ФН (2006) Может ли частота сердечных сокращений рассматриваться в качестве фактора риска сердечно-сосудистых заболеваний. In: Russian language Could heart rate be regarded as cardiovascular disease risk factor? https://cardiovascular.elpub.ru/jour/article/viewFile/1397/1064

27 People of Thai Heritage

Larry D. Purnell

27.1 Introduction

Siam, the land of the musical *The King and I*, is the former name of Thailand (meaning Land of the Free). Thailand is a country in Southeast Asia well known for its cuisine and exotic culture. A unified Thai kingdom was established in the mid-fourteenth century. Known as Siam until 1939, Thailand is the only Southeast Asian country never to have been colonized by a European power. Thailand today is composed of a unique blend of traditions reaching back to its origins as a mixture of Southeast Asian peoples, its background in Buddhism, and its profound influences inherited from the cultures of both India and China. For providers of health care to Thai patients, the beliefs and practices that stem from these combined traditions can present both opportunities and challenges.

This chapter is an update of a chapter written Ratchneewan Ross and Jeffrey Ross in a previous edition of the book.

L. D. Purnell (✉)
Excelsior College, Albany, NY, USA
e-mail: lpurnell@udel.edu

27.2 Overview, Inhabited Localities, and Topography

27.2.1 Overview

Thailand began a tradition of emulating Western political, economic, and cultural ideas in the late nineteenth century. In the closing decades of the twentieth century, Thailand—like several other Asian "economic tigers"—began a period of explosive economic growth. Thailand's growth continues today, but it has also been interrupted by periods of political conflict and economic instability. Thailand's bloodless coup of 2006 was followed in 2010 by alarming clashes of violence between "yellow shirt" and "red shirt" competing political factions. These conflicts have further strained Thailand's economic and cultural stability (Kane 2010). Indeed, the political opposition of these two factions can be said to reflect a deeper and broader cultural divide in Thailand between the urban elite of the city and the poor of the country, thus challenging Thailand's "national capacity for creative compromise" and—at least in some sense—contributing to a kind of "collective anxiety" among the Thai population (McCargo 2008).

Most visitors to Thailand continue to be impressed by the unique ways with which the Thai people manage a precarious balance between the contrasts of the old and the new, between the rich traditions of their past and the

© Springer Nature Switzerland AG 2021

L. D. Purnell, E. A. Fenkl (eds.), *Textbook for Transcultural Health Care: A Population Approach*,
https://doi.org/10.1007/978-3-030-51399-3_27

frenetic influences of modern economic competition, all amid a continually shifting global culture.

Thailand is located north of Malaysia, west of Cambodia, south and west of Laos, and east of Myanmar (formerly Burma). Further to the north lies China, now dramatically influencing Thailand's political and economic spheres. Thailand's land mass is (513,115 km^2, 198,115 mi^2) with a population of over 68,615,858 (CIA World Factbook 2019). The highest population density is found in and around Bangkok, the Capitol. Significant population clusters are found throughout large parts of the country, particularly north and northeast of Bangkok and in the extreme southern region of the country. Bangkok, once called the "Venice of the East" because of its historic canal system, is the vast and vibrantly pulsating hub of the country. More than anywhere else, it embodies the contrasts between the old and the new in the country.

27.2.2 Heritage and Residence

The earliest knowledge of what today is Thailand is shrouded in lost histories of the ancient peoples of Southeast Asia. New cultures arose as kingdoms shifted through the centuries. The *Dvaravati* (first century BC to the eleventh century AD) were strongly influenced by Indian culture so that even today the Rama legends of Indian mythology form an integral part of Thailand's belief system (Hoare 2004).

The first people culturally considered as "Thais" probably migrated from the south of China. Sukhothai, founded in the thirteenth century AD, is considered the first kingdom of Siam (or Thailand). Its most famous king was Ramkhamhaeng, who is credited with developing the first Thai alphabet. Sukhothai had a profound influence on the development of Buddhist theology and classical art in Thai culture (Hoare 2004). The Sukhothai period was eclipsed in 1350 by the extremely powerful kingdom of Ayutthaya on the Chao Praya River. Although Ayutthaya eventually met its tragic demise when the Burmese sacked the city in 1767, it still represents a magnificent blossoming of artistic and cultural expression in the history of Siam (Hoare 2004).

In 1932, Thailand appointed its first prime minister. Thereafter, the king no longer served in any critical decision-making capacities (Hoare 2004). Still, the lineage of Thai kingships continues, and the Thai people continue to love and deeply revere their king. This intimate relationship between royalty and the people is intertwined with Thai Buddhism and the Thai peoples' perception of their king as divinely ordained. The king is usually not directly involved in Thai politics, but if a strong moral issue arises, he generally helps in addressing the problem guided by his peace and wisdom (Hoare 2004). In 2006, the Thais celebrated their beloved King Rama IX's 60th anniversary. His monarchy is now the oldest in the world. Any criticism of the king and his family is not at all acceptable to Thais and is even forbidden by law. Yet, Thailand's present constitutional monarchy is a democratic form of government built around the actual governing authority of the prime minister and the parliament.

27.2.3 Reasons for Migration and Associated Economic Factors

Little information is available as to where Thais migrate except to the United States. Over 300,000 Thais live in the US, most of them in Los Angeles followed by New York City, Houston, and Chicago (Pew Research Foundation 2017). In general, Thais have continued in their migration to the US in search of better opportunities.

The first two Thai immigrants in the US were *Eng* and *Chang*, the famous Siamese twins who captured the world's attention because of their conjoined chests and whose career was a public exhibition. They settled in North Carolina in 1839, later changed their Thai last name to Bunker, and got married to American women (Oliver 2019).

During the Vietnam War, many Thai women married American GIs and immigrated to the US

(Bao 2005). Immediate family members of these American Thais often followed them and settled in the new country. Many Thai professionals such as physicians, pharmacists, and engineers have immigrated to the US and Canada to further their studies under scholarship programs; many of them have remained, finding professional careers in the United States and Canada (Association of Thai Professionals in America and Canada 2019).

27.3 Communication

27.3.1 Dominant Language and Dialects

The standard Thai dialect is derived from *Pali* and *Sanskrit* (ancient South Asian languages) and is the official language in Thailand. The Thai language is a fixed tonal language having five tones. Thus, the same phonetic sound can have different meanings depending on the tone. The written alphabet is a complicated system of 44 letters with over 33 vowels or vowel combinations.

English is used in international schools, tourist places, and sometimes among Thai elite society. Although English is taught in Thai schools, the English proficiency of Tsai people in general is not very high, especially when compared with certain other Southeast Asian countries such as Malaysia or Singapore. This may be due in part to Thailand's having never been colonized.

The north, northeast, and southern regions of Thailand are all areas with unique dialects of their own. The dialect in northern Thailand is Pasah Nua, literally "the northern language." Thais in the Northeast speak Pasah Isaan, "the northeast language," which is a mixture of Laotian and other dialects. Pasah Isaan usually sounds very foreign to the ears of people in other regions of Thailand. The dialect of southern Thailand is Pasah Dai, "the southern language," and is the fastest-sounding among the dialects. A recent trend, however, has been that many parents in the northern or northeastern regions choose not to teach their children their regional dialects, in part, because they believe that the dialects do not sound modern or cultured.

27.3.2 Cultural Communication Patterns

Age and status in Thailand contribute greatly to how Thais communicate with one another. According to Thai culture, a younger person is expected to show respect for an older person through his or her gestures and language. A Thai female uses the word "*Kah*" and a Thai male uses "*Kraab*" at the end of a sentence to add politeness in a conversation. Looking in a person's eyes and conversing quietly reflects respect and politeness. A distance of 1 1/2 to 2 ft between two speakers is preferable.

In terms of body language, kisses and hugs between a male and a female are not traditional in Thai culture. Thais usually greet each other with the ***"Wai"*** motion—putting the palms of both hands together in a prayer-like gesture and bowing the head slightly. This gesture is used by both men and women of all age groups. Respect for older people, an important aspect of Thai culture, is always signaled by a younger person gesturing with the "*Wai*" to the older person first.

27.3.3 Temporal Relationships

Traditional Thai families are nuclear in nature. Today, however, single families are becoming more common in Thailand. In any case, it is not uncommon for a single Thai to live with her or his sibling(s), cousin(s), aunt(s), uncle(s), grandparent(s), or parent(s). A friendship between two individuals who are not biologically related can often evolve into a family-like relationship. Thus, a Thai may become like a brother, a sister, an aunt, an uncle, a parent, or a grandparent to a friend.

Respect for seniority is crucial among Thais. Visiting and bringing along a present or giving money to elders during the Thai New Year is an important role obligation for younger Thais. When elders in a Thai family become too old to take care of themselves, younger members are morally required to care for them. Only in very rare circumstances do elderly Thais live alone.

27.3.4 Format for Names

Most Thais have long first and last names. A Thai is usually referred to by his or her first name, even in an official setting like school or work. The names usually have clear meanings in Thai. The first name is often given by a Buddhist monk or fortune-teller based on the date, day of the week, and time of a newborn's birth. But often parents name their children themselves. More recently, some parents have begun to give their children Western first names, such as John, Matthew, or Amy.

In general, when a woman marries, she usually takes her husband's last name. A couple's children also take their father's last name. A recent Thai law, however, regulates that a married woman does not have to use her husband's last name if she prefers not to, thus legally sealing an already existing cultural shift.

When Thai names are transcribed in English, the spelling is merely a phonetic translation from its spelling in the Thai alphabet. Because Thai is a tonal language, the correct pronunciation of names cannot be ascertained from their spelling in English. For health-care providers, the best course is to ask Thai patients how to pronounce their names and do the best one can in approximating it.

Importantly, almost all Thais have a short nickname used by their family and close friends and often by colleagues at work. Nicknames normally have no relationship with first names. They are often humorous to Thais themselves. Nicknames are usually either Thai or English words. They might be derived from names of colors, body types, fruits, or any number of other things. Health-care providers should feel free to ask their patients if they wish to be called by their nickname. The client may well prefer it. For medical record keeping, include both names.

27.4 Family Roles and Organization

27.4.1 Head of Household and Gender Roles

Gender is another important aspect in Thai families. A man is the head of the household in a traditional Thai family, usually being the breadwinner and managing important tasks. This view is reflected in an elder's teaching on a wedding day: "The man is the front step of an elephant. The woman is the hind step."

In most Thai families, responsibilities involving household chores and taking care of children belong to a woman. If a woman works outside the home, a maid is sometimes hired to help with the household chores and babysitting. Many Thai men have much more leisure time than Thai women, regardless of the employment status of a woman. However, more Thai families today have begun to divide household chores between men and women.

27.4.2 Prescriptive, Restrictive, and Taboo Behaviors for Children and Adolescents

Thai children are taught to respect elders. Talking back to elders is discouraged. The role of children as students in school is very important. Many Thai parents choose a career deemed suited to their child's abilities and characteristics. The degree to which children assist with household chores depends upon a family's economic status; the poorer the family, the more chores children do.

Thai female adolescents have traditionally been expected to protect their virginity until marriage. Dating with a chaperone present is preferable to parents. However, more and more Thai adolescents date on their own today. Social attitudes are changing rapidly in Thailand, and those of the youth culture are strongly influenced by global trends related to music, entertainment, and social mores. These are often challenging to older traditions and can conflict with those inherited through Buddhist theology.

27.4.3 Family Goals and Priorities

Children are the center of the family for Thais. Many Thai children, therefore, sleep with their parents from birth until some point in time before they reach adolescence. Thai parents do not feel comfortable leaving their infants in a separate bedroom. Often, children are spoon-fed by adults until they are 6–7 years old. This can sometimes appear unusual to Westerners.

Most Thai parents hope their children will go to college. They will pay whatever they can for tuition fees and support even through graduate school. Education is so vitally important for Thais that Westerners are often amazed when a Thai spouse will leave his or her partner or children behind for years to further studies aboard.

Marriages in Thailand used to be mainly arranged by the parents. Today, young Thais have more freedom to select a spouse. Nevertheless, sometimes parents may make the final decision as to whether or not a bride or groom is acceptable. In this context, younger Thais are clearly expected to care for older people, including older in-laws, when they are in need.

27.4.4 Alternative Lifestyles

Gays and lesbians are more accepted in Thailand than in the past. Before the mid 1980s, commercial lounges and bars were the main or the only places for gays and lesbians for social gatherings. Since the mid 1990s, Thai gays and lesbians have had more venues to meet and advance a positive lifestyle. These new places include launched boutiques, hotels, restaurants, karaoke clubs, pubs, and spas (Utopia 2019).

The first Thai lesbian organization was founded in Bangkok in 1986 by a popular Thai singer. Eight years later, the first Southeast Asian gay and lesbian center was established. The center is a resource for gays and lesbians to find books and presents. Both of the organizations have at least two common goals, which include a movement for lesbian and gay rights and efforts to combat HIV/AIDS (Utopia 2019). At present, gay marriage is not supported by Thai laws. However, civil unions are accepted since 2019 (Reuters 2019).

27.5 Workforce Issues

27.5.1 Culture in the Workplace

Most Thais usually try to avoid personal conflicts at work and are hard workers. Although the family is deemed very important for Thais, in many circumstances, especially for economic reasons, work comes before family. For instance, a husband and his wife in Thailand often work in different provinces. A good number of the Thai couples reunite once a month. Taking a leave from work for a major surgery or a death or dying of family members besides one's spouse, child, or parent may not be supported by Thai agencies.

In general terms, Thai Americans (and most likely in other countries) tend to socialize among themselves rather than mix with Americans or people from other cultures. Therefore, some Thai Americans may not deeply understand American culture. However, other Thai Americans relate well to their surrounding culture, especially first-generation American-born Thais who tend to help change or adaptation come more easily to their parents (Advameg Inc. 2019).

27.5.2 Issues Related to Autonomy

Like many other Asian Americans, Thai Americans respect their supervisors because seniority is strongly valued in their culture. Thus, they might not be assertive at work. Therefore, supervisors may be wise to provide open discussions and expression of opportunities for their Thai American colleagues.

English proficiency among some Thais is low. Therefore, with Thai Americans who are learning English as their second language, the language used in the workplace should be clear. Slang expressions should be avoided. If used, slang expressions need to be clarified.

27.6 Biocultural Ecology

27.6.1 Skin Color and Other Biological Variations

An estimated 75% of the population in Thailand are pure "Thai"; 14% are Chinese; and the rest (11%) are Malay, Lao, Mon, Cambodian, Vietnamese, Asian Indian, Caucasian, or hill-dweller tribes—Karen, Lisu, Ahka, Lahu, Mien, and Hmong (CIA World Factbook 2019).

Some Thais in northeast Thailand (*Isaan*) emigrated from Laos or Cambodia. In general, *Isaan* Thais have darker skin color (dark brown) than other Thais who live in the north and central regions. The facial profile of *Isaan* Thais is akin to that of Laotians, with a relatively flat nose and broad prominent cheekbones. Some Thais in the north immigrated to Thailand from China or Burma. They tend to have finer skin texture and lighter skin color than other Thais. Their noses are a little longer and their cheekbones narrower than those of *Isaan* Thais. Central Thais generally have medium skin color compared with that in the rest of the country. Their facial profile is a mixture of *Isaan* Thais and northern Thais. Southern Thais, some of whom migrated from Malaysia, are likely to have darker skin color. Their facial profile is similar to that of Malay.

Other Thais have combined Thai and Chinese, Vietnamese, Malaysian, Laotian, or other heritage, with skin color and facial profiles representing mixtures of such racial combinations. Overall, regardless of skin color or facial profile, the Thais' size and body structure are usually much smaller than those of Caucasians.

27.6.2 Diseases and Health Conditions

Thai scientists in collaboration with scientists from Riken Yokohama Institute in Japan and Yale University in the US successfully identified a genetic pattern common to Thais by analyzing blood samples from 280 Thais from all four regions of the country (National Center for Genetic Engineering and Biotechnology [BIOTEC] 2006). This breakthrough, hopefully, will help scientists to better understand Thais' responses to a variety of antigens, drug metabolism, and genetic disorders.

Glucose-6-phosphate dehydrogenase deficiency (G-6-PD) is the most common genetic disorder among humans. Sixty-five percent of Thai newborns' jaundice is caused by this deficiency (Nuchprayoon et al. 2002). Usually, the enzyme regulates how red blood cells function. When a person lacks the enzyme, her or his red blood cells can be hemolyzed by certain medications, foods, or infections. The condition is called "hemolytic anemia." In most cases, when the cause of the anemia is removed, symptoms disappear. In rare cases, people with G-6-PD deficiency have persistent anemia and need to be monitored on a regular basis (Nuchprayoon et al. 2002).

Thalassemia, another genetic disorder is prevalent among Thais. Thirteen percent of Thais have inherited this disorder, and 50% of those who are affected by the disorder come from *Isaan,* or the northeast of Thailand (Fucharoen et al. 2006). Symptoms among Thais with thalassemia range from asymptomatic to severe anemia (Fucharoen et al. 2006). When Thai patients show anemic symptoms, they should be tested for thalassemia and identified for care if necessary.

27.6.3 Variations in Drug Metabolism

Literature reporting some variations in drug metabolism between Thais and non-Thais is mostly associated with antiretroviral medications. For example, a study revealed that using indinavir/ritonavir dose (400/100 mg) as a combined antiretroviral drug among Thais is more preferable than using indinavir/ritonavir dose (600/100 mg) as used among Caucasians owing to the smaller body size of the Thais (Cressey et al. 2005). This lower-dose medicine results in fewer side effects and greater adherence for Thais than the higher-dose medicine. Although the lower-dose medication provided lower plasma concentrations among the Thai participants, the low dose seems to be effective as evidenced by a suppression of viral replication through 48-week follow-ups (Cressey et al. 2005). Therefore, when treating Thai patients, dosing recommendations derived from Caucasian patients may not be appropriate. As a general rule, a lower dose may be more beneficial for Thais, possibly resulting in fewer severe side effects and greater adherence to the medications.

27.7 High-Risk Behaviors

27.7.1 Health-Care Practices

The top 10 causes of death in Thailand are neoplasms, cardiovascular disease, communicable diseases (malaria and influenza), musculoskeletal disorders, diabetes, substance misuse, transport injuries, neurological disorders, diarrheal diseases, and chronic respiratory diseases (Global Health—Thailand 2018). In addition, food and waterborne diseases include bacterial diarrhea; vectorborne diseases include dengue fever, Japanese encephalitis, and malaria (CIA World Factbook 2019). The extent of these diseases among Thais upon immigration is unknown but health providers need to assess Thais for these conditions. According to the Thai Ministry of Public Health: Thailand (2018) the most significant major health problems among the Thais are (a) Low birth weight and perinatal asphyxia (b) HIV/AIDS, road traffic injuries, drug abuse, schizophrenia, and alcohol abuse, road traffic injuries, diabetes, and liver cancer, and (c) cerebrovascular diseases, emphysema, and diabetes.

The first patient with AIDS in Thailand, reported in September of 1984, was a Thai gay man who studied in the US and moved back to Thailand. Since that year, incidences of HIV infection have been reported throughout the country. HIV infection rates in Thailand peaked at 4% in 1991, with over 140,000 new cases in that year. Rates declined to 1.5% by 2003, partly due to the 100 percent condom use campaign promoted among high-risk groups by the Thai government (Ministry of Public Health 2005). The current rate in Thailand is 1.1% (CIA World Factbook, Thailand 2019).

A study in Thailand in 24 provinces found homosexual behavior at 0.3% among high school students, 2–3% among vocational school students, and 4.7% among men in the military, with only 50% of men having sex with men using condoms consistently (National AIDS Prevention and Alleviation Committee 2010). HIV rates among men who have sex with men are reported to be highest (17–31%) in large tourist cities like Bangkok, Chiang Mai, and Phuket, and this is becoming a great concern for health professionals and the Thai government (National AIDS Prevention and Alleviation Committee 2010).

27.8 Nutrition

27.8.1 Meaning of Food

"We should eat to live, not live to eat" is a famous saying not only in Latin but also in Thai, reflecting the central importance and meaning of food in the Thai culture. Many Thais live their lives by following such a saying.

In general, an individual portion of a Thai dish is about one-third to one-fifth of a typical U.S. dish in terms of volume. As a result, most Thais are slim owing to these smaller portions and also the types of food they eat. Thais believe that foods containing adequate essential nutrients help to maintain life and growth and delay illness later in life (Kosulwat 2002). A Thai balanced diet usually includes low-fat/low-meat dishes with a large percentage of vegetable and legumes. Rice and fish are main staples (Kosulwat 2002).

27.8.2 Common Foods and Food Rituals

In general, rice is the main source of carbohydrates in Thai dishes, but noodles are also found in many favorite recipes. Vegetables and meats are usually fried or grilled and prepared in many combined variations to supplement rice. Overall, pork or chicken is eaten more than beef. All meats are consumed more sparingly in proportion to vegetables when compared with a Western diet. Fish and other forms of seafood are also regularly enjoyed.

Communal eating is an essential part of the Thai culture. Friends and families eat seated together either on the floor or at a table. Either way, when rice is part of the meal, Thais begin with a large amount of rice on their plates and reach to central communal plates of combined meat and vegetables to add to their rice. This is done by all in a free fashion throughout the meal, with some families using a serving spoon to take from the communal dishes and others using their individual tablespoons. The tablespoons are the main instruments for eating, with the fork used only as a guide; knives are not often used because the meats in Thai recipes are usually pre-cut. Noodle recipes are much loved by Thais and prepared with the noodles already mixed in with meats and vegetables.

For all foods, seasonings are critical to the Thai artistry of accommodating different palettes. Fish and oyster sauces are very often combined with soy sauce as a basic starting point for many recipes. Thai chili pepper is the basic ingredient added to control the degree of spiciness in foods. Many Thais love very spicy food, but not all. *Tom-Yum* is a traditional spicy Thai soup that is gaining popularity worldwide. It has been found to have positive effects on people's health because of its ingredients, which include lemon grass, galangal roots, kaffir lime leaves, hot chilies, red onions, and garlic (Siripongvutikorn et al. 2005). The soup's antimicrobial effects come from its chilies, onions, and garlic (Siripongvutikorn et al. 2005). Onions and garlic can function against diabetes and hypercholesterolemia. Fresh garlic, used as an ingredient in *Som-Tum* and many other Thai dishes, has been identified as an antifungal, antiparasitic, and antiviral agent (Siripongvutikorn et al. 2005).

Som-Tum is a famous spicy Thai salad originating from the northeast of Thailand. Its ingredients include fresh shredded papaya, cut tomatoes, tamarind juice, fish sauce, salt, sugar, fresh crushed garlic, and hot chilies. Sometimes, cooked or raw fermented fish is added. *Som-Tum* is usually served with hot *sticky* (sweet) rice, which is a favorite in the Northeast. Sources of protein, such as Thai beef/pork jerky and grilled chicken are often served with *Som-Tum* and *sticky* rice. Overall, this course of *Som-Tum*, *sticky* rice, and sources of protein is considered an enjoyable delicacy by Thais in all areas of society.

In the past, many Thais became sick and died from eating raw fermented fish, which contains *Opisthorchis viverrini*, a liver fluke, found to cause cholangiocarcinoma in humans (Watanapa and Watanapa 2002). Today, because of increased health education provided by nurses and other health professionals, Thais are more knowledgeable about the dangers of eating raw fish. Nevertheless, some Thais may persist in eating raw fermented fish because of entrenched eating habits and their attraction to its taste and smell. An assessment regarding any preference for eating raw fermented fish could be helpful.

A study conducted in Thailand revealed that many healthy Thai dishes are being replaced by foods containing a high quantity of fat and meat, related to the country's evolution from an agricultural to a newly industrialized country. Food produced in Thailand is now more important for exportation purposes and the economy than for domestic consumption (Kosulwat 2002). Thai families have less time to cook. Many tend to eat at Western-style restaurants serving foods high in fats, meat, and sugar content. As a result, obesity rates among Thai children and adults have risen dramatically since the mid 1980s (Kosulwat 2002). A study revealed that Thai children with obesity have low self-esteem and are often ridiculed by their peers (Phakthoop and Ross 2006).

In a study among 102 Thais in the US, 79% changed their food habits when living in the US (Siripongvutikorn et al. 2005). They skip more meals and consume more Western foods and snacks such as white bread, salty items, fruit juice, soft drinks, and sweets. When they dine out, they tend to go to American or Chinese restaurants. Forty percent of the participants indicated that their diet has become less healthy owing to a lack of time for food preparation and the unavailability of some Thai ingredients and food choices (Siripongvutikorn et al. 2005). An analysis of this study, as based on the Food Guide Pyramid, reveals that most Thai participants living in the US consume enough fruits and vegetables; not enough bread and milk; and too much meat, fats, oils, and sweets. Health professionals should assess their Thai patients' food intake habits and encourage them to consume more fruits and vegetables. If needed, advice about an increase of bread and milk intake and limiting meat, fats, oils, and sweets should also be provided (Siripongvutikorn et al. 2005).

27.8.3 Dietary Practices for Health Promotion

For Thais, hot or warm foods or drinks are considered healthier than cold ones. This idea is based in part on a belief in "cold and hot" or "Yin and Yang," inherited from Thailand's profound Chinese influence. Many types of herbs are considered to promote health and work against cancer development. Some herbs are considered a panacea. Therefore, Thai dishes usually contain some kind of herbs, particularly garlic and hot chilies. Positive effects of some herbs have already been described.

27.8.4 Nutritional Deficiencies and Food Limitations

Iodine deficiency (IDD) used to be a major health concern in Thailand. In 1953, IDD was first identified in the northeastern and northern regions of Thailand, where there is no sea outlet. Aware of the problem, in 1965, the Thai government initiated a pilot project of salt iodization in a northern province. Owing to its success, the project has been further expanded. The first IDD survey, conducted until 1988, was completed in 15 provinces of two regions of Thailand, showing an IDD prevalence rate of 19.3%. In 1993, the salt iodization project was expanded nationwide, resulting in further success with an IDD rate of 1.3% in 2003. At present, the Thai government examines goiter rates among schoolchildren in 15 northeast and northern provinces and uses them as the Thai IDD indicator (Ministry of Public Health 2005).

Despite the salt iodization program's success, at the 2004 Review of Progress toward Sustainable Elimination of Iodine Deficiency held in Thailand, the Thai Ministry of Public Health indicated that only 51% of Thai households consumed enough iodized salt (Network for Sustained Elimination of Iodine Deficiency 2004). This was well below the international target of at least 90% set for the end of the year 2005. More than 34 million Thais do not consume enough iodized salt, and 375,000 newborns may suffer from IDD. However, the Thai Ministry of Industry and the U.S. Food and Drug Administration have begun working with salt producers to manage salt iodization programs. Together they brought the goiter prevalence rate in Thailand down to 2.2% in 2008 (as compared to rates for the same year in other countries: 19.4 in Australia, 17.9 in India, 30.0 in Turkey, and 14.5 in Switzerland) (Network for Sustained Elimination of Iodine Deficiency 2011).

In Thailand, only seven cases of anorexia nervosa have been reported (Jennings et al. 2006). However, evidence exists that young Thais in particular are increasingly becoming susceptible to developing eating disorders. A study among 101 Thais in Thailand, 110 Caucasian Australians, and 130 Asian Australians found that the Thai participants reported the highest scores on eating disorder attitudes and psychopathology (Jennings et al. 2006). Recently, pressure to be thin has become more extreme in Thailand than in Australia. The evidence suggests that eating disorders may not be limited to Westerners, as we used to believe. Such disorders will become more prevalent among Thais in the near future.

27.9 Pregnancy and Childbearing Practices

27.9.1 Fertility Practices and Views Toward Pregnancy

Thai women view pregnancy as a special time in their lives when they need extra care physically and emotionally (Nigenda et al. 2003). They acknowledge that this is a time when their moods can be unstable. Ideally, the age of 20 years is the optimal time for pregnancy owing to the women's physical and emotional maturity. Thai women want their husbands and their mothers to be supportive of their pregnancies. Some women state that the most common side effects of pregnancy are excessive white vaginal discharge, frequent urination, and morning sickness (Nigenda et al. 2003). Owing to modesty, especially during a vaginal examination, Thai women prefer female health-care providers over their male counterparts. They do not feel comfortable exposing their bodies to male providers (Nigenda et al. 2003).

27.9.2 Prescriptive, Restrictive, and Taboo Practices in the Childbearing Family

The descriptions in this section are based on literature review and the previous authors' experience working with pregnant and postpartum Thai women. During the childbearing period, Thai women basically receive advice from their mothers about what to do or not to do. Their mothers are the most significant persons who direct their practices during this time. Some of the practices presented herein are not stereotypical among all Thais; rather, they reflect some general practices or beliefs of some Thais in some particular areas of the country.

During pregnancy, the mothers of some pregnant Thai women may discourage their daughters from particular practices or behavior. For example, pregnant women are advised not to complain or get upset so that newborns will be happy and stay happy for the rest of their lives. They may also be advised not to sit on stairs or doorsills to avoid a difficult labor and delivery. When a pregnant mother blocks other people from going up and down stairs or in and out of a doorway, the unborn baby could be blocked inside the mother's uterus.

Astrology and animism play major roles in many Thais' lives. In general, Thai pregnant women are discouraged from visiting a hospitalized person (regardless of the kind of sickness), attending a funeral ceremony, or visiting a house where there has been a death (Kaewsarn et al. 2003a). Such practices are believed to prevent the pregnant woman and her unborn baby from catching any illness or getting haunted by a spirit or ghost.

Some women believe that eating eggs may result in having smelly newborns (Nigenda et al. 2003). Some avoid drinking coconut juice, believing that it can cause too much vernix caseosa (fat on the newborn's skin), whereas others drink a lot of the juice, believing that it will help their newborns to have smooth and beautiful skin texture. Some believe that drinking chocolate milk, eating chocolate, or drinking coffee will cause their newborns to have a darker skin color. Most Thais view lighter skin as more favorable.

Pregnant women from the central region of Thailand are often seen with a safety pin on their outfit over their belly. The pin works against a kind of ghost who always wants to steal the unborn baby from a mother's womb. Pregnant Thai women, especially those with Chinese descendants and their families, may ask their obstetric physicians to perform selective cesarean sections, believing that the date and time of their babies' births can greatly affect their children's future as based on the Chinese Zodiac calendar and fortune-telling (Ross et al. 2007a).

Like many other Southeast Asian women, postpartum Thai mothers practice the concept of "Yin" and "Yang" (cold and hot) (Kaewsarn et al. 2003a). After a child is born, the mother is left cold and wet. Therefore, the mother should gain some heat to dry out her body, especially her uterus (Kaewsarn et al. 2003a). To gain heat, some mothers practice *Yue Fai*, which literally means "being with fire." There are a couple of ways to perform *Yue Fai*. The new mother lies down either on a bed above a bonfire or on a wooden plank nearby. The fire is tended for as long as the mother is supposed to be near the fire, which may be from 1 to 30 days. Reasons given by Thai mothers for practicing *Yue Fai* include desiring an increase of milk, faster involution of the uterus, and prevention of illnesses and bone ache (Kaewsarn et al. 2003b). Some drawbacks of this ritual, include inconvenience, discomfort, and complications such as heat rashes, sweating, dehydration, and minor burns (Kaewsarn et al. 2003b). To be able to perform *Yue Fai,* space is needed and a family member must keep tending the fire. Without enough space and a 24/7 support person, *Yue Fai* is not possible. The practice of *Yue Fai* outside Thailand is unknown.

When *Yue Fai*, the ultimate practice for gaining heat during the postpartum period, is not possible, Thai mothers are advised to use a combination of practices, including a perineal heat light, a hot Sitz bath, sauna heat belts, and warm showers (Kaewsarn et al. 2003b). Warm and hot drinks and foods are consumed; ice chips or ice cubes are avoided.

In general, all Thai mothers are allowed by their mothers to drink warm and hot non-alcoholic liquids. However, there is no consensus about the types of protein, vegetables, and fruit the postpartum mother should consume. Many postpartum Thai women are not restricted to proteins, vegetables, and fruit, but some are.

Sources of protein include pork, chicken, fish, eggs, milk, catfish, internal organs, beef, water buffalo meat, and shrimp (Kaewsarn et al. 2003b). However, some mothers might be advised to not eat eggs, chicken, or buffalo meat, believing that the new mothers' perineum may not heal. For some postpartum Thai mothers, chicken is a taboo food for women after delivery: The belief is that a chicken likes to scratch the ground to look for food. Therefore, the chicken meat could scratch open the perineum.

Eggs are avoided by some, believing that they could cause a big scar on the perineum. Water buffalo meat is tough and cheap and, therefore, seen as unhealthy by Thais. Based on this belief, it is thought that the healing process of the new mother's perineum could be jeopardized by its consumption.

Vegetables eaten by postpartum mothers may include lettuce, banana flower, lemon grass, onion, ginger, cabbage, hairy melon, snake beans, chili, peppers, and bamboo shoots (Kaewsarn et al. 2003b). Acceptable fruits after the postpartum period may include oranges, bananas, tamarind, watermelon, jack fruit, and durian, an oval fruit with a hard-spiny rind. However, some women avoid durian because of its strong smell. For traditional Thai families, especially those from rural Thailand, the new mother might be restricted to a few items of food for the first few weeks. For example, she might be allowed to take only rice soup with salt without any protein or fruit. Some postpartum Thai women drink ***Ya Dong***, a Thai non-alcoholic or alcoholic drink infused with herbs. Herbs used in *Ya Dong* may include ginseng, galangal, peppermint, cinnamon, Spirulina, and plant roots. As perceived by many Thais, *Ya Dong* is famous for its medicinal qualities. The drink helps with blood production and drying out the uterus quickly.

Expecting Thai fathers, like those of many cultures, have begun to participate more in the childbearing experience. They tend now to desire more strongly to protect the unborn baby and to become more involved with the mother in preparing for postpartum care (Sansiriphun et al. 2010). Particular regional cultural practices may influence how some Thai fathers respond to society's changing expectations for them. Moreover, Buddhist theology has been found to be "embedded in the beliefs and strategies" of expecting Thai fathers (Sansiriphun et al. 2010).

27.10 Death Rituals

27.10.1 Death Rituals and Expectations

Because most Thais are Buddhists, only the funeral rites in connection with Buddhism are addressed here. Thai Buddhists believe that after a person dies, the person will be reborn somewhere else based on that person's ***Karma*** (Dhammanada 2002). "*Karma* means 'action' and..refers to the process by which a person's moral behavior or actions have consequences for the person's future, either in the present or later life" (Ross et al. 2007b, p. 4).

In general, Thai Buddhists follow the custom of cremating the bodies of the deceased because when the Buddha passed away, his body was cremated. According to the Buddha's teaching, a funeral ceremony should be simple. Unfortunately, many Thai Buddhists (and some other Buddhists) have transformed what was a traditionally simple cremation ceremony into one that is overly extravagant.

> *The consciousness or mental energy of the departed person has no connection with the body left behind. . . . A dead body is simply an old rotten simple house which the departed person's life occupied. The Buddha called it "a useless log." Many people believe that if the deceased is not given a proper burial or if a sanctified tombstone is not placed on the grave, then the soul of the deceased will wander to the four corners of the world and weep and wail and sometimes even*

*return to disturb the relatives. Such a belief cannot be found in Buddhism (*Dhammanada 2002, *p. 246).*

In the funeral ceremony, Buddhist monks are often invited to chant verses to the dead and the family. Food and candles are offered to the monks. Many Thai Buddhists believe that such chanting will benefit the spirit of the dead, regardless of Buddha's teaching about the unbound relationship between the body and the spirit. The ashes from the cremation are buried at a cemetery. Sometimes, a portion of the ashes is sprinkled in a river. If possible, when the family of the dead returns home after a sojourn away from Thailand, some of the ashes may be sprinkled again in a river or near the deceased's hometown.

27.10.2 Responses to Death and Grief

During the funeral ceremony, the family gets together. The sons of the deceased are expected to be ordained for a short period of time, ranging from a week to 3 months. The ordination is believed to help the dead go to heaven. Female relatives normally wail quietly. The family members pray quietly to the dead before the cremation to ask for forgiveness and wish the dead to be reborn in a happy and peaceful home. Often, in their prayer, family members wish for themselves to be reborn in the same family with the same relation to the dead in their next life.

27.11 Spirituality

27.11.1 Dominant Religion and Use of Prayer

Approximately 94.6% of the Thais are Buddhist; the rest are Muslim (4.6%), Christian (0.7%), and Hindu or other (0.1%) (CIA World Factbook 2019). In the US, over three million people are Buddhist, most coming from Asian countries, including Thailand (Eck 2001). Buddhism is an exceptionally tolerant religion with its roots in Hinduism. Although precepts grounded in Buddhism are fundamental to the spiritual makeup of most Thais, animistic beliefs generally have equal meaning and play a parallel role in their belief system.

Although not in agreement with all other religious beliefs, Thai Buddhists are free to incorporate any other religious values and animism to their beliefs and practices when deemed good. Most Thais in all socioeconomic strata to some degree incorporate animism, fortune-telling, and astrology. Studies have shown that ancient spirits are prayed to by many Thai patients or their caregivers. Fortune-telling plays a major role in how Thais deal with illnesses (Ross et al. 2007a; Rungreangkulkij and Chesla 2002).

Many families in Thailand have a *spirit house* where they believe that the ancient spirits of the land (*Pra Poom*) dwell: two little statues of the *Pra Poom* (one male and one female) are placed inside a unique little abode that rests on a post or column. This house is usually at least as high as the eye level of an adult so as to indicate respect to the family for the *Pra Poom*. Their abode can be either very simple or quite decorative, depending upon how much the family can afford, and faces either north or east (in the belief that these two directions are superior to the south and west). Miniature figures of a couple of horses and elephants are often placed in front of the *Pra Poom* figures to accompany them. Fresh flowers, food, and drink are placed in tiny plates, bowls, and cups as offerings. These may be placed everyday, or about once a month. The family members pray to the *Pra Poom* as often as they wish. Usually, the family prays and gives offerings to the *Pra Poom* more often when asking for blessings and faster healing of an ill family member. It is unknown how much of these practices are completed when Thais migrate to other countries.

27.11.2 Meaning of Life and Individual Sources of Strength

For most Thais, family support along with Buddhism is a crucial source of strength. Parents are obliged to care for their ill children, regardless of a child's age or type of illness (Rungreangkulkij and Chesla 2002). Children or the unborn babies of HIV-positive pregnant Thai women have been identified as a major source of strength for their mothers (Jirapaet 2001; Ross et al. 2007c).

In a study among Thai mothers of schizophrenic adult children, the mothers practiced ***Thum-jai*** as a way to cope with a situation perceived to be unchangeable (Rungreangkulkij and Chesla 2002). *Thum-jai* means "let it be" or "whatever will be, will be." By practicing *Thum-jai*, a person will be able to accept the reality of a challenge or problem and try to move on in his or her life with calmness and peace. The mothers in Rungreangkulkij and Chesla's study (Rungreangkulkij and Chesla 2002) stated that when their sick children misbehaved, they smoothed their own heart with "water." For the Thai, a metaphor of "water versus fire" indicates "calmness versus anger/frustration." The "fire" should be put out by "water" in a person's heart to defeat a crisis situation. The mothers in this study offered that calmness and gentle speech usually worked better than scolding in calming down their schizophrenic children (Rungreangkulkij and Chesla 2002).

27.11.3 Spiritual Beliefs and Health-Care Practices

Buddhism significantly pervades the life of many Thais, whether in Thailand or elsewhere (Burnard and Naiyapatana 2004). When coping with difficulties or illnesses, many Thai laypeople and health-care professionals follow Buddha's teaching (Ross et al. 2007b). Like most Buddhists, the ultimate goal for a Buddhist Thai is to reach *Nirvana*. This is the end of reincarnation or the cycle of rebirths. When there is no rebirth, there is no suffering. Either they are happy or suffering. "Peace" is the ultimate goal (Dhammanada 2002). Results from a study reflect this belief by reporting that the ultimate goal of HIV-positive postpartum Thai women (alongside goals for their children) is to live with their HIV infection in peace. The women thus stated that they followed the Buddha's teachings through their beliefs in *Karma,* the Five Precepts, and the Four Noble Truths in order to live in peace with HIV (Ross et al. 2007b).

Karma is strongly associated with belief about rebirth; thus, many Thai patients (or caregivers) believe that unwholesome *Karma* from their past life has caused them to become ill in the present life. They believe that the illness can be improved by following the Five Precepts so that their present or next life (or the lives of their loved ones) will be improved (Ross et al. 2007b). Merit making—a way to decrease selfishness and greed and a way to be hopeful for a better present and future life—is performed by many Thais. Merit making includes activities such as freeing animals or birds, donating money to the poor or a temple, offering food to monks, and tangibly helping those in need, emotionally, or financially (Ross et al. 2007b; Tongprateep 2000).

Five Precepts are comparable with half of the Christian Ten Commandments and stress abstinence from killing, stealing, lying, sexual misconduct, and illicit drugs and alcohol consumption (Smith 1994). A study with seven HIV-positive postpartum Buddhist Thai women revealed that the participants all decided to carry their pregnancies to term instead of ending them. They all stated that ending a pregnancy is a type of killing, which is considered a sin. Furthermore, all of the women in this study believed that such unwholesome action would follow them in their next reincarnation as bad *Karma* (Ross et al. 2007b). In another study, it was found that the Five Precepts are observed by older Thai people to help them feel happy and peaceful (Tongprateep 2000).

The Four Noble Truths reflect tenets about life, suffering, and the cessation of suffering. The First Noble Truth maintains that life is suffering, and that suffering as such is found in four unavoidable life moments—namely, birth, illness, aging, and death. The Second Noble Truth maintains that the cause of all suffering is Tanha, or personal desire. The Third Noble Truth is a belief that overcoming Tanha is attainable. The Fourth Noble Truth outlines paths to end suffering (Smith 1994).

A qualitative study reported that the Four Noble Truths were followed by HIV-positive pregnant Thai women to cope with their infection (Ross et al. 2007b). The women stated that they began dealing with their illness by accepting the truth that everyone dies anyway at some point in life (The First Noble Truth) and that their suffering came from their personal desire (The Second Noble Truth). In other words, their desire arose by thinking of themselves as a real existence in the world rather than as an illusion. "Self" is like a mirage, or an imagined being (Flanagan 2005). Therefore, when a person becomes selfless, the person is freed from suffering (Smith 1994). The women in the study tried to think that their body and soul were not theirs, but instead imagined elements. To overcome desire, the participants tried to focus on universal life (The Third Noble Truth) by thinking about their infants instead of themselves and by meditating and praying. They also tried to follow the Middle Way, or a path between the two extremes of self-pleasure and self-mortification (Dhammanada 2002), as a means to end their suffering. In accordance with the Buddha's teaching that any extremes of thoughts, behavior, or speech are not wholesome, they reported trying not to feel too badly about themselves in order to be peaceful.

Meditation and prayer are ways for many Thais to cope with an illness. Studies revealed that both Thai pregnant and nonpregnant women meditated and prayed to the Buddha and supreme beings in order to help them cope with HIV/AIDS (Dane 2000; Jirapaet 2001; Ross et al. 2007b). Meditation is a means for Thai older people to enhance their self-awareness, peace of mind, sleep, and physical health (Tongprateep 2000). For Thai older people in the US and probably elsewhere, meditation and prayer help them feel peaceful, perceive life as valuable, value tranquil relationships with family and friends, and experience meaning and confidence in death. For Thais, health and spirituality are intertwined and are important aspects of life (Pincharoen and Congdon 2003).

In conclusion, the spiritual concepts of Karma, Nirvana, the Five Precepts, the Middle Way, and the Four Noble Truths are all important for Buddhist Thais. Ideally, when health professionals are aware of Buddhist concepts in caring for their sick or healthy Thai patients, the quality of care can be significantly enhanced.

27.12 Health-Care Practices

27.12.1 Health-Seeking Beliefs and Behaviors

In Thailand, most Thais rely on government health-care facilities, especially in the northeast region, or *Isaan*. People in *Isaan* hold strong traditional beliefs and practices. They tend to be poorer and less educated than the rest of the country. In the *Isaan* area, statistical rates of gynecological problems are relatively low; yet, *Isaan* women's self-reports show high rates of gynecological complaints associated with vaginal discharge and pain "in the uterus" (Boonmongkon et al. 2001). This contrast is often explained by a lack of comprehension among some *Isaan* women who do not understand clearly the physiological changes of their menstrual cycle and the amount of vaginal discharge. Some of the pain "in the uterus" with which they are concerned may well be related to physiological pain during ovulation. These same beliefs most likely continue upon migration.

Boonmongkon et al. (2001) reported that *Isaan* women's complaints and concerns about vaginal discharge and pain in the uterus may have an extreme impact on their lives. The women believe that such problems will turn into cervical cancer. This belief causes them to visit health-care facilities often, self-treat by relying on small doses of inappropriate antibiotics, be unhappy with their sexual relationship with their husband, and suffer from worries of their "ailments" (Boonmongkon et al. 2001).

Most *Isaan* women in Boonmongkon and colleagues' study (Boonmongkon et al. 2001) believed that their sustaining problems of vaginal discharge and pain "in the uterus" stemmed from their inappropriate practices after postpartum or significant past events. For instance, over 25% of the women stated that their chronic symptoms resulted from their inadequate practices of "lying by fire" and this caused their uterus to stay wet. Examples of other past experiences that the women believed caused their sustaining gynecological problems include hard work in youth, abortion, pushing too hard during delivery, and sterilization. Some women indicated that they did not receive adequate information about their problems from health professionals but did not feel like asking questions for fear of being scolded. Therefore, health-care professionals should bear in mind that Thai women, especially from *Isaan,* may need more information regarding physiological changes related to their menstrual cycle and may need encouragement to ask any questions they have regarding their gynecological concerns.

27.12.2 Responsibility for Health Care

Health promotion and disease prevention behavior among Thais are very limited. Although all Thais are covered by some kind of health insurance in Thailand, including the Universal Coverage of Health Care Scheme (75-cent health care), only 5.3% of the population used health promotion services, which include immunization, prenatal care, family planning, postpartum care, yearly checkups, dental care, and some other services (Ministry of Public Health 2005). Among those who did use such services, one-third went to urban health centers, 28.7% to community hospitals, and 11.3% to general/regional hospitals. One-third of the services used were yearly checkups, and one-third included immunization (Ministry of Public Health 2005).

27.12.3 Folk and Traditional Practices

Folk practices are more common among less-educated, rural Thais. Many Thais believe that bad *Karma* and negative supernatural power causes mental illness. Therefore, folk therapies from traditional healers are the first resource for many Thai families. When such therapies do not seem to work, they go to contemporary medical facilities as their second resource. Folk therapies may include healing ceremonies, using shamans (as a mediator) to converse with supernatural beings (such as black magic, evil beings, and ancient/natural spirits), negotiating with them that the sick person might be released from their illness. In such ceremonies, holy water or oil is usually used to anoint the sick (Rungreangkulkij and Chesla 2002). Little is known amout Thai's use of folk practices upon emigration to other countries.

Khwan is a Thai concept about the "power inside," or the "life spirit." Thais believe that *Khwan* enters the newborn's anterior fontanel during delivery. *Khwan* is different from self-esteem; it is thought of as a "life force" that can vanish when people are in a stage of shock, mental illness, or far away from home (Burnard and Naiyapatana 2004). In a *Khwan* ceremony, fresh flowers are offered by the sick or the person who has lost her or his *Khwan.* A monk or an older person then ties a blessed white thin string around a person's wrist, believing that the blessed string will tie the *Khwan* again to the person's body (Burnard and Naiyapatana 2004).

27.12.4 Barriers to Health Care

For Buddhists, when one is too extreme in one's speech, thoughts, or behavior, it is considered unwholesome, as based on their belief in the Middle Way (Dhammanada 2002). In this sense, some Buddhist Thais may not seek health care until their symptoms become severe. In addition, stigmatization attached to mental illness and beliefs in animism and *Karma* tend to prevent some Thais from seeking professional help when mental health problems arise. Some may not seek assistance from health-care professionals until they realize that traditional healers, Shamans, cannot help them (Rungreangkulkij and Chesla 2002).

27.12.5 Cultural Responses to Health and Illness

Adhering to their belief in the Middle Way, many Thais may appear stoic in trying to withhold expressions of pain or suffering from their illness. Health-care professionals may need to rely more on nonverbal clues for pain or some psychological-emotional distress when assessing their Buddhist Thai patients.

Many Thais, and even some health professionals, equate depression with psychosis (Ross et al. 2007a). Thus, when clinical depression is diagnosed, health-care professionals should make extra efforts to encourage depressed Thai patients to get help and treatment, along with assuring them that depression and psychosis are different disorders.

27.12.6 Blood Transfusions and Organ Donation

No religious beliefs against blood transfusion exist for Thais. However, donating and receiving organs is another matter. Although acceptable among many Thais, belief in their rebirth might prevent some from donating their organs, believing that they might not have the organ when needed in the next life.

27.13 Health-Care Providers

27.13.1 Traditional Versus Biomedical Providers

Thais tend to consult their family and friends first when they feel ill or have medical problems. Thai women usually seek female providers for childbearing care and gynecological problems owing to their modesty and their culture. However, if female providers are not available, they are generally willing to accept male providers.

27.13.2 Status of Health-Care Providers

Respect for seniority is a strong cultural value among Thais. Thus, less-experienced health professionals are expected to respect those with more experience in the same profession. In general, when comparing physicians, head nurses, and junior nurses, Thai physicians receive the most respect, followed by head nurses and junior nurses. In some cases, very senior head nurses receive the same level of respect as physicians (Burnard and Naiyapatana 2004).

Based on a concept of "Thainess," as expressed by many Thais, and especially in terms of being Buddhist, Thai nurses reported that they often incorporate Buddhist ideas and beliefs in caring for their chronically ill patients (Burnard and Naiyapatana 2004; Sawatphanit et al. 2004).

Like many other patients in developing countries (Withell 2000), some Thai patients, and especially those of lower socioeconomic status, can be passive in voicing their needs and requesting care and services from health-care providers (Jirapaet 2001). Therefore, health professionals in countries outside of Thailand are advised to evaluate the level of passivity among their Thai patients so that their needs can be addressed with an eye toward optimal care.

Reflective Exercises

1. What are the three main dialects spoken in Thailand? How might you get an interpreter for each dialect?
2. What are nonverbal practices used by This?
3. Identify at least two genetic conditions common among Thais. What implications do they have for health and health teaching?
4. Identify common health conditions among Thais.

5. What is *Tom-Yum*? What are the ingredients and what are it health properties? For what conditions is it advantageous?
6. *Som-tum*? What are the ingredients and what are it health properties? For what conditions is it advantageous?
7. What foods might be avoided among some Thai postpartum women? Why are these foods avoided?
8. What is the dominant religion of Thais? What are its principles?
9. Identify several Thai folk practices. For what conditions are they used?
10. What are the roles for men and for women in traditional Thai culture?

References

Advameg Inc. (2019) Thai American. http://www.everyculture.com/multi/Sr-Z/Thai-Americans.html

Association of Thai Professionals in America and Canada (2019). http://www.atpac.org/about-atpac/

Bao J (2005) Merit-making capitalism: re-territorializing Thai Buddhism in Silicon Valley, California. J Asian Am Stud 8(2):115–142

Boonmongkon P, Nichter M, Pylypa J (2001) *Mot Luuk* problems in Northeast Thailand: why women's health concerns matter as much as disease rates. Soc Sci Med 53:1095–1112

Burnard P, Naiyapatana W (2004) Culture and communication in Thai nursing: a report of an ethnographic study. Int J Nurs Stud 41(7):755–765

Central Intelligence Agency (2019) The world factbook: Thailand. https://www.cia.gov/library/publications/the-world-factbook/geos/th.html

Cressey TR, Leenasirimakul P, Jourdain G, Tod M, Sukrakanchana P, Kunkeaw S, Puttimit C, Tod M, Jourdain G, Lallemant MJ (2005) Low-doses of indinavir boosted with ritonavir in HIV-infected Thai patients: pharmacokinetics, efficacy and tolerability. J Antimicrob Chemother 55(6): 1041–1044

Dane B (2000) Thai women: meditation as a way to cope with AIDS. J Relig Health 39:5–21

Dhammanada KS (2002) What Buddhists believe. Buddhist Missionary Society Malaysia, Kuala Lumpur

Eck DL (2001) A new religious America. http://pluralism.org/publications/new_religious_america/

Flanagan A (2005) Buddhism. http://buddhism.about.com/cs/ethics/a/BasicsKama.htm

Fucharoen G, Trithipsombat J, Sirithawee S, Yamsri S, Changtrakul Y, Sanchaisuriya K, Fuchareaon S (2006) Molecular and hematological profiles of hemoglobin EE disease with different forms of a-thalassemia. Ann Hematol 85(7):450–454

Global Health—Thailand (2018). https://www.cdc.gov/globalhealth/countries/thailand/default.htm

Hoare TD (2004) Thailand: a global study handbook. ABC CLIO, Santa Barbara CA

Jennings PS, Forbes D, McDermott B, Hulse G, Juniper S (2006) Eating disorder attitudes and psychopathology in Caucasian Australian, Asian Australian and Thai university students. Aust N Z J Psychiatry 40(2):143–149

Jirapaet V (2001) Factors affecting maternal role attainment among low-income, Thai, HIV-positive mothers. J Transcult Nurs 12:25–33

Kaewsarn P, Moyle W, Creedy D (2003a) Traditional postpartum practices among Thai women. J Adv Nurs 41:358–366

Kaewsarn P, Moyle W, Creedy D (2003b) Thai nurses' beliefs about breastfeeding and postpartum practices. J Clin Nurs 12:467–475

Kane S (2010) Thailand's political crisis: which color is conservative? SAIS Rev 30(1):105–108

Kosulwat V (2002) The nutrition and health transition in Thailand. Public Health Nutr 5(1A):183–189

McCargo D (2008) Thailand: state of anxiety. Southeast Asian Affairs 2008:333–356

Ministry of Public Health (2005) Thailand health profile 2001–2004. Printing Press, Express Transportation Organization, Bangkok

Ministry of Public Health (2018) Thailand health profiles: 2005–2007. http://www.moph.go.th/ops/thp/index.php?option=com_content&task=view&id=6&Itemid=2&lang=en

National AIDS Prevention and Alleviation Committee (2010) UNGASS Country progress report Thailand: reporting period January 2008–December 2009. http://www.unaids.org/en/dataanalysis/monitoringcountryprogress/2010progressreportssubmittedbycountries/thailand_2010_country_progress_report_en.pdf

National Center for Genetic Engineering and Biotechnology (2006) Gene sequence of Thais identified. http://www.ncbi.nlm.nih.gov/pubmed/16889678

Network for Sustained Elimination of Iodine Deficiency (2004) Optimal iodine nutrition in the Americas. http://www.iodinenetwork.net/About_Ameet_Lima.htm

Network for Sustained Elimination of Iodine Deficiency (2011) Iodine global scorecard 2008. http://www.iodinenetwork.net/documents/scorecard-2008.pdf

Nigenda G, Langer A, Kuchisit C, Romero M, Rojas G, Al-Osimy M (2003) Women's opinions on antenatal care in developing countries: results of a study in Cuba, Thailand, Saudi Arabia and Argentina. BMC Public Health 3:17. http://www.biomedcentral.com/1471-2458/3/17

Nuchprayoon I, Sanpavat S, Nuchprayoon S (2002) Glucose-6-phosphate dehydrogenase (G6PD) muta-

tions in Thailand: G6PD Viangchan (871G>A) is the most common deficiency variant in the Thai population. Hum Mutat 19(2):185
Oliver M (2019) Chang and Eng Bunker: the strange story of the original Siamese twins. https://allthatsinteresting.com/chang-eng-bunker-siamese-twins
Pew Research Center (2017) Thai population in the United States. https://www.pewsocialtrends.org/chart/thai-population-in-the-u-s/
Phakthoop M, Ross R (2006) Antecedents, consequences, and management of obesity: a descriptive qualitative study among obese children in Chonburi Province. J Facul Nurs 14:34–48
Pincharoen S, Congdon J (2003) Spirituality and health in older Thai persons in the United States. West J Nurs Res 25(1):93–108
Reuters (2019) Gay couples to live more freely with Thai civil unions. https://www.reuters.com/article/us-thailand-lgbt-lawmaking-feature/gay-couples-to-live-more-freely-with-thai-civil-unions-idUSKCN1PX00E
Ross R, Sawatphanit W, Suwansujarid T, Draucker CB (2007a) Life story and depression of an HIV-positive, pregnant Thai woman who was a former sex worker: case study. Arch Psychiatr Nurs 21(1):32–39
Ross R, Sawatphanit W, Suwansujarid T (2007b) Finding peace (*Kwam Sa-ngob Jai*): a Buddhist way to live with HIV. J Holist Nurs 25:228–235
Ross R, Sawatphanit W, Draucker C, Suwansujarid T (2007c) The lived experiences of HIV-positive, pregnant women in Thailand. Health Care Women Int 28(8):731–744
Rungreangkulkij S, Chesla C (2002) Smooth a heart with water: Thai mothers care for a child with schizophrenia. Arch Psychiatr Nurs 15:120–127
Sansiriphun N, Kantaruksa K, Klunklin A, Baosaung C, Jordan P (2010) Thai men becoming a first-time father. Nurs Health Sci 12(4):403–409
Sawatphanit W, Ross R, Suwansujarid T (2004) Development of self-esteem among HIV positive pregnant women in Thailand: action research. J Sci Technol Human 2(2):55–69
Siripongvutikorn S, Thummaratwasik P, Huang Y (2005) Antimicrobial and antioxidation effects of Thai seasoning, *Tom-Yum*. Soc Food Sci Technol 38:347–352
Smith H (1994) The illustrated world's religions: a guide to our wisdom traditions. HarperCollins, San Francisco
Tongprateep T (2000) The essential elements of spirituality among rural Thai elders. J Adv Nurs 31:197–203
Utopia (2019) Travel and resources: Thailand. http://www.utopia-asia.com/tipsthai.htm
Watanapa P, Watanapa WB (2002) Liver fluke-associated cholangiocarcinoma. Br J Surg 89:962–970
Withell B (2000) A study of the experiences of women living with HIV/AIDS in Uganda. Int J Palliat Nurs 6:234–244

28 People of Turkish Heritage

Marshelle Thobaben and Sema Kuguoglu

28.1 Introduction

A comparable diversity can be seen in the human history of Turkey where over the past 10,000 years various civilizations have risen and fallen due to invasions by newcomers, disease epidemics, and natural disasters such as earthquakes. It continues to be a land of educational, religious, and cultural diversity. What is presented about the Turkish culture in this chapter is based on studies from Turkey and on observations of and experiences with Turkish immigrants. Turkey is one of the 18 most populated countries in the world and has the third largest population in the Middle East, and largest in Europe.

28.2 Overview, Inhabited Localities, and Topography

28.2.1 Overview

Türkiye (Turkey), as it is written in Turkish, means "land of Turks." It is located in the Northern Hemisphere, almost equidistant to the North Pole and the equator. The shape of Turkey resembles a rectangle, stretching in the east–west direction for approximately 1565 km (972 miles) and in the north–south direction for nearly 650 km (404 miles). It is bordered by Georgia, Armenia, and Nahcivan (Azerbaijan) to the northeast; the Islamic Republic of Iran to the east; Iraq and Syria to the south; Greece and Bulgaria in the Thrace to the west; and Russia, Ukraine, and Romania to the north and northwest (through the Black Sea). The Anatolian peninsula is the westernmost point of Asia, divided from Europe by the Bosporus and Dardanelles straits. Thrace is in the western part of Turkey on the European continent.

Turkey has a diverse geography. It is only slightly larger than the state of Texas in the US with a total area of 783,562 square kilometers (486,882 sq. mi.). Its land area is 769,632 km^2 (478,227 sq. mi.) and water 13,930 km^2 (8565 sq. mi.). About 3% of Turkey lies in Southeastern Europe (Thrace) and the remainder in Southwestern Asia also called Anatolia or Asia Minor. The sea surrounds Turkey on three sides. The Mediterranean Sea turns into the Aegean Sea along the west coast of Turkey, facing Greece. In the northern part of the Aegean, Çanakkale Bogazi (the Dardanelles) give passage to the Marmara Denizi (Sea of Marmara), which then opens into the Black Sea through the Istanbul Bogazi (the Bosporus) (CIA World Factbook 2019).

M. Thobaben (✉)
Humbolt State University, Arcata, CA, USA
e-mail: Marshelle.Thobaben@humboldt.edu

S. Kuguoglu
Department of Nursing, Istanbul Medipol University, Istanbul, Turkey
e-mail: skuguoglu@medipol.edu.tr

© Springer Nature Switzerland AG 2021
L. D. Purnell, E. A. Fenkl (eds.), *Textbook for Transcultural Health Care: A Population Approach*,
https://doi.org/10.1007/978-3-030-51399-3_28

The first historical reference to the Turks appears in Chinese records dating back around 200 BC, which refer to tribes called the Hsiung-nu (an early form of the Western term *Hun*). They lived in an area bounded by the Altai Mountains, Lake Baykal, and the northern edge of the Gobi Desert. They are believed to have been the ancestors of the Turks. In AD 552 many ethnic Turks began to converge under the Gokturks, and later under the Uygurs of Turkistan, followed by the Mongols. In the tenth century, Turkey became fully Muslim and accepted the Arabic script. Under the influence of the Muslim religion, Turkish language and literature were developed, and the building of mosques, schools, and bridges began (CIA World Factbook 2019).

The Seljuk Turks defeated the Christian Byzantine Empire in 1071, resulting in the first of the Christian crusades against Muslims. The Seljuks contributed to medical science and established medical institutions and hospitals in most cities. When the Seljuk Empire collapsed at the end of the thirteenth century the Ottomans established rule and in 1453 claimed Constantinople as the capital, renaming it *Istanbul*. The modern Turkish State is a descendent of the Ottoman Empire. Based on a tolerance of differences among its subjects, the Ottoman Empire endured for 600 years and at its height stretched from Poland to Yemen and from Italy to Iran.

In 1876 a constitutional monarchy was established under a sovereign sultan, but separatist movements, their subsequent repression, and an emerging Turkish nationalism resulted in the "Young Turk" revolution of 1908 and the erosion of the sultan's powers. During this time, modest advances in women's rights began, including the unveiling of nurses in the Balkan Wars and more educational opportunities for women.

An armistice at the end of World War I left the Empire stripped of all but present-day Turkey, occupied by Greek, French, British, and Italian armies, and established independence for Armenia and autonomy for Kurds in eastern Anatolia. However, the Treaty of Lausanne in 1923 officially ended Allied occupation, partitioning Armenia between Russia and Turkey, reinstating the Kurds, and proclaiming an independent Republic of Turkey, with Ankara as its new capital.

Although Westernization had begun before independence, Turkey's president, Mustafa Kemal Atatürk, became synonymous with Westernization and secularism. During his presidency from 1923 to 1938, he initiated many reforms, including banning the fez, outlawing polygamy, instituting marriage as a civil contract, abolishing communal law for ethnic minorities, removing Islam as the state religion, promoting nationalism and pride, instituting educational and cultural reforms, making surnames obligatory, changing the weekly day of rest from Friday to Sunday, and electing 17 female deputies to the National Assembly. Atatürk died on November 10, 1938, but he is still revered as the father of Turkey, and his image can be found in most government and public offices. Turkey remained neutral in World War II, but the post war economy and Cold War politics prompted U.S. economic and military aid in 1947, forging the political ties that endure today. Despite three bloodless military coups in 1960, 1971, and 1980, Turkey was a multiparty democratic system, a Republican parliamentary democracy until 2017 when the voters approved constitutional amendments changing Turkey from a parliamentary to a presidential system (CIA World Factbook 2019; Information Please 2019).

Turkey joined the United Nations (UN) in 1945, became a member of the North Atlantic Treaty Organization (NATO) in 1952, an associate member of the European Community in 1964, and began accession membership talks with the European Union in 2005. Voters approved a referendum in September 2010 that made several constitutional changes including Parliament having increased oversight and diminishing the power of the judiciary and the military; additionally, it provided wider democratic freedoms for Turkey's citizens (CIA World Factbook 2019; Information Please 2019). Turkey remains strategically important to the West and is a strong ally of the US because of its geopolitical location and its cultural and religious ties.

28.2.2 Heritage and Residence

The first national recorded population of the Republic of Turkey was 13.6 million in 1927. The population in 2018 was 81,257,239 with 24% of the population age 14 years and younger, and 8% age 65 years and older. Roughly, 70–75% of the population is Turkish, 19% Kurdish, and 7–12% other minorities. Approximately, 75% of the population lives in cities, such as Istanbul, Ankara, Izmir, and the remainder in villages (CIA World Factbook 2019; Turkey Statistical Yearbook [TSY] 2019). The capital city of Turkey is Ankara, but the historic capital, Istanbul, remains the financial, economic, and cultural center of the country.

Until the 1950s most Turks were peasants living in isolated, self-sufficient villages with their extended family and practical folk-belief system. Depeasantization, migration, and urban settlement have continued, and today squatter housing districts populated by rural "immigrants" in major cities have resulted in permanent low-income neighborhoods juxtaposed against modern urban development. Changes in the social structure and people's expectations are also shifting. For example, older people's ability to live in their familiar housing environments, particularly, in large cities and metropolitan areas is forcing the government to change its policy and to strive to provide affordable housing and care centers for them (Turel 2009).

Over the past two decades, Turkey has been hit by several moderate to large earthquakes that resulted in a significant number of casualties and heavily damaged or collapsed buildings. This has been as a result of inadequate seismic performance of multistory reinforced concrete buildings, typically three to seven stories in height. A recent study indicates that a considerable portion of existing building stock may not be safe enough in Turkey (Inel et al. 2008).

As a result of extensive foreign trade, larger coastal cities are undergoing many changes, which have resulted in an urban environment with a dual character, representing the traditional old way of life and the ensuing new class. Every aspect of life and society is being affected, including changes in values, recreational activities, mass communication and media, and women's status.

Observations suggest that everyday practices of the people, as well as their folk beliefs, are truly changing. However, the Turks still depend on nuclear and extended family and friends for adjustment, job possibilities, and money.

28.2.3 Reasons for Migration and Associated Economic Factors

The U.S. Census Bureau (2017) reported 117,366 people of Turkish descent living in the US. However, according to conservative estimates, there are approximately 350,000–500,000 Americans of Turkish descent living across the US. The largest concentration of Turkish Americans is in New York/New Jersey and in California (TCA 2019).

The Turkish immigrant population in the US differs significantly from most of the Turkish population that inhabits Europe in terms of both demographic makeup and socioeconomic status and integration. A high proportion of Turks come from the elite and upper-middle classes, interspersed with smaller groups of middle-class students and skilled laborers who are supported privately or by the government.

Economic reasons, such as unemployment and poor salaries, are the major reasons Turks leave to work in other countries (İcduygu 2008). Although Turks have emigrated throughout the world, many have lived in Western Europe since the 1960s and in 1970s North Africa and the Middle East, largely as a result of "guest worker" programs. Since the 1990s, Turkish workers have also moved to the neighboring former communists' countries such as Russian Federation and Ukraine (İcduygu 2008). A large "Turkic belt" stretches from the Balkans across Turkey, Iran, Central Asia, the former republics of the Soviet Union, and deep into the borders of Mongolia. This belt includes many ethnic Turks who may share cultural, linguistic, religious, and certainly historical links with the people of Turkey.

Research studies have indicated that any concern about excessive "brain drain" from the immigration of some of Turkey's intellectual, academic, and other highly skilled professionals to the US is unfounded; it has not created a threat to Turkey's economic, scientific, social, and cultural development. Additionally, Turks living in the US, and elsewhere usually maintain strong bonds with their Turkish families and pass on their Turkish cultural values, traditions and language to their children born in the United States (Köser-Akçapar 2006). Turks who have lived or studied in the US generally have higher status and greater employment opportunities in Turkey.

28.2.4 Educational Status and Occupations

Education is highly valued in Turkey by all socioeconomic groups. Coeducational primary and secondary education is provided at no cost and is guaranteed under the Constitution. It consists of public and private school at all levels, ranking from pre-primary (1 year), primary (8 years), high school (4 years), and universities (4–6 years). In 1997, 5 years of compulsory primary school was extended to 8 years including the middle schools. Primary school starts at age of 7 and ends at 13. High schools were extended from 3 to 4 years in 2005. High school includes a number of options, including general, technical, trade, vocational, and theological training. Higher education institutions include universities, faculties, institutes, higher schools, vocational higher schools, conservatories, and research and application centers. Students who wish to pursue a university education must take a state examination that determines both their admission to the institution and their subject of study. In a recent study it was reported that only 22% of the students who took the nationwide competitive entrance examination were placed in a university program in Turkish universities. Turkey's university distance education program, one of the largest in the world, annually accepts only about 15% of students who apply (Tasçı and Oksuzler 2010).

A high level of education exists among people of Turkish descent living in the US. Significant numbers hold advanced degrees, and most are employed in professional, managerial, and technical occupations.

Turkey Statistical Yearbook (2019) reported that of the 48% of Turkey's working age population who participated in the labor force, 70.5% were men and 26% were women; 45.8% worked in urban areas (69.9% male and 22.3% female) and 52.7% in rural areas (72% male and 34.6 female). Of those workers employed in the agriculture sector, 46% were unpaid family workers; 76.9% of the unpaid family workers were female, while 23.1% were male. The unemployment rate was estimated to be 14% in 2009 (Turkey Statistical Yearbook (TSY) 2008).

Persons not in the labor force composed 52.1% of the working-age population. The main subgroups were persons who were busy with household chores (44.9%), students, and disabled and retired persons (TSY 2008). For cultural reasons, many women have continued to maintain their traditional roles and do not work outside the home because it interferes with their household responsibilities, including caring for their children, and it may require them to work with men from outside their immediate family.

28.3 Communication

28.3.1 Dominant Language and Dialects

A Uralic-Altaic language, Turkish is spoken by 90% of the population. The Turkish language has approximately 20 dialects, including Yakut, Chuvash, Turkoman, Uzbek, Kazakh, and the language of the Gagavuz people. Differences in some of the dialects are so great that they are considered separate languages.

Through the centuries, Turks borrowed from Arabic and Persian languages, and bits of "Turkified" French and English can also be found. Until 1928, Turkish was written in Arabic script, but under Atatürk's direction, a Turkish alphabet was developed based on Latin script.

The Turkish alphabet is much like the English alphabet, although it does not have a "w" or an "x," and additional sounds are symbolized by an "ı" without a dot; a "ğ," an "ö," and a "ü" with accents; and an "ş," and a "ç" with a cedilla, symbolizing "sh" and "ch," respectively. The Turkish language does not distinguish gender pronouns such as "he" from "she" or "her" from "his"; therefore, Turks learning English may inadvertently confuse these pronouns. However, Turkish does distinguish a formal from an informal "you," signifying the importance of status in Turkish society.

Typical of many Mediterranean cultures, speaking in loud voices is common; this may not signify anger, but rather excitement or deep involvement in a discussion. It can be common for more than one person to speak at the same time or to interrupt another person, which is not necessarily considered rude. However, someone of lower status should not interrupt someone of higher status.

28.3.2 Cultural Communication Patterns

In the Turkish's culture group affiliation is valued over individualism. In fact, identity may be determined by family membership, group, school, or work associations. An individual's behavior is expected to conform to the norms or traditions of the group, and Turks tend to be more people and relationship oriented than Americans. Although Turks may take longer than Americans to form friendships, these relationships last longer, formality is decreased significantly, and interdependence is encouraged as a source of strength. In this group-oriented culture, Turks generally do not desire much privacy and tend to rely on cooperation between family and friends, although competition between groups can be fierce.

Turks value harmony over confrontation. However, Turkish communication style is characteristic of Mediterranean cultures in which the outward show of feelings is less restrained. For women, expressions of anger are usually acceptable only within same-sex friendships and kinship networks or toward those of lower social status. Generally, women are not free to vent their anger toward their husbands or other powerful men.

Children are very accustomed to being held, hugged, and kissed by family and friends of the family. Touching, holding hands, and patting one another on the back are acceptable behaviors between same-sex friends and opposite sex partners. It is common to see same-sex friends, especially among the older generations, holding hands or linking arms while walking. Likewise, personal space is closer between same-sex friends and opposite-sex partners; physical proximity is valued as a sign of emotional closeness. Very strict Muslims generally do not shake hands or touch members of the opposite sex, especially, if they are not related. Health-care providers are usually looked upon as professionals and touch is allowed and expected when necessary.

Eye contact may be used as a way of demonstrating respect. When interacting with someone of higher status, a person is expected to maintain occasional eye contact to show attention; however, prolonged eye contact may be considered rude or interpreted as flirting.

Turkish people tend to dress formally; men wear suits rather than sports jackets and slacks on social occasions. Women tend to dress modestly and wear skirts and dresses rather than slacks. Black clothing accented with gold jewelry is quite popular. More traditional Muslim women may wear very modest clothing and cover their heads with a scarf, either black or a colorful print. However, styles continue to change, and denim jeans and casual dress are becoming common among young people for less formal occasions.

Turks tend to openly display emotions such as happiness, disgust, approval, disapproval, and sadness through facial expressions and gestures. Two unique gestures in Turkish culture include signals for "no" and signs for approval or appreciation. "No" is indicated by raising the eyebrows or lifting the chin slightly while making a snapping or "tsk" sound with the mouth. Appreciation may be expressed by holding the tips of the fingers and thumb together and kissing them. This signal is commonly used to express appreciation for food.

Various phrases are commonly used by Turks. *Allahaismarladik* (God watch over you) is said to someone leaving and is responded to with *gule gule* (go with smiles). *Ellerine saglik* (health to your hands) communicates appreciation for a good meal, and the cook responds with *afiyet olsun* (good appetite). *Cok yasa* (live long) is said after someone sneezes with a response of *sen de gor* (you see a long life, too). *Masallah* (God protect from the evil eye) is said, for example, when one has a healthy baby or when one has achieved something good, whereas ***insallah*** (God willing) is said when something is wished to happen.

Turkish people take pride in keeping their homes immaculately clean, and one is expected to remove one's shoes inside the home. Most hosts in Turkey and many in the US and elsewhere offer slippers to their guests. Whether wearing shoes or not, showing the sole of one's foot is considered to be offensive in Turkish culture. Women are expected to sit modestly with knees together and not crossed.

Tortumluoğlu et al. (2005b) conducted a qualitative study in a village in eastern Turkey by examining individual cultural communication characteristics. Comments from the participants included the following:

- According to our religion, men who are not our legal husband are not allowed to listen to our voices. A woman cannot speak out loud and cannot laugh in the community; she would be like bad woman (woman over 65 years old).
- We do the duties of the bride (act like a servant) for our husband's relatives and mother-in-law. We can never speak near them. To speak would be disrespectful (bride, 15 years old).
- If the person across from us is a woman, we hug and kiss, but if he is a man outside the family, we don't touch him. We will not eat at the same table with men outside the family; we won't be together with them at weddings; we won't sit next to them on the bus; we wouldn't go near them without covering our heads and bodies. We don't look them in the eye (women over 65 years old).
- We're uncomfortable being examined or given a shot by a man because they are strangers. It is very sinful to go to a man even our husband and open what is covered (female, age unknown).
- I would not take my wife, daughter, or daughter-in-law to a male doctor. I would not show them to unrelated men (man over 65 years old).
- Even if I knew my wife would die, I wouldn't take her to a male obstetrician (boy about 15 years old).

28.3.3 Temporal Relationships

Turks tend to have a relaxed attitude about time; social visits can begin late and continue well into the night. Whereas punctuality in social engagements is not highly important, in business relationships punctuality among Turkish Americans and Turks in other individualistic cultures is gaining in importance.

28.3.4 Format for Names

Turks value status and hierarchy. Demonstrating respect for those of higher status is mandatory and determines the quality of interactions with a person. A variety of titles are used to show respect and acknowledge status. Strangers are always greeted with their title, such as *Bey* (Mr.), *Hanim* (Mrs., Miss, or Ms.), *Doktor* (Dr.), or *Profesör (Professor)*. Members of the family are also addressed using specific titles that recognize relationships, such as *agabey* (older brother or older close male friend), *amca* (uncle or elderly male relative or stranger), *abla* (older sister or older close female friend), *teyze* (maternal aunt or older female relative or older female stranger), and *yenge* (wife of a brother or paternal uncle).

When friends or family members greet, it is customary for each to shake hands and to kiss one another on each cheek. Traditionally, when greeting someone of very high status or an elderly person, one might grasp his or her hand and kiss it and then bring it to touch one's forehead in a gesture of respect.

28.4 Family Roles and Organization

28.4.1 Head of Household and Gender Roles

In a very traditional Turkish home the father is considered the absolute ruler. The concept of *izin* (permission to leave to do something specific) captures this significance. In rural and traditional families, women may require *izin* from the head of household for doing simple things, such as shopping, traveling, or visiting their nurse midwife, physician, or dentist. The justification is that the one who earns the money may spend the money. The person who bestows *izin* is responsible for the protection of the *izinli* (person who requires the *izin*). *Izin* exhibits a structure of authority that is both hierarchical and patriarchal; therefore, women typically require *izin* more often than do men. A young wife (*gelin*) may require *izin* from her husband and from her mother-in-law. All are ultimately responsible to the *gelin*'s father-in-law, who is usually the absolute ruler of the traditional extended family (Tortumluoğlu et al. 2005a, b).

Less-traditional families show more equality between spouses, especially in nuclear families in which the wife is well educated and works outside the home. Yet remnants of traditional family structure prevail and the husband often takes on the role of ultimate decision maker, especially in matters of finance. Women may work full time outside the home in addition to assuming full responsibility for running the daily activities inside the home.

Modern Turkish women tend to be more Westernized than some of their Middle Eastern or Muslim counterparts. The first institution for higher learning for women in Turkey was established in 1910. In 1917, women earned the right to divorce and to reject polygamous marriage. Atatürk's new republic abolished the old legal system based on religion and secularization, giving women equal rights to education and no longer requiring them to wear veils and long overgarments. Legal marriage does not permit polygamy, although some may practice it outside the law. Women have had the right to vote since the early 1930s. In 1966, a charter of the International Labor Organization passed the equivalent of an Equal Rights Amendment, requiring equal wages to both sexes for work of an equal nature.

28.4.2 Family Goals and Priorities

A woman's age and the number, age, and gender of her living children can influence her status in the family and the community but varies depending upon such things as education, religious practice, socioeconomic level, urbanization, and professional achievement. Generally, a young *gelin* (woman aged 15–30 years) has the lowest status, middle-aged" woman (30–45 years) has intermediate status, a "mature" woman (45–65 years) has the highest status, and an "old age" woman (65 years or older) is highly respected but not very powerful. Working outside the home is associated with status positively in the urban context and negatively in the rural context. Professional employment and education raise the status of women. Thus, health-care providers may find significant variations regarding gender roles when working with Turkish patients.

28.4.3 Prescriptive, Restrictive, and Taboo Behaviors for Children and Adolescents

Children are held very dear in the Turkish family, and they are expected to act as young children, not small adults. They are accustomed to receiving attention from family, friends, and visitors. Kissing children and pinching their cheeks is quite common.

Once children enter school, they are expected to study hard, show respect, and obey their elders, including older siblings. This concept is referred to as ***hizmet*** (duty or service). As children age, they are socialized into more-traditional gender roles. Girls are expected to help care for younger siblings, to help at mealtimes, and to learn to cook. Traditionally, children are not allowed to act out or talk back to their superiors. Light corporal punishment is generally acceptable. Circumcision is a major rite of passage for a male child. This is a time of celebration within the extended family, and newly circumcised boys are honored with gifts. Traditionally, boys can be circumcised up to the age of about 12, although the modern trend is to perform the circumcision in the hospital shortly after birth.

Rankina and Aytaç's (2008) research found that the religiosity of the parents, the vast majority of whom were Muslim, had no effect on the schooling of Turkish children, whether male or female. In contrast, patriarchal family beliefs and practices discouraged the education of children, particularly girls. Their findings also showed a father's disapproval of daughters going out in public without a headscarf reduced the likelihood of girls finishing or going beyond primary school. Thus, family cultural traits may continue to represent a significant barrier to gender equality in education (Rankina and Aytaç 2008).

As children reach adolescence, they are expected to continue to work hard in school and show respect for superiors. The US and Western culture and lifestyles are exported to Turkey via the various social networks. O'Neil and Güler (2010) explored the meaning high school and university students attached to American popular culture and found no evidence that that American popular culture was in danger of overwhelming Turkish culture. Young adults like to move back and forth between indigenous and foreign products, including American ones, and as a result the researchers felt this continued to embody a multiplicity and hybridity that has characterized Turkish culture for centuries.

Young people in the urban areas may talk more about sex and engage more freely in sexual activity than previous generations; however, sexuality largely remains a taboo and is regarded as a forbidden topic for social and cultural reasons. Though not common among rural Turks, urban adolescents are beginning to date in pairs, in addition to the more traditionally accepted practice of group outings. However, sexual interaction is strongly discouraged among youth and the unmarried, especially young women. Virginity in unmarried women is a strong cultural value. According to a study conducted with university students in Turkey, 82.4% of female students and 86.5% of male students were virgins when they married, because of social rules and religious beliefs. Sixty-two percent of female students practiced sexual abstinence (Tortumluoğlu et al. 2006).

Parents are expected to provide sexual education within the family but often have insufficient knowledge on the subject. Kukulu et al. (2009) recommended that structured sex education that incorporated knowledge of specific aspects of the Islamic culture experience would help to promote healthy sexual behavior and decrease sexual myths, such as marrying a virgin increases sexual satisfaction.

Successful completion of high school or university education is a first step toward adulthood. Although education earns respect in the family, the concept of *hizmet* still applies. A further step for men is the completion of required military service (*askerlik*), the duration of which varies depending on the population and the needs in Turkey. In addition, employment and earning money are symbols of adulthood for both men and women.

Marriage is perhaps the most important developmental task for adulthood. Young people generally live in their parents' home until they are married, unless school or work necessitates other arrangements. This practice may be quite different among assimilated Turks in America and other countries. The Turkish word for marriage, ***evli***, translates to "with house." Family remains an important factor in marriage. Marrying into a "good family," having a high-status occupation, and achieving wealth are means of attaining higher social status for both the individual and the entire family. Family members' accomplishments raise the entire family's status, whereas failures have an equally broad effect. Thus, individuals must always consider what impact their actions will have on the family. Often, they consult parents or other family members before making major decisions.

Arranged marriages occurred most often among less-educated, older individuals. Family initiated marriages range from rare contractual agreements between parents to the relatively common introduction and gentle encouragement of a newly formed couple. The more traditional family will "choose" a spouse for a son by considering the individual's personality, talents, and appearance. For a daughter, it is more important to consider the individual and his family because she marries *into* the husband's family.

Elders are attributed authority and respect until they become weak or retired, at which time their authoritative roles diminish. However, respect always remains a factor.

Although financial independence is valued in Turkish culture, independence from the family is not encouraged. Adult children, especially men, remain an integral part of their parents' lives and parents expect their children to care for them in their old age which is regarded as normal, not as an added burden. Grandparents play a significant role in raising their grandchildren, especially if they live in the same home.

The extended family is very important. Even the apparent increase in nuclear households does not rule out the networks among closely related families. Whether or not they live under the same roof, a young family may still live under the supervision of the husband's parents or, at least, maintain an interdependent relationship. In many Turkish families, aunts, uncles, cousins, and in-laws form the extended family. Visits with local relatives are assumed and mandatory when traveling. Extended family members have a social relationship and may also play an authoritative role within the network. A cooperative relationship, which includes sharing child care, labor, and food, when necessary, and providing companionship, is essential between women in an extended family or neighborhood.

28.4.4 Alternative Lifestyles

Divorce is becoming more common, but it remains socially undesirable, especially for women, for whom remarriage opportunities may be limited to divorced or widowed men. Widows, however, are generally taken care of by their late husband's families and depending on their age and socioeconomic background may have the option to remarry. Premarital cohabitation and unwed motherhood are strongly discouraged, especially among more-traditional families, although living together before marriage is not uncommon in larger cities and among immigrant Turks.

Even though being a gay man or lesbian is not a crime or considered a disease, homosexuality is only beginning to be received "at a distance." In Oksal's (2008) study of familial patterns of attitudes toward lesbians and gay men, he found that young adults' attitudes toward lesbians and gay men were more liberal than those of their parents. However, on the whole, Turkish family members have quite negative attitudes toward homosexuality, most likely linked to religious beliefs. Most Turkish people are in agreement with Islamic values that regard homosexuality as a sin and unacceptable (Oksal 2008).

28.5 Workforce Issues

28.5.1 Culture in the Workplace

Because Turkey is a group-oriented culture, the Turkish workplace may be more team oriented than in the US and other individualistic cultures. Turkish relationship orientation may lead to dependence on personal contacts and networks to accomplish tasks, and from the American perspective, developing these relationships and networks may appear as nepotism or as too much socializing. In contrast, the Turkish immigrant employee may not feel a sense of belonging in a less relationship-oriented work milieu.

Hierarchical structure is highly pervasive throughout Turkish culture and the workplace is no exception. Turkish employees expect an authoritative relationship between superior and subordinates. However, indirect criticism is expected and appreciated in order to "save face." A Turk may be highly offended if openly criticized, especially in front of other people. They may be reticent about asking questions for fear of exposing a lack of knowledge. Yet, Turks may exhibit modesty when applying for a job or a promotion relying more on the recommendations of others than on pointing out their own strengths.

Because military service is mandatory for men who wish to maintain their Turkish citizenship (even those living abroad), young Turkish men who reside outside Turkey may need to take an extended leave to complete their military service.

28.5.2 Issues Related to Autonomy

Because most Turkish immigrants speak English, language barriers in the workplace may be only subtle. However, dealing with differences of opinion between parties of equal hierarchical level may present difficulty. Turks perceive that aggressive face-to-face confrontation may cause relationships to deteriorate; therefore, the dominant means of conflict resolution is collaboration reinforced by compromise and forcing. Compromise and avoidance behaviors are more likely among peers, whereas accommodation behaviors are used with superiors. Their way of handling differences of opinion is brisk and clear-cut when an authority relationship exists between the two parties.

Turkey is known for its high-power distance (the psychological and emotional distance between superiors and subordinates), respect for authority, centralized administration, and authoritarian leadership style. In Turkish culture a manager's authoritative control is often more important than the achievement of organizational goals.

28.6 Biocultural Ecology

28.6.1 Skin Color and Other Biological Variations

The Turkish population is a mosaic in terms of appearance, complexion, and coloration because of historical migration and inhabitance patterns. Appearances range from light-skinned with blue or green eyes to olive or darker skin tones with brown eyes. Mongolian spots, usually found at or near the sacrum, are common among Turkish babies and should not be confused with bruising. Racially, 75.6% of the men and 77.7% of the women are in the *brakisefal* (having a short, broad head) category, which is a shared symbol of the Dinaric Alpines (Gültekin and Koca 2003).

28.6.2 Diseases and Health Conditions

According a recent health survey (TSY 2008) that was sent to all settlements in the territory of the Republic of Turkey, 71.9% of men and 55.5% of women stated their general health status was good or very good; 75.2% of men and 58.8% of women living in urban areas stated their health was good or very good; while only 63.8% of men and 48.4% of women in rural areas stated their health status was good or very good (TSY 2008). Life expectancy at birth in 2018 was estimated to be 75.3 years for the total population (72.9 years for males and 75.3 years for females) (CIA World Factbook 2019).

The leading causes of death include major vascular diseases (ischemic heart disease, stroke), chronic obstructive lung disease and lung cancer in men, perinatal problems, lower respiratory infections, and diarrheal diseases. Injuries cause about 6–8% of deaths, although this may be an underestimate (Akgun et al. 2007). There is also a high prevalence of obesity, hypertension, and diabetes, especially in Turkish women.

Lactose intolerance rises among populations farther south and east in Europe. The Black Sea region tends to have a relatively high incidence of helminthiasis (intestinal worm). Endemic goiter associated with iodine deficiency, despite iodine prophylaxis (ID), still exists in 27.8% of the Turkish population. It has been eliminated in most of the urban population; however, it is prevalent in rural areas and in particular geographical regions (Erdoğan et al. 2009). Tuberculosis continues to be prevalent in the Aegean areas and in southeastern Anatolia.

Behçet's disease (BD) is a systemic inflammatory disorder of unknown etiology with a strong genetic component. It is characterized by recurrent attacks of oral aphthous ulcers, genital ulcers, skin lesions, uveitis or other manifestations affecting the blood vessels, gastrointestinal tract, and respiratory and central nervous system; the inflammatory lesions at particular sites, such as the eyes, brain, or major vessels can result in permanent tissue damage and cause chronic manifestations or even death (Gul 2007). It is prevalent in Japan and China in the Far East to the Mediterranean Sea, including countries such as Turkey and Iran, and usually starts in the second and third decade of life. The male-to-female ratio is approximately equal, although BD runs a more severe course in men and in those aged <25 years at onset (Gul 2007).

Beta thalassemia is the most common inherited blood disorder in Turkey and represents a major public health problem. It is characterized by reduced or absent beta globin gene expression. Beta thalassemia, alpha thalassemia, and sickle cell anemia are also the most common hemoglobinopathies in Turkey. Although the overall frequency of beta thalassemia in Turkey is 2%, there are significant regional differences. The incidence of beta thalassemia in the Denizli province is estimated between 2.6 and 3.7% (Bahadir et al. 2009).

The Turkish government has national healthcare prevention programs for communicable diseases, tobacco control, cardiovascular diseases, chronic respiratory diseases, cancer, and a newly developed Obesity Prevention and Control Program (Ministry of Health of Turkey 2010).

Because of the diversity in climate, topography, and culture in Turkey, it is essential to ascertain the specific geographic origin of a Turkish immigrant. Health-care providers may need to assess newer Turkish immigrants for tuberculosis, malaria, or other potential health problems found in Turkey.

28.6.3 Variations in Drug Metabolism

The literature reports no studies regarding variations in drug metabolism and interactions for Turks. Given the diversity of ethnicity, one cannot extrapolate data from other ethnic or minority groups and apply them to Turkish peoples. This is one area in which research is needed.

28.7 High-Risk Behaviors

Beser et al. (2007) studied health-promotion lifestyle profiles of 264 Turkish workers to determine the factors that affect their lifestyles. The research found that the workers did not have the desirable degree of health responsibility because they did not consider health controls as a necessity to lead a healthy life. If individuals can do their daily routines and if their health does not prevent them from going to work, they do not consider themselves ill. The workers obtained the highest scores on interpersonal support (family members support each other during difficult times) which is the hallmark of Turkish culture.

There is a significant risk among farmers exposed to exogenous carcinogens such as artificial fertilizers and insecticides. Cigarette smoking is widespread in Turkey and tends to start at an early age. Turkey, a major producer of tobacco in the world, has instituted very limited antitobacco activities. Passive smoke has been associated with an increased incidence of asthma and allergic diseases among Turkish children.

Despite stereotypes promoted in the American film *Midnight Express*, drug use is not common among mainstream Turks. They tend to consume less alcohol than Americans or Europeans, perhaps as a result of the Muslim culture that discourages more than moderate alcohol use. In general, it is more acceptable for men than women to drink alcohol; however, this is becoming less so as Turkey becomes more Westernized. According to the Who Health Organization (2019) in 2016 82.7% males, and 95.1% females were lifetime abstainers from alcohol (WHO 2018).

There is a high risk for sexually transmitted diseases in Turkey. The tendency of men in Turkey, which has a particularly young population, to view themselves as strong and immune to disease and the positive view of men in traditional

Turkish culture to have sexual relationships with more than one woman increases the danger for both the man and his wife.

28.7.1 Health-Care Practices

Health beliefs and behaviors of the Turkish immigrant may vary according to the variant cultural characteristics (see Chap. 1). Health-seeking behaviors promoted by the Turkish government include a strictly enforced law requiring the wearing of seat belts in motor vehicles. Helmet laws for motorcycle drivers have not been instituted.

Aerobics studios and athletic facilities exist in major urban areas, but the idea of cardiovascular fitness is relatively new, and Turks may be more likely to seek outdoor activities such as picnicking or, among men, playing soccer. However, because many Turks, especially in rural areas, do not have modern conveniences such as elevators in apartment buildings, automobiles, or clothes dryers, their daily life inherently requires more caloric output than life in the US, an important issue to keep in mind when adjusting to life in the US and other countries.

28.8 Nutrition

28.8.1 Meaning of Food

Turks take great pride in the fact that French, Chinese, and Turkish cooking are reportedly the three foremost cuisines in the world. Turkish cuisine is influenced by the many civilizations encountered by nomadic Turks over the centuries, as well as by a mixture of delicacies from different regions of the vast Ottoman Empire. Therefore, food choices are varied and tend to provide a healthy, balanced diet.

Food is a highly valued symbol of hospitality that communicates love and respect to those for whom it is prepared. Whereas a typical family dinner may be simple, guests are generally served a bountiful array of dishes—more food is always better. Tea and a snack are always on hand for visitors. Dinner guests may have difficulty finishing everything on their plates because hostesses may relentlessly offer to replace what has been eaten. Polite guests refuse the first offer, but the hungry need not worry—offers are made again and again.

Food is generally presented in an appetizing manner, and many foods have names intended to be enticing or, at least, entertaining. For example, *kadinbudu köfte* translates as "lady's thigh meatballs"; *imambayildi*, or "the priest fainted," is an eggplant dish with lots of garlic; and *asure*, or "Noah's pudding," is a dessert in which more than two of everything is included.

Turks typically eat their evening meal later than most Americans, at about 8 p.m., something health-care professionals may need to take into consideration when teaching Turkish patients about medication therapies.

28.8.2 Common Foods and Food Rituals

Turkish cooking is quite delicious, not terribly spicy, and prepared artfully and fastidiously, because Turkish appetites tend to be discriminating. Breakfast is typically a simple meal of white feta cheese (*beyaz peynir*), olives, tomatoes, eggs, cucumbers, toast, jam, honey, Turkish sausage (Turk *sucugu*), and Turkish tea. Hot midday or evening meals may include any of the following foods:

- *Çorba* (soups) range from light to substantial. *Meze* (hors d'oeuvres) include a great variety of small dishes, either hot or cold, such as *yaprak dolma* (sarma) (stuffed grape leaves in olive oil), olives, circassian or *çerkez tavuğu* (chicken with walnut sauce), *çiroz* (dried mackerel), *leblebi* (roasted chick peas), or *sigara böreği* (a savory cheese pastry fried until crispy). *Meze* may be accompanied by *rakí*, traditional anisette liquor distilled from grapes that is served with water over ice and drunk slowly. Sharing a glass of *rakí* is usually toasted with the phrase *Şerefe* (to your honor).
- Salads include lettuce, tomatoes, cucumbers, onions, and other raw vegetables with a dress-

ing of olive oil and lemon juice or vinegar. Olive oil and lemon are staples in Turkish culinary preparation.

- Turks generally prepare meat in small pieces in combination with vegetables, potatoes, or rice. Famous Turkish cuisine includes köfte, small spicy meatballs, and kebab, **skewered** beef or lamb and vegetables. Whereas poultry is less common, fish has a special place in Turkish cuisine because of its variety, freshness, and availability.
- Turkey is the birthplace of yogurt, which is an essential part of the Turkish diet and is generally served with hot meals rather than as cold breakfast food.
- With the abundant produce in Turkey, vegetables play a large role in the Turkish kitchen. Vegetables are served cooked or raw, hot or cold, as part of a stew or casserole, or stuffed (*dolma*) with meat, rice, and currants.
- Rice and *börek* are important parts of Turkish culinary tradition. *Börek* is made by wrapping *yufka* (thin sheets of flour-based dough) around meat, cheese, potato, or spinach and then frying or baking until the dough is flaky.
- Turkish desserts fall into four categories: rich and sweet pastry, such as *baklava*; puddings; *komposto* (cooked fruits); and fresh fruits. In fact, most meals conclude with fresh fruit and coffee or tea.
- Turkish *kahve*, from which the English word "coffee" is derived, is famous for its dark, thick, sweet taste. Cooked with a *cezve* (coffee pot), it is served in small, demitasse cups. Coffee grounds left in the cup can foretell one's future. Turkish *çay* (tea) is prepared using a two-tiered *çaydanlik* (teapot), allowing the hostess to serve guests according to their preference for *koyu* (dark) or *açík* (light) tea. *Ayran*, a mixture of yogurt and milk, is the national cold drink and is drunk by children and adults alike.

The Muslim religion requires abstinence from eating pork and drinking alcohol, but not all Muslims abstain, depending on their degree of religious practice. Given the diversity of food options for Turks in America and other countries, health-care providers need to provide dietary counseling according to the individual's unique food choices and practices.

The Islamic tradition of **Ramazan or Ramadan** is a month of fasting (*oruç tutmak*) observed by practicing Muslims throughout the world. During Ramazan, one is not allowed to eat or drink anything from sunrise to sunset as a test of willpower and as a reminder of the preciousness of the food provided by a gracious Allah (God). Many Muslims also stop smoking during this month. Delicious unleavened bread called *pide* is sold everywhere only during Ramazan. Observance of this tradition varies from some not observing it to others who strictly follow the ritual and do not bring *anything* to their mouth during daylight hours. Sunni Muslims, the majority of Muslims in Turkey, start practicing Ramazan at age 10 or 11, and some believe that women have the duty to fast even during pregnancy and the postnatal period. Generally, pregnant and postpartum women, travelers, and those who are ill are excused from fasting, but they may be required to make up lost time at a later date.

Ramazan is determined by the lunar calendar and, therefore, can take place at various times in the year. Typically, Turks who are fasting eat breakfast, or *sahur*, before dawn and before *ezan* (the call to prayer). The evening meal, *iftar*, is something all look forward to with great anticipation, and Turkish women who almost invariably do all the cooking create veritable feasts each night. This is a time to visit with friends and relatives, so dinner invitations abound during Ramazan. In a sense, Ramazan is a spiritual and physical cleansing that brings the community together.

Despite fasting many Turks actually gain weight during the month because of their inactivity during the day and eating well at the end of each day. Fasting also can cause a variety of digestive problems and may endanger the health of a pregnant or postnatal woman and her baby. Health-care professionals should provide factual information regarding these issues.

Another holiday based on Islamic practice is the *Kurban Bayram* or sacrificial holiday. In Turkey, an animal such as a goat or sheep may be butchered and the meat divided and distributed to the poor. New Year's Day is celebrated much like American's Christmas, with a large feast including a turkey dinner, the exchange of gifts, sometimes a tree, and socializing with family and friends.

28.8.3 Dietary Practices for Health Promotion

Traditional dietary practices linking food to health have carried through to the modern day, even among highly educated Turks. Molasses and *baklava*, *lokum* (Turkish delight), tahini, and honey and nuts and raisins are believed to increase strength and sexual vigor. Fruits, especially bananas, oranges, tangerines, and apples, are brought to convalescing people helping them to regain their strength and aid in the healing process. Milk, which is not commonly drunk by adults, is considered more medicinal than yogurt. Chicken soup is a common remedy for cold and flu symptoms. An ***ebe kadin or kadin ana***, a traditional midwife or healer in Turkey, relies on various herbs and home remedies to heal patients. *Ebegömeci*, a spinach-like leaf or herb, may be prepared for topical or oral use to treat inflammation, infection, and sometimes infertility. *Ihlamur* tea, *tarçin* (cinnamon), *kant* (hot sugar water), ginger, mint, and various roots are used separately or in various combinations to treat rheumatism, low blood pressure, intestinal gas, and colds and flu. Nettles may be used topically for rheumatism, arthritis, and varicose veins.

A folk remedy for diabetes involves boiling olive leaves and, after refrigerating, drinking the juice. *Lapa*, a watery rice mixture with a gruel-like texture, or a boiled potato may be used to treat diarrhea; this is followed by yogurt to replace the natural flora of the intestines. Health-care professionals need to ask Turkish patients if they are using folk dietary practices and may incorporate these into prescription therapies. Some traditional Turkish foods unavailable in most parts of the United States include *pastirma* (Turkish version of pastrami), *sucuk* (Turkish sausage), and various types of cheeses such as *kaşar*. *Yufka* is an essential ingredient in many Turkish recipes such as *börek*, although phyllo dough may be an adequate substitute.

28.8.4 Nutritional Deficiencies and Food Limitations

Population groups at greatest risk for malnutrition are preschool children, female adolescents, mothers, and the economically disadvantaged. Health-care providers may need to consider extensive nutritional assessments for more recent Turkish immigrants.

Another important health problem is rickets, caused by vitamin D deficiency. Rickets is seen more frequently in children under 2 years old and has a 6–20% prevalence, with the leading causes being not taking children outside into the sunshine and not feeding children sufficiently from the dairy food group. Cinar et al. (2006) reported that 62% of mothers who had children from 0 to 12 months old in Sakarya, Turkey, believed that sunlight was "harmful" for their children; however, the majority (80%) of mothers named one benefit a child received from intentional baby sunning. "Sun causes cutaneous diseases" was the most frequently cited harm (Cinar et al. 2006). Other prevalent nutritional problems in some communities in Turkey arc skin, mucosa, eye, and lip symptoms from riboflavin (vitamin B_2) and vitamin A deficiencies and bleeding gums from vitamin C deficiency.

Turks who are Muslim are forbidden to eat meat from a carcass, blood, pork, or the meat of animals sacrificed in the name of anyone other than Allah. The list of forbidden animals also includes those with tusks, wild game, and those torn apart by wolves, bears, dogs, squirrels, and foxes. The meat of birds that hunt with their claws is also forbidden by religious leaders. Additionally, the list of animals includes animals such as snakes, frogs, turtles, and crabs (Meals in Koran 2006).

Reflective Exercise

Mr. Yılmaz, a 61-year-old Turkish man, is brought to the emergency room by his wife. He complains he has had nausea and vomiting for 2 days and has been confusion. His blood glucose is 796 mg/dL. He was given intravenous regular insulin (Novolin R) to stabilize his blood sugar. He was unaware that he was a diabetic. He mentioned that he likes Turkish desserts, particularly, the rich and sweet pastry, such as *baklava* and puddings. He also mentioned that he practices folk remedies for illnesses. He will require teaching for his newly diagnosed diabetes.

1. What questions should the nurse ask Mr. Yilmaz about his cultural practices related to the traditional diet of Turkish people?
2. Mr. Yilmaz believes olive oil is all he needs to keep his diabetes under control. Considering this and other potential cultural beliefs, how should the nurse incorporate them into his diabetic regime?
3. Mrs. Yilmaz believes she cannot go against her husband's wishes even if he will not follow his diabetic regime. The nurse recognizes that for some Turkish women, the man has the final word. What advice should she be given?

28.9 Pregnancy and Childbearing Practices

28.9.1 Fertility Practices and Views Toward Pregnancy

In 1975 approximately five children were born to each Turkish woman, but by 2018 that figure dropped to 2 children (CIA World Factbook 2019). According to the 2008 *Turkey Demographic and Health Survey*, there is a tendency for women to have children early in the childbearing period (7 out of 10 births took place before the age 30); however, the 25- to 29-year age group had the highest age-specific fertility rate, which indicates there is a trend toward postponing childbearing until later years (Turkey Demographic and Health Survey 2009). Infant mortality rates were 16 deaths/1000 live births in 2015 (CIA World Factbook 2019). There is a diminishing trend in the fertility ratios with the beginning of widespread use of modern contraceptive methods, the births in the very young or advanced ages, and the births frequently are of the foremost causes of maternal death (Kara et al. 2010).

Motherhood is accorded great respect and pregnant women are usually made comfortable in any way possible, including satisfying their cravings. Pregnant women may continue their daily activities or work as long as they are comfortable. Education efforts have increased prenatal practices throughout Turkey. In urban areas, monthly prenatal visits are usually made with an obstetrician. In rural areas in which physicians may be scarce, midwives provide care to pregnant women. However, pregnancy is considered by a number of people as a shameful condition that ought to be concealed (Ayaz and Efe 2008).

28.9.2 Folk Practices for Fertility

In traditional Turkish culture, one of the most important desires of a married woman is to have a child. A woman who has not had a child is faced with social pressure and accusations and, thus, may try to use some traditional practices to increase fertility. Some women damage their bodies by using these practices; sometimes, the damage is permanent (Kayhan et al. 2006). Some of the traditional practices women use to increase fertility include burying the woman in sand, placing her on heat, taking her to thermal springs, applying a poultice, talking to a religious leader, going to a saint's gravesite, having an amulet written, putting a mixture inside the womb, eating meat and wheat brought back from the pilgrimage, sitting on the placenta of a newborn baby, boiling parsley and sitting over its steam, sitting over a milk steam, and going to the *hamam* (Turkish bath) (Kayhan et al. 2006).

28.9.3 Modern and Folk Practices for Preventing Pregnancy

According to 2008 Turkey Demographic and Health Survey, 73% of married women used some method of contraception. Modern contraception was used by 46% of the women while 27% used traditional methods (25% used withdrawal [coitus interruptus]). The most prevalent modern contraceptive methods used were IUDs (17%) and the pill (14%); additionally, female sterilization was used by 8% of the married women (Turkey Demographic and Health Survey 2009). Women who prefer the traditional method of coitus interruptus to prevent pregnancy believe it is free from side effects, a clean method, and unlike condom use, did not seriously affect the sexual pleasure gained by their partners (Ciftcioglu and Behice 2009).

Other folk practices and beliefs that relate to preventing pregnancy include (a) birth control pills cause cancer and make you fat; (b) an IUD can migrate to the stomach or invalidate ritual cleansing; (c) vasectomy ends a man's sexual life; and (d) using some methods is a sin (Kayhan et al. 2006). Others include vaginal douche, calendar method, putting lemon or alum on the sexual organs, drinking henna water, putting aspirin in the vagina, and putting honey and horseflies in food and eating it (Kayhan et al. 2006).

28.9.4 Modern and Folk Practices for Terminating Pregnancy

The risk of unwanted pregnancy increases with inadequate information about birth control, negligence, poverty, low level of education, and inappropriate protection methods (Kavlak et al. 2006). According to the law, married women may request to have an abortion with the consent of their husbands; single women who are older than 18 years can have an abortion at their own request; and single women less than 18 years can have an abortion with the consent of their parent Turkey Population Planning Law 1983). The rate of voluntary abortion is about 11% (Kavlak et al. 2006). The widespread use of abortion has had a significant effect on decreasing the fertility rate. It is common practice that women in families who have chosen to limit the number of children first seek to have an abortion and then they learn about methods to prevent pregnancy and begin to use them (Kavlak et al. 2006).

Traditional methods used to terminate pregnancy include pressing on the abdomen with a stone, mixing matches with trash and putting it in the womb, carrying heavy loads, aborting the child with a beetroot branch and chicken wing, boiling poison ivy leaves and standing in its heat (Kayhan et al. 2006).

28.9.5 Prescriptive, Restrictive, and Taboo Practices in the Childbearing Family

The pregnant Turkish woman is encouraged to keep her strength up by eating foods that are rich in nutrients; however, in poorer families, these nutrients may not be available. Many pregnant women take prenatal vitamins, drink a lot of milk, and apply salves such as Vaseline to avoid stretch marks. Light exercise, such as walking, is encouraged, but weather conditions often hamper such efforts because Turks generally tend to avoid wet or cold weather fearing its ill effects on one's health.

28.9.6 Modern and Folk Practices to Facilitate Childbirth and Postpartum Period

Most women prefer hospitals for physician-assisted child delivery; however, midwives are accepted in rural areas when a physician is not available. Particularly in rural areas, the more natural squatting or semi-sitting position is preferred to the supine position during delivery. Expressions of discomfort and pain are quite acceptable, although Laz women from the Black Sea area tend to be stoic.

Some traditional women believe they should not eat fish, sheep's heads, sheep's trotters, and rabbit meat during pregnancy because eating fish would cause the baby to be mute, not to develop bones, and to float like a fish. Eating sheep's heads or trotters is thought to cause the baby to have a runny nose and consumption of rabbit meat to cause a cleft lip (Ayaz and Efe 2008).

Folk practices used to make childbirth easier include unlocking places that are locked, untying the woman's hair ribbons, unbuttoning buttons, standing straight and turning so the child will move, drinking water that has been prayed over by religious leaders, enclosing the woman around her waist and rocking her three times, and putting her in a blanket and rocking her three times (Kayhan et al. 2006). Still others include jumping from a high point because it facilitates birth, bumping women in the back and shaking them since it makes birth easier, shaking women in a sheet to facilitate the birth process, and anointing the genitals to make the birthing process easier (Ayaz and Efe 2008).

It is becoming more acceptable and more common for the husband and other relatives to be present during the birthing process as it is with the immigrant Turkish population in the US and other European countries.

The postpartum period can last up to 40 days. During this time, a woman is under the effect of many supernatural powers. There is a folk saying that for 40 days the grave is open for the woman postpartum. She is not left alone during this time, which is called *lohusa*. At the end of the 40 days she returns to normal life. She is bathed with abundant water and prayers are read. The infant is also bathed in a similar manner. Eating boiled potatoes, thick rice soup, cola with aspirin, ground coffee–lemon mixture, apricot, roasted chickpeas, or olive paste and spreading herbal dough on the stomach are also widely used (Ogut and Gurkan 2005).

Light exercise is encouraged during the postpartum period, and bathing, an important part of the Muslim tradition, is strongly recommended. A special food called *loğusa serbeti* or *loğusalik* is served to the woman. *Loğusalik* is a sweet, sherbet-like foodstuff, prepared by dissolving *loğusalik* beads (available in stores in Turkey) in hot water. This high-carbohydrate mixture is said to increase the woman's strength. Postpartum women drink hot soups and other fluids such as milk, especially when breastfeeding. Most Turks realize the value of breastfeeding, which is practiced modestly.

28.9.7 Folk Practices for Newborns and Children

Newborns are treated as cherished gifts. Healthy babies are greeted with *Masallah* (may God bless and protect). The 2008 *Turkey Demographic and Health Survey* reported that nearly all infants are breastfed for the first months after birth (Turkey Demographic and Health Survey 2009). The rates of Cesarean delivery are increasing which is having an effect on the initiation and duration of breastfeeding (Cakmak and Kuguoglu 2007). There are many folk practices associated with newborns. One is to not give water or breastfeed a newborn until the call to prayer has been announced three times; the belief is that this makes the baby become patient, intelligent, and religious. The first milk (colostrum) from the mother's breast is considered impure so it is discarded and instead the baby is fed with sugary water, cow or goat's milk, honey, and butter (Ayaz and Efe 2008). Another traditional practice is to bathe a newborn immediately after birth in salt water to have healthy skin and to prevent sweat from being offensive when they grow older. It is also believed to prevent diaper rash as well as some diseases and to cause future injuries to heal rapidly. This practice is called salting or striking the child with salt. The infant may also have salt put in his or her mouth to prevent bad breath which can cause dehydration (Ayaz and Efe 2008).

When a newborn infant has jaundice, a variety of traditional practices may be used, including tying a yellow ribbon to the crib, dressing the infant in yellow clothes, having the infant drink her or his own urine, using a razor blade to cut between the infant's eyebrows and between his or her fingers, putting a gold coin on the infant, dressing the infant in yellow clothing, and bathing the infant with gold water and the yellow yolk of egg (Ayaz and Efe 2008).

When the umbilical cord is cut it is given a name; the umbilical cord's name is used to call the person when they die. The name is given by saying it three times into the right ear during the call to prayer.

A small blue bead called a ***nazar boncuk***, believed to protect the child from the "evil eye," is usually placed on the child's left shoulder. This practice is believed to protect the child from the evil angel whispering in the left ear, often portrayed in Christian religious art. However, it may be swallowed or aspirated by the infant. A child may be taken to a *hoca* religious leader to have an amulet written to recover from an evil eye.

Dressing the baby with a sand-filled diaper (holluk), which is spreading fine soil over the baby's diaper to absorb wetness and to keep the baby warm and comfortable,. It is done to prevent diaper rash and is believed it promotes harmony with nature; the earth is regarded as nutritious and a source of power. It can be harmful because it may cause parasitic infections and tetanus in the newborn, whose immune system is not fully developed (Ayaz and Efe 2008).

Other traditional practices are not breastfeeding a baby with diarrhea or feeding a baby with diarrhea a mixture of coffee and yogurt, which can be quite harmful because diarrhea is a common cause of infant mortality. Another is putting soap into the rectum when a child is constipated, which may harm the baby as the soap irritates the intestinal mucous membrane (Ayaz and Efe 2008).

Other folk practices include placing iron under the baby's mattress to protect against anemia and placing a red bow on the crib to distract any envy or negativity. In eastern Turkey, an infant may be put under soil based on the belief that keeping the infant warm will keep her or him as healthy as the soil. This practice can irritate the infant's skin and even result in death from tetanus.

Swaddling infants, a common practice, has benefits such as helping infants sleep longer, decreasing physiologic distress, improving neuromuscular development, soothing pain, and in excessively crying infants reducing crying and regulating temperature. However, it can also cause hyperthermia, hip dysplasia, and respiratory infections and increase the risk of sudden infant death syndrome with the combination of swaddling with the infant in a prone position, which makes it necessary to warn parents to stop swaddling if infants attempt to turn (Van Sleuwen et al. 2007). Herbal therapies are commonly given to children for respiratory and digestive problems (Ozturk and Karayagiz 2008). Health-care providers must teach mothers to use swaddling properly and with caution.

Health-care providers need to assess the use of prescriptive, restrictive, and taboo practices for pregnancy, labor and delivery, and postpartum because some women still carry out traditional practices that may adversely affect them or their infants. It is important to gain an understanding of these potentially harmful customs and cultural beliefs so that health education programs can be implemented that dissuade women from resorting to and continuing these practices.

Reflective Exercise

The nurse at a local clinic sees Mrs. Aydın who expresses concern that she has is not pregnant. She has been married a year. In traditional Turkish culture one of the most important desires of a married woman is to have a child. She is facing social pressure and accusations and, thus, want to try some traditional practices to increase her fertility.

1. What background information does the nurse need to know about the traditional fertility practices some Turkish women use to increase their possibility of getting pregnant before talking with Mrs. Demir?
2. When she is taking Mrs. Aydin history what questions should she about her cultural practices related pregnancy?
3. How should the nurse approach Mrs. Aydın to assess her knowledge of fertility and suggest her husband may need to be tested to see if he has an abnormal sperm production?

28.10 Death Rituals

28.10.1 Death Rituals and Expectations

When death occurs the deceased individual's family members cry in the most natural manner. Neighbors who hear about the death gather at the home of the deceased to share in the suffering of the family, to console them, and to help with the initial preparations. In the first week after a death, all close friends and relatives will help with the funeral arrangements, prepare food for the grieving family, assist at household chores, and deal with family and friends who come to pay their respects and be continuously by the family's side in full support (Cimete and Kuguoglu 2006).

Having prayers said is a common practice. In villages, townships, and small cities, a news reader goes from house to house to announce the death; placing a death notice in the newspaper is more common in large cities. Commercial funeral agencies in large cities make necessary preparations for the burial as well as preparing death notices. Some of the procedures done immediately after death deal directly with the corpse and others involve arranging the environment around the corpse. Turkish Muslims do not generally practice cremation because the body must remain whole. Frequently, the body is displayed in the home for a day or two; it is then placed in a coffin and taken to the ***cami*** (mosque) to be visited primarily by men. In rural areas, showing respect for the deceased by participating in the funeral procession is very important.

Religious or traditional reasons require the preparations for burial. These preparations include three important procedures: bathing, wrapping in a shroud, and saying funeral prayers outside a mosque. After someone dies, he or she is prepared quickly for burial. If a person dies in the morning, he or she is buried after the mid-afternoon prayers; a person who dies during the night is buried in the morning. The funeral may be delayed to await the arrival of distant relatives.

Common rituals after death are closing the eyes of the deceased, tying the chin, turning the head toward Mecca, putting the feet next to each other, putting the hands together on the abdomen, and removing clothing. In some places, the bed is changed; a knife, iron, or other metal object is placed on the abdomen of the deceased; the room in which the deceased is lying is cleaned and well lighted; and the Koran is read at the head of the deceased.

The majority of those in Anatolia wash the corpse before burial. Women wash a deceased woman, and men wash a deceased man. The people who do this procedure are professional cleansers, religious leaders, experienced people, the religious community, or in some cases, someone from the home of the deceased or a neighbor, or in some places, a person mentioned in the will. In large cities, the washing is done in funeral homes; in villages, however, a sheltered corner in someone's garden is used.

Shrouding is the second important procedure after the washing of the corpse and before burial. The fabric for the shroud is white and the number of pieces of cloth varies for men and women. Again, the majority of the people who live in Anatolia practice this procedure.

Funeral prayers are the third procedure. According to the Islamic religion, several conditions need to be met for funeral prayers to be said. After the funeral prayers are said, the corpse is taken to the cemetery in a coffin. The corpse is placed in the grave with the right side facing the direction of Mecca. When the body is placed inside the grave, a wooden board is leaned against one wall to protect the body from the dirt used to fill the grave. The corpse is generally placed in the grave without the coffin; however, it may also be buried in the coffin. Placing a gravestone with inscriptions to give the identity, gender, and fate of the deceased is very common.

After a funeral prayer at the mosque, the body is interred. For the first 7 days there will be continuous religious rites and a religious ceremony on the 7th, 40th, and 52nd days. Prayers are said and special food will be prepared and distributed to the guests. The first 7 days after death more prayers are said and *helva* (a sweet dessert) is served in honor of the deceased. The traditional mourning period is 40 days, during which time traditional women may wear black clothes or a black scarf.

The clothes and personal belongings of the deceased will be sent to the poor. A few belongings will be saved by close family members as memories of the deceased. Again, donations will be distributed to the poor or to religious organizations. The underlying belief is that food, money, and clothing are distributed to the poor so that the deceased will not be left hungry, naked, and cold (out in the open) in the other world, and that the prayers will help the deceased to be received into Heaven. Although all of these rituals are aimed at supporting the family who is suffering the loss, family members are not given the opportunity to reveal their emotions after a death and even further, they are kept occupied so that they remember the loss as little as possible (Cimete and Kuguoglu 2006).

28.10.2 Responses to Death and Grief

Although Muslim Turks believe in the afterlife, death is always an occasion of great sorrow and mourning. An expression of sympathy to one who has just lost someone to death is *Basiniz sağolsun* (may your head be healthy), hoping that one is not overwhelmed with grief. Mourning, synonymous with grief over the death of someone, is important and is done for a specific period of time for the purpose of adapting to the new situation and decreasing suffering.

There are no home care or hospice systems within the health system of Turkey. This may lead families to an isolated state and may keep them from advancing optimally through the stages of the grief process.

Reflective Exercise

A home health nursing made a visit to Mrs. Demir, an 81-year-old Turkish female who has been diagnosed with end-stage renal disease. She complained of progressively worsening shortness of breath, loss of appetite, fatigue, difficult sleeping and feeling nauseated. Her blood pressure was elevated, she had bilateral extremity swelling and her urinary output had decreased. During the home health nurse's visit, Mrs. Demir's friend says to her "Geçmiş olsun" ("May it be the past", which is often said when someone is sick. It recognizes their pain, but expresses hope it will soon be behind them). For a variety of reasons terminally ill patients are generally not told the severity of their conditions.

1. What questions should the nurse consider to find out the extent of Mrs. Demir's knowledge of her terminal disease?
2. Many believe that informing a patient of a terminal illness may take away the hope, motivation and energy that should be directed toward healing. How would the nurse determine Mrs. Demir's state of mind?
3. How would the nurse assess Mrs. Demir's anxiety related to her fear of dying and concern about those being left behind?
4. Mrs. Demir, a Muslim, may believe that no one can second-guess Allah, for who can know if Allah has a miracle in mind. How would the nurse assess if Mrs. Demir holds this belief and the effect on her treatment?

28.11 Spirituality

28.11.1 Dominant Religion and Use of Prayer

Turks are 99.8% Muslim, but freedom of religion is mandated by the Turkish secular state (CIA World Factbook 2019). Most are Sunni Muslims, with a minority from the Alevi Muslim group. Other religious minority groups include Jews (mostly Sephardic) and Christians. Proselytizing is illegal in Turkey.

Most Turks who emigrate to the West tend to be very moderate Muslims. Traditional *namaz* (prayer) is practiced five times each day and can take place in the *cami* or elsewhere, as long as one is facing ***kíble*** (the holy city of Mecca). A special small rug, called ***seccade***, is used for praying in places other than the *cami*. When entering the *cami*, shoes are always removed and women must cover their heads. Men and women go to separate parts of the *cami* for prayer. One prepares for prayer by a ritual cleansing called *abdest*, which at a minimum includes washing the face, ears, nostrils, neck, hands to the elbow, and feet and legs to the knee three times each. Sometimes washing facilities are available at the *cami*. Caregivers may need to make special arrangements and be sensitive to the need for Muslims to practice their religious obligations when they are in a health-care facility.

Tortumluoğlu et al. (2005b) conducted a qualitative study about religious characteristics in a village in Turkey and found that

- In the village, everyone knows the command of Allah. They do their ritual cleansing; say their prayers; fast; those who are able, make the pilgrimage; give alms; read the Koran; and do what the Koran says.
- In our village, a lot of people keep the fast for 3 months. During Ramazan, we also say the *teravi* prayer (an extra prayer in addition to those said five times a day). We go to the mosque and we both say our prayers in the designated area and also listen to the sermon.
- We do what Allah said and live according to our religion. We don't show ourselves to those outside the family. We read, pray, and fast, and if Allah allows, we will go to heaven in the afterlife.

28.11.2 Meaning of Life and Individual Sources of Strength

Turks rely on their religious beliefs and practices and their family and friends for strength and meaning in life. One's degree of religiosity influences the importance of prayer in giving meaning to life. A little known fact is that St. Nicholas was born and lived in Patara, in southern Turkey, in the fourth century AD where he became known as *Santa Claus*.

28.11.3 Spiritual Beliefs and Health-Care Practices

Religious beliefs intertwined with folk beliefs continue to influence Turkish lifestyle. Spiritual leaders or healers are sought most often for assistance with relationship or emotional problems and, less frequently, for physical problems. A *muska*, a paper inscribed by a *hoca* (spiritual teacher) with a prayer in Arabic, is wrapped in fabric and then hidden in the home or worn by the person seeking help. *Turbe* and *yatir* are the practice of going to the saints' graves to pray about wishes, mental or emotional problems, or fertility problems. ***Tesbih***, the small beads traditionally used for praying, now take a more-secular meaning and are often referred to as *worry beads*.

Health teaching strategies for Turks should include the recognition and prevention of dehydration, bloating, constipation, hypoglycemia, and fatigue during periods of Ramazan fasting. In addition, religious or folk items should not be removed from the health-care facility because they provide comfort for the client, and their removal may increase anxiety.

28.12 Health-Care Practices

28.12.1 Health-Seeking Beliefs and Behaviors

Most Turks rely on Western medicine and highly trained professionals for health and curative care. However, remnants of traditional beliefs continue to have an impact on health-care practices. Thus, health-care providers may wish to incorporate factual information regarding disease causation and treatment into patient education planning.

28.12.2 Responsibility for Health Care

Turkish children are routinely immunized against diphtheria, tetanus, whooping cough, measles, polio, hepatitis B, mumps, rubella, and TB. There has been significant improvement in the vaccination rates, between 2003 and 2008, the rate for children fully vaccinated rose from 54 to 74% (TDHS 2009).

For a variety of reasons terminally ill patients are generally not told the severity of their conditions. Many believe that informing a client of a terminal illness may take away the hope, motivation, and energy that should be directed toward healing, or it may cause the client additional anxiety related to the fear of dying and concern about those being left behind. Furthermore, no one can second-guess Allah, for who can know if Allah has a miracle in mind?

Turkey has one of the highest rates of consumption of over-the-counter antibiotics and painkillers; aspirin is commonly used as a panacea for a variety of ailments, including gastric upset. Turks, especially those who have difficulty affording the services of a physician, commonly consult a pharmacist before visiting a physician. Fever- and pain-reducing medicines and cough syrups are frequently purchased without professional medical consultation. Health professionals must assess Turkish patients for their use of over-the-counter medications to prevent conflicting or potentiating effects with prescription medications.

28.12.3 Folk and Traditional Practices

Turkey is a country where civilizations have been established since the ancient ages resulting in a rich folklore. There is a high prevalence of traditional health-care practices among Turks; these practices are so significant in parts of the culture that they cannot be ignored. They are very common, particularly in rural areas, because they often cannot access health services and do not have the financial resources to see physicians so must rely on the traditional health practices. However, such practices may be harmful to a person's health and may delay early treatment.

Engin and Pasinlioğlu's (2000) study, conducted in the center of Erzurum with infertile women, showed that 44.6% of women were assisted by untrained midwives, 57.8% used witch doctor medicine, and 39.1% were prayed over by religious people to treat their infertility.

The findings from Özyazıcıoğlu's (2000) study, conducted in the center of Erzurum, was that mothers with at least one child older than 12 months had a high rate of using traditional treatments for such health problems as the common cold and nasal congestion (26.39%), earache and ear drainage (17.22%), stomach ache (54.96%), constipation (24.88%), burns (24.39%), poisoning (23.08%), cuts and bleeding (67.56%), and fractures (41.67%). Tortumluoğlu et al. (2004) reported that 82.2% of older people in their study used traditional practices for burns, 76.7% for insect bites, 64.4% for the common cold, 63% for stomach problems, 63% for high fever, 63% for warts, 56.2% for sties, and 54.4% for constipation.

Ugulu and Baslar (2010) found that most people (68%) in their study from four Turkish cities continued to use traditional systems of health care including medicinal plants alone or in combination with other ingredients, such as flour, honey, and oil. Phyto preparations (salves, gels, creams), medicinal plants, are used for the treatment of various diseases of skin and mucous membranes.

Toprak and Demir found that the most common traditional methods for treating hypertension among their research subjects were eating yogurt with garlic (27.8%) and eating sour foods, such as lemon and grapefruit (25%). Resting, drinking *ayran* (a Turkish drink made with yogurt and water), and applying a cold bag to the head were the other methods used for coping with hypertension (Toprak and Demir 2007).

Kara (2009) found that many patients with end-stage renal disease undergoing hemodialysis also used herbs to treat their health problems. The majority received the information about which herbs to use from their families and friends. They did not disclose the use of herbal products to their physicians (Kara 2009). Additional traditional health-care practices to treat illness or symptoms include applying rubbing alcohol or a wet cloth to bring down a fever and warming the back to treat coughing. Health-care providers should be aware of the health risks caused by certain traditional health-care practices and educate the patients and families about the potential risks.

Turkey also encourages health tourism at their 1500 thermal spas. These spas treat conditions such as rheumatism, respiratory and digestive problems, diabetes, skin conditions, gallstones, female diseases, kidney and heart conditions, nerves, obesity, and hyperlipidemia.

The concept of the evil eye is prevalent in many cultures, including Turkish culture. Specific to health, it is a cultural inclination not to speak too well of one's health for fear that one may incur misfortune through others envy or *nazar*. Turkish patients, therefore, may be more inclined than other ethnic groups to complain about health. So pervasive is this concept that taxi drivers and medical doctors alike respect the ***nazar boncuk***, a blue bead used as protection from the evil eye. Some Turks may believe that excessive complaining may bring the benefit of closer medical attention. However, when describing an illness, a person avoids using oneself or another person as an example for fear of inviting the illness or condition upon that person.

Kolonya (cologne) is part of a traditional practice that crosses religious and secular lines. Originally derived from the religious value of cleanliness, cologne is sprinkled on the hands of guests before and after eating to provide cleanliness and a fresh lemon scent. Inhaling from a cloth or handkerchief doused with cologne may be used for relief from motion sickness. In the hospital, patients may offer cologne to a physician or nurse prior to examination. An essential part of hospitality, it also has some medicinal intent. In fact, cologne is approximately 70% alcohol and does have a bactericidal quality.

28.12.4 Cultural Responses to Health and Illness

An autonomy-centered approach in Turkish health care is relatively new. The Regulation on Patient Rights was enacted in 1998 but only recently have there been tangible steps toward its implementation in the health-care system. It is becoming more common for patients to want to know their diagnosis and express their wishes and expectations about their health care; traditionally, patients had to be in compliance with the traditions of the paternalistic medical model, which demanded compliance from physicians (if not obedience) without considering patients' opinions or wishes (Guven 2010).

Seriously ill people are expected to conserve their energy to allow their minds and bodies to fight their illnesses; thus, reducing their workload and avoiding unnecessary energy expenditure are acceptable. During hospitalization, ***refakatçí*** refers to the person who stays overnight with the client, providing emotional and physical support and comfort. A show of concern and *Şevkat* (compassion) for the client eases her or his fears and reduces loneliness. Family members may also attend to physical needs such as bathing. A balanced, healthy diet is considered essential to regaining one's health; thus, Turks frequently bring food from home for the patient.

Although the degree of pain expression varies according to regional origin, Turkish culture allows freedom to express pain, either through emotional outbursts or through verbal complaints. General observations about Turkish culture suggest that although stigma is attached to mental illness, many families seek treatment or care for the client at home. Mental health services are basically curative rather than preventive.

There is a concern that the risks for mental disorders and depression in older people will increase in the near future due to factors such as rapidly changing social structure, urban migration, and shifting to nuclear family life. Families will no longer be in a situation to care for older people, physically, psychologically, or economically (Nahcivan and Demirezen 2005).

Postpartum depression (PPD) is common among Turkish women, though the majority of women suffering from PPD receive no treatment and it may go undetected. There is an increased need to educate public and health-care providers about PPD and to develop nursing interventions to provide support to postpartum women (Dindar and Erdogan 2007).

28.12.5 Blood Transfusions and Organ Donation

According to Turkish law 2238, which went into effect in 1979, obtaining, storing, grafting, and transplanting organs and tissues for the purpose of treatment, diagnosis, and science are subject to regulations. In 1980, the Ministry of Religious Affairs stated that organ and tissue transplantation is permissible when it is done according to the following conditions:

- To save a patient's life and when no other alternative exists, as established by a licensed physician whose honesty is reliable.
- A dominant medical opinion is that the illness cannot be treated in another way.
- The donor's organ or tissue is taken while this procedure is being done.
- To prevent the disturbance in the peace and order of society, the donor must have given permission when healthy (prior to their death), or if no declaration was made while alive, the next of kin are willing.
- No payment of any kind can be received in exchange for the donated organ or tissue.
- The donee must be willing to have the transplantation (Turkish Transplantation Society 2006).

Muslims traditionally prefer that their body remain intact after death, a belief that can conflict with organ donation. Former Prime Minister and President Turgut Ozal and his wife promoted organ donation by publicly signing donor cards and encouraging others to do so. Ağartan et al. (2006) in their study reported that 39.8% of nurses were against organ and tissue transplantation because they believed it was not acceptable in the Islamic religion. Additionally, they would not want their bodies disturbed after death; 31.1% were afraid that their organs would be taken before they had died.

Blood transfusions are gaining acceptance. However, Turkish people generally prefer to receive blood from family members.

28.12.6 Barriers to Health Care

In general, women are responsible for the actual caregiving of the ill and the elderly in the home. However, in traditional households, the mother-in-law or father-in-law, depending on who controls the finances in the family, makes decisions about going to the physician. In many situations, the person who is respected as the most educated has primary input into decisions about health care. An additional barrier can exist for devout Muslim women when a female health-care provider is not available. An over reliance on folk and traditional practices can also be a barrier.

28.13 Health-Care Providers

28.13.1 Traditional Versus Biomedical Providers

Although Turkish people are inclined toward Westernized health-seeking behaviors, medical care in Turkey tends to be holistic. Great value is placed on emotional well-being, especially as it affects physical well-being. Emotional health is considered instrumental to the healing process. Physicians may be "adopted" as members of their patients' families, and it is common to give gifts (usually food) to physicians as an expression of gratitude. A Turkish physician would never refuse gifts or interpret them as a bribe for better care. When modern medicine is not available, accessible and affordable, or has not been effective, Turks may seek the care of a traditional healer.

Generally, physicians are viewed and respected as professionals, so caring for someone of the opposite sex is not an issue among most Turks. However, it is always advisable to ask patients their opinion or preference.

28.13.2 Status of Health-Care Providers

Physicians and to a lesser extent, nurses and midwives, have historically been held in very high esteem. Patients rarely question the authority of physicians, but the notion of obtaining a second opinion is gaining popularity.

The university qualifying examination system allows only the very top academic echelon of students to study medicine. Nursing master's programs began in 1968 and doctoral programs in 1972. A recent study identified problems nurses faced during their postgraduate education which included the lack of associate professors, lack of foreign language skills, not enough time to do research due to being overworked, physiological stress, conflict within the office, and economical problems (Canbulat et al. 2007).

The relationship between physicians and nurses is hierarchical. Currently, this situation is based more on educational level than on gender because a great number of women enter the medical profession. Most male nurses work in community settings. Neither physicians nor nurses share the same financial benefits of health-care professionals in the United States.

References

Ağartan E, Önder A, Memiş S, Baklaya N (2006) Hemşireler organ ve doku bağışı konusunda yeterince duyarlımı? Ulusal Hemşirelik Öğrencileri Kongresi. Kongre kitabı. Harran Üniversitesi, Şanlıurfa. [Are nurses sensitive about the donation of organs and tissue?] National Nursing Students Congress Book. Harran University, Sanlıurfa, p 214

Akgun S, Rao C, Yardim N, Basara BB, Aydin O, Mollahaliloglu S, Lopez AD (2007) Estimating mortality and causes of death in Turkey: methods, results and policy implications. Eur J Pub Health 17(6):593–599

Ayaz S, Efe S (2008) Potentially harmful traditional practices during pregnancy and postpartum. Eur J Contracept Reprod Health Care 13(3):282–288

Bahadir A, Öztürk O, Atalay A, Atalay E (2009) Beta globin gene cluster haplotypes of the beta thalassemia mutations observed in Denizli province of Turkey. Turk J Haematol 26(3):129–137

Beser A, Bahar Z, Buyukkaya D (2007) Health promoting behaviors and factors related to lifestyle among Turkish workers and occupational health nurses' responsibilities in their health promoting activities. Ind Health 45:151–159

Cakmak H, Kuguoglu S (2007) Comparison of the breastfeeding patterns of mothers who delivered their babies per vagina and via cesarean section: an observational study using the LATCH breastfeeding charting system. Int J Nurs Stud 44(7):1128–1137

Canbulat N, Demirgöz M, Cingil D, Saklı F (2007) A general overview of the nursing academicians in Turkey. Int J Hum Sci 4(1). http://www.insanbilimleri.com/ojs/index.php/uib/article/view/49

Central Intelligence Agency (CIA) CIA WorldFactbook: Turkey (2019). https://www.cia.gov/library/publications/the-world-factbook/geos/tu.html

Ciftcioglu S, Behice E (2009) Coitus interruptus as a contraceptive method: Turkish women's perceptions and experiences. J Adv Nurs 65(8):1686–1694

Cimete G, Kuguoglu S (2006) Grief responses of Turkish families after the death of their children from cancer. J Loss Trauma Int Perspect Stress Coping 11(1):31–51

Cinar ND, Tuncay MF, Topsever PT, Ucar F, Akgul S, Gorpelioglu S (2006) Intentional sun exposure in infancy in Sakarya, Turkey. Saudi Med J 27(8):1222–1225

Dindar I, Erdogan S (2007) Screening of Turkish women for postpartum depression within the first postpartum year: the risk profile of a community sample. Public Health Nurs 24(2):176–183

Engin R, Pasinlioğlu T (2000) Erzurum ve Yöresinde İnfertil Kadınların İnfertilite [Infertile women's traditional practices and religious beliefs about infertility in Erzurum and the surroundings]. Master's thesis, Ataturk University Health Sciences Institute, Erzurum, Turkey

Erdoğan MF, Ağbaht K, Altunsu T, Ozbağ S, Yücesan F, Tezel B, Sargin C, Ilbeğ I, Artik N, Kösc R, Erdoğan G (2009) Current iodine status in Turkey. J Endocrinol Investig 32(7):617–622

Gul A (2007) Standard and novel therapeutic approaches to Behçet's disease. Drugs 67(14):2013–2022

Gültekin T, Koca B (2003) Cumhuriyet döneminden günümüze ülkemizde gerçeklies¸tirilen ırk çalıs¸maları. [Race studies in our country since the constitution of the republic to nowadays]. J Anthropol 14:1–24

Guven T (2010) Truth-telling in cancer: examining the cultural incompatibility argument in Turkey. Nurs Ethics 17(2):159–166

İcduygu A (2008) Circular migration and turkey: an overview of the past and present—some demo-economic implications. Carim Analytic and Synthetic Notes 2008/10. Circular Migration Series. Demographic and Economic Module. http://cadmus.eui.eu/bitstream/handle/1814/8331/CARIM_AS%26N_2008_10.pdf?sequence=1

Inel M, Ozmen B, Bilgin H (2008) Re-evaluation of building damage during recent earthquakes in Turkey. Eng Struct 30(2):412–427

Information Please (2019) Turkey. https://www.infoplease.com/world/countries/turkey

Kara B (2009) Herbal product use in a sample of Turkish patients undergoing haemodialysis. J Clin Nurs 18:2197–2205

Kara M, Yilmaz E, Töz E, Avci I (2010) The contraceptive methods used in A_rı, Turkey. J Gynecol Obstet 20(1):10–13

Kavlak O, Atan S, Saruhan A, Sevil U (2006) Preventing and terminating unwanted pregnancies in Turkey. J Nurs Scholarsh 38(1):6–10

Kayhan S, Güzlek C, Özdemir G, İpsala E, Tortumluoğlu G (2006) Women's practices about conception, contraception, terminating of pregnancy, and facilitating of delivery in Çanakkale. Jinekoloji ve Obstetri Dergisi. Baskıda. J Obstet Gynecol 17(3). http://jinekoloji.turkiyeklinikleri.com/index.php?lang=tr

Köser-Akçapar S (2006) Do brains really go down the drain? Rev Eur Migr Int 22(3):79–107. http://remi.revues.org/index3281.html

Kukulu K, Gursoy E, Gulsen S (2009) Turkish university students' beliefs in sexual myths. Sex Disabil 27:49–59

Meals in Koran and eating in sects and tradition (2006). http://www.zpluspartners.com/kosherhalal2.pdf

Ministry of Health of Turkey (MHT) (2010) General Directorate of Primary Health Care, obesity prevention and control program of Turkey (2010–2014). Kurban Matbaacilik Yayincilik, Ankara

Nahcivan N, Demirezen E (2005) Depressive symptomatology among Turkish older adults with low incomes in a rural community sample. J Clin Nurs 14:1232–1240

O'Neil M, Güler F (2010) Strangers to and producers of their own culture: American popular culture and Turkish young people. Comp Am Stud 8(3):230–243

Ogut Y, Gurkan A (2005) A study on the traditional attitudes and applications related to diarrhea. III. Uluslararası—X. Ulusal hems¸irelik kongresi. III International—X National Nursing Congress , Eylül, İzmir, p 95

Oksal K (2008) Turkish family members' attitudes toward lesbians and gay men. Sex Roles 58:514–525

Ozturk C, Karayagiz G (2008) Exploration of the use of complementary and alternative medicine among Turkish children. J Clin Nurs 17:2558–2564

Özyazıcıoğlu N (2000) Erzurum il merkezinde 12 aylık çocuğu olan annelerin çocuk büyütmeye ilişkin yaptıkları geleneksel uygulamalar [Mothers' traditional practices about fostering 12-months-old child in Erzurum]. Master's thesis, Ataturk University Health Sciences Institute, Erzurum

Rankina B, Aytaç I (2008) Religiosity, the headscarf, and education in Turkey: an analysis of 1988 data and current implications. Br J Sociol Educ 29(3):273–287

Tasçı M, Oksuzler O (2010) Income differentials and education in Turkey: evidence from individual level data. Int Res J Financ Econ 51:119–131. http://www.eurojournals.com/irjfe_51_10.pdf

The Turkey Demographic and Health Survey–2008. (TDHS) (2009) Hacettepe University Institute of population studies. Tezcan S. Project Director. http://www.hips.hacettepe.edu.tr/eng/index.html

Toprak D, Demir S (2007) Treatment choices of hypertensive patients in Turkey. Behav Med 33:5–10

Tortumluoğlu G, Karahan DE, Bakır B, Türk R (2004) Kırsal alanda yağayan yağlıların yaygın görülen sağlık problemlerine yönelik yaptıkları geleneksel uygulamaların tanımlanması. Uluslararası İnsan Bilimleri Dergisi. [Defining Traditional Health İmplementation applied by old people living in rural area for their common health problem]. Int J Human Sci 13(1):1–16

Tortumluoğlu G, Bayat M, Sevig U (2005a) The evaluation of the individuals in health viewpoint by "Giger and Davidhizar's transcultural assessment model" (p. 141). III. Uluslararası—X. Ulusal Hemşirelik Kongresi. III International—X National Nursing Congress. Eylül, İzmir

Tortumluoğlu G, Bedir E, Sevig U (2005b) The evaluation of "The guide to definite cultural characteristics of Turkish society with a health viewpoint of individuals in Erzurum." III. Uluslararası—X. Ulusal hemşirelik kongresi. III International—X National Nursing Conference. Eylül, Yayınlanmamış yayın, İzmir

Tortumluoğlu G, Ersay A, Pamukçu K, Şenyüz P (2006) Farklı sağlık alanlarında eğitim gören yüksekokul öğrencilerinde cinsellik. [Sexual behaviors of college students in educating different health fields]. National nursing students congress book, vol. 5. Harran University, Sanlıurfa

Turel G (2009) Provision of housing and services for the elderly in Turkey. Beykent University. J Sci Technol 3(1):90–103

Turkey: Population Planning Law: No. 2827 (1983). http://www.hsph.harvard.edu/population/abortion/TURKEY.abo.htm

Turkey Statistical Yearbook. (TSY) (2008) Health Survey 2008. Publication Number 3452. Turkish Statistical Institute, Ankara, p 2

Turkey Statistical Yearbook. (TSY) (2019). https://www.statista.com/topics/1442/turkey/

Turkish Coalition of America (2019) The Turkish American community. http://www.tc-america.org/turkish-american-community/

Turkish Transplantation Society. (2006). http://www.tond.org.tr/tr/

U.S. Census Bureau: Statistical Abstract of the United States. (USCB) (2017) Place of birth for foreign born population in the United States. https://factfinder.census.gov/faces/tableservices/jsf/pages/productview.xhtml?src=bkmk

Ugulu I, Baslar S (2010) The determination and fidelity level of medicinal plants used to make traditional Turkish salves. J Altern Complement Med 16(3):313–322

Van Sleuwen B, Engelberts A, Boere-Boonekamp M, Kuis W, Schulpen T, L'Hoir M (2007) Swaddling: a systematic review. Pediatrics 120(4):1097–1106

WHO (2019) Turkey

World Health Organization (2018) Turkey. https://www.who.int/substance_abuse/publications/global_alcohol_report/profiles/tur.pdf

29 People of Vietnamese Heritage

Carol O. Long

29.1 Introduction

Vietnam is located at the extreme southeast corner of the Asian mainland, bordering the Gulf of Thailand, Gulf of Tonkin, and South China Sea, and alongside the countries of China to the north and Laos and Cambodia to the west. The capital is Hanoi, North Vietnam. With a population of over 97 million in a land mass of 127,330 sq. mi. (Central Intelligence Agency (CIA) 2020), it is relatively narrow in width, but its north–south length equals the distance from Minneapolis to New Orleans in the United States. This chapter follows the 12 domains of culture as depicted in the Purnell Model (See Chap. 2).

29.2 Overview, Inhabited Localities, and Topography

29.2.1 Overview

Vietnam consists largely of a remarkable blend of rugged mountains and the broad, flat Mekong and Red River deltas, which mainly produce rice. Other features are a long, narrow coastal plain and other river lowlands, where most ethnic Vietnamese live. Much of the rest of the country is covered with tropical forests. The climate in the southern region of Vietnam is tropical, and in the northern region is rainy, hot, and humid during monsoon season.

29.2.2 Heritage and Residence

The Vietnamese are an Asian racial group closely related to the Chinese. The population shares some characteristics with other Asian and Pacific Islander groups, yet many aspects of its history and culture are unique. Vietnam was under Chinese control from 111 BC to AD 939 (Vu 2016). At that time, a variety of Chinese beliefs and traditions were introduced to Vietnam, including religions and philosophies of Confucianism, Buddhism, and Taoism. In addition, the system of Chinese medicine was adopted widely. European merchants and missionaries arrived in Vietnam during the sixteenth century, and the French established a political foothold and instituted changes in government and education, including Western medical practices (Vu 2016). Chinese and French influences persist even today.

Vietnam alone has eight different ethnic groups, the majority (86%) of whom are Kinh or Viet. Current life expectancy at birth for females in Vietnam is 76.7 years, and for males,

This chapter is a revision written by Susan Mattson in the previous edition of the book.

C. O. Long (✉)
Fredericksburg, VA, USA

© Springer Nature Switzerland AG 2021
L. D. Purnell, E. A. Fenkl (eds.), *Textbook for Transcultural Health Care: A Population Approach*,
https://doi.org/10.1007/978-3-030-51399-3_29

71.4 years and 73.9 for the total population. The fertility rate is a low 1.79 children per female (CIA 2020). One factor in providing proper health-care to Vietnamese is understanding that they differ substantially between and among themselves, depending on the variant cultural characteristics of culture (see Chap. 1 in this book). This is in addition to understanding that Vietnamese people primarily have been emigrating to the US as well as other Asian countries, Canada, and France since 1975. Vietnamese, whether as immigrants or sojourners, have fled their country to escape war, persecution, or possible loss of life. Newer immigrants come to be with family members and seek a new life in the United States.

29.2.3 Immigration to the United States

Initial Vietnamese immigrants confronted a unique set of problems, including dissimilarity of culture, no family or relatives to offer initial support, and a negative identification with the unpopular Vietnam War. Many Vietnamese were involuntary immigrants, with their expatriation unexpected and unplanned; their departures were often precipitous and tragic. Escape attempts were long, harrowing, and for many, fatal. Survivors were often placed in squalid refugee camps for years.

The first wave of Vietnamese immigration began in April 1975, when South Vietnam fell under the Communist control of North Vietnam and the Viet Cong. At that time, many South Vietnamese businessmen, military officers, professionals, and others closely involved with America or the South Vietnamese government feared persecution by the new regime and sought to escape. Better-educated, first-wave immigrants from urban areas had professional, technical, or managerial backgrounds. American ships and aircraft rescued some; many were temporarily located in refugee camps in Southeast Asia, and then sent to relocation camps in the United States. The 130,000 Vietnamese refugees who arrived in the United States came mainly from urban areas, especially Saigon (Ho Chi Ming City), and consequently had some prior orientation to Western culture. Many spoke English or soon learned English in relocation centers. More than half were Christian. Sixty-two percent consisted of family units of at least five people, and nearly half were female. They were dispersed over much of the United States, often in the care of sponsoring American families. These first-wave immigrants adjusted well in comparison to the subsequent wave (Rkasnuam and Batalova 2014).

Soon after, further events in Vietnam triggered a second wave of immigration. Many Vietnamese grew disenchanted with Communism and their decreased living standard. Great numbers had been forced into labor in new countryside settlements, and young men were often fearful of being called to fight against China or in the new war with Cambodia. Some left by land across Cambodia or Laos, commonly joining refugees from those countries in an effort to reach Thailand. For more than a decade, over one to two million Vietnamese, known as the "boat people," departed Vietnam in small, often unseaworthy and overcrowded vessels in hopes of reaching Malaysia, Hong Kong, the Philippines, or another non-Communist ports (Rkasnuam and Batalova 2014). Half died during their journey. Many were forcibly repatriated to Vietnam or eventually returned voluntarily; others continued to languish in camps for years.

Less-educated, second-wave immigrants from more rural areas were fishermen, farmers, and soldiers and had only minimal exposure to Western culture. Most of the second-wave refugees represented lower socioeconomic groups and had less education and little exposure to Western cultures. Most did not speak English. This wave of Vietnamese included far more young men than women, children, or older people, which disrupted intact families and normal gender ratios. Many spent months or years in refugee camps under deplorable and regimented conditions. The United States passed the *Refugee Act of 1980* as an amendment to the *Immigration and Nationality Act* that was passed decades before in response to this second wave and widened the scope of resources available to assist refugees or individuals who fled their native country and could not return for fear of persecution and physical harm. When they finally arrived in the United States and Canada, many did not fit into American communities, did not learn English effectively, and remained unemployed or obtained menial jobs. Thus, the second-wave immigrants were significantly more disadvantaged.

The continuing persecution of individuals in Vietnam led to a third wave of immigration, beginning in 1979 with the creation of the *Orderly Departure Program*, which provided safe and legal exit for Vietnamese seeking to reunite with family members already in America. Former military officers and soldiers in prison or re-education camps were allowed to come the United States with their families, resulting in the immigration of 200,000 individuals by the mid-1990s.

The fourth wave of immigration was sparked by the *Humanitarian Operation Program of 1989* which permitted more than 70,000 current and former political prisoners to immigrate. Finally, the *Amerasian Homecoming Act of 1988* allowed the children of Vietnamese civilians and American soldiers to immigrate to the United States. Many of the Amerasian children were orphans who had lived on the street, received no formal education, and had been subjected to prejudice and discrimination in Vietnam (Amerasians without Borders 2019). Large scale immigration of Vietnamese to the US subsided in the mid-1990s, shifting to regular family-sponsored immigration methods (Zhou and Bankson 2000)

29.2.4 Today's Vietnamese Americans

Today there are an estimated 1.3–1.6 million Vietnamese immigrants in the United States. Most reside in California (40%), Texas (13%) and Washington and Florida (4% each) (Alperin and Batalova 2018; Pew Research Center 2012; Rkasnuam and Batalova 2014; U.S. Census 2011).

Vietnamese Americans are considered to be upbeat and optimistic. Most Vietnamese Americans age 18 and older today (84%) were born outside of the US. Nineteen percent of Vietnamese American adults age 25 and older have obtained a bachelor's degree and 7% an advanced degree compared to the US population of 18 and 10% (Pew Research Center 2012). Vietnamese Americans do well financially with an annual income of $53,400 compared to $66,000 (all Asians) and $49,800 for the total US population (Pew Research Center 2012).

Vietnamese place a high value on education and accord scholars an honored place in society. The teacher is highly respected as a symbol of learning and culture. In contrast to American schools' emphasis on experimentation and critical thinking, Vietnamese schools emphasize observation, memorization, and repetitive learning. This style of learning is still predominant in Vietnam, including the universities with schools of medicine and nursing. Most Vietnamese men and women in America are very educationally oriented and take full advantage of educational opportunities when possible. Educational level and occupation continue to vary depending upon their time of emigration.

29.3 Communication

29.3.1 Dominant Languages and Dialects

The official language of Vietnam is Vietnamese, or *tiěng Việt,* with English increasingly being favored as a second language, followed by French, Chinese, Khmer, and other mountain area languages (CIA 2020). Other languages spoken include Chinese, Khmer, Chan and others by tribes residing in mountainous regions. Ethnic Vietnamese speak a single distinctive language, with northern, central, and southern dialects, all of which can be understood by anyone speaking any of these dialects.

The Vietnamese language is tonal, resembling Chinese and contains many borrowed words, but someone speaking one of these languages cannot necessarily understand the other. All words in Vietnamese consist of a single syllable, although two words are commonly joined with a hyphen to form a new word. Verbs do not change forms, articles are not used, nouns do not have plural endings, and there are no prefixes, suffixes, definitives, or distinctions among pronouns. Contextually, the Vietnamese language is musical, flowing, and polytonal, with each tone of a vowel conveying a different meaning to the word. The language is spoken softly, and its monosyllabic structure lends itself to rapidity, but spoken pace varies according to the situation. Whereas grammar is mostly simple, pronunciation can be difficult for Westerners, mainly because each vowel can be spoken in five or six tones that may completely change the meaning of the word.

Vietnamese is the only language of the Asian mainland that, like English, is regularly written in the Roman alphabet since it was introduced by the French in the seventeenth century. Although the letters are the same, pronunciation of vowels may vary radically depending on associated marks indicating tone and accent, and certain consonant combinations take on unusual sounds.

Most Vietnamese refugees, even those who have been in the United States for many years, do not feel competent in speaking English. It is estimated that 31% of immigrants age 18 and older speak English very well compared to 53% collectively with all American-Asian groups (Pew Research Center 2012). Although many refugees eventually learn English, their skills may not be adequate in certain situations. The important subtleties in describing medical conditions and symptoms, or the more abstract presentation of ideas during psychiatric interviews may be particularly difficult. Health-care providers may need to watch patients for behavioral cues, use simple sentences, paraphrase words with multiple meanings, avoid metaphors and idiomatic expressions, ask for correction of understanding, and explain all points carefully. Another cause for concern are homonyms such as meat versus meet, here versus hear, their versus there, and past versus passed. Approaching patients in a quiet, unhurried manner, opening discussions with small talk, and directing the initial conversation to the oldest member of the group facilitate communication, if appropriate.

29.3.2 Cultural Communication Patterns

Traditional Vietnamese religious beliefs transmitted through generations produce an attitude toward life that may be perceived as passive. For example, whenever confronted with a direct but delicate question, many Vietnamese cannot easily give a blunt "no" as an answer because they feel that such an answer may create disharmony. Self-control, another traditional value, encourages keeping to oneself, whereas expressions of disagreement that may irritate or offend another person are avoided. Individuals may be in pain, distraught, or unhappy, yet they rarely complain except perhaps to friends or relatives. Expressing emotions is considered a weakness and interferes with self-control and they may be unaccustomed to discussing their personal feelings openly with others. Instead, at times of distress or loss, they often complain of physical discomforts such as headaches, backaches, or insomnia. Vietnamese tend to be very polite and guarded. Sparing one's feelings is considered more important than factual truth.

The strong influence of the Confucian code of ethics means that proper form and appearance are important to Vietnamese people and provide the foundation for nonverbal communication patterns. For example, the head is a sacred part of the body and should not be touched. Similarly, the feet are the lowest part of the body and to place one's feet on a desk is considered offensive to a Vietnamese person. To signal for someone to come by using an upturned finger is a provocation, usually done to a dog; waving the hand is considered more proper.

Hugging and kissing are not seen outside the privacy of the home. Men greet one another with a handshake but do not shake hands with a woman unless she offers her hand first. Women do not usually shake hands. Two men or two women can walk hand in hand without implying sexual connotations. However, for a man to touch a woman in the presence of others is insulting.

Even if someone learns how to pronounce and translate Vietnamese, problems may remain with respect to intended meaning of various words.

For example, the word for "yes," rather than expressing a positive answer or agreement, may simply reflect an avoidance of confrontation or a desire to please the other person. The terms "hot" and "cold," rather than expressing physical feelings associated with fever and chills, may actually relate to other conditions associated with perceived bodily imbalances. Various medical problems might be described differently from what a Westerner might expect; for example, a "weak heart" may refer to palpitations or dizziness, a "weak kidney" to sexual dysfunction, a "weak nervous system" to headaches, and a "weak stomach or liver" to indigestion.

Looking another person directly in the eyes may be deemed disrespectful. Women may be reluctant to discuss sex, childbearing, or contraception when men are present and demonstrate this unwillingness by giggling, shrugging their shoulders, or averting their eyes. Negative emotions and expressions may be conveyed by silence or a reluctant smile. A smile may express joy, convey stoicism in the face of difficulty, indicate an apology for a minor social offense, or be a response to a scolding to show sincere acknowledgment for the wrongdoing or to convey the absence of ill feelings. Vietnamese prefer more physical distance during personal and social relationships than some other cultures, but extended Vietnamese families of many individuals live comfortably together in close quarters, carrying on the tradition of their home country.

29.3.3 Temporal Relationships

Religious practices and family tradition place emphasis on continuity, cycles, and worship of ancestors. Traditional Vietnamese may be less concerned about the precise schedules than are European Americans and other individualistic cultures, concentrating on the present and to some extent, on the future.

Asian people are known to arrive late for appointments. Noncompliance in keeping appointments may relate to not understanding oral or written instructions or to not knowing how to use the telephone. One other aspect of time involves the concept of age. Vietnamese people pay much less attention to people's precise ages than do Americans. Actual dates of birth may pass unnoticed, with everyone celebrating their birthdays together during the Lunar New Year (*Tet*) in January or February. In addition, a person's age is calculated roughly from the time of conception; most children are considered to be already a year old at birth and gain a year each Tet. A child born just before Tet could be regarded as 2 years old when only a few days old by American standards. Because the practice of determining age is so different in Vietnam, many immigrants who do not know their exact birth date are often assigned January 1 for official records.

When a friend is invited on an outing, the bill is paid for by the person offering the invitation. When giving gifts, the giver often discounts the item, even though it may be of great value. The recipient of a gift is expected to display significant gratitude, which sometimes lasts a lifetime. Some may be reluctant to accept a gift because of the burden of gratitude. Vietnamese-Americans may refuse a gift on the first offer, even if they intend to accept it eventually, so as not to appear greedy.

29.3.4 Format for Names

Most traditional Vietnamese names consist of a family name, a middle name, and a given name of one or two words, always written in that order. There are relatively few family names, with Nguyen (pronounced "nwin") and Tran accounting for more than half of all Vietnamese names. Other common family names are Cao, Dinh, Hoang, Le, Ly, Ngo, Phan, Vu/Vo, and Pho. Additionally, there is little diversity in middle names, with Van being used regularly for men and Thi (pronounced "tee") for women. Given names frequently have a direct meaning, such as a season of the year or an object of admiration. Family members often refer to offspring by a numerical nickname indicating their order of birth.

This practice may increase the difficulty of modern record-keeping and identification of specific individuals. Therefore, use the family name in combination with the given name. Indeed, Vietnamese refer to one another by given name in both formal and informal situations. For example, a typical woman's name is Tran Thi Thu, which is how she would write or give her name if requested. She would expect to be called simply Thu or sometimes *Chi* (sister) Thu by friends and family. In other situations, she would expect to be addressed as *Cô* (Miss) or *Ba* (Mrs.) Thu. If married to a man named Nguyen Van Kha, the proper way to address her would be as Mrs. Kha, but she would retain her full three-part maiden name for formal purposes. The man would always be known as Kha or *Ong* (Mr.) Kha. Some Vietnamese American women have adopted their husband's family name. Children always take the father's family name.

29.4 Family Roles and Organization

29.4.1 Head of Household and Gender Roles

The traditional Vietnamese family is strictly patriarchal and is almost always an extended family structure, with the man having the duty of carrying on the family name through his progeny. Some families who are not accustomed to female authority figures may have difficulty relating to women as professional health-care providers, although this is changing in Vietnam. With the move into Western society, the father may no longer be the undisputed head of the household, and the parents' authority may be undermined. Immigrant Vietnamese families frequently experience role reversals, with wives or children adapting more easily than men.

A Vietnamese woman lives with her husband's family after marriage but retains her own identity. Within the traditional family, the division of labor is gender related: the husband deals with matters outside the home, and the wife is responsible for the actual care of the home, and often makes health-care decisions for the family. While many Vietnamese and Vietnamese American women work outside the home, they also continue as the primary caretaker of the home. Although her role in family affairs increases with time, a Vietnamese wife is expected to be dutiful and respectful toward her husband and his parents throughout the marriage.

Vietnamese immigrants of all subgroups have experienced degrees of reversal of the provider and recipient roles that existed among family members in Vietnam. "Women's jobs," such as hotel maid, sewing machine operator, and food-service worker, are more readily available than male-oriented unskilled occupations; today more men are employed in these jobs. Role reversals between parents and children are also common because children often learn the English language and American customs more rapidly than their parents and therefore, may be able to find employment more quickly. Vietnamese families in the United States experience a greater tendency toward nuclearization, growth in spousal interaction and interdependency, more-egalitarian spousal relations, and shared decision-making than their traditional counterparts.

29.4.2 Prescriptive, Restrictive, and Taboo Behaviors for Children and Adolescents

Traditionally, children are expected to be obedient and devoted to their parents, their identity being an extension of the parents. Children are obliged to do everything possible to please their parents while they are alive and to worship their memory after death. The eldest son is usually responsible for rituals honoring the memory and invoking the blessings of departed ancestors. This pattern may be ingrained from early childhood.

Vietnamese children are prized and valued because they carry the family lineage. For the first 2 years, their mothers primarily care for them; thereafter, their grandmothers and others take on much of the responsibility. Working women may rely on family members to care for their children. Parents usually do not discipline or place extensive limits on their children at a young age. Generally, Vietnamese do not use corporal punishment such as spanking; rather, they speak to the children in a quiet, controlled manner.

Young people are expected to continue to respect their elders and to avoid behavior that might dishonor the family. As a result of their exposure to Western cultures, a disproportionate share of young people have difficulty adapting to this expectation. A conflict often develops between the traditional notion of filial piety, with its requisite subordination of self and unquestioning obedience to parental authority, and the pressures and needs associated with adaptation to their new life. Ironically, successful relationships with Americans at school have placed Vietnamese adolescents at risk for conflicts with their parents. Conversely, failure to form such relationships with their American peers has sometimes appeared to be a precursor of emotional distress. Parents do, however, show relative approval for adolescent freedom of choice regarding dating, marriage, and career choices.

The extreme bipolarities of the adaptation of Vietnamese youth are sometimes overemphasized. Members of one group, usually the children of the first-wave refugees, are often portrayed as academic superstars. At the other end of the social spectrum are the criminal and gang elements, who often direct their activities against other Asian immigrants. Most Vietnamese adolescents, however, fall between these two extremes and have the same pressures and concerns as other youths.

29.4.3 Family Goals and Priorities

The traditional Vietnamese family is perhaps the most basic, enduring, and self-consciously acknowledged form of national culture among refugees, providing lifelong protection and guidance to the individual. The family, usually large, patriarchal, and extended, includes minor children, married sons, daughters-in-law, unmarried grown daughters, and grandchildren under the same roof. Other close relatives may be included within the extended family structure. The family is explicitly structured with assigned priorities, identifying parental ties as paramount. A son's obligations and duties to his parents may assume a higher value than those to his wife, children, or siblings. Sibling relationships are considered permanent. Vietnamese self is defined more along the lines of family roles and responsibilities and less along individual lines. These mutual family tasks provide a framework for individual behavior, promoting a sense of interdependence, belonging, and support. The traditional family has been altered as a consequence of Western influence, urbanization, dual-family member employment, and ongoing acculturation and still the effects of the war-induced absence of men. Nevertheless, many Vietnamese continue to uphold this social form as the preferable basis of social organization upon migrating.

Traditionally, older people are honored and have a key role in transmitting guidelines related to social behavior, preparing younger people for handling stressful life events, and serving as sources of support in coping with life crises. Older people are usually consulted for important decisions. Addressing a client in the presence of an older person, whether they speak English or not, instead of the elder, may be interpreted as disrespectful to the family. Homesickness and bewilderment are especially acute in older refugees when confronted with the strange Western culture and despair about the future. Accustomed to considerable respect and esteem in their homeland, they may feel increasingly alienated and alone as the younger generations adopt new values and ignore the counsel and values of the elders. Living within the family unit facilitates the social adjustment of older refugees into their new society and culture.

Traditional Vietnamese are class conscious and rarely associate with individuals at different levels of society. Traditional respect is accorded to people in authoritative positions who are well educated or otherwise successful or who have professional titles. Two concepts govern the gain and loss of prestige and power, thereby maintaining face: *mien*, based on wealth and power, and *lien*, based on demonstration of control over and responsibility for moral character. For example, to smile in the face of adversity is to maintain *lien* and is considered of great importance.

29.4.4 Alternative Lifestyles

The complex extended Vietnamese family is extremely vulnerable to change. Many young people, frequently unmarried couples, seek their own living accommodations away from the control of older generations. Unattached male refugees may join *pseudofamilies*, households made up of close and distant relatives and friends who share accommodations, finances, and companionship. These families form an important source of social support in the refugee communities. Because of the high regard for chastity placed on Vietnamese adolescents, the number of single-parent households is low, as is the divorce rate.

During the Vietnam War, LGBT (lesbian, gay, bisexual and transgender) Asians and Americans protested the war. At the same time, gay liberation groups took hold which challenged labelling and the inference of being un-American-like and racism emerged as a confounding variable during this time (McConnell and Holloway n.d.). For people living in Vietnam today, differing sexual orientation is difficult for those to face because being gay or lesbian brings shame upon the family, causing many gays and lesbians to remain closeted. As in the United States and elsewhere, Vietnamese society is undergoing changes in the rights of gay, lesbian, transgender, and queer people; however, current political and family norms, through the process called *outlawing,* have excluded the rights of these individuals (Horton and Rydstrom 2019).

29.5 Workforce Issues

29.5.1 Culture in the Workplace

First-wave immigrants adjusted well to the American workplace, and within a decade, their average income equalled that of the general US population. Many later immigrants, who had less education and did not know English, ended up working in lower-paying jobs. However, some learned English and opened their own businesses and prospered.

Traditionally, priority is given to the concerns of the family rather than to those of the employer. However, this emphasis is not a detriment to productivity in work habits, because a good work record and steady pay bring honor and prosperity to the family. Most are highly adaptable and adjust their work habits to meet requirements for successful employment.

Most Vietnamese respect authority figures with impressive titles, achievement, education, and a harmonious work environment. They may be less concerned about such factors as punctuality, adherence to deadlines, and competition. Other traditions include a willingness to work hard, sacrifice current comforts, and save for the future to ensure that they assimilate well into the workforce. Eighty-three percent believe they can get ahead if they work hard and close to half believe their children will be better off than them in the future (Pew Research Center 2012). Many seek the same material, financial, and status rewards that beckon native-born Americans.

29.5.2 Issues Related to Autonomy

Confucianism and its stress on the maintenance of formal hierarchies within governmental, religious, and educational institutions; commercial establishments; and families have heavily influenced the Vietnamese outlook. This cultural background results in conformity and reluctance to undertake independent action. At the same time, the cultural outlook of company and family values superseding personal values creates a cohesive work group. Moreover, because many fear losing their job if they speak out about inequities; they are likely to be taken advantage of by some more-unscrupulous employers.

29.5.3 Language Barriers

Vietnamese Americans quickly learn vocabulary for pragmatic communication but may have difficulty with complex verbal skills. Values related to their own culture discourage disclosure of inner thoughts and feelings. These barriers may adversely affect employment opportunities and limit their ability to communicate needs relative to social, psychological, and economic matters. Employers may need to allow extra time and provide visually oriented instructions and programs that enhance communications to promote increased harmony in the workplace.

29.6 Biocultural Ecology

29.6.1 Skin Color and Other Biological Variations

Vietnamese are members of the Mongolian or Asian race. Although their skin is often referred to as "yellow," it varies considerably in color, ranging from pale ivory to dark brown. Mongolian spots, bluish discolorations on the lower back of a newborn child, are normal hyperpigmented areas in many Asians and dark-skinned races.

To assess for oxygenation and cyanosis in dark-skinned Vietnamese, the health-care provider must examine the sclerae, conjunctivae, buccal mucosa, tongue, lips, nailbeds, palms of the hands, and soles of the feet. These same areas should be observed for adverse reactions during blood transfusions, giving special attention to diaphoresis on the forehead, upper lip, and palms, which may signify impending shock.

One of the first signs of iron deficiency anemia is pallor, which varies with skin tones. Dark skin loses the normal underlying red tones, so that Vietnamese patients with brown skin will appear yellow-brown. Petechiae and rashes may be hidden in dark-skinned individuals as well, but these can be detected by observing for patches of melanin in the buccal mucosa and on the conjunctivae. Jaundice can be observed in dark-skinned Vietnamese as a yellow discoloration of the conjunctiva. Because many dark-skinned individuals have carotene deposits in the subconjunctival fat and sclera, the hard palate should also be assessed.

The Vietnamese are usually small in physical stature and light in build relative to most Europeans and European Americans. Adult women average 5 ft tall and weigh 80–100 lbs. Men average a few inches taller and weigh 110–130 lbs. Although Roberts et al. (1985) reported no significant difference in birth weight between refugee babies and those of other parents, Vietnamese children are small by American standards, not fitting the published growth curves. The study by Vangen et al. (2002) of the birth weights for Vietnamese, Pakistani, Norwegian, and African American babies found that the mean birth weights were largely unrelated to perinatal mortality, which was lowest for the Vietnamese. They concluded that the differences in perinatal mortality between ethnic groups were not explained by differences in mean birth weight. Paradoxical differences in birth weight–specific mortality rates could be resolved by adjustment to a relative scale. Thus, growth charts commonly used in America cannot provide adequate assessments for evaluating the physical development of Vietnamese children. Other parameters such as parental height and weight, apparent state of health, the energy level of the child, and progressive development over time need to be considered. In addition, the difference between refugee status of the seventies until immigration during modern times may erase these older findings.

Typical physical features of the Vietnamese include almond-shaped eyes, sparse body hair, and coarse head hair. Vietnamese also have dry earwax, which is gray and brittle. People with dry earwax have few apocrine glands, especially in the underarm area, and thus produce less sweat and associated body odor. Asians generally have larger teeth than Europeans and Americans, creating a normal tendency toward a prognathic profile. In addition, there may be a torus, bony protuberance, on the midline of the palate or on the inner side of the mandible near the second premolar (Garcia-Garcia et al. 2010).

Betel nut pigmentation may be found in some Vietnamese adults, resulting from the practice of chewing betel leaves (*chau*). This practice is common among older women and has a narcotic effect on diseased gums. Some older women lacquer their teeth, believing that it strengthens the teeth and symbolizes beauty and wealth.

29.6.2 Diseases and Health Conditions

Vietnamese Americans are less likely to report needing mental health-care and also less likely to discuss such issues with a professional. Still today, Asian Pacific Islander older adults still have difficulty talking with a professional and that is the reason for not seeking treatment (Sorkin et al. 2016). Many of the problems are believed to be related to the Vietnam War and leaving the country in 1975, which led to physical problems, psychological stress, and depression. "They already had pre-war trauma, and they come to the US; it's a new country, a new language, and they have to find jobs. Thirty years after the war, there are still having problems (Sorkin et al. 2008). Mental-health research has indicated that Vietnamese refugees have disturbingly high rates of depression, generalized anxiety disorders, and post-traumatic stress associated with military combat, political imprisonment, harrowing events during escapes by sea, and brutal pirate attacks. Chronic personal and emotional problems often stem from post-traumatic stress experiences in this population (Kim et al. 2019). In addition, acculturation may influence the stigmatization of getting help for personal problems (Do et al. 2018).

Related to mental health concerns, Kinzie and Manson (1982) and Buchwald et al. (1993) developed a Vietnamese Depression Scale (VDS), which uses terms that allow an English-speaking practitioner to make a cross-cultural assessment of the clinical characteristics of depressed Vietnamese patients. Health-care providers working with Vietnamese patients may find this scale useful when providing mental-health services. Recent research indicated that a high prevalence of depressive symptoms are evident among Vietnamese Americans associated with increased perceived discrimination (Chau et al. 2018).

Of immediate concern to health-care providers working with Vietnamese refugees is the treatment of infectious diseases that jeopardize both the refugee and the resident population. Some refugees suffer from malaria, intestinal parasites, worms associated with poorly cooked foods, and other problems associated with the tropics and rural areas, crowded cities, and rural areas, including tuberculosis, hepatitis B, and anemia, mostly present in children. Waterborne diseases include bacterial diarrhea, hepatitis A, and typhoid fever. Vectorborne diseases include dengue fever, malaria, and Japanese encephalitis. Hepatitis B virus vaccination is recommended for all newborn refugee children.

To determine the presence of parasites, health-care providers must assess for symptoms of anemia, lassitude, failure to thrive, abdominal pain, weight loss, and skin rashes. In the first two waves of refugees, major health problems also included skin infections caused by fungus, impetigo, scabies, and lice (7–15%); infections of the upper respiratory tract and otitis media (20%); anemia including parasitic iron deficiency (16–40%), with a higher occurrence in young children; hemoglobin disorders (30%); chronic diseases (10%); and malnutrition and poor immunization status (Ross 1982). Screening for erythrocytic microcytosis is also important, the condition which can lead to an incorrect diagnosis of iron deficiency and inappropriate treatment with iron. Microcytosis may also be evident as Southeast Asians is most likely a reflection of the presence of thalassemia or of hemoglobin E trait, conditions that are usually harmless and need no treatment. These disorders should be suspected in people with findings consistent with tuberculosis but with a negative purified protein derivative response (Ross 1982).

The health-care provider should consider screening newer refugees and immigrants from Vietnam for nutritional deficits; hepatitis B; tuberculosis (as differentiated from melioidosis and paragonimiasis); parasites such as roundworm, hookworm, filaria, flukes, amoebae, and giardia; malaria; HIV; Hansen's disease; and post-traumatic stress disorder. Recommended laboratory and other tests for refugees include a nutritional assessment, stool for ova and parasites, hemoglobin and hematocrit, and a chest radiograph for tuberculosis. Current research indicates low screening, treatment and immunization rates among newly arrived immigrants indicating the need for following protocols and creating better methods to assure that these health risks are identified (Waldorf et al. 2014). In addition, moderate to severe dental problems may occur in newer immigrants, especially children. Clinical practice guidelines and materials from the federal government are helpful in health screening and for concurrent medical conditions or diseases for Vietnamese Americans children and adults (CDC 2008; USDHHS 2013).

29.6.3 Variations in Drug Metabolism

Little pertinent drug research exists specifically on Vietnamese-Americans and research in this area continues to evolve. Clinical studies comparing other Asians with European Americans provide some idea of what might be expected. For example, the Chinese are twice as sensitive to the effects of propranolol on blood pressure and heart rate; experience a greater increase in heart rate from atropine; require lower doses of benzodiazepines, diazepam, and alprazolam because of their increased sensitivity to the sedative effects of these drugs; require lower doses of imipramine, desipramine, amitriptyline, and clomipramine; and are less sensitive to cardiovascular and respiratory side effects of analgesics (e.g., morphine) but are more sensitive to their gastrointestinal side effects. Asians require lower doses of neuroleptics (e.g., haloperidol) (Levy 1993).

Lin and Shen (1991) expressed concern about the lack of research on pharmacotherapy specifically related to major depressive and post-traumatic stress disorders in Southeast Asian refugees. They suggested that drug metabolism is comparable with that of other Asian groups with important common traits such as genetic, cultural, and environmental influences. Asian diets, for example, are similar in their higher carbohydrate-to-protein ratio, which significantly influences the metabolism of some commonly prescribed drugs. Also, because most Asians come from areas with similar degrees of socioeconomic development, exposure to various enzyme-inducing agents, such as industrial toxins, is likely to be similar. Conversely, the exposure of the refugees to war, trauma, starvation, and other adverse conditions could have an effect on the enzyme systems governing psychotropic medications. One precaution involves the continued extensive use of traditional herbal medicines by refugees. Some of these herbal drugs have active pharmacologic properties that may interact with psychotropic drugs. For example, some may cause atropine psychosis when ingested concomitantly with tricyclic antidepressants or low-potency neuroleptics.

Significantly lower dosages of psychotropic medications are prescribed in Asian countries than are common in Western countries. Low doses of antidepressant medications are often effective. Weight standards for neuroleptic dose ranges are significantly lower in Asians than among White Americans (Levy 1993). Because Vietnamese are considerably smaller than most White Americans, medication dosages may need to be reduced. Vietnamese generally consider American medicines more concentrated than Asian medicines; thus, they may take only half of the dosage prescribed.

29.7 High-Risk Behaviors

Adult cigarette smoking has steadily declined since the 1964 seminal report by the Surgeon General linking smoking to bad health. However, different ethnic groups are more likely to smoke versus others. Smoking prevalence among Asians was 10.9% during 2010–2013. However, Vietnamese reported 24.4% for men and 7.9% for women or 16% overall (Martell et al. 2016).

Researchers investigating alcohol use among Asia Americans has been largely understudied, yet it has been assumed that alcohol use has been low. Iwamoto et al. (2016) reported an increase in alcohol use in this demographic, most significantly among 18–25-year olds. Suggestive causes may be the effects of acculturation. Higher levels of depression, distress, and discrimination may be important factors. Yet both Chinese and Vietnamese Americans drink less than their Asian counterparts (Japanese, Korean, multi-Asian

Americans). Thus, much more research on this topic is necessary. Finally, many Asians are slow metabolizers of alcohol. Thus, Asians are more sensitive than European Americans and White groups to the adverse effects of alcohol as expressed by facial flushing, palpitations, and tachycardia. These physiological effects may decrease the propensity for drinking alcohol.

Two gastrointestinal problems relate to high-risk behavior. High rates of gastrointestinal cancer may be due to asbestos that is used in the process of "polishing" rice in some parts of the world. Colorectal cancer is the fourth most common cancer in the United States, and the third most common among Vietnamese adults in California. Yet Vietnamese Americans have lower rates of screening for colorectal cancer compared to other Asian Americans and Whites (Medical News Today 2010). Trichinosis risk is 25 times greater in Southeast Asian refugees than in the general population. This increased risk is related to undercooking pork and purchasing pigs directly from farms.

29.8 Nutrition

29.8.1 Meaning of Food and Rituals

Meals are an important time to the Vietnamese, allowing the entire family to come together and share a common activity. Preparation is precise and may occupy much of the day. Celebrations and holidays involve elaborately prepared meals.

Because of their size, the normal daily caloric intake of the Vietnamese is approximately two-thirds that of average Americans. Rice is the main staple in the diet, providing up to 80% of daily calories. Other common foods are fish (including shellfish), pork, chicken, soybean curd (tofu), noodles, various soups, and green vegetables. Preferred fruits are bananas, mangoes, papayas, oranges, coconuts, pineapples, and grapefruits. Soy sauce, garlic, onions, ginger root, lemon, and chili peppers are used as seasoning.

The Vietnamese eat almost exclusively white or polished rice, disdaining the more nutritious brown or unpolished variety. Rice and other foods are commonly served with *nuoc mam*, a salty, marinated fish oil sauce. A meal typically consists of rice, *nuoc mam* and a variety of other seasonings, green vegetables, and sometimes meat cut into slivers. Chicken and duck eggs may be used. The Vietnamese prefer white bread, particularly French loaves and rolls, and pastry. A regular dish is *pho*, a soup containing rice noodles, thinly sliced beef or chicken, and scallions.

Other Vietnamese dishes resemble Chinese foods commonly seen in the United States. Some of these include *com chien* (fried rice) and *thit bo xau ca chua* (beef fried with tomatoes). Perhaps the favorite of Americans is *cha gio* (pronounced "cha-yuh"), a combination of finely chopped vegetables, mushrooms, meat or bean curd, rolled into delicate rice paper and deep fried. If fried, it is also called a "spring" roll, while if left uncooked (the rice paper), it is a "summer" roll. It is served as part of elaborate meals or during celebrations; proper preparation may require many hours.

Vietnamese eat three meals a day: a light breakfast, a large lunch, and dinner, with optional snacks. Meals are served communal style, with food being placed in the center of the table or passed around, with everyone taking what they wish. If in a restaurant, the various dishes are often brought out when they are prepared, not necessarily all at once. Children wait for their elders to pass each dish. Chopsticks and sometimes spoons are used for eating. Knives are seldom necessary at the table, because meat and vegetables are usually cut into small pieces before serving. Stir frying, steaming, roasting, and boiling are the preferred methods of cooking. Hot tea is the usual beverage.

29.8.2 Dietary Practices for Health Promotion

A predominant aspect of the traditional Asian system of health maintenance is the principle of balance between two opposing natural forces, known as *am* and *duong* in Vietnamese. As with medicines, these forces are represented by foods that are considered hot (*duong*) or cold (*am*). The terms have nothing to do with temperature and are only partly associated with seasoning. Rice, flour, potatoes, most fruits and vegetables, fish, duck, and other things that grow in water are considered cold. Most other meats, fish sauce, eggs, spices, peppers, onions, candies, and sweets are hot. Tea is cold, coffee is hot, water is cold, and ice is hot.

Illness or trauma may require therapeutic adjustment of hot–cold balance to restore equilibrium. Hot foods and beverages, used to replace and strengthen the blood, are preferred after surgery or childbirth. During illness, certain foods are consumed in greater quantity, such as a light rice gruel (*chao*) mixed with sugar or sweetened condensed milk, and a few pieces of salty pork cooked with fish sauce. Fresh fruits and vegetables are usually avoided, being considered too cold. Water, juices, and other cold drinks are restricted. Nutritional counseling should take into consideration these factors and other aspects of the usual Vietnamese diet, because advice to simply eat certain kinds of American foods may be ignored.

The evaluation of food practices among various populations continues to be studied. A study conducted in California for Asian Americans of the 5-day requirement of fruits in vegetables found that Vietnamese Americans who were female were less likely to meet this nutrition standard. Vietnamese Americans overall were unlikely to meet these standards compared to their counterparts born in the US (Sarwar et al. 2015). This has implications for health teaching as higher vegetable and fruit intake is associated with lower risk for cardiovascular disease and cancer.

29.8.3 Nutritional Deficiencies and Food Limitations

The traditional Vietnamese diet is basically nutritious with high use of fruit comparing favorably with U.S. federal guidelines for a diet low in fat and sugar, high in complex carbohydrates, and moderate in fiber. However, the prevalence of anemia in children may be associated with an iron deficiency. Many pregnant women have thalassemia β which may be genetically transmitted to their children (Nguyen 2015). The Vietnamese diet may also be deficient in calcium and zinc but exceedingly high in sodium, with implications relevant to hypertension.

Before 1975, immigrants encountered difficulty in preparing traditional dishes, especially in areas with no established Vietnamese community. Even then, the determined housewife could assemble most necessary ingredients through judicious selections at ethnic American, Chinese, Korean, and Indian groceries. Today, nearly all common Vietnamese foods are available at reasonable cost, except perhaps for certain native fruits and vegetables. Vietnamese dishes are often built around rice, broth, and layers of flavor. In addition, Vietnamese Americans have changed their diet to a degree, often increasing their fat intake.

Understanding nutritional practices in Vietnam can augment some understanding of Vietnamese Americans. Immigrants may eat fewer grains, fruits and vegetables while consuming more fat, cholesterol, sugared drinks, and fast foods. Most Vietnamese adults and many children have lactose intolerance, which may cause problems in schools, other institutional settings, and adoptive families. Calcium intake may be poor and lactose intolerance hard to manage if appropriate substitute foods are unfamiliar or difficult to obtain. Health-care providers may need to encourage the use of substitute milk products that are based on soybeans. Additional challenges are a different food measurement system (metric vs. apothecary), working women and distaste for food that is not fresh (UNE Applied Nutrition Program 2019).

Reflective Exercise #1

Miss Thu, a very thin recent 20-year-old Vietnamese immigrant presents to the urgent care clinic with gastrointestinal distress. The on-duty nurse takes her nutritional history. Miss Thu had visited with some friends at an Italian restaurant where she ate Fettuccini alfredo, the only food at the meal that was new to her. She was concerned that she had food poisoning, even though her friends ate the same food and had no problems. The nurse suspects that she has a lactose intolerance.

1. What is the immediate treatment for lactose intolerance?
2. Is lactose intolerance an immediate threat for Miss Thu?
3. What recommendations can the nurse make on a short-term basis upon discharge?
4. Would you refer Miss Thu to a nutritionist for dietary counseling?
5. To get her calcium needs met, what foods should the nutritionist encourage her to avoid?

29.9 Pregnancy and Childbearing Practices

29.9.1 Fertility Practices and Views Toward Pregnancy

Capturing fertility rates, practices, including contraception, abortion and fertility rates of immigrant Vietnamese are relatively unknown. In Vietnam, the birth rate is 15.2 births per 1000 population (CIA Factbook 2020). Demographic information about Asian-American mothers indicates that Vietnamese Americans accounted for 7.9% of the births to non-Hispanic Asian women in 2016 with the state of Louisiana in the US claiming the highest percentage of Vietnamese American women giving birth (32.5%). Of the women born outside of the US, Vietnamese were least likely to be married and have not completed college, but most likely to receive prenatal Special Supplemental Nutrition Program for Women, Infants, and Children (WIC) (Driscoll 2018).

After arriving in the US, women often desire information on contraception but may be afraid to ask. The problem stems in part from their cultural background and emphasis on premarital modesty and virginity. However, when contraception is addressed and information made available Vietnamese women choose some method of contraception. Providers should avoid forceful family-planning indoctrination on the first encounter, but such information is usually well received on subsequent visits.

29.9.2 Prescriptive, Restrictive, and Taboo Practices in the Childbearing Family

Traditionally, prescriptive food practices for a healthy pregnancy include noodles, sweets, sour foods, and fruit but avoidance of fish, salty foods, and rice. After birth, to restore equilibrium and provide adequate warmth to the breast milk, women consume soups with chili peppers, salty fish and meat dishes, and wine steeped with herbs. In addition to hot *(duong)* and cold *(am)*, foods are classified as tonic and wind. Tonic foods include animal protein, fat, sugar, and carbohydrates; they are usually also hot and sweet. Sour and sometimes raw and cold foods are classified as antitonic. Wind foods, often classified as cold, include leafy vegetables, fruit, beef, mutton, fowl, fish, and glutinous rice. It is considered critical to increase or decrease foods in various categories to restore bodily balances upset by unusual or stressful conditions such as pregnancy. Whereas the balance of foods may be followed, the terminology is not consistently used.

During the first trimester, the expectant mother is considered to be in a weak, cold, and antitonic state. Therefore, she should correct the imbalance by eating hot foods such as ripe mangoes, grapes, ginger, peppers, alcohol, and coffee. To provide energy and food for the fetus, she is prescribed tonic foods, including a basic diet of steamed rice and pork. Cold foods, including mung beans, green coconut, spinach, and melon, and antitonic foods, such as vinegar, pineapple, and lemon are avoided during the first trimester.

In the second trimester, the pregnant woman is considered to be in a neutral state. Cold foods are introduced, and the tonic diet is continued.

During the third trimester, when the woman may feel hot and suffer from indigestion and constipation, cold foods are prescribed and hot foods are avoided or strictly limited. Tonic foods, which are believed to increase birth weight, are restricted to reduce the chances of a large baby, which would make birthing difficult. Wind foods are generally avoided throughout pregnancy, because they are associated with convulsions, allergic reactions, asthma, and other problems. This regimen may appear more complex and restrictive than it actually is in practice. Most women use it only as a general guide, commonly restricting, rather than totally abstaining from, the proscribed foods. A great variety of food, including rice, many kinds of vegetables and fruits, various seasonings, and certain meats and fish, is generally permissible throughout pregnancy.

Intensive prenatal care is not the norm in Southeast Asia. Many women do not seek medical attention until the third trimester because of cost, fear, or lack of perceived need. For obstetric and gynecological matters, Vietnamese women feel more comfortable with a female physician or midwife.

Traditionally, Vietnamese women maintain physical activity to keep the fetus moving and to prevent edema, miscarriage, or premature delivery. Prolonged labor may result from idleness; an undesirable large baby may result from afternoon napping. Additional restrictive beliefs include avoiding heavy lifting and strenuous work; raising the arms above the head that pulls on the placenta causing it to break; and sexual relations late in pregnancy that may cause respiratory stress in the infant. In Vietnam, many consider it taboo for pregnant women to attend weddings or funerals. However, they often look at pictures of happy families and healthy children, believing that it helps give birth to healthy babies.

In Vietnam, some rural children are delivered in a screened-off portion of the home or in a special birth house by certified midwives; more frequently though, more are born in hospitals with Western-trained physicians or midwives in attendance, especially in the cities and towns, although they may receive their prenatal care in the rural clinics. Southeast Asians generally dislike invasive procedures, such as episiotomies, cesarean sections, circumcisions, nasal oxygen, and intravenous fluids. However, unlike some women of other ethnic groups, Vietnamese women may ask for anesthesia during labor and delivery and epidurals are becoming popular if the woman can pay. Otherwise, once in labor, the Vietnamese woman tries to maintain self-control and may even smile continuously. Her period of labor is usually short, and there may be no warning of impending delivery. Although a special bed may be available, the mother may prefer walking around during labor and squatting during the birth process. This position is less traumatic than others for both mother and baby and results in fewer and less-serious lacerations. This is a deviation from normal birth practices in the US and may need to be discussed with the attending physician or midwife prior to birth.

Because the head is considered sacred, neither that of the mother nor that of the infant should be touched or stroked. Removal of vernix from the infant's head can cause distress. The practice of inserting intravenous devices into infants' scalps can be particularly stressful to Vietnamese families. Health-care providers need to stress the importance and necessity of this invasive procedure and select other venous routes if possible.

Customary practices include clearing the neonate's throat using the finger, cutting the umbilical cord with a nonmetal instrument, quickly burying the placenta to protect the infant's health and ritually cleaning the mother in a manner that does not involve actual bathing with water.

Because body heat is lost during delivery, Vietnamese women avoid cold foods and beverages and increase consumption of hot foods to replace and strengthen their blood. Ice water and other cold drinks are usually not welcome; thus, the usual practice of offering a newly delivered mother a cold drink should be replaced with something hot—either water or tea is usually available. This can accomplish the nurse's goal of replacing fluids and maintain the patient's cultural heritage. Most raw vegetables, fruits, and sour items are taken in lesser amounts. Prescriptive foods include steamed rice, fish sauce, pork, chicken, eggs, soups with chili or black peppers, other highly seasoned and salty items, wine, and sweets.

Because water is cold, women traditionally do not fully bathe, shower, or wash their hair for a month after delivery. Some Vietnamese women have complained that they were adversely affected by showering shortly after delivery in American hospitals. Others, however, have welcomed the opportunity to shower and seem willing to give up other traditional practices. Postpartum women also avoid drafts and strenuous activity; wear warm clothing; stay in bed, indoors, or both for about a month and avoid sexual intercourse for months. In the past, postpartum women remained in a special bed above a slow-burning fire. This practice still continues with the use of hot-water bottles or electric blankets.

Other women in the family assume responsibility for the baby's care. In Vietnam, husbands would never be present at their child's delivery. For Vietnamese in the US, this varies and some men do attend deliveries. The mother's inactivity and dependence on others may be incorrectly interpreted by health-care workers as apathy, depression, or lack of attachment to the baby. A newborn is often dressed in old clothes; it is considered taboo to praise the child lest jealous spirits steal the infant. The mother may be reluctant to cut the child's hair or nails for fear that this might cause illness. The infant is generally maintained on a diet of milk for the first year, with the introduction of rice gruel at around 6 months. There is little formal toilet training; the child usually learns by imitating an older child. One can see mothers holding their naked babies away from them to urinate, and "whispering" in their ears to stimulate a bowel movement. The child is then cleaned and returned to his or her usual clothing.

Breastfeeding is customary in Vietnam, but since resettlement, some variations on this practice have been instituted. Some Southeast Asian women discard colostrum and feed the baby rice paste or boiled sugar water for several days. This does not indicate a decision against breastfeeding. After the milk comes in, both mother and baby benefit from the hot foods consumed by the mother for the first month. Then, however, a conflict arises: the mother believes that hot foods benefit her health but that cold foods ensure healthy breast milk. Having the mother change from breastfeeding to formula can easily solve this dilemma; however, it is counterproductive to the medical and nursing community's efforts to promote breastfeeding during the baby's first year. If the mother cannot afford formula, she may use fresh milk or rice boiled with water, which may result in anemia and growth retardation. Some health-care professionals concerned about these developments and their impact on the infant's health have recommended educational programs that might restore conditions conducive to traditional breastfeeding.

Finally, one recent study explored Vietnamese American mothers and their experiences after birth related to traditions, post-partum depression, and seeking help. The results indicated that half of the sample of 15 women had sadness during the post-partum period and expressed the need to maintain strong cultural connections and value (Park et al. 2017). Some reported feelings of isolation and sadness but not of significant concern to seek out professional help. The importance of post-partum cultural traditions warrants important consideration for new mothers of Vietnamese heritage.

29.10 Death Rituals

29.10.1 Death Rituals and Expectations

Vietnamese accept death as a normal part of the life process. The traditional stoicism of the Vietnamese, the influence of Buddhism with its emphasis on cyclic continuity and reincarnation, and the pervading association of current activities with ancestral spirits and burial places contribute to attitudes toward death. Other Vietnamese, who may prefer cremation, will donate body parts under certain circumstances.

Most Vietnamese have an aversion to hospitals and prefer to die at home. Some believe that a person who dies outside the home becomes a wandering soul with no place to rest. Family members think that they can provide more comfort to the dying person at home. Ancestors are commonly honored and worshipped and are believed to bestow protection on the living.

Generally, Southeast Asians tend not to want to artificially prolong life and suffering, but it may still be difficult for relatives to consent to terminating active intervention, which might be viewed as contributing to the death of an ancestor who would shape the fates of the living (Muecke 1983). Asian Americans are also less likely than White Americans to complete an advance directive (Sun et al. 2017). Researchers created and tested a church-based and culturally appropriate education intervention to improve the knowledge of advance directives for Vietnamese Americans in California. Pre and post-test results indicated improved intentions, a change in belief and attitudes about medical decision-making, and also the completion of advance directives for churchgoers in this pilot study (Sun et al. 2017). However, sensitivity to the cultural values of Vietnamese cannot be underestimated. Varied immigration experiences and levels of acculturation play out in the care of the seriously ill with a high level of family involvement. Lack of culturally sensitive approaches in end of life care for this population, communication challenges and attention to advance care planning, can contribute to less than optimal end-of-life care (Tran et al. 2019).

Few Vietnamese families consent to autopsy unless they know and agree with the reasons for it. Older Vietnamese, on realizing the inevitability of death, sometimes purchase coffins in advance, display them beneath the household altar, and choose burial sites with a favorable position. Although Vietnamese custom is associated with proper burial practices and maintenance of ancestral tombs, cremation is an acceptable practice to some families.

29.10.2 Responses to Death and Grief

Vietnamese families may wish to gather around the body of a recently deceased relative and express great emotion. Traditional mourning practices include wearing white clothes for 14 days, the subsequent wearing of black armbands by men and white headbands by women, and the yearly celebration of the anniversary of a person's death. In Vietnam, death rituals involve a display of personalized banners and inscriptions on objects and placards which signify the mourner's grief (Shovet 2018). Such observances, together with ritual cleaning and worship at ancestral graves, help reinforce family ties and are deeply woven into Vietnamese culture. Departure from Vietnam has greatly curtailed the observance of these practices, leaving a painful void for many.

Priests and monks should be called only at the request of the client or family. Clergy visitation is usually associated with last rites by the Vietnamese, especially those influenced by Catholicism, and can actually be upsetting to hospitalized patients. Sending flowers may be startling, because flowers are usually reserved for the rites of the dead.

29.11 Spirituality

29.11.1 Dominant Religions

According to the CIA, religions practiced by those living in Vietnam are Buddhism (7.9%), Catholic, (6.6%), Hao Hao (1.7%), Cao Dai (0.9%), Protestant (0.9%), Muslim (0.1%), and none (81.8%) (CIA 2020). Some are driven by the desire to learn, to relieve suffering, to produce beauty, to assist the progress of civilization, and to gain strength from participating in ethnic community activities. Other religions include Taoism and Confucianism, which are basically offshoots and combinations of the major faiths (see Table 29.1). Animism is found mainly among the highland tribes. Many Vietnamese believe that deities and spirits control the universe and that the spirits of dead relatives continue to dwell in the home.

Although some Vietnamese refugees are Catholic, or have converted to other branches of Christianity, many follow Buddhist concepts. Buddhism on the whole is best understood not as a religion in the Western sense but more of a philosophy of life and impacts profoundly on the health-care beliefs and practices of the

Table 29.1 Religions and spiritual beliefs

Eastern philosophies	Spiritual beliefs
Buddhism	Central to Buddhism is the concept of following the correct path of life, thus eliminating suffering that is caused by desire. Another tenet is that the world is a cycle of ordeals: to be born, grow old, fall ill, and die. In addition, people's present lives predetermine their own and their dependents' future lives
Confucianism	Confucianism stresses harmony through maintenance of the proper order of social hierarchies, ethics, worship of ancestors, and the virtues of chastity and faithfulness
Taoism	Taoism teaches harmony, allowing events to follow a natural course that one should not attempt to change. This may be perceived as a passive response to spirituality whereas Westerners maintenance of self-control is dominant

Vietnamese. If one lives in adherence to the Buddhist path, one can expect less suffering in future existences. Buddhism stresses disconnection to the present, especially materialism and self-aggrandizement. Thus, pain and illness are sometime endured and health-seeking remedies delayed because of this belief in fate. Similarly, preventive health-care has little meaning in this philosophy.

Respect for and veneration of ancestors is associated with Buddhism and Confucianism. The prospect of burial away from ancestral burial sites may be a source of significant distress to older Vietnamese and particularly those who now live in the US.

While most Vietnamese who practice a religion are Buddhists, some almost never visit temples or perform rituals. Others, both Buddhist and Christian, may maintain a religious altar in the home and conduct regular religious observances. In cases of severe illness, prayers and offerings may be made at a temple.

The family is the fundamental social unit and the primary source of cohesion and continuity. Whereas the wish to bring honor and prosperity to the family remains a dominant force for most Vietnamese, some find meaning in life from the practice of Buddhism or other religions. A tenet of Buddhism holds that the family unit is more important than the individual, with less emphasis on the "self." Accordingly, health-care decision making is frequently a family matter. Correspondingly, the family is typically involved in health-care treatments.

29.12 Health-Care Practices

The Vietnamese approach to health-care is one of ambivalence. Many Vietnamese immigrants are accustomed to depending on the family unit and traditional means of providing for health needs. They may be distrustful of outsiders and Western methods. Most are familiar with immunizations and diagnostic tests; they do want to avoid health problems and are anxious to follow reasonable procedures. Newly arrived refugees and immigrants are less likely to seek Western health-care; however, once established, Vietnamese are the most likely of Southeast Asians to seek care and to do so earlier (Strand and Jones 1983). Most Southeast Asian refugees want to go to a physician for an illness, but they rarely seek care when they are asymptomatic (for screening and prevention services), and few are familiar with the appointment system. Some regard the most-convenient physician as the closest one not requiring an appointment and accepting medical coupons, which usually translates into a hospital emergency room (Muecke 1983).

29.12.1 Health-Seeking Beliefs and Behaviors

In contrast, the Vietnamese family may not seek outside assistance for illness until it has exhausted its own resources. The family may try various home remedies, allowing the condition to become serious before seeking professional assistance. Once a physician or nurse has been consulted, the Vietnamese are usually quite cooperative and respect the wisdom and experience of health-care professionals. Hospitalization is viewed as a last resort and is acceptable only in case of an emergency when everything else has failed. With respect to mental health, Vietnamese do not easily trust authority figures, including treatment staff, because of their refugee experiences.

The diagnosis of illness is frequently understood in three different, although overlapping models. The first, the least common, could be considered supernatural or spiritual, where illness can be brought on by a curse or sorcery or failure to observe a religious ethic or belief. Traditional medical providers are common, both in the US and Vietnam; some are specialists in the more magico-religious realm and may be called upon to exorcise a bad spirit via chanting, a potion, or consultation from an ancient Chinese text.

The use of amulets and other forms of spiritual protection is also commonly employed. For example, babies and children often wear *bua*, an amulet of cloth containing a Buddhist verse, or that has been blessed by a monk. It is worn on a string around the wrist or neck.

As mentioned earlier in the chapter, Vietnamese traditionally do not have a concept of mental illness as discrete form somatic illness, and thus rarely use Western-based psychological and psychiatric services. Instead, most mental health issues such as depression or anxiety fall into this spiritual health realm and are treated appropriately. Similarly, somatization is common, and treatments overlap with Western treatments and metaphysical interventions described below.

Second, a widespread belief is that the universe is composed of opposing elements held in balance; health is a state of balance between these forces, know as *am* and *duong,* based on the more familiar concepts of the Chinese *yin* and *yang*. In health, these concepts are frequently translated as "hot" and "cold," although they do not necessarily refer to temperature. Illness results when there is an imbalance of the "vital" forces; the imbalance can be a result of a physiological state, such as pregnancy or fatigue, or it can be brought on by extrinsic factors like diet or overexposure to "wind," one of the body forces or humors first described by Galen. Balance can be restored by a number of means, including diet changes to compensate for the excess of "hot" or "cold" Western medicines and injections, traditional medicines, herbs, and medical practices. Naturalistic explanations for poor health include eating spoiled food and exposure to inclement weather. The natural element known as *cao gio* is associated with bad weather.

Third, most Vietnamese Americans also recognize the more Western concept of disease causation such as the germ theory. A widespread understanding is that disease can come from contaminants in the environment, even if full concepts of microbiology or virology are not grasped. Thus, through decades of French occupation and more recently the American influence, even the most rural Vietnamese has come to know the power of antibiotics.

When Vietnamese enter the modern healthcare setting, they do so frequently with the goal to relieve symptoms; in general, the patient expects a medicine to cure the illness immediately. When something is not prescribed initially, the patient is likely to seek care elsewhere, either directly from a Vietnamese pharmacist or specialized "injectionists." Newly arrived immigrants are used to receiving the medication directly from the doctor; the concept of a "prescription" written on a piece of paper to take to a pharmacy to be "filled" is foreign to them. They may feel that this piece of paper contains instructions for the patient, and not follow through with obtaining the medication.

Vietnamese frequently discontinue medicines after the symptoms disappear; similarly, if symptoms are not perceived, there is no illness. Thus, preventive, long-term medications like antihypertensives must be prescribed with culturally sensitive education. It is quite common for Vietnamese patients to amass large quantities of half-used prescription drugs, even antibiotics, many of which are shared with friends and may be sent back to family in Vietnam. Additionally, Vietnamese commonly believe that Western pharmaceuticals are developed for Americans and Europeans, and hence dosages are too strong for more slightly built Vietnamese, resulting in self-adjustment of dosages.

The Vietnamese hold great respect for those with education, especially physicians. The doctor is considered the expert on health; diagnosis and treatment should happen at the first visit with little examination or personally invasive laboratory or other diagnostic tests. Commonly, laboratory procedures involving the drawing of blood are feared and resisted by Vietnamese, who believe the blood loss will make them sicker and that the body cannot replace what was lost. Surgery is especially feared for this reason. Overall, as health is believed to be a function of balance, surgery would be considered an option of last resort, as the removal of an organ would alter the internal balance.

Vietnamese view health and illness from a variety of different perspectives, sometimes simultaneously. It is not uncommon for a sick person to interpret their illness as an interaction of spiritual factors, internal balance inequities, and even an infective process. They will thus combine diagnostic and treatment elements from all three models in order to get the maximum health benefits (Rasbridge 2004).

The belief that life is predetermined is a deterrent to seeking health-care. For many Vietnamese, diagnostic tests are baffling, inconvenient, and often unnecessary. Procedures such as circumcision or tonsillectomy, which biomedicine considers simple, are generally unknown to the Vietnamese. Invasive procedures are frightening. The prospect of surgery can be terrifying. The fear of mutilation stems from widespread beliefs among non-Christians that souls are attached to different parts of the body and can leave the body causing illness or death. Loss of blood from any route is feared, and the Vietnamese may refuse to have blood drawn for laboratory tests. The client may complain, though not to the health-care worker, of feeling weak for months. A Vietnamese client in America may feel that any body tissue or fluid removed cannot be replaced, and the body suffers the loss in this life as well as into the next.

The concept of long-term medication for chronic illnesses and acceptance of unpleasant side effects and increased autonomic symptoms, which are standard components of modern Western medicine, are not congruent with traditional notions of safe and effective treatment of illnesses.

29.12.2 Responsibility for Health-Care

In Vietnam, the family is the primary provider of health-care, even in hospitals. This practice survives because of tradition and a shortage of professional personnel. Their own families attend hospitalized patients' day and night. The importance of involving family members, including elder family members or clan leaders, in all major treatment decisions regarding physical and mental health must be stressed.

Health-care in Vietnam is crisis-oriented, with symptom relief as the goal. Vietnamese typically deal with illness by means of self-care, self-medication, and the use of herbal medicines. Facsimiles of Western prescription drugs are sold over the counter throughout Southeast Asia, which may explain the increasing resistance of bacteria to several readily available antibiotics.

Many Vietnamese believe that Western medicine is very powerful and cures quickly, but few understand the risks of overdosages or underdosages. Patients being treated for depression who fail to take their antidepressants evidence improvement after receiving instructions for taking their medication. Vietnamese patients may not follow prescribed schedules of medication for the treatment and prevention of tuberculosis. Extensive education, repetition of instructions, and home visitations are necessary.

Vietnamese-American women are twice as likely to receive a cervical cancer diagnosis compared to their White or Hispanic counterparts (Nguyen-Truong et al. 2017). The prevalence of the disease is the result of lack of education, reluctance to seek early treatment, fear that nothing can be done, low utilization of annual Pap smears, and failure to follow up on abnormal Pap smears. Some evidence also implicates human papillomavirus (HPV), a sexually transmitted etiological factor, in the pathogenesis of cervical cancer. Cancer and other problems common to Vietnamese people may also be associated with the widespread application of chemical agents during the Vietnam War.

Additionally, the problem has been associated with lack of organized language services and, thus, a failure by women to comprehend the severity of the situation and the potential for recovery if regular treatment begins early enough. To increase follow-up visits and care, it may be necessary to carefully explain the problems that may result if they do not follow up after an abnormal Pap smear. Women should understand that lack of symptoms or pain may be only temporary and that experiences of acquaintances may not apply to them. A community-based cervical cancer education focused on improved knowledge dissemination about cervical cancer and screening can reduce perceived barriers related to screening and promote regular cancer screening and follow-up treatment (Fang et al. 2019).

29.12.3 Folk and Traditional Practices

Folk and traditional practices may be evident for Vietnamese Americans. The forces of *am* (cold) and *duong* (hot) are pervasive forces in the practice of Traditional Vietnamese Medicine. *Am* represents factors that are considered negative, feminine, dark, and empty, whereas *duong* represents those that are positive, masculine, light, and full. These terms are applied to various parts, organs, and processes of the body. For example, the inside of the body is *am*, and the surface is *duong*. The front part of the body is *am*, and the back is *duong*. The liver, heart, spleen, lungs, and kidneys are *am*, and the gallbladder, stomach, intestines, bladder, and lymph system are *duong*. *Am* stores strength, and care must be taken not to use it up too quickly. *Duong* protects the body from outside forces, and if it is not cared for, the organs are thrown into disorder. Proper balance of these two life forces ensures the correct circulation of blood and good health. If the balance is not proper, life is short.

Diseases and other debilitating conditions result from either cold or hot influences. For example, diarrhea and some febrile diseases are due to an excess of cold, whereas pimples and other skin problems result from an excess of hot. Countermeasures involve using foods, medications, and treatments that have properties opposite those of the problem and avoiding foods that would intensify the problem. Asian herbs are cold, and Western medicines are hot. A widely held belief among Vietnamese refugees is that Asian medicine relieves symptoms of a disease more quickly than Western medicine but that Western medications can actually cure the illness. Many prefer Asian methods for children. Reliance on traditional folk medicine is declining in the United States, partly because of the unavailability of suitable shamans and traditional herbs. The following are common treatments practiced in Vietnam and continued to some degree in the US (Table 29.2).

Researchers are exploring the use of herbs in the control of diabetes, long-established in Traditional Vietnamese Medicine, exploring the use 20 species of herbs that suggest clinical application for diabetes management is promising with further exploration (Le et al. 2019).

Like Traditional Chinese Medicine, Vietnamese may use animal products for health purposes. Two additional practices in Vietnam are consuming gelatinized tiger bones to gain strength and taking powdered rhinoceroses' horn to reduce fever or as an aphrodisiac (Hsu 2017). Many of the products contain ingredients from endangered, threatened, or protected species. The use of Traditional Vietnamese Medicine continues to be an important part of health-care for Vietnamese and requires further study to better understand the health-care implications (Nguyen et al. 2016).

29.12.4 Barriers to Health-Care

Barriers to adequate health-care for Vietnamese include

1. Subjective beliefs and the cost of health-care
2. Lack of access to a primary health-care provider
3. Differences between Western and Asian health-care practices
4. Caregivers' judgment of Vietnamese as deviant and unmotivated because of noncompliance with medication schedules, diagnostic

Table 29.2 Common health treatments

Common treatments	Purpose
Acupuncture, acupressure, and acumassage	Relieve symptomatic stress and pain
Balms and oils such as Red Tiger balm	These topical materials are applied to affected areas for relief of bone and muscle ailments
Be bao or bat gio	Also knows as "skin pinching", is a treatment for headache or sore throat. The skin of the affected area is repeatedly squeezed between the thumb and the forefinger of both hands, as the hands converge toward the center of the face. The objective is to produce ecchymoses or petechiae
Cao gio (or coining)	Literally meaning "rubbing out the wind," is used for treating colds, sore throats, flu, sinusitis, and similar ailments. An ointment or hot balm oil is spread across the back, chest, or shoulders and rubbed with the edge of a coin (preferably silver) in short, firm strokes. This technique brings blood under the skin, resulting in dark ecchymotic stripes, so the offending wind can escape. Health-care professionals must be careful not to interpret these ecchymotic areas as evidence of child abuse. However, dermabrasion may provide a portal for infection
Eating organ meats	Food such as liver, kidneys, testes, brains, and bones of an animal is said to increase the strength of the corresponding human part
Giac	(or cup suctioning) another dermabrasive procedure, is used to relieve stress, headaches, and joint and muscle pain. A small cup is heated and placed on the skin with the open side down. As the cup cools, it contracts the skin and draws unwanted hot energy into the cup. This treatment leaves marks that may appear as large bruises
Herbal teas, soups, and	Other concoctions are taken for various problems, generally in the sense of using cold measures to overcome hot illnesses
Moxibustion	Is used to counter conditions associated with excess cold, including labor and delivery. Pulverized wormwood or incense is heated and placed directly on the skin at certain meridians
Xong	Also known as "steaming" relieves motion sickness or cold-related problems. Herbs or an agent such as Vicks® VapoRub is put into boiling water, and the vapor is inhaled. Small containers of aromatic oils or liniments are sometimes carried and inhaled directly

tests, follow-up care, and their failure to keep appointments
5. Inability to communicate effectively in the English language by recent immigrants who lack confidence in their ability to communicate their needs, failure of providers to communicate adequately, or lack of an interpreter
6. Avoidance of Western providers out of fear that traditional methods will be criticized
7. Fear of conflicts and ridicule resulting in loss of face
8. Lack of knowledge of the availability of resources

Additional barriers exist for Vietnamese people when seeking mental-health-care. These include fear of stigmatization, difficulty locating agencies that can provide assistance without distorted professional and cultural communication, and reluctance to express inner feelings (Do et al. 2018).

29.12.5 Cultural Responses to Health and Illness

Fatalistic attitudes and the belief that problems are punishment may reduce the degree of complaining and expression of pain among the Vietnamese who view endurance as an indicator of strong character. One accepts pain as part of life and attempts to maintain self-control as a means of relief. A deep cultural restraint against showing weakness limits the use of pain medication. However, the sick person is allowed to depend on family and receives a great deal of attention and care.

Many Vietnamese believe that mental illness results from offending a deity and that it brings disgrace to the family and, therefore, must be concealed. A shaman may be enlisted to help, and additional therapy is sought only with the greatest discretion and often after a dangerous delay. Emotional disturbance is usually attributed to possession by malicious spirits, the bad luck of familial inheritance, or for Buddhists, bad karma accumulated by misdeeds in past lives. The term *psychiatrist* has no direct translation in Vietnamese and may be interpreted to mean nerve physician or specialist who treats crazy people. The nervous system is sometimes seen as the source of mental problems—neurosis being thought of as "weakness of the nerves" and psychosis as "turmoil of the nerves."

Physically disabled people are common and readily seen in Vietnam. Some are veterans or survivors of the Vietnam War and others have been affected by congenital disabilities (often from environmental toxins) or birth injuries. To the extent that resources allow, they are treated well and cared for by their families and the government. In contrast, a mentally disabled person may be stigmatized by the family and society and can jeopardize the ability of relatives to find marriage partners. The mentally disabled are usually harbored within their families unless they become destructive; then, they may be admitted to a hospital.

Mental health illness can cause a significant stigma in the Vietnamese culture. How might a Vietnamese patient explain a mental illness?

29.12.6 Blood Transfusions and Organ Donation

Historically, Vietnamese believe that the body must be kept intact even after death, they may be averse to blood transfusions and organ donation. Trends have changed in Vietnam as the need for blood has prompted more people to donate blood to meet the demand (Vietnam makes blood donation mandatory in controversial draft law, January 9, 2017; A new chapter in Viet Nam quest for save blood, January 18, 2017). However, some staff in a rural hospital in Vietnam donated blood after learning that the body replenished its blood supply. The smaller size of Vietnamese adults can be a challenge for blood donations as women must weight 42 kg and over and 45 kg and above for men and an acceptable haemoglobin level.

Many Vietnamese, even those whose families have long been Christian, may object to removal of body parts or organ donation. However, like blood donations, organ donations are increasing in as they are considered acceptable in Vietnam and as technology for organ transplantation continues to improve (Tran et al. 2018).

29.13 Health-Care Providers

29.13.1 Traditional Versus Biomedical Providers

Four kinds of traditional and folk providers exist in Vietnam. The first group includes Asian physicians who are learned individuals and employ herbal medication and acupuncture. The second group consists of more informal folk healers who use special herbs and diets as cures based on natural or pragmatic approaches. The secrets of folk medicine are passed down through the generations. The third group includes various forms of spiritual healers, some with a specific religious outlook and others with powers to drive away malevolent spirits. The fourth group is made up of magicians or sorcerers who have magical curative powers but no communication with the spirits. Many Vietnamese consult one or more of these healers in an attempt to find a cure.

Whereas many Vietnamese have great respect for professional, well-educated people, they may be distrustful of outside authority figures. Most Vietnamese have come to America to escape oppressive authority. Refugees generally expect health-care professionals to be experts. A common suspicion is that divulging personal information for a medical history could jeopardize their legal rights. Respect and mistrust are not mutually exclusive concepts for Vietnamese seeking care from Western providers. Because of the need to build trust with a Vietnamese client, it is particularly important to acknowledge and support traditional belief systems.

Traditional Asian male providers do not usually touch the bodies of female patients and sometimes use a doll to point out the nature of a problem. Whereas most Vietnamese may no longer insist on the use of this practice, adults, particularly young and unmarried women, are more comfortable with health-care providers of the same gender. Pelvic examinations on unmarried women should not be made on the first visit or without careful advance explanation and preparation. When such an examination is necessary, the woman may want her husband present. If possible, the practitioner and an interpreter should both be female. Women may not want to even discuss sexual problems, reproductive matters, and birth control techniques until after an initial visit and after confidence has been established in the practitioner.

29.13.2 Status of Health-Care Providers

Because of the shortage of physicians in Vietnam, medical assistants, nurses, village health-care workers, self-trained individuals, and injectionists practice Western medicine. Paralleling these approaches are the traditional systems of Asian and folk medicine. Traditional healers often provide Vietnamese with necessary social support that may be lacking with Western providers. However, all are respected and have high status and may be used concurrently or separately, according to the illness and varying beliefs of each individual.

References

A new chapter in Viet Nam quest for safe blood (2017) Viet Nam News. https://vietnamnews.vn/society/health/349873/a-new-chapter-in-viet-nam-quest-for-safe-blood.html#l5vzqudBzFX3bqaR.97

Alperin E, Batalova J (2018) Vietnamese immigrants in the United States. https://www.migrationpolicy.org

Amerasians without Borders (2019). https://amerasianswithoutborders.us/

Buchwald D, Manson SM, Dinges NG, Kean EM, Kinzie JD (1993) Prevalence of depressive symptoms among established Vietnamese refugees in the United States. J Gen Intern Med 8(2):76–81

Centers for Disease Control and Prevention (CDC) (2008) Promoting cultural sensitivity: a practical guide for tuberculosis programs that provide services to persons from Vietnam. U.S. Department of Health and Human Services, Atlanta GA

Central Intelligence Agency (CIA) (2020) East Asia/Southeast Asia: Vietnam. Accessed July 6, 2020 from https://www.cia.gov/library/publications/resources/the-world-factbook/geos/print_vm.html

Chau V, Bowie JV, Juon H-S (2018) The association of perceived discrimination and depressive symptoms among Chinese, Korean, and Vietnamese Americans. Cult Divers Ethn Minor Psychol 24(3):389–399

Do M, McCleary J, Nguyen D, Winfrey K (2018) Mental illness, public stigma, culture and acculturation among Vietnamese Americans. J Clin Transl Sci 2(S1):17–19

Driscoll AK (2018) Asian-American mothers: demographic characteristics by maternal place of birth and Asia subgroup, 2016. Natl Vital Stat Rep 67(2):1–12

Fang CY, Lee M, Feng Z, Tan Y, Levine F, Nguyen C, Ma GX (2019) Community-based cervical cancer education: changes in knowledge and beliefs among Vietnamese American women. J Community Health 44(3):525–533

Garcia-Garcia AS, Martinez-Gonzalez J-M, Gomez-Font R, Soto-Rivadeneira A, Oviedo-Roldan L (2010) Current status of torus Palatinus and torus Mandibularis. Med Oral Patol Oral Cir Bucal 15(2):e353–e360

Horton P, Rydstrom H (2019) Reshaping boundaries family politics and GLBTQ resistance in urban Vietnam. J GLBT Fam Stud 15(3):290–305. https://doi.org/10.1080/1550428X.2018.1518739

Hsu J (2017) The hard truth about the rhino horn "aphrodisiac" market. Sci Am 2019. https://www.scientificamerican.com/article/the-hard-truth-about-the-rhino-horn-aphrodisiac-market/

Iwamoto DK, Kaya A, Grivel M, Clinton L (2016) Under-researched demographics: heavy episodic drinking and alcohol-related problems among Asian Americans. Alcohol Res 38(1):17–25

Kim I, Keovisai M, Kim W, Richards-Desai S, Yalim AC (2019) Trauma, discrimination, and psychological distress across Vietnamese refugees and immigrants: a life course perspective. Community Ment Health J 55(3):385–393

Kinzie JD, Manson SM (1982) Development and validation of a Vietnamese-language depression rating scale. Am J Psychiatr 139(10):1276–1281

Le Q, Lay H, Wu M (2019) Herbs for the management of diabetes mellitus in traditional Vietnamese medicine to manage diabetes mellitus. J Appl Biopharm Pharmacokinet 7:1–7

Levy RA (1993) Ethnic and racial differences in response to medicines: preserving individualized therapy in managed pharmaceutical programmes. Pharmaceutic Med 7:139–165

Lin K, Shen WW (1991) Pharmacotherapy for southeast Asian psychiatric patients. J Nerv Ment Dis 179(6):346–350

Martell BN, Garrett BE, Caraballo RS (2016) Disparities in adult cigarette smoking—United States, 2002–2005, 2010–2013. Morbid Mortal Wkly Rep 65(30):753–758

McConnell K, Holloway K (n.d.) Asian, American, and Queer. Dorothy F. Schmidt College of Arts and Letters, p 70–79. http://www.fau.edu/artsandletters/

Medical News Today (2010) Huge health disparities revealed among Asian-Americans, Native Hawaiians, Asian immigrants. http://www.medicalnewstoday.com//articles/182881.php

Muecke MA (1983) Caring for southeast Asian refugees in the American health care system. Am J Public Health 73(4):431–438

Nguyen HN (2015) Thalassemia in Vietnam. Ann Transl Med 3(S2):AB035

Nguyen LT, Kaptchuk TJ, Davis RB, Nguyen G, Pham V, Tringale SM et al (2016) The use of traditional Vietnamese medicine among Vietnamese immigrants attending an urban community health Center in the United States. J Altern Complement Med 22(2):145–153

Nguyen-Truong CKY, Hassounek, D, Lee-Lin F, Hsiao C, Le TV, Tang J et al (2017) Health care providers' perspectives on barriers and facilitators to cervical cancer screening in Vietnamese American women. J Transcult Nurs 29(5):441–448.

Park VMT, Goyal D, Nguyen T, Lien H, Rosidi D (2017) Postpartum traditions, mental health, and help-seeking considerations among Vietnamese women: a mixed methods pilot study. J Behav Heal Serv Res 44(3):428–441

Pew Research Center. Social and Demographic Trends (2012) Vietnamese Americans. https://www.pewsocialtrends.org/rise-of-asian-americans-2012-analysis/vietnamese/. Accessed 30 Sep 2019

Rasbridge L (2004) Vietnamese. In: Kemp C, Rasbridge L (eds) Refugee and immigrant health: a handbook for health professionals. University of Cambridge Press, Cambridge, pp 346–358

Rkasnuam H, Batalova J (2014) Vietnamese immigrants in the United States. Migration Policy Institute. www.migrationpolicy.org

Roberts NS, Copel JA, Bhutan Y, Otis S (1985) Intestinal parasites and other infections during pregnancy in southeast Asian refugees. J Reprod Med 30(10):720–725

Ross TF (1982) Health care problems of southeast Asian refugees. West J Med 136(1):35–43

Sarwar E, Arias D, Becerra BJ, Becerra MB (2015) Sociodemographic correlates of dietary practices among Asian-Americans: results from the California health interview survey. J Racial Ethn Health Disparities 2:494–500

Shovet M (2018) Two deaths and a funeral: ritual inscriptions' affordances for mourning and moral personhood in Vietnam. Am Ethnol 45(1):60–73

Sorkin D, Tan A, Hays R, Mangione C, Ngo-Metzger Q (2008) Self-reported health status of Vietnamese and non-Hispanic white older adults in California. J Am Geriatr Soc 56(8):1543–1548

Sorkin DH, Murphy M, Nguyen H, Biegler KA (2016) Barriers to mental health care for an ethnically and racially diverse sample of older adults. J Am Geriatr Soc 64(10):2138–2143

Strand PJ, Jones W (1983) Health service utilization by Indochinese refugees. Med Care 21(11):1089–1098

Sun E, Bui Q, Tsoh JY, Gildengorin G, Chan J, Cheng J et al (2017) Efficacy of a church-based, culturally tailored program to promote completion of advance directives among Asian Americans. J Immigr Minor Health 19(2):381–391

Tran SN, Thu TN, Thai SM, Nguyen HT, Bui TV, Lo-Cao EAY et al (2018) Current status of organ donation for transplantation in Vietnam. Transplantation 102:S382

Tran QNH, Dieu-Hien H, King IN, Sheehan K, Iglowitz M, Periyakoil VS (2019) Providing culturally respectful care for seriously-ill Vietnamese Americans. J Pain Symptom Manag 58(2):344–354

U.S. Census (2011) The Vietnamese population in the United States: 2010. http://www.vasummit2011.org/docs/research/The%20Vietnamese%20Population%202010_July%202.2011.pdf. Accessed 15 Sep 2019

U.S. Department of Health and Human Services (USDHHS), Centers for Disease Control and Prevention (CDC), National Center for Emerging and Zoonotic Infectious Diseases, Division of Global Migration and Quarantine (2013). Guidelines for evaluation of the nutritional status and growth of refugee children during the domestic medical screen examination. www.cdc.gov

UNE Applied Nutrition Program (2019) Food and culture fact sheet Vietnam. In: All student-created educational resources, p 127. https://dune.une.edu/an_studedres/127

Vangen S, Stoltenberg C, Rolv Skjaerven R, Magnus P, Harris JR, Stray-Pedersen B (2002) The heavier the better? Birthweight and perinatal mortality in different ethnic groups. Int J Epidemiol 31:654–660

Vietnam makes blood donation mandatory in controversial draft law (2017) Troi Tre News. https://tuoitrenews.vn/society/38934/vietnam-makes-blood-donation-mandatory-in-controversial-draft-law

Vu T 2016 State formation on Chinas southern frontier: Vietnam as a shadow empire and hegemon, 40 HumaNetten, NR 37, Hösten. https://open.lnu.se/index.php/hn/article/view/388/335

Waldorf B, Gill C, Crosby SS (2014) Assessing adherence to accepted National Guidelines for immigrant and refugee screening and vaccines in an urban primary care practice: a retrospective chart review. J Immigr Minor Health 16(5):839–845

Zhou M, Bankston CL (2000) Straddling two social worlds: the experience of vietnamese refugee children in the United States. ERIC Urban Diversity Series No. 111, ERIC Clearinghouse on Urban Education, Institute for Urban and Minority Education

Glossary

A

Aagwachse Amish folk illness, referred to in English as liver-grown, with symptoms of abdominal distress believed to be caused by too much jostling, especially occurring in infants during buggy rides.

Abnemme Amish folk illness characterized by "wasting away"; usually affects infants or young children who seem to be too lean and not active.

Abwaarde Amish term for ministering to someone by being present and serving when someone is sick in bed.

Acculturate To modify or give up traits from the culture of origin as a result of contact with another culture.

Achegewe Amish term for warm hands.

Allah Greatest and most inclusive of the names of God. Arabic word used to describe the God worshipped by Muslims, Christians, and Jews.

Am Pervasive force in Vietnamese traditional medicine, associated with cold conditions and things that are dark, negative, feminine, and empty.

Anabaptist Adherent of the radical wing of the Protestant Reformation who espouses baptism of adult believers.

Anatolia Geographic and historical term denoting the westernmost protrusion of Asia, comprising the majority of the Republic of Turkey.

Antyesti Hindu equivalent of last rites.

Apocopation Dropping the first vowel when one word ends with a vowel and the next word begins with a vowel.

Arabic Semitic language of the Arabs.

Ashkenazi Descended from Eastern Europe and Russia.

Assimilate To gradually adopt and incorporate the characteristics of the prevailing culture.

Attitude State of mind or feeling with regard to some matter of a culture.

!Ay bendito! Frequently used Puerto Rican phrase expressing astonishment, surprise, lament, or pain.

Ayurveda Traditional Asian Indian medicine.

B

Barrenillos Spanish term for obsessions.

Baten Iranian term for inner self.

Be bao or bat gio Vietnamese folk practice in which the skin is pinched in order to produce ecchymosis and petechiae; practiced to relieve sore throats and headaches.

Boat people Haitian or Cuban immigrants who arrive in small boats; usually of undocumented status.

Boricua Puerto Rican term used with great pride; name given to Puerto Rico by the Taino Indians.

Botanica Traditional Cuban or other Spanish store selling a variety of herbs, ointments, oils, powders, incenses, and religious figurines used in Santería.

Brauche Folk healing art common among Pennsylvania Germans.

Braucher Amish practitioner of brauche, a folk healer.

© Springer Nature Switzerland AG 2021
L. D. Purnell, E. A. Fenkl (eds.), *Textbook for Transcultural Health Care: A Population Approach*,
https://doi.org/10.1007/978-3-030-51399-3

Bris or brit milah Ritual circumcision of a male Jewish child.

Bureau of Indian Affairs (BIA) Federal agency responsible for ensuring services to Native Americans, Alaskan Indians, and Eskimo tribes.

C

Caida de la mollera Condition of fallen fontanelle, believed to occur because the infant was withdrawn too harshly from the nipple; common among some Spanish-speaking populations.

Cami Turkish word for mosque.

Cao gio Vietnamese practice of placing ointments or hot balm oil across the chest, back, or shoulders and rubbing with a coin; used to treat colds, sore throats, flu, and sinusitis.

Carı˘noso(a) Hispanic term for caring, in both verbal and nonverbal communications.

Catimbozeiros Portuguese word for sorcerer; can be a folk practitioner.

Celtic Belonging to a group of Indo-European languages: Irish, Welsh, or Breton.

Chasidic (or Hasidic) Ultra-Orthodox Jewish sect.

Cheshm-i-bad Iranian term for evil eye.

Cho Haitian word for cold.

Chondo- Korean naturalistic religion that combines Confucianism, Buddhism, and Daoism.

Choteo Cuban term for a lighthearted attitude, involving teasing, bantering, and exaggeration.

Collectivism Moral, political, or social outlook that stresses human interdependence.

Comadre Portuguese term for godmother.

Community Group of people having a common interest or identity; goes beyond the physical environment to include the physical, social, and symbolic characteristics that cause people to connect.

Compadrazgo Spanish term for a system of personal relationships in which friends or relatives are considered part of the family whether or not there is a blood relationship.

Compadre Portuguese term for godfather.

Confianza Hispanic term for trust developed between individuals; essential for effective communication and interpersonal interactions in health-care settings.

Conservative Jewish term for the religious group between Reform and Orthodox in terms of religious practice.

Contadini Italian term for peasants.

Cornicelli Italian charm with little red horns worn for good luck.

Creole Rich and expressive language derived from two other languages, such as French and Fon, an African tongue.

Cultural awareness Appreciation of the external signs of diversity such as the arts, music, dress, and physical characteristics.

Cultural competence Having the knowledge, abilities, and skills to deliver care congruent with the patient's cultural beliefs and practices. (See Chap. 1 for a more extensive definition.)

Cultural humility Focuses on the process of intercultural exchange, paying explicit attention to clarifying the professional's values and beliefs through self-reflection.

Cultural imperialism Practice of extending the policies and practices of one organization (usually the dominant one) to disenfranchised and minority groups.

Cultural imposition Intrusive application of the majority cultural view onto individuals and families.

Cultural leverage A process whereby the principles of cultural competence are deliberately invoked to develop interventions.

Cultural relativism Belief that the behaviors and practices of people should be judged only from the context of their cultural system.

Cultural safety Expresses the diversity that exists within cultural groups and includes the social determinants of health, religion, and gender in addition to ethnicity.

Cultural sensitivity Having to do with personal attitudes and not saying things that may be offensive to someone from a cultural or ethnic background different from the health-care provider's background.

Culture Totality of socially transmitted behavior patterns, arts, beliefs, values, customs, lifeways, and all other products of human work and thought characteristics of a population of people that guides their worldview and decision-making. Patterns may be explicit or implicit, are primarily learned and transmitted within the family, and are shared by the majority of the culture.

Curandeiro Portuguese folk practitioner whose healing powers are divinely given.

Curandero Traditional folk practitioner common in Spanish-speaking communities; treats traditional illness not caused by witchcraft.

D

Daadihaus Amish grandparents' cottage adjacent to farmhouse.

Dan wei Functional unit of Chinese society; work unit or neighborhood unit responsible to and for the Chinese people's way of life.

Dao Balance between yin and yang.

Dayah Arab midwife.

Decensos Spanish term for fainting spells.

Demut German term for humility, a priority value for the Amish, the effects of which may be seen in details such as the height of the crown of an Amish man's hat, as well as in very general features such as the modest and unassuming bearing and demeanor usually shown by Amish in public. This behavior is reinforced by frequent verbal warnings against its opposite, hochmut, pride or arrogance, which is to be avoided.

Deitsch/Duetsch Pennsylvania German (sometimes incorrectly anglicized as Pennsylvania Dutch); American dialect derived from several uplands and Alemannic German dialects, with an admixture of American English vocabulary.

Docte fey Haitian word for leaf doctor.

Docte zo Haitian word for bonesetter.

Doule Haitian word for pain.

Duong Vietnamese force used in traditional health practice, associated with things positive, masculine, light, and full.

E

Eid Arabic, Iranian, and Somali term for celebration of a feast—for example, Eid Gorgan (day/feast ending pilgrimage to Mecca); Eid-al-Fitr (last day of the month of Ramadan).

El ataque/ataque de nervios Hyperkinetic spasmodic activity common in Spanish-speaking groups. The purpose is to release strong feelings or emotions. The person requires no treatment, and the condition subsides spontaneously. It is an expression of deep anger or depression.

Empacho Condition common among some Spanish-speaking populations; believed to be caused by a bolus of food stuck in the gastrointestinal tract. Massage of the abdomen is believed to relieve the condition.

Enculturation A natural conscious and unconscious conditioning process of learning accepted cultural norms, values, and roles in society.

Endropi Greek term for shame.

Espiritista (espiritualista) Spanish or Portuguese folk practitioners who receive their talent from "God"; treat conditions believed to be caused by witchcraft.

Estampitas Spanish for little statues of saints.

Ethic of neutrality Avoiding aggression and assertiveness, not interfering with others' lives unless asked to do so, avoiding dominance over others, and avoiding arguments and seeking agreement.

Ethnic group Group of people who have had experiences different from those of the dominant culture in status, background, residence, religion, education, or other factors that functionally unify the group and act collectively on one another. Pertains to a religious, racial, national, or cultural group.

Ethnocentrism Universal tendency for human beings to think that our own ways of thinking, acting, and believing are the only right, proper, and natural ones and to believe that those who differ greatly are strange, bizarre, or unenlightened.

F

Familism Social pattern in which family solidarity and tradition assume a superior position over individual rights and interests.

Falasha Black Jews originating from Africa.

Fam saj Haitian Creole word for lay midwife.

Farsi The national language of Iran.

Fatalism Acceptance that occurrences in life are predetermined by fate and cannot be changed by human beings.

Freindschaft Amish three-generational extended family network of relationships.

Fret Haitian word for cold.

G

Gaelic The language spoken in Ireland.

Garm Iranian term for hot.

Garmie Iranian digestive problem caused from eating too much hot food.

Gelassenheit Amish term for submission, yielding, surrender of self and ego to the higher will of the group or deity.

Gemeinschaft German word for community.

Generalization Reducing numerous characteristics of an individual or group of people to a general form that renders them indistinguishable. Generalizations have to be validated by the individual.

Giac Vietnamese dermabrasive procedure performed with cup suctioning.

Giagia Greek term for grandma.

Global society Seeing the world as one large community as people travel and interact.

Great Eid Islamic feast of 4 days.

Guanxi Chinese term defining how relatives are expected to help one another through connections, used by Chinese society in a manner similar to the use of money in other cultures.

Gullah Creole language spoken by African Americans who reside on or near the islands off the coasts of Georgia and the Carolinas.

H

Hasidic (or Chasidic) Ultra-Orthodox Jewish sect.

Hadith Oral tradition of the Prophet Muhammad; collection of words and deeds that form the basis of Muslim law.

Hajji (Hajj an haji) Annual pilgrimage to Mecca.

Halal The lawful—that which is permitted by Allah; also, the term used to describe ritual slaughter of meat.

Hanbang Traditional Korean medical-care system.

Hanui Korean word for oriental medicine doctor.

Hanyak Korean traditional herbal medicine used to create harmony between oneself and the larger cosmology; a healing method for body and soul.

Haram The unlawful—that which is prohibited by Allah; anyone who engages in what is prohibited is liable to incur punishment hereafter (as well as legal punishment in countries that incorporate Islamic law into legal codes).

Hasidic Jewish ultra-Orthodox sect.

Hebrew Language of Israel and of Jewish prayer.

Hejab Iranian term for any behavior that expresses modesty in public—for example, in women, modest attire (loose dress or head scarf) or shy, self-limiting behavior in relating to the other gender.

Hijab Modest covering of a Muslim woman; conceals the head and the body, except for the hands and face, with loosely fitting, nontransparent clothing.

Hilot Filipino folk healer and massage therapist.

Hindi Primary language of India.

Hispanic American of Spanish or Latin American origin.

Hochmut Amish term for pride and arrogance.

Hogan Earth-covered Navajo dwelling.

Honor Spanish term for goodness or virtue; can be diminished or lost by an immoral or unworthy act.

Hot-and-cold theory Hispanic concept that illness is caused when the body is exposed to an imbalance of hot and cold; foods are also classified as hot or cold.

Hwa-byung Korean traditional illness that occurs from repressing anger or other strong emotions.

Hwangap Significant celebration in Korean society—at the age of 60, a person starts the calendar cycle over again.

I

Ideology Thoughts and beliefs that reflect the social needs and aspirations of an individual or an ethnocultural group.

Imam Muslim leader of the prayer; usually the most learned member of the local Islamic community.

Indian Health Service Federal agency that has the responsibility for providing health services to Native Americans.

Indids Asian Indians who have a light brown skin color.

Individualism Term used to describe a moral, political, or social outlook that stresses human independence and the importance of individual self-reliance, and freedom.

Individuality The sense that each person has a separate and equal place in the community and where individuals who are considered "eccentrics or local characters" are tolerated.

Indochinese Individuals originating from Vietnam, Cambodia, or Laos.

Insallah or Insh'Allah Arabic and Turkish word for "if God wills."

Islam Monotheistic religion in which the supreme deity is Allah; according to Muslim belief, God imparted his final revelations—the Holy Qur'an—through his last prophet, Mohammed, thereby completing Judaism and Christianity.

J

Jerbero Spanish folk practitioner who specializes in treating health conditions through the use of herbal therapy.

Jinn Arabic term for demons.

Judaism Refers to a religion, people, and a culture.

K

Kaddish Jewish prayer said for the dead.

Kapwa A word meant to bring Filipinos together, **means** "shared identity," "equality," and "being with others." To put it simply, **kapwa** is the obligation towards our fellow man.

Karma Hindu term for actions performed in the present life and the accumulated effects from past lives.

Kashrut or kashrus Jewish laws that dictate which foods are permissible under religious law.

Ki Korean term for the energy that flows through living creatures.

Koran See Qur'an.

Kosher Kashrut laws in the Jewish religion.

Koumbari Greek term for coparents.

Kut Korean shamanistic ceremony to eliminate the evil spirits causing an illness.

L

La gente de la raza Phrase denoting a genetic determination to which all Spanish-speaking people belong, regardless of class differences or place of birth.

Latino(a) Person from Latin America.

Laying on of hands Spiritual practice of placing one's hands on an individual for the purpose of healing.

Lien Vietnamese concept that represents control over and responsibility for moral character.

M

Maalesh Arabic term meaning never mind, it doesn't matter; substantial efforts are directed at maintaining pleasant relationships and preserving dignity and honor; hostility in response to perceived wrongdoing is warded off by an attitude of maalesh.

Machismo Sense of masculinity that stresses virility, courage, and domination of women; includes the need to display physical strength, bravery, and virility.

Madichon Haitian term used when children are disrespectful; it means that their future will be marred by misfortune.

Maghi Italian word for witch.

Magissa Greek folk healer.

Mal ojo Spanish term for the evil eye, a hex condition with unspecific signs and symptoms believed to be caused by an older person admiring a younger person; condition can be reversed if the person doing the admiring touches the person being admired.

Marielitos Cuban immigrants who arrived in 1980 on a massive boatlift from Muriel Harbor, Cuba, to Key West, Florida.

Masallah Turkish term for God bless and protect.

Matiasma Greek term for the evil eye.

Mestizo(a) Person of mixed Spanish and Native American heritage.

Mezuzah Container with biblical writings; placed on the doorpost of homes or hung around the neck on a necklace.

Mien Vietnamese concept based on wealth and power.

Mohammed Prophet of God and founder of Islam.

Mohel Ritual circumciser in the Jewish faith.

Moreno Portuguese Brazilian individual who has black or brown hair and dark eyes.

Moslem See Muslim.

MosqueMuslim place of worship.

Moxibustion Vietnamese health-care practice in which pulverized wormwood is heated and placed directly on the skin at specified meridians to counter conditions associated with excess cold.

Mukrah Arabic term for undesirable but not forbidden.

Mulatto Person of mixed European and African heritage.

Mundang Korean folk healer who has special abilities for communicating with the spirits and in treating illnesses after all other means of treatment are exhausted.

Muslim Person who follows the Islamic faith, the world's second-largest religion.

N

Naharati Iranian term for generalized distress.

Navajo neuropathy Neurological condition confined to Navajo Indians; characterized by a complete absence of myelinated fibers resulting in short stature, sexual infantilism, systemic infection, hypotonia, areflexia, loss of sensation in the extremities, corneal ulcerations, acral mutilation, and painless fractures.

Nazar Turkish term for envy.

Nazar boncuk Small blue bead used among Turkish people to protect a child from the evil eye.

Nervioso(a) Hispanic term used to describe signs and symptoms of nervousness, anxiety, sadness, and grief.

Nevra Greek folk illness.

Niuyoricans Puerto Ricans born in New York.

Nihon/Nippon Japanese name for Japan.

O

Old Order Amish Most conservative and traditionalist group among the followers of Jacob Ammann; today simply called Amish, but technically known as Old Order Amish Mennonite to distinguish them from other related Amish and Mennonite groups.

Oppression Haitian ailment related to asthma; describes a state of anxiety and hyperventilation.

Ordnung Codified rules and regulations that govern the behavior of a local Amish church district, or congregation; local consensus of faith and practice; also the German term for order.

Orishas Gods or spirits in Santería.

Orthodox Traditional Judaism.

P

Padrone or capo di famiglia Italian word for master, head of the family.

Pappous Greek word for grandfather.

Parve Jewish term used for foods that are neutral and can be eaten with meat or milk products.

Pasah Dai Dialect in southern Thailand.

Pasah Isaan Dialect in northeastern Thailand.

Pasah Nua Dialect in northern Thailand.

Personalismo Spanish word for emphasis on intimate, personal relationships as more important than impersonal, bureaucratic relationships.

Philptimo Greek term for respect.
Phylacto Greek amulet worn to ward off envy.
Pikirist Haitian word for injectionist.
Pogrom Organized persecution or massacre of a minority group.
Ponos Greek word for pain.
Practika Greek herbal remedies.
Pseudofamilies Vietnamese households made up of close and distant relatives and friends that share accommodations, finances, and fellowship.
Pu tong hua Recognized language of China.

Q

Qi One of five substances or elements of traditional Chinese medicine; encompasses the foundation of the energy of the body, environment, and universe; includes all sources and expenditures of energy.
Quinceñera A Hispanic/Latino girl's 15th birthday that celebrates her passage into womanhood.
Qur'an or Koran Muslim holy book; believed by Muslims to contain God's final revelations to humankind.

R

Rabbi Jewish religious leader.
Ramadan or Ramazan The 9th month of the Islamic year during which Muslims are required to fast during daylight hours for 30 days.
Reconstructionism Mosaic of the three main branches of Judaism; is an evolving religion of the Jewish people; seeks to adapt Jewish beliefs and practices to the needs of the contemporary world.
Reform Liberal or Progressive Judaism.
Remedios caserios Portuguese (Brazilian) home medicine or remedy.
Remedios populares Portuguese (Brazilian) folk medicine practices.
Respeto Hispanic term denoting respect; refers to the qualities developed toward others such as parents, the elderly, and educated people who are expected to be honored, admired, and respected.
Restavec Haitian term to denote children who are sent to live with a nonparent family for the purposes of improving their lives economically.
Rezadeiras Brazilian spiritual leaders.

S

Sabra Jew who was born in Israel.
Santería 300-year-old Afro-Cuban religion that syncretizes Roman Catholic elements with ancient Yoruba tribal beliefs and practices.
Santero Practitioner of Santería.
Sard Iranian term for cold.
Sardie Iranian digestive problem; occurs from eating too much cold food.
Sephardic Jewish term for being descended from Spain, Portugal, the Mediterranean, Africa, or Central or South America.
Severe combined immune deficiency syndrome Immune deficiency syndrome (unrelated to AIDS), characterized by a failure of antibody response and cell-mediated immunity.
Sheikhs The most learned individuals in an Islamic community.
Simpatia Spanish term for smooth interpersonal relationships; characterized by courtesy, respect, and the absence of harsh criticism or confrontation.
Small Eid Islamic holy feast of 3 days.
Sobador Spanish folk practitioner, similar to a chiropractor, who treats illnesses and conditions affecting the joints and musculoskeletal system.
Spanglish Sentence structure that includes both English and Spanish words.
Stereotyping Oversimplified conception, opinion, or belief about some aspect of an individual or group of people.
Subculture Group of people who have had experiences different from those of the dominant culture in status, ethnic background, residence, religion, education, or other factors

that functionally unify the group and act collectively on one another.

Susto "Magical fright," a condition believed to be caused by witchcraft; symptoms can be quite varied and include both mental and physical concerns.

Synagogue, temple, or shul Jewish house of worship.

T

T'ai chi Chinese system of exercise for mind and body control.

Ta'arof Iranian ritual expressing courtesy.

Tae-kyo Korean term, literally fetus education, with the objective being health and well-being of the fetus and the mother through art, beautiful objects, and a serene environment.

Tae-mong Korean term signifying the beginning of pregnancy; the pregnant woman dreams of conception of the fetus.

Tagalog Filipino national language.

Tesbih Turkish small beads traditionally used for praying now take a more secular meaning and are often referred to as worry beads.

Tet Asian Lunar New Year; celebrated in January or February.

Torah Five books of Moses; referred to in the Jewish faith.

Treyf Jewish term for forbidden or unclean.

Tribe Native American social organization comprising several local villages, bands, districts, lineages, or other groups who share a common ancestry, language, and culture.

Tridosha Theory that the body is made up of five elements: fire, air, space, water, and earth.

Tudo bom Portuguese word for great, often said in a stoical sense.

Two spirit Term used among AI/AN populations to connote diverse gender and sexual identities.

V

Variant cultural characteristics Determine a person's adherence to beliefs and values of his or her dominant culture. Includes nationality, race, color, gender, age, religious affiliation, educational status, socioeconomic status, occupation, military experience, political beliefs, urban or rural residence, enclave identity, marital status, parental status, physical characteristics, sexual orientation, gender issues, and reason for migration (sojourner, immigrant, or undocumented status).

Velorio Spanish term for a wake; a festive occasion following the burial of a person.

Vendouses Greek practice of cupping.

Verguenza Spanish term for a consciousness of public opinion and the judgment of the entire community.

Via vecchia Italian for old way.

Viddui Jewish personal confession recited when death is imminent.

Visiting High-frequency custom of family-to-family home visits that help to maintain kinship and church ties and the flow of information within the Amish community.

Voudou or voodoo Vibrant religion born from slavery and revolt; the term means sacred in the African language of Fon.

W

Wake Watch over a deceased person before burial; usually accompanied by a celebration, which may include feasting.

Warm hands Healing art related to therapeutic touch; regarded by Amish as a gift to be applied for the good of others in need of healing; a form of brauche.

Worldview Way an individual or group of people look upon their universe to form values about their life and the world around them.

Y

Yang In Chinese belief system, one of the two opposing principles of the balance of life; can be either a single phenomenon or a state of being of a phenomenon. See yin.

Yarmulke Jewish head covering worn by men.

Yerbero See jerbero.

Yiddish Language often spoken by elderly Jews.

Yin In Chinese belief system, one of the two opposing principles of the balance of life; can be either a single phenomenon or a state of being of a phenomenon. See yang.

Z

Zaher Iranian term for public persona.

Zhong guo The Chinese name for China and means "middle kingdom."

Zong Vietnamese herbal preparation; relieves motion sickness or cold-related problems.

Printed in the United States
By Bookmasters